Volume (gas)

1 cm³	=	0.061 in³
1 m³	=	35.314 ft³
1 m³	=	1.308 yd³
1 ft³	=	0.0283 m³
1 in³	=	16.387 cm³

Volume (liquid)

1 mL	=	0.271 dr (dram)
1 mL	=	0.0338 fl oz
1 L	=	33.815 fl oz
1 L	=	2.113 pt
1 L	=	1.057 qt
1 L	=	0.264 gal
1 dr	=	3.697 mL
1 fl oz	=	29.573 mL
1 pt	=	473.166 mL
1 qt	=	946.332 mL
1 gal	=	3.785 L

Length

1 mm	=	0.0394 in
1 mm	=	0.00328 ft
1 mm	=	0.0011 yd
1 cm	=	0.3937 in
1 cm	=	0.0328 ft
1 m	=	39.37 in
1 m	=	3.281 ft
1 m	=	1.0936 yd
1 in	=	25.4 mm
1 in	=	0.0833 ft
1 in	=	0.0254 m
1 ft	=	304.8 mm
1 ft	=	30.48 cm
1 ft	=	0.3048 m
1 yd	=	914.40 mm
1 yd	=	91.44 cm
1 yd	=	0.9144 m

Pressure

1 atm	=	14.696 lb/in²
1 atm	=	2116.2 lb/ft²
1 atm	=	33.93 ft of water
1 atm	=	1.0133 bars
1 atm	=	29.921 in of mercury
1 atm	=	760.0 mm of mercury
1 cm of Hg	=	5.352 in of water
1 cm of Hg	=	0.1934 lb/in²
1 in of Hg	=	70.727 lb/ft²
1 in of Hg	=	13.60 in of water
1 in of Hg	=	345.32 kg/cm²
1 kg/m²	=	0.00142 lb/in²
1 kg/m²	=	0.205 lb/ft²
1 kg/cm²	=	0.968 atm
1 g/mL	=	0.0361 lb/in³
1 ft of water	=	0.0295 atm
1 lb/in²	=	70.31 g/cm²

HUMAN ANATOMY AND PHYSIOLOGY

HUMAN ANATOMY AND PHYSIOLOGY

ANTHONY J. GAUDIN, PH.D.

California State University, Northridge

KENNETH C. JONES, PH.D.

California State University, Northridge

with

James G. Cotanche, M.D.

University of Massachusetts, Amherst

Josephine Ryan, D.N.Sc.

University of Massachusetts, Amherst

HARCOURT BRACE JOVANOVICH, PUBLISHERS

and its subsidiary ACADEMIC PRESS

San Diego   New York   Chicago   Austin   Washington, D.C.
London   Sydney   Tokyo   Toronto

*To
Susanne
and
Shirley*

Cover illustrations •	Keith Kasnot
Text illustrations •	Keith Kasnot
•	Darwen and Vally Hennings
•	Sandra McMahon
•	Signature Design Associates

Copyright © 1989 by Harcourt Brace Jovanovich, Inc.

All rights reserved. No part of this publication may be reproduced or transmitted in any form or by any means, electronic or mechanical, including photocopy, recording, or any information storage and retrieval system, without permission in writing from the publisher.

Requests for permission to make copies of any part of the work should be mailed to: Permissions, Harcourt Brace Jovanovich, Publishers, Orlando, Florida 32887

ISBN: 0-15-539705-2

Library of Congress Catalog Card Number: 88-80612

Printed in the United States of America

Copyrights and Acknowledgments and Illustration Credits appear on pages 834–836, which constitute a continuation of the copyright page.

Preface to the Instructor

Our philosophy about the function of a textbook has evolved over 25 years of teaching. In that time, we have developed a strong identification with our role as educators and as guides through the learning process. It is clear to us that a textbook should firmly support us in this responsible role. Ultimately, the student is accountable for how well he or she learns a subject, but as instructors, we play the most direct and important role in helping the student meet that responsibility. It is our task to convey information clearly and in a logical order, to identify and explain the relationships among different elements of that information, to put the importance of a subject into perspective, to instill in the student an appreciation for the subject, and to develop an ability and interest that will enable him or her to pursue the subject long after the course is over. To us, a textbook is an indispensable aid in that process, an adjunct that will provide reinforcement, help stimulate interest, and provide information that we do not have time to discuss. The ideal textbook should make our job easier and our efforts more effective.

We have written this book with the assumption that the students in your class are generally freshmen and sophomores and are taking your course in human anatomy and physiology to fulfill a requirement for entry into a related field. As such, most of them are not biology majors and may have had only minimal background in chemistry and biology. Some students may expect the subject to be particularly difficult and view it as a potential obstacle in their progress to the profession they have chosen. The students in your course are generally bright, alert, excited about life, and challenged by new ideas, but there is probably considerable outside competition for their time and interest, which makes your job more difficult.

This text has been designed to serve as the nucleus of a comprehensive package of instructional materials. The book itself contains numerous aids that will help to stimulate interest, to enable students to assess their progress independent of class examinations, and to recognize the relevance of the information to their lives. Elements are placed where they best support the text so as to have the greatest pedagogical value. Ancillary materials are available for the students and instructor that are designed to provide additional exposure through alternative media.

The Book

The text is organized into seven units arranged in a logical order that corresponds to the way the course is usually taught. Unit I deals with the ideas and principles that underlie the material in subsequent chapters, including an overview of principles of anatomy, physiology, chemistry, cell structure and function, and the organization of tissues. Unit II includes chapters that deal with systems involved in support and movement. In this unit are chapters on skin, skeletal system, articulations, muscles, and muscle function. Unit III contains chapters that describe cellular and chemical control and integration of body processes. Chapters in this unit deal with the organization of the nervous system, generation and transmission of nerve impulses, special senses, and hormonal control. Unit IV deals with the metabolic processes, including chapters on digestion, general metabolism, and the structure and function of the respiratory system. Unit V describes the structure and function of blood and the cardiovascular and lymphatic systems and their role in transport. Unit VI deals with homeostatic control mechanisms and includes chapters on the urinary system and the composition and control of body fluids. Unit VII contains chapters that describe the structure and function of the reproductive systems, fertilization, embryonic and fetal development, and the structure and mechanism of gene action.

Each chapter includes several pedagogical elements to help the student learn the material, assess his or her progress, and understand the relationship of the material to health and medicine. For example, each chapter begins with a brief introduction to the material that follows and a list of important concepts to be covered. Each chapter is divided into major sections, each of which begins with a brief listing of the objectives of that section, which alerts the student to salient points to look for in the section. Technical terms are identified in boldface type and pronunciations are provided where appropriate. Etymologies for many technical terms appear in the upper page margin to help broaden the student's knowledge of the bases for the terminology. Numerous clinical notes and essays are included in each chapter (over 220 in the book) to help the student understand the clinical relevance of the material. Descriptions of the embryological development of organ systems are placed in boxed passages in appropriate chapters for those who wish to use an embryological approach to teaching this subject. Each chapter is summarized in a study outline written in a narrative style. An objective examination, usually with about 30 to 50 questions, is included at the end of each chapter. Answers to these are provided in an appendix. Questions for discussion or essay examination are also provided, some of which cover the clinical application material.

The backmatter includes an extensive glossary to which the student can refer for help on technical terms. There is also an appendix that lists and discusses areas of employment in allied health fields and one on Latin and Greek word parts for building terminology. The endpapers include a list of common pharmacological and medical abbreviations and tables of unit conversion factors and equivalents.

Ancillary Materials
Several items are available to help round out the package for the student. A Study Guide, written by Dr. Billie Perkins, is keyed directly to the text. It provides additional discussion, exercises, and questions. It is a valuable supplement to the text for the student. Computer-aided learning software is also available for students' self-study, as is the software-supplemented text *Medical Terminology for the Health Professions* by Bryan C. Smith and Bentley E. Smith. To help engage students as they learn anatomy, also available is *Coloring Atlas of Human Anatomy* by Edwin Chin, Jr., and Marvin M. Shrewsbury.

To help you, the instructor, the package includes an Instructor's Manual, which contains lecture outlines keyed to the text, additional examination questions and answers, and references to additional sources of material and information. Overhead color transparencies and color slides of 225 figures contained in the text are provided, as well as a software test bank and examination-generating computer program for PC-DOS, MS-DOS, Macintosh, and Apple II computers. As a special supplement, we are offering to adopters a 50-minute videotape consisting of two 25-minute segments featuring the dissection of the human heart and the digestive system, respectively. The videotape script was written by Anthony Gaudin and Garry Porter.

An Invitation
In short, along with the staff at HBJ, we have assembled as complete a package as possible to help you motivate and teach your students human anatomy and physiology. We feel that this package is a powerful beginning, and we hope to improve upon it as time goes on. We invite your comments, corrections, and suggestions on how we can do so. Please do not hesitate to contact us directly at the Department of Biology, California State University, Northridge, 18111 Nordhoff Street, Northridge, California 91311, or through Ms. Cathleen Petree, Acquisitions Editor, Harcourt Brace Jovanovich, 1250 Sixth Avenue, San Diego, California 92101.

Acknowledgments
Scores of people have been involved in the production of this text. Each one has made a unique contribution to improving the quality of the book. We would first like to thank James Cotanche, M.D., and Josephine Ryan, D.N.Sc., for providing the clinical material incorporated in each chapter. Their expertise, their choices of examples, and their skill at explaining complex material have provided a valuable addition to this text. We would also like to thank Robert Neveln, M.D., and Frederick Soldau, D.D.S., for their valuable consultation on the clinical material.

We are very grateful to Kathleen French and William Kristan, University of California, San Diego, and Annalisa Berta, San Diego State University, for their expert suggestions and improvements to the art program.

We thank Shirley Jones for spending so many long and late hours typing much of the manuscript and all of the glossary.

Editors Bill Bryden and Dick Morel gave us valuable advice in the formative stages of the book's development, and we thank them for their encouragement and assistance.

As their successor, Acquisitions Editor Cathleen Petree has done an outstanding job in guiding the development of this text and in overseeing its completion. She is innovative, energetic, tactful, and capable. We are fortunate to have her as Acquisitions Editor.

We thank Candace Young, Art Editor, and Vicki Kelly, her assistant, for their perserverance in finding and obtaining the photographic illustrations in the text. Candace's ability to organize and coordinate such a complex task borders on the awe-inspiring.

Don Fujimoto, Art Director, was responsible for the design of the book, and has done a beautiful job. He also commissioned and oversaw the many artists who have been involved in this book. Thanks to him, the art is attractive and effectively supports the text. We would like to thank the artists responsible for the outstanding artwork in the book: Keith Kasnot, Darwen and Vally Hennings, Sandra McMahon, and Signature Design Associates.

Lynne Bush, Production Manager; Karen Denhams, Production Editor; Amy Krammes, Promotion; and Nancy Evans, Marketing Manager, have played key roles in developing, producing, and marketing the text and supplementary materials. Without exception, they have been dedicated, competent, efficient, and a pleasure to work with, and we are grateful to them.

However, of all the people involved in the production of this book, Kathy Walker must be singled out for special thanks. As Manuscript Editor, she has been the major force in assuring consistency in style, organization, and format of the text; in assuring that illustrations and text match; and in coordinating the myriad of activities associated with a project of this magnitude. In the course of her involvement with this book, she has learned considerable biology and has made many valuable suggestions regarding the book's content. Through all of this, her attention to detail, her patience, and her commitment to the project have never waned. She has become a good friend, and we are deeply indebted to her.

Reviewers

In the course of this book's development, the following reviewers read, corrected, and commented on individual chapters or on the entire manuscript. They made many useful suggestions on the book's content and organization and on ways to improve the clarity or accuracy of various passages. We are grateful for their help and encouragement. We often incorporated their comments. Nevertheless, there were times when we disagreed on the best way to present a subject or on an interpretation of a fact, and we take responsibility for any errors, omissions, or lack of clarity that remain.

Robert Allen	De Anza College, CA
James Averett	Nassau Community College, NY
John Baust	University of Houston
Mary Jane Burge	Cuyahoga Community College (Metro), OH
Ken Bynum	University of North Carolina, Chapel Hill
Lou Ann Clark	Lansing Community College, MI
John Cunningham	Keene State University, NH
Russell P. Davis	University of Arizona
Gilbert Desha	Tarrant County Community College (South), TX
Mary Desmond	Villanova University, PA
R. Scott Dunham	Illinois Central College
Charles Ellis	formerly of Iowa State University
Ira Fowler	University of Kentucky
J. E. Kendrick	University of Wisconsin, Madison
Robert J. Laird	University of Central Florida
Tom Leslie	Saddleback College, CA
Harvey Liftin	Broward Community College, FL
Chris Loretz	State University of New York at Buffalo
James Love	College of Du Page, IL
Elden Martin	Bowling Green State University, OH
Dennis Meerdink	University of Arizona
Robert E. Nabors	Tarrant County Community College (South), TX
Richard Northrup	Boston College
Ethel Sloane	University of Wisconsin, Milwaukee
John Venable	Colorado State University, Ft. Collins
Jacob Wiebers	Purdue University, IN
Don Wheeler	Cuyahoga Community College (Metro), OH

Preface to the Student

Prefaces presumably are intended to be read first, since they are always placed at the beginning of a book. However, this preface should be read after you have spent some time browsing through the pages that follow. As you do so, notice the way the book has been organized and the many elements that we have included to help you in your study of human anatomy and physiology.

In writing this book, we have been guided by the experiences of students whom we have taught over the past two decades. Our classroom and laboratory encounters showed us a need for a text that not only presents and organizes information, but also motivates and helps the student in a significant way to understand the subject, its implications, and its relevance. Our experience has been that many students confuse reading with studying, and often can't understand why reading a chapter one or two times is not sufficient for a thorough understanding of a subject. Unless one is an exceptional student, reading alone will not provide the insights and understanding that a deeper and more intensive examination will provide. This is as true in anatomy and physiology as it is in any subject.

By the time you have completed your study of this text, you will realize that the human body is exceedingly complex in structure and function, yet not so complex as to defy analysis. Regardless of the detail in which any of the topics are described, you should be aware that much more could have been written and that our coverage can only be considered an introduction to human anatomy and physiology. By the time you read this, even more will be known about many of the mechanisms, processes, and structures we have described. Like any science, human anatomy and physiology is a dynamic field, constantly growing and changing in the light of new information. We hope that this introductory textbook will provide you with a foundation for understanding and appreciating new discoveries as they develop.

Organization and Elements
The text is organized into seven logical groupings of chapters. Unit I, not surprisingly, focuses on fundamental principles that underlie the subject matter covered in the remaining units. In Chapter 1, you will be introduced to some of the basic terminology of anatomy and physiology and be given an overview of the subjects. Chapters 2, 3, and 4 deal with the fundamental principles of chemistry, cell structure and function, and the organization of cells into tissues. These subjects are the foundation of everything that follows in the text. Units II through VI deal with anatomical structures and physiological processes involved in the development and maintenance of the body, including the organs used for support and movement, integration and control of body processes, supply of nutrients and oxygen, transport and distribution, waste excretion, and maintenance of the internal environment. Unit VII describes the structures and mechanisms of reproduction, development, and heredity.

Each chapter begins with a brief introduction and a list of the important concepts in the chapter. The chapter itself is divided into sections that logically break down the subject into major subdivisions, with each section preceded by a short list of learning objectives. Within each section, key words are introduced in boldface type and are accompanied by the pronunciation. Etymologies (word origins) for selected terms are provided in the upper margin on the page where the term is introduced. A narrative outline at the end of each chapter summarizes the chapter according to primary and secondary headings. A set of objective and essay questions is provided at the end of each chapter, with answers to the objective questions appearing in an appendix at the end of the text.

Supplementing the information in each chapter are "boxed" sections that describe several clinical conditions related to the chapter topic. You may find that this material heightens your interest in the subject and provides clinical relevance by describing what happens to the body when something malfunctions. In addition, embryologic development is included in boxes in those chapters that deal with anatomical structures and systems.

How to Use the Text
When beginning each chapter, read the introduction carefully, being sure that you grasp the general significance of the subject and the points to be covered. As you read the chapter, stop occasionally and reflect back on the introduction to remind yourself of the importance and role of the information you are reading. As you begin each section, read the objectives carefully, noting terms that are new and unfamiliar and the purpose of each objective. It might be useful to jot them down so they will be immediately available as you read the section that follows. As you read the section, identify the points that relate to the objectives. When you finish the section, stop, think about the objectives, and see if you can meet them. Don't be afraid to go back and dig out the answer if you can't remember certain points.

After finishing the chapter, slowly read and think about each statement in the Study Outline. Since it is a summary, much of the detailed information is omitted. Try to remember that information, and again, don't hesitate to go back into the chapter to refresh your memory on points you can't recall. Ask yourself how each statement fits into the subject as a whole and what objectives it meets.

Then, to make certain you have grasped the specific information in the chapter, answer the questions in the objective examination. The answers are provided in the back of the book, but wait until you have finished the entire exam before referring to them. For questions that you answer incorrectly, go back into the chapter and find the statements that provide the basis for the correct answer. Perhaps the best indicator of how well you understand the material is whether you can explain a subject in your own words. For that purpose, several essay questions have been provided for you to test your ability to explain a principle or to predict an outcome on the basis of the information learned in the chapter. For some students, discussion with other students in the class is a helpful way to develop the intellectual involvement on which understanding is based. Thus, we encourage small group study. The essay questions can provide a basis for those discussions.

In addition to the learning aids included in the text, a Study Guide has been prepared that is keyed to the text and that provides additional questions, exercises, applications, and activities. Computer software, an anatomy coloring book, and a text on medical terminology for the health professions are also available to help you throughout the course.

One Final Note
We are anxious that this text be as accurate, understandable, and useful as a textbook can possibly be. If you have suggestions on how we might improve it in any way, please do not hesitate to contact us. We can be reached by writing to the Department of Biology, California State University, Northridge, 18111 Nordhoff Street, Northridge, California 91330. We will appreciate your suggestions.

Guide to Pronunciation
Pronunciation of technical terms in this textbook is based on the system used in *Taber's Cyclopedic Medical Dictionary*, 15th edition, published by F. A. Davis Company, Philadelphia. Two *diacritics* (marks over the vowels) are used. The *macron* ¯ indicates the long sound of a vowel, as a in fāte, e in bēlief, i in bīcycle, o in ōvary, and u in ūnion. The *breve* ˘ indicates the short sound of a vowel, as a in ăpple, e in ĕlbow, i in ĭs, o in hŏt, and u in sŭpper. Syllables are separated from one another by a bullet ·, unless they are accented. A syllable that receives primary emphasis is indicated with a ′ and one that receives less stress is indicated with a ″. Thus, the pronunciation of *hypothalamus*, for example, would be indicated as hī″pō·thăl′ă·mŭs.

Brief Contents

Unit I
Organization of the Human Body 2

1	Introduction to the Human Body	4
2	The Chemical Basis of Life	18
3	Cell Structure and Organization	46
4	Tissues	80

Unit II
Support and Movement Systems 98

5	The Integumentary System	100
6	Development and Physiology of the Skeletal System	114
7	Anatomy of The Skeletal System	130
8	Articulations: The Joints Between Bones	164
9	Muscles: Cellular Organization and Physiology	180
10	Musculature of the Human Body	206

Unit III
Nervous and Chemical Integration and Control Systems 258

11	The Nervous System: Neurons and Impulses	260
12	The Central Nervous System	290
13	The Peripheral Nervous System	322
14	The Autonomic Nervous System	342
15	Functional Aspects of the Nervous System	356
16	Special Sense Organs	376
17	The Endocrine System	406

Unit IV
Metabolic Processes: Digestion and Respiration 438

18	The Digestive System	440
19	Metabolic Processes	478
20	The Respiratory System	504

Unit V
Transport and Circulatory Systems 536

21	The Blood	538
22	The Cardiovascular System: The Heart	558
23	The Cardiovascular System: Blood Vessels	584
24	The Lymphatic System and Immunity	618

Unit VI
Homeostatic Control of Body Fluids 644

25	The Urinary System	646
26	Composition and Control of Body Fluids	680

Unit VII
Reproduction: The Continuity of Life 696

27	Reproductive Organs and Their Functions	698
28	Development and Inheritance	732

Unit I

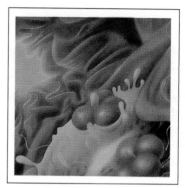

Organization of the
Human Body   2

Detailed Contents

1 Introduction to the Human Body	**4**
The Fields of Anatomy and Physiology	5
Structural Plan of the Human Body	7
Atoms and Molecules	7
Cells	7
Tissues	7
Organs	7
Systems	7
Anatomic Regions and Positions	7
The Characteristics of Life	10
Organizational Complexity	10
Metabolism	10
Homeostasis	12
Irritability	16
Reproduction	16
2 The Chemical Basis of Life	**18**
The Nature of Matter	19
Chemical Elements	19
Atoms	19
Isotopes	20
Atomic Mass	21
Electron Orbitals	21
Ions	23
Chemical Bonds	23
Ionic Bonds	23
Covalent Bonds	23
Hydrogen Bonds	25
Molecules	25
Chemical Reactions	26
Rearrangement Reactions	26
Addition Reactions	28
Cleavage Reactions	28
Transfer Reactions	28
Oxidation-Reduction Reactions	28
Metabolic Reactions	28
Factors Affecting Reaction Rate	29
Concentration	29
Enzymes	29
Hydrogen Ion Concentration	30
Acid and Bases	31
Buffers	32
Categories of Organic Compounds	33
Carbohydrates	34
Lipids	35
Fatty Acids	35
Neutral Fats	35
Phospholipids	36
Steroids	36
Proteins	37
Nucleic Acids	37
3 Cell Structure and Organization	**46**
Cellular Basis of Life	47
Plasmalemma	47
Organization	47
Junctions	47
Movement Through the Cell Membrane	50
Effects of Size, Charge, and Solubility	50
Osmosis	51
Facilitated Diffusion	54
Active Transport	54
Mechanisms of Transport	55
Exocytosis and Endocytosis	55
Contents of the Cytoplasm	57
Endoplasmic Reticulum and Ribosomes	57
Golgi Apparatus and Lysosomes	57
Peroxisomes	59
Mitochondria	59
Cytoskeleton	61
Cilia and Flagella	63
The Nucleus and Its Contents	64
Nuclear Envelope	64
Nucleolus	64
Chromatin	65
Cell Division	66
Cytoplasmic and Nuclear Division	66
Mitosis	67
Prophase	67
Metaphase	67
Anaphase	68
Telophase	68
DNA Function	68
Replication	68
Transcription	70
Translation	70

4 Tissues	80
Epithelial Tissue	81
Simple Epithelial Tissue	82
Pseudostratified Epithelial Tissue	83
Stratified Epithelial Tissue	83
Glandular Epithelium	84
Functional Classification of Exocrine Glands	84
Structural Classification of Exocrine Glands	85
Connective Tissue	86
Embryonic Versus Adult Connective Tissue	87
Adult Connective Tissue Proper	87
Loose Connective Tissue	88
Dense Connective Tissue	90
Blood	90
Cartilage	90
Bone	90
Muscle Tissue	91
Smooth Muscle	91
Skeletal Muscle	91
Cardiac Muscle	92
Nervous Tissue	92

Unit II

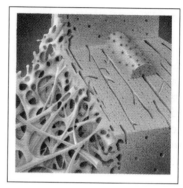

Support and Movement Systems 98

5	**The Integumentary System**	**100**

Epidermis	101
Epidermal Layers	101
Pigmentation	103
Dermis	105
Epidermal Derivatives	106
Nails	106
Hair	107
Glands of the Skin	108
Functions of the Skin	109
Protection	109
Excretion	110
Temperature Regulation	110
Production of Vitamin D	110
Sensations	110

6	**Development and Physiology of the Skeletal System**	**114**

Functions of the Skeletal System	115
Histology of Bone	115
Classification Based on External Appearance	115
Classification Based on Internal Structure	115
Development of the Skeleton	119
Development of Cartilage	119
Development of Bone	119
Intramembranous Ossification	120
Endochondral Ossification	121
Physiological Maintenance of the Skeleton	123
Osteogenesis and Osteoclasis	123
Remodeling of the Skeletal System	124

7	**Anatomy of the Skeletal System**	**130**

Divisions of the Skeleton	131
Skull	131
Cranial Group	133
Frontal	133
Parietals	133
Temporals	133
Occipitals	136
Sphenoid	138
Ethmoid	139
Facial Group	140
Inferior Nasal Conchae	140
Vomer	140
Lacrimals	140
Nasals	140
Maxillae	140
Palatines	141
Zygomatics	141
Mandible	141
Sutures and Wormian Bones	142
Hyoid Bone	142
Vertebral Column	143
Curves of the Vertebral Column	143
A Typical Vertebra	143
Types of Vertebrae	143
Cervical Vertebrae	143
Thoracic Vertebrae	145
Lumbar Vertebrae	145
Sacrum	145
Coccyx	147
Thorax	147
Sternum	147
Ribs	147
Pectoral Girdle	148
Clavicle	148
Scapula	148
Upper Extremity	149
Humerus	149
Ulna and Radius	150
Manus	150
Carpus	150
Metacarpus	152
Phalanges	152
Pelvic Girdle	152
Os Coxae	152
Differences Between Female and Male Pelvic Bones	154
Lower Extremity	154
Femur	154
Patella	155
Tibia	155
Febula	155
Pes	155
Tarsus	155
Metatarsus	155
Phalanges	155
Arches of the Foot	155

8 Articulations: The Joints Between Bones — 164

- Articulations — 165
  - Functional Classification — 165
  - Structural Classification — 165
    - Fibrous Joints — 165
    - Cartilaginous Joints — 166
    - Synovial Joints — 167
  - Movements of Synovial Joints — 168
    - Gliding Joints — 168
    - Hinge Joints — 168
    - Ellipsoidal Joints — 168
    - Pivot Joints — 169
    - Ball-and-Socket Joints — 169
    - Saddle Joints — 170
  - Movements of the Skeleton — 170
  - The Knee: A Special Joint — 174

9 Muscles: Cellular Organization and Physiology — 180

- Organization of Skeletal Muscle — 181
  - Motor Units — 183
  - Structure of the Skeletal Muscle Fiber — 183
  - Sarcomeres — 184
    - Molecular Organization of Filaments — 186
    - Membranes of the Sarcomere — 187
- Muscle Contraction — 188
  - Initiation of Contraction — 189
    - Removal of Acetylcholine — 189
    - All-or-None Effect — 190
  - Mechanism of Contraction — 190
  - Relaxation — 192
- Physiology of Muscle Contractions — 192
  - Twitches — 193
  - Energy Supply for Contraction — 195
    - Fast Twitch Fibers — 195
    - Slow Twitch Fibers — 196
  - Oxygen Debt — 196
  - Heat — 197
  - Chemical Waste Products — 197
- Cardiac Muscle — 197
- Smooth Muscle — 198
- Effects of Training on Muscles — 199
  - Strength and Endurance — 200
  - Isometric and Isotonic Contractions — 200

10 Musculature of the Human Body — 206

- Muscles and Their Movements — 207
  - Naming the Muscles — 207
  - Shapes of Skeletal Muscles — 207
  - Origin, Insertion, and Action — 207
  - Levers — 209
- Muscles of the Head and Neck — 214
  - Muscles of Facial Expression — 214
  - Muscles That Move the Eyeball — 216
  - Muscles of Mastication — 218
  - Muscles of the Tongue — 219
  - Muscles of the Neck — 220
- Muscles of the Trunk — 222
  - Muscles of the Vertebral Column — 222
  - Muscles of the Thoracic Wall — 224
  - Muscles of the Abdominal Wall — 226
  - Muscles of the Pelvic Floor — 228
- Muscles of the Pectoral Girdle and Upper Extremity — 230
  - Muscles That Move the Scapula — 230
  - Muscles That Move the Humerus — 232
  - Muscles That Move the Forearm — 234
  - Forearm Flexors of the Wrist and Hand — 236
  - Forearm Extensors of the Wrist and Hand — 238
  - Intrinsic Muscles of the Hand — 240
- Muscles of the Lower Extremity — 242
  - Muscles That Move the Thigh — 242
  - Muscles That Move the Leg — 246
  - Muscles That Move the Foot and Toes — 248
  - Intrinsic Muscles of the Foot — 250

Unit III

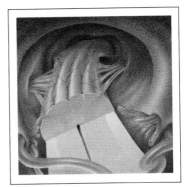

Nervous and Chemical Integration and Control Systems 258

11 The Nervous System: Neurons and Impulses — 260

- Organization of the Nervous System — 261
  - The CNS and the PNS — 261
  - Organization of a Nerve — 261
- Cells of the Nervous System — 262
  - Neurons — 262
  - Neuroglia — 264
    - Neuroglia of the CNS — 265
    - Neuroglia of the PNS — 265
- Membrane Potential — 266
  - Electric Potentials — 266
  - Resting Potentials — 267
- An Action Potential — 269
  - Depolarization — 269
  - Repolarization — 269
- Propagating an Action Potential — 270
  - Initiating an Impulse — 271
  - Speed of an Impulse — 271
  - Volleys — 274
- Transmission from Neuron to Neuron — 274
  - Communication Across a Synapse — 276
  - Release of Neurotransmitters — 277
  - Mode of Action of Neurotransmitters — 277
  - Removal of Neurotransmitters — 277
  - Electrical Synapses — 278
- Postsynaptic Membrane Potentials — 278
  - Excitatory Postsynaptic Potential — 278
    - Spatial Summation — 278
    - Temporal Summation — 278
  - Inhibitory Postsynaptic Potential — 278
- Integrating the Stimuli — 279
  - Divergence and Convergence — 279
  - Encoding — 280
- Neurotransmitters — 281

12 The Central Nervous System — 290

- Telencephalon (Cerebrum) — 291
  - Cerebral Cortex — 291
  - Functional Areas of the Cerebral Cortex — 295
    - Sensory Areas — 295
    - Motor Areas — 296
    - Association Areas — 296
  - White Matter of the Cerebrum — 296
  - Gray Matter of the Cerebrum — 297

Diencephalon	299
Thalamus	299
Hypothalamus	300
Epithalamus	301
Limbic System	301
Mesencephalon	302
Corpora Quadrigemina	303
Cerebral Peduncles	303
Metencephalon	303
Cerebellum	303
Pons	305
Myelencephalon	305
Medulla Oblongata	305
Reticular Formation	306
Ventricles and Meninges of the Central Nervous System	307
Ventricles of the Brain	307
Meninges of the Brain and Spinal Cord	309
Dura Mater	309
Pia Mater	309
Arachnoid	310
Blood Supply	310
Spinal Cord	311
Anatomy of the Spinal Cord	311
Gray and White Matter of the Spinal Cord	312
Ascending and Descending Fasciculi	313
Spinal Cord Reflexes	314
Monosynaptic Reflexes	314
Polysynaptic Reflexes	317

13 The Peripheral Nervous System — 322

Cranial Nerves	323
Olfactory Nerves (I)	324
Optic Nerves (II)	325
Oculomotor Nerves (III)	326
Trochlear Nerves (IV)	324
Trigeminal Nerves (V)	328
Abducens Nerves (VI)	329
Facial Nerves (VII)	329
Vestibulocochlear Nerves (VIII)	329
Glossopharyngeal Nerves (IX)	330
Vagus Nerves (X)	331
Accessory Nerves (XI)	332
Hypoglossal Nerves (XII)	332
Spinal Nerves	333
Anatomy of Adult Spinal Nerves	333
Peripheral Regions Innervated by Spinal Nerves	334
Cervical Plexus	334
Brachial Plexus	336
Lumbosacral Plexus	338

14 The Autonomic Nervous System — 342

Distribution of Neurons	343
Anatomy of the Autonomic Nervous System	343
Sympathetic System	344
Parasympathetic System	344
Physiology of the Autonomic Nervous System	347
Neurotransmitters	347
Membrane Receptors	347
Control Through Dual Innervation	349
Sensory Input into the Autonomic Nervous System	349
Conscious Control of the Autonomic Nervous System	349
Biofeedback	351
Forms of Meditation	351

15 Functional Aspects of the Nervous System — 356

Sensory Reception	357
An Overview of Sensory Reception and Perception	357
Classification of Senses	357
Classification of Receptors	357
Stimulation of Receptors	357
Adequate Stimulus	357
Receptor Potentials	357
Adaptation and Afterimage	358
Projection	358
General Sense Receptors	359
Mechanoreceptors	359
Touch Receptors	359
Pressure Receptors	360
Thermoreceptors	360
Nociceptors	360
Proprioceptors	360
Sensory Pathways	361
Touch and Pressure	362
Proprioception	362
Temperature and Pain	363
Motor Pathways	364
Reflex Motor Pathways	364
Voluntary Motor Pathways	364
Pyramidal Pathways	365
Extrapyramidal Pathways	366
The Diver	366

Integrative Functions of the Brain	367
Consciousness Versus Unconsciousness	367
Sleep	369
Memory	370
Short-Term Memory	370
Long-Term Memory	371
Learning	372

16 The Special Sense Organs — 376

Olfaction: The Sense of Smell	377
Stimulation of Olfactory Receptors	377
Olfactory Adaptation	378
Olfactory Pathway	378
Gustation: The Sense of Taste	379
Gustatory Cells	379
Gustatory Pathway	380
Anatomy of the Eye	380
The Eyeball	380
Fibrous Tunic	380
Vascular Tunic	380
Retina	382
Lens	383
Chambers of the Eyeball	385
Accessory Structures of the Eye	385
Eyebrows	385
Eyelids	385
Lacrimal Gland	386
Physiology of Vision	387
Formation of the Image	387
Visual Accommodation	387
Convergence and Binocular Vision	390
Depth of Field	391
Biochemistry of Vision	391
Visual Pigments in Rods	391
Visual Pigments in Cones	392
Visual Pathways to the Brain	393
Anatomy of the Ear	393
External Ear	393
Middle Ear	393
Internal Ear	396
Semicircular Canals and Vestibule	396
Cochlea	396
Innervation	398
Physiology of Hearing	398
Sound	398
Transmission of Sound to the Inner Ear	398
Functions of the Cochlea	398
Sensory Pathways for Hearing	399
Physiology of Equilibrium	400
Static Equilibrium	400
Dynamic Equilibrium	401
Nerve Pathways in Equilibrium	401

17 The Endocrine System — 406

Hormones	407
Chemical Composition of Hormones	407
Concentration of Hormones	407
Source of Hormones	407
Mechanism of Action	408
Adenylate Cyclase	408
Steroid Hormones	410
Regulation of Hormone Activity	411
Hypothalamus and Pituitary Gland	412
Hormones of the Posterior Pituitary Gland	413
Antidiuretic Hormone	413
Oxytocin	414
Hormones of the Anterior Pituitary Gland	415
Prolactin	416
Growth Hormone	417
Melanocyte-Stimulating Hormone	417
Adrenocorticotrophic Hormone (ACTH)	417
Thyroid-Stimulating Hormone (TSH)	417
Gonadotrophins	417
Thyroid and Parathyroid Glands	418
Anatomy of the Thyroid Gland	418
Thyroid Hormones	418
Effects of Thyroid Hormones	419
Calcitonin	420
Parathyroid Glands	420
Adrenal Glands	422
Adrenal Cortex	422
Hormones of the Zona Glomerulosa	423
Hormones of the Zona Fasciculata	424
Hormones of the Zona Reticularis	424
Adrenal Medulla	424
Pancreas	426
Insulin	426
Glucagon	428
Somatostatin	429
Other Endocrine Glands	429
Gonads	429
Thymus Gland	429
Pineal Gland	431
Sources of Additional Hormones	431
Prostaglandins	431

UNIT IV

Metabolic Processes: Digestion and Respiration 438

18	**The Digestive System**	**440**
	Histology	441
	Mouth	443
	Tongue	444
	Gingivae and Teeth	444
	Salivary Glands	447
	Digestion in the Mouth	448
	Pharynx and Esophagus	449
	Pharynx	449
	Esophagus	449
	Swallowing	449
	Buccal Stage	449
	Pharyngeal Stage	449
	Esophageal Stage	449
	Stomach	451
	Digestion in the Stomach	451
	Source and Composition of Gastric Juice	452
	Cardiac and Gastric Glands	452
	Pyloric Glands	453
	Control of Gastric Gland Secretion	454
	Cephalic Phase	454
	Gastric Phase	454
	Intestinal Phase	454
	Control Of Stomach Emptying	454
	Neural Control	454
	Hormonal Control	455
	Small Intestine	455
	Digestion in the Small Intestine	457
	Neutralization of Chyme	458
	Pancreatic Enzymes	458
	Protein Digestion	459
	Carbohydrate Digestion	459
	Fat and Nucleic Acid Digestion	459
	Role of Bile	459
	Intestinal Enzymes	460
	Carbohydrate Digestion	460
	Peptide Digestion	460
	Fat and Nucleic Acid Digestion	460
	Absorption	460
	Absorption of Carbohydrate	461
	Absorption of Amino Acids	462
	Absorption of Fat	462
	Absorption of Water and Minerals	463
	Large Intestine	464
	Digestion in the Large Intestine	466
	Intestinal Flora	466
	Defecation	467

Accessory Structures	468
Liver	468
Gallbladder	470
Pancreas	470
Membranes of the Abdominal Cavity	471

19 Metabolic Processes — 478

Overview of Metabolism	478
Carbohydrate Metabolism	479
Metabolism of Glucose	480
Glucose Storage	480
Glucose Release	480
Control of Glycogen Synthesis and Degradation	480
Catabolism of Glucose	480
Glycolysis	481
Fate of Pyruvic Acid	482
Krebs Cycle	482
Respiratory Chain	483
Oxidative Phosphorylation	484
An Overview of Cellular Respiration	485
Glucose as a Source of Carbon	485
Lipid Metabolism	487
Transport of Lipids	487
Storage of Fats	487
Metabolism of Fats	488
Catabolism of Fats	488
Anabolism of Fats	490
Protein and Amino Acid Metabolism	491
Cofactors in Metabolism	492
Minerals	492
Vitamins	492
Water-Soluble Vitamins	493
Fat-Soluble Vitamins	493
Heat as a By-Product of Metabolic Reactions	493
Temperature	494
Production and Distribution of Heat	494
Regulation of Body Temperature	494
Control by the Hypothalamus	494
Mechanisms for Warming the Body	495
Mechanisms for Cooling the Body	496
Heat Loss	496
Basal Metabolic Rate	497

20 The Respiratory System — 504

Organization of the Respiratory Tract	505
Upper Respiratory Tract	506
Nasal Cavity	507
Pharynx	508
Larynx	509
Voice and Singing	509
Lower Respiratory Tract	510
Gross Morphology of the Lungs	510
Pulmonary Tree	512
Conducting Division	513
Respiratory Division	513
Structure of the Alveolar Wall	514
Breathing Mechanics	516
Inspiration and Expiration	516
Respiratory Volumes	517
Properties of Gases	519
Gas Pressure	519
Partial Pressure	519
Ideal Gas Laws	519
Solubility of Gases in Liquids	520
Gas Exchange in Alveoli	521
Factors Affecting Gas Exchange	521
Oxygen Transport	521
Structure of Hemoglobin	523
Oxygen Binding by Hemoglobin	523
Factors Affecting Oxygen Release	524
Carbon Dioxide and Hydrogen Ion Transport	524
Regulation of Respiration	525
Neural Control of Breathing	526
Peripheral Receptors	526
Chemical Regulation of Breathing	527
Peripheral Chemoreceptors	527
Medullary Chemoreceptors	527
Interaction Between Medullary and Peripheral Chemoreceptors	528

Unit V

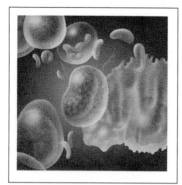

Transport and Circulatory Systems 536

21 The Blood — 538

- Functions, Composition, and Production of Blood — 539
  - Functions of Blood — 539
  - Composition of Blood — 539
    - Plasma — 539
    - Formed Elements — 541
  - Production of Blood Cells — 541
- Erythrocytes — 543
  - Structure — 543
  - Function — 543
  - Production — 544
  - Replacement — 545
  - Control of Erythrocyte Production — 545
  - Blood Groups — 546
- Leukocytes — 546
  - Structure — 546
  - Function — 547
  - Production — 548
  - Replacement — 548
  - Control of Leukocyte Production — 548
- Thrombocytes — 549
  - Function — 549
  - Production — 549
- Hemostasis — 550
  - Vascular Constriction — 550
  - Platelet Plug Formation — 550
  - Coagulation — 550
    - Intrinsic Pathway — 552
    - Extrinsic Pathway — 553
    - Enhancement of Clot Formation — 553
  - Factors Affecting Clotting Time — 553
  - Limiting Growth of the Clot — 553
  - Retraction and Dissolution of the Clot — 553
  - Preventing Abnormal Clotting — 553

22 The Cardiovascular System: The Heart — 558

- Anatomy of the Heart — 559
  - Pericardium — 559
  - Muscular Walls of the Heart — 561
  - Chambers and Valves of the Heart — 561
    - Right Atrium — 562
    - Right Ventricle — 564
    - Left Atrium — 564
    - Left Ventricle — 566

Cardiac Intrinsic Blood Supply	566
Cardiac Physiology	567
Myocardium	568
Conduction System of the Heart	568
Cardiac Contraction	568
Electrocardiography	570
Cardiac Cycle	572
Mechanical Changes During the Cardiac Cycle	572
Pressure Changes During the Cardiac Cycle	573
Heart Sounds	574
Cardiac Output	574
End-Systolic Volume	575
End-Diastolic Volume	575
Starling's Law of the Heart	576
Physiologic Control of Heart Rate	576
Neural Control	576
Additional Factors Regulating Heart Rate	577
Temperature	578
Chemicals	578
Psychological Factors	578
Age and Sex	578

23 The Cardiovascular System: Blood Vessels   584

Blood Vessels	585
Arteries	585
Arterioles	586
Capillaries	586
Venules	586
Veins	587
Anatomy of the Circulatory Pathways	587
Pulmonary Circulation	589
Systemic Circulation: Arteries	589
Aorta	591
Arteries of the Head and Neck	591
Arteries of the Upper Extremities	591
Arteries of the Thorax	591
Arteries of the Abdomen	591
Arteries of the Lower Extremities	591
Systemic Circulation: Veins	596
Veins of the Head and Neck	597
Veins of the Upper Extremities	600
Veins of the Thorax	600
Veins of the Abdominal Cavity	601
Hepatic Portal System	601
Veins of the Lower Extremities	602
Fetal Circulation	603
Arterial Circulation	605
Factors Affecting Blood Pressure	606
Blood Volume	606
Normal Blood Pressure	606
Peripheral Resistance	606
Blood Viscosity	607
Elasticity of the Arteries	607
Control of Blood Pressure	607
Neural Control of Blood Pressure	607
Pressoreceptors and Chemoreceptors	608
Cerebral Control Centers	608
Chemical Control of Blood Pressure	608
Local Control of Blood Pressure	609
Capillary Exchange	610
Hydrostatic Pressure	610
Osmotic Pressure	611
Venous Circulation	611
Venous Valves	611
Muscular Activity	611
Velocity of the Blood	611
Breathing Activities	612
Blood Reservoirs	612

24 The Lymphatic System and Immunity   618

Lymphatic System	619
Lymphatic Vessels	619
Lymph Nodes	620
Lymph Circulation	621
Pathway	621
Mechanisms	621
Lymphatic Organs	622
Spleen	623
Tonsils	624
Thymus Gland	624
Nonspecific Defense Against Disease	624
Skin and Mucous Membrane	625
Mechanical Processes	625
Chemical Processes	625
Nonspecific Chemicals	625
Interferon	625
Complement	626
Properdin	626
Phagocytosis	626
Inflammation	627

Specific Defense Against Disease	627
Antigens	627
Structure	627
Activity	627
Antibodies	628
Structure	628
Activity	628
Cellular and Humoral Immunity	629
Cellular Immunity and T cells	629
T Cell Lymphocytes	629
Cellular Immune Response	631
Natural Killer Cells	632
Humoral Immunity and B Cells	632
B Cell Lymphocytes	632
Humoral Immune Response	632
Organ Transplants	633
Immunity and Vaccination	634
Induced Immunity	634
Natural Immunity	634
Active and Passive Immunity	635
Monoclonal Antibodies	635
Blood Groups	636
ABO Blood Groups	636
Transfusions	637
Rh Blood Group	637

UNIT VI

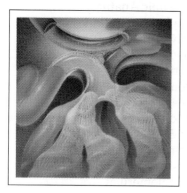

Homeostatic Control of Body Fluids   644

25 The Urinary System — 646

- Anatomy of the Kidneys — 647
  - External Structure — 647
  - Internal Organization — 649
  - Microscopic Anatomy — 650
- The Nephron — 651
  - Renal Corpuscle — 651
    - Bowman's Capsule — 652
    - Glomerulus — 652
    - Glomerular Basement Membrane — 653
  - Tubular Portion of a Nephron — 653
  - Location and Dimensions of Nephrons — 653
  - Blood Supply to a Nephron — 654
    - Distribution Within the Kidney — 654
    - Passage of Blood Through a Glomerulus — 654
    - Juxtaglomerular Apparatus — 654
    - Blood Distribution after Leaving a Glomerulus — 655
  - Nerve and Lymphatic Supply to a Nephron — 656
- Physiology of the Kidneys — 657
  - Filtration — 657
    - Glomerular Blood Pressure — 657
    - Measurement of Glomerular Filtration Rate — 658
  - Function of the Tubular Portion of a Nephron — 659
    - Reabsorption — 659
    - Secretion — 661
  - Concentrating the Urine — 662
    - Countercurrent Concentrating Mechanism — 662
    - Role of the Vasa Recta — 664
    - Regulation of Urine Concentration and Water Excretion — 665
- Homeostatic Functions of the Kidneys — 665
  - Renal Regulation of Sodium Content — 666
  - Control of Blood Pressure — 667
  - Renal Regulation of pH — 667
  - Potassium Absorption and Secretion — 668
  - Regulation of Red Blood Cell Production — 668
  - Vitamin D Activation — 668
- Conduction, Storage, and Elimination of Urine — 669
  - Conduction through the Ureters — 669
  - Urinary Bladder — 670
  - Urethra — 670
  - Micturition — 672
- Composition of Urine — 673

26 Composition and Control of Body Fluids — 680

- Regulating Water Content — 681
  - Extracellular and Intracellular Fluids — 681
  - Water Intake — 681
  - Water Output — 681
- Composition of Body Fluids — 682
  - Electrolytes — 682
  - Major Cations in Body Fluid — 683
    - Sodium — 683
    - Potassium — 684
    - Calcium — 685
    - Magnesium — 686
    - Other Cations — 687
  - Major Anions in Body Fluids — 687
- Acid-Base Balance — 687
  - pH Buffers — 687
    - Carbonic Acid–Bicarbonate Buffer — 687
    - Phosphate Buffers — 687
    - Hemoglobin-Oxyhemoglobin Buffer — 687
    - Protein Buffer — 687
  - Respiratory Ventilation — 689
  - Kidney Control — 689
- Acid-Base Imbalance — 689
  - Acidosis — 689
  - Alkalosis — 690

Unit VII

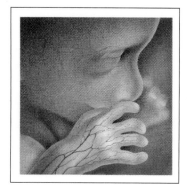

Reproduction: The Continuity of Life 696

27	**Reproductive Organs and Their Functions**	**698**
	Anatomy and Function of the Male Reproductive Tract	699
	Scrotum	699
	Testes	699
	Epididymis	702
	Ductus Deferens and Ejaculatory Duct	703
	Urethra	703
	Production and Ejaculation of Sperm	703
	Hormonal Control of Sperm Production	705
	Testosterone	705
	Inhibin	706
	Accessory Glands	706
	Composition of Semen	706
	Penis	707
	Anatomy and Function of the Female Reproductive Tract	708
	Ovaries	709
	Oviducts	709
	Uterus	711
	Cervix	713
	Vagina	713
	External Genitalia (Vulva)	714
	Mammary Glands	715
	Production and Release of Eggs	717
	Development of Follicles	717
	Fate of Follicles	717
	Hormonal Control of Egg Production	719
	Menstrual Cycle	719
	Menopause	722
	Coitus	722
	Male Sexual Response	722
	Arousal and Erection	722
	Plateau	722
	Orgasm and Ejaculation	722
	Resolution	722
	Female Sexual Response	722
	Arousal	722
	Plateau	723
	Orgasm	724
	Resolution	724
	Fertilization	724
	Capacitation	724
	Fusion of Sperm and Egg	726

28 Development and Inheritance		**732**
Embryonic Development		733
Cleavage		733
Implantation		733
Primary Germ Layers		734
Embryonic Membranes		735
Amnion		735
Yolk Sac		735
Allantois		738
Chorion		738
Placenta and Umbilical Cord		738
Fetal Development		741
Twins		741
Pregnancy and Childbirth		743
Diagnosis of Pregnancy		743
Duration of Pregnancy		744
Changes in Physiology During Pregnancy		745
Cardiovascular Changes		745
Pulmonary Changes		745
Renal Changes		745
Digestive System Changes		745
Endocrine Changes		747
Labor and Delivery		747
Meiosis		749
Meiosis in Males		749
Meiosis in Females		751
Inheritance		753
Genotype and Phenotype		753
Distribution of Genes		753
Multiple Alleles		755
Segregation and Independent Assortment of Genes		755
Chromosomal Linkage		756
RFLP-Based Mapping of Chromosomes		758
Sex Linkage		759
Genetic Disorders due to Chromosomal Abnormalities		761
Chromosomal Nondisjunction		761
Down Syndrome		762
Other Examples of Autosomal Nondisjunction		763
Other Chromosomal Abnormalities		763
Prenatal Detection of Genetic and Chromosomal Abnormalities		764

Appendix I Answers to Self-Tests of Chapter Objectives	772
Appendix II Allied Health Careers	776
Appendix III Word Parts for Building Anatomical and Medical Terminology	783
Glossary	787
Copyrights and Acknowledgments	834
Index	837

List of Clinical Application Boxes

UNIT I
Organization of the Human Body

1 Introduction to the Human Body
- An Introduction — 5
- Measuring the Body — 9
- Autopsy — 12
- The Clinic: Radiographic Anatomy — 14

2 The Chemical Basis of Life
- Photon Imaging — 20
- Radioisotope Scanning — 22
- PET and SPECT: The Future — 24
- Serum Electrolytes — 31
- Saturated Fats — 35
- The Clinic: Vitamins — 41

3 Cell Structure and Organization
- Renal Dialysis — 52
- Genetic Engineering — 70
- The Clinic: Neoplasms: Cells out of Control — 74

4 Tissues
- Scarring — 84
- Abscess — 90
- Ligament Repair — 91
- The Clinic: Collagen Vascular Diseases — 93

Unit I Case Study: A Patient with Hodgkin's Disease — 96

UNIT II
Support and Movement Systems

5 The Integumentary System
- Common Skin Problems — 102
- Viral Skin Infections — 104
- Eczema — 105
- Suntanning — 106
- Burns — 109
- The Clinic: Dermatology and Skin Disorders — 111

6 Development and Physiology of the Skeletal System
- Fractures — 116
- Torn Cartilage — 119
- Fibrous Dysplasia — 121
- Myositis Ossificans — 121
- Epiphyseal Fractures — 123
- Osteomyelitis — 124
- Bone Ulcers — 124
- The Clinic: Diseases of the Skeletal System — 125

7 Anatomy of the Skeletal System
- Craniosynostosis — 133
- Mastoiditis — 134
- Fracture of the Cribriform Plate — 140
- Fracture of the Nose — 140
- Zygomatic Fracture — 141
- Dislocation of the Mandible — 142
- Whiplash — 144
- Abnormalities of the Feet — 156
- The Clinic: Disorders of the Spine — 158

8 Articulations: The Joints Between Bones
- Gout and Pseudogout — 166
- Synovitis — 167
- Bursitis — 168
- Hip Prosthesis — 169
- Rheumatoid Arthritis — 170
- Sports Injuries — 174
- Stability of Joints — 176
- The Clinic: Disorders of the Joints — 177

9 Muscles: Cellular Organization and Physiology
- Tendonitis — 186
- Drug-induced Neuromuscular Junction Block — 190
- Calcium Channel Blockers — 191
- Rigor Mortis — 192
- Abnormal Muscle Contractions — 193
- Fast and Slow Twitch Fibers — 196
- Myoglobinuria — 196
- Serum Creatinine — 197
- Charley Horse — 200
- The Clinic: Trends in Body Building — 201

10 Musculature of the Human Body
- Musculotension Headaches — 214
- Torticollis — 220
- Diaphragmatic Hernias — 224
- Iliotibial Band Friction Syndrome — 242
- Piriformis Syndrome — 242
- Muscle Tears or Pulls — 246
- Torn Plantaris Muscle — 248
- The Clinic: Muscular Diseases — 252

Unit II Case Study: Overuse Syndromes — 256

UNIT III
Nervous and Chemical Integration and Control Systems

11 **The Nervous System: Neurons and Impulses**
- Multiple Sclerosis — 265
- Endorphins and Enkephalins — 281
- Huntington's Disease — 283
- Beta-Blockers — 284
- The Clinic: Diseases of the Nervous System — 286

12 **The Central Nervous System**
- Lobotomy — 296
- Alzheimer's Disease: The False Epidemic? — 298
- Headaches — 299
- Epilepsy — 304
- Brain Death — 305
- Hydrocephalus — 307
- Spinal Tap — 311
- Stroke — 314
- Spinal Cord Injury — 315
- Reflexes — 316
- The Clinic: Examination of the Nervous System — 318

13 **The Peripheral Nervous System**
- Optic Chiasma and Pituitary Tumors — 326
- "Eye Signs" and Diagnosis — 327
- Trigeminal Neuralgia — 328
- Anesthesia — 329
- Bell's Palsy — 330
- Electrodiagnosis — 336
- Sciatica and Ruptured Disks — 338
- The Clinic: Peripheral Neuropathy — 339

14 **The Autonomic Nervous System**
- Substance Abuse — 350
- The Clinic: Influencing the ANS with Drugs — 352

15 **Functional Aspects of the Nervous System**
- Referred Pain — 360
- Control of Pain — 364
- Extrapyramidal Disorders — 366
- Jet Lag — 367
- Sleep Disorders — 369
- The Clinic: Developmental Dyslexia — 373

16 **The Special Sense Organs**
- Anosmia — 379
- Testing Gustation — 380
- Corneal Transplants — 382
- The Eye: Window to the Body — 384
- Contact Lenses — 385
- Diseases Affecting the Eye — 387
- Visual Disorders — 389
- Strabismus — 390
- Color Blindness — 393
- Disturbances to Equilibrium — 402
- The Clinic: Hearing Tests and Disorders — 403

17 **The Endocrine System**
- Diabetes Insipidus — 415
- Growth Abnormalities — 417
- Aldosteronism — 424
- Diseases of the Adrenal Cortex — 425
- Pheochromocytoma — 426
- Hyperinsulinism — 427
- The Clinic: Diabetes Mellitus — 432

Unit III Case Study: Paralysis — 437

UNIT IV
Metabolic Processes: Digestion and Respiration

18 **The Digestive System**
- Teeth and Gum Disorders — 446
- Peptic Ulcer Disease — 452
- Gastritis — 454
- Gastrointestinal Diagnostic Techniques — 456
- Gallstones — 460
- Lactose Intolerance — 460
- Appendicitis — 465
- Ulcerative Colitis — 467
- Dysentery — 467
- Cirrhosis of the Liver — 470
- Hepatitis — 471
- The Clinic: Disorders of the Gastrointestinal Tract — 472

19 **Metabolic Processes**
- Obesity — 488
- Anorexia Nervosa — 489
- Heat Stroke — 496
- Fever — 497
- The Clinic: Diets — 498

20 The Respiratory System

Upper Respiratory Infection	507
Sinusitis	508
Lung Cancer	511
Pneumonia	513
Tuberculosis	514
Atelectasis	515
Pulmonary Embolism	517
Diving Illness	525
Sudden Infant Death Syndrome	526
Hyperventilation and Hypoventilation	527
Altitude Sickness	529
The Clinic: Diseases of the Respiratory System	530
Unit IV Case Study: The Strange Asthmatic	535

Unit V
Transport and Circulatory System

21 The Blood

Malaria	544
Reticulocyte Counts	544
Anemia	546
Complete Blood Count	548
Sedimentation Rate	551
The Clinic: Diseases of the Blood	554

22 The Cardiovascular System: The Heart

Open Heart Surgery	562
Fainting	570
Office Evaluation of the Cardiac Patient	574
Special Cardiac Diagnostic Procedures	575
Congestive Heart Failure	576
The Clinic: Diseases of the Heart	579

23 The Cardiovascular System: Blood Vessels

Thrombosis and Embolism	588
Evaluation of Blood Vessel Disorders	593
Hypertension	607
Aneurysm	607
Shock	609
Varicose Veins	611
The Clinic: Diseases of the Blood Vessels	613

24 The Lymphatic System and Immunity

Mastectomy	622
Splenectomy	623
Hypochondriasis	625
AIDS	630
Eradication of Smallpox	635
Hemolytic Disease of the Newborn	636
The Clinic: Immunological and Allergic Disorders	638
Unit V Case Study: "Heart Patient"	643

Unit VI
Homeostatic Control of Body Fluids

25 The Urinary System

Urinary Tract Imaging Techniques	649
Renal Failure	658
Syndromes Associated with Renal Disease	660
Urinary Calculi	670
Urinalysis	673
The Clinic: Urinary System Disorders	675

26 Composition and Control of Body Fluids

Body Fluid and Electrolyte Imbalance: An Example	684
Repair of Fluid and Electrolyte Imbalance: An Example	685
The Clinic: Diseases Affecting the Body Fluids	691
Unit VI Case Study: Diabetes and Fluid Imbalance	694

Unit VII
Reproduction: The Continuity of Life

27 Reproductive Organs and Their Functions
Sterilization	704
Endometriosis	713
Cancer of the Cervix	714
Breast Feeding	716
Premenstrual Tension Syndrome	720
Dysmenorrhea	720
Sexually Transmitted Diseases	723
Safe Sex	724
Contraceptive Devices	725
The Clinic: Disorders of the Reproductive System	728

28 Development and Inheritance
Artificial Reproduction	734
Amniocentesis	738
Ultrasonography of the Uterus	739
Abortion	743
Complications During Pregnancy	744
Natural Childbirth Preparation	746
Complications of Labor and Delivery	747
The Clinic: Advances in Molecular Genetics	766

Unit VII Case Study: A Problem of Reproductive Failure — 771

HUMAN ANATOMY AND PHYSIOLOGY

Unit I

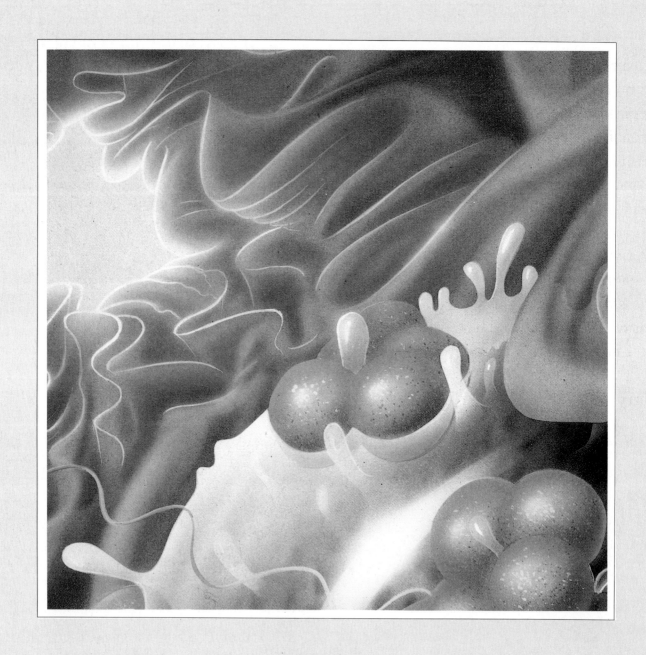

Organization of the Human Body

Unit I is an introduction to the study of human anatomy and physiology. It begins with an overview of the organization of the body. Essential terminology is introduced and defined. Next, we examine the body from the perspective of three different levels of organization. First, we study atoms and molecules, the primary level of organization. Next, we examine cells, the smallest organization level that shows all the characteristics of life. The unit ends with a description of how cells are organized into tissues that carry on the functions of the body.

1
Introduction to the Human Body

2
The Chemical Basis of Life

3
Cell Structure and Organization

4
Tissues

1
Introduction to the Human Body

The human body is a magnificent structure. A human being is an intricate organism composed of many parts that cooperate in all the activities we associate with life. This chapter introduces you to the study of the human body. In this chapter, we will consider

1. the characteristics of life and how each relates to human life,
2. the principle of homeostasis and how feedback mechanisms operate in maintaining a stable internal environment,
3. the different levels of anatomic organization within the body,
4. the major regions and cavities of the body and the major structures in each, and
5. the major terms that describe positions and dimensional perspectives used in studying human anatomy.

anatomy: (Gr.) *anatome*, from *ana*, up + *tome*, cutting
physiology: (Gr.) *physis*, nature + *logos*, to study
histology: (Gr.) *histos*, tissue + *logos*, to study
radiography: (L.) *radius*, ray + (Gr.) *graphein*, to write

The Fields of Anatomy and Physiology

After studying this section, you should be able to:

1. Define anatomy and physiology and tell how the two differ in their approaches to the study of the human organism.
2. List several subfields of both anatomy and physiology.

A comprehensive study of the human body requires information and techniques from many branches of science. Two of the most important branches are **anatomy** and **physiology.** Anatomy is the study of body structure, while physiology concerns body performance during life functions. The study of anatomy may be undertaken in several different ways. **Gross anatomy** is the study of the major structures of the body, while **microscopic anatomy,** also called **histology** (hĭs·tŏl′ō·jē), deals with those details of structure that require the use of a microscope. **Regional anatomy** involves studying specific regions of the body, such as the head, trunk, or limbs. **Systemic anatomy** deals with the functions of the body, such as digestion or reproduction. **Developmental anatomy** traces changes that occur in a maturing embryo and fetus.

Perfection of highly specialized devices have opened new approaches to the study of anatomy. **Radiographic anatomy** involves the study of **radiographs** (also called roentgenograms or **X rays**) that are produced when X rays pass through the body and expose a plate of photographic film (Figure 1-1a) (see The Clinic: Radiographic Anatomy). **Computerized axial tomography (CAT)** is a more specialized type of radiographic technique. In this procedure a series of X rays produce a radiograph of a specific level of the body, providing internal perspectives previously unavailable

CLINICAL BOXES

AN INTRODUCTION

Many of you are taking this human anatomy and physiology course because you plan to pursue careers in the health fields. Throughout the chapters of this anatomy and physiology text you will notice many boxed passages of clinical significance. These boxes offer you various medical applications of the text material you will be studying. The applied material will help you to understand how important the knowledge of a properly functioning system is to the diagnosis and treatment of an improperly functioning system.

The boxes are grouped into three kinds of clinical applications. The brief "Clinical Notes" offer quick insights into common medical applications of the systems being discussed in the text. Another type of box, entitled "The Clinic," falls at the end of every chapter and presents an important disorder particularly relevant to the chapter or serves to summarize disorders that can occur in a system. The disorders are usually organized into six categories: development or heredity, infection, trauma or injury, immunology or allergies, tumors and cancer, and aging. Finally, the third kind of medical application box is called a "Case Study." One occurs at the end of each unit and serves to put the textual material into action. These case studies are based on actual clinical situations and may well be similar to ones you will encounter in your future careers.

to physicians. Such views of internal anatomy are called **CAT scans** (Figure 1-1b).

Just as anatomy has subfields or specialties, so does physiology. Physiologists are concerned with the chemical and physical reactions that occur in the body. At the lowest level is **cell physiology,** which investigates individual cells and their special activities vital to life and proper functioning of the whole body. Additional study specialties in physiology include **renal physiology** (kidney function),

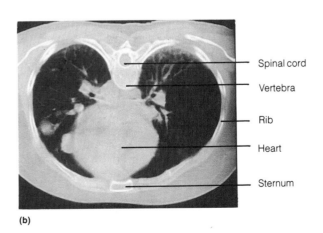

Figure 1-1 (a) An X ray of the thoracic area. (b) A CAT scan showing a transverse section through the same region of the body.

kinesiology:	(Gr.) *kinesis*, movement + *logos*, to study
cardiology:	(Gr.) *kardia*, heart + *logos*, to study

neurophysiology (nervous system function), **kinesiology** (kĭ·nē″sē·ŏl′ō·jē) (muscle function), **cardiology** (kăr·dē·ŏl′ō·jē) (heart function), and **respiratory physiology** (lung function), among others.

Although anatomy and physiology are separate disciplines, the two are intimately associated. Each organ in the body has a specific anatomical form and shape that is often directly related to its function. Likewise, the function of an organ may be determined by its anatomy. For example, the bulbous form and shape of the stomach indicates its function as a storage space for food, and the physiology of its individual cells is influenced by their positions within the organ. In a similar manner, anatomical differences between the stomach and the skeleton, for example, are due to the skeleton's function of structural support, rather than storage and digestion of food.

Figure 1-2 The human body is organized at several levels of increasing structural complexity. Here we see how complexity increases from the molecular to the systemic level.

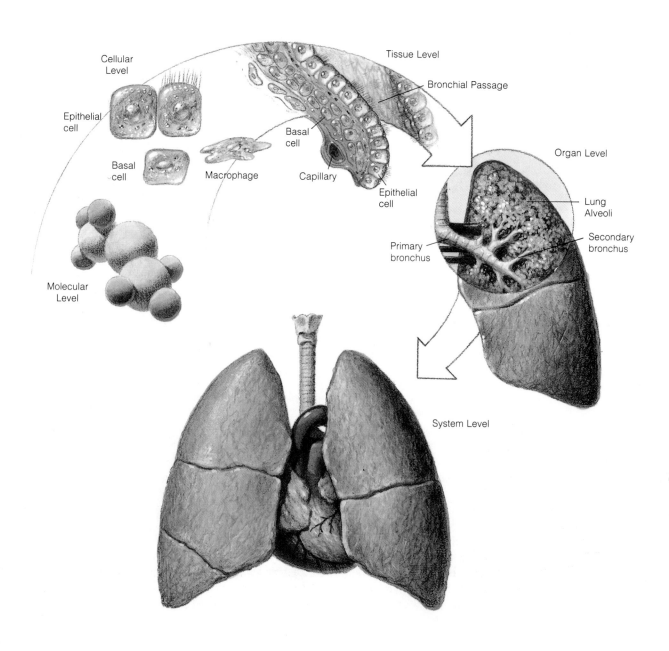

organelles: (L.) *organum,* instrument + *-ellus,* diminutive

Structural Plan of the Human Body

After studying this section, you should be able to:
1. List the five principal levels of organizational complexity found in the human body.
2. Describe a typical organ and indicate its organizational levels.

The structural plan of the human body as an organism involves a high degree of organization at five principal levels of complexity: molecules, cells, tissues, organs, and systems (Figure 1-2). These five levels are discussed below.

Atoms and Molecules

The lowest level of organization involves the atoms and molecules that compose the basic chemical substances of the body. These substances range from small, simple molecules of oxygen and water to protein molecules so large and complex that they will not dissolve in the blood. Body chemistry can be very complicated, and Chapter 2 introduces atoms and molecules and the mechanics of chemical reactions.

Cells

Atoms and molecules combine to form the next level of organization: cells. The **cells** of the human body are the fundamental structural and functional units of life. They are the smallest units that show the characteristics of life. Typically they are made up of smaller components called **organelles** (or″găn·ĕls′) (see Chapter 3), where many of the chemical reactions of life take place. Ultimately these reactions can be carried out only because the organelles have precise organizational patterns at the chemical level.

Tissues

Similar cells are grouped together to form the next higher organizational level: tissues. A **tissue** is a group of similar cells that perform a certain function. The lungs, for example, are simple in outward appearance, yet are made up of glandular, muscular, vascular, nervous, and protective tissues, each of which is responsible for a particular function. Rings of cartilage support the bronchial passages and keep them from collapsing. Glandular cells secrete mucus that traps dust particles, microbes, and other solids that come into the respiratory passages with each breath. Tiny hairs (cilia) on other cells move the trapped particles out of the lungs up to the back of the throat, from where it can be expelled. Smooth muscles under the direction of nerves in the bronchial tubes can constrict or dilate these passages. A network of blood vessels bring blood close to the air in the respiratory sacs. This entire assemblage is held together both externally and internally by tough connective tissues, and the lungs are encased in a thin, slippery membrane that eases the movements made with each breath.

Organs

Tissues in turn are grouped into structurally and functionally integrated units called **organs.** Each organ in the body performs specific functions. For example, the lungs (Figure 1-2), organs of the respiratory system, are responsible for the exchange of oxygen and carbon dioxide between the blood and the air in the respiratory sacs. But the lungs are only one component of the respiratory system, along with the larynx, trachea, and bronchial tubes.

Systems

The human body is organized into a number of systems. A **system** is a group of organs that work together to perform specific functions. For example, the **respiratory system** exchanges oxygen and carbon dioxide between the blood and the atmosphere, and the **circulatory system** distributes oxygen, nutrients, and other chemicals to all parts of the body. The **skeletal system** supports the body, and an associated **muscular system** provides for movement. The human body is composed of 10 such systems, which, in addition to those already listed, include the digestive, nervous, endocrine, lymphatic, urinary, and reproductive systems (Figure 1-3).

Anatomic Regions and Positions

After studying this section, you should be able to:
1. Name the three major regions of the body.
2. Describe the organization and contents of the two major cavities of the body.
3. Enumerate the terms used to describe sections cut through the body.

In addition to the three major regions of the body—the **head,** the **trunk,** and the **appendages**—two major cavities are formed during embryonic development: a **posterior cavity,** sometimes called the **dorsal cavity,** which houses the brain and spinal cord, and an **anterior cavity,** sometimes

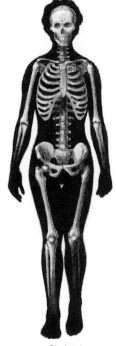
Skeletal system

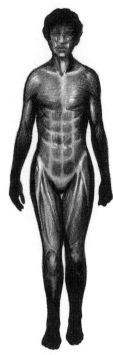
Muscular system

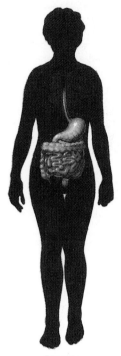
Digestive system

Respiratory system

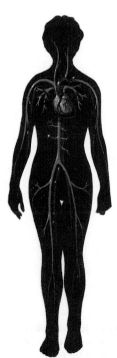
Circulatory system

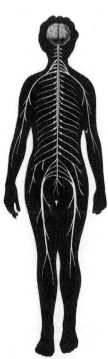
Nervous system

Endocrine system

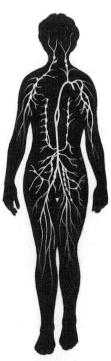
Lymphatic system

pericardium: (Gr.) *peri*, around, near + *kardia*, heart + (L.) *ium*, quality or nature of
mediastinum: (L.) *mediastinus*, medial
anthropometry: (Gr.) *anthropos*, man + *metron*, to measure

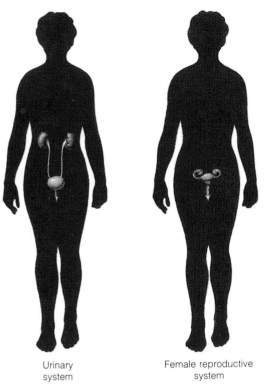
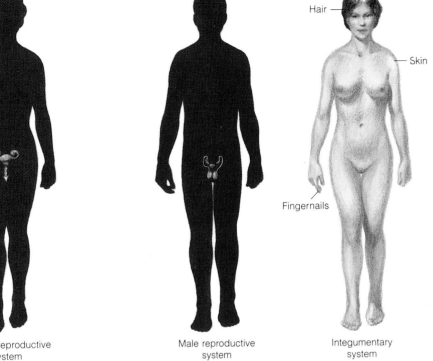

Figure 1-3 Organ systems of the body. The organ systems of the body are arranged into functional units that specialize in different activities, such as support, movement, internal transport, gaseous exchange, and reproduction.

called the **ventral cavity,** which contains most of the soft organs of the body (Figure 1-4). The posterior cavity is subdivided into two cavities, **cranial** and **vertebral.** The anterior cavity is partitioned by a muscular **diaphragm** forming the **thoracic** and **abdominopelvic** (ăb·dŏm″ĭ·nō·pĕl′vĭk) **cavities.** During formation of organs, several membranes subdivide the thoracic cavity into three more discrete regions: the **pericardial** (pĕr·ĭ·kăr′dē·ăl) **cavity,** which contains the heart; the **pleural cavities,** which contain the lungs; and a tissue-filled "cavity" called the **mediastinum** (mē″dē·ăs·tĭ′nŭm), which contains and holds in place the thymus gland, the esophagus, trachea, and several large blood vessels. An indistinct **pelvic cavity** is often described as an extension of the abdominal cavity. Its boundaries can be delineated by drawing an imaginary line from the right to the left iliac crests. The pelvic cavity is the area of the abdominal cavity below the pelvic crests, extending down into the confines of the pelvic skeleton.

Physicians, nurses, and other health practitioners assume that the abdominal area has **four quadrants** (right upper,

CLINICAL NOTE

MEASURING THE BODY

The science of body measurement, called **anthropometry** (ăn·thrō·pŏm′et·rē), is a very old one. In ancient times various measurements, particularly of the head, were used to classify the different races of human beings. Later, extensive studies of weight and stature were done to show the rate of development of children and to show that certain environmental differences (e.g., rural versus urban living conditions) influenced the rates of growth in children as much as genetic factors. In more recent times there were attempts to use body measurement as a means of identification of specific individuals, usually criminals, by making several measurements, such as the distance between the pupils of the eyes and the width, length, and circumference of the head. These measurements were eventually replaced with the discovery and use of modern finger printing methods.

Today there is renewed interest in anthropometry because of the popularity of sports and athletics. Many specialists in sports medicine are doing research on the characteristics that determine what types of bodily configurations and physical attributes make the best athletes for specific sports. They are also measuring the percentage of body fat, amounts of various ions, joint flexibility, and types of muscle fibers of world class athletes in hopes of predicting the potential of athletic candidates.

sagittal: (L.) *sagitta*, arrow
metabolism: (Gr.) *metabole*, change + *-ismos*, denoting condition

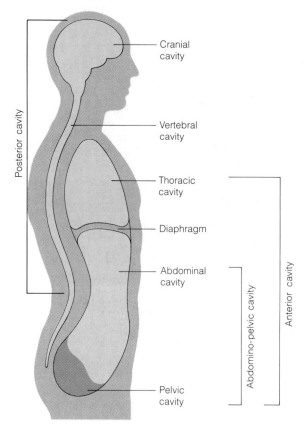

Figure 1-4 The human body is divided into two major cavities. The posterior cavity is subdivided into cranial and vertebral cavities, while the anterior cavity is subdivided into thoracic and abdominopelvic cavities.

left upper, right lower, and left lower) or **nine regions**, which include the right hypochondriac, epigastric, left hypochondriac, right lumbar, umbilical, left lumbar, right iliac, hypogastric, and left iliac regions (Figure 1-5). In practice, they locate specific organs within each of these quadrants or regions.

Several terms describe the positions of different parts of the body in relation to body **planes** and **sections** (Figure 1-6 and Table 1-1). You will use these terms throughout your study of anatomy. Learning them will make your study of anatomy easier. In addition to regarding diagrams and drawings with a three-dimensional perspective, you will also learn internal relationships by studying sections cut through body parts in different planes. A **frontal (coronal** (kō·rō′năl) section (Figure 1-6) divides the body into **anterior (ventral)** and **posterior (dorsal)** portions. **Sagittal** (săj′ĭ·tăl) sections divide the body into right and left halves; a **midsagittal** section passes through the midline of the body, and a **parasagittal** section passes through any area to the right or left of the midline. A **transverse** section (or cross-section) is perpendicular to the long axis of the body and divides the body into **superior** and **inferior** portions. An **oblique** section is one that is cut at any angle other than 90°.

The Characteristics of Life

After studying this section, you should be able to:

1. List the five most commonly recognized characteristics used to identify living things.
2. Cite an example of how homeostatic mechanisms maintain a steady state in the internal environment.

Living organisms exhibit a number of characteristic qualities. The following five are the most commonly recognized characteristics usually ascribed to living organisms and used to identify them.

Organizational Complexity

Although it is not a unique characteristic of life, all living organisms do show varying degrees of organizational complexity. Chemical crystals, among many nonliving things, also show intricate and complex organization. As discussed previously in this chapter, many levels of complexity can be recognized in the human organism, from atoms at the lowest level to the entire body at the highest level.

Metabolism

Metabolism (mĕ·tăb′ō·lĭzm) refers to the constant exchange of energy and materials between an organism and its environment. When organisms eat, they take in fuel in the form of food to obtain the energy necessary to power bodily functions. As the fuel is reduced to smaller fragments, energy is released. It is either used immediately, perhaps to provide body heat, or is stored in various chemical compounds for later use. However, not all food is used for heat production or for energy storage. Some of it provides the energy to synthesize new materials for the growth and production of new tissues, and the remainder is discarded and excreted as waste.

Figure 1-5 (Top, opposite) Spatial regions of the abdominal cavity. Either (a) four quadrants or (b) nine regions are recognized, allowing a more precise indication of the areas within which organs are found.

Figure 1-6 (Bottom, opposite) The directional terms used in describing the human body (a) from a back perspective, (b) from a side perspective, and (c) the anatomical planes. See the text and Table 1-1 for definitions of these terms.

The Characteristics of Life 11

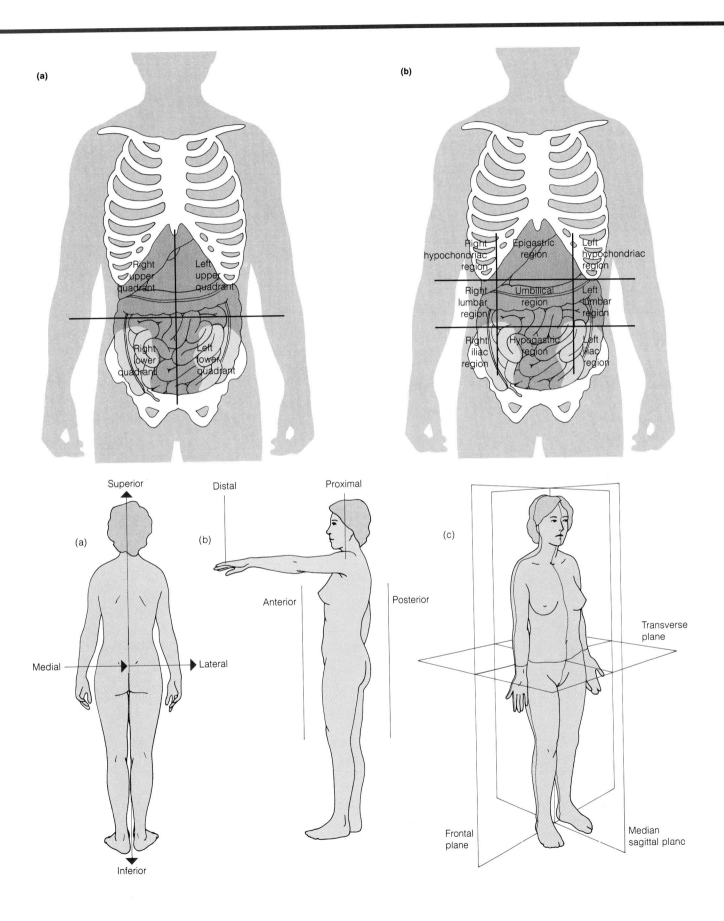

homeostasis: (Gr.) *homoios*, alike + *stasis*, standing
autopsy: (Gr.) *autos*, self + *opsis*, to view (to see for oneself)
pathology: (Gr.) *pathos*, suffering, disease + *logos*, to study

Table 1-1 ANATOMICAL TERMS

Term	Description
Anterior	Toward the front of the body
Posterior	Toward the rear of the body
Medial	Toward the midline of the body
Lateral	Toward the sides of the body
Central	Toward the midline of the body or a specific organ
Peripheral	Toward the outer edge of the body or a specific organ
Superior (cranial)	Toward the head
Inferior (caudal)	Toward the feet
Proximal	Used primarily with reference to limbs; closer to the attached end of a limb or the midline of the body
Distal	Used primarily with reference to limbs; closer to the free end of a limb or away from the midline of the body
Superficial (external)	Closer to the surface of the body or a specific organ
Deep (internal)	Closer to the midline of the body or a specific organ
Frontal section	Section dividing the body into anterior and posterior portions
Coronal section	Section dividing the head into anterior and posterior portions
Sagittal section	Section dividing the body into right and left portions
Transverse section	Section dividing the body into superior and inferior portions

CLINICAL NOTE

AUTOPSY

An **autopsy** (aw'tŏp·sē), or postmortem examination, is a very thorough gross and microscopic examination of all of the organs and tissues of a body after death has occurred. This examination is very useful in determining the exact cause of death and discovering other conditions that may have contributed to the patient's illness. It is helpful in assisting physicians to understand the processes of disease. An autopsy may also be very important to the family in explaining the final illness and in solving legal matters that sometimes complicate the loss of a family member.

Most autopsies are done by permission of the family, but deaths that result from trauma or that occur under suspicious circumstances or without recent medical attendance may be declared coroner's cases and the examination ordered by local or state officials. **Coroners** are officials who are responsible for investigating accidental or suspicious deaths. Frequently the coroner is a specially trained physician called a **medical examiner.** Autopsies are usually performed by specially trained physicians called **pathologists** who are skilled at recognizing disease processes by careful examination of organs and microscopic examination of tissues and cells.

The examination is performed with respect and privacy similar to a surgical operation. The entire body and each organ and system is carefully examined, measured, weighed, and described. Small representative pieces of each organ are removed and prepared for microscopic examination. A final complete report is written that includes a discussion of the clinical course of the patient, the gross and microscopic description of all of the organs, and finally, a summary of the cause of death and all secondary diagnoses. The final report is a legal document that is filed in the patient's record and retained by the hospital or coroner's office.

Metabolic processes are often divided into two categories that can occur simultaneously. **Anabolic** (ăn″ă·bŏl′ĭk) processes require energy and build simple molecules into more complex compounds and tissues in the body; **catabolic** (kăt″ă·bŏl′ĭk) processes reduce the size of complex molecules and tissues into simpler compounds with the release of energy. For example, anabolic processes exceed catabolic processes in children as they grow, until they reach adult size. In an adult of stable size and weight, the two processes roughly balance one another and there is neither a net gain nor a loss.

Homeostasis

In 1852 the French physiologist Claude Bernard observed that cells and tissues of the human body were able to maintain an environment, called the **extracellular fluid (ECF)** (or **tissue fluid**) that is different from their external environment. Only after the 1920s, however, was the word **homeostasis** (hō″mē·ō·stā′sĭs) coined by the American physiologist Walter Cannon as a label for the process. Homeostasis is the process by which the body maintains a stable environment for its cells and tissues, in spite of changes in the external environment. Later research emphasized that most of the body's organ systems function to maintain this steady state within the internal environment.

Homeostatic mechanisms are numerous in the body. They regulate such things as body temperature, blood pressure, nutrient levels in the blood, and breathing rate, as well as many other body functions. The control of body temperature at a fairly constant level despite environmental temperature changes, changes in food and fluid intake, and changes in exercise level is a good example of many systems working together to maintain homeostasis (see Chapter 19 Clinical Note: Fever). Most of these mechanisms involve self-regulating control called **feedback.** A feedback mechanism is any system in which results produced by the mechanism are used in self-regulation.

The body's ability to regulate the concentration of calcium in the blood is another important example of a homeostatic mechanism found in the body. Figure 1-7 is a diagrammatic representation of the regulation of blood calcium. The maintenance of such a concentration very close to 10 milligrams (mg) of calcium per 100 milliliters (ml) of blood is essential for proper functioning of nerves and muscles and for the efficient formation of blood clots to stop bleeding. Significant increases or decreases in calcium concentration can have serious results. Unusually low calcium levels delay the formation of blood clots, increase the activity of nerves and muscles, and may even cause muscular convulsions. Excessive levels of calcium decrease muscle activity, particularly in the muscles of the heart.

The homeostatic mechanism regulating calcium concentration involves secretions from the **parathyroid** and **thyroid** glands located in the anterior region of the neck (see Chapter 17). Certain parathyroid gland cells are sensitive to the concentration of blood calcium. A decrease in this concentration stimulates these cells to increase output of the substance **parathyroid hormone.** This hormone is carried in the blood to all parts of the body where it has three effects that increase the level of blood calcium. First, it stimulates small amounts of bone tissue to dissolve. (About 99% of the body's calcium is stored in bone tissue.) The dissolved calcium enters the blood, raising the concentration. Second, parathyroid hormone converts vitamin D into a compound that stimulates the absorption of calcium present in food in the intestine. Third, parathyroid hormone causes the kidneys to remove less calcium from the blood during the formation of urine. All three effects of parathyroid hormone raise the concentration of calcium in the blood.

Once calcium levels have risen, the feedback mechanism of this control system commences. The increasing level of blood calcium decreases the stimulation of the parathyroid cells, which lowers the output of parathyroid hormone, thereby reducing the amount of calcium entering the blood. Because the feedback information *reduces* the output of parathyroid hormone, it is a **negative feedback mechanism.** Feedback from certain other mechanisms in the body *stimulate* the output of certain glands rather than decreasing it (as in the present case) and are thus called **positive feedback mechanisms.**

As indicated in Figure 1-7, an abnormal rise in the blood calcium level stimulates certain thyroid gland cells to produce more of a substance called **calcitonin** (kăl″sĭ·tō′nĭn). Figure 1-7 shows that calcitonin decreases the concentration of blood calcium. It does so by stimulating certain bone cells

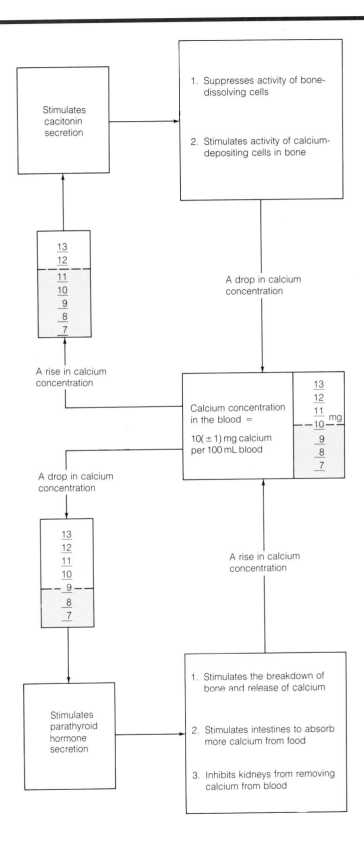

Figure 1-7 The homeostatic regulation of calcium concentration in the blood plasma. Note that changes away from the optimal level of 10 mg of calcium per 100 ml of blood initiate feedback mechanisms that return the calcium level to optimal.

tomogram: (Gr.) *tome*, cutting + *graphein*, to write

THE CLINIC

RADIOGRAPHIC ANATOMY

Radiographic anatomy is a special branch of anatomy that studies the internal body structures by passing electromagnetic waves through the body onto a photographic plate and then interpreting the resulting picture. It is sometimes called **roentgenology** (rĕnt″gĕn·ŏl′ō·jē) in honor of the German physicist Wilhelm Konrad Roentgen who discovered X rays in 1895 while he was experimenting with passing an electric current through gases. He attempted to pass a current through a vacuum tube and noted a strange fluorescence on a nearby barium-coated screen. He called the newly discovered form of radiation **X ray.** Later he discovered that the new ray would pass through the human body and expose a photographic plate.

Radiographic Techniques

The X ray or roentgenogram is a two-dimensional photograph of the structures of the body made by passing X rays through the body onto a photographic plate (film). Some parts of the body, such as bone, absorb more of the X rays than other parts, so certain structures and conditions are made visible (see Figure 1-1a). The X ray is one of the most commonly used diagnostic techniques. Physicians who specialize in interpretation of the films are called **radiologists.**

A **xeroradiograph** is a type of X ray made using photographic plates covered with an electrically charged powdery substance held between two metal plates. The X rays alter the charge on the substance to varying degrees depending on the tissues they penetrate (see Figure 6-3). Xeroradiographs are particularly useful for viewing soft tissues, which often do not show up well on normal X rays.

The **tomogram** is an X-ray technique that provides a way of obtaining finer detail of a specific area where an abnormality may have been seen on the routine X ray. Tomography involves taking multiple pictures of the area, each at a slightly different level, like slicing bread.

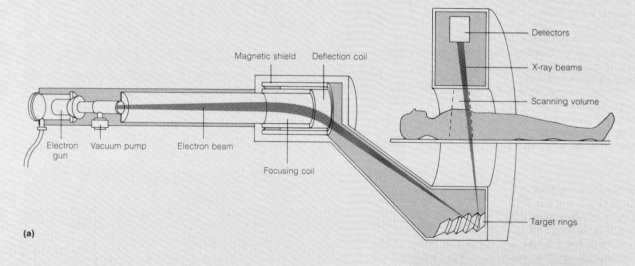

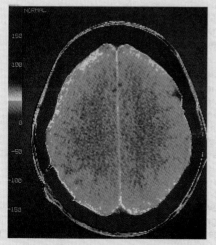

Figure 1 (a) CAT scan apparatus. A scanning electron beam produces X rays that pass through a selected plane of the body from many different angles within a fraction of a second. This can provide a real-time image of objects in rapid motion, such as the heart. (b) CAT scan image of a normal brain, transverse section.

Contrast radiography is a technique of introducing a radiopaque substance (one that does not allow X rays to pass through it) into the body to delineate the internal structures better. One such technique is the gastrointestinal series in which a drink containing barium is swallowed by the patient prior to being X rayed to outline the lining of the gastrointestinal tract. Another example of contrast radiography is the **myelogram** in which an iodine-containing dye is injected into the spinal canal to demonstrate any abnormalities, such as a herniated intervertebral disk. When a radiopaque dye is injected into a joint to visualize the internal structure, the examination is called an **arthrogram.**

Another development in radiographic technology is the **computerized axial tomogram (CAT scan).** This system employs a series of X-ray sources ringing the body with metering devices on the opposite sides (Figure 1a). In this manner the actual amount of X rays absorbed by the structure can be measured and fed to a computer that reconstructs the image of a cross-section of the body on a television monitor (Figure 1b). Multiple slices (tomograms) can be obtained to aid in visualizing large cross-sections of the body such as the head or chest. These transverse pictures have stimulated a renewed interest in the study of transverse sectional anatomy.

Continuing developments have led to the **dynamic spatial reconstructor (DSR)** which uses 28 revolving X-ray guns and is capable of producing moving, three-dimensional, life-sized images in any plane. It can also produce enlargements, and it features instant replay and stop action viewing.

Sonographic Techniques

The **sonogram** is a method of viewing the internal anatomy of the body by using high frequency sound waves (ultrasound) (see figure in Chapter 28 Clinical Note: Ultrasonography of the Uterus). Sonograms are very useful in the examination of the heart and the pregnant abdomen because the sound waves are painless and apparently harmless and because the examinations may be repeated frequently to demonstrate any changes in the organs examined. Sonograms are also relatively inexpensive.

Nuclear Imaging

The newest development in radiographic anatomy is **nuclear magnetic resonance (NMR).** The NMR systems use low energy radio waves and a strong magnetic field. The patient is placed in a tunnel-shaped magnet 3000 times the strength of the earth's magnetic field (Figure 2a). The magnet is used to align the protons in the atoms of the cells. Then short bursts of radio waves are introduced causing the protons to wobble. When the frequencies of the wobbling motion and the radio waves coincide, resonance has been achieved. The radio signal is then stopped, and the protons realign with the magnetic field and in so doing emit radio waves that can be measured. Since every chemical substance in the body has its own resonance properties, the NMR reveals chemical and structural information indicating and pinpointing concentrations of specific chemicals. The information from the multiple bursts of radio waves is processed by a computer and displayed on a television monitor (Figure 2b). The NMR promises to provide clearer images of cross-sectional anatomy without the use of X rays or contrast media. It will also provide information about changes within structures and possibly subtle information on chemical imbalances. Since NMR units are very costly, an NMR examination costs about twice that of a conventional CAT scan.

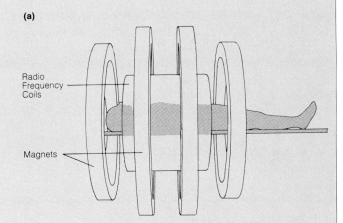

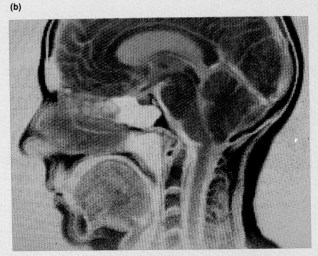

Figure 2 (a) Nuclear magnetic resonance (NMR) imaging apparatus. Radio frequency coils produce radio waves, and a powerful magnet is used to set up a gradient magnetic field. (b) NMR image of the head, sagittal section.

to remove calcium from the blood by producing new bone tissue. As the calcium level in the blood drops, the other negative feedback mechanism just described eventually reduces the production of calcitonin. Thus, interplay between these two systems maintains a steady state in the calcium concentration in the blood.

Control of calcium is only one of many homeostatic mechanisms. The principle of homeostasis is vital to the study of anatomy and particularly physiology. It provides a conceptual framework for understanding the many structural and functional adaptations in the human body.

Irritability

Irritability is the term used by biologists to denote the ability of organisms to respond to internal and external stimuli. Specifically, sense organs, muscles, and nerves in the human body react in a coordinated fashion in response to stimuli. For example, if a person is walking along the street and sees a speeding car approach, sense organs and muscles respond to the stimulus (the car) and cooperate to move the individual away from the oncoming vehicle.

As we shall see in later chapters (in Unit III), the **nervous system** and the **endocrine** (ĕn′dō·krīn) **system** regulate responses to both external and internal stimuli.

Reproduction

Another important characteristic of life is **reproduction.** In humans, reproduction occurs when specialized cells of a male and female unite to produce a new individual. In Chapters 27 and 28 we will study specific organs that produce these cells and the processes involved in their union. We will also study how the new individual develops and what determines hereditary linkage between parents and children.

Now that we have presented the overall organization of the body, we can begin a detailed study of human anatomy and physiology. Chapters 2 and 3 concern molecular and cellular organization. Chapter 4 deals with a more complex level of organization: the combination of cells into tissues.

STUDY OUTLINE

The Fields of Anatomy and Physiology (pp. 5–6)
  Anatomy is the study of body structure and involves all levels of organization from the gross characteristics of organs to the microscopic examination of cellular details. Anatomists often make use of developmental information and use *X rays or radiographs* in studying internal details of organs.

  Physiology is the study of body function and may involve individual cells, organs, or an entire system.

Structural Plan of the Human Body (p. 7)
  The human body is organized at five principal levels: molecules, cells, tissues, organs, and systems.

  Atoms and Molecules Atoms and molecules represent the lowest level of organizational complexity in the body.

  Cells Cells are the fundamental structural and functional units of life and are the smallest units that show the characteristics of life.

  Tissues A *tissue* is a group of similar cells that perform a certain function. There are, for example, glandular, muscular, vascular, nervous, and protective tissues.

  Organs Tissues are grouped into structurally and functionally integrated units called *organs*, such as the lungs, stomach, and liver.

  System A *system* is a group of organs that work together to perform specific functions.

Anatomic Regions and Positions (pp. 7–10)
  The human body has three major regions, *head, trunk,* and *appendages,* and two major cavities, *thoracic* and *abdominopelvic.* The major planes of the body are *frontal, sagittal,* and *transverse.*

The Characteristics of Life (pp. 10–16)
  Living organisms exhibit five characteristics not found as a group in nonliving things.

  Organizational Complexity The levels of organizational complexity in humans are molecules, cells, tissues, organs, and systems.

  Metabolism Metabolism is the constant interchange of matter and energy between an organism and its environment. *Anabolic* processes incorporate materials and energy into the structure of the body, while *catabolic* processes break down materials with the release of energy.

  Homeostasis Homeostasis is the process by which the body maintains a stable environment for its cells and tissues. It involves self-regulating control mechanisms that are governed by *feedback* from the results of the process.

  Irritability Irritability is the phenomenon of organisms responding to internal and external stimuli.

  Reproduction Reproduction is the ability to produce new individuals.

SELF-TEST OF CHAPTER OBJECTIVES

True-False Questions
1. Positive feedback mechanisms operate by stimulating a mechanism, whereas negative feedback mechanisms inhibit an aspect of a homeostatic device.

2. Irritability is defined as the ability of the body to break large molecules down into smaller ones.
3. The term *posterior* refers to the upper part of the body.
4. Anabolic reactions in the body build up cells and molecules, whereas catabolic reactions break down cells and molecules.
5. The abdominal cavity is described as a posterior cavity in the body.
6. The highest organizational level in the body is the organ level.
7. Reproduction is an important characteristic of life.
8. The mediastinum is located in the abdominal cavity.
9. The nervous system allows the body to exercise the life characteristic of irritability.
10. A sagittal section would divide the body into right and left portions.

Matching Questions
Match the anatomical terms with the correct definitions:

11. medial
12. posterior
13. distal
14. anterior
15. proximal

a. the front side of the body
b. the free end of a limb
c. toward the midline of the body
d. toward the rear of the body
e. toward the attached end of a limb

Match the anatomical terms with the correct definition:

16. transverse section
17. frontal section
18. parasagittal section
19. midsagittal section

a. divides the body into equal right and left portions
b. divides the body into superior and inferior portions
c. divides the body into unequal right and left portions
d. divides the body into anterior and posterior portions

Multiple-Choice Questions
20. The constant exchange of energy and materials between an individual and its environment is called
    a. homeostasis
    b. metabolism
    c. feedback mechanism
    d. irritability
21. Which of the following represents the highest level of organizational complexity found in the human body?
    a. molecules
    b. tissues
    c. systems
    d. organs
22. The process used by the body to maintain a stable environment for its cells is called
    a. metabolism
    b. irritability
    c. anabolism
    d. homeostasis
23. Which of the following organs is not located in the abdominal cavity?
    a. liver
    b. lungs
    c. stomach
    d. intestine
24. Which of the following is an example of an anabolic mechanism in the body?
    a. weight loss
    b. weight gain
    c. digestion of foods in the digestive system
    d. destruction of old body cells
25. The mechanism used by the body to maintain the concentration of calcium in the blood at a very stable level is an example of
    a. metabolism
    b. anabolism
    c. catabolism
    d. homeostasis

Essay Questions
26. List the five major characteristics usually attributed to and used to identify living organisms.
27. Define homeostasis and give an example of this process in humans.
28. What is the difference between a positive and a negative feedback mechanism?
29. List the major planes along which sections can be cut through the body.
30. Beginning at the molecular level, describe the increasingly complex levels of organization in the body.
31. Name the 10 organ systems of the human body.
32. Define the term *metabolism* and distinguish between anabolism and catabolism.

Clinical Application Questions
33. Compare and contrast computerized axial tomography with nuclear imaging.
34. How does the study of anthropometry relate to sports medicine?
35. How does a pathologist's work differ from that of a coroner?

2
The Chemical Basis of Life

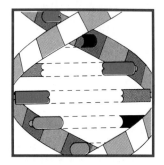

Although many of the substances found in living organisms are unique to life, the ingredients from which they are formed and the processes that form them are not. Thus, it is appropriate to begin an examination of anatomy and physiology with a study of the structure of matter in general and a look at some of the specific processes used in the production of different forms of matter. To understand these processes, it is first necessary to understand some of the chemical structures and principles involved. The purpose of this chapter is to introduce them to you. In it, we will consider

1. the nature and fundamental structure of all matter,
2. the forces that hold matter together and give it form,
3. how matter interacts to produce new forms,
4. the way large chemical products are formed by combining smaller subunits, and
5. classes of chemical substances found in biological matter.

The Nature of Matter

After studying this section, you should be able to:

1. Define an element.
2. Recognize the symbols that represent elements important to physiology.
3. Define an atom and identify the three primary particles from which atoms are formed.
4. Describe the structure of an atom and explain how the atoms of different elements vary.
5. Describe what an ion is and how ions are formed.
6. Define a chemical bond and distinguish among covalent, ionic, and hydrogen bonds.
7. Explain what a molecule is and describe how larger molecules are formed by combining smaller subunits.

Matter is anything that occupies space and can be perceived by one or more of the senses. Over the decades, chemists and physicists have determined that all matter consists of a relatively few elementary materials, combined in countless ways to produce the nearly infinite variety of substances that make up the universe. Included within this variety of substances is the matter found in living organisms. While representing a tiny fraction of the forms in which matter exists, biological matter is nevertheless a complex mixture of substances combined in thousands of ways to produce those structures that make life possible.

Chemical Elements

In 1803 a British chemist named John Dalton advanced the theory that all matter consists of tiny, indivisible particles that he termed **atoms**. According to Dalton's theory, different types of atoms exist, each type constituting the fundamental particles of various **chemical elements**. All the atoms of one element were considered by him to be identical to one another but different from the atoms of any other element. He believed that in a chemical reaction, atoms of the same or different elements interact to form combinations of atoms. Dalton theorized that most matter consists of atoms of different elements combined in various ways.

Today we recognize that Dalton's ideas about the atomic nature of chemical elements are basically correct. During the nearly two centuries since he formulated his theory, more than 100 different elements have been identified. Of these, about one-third are found in the human body (Table 2-1). Of this third, four constitute over 99% of the total weight of the human body: hydrogen (63%), oxygen (25.5%), carbon (9.5%), and nitrogen (1.4%). The remainder range in quantity from a few tenths of a percent to amounts that are so small that they are difficult to measure accurately.

For convenience, each element is designated by a one- or two-letter abbreviation called a **chemical symbol**. For example, hydrogen is designated by H, oxygen by O, and carbon by C. In some cases, the symbol derives from the Latin name of the element rather than the English name. Sodium, for example, is designated by Na for the Latin *natrium*, and potassium by K for its Latin name *kalium*.

Atoms

All atoms, except certain atoms of the element hydrogen (H), are composed of the same three kinds of basic units, particles called **protons, neutrons,** and **electrons.** The protons and neutrons are located in a central region of the atom called the **nucleus** (Figure 2-1). Electrons travel in the space surrounding the nucleus, moving around the nucleus at nearly the speed of light. Protons and electrons each possess an **electric charge**. Although these charges are of equal magnitude, they have opposite properties, defined as **positive** in the proton and **negative** in the electron. Neutrons have no electric charge. The number of protons in an atom equals the number of electrons, so an atom possesses the same number of positive and negative charges, and there is no net charge on the atom. The atom is thus **electrically neutral**. Protons and neutrons have about the same mass, each possessing 1,840 times as much mass as an electron.

Table 2-1 THE MOST ABUNDANT CHEMICAL ELEMENT OF THE BODY

Element	Symbol	Atomic Number	Atomic Mass	Percentage of Body Mass
Oxygen	O	8	16.00	65
Carbon	C	6	12.01	18
Hydrogen	H	1	1.01	10
Nitrogen	N	7	14.01	3
Calcium	Ca	20	40.08	1.5
Phosphorus	P	15	30.97	1
Potassium	K	19	39.10	0.4
Sulfur	S	16	32.06	0.3
Sodium	Na	11	22.99	0.2
Magnesium	Mg	12	24.31	0.1
Chlorine	Cl	17	35.45	0.1
Iron	Fe	26	55.85	Trace
Iodine	I	53	126.90	Trace
Manganese	Mn	25	54.94	Trace
Copper	Cu	29	63.55	Trace
Zinc	Zn	30	65.37	Trace
Cobalt	Co	27	58.93	Trace
Fluorine	F	9	19.00	Trace
Molybdenum	Mo	42	95.94	Trace
Selenium	Se	34	78.96	Trace

isotope: (Gr.) *isos*, equal + *topos*, place

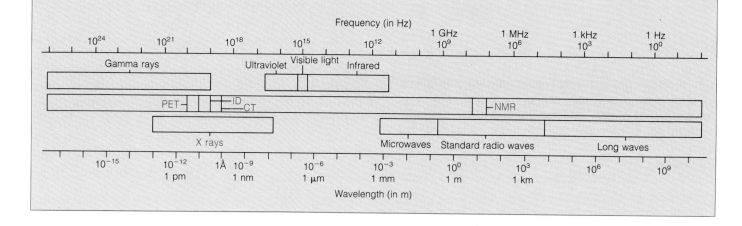

CLINICAL NOTE

PHOTON IMAGING

Several of the clinical notes in Chapters 1 and 2 deal with various techniques for obtaining information on internal structure and chemical reactions occurring in the body. Most of these techniques make use of **photon imaging.**

The modern theory of the fundamental structure of the universe describes two basic classes of particles: **structural units,** such as protons, neutrons, and electrons, and **intermediary units,** which travel among the structural units and provide the energy for physical or chemical reactions. **Photons** are intermediary particles that travel between charged structural particles, such as protons and electrons.

Photons are generated when a charged particle goes from a higher to a lower energy state. **Electromagnetic radiation** is made up of photons, which are classified according to their **oscillation frequency** or **wavelength** (see figure). Photons with wavelengths most familiar to us are visible light. Less familiar photons include X rays, radio waves, and gamma rays. Visible light photons tend to bounce off solid structures, which allows us to see and take pictures of the surface of solid objects. However, X-ray, radio wave, and gamma ray photons will pass through most solid objects, including the human body, and are therefore useful in studying the internal structure of the body. Thus, all these techniques are types of photon imaging.

Photons in several regions of the electromagnetic spectrum are suitable for medical imaging. The graph shows the range of wavelengths (and frequencies) used in X rays, PET, ID (iodine dichromotography), CAT (or CT) scans, and NMR imaging.

The atoms of one element differ from the atoms of all other elements in the number of protons and electrons they possess. Each element has a particular **atomic number** defined as the number of protons present in the nucleus of the atoms of that element. For example, carbon atoms contain 6 protons and have an atomic number of 6. Sodium, with 11 protons in its atoms, has an atomic number of 11.

Isotopes

Unlike the number of protons and electrons, which are characteristic of the atoms of a given element, the number of neutrons in the nucleus may vary. Atoms of oxygen, for example, have eight protons and eight electrons but may have eight, nine, or ten neutrons. Similarly, atoms of carbon always have six protons and six electrons but may have anywhere from four to ten neutrons. Such different forms of an element are called **isotopes** (ī′sō·tōps). The total number of protons and neutrons in a particular isotope is referred to as the **mass number** of the isotope. For example, hydrogen atoms have mass numbers of 1, 2, or 3, which corresponds to isotopes with one proton and zero, one, or two neutrons. In the case of hydrogen, the most common isotope by far is the one that lacks a neutron. The other two isotopes are so unusual that they are given specific names—the isotope with one neutron is called **deuterium** (dū·tē′rē·ŭm) and the isotope with two neutrons is called **tritium** (trĭt″ē·ŭm). Isotopes of other elements are often designated by the elemental name followed by the mass number of the isotope. For example, the isotope of carbon that possesses eight neutrons and six protons is designated carbon-14. The mass number of an isotope is indicated by a superscript preceding the symbol; thus, carbon-14 is ^{14}C.

Certain isotopes of many elements are unstable and disintegrate spontaneously. This tendency to disintegrate is called **radioactivity.** When an atom disintegrates, electrons,

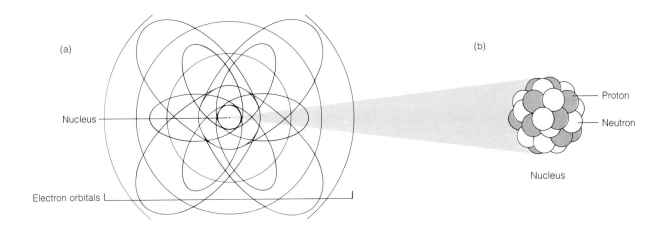

Figure 2-1 Structure of an atom. (a) An atom consists of a central nucleus, containing the atom's protons and neutrons, surrounded by orbitals in which the electrons travel. (b) Although it contains almost all of the mass of the atom, the diameter of the nucleus is tiny compared with the overall diameter of the atom.

nuclear particles, and energy that had been present in the atom are released as **radiation** (see Clinical Note: Radioisotope Scanning). Radioactivity is often characteristic of isotopes in which the number of neutrons is greater than the number of protons, as, for example, in tritium and carbon-14. It is not necessarily true, however, that such isotopes are radioactive, since oxygen-18, with eight protons and ten neutrons, and carbon-13, with six protons and seven neutrons, are stable. In addition, all the elements with atomic numbers greater than 84 are radioactive, but these elements are not normally present in biological material.

Atomic Mass

The **atomic mass** of an element is the sum of the protons and neutrons in the nucleus of an atom of the element. A unit of atomic mass is the **dalton**, defined as the weight of a single hydrogen atom (1.66×10^{-24} grams). Since elements generally consist of mixtures of isotopes, no single atom's sum of protons and neutrons represents all the atoms of the elements. Atomic mass is therefore the *average* mass of the isotopes present in a naturally occurring mixture of isotopes, taking into account the proportion of each isotope in the mixture. Chlorine, for example, has an atomic mass of 35.453 because it consists of two different isotopes, one with an atomic mass of 35, which comprises about 76% of the total atoms, and the other with an atomic mass of 37, which comprises about 24% of the total atoms. In many cases, one isotope is much more common than the rest so that the atomic mass of the element is nearly the same as the atomic mass of the major isotope. In carbon, for example, nearly 99% of the atoms have six protons and six neutrons, with an atomic mass of 12, and only slightly more than 1% have six protons and seven neutrons, with an atomic mass of 13. (The remaining five isotopes of carbon together account for less than 0.01% of the total.) As a result, the atomic mass of a naturally occurring mixture of carbon atoms is 12.011. The average mass of an element is also referred to as its **atomic weight.**

Electron Orbitals

As the electrons of an atom travel around the nucleus, they travel in specific regions of space referred to as **orbitals.** The many orbitals in which electrons can travel are organized into groups called **shells,** which can be further divided into **subshells** (Figure 2-2). Electrons present in the orbitals of a particular shell exist at a particular **energy level.** Electrons in the orbital closest to the nucleus have the lowest energy level, whereas electrons in outer orbitals have higher energy levels.

The number of orbitals in which electrons actually travel depends on the number of electrons in the atom. In the simplest atom, that of hydrogen, there is only one electron and it travels in a spherical orbital that surrounds the nucleus. In more complex atoms—that is, those with a greater number of electrons—as many as two electrons travel in this same orbital and any additional electrons travel in orbitals corresponding to successively higher energy levels. In the largest atoms there may be as many as seven different shells, each containing electrons traveling in several different orbitals (Figure 2-2).

Electrons that travel in the highest energy orbitals of an atom are primarily responsible for the chemical properties of the atom. Because elements differ in the number of electrons present in their atoms, no two elements have electrons at the same highest energy level. Consequently, their chemical properties are different. That is, if two atoms had exactly

22 Chapter 2 The Chemical Basis of Life

radioisotope: (L.) *radius*, ray + (Gr.) *isos*, equal + (Gr.) *topos*, place

---

CLINICAL NOTE

RADIOISOTOPE SCANNING

Some isotopes are unstable and tend to decay or change their nuclear structure to a more stable condition. These are considered to be radioactive. When certain isotopes decay, they give off **gamma rays,** which are similar to X rays, which pass through the body and may be detected. These are called **radioisotopes** (rā″dē·ō·ī′sō·tōps).

Medicine has made use of these unstable isotopes to help in the diagnosis and treatment of disease. The radioisotopes used in medicine are produced artificially by the use of particle accelerators such as the cyclotron. Minute doses of the radioisotopes are introduced into the body and the gamma rays they emit are detected by scintillation counters or scintillation cameras.

The radioisotopes may be used in the form of simple salts of the element, or they may be combined with organic compounds. Since the thyroid is the only tissue in the body to metabolize iodine, iodine becomes concentrated in this gland. Minute doses of iodine-131 (^{131}I), a radioisotope of iodine, can be injected intravenously, and the thyroid can be scanned with a scintillation camera to obtain a picture of the gland. The technique is especially useful in the diagnosis of thyroid tumors since most cancerous tumors do not metabolize the iodine and therefore give a negative or "cold" image on the scan. In contrast, non-cancerous nodules are usually overactive and incorporate more of the radioisotope, producing a positive or "hot" image.

There are many radioisotopes that can be used to examine many different organ systems. Technetium (^{99}Tc) is used for scanning the liver and spleen. ^{99}Tc-labeled polyphosphate or strontium-85 (^{85}Sr) are used in bone scans. **Total body scanning** using gallium citrate (^{67}Ga) has been useful in detecting tumor spread and hidden areas of infection.

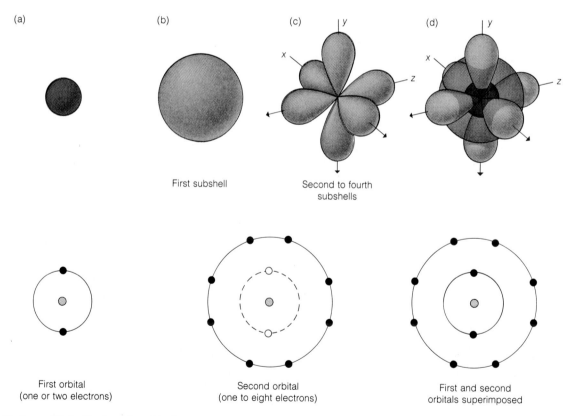

Figure 2-2 Electron orbitals. Electrons travel in the space around the nucleus of an atom in volumes referred to as orbitals. Orbitals are grouped into principal energy levels which, except for the first, are further subdivided into subshells. (a) The lowest energy orbital consists of a single, spherical volume which contains one electron in hydrogen atoms and two electrons in all other atoms. (b) In atoms with more than two electrons, one or two electrons are in a spherical subshell of the second energy level. (c) Up to six additional electrons are in sausage-shaped subshells that lie at right angles to one another. (d) The first and second energy levels are shown superimposed. The lower drawings illustrate a shorthand way of depicting electron distribution.

covalent: (L.) *co*, share + *valentia*, vigor, strength

the same number of electrons, those electrons would be distributed identically. The atoms would then have the same chemical properties and would be atoms of the same element.

Ions

An electron in the orbital with the highest energy level is sometimes able to leave an atom and go to a neighboring atom, which incorporates it into its own highest available energy level orbital. When this happens, the atom that lost the electron is left with more protons than electrons and thus has a net positive charge. The atom that gains the electron then has more electrons than protons and thus has a net negative charge. Such electrically charged atoms are called ions (Figure 2-3). As we shall see, electrical attractions between ions are the basis for many important physiological phenomena.

Chemical Bonds

Whether or not one atom will react with another depends largely on the number of electrons in their respective highest energy level shells. For example, hydrogen and helium have only one shell, and two electrons in this shell make an atom relatively inert. Helium possesses two electrons in this shell and is thus inert. Hydrogen, on the other hand, has one electron in this shell and is thus quite reactive. Reactivity of the remaining elements works in much the same way. According to the **octet rule,** any atom that has more than one shell will be generally inert if there are eight electrons in its highest energy shell. If there are fewer than eight electrons in this shell, they will react in a way that causes them to have eight electrons in that shell. There are two major ways in which an atom can acquire a filled high energy shell: by *ionizing*, that is, by gaining or losing enough electrons to have a filled shell, or by *sharing electrons* with another atom.

Ionic Bonds

We have already seen that an ion is formed when an atom has a greater or lesser number of protons than it has electrons. Depending on the number of electrons present, an atom can "obey" the octet rule either by gaining enough electrons to have a total of eight or by giving up the electrons in the outermost shell so that the next lower energy shell becomes the highest one. Gaining electrons will produce a *negative ion* and losing electrons will produce a *positive ion*.

Positively and negatively charged ions can be held together by the *electrostatic* forces that particles of opposite charge exert on one another. Sodium chloride, which is common table salt, is an example of a substance formed by ionic

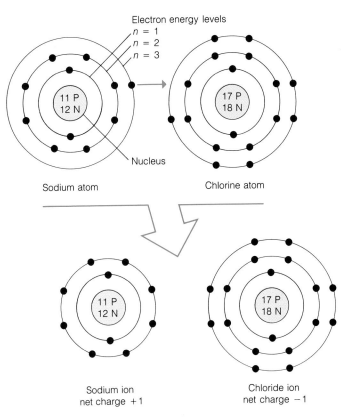

Figure 2-3 Ionization. Ions are formed when electrons are lost or gained by atoms or molecules. This diagram shows how an electron might be lost from the outermost orbital of a sodium atom to the outermost orbital of a chlorine atom. The sodium ion that results has one more proton than it has electrons and has a net charge of +1. The chloride ion that results has one more electron than it has protons and has a net charge of −1.

bonds between sodium and chloride ions, designated as Na^+ and Cl^-. A crystal of table salt consists of sodium and chloride ions that are held together in a three-dimensional **crystalline lattice** formed as each ion attracts ions of opposite charge. The electrostatic force that holds ions of opposite charge together is called an **ionic bond** (Figure 2-4).

Covalent Bonds

Bonds that result when electrons are shared are called **covalent bonds** (Figure 2-5). In a covalent bond, a pair of electrons, one from each atom, travels in the space between the atoms. For example, in water (H_2O) one pair of electrons travels in the space between the oxygen atom and one of the hydrogen nuclei, and another pair of electrons travels in the space between the oxygen atom and the second hydrogen nucleus. In each bond, one electron is contributed by the hydrogen atom and the other by the oxygen

24 Chapter 2 The Chemical Basis of Life

tomography: (Gr.) *tome*, incision + *graphein*, to write

CLINICAL NOTE

PET AND SPECT: THE FUTURE

A very recent development in radiographic imaging of the body that produces images of great clarity and accurately identifies chemical and physiologic changes is **positron emission tomography** (tō·mŏg′ră·fē) or **PET.** PET uses a combination of photon imaging and radioisotope scanning. The PET image is produced by gamma rays released from specially constructed radioisotopes such as ^{13}N, ^{11}C, or ^{18}F. These highly unstable isotopes are combined with various substrates used in normal cellular metabolism. The unstable **substrate** (or metabolic compound) is injected into the patient and enters the normal metabolic pathway. The unstable isotope releases gamma rays that can be detected and used to form an image that gives us information about where these substrates are used in the body (see figure).

PET has been used to study the glucose metabolism of the brain, and it has also been useful in the study of blood flow and blood volume in the brain. PET has been useful in research on several neurologic conditions, such as epilepsy, Alzheimer's disease, language and visual impairments, and strokes. PET has also been used to study blood flow in the heart and the evolution of heart attacks. One of the disadvantages of PET is that the radioisotopes used are so unstable that they decay within a matter of minutes. This requires the close proximity of a cyclotron to produce the isotopes used.

Single photon emission computed tomography, or **SPECT,** is a less complicated, less expensive technique similar to PET that demonstrates brain blood flow using more stable isotopes and readily available detectors. SPECT is useful in diagnosing patients with stroke, brain tumors, epilepsy, and senile dementia.

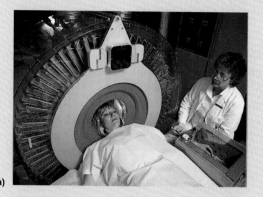

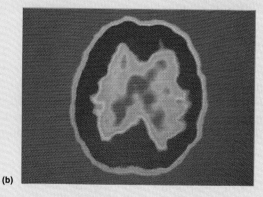

PET imaging. (a) Patient undergoing PET scan. (b) PET image of a normal brain, transverse section.

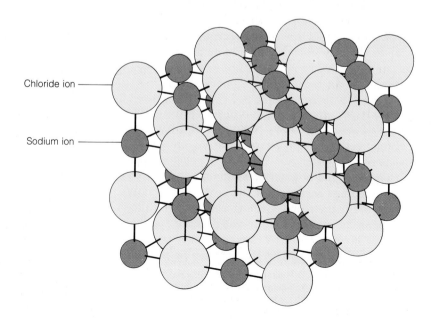

Chloride ion

Sodium ion

macromolecule: (Gr.) *makros*, large, long + (L.) *molecula*, diminutive of *moles*, mass, bulk
polymer: (Gr.) *polus*, much, many + *meros*, part

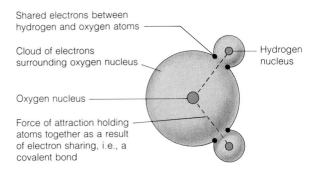

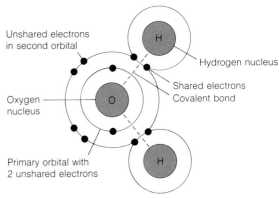

Figure 2-5 Covalent bonds. Covalent bonds are formed when two atoms share electrons. These diagrams of a water molecule show how an oxygen atom shares electrons with two hydrogen atoms. Instead of traveling in orbitals around the nucleus as they would in individual atoms, the shared electrons in water spend most of their time in the space between the nuclei.

Hydrogen Bonds

The unique structure of a hydrogen atom, possessing a single electron that travels in a single electron shell, imparts an unusual character to the atom. When a hydrogen atom forms a covalent bond with an oxygen or nitrogen atom, the shared electrons spend most of the time in the space between the nuclei, exposing the hydrogen nucleus to the neighboring space. This does not occur in the covalent bonds of other elements because electrons in other orbitals continue to encircle the nucleus. In the case of hydrogen, the positive charge of the exposed proton radiates outward.

If an atom of nitrogen or oxygen in a neighboring molecule is brought near, an electrostatic force develops between the proton in the hydrogen nucleus and the electrons of the neighboring atom. The bond that develops is a **hydrogen bond** (Figure 2-6). Although relatively weak compared with ionic or covalent bonds, hydrogen bonds are extremely important in determining how molecules react with one another, and when present in large numbers, can add up to a sizable force. Ice, for example, occurs when the hydrogen and oxygen atoms of water molecules are attracted to one another. In ice, the water molecules become arranged in a crystalline lattice held together by hydrogen bonds (Figure 2-6c).

Hydrogen bonds can also form between hydrogen atoms and nitrogen or oxygen atoms in the same molecule, where they help to determine the specific three-dimensional shape of the molecule. Most of the large molecules described later in this chapter owe their specific three-dimensional shape in part to intramolecular hydrogen bonding.

atom. As a result, the oxygen atom has eight electrons in its highest energy shell, six of its own and two contributed by the hydrogen atoms. Likewise, each of the hydrogen atoms has two electrons in its shell, one of its own and one "borrowed" from the oxygen atom. The highest energy shells of all three atoms are therefore "filled," and the sharing of electrons acts as a bond holding the atoms together.

Sometimes atoms are held together as a result of the sharing of two pairs of electrons. When this is true, the covalent bond formed is referred to as a **double bond.** **Triple bonds** also occur occasionally as a result of the sharing of three pairs of electrons by two atoms.

Molecules

Most of the mass of a cell is made up of atoms bound together into groups of atoms called **molecules.** Some molecules contain as few as two atoms, as in the molecules of oxygen and nitrogen in the air we breathe, whereas other molecules may contain atoms numbering in the billions. A molecule of DNA, for example, typically has 100 billion or more individual atoms. Such large molecules are often referred to as **macromolecules.** Macromolecules are made up of large numbers of smaller molecules joined together by bonds. Such macromolecules are also referred to as **polymers,** and their formation through the successive addition of smaller subunits is called **polymerization.**

The kinds and numbers of atoms present in a molecule are indicated by the chemical symbols of the atoms, each followed by a subscript indicating the number of atoms of that element if more than one atom is present. For example, the chemical formula for water is H_2O, indicating that each molecule consists of two hydrogen atoms and one oxygen

Figure 2-4 (left) Ionic bonds. The attractive force that holds ions of opposite charge together is an ionic bond. Because the electric charge of an ion radiates in all directions, ions can form bonds with several ions of opposite charge simultaneously. As ionic bonds are formed, the ions become arranged in a crystalline lattice. Here, sodium (Na^+) and chloride (Cl^-) are shown in a lattice of ions present in common table salt.

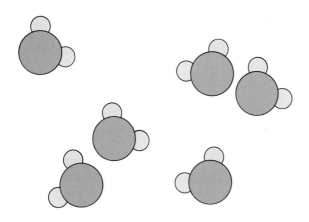

(a) Vapor

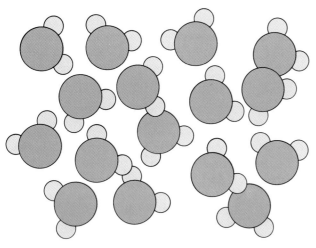

(b) Liquid water

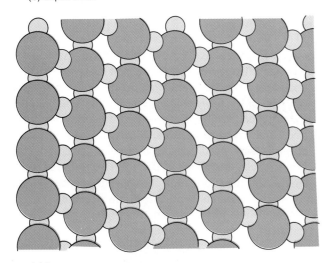

(c) Ice

atom. Glucose, an important substance in physiology, has the formula $C_6H_{12}O_6$, indicating that each molecule consists of six carbon atoms, twelve hydrogen atoms, and six oxygen atoms.

Chemical Reactions

After studying this section, you should be able to:

1. Define a rearrangement reaction and explain isomerization.
2. Describe reactions that involve the exchange of atoms between molecules.
3. Explain what occurs during an oxidation-reduction reaction.

Chemical reactions are the interactions that occur between molecules and that result in changes in their structure (Figure 2-7). The following are a few of the more common types of reactions on which the chemistry of life is based.

Rearrangement Reactions

A **rearrangement reaction** is a chemical reaction in which a molecule changes from one three-dimensional configuration to another. Figure 2-7a shows an example of rearrangement involving a **conformational change** in a molecule of glucose. All the individual bonds remain, but the atoms change their spatial relationships with one another. As a result of the rearrangement, the molecule changes its shape. The new form is still glucose, but because of its new shape, it has somewhat different chemical properties than it had originally.

Figure 2-6 (left) Hydrogen bonds in water. (a) Water molecules in a gaseous state show little attraction for one another, resulting in a random orientation of the molecules with respect to one another. (b) In liquid water, the molecules form transient hydrogen bonds between oxygen and hydrogen atoms of neighboring molecules. As water gets colder, the bonds become longer lasting and the water molecules become arranged less and less randomly. (c) In ice, hydrogen bonds have locked the water molecules into a crystalline lattice, and there is relatively little breaking and reforming of the bonds.

Figure 2-7 (opposite right) Representative types of reactions. (a) Conformational change. Atoms remain bound to one another, but angles between atoms change. (b) Isomerization. The same atoms remain in the molecule, but new bonds are formed and original bonds are broken. (c) Addition reaction. Atoms of different molecules are combined into a new molecule. (d) Cleavage reaction. A molecule is cleaved into two smaller molecules. (e) Transfer reaction. A group of atoms is removed from one molecule and added to another.

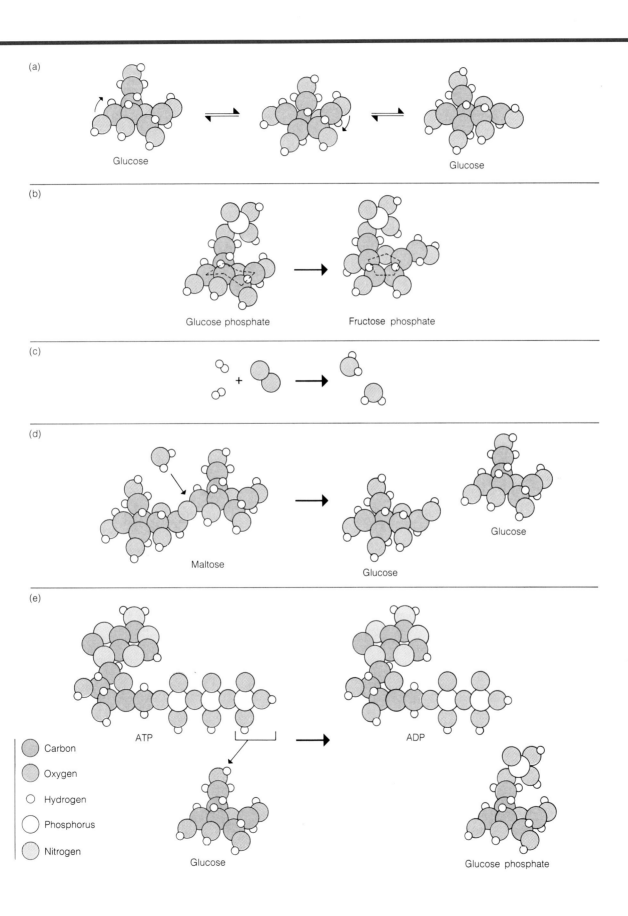

isomer: (Gr.) *isos*, equal + *meros*, part

A more complicated change is illustrated in Figure 2-7b. In this case, the atoms again change their spatial relationships, but here the changes also involve the breaking of certain bonds and the formation of new ones. The resulting changes in the chemical properties of the molecule are sufficient to identify it as a different substance and to give it a new name. In this case, a molecule of *glucose phosphate* is rearranged to a molecule of *fructose phosphate*. All the same atoms are still present, but they have been rearranged as a result of the formation of new bonds. This type of rearrangement is referred to as **isomerization** (ī·sŏm″ĕr·ī·zā′shŭn), and the two different molecules are referred to as **chemical isomers** (ī′sō·mĕrs).

Addition Reactions

An **addition reaction** occurs when atoms or molecules come together to form a new molecule (Figure 2-7c). For example, when hydrogen is burned it reacts with oxygen to form water. Hydrogen and oxygen molecules each contain two atoms held together by covalent bonds (that is, H_2 and O_2). In this reaction, two molecules of hydrogen and one molecule of oxygen combine to produce two molecules of water. The reaction occurs as a result of the collision of the two different kinds of molecules, which brings them close enough together to be able to share electrons. When this happens, new covalent bonds are formed and new molecules are produced.

Cleavage Reactions

A **cleavage reaction** is one in which a molecule is split into two smaller molecules. For example, sucrose is split in half when it is digested in the small intestine, producing a molecule of glucose and a molecule of fructose from each sucrose molecule that is split (Figure 2-7d). This kind of reaction characterizes many of the reactions of digestion, the process by which food is broken down in the digestive tract to products small enough to be absorbed by cells that line the tract. Cleavage reactions usually also involve a molecule of water, with a hydrogen atom from the water being added to one product and the remaining —OH (called a **hydroxyl group**) added to the other product.

Transfer Reactions

A **transfer reaction** is a chemical reaction in which an atom or a group of atoms is transferred from one molecule to another. Sometimes the transfer involves an *exchange* of atoms, in which case the reaction is referred to as a **replacement** or **exchange reaction**. One common example of this type of reaction in the human body involves the transfer of a phosphate group from a molecule of **adenosine triphosphate (ATP)** to glucose (Figure 2-7e). A phosphate group is a group of atoms consisting of one phosphorus atom and four oxygen atoms. ATP is made up of a molecule of **adenosine** and three phosphate groups. Under suitable conditions, the covalent bond that binds one of the phosphate groups to the ATP molecule is cleaved, and the phosphate group is transferred to the glucose molecule. The adenosine triphosphate has been converted to **adenosine diphosphate (ADP)**, which has only two phosphate groups. The glucose molecule with a phosphate group attached is called **glucose phosphate**.

Oxidation-Reduction Reactions

Some chemical reactions involve the transfer of one or more electrons from one molecule to another. Such reactions are called **oxidation-reduction reactions.** The molecule that gains the electron is termed an **electron acceptor,** and the one that loses it is an **electron donor.** If either of the molecules is electrically neutral prior to the reaction, it becomes an ion because it is left with a different number of electrons than protons. The electron acceptor is said to be **reduced** as a result of receiving the electron, and the electron donor is **oxidized** as a result of losing the electron (Figure 2-8). Oxidation-reduction reactions are especially important in the reactions that cells use to obtain energy from chemical fuels.

Metabolic Reactions

After studying this section, you should be able to:

1. Define metabolism and explain how several reactions can form a metabolic pathway.
2. Distinguish between anabolic and catabolic reactions.
3. Explain the concept of reaction rate, and describe how temperature and concentration can affect reaction rate.
4. Explain what an enzyme is and how it speeds up a reaction.

There are on the order of tens of thousands of different chemical reactions that go on in a living system such as a human body. These reactions are usually associated with one another in **metabolic pathways,** sequences of reactions in which the products of one reaction are the starting material of other reactions. In this way, chemical fuels are broken down in processes that provide energy, substances are formed that are used as building materials for cells and tissues, and waste molecules are produced and discarded. At the cellular level, life is the result of the coordination and regulation of these reaction pathways.

anabolism: (Gr.) *ana*, up + *bole*, throw + *ismos*, denoting condition
catabolism: (Gr.) *kata*. down + *bole*, throw + *ismos*, condition

 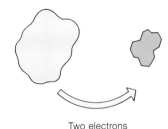  

Reduced compound A   Oxidized compound B   Two electrons   Oxidized compound A   Reduced compound B

Figure 2-8 Oxidation-reduction reactions. In an oxidation-reduction reaction, one or more electrons pass from one atom or molecule to another. The electron donor is oxidized in the process, and the molecule receiving the electron is reduced.

There are two general categories of metabolic pathways: those that form larger compounds from smaller ones and those that break down larger molecules into smaller ones. The synthesis of larger compounds from smaller ones is called **anabolism** (ă·năb´ō·lĭzm). These reactions produce the more complex molecules (such as fats, carbohydrates, and proteins) described later in this chapter. The breakdown of larger molecules into smaller ones is termed **catabolism** (kă·tăb´ō·lĭzm). These reactions generally produce molecules that can be used as chemical fuels or that are discarded as waste products. (See Chapter 19 for a more detailed discussion of metabolic processes.)

Factors Affecting Reaction Rate

A **reaction rate** is the speed at which products are produced in a reaction or, conversely, the speed at which the reacting components are used up. Consider, for example, an addition reaction that involves the combination of two different molecules. In this case, the reaction only occurs if the reacting molecules come into contact with one another. The rate of this reaction depends on how often the molecules collide and the force with which they collide when they do. Both of these factors can be influenced by changing the number of molecules and the speed at which they travel. Increasing the concentration of a solution means there are more molecules, and thus collisions are more likely. Increasing the temperature means increasing the speed of the molecular movement, and thus collisions are both more likely and more forceful. As we shall see in the next section, the concentration of hydrogen ions (pH) also affects reaction rates.

Concentration

Concentration refers to the amount of a substance within a given volume, such as ounces per gallon, grams per liter (g/L), or milligrams per milliliter (mg/mL). A useful way to express concentration, at least for chemical purposes, is by the number of atoms or molecules of a substance within a given volume. Generally, however, the number of atoms or molecules in a customary unit of volume, such as a liter, is so large that it is more convenient to deal with a unitless number called a **mole.**

A mole (abbreviated *mol*) is equal to 6.022×10^{23} of anything, regardless of whether we are speaking of atoms or elephants. That is, 6.022×10^{23} elephants is a mole of elephants and 6.022×10^{23} oxygen molecules is a mole of oxygen molecules. This number is called **Avogadro's number,** which is defined as the number of atoms present in exactly 12 grams (g) of carbon-12. Since the atomic mass of an element is proportional to the average number of protons and neutrons in the nuclei of its atoms, there is also 1 mol of atoms in 1.008 g of hydrogen (atomic mass equals 1.008) and in 196.967 g of gold (atomic mass equals 196.967).

The same principle applies to **molecular weight,** which is simply the sum of the atomic weights of the atoms present in a molecule. Like atomic weight, molecular weight is a unitless measurement. For example, the molecular weight of water equals 18.010 (the atomic weights of 2 hydrogen atoms plus 1 oxygen atom equals 18.010), and the molecular weight of table sugar equals 342.299, which is the combined atomic weights of 12 carbon atoms, 22 hydrogen atoms, and 11 oxygen atoms. Thus, 1 mol of water weighs 18.010 g and 1 mol of table sugar weighs 342.299 g, each mole containing 6.022×10^{23} molecules.

Enzymes

Life is a collection of thousands of chemical reactions, many of which have been duplicated in the laboratory under special controlled conditions. One difference between a reaction that is duplicated in a laboratory and one that occurs in life is that the laboratory reaction goes much slower than it does in the organism. This is because living systems possess **enzymes** (ĕn´zīms), molecules that cause reactions to proceed faster than the reactions would occur in the absence of the enzyme. The speeding up of a reaction by a substance

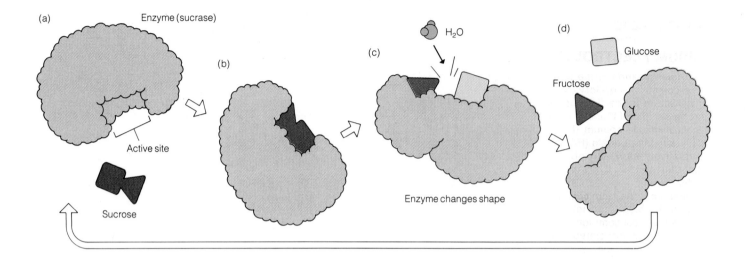

Figure 2-9 Cleavage of sucrose to glucose and fructose. (a) Sucrose conforms to the shape of an active site on the surface of a sucrase molecule. (b) Sucrose binds to the enzyme at the active site, (c) inducing a change in shape in the sucrase molecule. This induces the cleavage of sucrose in a reaction involving a water molecule. (d) Glucose and fructose are released, and the enzyme is available to catalyze another reaction.

that is not consumed in the reaction is a process called **catalysis** (kă·tăl´ĭ·sĭs), and an enzyme is considered to be a **biological catalyst.**

Consider, for example, the reaction described earlier in which sucrose is cleaved to form glucose and fructose (see Figure 2-7d). Sucrose molecules exist in many different conformations when dissolved in water. Bonds of the molecule stretch and bend in many directions, causing it to change rapidly from one configuration to another. Most of these configurations are stable, but in some, a strain is placed on the bond that connects the glucose and fructose portions, causing the bond to break. The rate of cleavage is low because few molecules are in this unstable configuration at any instant, and those that are may quickly revert to a stable state before cleavage occurs.

In general, enzymes speed up reactions by first combining with specific molecules, called **substrate molecules,** that bind to specific **active sites** on the surface of the enzyme. Combination of the substrate molecule with an active site causes the substrate to change its shape to an unstable, reactive conformation. The substrate molecule then reacts and is converted to the product of the reaction. The product does not bind tightly to the enzyme the way the substrate did and is immediately released. This frees the enzyme to allow it to combine with another substrate molecule and repeat the reaction.

In the enzymatic cleavage of sucrose, a sucrose molecule acts as a substrate for a particular enzyme, combining with the enzyme at specific sites on the enzyme surface (Figure 2-9). This attachment causes a change in the shape of the sucrose molecule to a form that readily cleaves to produce glucose and fructose. Neither glucose nor fructose is attracted to the enzyme the way sucrose was, so both are released and the enzyme is free to repeat the reaction with another sucrose molecule. In the small intestine, where sucrose is broken down during digestion, this reaction occurs perhaps a half a million times faster than it would if the enzyme were not present, providing a rapid means for the conversion of sucrose to a form that can be absorbed by the body.

Hydrogen Ion Concentration

After studying this section, you should be able to:

1. **Define pH.**
2. **Explain how the pH of a solution changes in relation to changes in the hydrogen ion concentration of the solution.**
3. **List the pH values of several common substances.**
4. **Define and distinguish between acids and bases.**
5. **Explain what pH buffers are and why they are important in metabolism.**
6. **List three buffer mechanisms important in physiology.**

Hydrogen ions (H^+) are a particularly important substance in the fluids of the body. If the hydrogen ion concentration is too low or too high, some reactions will not occur

electrolyte: (Gr.) *elektron,* amber + *lytos,* soluble

> **CLINICAL NOTE**
>
> **SERUM ELECTROLYTES**
>
> The determination of the **serum electrolytes** is one of the most common laboratory examinations performed in medicine. Serum is the fluid portion of whole blood removed after a sample of blood is allowed to clot. The serum electrolytes include sodium (Na^+), potassium (K^+), chlorides (Cl^-), and bicarbonate (HCO_3^-). Testing for these four ions is relatively easy and gives a great deal of information about the state of hydration and acid-base balance of the body.
>
> Sodium is the main **cation** in the extracellular fluids (the fluid in the blood and intercellular spaces), while potassium is the most common cation within the cells. Chlorides are the **anions** associated with the sodium and potassium. The bicarbonate ion is part of the acid-base buffering system of the body.
>
> The serum electrolytes are sampled in patients for whom there has been severe restriction of fluid intake or severe loss of bodily fluids, as in prolonged vomiting and diarrhea, and in diabetic patients who are losing large amounts of body fluids through the kidneys to help remove the excessive sugar from the body. They are also sampled in seriously burned patients who can lose large amounts of fluids through the burn. Regular sampling of the serum electrolytes allows the calculation of actual amounts of fluid and ions lost so that suitable intravenous fluids can be administered to restore the fluid and ion balance as well as to adjust the pH of the body. Knowledge of serum electrolyte levels can also provide initial clues to disease of the kidneys and the endocrine system.
>
> Since potassium is the main intracellular cation, changes in potassium concentration can produce severe toxic effects on the heart muscle and nervous system. In patients with diseases that may affect the level of potassium, this ion must be regulated very carefully. It is important to be sure that the kidneys are functioning properly before giving patients intravenous potassium since the kidneys are the main regulating organ of potassium levels. The heart muscle and nerves are so sensitive to the level of potassium that an **electrocardiogram** (EKG) is sometimes used to detect changes in the potassium level. The probability of heart failure is high among anorexics and bulimics who upset their serum electrolytes through their restrictive and/or purging behaviors (see Chapter 19 Clinical Note: Anorexia Nervosa).

Table 2-2 RELATIONSHIP OF pH TO H^+ CONCENTRATION

pH	H^+ Concentration (mol/L)	
1	0.1	Acidic
2	0.01	
3	0.001	
4	0.0001	
5	0.00001	
6	0.000001	
7	0.0000001	Neutral pH
8	0.00000001	
9	0.000000001	
10	0.0000000001	
11	0.00000000001	
12	0.000000000001	
13	0.0000000000001	
14	0.00000000000001	Basic

properly or may not occur at all, and serious problems can develop. The concentration of H^+ in water is expressed as **pH;** the pH of a solution is defined as the negative logarithm of the moles of hydrogen ion in 1 liter (L) of solution. Pure water has a concentration of 0.0000001 mol of hydrogen ions/L (corresponding to 6.022×10^{16} ions/L or 0.0000001 g/L). This number is the same as $1/10^7$, or 10^{-7} mol/L. The logarithm of this concentration (the number that is written as a superscript) is -7, so the **negative** logarithm is $+7$, and the pH is thus 7. A solution at pH 8 would have 0.00000001 mol of hydrogen ions/L, one-tenth as much as a solution at pH 7, and a solution at pH 6 would have ten times as much (0.000001 mol/L). In other words, the *higher* the pH value of a solution, the *fewer* hydrogen ions present (Table 2-2). The pH of arterial blood and tissue fluids is normally about 7.40, whereas the pH of blood in the veins is normally about 7.38. Consequently, these fluids usually contain slightly less H^+ than is present in pure water. These pH values may not seem to be significantly different from that of pure water, but small differences can have profound effects on the ability of enzymes to catalyze reactions. As a result of small but uncontrolled changes in pH, many enzymatically controlled reactions essential to life can go awry and result in death.

Acids and Bases

Chemicals that ionize and release hydrogen ions in water are called **acids.** If most or all of the molecules of the acid ionize, it is considered to be a strong acid. Hydrochloric acid, a substance released into the stomach to aid in digestion, is an example of a strong acid since in water it ionizes completely to chloride ion (Cl^-) and hydrogen ion (H^+) (Figure 2-10a). Weak acids do not ionize completely. Acetic acid, the compound that gives vinegar its sour taste, only ionizes about 0.002% when mixed with water (Figure 2-10b).

Chemicals that reduce the concentration of hydrogen ions are called **bases.** Many bases act by removing hydrogen ions from water. For example, ammonia (NH_3) combines with dissolved hydrogen ions to produce ammonium ion (NH_4^+), lowering the amount of dissolved hydrogen ions in the process. Other bases ionize to produce **hydroxyl ions** (OH^-), which combine with hydrogen ions to form water.

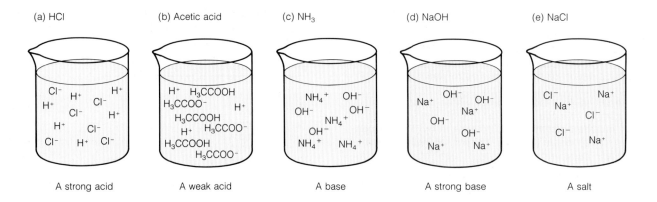

Figure 2-10 Acids, bases, and salts. (a) A strong acid is a substance that ionizes completely in water, releasing H^+ as one of its products. (b) Weak acids, such as acetic acid, (CH_3COOH) also release H^+, but not all the molecules are ionized at the same time. (c) Molecules that remove H^+ from solution, such as NH_3 does when it combines with H^+, are bases. (d) Some bases reduce H^+ by releasing OH^-, which combines with H^+ to form water. Similarly to acids, strong bases ionize completely and weak bases ionize partially. (e) Salts are formed by reactions between acids and bases.

This reduces the hydrogen ion concentration and raises the pH of the solution, that is, makes it more basic. As with acids, strong bases ionize completely and weak bases ionize partially (Figures 2-10c and d).

Solutions can also be classified as acidic or basic in relation to one another. Egg white, for example, has a pH of about 8, and although it is basic when compared with pure water, it is considerably more acidic than milk of magnesia at about pH 10.5 (Table 2-3). Tomato juice, with a pH of about 4.2, is acidic compared with water but is basic relative to lemon juice, which has a pH of about 2.3.

Acids and bases react with one another to produce **salts** (Figure 2-10e). For example, sodium hydroxide (NaOH), a strong base, ionizes to Na^+ and OH^- in water. If HCl is also present, it will ionize to H^+ and Cl^-. Hydrogen and hydroxyl ions produced from the HCl and NaOH will combine to form water, leaving the sodium and chloride ions in solution as a salt.

Buffers

Body tissues produce considerable amounts of hydrogen ions, yet tissues are so sensitive to changes in pH that levels lower than about 7.0 and higher than 7.8 are lethal. Death would occur quickly if there were no mechanisms for removing excess hydrogen ions or adding hydrogen ions when their concentration becomes low. The primary mechanisms for adding or removing hydrogen ions are chemical systems called **pH buffers**. A pH buffer is a substance that reduces the effect of added acids or bases by absorbing or releasing hydrogen ions.

The most important pH buffer in the body is one that involves the chemical interaction of carbonic acid (H_2CO_3) with bicarbonate ion (HCO_3^-). Carbonic acid is a weak acid, ionizing in water according to this reaction:

$$H_2CO_3 \rightleftharpoons HCO_3^- + H^+$$

This equation is a chemist's shorthand way of saying that carbonic acid splits into bicarbonate ions and hydrogen ions. The arrows pointing in both directions indicate that the reaction is **reversible**, meaning that bicarbonate and hydrogen ions can also combine to make carbonic acid. Substances that are involved in a reaction are referred to as **reactants**, and the substances produced by the reaction are called **products**. Thus, in a reversible reaction, the reactants of the reaction

Table 2-3 pH OF SOME COMMON FLUIDS

Substance	pH
Stomach digestive juice	1.0–3.0
Lemon juice	2.2–2.4
Vinegar	2.4–3.4
Apple cider	2.9–3.3
Grapefruit	3.0–3.3
Orange juice	3.0–4.0
Tomato juice	4.0–4.4
Urine	4.8–8.4
Milk	6.3–6.6
Saliva	6.5–7.5
Drinking water	6.5–8.0
Bile	6.8–7.0
Pancreatic digestive juice	7.1–8.2
Blood plasma	7.3–7.5
Spinal fluid	7.3–7.5
Semen	7.4–7.5
Egg white	7.6–8.0
Milk of magnesia	10.0–11.0
Lime, water saturated	12.0

that proceeds in one direction are the products of the reaction that proceeds in the opposite direction.

If we mixed the components of this reaction together, they would react according to this equation, and concentrations of each of the components would quickly stabilize and remain constant. At that point, the reaction would be in **equilibrium.** If we measured the concentration of each of these components at various times, we would always get the same results. Just because the concentrations remain constant, however, does not mean that the chemical reactions have ceased. It means that the reactions continue to occur in both directions at rates that exactly balance one another. The concentration of each of the components is stable because each compound is being produced as fast as it is being used.

In any reaction at equilibrium at a specific temperature, the ratio of the concentration of the products of the reaction to the concentration of the reactants is a constant numerical value called the **equilibrium constant.** When more than one reactant or product is involved, the concentration of the reactants or products is obtained by multiplying the individual concentrations together. To obtain the equilibrium constant of the carbonic acid–bicarbonate reaction, for example, the concentration of hydrogen ions is multiplied by the concentration of bicarbonate ions, which is then divided by the concentration of carbonic acid. (Note that concentrations of substances in moles per liter are indicated by square brackets.)

$$\frac{[H^+] \times [HCO_3^-]}{[H_2CO_3]} = \text{Equilibrium constant}$$

We need not concern ourselves here with the specific value of the equilibrium constant, only with the concept that the ratio of the concentrations will equal that constant value.

Suppose we have a solution that contains these components in equilibrium and we add a small amount of hydrochloric acid to it. In water, HCl ionizes completely to H^+ and Cl^-, thus, with the addition of HCl, the total amount of H^+ is increased by the amount of HCl added to the solution. Since we have added H^+ to our solution, the ratio between the concentration of the products and the concentration of the reactants will change because the amount of H^+ has changed. As a result, the ratio of products to reactants will no longer equal the equilibrium constant.

The equilibrium constant will be restored if either the concentration of HCO_3^- decreases or the concentration of H_2CO_3 increases. In fact, both things happen—some of the H^+ combines with HCO_3^- and produces more H_2CO_3. This results in a simultaneous decrease both in HCO_3^- and in some of the newly added H^+. The reaction is therefore "driven to the left" until equilibrium is reestablished. In a buffered reaction like this, where a molecule can gain or lose a hydrogen ion, the form of the molecule that possesses the hydrogen ion is an acid, and its form when it lacks the hydrogen ion is called the **conjugate base** of the acid.

The important point to realize is that some of the newly added H^+ has been absorbed by HCO_3^- to produce H_2CO_3; that is, the effect on the pH of the solution from adding HCl has been "buffered" by HCO_3^-.

Buffer systems also reduce the effect that added bases have on the pH of a solution. Let us suppose that we have a solution that contains H^+, HCO_3^-, and H_2CO_3 in equilibrium and that we add a small amount of NaOH to the solution. As NaOH (a base) ionizes to sodium ions and hydroxyl ions, hydrogen ions are removed as they combine with the hydroxyl ions, causing the pH to rise.

As H^+ is removed, the ratio of the concentrations of the reaction components changes and no longer equals the equilibrium constant. The ratio is restored to the equilibrium constant in just the opposite way from what we saw following the addition of HCl. That is, enough H_2CO_3 ionizes to H^+ and HCO_3^- to reestablish the equilibrium ratio. In this case, the reaction is "driven to the right." H^+ is added to the solution, compensating for the H^+ that combined with OH^-. In other words, the effect on the pH of the added NaOH has been "buffered" by the H_2CO_3.

There are several other buffer mechanisms that function to maintain the pH of body fluids within necessary limits, and each one works in a manner similar to the carbonic acid–bicarbonate system. These include, in particular, phosphate buffers and the buffering effect of proteins present in the blood, tissue fluids, and cell interiors. In each case, the molecules that are involved are present in an equilibrium mixture of the acid and its conjugate base. Addition or removal of hydrogen ions to the fluid is then followed by absorption or release of hydrogen ions by the buffer molecules until the equilibrium constant is reestablished.

Categories of Organic Compounds

After studying this section, you should be able to:

1. **Define carbohydrates and explain how carbohydrate subunits are used as building blocks in the formation of polysaccharides.**
2. **Distinguish between the structures and functions of starch, cellulose, and glycogen.**
3. **Define lipids and describe the different categories of lipids.**
4. **Describe how subunits are arranged to form different types of lipids.**

carbohydrate: (L.) *carbo*, coal + (Gr.) *hydros*, water
monosaccharide: (Gr.) *monos*, single + (L.) *saccharum*, sugar

5. Describe the structure common to amino acids and explain how amino acids are joined to form proteins.
6. List the subunits that make up nucleic acids and describe the differences in the structures of RNA and DNA.

Life, as we know it, is only possible because of the unique properties of the carbon atom. Compounds that contain carbon are generally referred to as **organic compounds,** in reference to a view once held that carbon-containing compounds were only produced by living things. Because of their size and their distribution of electrons, carbon atoms form relatively stable covalent bonds with one another and with many other elements, especially hydrogen, oxygen, nitrogen, and sulfur. In addition, carbon atoms can form branched and unbranched chains and rings of different lengths that can include atoms of other elements. Because of these properties, carbon compounds are found in an enormous diversity in living matter.

Most organic compounds we encounter in biology fall into one of four general categories of substances: carbohydrates, lipids, proteins, and nucleic acids.

Carbohydrates

A **carbohydrate** is a compound that consists of carbon, hydrogen, and oxygen in an approximate ratio of 1:2:1 (Figure 2-11). Many carbohydrates are macromolecules built up from large numbers of fundamental units called **monosaccharides** (mŏn·ō·săk´ă·rīds). Monosaccharides are simple sugar molecules, usually containing three to six carbon atoms. For example, glyceraldehyde ($C_3H_6O_3$) is a **triose** since it has three carbon atoms, and ribose ($C_5H_{10}O_5$) is a **pentose** since it has five carbon atoms. The most prevalent monosaccharide by far is the six-carbon sugar **glucose** (glū´kōs). Monosaccharides with six carbon atoms such as glucose ($C_6H_{12}O_6$) are also called **hexoses.** Several structural isomers of glucose are also common, including the hexoses **fructose** (frūk´tōs) and **galactose** (gă·lăk´tōs).

Monosaccharides may be joined to form **disaccharides,** molecules that consists of two monosaccharides. Common disaccharides include **sucrose,** or table sugar, which consists of a molecule of glucose bound to a molecule of fructose; **lactose,** or milk sugar, which consists of a glucose molecule bound to galactose; and **maltose,** which consists of two glucose molecules. Much of the sugar we eat consists of disaccharides, which are broken down into their monosaccharide subunits in the digestive tract.

Monosaccharides may be linked together to form larger carbohydrate polymers called **polysaccharides.** Examples of polysaccharides include starch and cellulose, each of which is

Figure 2-11 Carbohydrates. Carbohydrates are molecules that consist of monosaccharides, which exist as chains of carbon atoms with hydrogen (H) and hydroxyl groups (—OH) attached to each carbon. Carbon, hydrogen, and oxygen exist in the ratio of 1:2:1 in monosaccharides. Disaccharides are formed when two monosaccharides bond together. The reaction is accompanied by loss of a water molecule, so the 1:2:1 ratio is only approximate in carbohydrates that contain more than one monosaccharide.

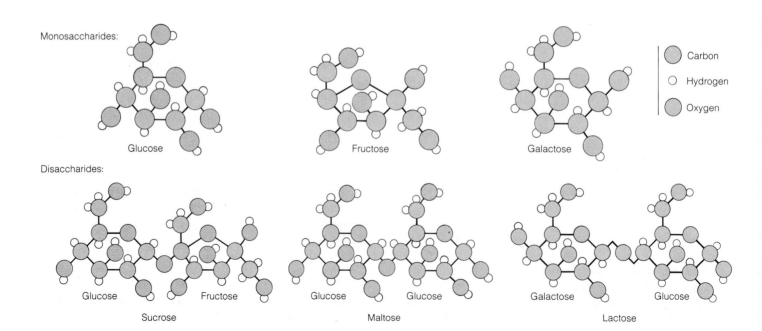

glycogen: (Gr.) *glykys*, sweet + (L.) *genitus*, producing

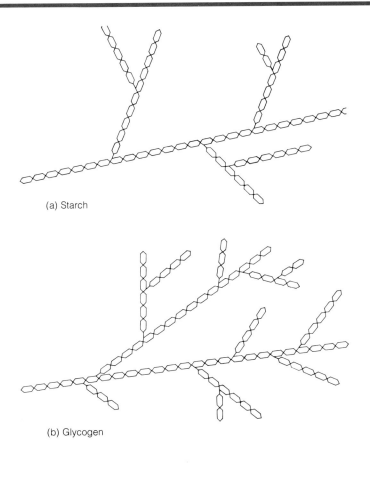

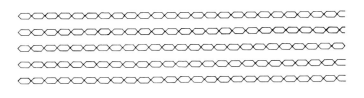

(a) Starch

(b) Glycogen

(c) Cellulose

Figure 2-12 Polysaccharides. (a) Starch and (b) glycogen are examples of polysaccharides formed from thousands of individual glucose units covalently bound in branching chains. (c) Cellulose also consists of thousands of glucose molecules, but the molecules are linked differently than they are in starch or glycogen.

CLINICAL NOTE

SATURATED FATS

Most **saturated fats** in the human diet are from animal sources, including milk, cheese, cottage cheese, lard, butter, lamb, and pork. Coconut oil, a major ingredient in nondairy creamers, is a common source of nonanimal saturated fats. Consumption of excessive amounts of saturated fats can lead to increased chance of heart disease and a buildup of cholesterol deposits in the arteries (see Chapter 23).

cotton or paper, which are virtually 100% cellulose, the glucose subunits remain unavailable because we lack an enzyme for the breakdown of cellulose. Nevertheless, cellulose is valuable because it provides the fiber necessary in a well-balanced diet. In humans and other animals, the most common polysaccharide is **glycogen,** a molecule much like starch in composition. Glycogen is the form in which animal cells store their glucose and one that can be broken down readily when energy needs to be made available.

Lipids

Lipids are fatty and oily compounds that are generally insoluble in water and soluble in certain organic solvents. They are especially soluble in chloroform and benzene. The most common compounds in this class of substances are the fatty acids, neutral fats, phospholipids and steroids.

Fatty Acids

Fatty acids are long chains of carbon atoms, each with one or two hydrogen atoms attached. The molecule terminates in a carboxyl group ($-COOH$) at one end and a methyl group ($-CH_3$) at the other end. Most fatty acids have an even number of carbon atoms in the chain, typically containing 16 or 18 carbon atoms. **Saturated** fatty acids have two hydrogen atoms attached to each carbon atom in the chain, whereas in **unsaturated** fatty acids, certain pairs of carbon atoms are joined by a double bond. Palmitic acid, shown in Figure 2-13a, is an example of a saturated fatty acid common to animal and vegetable fats (see Clinical Note: Saturated Fats). Oleic acid (Figure 2-13b) is an example of an unsaturated fatty acid that contains one pair of double-bonded carbon atoms. Oleic acid is found in olive oil.

Neutral Fats

Neutral fats are formed by combining one to three fatty acids with the three-carbon compound **glycerol** (glĭs´ĕr·ŏl).

formed from many thousands of individual glucose molecules (Figure 2-12). Although these polysaccharides are both polymers of glucose, they are different from one another in the way the glucose units are linked together. **Starch** is an important food source since when it is digested it releases glucose. **Cellulose,** in contrast, is indigestible; if we eat

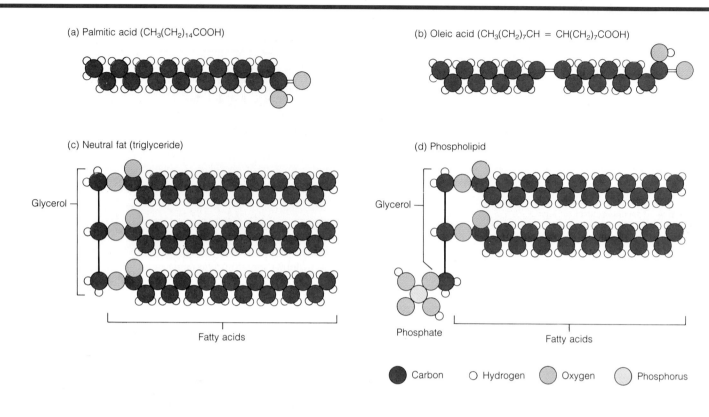

Each of the fatty acids is attached by a dehydration reaction that occurs between the carboxyl group of the fatty acid and a hydroxyl group on the glycerol molecule (Figure 2-13c). Since there are several kinds of fatty acids, differing in the number of carbon atoms in the chain and the degree to which they are saturated, many different kinds of neutral fats can be made. Neutral fats are the form in which much of the body's energy reserve is stored. Layers of fatty tissue also serve an important insulating and cushioning role, protecting the body from external temperature extremes and protecting internal organs from physical shock. Neutral fats are also known as glycerides; mono-, di-, and triglycerides have one, two, and three fatty acids attached to glycerol, respectively.

Phospholipids

Phospholipids (fŏs″fō·lĭp′ĭds) are similar to neutral fats in that they consist of fatty acids attached to a glycerol molecule. They differ, however, in that there are only two fatty acids attached, the third position being occupied by a phosphate group (Figure 2-13d). In many of the phospholipids,

Figure 2-13 Glycerides and fatty acids. Glycerides are compounds formed by joining glycerol and fatty acids. Fatty acids are chains of carbon atoms with associated hydrogen atoms and a —COOH group at one end. (a) In saturated fatty acids, such as palmitic acid, the carbon atoms are attached by single covalent bonds. (b) In unsaturated fatty acids, such as oleic acid, one or more pairs of carbon atoms are joined by double covalent bonds. (c) Neutral fats (triglycerides) consist of a glycerol molecule to which are attached three fatty acids. (d) A phospholipid, common in cell membranes, consists of a glycerol molecule to which are attached two fatty acids and a phosphate group.

the phosphate group has one of several additional groups of diverse structures attached to it. Phospholipids are important constituents of cell membranes.

Steroids

A **steroid** is a lipid that lacks a subunit structure, consisting of a single relatively large but compact molecule. This is a diverse group of compounds having many different functions. Testosterone and estrogen, for example, are steroids that regulate sexual development and behavior in males and

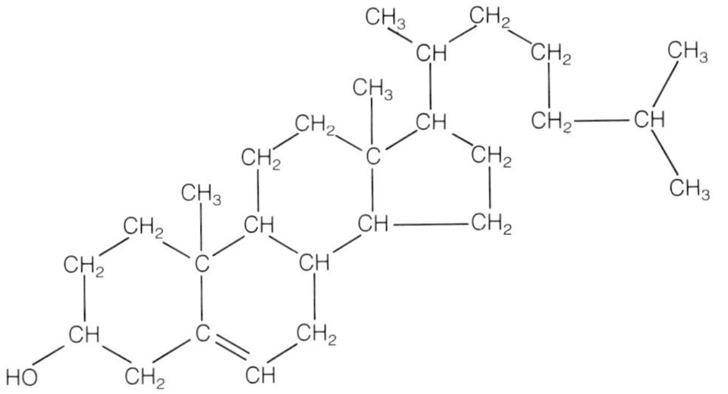

Figure 2-14 The cholesterol molecule. This is an example of a steroid, a group of lipids responsible for diverse functions.

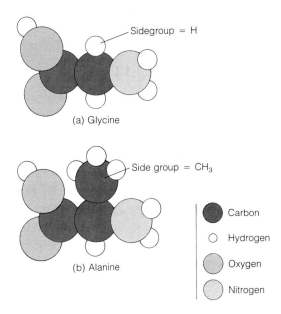

(a) Glycine

(b) Alanine

Figure 2-15 Amino acids. Amino acids are characterized by a carboxyl group (—COOH) and an amino group (—NH$_2$) attached to a central carbon atom. A hydrogen atom and a variable side group are also attached to that carbon atom. Each amino acid differs from the others in the nature of the side group. (a) In glycine, the side group is a hydrogen atom, whereas in (b) alanine the side group is a carbon atom with three attached hydrogen atoms. Twenty different amino acids form the building blocks for proteins (see Figure 2-17).

females. Perhaps the best known steroid is **cholesterol** (kō·lĕs′tĕr·ŏl), a basic steroid found in cell membranes and, when deposited in the circulatory system, is instrumental in causing heart and circulatory disease (Figure 2-14). Other steroids help regulate the concentration of salt in body fluids.

Proteins

Proteins are nitrogen-containing molecules made of subunits called **amino acids** that are linked together in long chains. The chain assumes a specific configuration that gives each protein its characteristic shape. Proteins serve a myriad of roles in life: providing a scaffold that gives cells their shape, serving as regulators of developmental processes, regulating the movement of materials into and out of cells, providing the mechanism for muscular movement, protecting the body against infection, and serving as enzymes that catalyze the thousands of reactions that go on in a cell.

Amino acids have an **amino group** (—NH$_2$) and a carboxyl group (—COOH) attached to a central carbon atom (Figure 2-15). The central carbon atom also has a hydrogen atom and a "side group" attached to it. There are 20 different amino acids, all differing from one another in the nature of this side group. In the simplest case, the amino acid **glycine** (glī′sēn), the side group is a single hydrogen atom. In the amino acid **alanine** (ăl′ă·nēn), the side group is a carbon atom with three hydrogen atoms (—CH$_3$, a methyl group). In tryptophan, one of the more complex amino acids, the side group is a collection of eight carbon atoms, six hydrogen atoms, and one nitrogen atom.

In a protein, the amino acids are joined by **peptide bonds**, which are covalent bonds formed by a dehydration reaction between the amino group of one amino acid and the carboxyl group of the next (Figure 2-16). Since there are 20 different amino acids to choose from, the variety of sequences possible for a protein, even those that contain only a few amino acids, is quite large. Most proteins, in fact, have sequences of 100 or more amino acids, so the number of possible variations is enormous.

There are four aspects to the structure of a protein: primary, secondary, tertiary, and quaternary structures. The **primary** structure of a protein resides in its specific amino acid sequence (Figure 2-17a). Changes in this sequence can profoundly affect the remaining aspects of the protein's structure, to the point where the protein loses its biological activity. **Secondary** structure results from hydrogen bonds that form between the oxygen atom in the carboxyl group of one amino acid and the hydrogen attached to the amino group of the amino acid three positions down the chain. The corresponding atoms of successive amino acids are brought into close enough proximity by these hydrogen bonds to cause the chain to assume a coiled configuration called an **alpha helix** (Figure 2-17b). Not all portions of a chain are necessarily in this configuration, however, and the presence of certain amino acids can interfere with the

38   Chapter 2 The Chemical Basis of Life

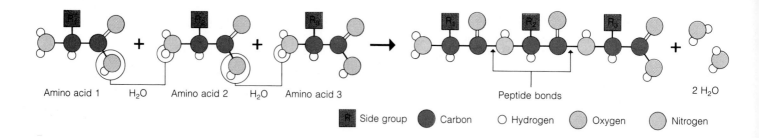

formation of a helix by causing a change in the direction of the chain.

Tertiary structure refers to the specific three-dimensional configuration assumed by a chain of amino acids that results from the combined effect of the amino acids that comprise its primary structure (Figure 2-17c). Such a specific three-dimensional shape is the result of many forces that can develop between different parts of the molecule as they are brought close together by the folding back and forth of the chain. Specific tertiary structure is partly stabilized by covalent bonds that form between two cysteine amino acids, perhaps widely separated in the chain, that are brought together. Cysteine contains a **sulfhydryl** (sŭlf·hī′drĭl) group (—SH) at the end of its side chain. When two such groups are brought together, the hydrogen atoms are removed and

Figure 2-16 A peptide bond. Amino acids are joined by peptide bonds, which are covalent bonds that link the carboxyl group of one amino acid to the amino group of another. Formation of the bond is accompanied by the formation of a water molecule.

Figure 2-17 Protein structure. There are three, and often four, aspects to the structure of a protein. (a) Primary structure is the specific sequence of amino acids in the polypeptide chain. (b) Chains of amino acids often assume a helical form, referred to as secondary structure in the protein. Dotted lines are hydrogen bond. (Some hydrogen atoms are not shown for simplification.) (c) Tertiary structure refers to the overall three-dimensional structure of the polypeptide that results from its primary and secondary structure. (d) When a protein consists of two or more polypeptide chains, it possesses quaternary structure.

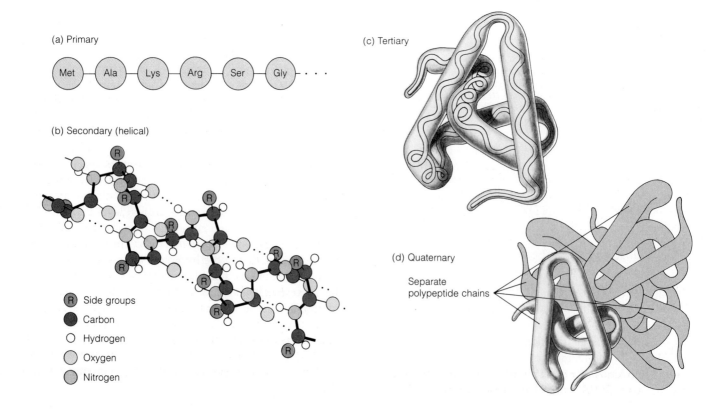

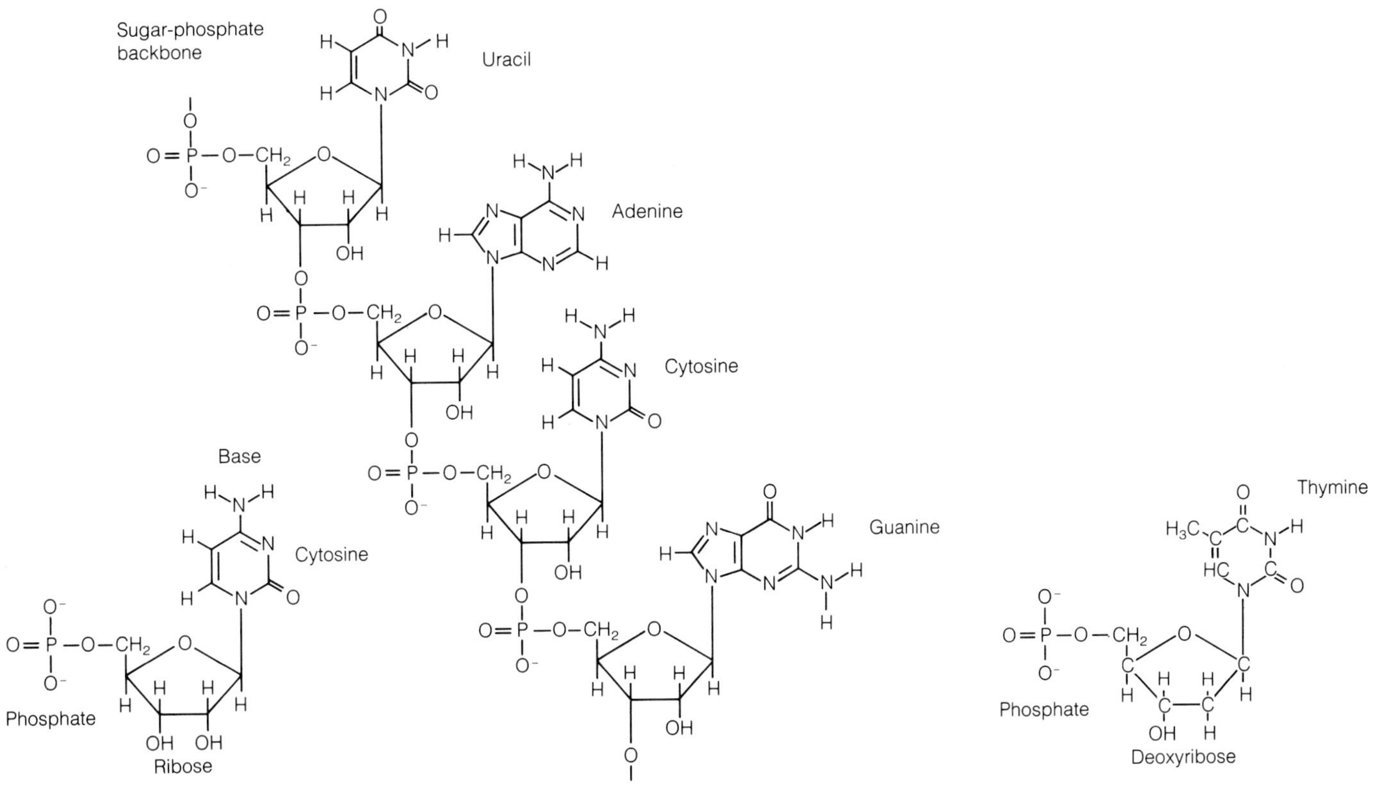

Figure 2-18 Nucleotides of RNA and DNA. (a) Ribonucleotides. Those that form RNA contain ribose, a five-carbon sugar, to which are attached a phosphate group and one of four organic bases, adenine, uracil, guanine, or cytosine. (b) Polynucleotides. RNA consists of chains of ribonucleotides joined by covalent linkages between the phosphate group of one ribonucleotide and the ribose group of the next. (c) A deoxyribonucleotide. These subunits of DNA contain deoxyribose in place of ribose and thymine in place of uracil.

a covalent bond forms between the sulfur atoms, tying the two regions of the chain together at that point.

Many proteins actually consist of more than one chain of amino acids. When a protein consists of two or more chains of the same or different primary structure, the protein possesses a **quaternary** structure (Figure 2-17d). The same factors that determine the structure of proteins with single chains also exist in proteins of this type and here again determine the overall three-dimensional shape of the molecule on which its properties are based.

Nucleic Acids

There are two kinds of **nucleic acids**—**ribonucleic** (rī″bō·nū″klē´ĭk) **acid (RNA)** and **deoxyribonucleic** (dē·ŏk″sē·rī″bō·nū·klē´ĭk) **acid (DNA)**—each of which is a polymer made up of subunits called **nucleotides** (nū´klē·ō·tīds). In RNA each nucleotide is composed of **ribose** (rī´bōs), a pentose type sugar, to which a phosphate group is attached, along with one of four different **organic bases** (Figure 2-18a). The main organic bases of RNA are **adenine** (ăd´ĕ·nēn), **uracil** (ū´ră·sĭl), **guanine** (guă´nēn), and **cytosine** (sī´tō·sĭn), although derivatives of these bases exist in small amounts in certain kinds of RNA molecules. Ribose-containing nucleotides are called **ribonucleotides.** Nucleotides are joined by covalent bonds that connect the phosphate group of one nucleotide to the ribose portion of the next nucleotide to form **polynucleotides** (pŏl″ĭ·nū´klē·ō·tĭd) (Figure 2-18b). Because there are four different kinds of nucleotides, many different sequences of nucleotides are possible, each resulting in a different RNA molecule.

DNA subunits differ from those of RNA in possessing the sugar **deoxyribose** instead of ribose and the base **thymine** (thī´mĭn) instead of uracil (Figure 2-18c). Deoxyribose differs from ribose by having one less oxygen atom. Thymine differs from uracil in having an additional methyl group (—CH_3). A small amount of thymine is also present in certain kinds of RNA molecules in addition to greater amounts of uracil.

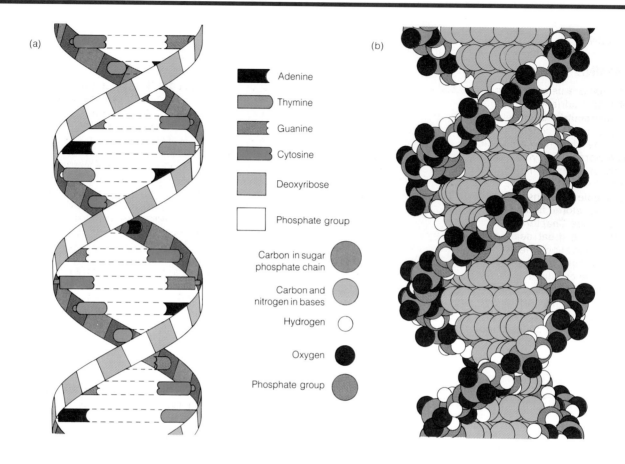

Figure 2-19 DNA. DNA usually consists of two chains of deoxyribonucleotides wrapped around one another to form a double-stranded helical molecule. (a) Schematic drawing shows how the two strands are held in association with one another by hydrogen bonds that form between adenine (A) and thymine (T) subunits and between guanine (G) and cytosine (C) subunits that lie opposite one another in the two strands. (b) Space-filling model shows a more representative view of the three-dimensional nature of the complex DNA molecule.

Although there is considerable similarity between the nucleotides of DNA and RNA and the way they are attached to one another, a significant difference does exist between the two molecules at a higher level of organization. While the RNA molecule consists of one long single strand, generally containing several hundred nucleotides at most, DNA consists of a long double strand of nucleotides, numbering on the order of 100 million or more in each strand. The two strands are twisted around one another to form a **double helix,** and they are held together by hydrogen bonds that form between adenine and thymine and between guanine and cytosine (Figure 2-19). The DNA molecule stores genetic information as specific sequences of such nucleotide pairs. The manner in which this information is used by the cell is described in more detail in Chapter 3 in the section on DNA function.

Certain nucleotides serve important roles quite apart from their role as components of nucleic acids. Guanosine monophosphate (GMP) is a guanine-containing nucleotide that plays a role in energy metabolism. Adenosine triphosphate (ATP), discussed earlier, is an especially important source of readily available energy in many cell reactions. Because of the importance of these compounds, considerable attention will be devoted to them in later chapters.

In this chapter we have seen how molecules can react to form new substances through internal rearrangements and exchange of atoms. We have also demonstrated the importance of organic compounds to living organisms. In the next chapter we will study cells, the smallest organizational units possessing all characteristics of life. In that chapter you will see how the chemical reactions studied here are responsible for the life functions of cells.

THE CLINIC

VITAMINS

Foods that are used to nourish the body are called **nutriment**. **Macronutrients** are carbohydrates, fats, and proteins, while **micronutrients** include vitamins and minerals. **Vitamins** are usually consumed in the diet in quantities less than 1 g per day. They are absorbed unchanged and serve as cofactors to enzymes, substances that increase the velocity of chemical reactions in the body. Vitamins are classified as water soluble (the B vitamins and vitamin C) or fat soluble (vitamins A, D, and K). The B vitamins (except B_{12}) are not stored in the body and need to be replenished each day (see Chapter 19).

There is a great deal of misunderstanding about the use of vitamin supplements among today's consumers, and millions of dollars are spent annually for vitamin pills. In reality, if we consume an adequate nutritious diet there is little need for vitamin supplements. Some fitness enthusiasts embrace a program of megadoses of vitamins which may indeed be toxic.

Specific vitamin deficiencies have been described in the past and related to specific disease syndromes. These are very rare in modern industrial nations today. Lack of vitamin C was a scourge of sailors in the days of sailing ships since they had no source of fresh fruits and vegetables on long voyages. Many developed **scurvy**, a disease producing skin ulcerations, ulcers of the mouth, and bleeding problems that resulted in many deaths. A British naval surgeon discovered that the disease could be prevented with daily doses of citrus juices (usually limes), thus coining the term "limies" for British sailors. **Beriberi** was caused by a lack of vitamin B_1 and resulted in nerve pains, heart failure, and mental illness. **Pellagara**, or deficiency of niacin, caused skin rashes, a painful tongue, gastrointestinal problems, and mental disorders.

These diseases are rarely seen today. What we do see on occasion are multiple vitamin deficiencies in patients with severe alcoholism who drink but do not eat. Famine in some Third World countries produces starvation, a general lack of all nutrients.

STUDY OUTLINE

The Nature of Matter (pp. 19–26)

Matter is anything that occupies space and can be perceived by one or more of the senses. It consists of relatively few components combined into many different forms.

Chemical Elements Matter is made up of specific *chemical elements*, about 100 of which have been identified. About one-third of those are important to life. Each element is identified by a symbol that is unique to that element. The symbol usually consists of the first one or two letters of the English or Latin name of the element.

Atoms The basic unit of an element is the *atom*. Atoms consist of subunits—*protons, neutrons,* and *electrons*—which are combined in specific numbers and ratios unique to the atoms of that element. The number of protons in the nucleus of the atoms of an element constitute that element's *atomic number*.

Protons and neutrons of an atom are organized into a central *nucleus*. Electrons travel in the space surrounding the nucleus. Protons have a positive electric charge, electrons a negative charge, and neutrons no charge. The number of protons is the same as the number of electrons, so there is no net electric charge in an atom.

Isotopes Forms of an element that differ in their number of neutrons are called *isotopes*; isotopes of a given element generally have the same chemical properties. The total number of protons and neutrons in an isotope is the *mass number*. Some isotopes are unstable, disintegrating spontaneously into fragments of the atom and releasing energy and particles. Unstable elements are *radioactive*.

Atomic Mass The sum of the protons plus the average number of neutrons in the atoms of an element is the *atomic mass*. The unit of atomic mass is the *dalton*, stipulated as the weight of a single hydrogen atom. The average mass of an element, obtained when the different isotopes are considered, is the *atomic weight* of the element.

Electron Orbitals Electrons travel in the space around a nucleus in specific regions called *orbitals*. Orbitals are organized into groups called *shells*, each of which represents particular *energy levels* for the electrons that travel in it. Electrons in the outermost or highest energy level are primarily responsible for the chemical properties of the atom.

Ions Electrons can be added to or subtracted from the outermost orbitals of most atoms. When this happens there is no longer a balance between the electrons and protons of the atom, and a net electric charge results. Such a charged atom is an *ion*.

Chemical Bonds The association of two atoms or ions with one another is often stabilized by forces of attraction called *bonds*.

Negatively and positively charged ions may be held together by the force that oppositely charged particles exert on one another. Such a force is an *ionic bond*.

Atoms may be held together by a force that develops as a result of the sharing of electrons by two different atoms. Such a force is a *covalent bond*.

Because of the unique structure of a hydrogen atom, containing a single electron shell, formation of a covalent bond with an oxygen or nitrogen atom causes the hydrogen atom's electrons to be distributed generally in the space between the two atoms. This leaves the proton exposed. The positive charge of the proton and the negatively charged surface of neighboring atoms exert a force on one another. This is a *hydrogen bond.*

Molecules Two or more atoms held together by covalent bonds constitutes a *molecule.* Large molecules, called *macromolecules,* are often formed by the accumulation of smaller molecules in a process called *polymerization.* Molecules are designated by the chemical symbols for the atoms in it followed by the number of that atom, such as $C_6H_{12}O_6$ for glucose.

Chemical Reactions (pp. 26–27)
Interactions that occur within and between molecules that result in a change in their structure are called *chemical reactions.*

Rearrangement Reactions A *rearrangement reaction* is a change in the spatial relationship of the atoms within a molecule.

Addition Reactions The combining of atoms, or groups of atoms, to form new molecules is an *addition reaction.*

Cleavage Reactions In a *cleavage reaction,* the atoms of a molecule are divided into two groups, each of which forms a new molecule.

Transfer Reactions In a *transfer reaction,* an atom or group of atoms leaves one molecule and becomes attached to another group of atoms.

Oxidation-Reduction Reactions An *oxidation-reduction reaction* is one in which one or more electrons is transferred from one atom or molecule to another.

Metabolic Reactions (pp. 27–30)
The total of the reactions of life are *metabolic reactions.* These reactions are frequently associated with the formation of larger molecules, *anabolism,* or the disassembly of such larger molecules, *catabolism.*

Factors Affecting Reactions Rate Reactions occur at rates that are dependent generally on conditions that affect the likelihood of reacting molecules coming into contact with one another. Concentration, temperature, and pH are especially important in determining the rate at which a reaction occurs.

Concentration Concentration is the amount of a substance within a given volume, expressed in weight per volume, or in numbers of molecules of a substance per volume. A *mole* is an amount of a substance that contains the same number of units as present in exactly 12 grams of carbon-12. This number, called *Avogadro's number,* is 6.02×10^{23}. The weight of a mole of a substance is equal to the sum of the atomic weights of the atoms in the substance (its *molecular weight*) expressed in grams.

Enzymes The rate at which a metabolic reaction occurs is almost always controlled by an *enzyme.* Enzymes are usually proteins that interact with the components of a reaction in such a way as to increase the likelihood of a reaction occurring. Most enzymes speed up, or *catalyze,* one specific reaction or class of reactions.

Hydrogen Ion Concentration (pp. 30–33)
The concentration of hydrogen ions is particularly important in determining the rate at which a reaction occurs. Hydrogen ion concentration is expressed as *pH,* defined as the negative logarithm of the hydrogen ion concentration expressed in moles per liter of water.

Acids and Bases Solutions can be classified as *acids* or *bases* depending on their pH. A solution with a pH less than 7.0 is an acid, one with a pH greater than 7.0 is a base, and one with a pH of 7.0 is considered *neutral.*

Weak and strong acids and bases are characterized by the degree to which they ionize. Strong acids and bases ionize completely, while weak acids and bases ionize incompletely.

Salts are produced by the reaction of acids and bases with one another.

Buffers Chemicals that tend to absorb or release hydrogen ions in response to changes in pH are called *buffers.* Buffer systems reduce the effect that added acids and bases have on the pH of a solution. Sufficient ions are absorbed or released to maintain an *equilibrium* concentration. An *equilibrium constant* is a numerical ratio of the concentration of reactants and products of a chemical reaction.

Categories of Organic Compounds (pp. 33–40)
Organic compounds consist of molecules that contain carbon. There are many different forms of organic molecules, but most can be classified into one of four general categories.

Carbohydrates *Carbohydrates* consist of carbon, hydrogen, and oxygen atoms in the approximate ratio of 1:2:1. The basic unit of a carbohydrate is a *monosaccharide,* which can combine with other monosaccharides to form larger molecules. *Disaccharides* are formed from two monosaccharides, and *polysaccharides* may contain thousands of monosaccharide subunits.

Lipids Fatty and oily compounds that are relatively insoluble in water and soluble in certain organic solvents are called *lipids.* The most common lipids are fatty acids, neutral fats, phospholipids, and steroids.

Fatty acids are long chains of carbon atoms, each carbon having one or two hydrogen atoms attached. A carboxyl group (—COOH) is present at one end. In saturated fatty acids, all the carbon atoms are bound to neighboring carbon atoms by single bonds. In unsaturated fatty acids, two or more adjacent carbon atoms are bound by a double covalent bond.

Neutral fats (or triglycerides) are combinations of three fatty acids with a glycerol molecule. *Phospholipids* are similar to neutral fats except that one of the fatty acids is replaced by a phosphate group. Sometimes the phosphate group has still another molecule attached to it. *Steroids* such as cholesterol are lipids that lack a subunit structure. They consist of as many as 30 carbon atoms that form rings of carbon atoms within the molecule.

Proteins Nitrogen-containing molecules composed of amino acid subunits are *proteins*. The amino acid subunits form chains that assume specific three-dimensional shapes that are important to the protein's function.

Amino acids have an amine group and a carboxyl group separated by a single carbon atom. The intervening carbon atom has, in addition, a hydrogen atom and a variable side chain attached. Amino acids differ from one another in the nature of the side chain.

Amino acids are attached to one another by *peptide bonds* that form between the amine group of one amino acid and the carboxyl group of the next.

Primary, secondary, tertiary, and *quaternary* aspects of the structure of a protein relate to amino acid sequence, conformation of the chain that results from hydrogen bonds that form between amino acids in the chain, the overall three-dimensional shape of the protein, and, when present, the association of two or more polypeptide chains in a single protein molecule.

Nucleic Acids Polymers that may consist of many thousands of nucleotide subunits are called *nucleic acids*. Two types of nucleic acids exist: ribonucleic acid (RNA) and deoxyribonucleic acid (DNA).

Ribonucleic acid (RNA) is composed of a single chain of *nucleotides*. Each nucleotide consists of a phosphate portion, a five-carbon sugar (ribose), and a nitrogenous base. Nucleotides differ from one another in the nature of the base, which may be *adenine, uracil, guanine,* or *cytosine*.

Deoxyribonucleic acid (DNA) is composed of a double chain of nucleotides wound around one another in a double-stranded helix. Each chain consists of a phosphate portion, a five-carbon sugar (deoxyribose), and a nitrogenous base. In DNA, uracil is absent, and thymine is present as one of the four bases. The two chains are held together by hydrogen bonds that form between adenine and thymine and between guanine and cytosine in corresponding positions in the two chains.

Adenosine and *guanosine* phosphate molecules are nucleotides that are important in energy metabolism.

SELF-TEST OF CHAPTER OBJECTIVES

True-False Questions
1. Atoms of a given element may not have the same number of protons, neutrons, and electrons.
2. Neutrons, protons, and electrons have the same masses.
3. The outermost electron orbital in an atom has the lowest energy level.
4. Chlorine atoms ionize by losing the seven electrons of their outermost orbital.
5. The forces that hold water molecules together in ice are called ionic bonds.
6. Electrons shared between atoms create forces of attraction called covalent bonds.
7. The greater the hydrogen ion concentration, the higher the pH.
8. The pH of a buffered solution does not change with the addition of acid.
9. Phosphate groups are found in neutral fats.
10. Thymine is one of the four main organic bases of DNA, but not of RNA.

Multiple-Choice and Fill-In Questions
11. Atoms of which of the following have the greatest number of protons?
    a. hydrogen   c. oxygen
    b. carbon   d. chlorine
12. Which of the following particles that makes up an atom has no electric charge?
    a. proton
    b. neutron
    c. electron
13. Which of the following lists is in order of increasing size?
    a. proton, nucleus, orbital   c. electron, orbital, nucleus
    b. nucleus, orbital, electron   d. nucleus, neutron, orbital
14. An atom that possesses an electric charge as a result of gaining or losing an electron is called a(n)
    a. positron   c. molecule
    b. ion   d. nucleus
15. Forces of attraction between atoms in a molecule are called _____.
16. The attraction that results from the sharing of electrons between two atoms is called a(n) _____ _____.
17. The attraction that exists between positively and negatively charged ions is called a(n) _____ _____.
18. The conversion of a molecule of glucose to a molecule of fructose involves
    a. the rearrangement of the atoms of the molecule
    b. addition of atoms from another molecule to glucose
    c. splitting of a molecule of glucose
    d. removal of electrons from glucose
19. The production of glucose phosphate from ATP and glucose involves
    a. the transfer of a phosphate group from a molecule of ATP to a molecule of glucose
    b. cleavage of the glucose molecule into two equal halves
    c. the transfer of electrons
    d. the addition of ATP to glucose
20. Which of the following has the highest pH?
    a. a solution with 0.000001 moles of H^+ per liter H_2O
    b. a solution with 0.0000001 moles of H^+ per liter H_2O
    c. a solution with 0.00000001 moles of H^+ per liter H_2O
21. Which of the following is the most acidic?
    a. a solution with 0.000001 moles of H^+ per liter H_2O
    b. a solution with 0.0000001 moles of H^+ per liter H_2O
    c. a solution with 0.00000001 moles of H^+ per liter H_2O

22. In a solution of acetic acid in water, some of the acetic acid molecules ionize to produce a mixture of acetate ions, hydrogen ions, and acetic acid. What effect will the addition of more acetic acid to the solution have on the pH of the solution?
    a. raise it
    b. lower it
23. Acetate ions can absorb H^+. Acetate ions can therefore act as a(n) _____ _____.
24. Which of the following yields only glucose as a result of a cleavage reaction?
    a. maltose   c. lactose
    b. sucrose   d. galactose
25. Which of the following is not one of the major bases of DNA?
    a. adenine   c. cytosine
    b. guanine   d. uracil
26. The chemical formula for both glucose and fructose is $C_6H_{12}O_6$. These two molecules can be combined to produce a molecule of sucrose, which has the chemical formula $C_{12}H_{22}O_{11}$. The reaction that joins glucose and fructose is apparently an example of
    a. an oxidation-reduction reaction
    b. a dehydration reaction
    c. an isomerization reaction
    d. a polymerization reaction
27. Maltose is formed from two molecules of glucose, thus maltose is an example of a
    a. monosaccharide   c. peptide
    b. disaccharide   d. polysaccharide
28. If the atoms of a particular element each have 17 electrons and the mass of the most common isotope is 35, how many neutrons are present in that isotope?
    a. 17   c. 20
    b. 18   d. 35
29. Atom A has a mass of 14 and an atomic number of 6. Atom B has a mass of 15 and an atomic number of 7. Atom C has a mass of 14 and an atomic number of 7. Atom D has a mass of 13 and an atomic number of 6. Which of these atoms are isotopes of one another?
    a. A and B, C and D   c. A and C, B and D
    b. A and D, B and C   d. A and C only
30. Hemoglobin is a protein in blood cells that is responsible for carrying oxygen from the lungs to the tissues. Hemoglobin consists of four chains of amino acids, thus it is an example of a protein that
    a. lacks tertiary structure
    b. lacks secondary structure
    c. possesses quaternary structure
    d. lacks quaternary structure
31. One liter of the air we breathe (at sea level and at 25°C) usually contains about 5.4×10^{21} molecules of oxygen. About how many moles of oxygen are usually present per liter of air?
    a. 1.0   c. 0.01
    b. 0.1   d. 0.001
32. When someone diets successfully, stored fat molecules are broken down and used for fuel and for the synthesis of new substances. Breakdown of fat molecules is an example of which of these reactions?
    a. anabolic   c. dehydration
    b. catabolic   d. depolymerization
33. Which of the following associations of compound and function is not correct?
    a. glycogen, energy storage
    b. phospholipid, membrane structure
    c. fatty acid, enzyme catalysis
    d. DNA, information storage
34. How many electrons are shared between two atoms held together by a triple covalent bond?
    a. 2   c. 6
    b. 4   d. 8
35. In which of the following compounds would you not find nitrogen?
    a. protein   c. RNA
    b. DNA   d. starch

Essay Questions

36. If a protein contains more than 100 amino acids, it could conceivably exist in an astronomical number of different amino acid sequences (more than 10^{130}). Yet the number of different sequences actually produced by living organisms is estimated to be only a few tens of thousands. Speculate on why so relatively few different proteins are produced.
37. Describe how an enzyme is able to speed up one chemical reaction and yet have no effect on other chemical reactions.
38. Draw a model of a hydrogen atom covalently bound to an oxygen atom and forming a hydrogen bond to a nitrogen atom.
39. Describe four aspects of protein structure and cite the kinds of bonds involved in establishing each aspect.
40. Show how monosaccharides can be assembled to form successively more complex molecules.
41. Describe how an oxidation-reduction reaction occurs, identifying the molecules that are oxidized and reduced in the process.
42. List four differences in the structure of RNA and DNA.
43. List and describe factors that influence the rate of a chemical reaction, describing how each factor influences the rate.
44. Hydrogen bonds can be disrupted by changes in pH. Enzymes are inactivated at improper pH levels. Speculate on the relationship between these two statements.
45. Explain why an increase in hydrogen ion concentration is associated with a decrease in pH.
46. Given that in a chemical reaction at equilibrium, the ratio of products to reactants is constant, describe what the effect will be of adding products to the mixture.
47. Different isotopes of an element generally have the same chemical properties. Describe the difference in structure of isotopes of an element and explain why the chemical properties of the isotopes will be the same in spite of differences in structure.
48. Describe the structure of a water molecule and explain what happens when liquid water becomes ice.

49. Atoms of the element silicon contain 14 protons and have an atomic weight (in the most common isotope) of 28. The most common isotope of carbon has 6 protons and an atomic weight of 12. Compare these atoms with respect to numbers of protons, neutrons, electrons, and electron orbitals.
50. Oleic acid is a fatty acid that contains 18 carbon atoms, 34 hydrogen atoms, and 2 oxygen atoms. Stearic acid also contains 18 carbon atoms and 2 oxygen atoms but 36 hydrogen atoms. Explain which of these molecules is unsaturated and explain the basis for the fewer number of hydrogen atoms in oleic acid compared with stearic acid.

Clinical Application Questions
51. Briefly describe what photons are and various techniques that use photons to form images of internal body structures.
52. Define *radioisotopes* and explain how they are used in diagnosing and treating certain diseases.
53. What do PET and SPECT stand for? Explain how these techniques are being used now (and may be used in the future) to study the body in great detail.
54. Define the term serum electrolytes and explain why the concentration and proper balance of serum electrolytes are important to life.
55. Distinguish between macronutrient and micronutrients and cite examples of diseases that might develop as a result of a micronutrient deficiency.

3
Cell Structure and Organization

One of the fundamental principles of biology is that organisms consist of living subunits called **cells.** Collectively, cells form the structure of an organism and perform the chemistry that underlies the organism's physiology. To understand fully the principles of anatomy and physiology, it is thus necessary to understand what cells are and how they function. Consequently, in this chapter we will consider

1. the cellular basis of life,
2. the organization and structure of typical cells,
3. the functions performed by specific structures within cells,
4. the mechanism for the production of new cells, and
5. how hereditary information stored in a cell is used.

cytology: (Gr.) *kytos*, cell + *logy*, study of
cytoplasm: (Gr.) *kytos*, cell + *plasma*, substance
plasmalemma: (Gr.) *plasma*, substance + *lemma*, skin, husk
nucleoplasm: (L.) *nucleus*, kernel + *plasma*, substance

Cellular Basis of Life

In the late nineteenth century two scientists named Matthias Schleiden and Theodor Schwann independently pointed out that virtually all organisms consist of cells and that these cells are responsible for the properties of the organism itself. Schleiden and Schwann had drawn their conclusions from the work and observations of numerous investigators who had preceded them and who had already laid the foundation for a branch of biology called **cytology** (sī·tŏl´ō·jē). Cytology, the study of cell structure and function, today remains an area of intensive and productive research.

The generalizations made by Schleiden and Schwann are no less applicable to humans than they are to other organisms. There are probably on the order of 10^{13}, or one hundred trillion, cells in the adult human body, ranging in size from tiny red blood cells, disks 7.5 micrometers (μm) in diameter by 2 μm thick, to cells of the nervous system, which may be more than 1 m long. The combined activity of these cells makes possible the phenomena that we associate with life—the ability to move, to reproduce, to react to the environment, to breathe, and indeed, even to think. Consequently, any thorough knowledge of the anatomy and physiology of the human organism requires a knowledge of the structure of cells and the roles they play in the life of the individual.

Most cells consist of two major divisions: the **cytoplasm** (sī´tō·plăzm) and the **nucleus** (nū´klē·ŭs). The cytoplasm is a complex mixture of structures enclosed within a membrane called the **plasmalemma** (plăz˝mă·lĕm´ă) or **cell membrane.** The nucleus is embedded in the cytoplasm and contains a mixture of substances called **nucleoplasm** (nū´klē·ō·plăzm˝). It is also bounded by a membrane called the **nuclear envelope.** Figure 3-1 shows a thin section of a cell under an electron microscope and a drawing of a composite cell containing all of the major cellular subunits.

Plasmalemma

After studying this section, you should be able to:

1. Describe the function of the cell membrane.
2. List the chemical components of a cell membrane and draw a diagram showing the membrane's organization.
3. Explain the basis for the term *fluid mosaic* in relation to the organization of the membrane.
4. List and describe the structure and function of three types of specialized regions of cell membranes formed where cells come in contact with one another.

The plasmalemma, or cell membrane, plays an important role in regulating the movement of many substances into and out of the cytoplasm, since all the substances that enter or exit the cell must pass through it. The smallest molecules, such as water, oxygen, and carbon dioxide, are relatively free to diffuse through the membrane, but larger molecules, such as glucose or amino acids, are not. It is the passage of these larger molecules that is controlled. Because some substances can pass through the membrane and others cannot, the membrane is said to be **selectively permeable.**

Organization

The plasmalemma is made of lipids, proteins, and carbohydrates organized into a thin covering only about 8 to 10 nanometers (nm) thick. The lipid portion comprises about half the mass of the membrane and consists primarily of phospholipid, with cholesterol and other lipids also present in smaller amounts. The lipid molecules are arranged in a double layer called a **lipid bilayer,** which forms about half the thickness of the membrane. In each layer, the lipid molecules are oriented with their chains of fatty acids directed toward the center of the bilayer and with the glycerol portions lying on either surface. The double nature of the bilayer is readily observed in electron micrographs as a pair of densely stained lines, separated by a more lightly stained region (Figure 3-2).

The proteins of the membrane are usually embedded in the lipid bilayer. Some protein molecules are embedded entirely in one of the two lipid layers, while others extend through both layers and are in contact with the fluid on both sides of the membrane. The carbohydrate portion of the membrane consists of individual sugar molecules or short chains of sugar molecules that are attached to some of the protein and lipid molecules of the membrane. These sugars are found on the exterior of the membrane.

An important feature of the membrane is that the individual lipid and protein molecules are not covalently bound together. Instead, the lipid portion acts as a two-dimensional fluid in which individual molecules are free to move about and in which protein molecules essentially float. The cell membrane is often referred to as a **fluid mosaic,** reflecting both its fluid nature and the mosaic of proteins embedded within it (Figure 3-3).

Junctions

In some areas of the plasmalemma, cells come into contact with one another and form specialized structures called **junctions.** There are three categories of junctions named on the basis of the functions they perform: communicating junctions, impermeable junctions, and adhering junctions.

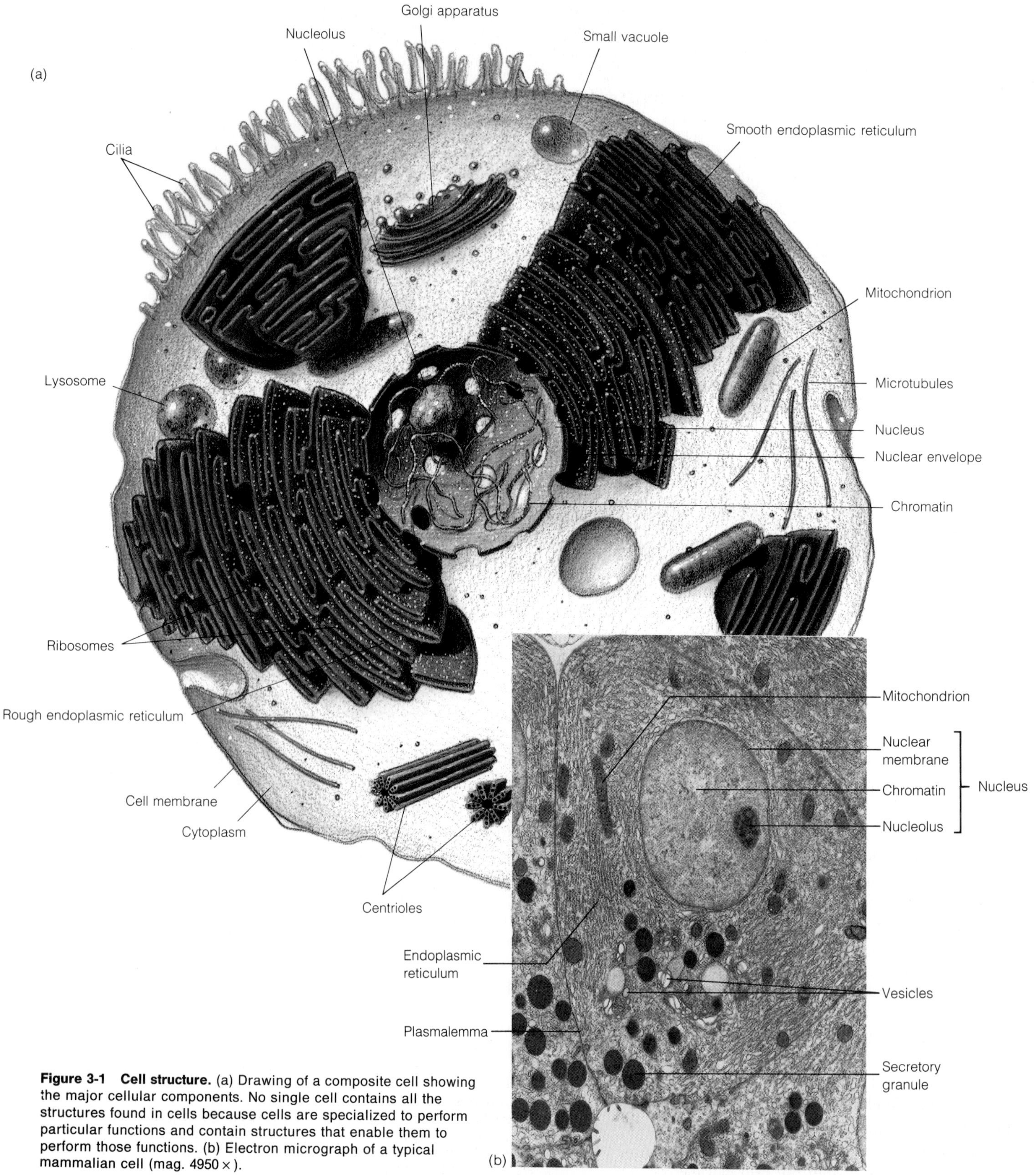

48 Chapter 3 Cell Structure and Organization

Figure 3-1 Cell structure. (a) Drawing of a composite cell showing the major cellular components. No single cell contains all the structures found in cells because cells are specialized to perform particular functions and contain structures that enable them to perform those functions. (b) Electron micrograph of a typical mammalian cell (mag. 4950×).

Plasmalemma

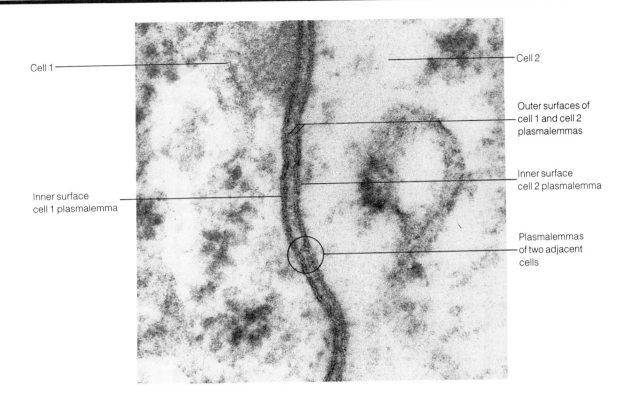

Figure 3-2 Plasmalemma. This electron micrograph shows the plasmalemmas (or cell membranes) of two adjacent cells. Each appears as a pair of parallel dark lines separated by a lighter region. The dark lines correspond to water soluble heads of phospholipids and other lipids within the inner and outer surfaces of the membrane. The light space between them corresponds to a region where the water insoluble tails of the fatty acids meet in the interior of the membrane. A plasmalemma is typically about 5 nm wide (mag. 162,648×).

Communicating junctions are formed when the membranes of two adjacent cells lie parallel and especially close together. Viewed through an electron microscope, the two membranes appear to form a single, multilayered structure with a thickness of about 14 nm. Within this region the membranes are separated from each other by a gap of only about 2 to 4 nm (Figure 3-4a). Structures called **connexons** bridge this gap and form channels that provide a direct connection between the cytoplasm on either side of the junction. Communicating junctions are points in the membrane where small molecules can move directly from one cell to the next, allowing adjacent cells to exchange substances. Communicating junctions are found in most types of cells, but they are *not* found in blood cells, certain muscle cells, and cells that line glands and hollow tubular organs such as the small intestine. The most common form of communicating junction is called a **gap junction**, in reference to the narrow space that lies between the two cell membranes.

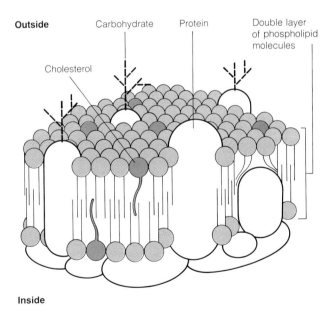

Figure 3-3 Fluid mosaic structure of a cell membrane. The proteins are embedded in a fluid lipid bilayer, some penetrating both surfaces, others embedded in one surface or the other. In a plasmalemma most proteins and many lipids in the outer surface have chains of sugar molecules (carbohydrates) attached. The carbohydrates in the membrane can represent as much as 10% of the weight of the membrane.

desmosome: (Gr.) *desmos*, band + *soma*, body

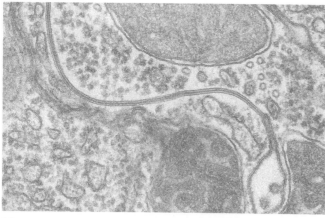

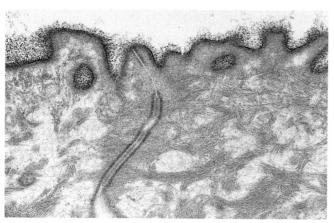

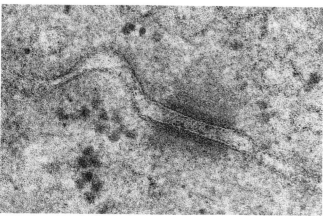

Figure 3-4 Junctions in cell membranes. (a) One type of communicating junction is called a gap junction (mag. 29,628×). Communicating junctions are points where small molecules can pass directly from one cell to another. (b) An impermeable or tight junction (this one from frog skin) seals the regions where adjacent cells are in contact and prevents the flow of materials between the cells (mag. 31,274×). (c) An adhering junction or desmosome (this one from rat liver) is a point where membranes of neighboring cells are attached to one another (mag. 139,305).

Impermeable junctions are often found between cells of glands and hollow organs where they seal the contacting surfaces of neighboring cells. These junctions hold the cells together and prevent the diffusion of substances through the wall of the organ, requiring instead that materials pass through the cells. This enables the cells to regulate the passage of materials into or out of the organ. Impermeable junctions are also referred to as **tight junctions** (Figure 3-4b).

In **adhering junctions**, the plasmalemmas of adjacent cells lie parallel to one another, separated by a space of about 25 nm (Figure 3-4c). Fine filaments lie across the junction serving to hold the cells together. Adhering junctions are also referred to as **desmosomes** (dĕs′mō·sōms). In **spot desmosomes**, adhering junctions appear as spots along the membrane of neighboring cells where they essentially rivet the cells together. **Belt desmosomes** occupy more of the membrane surface, often encircling the cell and attaching it to all of its neighbors. Adhering junctions and their associated filaments are particularly abundant in the cells of skin and other tissues that must withstand considerable mechanical stress without separating from one another.

Movement Through the Cell Membrane

After studying this section, you should be able to:
1. Describe the factors that influence the ability of a substance to pass through a cell membrane.
2. Define osmosis and describe the molecular mechanism for it.
3. Distinguish between the mechanisms of facilitated diffusion and active transport as means of passage of smaller molecules through a cell membrane.
4. Define and distinguish between endocytosis and exocytosis as means of transporting larger molecules through a cell membrane.

Materials must be able to pass through the cell membrane if the cell is to use them or if (in the case of waste products) the cell is to rid itself of them. Many different kinds of substances pass through a membrane, ranging in size from water molecules and molecules of dissolved oxygen and carbon dioxide to sugar and amino acids. Even proteins and other large molecules are able to enter some cells.

Effects of Size, Charge, and Solubility

The ease with which a substance passes through the membrane depends on three factors: its molecular size, the electric charge on the molecule, and the solubility of the substance in the membrane. In general, relatively small uncharged molecules are able to pass through easily; their small size counteracts their insolubility in lipids. For example, water is relatively insoluble in lipids, yet it passes through the membrane rapidly, as do oxygen and carbon dioxide molecules. These molecules are able to pass between the larger molecules that make up the membrane. Larger molecules and ions of individual molecules are less free to pass through the membrane because of their size and/or their insolubility in the lipid portion. Movement of some of these substances requires special transport molecules in the membrane to carry them across. Still other substances may not be able to pass through at all because the membrane lacks specific carriers for them (see the section on Mechanisms of Transport).

The driving force behind the movement of any substance through a membrane is its **kinetic energy.** A substance that possesses kinetic energy is in constant motion—the more kinetic energy it has, the more motion. At the molecular level in a liquid or gas, this motion takes the form of individual molecules flying in random directions through space, tumbling as they go, and colliding with one another and with the walls of the container that holds them. In a solid, individual molecules are not so free to move about, but are instead more or less locked in place. Even in these molecules, however, there can be considerable vibrational motion depending on the kinetic energy possessed by the molecules.

The kinetic energy of matter is responsible for a very important and fundamental phenomenon called **diffusion.** Diffusion is the random movement of atoms and molecules from regions where they are most highly concentrated to regions where they are less concentrated. If a gaseous or liquid substance, for example, is more concentrated in one region of a container than in another, random collisions of the molecules with one another and with the walls of the container will result in a diffusion of the substance away from the center of concentration into the surrounding regions. In time, the substance becomes uniformly distributed throughout the container, and although the molecules are all still in motion, there will be an equal concentration throughout the container (Figure 3-5).

Osmosis

The movement of water through a selectively permeable membrane is a special kind of diffusion called **osmosis** (ŏz·mō´sĭs). In osmosis, the rate of diffusion through a membrane is influenced by the presence of dissolved materials in the water, which may be unable to pass through the membrane. To understand the influence that these substances can have, suppose we consider a container of pure water divided into two compartments by a selectively permeable

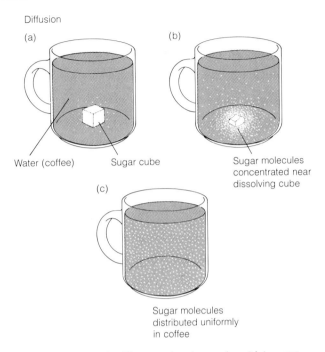

Figure 3-5   Diffusion. (a) When a cube of sugar is put into a cup of coffee, it begins to dissolve. (b) As the cube dissolves, sugar molecules are concentrated in the region of the dissolving cube. Random collisions between water and sugar molecules cause a net movement of sugar molecules away from the cube. (c) At equilibrium, long after the sugar is completely dissolved, the sugar molecules are uniformly distributed throughout the liquid. Stirring speeds up the mixing of the dissolved sugar in the coffee.

membrane, such as a piece of plastic wrap. This material serves quite well for our example, for although water will not flow as a liquid through this material, individual water molecules will diffuse through its pores (Figure 3-6).

Water molecules on either side of the membrane have the same kinetic energy and are equally likely to pass through it from one side to the other. As a result, the rate at which molecules pass through the membrane is the same in both directions. Since there is no difference in the rate of flow in the two directions, the water volume stays the same on both sides of the membrane, and nothing appears to be happening.

Now suppose that we add a small amount of sugar to the water on one side, making, for example, a 5% solution of glucose. The pores in the membrane are too small to allow glucose molecules to pass through, so the glucose stays on the side where it was added.

The glucose has two effects on the water in which it is dissolved. One effect is simply to reduce the concentration of water molecules, since the dissolved glucose molecules occupy space that would otherwise be occupied by water. A second effect is to reduce the activity of many of the water molecules. Each sugar molecule is, in effect, surrounded by a "cloud" of water molecules that are attracted to the

dialysis: (Gr.) *dia*, through + *lysis*, dissolution
hypotonic: (Gr.) *hypo*, under + *tonos*, tension
hypertonic: (Gr.) *hyper*, over + *tonos*, tension
isotonic: (Gr.) *isos*, equal + *tonos*, tension

CLINICAL NOTE

RENAL DIALYSIS

The kidneys are vital organs of the human body that play the major role in maintaining the body's water and electrolyte balance and in excreting the waste products of metabolism. When the kidneys fail, life is threatened by retention of waste products and fluid and an inability of the body to maintain its acid-base and electrolyte balance. Kidney (renal) failure is a common medical problem. **Renal dialysis** (dī·ăl´ĭ·sĭs) is a method of helping patients whose kidneys have failed by taking advantage of the processes of **osmosis** and **diffusion**. In dialysis, diffusable substances move across a semipermeable membrane from an area of greater concentration to an area of lower concentration.

There are two types of dialysis. **Extracorporeal hemodialysis** is a technique used to treat renal failure in which the blood from the patient is circulated through a series of fine tubules made of a semipermeable membrane (such as cellophane) immersed in a bath of a special solution called the **dialysate**. The waste products and elevated salts are removed from the blood by diffusion across the membrane. If necessary, extra fluid can be removed by osmosis with the use of a **hypertonic dialysate** prepared by adding a special sugar to the solution. The cleansed blood is then returned to the body. Patients who receive renal dialysis usually have three or four treatments per week. There are many complications with renal dialysis, and it is very expensive since it is usually performed in the hospital or a dialysis center and it involves the use of expensive equipment (see figure).

Peritoneal dialysis is a less complicated procedure that can be performed by the patient at home. This technique takes advantage of the large **peritoneal membrane** that lines the abdominal cavity and part of the intestines. A reusable plastic tube is inserted through the abdominal wall and several liters of sterile dialysate are allowed to run into the abdominal cavity. After a suitable period of time, the fluid is drained away, removing the waste products, salts, and excess water. The patient can repeat the process several times a day if necessary. Peritoneal dialysis is not as efficient as extracorporeal dialysis, but it is less complicated, can be done at home, and is therefore less expensive.

These are two important techniques for which medical science uses the natural processes of diffusion and osmosis to help solve a very difficult medical problem.

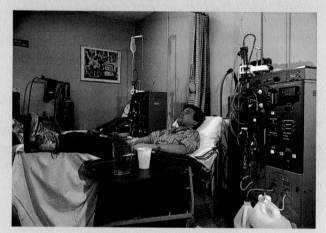

Renal dialysis. A patient undergoing treatment on a renal dialysis machine.

---

glucose molecule and held by hydrogen bonds. Although all the original water molecules are still present, their kinetic motion is reduced as a result of this attraction (Figure 3-7).

Since the rate of passage of water molecules through the membrane depends on their concentration and kinetic motion, the rate of flow of water from the glucose-containing side to the pure water side of the membrane is reduced. As a result, a difference in the rate of flow in the two directions develops: the rate of flow from the side containing the pure water stays the same and the rate of flow from the side where the sugar was added decreases. There is now a **net difference** between the two rates: more water flows from the side with the pure water into the side where the sugar is dissolved than flows in the opposite direction. As a result, the level of the sugar solution rises and the level of pure water falls.

We would observe similar results if we dissolved unequal amounts of sugar in the two sides of the container. In this case, the decrease in flow will be greater from the side with the highest concentration of sugar because more sugar molecules are present to reduce the activity of the water molecules on that side. Flow is reduced from both sides, but unequally, so that more water flows from the side with the lower sugar concentration than from the side with the higher sugar concentration.

When two solutions contain different concentrations of dissolved material, the solution with the lower concentration is said to be **hypotonic** (hī·pō·tŏn´ĭk) to the other. Conversely, the solution with the higher concentration is said to be **hypertonic** (hi˝pĕr·tŏn´ĭk) to the other. If the concentrations are equal, the two solutions are said to be **isotonic** (ī˝sō·tŏn´ĭk). From this discussion, it should be apparent that the net flow of water through a selectively permeable membrane is from a hypotonic solution to a hypertonic one, and that when the two solutions are isotonic, the flow rates are balanced and there is no net flow.

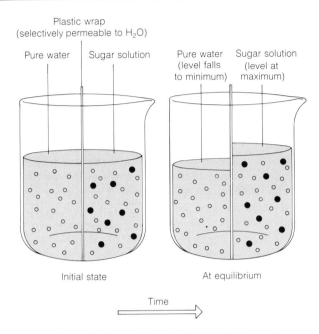

Figure 3-6 Osmosis. If a container is divided into two sections by plastic wrap (or any other selectively permeable membrane) and pure water is placed in one side and sugar solution in the other, a net flow of water will occur from the pure water into the sugar solution. As the volume of the sugar solution increases, more water will flow back into the sugarless side (due to hydrostatic pressure) until, at equilibrium, flow rates in the two directions are equal.

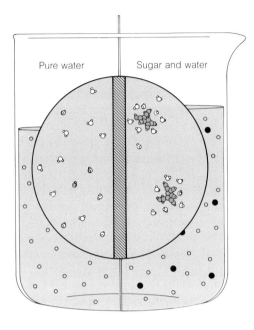

Figure 3-7 The attraction of water molecules to dissolved sugars and other substances is created by electromagnetic forces between locally positive and negative regions in the molecules. This causes water molecules to form clouds (shown in magnification) around sugar molecules, reducing the activity of the water molecules.

The same principle applies to the flow of water through the plasmalemma. On one side of the cell's membrane is its cytoplasm, a complex mixture of dissolved chemicals and various cell components. On the other side, separated from the cytoplasm by the membrane, is the water that bathes the cell. Individual water molecules pass through the membrane from both sides because the water molecules on both sides are all in motion and the membrane is permeable to them. However, many substances dissolved in the water are not able to diffuse through the membrane because the membrane is impermeable to them. If the solution outside the cell is hypotonic to the cell contents, there will be a net flow of water into the cell. In contrast, if the cell is in a hypertonic solution, water will flow out (Figure 3-8).

Cells are usually bathed in a solution of sugars, amino acids, proteins, and other substances. This external fluid is nearly isotonic to the cell's interior, so there is usually little difference in the flow rate into and out of the cell. However, cells can often control the flow of water through their membrane by transporting dissolved materials into or out of the cytoplasm. Under these conditions, cells can gain or lose water as necessary to maintain a suitable concentration of water in their interiors.

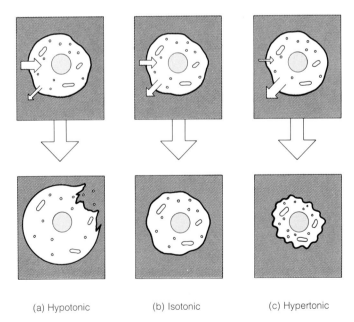

(a) Hypotonic   (b) Isotonic   (c) Hypertonic

Figure 3-8 Flow of water through the plasmalemma. (a) If a cell is placed in pure water (hypotonic to the cell), more water will flow in than out, and the cell will swell, perhaps to the point of bursting. (b) If placed in an isotonic solution, there is no difference in flow rates in or out of the cell, and the cell maintains its size. (c) When placed in a solution hypertonic to the cell, water will flow out faster than it flows in, and the cell will shrink.

Facilitated Diffusion

There are several ways a cell can transport larger substances across its membrane. One mechanism is called **facilitated diffusion**, which is the movement of molecules through the membrane with the assistance of specialized "carrier" molecules located in the membrane. Carrier molecules are proteins that combine with a particular kind of molecule in the solution on one side of the membrane, transport it through the membrane to the opposite side, and then release it (Figure 3-9a).

The rate of movement of a substance by this mechanism will still depend in part on the relative concentration of the substance on the two sides of the membrane. As in simple diffusion, any net flow will be from the side of the membrane where the concentration of the substance is highest to the side where it is lowest. The difference between this mechanism and simple diffusion, however, is that the membrane must contain the appropriate carrier molecules for each substance to be transported. The rate of flow will thus depend on both the difference in concentration across the membrane and the availability of carrier molecules in the membrane.

Active Transport

Active transport is the movement of molecules through the membrane from a region of low concentration into a region of higher concentration. Active transport is similar to facilitated diffusion in that it relies on membrane proteins for the transport of substances that would otherwise be unable to enter a cell. An important difference, however, is that active transport requires energy, and in order for substances to be transported, the cell must expend energy. ATP is the main source of energy used for this process. The use of ATP makes it possible for the cell to transport substances, regardless of the concentration, on the two sides of the membrane. A substance can be transported into a region where it is already in high concentration, something that could not be done if simple or facilitated diffusion were involved (Figure 3-9b).

Figure 3-9 (a) Facilitated diffusion involves movement from an area of high concentration to an area of low concentration using specialized carrier molecules. (b) Active transport involves movement against a concentration gradient, using ATP as a source of energy as well as specialized carrier molecules.

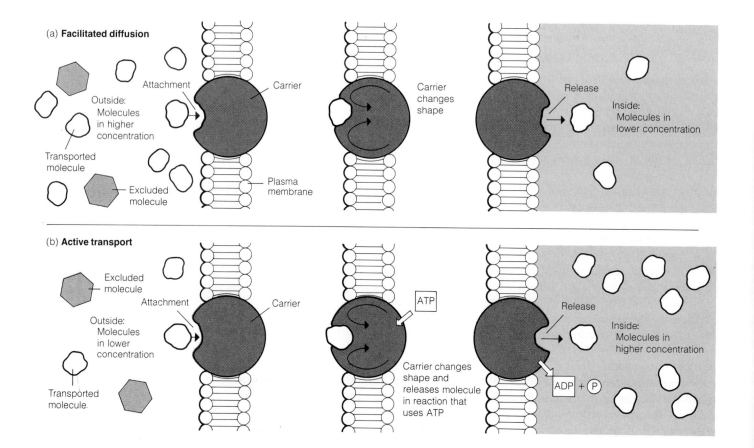

exocytosis: (Gr.) *exo*, without + *kytos*, cell + *osis*, a condition
endocytosis: (Gr.) *endon*, within + *kytos*, cell + *osis*, a condition
pseudopodia: (Gr.) *pseudos*, false + *podos*, foot + (L.) *ium*, denoting a quality or nature

---

Active transport is a primary mechanism for transporting glucose and other potential fuel molecules into a cell. It is also the principal means by which nutrients are removed from the intestine during digestion and transported to the circulatory system. Certain cells of the kidney use active transport in regulating the concentration of various salts in the blood. Similarly, the excretion of waste products by cells is often accomplished by this mechanism.

Mechanisms of Transport

The specialized carrier proteins that act in facilitated diffusion and active transport are called **permeases** (pĕr′mē·āz″ĕs). The manner in which a permease transports a substance through the cell membrane is not known, but two hypotheses have been proposed to account for their action. According to the **fixed pore** hypothesis, the permease provides a passageway through the lipid. The inner portions of the protein are soluble to the substance, and the substance essentially diffuses through the permease itself from one side of the membrane to the other. According to the **conformational change** hypothesis, a molecule of the substance combines with a permease molecule and causes the permease to change its shape in such a way that it essentially carries the attached molecule to the opposite side where it is released (Figure 3-10).

Exocytosis and Endocytosis

While diffusion and active transport provide a means of transporting relatively small molecules across the membrane, they are not adequate mechanisms for the transport of larger molecules. Yet proteins and large particles do pass into and out of many cells. For example, cells of the digestive tract secrete enzymes used in the digestion of proteins, carbohydrates, and nucleic acids. Likewise, cells of the pituitary gland secrete several regulatory proteins that are carried to other glands and tissues in the body. Also, certain cells of the kidney take in small proteins as the proteins are recovered from urine as it is being produced by the kidney.

When relatively large substances are secreted, they are generally packaged in small membranous spheres called **vesicles** and transported to the cell exterior in a process called **exocytosis** (ĕks″ō·sī·tō′sĭs). These vesicles are produced by the Golgi body and migrate to the plasmalemma (see the section on Golgi Apparatus and Lysosomes). As the vesicles fuse with the plasmalemma, their contents are released to the outside of the cell (Figure 3-11).

Passage is not always outward, however, and many cells are able to obtain large particles from the environment by engulfing them with protoplasm in a process called **endocytosis** (ĕn″dō·sī·tō′sĭs). Specialized cells such as leukocytes

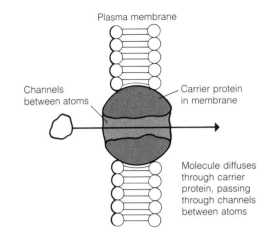

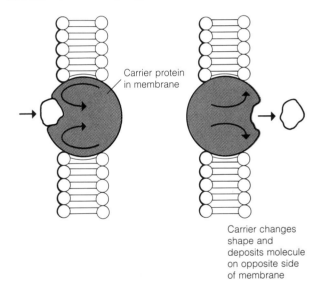

Figure 3-10 Alternate hypotheses for the movement of molecules through a membrane. (a) According to the fixed pore hypothesis, substances diffuse through spaces within the carrier molecule. (b) According to the conformational change model, attachment of a molecule to an active site in the carrier causes a change in the shape of the carrier. The change in shape carries the substance from one side of the membrane to the other.

(lū′kō·sīts) (the white cells of the blood) and **Kupffer** (kūp′fĕr) cells (certain cells of the liver) are able to engulf whole bacteria by extending **pseudopodia** (sū″dō·pō′dē·ă), which are long cytoplasmic projections, around the bacterium until it is surrounded by the cell's protoplasm and enclosed within a membrane (Figure 3-12). Endocytosis can

pinocytosis: (Gr.) *pinein*, to drink + *kytos*, cell + *osis*, a condition
phagocytosis (Gr.) *phagein*, to eat + *kytos*, cell + *osis*, a condition

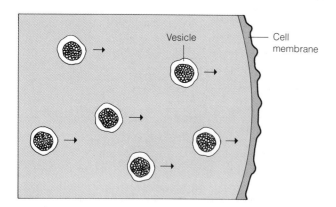

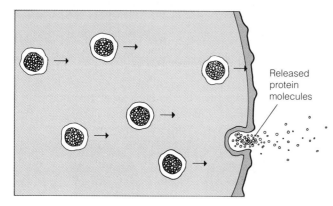

Figure 3-11 Exocytosis. Vesicles that contain substances to be secreted by the cell migrate in the cytoplasm to the cell membrane. Fusion of the vesicle with the cell membrane opens the vesicle to the cell exterior, and the vesicular contents are then released.

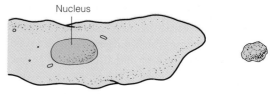

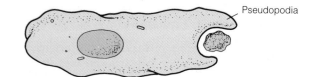

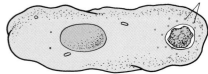

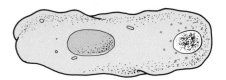

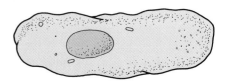

Figure 3-12 Endocytosis. Cells take in liquid droplets (pinocytosis) or solid particles (phagocytosis) by surrounding them with cytoplasm. Vesicles that contain digestive enzymes (lysosomes) then fuse with the membrane surrounding the droplet or particle and release their enzymes.

also involve liquid materials in the environment, such as fat droplets and aqueous solutions. When the substance engulfed is contained in relatively small vesicles, as is usually the case with liquids, the process is called **pinocytosis** (pī″nō·sī·tō′sĭs), meaning "cell drinking." When the engulfed substance is enclosed in a larger vesicle, as is usually the case with solids, the process is called **phagocytosis** (făg″ō·sī·tō′sĭs), meaning "cell eating." Pinocytosis is common to most cells since they are continually ingesting fluids. Phagocytosis, on the other hand, is usually performed by specialized cells.

Like active transport, pinocytosis and phagocytosis are energy-requiring processes that can move materials into the cell against a concentration gradient. Unlike active transport, however, they appear not to be selective processes and are stimulated by many different kinds of substances.

Many substances enter a cell by **receptor-mediated endocytosis**, an energy-requiring process that enables a cell to transport a substance against a concentration gradient. The process begins as a substance combines with specific **receptors** in the cell membrane. The receptor-substance complexes become concentrated in shallow pits in the membrane which invaginate and become vesicles that are released in the cytoplasm. The receptor molecules and the substances bound to them then separate, and small receptor-rich vesicles bud off and return to the membrane where the process can be repeated. The remaining receptor-free vesicles, still containing the transported substance, fuse to form larger vesicles called **endosomes.**

endoplasmic reticulum: (Gr.) *endon*, within + *plasma*, substance + *reticulum*, little net
lysosome: (Gr.) *lysis*, dissolving + *soma*, body

Contents of the Cytoplasm

After studying this section, you should be able to:

1. Define an organelle.
2. Describe the organization of the endoplasmic reticulum, including the way it divides the cytoplasm into cytosol and luminal portions.
3. Explain the basis for the difference in appearance of smooth and rough portions of the endoplasmic reticulum.
4. Describe the appearance and function of ribosomes.
5. Describe the organization of a Golgi apparatus, explaining how vesicles are derived from it.
6. Draw a mitochondrion in cross-section, illustrating the inner and outer membranes.
7. List and describe the filaments responsible for the form of a cell and movement of organelles within a cell.
8. Describe the structure of flagellae and cilia and explain their roles.

For a long time the cytoplasm was thought to be a relatively amorphous granular substance with a semifluid to gelatinous consistency. With the development of the electron microscope, however, it became apparent that the cytoplasm is an extremely complex structure, containing an elaborate network of membranes in which are suspended large numbers of subcellular structures. The network of membranes is called the **endoplasmic reticulum** (ĕn″dō·plăz′mĭk rĕ·tĭk′ū·lŭm) (or **ER**), and the suspended structures are called **organelles** (Figure 3-13).

Endoplasmic Reticulum and Ribosomes

The structure of the membranes of the endoplasmic reticulum is similar to that of the plasmalemma in that it involves a double layer of lipid molecules and associated proteins. The differences between the two kinds of membranes lie mainly in the specific lipids and proteins that each contains and in the functions they perform for the cell. While proteins of the plasmalemma are concerned with the transport of substances across the cell membrane, proteins of the ER are largely responsible for the reactions that occur within the cytoplasm itself.

The membranes of the ER twist and turn in such a way that they form a network of channels and caverns that extend throughout the cytoplasm (Figure 3-14). This network is considerably more complex in chemically active cells than it is in relatively inactive ones. The network of membranes divides the cytoplasm into two phases: a **luminal** (lū′mĭ·năl) **phase,** the fluid contained within the chambers and channels of the ER, and a **cytosol** (sī′tō·sŏl) **phase,** the portion of the cytoplasm that surrounds the chambers and channels. Ribosomes and other organelles of the cytoplasm are generally contained within the cytosol.

Examination of electron micrographs of ER reveals two distinct forms: one that appears relatively smooth, called the **smooth ER,** and one that appears relatively rough, called the **rough ER.** The appearance of rough ER results from the presence of many tiny organelles called **ribosomes** associated with that portion of the ER. Ribosomes are involved in protein synthesis (see the section on DNA Function), and rough ER is the region where protein synthesis occurs. Cells that produce large amounts of protein contain correspondingly large amounts of rough ER.

Each ribosome consist of one large and one small subunit (Figure 3-15). Together, the subunits contain 4 different RNA molecules and over 70 different protein molecules. Partial assembly of ribosomal subunits takes place in the **nucleolus,** a structure within the nucleus, and final assembly of functional ribosomes takes place in the cytosol. A complete ribosome is about 20 nm by 30 nm in size.

The number of ribosomes present in a cell usually reflects the amount of protein synthesis going on, and the number may change dramatically as the activity of the cell changes.

Golgi Apparatus and Lysosomes

There are places in the ER where membranous, flattened sacs form several closely spaced parallel membranes; these comprise the **Golgi** (gŏl′jē) **apparatus.** Membranes of the Golgi apparatus are similar in appearance to the membranes of smooth ER (Figures 3-16a and b). Substances produced in other regions of the cell (such as proteins produced in the rough ER) pass through channels in the luminal phase of the ER and into the Golgi apparatus. Once within the Golgi apparatus, the substances pass through the sacs of the Golgi and accumulate at the end of a channel. The end then pinches off, leaving the substance enclosed in a membrane-bound "bag" (Figure 3-16c). The bag, now called a vesicle, usually migrates to the periphery of the cytoplasm where it fuses with the plasmalemma, depositing its contents outside the cell (see Figure 3-11).

Some vesicles carry chemicals used for the digestion of proteins, lipids, and other large molecules within the cell itself. Such vesicles are called **lysosomes** (see Figure 3-12). Proteins in the lysosomes are used by the cell to digest "foreign" proteins, such as the proteins of an infecting bacterium. White blood cells and certain cells of the liver, spleen, and lungs capture and digest bacteria and foreign proteins in the blood in this way. The contents of lysosomes are

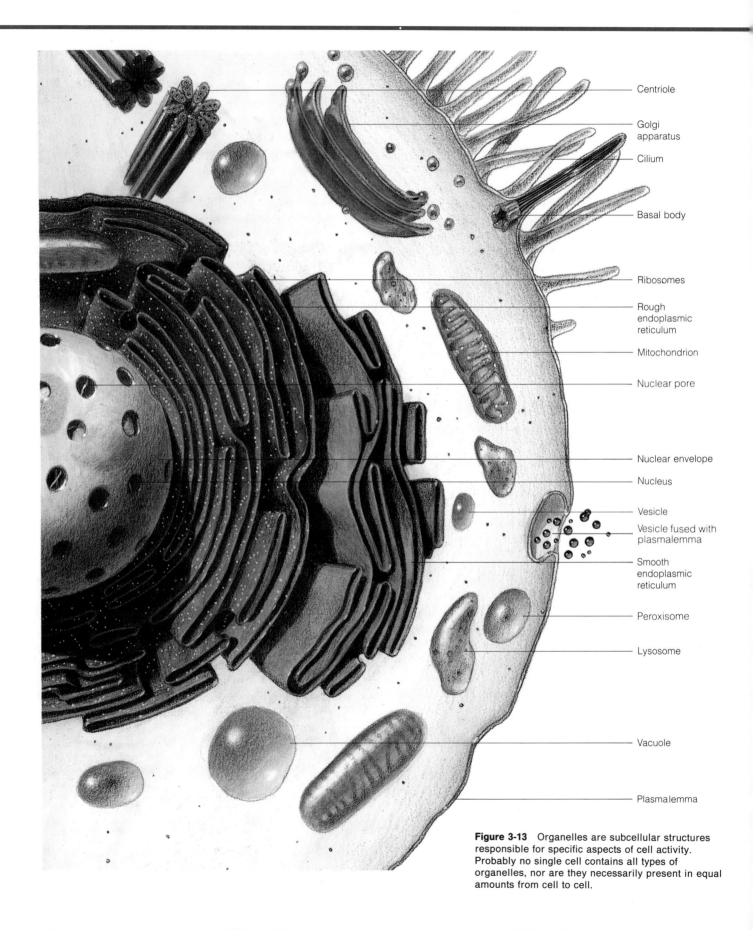

Figure 3-13 Organelles are subcellular structures responsible for specific aspects of cell activity. Probably no single cell contains all types of organelles, nor are they necessarily present in equal amounts from cell to cell.

| peroxisome: | (Chem.) peroxide + (Gr.) *soma*, body |
| mitochondrion: | (Gr.) *mitos*, thread + *chondrion*, grit |

usually maintained at about pH 5, which is considerably more acidic than the cytoplasm outside the vesicle. The digestive proteins inside the lysosome require acidic conditions to function. This protects the cell against self-digestion in the event that the contents of the lysosome leak into the cytoplasm.

Peroxisomes

Peroxisomes (pĕ·rŏks´ĭ·sōms) are special vesicles that contain enzymes important to the removal of many toxic substances and to the breakdown of fatty acids. These reactions are accomplished by the oxidization of certain compounds and by the transfer of the released electrons and hydrogen atoms to molecular oxygen. The oxygen molecules are converted to hydrogen peroxide, H_2O_2. Hydrogen peroxide, in turn, is broken down to water and molecular oxygen, a reaction that is accompanied by the oxidation of several other substances that would be toxic to the cell if allowed to accumulate. Because of these reactions, peroxisomes are often the site of much of the oxygen utilization that occurs in a cell. Peroxisomes appear to be produced by direct budding off from regions of the smooth ER rather than being produced by the Golgi apparatus.

Mitochondria

Mitochondria (mīt˝ō·kŏn´drē·ă) are organelles that play the major role in converting the chemical energy of carbohydrates, proteins, and lipids into forms that can be used more directly by the cell. The number of mitochondria present in a cell depends on the cell's metabolic activity. Cells that are the most active generally possess the greatest number of mitochondria. Considerable attention will be given to

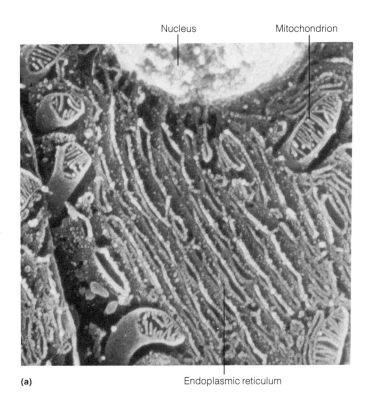
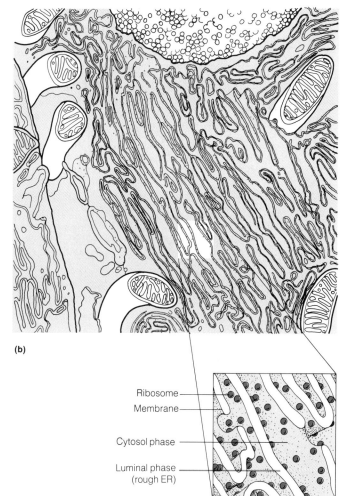

Figure 3-14 Endoplasmic reticulum (ER). (a) Scanning electron micrograph of ER (mag. 20,825×). The membranes of the ER have been cut transversely. (b) Interpretative drawing of the EM. The ER is a network of membranes in the cytoplasm that separates the cytoplasm into luminal and cytosol phases, which are the fluid-filled channels and their surrounding cytoplasm, respectively.

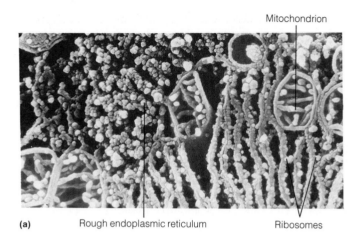

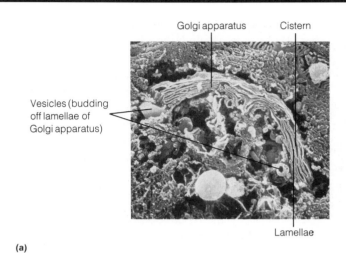

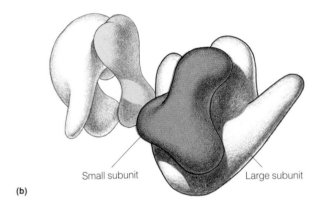

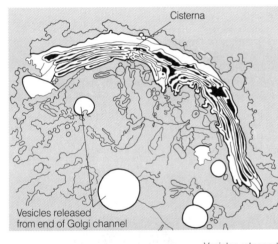

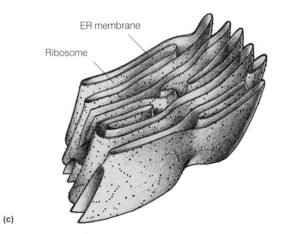

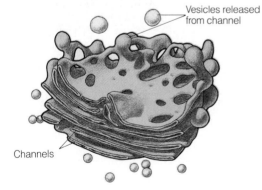

Figure 3-15 Ribosomes consist of large and small subunits which, when assembled, are involved in the synthesis of proteins. (a) Scanning electron micrograph of ribosomes, which show up as small white balls on the reticulum walls (mag. 40,218). (b) Diagrammatic model of a bacterial ribosome. Ribosomes of higher organisms are thought to be generally similar. (c) In rough ER, ribosomes lie on the cytosol side of the ER membrane.

Figure 3-16 A well-developed Golgi apparatus is characteristic of cells that manufacture and secrete substances. The membrane of the Golgi apparatus is continuous with the membrane of the endoplasmic reticulum. (a) Scanning electron micrograph showing vesicles budding off the ends of the Golgi channels (mag. 13,199×). (b) Interpretative drawing of the SEM. (c) Perspective drawing of a Golgi apparatus.

mitochondrial function in Chapter 19 (Metabolic Processes), and the discussion here will be confined to their structure.

Mitochondria range in size from about 1 to 10 μm long and from about 0.3 to 1 μm in diameter. They range in shape from spherical or sausage-shaped structures to elaborately branched structures. A mitochondrion consists of two membranous bags, one contained within the other, and each possessing an intricate arrangement of specialized lipids and proteins necessary for the mitochondrion to function. The inner bag is extensively folded, with the infoldings, called **cristae**, extending into the central cavity of the mitochondrion (Figure 3-17). The space between the membranes as well as the central cavity is filled with a fluid called the **matrix**.

The inner and outer membranes are distinct from one another in the proteins and lipids they contain and in the functions they perform. The outer membrane contains enzymes used in the chemical breakdown of carbohydrates, and the inner membrane proteins are involved in converting energy in the bonds of those breakdown products to ATP.

Cytoskeleton

The cell possesses a complex internal skeleton made up of a tiny but elaborate network of long slender tubes and filaments. This network is used by the cell to maintain a specific shape, to distribute materials throughout its cytoplasm, and to control those rearrangements of the cytoplasm that result in cell movement.

Even the largest of these tubes is too narrow to be seen by ordinary light microscopy, and special fluorescence techniques must be used to make them visible (Figure 3-18a). These tubes are called **microtubules** (mī″krō·tū′būls). Each one is formed from about 13 small filaments, each about 5 nm in diameter, arranged in a cylindrical pattern to form a hollow tubule about 20 to 30 nm in diameter (Figure 3-18b).

Microtubules extend throughout the cell from a region surrounding the cell nucleus to the cell perimeter. The shape of a cell is determined by the arrangement of microtubules. Microtubules (and filaments, discussed next) are also involved in the transport of organelles within the cytoplasm, providing avenues along which the organelles travel. The mechanism for this movement (Figure 3-19) may be similar to mechanisms that function in muscles to cause contraction (see Chapter 9).

Filaments are solid cylinders that differ from microtubules in their structure and in the kinds of proteins from which they are made. There are at least three different kinds of filaments found in cells. **Intermediate filaments** are about 10 nm in diameter and play a major role in anchoring the

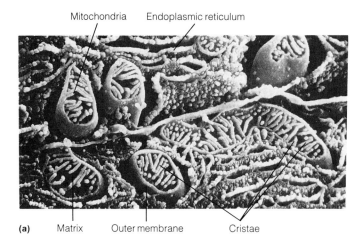
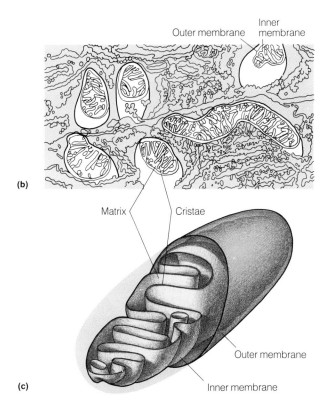

Figure 3-17 The mitochondrion is responsible for energy metabolism in a cell. They are especially prevalent in cells that use large amounts of chemical energy. (a) Scanning electron micrograph showing internal structure of mitochondrion (mag. 24,187 ×). (b) Interpretative drawing of the SEM. (c) Perspective drawing of a mitochondrion.

62 Chapter 3 Cell Structure and Organization

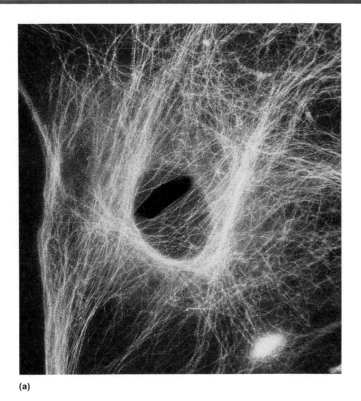

(a)

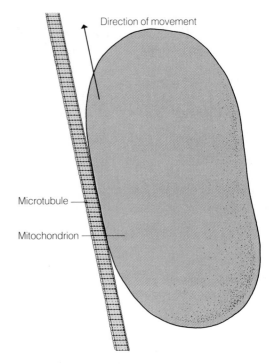

Figure 3-19 Movement of a mitochondrion along a microtubule. Proteins in the mitochondrial membrane interact with proteins in the microtubule in a way that pulls the mitochondrion along the tubule.

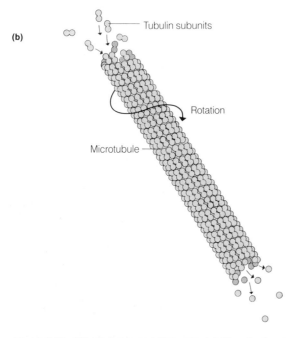

Figure 3-18 Microtubules and filaments. (a) Visualization by fluorescence microscopy. (b) The organization of a microtubule. It is formed from individual protein subunits (tubulin molecules) and is constantly synthesized in a treadmill fashion, as tubulin units are added at one end and removed at the other.

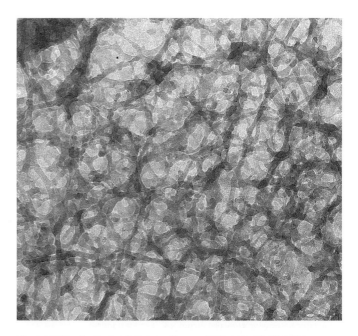

Figure 3-20 Ground substance. At very high magnification the cytoplasm can be seen to contain an elaborate network of very fine filaments called the microtrabecular lattice (mag. 87,450×).

axoneme: (Gr.) *axon*, axle + *nema*, thread

nucleus in the cytoplasm. **Microfilaments** are 5 to 6 nm in diameter, and they form a meshwork located predominantly to one side of the cytoplasm. In a migrating cell these filaments are oriented in a direction parallel to the direction in which a cell is moving; for this reason, microfilaments are thought to play a role in cell movement.

At the smallest end of the scale, and close to the limit of resolution of even the most powerful electron microscopes, are extremely fine filaments that together form an elaborate network termed the **microtrabecular** (mī″krō·tră·bĕk′ū·lăr) **lattice.** These filaments range from 3 to 6 nm in diameter and are distributed throughout the cytosol to form a **ground substance** (Figure 3-20). Microtrabeculae appear to support ribosomes that are involved in protein synthesis.

Cilia and Flagella

Microtubules are also found in special organelles called **cilia** (sĭl′ē·ă) and **flagella** (flă·jĕl′ă). Cilia are small, hairlike projections on the surfaces of cells that line certain passages in the respiratory and reproductive systems. The wavelike motion of the cilia serves to transport materials along the passageway. This is the mechanism by which dust and other foreign particles are carried from the nostrils after having been filtered from the air passing through the nostrils. In the female reproductive tract the wavelike motion of cilia carries the egg through a tube called the oviduct to the uterus.

Flagella are the long, whiplike tails that enable sperm to swim through their liquid environment. They are longer than cilia, but their movement appears to be due to a similar mechanism.

Cilia and flagella contain a central core called an **axoneme** (ăks′ō·nēm), which is responsible for the bending of the organelle. The axoneme consists of a cylinder of nine microtubules that surrounds a central pair of microtubules. It originates at a structure called a **basal body** located in the cytoplasm of the cell. Each microtubule is attached to the basal body and extends through the entire length of the cilium or flagellum (Figure 3-21).

Figure 3-21 Fine structure of cilia and flagella. (a) Each cilium and flagellum has an axoneme made up of a central pair of microtubules. (b) and (c) These are surrounded by a cylinder of nine "doublet" microtubules held in place by arms that extend from one doublet to the next. The structure of the arms allows the cilium or flagellum to bend without losing the cylindrical organization of the axoneme. (d) Electron micrograph of cross-section through a flagellum (mag. 176,757).

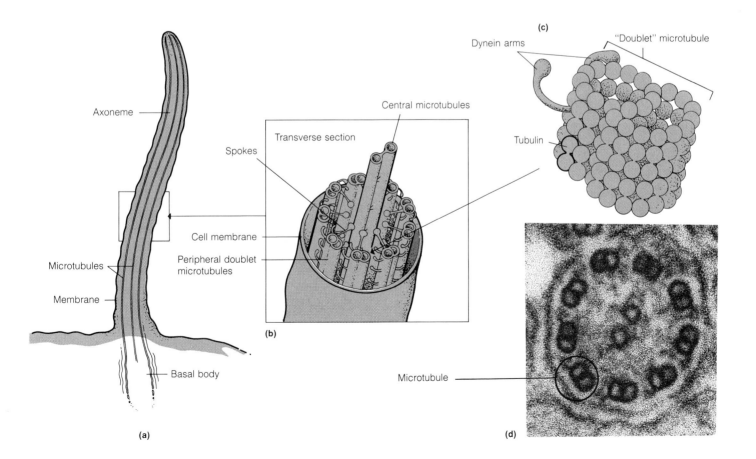

perinuclear: (Gr.) *peri*, around + (L.) *nucleus*, kernel

The Nucleus and Its Contents

After studying this section, you should be able to:

1. Describe the structure of the nucleus.
2. Describe the organization of the two membranes that surround the nucleus.
3. Cite the functions of the nucleolus.
4. Identify and list the components of chromatin.
5. Describe the subunit structure of chromatin.

The nucleus contains the cell's genetic material. It is usually the largest organelle of the cell, typically measuring about 0.5 µm in diameter, although this measurement varies considerably with cell size and type (Figure 3-22). The nucleus may be found suspended in the central region of the cell or over to one side, usually in a region of relatively high metabolic activity. Depending on the kind of cell in which it is found, the nucleus varies in shape. In some cells it is roughly spherical, and in others it is shaped more like an elongated sphere or is sometimes branched or lobed. In leukocytes (white blood cells) the nucleus is often horseshoe shaped.

Although most cells possess a single nucleus, there are some cells that lack a nucleus and others that contain more than one nucleus and sometimes many nuclei. For example, mature red blood cells lack a nucleus, whereas many liver and cartilage cells contain two nuclei, and skeletal muscle cells contain numerous nuclei.

Nuclear Envelope

The nucleus is bounded by two distinct membranes which together comprise the **nuclear envelope.** Between the two membranes is an area called the **perinuclear space** (Figure 3-23). Each membrane of the nucleus is produced from membranes of the endoplasmic reticulum immediately after cell division. In some cases the outer membrane appears to be continuous with membranes of the ER, forming a pathway from the perinuclear space to the lumenal phase of the cytoplasm.

Nucleolus

The **nucleolus** (nū·klē′ō·lŭs) is an inclusion of the nucleus that appears as a dark, roughly spherical structure not bounded by a membrane. Nucleoli consist of about 3 to 10% RNA, about the same amount of DNA, and as much as 90% or more protein. High resolution electron micrographs reveal distinct granular, fibrillar, and amorphous regions in the nucleolus, but the significance of these differences in internal

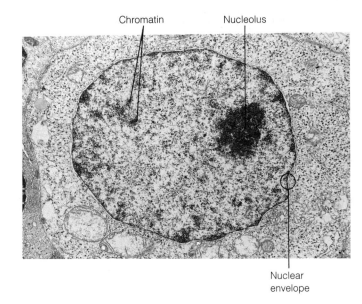

Figure 3-22 The nucleus. This electron micrograph shows the double membrane bounding the nucleus, the outer one of which is continuous with the endoplasmic reticulum (mag. 1500×). Nuclear pores lie at junctions where the two membranes come together. The grainy material within the nucleus is chromatin, which contains the genetic material of the cell. The nucleolus is an irregular structure in which ribosomal subunits are assembled.

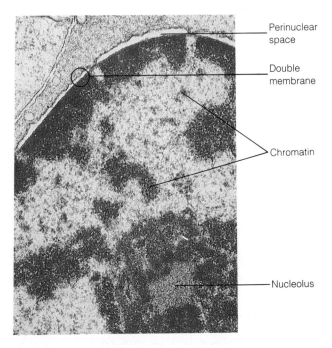

Figure 3-23 Perinuclear space. The space between the two membranes that surround the nucleus is the perinuclear space. Substances that pass from the nucleus into the perinuclear space then pass into the luminal phase of the cytoplasm.

chromatin: (Gr.) *chroma*, color
nucleosome: (L.) *nucleus*, kernel + (Gr.) *soma*, body

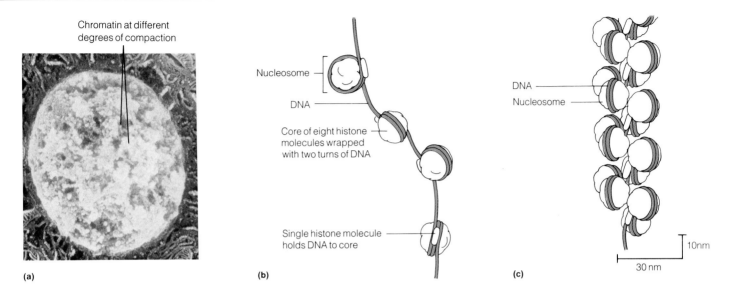

Figure 3-24 Molecular organization of chromatin. (a) This scanning electron micrograph shows that chromatin is a granular mass that lies suspended throughout the nucleus. Chromatin has a granular appearance due to differences in the degree of condensation in different portions (mag. 12,500×). (b) The fundamental unit of chromatin is the nucleosome, which consists of an assemblage of nine histone molecules wrapped in two turns of DNA. (c) Nucleosomes form solenoids by coiling to produce a chromatin fiber about 30 nm in diameter.

structure is unknown. The nucleolus forms at a region in the nucleus called the **nucleolus organizer.** The nucleolus organizer produces large quantities of RNA that are collected in the nucleolus where they are modified and partially assembled into ribosomal subunits.

Chromatin

Most of the interior of the nucleus consists of a mass of material called **chromatin** (krō′mă·tĭn), so named because it stains brightly with certain dyes used in light microscopy. In the unstained cell, chromatin has a characteristic grainy appearance (Figure 3-24a). Chemically, it consists almost entirely of molecules of protein and nucleic acid arranged in long, slender fibrils that are bathed in the nucleoplasm. The nucleic acid portion of chromatin consists of both DNA and RNA, the DNA being present in extremely long molecules. In humans the average DNA molecule has about 130 million nucleotide pairs. If laid end-to-end, the DNA in the 46 chromosomes of a human cell would be over a meter long, all in a nucleus about 5 millionths of a meter in diameter! This DNA contains the genetic information of the cell, that is, the information used for the synthesis of enzymes and other proteins that regulate the chemical activities of the cell.

The RNA component of chromatin is a heterogeneous mixture of molecules that range in length from molecules with as few as eight or ten nucleotides to molecules containing many tens of thousands of nucleotides. RNA molecules are made in the nucleus using portions of the DNA as a template to specify the order of nucleotides assembled. Many of these RNA molecules are destined to travel to the cytoplasm where they will be used for the synthesis of protein. Some of them, specifically the RNA molecules of ribosomes, are first assembled with proteins into ribosomal subunits and are then transported into the cytoplasm.

The proteins of chromatin are classified into two general categories called **histones** (hĭs′tōns) and **nonhistone chromosomal proteins** (abbreviated NHCP). Histones are closely associated with DNA to form chromatin subunits called **nucleosomes** (nū′klē·ō·sōms). A typical nucleosome contains nine histone molecules surrounded by two turns of DNA. Nucleosomes have a diameter of about 10 nm. They give the appearance of many beads on a string when viewed through an electron microscope (Figure 3-24b).

These strings of beads, in turn, may be organized into coils of nucleosomes, producing higher levels of organization called **solenoids** (sō′lĕ·noyds) (Figure 3-24c). In a solenoidal region of the chromatin, the fibril formed by the coil of nucleosomes is about 30 nm in diameter. Variations in the degree of aggregation in different regions of the chromatin are responsible for its grainy appearance.

Nonhistone chromosomal proteins include a diverse group of proteins engaged in the many biochemical activities that go on in chromatin. Included within this group are the enzymes and other proteins that are involved in replicating DNA, synthesizing RNA, regulating the action of the genes present in the DNA, and probably regulating the degree of aggregation of the chromatin.

cytokinesis: (Gr.) *kytos*, cell + *kinesis*, motion
karyokinesis: (Gr.) *karyon*, nucleus + *kinesis*, motion

Cell Division

After studying this section, you should be able to:

1. Distinguish between cytoplasmic and nuclear division.
2. Explain the role of mitosis as a mechanism for distributing duplicated DNA to new cells.
3. Diagram a typical chromosome as observed with a light microscope during mitosis.
4. Distinguish between chromatids and chromosomes.
5. Describe the structure of a mitotic spindle and explain its role in chromosome distribution during mitosis.
6. List the five stages of mitosis and describe the appearance and behavior of chromosomes associated with each stage.

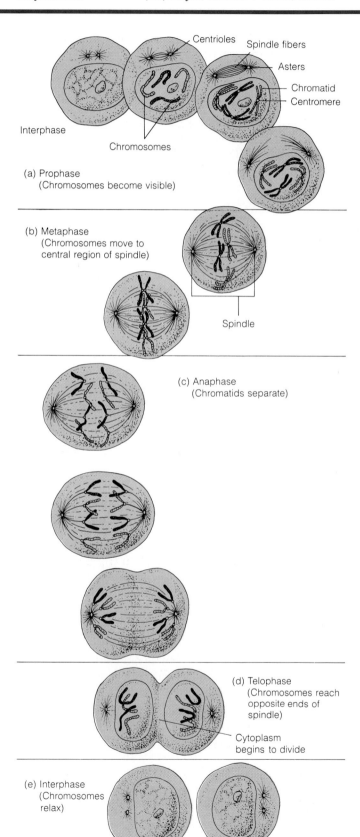

(a) Prophase (Chromosomes become visible)

(b) Metaphase (Chromosomes move to central region of spindle)

(c) Anaphase (Chromatids separate)

(d) Telophase (Chromosomes reach opposite ends of spindle) Cytoplasm begins to divide

(e) Interphase (Chromosomes relax)

The life of a new individual begins at fertilization, when a sperm from a man fuses with an egg from a woman. The product of that fusion is a single cell called a **zygote** (zī′gōt) (see Chapter 28). That single cell then initiates a series of **cell divisions** which, in time, result in a human adult having within his or her body several hundred trillion cells. Cell division continues even after maturity to replace existing cells that die.

Cytoplasmic and Nuclear Division

Cell division consists of two phases, one consisting of the division of the cell's cytoplasm into two portions, called **cytokinesis** (sī″tō·kĭ·nē′sĭs), and another that results in the division of the nucleus, called **karyokinesis** (kăr″ē·ō·kĭn·ē′sĭs). One nucleus produced by karyokinesis is distributed to each of the portions of the cytoplasm produced by cytokinesis.

Sometime prior to cell division, the DNA in the nucleus is replicated, exactly doubling the amount of DNA present. This doubled quantity of DNA is then distributed equally by cell division to each of the new cells. This results in two cells each with the same amount of DNA that had been present prior to doubling. Because the nucleotide sequence is accu-

Figure 3-25 Mitosis. (a) In prophase, chromosomes become visible as they shorten, thicken, and divide into two chromatids that remain attached to one another at the centromere. The centriole divides, and the two products move to opposite sides of the nucleus. A spindle forms and spans the cell. (b) Metaphase begins when the chromosomes have reached a plane in the central region of the cell. (c) Chromatids that comprise a chromosome separate and move to opposite ends of the spindle during anaphase. (d) Telophase is marked by the accumulation of the chromosomes at opposite ends of the spindle. The cytoplasm begins to divide. (e) After nuclear and cytoplasmic division, the chromosomes relax in the newly formed nuclei and each cell enters interphase.

mitosis: (Gr.) *mitos*, thread
chromosome: (Gr.) *chroma*, colored + *soma*, body
centromere: (L.) *centrum*, center + (Gr.) *mere*, part

rately duplicated when DNA is replicated, each of the cells also has the same genetic information as the cell from which it was produced.

Mitosis

Cytokinesis and karyokinesis together comprise **mitosis** (mī·tō′sĭs), the process that a cell uses to distribute its duplicate copies of DNA. Mitosis consists of four successive stages: prophase, metaphase, anaphase, and telophase. A nucleus that is not undergoing mitosis is said to be in **interphase**.

Prophase

Prophase begins as the granular contents of the nucleus transform into a tangled mass of long, slender threads called **chromosomes**. The transition results from the progressive contraction of the fine chromatin filaments into thicker and thicker fibers until they are thick enough to be seen as chromosomes (Figure 3-25a). As the process continues, it becomes apparent that each chromosome consists of two longitudinal halves, called **chromatids** (krō′mă·tĭds) that are connected to one another at a constricted region of the chromosome called the **centromere** (sĕn′trō·mēr).

After the chromosomes have fully contracted, each one displays a characteristic size and a centromere that is located at a characteristic point. The lengths of a chromosome on either side of the centromere are referred to as **chromosome arms**, and depending on where the centromere is located, the arms may be equal or unequal in length. In addition, special staining techniques can be used that give each chromosome a characteristic banded appearance, enabling each chromosome to be identified on the basis of size, location of the centromere, and banding pattern (Figure 3-26).

Simultaneous with the appearance of the chromosomes, the nuclear membrane disintegrates and a network of fine fibrils develops in the cell. Each fibril in the network consists of bundles of microtubules. Some fibrils stretch from one side of the cell to the other. Other fibrils extend from the centromere of each chromatid to one or the other side of the cell. Collectively, the fibers form a **spindle**.

Metaphase

The second phase of mitosis, **metaphase**, is marked by the migration of the chromosomes to a region midway between

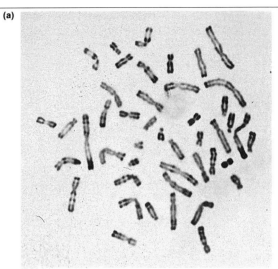
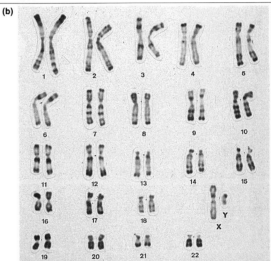
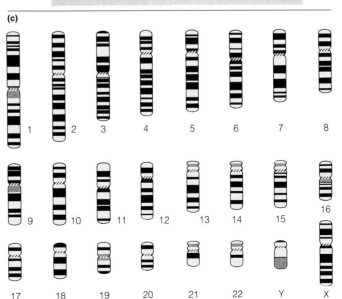

Figure 3-26 Chromosomes. (a) Photomicrograph showing stained chromosomes from a human male at metaphase of mitosis. (b) The same chromosomes arranged according to size, centromeric location, and similarity of appearance. Note that all but the last two can be arranged in similiar pairs. In a female there would be two X chromosomes and a Y chromosome would be absent. (c) Schematic representation of one chromatid from each pair in part (b).

the ends of the spindle. By now the spindle is well developed and the nuclear membrane is gone (Figure 3-25b).

Anaphase

In **anaphase,** the third phase of mitosis, the centromere of each chromatid splits and the sister chromatids move toward opposite ends of the spindle (Figure 3-25c). The exact mechanism for the movement of the chromatids is unknown, but one possible explanation is that a shortening of the spindle fibers pulls the chromatids apart.

Telophase

During **telophase,** the final phase of mitosis, the chromatids, now referred to as chromosomes, are clumped together at opposite ends of the spindle (Figure 3-25d). Following telophase, the chromosomes relax and return to the long, slender, stranded state characteristic of the nondividing nucleus during interphase (Figure 3-25e). At the same time, the spindle disintegrates and nuclear membranes form around each mass of chromosomes. The cytoplasm surrounding the two new nuclei divides in two, and two new **daughter cells** are formed.

DNA Function

After studying this section, you should be able to:

1. Generally describe the process of DNA replication.
2. Describe how information is stored in DNA.
3. Define transcription and explain how a sequence of nucleotides in DNA is used to specify a sequence of nucleotides in RNA.
4. List the three types of RNA used in protein synthesis and describe their function.
5. Define translation and explain the events that occur as a ribosome moves along a messenger RNA molecule and amino acids are incorporated into a polypeptide.

Replication

Mitosis thus provides a mechanism for the packaging and distribution of newly formed DNA that has been produced prior to the mitotic division. DNA is duplicated by a process called **replication,** in which each of the two strands of the DNA double helix serves as a **template** for the synthesis of a new strand. In this process enzymes responsible for synthesizing the DNA pass along the strands, causing the addition of a nucleotide in the new chain at a position opposite each nucleotide in the template. This is a complex process involving enzymes and other proteins that cause the strands to unwind and separate, as well as enzymes that catalyze the addition of nucleotides to the growing end of each new strand. The enzyme that catalyzes the addition of each new nucleotide is called **DNA polymerase.** Because of the property of complementary base pairing, a thymine-containing nucleotide is normally incorporated opposite an adenine-containing nucleotide, adenine opposite thymine, guanine opposite cytosine, and cytosine opposite guanine. The result is two double-stranded molecules where one had been before. (See the section on Nucleic Acids in Chapter 2 for an overview of DNA and RNA.)

A DNA molecule is enormously long compared with its width, and an average DNA molecule in a human cell contains over 130 million nucleotide pairs (Figure 3-27a). Prior to DNA replication, each chromosome contains a single DNA molecule joined with histone molecules to form nucleosomes and with other proteins and RNA to form a complex aggregate of molecules that together comprise the chromosome. Replication occurs at several points along a DNA molecule at the same time. The strands are separated from one another at these points to form many **replication bubbles** within which each of the separated strands serves as a template for the synthesis of a new strand (Figure 3-27b). Each end of a replication bubble consists of a **replication fork,** the Y formed where the two strands of the original molecule are separating (Figure 3-27d).

As each new strand grows longer with the addition of nucleotides to each of its ends, the replication bubble also grows longer, increasing in length as the replication forks move away from the point where replication had been initiated. The bubbles continue to grow until the replication forks of neighboring replication bubbles meet and merge. When all the bubbles have merged, replication of the original DNA molecule is complete, and there are two newly formed DNA molecules where one had been before (Figure 3-27c).

The significance of this process is that each of the DNA molecules has the same sequence of nucleotide pairs as the other and as the original molecule from which it was produced. Since genetic information is encoded in this sequence of nucleotide pairs, the two new DNA molecules have the same information as each other and as the molecule from which they were produced. These molecules, along with the proteins and other substances with which they are associated, comprise the sister chromatids observed during mitosis. Mitosis is thus a mechanism for separating DNA molecules produced by replication and placing them in separate cells.

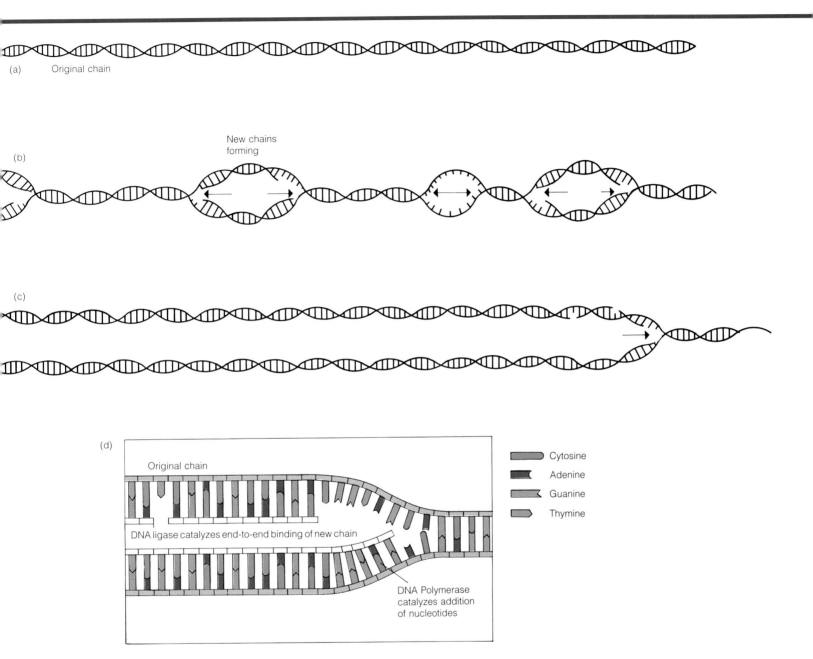

Figure 3-27 DNA replication. When DNA replicates, two double-stranded molecules with the same nucleotide sequence as the original one are produced. This maintains the genetic information in the molecules, and they are distributed to separate cells at the time of cell division. (a) The average chromosome contains about 130 million base pairs of DNA in a single long, narrow double-stranded molecule. (b) Replication begins at many different sites proceeding in both directions from each site to form replication bubbles. (c) Growth continues outward until neighboring bubbles fuse and produce two separate DNA molecules, each identical in sequence to the original molecule. (d) Detail of the replication fork. Nucleotides are incorporated simultaneously into both strands at the growing fork, moving to the right in a continuously extending strand on the lower side and to the left in a series of short fragments on the upper strand.

CLINICAL NOTE

GENETIC ENGINEERING

Over the past 10 to 15 years such dramatic technological developments in biology have occurred that they have often been referred to as a "biological revolution." These developments have enabled investigators to remove DNA from an organism, identify specific genes or other regions in the DNA, isolate them, insert them into bacteria or other microorganisms, and induce those microorganisms to produce more of the DNA. This complex process is called **DNA cloning.** In addition, by carefully controlling the manner in which the DNA is introduced, the microorganism can often be induced to produce the chemical product of the gene, regardless of the source of DNA introduced into it. This makes it possible, for example, to use bacteria such as *Escherichia coli* to produce human enzymes and other proteins.

The techniques employed generally involve combining DNA from one source with DNA from another source. DNA molecules that have been produced by combining DNA from two or more sources is referred to as **recombinant DNA.** The techniques used to produce recombinant DNA molecules are collectively referred to as **genetic engineering.**

The techniques of genetic engineering are extremely powerful. They enable investigators, for example, to isolate a single gene out of the tens of thousands of genes present in the cells of a human or other mammal. This amounts to specifically identifying and isolating a fraction of the DNA of a cell that corresponds to about 0.0003% of the total. This is akin to stretching a rope from Los Angeles to San Francisco and locating a particular stretch of it about 7 feet long while flying over it in the dark!

The ability to find and isolate a particular portion of human DNA and then to introduce it into a microorganism makes it possible to produce otherwise extremely rare proteins efficiently and in large quantity. This has already resulted in the production of many proteins of pharmaceutical value, such as human insulin, growth hormone, and hormones that stimulate the production of blood cells. Conventional techniques of isolation, purification, and recovery of these hormones from animal sources can be extremely expensive, and the hormones may not always be suitable for use in humans. Substances produced by recombinant DNA-based procedures, in contrast, are cheaper, relatively clean to begin with (so they can be purified more thoroughly), and free of the immune response often induced by substances isolated from animal sources.

Genetic engineering is also providing a clearer understanding of the molecular basis of many hereditary illnesses. Genes for many blood diseases, color blindness, diseases of the immune system, hormone deficiencies, digestive disorders, and hereditary nervous system disorders, to name just a few, have been isolated and their differences from normal forms of the corresponding genes determined. Knowledge of the molecular bases for many genetic diseases gained in this way will help researchers develop strategies for managing those diseases.

Transcription

The primary function of the information stored in DNA is to specify the structure of proteins, that is, to specify the order in which amino acids are incorporated into a protein at the time of its synthesis. This is accomplished by two processes called transcription and translation.

Transcription is a process in which DNA directs the synthesis of RNA molecules that are then used for protein synthesis. This process is somewhat like replication in that the nucleotides of the strand of DNA specify, in a complementary manner, the nucleotides to be incorporated into the new RNA molecule. In this case, however, an adenine base in the DNA strand serves as a template for the incorporation of a uracil-containing nucleotide in the RNA molecule (Figure 3-28).

Transcription begins at specific points along a DNA molecule as molecules of an enzyme called **RNA polymerase** travel along the DNA. As the polymerase passes along the DNA, the strands of the double helix unwind ahead of it and come back together into a double helix behind it. Within the short bubble formed by this unwinding and rewinding, the RNA polymerase catalyzes the addition of ribonucleotides to the end of a growing RNA molecule. In time, the RNA polymerase reaches a sequence of deoxyribonucleotides that signals the end of the region to be transcribed, and the newly formed RNA molecule and the RNA polymerase are released from the DNA molecule.

Following its synthesis, the RNA molecule is subjected to several chemical modifications collectively called **RNA processing,** after which it passes from the nucleus into the cytoplasm.

Translation

Translation is the process by which information transcribed into RNA is used to make proteins. There are three different kinds of RNA involved in the process: **messenger RNA (mRNA), transfer RNA (tRNA),** and **ribosomal RNA (rRNA).** Of these, mRNA carries information from the nucleus to the cytoplasm. Information encoded in the sequence of nucleotides in mRNA specifies the sequence of amino acids to be incorporated into a protein. Molecules of tRNA also travel to the cytoplasm where they associate with specific amino acids. There are several dozen different tRNA species, each one of which combines with a particular amino acid. Their role is to carry these amino acids to the site of protein synthesis in the cytoplasm. These sites of protein synthesis are provided by the ribosomes, complex structures which contain many protein molecules in addition to the rRNA molecules (Figure 3-29).

Figure 3-28 Transcription. Genetic information stored in DNA is transcribed into RNA molecules which are transported to the cytoplasm.

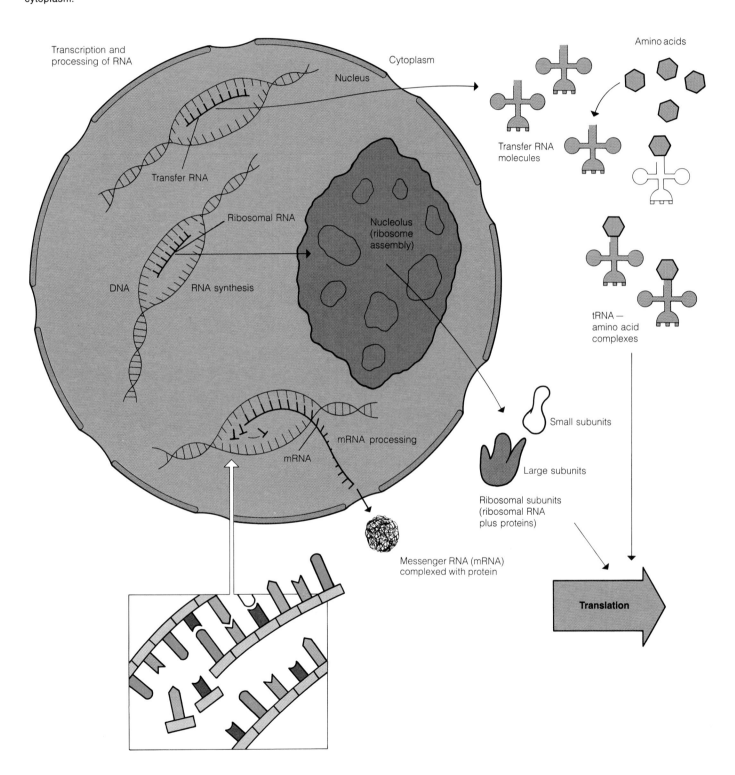

Figure 3-29 Translation. In the cytoplasm, the information transcribed into a messenger RNA molecule is used to specify the sequence of amino acids to be incorporated into a protein.

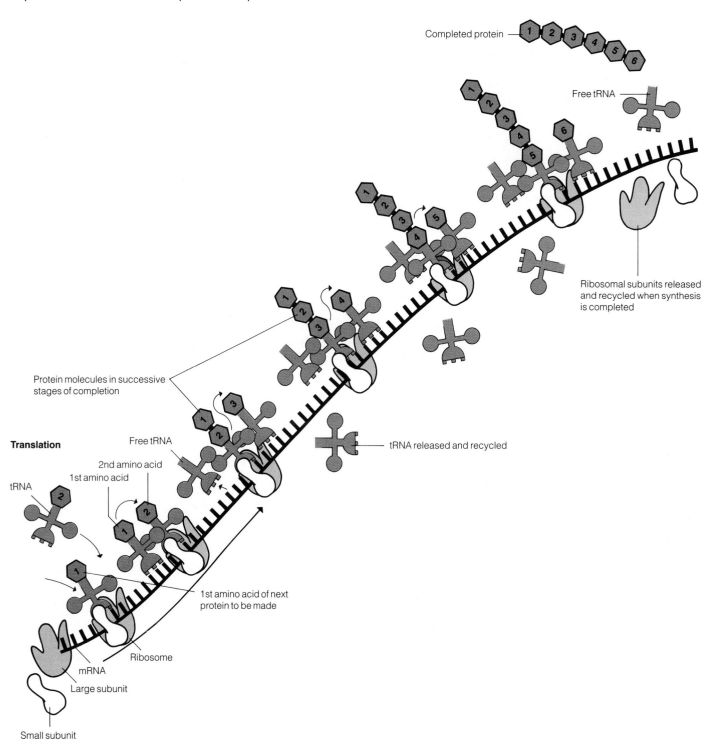

polysome: (Gr.) *poly*, many + *soma*, body

Once an mRNA molecule reaches the cytoplasm, a ribosome attaches to one end of it and begins to travel along its length. As the ribosome travels along the mRNA molecule, tRNA molecules, which have previously combined with amino acids, sequentially associated with the mRNA molecule and deliver their respective amino acids to the growing polypeptide.

Translation involves three distinct phases: **initiation, elongation,** and **termination.** Initiation is marked by the binding of an mRNA molecule to a ribosome and a tRNA–amino acid complex. Once this binding is complete, elongation occurs as the ribosome moves ahead and provides a site for a second tRNA–amino acid complex to associate with the ribosome and mRNA. With the association of the second tRNA, the bond holding the first amino acid and tRNA together cleaves, and a peptide bond is formed between the two amino acids. The second amino acid is still attached to its tRNA. The ribosome continues to travel along the mRNA molecule and provides a site for a third tRNA–amino acid complex to move into position. Again, the bond between the previous amino acid and its tRNA cleaves, and another peptide bond forms, this time between the second amino acid and the third. This sequence of events continues over and over again as the ribosome continues along the mRNA, with the strand of amino acids increasing one at a time until synthesis of the polypeptide is complete and termination occurs.

Appropriate amino acids are selected during elongation on the basis of the sequence of nucleotides present in the mRNA. Information in the mRNA is arranged in successive groups of three nucleotides called **codons** (kō′dŏns). Each codon is complementary to a sequence of three nucleotides in a tRNA called an **anticodon,** so that each codon in the mRNA specifies a tRNA with a complementary anticodon. Since each tRNA species is associated with a specific amino acid, selection of a tRNA with the correct anticodon simultaneously selects for a tRNA molecule with a particular amino acid. Termination occurs when the ribosome reaches a specific nucleotide sequence on the mRNA that causes the ribosome and the complete polypeptide chain to fall off the mRNA molecule.

Molecules of mRNA are generally long enough for several ribosomes to move along their length at the same time; thus, several proteins can be synthesized simultaneously. A mRNA molecule with several ribosomes attached is called a **polysome** (pŏl′ĭ·sōm).

From this brief overview of cell structure, it should be evident that the cell is an enormously complex structure. Keep in mind as you proceed through the chapters that follow that the structure of the cell is responsible for the cell's ability to perform various functions, and the functions performed by a cell are the foundation on which the physiology of the individual rests.

STUDY OUTLINE

Cellular Basis of Life (p. 47)
Like that of other organisms, the human body is made up of cells. Specialization of these cells makes possible the various aspects of life.

Plasmalemma (pp. 47–50)
Cells are surrounded by a membrane called the *plasmalemma,* which controls entrance and exit of materials to and from the cell.

Organization The plasmalemma consists of proteins, lipids, and carbohydrates organized into a thin film that surrounds the cell.

The lipid portion produces a double-layered structure in which are embedded the proteins.

Carbohydrates are attached primarily to the proteins, projecting outward from the outer surface.

The lipid molecules form a fluid layer in which proteins are embedded. The cell membrane is a mosaic of proteins floating in a fluid layer of lipids.

Junctions Plasmalemmas of adjacent cells come into contact with one another and form *junctions,* of which there are three types.

Communicating junctions are points in the membranes of adjacent cells where there is a direct connection between the cytoplasms of the cells. The most common form is called a gap junction.

Impermeable junctions, or *tight junctions,* seal the contacting surfaces of adjacent cells, preventing the passage of material.

Adhering junctions, or *desmosomes,* hold the plasmalemmas of adjacent cells together, preventing the cell membranes from sliding past one another.

Movement Through the Cell Membrane (pp. 50–56)
Effects of Size, Charge, and Solubility Movement of materials through a membrane relies on the *kinetic energy* (energy of movement) of the material and the permeability of the membrane to the substance.

Diffusion is the spontaneous movement of materials from a region of high concentration to a region of low concentration. Diffusion results from the kinetic energy of the molecules of the substance.

lipoma	(Gr.) *lipos*, fat + *oma*, tumor
adenoma	(Gr.) *aden*, gland + *oma*, tumor
neurofibroma	(Gr.) *neuron*, nerve + (L.) *fibra*, fiber + (Gr.) *oma*, tumor
metastasis	(Gr.) *meta*, beyond + *stasis*, stand
sarcoma	(Gr.) *sarx*, flesh + *oma*, tumor

THE CLINIC

NEOPLASMS: CELLS OUT OF CONTROL

Neoplasms, usually called **tumors,** are abnormal aggregations of cells growing out of control. Cells in neoplasms are usually less differentiated than normal cells; that is, they have lost some or all of the highly specialized features and functions characteristic of the normal tissue cells from which they grew. These new growths are either **benign** (noncancerous) or **malignant** (cancerous). The cells in benign neoplasms tend to be made up of cells that are more differentiated than the cells of malignant tumors.

Benign Tumors

Benign tumors grow more slowly than malignant neoplasms, and they are frequently surrounded by a fibrous coating. They remain in a localized area and do not spread to other parts of the body. Because benign tumors are relatively well defined, they are usually easily removed by surgery.

Although the word *benign* implies "not harmful," local benign growths can damage the body by causing pressure on adjacent organs. In addition, some benign neoplasms are sufficiently well differentiated to produce excessive amounts of hormones or enzymes that can cause abnormal symptoms in patients. For example, benign tumors of the thyroid gland can cause severe symptoms because of their production of excessive amounts of the normal thyroid hormone thyroxine. This results in hyperthyroidism, which is manifested by rapid heart rate, weight loss, protruding eyeballs, and at times heart failure. Benign neoplasms in the pancreas can increase the production of insulin to abnormal levels causing imbalance in the metabolism of glucose, the main sugar in the blood (see Chapter 17). Surgical removal of tumors such as these can result in a cure for the patient.

Benign tumors are named for the tissue from which they arise with the suffix *-oma* added. For example, **lipomas** (lĭ·pō′măs) are benign tumors of fatty tissues that often occur just beneath the skin causing soft, semimobile, painless lumps anywhere on the body. **Adenomas** (ăd″ĕ·nō′măs) are benign growths of glandular tissues such as thyroid adenomas.

Multiple neurofibroma (nū″rō·fī·brō′mă) is a hereditary disease that results in many benign tumors of fibrous cells and nerve cells throughout the body causing severe deformity and symptoms from the pressure of the expanding tumors. This disease has recently attained attention as "Elephant Man disease," but it has long been known to physicians as von Recklinghausen's disease.

Malignant Tumors: Cancer

Probably no other disease strikes as much fear in the hearts of people than cancer. **Cancer** is the second leading cause of death in the United States at the present time, surpassed only by heart disease. The number of deaths due to cancer in this country has doubled every 30 years during this century and now stands at 400,000 deaths per year. About one person out of every four in the United States will have cancer within his or her lifetime, and one in six will die from the disease.

Cancers are new growths that are usually less well differentiated in their cell types than benign tumors and tend to spread to other areas of the body and produce new growths there. The capacity for malignant neoplasms to spread randomly to other parts of the body most clearly differentiates malignant from benign growths. Cancer cells may spread locally by fingers or sheets of invasive tissue into adjacent areas, or cells may break away from the original site and travel through the bloodstream or lymphatic system (see Chapter 24) to a distant site where they may begin to grow and form secondary tumors. The spread of cancer cells from the original (primary) site to distant areas is called **metastasis** (mĕ·tăs′tă·sĭs). Sites of metastasis may in turn metastasize to further areas. Cancers often have a preferred site of metastasis, for example, cancer of the breast frequently metastasizes to bone, lung, or brain tissues. Malignant neoplasms generally grow rapidly and recruit a vascular system to support their growth. Frequently, a malignant tumor will grow so rapidly that it appears to outstrip its vascular supply, and areas of dead tissue appear in malignant growths.

Cancers may be classified into two basic types: those originating from epithelial tissues are called **carcinomas** and those from connective tissues are called **sarcomas.** The classification of malignant neoplasms, however, is very complex.

The damage caused by cancer is due to the widespread destruction of normal cells and organs by pressure from the expanding tumors and by competition for nutrients and metabolites by the cancerous cells. Some cancers also cause inflammatory reactions that result in symptoms. Malignant neoplasms have a variety of systemic effects, including **cachexia** (severe wasting of the individual), loss of normal function of organs, and pain. The pain often associated with cancer is due to pressure of the expanding tumor on unyielding tissues such as bone, pressure on adjacent nerves, and sometimes toxic effects.

Causes of Cancer Cancer is not a single disease with a single cause, but is an abnormal growth of cells in response to many environmental and hereditary factors such as **carcinogenic** (cancer-causing) chemicals, viruses, and ionizing radiation (X rays and sunlight). However, the exact mechanisms that cause runaway cell division remains unknown. It has been estimated that up to 80% of cancers in the United States may be due to environmental factors. The single major cause of preventable cancer is the use of tobacco, particularly from smoking cigarettes. The death rate from cancer of the lungs now exceeds 110,000 per year (see Chapter 20). The use of tobacco is implicated in 98,000 of these deaths per year. About 5% of cancer deaths in this country are related to the excessive use of alcohol. The occupational exposure to chemicals and ionizing radiation is responsible for another 4% of cancer deaths. Infectious agents have also been implicated in a small

oncologist   (Gr.) *ogkos,* mass + *logos,* to study

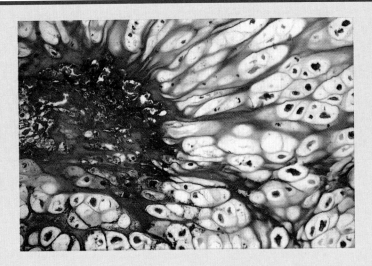

This photomicrograph of cartilage tissue shows the diastrous effect of cancer on our cells. This tissue is affected by chondrosarcoma.

percentage of cancers. There is strong evidence linking the hepatitis B virus with cancer of the liver, especially in Asia and Africa. The **Epstein-Barr virus** (which causes infectious mononucleosis in this country) is associated with several types of cancer (see Chapter 20). Cancer of the bladder has been caused by infestation of a parasitic worm in tropical countries. Heredity can also play a role in the occurrence of cancer. Daughters of women who have cancer of the breast are more likely to develop similar cancers. Male relatives of men with cancer of the colon are predisposed to developing colonic cancers. The incidence of particular cancers is different for men and women. For instance, men are almost twice as likely to get cancer of the urinary tract than are women.

Treatment and Management Physicians who specialize in the diagnosis and treatment of cancers are called **oncologists.** Oncology is so complex and diverse that it involves nearly every aspect of the health sciences.

Of course, the best form of treatment is prevention through awareness of the potential dangers of our lifestyles. Giving up cigarette smoking is an obvious preventative measure. If we were to become a nonsmoking society, we would avoid 98,000 tobacco-related deaths from lung cancer each year in addition to other smoking-related diseases of the heart and lungs. It is also important that we are aware of the dangers of exposure to known or suspected carcinogens in the home and workplace. Early detection is the second line of defense. Many cancers can be detected long before they metastasize and can be successfully eradicated before that first cell breaks away and spreads to distant sites. Many effective methods of early detection are simple and inexpensive. These include the **Papanicolaou test** (the "pap" smear) for cervical cancer in women, stool examination for blood to detect cancer of the colon, and self-breast examination and mammography for breast cancer (see Chapter 27).

Treatment for advanced cancer is much more complicated and involves surgery, chemotherapy (drugs), and radiation treatments. Chemotherapy and radiation therapy are aimed at disrupting the cell division process of the cancer cells and killing them. These two treatments tend to affect rapidly dividing cancer cells and fortunately have less effect on normally dividing cells. However, body hair and other rapidly growing tissues are sloughed off during treatment for cancer. Surgical excision combined with radiation and chemotherapy have produced successful cures for many cancers.

Immunotherapy is a more recent method of cancer treatment. Various techniques have been developed in an attempt to enhance the body's own immune system to resist the effects of cancer (see Chapter 24). Because of the complexity of the body's immune system, there has not been spectacular progress in this form of cancer therapy, but a few techniques deserve mention. **Interferon** is a substance released by cells that have been invaded by viruses. It has nonspecific effects on the immune system, suppressing some parts of the system while activating others. Interferon also inhibits the synthesis of DNA, which may directly inhibit cell growth in rapidly growing cancer cells. At the present time, it has not proven to be very useful.

The use of **monoclonal antibodies** is a more specific approach to cancer therapy. The advent of recombinant DNA technology has made it possible to produce many genetically identical offspring (clones) of a single cell. Each monoclonal cell can be induced to produce an antibody specific to certain components of tumor cells. In this manner, a substance can be produced that attacks only the tumor cells, thus avoiding damage to the normal healthy cells. Progress in this technology holds great hope for more effective and less harmful treatment of malignant neoplasms.

Osmosis The diffusion of water through a selectively permeable membrane is called *osmosis*. The rate of diffusion of water depends on the activity of the water molecules. Differences in the activity of water on the two sides of a *selectively permeable* membrane results in a net flow of water through the membrane.

When the solute concentrations on the two sides of a membrane are not the same, the side with the lower solute concentration is *hypotonic* to the other solution; the solution with the higher solute concentration is *hypertonic*. If the solute concentrations are the same, the solutions are *isotonic* to one another. Net water flow is from the hypotonic solution to the hypertonic solution.

Facilitated Diffusion Molecules that are too large to pass through a membrane by simple diffusion are often able to pass through with the help of carrier molecules. In *facilitated diffusion* net movement of the molecules transported by the special carrier molecules is from the side of higher concentration to the side of lower concentration. Chemical energy is not expended by the cell in the process.

Active Transport The movement of materials through the membrane from a region of low concentration into a region of high concentration is *active transport*. It utilizes special carrier molecules and requires the expenditure of chemical energy by the cell.

Mechanisms of Transport Carrier molecules used in facilitated diffusion and active transport are *permeases*. They may aid in the transport of materials by allowing the substances to simply diffuse through them or by undergoing a conformational change after combining with the substance. The conformational change causes the transported molecule to be carried across the membrane.

Exocytosis and Endocytosis Movement of materials through the membrane can also be effected by engulfment of material outside the cell by the cell membrane (*endocytosis*) or by the fusion of vesicles within the cell with the cell membrane (*exocytosis*). Uptake of solid matter by a cell is called *phagocytosis*; uptake of liquids is *pinocytosis*.

Contents of the Cytoplasm (pp. 57–63)
Cytoplasm is a complex mixture of chemicals and *organelles* specialized to perform particular tasks for the cell.

Endoplasmic Reticulum and Ribosomes A network of membranes widely distributed in the cytoplasm is called the *endoplasmic reticulum*, which divides the cytoplasm into *luminal* and *cytosol* phases. The endoplasmic reticulum has a rough appearance in regions where there are many ribosomes and a smooth appearance in regions where there are fewer ribosomes.

Ribosomes are structures assembled in the nucleolus that are responsible for protein synthesis in the cytoplasm.

Golgi Apparatus and Lysosomes The membranous structure responsible for the production of vesicles is a *Golgi apparatus*. Vesicles bud off from sacs of the Golgi apparatus, enclosing dissolved materials that had been in the fluid of the Golgi apparatus.

Lysosomes are vesicles that carry digestive chemicals used for digestion of macromolecules.

Peroxisomes Vesicles that carry enzymes that catalyze the oxidation of many toxic substances are called *peroxisomes*. They may be produced directly from the endoplasmic reticulum rather than from the Golgi apparatus.

Mitochondria The organelle responsible for the production of ATP, the primary chemical energy source for many cell processes, is the *mitochondrion*. It consists of an inner and outer membrane, a folded *cristae*, and a central fluid *matrix*. The number of mitochondria in a cell is indicative of the amount of energy-requiring reactions occurring in the cell.

Cytoskeleton The shape of a cell, its movement, and the transport of organelles within it depend on the cell's *cytoskeleton*. Cytoskeletons consist of elaborate networks of *microtubules* and *microfilaments* of various sizes and structures.

Cilia and Flagella *Cilia* are small, hairlike appendages on the surfaces of cells that line many tubular organs. Wavelike undulations of the cilia set up currents that help propel material through the tube. *Flagella* are long, whiplike tails that propel sperm. Movement of cilia and flagella is caused by a cylinder of microtubules at the base of the structure called an *axoneme*. A *basal body* lies at the base of the axoneme.

The Nucleus and its Contents (pp. 64–65)
The *nucleus* is a large, usually spherical organelle that houses the cell's genetic material.

Nuclear Envelope The outermost surface of the nucleus, or *nuclear envelope*, consists of a double membrane, each separated from the other by a *perinuclear space*. The outermost membrane of the two is continuous with the endoplasmic reticulum.

Nucleolus The *nucleolus* is an irregularly shaped, amorphous structure in the nucleus in which ribosome subunits are formed. It is not surrounded by a membrane.

Chromatin The granular contents of a nucleus is the *chromatin*, which consists primarily of DNA and associated proteins. RNA is also present in chromatin and represents molecules that will be transported to the cytoplasm where they will be involved in protein synthesis. Proteins in chromatin are principally *histones*, molecules that associate with DNA to form nucleohistones, and a diverse group of *nonhistone* chromosomal proteins.

Cell Division (pp. 66–68)
The process by which new cells are formed.

Cytokinesis and Karyokinesis Cell division consists of two phases: *cytokinesis*, the division of the cytoplasm, and *karyokinesis*, the division of the nucleus. The two phases together comprise *mitosis*.

Phases of Mitosis *Interphase* is the period of a cell cycle during which mitosis is not occurring.

Prophase is the period in mitosis when *chromosomes* become visible as long, slender threads. Each chromosome consists of two *chromatids*, held together at the chromosome's *centromere*. The nuclear envelope disintegrates and a *spindle* apparatus forms.

Metaphase is the phase in mitosis when the chromosomes line up in the central region of the spindle.

Anaphase is the phase when chromatids separate from one another and are pulled to opposite ends of the spindle.

Telophase is the time in mitosis after the chromatids (now called chromosomes) have reached the ends of the spindle. The chromosomes relax, new nuclear membranes form, the cytoplasm divides, and the two newly formed cells enter interphase.

DNA Function (pp. 68–73)

Replication DNA is duplicated by a process called *replication*, in which each strand serves as a *template* for the synthesis of a new strand. Although many enzymes and proteins are involved, the specific enzyme that catalyzes the addition of nucleotides is *DNA polymerase*.

Replication is initiated at many points along a DNA molecule, forming *replication bubbles* that grow outward from the initiation site until they merge with the neighboring bubbles.

Transcription *Transcription* is a process in which RNA is made using DNA as a template. Transcription is catalyzed by RNA polymerase molecules that begin catalyzing RNA synthesis at specific points along the DNA molecule. The RNA molecules formed are usually modified further to produce the functional RNA molecule.

Translation *Translation* is the process in which proteins are made using information transcribed into RNA molecules. Three kinds of RNA molecules used in translation are *messenger RNA* (mRNA), *transfer RNA* (tRNA), and *ribosomal RNA* (rRNA).

Messenger RNA carries information to the sites of protein synthesis as a sequence of nucleotides transcribed from DNA.

Transfer RNA molecules combine with specific amino acids and carry them to messenger RNA molecules where they are incorporated into polypeptides.

Translation consists of three phases: *initiation, elongation,* and *termination*. Appropriate amino acids are selected on the basis of complementarity of *codons* in the messenger RNA and *anticodons* in the transfer RNA to which a specific amino acid is attached.

Several ribosomes may be associated with a messenger RNA molecule at the same time, forming *polysomes*.

SELF-TEST OF CHAPTER OBJECTIVES

True-False Questions
1. The mitochondrion has a single membrane surrounding it.
2. The central portion of a membrane is the most water soluble.
3. Substances can be transported across a membrane from a region where they are in low concentration to a region where they are in high concentration by facilitated diffusion.
4. Osmosis refers specifically to the movement of water through a selectively permeable membrane.
5. The dissolved solutes inside a red blood cell have the same osmotic properties as a solution of 0.9% table salt. If that cell is placed in pure water, the cell will lose water by osmosis.
6. Cells in a 1% solution of sodium ions that are maintaining a 2% internal concentration are probably doing so by facilitated diffusion.
7. Rough endoplasmic reticulum is rough in appearance because of the presence of large numbers of mitochondria.
8. Material leaving the nucleus through the nuclear pores enters the luminal phase of the cytoplasm.
9. Ribosomes are partially assembled in a part of the nucleus called the nucleolus.
10. Chromosomes are condensed forms of the fibrils of chromatin present in the interphase nucleus.

Matching Questions
Match the structure on the left with its probable function on the right.

11. communication junctions
12. impermeable junctions
13. adhering junctions

a. prevents diffusion of materials into spaces between cells
b. permits the movement of materials directly from cell to cell
c. imparts strength to tissue

Indicate which of the chemical substances on the right can be found in the following cell structures:

14. chromatin
15. plasmalemma
16. ribosomes
17. nucleolus

a. neutral fats
b. phospholipids
c. DNA
d. carbohydrate
e. steroids
f. RNA
g. protein

Multiple-Choice Questions
18. When a mitochondrion migrates through the cytoplasm of a cell, which of the following is probably involved?
    a. endoplasmic reticulum
    b. microtubules
    c. microfilaments
    d. microtrabecular lattice
19. There is 0.1 g of dissolved sugar per 1 mL of water on one side of a membrane (side A) and 0.2 g/mL of water on the other side (side B). Molecules of the sugar will not pass

through the membrane, but they attract water molecules and hold them in a "cloud" around themselves. Water passes through the membrane readily. In which direction does the water pass most rapidly?
   a. from side A to side B
   b. from side B to side A
   c. both directions at equal rates
   d. Neither. There is no flow in either direction.
20. Which of the following is responsible for the movement of cilia?
   a. microtubules   c. microfilaments
   b. intermediate filaments   d. ground substance
21. Which of the following is not bounded by a membrane?
   a. cytoplasm   c. mitochondrion
   b. ribosome   d. nucleus
22. Which of the following cells would you expect to have the most extensively developed Golgi apparatus?
   a. an epithelial cell in the skin
   b. a cell in the intestine that secretes digestive proteins
   c. a nerve cell
23. At what stage of mitosis do sister chromatids move toward the opposite ends of the spindle?
   a. interphase   c. metaphase
   b. prophase   d. anaphase
24. Which of the following is not a component of chromatin?
   a. protein   c. carbohydrate
   b. DNA   d. RNA
25. Which of the following forms of RNA carries amino acids to sites of protein synthesis in the cytoplasm?
   a. messenger RNA   c. ribosomal RNA
   b. transfer RNA   d. nuclear RNA
26. Which of the following would pass through the plasmalemma most readily?
   a. water molecule   c. amino acid molecule
   b. glucose molecule   d. protein molecule
27. Which of the following processes requires the expenditure of energy by a cell?
   a. simple diffusion   c. active transport
   b. facilitated diffusion   d. all of these
28. Which of the following is continuous with the perinuclear space?
   a. cytosol   c. mitochondrial matrix
   b. luminal phase   d. nuclear matrix
29. From which of the following structures are vesicles produced?
   a. mitochondria   c. Golgi apparatus
   b. nucleus   d. endoplasmic reticulum
30. Which of the following describes the function of a nucleolus?
   a. stores genetic information for protein synthesis
   b. stores and assembles ribosomal subunits
   c. produces ATP
   d. transports materials from the nucleus
31. Where in the cell would histones be found in greatest concentration?
   a. plasmalemma   c. nucleolus
   b. cytoplasm   d. chromatin
32. Which of the following lists the stages of mitosis in the correct order?
   a. prophase, metaphase, telophase, anaphase
   b. prophase, metaphase, anaphase, telophase
   c. anaphase, metaphase, prophase, telophase
   d. metaphase, prophase, telophase, anaphase
33. Where would glycoproteins most likely be found in a cell?
   a. on the outer surface of the plasmalemma
   b. on the inner surface of the plasmalemma
   c. on the cytosol surface of the endoplasmic reticulum
   d. on the luminal surface of the endoplasmic reticulum
34. Two cells, designated A and B, lie in contact with one another. Water flows by osmosis from cell A to cell B. Therefore,
   a. cells A and B are isotonic
   b. cell A is hypertonic to cell B
   c. cell A is hypotonic to cell B
   d. cell A's turgor pressure is greater than cell B's
35. Where are basal bodies found in a cell?
   a. at the bottom of the cell
   b. at the base of a flagellum or cilium
   c. combined with the acidic bodies in the nucleus
   d. in the nucleolus

Essay Questions

36. Explain in general the function served by organelles within a cell.
37. Draw a cross-sectional view of a cell, including and identifying the (a) plasmalemma, (b) endoplasmic reticulum, (c) vesicles, (d) Golgi apparatus, (e) mitochondria, (f) ribosomes, (g) nucleus, and (h) nucleolus.
38. Draw a cross-sectional view of the plasmalemma, labeling the major features.
39. Describe the mechanism by which each of the following substances can enter a cell: (a) H_2O, (b) glucose, and (c) a protein.
40. Describe three types of cell junctions and describe the function of each.
41. Explain this statement: the rate of flow of water through a cell membrane depends in part on the activity of the water molecules.
42. Suppose pure water lies on either side of a cell membrane, but the water on one side is warmer than the water on the other side. Would there be a net flow of water through the membrane? Why or why not?
43. Draw a diagram illustrating the difference in the mechanisms of facilitated and active transport.
44. Diagram the passage of a protein molecule synthesized on a portion of the rough endoplasmic reticulum to its secretion at the plasmalemma.
45. List and describe the components of a cytoskeleton.
46. Diagram mitosis in a cell that has four chromatids. Identify the five stages of mitosis, the spindle and spindle fibers, and the chromosomes and chromatids.
47. Draw a typical chromosome as it might appear at metaphase of mitosis.

48. Describe the function of a lysosome in the degradation of a substance taken into the cell by phagocytosis.
49. An amino acid is coded by a sequence of three nucleotides in messenger RNA. How many such combinations are possible?
50. Diagram the processes of transcription and translation, showing how information stored in a DNA molecule in a nucleus can be used to specify an amino acid sequence in a protein produced in the cytoplasm.

Clinical Application Questions

51. How does renal dialysis make use of the natural processes of diffusion and osmosis to filter the blood?
52. Briefly describe the potential benefits that the application of genetic engineering techniques can have on human physiology.
53. Explain the difference between benign tumors and malignant tumors.
54. List some of the major causes of cancer and the treatments in use today.

4
Tissues

In Chapters 2 and 3, the basic building blocks of the human body, namely molecules and cells, were described. In those chapters we stressed the nature and importance of structure and functions of individual cells. But with few exceptions—certain white blood cells, sperm cells, and egg cells, for example—cells do not function as individual entities. Instead, the body is composed of cells bound together into groups called **tissues** that act together to perform one or more specific tasks. The present chapter is concerned with this level of organization.

Four basic types of tissues exist: epithelial tissue, connective tissue, muscle tissue, and nervous tissue. The relationship between cells and these tissues is analogous to that between bricks and a brick building. The cells of a tissue are held together by extracellular material. It ranges from a simple extracellular fluid to a rigid mineral deposit found in bone.

In this chapter we will discuss

1. the definition of a tissue and indicate its position in the organizational levels of the body,
2. the four basic types of tissues found in the human body,
3. epithelial tissues and the functions they perform,
4. connective tissues and the difference between loose and dense connective tissues,
5. bone and cartilage,
6. the differences among smooth, skeletal, and cardiac muscle tissues, and
7. nervous tissue.

Epithelial Tissue

After completing this section, you should be able to:

1. Differentiate among simple, pseudostratified, and stratified epithelia.
2. Indicate the major areas in the body where epithelia are found.
3. Describe the functions of different types of epithelia.
4. Describe the differences between exocrine and endocrine glands.

Epithelia (ĕp″ĭ·thē′lē·ă) are composed of layers of cells packed together closely. They form the outer covering of the body and of individual organs and the inner lining of various spaces in the body, such as the thoracic and abdominal cavities. They also form tubes, such as the digestive and respiratory tubes, ducts of glands, and the inner lining of the blood vessels. These tissues function as protective layers and also secrete mucus to lubricate and moisten some exposed surfaces. Specialized types of epithelia are modified to form parts of certain sense organs and also to form glands that produce and secrete specialized chemical products.

One surface of an epithelium is usually free and exposed to either air or a fluid. The opposite surface of the tissue rests on a **basement membrane** (Figure 4-1), which holds the epithelial cells together and binds them to underlying connective tissue. The basement membrane is composed of materials secreted by both deep epithelial cells and adjacent connective tissue cells. Together these secretions cement the two tissues together.

Epithelia are classified on the basis of the functions they perform, the shapes of the surface cells, and the manner in which they form layers. Functionally, epithelia serve either to cover and line various organs and cavities or to secrete cellular products. Structurally, epithelia can be classified into three major groups: simple, pseudostratified, and stratified. **Simple epithelium** is composed of a single layer of cells (Figures 4-2a–c). **Pseudostratified epithelium** also consists of a single layer, but since some of the cells in the tissue are shorter than others and do not extend entirely from the basement membrane to the surface, the tissue appears to be composed of more than one layer of cells (Figure 4-2d). **Stratified epithelium** is composed of several to many layers of cells (Figures 4-2e–h).

Figure 4-1 Epithelial tissues are exposed on one surface, and the opposite surface is firmly attached to a basement membrane which in turn attaches to underlying connective tissue. (a) Diagram of a typical epithelial cell from a salivary gland. (b) An electron micrograph of an actual salivary gland cell (2000×).

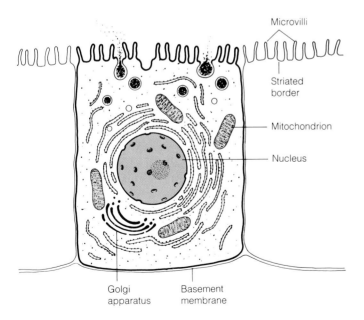

(a)

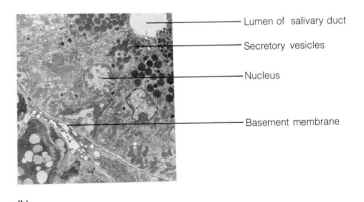

(b)

82 Chapter 4 Tissues

squamous: (L.) *squama*, scale
epithelium: (Gr.) *epi*, on + *thele*, nipple
endothelium: (Gr.) *endo*, within + *thele*, nipple
mesothelium: (Gr.) *mesos*, middle + *thele*, nipple

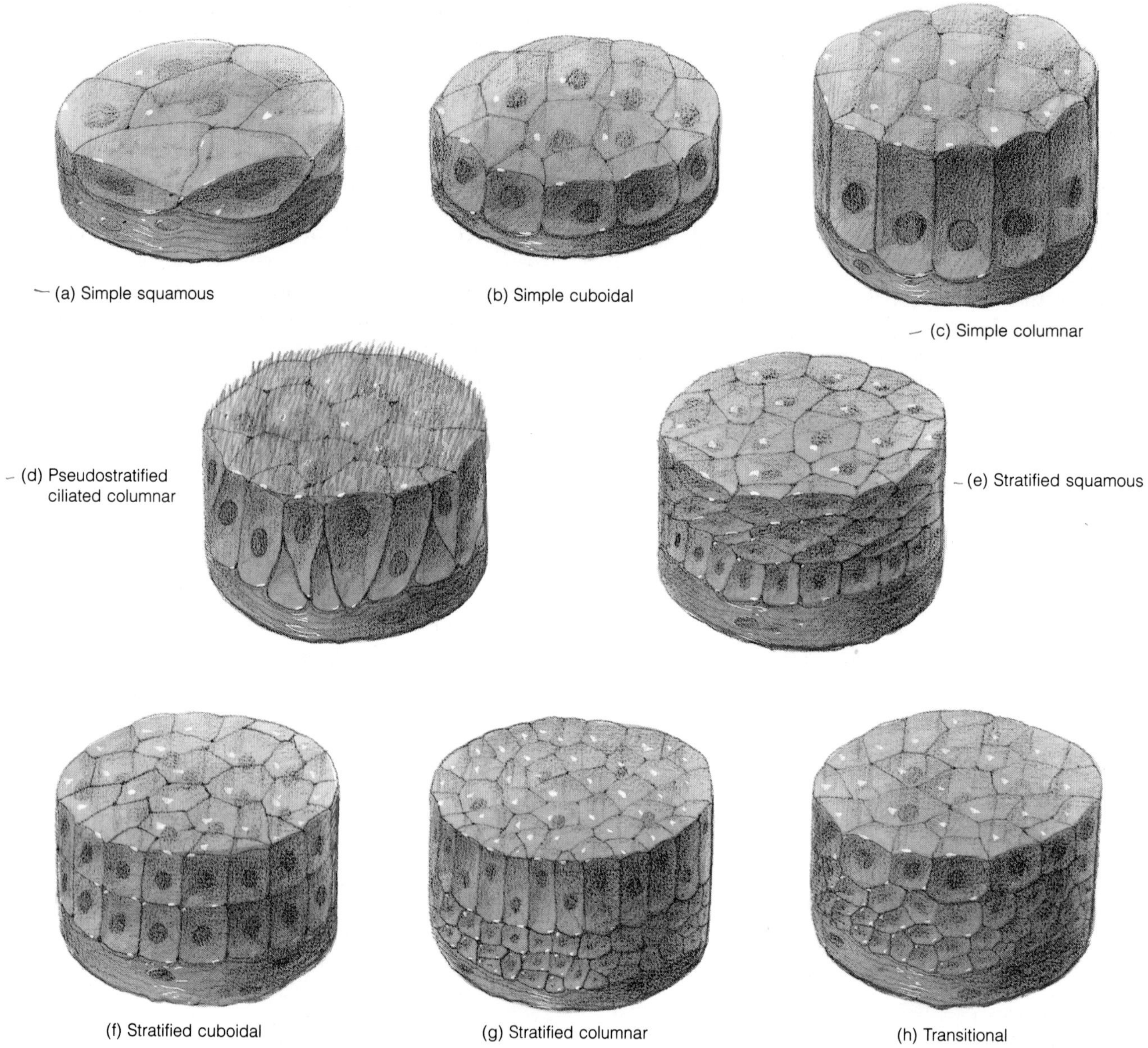

(a) Simple squamous

(b) Simple cuboidal

(c) Simple columnar

(d) Pseudostratified ciliated columnar

(e) Stratified squamous

(f) Stratified cuboidal

(g) Stratified columnar

(h) Transitional

Figure 4-2 Epithelial tissues range in complexity from flat sheets of single cells to thick layers composed of cells of a variety of sizes and shapes. Specific locations in the body of each of these tissue types are given in the text.

Simple Epithelial Tissue

A simple epithelium is composed of cells of three basically different shapes: flattened (or squamous) cells, cuboidal cells, and elongate (or columnar) cells.

Simple squamous (skwā′mŭs) **epithelium** is composed of a single layer of flattened cells that fit together like pieces in a jigsaw puzzle (Figure 4-2a). This type of epithelium lines the interior of the blood and lymphatic vessels, as well as the heart, where it is called **endothelium** (ĕn″dō·thē′lē·ŭm). It also lines thoracic and abdominal cavities and covers the organs located in these areas. Here it is called **mesothelium** (mĕs″ō·thē′lē·ŭm). Additionally, it lines the smallest respiratory ducts and air sacs of the lungs, the smallest ducts of certain glands, portions of the ducts that filter wastes from the blood in the kidney, and the tympanic cavity and membranous labyrinth in the middle and inner ear. The thinness of this tissue allows many materials to pass through it in either direction by diffusion, osmosis, or filtration.

Simple cuboidal epithelium consists of a single layer of small, cube-shaped cells (Figure 4-2b). It is found in the retina, on the anterior surface of the lens, and the posterior surface of the iris. Cuboidal epithelium is also found in several small networks of blood vessels in the brain, the outer covering of the ovary, the lining of the adult kidney tubules, and part of the secretory portions and ducts of many glands, such as the thyroid and sweat glands. Simple cuboidal epithelium functions in the active transport of materials.

Simple columnar epithelium is similar to cuboidal epithelium except that columnar cells are taller than they are wide (Figure 4-2c). As you might expect, some cells in columnar epithelium are intermediate between typical cuboidal and typical columnar cells. Simple columnar epithelium is found lining much of the respiratory tract, stomach and intestines, gallbladder, and uterus and uterine tubes; it also occurs in some digestive glands.

Columnar cells often bear cilia on their exposed surfaces. In the digestive tract and the upper respiratory tract, some of the columnar cells, known as **goblet cells,** are specialized to produce mucus that accumulates at their free ends, resulting in a swollen appearance. Mucus released from goblet cells forms a protective lubricant layer between the cells of the digestive tract and the food that passes through it. In the respiratory tract, dirt particles and other foreign materials are trapped in this mucus and, by coordinated wavelike movements of cilia, are moved up to the throat and mouth where they can be eliminated either by swallowing or expectoration. Waves of ciliary movement account for the movement of reproductive cells within the tubes of the female reproductive tract.

Pseudostratified Epithelial Tissue

Pseudostratified columnar epithelium is so named because it appears to be composed of several layers of cells, but it actually consists of a single layer of cells of varying heights, all of which are attached to the basement membrane (Figure 4-2d). Pseudostratified epithelium is often ciliated, and in the respiratory tract it contains goblet cells. It lines the male urethra (the tube that carries urine out of the urinary bladder) and the large ducts of many excretory glands. Ciliated pseudostratified columnar epithelium lines the nasal cavities, the larger tubes of the upper respiratory tract, the auditory tubes of the ear, and certain male reproductive ducts.

Stratified Epithelial Tissue

Up to this point we have examined epithelial tissues composed of a single layer of cells. These single layers can withstand little trauma or abuse because they are so thin. **Stratified epithelium** is thicker and consists of multiple layers, thus it can withstand a variety of stresses including stretching and friction. Cells found in the upper layers of stratified tissues have essentially the same shape as those that comprise simple epithelia: squamous, cuboidal, and columnar. These cells provide the basis for the classification of stratified epithelium.

Stratified squamous epithelium consists of several layers of cells of varying shapes (Figure 4-2e). The deepest or **basal layer** of cells, those attached to the basement membrane, varies from cuboidal to columnar, and these deep layers may be folded over one another. Cells in the basal layer undergo cell divisions mainly at night, producing a continuous supply of new cells. As the cells age, they are pushed closer and closer to the surface by newly produced cells of the basal layer. As they approach the surface, they become flatter, eventually assuming a scalelike (squamous) appearance when they reach the surface. Cells thus displaced from the basal layer are moved farther and farther from the nutritive blood vessels in the underlying connective tissue, and their metabolic activity is reduced and may even cease. Eventually, they slough off and are replaced by cells from deeper layers.

Stratified squamous epithelium is found in two major regions. One of these regions is the outer layer of skin that covers the body. Exposure to the sun and air causes drying, promoting the cells to produce a protective, waterproof protein called **keratin** (kĕr´ă·tĭn) which replaces the living contents of the cells. This results in the formation of a layer of dead cells that is extremely tough, waterproof, and very resistant to invasion by bacteria. This is **keratinized squamous epithelium.** Whether keratinized or not, the outer layers of this tissue are continuously worn away and replaced from beneath.

Stratified squamous epithelium also occurs in moist areas lining the upper and lower ends of the digestive tube, the larynx, the cornea, and parts of the urethra and vagina. In these places the surface cells usually remain alive, retain their cytoplasm and nuclei, and produce only small amounts of keratin. Such moist tissues form a **nonkeratinized squamous epithelium.**

Stratified cuboidal epithelium consists of two or more layers of cuboidal cells (Figure 4-2f) and is found in the ducts of adult sweat glands and the mucous membrane that lines the eyelids. This tissue functions primarily as a protective layer of cells, but it may also secrete mucus.

Stratified columnar epithelium is difficult to distinguish from pseudostratified columnar epithelium. The chief distinguishing characteristic of this sparsely distributed tissue is the elongation of its surface cells (Figure 4-2g). Stratified columnar epithelium is very localized in occurrence, being found in parts of the upper respiratory tract, male urethra, pharynx, larynx, conjunctiva, and parts of the

keloid: (Gr.) *kele*, tumor + *eidos*, form
endocrine: (Gr.) *endo*, within + *krinein*, to separate
exocrine: (Gr.) *exo*, outside + *krinein*, to separate

excretory ducts of mammary and salivary glands. It functions primarily as a protective layer of cells that may also secrete mucus.

Transitional epithelium, also called **urothelium** (ū″rō·thē′lē·ŭm), is unique. It was originally designated as "transitional" because some thought it provided a transition in cell types between stratified nonkeratinized squamous epithelium and stratified columnar epithelium. Actually, the appearance of this tissue changes under different conditions of stretching (Figure 4-2h). It is found only in the urinary bladder and in the tubes leading into and out of the bladder, hence the name "urothelium." These organs are subject to much stress as the regular accumulation of urine causes them to fill and increase their size many times compared to when they are empty. In an empty bladder, cells of the interior lining tend to pile up on one another, producing a many-layered tissue that resembles stratified squamous epithelium, except that the surface layer consists of large, rounded cells instead of flattened, scalelike cells. As the bladder fills with urine, the cells flatten out and slide across one another, producing a thin, but still stratified layer of cells only two or three cells thick with a squamous surface layer. When the bladder is emptied, the tissue returns to its thick, stratified condition.

Glandular Epithelium

Recall that some epithelial cells function as glands rather than as protective layers of tissue. A **gland** is any cell or group of cells that secretes specialized materials. There are two basic types: endocrine glands and exocrine glands. **Endocrine glands** empty their secretions directly into the blood vessels of the gland, and the circulatory system serves to transport the endocrine secretions, or **hormones**, throughout the body. The thyroid gland, located on either side of the larynx and trachea, is an example of an endocrine gland. Its hormones help in the regulation of metabolism, growth, and development. The functions of endocrine glands will be discussed more fully in Chapter 17 on the Endocrine System.

Exocrine glands produce secretions that flow through a duct to the specific area where the secretion is used, either the body surface or the lumen of a hollow body cavity. Sweat glands secreting onto the skin and salivary glands secreting into the mouth are typical exocrine glands.

Functional Classification of Exocrine Glands

Exocrine glands can be classified into three functional groups depending on their mode of secretion: holocrine, apocrine,

---

CLINICAL NOTE

SCARRING

The human body has great recuperative powers. The ability of body tissues to recover from injuries depends in part on the specialization of the cells that make up the injured tissue. Highly specialized cells of the central nervous system have little ability to recover from serious injury. In contrast, the skin readily repairs scrapes and cuts. The repair process usually involves replacing the injured cells with fibrous tissues rather than functioning specialized cells. The fibrous tissue is produced by **fibroblasts** in a haphazard manner resulting in a tough matrix we recognize as **scar tissue.** Scar tissue has little function other than to hold the tissues together or to fill a gap where large defects have occurred. Scar tissue has little elasticity and may cause loss of function in tissues such as muscle.

Excessive production of scar tissue can produce an unsightly heaped up mass on the skin called a **keloid** (see figure). The production of keloids seems to be a characteristic of some individuals and is quite common in people of the black race.

Scars can also occur internally, where they may form strong bands of fibrous tissue from one organ to another. These are known as **adhesions.** Adhesions in the abdominal cavity can result in obstruction of abdominal organs, requiring a surgical release (lysis) of the offending bands.

Surgeons are very meticulous in the handling of tissues to limit further injury. They remove damaged tissue, close spaces, and align layers of tissues in an attempt to prevent excessive scar tissue formation.

A Nuer tribesman from Sudan decorated with facial keloids.

holocrine: (Gr.) *holos*, whole + *krinein*, to separate
apocrine: (Gr.) *apo*, away from + *krinein*, to separate
merocrine: (Gr.) *meros*, part + *krinein*, to separate
alveolar: (L.) *alveus*, cavity
acinar: (L.) *acinus*, grape, berry

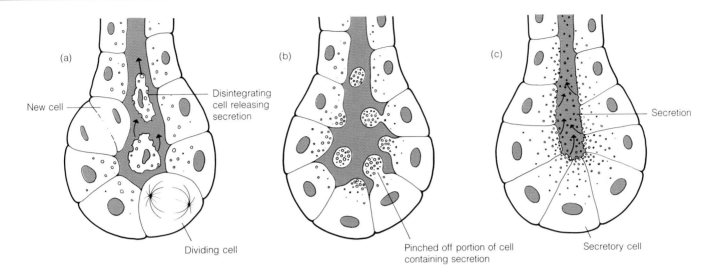

Figure 4-3 The functional types of exocrine glands are based on the mode of secretion: (a) holocrine, secretion of whole cells; (b) apocrine, secretion of pinched off parts of cells; (c) merocrine, direct secretion by diffusion or active transport.

and merocrine (Figure 4-3). **Holocrine** (hōl'ō·krĭn) **glands** secrete whole cells into ducts. The cells later disintegrate and release their contents as a glandular secretion. Oil (sebaceous) glands of the skin are classified as holocrine. **Apocrine** (ăp'ō·krĭn) **glands** accumulate cell secretions in the end of the cell adjacent to the surface, which is then pinched off and released into ducts. The cell thus loses a small quantity of cytoplasm and cell membrane along with the secretion. Mother's milk is produced in this manner by the mammary glands. **Merocrine** (mĕr'ō·krĭn) **glands,** such as mucous glands, salivary glands, and the pancreas, do not lose any cytoplasm during secretion. Instead, the secretion passes through the cell membrane, either by diffusion or by active transport, and is delivered directly to the duct.

Structural Classification of Exocrine Glands

Exocrine glands can be further classified by the complexity of their ducts and secretory portion. They are considered to be either **unicellular** or **multicellular.** Goblet cells are examples of unicellular glands. Table 4-1 provides a summary of glandular types.

Multicellular glands are of greater variety and are named according to whether the secretory duct is unbranched (**simple**) or branched (**compound**) (Figure 4-4). They are further differentiated on the bases of whether the secretory cells are arranged in a tubular configuration (**tubular**), concentrated in a saclike cluster (**alveolar** [ăl·vē'ō·lăr] or **acinar** [ăs'ĭ·năr]), or configured as a mixture of both (**tubuloalveolar** or **tubuloacinar**).

Simple tubular glands may contain a single, straight secretory portion, as do some of the digestive glands in

Table 4-1 SUMMARY CLASSIFICATION OF THE TYPES OF GLANDS FOUND IN THE BODY

Functional Classification

Endocrine
Exocrine
  Holocrine
  Apocrine
  Merocrine

Structural Classification of Exocrine Glands

Unicellular
Multicellular
  Simple
    Simple tubular
      Unbranched
      Branched
    Simple alveolar (acinar)
      Unbranched
      Branched
  Compound
    Compound tubular
    Compound alveolar
    Compound tubuloalveolar

stroma: (Gr.) *stroma.* bedding
parenchyma: (Gr.) *para,* beside + *en,* in + *chyma,* something infused

the intestinal wall, or an elongate, coiled secretory portion, as do ordinary sweat glands (Figure 4-4, top). Several simple glands have branched, tubular secretory portions that empty through a single excretory duct. **Simple alveolar glands** can be unbranched, such as those found in the male reproductive system, or branched, such as sebaceous glands that empty oil into hair follicles.

Compound glands are of three types (Figure 4-4, bottom). **Compound tubular glands** consist of several tubular glands that utilize several ducts in emptying their secretions. These glands are found in the kidneys, liver, testes, and other reproductive glands. **Compound alveolar glands** have essentially the same structural plan, except that the secretory portions are alveolar instead of tubular. Such glands are found in mammary glands, salivary glands (submandibular and sublingual), and some glands in the respiratory tract. **Compound tubuloalveolar glands** are in the pancreas, the larger glands of the respiratory tract, and parts of the digestive tract. In these glands the secretory portions are variable. As their name implies, they usually include a combination of tubular and alveolar regions.

Figure 4-4 Multicellular exocrine glands are classified according to structure into simple (top row) and compound types (bottom row). The active secretory portion of each gland is indicated in orange.

Connective Tissue

After completing this section, you should be able to:

1. Define connective tissue and describe how it differs from epithelial tissue.
2. Distinguish between embryonic and adult connective tissues.
3. Describe loose connective tissue and indicate where it is found.
4. Describe dense connective tissue and indicate where it is found.
5. Enumerate the three different kinds of cartilage.

The major function of **connective tissue** is to connect, hold together, and support different parts of the body. It provides support and protection for organs, and one particular connective tissue—blood—transports materials from one region of the body to another. Connective tissues in an organ are often referred to as the **stroma** (supporting tissue) in contrast to the **parenchyma** (păr·ĕn′kĭ·mă) (functioning part) of the organ. Compared to epithelial tissues, cells of connective tissues are less densely packed and are separated from one another by their secretions called the **ground substance** or **matrix**. The ground substance may take one of

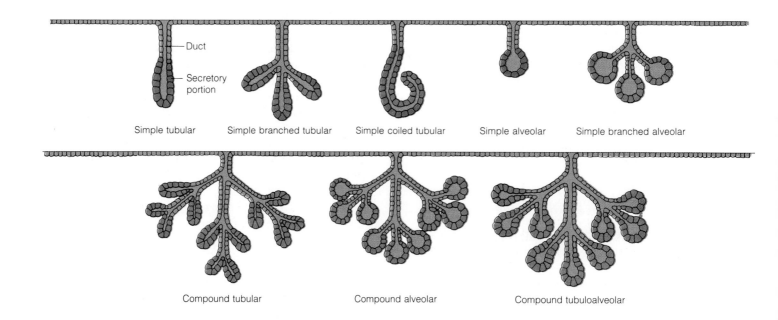

collagenous: (Fr.) *collagene*, from (Gr.) *kolla*, glue
mesenchyme: (Gr.) *mesos*, middle + *en*, in + *chyma*, something infused
fibroblast: (L.) *fibra*, fiber + (Gr.) *blastos*, bud, germ

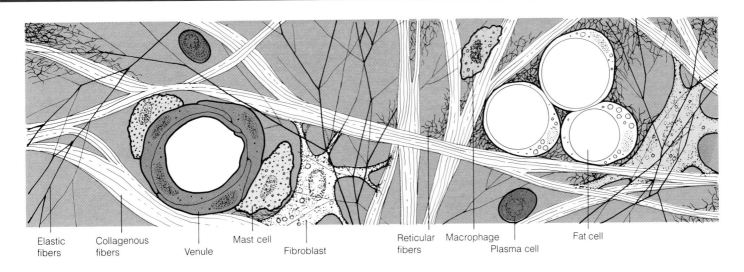

Figure 4-5 Loose connective tissue. This diagram illustrates the types of cells and fibers normally present in this tissue.

three forms: a free-flowing liquid such as blood plasma, a soft semisolid such as the cartilage of the nose, or a solid such as bone.

Three types of proteinaceous fibers produced by connective tissue cells have been identified in connective tissues (Figure 4-5). **Collagenous** (kŏl·lăj′ĕ·nŭs) **fibers** (white fibers) are unbranched, colorless fibers that are usually deposited in bundles to form wavy, parallel groups. **Elastic fibers** are branching, yellowish fibers that are deposited singly and stretch readily under tension. The individual fibers are straight rather than wavy. Although both types of fibers impart a certain amount of resilience to tissues, collagenous fibers are stronger than elastic ones and can withstand far greater tension before they break. **Reticular fibers** are inelastic, immature, collagenous fibers that add strength to the tissue. They are branched and colorless and also contain some glycoproteins.

Embryonic Versus Adult Connective Tissue

Certain connective tissues are found primarily in the embryo and fetus and are referred to as embryonic connective tissues. The developing human embryo has two types of connective tissue: mesenchyme and mucous (Figure 4-6). **Mesenchyme** (mĕs′ĕn·kīm) **tissue** is a loose, delicate, spongy tissue that is composed of irregularly shaped **mesenchymal cells** scattered through a fluid ground substance that often contains fine, reticular fibers. The ground substance of adults has a more gelatinous consistency than that of fetuses. Mesenchyme develops into many other tissues, including mucous tissue, bone, blood, blood vessels, muscles, and fat. Adult connective tissues, chiefly blood vessels, normally contain scattered mesenchyme cells that are not organized into a discrete tissue. These cells can transform themselves into fibroblasts and can secrete collagenous fibers. This activity occurs during the natural healing of a wound, when the collagenous fibers serve to fill a break in a tissue and knit the two severed ends together.

Mucous tissue develops from mesenchyme, and the two appear similar except that as the embryo develops, mucous tissue cells become larger and more elongate and develop cytoplasmic extensions that protrude from different parts of the cell (Figure 4-6b). These cells are called **fibroblasts** (fī′brō·blăsts) because they produce collagenous fibers. Fibroblasts are present in adult connective tissues as well. Mucous tissue also contains certain blood cells in its ground substance. Mucous tissue (called Wharton's jelly) is also found in the fetal umbilical cord where it provides support for the umbilical blood vessels (see Chapter 28).

Embryonic tissues evolve and differentiate by the time of birth into four types of adult connective tissues: connective tissue proper, blood, cartilage, and bone. This classification is based on the relative amount of cells, extracellular matrix, and fibers present. Each of these is discussed below.

Adult Connective Tissue Proper

Two major types of connective tissue proper exist. The first is a tissue with relatively few loosely arranged fibers and is referred to as **loose connective tissue.** The other connective tissue contains many fibers arranged in compact groups and is referred to as **dense connective tissue.**

areolar: (L.) *areola*, a small space
macrophage: (Gr.) *makros*, large + (Gr.) *phagein*, to eat
lipocyte: (Gr.) *lipos*, fat + *kytos*, cell
adipose: (L.) *adipis*, fat
reticular: (L.) *reticulum*, network

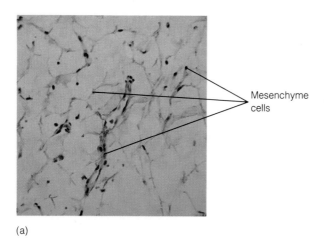

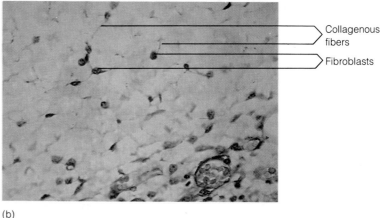

Figure 4-6 Embryonic connective tissues. (a) Mesenchyme tissue is found in many places in a developing embryo (mag. 972×). (b) Mucous tissue develops from mesenchyme tissue in the embryo and is also found in the umbilical cord (mag. 486×).

Loose Connective Tissue

Loose connective tissue (Figure 4-7a), also called **areolar** (ă·rē′ō·lăr) **tissue,** is the most widespread of all the connective tissues. The term *areolas*, which means "spaces," was used by early anatomists to describe the artificial air spaces that were created in the loose connective tissue of an animal during the skinning process. Found throughout the body, it serves as a loose packing for practically all organs and tissues. It is particularly abundant under the skin in a layer of tissue called the **subcutaneous** (sŭb″kū·tā′nē·ŭs) layer. It also surrounds all blood vessels. The ground substance separating these cells is an amorphous fluidlike or jellylike material (Figure 4-7a).

Loose connective tissue contains a variety of cells, the most common of which are fibroblasts and macrophages (Figure 4-5). Since fibroblasts produce collagenous fibers, they add strength to the tissue. **Macrophages** (măk′rō·fāj·ĕs) are cells that serve a defense function by engulfing and destroying bacteria, dead cells, debris, and other foreign materials. **Plasma cells,** which produce antibodies, are also often found in loose connective tissue.

Mast cells are cells distributed throughout loose connective tissue and are found in greatest abundance around small blood vessels (Figure 4-5). Mast cells produce *heparin* (hĕp′ă·rĭn), a chemical that prevents coagulation of blood, and *histamine* (hĭs′tă·mĭn), a chemical released when certain types of tissue damage occur and following exposure to certain *antigens*, such as dust or pollen (see Chapters 21 and 24). When released from mast cells, histamine causes enlargement (dilation) of the local blood vessels and the subsequent loss of fluid from these vessels into the surrounding tissues.

Fat cells, or **lipocytes,** occur normally in loose connective tissue (Figure 4-5). Each fat cell develops from a mesenchyme cell by the accumulation within the cytoplasm of small droplets of fat that increase in size and usually combine into a large, single droplet. This large droplet displaces cytoplasm to the cell edges. In a mature fat cell, only a thin layer of cytoplasm remains stretched around a large, central fat droplet resembling a ringlike signet structure.

Loose connective tissue contains all three types of fibers. Collagenous fibers are most abundant; elastic fibers are present throughout but in smaller numbers; and reticular fibers surround each cell.

Adipose tissue is a special type of loose connective tissue in which fat cells have accumulated to the point where they outnumber other cells (Figure 4-7b). Under microscopic examination adipose cells appear hollow because the fat droplets stored in living tissue are dissolved by the chemicals routinely used to preserve and stain the tissue for laboratory examination. Adipose tissue totals about 10% of body weight in a healthy person and is most abundant in the subcutaneous layer of skin, bone marrow, membranes of the abdominal and thoracic cavities, and around the kidneys. Fat serves as a layer of insulation under the skin. It also serves as a shock absorber around body organs and as an energy reserve.

Reticular tissue, another type of loose connective tissue, looks like mesenchyme (Figure 4-7c). It consists of **reticular cells** (similar to mesenchyme cells) and numerous interconnected reticular fibers arranged in an open latticework. The spaces between the lattice are filled by the cells of the particular organ where it is found. Reticular tissue functions in the body as a structural support for the tissues of such organs as the liver and the bone marrow.

Connective Tissue 89

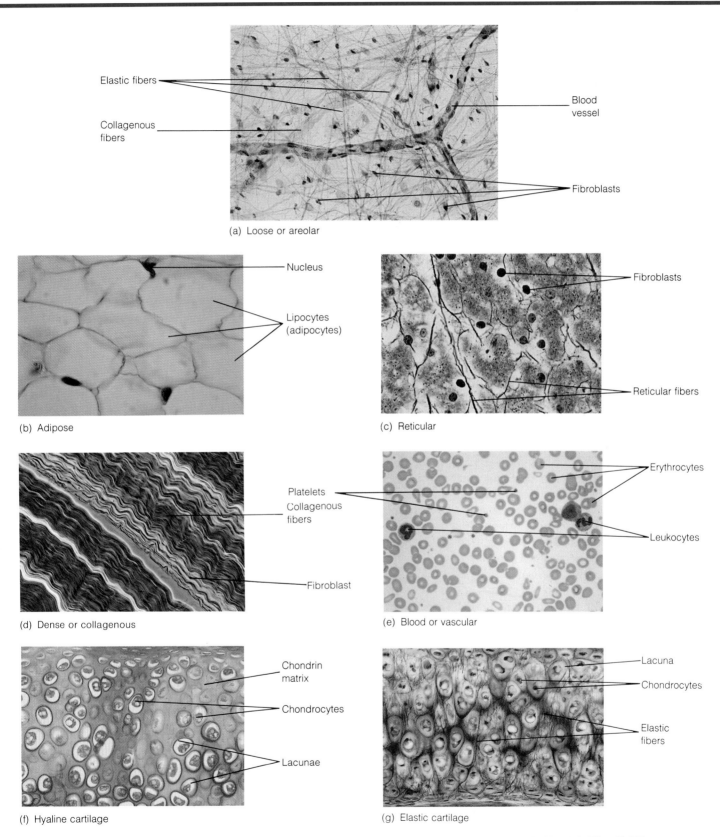

Figure 4-7 Types of adult connective tissues. Magnifications are as follows: (a) 176×; (b) 2095×; (c) 335×; (d) 209×; (e) 420×; (f) 335×.

aponeurosis: (Gr.) *apo*, off, away from + *neuron*, sinew
chondrocyte: (Gr.) *chondros*, grain + *kytos*, cell
hyaline: (Gr.) *hyalos*, glass
perichondrium: (Gr.) *peri*, around + *chondros*, cartilage

CLINICAL NOTE

ABSCESS

An **abscess** is a collection of pus that occurs at the site of an infection. It contains semifluid debris composed of decayed white blood cells and dead or decayed tissue cells. The abscess usually becomes walled off by connective tissues, and thus a barrier is formed so that further spread of the infection is prevented. Repair can only occur after the debris has been cleared and may occur when the abscess discharges its contents or is drained by a surgeon. If the abscess is not drained, healing is delayed until all the contents have been digested and reabsorbed. Abscesses may occur in the skin and other tissues.

Dense Connective Tissue

Dense connective tissue can be either regularly or irregularly arranged. Two types of **regularly arranged dense connective tissue** are recognized: one in which collagenous fibers are most numerous, another in which elastic fibers are most numerous. The former type is found in **tendons** (Figure 4-7d) (tissue that attaches muscles to bones), most **ligaments** (tissue that holds bones together at a joint), and **aponeuroses** (ăp″ō·nū·rō′sēs) (tissue that connects muscles to other muscles or bones). Ligaments also contain some elastic fibers. Strength is at a premium in these situations, and the thick bundles of collagenous fibers serve this function well (see Clinical Note: Ligament Repair). These tissues are often referred to collectively as **white fibrous connective tissue.** The second type is called **regularly arranged elastic tissue** and is found in vocal cords, certain ligaments of the vertebrae, and the penis. The abundance of elastic fibers in this tissue gives it a yellowish color and the ability to stretch under tension and return to its original shape when the tension is released.

Irregularly arranged dense connective tissue can also be differentiated into two types. Collagenous fibers predominate in the tough, fibrous **capsules** around such organs as the liver, spleen, testes, kidneys, heart, and lymph nodes. Elastic fibers predominate in the connective tissue portions of the walls of the arteries and the walls of the bronchial tubes and lungs.

Blood

Blood, or **vascular tissue,** is recognized as one of the four major types of connective tissues (Figure 4-7e). It is composed of a number of cell types that are dispersed in a fluid ground substance called **plasma.** Blood functions primarily as the transport system of the body. Because of the numerous, highly specialized cells in this tissue, we will postpone a detailed description of the morphology and functions of blood until Chapter 21.

Cartilage

Cartilage and bone are two types of connective tissue that are intimately associated embryologically, structurally, and functionally. Both function as support structures in the skeletal system. The extent of their interrelationship will emerge as we describe their formation and structure. Cartilage tissue consists of cartilage cells, called **chondrocytes,** (kŏn′drō·sīts) distributed in an amorphous ground substance **(chondrin)** that may contain fibers. Three types of adult cartilage are hyaline, elastic, and fibrocartilage. **Hyaline** (hī′a·lĭn) **cartilage** is translucent and has a bluish-white color reminiscent of pearls. It is found at the ends of long bones and it forms the cartilage of the ribs and nose. It also forms practically all of the skeleton in a developing fetus. Microscope preparations of hyaline cartilage (Figure 4-7f) show large cartilage cells embedded in an amorphous ground substance. Fine collagenous fibers are also present.

Elastic cartilage contains elastic fibers that have been deposited in the ground substance (Figure 4-7g). It is found in some of the cartilages of the upper respiratory tract, and it forms the external ear. A dense layer of connective tissue called the **perichondrium** surrounds most cartilage, except where it articulates with another cartilage or with a bone (see Chapter 6).

Fibrocartilage contains dense deposits of collagenous fibers. Only small amounts of amorphous ground substance surround the chondrocytes. Fibrocartilage provides strong support and is tough enough to withstand much physical abuse. It is found in the intervertebral disks of the vertebral column, in the pubic symphysis, in the cartilage disks (menisci) of the knee, and as a component of some ligaments and tendons. Fibrocartilage is never found in isolated circumstances; it is always adjacent to and gradually merges and changes into either hyaline cartilage or dense connective tissue. Fibrocartilage lacks a perichondrium.

Cartilage tissue lacks an internal network of blood vessels and lymphatics. Consequently, the chondrocytes do not have a direct source of blood plasma for oxygen, nutrition, or waste removal. These needs are satisfied by diffusion of materials from surface blood vessels through the ground substance. This accounts for the prolonged healing period required to repair damaged cartilage, ligaments, and tendons.

Bone

Bone, or **osseous tissue,** is one of the most specialized of all the connective tissues. It contains living cells called

osteocyte: (Gr.) *osteon*, bone + *kytos*, cell

CLINICAL NOTE

LIGAMENT REPAIR

The recent wave of interest in sports and physical fitness has also resulted in an epidemic of sports-related injuries. Running in particular is a major cause of ligament damage. One of the most disabling injuries is a complete tear of a major **ligament**. Ligamentous tissue does not tend to repair itself well because of the nature of its structure and because it usually needs to be repaired under specific tension to restore the stability of the joint. Sports medicine physicians have found that primary repair of ligaments must be accomplished within 7 to 10 days of the injury or the ligament will form disorganized scar tissue and a poor result. If the injury is seen late, a **secondary reconstruction** may be necessary which makes use of transplanted ligaments from other structures or even cadaver (deceased bodies) ligaments. There have been many attempts to develop artificial ligaments out of coated carbon fibers and manmade materials with only moderate success. Reconstruction operations are major procedures and require a prolonged period of rehabilitation.

To avoid ligament damage, practice thorough stretching exercises prior to any sports activity and possibly shift more toward less stressful aerobic exercise, such as bicycling or swimming.

osteocytes surrounded by fibers and a ground substance like that of other connective tissues. It differs, however, because the ground substance contains sufficient deposits of inorganic salts to make bone a heavy, rigid substance. Because of its many specializations, descriptions of this tissue will be deferred until Chapter 6, as part of an introduction to the skeletal system.

Muscle Tissue

After completing this and the following section, you should be able to:

1. Describe the three types of muscle tissue, and explain how they differ from each other.
2. List the major types of cells found in nervous tissue.

Muscle is one of the most specialized tissues. Its major specialized function is contraction, and this activity will be described in detail in Chapter 9 on Muscles: Cellular Organization and Physiology. These contractions produce movements of body parts and the movement of fluids and solids within the body. There are three basic muscle types: smooth, skeletal, and cardiac.

Smooth Muscle

Smooth muscle has the least complex histological structure of the three types. It is composed of elongate, tapering, isolated, or small groups of cells (Figure 4-8a). Each cell has one centrally located nucleus. The cells vary greatly in length. Smooth muscle is found in the deep dermal layers of the skin; in the walls of the digestive, respiratory, reproductive, and urinary tracts; in the walls of arteries, veins, and larger lymphatic vessels; and in the iris of the eye. Smooth muscle tissue is also referred to as *involuntary muscle* because most smooth contractions occur without voluntary control. It is also called *visceral muscle* because it is found in the walls of internal (visceral) organs.

Skeletal Muscle

Skeletal muscle makes up the "meat" or "flesh" of the body. In contrast to smooth muscle, it is composed of elongate, multinucleate, cylindrical structures called fibers (not to be confused with the fibers described in the ground substance of connective tissue or nerve fibers) that bear alternating dark and light cross-markings or **striations** (Figure 4-8b). The multinuclear nature of the elongate fibers results from the fusion of several developing cells during the embryonic differentiation of the tissue. The striations are caused by the presence of intracellular fibrils, which will be discussed in detail in Chapter 9. Skeletal muscle is also referred to as *voluntary muscle* because its contractions are under voluntary control.

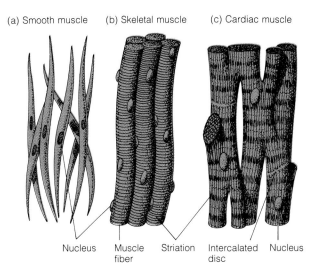

Figure 4-8 The three types of muscle tissue. Each is capable of contracting and causing the movement of body parts and materials within the body. (a) Smooth muscle. (b) Skeletal muscle. (c) Cardiac muscle.

Nervous Tissue

Nervous tissue is another specialized connective tissue. Because the structure and function of nervous tissue are complex, a more detailed discussion of it can be found in Chapter 11 where we will describe the many variations of nerve cells and how each type contributes to the functioning of the entire nervous system.

Nervous tissue is composed of two types of cells: **neurons** and **neuroglia.** Neurons (Figure 4-9) are capable of two extremely important functions: they respond to stimuli in their environment and they conduct impulses from one cell to another, sometimes over great distances. Neurons possess elongate cytoplasmic extensions from the cell body called **dendrites** and **axons,** which receive and transmit impulses, respectively. Neuroglia cells also possess cytoplasmic extensions, but rather than conducting impulses, they act as the connective tissue of the nervous system.

Tissues provide the structural framework within which individual cells perform their functions. The separation of tissues into units that perform different functions, such as the secretion of products, movement, and the transfer of information, allows a more efficient division of labor within the body. The development of connective tissues that support and tie the parenchyma cells together maintains the integrity of the body as a whole.

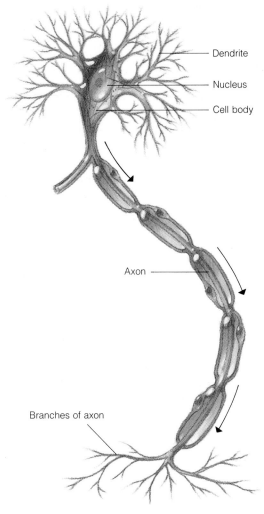

Figure 4-9 Typical neuron. Neurons are nerve cells specialized to respond to stimuli and transmit information. Arrows indicate the direction of nerve impulse flow.

Cardiac Muscle

Cardiac muscle is composed of branched cells called cardiac monocytes that are smaller than skeletal muscle monocytes. The striae in the former are much "weaker" than the latter (Figure 4-8c). Cardiac muscle cells are separated from one another at their ends by irregular transverse thickenings of their plasma membranes called **intercalated disks.** These disks strengthen the cardiac muscle tissue and provide for communication between adjacent cells. The details of their physiological importance in the functions of the heart will be discussed in Chapter 22 on the heart. Cardiac muscle contains myofibrils identical to those found in skeletal muscle, except the striations produced are not as distinct. Cardiac muscle is found only in the heart and at the very base of the large blood vessels attached to the heart.

STUDY OUTLINE

Epithelial Tissues (pp. 81–86)

A *tissue* is a group of cells that act together to perform one or more specific tasks. *Epithelial tissues* are composed of layers of cells packed together. They cover the body, individual organs, and line internal spaces. They physically protect organs and secrete mucus.

  Simple Epithelial Tissue Simple epithelium is composed of a single layer of cells. *Simple squamous epithelium* is composed of flattened cells, *simple cuboidal epithelium* of cuboidal cells, and *simple columnar epithelium* of columnar cells. Columnar cells may be modified into *goblet cells* that produce and secrete mucus.

  Pseudostratified Epithelial Tissue Pseudostratified epithelium consists of a single layer of cells of varying heights, giving the superficial appearance of multiple layers.

  Stratified Epithelial Tissue Stratified epithelium consists of multiple layers of cells. Cell shapes may be flattened (squamous), cuboidal, or columnar.

  Stratified squamous epithelium consists of several layers of flattened cells. Cells in the deepest or *basal layer* may be

scleroderma: (Gr.) *skleros*, hard + *derma*, skin
dysphagia: (Gr.) *dys*, painful, difficult + *phagein*, to eat
dermatomyositis: (Gr.) *dermatos*, skin + *mys*, muscle + *itis*, inflammation
polyarteritis: (Gr.) *poly*, many + *arteria*, artery + *itis*, inflammation
ischemia: (Gr.) *ischein*, to hold back + *haima*, blood

THE CLINIC

COLLAGEN VASCULAR DISEASES

The **collagen vascular diseases** are a group of bizarre diseases usually of unknown cause that are grouped together because of their effects on connective tissues. Since connective tissues are widely distributed throughout the body, the collagen vascular diseases affect many organ systems. Although the true cause of these diseases is largely unknown, most of them are thought to be **autoimmune** disorders. Autoimmune disorders are those in which the body produces **antibodies** against itself. Often these abnormal antibodies can be detected and identified in the patient's blood, which may help in identifying the disease. Several specific diseases have been identified.

Systemic Lupus Erythematosus (SLE)

This is an inflammatory disease of the connective tissues seen mostly in young women. It may effect any system in the body. General symptoms such as prolonged fever, weakness, and joint pains occur. Sometimes a peculiar **butterfly rash** is present over the bridge of the nose and extending onto the cheeks in a butterfly pattern. The disease may be precipitated or made worse by exposure to excessive sunlight or certain drugs such as sulfa drugs or barbiturates. SLE has such diffuse symptoms that it may not be readily diagnosed. Such a delay in diagnosis and treatment may cause a great deal of anxiety and stress in these unfortunate patients. Treatment with cortisone drugs provides marked relief of the symptoms.

Progressive Systemic Sclerosis (Scleroderma)

This is a chronic disease characterized by diffuse **fibrosis** (scar tissue formation) and vascular abnormalities in the skin, joints, and gastrointestinal tract. The disease is more common in women. The symptoms include thickening of the skin on the extremities, sometimes with calcium deposits in the skin; **dysphagia**, which is difficulty in swallowing due to fibrosis of the esophagus; and **Raynaud's phenomenon**. Raynaud's phenomenon is a cadaverous, painful blanching of the fingers and toes when exposed to cold. There is no specific treatment for this disease and it is usually slowly progressive.

Dermatomyositis (Polymyositis)

This is another connective tissue disease that causes inflammation and degeneration (wasting) of the skin and muscles. The skin changes may include a peculiar violet discoloration around the patient's eyes and rashes on the fingers. There is also progressive weakness of the muscles of the trunk, shoulder girdle, and hips. Twenty percent of patients with this disease have been found to develop cancer during the course of the illness. Treatment with cortisone and aggressive treatment of associated cancer help provide relief.

Polyarteritis Nodosa (Periarteritis nodosa)

This is another member of this strange group of connective tissue diseases that is characterized by inflammation and necrosis (death) of small- to medium-sized arteries throughout the body resulting in localized **ischemia** (ĭs·kē′mē·ă), or loss of blood supply, to the tissues supplied by these arteries. This disease is three times more common in men than women. Its symptoms vary widely and may mimic many other diseases. They include persistent fever, loss of weight, abdominal pain, and progressive renal failure. Unlike the other collagen diseases, abnormal antibodies are rarely found in polyarteritis. The diagnosis is confirmed by biopsy to demonstrate the inflamed small arteries. This disease is frequently fatal in spite of treatment with cortisone-like drugs.

---

cuboidal to columnar. As the outer layer of skin, this tissue contains a protective, waterproof protein called *keratin*.

Stratified cuboidal epithelium is found in the ducts of sweat glands and in the lining of the eyelids.

Stratified columnar epithelium is a protective tissue that also secretes mucus.

Transitional epithelium (*urothelium*) lines the urinary bladder and its tubes and is specialized to withstand the stress of stretching associated with the regular filling and emptying of the bladder.

Glandular Epithelium A *gland* is any cell or group of cells that secretes specialized materials. The secretions of *exocrine glands* are carried through ducts to specific areas, whereas *endocrine glands* empty their secretions directly into the blood vessels of the gland.

In a functional classification of exocrine glands, three types are differentiated. *Holocrine glands* secrete whole cells into ducts. *Apocrine glands* accumulate cell secretions in the end of the cell adjacent to the surface, which is then pinched off and released into ducts. *Merocrine glands* deliver only the secretion to the duct.

Exocrine glands can also be differentiated according to a structural classification. Exocrine glands can be unicellular or multicellular. Multicellular glands can be unbranched (simple) or branched (compound), and the secretory cells can be arranged in a tubular or saclike (alveolar) configuration. Combinations of these arrangements produce the following configurations: simple tubular, simple alveolar, compound tubular, compound alveolar, and compound tubuloalveolar.

Connective Tissues (pp. 86–91)

Connective tissues connect, hold together, protect, and support body parts. Their cells are separated from one another by secretions called the *ground substance* or *matrix*. These tissues may contain *collagenous, elastic, or reticular fibers* that add strength and elasticity to the tissue.

Embryonic Versus Adult Connective Tissue Embryonic connective tissues include *mesenchyme* and *mucous tissue*. Both contain irregularly shaped *mesenchymal cells* and *fibroblasts* and are the precursors of adult connective tissues. Four types of adult connective tissues are identifiable: connective tissue proper, blood, cartilage, and bone.

Adult Connective Tissue Proper There are two principal types of connective tissue proper: loose connective tissue and dense connective tissue.

Loose connective tissue (*areolar tissue*) is the most widespread of all connective tissues. Found throughout the body, it serves as a loose packing for practically all organs and tissues. It contains a variety of cells including fibroblasts, *mast cells*, *macrophages*, *plasma cells*, and *fat cells* (*lipocytes*). *Adipose tissue* and *reticular tissue* are forms of loose connective tissue.

Two types of *dense connective tissue* are recognized based on the internal arrangement of the fibers present. *Regularly arranged* tissues include *ligaments, tendons,* and *aponeuroses*. *Irregularly arranged* tissues are found in *capsules* surrounding organs, in the lungs, and in the walls of arteries. Both types of dense connective tissue are noted for their strength and elasticity.

Blood Blood (*vascular tissue*) is composed of several cell types dispersed in a fluid *plasma*. It functions as the transport system of the body.

Cartilage Cartilage is one of the support systems of the body. Three types of adult cartilage are *hyaline, elastic,* and *fibrocartilage*.

Bone Bone (*osseous tissue*) is another support system of the body. It contains large amounts of rigid inorganic salts.

Muscle Tissue (pp. 91–92)

Muscle is among the most specialized of tissues. Its most specialized function is contraction.

Smooth Muscle The least complex muscle is *smooth muscle*, which consists of individual cells. This type is also referred to as *involuntary muscle* because most smooth contractions occur without voluntary control.

Skeletal Muscle *Skeletal muscle* is composed of elongate, multinucleate, cylindrical fibers that bear striations. Contractions of these muscles are under voluntary control.

Cardiac Muscle A third type of muscle is composed of branched cells and is found in the heart, thus it is called *cardiac muscle*. The contractions of cardiac muscle are not under voluntary control.

Nervous Tissue (p. 92)

Nervous tissue is composed of specialized cells that respond to stimuli in their environment, and they conduct impulses from one cell to another.

SELF-TEST OF CHAPTER OBJECTIVES

True-False Questions
1. Connective tissues are composed of layers of cells that are packed closely together.
2. The exposed surface of a connective tissue is covered by a basement membrane.
3. Neuroglia is an integral part of nervous tissue.
4. Merocrine glands release whole cells as their secretion.
5. The human body is composed of two major types of tissues—epithelial and connective.
6. The functionally active cell in bone tissue is called a chondrocyte.
7. Stratified epithelial tissue differs from simple epithelial in that the former consists of several layers of cells, whereas the latter consists of only one layer of cells.
8. The walls of the heart are composed of smooth muscle fibers.
9. Cartilage tissue possesses an abundant blood supply.
10. Both ligaments and tendons are classified as varieties of dense connective tissue.

Matching Questions
Match the specific epithelial tissue on the left with the correct descriptive statement:

11. stratified squamous epithelium
12. simple squamous epithelium
13. pseudostratified epithelium
14. glandular epithelium
15. transitional epithelium

a. the thinnest epithelial tissue
b. lines the urinary system
c. forms the outer skin
d. forms endocrine glands
e. lines the nasal cavities

Match the item on the left with the appropriate descriptive statement:

16. fibrocartilage
17. neuroglia
18. matrix
19. mesenchyme
20. adipose

a. an embryonic tissue
b. forms the intervertebral disk
c. intercellular material of connective tissue
d. connective tissue of the nervous system
e. a loose adult connective tissue

Multiple-Choice Questions
21. Most of the inner lining of the stomach and intestine consists of
    a. simple cuboidal epithelium
    b. stratified squamous epithelium
    c. pseudostratified squamous epithelium
    d. simple columnar epithelium
22. Glands such as the mammary glands that release part of their cytoplasm along with their chemical secretion are known as
    a. merocrine glands   c. holocrine glands
    b. apocrine glands   d. acinar glands
23. Unbranched, colorless fibers that are found in wavy, parallel bundles are called
    a. collagenous fibers   c. elastic fibers
    b. connective fibers   d. fibroblasts

24. Which of the following is *not* a loose connective tissue?
    a. areolar tissue   c. cartilage
    b. adipose tissue   d. reticular tissue
25. Striated muscle tissue is also referred to as
    a. voluntary muscle   c. involuntary muscle
    b. smooth muscle   d. visceral muscle
26. Which connective tissue composes the tendons that connect muscles to bone?
    a. dense regular   c. dense irregular
    b. loose regular   d. loose irregular
27. Which type of cartilage forms the intervertebral disks of the spine?
    a. fibrocartilage   c. elastic cartilage
    b. hyaline cartilage   d. osteocartilage
28. Cardiac muscle cells are connected at their ends by
    a. fibrocartilage   c. intercalated disks
    b. elastic cartilage d. dense irregular connective tissue
29. Mesenchyme tissue is classified as
    a. loose epithelial tissue   c. mucous tissue
    b. embryonic tissue   d. ground substance
30. Which of these tissues forms a prominent component of ligaments?
    a. adipose tissue
    b. irregularly arranged dense connective tissue
    c. urothelium
    d. regularly arranged dense connective tissue

Essay Questions
31. Describe the major differences between dense and loose connective tissues.
32. Which structural and functional characteristics distinguish epithelia from connective tissues?
33. How does the microscopic appearance of adipose tissue differ from that of loose connective tissue?
34. How does muscle found in the visceral organs of the body differ in general structure and function from muscle tissue that moves the skeleton?
35. How does embryonic connective tissue differ from adult connective tissue?
36. Describe the four types of simple epithelia in terms of shapes of the cells and general functions.
37. What characteristics of blood would prompt you to classify blood as a connective tissue?
38. How do endocrine and exocrine glands differ in function?
39. Describe the structural complexity found in exocrine glands.
40. Blood vessels are not found in the layers of epithelial cells that line body cavities, such as the intestine and abdomen. Lacking direct contact with blood, how do these cells obtain oxygen and nourishment and get rid of wastes?

Clinical Application Questions
41. Discuss the difference between keloids and adhesions.
42. Define the term abscess.
43. Compare and contrast the collagen vascular diseases systemic lupus erythematosus and dermatomyositis.

UNIT I CASE STUDY

A PATIENT WITH HODGKIN'S DISEASE

The medical evaluation of patients with serious diseases often requires the use of many different techniques and tools to determine not only the nature of the ailment but also its extent. It is important for the physician to be aware of the tests available, how they are performed, what they show, and the cost to the patient. A good way to show some of the concepts introduced in the chapters of this unit is to examine the work-up of a young woman with **Hodgkin's disease.**

Hodgkin's disease is a fairly common malignant tumor of the lymph system. It occurs most frequently in children and young women. If the disease is discovered in its early stages when it is localized to a single lymph node area, it can be treated successfully with local X-ray treatment. The initial symptoms of Hodgkin's disease may be few and so the physician must be alert to make an early diagnosis. Once the disease has been diagnosed, usually by **biopsy** (surgically removing a piece of a suspect tissue and examining it under the microscope for characteristic cells), the disease must be **staged.** Staging is the process of determining the full extent of the disease. Once the stage of the disease has been determined, a treatment program is selected based on the recorded studies of many patients with similar stages of the disease and the results of many therapeutic trials.

Here is a simple staging classification for Hodgkin's disease:

STAGE I	Limited to a single nodal area
STAGE II	Involving two or more nodal areas on the same side of the diaphragm
STAGE III	Involving two or more nodal areas on both sides of the diaphragm
STAGE IV	Involving one or more extra lymphatic organs or tissues with or without lymph node involvement (see figure).

The patient in this case study is a 20-year-old woman who complains of feeling unduly tired, along with exhibiting a low grade fever and a lump in the left side of her neck. These symptoms have bothered her for several weeks. She has felt fairly well otherwise and has continued with her usual activities and work.

A general physical examination is normal except for a temperature of 99.8°F, a mild pallor (paleness) of her skin, and a 2 × 3-cm firm, rubbery lump in the left side of her neck. The lump is painless and appears to be attached to the deeper tissues beneath. There is no evidence of any local infection in the nearby organs, such as the tonsils or teeth. A complete blood count shows only a mild anemia (low red blood cell count). A routine chest X ray is also normal, as are selected blood chemistries for liver or kidney disease.

Suspecting that the woman's problem is Hodgkin's disease, her doctor refers her to a surgeon for an excisional biopsy of the node in the neck. The removed tissue is sent to the laboratory where it is processed and examined by a **pathologist** (a physician trained to recognize diseases by examining the various cells that are present in a tissue sample). The pathologist reports that the tissue is a lymph node containing Hodgkin's disease of the nodular sclerotic variety. There are four varieties of Hodgkin's tissues—**nodular sclerosis** is one of the least aggressive types. We now know what disease the patient has, but we still need to determine how extensive it is.

We begin the staging process with a CAT scan of the chest and abdomen since the CAT scan is noninvasive (nonsurgical)

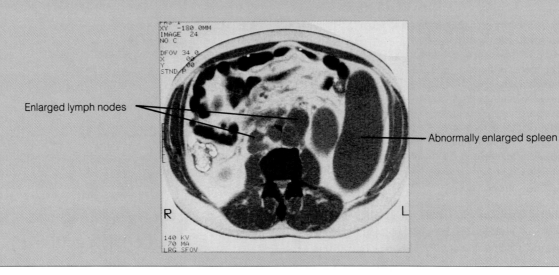

This CAT scan image through the thoracic region shows evidence of Stage IV of Hodgkin's disease. Note the enlargement of the lymph nodes and spleen.

and will give us a good, detailed view of the organs of the chest and abdomen. The CAT scan of our patient shows no suggestion of tumor in the abdomen or chest. Although this negative examination is encouraging, it cannot demonstrate minute spread of the tumor to distant lymph node areas. A more detailed examination of the lymph system below the diaphragm can be obtained with a **bipedal lymphangiogram** (lĭm·făn′jē·ō·grăm). This is a mildly invasive procedure in which a radiopaque dye is injected directly into the fine lymph channels in each foot (thus bipedal). The dye will travel up the lymph system in each leg and into the abdomen where it will outline all of the lymph nodes and channels. The dye is retained in them for a long enough period of time so that an X-ray examination can provide a detailed outline of the entire lymph system. The examination on our patient again demonstrates no evidence of the disease. If the lymphangiogram had been found to be abnormal, we would have proceeded with a **diagnostic laporotomy.** This is an extensive surgical exploration of the entire abdominal contents in which biopsies are performed on all suspicious areas as well as the liver and bone marrow. The spleen is also removed during such surgery because it is the most common site of spread and is difficult to irradiate without damaging the nearby kidneys. All of the tissues are examined by the pathologist in search of **metastatic disease** (secondary growth of the malignancy in a new location that arises from the primary growth). There are other techniques used to search for metastatic disease, such as **radionucleotide scanning**, which uses radioactive gallium. Perhaps in the near future some of the newer **photon imaging** techniques will provide us with all the information we need with a single scan.

Fortunately for our patient, the disease was discovered early and classified as a Stage I disease. Stage I Hodgkin's disease can be treated with local X-ray irradiation, which has a 90% assurance of complete cure.

This relatively simple case study shows how multiple techniques may be required to evaluate a patient's illness.

Unit II

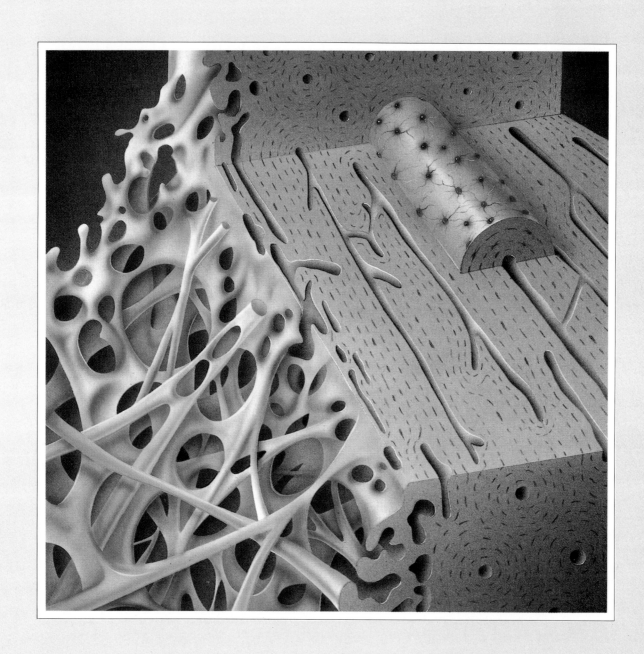

Support and Movement Systems

Unit II describes those systems of the body that provide support and movement. We begin with an examination of the integumentary system, which consists largely of the skin. The skin is one of the largest organs in the human body, and its condition often reflects the internal state of affairs. Next, we look at the skeleton, beginning at the microscopic level and proceeding to the anatomy of the bones and joints. The unit is completed with an examination of muscles, again proceeding from a microscopic perspective to a gross anatomical view of the major muscles of the body.

5
The Integumentary System

6
Development and Physiology of the Skeletal System

7
Anatomy of the Skeletal System

8
Articulations: The Joints Between Bones

9
Muscles: Cellular Organization and Physiology

10
Musculature of the Human Body

5
The Integumentary System

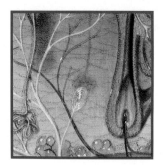

The integumentary system is composed of the **skin (integument)** and several specialized structures derived from the skin, including hair, nails, and several kinds of glands. The skin is composed of two layers. The outer layer is called the *epidermis* and the deeper layer is the *dermis*.

The skin covers practically the entire body and is one of the largest organs we possess. It is continuous with mucous membranes lining the mouth, anus, nasal cavity, ears, and eyes. In an average adult, it weighs about 4.5 kg and has a surface area of nearly 2 m^2, approximately half that of a ping-pong table. It varies in thickness from about 0.5 mm in the abdominal region to 3 mm in areas of maximum thickness, such as the palms of the hands and the soles of the feet.

In this chapter we will examine

1. the components of the integumentary system,
2. the differences between the epidermis and dermis of the skin,
3. the causes and control of skin pigmentation,
4. the growth processes that produce hair and nails,
5. the types of glands found in the skin, and
6. the numerous functions of the skin.

integument:	(L.) *integumentum*, covering		granulosum:	(L.) *granulum*, granule
epidermis:	(Gr.) *epi*, upon + *derma*, skin		germinativum:	(L.) *germinare*, to sprout
basale:	(L.) *basis*, base		keratohyalin:	(Gr.) *keras*, horn + *hyalos*, glass
spinosum:	(L.) *spinosus*, spiny		eleiden:	(Gr.) *elaion*, oil

Epidermis

After completing this section, you should be able to:

1. List the five epithelial layers found in the epidermis.
2. Discuss why the deepest two strata of the epidermis are referred to collectively as the stratum germinativum.
3. Describe the function of keratohyalin and hyalin in the skin.
4. Describe the role of melanin and carotene in determining skin color.

Epidermal Layers

The **epidermis** consists of stratified squamous epithelium (Figure 5-1). Recall that when you studied this type of tissue in the preceding chapter (see Figure 4-2), you learned that the deep cells in stratified squamous epithelium are constantly dividing, producing new cells that undergo change as they are pushed to the surface. These changes are so pronounced in the skin that microscope preparations reveal several prominent layers. The maximum number of layers is visible in the thick skin of the palms and soles. A study of one of these regions provides a good starting point for our examination of the integumentary system. Here, we find five prominent epidermal layers or strata: *basale, spinosum, granulosum, lucidum,* and *corneum.*

The epidermis lacks a system of blood vessels, therefore, all nutrients for epidermal cells must pass from vessels in the deeper dermis, either by passive or active transport, to epidermal cells that are several layers removed from the vessels. Waste materials move in the opposite direction.

The **stratum basale** (bā·sā′lē) is the deepest of the five layers (Figure 5-1). It consists of a single layer of columnar cells firmly attached to the underlying dermis. Stratum basale cells provide the source of cells in the other four layers. Frequent divisions of cells in the stratum basale (which occur mainly during sleep) produce daughter cells that subsequently are pushed into the next outer layer.

The **stratum spinosum** (spī·nō′sŭm) is the layer external to the stratum basale (Figure 5-2). It is so named because, when viewed under the microscope, the cells of this layer look like spiny sea urchins attached to one another. Cells here also undergo frequent divisions, and consequently, this layer and the stratum basale are together often called the **stratum germinativum** (jĕrm″ĭ·nă·tĭ′vŭm), a Latin term that reflects their role in producing the epidermis. Cells near the base of this layer are polygonal but become flattened as they are pushed toward the next outer layer.

The **stratum granulosum** (grăn″ū·lō′sŭm) is so named because the cells in this layer contain granules of a substance called **keratohyalin** (kĕr″ă·tō·hī′ă·lĭn). These cells convert keratohyalin into another protein, called **eleiden** (ĕ·lē′ĭ·dĭn), which is further processed into **keratin**. Recall from Chapter

Figure 5-1 The skin is divided into two major regions: a superficial epidermis, and a deeper dermis. The combined structure is underlain by thick, subcutaneous connective tissue.

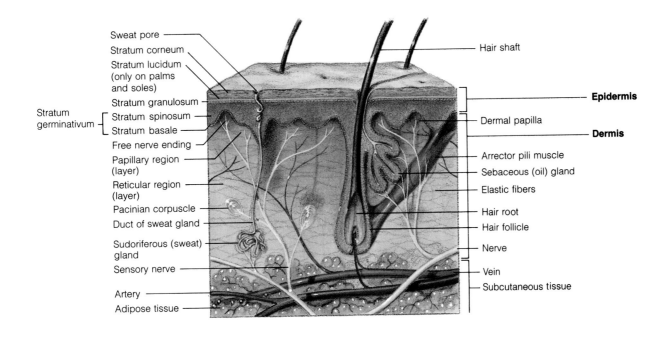

psoriasis: (Gr.) *psoriasis*, an itching

CLINICAL NOTE

COMMON SKIN PROBLEMS

Here are some common skin problems you may have heard about:

Psoriasis (sō·rī´ă·sĭs) is a dry, silvery, scaled rash that frequently occurs in patches on symmetrical extensor areas of the body such as the knees, elbows, base of the spine, and scalp (Figure 1a). Psoriasis also causes minute pits in the nails. It may also appear in areas of trauma to the skin, where it is called the **Koebner phenomenon.** The cause of psoriasis is not known. It can usually be controlled by application of various tar and cortisone creams and by exposure to ultraviolet light.

Cold sores are grouped, painful vesicles that may appear on the lips or genitalia and are caused by an infection with the **herpes simplex** virus (Figure 1b). This group of viruses has the nasty habit of invading a local nerve ganglion where it may lie dormant only to reactivate at some later time and produce a new crop of lesions. **Herpes genitalis** is a particularly severe form of the disease that produces painful lesions on the genitalia. It is a sexually transmitted disease that has reached epidemic proportions.

Hives, also called **urticaria** (ŭr·tĭ·kā´rē·ă) or wheals, are itchy, raised, flat papules usually caused by an allergic reaction that releases large amounts of **histamine** into the skin (Figure 1c). They can result from a reaction to a drug, a systemic infection, or even stress. Hives can be controlled with antihistamine drugs, cortisone, or adrenaline.

Warts, or **verrucae** (vĕ·ru´kē) are single or grouped, rough, raised tumors of the skin caused by the **papovaviruses.** They can be removed by using one of a number of destructive chemicals, such as liquid nitrogen, nitric acid, or salicylic acid.

Shingles (herpes zoster) is a viral infection caused by the chicken pox virus that produces a band of lesions similar to cold sores in the area of a superficial nerve. Unlike herpes simplex, zoster tends to be self-limited and does not usually recur. It can, however, result in a painful and persistent neuralgia (sharp pain along the nerve), particularly in elderly patients.

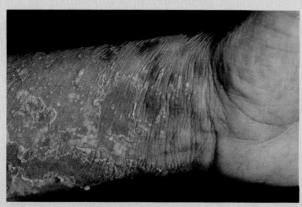

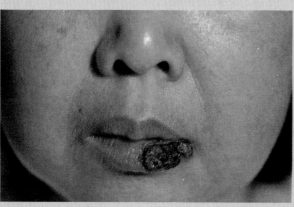

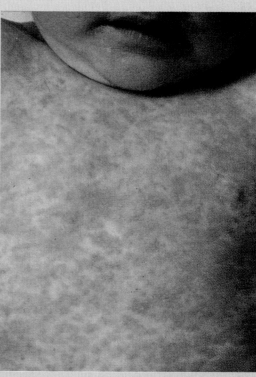

Figure 1 Skin patches showing (a) psoriasis, (b) cold sores (herpes simplex), and (c) hives (urticaria).

lucidum: (L.) *lucidus*, lucid
corneum: (L.) *corneus*, horny
melanin: (Gr.) *melas*, black
melanocytes: (Gr.) *melas*, black + *kytos*, cell

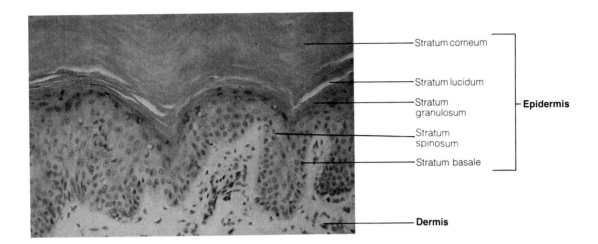

Figure 5-2 Structure of the skin. This photomicrograph shows the skin in the palm of the hand (mag. 1120×).

4 that keratin serves as a waterproofing material for the skin. This layer is most highly developed in the palms and soles, where it is four or five cell layers thick (Figure 5-1). It is only one or two cell layers thick or absent elsewhere. Cells moving into this keratinized layer become more flattened as they are pushed outward and farther away from their food supply. Lack of nutrients and the accumulation of wastes causes degeneration of their nuclei and cytoplasm, and eventually the cells die.

The **stratum lucidum** (lū′sĭ·dŭm) is a thin layer of dying and dead cells whose interiors are becoming more filled with eleiden. Eleiden is translucent in appearance, and the name *lucidum* refers to the clear appearance of this layer in microscopic sections (Figure 5-2). The stratum lucidum is present only on the palms and soles.

The **stratum corneum** (kōr′nē·ŭm) forms the outer exposed layer of epidermis (Figure 5-2). This layer is about 25 to 30 cell rows thick. As cells are pushed into this layer, they become extremely flattened squamous epithelial cells. The stratum corneum consists solely of dead cells whose cytoplasm has been replaced by keratin. The stratum corneum is thickest on the palms and soles (where it can be more than 100 cell rows thick) and in areas subject to continued mechanical stress from rubbing or friction (for example, **calluses** on the skin). Cells of the stratum corneum are constantly being sloughed off and replaced by cells originating from the stratum basale.

Pigmentation

Skin color in humans ranges from pinkish ivory to dark brown and is due to the relative abundance of two pigments, a yellowish one called **carotene** and a dark brown one called **melanin**. The effects of both pigments are also modified by the reddish coloration of blood in the capillaries of the underlying dermis.

Carotene enters the body chiefly as a component of yellow and red vegetables, such as carrots and tomatoes. Melanin is not eaten as food but is manufactured by specialized cells called **melanocytes** (mel′ăn·o·sīts) which are distributed throughout the stratum basale, except in the palms and soles. Numerous cytoplasmic projections extend from the melanocytes in between the cells of the stratum spinosum. Cells in the stratum germinativum actively accumulate melanin from the cytoplasmic projections through phagocytosis, causing the cells to darken. In Caucasian races, as the cells are pushed into the stratum corneum, the melanin breaks down and disappears and the surface layers are clear to yellowish (from the remaining carotene). In brown and black racial groups, the melanin does not disappear. **Freckles** are isolated areas of concentrated melanocytes and/or increased melanin production.

The amount of melanin produced in the skin is governed by heredity and environmental conditions. Even though all racial stocks possess a similar number of melanocytes, they produce different amounts of melanin. In fact, in all races there are some individuals, called **albinos,** who lack the ability to produce any melanin at all. In addition to having pink skin, albinos lack melanin in the hair and eyes. The genetic basis for these differences in melanin production will be discussed in Chapter 28.

The physiological basis for melanin production is poorly understood at present. However, its production is enhanced

rubeola: (L.) *ruber*, red
rubella: (L.) *ruber*, red
varicella: (L.) *varicella*, a small spot

CLINICAL NOTE

VIRAL SKIN INFECTIONS

Two kinds of measles as well as chicken pox are caused by viruses that infect the skin.

Rubeola

The familiar red measles, or **rubeola,** is a highly communicable disease that makes itself manifest by fever, watery eyes, a slight cough, and characteristic Koplik's spots on the inside of the cheek. After these symptoms, the red blotchy rash appears on about the third day, first on the face and then all over the body (Figure 1a). The rash lasts about three to seven days. The viral infection is passed through the respiratory secretions of persons who are infected about four days before and four days after the rash appears. Very rarely, pneumonia or encephalitis can occur as complications of this infection.

The care necessary for a child with rubeola includes decreasing the fever with antipyretics (nonaspirin) and cool baths. The child should be kept quiet and allowed to rest as much as possible. In the United States children are required to receive active immunization with a single dose of the live attenuated virus after the age of 15 months and before entering school.

Rubella

Also known as German measles, **rubella,** is a fairly mild but highly communicable viral disease with a characteristic spotty rash. The symptoms prior to the rash include cold symptoms, runny nose, inflamed eyes, fever, headache, and a general feeling of being unwell. This is the type of measles that it is important for pregnant women to avoid. An infection in a pregnant woman may lead to an infection in the fetus, resulting in a variety of malformations in the baby.

A person with rubella should be kept comfortable and be given cool baths to reduce the fever. He or she does need to be isolated to limit the communicability. As for rubeola, children should receive active immunity with a single dose of the live attenuated virus after the age of 12 months.

Chicken Pox

Varicella, commonly known as **chicken pox**, is a highly communicable disease of childhood. The period of communicability is from one to two days before the rash appears until all the lesions have crusted over. The incubation period for this viral infection is usually 13 to 17 days.

The fever accompanying this disorder begins abruptly; the child may appear mildly out of sorts, and then the characteristic skin eruptions appear. At first they are red raised spots that then become blisterlike for about four days (Figure 1b). At the end of this time they crust over and leave a scab. The lesions do not all appear at once but occur in successive "crops." While the lesions may occur anywhere on the body, including the scalp, they are more likely on the covered areas of the body, such as the chest, belly, and back.

The causative virus is the varicella-zoster virus, a herpesvirus that gains entrance to the body through the respiratory tract. After the person recovers from the acute viral infection, it is believed that the virus remains dormant in the body in the dorsal root ganglia of the spinal cord.

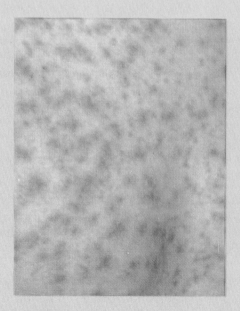 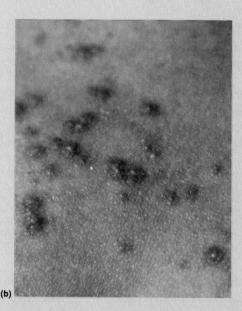

(b)

Figure 1 Skin patches showing skin rashes typical of (a) rubeola and (b) chicken pox.

tyrosinase: (Gr.) *tyros*, cheese + *ase*, suffix denoting an enzyme
dermis: (Gr.) *derma*, skin
corium: (L.) *corium*, leather
eczema: (Gr.) *ekzein*, to boil out
dermatitis: (Gr.) *derma*, skin + *itis*, inflammation

CLINICAL NOTE

ECZEMA

Eczema (ĕk′zĕ·mă) is a superficial inflammation of the skin: a **dermatitis**, which means an inflammation of the skin that may be either acute or chronic. Included under this heading is **contact dermatitis**, an acute or chronic inflammation of the skin due to external factors, and **atopic dermatitis**, a chronic, superficial, itchy inflammation of the skin in persons who have other atopic disorders, such as hay fever and asthma. Persons with atopic dermatitis just seem to have more sensitive skin that is dry and more inclined to be itchy.

Eczema appears as reddened, swollen, oozing, crusty, and scaly lesions that are sometimes blistered. Affected areas are itchy. These symptoms are caused by the skin's limited response to a variety of noxious environmental stimuli, such as excretions (urine from wet diapers), plant oils (poison ivy, sumac, or oak), and irritating chemicals (detergents, solvents, acids, or alkalis). The skin reacts with vasodilatation (which makes the skin reddened), swelling of the dermis, exudation of fluid accumulating between the cells, breakdown of the epidermal cells, erosion, crusting, and scabbing. Thickening and scaling can occur as a result of repair (see figure).

Eczema is managed by identifying (when possible) the offending noxious stimulant and removing it. The person may be given drugs to decrease the itching (usually antihistamines) and to decrease the inflammatory response (usually topical corticosteroids). The person should drink plenty of fluids, and soothing saline dressings will provide comfort and relief.

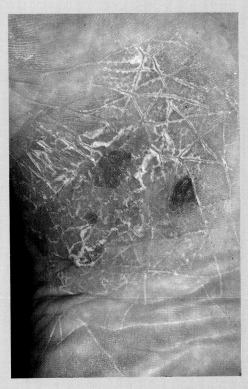

The skin on the sole of this foot is affected by eczema.

---

by exposure to X rays and ultraviolet light, two potentially harmful components of sunlight. The process involves the enzyme **tyrosinase** (tī·rō′sĭn·ās) converting the amino acid **tyrosine** (tī′rō·sēn) into melanin. Excess melanin thus produced accumulates in the stratum germinativum, causing a darkening or "tanning" of the skin. Generally speaking, the more a person is exposed to the sun, the darker the skin becomes (although a prolonged overexposure can cause a severe burn). The buildup of melanin blocks the penetration of ultraviolet light, thereby maintaining a more stable environment for cells of the dermis, another example of homeostasis in action.

In addition to stimulating melanin production, the ultraviolet rays in sunlight cause the conversion of certain chemicals in the skin to vitamin D, which is essential for the proper metabolism of calcium and phosphate in the skeleton. Children who are deficient in vitamin D develop the disease *rickets*, which is characterized by demineralization and softening of the bones. Although a certain amount of exposure to sunlight is necessary for vitamin D production, overexposure can have detrimental effects. Certain skin disorders, particularly tumors and skin cancer, are more prevalent in groups of people who experience prolonged and excessive exposure to sunlight or the ultraviolet rays used in tanning salons (see Clinical Note: Suntanning).

Dermis

After completing this section, you should be able to:

1. Describe the two strata of the dermis.
2. Distinguish between the dermis and the subcutaneous layer.

The **dermis** (sometimes called the **corium**) is the deeper layer of the skin. It is composed of dense, irregularly arranged connective tissue, which contains both collagenous

hypodermis: (Gr.) *hypo*, under + *derma*, skin
subcutaneous: (L.) *sub*, under + *cutis*, skin
reticular: (L.) *reticulum*, net
papillary: (L.) *papilla*, nipple

> **CLINICAL NOTE**
>
> **SUNTANNING**
>
> Since 1930, the incidence of skin cancer in the United States has increased tenfold, from 1 in 1500 people to 1 in 150 people. The chief cause is overexposure to the sun. Dermatologists believe that exposure to the sun creates cumulative and irreparable damage to the skin.
>
> The quest for the perfect tan has created 400,000 new cases of **basal cell carcinoma** a year. Basal cell carcinoma is one of three forms of skin cancer. The other two are **squamous cell carcinoma** and **malignant melanoma** (cancer of melanocytes). Basal cell carcinoma appears most frequently on the face and shoulders or chest as small, sometimes crusty nodules. Basal cell and squamous cell carcinomas are 100% reversible if treated immediately, usually by a curettage and electrodessication process. A sharp blade is used to scrape off the lesion, and an electric needle is then applied to seal the area. This process takes less than 30 minutes on an outpatient basis.
>
> Malignant melanomas, on the other hand, are far more serious. Melanomas radiate quickly to other organs. A mole that suddenly changes color or shape or that opens up is an initial sign of melanoma and should be checked immediately. Dermatologists believe that even just one very painful blistering case of sunburn during childhood or early adolescence can double the adult risk for melanoma.
>
> Melanocytes release the skin pigment melanin when exposed to the sun. Melanin absorbs ultraviolet rays from the sun and disperses them; these rays disrupt the molecular DNA structure of skin cells. There are two bands of ultraviolet rays: UVA, a long band, and UVB, a short band. UVB is more dangerous to the skin cells and, ironically, the more effective ray for tanning. UVB is most available at midday between 10 A.M. and 2 P.M., whereas UVA occurs more moderately throughout the day. UVA rays damage the collagen and elastin cells and are thought to be responsible for the characteristics of aging, such as wrinkles and skin dryness.
>
> Dermatologists have categorized the skin's sensitivity to sunburn into six pigment types, from the light-complexioned people with blue eyes and blond or red hair at highest risk, to black skin at lowest risk. Immunity to sunburn does not mean immunity to skin cancer, however. Blacks do develop skin cancer, but ten times less frequently.
>
> The Food and Drug Administration has set sun protection factor (SPF) standards according to the skin pigment types from 2 to 15 for the $350,000,000 per year suntan lotion industry. The time and frequency of application of the sunscreen are as important as the SPF number. Multiply the SPF number by the normal time spent in the sun until the skin reddens without any sunscreen, and the result is the time you are protected by using the sunscreen. Look for these active ingredients in the lotion: benzophenone and anthranilates (anti-UVA) and PABA and cinamates (anti-UVB).

and elastic fibers. The dermis contains blood vessels, nerves, hair follicles, sebaceous and sweat glands, and smooth muscle fibers. The dermis is connected to underlying organs and muscles by the **hypodermis,** or **subcutaneous layer** (also called **superficial fascia**). This layer often has extensive deposits of adipose tissue.

The dermis can be subdivided into two strata, the reticular and papillary regions (Figures 5-1 and 5-2). The **reticular region** is deeper and thicker than the papillary region, forming the base of the dermis. Collagenous and elastic fibers in this layer tend to be oriented parallel to the surface of the skin, thereby imparting considerable strength and elasticity to the region.

The **papillary** (păp´ĭ·lăr·ē) **region** is thinner and more superficial than the reticular region and is attached to the stratum basale of the epidermis. This region is so named because it contains numerous folds and ridges, called **papillae,** that protrude into the epidermis. These folds and ridges are clearly visible on the back of your hand and your palm and as your "fingerprints" at the ends of the fingers. Some of the papillae contain loops of capillary blood vessels, and others contain special nerve endings sensitive to touch.

Epidermal Derivatives

After completing this section, you should be able to:

1. **Enumerate and discuss the specialized derivatives that develop from the epidermis.**
2. **Describe the similarities in development of nails and hair.**
3. **Describe how a sebaceous gland differs from a sweat gland.**

Nails

Nails are flattened, keratinized structures found on the superior surfaces at the terminal ends of fingers and toes. They function as protective coverings for the digits, and they facilitate manipulation of tiny objects. Nails begin to grow on the fetus before the fourth month of pregnancy.

When fully formed, a nail possesses several anatomic regions. Nail growth occurs from the **nail matrix,** a layer of rapidly dividing epidermal cells (Figure 5-3). The cells pro-

hyponychium: (Gr.) *hypo*, under + *onyx*, nail
eponychium: (Gr.) *epi*, upon + *onyx*, nail
follicle: (L.) *folliculus*, small bag

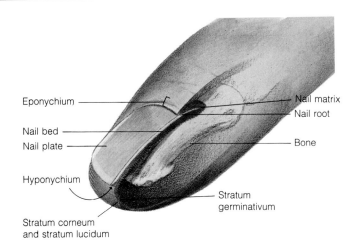

Figure 5-3 Structure of a fingernail.

duced by the nail matrix become highly cornified and scale-like and form the **nail plate** that grows out over the end of the digit. The nail plate is about 0.5 mm thick and consists of extremely flattened, fused squamous epithelial cells. Growth of the nail plate occurs at the nail matrix and continues throughout life. The rate of growth is about 1 mm per week, and at this rate it takes five to six months to replace a fingernail that is lost. Toenails grow at a somewhat slower rate, and replacement takes six to eight months.

The stratum germinativum of the fingertip extends under the nail plate where it forms the superficial layer of the **nail bed.** The stratum corneum of the fingertip only extends under the free edge of the nail and forms the **hyponychium** (hī·pō·nĭk′ē·ŭm). The epidermis surrounding the surface of the nail plate forms a nail wall from which the stratum corneum extends over the nail plate as the **eponychium** (ĕ·pō·nĭk′ē·ŭm) or cuticle (Figure 5-3). The base of the nail plate, called the **nail root**, contains recently formed nail material that has not completely dried. This area is visible above as a crescent white area called the **lunula.** The rest of the nail becomes transparent as it dries out, allowing the pinkish color of blood vessels in the nail bed to show through.

Hair

Hairs are elongate, threadlike growths of highly keratinized cells from the epidermis. They range in diameter from 0.05 mm to 0.5 mm and in length from less than 1 mm to nearly 2 m. In humans, hair is sparsely distributed over most of the body, and its function is reduced to that of physical protection and partial shading from the sun's rays (scalp) and protection from foreign particles (nostril hairs, eyelids, and eyebrows). Hair is absent from the palms, soles, lips, anus, and urinary and genital openings.

Hairs begin to form in the fetus during the third month of pregnancy, developing from inward growths of epidermis, called **hair follicles,** that invade the deeper dermis. The deep end of a follicle is an enlarged **hair bulb** (Figure 5-4), whose basal end is indented by a **papilla** composed of dermal connective tissue and blood capillaries from the dermis. Growth of a hair originates in a central area of the bulb called the **matrix,** which is an active area of dividing cells similar to the nail matrix. As cells are produced in the matrix, they are pushed outward, forming the threadlike hair that eventually projects from the open end of the follicle.

Longitudinally, a hair consists of a **root** that lies embedded in the follicle and a free **shaft.** Living cells produced at the matrix accumulate keratin and die as they are pushed up through the root, thus the shaft is composed of dead cells. Internally, a hair consists of three concentric layers: a central **cortex** surrounded by a **medulla** and a **cuticle** (Figure 5-5). All three layers develop from the matrix. The epithelial cells of the follicle surrounding the hair root can be divided into two sheaths. The **external root sheath** develops from epidermal strata that have grown into the skin surrounding the matrix. The **internal root sheath** develops from cells of the matrix. In turn, both sheaths of the follicle are surrounded by connective tissue of the dermis.

Hair color is due to the presence of melanin in the cortex and medulla of the hair shaft. In addition to the standard black, melanocytes in the hair bulb produce two color variants of melanin, brown and yellow. Blond, light colored, and red hair has a high proportion of the yellow variant.

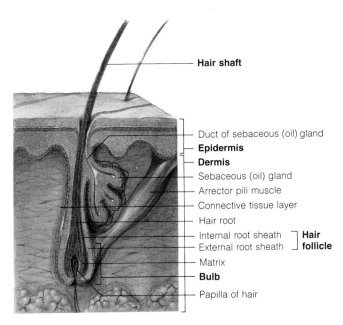

Figure 5-4 Hair. Notice how the layers of the epidermis extend down into the dermis at the site of growth of a hair.

sebaceous:	(L.) *sebum*, tallow, grease + *aceus*, having the appearance of
sudoriferous:	(L.) *sudor*, sweat + *ferre*, to carry
eccrine:	(Gr.) *ekkrinein*, to secrete

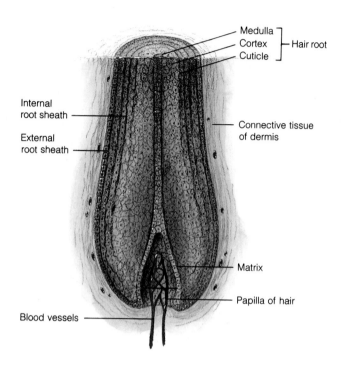

Figure 5-5 Longitudinal section through a hair follicle. Note the concentric layers that make up a fully formed hair.

Brown and black hair possesses more of the brown and black melanin. Hereditary factors influence which variant predominates, and natural bleaching by the sun can also occur. An almost infinite number of combinations of the pigment varieties is possible, and hair color often varies over different parts of the body or even just throughout the head or beard. Hair turns gray when the melanocytes in the hair bulb stop producing melanin, and white hair is the result of air present in the central medulla of the hair shaft.

Attached to a hair follicle is a slip of smooth muscle, **the arrector pili muscle** (Figure 5-4), which serves to erect the hairs, causing the familiar "gooseflesh" or "goosebumps" associated with shivering. In fur-bearing mammals, erection of hairs increases the insulation provided by the coat, but in humans, the hair is not thick enough to be effective in temperature control.

Each hair in the body has a definite growth pattern and a life span that ranges from about three months for eyelashes to about four years for hair in the scalp. At the end of its particular life span, the root of a hair separates from its matrix and remains in the follicle until it falls out or is pulled out. The follicle then shrinks and enters a period of rest. Following its rest period, the follicle will either degenerate completely or enlarge, form a new hair bulb, and produce another hair.

Glands of the Skin

The integumentary system contains two kinds of glands that are derivatives of the skin: sebaceous and sweat glands. **Sebaceous** (sĕ·bā'shŭs), or **oil glands**, are simple, unbranched or branched, alveolar, holocrine glands usually connected to hair follicles (Figure 5-4). The few that are located on the lips, the glans penis, and labia minora empty directly onto the surface of the skin, since hairs are absent on these structures. Sebaceous glands are most abundant in the scalp and other areas of the skin covered with hair. They are not found on the palms and soles. These glands produce an oily substance called **sebum,** which contains fats, cholesterol, and cellular remnants. It coats the surface of the skin and the shafts of hairs where it prevents excess water loss, lubricates and softens the stratum corneum, and softens the hair. It is also mildly toxic to certain bacteria.

Two types of **sweat** (or **sudoriferous**) **glands** are found in human skin. The majority are ordinary sweat glands called **eccrine** (ĕk'rĭn) **glands.** They are simple, coiled, tubular

ceruminous: (Med. L.) *cerumen*, from (L.) *cera*, was + *ous*, relating to

> ## CLINICAL NOTE
>
> ### BURNS
>
> **Burns** are injuries to the skin produced by exposure to hot liquids or steam, flame or hot surfaces, corrosive chemicals, or electric currents. Electrical burns may involve a much larger area of burn than is immediately apparent and may require close observation.
>
> Burns are characterized by the depth of the burn, the area (extent) of the burn, and the causative agent. The depth of a burn is described in degrees. **First-degree burns** are the least serious and involve only the superficial layers (stratum corneum, granulosum, and spinosum) of the epidermis and result in redness and mild pain. They usually heal completely in three to four days and do not result in scars. **Second-degree burns** involve the deeper layers of the epidermis and the upper papillary layer of the dermis and result in blistering of the skin. They are quite painful and result in the loss of some of the superficial layers of the skin. They should be protected with a coating of an antibiotic cream such as Silvadene to prevent infection.
>
> **Third-degree burns** involve the full thickness of the skin, the skin derivatives, and at times, even deeper structures. These burns are usually painless since the nerve endings have been destroyed. They are chalky white or charred black in color. Third-degree burns are very serious, resulting in full thickness loss of the skin and may require skin grafting; they usually result in severe scar formation. An extensive third-degree burn can result in the loss of tremendous amounts of serum and extracellular fluids into the burn and can require the replacement of large amounts of fluids. A recently developed technique to grow large sheets of new skin by tissue culturing small samples of the patient's own skin promises to greatly improve the treatment of these serious burns.
>
> The area of the burn is denoted as the percentage of the body burned, taking into account special areas such as the face, hands, and genitalia. A quick method for estimating the percentage of burns is to use the "rule of nines" (see figure).
>
>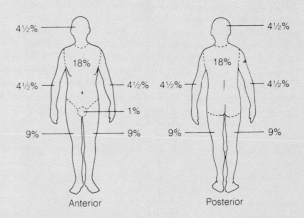
>
> The "rule of nines" is used to estimate the extent of burns covering the surface area of the body.

glands (Figure 5-1). These are found throughout the skin except for the lips, glans penis, and eardrum. Several million are distributed over the surface of an adult, and they are most numerous in the palms and soles. (No wonder it is so easy to get "sweaty palms" under certain conditions!) Ordinary sweat glands secrete an oily substance that contains water and many ingredients of urine. Sweat glands assist the kidneys in excreting wastes, and evaporation of sweat helps cool the body. The role of the sweat glands in temperature regulation is discussed in Chapter 19.

Certain specialized branched sweat glands are found in the **axilla** (armpit) and in the external auditory canal. During secretion, these glands pinch off small bits of cytoplasm near their surfaces and hence are apocrine in nature. The **ceruminous** (sĕ·rū′mĭ·nŭs) **glands** of the ear secrete a yellow substance called **cerumen** ("ear wax") that is rich in oils.

Mammary glands are specialized glands that produce milk. Their nature and function will be discussed along with the female reproductive system in Chapter 27.

Functions of the Skin

After completing this section, you should be able to:

1. List the major functions of the skin.
2. Describe how the skin provides both physical and chemical protection against invasion by microorganisms.
3. Discuss the importance of vitamin D_3.
4. Describe the role of sweat glands in controlling body temperature.

The skin has a variety of important functions, all of which contribute to homeostasis of the body.

Protection

The skin forms the body's first line of defense in many respects. The several layers of dead skin cells covering the body, reinforced by the presence of keratin, eleiden, and melanin, provide physical protection against mechanical injury, excess ultraviolet radiation, heat, and the harmful effects of many dangerous chemicals. In addition, pathogenic bacteria and other microbes are retarded in their ability to penetrate to the deeper living cells where they can thrive. The skin is even able to respond to continuous physical stress by producing a thickened area, such as a callus.

The skin responds chemically in protecting the body. Externally, the acidic nature of sweat inhibits the reproduction of many types of bacteria and kills others. Internally, the production of defense chemicals called antibodies destroys pathogens that invade the skin.

Excretion

The sweat glands discharge varying amounts of secretions, depending on internal and external temperatures. Sweat contains salts and nitrogenous wastes that are excreted in the watery solution of sweat.

Temperature Regulation

A major homeostatic function of the skin is the control of body temperature (see more on this in Chapter 19). A rise in body temperature causes arterioles in the skin to increase in diameter (dilate), bringing more heat to the surface. The sweat glands respond by producing greater quantities of sweat, which flows onto the skin. Evaporation of water in the increased amount of sweat carries heat away from the skin, lowering body temperature. A drop in either external air temperature or internal body temperature results in constriction of the skin arterioles, conservation of internal body heat, and a reduction in the production of sweat.

Production of Vitamin D

One of the effects of the exposure of the skin to the ultraviolet radiation in sunlight is the production of vitamin D_3 in skin cells. This vitamin is important in the absorption of calcium and phosphorus from foods in the intestine. Insufficient production of vitamin D_3 interferes with the proper metabolism of these minerals and with the development and maintenance of healthy bones and teeth.

Sensations

The skin also contains a great number of nerve endings and specialized sensory receptors. These cells respond to such stimuli as fine touch, deep pressure, heat, cold, and pain. The information they gather is relayed to the central nervous system, alerting the body's control centers of external and internal conditions. These signals are used to determine responses that guide body reactions.

In this chapter we have studied the structure of the skin and how its metabolic activities promote survival. In the following chapter we shift our focus of study from the surface of the body to an internal support system, the skeleton. In addition to protecting soft organs and providing support against gravity, the skeletal system is organized into a jointed system of levers that are used to accomplish locomotion.

STUDY OUTLINE

Epidermis (pp. 101–105)

Epidermal Layers The epidermis consists of stratified squamous epithelium. It lacks a system of blood vessels and relies on the transport of materials from vessels in the underlying dermis. It consists of as many as five layers (from deepest to outermost):

The *stratum basale* is the deepest layer and produces the cells of the other four layers through frequent cell divisions.

The *stratum spinosum* is a layer of young cells also involved in frequent cell divisions.

The *stratum granulosum* contains *keratohyalin*, a substance that produces *keratin*, a waterproofing agent for the skin. Cells in this layer flatten and eventually die as they lose nutrients.

The *stratum lucidum* is a thin layer of dead cells found only in the palms and soles. Its cells contain *eleiden*.

The *stratum corneum* forms the outer exposed layer of the epidermis and has keratin in place of cytoplasm in its cells.

Pigmentation Skin color is due to the relative abundance of two pigments, yellowish *carotene* and dark brown *melanin*. Carotene is acquired with yellow and red vegetable foods, while melanin is manufactured by special cells called *melanocytes*. Heredity and environmental conditions regulate the amount of melanin produced.

Dermis (pp. 105–106)

The *dermis* (or *corium*) is the deepest layer of the skin. It consists of two subdivisions. The *reticular region* is a thick, deeper region containing collagenous and elastic fibers that impart strength and elasticity to the tissue. The *papillary region* is more superficial and is attached to the stratum basale of the epidermis.

Epidermal Derivatives (pp. 106–109)

Nails Nails are flattened, keratinized structures on the superior surfaces at the terminal ends of the fingers and toes.

dermatology: (Gr.) *derma*, skin + *logos*, to study
impetigo: (L.) *impetere*, to attack

THE CLINIC

DERMATOLOGY AND SKIN DISORDERS

The branch of medicine that studies the diseases of the skin is called **dermatology**. The study of dermatology is important because many internal diseases are reflected in changes in the skin which may be seen and identified. In the past, dermatology has been closely related to the study of the sexually transmitted disease **syphilis** because of the multiple skin changes seen in this disease. Early physicians specializing in this area of medicine were certified as dermatologists and syphilologists.

Early studies in dermatology were largely that of describing the various skin rashes. **Lesions** were classified into descriptive types, such as **pustular** (containing pus), **vesicular** (containing clear fluids), **macular** (flat discolored spots), and **papular** (raised lesions). Diseases were given descriptive names, such as **erythema** (ĕr″ĭ·thē′mă) **multiforme** (a red rash of many forms), **erythema nodosum** (a red, nodular rash), and **erythema marginatum** (a red rash with sharp borders). Today, dermatology includes the study of the underlying cellular changes that result in skin diseases, but it still retains much of the old descriptive terminology.

Skin diseases have a multitude of causes which can be grouped into the following categories. Some examples are given for each type.

Developmental or Hereditary

Ichthyosis (ĭk″thē·ō′sĭs) is dry, scaling skin ("fish scale" skin) that is inherited. **Xeroderma pigmentosum** (zē″rō·dĕr′mă pĭg″mĕn·tō′sŭm) is a rare, disfiguring hereditary disease in which skin cancers frequently develop.

Infectious Diseases

There are many common diseases caused by infection. The common skin eruptions, including red measles (rubeola), German measles (rubella), and chicken pox are caused by viral infections (see Clinical Note: Viral Skin Infections). Bacterial infections of the skin are called **pyodermas** (pī·ō·dĕr′măs). **Impetigo** is a common pyoderma caused by the hemolytic streptococcus (a bacteria) which produces patches of weeping, crusted, red lesions. It is easily treated with antibiotics. Infection of the skin with fungal organisms are called **dermatophytoses** (dĕr″mă·tō·fī·tō′sēs). **Tinea circinata** (ring worm), **tinea cruris** (jock itch), and **tinea pedis** (athlete's foot) are all fungal infections of the skin. **Acne** is a more complicated problem involving infection, hormonal effects, keritinization, and sebum production on the skin.

Allergy and Hypersensitivity

There are many skin problems related to allergy and hypersensitivity. The skin is a highly reactive organ, and medicine makes use of this property to identify the cause of many allergic conditions by the use of **scratch** or **patch testing**. Suspected **allergens** are applied to the skin on patches or scratched into the skin using fine needles; the doctor then notes whether the skin reacts to the allergen by forming a red, raised spot (a **wheal**). The size of the wheal is directly proportional to the degree of hypersensitivity. Allergic people can also be desensitized by injecting serially stronger doses of the allergen into the skin—**allergy shots**. Some diseases of the skin are due to allergic reactions such as hives, poison ivy dermatitis, and eczema.

Tumors and Cancers

There are many tumors of the skin, both **benign** (innocent) and **malignant** (cancerous). Moles, birthmarks, and freckles are examples of benign skin tumors. Many skin cancers are related to excessive exposure to sunlight, particularly in blue-eyed, fair-skinned, blond people.

Aging of the Skin

Aging causes many changes in the skin and its derivatives: graying and thinning of the hair, thinning and wrinkling of the skin itself, and drying of the skin because of a decrease in the production of oils and sweat. Aging also causes the appearance of light-brown patches on the backs of the hands called **liver spots** and elevated, greasy, brown patches on the trunk called **seborrheic keratoses** (sĕb″ō·rē′ĭk kĕr′ă·tō″sēs).

A *nail plate* grows from a *nail matrix* and overlies the *nail bed*.

Hair Hairs are elongate, threadlike growths of highly keratinized cells from the epidermis. Hairs grow from *hair follicles*, epidermal structures that invade the dermis. Hair color is due to the presence of black, brown, and yellow variants of melanin. Hairs may be erected by an attached smooth muscle, the *arrector pili muscle*.

Glands of the Skin Two kinds of glands are present in the skin.

Sebaceous (oil) glands are simple, unbranched or branched, alveolar, holocrine glands usually connected to hair follicles. These glands produce *sebum*, an oily coating that lubricates, softens, and helps waterproof the skin and hair.

Sweat (sudoriferous) glands produce sweat, and *ceruminous glands* produce ear wax.

Functions of the Skin (pp. 109–110)

Protection The skin protects the body against mechanical injury and stress, ultraviolet radiation, heat, dangerous chemicals, bacteria, and other pathogens.

Excretion The skin excretes unwanted wastes in the form of sweat.

Temperature Regulation The skin cools the body through evaporation of sweat and warms it through conservation of internal body temperature. This is a major homeostatic function of the skin.

Production of Vitamin D The skin produces vitamin D, a vitamin important to the metabolism of calcium and phosphorus.

Sensations A myriad of nerve endings in the skin respond to external stimuli, allowing our central nervous system to determine responses that guide body reactions.

SELF-TEST OF CHAPTER OBJECTIVES

True-False Questions

1. The most superficial layer of the epidermis is called the stratum lucidum.
2. The cells of the skin are active in the formation of vitamin D.
3. The papillary region is the more superficial of the two subdivisions of the dermis.
4. Blood vessels, nerves, and connective tissue are found in the dermis but not in the epidermis.
5. Melanocytes found in the epidermis are cells that produce a brown pigment called melanin.
6. A scar is composed of epithelial parenchyma and is used to repair an injury to the skin with functional epithelial tissue.
7. Nails and hair are similar in that they are formed from dead epidermal cells that are packed together tightly.
8. Sebaceous (oil) glands are apocrine glands.
9. The acid nature of sweat inhibits the growth of bacteria on the skin.
10. The external root sheath of a hair develops from the matrix.

Matching Questions
Match the epidermal layers with the correct descriptive statement on the left:

11. has cells that look like spiny pin cushions
12. found only in the palms of the hands and the soles of the feet
13. has cells that contain granules of keratohyalin
14. the deepest layer of the epidermis
15. the layer of epidermis exposed to the air

a. stratum lucidum
b. stratum basale
c. stratum spinosum
d. stratum corneum
e. stratum granulosum

Match the epidermal glands with the correct descriptive statement on the left:

16. function as merocrine glands
17. usually connected to hair follicles
18. ceruminous glands of the ear are included here
19. are usually simple, coiled, tubular glands
20. are absent from the palms and soles

a. sebaceous glands
b. sweat glands

Multiple-Choice Questions

21. Which of the following is *not* a function of the skin?
    a. excretion of waste materials
    b. production of vitamin C
    c. detection of external stimuli
    d. helps to control body temperature
    e. production of vitamin D
22. The cells that eventually form a fingernail are produced by the
    a. nail root   d. hyponychium
    b. lunula   e. nail matrix
    c. nail plate
23. Which of the following is the thicker, deeper layer of the dermis?
    a. reticular region   d. hypodermis
    b. papillary region   e. subcutaneous layer
    c. corium
24. The internal root sheath develops from the
    a. matrix   d. stratum lucidum
    b. stratum granulosum e. stratum corneum
    c. stratum spinosum
25. "Goosebumps" are caused by the action of the
    a. hair matrix   d. arrector pili muscle
    b. internal root sheath e. cuticle
    c. external root sheath
26. The stratum germinativum consists of the
    a. stratum lucidum and stratum basale
    b. stratum basale and stratum spinosum
    c. stratum spinosum and stratum lucidum
    d. stratum spinosum and stratum corneum
    e. stratum basale and stratum corneum
27. The major waterproofing material of the skin is
    a. eleiden   d. melanin
    b. cerumen   e. keratin
    c. sebum
28. Which of the following glands coats hair with oil?
    a. eccrine   d. ceruminous
    b. sweat   e. sudoriferous
    c. sebaceous
29. The white crescent area visible at the base of a fingernail is the
    a. lunula   d. cuticle
    b. nail root e. hyponychium
    c. nail bed

30. The active growing part of a hair is the
    a. root
    b. internal root sheath
    c. medulla
    d. cortex
    e. matrix

Essay Questions
31. How do cells in the stratum spinosum of the epidermis receive nutrients?
32. Describe the relationship of keratohyalin and hyalin in the skin. What is their function?
33. Where is melanin produced in the skin? How does it get into additional skin cells?
34. Describe the major differences between the epidermis and dermis.
35. Which stratum of the epidermis is responsible for the growth of both nails and hair?
36. Describe the chemical differences between blond and black hair.
37. Compare and contrast sebaceous and ceruminous glands in terms of structure, location, and secretions.
38. List the functions of the skin.
39. Describe the internal structure of a hair.

Clinical Application Questions
40. Compare and contrast psoriasis versus cold sores.
41. Describe the difference among first-, second-, and third-degree burns.
42. How does rubella (German measles) differ from rubeola (red measles)?
43. Compare communicability of chicken pox and eczema.
44. Why is suntanning not really healthy?
45. List some of the many common skin disorders.

6

Development and Physiology of the Skeletal System

The human skeletal system is composed primarily of two kinds of tissues: cartilage and bone. Having a relatively rigid composition compared to other tissues, cartilage and bone perform four essential functions: *support, protection, formation of blood cells,* and *mineral storage*. The skeletal system also functions with the muscular system to produce many kinds of movement and locomotion. While the next chapter covers skeletal anatomy, this chapter is a study of the basic structure and functions of bone tissue. Here we will examine

1. the five major functions performed by the skeletal system,
2. the differences between cartilage and bone and the function of the specialized cells associated with each,
3. the development of bone and the differences between intramembranous and endochondral bone formation,
4. the processes of osteogenesis and osteoclasis,
5. factors that affect the physiological maintenance of the skeleton,
6. the feedback mechanisms involving calcium deposition in the skeleton and release from the skeleton,
7. several major bone diseases, and
8. the steps involved in the healing of a fractured bone.

hemapoiesis:	(Gr.) *haima*, blood + *poiein*, to make	osteocyte:	(Gr.) *osteon*, bone + *kytos*, cell
articulations:	(L.) *articulatus*, jointed	sesamoid:	(L.) *sesamon*, sesame + *eidos*, form
histology:	(Gr.) *histos*, tissue + *logy*, study of		

Functions of the Skeletal System

After completing this section, you should be able to:

1. List the major functions of the skeleton.
2. Indicate soft body organs that receive protection from the skeleton.
3. Discuss the role of the skeleton as a storage place for minerals.

Support is an important role of the skeleton. It bears the weight of all other tissues in the body and is responsible for our ability to stand erect. In addition to support, the skeleton protects several areas of the body. The bones of the cranium, for instance, protect the brain from injury. The vertebral column (backbone) surrounds and protects the spinal cord. The rib cage protects the lungs and heart located within the thoracic cavity, and bones of the pelvic girdle provide some protection for soft organs in the pelvic cavity.

Formation of blood cells is another important function of the skeletal system. In adult humans, the **red marrow** found in the interior (chiefly the ends) of certain bones produces all the red blood cells and many white ones. This process, known as **hemapoiesis** (hĕm″ă·poy·ē′sĭs) or **hematopoiesis** (hĕm″ă·tō·poy·ē′sĭs), will be discussed in detail in Chapter 21.

The skeletal system is the body's reservoir of minerals. Calcium, phosphorus, magnesium, potassium, carbonates, and other materials are removed from the foods we eat and stored as salts in our bones. Precise homeostatic mechanisms regulate the release or storage of these minerals. When some of the minerals are needed for chemical or physiological processes in another body tissue, they are released into the blood where they are available for use in many metabolic activities.

Finally, the skeletal system plays a large role in body movement and locomotion. Many bones are connected by joints or **articulations,** some of which are movable. Muscles stretch from one bone to another across a joint; when the muscles shorten, they produce movement of the skeletal parts. A detailed examination of articulations will be presented in Chapter 8, and the study of musculature is in Chapter 10.

In this chapter the development and physiology of cartilage and bones are examined. The **histology,** or microscopic structure, of cartilage was discussed in detail in Chapter 4. In the next chapter the specific bones of the human skeleton, their arrangement, and their anatomical landmarks will be presented.

Histology of Bone

After completing this section, you should be able to:

1. Describe the basic histology of bone tissue.
2. List five major types of bone based on external appearance.
3. Describe the major differences between spongy and compact bone.
4. Discuss the importance of the osteonal canal system in the physiologic maintenance of bone.

Bone is one of the most specialized of the connective tissues. Mature bone contains living cells called **osteocytes** (ŏs′tē·ō·sīts″) surrounded by a matrix of fibers and ground substance as do all other connective tissues. In the case of bone, collagen is the fiber type employed. Bone differs from other connective tissues, however, in that the matrix contains sufficient deposits of inorganic salts to make bone a heavy, rigid substance. The salt deposits are chiefly calcium phosphate (85%) and calcium carbonate (10%) that coat the collagen fibers. Furthermore, these inorganic salts give the matrix the physical state of a solid, thereby rendering diffusion through the matrix impossible and requiring an alternative method for the delivery of nutrients and the elimination of wastes. This alternative method is physically represented by an extensive network of interconnecting canals and passageways, the osteonal system, to be described later.

Classification Based on External Appearance

Based on external appearance, bones may be classified into five categories. **Flat bones** form such structures as the ribs, and the frontal and parietal bones of the cranium. **Long bones** (Figure 6-1a) make up most of the upper extremity (the humerus of the arm and radius and ulna of the forearm) and the lower extremity (the femur of the thigh and tibia and fibula of the shin). **Short bones** are found in the wrist and ankle. **Irregular bones** come in a wide variety of sizes and shapes and include such bones as the vertebrae and the pelvic bone. **Sesamoid** (sĕs′ă·moyd) **bones,** such as the patella (kneecap), are floating bones that do not attach solidly to other bones. All of these bones are discussed in detail in the next chapter.

Classification Based on Internal Structure

In addition to this classification based on shape, we recognize two types of bone, based on the relative solidity of the tissue: spongy bone and compact bone.

CLINICAL NOTE

FRACTURES

The most common injury to a bone is a **fracture** or break in the continuity of a bone. Fractures cause disability, require care during convalescence, and present the most common problem to affect the skeletal system. Few people go through life without breaking at least one bone.

Types of Fractures

There are many ways of describing fractured bones. The name given to a fracture defines its nature. Here is a list of fracture types following the usual classification. (Eight of these fractures are shown in the X rays below.)

Simple or Closed Fracture A bone is broken but the two ends of the bone remain in their normal position and do not puncture the skin.

"Greenstick" Fracture A bone is broken but the cortex on one side is still intact until part of the fibers break, similar to bending a green stick. This type of fracture occurs most frequently in young people whose bones are not completely calcified.

Comminuted Fracture There are several parts to a comminuted fracture, not just a single fracture line across the bone. A fragment of the bone may be isolated from the ends of the bone and is called a "butterfly."

Spiral Fracture The fracture line is not straight across the bone but "spirals" down the shaft. It is the result of torsional or twisting forces.

Displaced Fracture The ends of the broken bone are not in the usual alignment. They may be at angles or the fractured ends of the bone may even pass by one another, thus shortening the limb.

Compound or Open Fracture This fracture is associated with a break in the skin, thus exposing the bone to infection. A compound fracture may occur when the ends of the bone actually protrude through the skin, causing a laceration, or it may occur when a blow causes a break in the skin as well as in the bone. In either case, the wound is assumed to be contaminated. The injury is then treated with much more care, such as thorough cleaning and antibiotic treatment, to prevent infection in the bone.

Stress Fracture This is a common problem for athletes of endurance sports. Although the exact mechanism of the stress fracture is unknown, it is probably a local failure of bone to adapt to repeated abnormal stress. Stress fractures were first reported from the military and were called "march fractures." They occur most frequently in neophyte runners or in more experienced runners who are advancing their distance too rapidly. Dancers and gymnasts

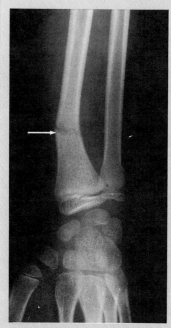
(a) Greenstick fracture

(b) Comminuted fracture

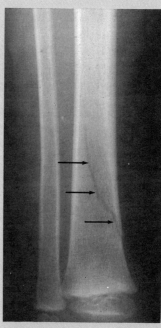
(c) Spiral fracture

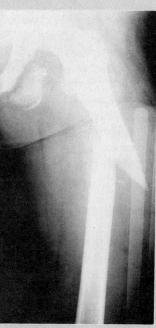
(d) Displaced compound fracture

can also develop stress fractures. The typical stress fracture occurs as localized pain and swelling in a bone of the lower leg or foot after prolonged running or other exercise. The pain usually prevents the athlete from running or dancing but usually does not interfere with normal activity such as walking. The fracture line does not appear on an X-ray examination until about three weeks after the onset of symptoms and then may show only small amounts of subperiosteal (beneath the periosteum) calcification as the bone attempts to heal the fracture. Usually stress fractures heal spontaneously after about six weeks of rest.

Colles' Fracture This is a break of the distal radius during a fall on outstretched arms. The fragment nearest the hand is displaced backward.

Pathologic Fracture The bone is weakened by disease conditions such as osteomalacia, osteoporosis, or rickets (see The Clinic: Diseases of the Skeletal System).

Repair of a Fracture

The repair of a broken bone is a very complex process. Immediately after a fracture occurs, the bone fragments must be aligned and then held immobilized in a cast. The length of time required for a fracture to heal depends on the type of fracture and on the kind of bone that has sustained the injury. (The navicular carpal and the fifth metatarsal often take a long time to heal.) Larger bones take longer to heal.

An unusual form of osteogenesis occurs when a bone is broken and healing takes place. Many of the blood vessels in the soft tissue surrounding the fracture site or in the periosteal membrane are broken, and a puddle of blood **(hematoma)** surrounds the fracture site. Calcium in the form of osteogenic cells is laid down in this hematoma in about two to three weeks following a fracture to form a collar that fuses the separate bone fragments. This healing bone is called a **callus,** and it acts to splint the fracture from the outside like an internal cast while the broken spicules of bones are reabsorbed by osteolytic cells from the inside to form compact bone. The compact bone strengthens the fracture site and renders the callus unnecessary. When the internal structure of the bone is well reestablished, the callus is reabsorbed. Frequently, the healing is so complete that the fracture site is no longer visible in a year or so following a fracture.

On rare occasions, fractures fail to heal, usually because of the interference of soft tissue between the ends of the broken bone. This is called an ununited fracture or a **nonunion.** Complications of this sort may require the surgical placement of a bone graft, or it can be treated by applying an electric current by means of implanted electrodes or coils placed onto the cast around the bone. The electric current stimulates the osteoblasts to produce more bone to heal the fracture.

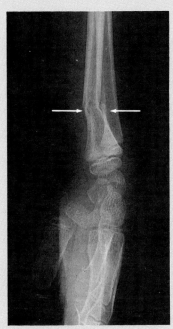
(e) Colles' fracture

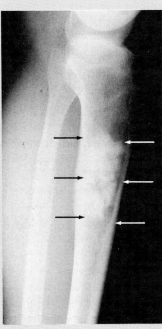
(f) Pathological fracture

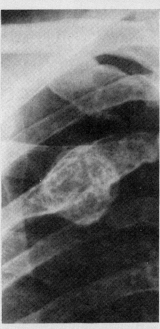
(g) Callus (on a rib)

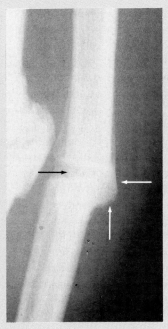
(h) Nonunion

118   Chapter 6   Development and Physiology of the Skeletal System

trabeculae:	(L.) *trabecula*, little beam
canaliculus:	(L.) *canaliculus*, small channel
endosteum:	(Gr.) *endon*, within + *osteon*, bone
periosteum:	(Gr.) *peri*, around + *osteon*, bone

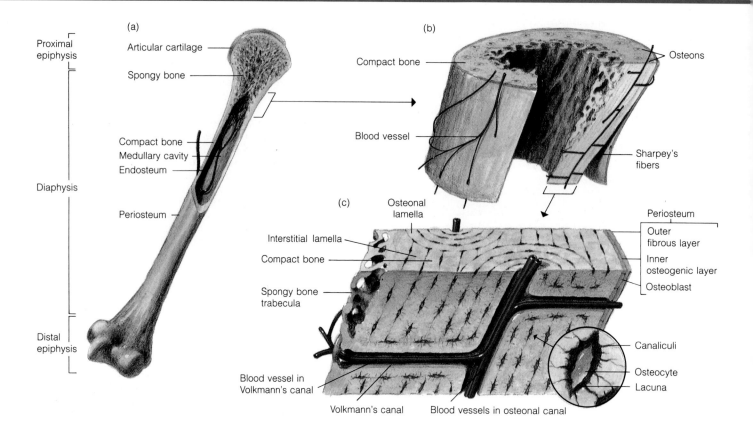

Figure 6-1 The internal structure of a fully formed long bone. (a) The entire bone. (b) Section through the upper end of the diaphysis (shaft) of a long bone. Note the junction between the osteons of the compact bone on the exterior and the trabecular bone on the interior. The periosteum is shown folded back to expose the Sharpey's fibers. (c) Enlarged view of the osteonal systems in compact bone.

Spongy bone (also called **cancellous** or **trabecular bone**) is found in the interior of flat bones and at the ends of long bones. It is composed of a network of branching, interconnected sheets and bars called **trabeculae** (tră·běk´ū·lē) (Figure 6-1). This internal structure creates a series of interconnected spaces that are filled with a vascular tissue called **marrow**, a functional part of the circulatory system producing both red and certain white blood cells.

Compact bone (also known as **dense** or **lamellar bone**) forms both the shaft of long bones and also a thin outer surface layer that covers the exterior of all bones. It is composed of concentric layers called **lamellae** (lă·měl´lē) of matrix deposited mostly around blood vessels that pass through the bone (Figure 6-1). A blood vessel and the series of lamellae and osteocytes that surround it form an **osteon** (also known as an **haversian system**). The minute blood vessels (capillaries) and nerves of osteons extend throughout bony tissue in longitudinally oriented **osteonal canals** (also called **haversian canals**) and transversely oriented **communicating canals** (also called **perforating** or **Volkmann's canals**).

The system of interconnecting canals in compact bone is so extensive that each osteocyte is no more than 0.5 mm away from a capillary. This is important because living osteocytes trapped in a mineralized matrix must have a source of nutrients and a route for waste disposal if they are to remain alive. An extensive network of microscopic canals called **canaliculi** (kăn˝ă·lĭk´ū·lī) (Figure 6-1c) allows the passage of materials between capillaries and osteocytes and thus aids the exchange of nutrients and wastes.

The osteonal canals of compact bone and marrow spaces of spongy bone are lined by a thin membrane called the **endosteum** (ĕn·dŏs´tē·ŭm), which functions in formation, repair, and maintenance of bone tissue. Interconnections between the marrow spaces and osteonal canals permit the endosteum to be continuous with the **periosteum** (pĕr·ē·ŏs´tē·ŭm), the tough sheath of fibrous connective tissue that adheres closely to the external surface of the bone. Short bundles of collagenous fibers called **Sharpey's fibers** connect the periosteum to minute pores in the surface of the bone.

chondroblast: (Gr.) *chondros*, cartilage + *blastos*, bud, germ
chondrin: (Gr.) *chondros*, cartilage
chondrocyte: (Gr.) *chondros*, cartilage + *kytos*, cell
perichondrium: (Gr.) *peri*, around + *chondros*, cartilage

meniscus: (Gr.) *meniskos*, crescent
arthroscope: (Gr.) *arthron*, joint + *skopein*, to examine

Development of the Skeleton

After completing this section, you should be able to:

1. Describe the development of both cartilage and bone.
2. Distinguish between intramembranous and endochondral ossification.
3. Describe primary and secondary ossification centers.

In addition to bone, a significant amount of cartilage is found in the skeletal system. In fact, most of the fetal skeleton and much of an infant's skeleton is cartilaginous. Unlike rigid bone, cartilage is flexible. Its matrix does not contain the mineral salts found in bone, but the presence of collagenous or elastic fibers imparts resilience and elasticity to cartilaginous portions of the skeleton. Refer back to Chapter 4 for the details of the histology of cartilage.

Development of Cartilage

Cartilage develops in the embryo when groups of young cartilage cells called **chondroblasts** (kŏn′drō·blăsts) begin secreting chondrin and producing collagenous fibers. **Chondrin** is the ground substance of cartilage. It is a complex material composed of protein and carbohydrates, and it is chondrin that gives cartilage its characteristic "rubbery" consistency. All cells begin life in an immature condition, and internal changes cause them to develop their adult characteristics. As young cartilage cells become surrounded by the ground substance and fibers they are producing, their ready access to nutrients is restricted, and their secretory activities are curtailed. This cessation of chondrin production signals the maturation of cartilage cells, and they are now referred to as mature **chondrocytes.**

Continued growth of the cartilaginous structure involves the development of new chondroblasts, hence cartilage, from the internal surface of the **perichondrium,** the fibrous sheath that surrounds most cartilaginous structures.

Growth of cartilage around its exterior surface is termed **appositional growth.** Cartilage can also grow by destruction of interior material and by addition of new chondrin by chondroblasts. This method is called **interstitial growth,** and it is characteristic of remodeling activities and endochondral bone growth, to be discussed under the section Endochondral Ossification.

Development of Bone

We have described the two kinds of bone: spongy or cancellous bone and compact or lamellar bone. The major differences between the two types lie mainly in their gross

---

CLINICAL NOTE

TORN CARTILAGE

Many of the synovial joints such as the knee contain a cartilaginous cushion or **meniscus.** A shearing or rotational injury to the joint can cause a tear in the meniscus that disrupts the smooth functioning of the joint. The knee joint contains two menisci which are frequently injured in athletes. The meniscus on the medial side of the knee is the most vulnerable and is torn 10 times more frequently than the lateral meniscus. Symptoms of a torn meniscus are pain at the joint line, cracking sounds, and **locking,** which is the inability to fully extend the knee.

Treatment of a torn meniscus is surgical removal of the loose piece of cartilage. Most of the surgery can now be performed by inserting a small telescope-like instrument called an **arthroscope** through a small incision on one side of the knee and small biting forceps through the opposite side of the knee (see figure). This technique, which uses fiberoptics, allows the surgeon to visualize the torn cartilage through the arthroscope and remove the loose piece with the forceps. The recovery period is brief, and the normal portion of the meniscus is preserved.

One of the most severe knee injuries is a complete tear of the medial collateral and anterior cruciate ligaments along with a torn medial meniscus. It is called the "terrible triad." Because cartilaginous structures lack an internal network of blood vessels resulting in poor healing, these types of injuries can be devastating to athletes (see Chapter 8 Clinical Note: Sports Injuries).

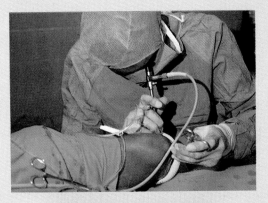

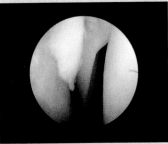

The arthroscope. (a) Arthroscopic surgery being performed on knee. (b) Arthroscopic view inside knee joint showing joint surfaces with cartilage meniscus.

endochondral: (Gr.) *endon*, within + *chondros*, cartilage
osteoblast: (Gr.) *osteon*, bone + *blastos*, bud, germ

morphology. In a developing embryo, both types can be formed in two different ways: directly from a sheet of mesodermal mesenchyme in a process called **intramembranous ossification** or by replacement of previously formed cartilage tissue through **endochondral ossification.**

Intramembranous Ossification

Intramembranous ossification is the method by which the flat bones of the skull and the clavicle (collar bone) develop (Figure 6-2). The process begins in a sheet of mesenchyme that has developed a rich supply of blood vessels. Some of the mesenchymal cells differentiate into **osteoblasts,** which are young bone cells capable of substantial metabolic activity. The cells then enlarge and begin to secrete collagen fibers and **glycosaminoglycans** (glī′kō·să·mē″nō·gly′kăns), the organic components that make up the matrix of bone. The collagen fibers tend to be oriented along the lines of stress in a developing bone (Figure 6-3). Simultaneously, the mineral components of bone dissolved in the fluid of the vascular spaces diffuse into the interstitial spaces. The enzyme **alkaline phosphatase,** secreted by the maturing osteoblasts, alters the pH of the interstitial spaces, causing the dissolved mineral components to precipitate out of solution around the collagen fibers, and a solid bone forms. In intramembranous

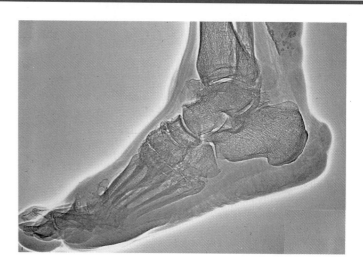

ossification, net diffusion toward the interstitial space is a dynamic process in which minerals lost to precipitation are replaced.

As progressively more organic matrix is secreted and more solid bone forms, the osteoblasts become trapped in small spaces **(lacunae).** Gradually, these cells curtail their secretory activities, and are now referred to as **osteocytes.** Unlike chondrocytes, osteocytes are not completely isolated from one another. Fine protoplasmic extensions, called cell junctions, maintain connections between neighboring cells. As minerals are deposited in the matrix around cellular extensions, the microscopic canals or sleeves called canaliculi are formed. Progressive mineral deposition around the osteoblasts and osteocytes produces the network of trabeculae characteristic of spongy bone (Figure 6-1).

Osteoblasts are incapable of mitosis, and current evidence indicates that they are produced by **osteoprogenitor cells** located on the inner periosteum and among the endosteal cells. Continued production of osteoblasts at the surface of the developing bone maintains layers of active cells that continue to add new bone. Eventually, the spaces at the surfaces become filled with the concentric lamellae of osteons, producing two layers of compact bone enclosing an inner area of spongy bone (Figure 6-4). The connective tissue surrounding the developing bone is transformed into the periosteum. As discussed in the section describing the histology of bone, an internal extension of this tissue into the interior cavities of the bone via the communicating canals forms the endosteum. Some osteoblasts associated with the periosteum

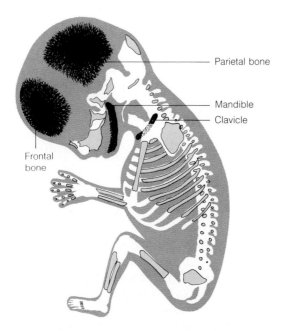

Figure 6-2 Intramembraneous bones (in black) versus endochondral bones in a developing fetus. Major intramembraneous centers in the flat bones of the skull and clavicle are shown here.

dysplasia:	(Gr.) *dys*, pain, difficulty + *plassein*, to form	myositis:	(Gr.) *mys*, muscle + *itis*, inflammation
diaphysis:	(Gr.) *dia*, through + *phyein*, to grow	ossificans:	(L.) *os*, bone + *facere*, to make
		epiphyses:	(Gr.) *epi*, on + *phyein*, to grow

> **CLINICAL NOTE**
>
> ### FIBROUS DYSPLASIA
>
> **Fibrous dysplasia** affects the development of one or several bones. Defective bone formation is caused by an abnormal fibrous tissue that displaces the normal cancellous bone and bone marrow. X rays show a degenerative cystlike appearance associated with cortical thinning and softening of the bone. Skeletal lesions are associated with the bony deformity and can lead to fractures. Fibrous dysplasia accompanied by early sexual development and abnormal skin pigmentation make up a clinical disorder known as **Albright's syndrome**.

> **CLINICAL NOTE**
>
> ### MYOSITIS OSSIFICANS
>
> **Myositis ossificans** is an unusual deposition of calcium within injured muscle tissue. It occurs most frequently in the large muscle mass of the anterior thigh following repeated blows to the area that can occur in blocking sports such as football. Massive swelling usually occurs in the involved muscle because of blood flow into the muscle; the player suffers severe pain and is unable to flex the knee. After about three weeks the accumulation of fine calcium deposits in the area of injury will be apparent on an X ray. The calcium deposit slowly condenses over a period of weeks and moves toward the underlying bone (the femur), eventually attaching to the bone. Within several months this deposit will appear as a spike or horn arising from the bone, gradually assimilating into the bone and disappearing after six months to one year. Some physicians postulate that the periosteum covering the bone is damaged in the initial injury and releases fibroblasts, which retain their embryonic potential to form bone within the damaged muscle and are thus responsible for this unusual calcification.

and endosteum retain their embryonic nature for possible future use in the forming and remodeling of mature bone.

Endochondral Ossification

Endochondral ossification is a slightly more complicated process than intramembranous ossification, but the results are essentially the same. Practically all bones develop by this method (see Figure 6-2). It is easiest to visualize in the development of a long bone of the arm or leg. In this process bones are first formed completely in cartilage. Once formed, the cartilage model resembles the adult structure and may increase in size. During the following discussion frequent reference to Figure 6-5 will be helpful.

During early life *in utero*, the ossification process begins when osteoprogenitor cells in the perichondrium surrounding the **diaphysis** (dī·ăf´ĭ·sĭs), or shaft, of the cartilage model enlarge, become osteoblasts, and begin forming a cylinder of compact bone around the cartilage model (Figure 6-5a–c). Once bone formation commences, the *perichondrium* is referred to as a *periosteum*. At about the same time, the interior of the shaft begins to degenerate and spaces appear within the cartilage. This degeneration seems to occur as calcium deposits accumulate in the tissue and impede the diffusion of nutrients and wastes, resulting in the death of many chondrocytes and a breakdown of the matrix. Blood vessels grow from the periosteum into the newly formed bony cylinder at the diaphysis (Figure 6-5a). They also penetrate the degenerating cartilage, further enlarging the cavities already present. These cavities thus become filled with progressively more blood vessels and embryonic connective tissue cells, some of which become osteoblasts that begin laying down spongy bone. The initial collection of small areas of internal calcification is the **primary ossification center** (Figure 6-5c) and the interconnected spaces are the **primary marrow spaces** (Figure 6-5).

Thus far, the interior of the cartilage model is being degraded and hollowed out and spongy bone is being deposited internally, while a cylinder of compact bone has formed around the diaphysis. These two activities are sufficient to account for growth in diameter of the bone. An additional series of events accomplishes lengthwise growth.

Secondary ossification centers form in the ends, or **epiphyses** (ĕ·pĭf´ĭ·sēs), of a long bone in the same way as the primary center (Figure 6-5e and f). Secondary centers first appear during fetal life in some bones and in early infancy and childhood in others. The activity of secondary ossification centers results in the replacement of cartilage in the epiphyses, with the exception of two areas. First, an

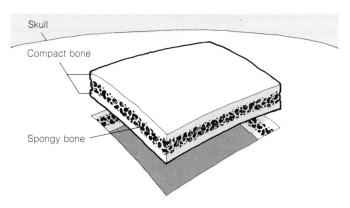

Figure 6-4 Flat bones of the skull have an internal network of spongy bone sandwiched between two layers of compact bone.

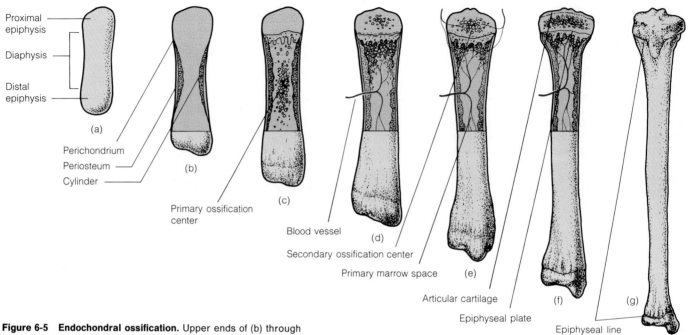

Figure 6-5 Endochondral ossification. Upper ends of (b) through (f) show cut-away views. During this type of bone formation a cartilage model forms first. The model is destroyed from the inside out, while bone matrix replaces the cartilage. (a) Cartilage model. (b) Cylinder formation. (c) Development of primary ossification center. (d) Introduction of blood vessels. (e) Marrow cavity formation along with thickening and lengthening of the cylinder. Appearance of secondary ossification centers. (f) Continued growth of secondary ossification centers. (g) Formation of the epiphyseal lines.

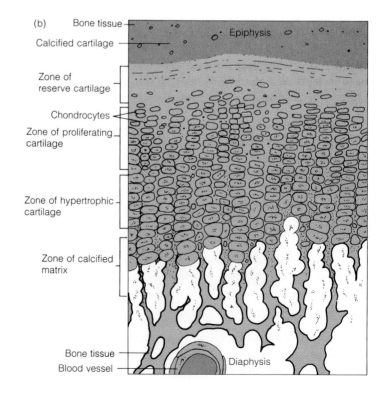

Figure 6-6 Epiphyseal plate. (a) X ray of a child's knee joint showing epiphyseal plate. (b) Enlarged view of section through epiphyseal plate of tibia (articulating surface is up). The plate is characterized by a series of zones that range from a layer of newly formed reserve cartilage on the outer surface, to older hypertrophic cartilage on the inner surface being replaced by bone.

metaphysis: (Gr.) *meta*, beyond + *phyein*, to grow
osteogenesis: (Gr.) *osteon*, bone + *gennan*, to produce
osteoclasis: (Gr.) *osteon*, bone + *klasis*, breaking

> **CLINICAL NOTE**
>
> **EPIPHYSEAL FRACTURES**
>
> The **epiphyseal plate** in children is weaker than the surrounding bone. Fractures can occur easily throughout this weaker area, or the whole epiphyseal plate may slip or be displaced. Some fractures within this actively growing area can result in a cessation of bone growth in the epiphysis and can cause a shortening or deformity in the mature bone. This problem has been of great concern among some physicians especially in light of the recent increased participation of children in contact sports. Statistics at the present time, however, have failed to support this concern.

articular cartilage remains as a protective cap over the articular end of the bone, and second, a transverse plate of cartilage, called the **metaphysis** (mĕ·tăf′ĭ·sis) or **epiphyseal plate,** remains between the epiphysis and diaphysis (Figures 6-5f and 6-6). The epiphyseal plate continues to produce new cartilage along the epiphyseal surface of the plate, while cartilage is destroyed and replaced by bone and marrow along the diaphyseal surface. Thus, the two epiphyseal plates move away from one another toward opposite ends of the bone and the diaphysis lengthens.

The continual production and replacement of epiphyseal cartilage results in lengthwise growth of the bone for the first 16 to 18 years of life. Between the ages of 20 and 25, the epiphyseal cartilages in the long bones are completely replaced by bone and growth ceases. The **epiphyseal line** on adult bones marks the last position of the metaphysis in a long bone (Figures 6-5g and 6-6). The spongy bone originally formed in the diaphysis is gradually removed, producing a **secondary marrow cavity** or **medullary cavity** surrounded by a cylinder of compact bone. In long bones the spongy bone remains only in the epiphyses.

Physiological Maintenance of the Skeleton

After completing this section, you should be able to:

1. Distinguish between osteogenesis and osteoclasis.
2. List three factors that regulate and influence the rate at which bone is formed and destroyed.
3. Discuss the role of osteogenesis and osteoclasis in remodeling the skeleton for changing needs.

Although the specialized and complicated processes of skeletal growth cease for the most part early in adult life, the skeletal system remains a dynamic part of the body, undergoing constant architectural changes and modifications as needed as well as constant slow turnover through reabsorption and replacement of existing bone. Physiologic maintenance of the skeletal system involves a balance between the deposition of bone, **osteogenesis,** and the destruction of bone, **osteoclasis.** Each process is regulated by several factors.

Osteogenesis and Osteoclasis

Osteogenesis is performed by osteoblasts and osteocytes, and this activity is dependent upon three important factors. First, sufficient amounts of calcium salts must be available to the osteoblasts and osteocytes. Osteogenesis begins during fetal life, and it is especially important that a pregnant woman maintain correct nutrition to supply her developing child with enough mineral salts for the osteogenic activities of the osteoblasts to proceed properly. After birth, infants, growing children, and adults to a lesser extent, need a constant supply of calcium salts to ensure the continued formation of their bones.

The second factor influencing osteogenesis is the presence of vitamins A, C, and D. (See Chapter 19 for a more complete discussion of these and other vitamins.) Each appears to function differently, but all three are important. Vitamin D must be present in cells of the intestinal tract for them to absorb calcium present in food. No matter how much calcium is eaten, if the body cells are low in vitamin D, the calcium will simply pass through the intestine and be excreted as waste. Vitamin C promotes the production of collagenous fibers, which form the framework for the deposition of calcium salts in a developing bone. Vitamin A appears to stimulate the activity of the osteoblasts.

The third factor influencing the maintenance of the skeleton is the activity of osteoclasts. **Osteoclasts** (not to be confused with *osteoblasts*) are specialized, multinucleate cells of uncertain origin (Figure 6-7). They may be formed by the fusion of several smaller cells, but their life cycle is poorly understood. Osteoclasts actively degrade bone by secreting enzymes that dissolve the mineral salts and transform them into soluble ions that enter the blood. These ions are important in several metabolic processes, including muscle contraction, nerve impulse transmission, energy transfer reactions involving ADP and ATP, and the maintenance of a proper acid-base balance in the blood. Bone tissue serves as a storage site for these salts.

Osteoclast activity is influenced by the presence of two hormones: calcitonin and parathormone. **Calcitonin** is manufactured in the parafollicular cells of the thyroid gland (see Chapter 17), where it is released into the blood and carried to all parts of the body. One of its effects is to inhibit destruction and reabsorption of the mineral components of

> **CLINICAL NOTE**
>
> ## OSTEOMYELITIS
>
> **Osteomyelitis** (ŏs″tē·ō·mī″ĕl·ī′tĭs) is an infection of bone most commonly caused by the bacteria *Staphylococcus aureus*. On rare occasions other organisms such as fungi, viruses, and rickettsiae can infect bone. Infection within the bone can occur by three main routes: direct trauma (open fracture, bullet wound, and so on), the spread of infection from adjacent tissues, and infection carried by the blood stream. In children the most commonly affected bones are the long bones (femur and tibia), which have a rich vascular marrow. The vertebrae are more commonly affected in adults since the marrow of their long bones has already been replaced by fatty tissue. Treatment includes antibiotic drugs given intravenously and minor surgery to drain the infected area.

> **CLINICAL NOTE**
>
> ## BONE ULCERS
>
> People with poor blood circulation (such as diabetics) and atherosclerosis patients can develop infected **ulcers** that may spread to the bone. In the acute state, there is formation of **edema** (ĕ·dē′mă) (a condition in which the body tissues contain an excessive amount of fluid) within the bone resulting in increased pressure. This pressure decreases the blood supply and leads to abscess formation and bone destruction. Treatment includes bed rest and usually six to eight weeks of intensive antibiotic therapy.

the bony matrix by osteoclasts. Since both reabsorption and deposition of bone are going on continuously, inhibition of reabsorption will cause blood calcium levels to drop as more calcium is held in the mineral deposits of solid bone.

Parathyroid hormone, or **parathormone**, is secreted by the parathyroid glands (see Chapter 17). These glands are sensitive to the level of calcium ions in the blood, and when the blood calcium level drops, they secrete additional amounts of parathormone. Parathormone stimulates osteoclastic activity, which raises the blood calcium level, which in turn decreases the activity of the parathyroid glands. Parathormone also inhibits the kidneys from removing calcium from the blood and excreting it into the urine. Recall that blood calcium regulation is also outlined in Chapter 1 as an example of a negative feedback mechanism.

Remodeling of the Skeletal System

The interplay of osteogenic and osteoclastic activities in the skeletal system results in a constant turnover of bone materials. In fact, it is estimated that the entire skeletal system is replaced approximately 10 times in an individual's lifetime. For the most part, this turnover of bone in the skeletal system involves interior regions not involved in supporting the mass of the body on a daily basis. As previously mentioned, the position and pattern of osteons within the bone produce an internal "grain" that is similar to the grain of a piece of wood (see Figure 6-3). This grain results from the orientation of the internal cells and collagenous fibers, which determines the strength of bone. As an individual grows, osteoclasts gradually degrade bones whose sizes, shapes, and internal grains are appropriate to the size and weight of a child, while osteoblasts lay down collagenous fibers and mineral

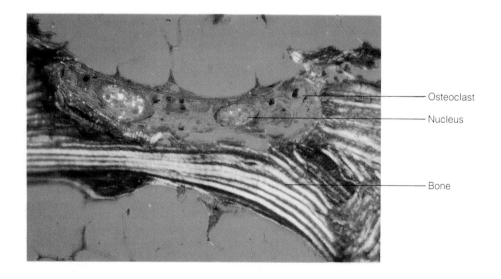

Figure 6-7 Osteoclast. This large sausage-shaped cell resting on the laminated bone tissue is actively dissolving the bone.

salts in new patterns to fit the size, weight, and strength requirements of a larger body.

Two fundamental aspects concern the process of bone remodeling in a growing individual. First, each bone undergoes changes, primarily associated with an increase in size. Second, the entire skeleton undergoes modifications to accommodate the shifting relationship of bones to one another and to meet the new stresses imposed on the body as a whole.

Internal remodeling in a bone begins almost immediately after it is first formed. Spongy bone laid down in the interior of both membranous and endochondral bones is ill-suited to withstand significant stresses, and its replacement begins early in life. This activity is particularly evident in diaphyses. Spongy bone of the central marrow cavity is removed early to accommodate the presence of the marrow tissue, and in a fully formed bone, it is found only at the epiphyses. Growth in diameter of a long bone is characterized by a steady removal of compact osteons from the interior of a diaphysis as new lamellae are added on the outside. This allows an increase in size and strength by spreading the weight borne by the bone over a larger cross-sectional diameter. This type of remodeling is characteristic of young, growing individuals.

Another type of skeletal remodeling related to age and lifestyle occurs throughout the life of an individual. Even after the epiphyseal plate ossifies and lengthwise growth has ceased, a bone may not be able to cope with all the situations that arise in an individual's lifetime. As a person proceeds from early adulthood through middle and old age, changes occur in weight and in the size and strength of muscles. Such changes put different stresses on the skeletal system, and the bones respond by remodeling themselves to meet the new stresses. Osteoclasts dissolve old osteons and osteoblasts build new ones, changing the internal grain of the bone to match the new lines of stress. This kind of activity is magnified in individuals who change their pattern of physical activity. For example, if one begins a program of jogging or weight lifting, stresses produced by the increased use of certain muscles stimulates remodeling activity in the bones. Unfortunately, one of the characteristics of aging is the progressive loss of the body's ability to respond to changing needs. Nevertheless, some remodeling activity continues throughout the life of the individual.

In this chapter we studied the histology and development of cartilage and bone. We also saw how the skeleton undergoes remodeling throughout life to meet different needs at different life stages. In the next chapter we will examine the detailed anatomy of the adult skeleton, emphasizing the anatomical landmarks associated with articulations and muscle attachments.

THE CLINIC

DISEASES OF THE SKELETAL SYSTEM

As you have seen, the skeletal system is not a static repository of minerals, but a dynamic, living, growing system of the body. Like all other major systems of the body, there is a multitude of medical conditions that affect the skeletal system. The following are just a few of the many clinical problems.

Developmental and Hereditary Diseases
Craniofacial Abnormalities Congenital facial deformities occur because of the maldevelopment of the first and second **visceral arches** (remnants of fetal gill-like structures) that form the facial bones and ear parts during fetal development. Two craniofacial birth defects occur when the lip or the palate do not fuse in the midline. These errors in human development are called **cleft lip** and **cleft palate.** Cleft lip is a cosmetic problem that can be repaired surgically before the child is three months old. Cleft palate is a much more serious disorder that interferes with feeding.

Babies with cleft palate have difficulty nursing and sucking, and food may escape through the cleft in the palate into the nose. A cleft palate infant requires a special nipple or feeding device, such as a dropper or spoon. The mother is encouraged to feed the child in this manner immediately after birth so that she can learn to participate. Reconstructive surgery of cleft palate is performed at about 18 months of age before the child speaks.

Surgical correction for a moderate cleft lip or cleft palate is usually successful. The child will have few speech problems and will not suffer much from self-image problems related to distorted facial features. However, a cooperative team of nurses, dentists, surgeons, pediatricians, and speech therapists is necessary for good progress.

Another developmental deformity is called **mandibulofacial dystocia** (măn·dĭb″ū·lō·fā′shĭ·ăl dĭs·tō′sĭ·ă). This results in a small mandible or jaw with external and middle ear abnormalities. As with the cleft palate disorders, surgical correction helps to repair some of the defects.

Abnormalities of the Feet Common developmental foot abnormalities include clubfeet, webbed toes, and pes planus (flat feet) (see Chapter 7 Clinical Note: Abnormalities of the Feet). Many of these conditions can be treated by serial casting. The more generalized defects of bone development often produce serious diseases.

Chondrodystrophies These are diseases involving conversion of cartilage to bone. The best known of this group is **achondroplasia,** which results in dwarfism usually with a normal size trunk and head but with very short extremities. There is no known treatment.

Osteogenesis Imperfecta Also known as brittle bone disease, osteogenesis imperfecta means "poorly developed bones." The victims of this affliction are unfortunate children having a defect in their bone development that results
(continued)

osteomalacia: (Gr.) *osteon*, bone + *malakia*, softening

THE CLINIC (continued)

DISEASES OF THE SKELETAL SYSTEM

in multiple fractures of their bones, even at birth. They suffer many fractures without trauma throughout their lives resulting in many deformities.

Metabolic Diseases of the Skeletal System

Immunologic and allergic diseases do not seem to affect bone. However, there are many metabolic and nutritional diseases of the skeletal system.

Rickets and Osteomalacia These are metabolic diseases caused by a deficiency of vitamin D. Clinical **rickets** in children one to two years of age (Figure 1) is produced by the combination of inadequate exposure to sunlight and a deficiency of vitamin D intake. Vitamin D is necessary for the absorption of calcium in the small intestine. If calcium is not absorbed, blood levels of calcium fall, leading to the withdrawal of calcium from bones so as to maintain blood calcium levels. Infants with rickets are restless and irritable because of bone pain. There is reduced calcification of the skull **(craniotabes),** creating a soft and crackly skull. An enlargement of the epiphyses of the long bones produces a beaded row along the ribs called the **rachitic rosary.** Crawling and walking are delayed, and when they do occur, bowed legs and knocked knees result because of the weakened bones. Occasionally seen in rickets patients is **achondroplasia,** a deformation of the cartilage at the epiphyses of long bones.

Osteomalacia is the counterpart of rickets in adults. Poor calcium absorption leads to decreased calcification of bone and thinning of bony cortices. Thinning and softening of bone may initially cause stiffness and pain. Ultimately, skeletal deformities occur, resulting in bowing and grotesque curvatures of the spine and long bones. Most commonly, the cause is related to intestinal malabsorption of calcium. Diets low in calcium and vitamin D also predispose an adult to this condition. Heavy loss of calcium during pregnancy and lactation frequently lead to the initial symptoms of the disease in susceptible individuals. Both rickets and osteomalacia may be treated by adequate calcium and phophorus intake and supplementary vitamin D intake.

Hypervitaminosis D Excessive amounts of vitamin D can be toxic enough to produce kidney failure, elevated serum calcium, and **metastatic calcifications** (calcium deposits in soft tissues).

There are also many other bone diseases related to changes in calcium or phosphorus metabolism, decreased vitamin C intake, and hyper- or hypoparathyroidism.

Tumors of Bone

Bone tumors are usually very painful since they expand in a closed space. They also weaken the bone and are responsible for nontraumatic (pathologic) fractures. There are many varieties of bone tumors. Fortunately, they occur rarely.

Osteochondroma The most common benign bone tumor is **osteochondroma.** On X rays these are the peculiar looking hornlike projections (exostoses) from a bone that have cartilagenous caps and look like mushrooms. They occur most frequently in adolescents aged 10 to 20. The exostoses may be singular or multiple. There is a strong familial trait in the multiple form, and about 10% undergo malignant change.

Malignant Myeloma Malignant bone tumors may be primary or metastatic (moved from one site to another) from another source, such as cancer of the breast or prostate gland. The most common primary bone tumor is **malignant myeloma,** which is in reality a tumor of the bone marrow. This disease tends to be multicentric (begins in multiple sites) and frequently will completely fill the marrow cavity. Diagnosis is by a bone marrow biopsy, a simple surgical procedure performed by pushing a special needle through the cortex of the bone into the marrow cavity to aspirate the marrow cells, which can then be examined like a blood smear under the microscope.

Osteogenic Sarcoma **Osteogenic sarcoma** is the second most common malignant bone tumor. It is also most common in adolescents aged 10 to 20. About half of these tumors occur around the knee. The five year survival rate is low, about 20%. Increasing evidence claims that the survival rate is dependent upon host tumor factors. Immunotherapy may help to increase the survival rate in the future.

Figure 1.   Rickets in a child, caused by vitamin D deficiencies. Note the extreme bowleggedness.

osteoporosis: (Gr.) *osteon*, bone + *poros*, passage + *osis*, condition

Aging
The aging process results in many changes to the skeletal system, such as wear and tear (degenerative arthritis), shortening of stature, and a flexed posture.

Osteoporosis Osteoporosis is an absolute decrease in total bone mass usually affecting postmenopausal women and the elderly (Figure 2). Older people in the United States suffer 180,000 to 200,000 hip fractures per year; 80% of them are related to osteoporosis. Mortality rates from hip fractures during the year following the fracture run 15 to 30% (27,000 to 60,000). Many of the survivors never regain independent living status. Osteoporosis also results in frequent wrist fractures and compression fractures of the spine. Osteoporosis-related fractures are estimated to cost 3.8 billion dollars annually.

Osteoporosis represents an imbalance between bone formation and bone resorption. Normally, bone resorption is balanced by bone formation. Old age, failure of the gonads (e.g., loss of estrogen after menopause in women), excessive levels of endogenous or exogenous cortisone, and poor calcium and vitamin D intake all cause an excess of bone resorption over formation, eventually resulting in osteoporosis. Also, when bones bear no weight, such as occurs when a person is immobilized by sickness, accident, or disability or simply leads a sedentary lifestyle, then bone resorption exceeds absorption and osteoporosis can result. Osteoporosis occurs primarily in trabecular bone, but long bones such as the femur are also affected.

Vertebral collapse as a result of osteoporosis can occur in the elderly and is accompanied by acute back pain following sudden, and perhaps trivial, bending or lifting movements. Several such events may lead to curvature of the spine or **kyphosis** or **dowager's hump** (Figure 2b). Kyphosis can compromise the elderly person's respiratory ability and may cause compression of nerves in the lumbar area.

At the present time we do not have a simple method for determining bone density and identifying osteoporosis early. Thus, current treatment of osteoporosis consists of estrogen therapy in postmenopausal women to limit bone resorption, vitamin D to promote calcium absorption from the gastrointestinal tract, and extra calcium included in the diet or taken as a supplement. However, new studies show that there may be no relationship between calcium intake and bone density within populations. This is a consensus of a National Institutes of Health panel, despite the popular press, which has touted the effectiveness of calcium supplements consumption. Consequently, the sales of calcium products have increased explosively. Studies have shown that weight-bearing exercise can slow bone loss. Also, one can decrease the risk factors for osteoporosis by not smoking and by limiting alcoholic intake. Prevention at an early age is the first line of defense against osteoporosis.

Osteitis deformans (Paget's disease) A slowly progressive bone disorder, **osteitis deformans** is characterized by an osteolytic phase followed by an osteoblastic phase that results in gross deformity of the bones. It is found in about 10% of people in their 80s and is more common in males. The disease usually affects the skull, pelvis, spine, femur, or tibia and produces deformity, bone pain, and pathologic fractures. The cause is as yet unknown.

Figure 2 Osteoporosis. (a) Photograph of a bone affected by osteoporosis; note the high degree of porosity. (b) Woman suffering from osteoporosis exhibiting dowager's hump. Note the X ray of her vertebrae.

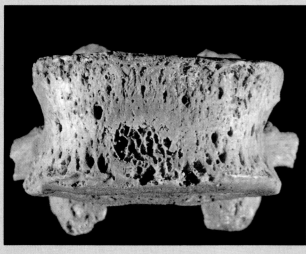
(a)

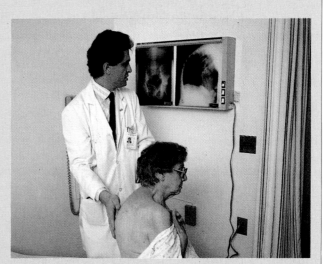
(b)

STUDY OUTLINE

Functions of the Skeletal System (p. 115)
The skeleton has five major functions in the human body: support, protection, formation of blood cells (*hemopoiesis*), mineral storage, and body movement.

Histology of Bone (pp. 115–118)
Bone is a specialized connective tissue. It contains living *osteocytes* surrounded by a matrix of collagen fibers that are coated with salts, chiefly calcium phosphate (85%) and calcium carbonate (10%).

Classification Based on External Appearance Based on external appearance, bones can be classified into five categories: flat (e.g., ribs), long (e.g., femur), short (e.g., wrist), irregular (e.g., vertebrae), and sesamoid (e.g., patella).

Classification Based on Internal Structure Based on internal structure, bone can be classified as spongy or compact. *Spongy* (*trabecular*) bone is found in the interior of flat bones and at the ends of long bones, and it is composed of a network of branching interconnected sheets and bars (*trabeculae*). Interconnected spaces are filled with a vascular tissue (*marrow*) and lined with *endosteum*, a thin membrane connected to the *periosteum*. *Compact* (*lamellar*) bone forms the exterior of bones and the shaft of long bones. It is composed of concentric layers (*lamellae*) of matrix deposited mostly around blood vessels that pass through the bone.

Development of the Skeleton (pp. 119–123)
Development of Cartilage Cartilage develops when young cartilage cells called *chondroblasts* produce collagenous fibers along with *chondrin*, the rubbery ground substance of cartilage. *Appositional growth* occurs at the surface of a cartilage structure as new chondroblasts are produced by the surrounding *perichondrium*. *Interstitial growth* occurs when the interior of a cartilage structure is destroyed and new chondrin is produced by chondroblasts.

Development of Bone Bone may develop directly from a sheet of mesodermal mesenchyme (intramembranous ossification) or by replacement of previously formed cartilage tissue (endochondral ossification).

During *intramembranous ossification*, mesenchymal cells differentiate into *osteoblasts*, enlarge, and begin to secrete collagen fibers, *glycosaminoglycans*, and *alkaline phosphatase*. Dissolved minerals in the interstitial fluid precipitate out of solution around the collagen fibers and a solid bone forms. Osteoblasts continue to be produced by *osteoprogenitor cells* on the inner periosteum and in the endosteum. The clavicle and flat bones of the skull form in this manner.

During *endochondral ossification*, a preformed cartilage model resembling the adult bone is replaced by bone. First a cylinder of bone is formed around the exterior of the bone. The interior of the cartilage model disintegrates and osteoblasts invade the structure and begin laying down bone, forming a *primary ossification center*. Secondary ossification centers form in the ends of long bones and at the edges of irregular bones. An *epiphyseal plate* remains between the ossification centers, and continued enlargement of the bone involves production of cartilage along the exterior surface of the plate, while cartilage is destroyed and replaced by bone at the interior surface of the plate.

Physiological Maintenance of the Skeleton (pp. 123–125)
Osteogenesis and Osteoclasis Physiological maintenance of the skeleton involves a balance between the deposition of bone, *osteogenesis*, and the destruction of bone, *osteoclasis*. Each process is regulated by several factors, including access to sufficient amounts of calcium salts and the presence of vitamins A, C, and D. *Calcitonin* is a thyroid hormone that inhibits the activity of *osteoclasts*, while *parathormone* stimulates osteoclastic activity.

Remodeling of the Skeletal System Remodeling of bone occurs throughout life. Each bone undergoes changes associated with an increase in size during the growth years and as one ages and changes lifestyle.

SELF-TEST OF CHAPTER OBJECTIVES

True-False Questions
1. Osteoclasis is the destruction of bone, whereas osteogenesis is the buildup of bone.
2. Parathyroid hormone stimulates osteoclastic activity in the breakdown of bone.
3. Chondrocytes are actively involved in secreting a chondrin matrix in the developing embryo.
4. During intramembranous ossification, osteoblasts may secrete the mineral matrix of bone in the form of concentric compact layers.
5. Cartilage possesses a more extensive blood supply than does bone.
6. Hemapoiesis is a function of the cartilaginous parts of the skeletal system.
7. One of the effects of the hormone calcitonin is to inhibit the activities of osteoclasts.
8. Mineral storage is one of the chief functions of bone.
9. Compact bone is composed primarily of trabeculae, and spongy bone is composed primarily of lamellae.
10. Osteonal canals are characteristic of spongy bone rather than compact bone.

Matching Questions
Match the cells on the left with appropriate tissues on the right:

11. osteoclasts
12. fibroblasts   a. bone
13. chondroclasts   b. cartilage
14. osteocytes   c. mesenchyme
15. osteoblasts

Match the cells on the right with the statements on the left:
16. actively absorb bone
17. actively produce bone
18. are stimulated by parathormone
19. are abundant in fetal bones
20. large, multinucleate cells

a. osteoblasts
b. osteoclasts

Multiple-Choice Questions
21. Which of the following vitamins does not seem essential for the formation of bone?
    a. vitamin A   c. vitamin C
    b. vitamin B   d. vitamin D
22. Which of the following vitamins promotes the absorption of calcium in the intestinal tract?
    a. vitamin A   c. vitamin C
    b. vitamin B   d. vitamin D
23. The hormone that stimulates osteoblasts and osteocytes to remove calcium ions from the blood and to secrete the ions as the mineral matrix of bone is
    a. calcitonin   c. vitamin A
    b. parathyroid hormone   d. vitamin C
24. Which of the following is a function of bone and not cartilage?
    a. movement   c. mineral storage
    b. protection   d. support
25. The tissue that surrounds a cartilaginous structure is called
    a. mesenchyme   c. elastic cartilage
    b. perichondrium d. periosteum
26. The tissue that surrounds a bony structure is called
    a. mesenchyme   c. elastic cartilage
    b. perichondrium d. periosteum
27. Cells found on the interior of the perichondrium that become osteoblasts are
    a. osteoprogenitor cells   c. chondroblasts
    b. chondroclasts   d. osteocytes
28. Vitamin D
    a. promotes osteoclast activity
    b. inhibits calcium absorption in the intestine
    c. stimulates calcium absorption in the intestine
    d. stimulates osteoblast activity
29. The end of a long bone is called the
    a. diaphysis   c. epiphyseal plate
    b. epiphysis   d. diaphyseal plate
30. The central shaft of a long bone is called the
    a. epiphysis   c. diaphyseal plate
    b. epiphyseal plate   d. diaphysis

Essay Questions
31. Imagine that the human skeleton is made entirely of cartilage and that it contains no bone. What functions could not be performed by this cartilage skeleton that are presently performed by our bony one?
32. Discuss the differences between osteoblasts and osteoclasts.
33. Describe the development of bone and differentiate between intramembranous and endochondral bone formation.
34. Discuss physiological factors that influence and control osteogenesis and osteoclasis.
35. Why is the relationship between calcium deposition and parathormone secretion called a negative feedback mechanism?
36. Discuss some of the reasons why the skeleton undergoes constant remodeling.

Clinical Application Questions
37. Describe the process of arthroscopy. For what types of injuries is it used?
38. Distinguish between a closed fracture and a compound fracture of a bone.
39. Outline the steps involved in the healing of a bone fracture.
40. Why does injured bone heal faster and more completely than injured cartilage?
41. Compare and contrast osteomyelitis and osteomalacia.
42. Describe osteoporosis. What are the methods currently used to avoid this condition?

7
Anatomy of the Skeletal System

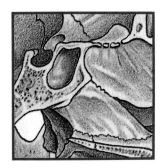

In the preceding chapter the embryology, physiology, and histology of the skeletal system were studied. In this chapter the detailed anatomy of the bones in the fully developed adult will be examined. We will see that anatomical features of the skeleton are associated primarily with two of its major functions: protection of vital organs (such as the brain, spinal cord, thoracic, and pelvic viscera) and support of the body in an upright position. Many anatomical landmarks also serve as places of attachment for muscles and ligaments and as channels for blood vessels and nerves. In this chapter we will consider

1. the differences between the axial and appendicular skeleton,
2. the features of the skull and the differences between the cranial and facial groups of bones,
3. the bones of the vertebral column,
4. the pectoral and pelvic girdles, their individual components, and how they function in the skeleton, and
5. the bones of the upper and lower extremities.

Divisions of the Skeleton

Bones of the skeletal system can be generally divided into several groups, depending on various criteria. For example, we can distinguish between flat bones and long bones, and bones of intramembranous ossification or bones of endochondral ossification. Anatomically, the 206 bones of the body are divisible into the axial skeleton and the appendicular skeleton (Figure 7-1 and Table 7-1).

The **axial skeleton,** so named because it forms the major axis of the body, consists of the skull, hyoid bone (at the base of the tongue), vertebral column, and rib cage. The **appendicular** (ăp″ĕn·dĭk′u·lăr) **skeleton** consists of the bones of the appendages, including the upper extremity (arm and forearm) and the lower extremity (thigh and leg), and their points of attachment to the axial skeleton. The upper extremities are attached to the upper part of the rib cage by the shoulder or **pectoral girdle,** and the lower extremity is attached to the vertebral column by the hip or **pelvic girdle.** Table 7-1 is a list of the specific bones within each division of the skeleton. Figure 7-1 shows the anatomical relationships of the bones to one another.

Table 7-2 is a list of the terms most frequently used in describing surface features and anatomical landmarks on bones.

Skull

After completing this section, you should be able to:

1. List the 28 bones in the skull.
2. Distinguish between the two groups of bones, cranial and facial, that make up the skull.
3. Identify the major sutures of the skull and name the bones they lie between.
4. Define wormian bones and indicate their usual location in the skull.

The **skull** consists of 28 bones that are arranged into a cranial group and a facial group. The **cranial group** contains the paired parietals and temporals and the solitary frontal, occipital, ethmoid, and sphenoid bones. These bones serve primarily to surround and protect the soft tissues of the brain. The tiny bones (ear ossicles) that transfer the vibrations of sound waves from the tympanum to the inner ear are discussed in Chapter 16. The 14 bones of the **facial group** include the nasals, maxillae, lacrimals, inferior nasal conchae, vomer, palatines, zygomatics, and mandible. The facial group is involved primarily with the protection and

Table 7-1 BONES OF THE BODY

Axial Skeleton	Number of Bones	Appendicular Skeleton	Number of Bones
Skull		Pectoral girdle	
Frontal	1	Clavicle	2
Parietals	2	Scapula	2
Temporals	2	Upper extremity	
Occipital	1	Humerus	2
Sphenoid	1	Ulna	2
Ethmoid	1	Radius	2
Nasals	2	Carpals	16
Maxillae	2	Metacarpals	10
Lacrimals	2	Phalanges	28
Inferior nasal		Pelvic girdle	
conchae	2	Os coxae	2
Vomer	1	Lower extremity	
Palatines	2	Femur	2
Zygomatics	2	Patella	2
Auditory		Tibia	2
ossicles*	6	Fibula	2
Mandible	1	Tarsals	14
Hyoid	1	Metatarsals	10
Vertebrae	25–26	Phalanges	28
Sternum	1		
Ribs	24		

Total number of bones = 80 + 126 = 206

* Auditory ossicles are discussed in Chapter 16.

Table 7-2 TERMS USED TO DESCRIBE SURFACE FEATURES OF THE SKELETON

Term	Description
Projections from a Bone	
Crest	A bony ridge
Condyle	A rounded, bony projection possessing a smooth surface for articulating with another bone
Epicondyle	A roughened projection near a condyle that is a place of attachment for tendons and ligaments
Head	The enlarged, often rounded end of a bone
Line	A low, bony ridge
Process	A generalized term designating a wide range of projections
Ramus	An elongate, sometimes columnar process
Shaft	The elongate middle portion of a long bone
Spine	A projection, often narrow and pointed
Trochanter	A large, blunt process
Tubercle	A small, usually rounded process
Tuberosity	A large process
Indentations into the Surface of a Bone	
Facet	A smooth, flat surface for articulation, usually rounded or oval
Fissure	A narrow, slitlike opening into or through a bone
Foramen	A hole in a bone
Fossa	A depression in a bone
Fovea	A shallow depression in a bone
Meatus	A deep canal penetrating a bone
Neck	A shallow groove separating the head from the shaft of a long bone
Sulcus	A narrow grove in a bone

132 Chapter 7 Anatomy of the Skeletal System

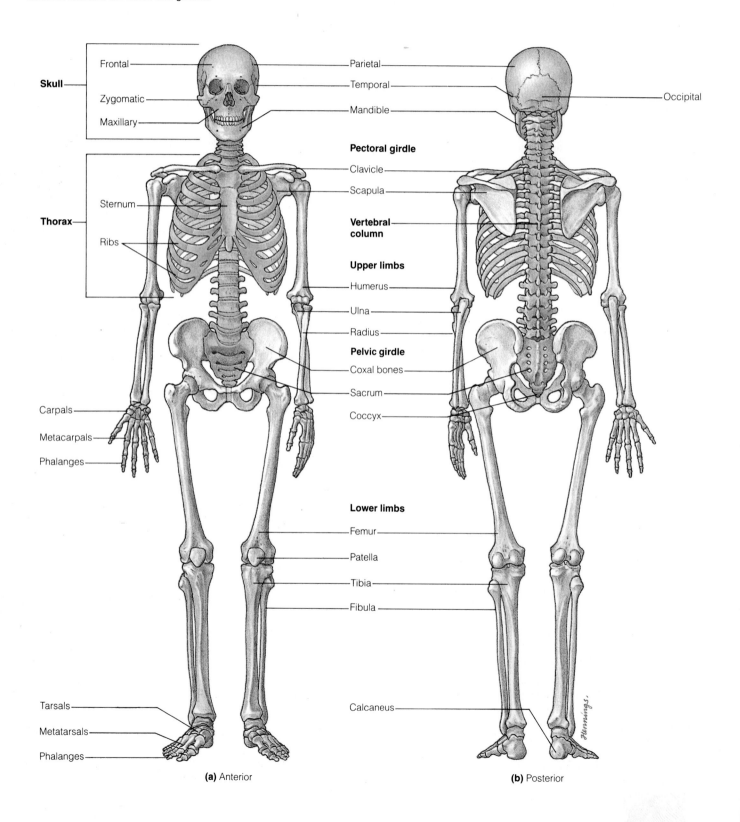

Figure 7-1 These anterior and posterior views of the skeleton show the two skeletal components: the axial skeleton forms the major axis of the body (highlighted), and the appendicular skeleton includes the limbs and girdles.

craniosynostosis:	(Gr.) *kranion*, skull + *syn*, together + *osteon*, bone + *osis*, condition	glabella:	(L.) *glaber*, bald
fontanel:	(Fr.) *fontanelle*, little fountain	parietal:	(L.) *paries*, wall

> **CLINICAL NOTE**
>
> **CRANIOSYNOSTOSIS**
>
> An abnormal condition that results in a premature fusion of the cranial bones is called **craniosynostosis** (krā″nē·ō·sĭn″ŏs·tō′sĭs). Normally, the growth of the developing brain tends to keep the sutures open and the bones discrete. In craniosynostosis, however, a spontaneous premature closure occurs involving part or all of the sutures. Brain growth becomes retarded by this constriction and the skull may become deformed. Neurosurgical procedures to separate the skull bones have been successful in allowing normal development.

functions of the eyes and nose and the functions of the mouth. The hyoid bone found inside the region of the mandible, at the base of the tongue, makes 29 bones total.

Cranial Group

During embryonic and fetal life, the skull bones in the cranial group are only loosely attached to one another (see Clinical Note: Craniosynostosis). This is especially true for those bones that form the **calvarium** (kăl·vā′rē·ŭm) (roof of the skull). The large membranous openings present between the cranial bones of the fetus are referred to as **fontanels** (fŏn″tă·nĕls′) (Figure 7-2), and they permit these bones to move and adjust their position, facilitating the passage of the infant's head through the birth canal. The bones of the skull develop individually and are held together by immovable joints called **sutures**.

The transition from a fetal and infant skull to that of an adult skull involves growth and fusion of the cranial bones into a solid unit that provides protection for the brain. The fontanels are normally filled in by age two years. It takes several additional years for the sutures to fully form as the skull grows to accommodate growth of the brain.

Several **foramina** (fōr·ă′mĭ·nă), holes that pierce the skull allow passage of blood vessels and nerves. Table 7-3 is a summary review of these foramina, their locations, and the structures passing through them.

Frontal

The **frontal** is a flat bone that forms the forehead, the roof of the **orbits** (eye sockets), and the anterior boundary of the cranial cavity (Figure 7-3). It forms from two ossification centers in the developing fetus that begin to fuse shortly after birth. Fusion is usually complete by the sixth year of life. The central flattened region of the frontal bone is called the **frontal squama** (skwā′mă). Just inferior to the squama are two horizontally directed brow ridges known as the **superciliary arches** or **supraorbital ridges.** The eyebrows grow from the skin covering these two ridges. Between the superciliary arches is a flattened and smooth depression called the **glabella**. A **supraorbital foramen** pierces the frontal bone above each orbit (Figure 7-4) and provides a passageway for blood vessels and nerves passing to the skin above the eyelids. The supraorbital foramen sometimes forms so close to the inferior edge of the frontal that it develops as a notch instead of a completely enclosed hole. The inferior projections at the sides of the frontal are the **zygomatic** (zī″gō·mă′tĭk) **processes** (Figure 7-5), so named because they make contact with the zygomatic bones. Two cavities, the **frontal sinuses,** lie within the frontal bone (see Figure 7-13).

Parietals

The **parietals** (pă·rī′ĕ·tăls) are paired flat bones just posterior to the frontal bone. Together they form most of the roof and sides of the skull (Figure 7-5). Each parietal bone has only two noteworthy landmarks: **superior** and **inferior temporal lines,** which are low ridges on the side of the head used as attachment points for a large chewing muscle, the temporalis.

Temporals

The paired **temporal** bones lie inferior to the parietal bones at the lateral borders of the skull (Figures 7-5 and 7-6). Each

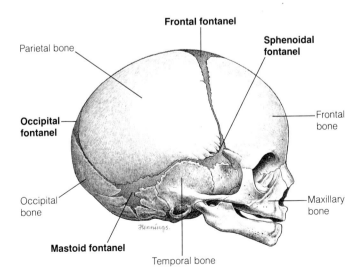

Figure 7-2 Lateral view of a fetal skull. Note the presence of large fontanels between the cranial bones.

petrous: (Gr.) *petros*, stone
mastoid: (Gr.) *mastos*, breast + *eidos*, form, shape
curettage: (Fr.) *curetage*, scraping

consists of several regions. The **squamous portion** is the flattened superior region that makes broad contact with the parietal. The zygomatic process projects anteriorly from near the midpoint of the lateral surface of the temporal and makes contact with the zygomatic bone. The cheekbone is referred to as the **zygomatic arch** and includes the zygomatic process of the temporal bone and the temporal process of the zygomatic bone. At the base and inferior surface of each zygomatic process, a shallow depression, the **mandibular fossa** (fŏs´ă) provides a point of articulation for the mandible. Another distinct portion of the temporal bone is the wedge-shaped **petrous** (pĕt´rŭs) **portion** (see Figure 7-8,) which extends internally from the squamous portion of the temporal bone into the cranial cavity. The petrous portion contains the organ for hearing (cochlear apparatus) and the balance organs (vestibular apparatus) (see Chapter 16). The **mastoid portion**, which contains the **mastoid sinuses** or **air cells** and a prominent **mastoid process**, forms the inferior and posterior part of the temporal bone. In lateral view the large **external acoustic** (or **auditory**) **meatus** is visible (Figures 7-5 and 7-6). This canal provides access to the tympanic membrane, which vibrates in response to sound waves. The inferiorly directed **styloid process** is also visible in this same view. Both the mastoid process and the styloid process are points for muscle attachments (see Chapter 10).

CLINICAL NOTE

MASTOIDITIS

Acute mastoiditis has historically been a very serious infection of the mastoid air cells of the temporal bone. Mastoid air cells or sinuses do not drain as do the paranasal sinuses. The infection is transmitted from middle ear infections, but thanks to the early use of antibiotics in the treatment of these infections, we seldom see cases of mastoiditis today. Many serious complications of mastoiditis once occurred, including erosion through the bone into the skull, resulting in meningitis. The treatment was a surgical drainage of the infection through the skin behind the ear and scraping out **(curettage)** all the air cells. This treatment has now been replaced by antibiotic therapy.

Table 7-3 A SUMMARY OF THE MAJOR FORAMINA OF THE SKULL

Name	Location	Structures Passing Through
Carotid canal	Petrous portion of temporal	Internal carotid artery
Foramen lacerum	Between sphenoid, temporal, and occipital bones	Branch of ascending pharyngeal artery
Foramen magnum	Occipital bone	Spinal roots of the accessory nerve (CN XI)
Foramen ovale	Greater wing of sphenoid	Mandibular branch of trigeminal nerve (CN V)
Foramen rotundum	Greater wing of sphenoid	Maxillary branch of trigeminal nerve (CN V)
Foramen spinosum	Posterior angle of sphenoid	Middle meningeal blood vessels
Greater palatine foramen	Posterior edge of palatine	Greater palatine nerve and descending palatine blood vessels
Hypoglossal canal	Next to occipital condyles	Hypoglossal nerve (CN XII) and ascending pharyngeal artery
Incisive foramen	Palatine process of maxilla, posterior to incisor teeth	Nasopalatine nerve and descending palatine blood vessels
Inferior orbital fissure	Within orbit, between maxilla and greater wing of sphenoid	Maxillary branch of trigeminal nerve (CN V), and infraorbital blood vessels
Infraorbital foramen	Maxilla, just below orbit	Infraorbital nerves and blood vessels
Internal acoustic meatus	Petrous portion of temporal	Facial (CN VII) and vestibulocochlear (CN VIII) nerves and labyrinthine artery
Jugular foramen	Between occipital and petrous portion of temporal bone	Internal jugular vein and glossopharyngeal (CN IX), vagus (CN X), and accessory (CN XI) nerves
Lacrimal canal	Lacrimal bone	Lacrimal duct
Lesser palatine foramen	Posterior edge of palatine	Lesser palatine nerves
Mandibular foramen	Medial surface of ramus of mandible	Inferior alveolar nerves and blood vessels
Mastoid foramen	Mastoid process of temporal bone	Blood vessels
Mental foramen	Anterior, external surface of mandible	Mental nerves and blood vessels
Olfactory foramen	Cribriform plate of ethmoid	Olfactory nerve (CN I)
Optic canal	Lesser wing of sphenoid	Optic nerve (CN II) and ophthalmic branch of internal carotid artery
Stylomastoid foramen	Base of temporal bone	Facial nerve (CN VII) and stylomastoid artery
Superior orbital fissure	Between greater and lesser wings of sphenoid	Occulomotor (CN III), trochlear (CN IV), ophthalmic branch of trigeminal (CN V), and abducens (CN VI) nerves and ophthalmic vein
Supraorbital foramen	Frontal bone, above orbit	Supraorbital nerve and blood vessels
Zygomaticofacial foramen	Zygomatic bone	Zygomaticofacial nerves and blood vessels

Skull   135

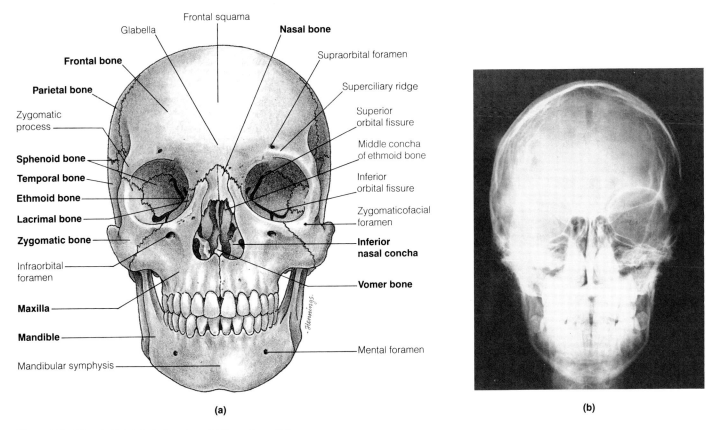

Figure 7-3 The skull. (a) Anterior view. (b) X ray from the same perspective.

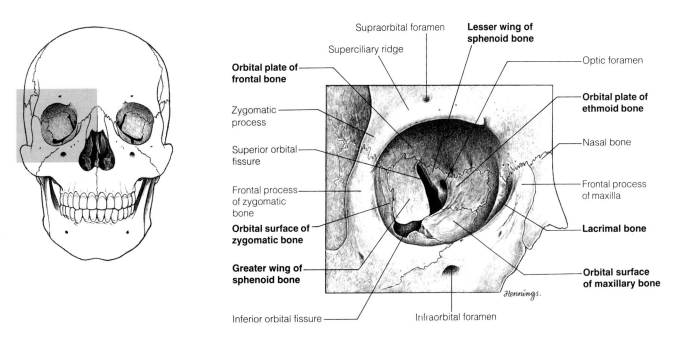

Figure 7-4 Anterior view of the orbital bones.

occipital: (L.) *occiput*, back of the head

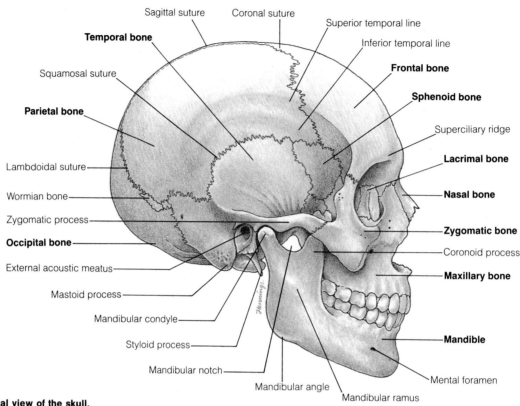

Figure 7-5 Lateral view of the skull.

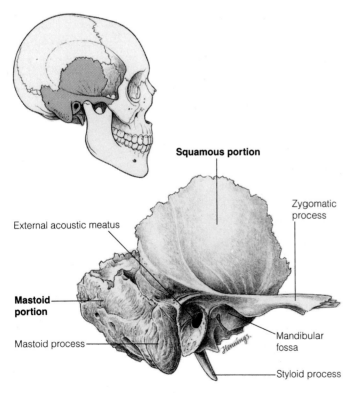

Figure 7-6 Lateral view of the temporal bone.

An inferior view of the temporal bone reveals three of the foramina that pierce this bone (Figure 7-7). The more anterior **carotid foramen** (or **canal**) provides an opening to a passageway for the internal carotid artery, which brings blood to the brain. The **jugular foramen** allows passage of the internal jugular vein, which drains blood from the brain, and three of the twelve large cranial nerves: the glossopharyngeal (cranial nerve IX), the vagus (CN X), and the spinal accessory (CN XI). (The cranial nerves and their functions are discussed in detail in Chapter 13.) The **stylomastoid** (stī″lō·măs′toyd) **foramen** is located immediately posterior to the styloid process and provides passage for the stylomastoid artery and CN VII, the facial nerve. Internal views of the temporal bone (Figures 7-8 and 7-9) reveal the **internal acoustic** (or **auditory**) **meatus**, through which pass the facial and auditory nerves and the labyrinthine artery.

Occipital

The **occipital** (ŏk·sĭp′ĭ·tăl) **bone** forms the posterior base of the skull (Figures 7-8 and 7-9). It is roughly oval and contains a large **foramen magnum**, through which the spinal cord passes. Posterior to the foramen magnum is a large flattened **squama**, and anteriorly is the **basilar portion** of

Skull 137

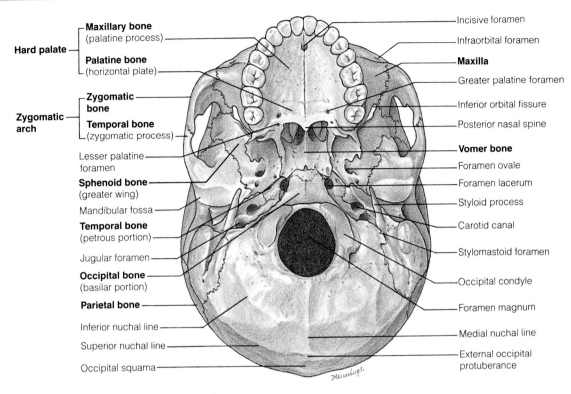

Figure 7-7 Inferior view of the skull.

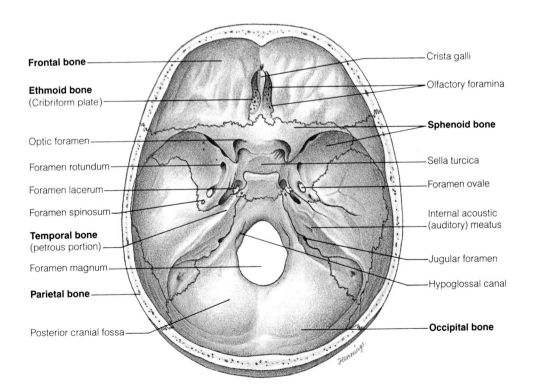

Figure 7-8 Superior view of the floor of the cranial cavity.

hypoglossal: (Gr.) *hypo*, under + *glossa*, tongue
sphenoid: (Gr.) *sphen*, wedge + *eidos*, form, shape
pterygoid: (Gr.) *pterygion*, wing + *eidos*, form, shape
sella turcica: (L.) *sella*, saddle + *turcica*, Turkish

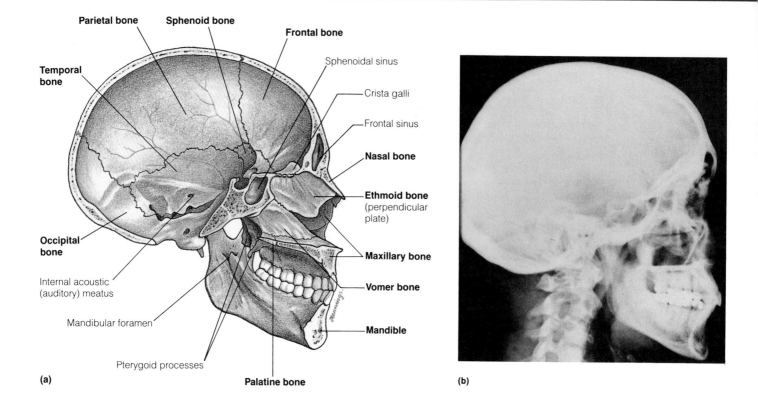

Figure 7-9 The skull. (a) Sagittal section. (b) X ray from the same perspective.

the occipital. On either side of the foramen magnum are two **lateral portions** whose chief anatomical features are the large **occipital condyles** (kŏn′dĭls) (Figure 7-7), which provide the only articulation (joint) between the skull and the vertebral column. The squama is marked by **superior, inferior,** and **medial nuchal** (nŭ′kăl) **lines,** which are slightly raised ridges that are points of attachment for certain neck muscles, and a prominent **external occipital protuberance** that serves the same function. In addition to the large foramen magnum, two smaller holes pierce the occipital bone. These are the openings to the **hypoglossal** (hī″pō·glŏs′ăl) **canals,** which allow passage of the hypoglossal cranial nerve (CN XII).

Sphenoid

The **sphenoid** (sfē′noyd) bone is wedged into the base of the cranial region (Figures 7-8 and 7-10). This bone makes contact with all the skull bones described thus far. When viewed either from above or below, it superficially resembles a butterfly or bat in flight. The central, roughly rectangular body of the sphenoid bears four pairs of winglike projections. The smaller, anterior pair are the **lesser wings,** and the larger, posterior pair are the **greater wings.** Together, all the wings form much of the central floor of the cranial cavity. An additional pair of **pterygoid** (tĕr′ĭ·goyd) **processes** are suspended from both sides of the inferior part of the sphenoid body (Figure 7-10b). These processes form part of the posterior walls of the nasal cavities, and they are attachment points for muscles used in chewing. The superior side of the body of the sphenoid bears a deep depression, the **sella turcica** (sĕl′ă tŭr′sĭ·kă) ("Turkish saddle"), which surrounds the pituitary gland. Several holes pierce the sphenoid bone; the largest of these is the **superior orbital fissure,** which allows passage of the cranial nerves that control eye movement. Medial to the elongate superior orbital fissure is the **optic foramen** through which the optic nerve passes. Three foramina are found at the base of each greater wing. In order, anterior to posterior, they are the **foramen rotundum** and **foramen ovale,** which allow passage of branches of one of the cranial nerves, and the **foramen spinosum,** which allows passage of blood vessels (Figure 7-10). The **foramen lacerum** (Figure 7-8) lies just posterior to the sphenoid and allows passage of a branch of the internal carotid artery.

ethmoid: (Gr.) *ethmos*, sieve + *edios*, form, shape
cribriform: (L.) *cribrum*, sieve + *forma*, shape
sinus: (L.) *sinus*, curve, hollow

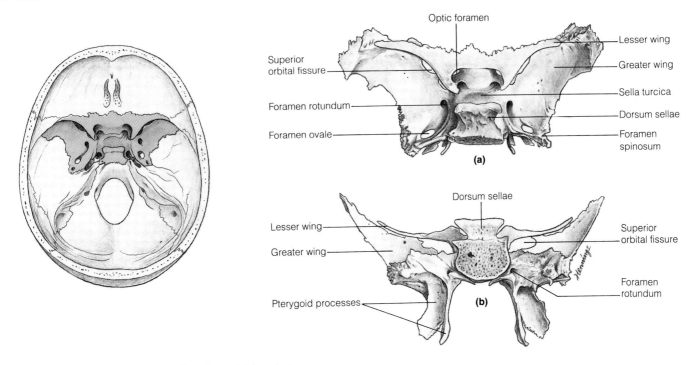

Figure 7-10 The sphenoid bone. (a) Superior and (b) posterior views.

Ethmoid

The **ethmoid** (Figures 7-3 and 7-11) is the chief bone of the nasal region. It consists of four principal parts: a **horizontal** or **cribriform** (krĭb´rĭ·form) **plate,** a medial **perpendicular plate,** and two **lateral masses.** The cribriform plate develops between the embryonic frontal bones and helps to form the anterior floor of the cranial cavity. It also forms the roof of the nasal cavity. The cribriform plate is pierced by many foramina through which pass the endings of the olfactory nerves, which determine our sense of smell. A small triangular process, called the **crista galli** (krĭs´tă găl´e), projects from the middle of the cribriform plate. The perpendicular plate hangs vertically from the cribriform plate and, along with the vomer, forms the **nasal septum** which divides the nasal cavity into two compartments (Figure 7-9). Deviation of the nasal septum can cause nasal congestion leading to infection and inflammation.

The lateral masses contain several **sinuses** and, consequently, are spongy in appearance. They project from the lateral edges of the cribriform plate and help to form much

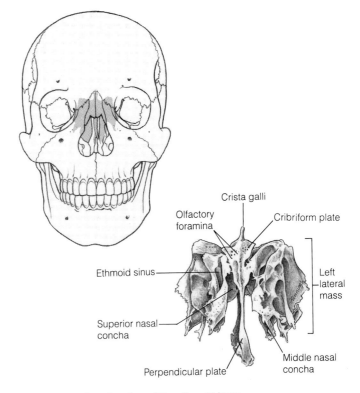

Figure 7-11 Anterior view of the ethmoid bone.

vomer: (L.) *vomer*, plowshare

> **CLINICAL NOTE**
>
> **FRACTURE OF THE CRIBRIFORM PLATE**
>
> The **cribriform plate** is a very thin bone that can be subject to fracture when there is severe injury to the face and nose, as can happen in an automobile accident. This injury can result in leakage of **cerebrospinal fluid** from the nose and a retrograde infection in the skull, which can lead in turn to meningitis. A fracture through the cribriform plate can also result in damage to the olfactory nerves and can cause permanent loss of smell (**anosmia**).

> **CLINICAL NOTE**
>
> **FRACTURE OF THE NOSE**
>
> A fracture of the nasal bones due to a direct blow to the front of the face is a common injury. This painful injury requires little treatment if the bones are not severely displaced. Displaced fractures must be **reduced** (restored to their original place) for cosmetic purposes and to restore the open nasal passages. A serious complication is a fracture or displacement of the nasal septum; this requires surgical correction.

---

of the medial walls of the orbits. The interior surfaces of the lateral masses are developed into two curved projections—**superior** and **middle nasal conchae** (kŏng′kā)—which extend into the nasal cavity and provide an underlying foundation for the mucous membrane covering them, which cleans, warms, and moistens air passing to the lungs.

Facial Group

There are almost twice as many facial bones as cranial bones. The facial group of skull bones is positioned at the front of the skull and is involved with the sense organs found there. Facial bones house and protect the soft organs that accomplish vision, smell, and taste.

Inferior Nasal Conchae

The **inferior nasal conchae** are paired, spiral-shaped bones on the inferior, lateral walls of the nasal cavities, attached to the maxillae (Figure 7-3). They provide the same function as the superior and middle conchae.

Vomer

The **vomer** is a thin, triangular bone forming the lower part of the nasal septum (Figures 7-7 and 7-9). This bone was named for the Latin "plowshare" because its shape resembles a plow. The vomer articulates with the sphenoid and perpendicular plate of the ethmoid above and with the palatine plate of the maxillae below.

Lacrimals

The **lacrimals** (lăk′rĭm·ăls) are the smallest bones in the face. They are rectangular bones that lie in the anterior medial corners of the orbits (Figures 7-3 and 7-5). Each bears a deep, rounded depression that serves to form most of the borders of the **lacrimal canal,** which leads the **lacrimal (tear) duct** into the nasal cavity. Tears produced in the lacrimal gland above the eye flow across the surface of the eye into the nasal cavity through the lacrimal duct (see Figure 16-7 in Chapter 16).

Nasals

The **nasals** are small, rectangular bones that form the bony bridge of the nose (Figure 7-3). They are wedged between the frontal bone and the maxillae. The cartilaginous portion of the nose is attached to the lower edges of the nasals (see Clinical Note: Fracture of the Nose).

Maxillae

The **maxillae** (măk·sĭl′ē) are paired, united bones that form the upper jaw, the roof of the mouth, the floor and lateral walls of the nasal cavities, and part of the floor of the orbits (Figures 7-3 and 7-12). Each maxilla (or maxillary bone) is composed of a body and four processes. The centrally located body contains a large **maxillary sinus** (Figure 7-13) that, like other sinuses in the skull, is connected to the nasal cavity. The **frontal process** extends superiorly from the body and lies between the lacrimal and nasal bones (Figure 7-4). The **zygomatic process** of the maxilla extends laterally from the body and articulates with the zygomatic bone. This process contains the **infraorbital foramen,** which allows passage of small nerves and arteries to tissues at the front of the face. The **alveolar process** forms a U-shaped inferior projection from the body of the maxillae. It bears the teeth of the upper jaw in small sockets called **alveoli** (ăl·vē′ō·lī). The **palatine process** (Figure 7-7) extends horizontally from the interior of the alveolar process and forms most of the **hard palate** (the bony roof of the mouth). The **incisive foramen** or **canal** (Figure 7-7) lies at the anterior end of the palatine process and allows passage of the nasopalatine nerve and branches of the descending palatine blood vessels. Sometimes, the two halves of the palatine process fail to fuse along the midline during development, producing a malformation called a *cleft palate* (see Chapter 6, The Clinic: Diseases of the Skeletal System).

CLINICAL NOTE

ZYGOMATIC FRACTURE

A fracture of the zygomatic (cheek) bone is a fairly common injury because of its prominent position on the face. A nondisplaced fracture is not a serious problem. A displaced fracture, however, may result in the dislocation of the floor of the orbit around the eye, thereby producing double vision. Another complication is the entrapment of the inferior rectus muscle of the eye, thus limiting upward gaze. These complicated fractures require surgical repair.

Palatines

The **palatine** (păl´ă·tīn) bones are paired, L-shaped bones that are wedged between the maxillae and the sphenoid bone. The horizontal plate of each fuses with the other and forms the rear of the hard palate (Figure 7-7). The vertical portions contribute to the walls of the nasal cavity. The **greater** and **lesser palatine foramina** pierce the palatines and allow passage of nerves and blood vessels.

Zygomatics

The **zygomatics**, or cheek bones, form the prominent projections of the cheeks (Figure 7-5) and much of the

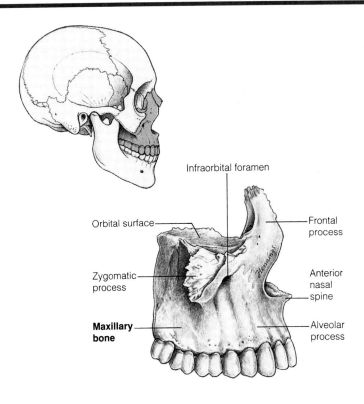

Figure 7-12 Lateral view of the maxilla.

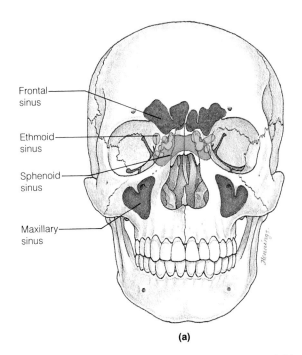

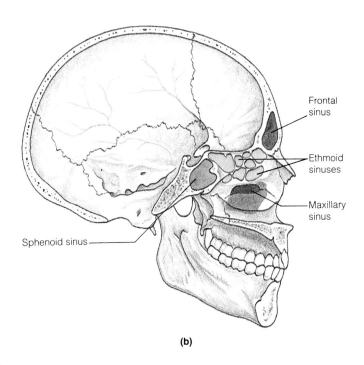

Figure 7-13 **Sinuses of the skull.** (a) Anterior view and (b) sagittal section.

symphysis: (Gr.) *symphysis*, a growing together
coronoid: (Gr.) *koronis*, something curved + *eidos*, form, shape
hyoid: (Gr.) *hyoeides*, U-shaped

CLINICAL NOTE

DISLOCATION OF THE MANDIBLE

On occasions, the mandibular condyle can dislocate or slip out of its place on the mandibular fossa of the temporal bone, causing a dislocated jaw. This usually occurs when a person yawns vigorously or sometimes following a blow to the jaw. This causes the jaws to lock in the open position so the person is unable to speak or chew. Reduction (repositioning) is a simple but hazardous procedure. The physician wraps his thumbs with protective towels and inserts them along the lower molars pressing down to unlock the mandible. The mandible snaps into place, restoring the patient to normal. The mandible may also fracture from a direct blow, which requires surgical alignment and wiring of the teeth for several weeks.

---

lateral floor of the orbits (see Clinical Note: Zygomatic Fracture). Each zygomatic bone has four prominent processes named for the areas into which they project: the **orbital, frontal, maxillary,** and **temporal processes.** The latter joins with the zygomatic process of the temporal bone to form the **zygomatic arch** (Figure 7-7). The **zygomaticofacial** (zī″gō·mǎt″ĭ·kō·fā′shǎl) **foramen** allows passage of the zygomaticofacial nerve and associated blood vessels (Figure 7-3).

Mandible

The **mandible** forms the lower jaw (Figures 7-3 and 7-5). During development the paired bones of the mandible fuse along a line called the **mandibular symphysis** (sĭm′fĭ·sĭs). Each side consists of a horizontal body and a vertical **ramus.** The body bears the teeth of the lower jaw inserted in the **alveolar process,** that portion of the bone that contains the sockets (alveoli) for teeth. A **mental foramen** at the anterior end allows passage of the mental nerve to the external surface, where it innervates the lower lip and tissues of the lower jaw. Each ramus bears two processes at its superior extremity: the **mandibular condyle** (kŏn′dīl) (or **condyloid process**), which articulates with the skull, and the **coronoid process,** which provides an expanded area for attachment of muscles used in biting and chewing (see Clinical Note: Dislocation of the Mandible). Each ramus also contains a **mandibular foramen** on its medial surface (Figure 7-9), which allows passage of the inferior alveolar branch of the trigeminal nerve (V) and its associated blood vessels.

Sutures and Wormian Bones

The bones of the skull articulate with one another through immovable connections called **sutures.** Most sutures are named after the two bones involved; for example, the **sphenofrontal suture** connects the sphenoid and frontal bones, and the **nasomaxillary suture** connects the nasal and maxillary bones. Several of the longer, more prominent sutures have been given names that are not related to the bones involved. The **coronal suture** connects the frontal bone with the parietals, and the **sagittal suture** connects the parietal bones along the top of the skull. The **lambdoidal suture** connects the occipital and parietal bones, and the **squamosal suture** connects the squamous portion of the temporal with the parietal (Figure 7-5).

Wormian bones are small bones that are often found along the sutures, particularly the squamosal and lambdoidal sutures (Figure 7-5). They are formed by interconnecting branches of the sutural line.

Hyoid Bone

The U-shaped **hyoid bone** (Figure 7-14) is held between the mandible and larynx (Adam's apple) by ligaments and muscles. It does not articulate solidly with any other bone and is thus often referred to as a **floating** or **sesamoid**

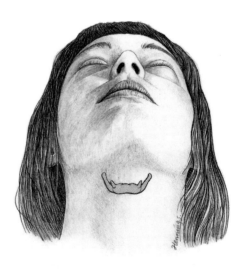

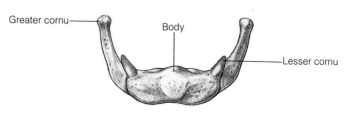

Figure 7-14 Anterior view of the hyoid bone.

(sĕs′ă·moyd) bone. The hyoid is composed of a horizontal body attached to two pairs of processes that resemble the horns of a bull. The **lesser cornua** (horns) are small projections at the lateral ends of the body, and the **greater cornua** extend posteriorly from the body. The hyoid serves as the place of attachment for several small muscles that extend into the tongue and along both sides of the larynx. These muscles are used in speaking, swallowing, and moving the tongue.

Vertebral Column

After completing this section, you should be able to:

1. Describe a typical vertebra.
2. Distinguish between the five types of vertebrae.
3. Describe the curves of the vertebral column.
4. Describe the thorax and identify its skeletal components.

The **vertebral column**, or **spine**, is composed of individual **vertebrae** arranged in a longitudinal series (Figure 7-15). It forms the vertical axis of support for the head and upper part of the body and protects the spinal cord. The spine measures approximately 70 cm (28 in.) in length in adults. Ribs and muscles attach to the vertebral column, and contraction of these muscles permits the spine to bend in several directions. Most of the vertebrae are separated from one another by oval **intervertebral disks** composed of fibrocartilage. The intervertebral disks serve as shock-absorbing pads that cushion the forces exerted along the column as one walks, runs, and jumps.

The vertebral column is divided into five distinct regions. At the superior end, seven **cervical vertebrae** form the neck. Just inferior to these are twelve **thoracic** (thō·răs′ĭk) **vertebrae** to which the ribs are attached. Five **lumbar vertebrae** are located below the thorax in the lower back region. Inferior to the lumbar region is the **sacrum**, which develops as five separate sacral vertebrae that later fuse into a single structure. Three, four, or five (usually four) vestigial **coccygeal** (kŏk·sĭj′ē·ăl) **vertebrae** are located at the inferior end of the column. These usually fuse into one or two bones called the **coccyx** (kŏk′sĭks).

The Clinic at the end of this chapter discusses the most common disorders of the spine.

Curves of the Vertebral Column

The vertebral column of an infant is curved anteriorly due to the position of the fetus in its mother's uterus (see Figure 6-2). This curve changes slightly a few months after birth when the infant gains sufficient strength to hold its head up. At this time the upper end of the cervical region begins to bend posteriorly in what is called the **cervical curve** (Figure 7-15). This curve allows the infant to balance the weight of its head upright over the vertebral column. (Most people have watched with fascination the bobbing and swaying of an infant trying to balance its head.) Later, when the child begins to walk, another curve (concave posteriorly) develops in the lumbar region. This **lumbar curve** allows the child to balance the weight of the upper body over its legs. In all, four curves are recognized at the final stages of development (from top to bottom): cervical, **thoracic**, lumbar, and **sacral** curves (Figure 7-15). Curves in the vertebral column help balance the weight of the upper body over the legs; along with the intervertebral discs they absorb shocks during walking, running, and jumping.

A Typical Vertebra

Except for certain regional differences we will consider later, most vertebrae have the same basic structure (Figure 7-16). The **body** is the oval, anterior region that bears the weight of the vertebrae above it. Attached to the body is the posterior **vertebral arch (neural arch)**, which together with the body forms a large **vertebral foramen (neural canal)** through which the spinal cord passes. The vertebral arch is attached to the body laterally by two **pedicles**. The roof of the vertebral arch is formed by the **laminae**, which unite and project posteriorly as the **spinous process**. Three paired processes extend from the laminae. The **superior articular processes** bear surfaces that articulate with the vertebra above, and the **inferior articular processes** (see Figure 7-19) articulate with the vertebra below. **Transverse processes** extend laterally from the base of the laminae providing surfaces for muscle attachments and articulations for ribs in the thoracic region. Indentations on the superior and inferior sides of the pedicles form **vertebral notches**. Superior and inferior notches of adjacent vertebrae fit together to form **intervertebral foramina** through which pass the spinal nerves.

Types of Vertebrae

In addition to the general characteristics just described, vertebrae possess certain structural features peculiar to their region and function. The three types are cervical, thoracic, and lumbar vertebrae.

Cervical Vertebrae

Vertebrae in the neck, or **cervical vertebrae** (numbered C1 through C7), possess smaller bodies and larger arches than do vertebrae in other regions. In addition, the transverse process of cervical vertebrae are pierced by **transverse foramina,** which allow passage of the vertebral artery and vein (Figure 7-17).

Because of their unique association with the skull and with movement of the head, the first two cervical vertebrae are different from the others. The first vertebra (C1) called the **atlas** (named after the mythological figure who supported the earth, analogous to the head, on his shoulder) is compact (Figure 7-18). It articulates with the occipital condyles of the skull and allows the up and down nodding movement of the head, as when you say yes. The atlas also lacks a spinous process and a body. During the development of the verteral column, the ossification center for the body of

> **CLINICAL NOTE**
>
> **WHIPLASH**
>
> **Whiplash** is a flexion-extension injury of the cervical vertebrae commonly incurred in a rear-end car collision. Fractures or dislocation of the vertebrae causing neurologic damage may occur along with whiplash. Fatalities occasionally result from this type of accident when the odontoid process of the C2 vertebra (the axis) are fractured and forced into the medulla oblongata of the brain.

Figure 7-15 The vertebral column. (a) Lateral view. (b) X ray of the cervical area of the vertebral column.

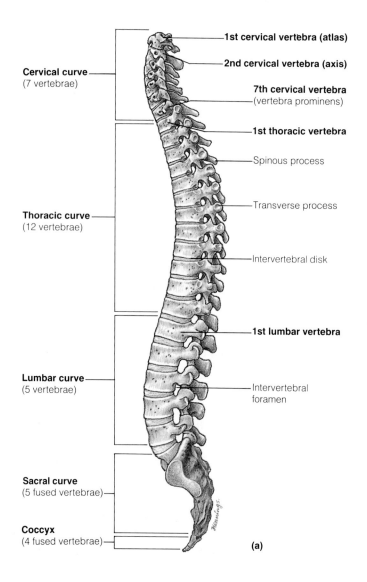
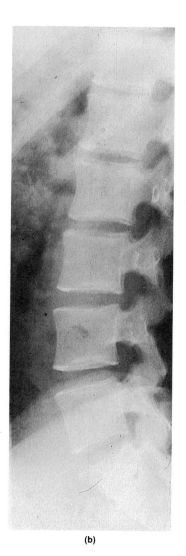

odontoid: (Gr.) *odous*, tooth + *eidos*, form, shape

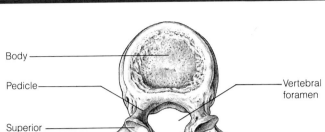
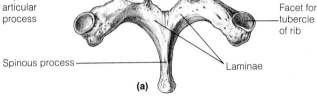

Figure 7-16 A typical thoracic vertebra. (a) Superior and (b) lateral views.

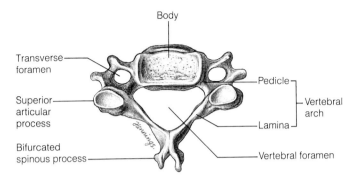

Figure 7-17 Superior view of a cervical vertebra.

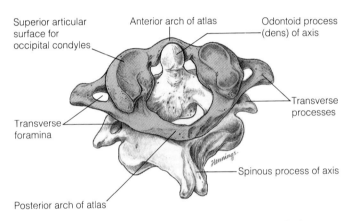

Figure 7-18 Oblique views of atlas (in color) and axis cervical vertebrae.

the atlas becomes separated, and it fuses with the body of the second cervical vertebra (C2) called the **axis** (Figure 7-18). This results in the growth of a toothlike **odontoid** (ō·dŏn′toyd) **process (dens)** on the superior surface of the body of the axis. The odontoid process extends into the space left in the atlas and provides a pivot around which the head and the atlas rotate. This motion is involved in moving the head from side to side, as when you shake your head to say "no."

The seventh cervical vertebra (called the **vertebra prominens** because it can be felt very easily at the base of the neck) has a prominent spinous process. Its transverse foramina are often reduced or even absent, and consequently the vertebral arteries and veins usually pass in front of the transverse processes and enter C6.

Thoracic Vertebrae

The **thoracic vertebrae** come closest to the description given for typical vertebrae. They are numbered T1 through T12. They do not have the transverse foramina found in the cervical vertebrae. Six **facets** mark where articulations are made with the ribs on most of the 12 vertebrae (Figure 7-16). Around the body of each vertebra, these facets (also called **demifacets**) articulate with the heads of the ribs. Facets on the transverse processes (except for those on the 11th and 12th vertebrae) articulate with the rib tubercles.

Lumbar Vertebrae

The **lumbar vertebrae** (numbered L1 through L5) also fit the description of a typical vertebra. They have massive, blunted bodies and thick spinous processes (Figure 7-19). The transverse and articulating processes of lumbar vertebrae also tend to be relatively short and blunt. They lack facets and demifacets.

Sacrum

The **sacrum** is produced by fusion of five **sacral vertebrae** numbered S1 through S5 (Figure 7-20). Evidence of this fusion is visible as **transverse lines** on the anterior surface of

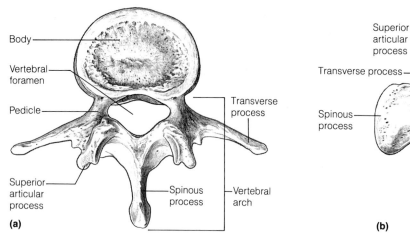

Figure 7-19 A lumbar vertebra. (a) Superior and (b) lateral views.

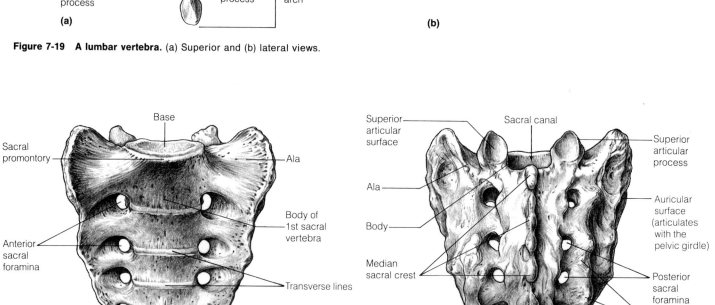

Figure 7-20 The sacrum and coccyx. (a) Anterior and (b) posterior views.

the sacrum. The **sacral foramina** that penetrate the sacrum on both sides of the transverse lines also indicate areas of fusion. The anterior surface of the sacrum is concave and the posterior surface is convex (more predominantly in females than males). The superior region, the **base,** forms the articulation with the fifth lumbar vertebra. On either side of the base is a **superior articular process** and an **ala** (wing). Just below the ala are two **auricular surfaces** that articulate with the pelvic girdle. Anteriorly, the **sacral promontory** projects just below the base, and the lower end forms the **apex.** Posteriorly, several anatomical landmarks can be recognized. The **median sacral crest** marks the spinous processes of the fused sacral vertebrae, and the **sacral canal** represents the combined vertebral foramina. The inferior opening of the sacral canal is the **sacral hiatus.** On either side of the median crest is a **lateral sacral crest.**

manubrium: (L.) *manubrium,* handle
xiphoid: (Gr.) *xiphos,* sword + *eidos,* form, shape

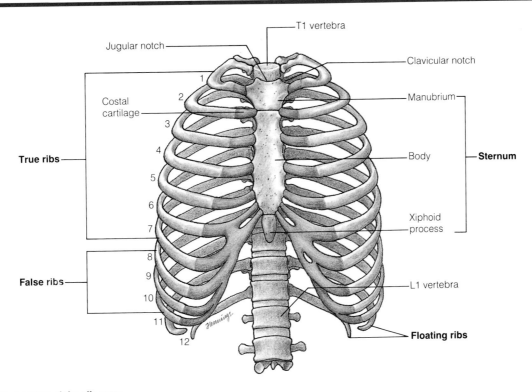

Figure 7-21 Anterior view of the rib cage.

Coccyx

The **coccyx** undoubtedly represents the vestigial remnant of a tail that was functional in distant human ancestors. However, it functions now primarily as a place for muscle attachments. The coccyx is formed from the fusion of three to five rudimentary **coccygeal vertebrae,** which lack most of the structures previously discussed (Figure 7-20). Various patterns of fusion occur, and in adults, the coccyx usually consists of one or two fused units.

Thorax

The **thorax,** or chest region, is that portion of the trunk associated with the thoracic vertebrae. The anterior and lateral skeletal margins of the thorax are the sternum and rib cage, respectively.

Sternum

The **sternum,** or breastbone, lies in the anterior midline of the thorax and consists of three recognizable units: manubrium, body, and xiphoid process (Figure 7-21). The **manubrium** (mă·nū'brē·ŭm) is roughly triangular in shape, with the base oriented superiorly. It bears a single **jugular notch** medially and two **clavicular notches** laterally. The manubrium articulates with the first and second ribs. The **body** is rectangularly tapered at its caudal end. It articulates with the costal cartilages of the second through seventh ribs. The **xiphoid** (zĭf'oyd) process is a small, inferior projection. It is variable in shape and may be perforated or forked. It serves as a place of attachment for some abdominal muscles.

Ribs

The **rib cage** is formed by the sternum plus 12 pairs of ribs and **costal cartilages** (Figure 7-21). The first 7 pairs of ribs, called **true ribs,** make direct contact with the sternum through their costal cartilages. The 8th, 9th, and 10th pairs attach to the costal cartilage of the 7th rib and are called **false ribs.** The 11th and 12th pairs are false ribs that do not attach to the sternum at all, and for this reason they are known as **floating** or **vertebral ribs.** These last two pairs are held in position by the muscles of the lateral thoracic wall. Occasionally, an extra rib develops, either in association with the 7th cervical vertebra or the 1st lumbar vertebra.

Figure 7-22 shows the structure of a typical rib. Posteriorly, it bears an expanded **head** that contains two **articular facets.** These facets make contact with the bodies of two

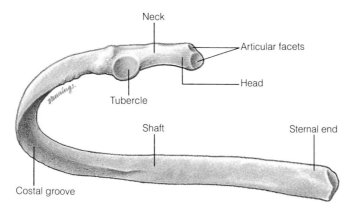

Figure 7-22 Anterior view of a typical rib.

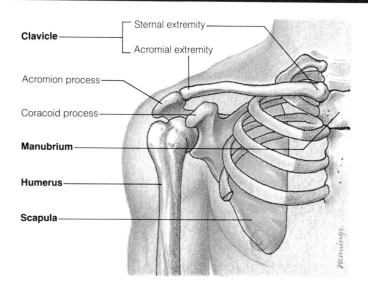

Figure 7-23 Anterior view of the pectoral (shoulder) girdle.

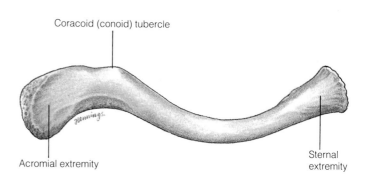

Figure 7-24 Anterior view of the clavicle.

adjacent thoracic vertebrae. Ribs 1, 10, 11, and 12 articulate with only one vertebral body and, consequently, have only one articular facet. A slightly constricted area called the **neck** is just lateral to the head. A prominent **tubercle** (tū′bĕr·kl) projects from the posterior surface of the rib and marks the beginning of the **body** of the rib, which curves anteriorly until it reaches the costal cartilage. The tubercle bears an **articular facet** that contacts the transverse process of a thoracic vertebra (except for the 11th and 12th ribs). Lateral to the tubercle, the curved body of the rib contains a sharp **angle**. A **costal groove**, which houses intercostal vessels and nerves, extends inferiorly along the medial anterior surface of the body.

Pectoral Girdle

After completing this section, you should be able to:

1. Identify the bones of the pectoral girdle.
2. Describe the major anatomic landmarks on the clavicle and scapula.

The arm is attached to the axial skeleton via the **pectoral** (or shoulder) **girdle** (Figure 7-23). It consists of two paired bones: the clavicles and scapulae.

Clavicle

The **clavicles** (klăv′ĭ·kls), or collarbones, have the shape of a stretched S (Figure 7-24). Each one articulates with the manubrium medially and the scapula laterally. The medial end of the clavicle, the **sternal extremity**, is thick and blunt. It appears circular in cross section. The **acromial extremity**, the lateral end of the clavicle, is flattened. A low projection called the **coracoid** (or **conoid**) **tubercle** on the inferior surface near the acromial end provides a place for attachment of ligaments that hold the clavicle to the coracoid process of the scapula.

Scapula

The **scapulae** (skăp′ū·lē), or shoulder blades, form the major bones of the pectoral girdle. They are found at the craniolateral corners of the thorax, on the posterior surface. Each is triangular (Figure 7-25) and has several prominent processes that serve as attachments for muscles and ligaments. The three **borders** of each scapula are named for their positions: **superior**, **lateral** (or **axillary**), and **medial** (or **vertebral**). A **scapular notch** is present in the superior border. The junction of the lateral and superior borders is marked by a

glenoid:	(Gr.) *glene*, socket + *eidos*, form, shape
acromion:	(Gr.) *akros*, summit + *osmos*, shoulder
bicipital:	(L.) *biceps*, having two heads
trochlea:	(Gr.) *trochilia*, pulley
capitulum:	(L.) *capitulum*, small head

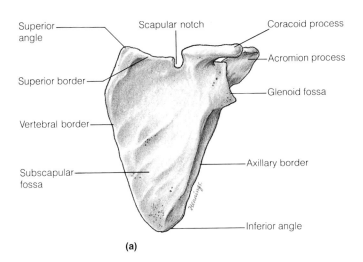
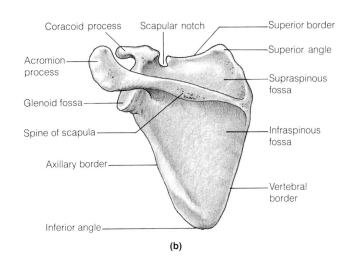

Figure 7-25 The scapula. (a) Anterior and (b) posterior views.

shallow, oval depression, called the **glenoid fossa** (glĕ′noyd fŏs′ă), which provides an articulation point for the humerus. Because a scapula is triangular, three angles—the **superior, lateral,** and **inferior**—are also recognized. The anterior surface of the scapula is concave; this area is the **subscapular fossa** (sŭb·skăp′ū·lăr fŏs′ă). The bone's posterior surface bears a transversely oriented **spine** dividing the surface into **supraspinous** and **infraspinous fossae,** areas for muscle attachments. The distal end of this spine expands into a flattened **acromion** (ă·krō′mē·ŏn) that articulates with the acromial extremity of the clavicle. The **coracoid process** projects anteriorly from the superior border. The name is derived from the Greek *korakaeides*, meaning "ravenlike," because its shape resembles a raven's beak.

Upper Extremity

After completing this section, you should be able to:

1. Identify the major segments of the upper extremity.
2. List the bones of the wrist and indicate how they articulate.

The upper extremity includes the arm, forearm, wrist, and hand. (The upper arm in everyday language is considered by anatomists to be the "arm," as distinguished from the "forearm.") The upper extremity attaches to the rib cage by means of the scapula and clavicle. The upper extremity is used primarily to manipulate objects in the environment.

Humerus

The **humerus,** the arm bone, is the largest bone of the upper extremity (Figure 7-26). Recall from Chapter 6 that like most long bones, it has a diaphysis and two epiphyses. The proximal epiphysis, known as the **head,** articulates with the scapula at the glenoid fossa. Interestingly, two necks are on the humerus. The **anatomic neck** is the groove just distal to the head, and the **surgical neck,** which (unlike the anatomical neck) fractures easily, is at the proximal end of the diaphysis. Between the two necks are two large projections, the **greater** and **lesser tubercles,** which serve as places for muscle attachment. The **intertubercular sulcus** (also called the **bicipital groove**) passes longitudinally between the greater and lesser tubercles. The diaphysis of the humerus bears the **deltoid tuberosity** (tū·bĕr·ŏs′·ĭ·tē), which is one of the attachments of the deltoid muscle of the shoulder.

The distal end of the humerus bears a **lateral** and a **medial epicondyle** (ĕp·ĭ·kon′dīl), both anchor muscles. The distal epiphysis of the humerus has two specialized condyles. The more medial **trochlea** (trōk′lē·ă) articulates with the ulna, one of the forearm bones; the more lateral **capitulum** (kă·pĭt′ū·lŭm) is a rounded projection where the radius (the other forearm bone) articulates. The humerus also has

150 Chapter 7 Anatomy of the Skeletal System

styloid: (Gr.) *stylos*, pillar + *eidos*, form, shape
triquetrum: (L.) *triquetrum*, triangle
pisiform: (L.) *pisum*, pea + *forma*, shape
trapezium: (Gr.) *trapezion*, small table
hamate: (L.) *hamatus*, hooked

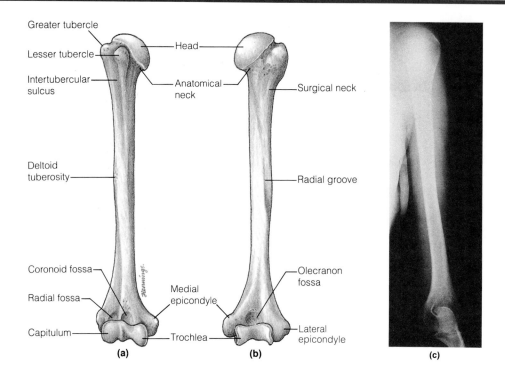

Figure 7-26 The humerus. (a) Anterior and (b) posterior views. (c) X ray of the humerus.

depressions to accommodate parts of the forearm bones when the elbow opens and closes. Anteriorly, a medial **coronoid fossa** articulates with the coronoid process of the ulna, and a lateral **radial fossa** meets the head of the radius. Posteriorly, a large **olecranon** (ō·lĕk′răn·ŏn) **fossa** accommodates the **olecranon process** when the forearm is fully extended.

Ulna and Radius

As already mentioned, the forearm includes two bones: the ulna and the radius. The **ulna** is longer than the radius and lies medial to it when the hand is palm up (Figure 7-27). The large **olecranon process** forms the elbow projection at the proximal end. The **trochlear** (or **semilunar**) **notch** on the proximal anterior surface articulates with the trochlea of the humerus (Figure 7-26). Just distal to the trochlear notch is the prominent **coronoid process.** It bears a rounded **radial notch** on its lateral side into which the head of the radius fits. A small **ulnar tuberosity** projects from the shaft of the bone just distal to the coronoid process on the anterior surface. The distal end of the ulna is a rounded **head** around which the distal end of the radius bone rotates. A narrow **styloid process** projects medially from the head. The sharp edges of both the ulna and radius that face each other are known as **interosseous** (ĭn″tĕr·ŏs′ē·ŭs) **borders.**

The **radius** is the shorter of the two forearm bones (Figure 7-27). Its head at the proximal end is shaped like a small flat knob. A depression on the proximal surface of the head articulates with the capitulum of the humerus (Figure 7-26). The constriction just distal to the head is the **neck.** A prominent **radial tuberosity** projects medially from the shaft distal to the neck. The distal end of the radius has a tapering styloid process, a small **ulnar notch** that fits the head of the ulna, and a large concave area that articulates with the bones of the carpus (wrist).

Manus

The **manus** (hand) is composed of the carpus, metacarpus, and phalanges.

Carpus

Eight small bones called **carpals** are arranged in two rows of four each to form the **carpus** (wrist) (Figure 7-28). They connect the forearm and hand. From lateral to medial, the proximal row contains the **scaphoid** (or **navicular**), **lunate**, **triquetrum** (trī·kwē′trŭm), and **pisiform,** while the distal row contains the **trapezium** (tră·pē′zē·ŭm) (or **greater multangular**), **trapezoid** (or **lesser multangular**), **capitate,** and **hamate.**

Upper Extremity 151

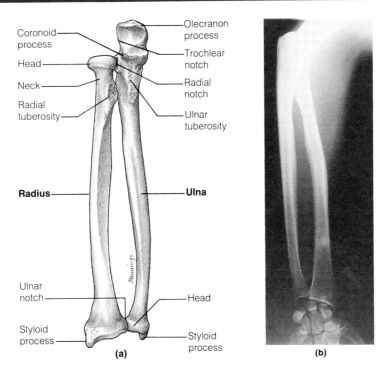

Figure 7-27 **The radius and ulna.** (a) Anterior view. (b) X ray from the posterior perspective.

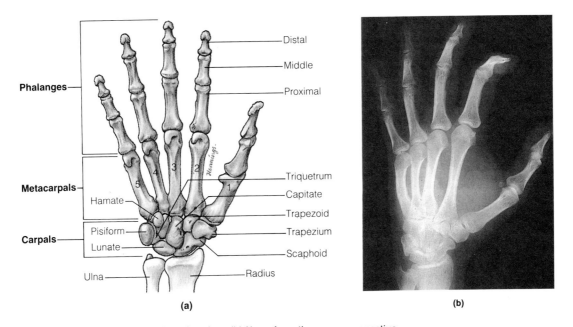

Figure 7-28 **Bones of the wrist and hand.** (a) Anterior view. (b) X ray from the same perspective.

metacarpus: (Gr.) *meta*, beyond + *karpos*, wrist
phalanges: (Gr.) *phalangx*, battleline
acetabulum: (L.) *acetabulum*, small cup for vinegar
pelvis: (L.) *pelvis*, basin

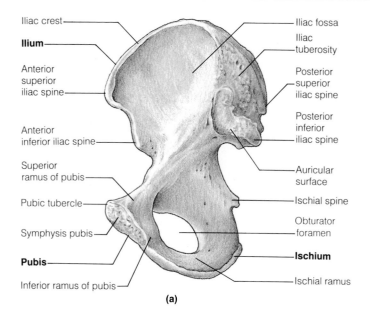
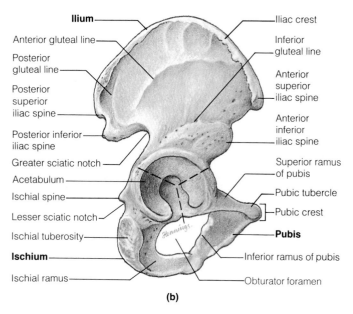

Figure 7-29 **Right os coxa.** (a) Medial and (b) lateral views.

Metacarpus

The **metacarpus** (Figure 7-28) is composed of five elongate **metacarpals** that form the palm of the hand. Each contains a diaphysis and prominent epiphyses characteristic of long bones, along with a proximal base, shaft, and distal head. The distal epiphyses form the knuckles where the fingers join the palm.

Phalanges

The **phalanges** (fă·lăn′jēz) are digital bones that form the fingers (as well as the toes). Each has a base, shaft, and head (Figure 7-28). The thumb (or **pollex**) has two phalanges and the remaining fingers each have three attached end to end. The distal phalanx of each finger is tapered where it supports the pad of connective tissue at the fingertip.

Pelvic Girdle

After completing this section, you should be able to:

1. Describe the two bones of the pelvic girdle.
2. List the differences that distinguish a male pelvic girdle from that of a female.

Os Coxae

A pair of curved bones called the **os coxae** (ŏs cŏx′ē) or **coxal bones,** commonly known as the hipbones or pelvic bones, form the **pelvic girdle** (Figure 7-29). This girdle articulates solidly with the sacrum. It supports the weight of the upper body and distributes it to the legs. Each os coxa develops from ossification centers of three embryonic bones that fuse in the adult. The names of the three bones are retained for the corresponding regions in the adult pelvic girdle. These three areas include the superiorly located ilium, the ischium at the posterior corner, and the pubis at the anterior inferior corner. The **acetabulum** (ăs″ĕ·tăb′ū·lŭm), a large depression on the lateral surface of the coxal bone, indicates the point of fusion of the three bony elements. It also serves as the area of articulation with the femur. Embryologically, the acetabulum forms as a result of physical pressure from the head of the femur bearing on the pelvic bones. In abnormal cases with little or no opposition, the acetabulum is shallow and can dislocate relatively easily.

Several anatomic regions are recognized in association with the pelvic girdle (Figure 7-30). The os coxae and sacrum enclose a space called the **pelvis** (Latin for "basin"). The **brim** of the pelvis is formed by a circular line passing along the upper edge of the pubic bones and the sacral promontory. The space above the brim is the **greater** (or **false**) **pelvis** and the space below is the **lesser** (or **true**) **pelvis.** The true pelvis has bony walls encircling it, while the false pelvis has only posterior and lateral walls formed by the sacrum and paired ilia. The muscular abdominal wall forms the anterior border of the false pelvis. Because the pelvic brim encircles

pubis: (L.) *pubes*, mature

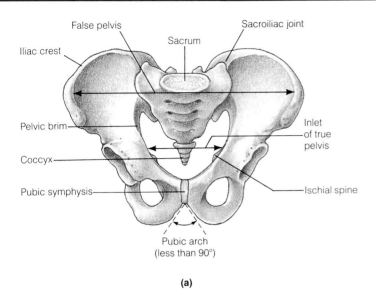

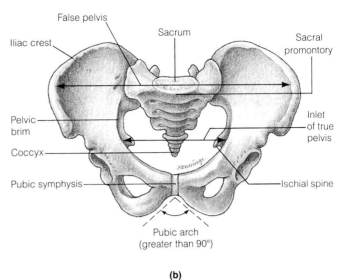

Figure 7-30 The pelvis. Anterior view of (a) male and (b) female pelvises. (c) X ray from the same perspective of the female pelvis.

the superior entrance into the true pelvis, it is named the **pelvic inlet**. The **pelvic outlet** is the inferior opening of the pelvis, and it is bounded by the coccyx posteriorly and the ischia and pubic bones anterolaterally.

The **ilium** (Figure 7-29) is the largest of the three bone regions of the os coxa. Its superior border is the **iliac crest**, whose anterior and posterior ends are the **anterior superior iliac** and **posterior superior iliac spines**, respectively. The iliac crest serves as a muscle attachment. In addition, **anterior inferior iliac** and **posterior inferior iliac spines** are recognized. The **greater sciatic** (sī·at'ĭk) **notch** on the posterior border accommodates passage of the largest nerve in the body, the sciatic nerve (see Chapter 13), which courses from the spinal cord into the thigh. The iliac lateral surface bears three low ridges: the **anterior, posterior,** and **inferior gluteal lines,** which demarcate the margins of attachment of large muscles. The internal surface of the ilium is broadly concave and forms the **iliac fossa**. The posterior border of the ilium enlarges into a roughened **auricular surface** that articulates with the sacrum. The **iliac tuberosity** projects from this surface.

The **ischium** (Figure 7-29) forms the lower, posterior region of the os coxae. Its outer border bears a conical **ischial spine** projecting posteriorly; a large, expanded **ischial tuberosity** projects inferiorly. These processes are separated by the **lesser sciatic notch**. Anyone who has sat for a long time on a hard bench has become painfully aware of the *exact*

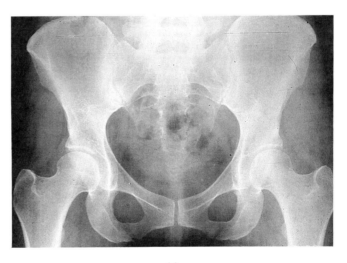

location of the ischial tuberosities! The lower process of the ischium is the **ramus**.

The **pubis** (Figure 7-29), the smallest of the three coxal regions, forms the anterior inferior portion of the os coxa. It consists of two narrow columns, the **superior** and **inferior rami**, connected by a body. The upper border of the superior ramus bears a **pubic crest**. The pubic bones articulate with one another anteriorly at the **pubic symphysis**. Together the pubis and the ischium border the large opening called the **obturator foramen**.

fovea: (L.) *fovea*, a small depression
trochanter: (Gr.) *trochanter*, runner

Differences Between Female and Male Pelvic Bones

The female pelvic bones are indirectly involved in pregnancy and childbirth. They are wider, lighter in weight, and less thick than in males (Figure 7-30), and the female acetabulum is smaller and directed more anteriorly than in males of comparable size. The spines of females are low and blunt, the sciatic notches are shallow, and the obturator foramen is smaller and more oval. The pubic arch forms an angle of less than 90° in males, but an angle of more than 90° in females. The female ilia are wider and offer greater support for a developing fetus. The inlet of the true pelvis is narrow and heart shaped in males, but circular and wide in females (Figure 7-30) to better facilitate the passage of the child during birth. The outlet of the true pelvis is small in males but wide in females, and the coccyx is flexible in females.

Lower Extremity

After completing this section, you should be able to:

1. List the bones of the lower extremity.
2. Describe the similarities and differences between the hand and foot.
3. Describe the arches of the foot.

The lower extremity includes the thigh, knee, leg, ankle, and foot. (The lower leg in everyday language is considered by the anatomist to be the "leg," as distinguished from the thigh.) The lower extremity is attached to the pelvic girdle, and its primary purpose is for mobility.

Femur

The **femur**, or thigh bone, is the longest bone in the body (Figure 7-31). It articulates with the pelvic girdle at the acetabulum and, at its distal end, with the tibia (see Figure 7-32). The femur has a well-marked diaphysis and epiphyses typical of long bones. The proximal end bears a large, rounded head, which is held by the neck at an approximate 120° angle to the shaft. The head has a small depression, the **fovea centralis** (fo′vē·ă sĕn·tră′lĭs) which houses a ligament that helps hold the femur in place. Just distal to the neck are two large prominences that are places for muscle attachment, the more superior **greater trochanter** (trō·kăn′tĕr) and the **lesser trochanter**. These two are joined by an **intertrochanteric line** anteriorly and an **intertrochanteric crest** posteriorly. An **intertrochanteric fossa** lies medial to the crest. The posterior surface of the femur's shaft bears a ridge, the **linea aspera** (lĭn′ē·ă ăs·pĕr′ă) whose superior end expands into a

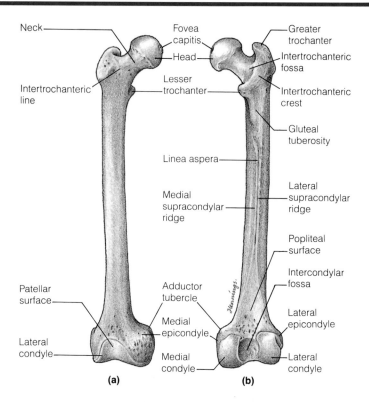

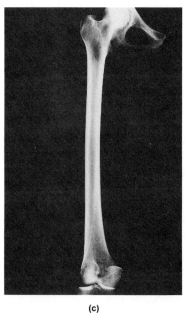

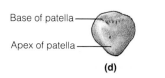 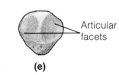

Figure 7-31 The femur and patella. (a) Anterior and (b) posterior views of the femur. (c) X ray of a femur. (d) Anterior and (e) posterior views of the patella.

patella: (L.) *patella*, small pan

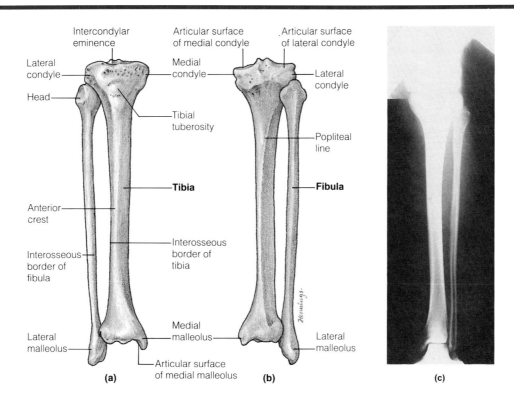

Figure 7-32 The tibia and fibula. (a) Anterior and (b) posterior views. (c) X ray from the posterior perspective.

gluteal tuberosity. Both serve as muscle attachments. The femur's distal end bears two large, rounded **condyles, lateral** and **medial,** that articulate with the tibia. The lateral and medial surfaces of the distal end expand into roughened **lateral** and **medial epicondyles,** which are sites where muscles are fastened. The anterior surface between the condyles is modified into a wide, shallow depression called the **patellar surface;** posteriorly a deep **intercondylar fossa** is present.

The femur angles toward the midline of the body at the knees. This distributes an individual's weight toward the body's center of gravity and makes it easier to stand and walk erect. Sometimes the angle is extreme and results in a condition known as "knock-knees."

Patella

The **patella,** or kneecap, is a floating sesamoid bone embedded in a tendon and does not make solid contact with other bones. It protects the knee joint and increases the leverage of the large anterior thigh muscles. The kneecap is small and triangular (Figures 7-31d and e), with its pointed end **(apex)** distal and its broad base proximal. On its posterior surface are **articular facets** for the medial and lateral condyles of the femur.

Tibia

The **tibia,** or shinbone is the larger of the two bones of the leg (Figure 7-32). It lies medial to the fibula, the other leg bone. The tibia bears most of the weight of the body. The proximal end of the tibia expands and has **lateral** and **medial condyles** for articulation with the femur. An **intercondylar eminence** projects up between the two condyles. The anterior surface of the shaft has a roughened **tibial tuberosity** just below the condyles, and an **articular facet** articulates with the head of the fibula. The tibia's distal end is modified into a large concave surface that forms part of the ankle joint. The **medial malleolus** (măl·ē′ō·lŭs) extends from the medial side of this concavity. Opposite the medial malleolus is a small notch where the fibula articulates.

Fibula

The **fibula,** the other lower leg bone, is narrower than the tibia (Figure 7-32). Its proximal end expands into a head that articulates with the proximal end of the tibia near the lateral condyle. The distal end expands into the **lateral malleolus,** which helps form the ankle. The shaft of the fibula is often triangular in cross section and is usually twisted.

talipes: (L.) *talus*, ankle + *pes*, foot
navicular: (L.) *navicula*, little boat

CLINICAL NOTE

ABNORMALITIES OF THE FEET

Flatfoot is a condition resulting from loss of the longitudinal arch during weight bearing (while standing). The arch is collapsed because of weakening of the longitudinal ligaments. This condition requires no treatment other than arch support for symptoms of strain.

Pigeon toe (or **congenital metatarsus varus**) is a relatively common disorder in which the forefoot is turned in while the hindfoot is in a neutral position. If the deformity is not corrected, the child will toe-in while walking. Treatment involves passive exercises and casting, with complete resolution usually achieved in three to six weeks.

Congenital inversion of the foot is a common inherited condition that is also called **hyperpronation** of the foot. For the nonathlete it is of little consequence, but with the increasing participation in running sports, hyperpronation causes (or at least contributes to) many painful running injuries. It contributes to such problems as unstable ankles, shin splints, chondromalacia of the patella and even low back pain. Hyperpronation can be prevented with the use of proper footwear and arch supports (see Unit II Case Study: Overuse Syndromes).

Talipes (tăl′ĭ·pez) is the medical term for **clubfoot** (see figure). This is a congenital deformity of the ankle (talus) and foot (pes). The deformity may involve one or both feet and is corrected by splinting or casting in such a way as to stretch and manipulate the foot into a functional position.

A **bunion** is an abnormal protrusion of the head of the metatarsus below the big toe. It may be a congenital defect, but is often caused by the wearing of high-heeled, pointy-toed shoes. The high heels drive the foot forward into the toe box of the shoe, and in the narrow space the great toe is driven toward the outside of the foot. Chronic irritation leads to inflammation and the buildup of bone. Prevention and treatment consists of wearing properly fitted flat or low-heeled shoes.

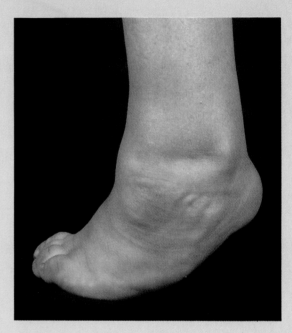

A foot deformed by talipes (clubfoot).

Pes

The **pes**, or foot, is composed of the tarsus metatarsus, and phalanges (see Clinical Note: Abnormalities of the Feet).

Tarsus

The **tarsus** contains seven bones called **tarsals** of different sizes and shapes (Figure 7-33). The **talus** is the primary bone of the ankle joint. It lies distal to the tibia and fibula between the two malleoli (Figure 7-32). The talus bears the weight of the erect body and transfers this weight to other bones of the foot. The **calcaneus** (kăl·kā′nē·ŭs), or heel bone, is the largest tarsal bone. The **navicular** lies anterior to the talus, and the **cuboid** lies anterior to the calcaneus. Three wedge-shaped bones—the **medial, intermediate,** and **lateral cuneiforms**—lie anterior to the navicular and medial to the cuboid.

Metatarsus

The **metatarsus** consists of five **metatarsal** bones (Figure 7-33). All metatarsals have well-marked diaphyses and epiphyses and articulate with the cuboid and cuneiform bones.

Phalanges

The **phalanges** form the digits of the foot and are similar to those of the hand (Figure 7-33). The **hallux** (big toe) is composed of two phalanges, and each remaining toe has three (comparable to the hand). The distal phalanx of all five toes is tapered distally, as in the fingers.

Arches of the Foot

In most adults the foot has three **arches** (Figure 7-34), two longitudinal and one transverse. The arches are maintained

Lower Extremity   **157**

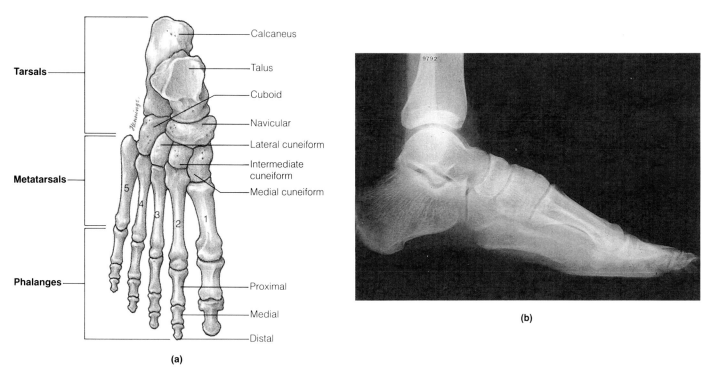

Figure 7-33   Bones of the foot. (a) Superior view of the foot. (b) X ray showing a lateral view of the foot.

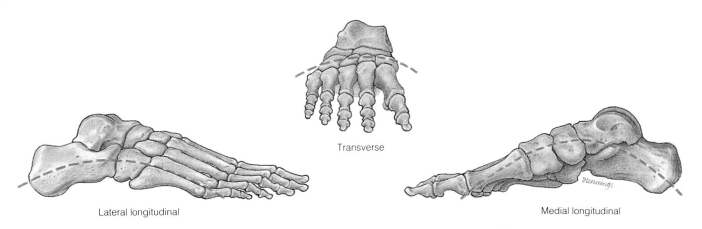

Figure 7-34   Arches of the foot.

by ligaments and muscles, and they give the foot resilience in bearing the body's weight when walking or running. The **medial longitudinal arch** extends from the calcaneus (heel bone) through the talus, navicular, cuneiforms, and to the metatarsals of toes one, two, and three. The **lateral longitudinal arch** extends from the calcaneus through the cuboid and to the metatarsals of toes four and five. The **transverse arch** extends across the foot from the calcaneus through the cuboid, navicular, and cuneiforms. In some individuals the foot ligaments and muscles develop improperly or may become weakened in their ability to maintain the arches. Such a condition is referred to as "flatfoot" (see Clinical Note: Abnormalities of the Feet). Congenital (inherited) flatfoot may be only mildly troublesome, but when it results from weakening of the structural elements of the foot, it is often accompanied by pain and discomfort.

hyperlordosis: (Gr.) *hyper*, above + *lordosis*, bending
kyphosis: (Gr.) *kyphosis*, hunchback
scoliosis: (Gr.) *skoliosis*, curvature condition

THE CLINIC

DISORDERS OF THE SPINE

There are many problems associated with the axial skeleton and with the spine in particular. The spine is a very complex structure and is responsible for the major weight bearing and movement of the body. Pain in the cervical spine and lower back are very common complaints and cause a great deal of disability in many patients. The following are some of the more common problems associated with the spine.

Abnormal Spinal Curves

Abnormalities of the normal spinal curves are of interest to the physician since they may be indications of more serious diseases. The spine can be abnormally curved in a variety of ways. Straightening of the normal cervical curve from a lateral view as seen on an X ray may be a clue to injury to the soft tissues of the neck and may suggest muscle spasm; this is sometimes referred to as **military spine.** The exaggerated dorsal curving of the thoracic spine in young people, known as **Scheuermann's disease,** is sometimes caused by an inflammation of the epiphysis of the vertebrae, resulting in upper back pain. The same exaggerated dorsal curving of the thoracic spine in elderly women results from compression fractures of the vertebrae due to osteoporosis (see Chapter 6, The Clinic: Diseases of the Skeletal System) and is called a **dowager's hump.** An isolated hump that was once seen in children with tuberculosis of the spine was called a **gibbous.**

The increased incurving of the lumbar spine is called **hyperlordosis** (or swayback) and is one of the many causes of chronic low back pain. Poor posture, obesity, and pregnancy may be contributing factors. Flattening of the sacral curve can obstruct the normal passage of the child through the birth canal and cause problems with delivery. **Kyphosis** is an exaggeration of the posterior curve of the thoracic spine. It may be associated with poor posture (as is indicated by rounded shoulders), congenital deformities, or diseases that cause bone destruction. Usually, there is no permanent loss of function, but if back pain becomes severe, surgical correction may be advised.

Scoliosis (skō′lē·ō′sis) is the side-to-side and associated rotation of the spine (Figure 1). It affects 2% of all people in the United States over the age of 14. Only about one-tenth of these require treatment. The most common type of scoliosis is the developmental or inherited type, which is most commonly seen in teenaged girls and is often entirely asymptomatic. It is initially not noticeable under clothing and symptoms of pain are uncommon. It is very important that scoliosis be identified early in its course so that patients can be followed to detect any signs of serious progression of the defect. A high proportion of cases of this type of scoliosis are mild and self-limited in that the progression of the curvature often ends when bone growth is completed in the early 20s. Minor scoliosis requires no treatment other than observation and perhaps exercises to maintain mobility of the spine. A few cases, however, tend to be progressive and if not treated can lead to disfigurement and compromise of the heart and lungs in adult life.

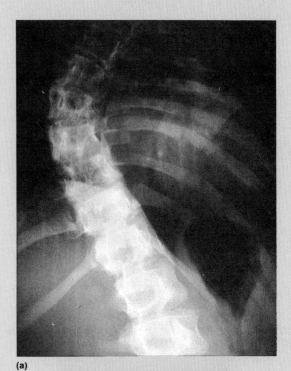

(a)

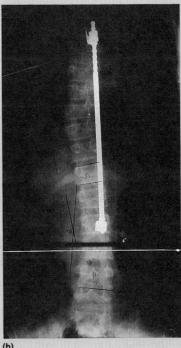

(b)

Figure 1 Scoliosis. (a) X ray showing characteristic curved spine caused by scoliosis. (b) X ray showing correcting pin in place to straighten spine.

spondylolysis: (Gr.) *spondylos*, vertebra + *lysis*, dissolution

This more progressive type requires treatment with prolonged bracing or surgical correction with pins. Other more rare causes of scoliosis include congenital defects of the vertebrae, such as **hemivertebra** (in which only half of the vertebra forms), and tumors, such as **neurofibromatosis** (elephant man's disease). Both of these conditions cause the spine to tilt. Poliomyelitis was also once a common cause of scoliosis due to the muscle paralysis associated with this disease. A recent development in treating scoliosis has been the use of electrical stimulation of the muscles at selected areas of the curve, usually administered during sleep, to help correct the abnormal curvatures. Most school systems today have a required examination that detects early scoliosis so that it can be treated.

Trauma to the Spine

The **intervertebral disks** are cushions of fibrocartilage that are positioned between the bodies of the vertebrae located between the axis and the sacrum. A disk consists of a soft, gelatinous center **(nucleus pulposus)** surrounded by a tough coating of fibrocartilage **(annulus fibrosis)**. A common cause of severe and persistent low back pain and associated leg pain called **sciatica** (sī·ăt′ĭ·kă) is a rupture or herniation of one of these intervertebral disks. In a younger patient the rupture is usually associated with an injury, such as when the spine is forcefully flexed to lift excessive weight or following a rear-end car collision. The excessive flexion force on the disk causes a tear in the annulus fibrosus and partial extrusion of the nucleus pulposus. The tear in the annulus fibrosus usually occurs in the posterior aspect of the annulus. If the tear is incomplete, there may be just a bulging of the nucleus; this is referred to as a **herniated intervertebral disk.** This injury creates pressure on the spinal cord itself or on one of the intervertebral nerve roots and can cause pain or muscle weakness in the area that is innervated.

Patients with damaged intervertebral disks usually complain of back pain and **referred pain** (pain felt in an area far from the injured area) down the back of the leg. Any movement that tends to stretch the entrapped nerve root such as straight leg raising or prolonged sitting will reproduce the pain. Although the most common site for disk injury is in the lower lumbar spine, it can also occur in the cervical spine and produce symptoms of pain, numbness, and weakness in the arm or hand. The treatment of a ruptured intervertebral disk usually consists of rest, pain medication, and specific exercise programs to relieve the nerve root pressure and to strengthen the back muscles. A few patients with severe symptoms or those who do not respond to the conservative treatment may require surgical removal of the herniated disk material. Another technique consists of injecting an enzyme called chymopapain into the nucleus pulposus; the enzyme causes the gelatinous material to liquify, thereby relieving the pressure. The injection is made through the skin with X-ray guidance and avoids open surgery.

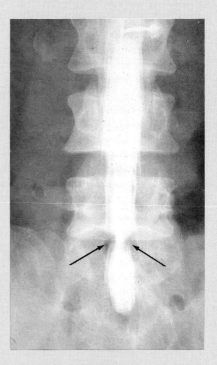

Figure 2 X ray showing ruptured disk.

Another common cause of low back pain in young adults is called **spondylolysis** (spŏn′dĭ·lŏl′ĭ·sĭs), or ruptured disk, which is a tear in the annulus fibrosis that allows the nucleus pulposus to protrude (Figure 2). It is caused by a defect in the pars interarticularis of the L4 or L5 vertebra. This defect may be a failure of the pars to calcify or may be caused by a stress fracture. The protruding material may press on nerve roots where they exit the spinal cord causing symptoms of pain, tingling, or loss of function. The most common cause is a sudden lifting or jerking movement. If the defect occurs on both sides of the vertebrae, it can result in instability with forward slipping of the L5 body onto the S1 body; it is then called **spondylolisthesis.** This problem is seen frequently in gymnasts and interior linemen in football. It causes low back pain and sciatic nerve pain with hyperextension of the spine. In both spondylolysis and spondylolisthesis prolonged bed rest helps to reduce pressure on the disk to allow healing.

Fractures of the Spine

Spinal fractures may occur when excessive force is applied to the spine in flexion, extension, or along the long axis of the spine. The most common spinal fracture is a compression fracture of the vertebra body. This type of fracture

(continued)

quadriplegia: (L.) *quattuor*, four + (Gr.) *plege*, stroke

THE CLINIC (continued)

DISORDERS OF THE SPINE

may occur when a patient lands heavily on the feet or buttocks from a height. It usually involves the lower thoracic or upper lumbar vertebrae and seldom results in neurologic complications. A similar type of compression fracture may occur without serious trauma in patients with tumors involving the vertebrae or in the elderly with osteoporosis.

A more serious spinal fracture can occur when there is forced extension or flexion of the spine that results in the fracture of the posterior elements of the vertebrae. This type of fracture may result in instability of the spine with dislocation of the vertebrae and damage to the spinal cord. One cause is diving into shallow water and hitting the bottom of the pool, which causes damage to the spinal cord at the C5 to C6 level. A fall on the stairs or on slippery surfaces may result in the patient striking the chin and forcefully extending the neck, thus damaging the posterior elements of the cervical vertebrae. Automobile accidents are the largest cause of spinal fractures and usually result in injury to the cervical or upper lumbar spine. Fractures may also occur in the dorsal or lateral processes of a vertebra from a direct blow; these are painful injuries but are usually of no serious consequence.

The ultimate outcome of these injuries depends on the extent of injury to the spinal cord. A simple compression fracture can be treated with bed rest and bracing until the fracture heals. Cervical or lumbar fractures complicated by spinal cord damage may result in sudden death, such as the "hangman's fracture" of the C2 vertebra. Permanent paralysis of all four extremities **(quadriplegia)** can be the result of a C5 to C6 injury. Paralysis of just the legs **(paraplegia)** can occur in a lumbar spine injury.

Diseases and Developmental Abnormalities

Paget's disease is a common chronic skeletal disease of unknown cause affecting about 2% of the population over age 40. The disease affects local areas of bone, especially the pelvis, femur, skull, and tibia, but it may be more widespread. Histologically, there is an initial increased absorption of bone by osteoclasts followed by the replacement of normal marrow with vascular fibrous connective tissue. The resorbed bone may then be replaced by disorganized, coarse trabecular bone. This gives the bone its characteristic mosaic pattern on an X ray. Many affected patients have no complaints related to the disease and are only diagnosed because of a routine X-ray examination. Occasionally, the person will report pain and swelling or will notice a deformity of a long bone. Usually it is a chronic disorder requiring no treatment. Complications, including paralysis, can result from overgrowth of pagetic bone at the base of the skull **(platybasia)** due to compression of the brain stem.

In this chapter we studied the anatomy of the skeletal system. You learned the anatomical landmarks associated with articulations between bones, with attachments of muscles, and with passageways for blood vessels and nerves. In the next chapter we will examine the manner in which bones articulate with one another.

STUDY OUTLINE

Divisions of the Skeleton (p. 131)

The 206 bones of the body are organized into the axial skeleton and the appendicular skeleton. The *axial skeleton* consists of the skull, hyoid bone, vertebral column, and rib cage. The *appendicular skeleton* includes the bones of the appendages, including the upper extremity (arm and forearm) and the lower extremity (thigh and leg). The upper extremities are attached to the rib cage by the pectoral girdle and the lower extremities are attached to the vertebral column by the pelvic girdle.

Skull (pp. 131–142)

The *skull* consists of 28 bones arranged into *cranial* and *facial groups*, plus the hyoid bone, making a total of 29. During embryonic and fetal life skull bones are separated from each other by *fontanels*.

Cranial Group These bones serve primarily to surround and protect the soft tissues of the brain. The cranial group includes the *frontal*, *parietals*, *temporals*, *occipital*, *sphenoid*, and *ethmoid*.

Facial Group The facial group is concerned primarily with the protection and functions of the eyes and nose and the functions of the mouth. Bones included here are the *inferior nasal conchae*, *vomer*, *lacrimals*, *nasals*, *maxillae*, *palatines*, *zygomatics*, and *mandible*.

Sutures and Wormian Bones The bones of the skull articulate through immovable joints called *sutures*. The *coronal*, *sagittal*, *lambdoidal*, and *squamosal* sutures are prominent anatomical landmarks in the skull. *Wormian bones* are small bones that are found at many locations along the sutures, particularly the squamosal and lambdoidal sutures.

Hyoid Bone (pp. 142–143)

The U-shaped *hyoid* is a sesamoid bone held between the mandible and larynx by ligaments and muscles. It serves as the place of attachment for several small muscles used in speaking, swallowing, and moving the tongue.

Vertebral Column (pp. 143–147)

The *vertebral column* (or *spine*) is composed of individual *vertebrae* arranged in a longitudinal series. It forms the vertical axis of support for the head and upper part of the body and protects the spinal cord. The vertebrae are separated from one another by oval *intervertebral disks*. The vertebral column is divided into five distinct regions: *cervical, thoracic, lumbar, sacral,* and *coccygeal.*

> ***Curves of the Vertebral Column*** The vertebral column of an infant is curved anteriorly due to the position of the fetus in the mother's uterus. As the child grows and begins to walk, four curves develop in the spine: *cervical, thoracic, lumbar,* and *sacral*. These curves help balance the weight of the upper body over the legs and help absorb shocks during walking, running, and jumping.
>
> ***A Typical Vertebra*** Most vertebrae have the same basic structures: *body, vertebral arch, vertebral foramen, spinous process, transverse processes,* and *articular processes.*
>
> ***Types of Vertebrae*** There are three types of vertebrae with different shapes and functions. Seven *cervical vertebrae* comprise the spine of the neck. The first cervical, the *atlas*, articulates with the skull, while the second, the *axis*, provides for rotation of the head. All cervical vertebrae possess *transverse foramina*.
>
> Twelve *thoracic vertebrae* provide articulation with the ribs.
>
> Five *lumbar vertebrae* make up much of the abdominal part of the spine. They have massive, blunted bodies and thick, spinous processes.
>
> ***Sacrum*** The *sacrum* consists of five fused sacral vertebrae. The vertebral canals of the individual vertebrae are fused into a *sacral canal*.
>
> ***Coccyx*** The *coccyx* is formed from the fusion of three to five rudimentary *coccygeal vertebrae*, which lack most of the features of other vertebrae.

Thorax (pp. 147–148)

The *thorax*, or chest region, is that portion of the trunk associated with the thoracic vertebrae. It includes the sternum and rib cage.

> ***Sternum*** The *sternum* lies in the anterior midline of the thorax and consists of three units: *manubrium, body,* and *xiphoid process*.
>
> ***Ribs*** The *rib cage* consists of the sternum and 12 pairs of ribs and *costal cartilages*. Ribs 1 through 7 are called *true ribs*, 8 through 10 *false ribs*, and 11 and 12 *floating ribs*.

Pectoral Girdle (pp. 148–149)

The *pectoral girdle* consists of the clavicles and scapulae and serves to attach the arm to the axial skeleton.

> ***Clavicle*** The *clavicles* (collarbones) are S-shaped.
>
> ***Scapula*** The *scapulae* (shoulder blades) form the major bones of the pectoral girdle. Each is triangular and has several prominent processes that serve as attachments for muscles and ligaments.

Upper Extremity (pp. 149–152)

The upper extremity consists of the arm, forearm, wrist, and hand.

> ***Humerus*** The *humerus* is the large bone of the arm. It articulates with the scapula at the proximal end (shoulder) and with the ulna and radius at the elbow.
>
> ***Ulna and Radius*** The *ulna* and *radius* are the bones of the forearm.
>
> ***Manus*** The *manus*, or hand, consists of the *carpus, metacarpus,* and *phalanges*. The carpus is composed of eight small bones (*carpals*) arranged in two rows of four each. The metacarpus is composed of five elongate *metacarpals* that form the palm of the hand. The phalanges are digital bones that form the fingers.

Pelvic Girdle (pp. 152–154)

The *pelvic girdle* is composed of paired pelvic bones called *os coxae* or *coxal bones*.

> ***Os Coxae*** The os coxae articulate solidly with the sacrum and support the weight from the upper body. The os coxae and sacrum enclose a space called the *pelvis*. Each coxal bone develops from three embryonic regions that fuse into one bone in the adult. Three bone regions of the os coxae are the *ilium, ischium,* and *pubis*.
>
> ***Differences Between Female and Male Pelvic Bones*** The female pelvic bones are wider, lighter in weight, and less thick than those of males, and the female acetabulum is smaller and directed more anteriorly than in males of comparable size. The pubic arch forms an angle of less than 90° in males, but an angle of more than 90° in females. The inlet of the true pelvis is narrow and heart shaped in males, but circular and wide in females.

Lower Extremity (pp. 154–157)

The lower extremity includes the thigh, knee, leg, ankle, and foot.

> ***Femur*** The *femur* (thigh bone) articulates with the pelvic girdle at the acetabulum and is the longest bone in the body.
>
> ***Patella*** The *patella* (kneecap) is a floating sesamoid bone that protects the knee joint and increases the leverage of the large anterior thigh muscles.
>
> ***Tibia*** The *tibia* (shinbone) is the larger of the two bones of the lower leg.
>
> ***Fibula*** The *fibula* is the smaller of the bones of the lower leg.
>
> ***Pes*** The *pes* (foot) is composed of *tarsus, metatarsus,* and *phalanges*. Seven *tarsal* bones comprise the ankle region. Five *metatarsals* comprise the arch of the foot, and the phalanges comprise the toes. Three resilient, shock-absorbing arches—the *medial longitudinal, lateral longitudinal,* and *transverse*—are found in the foot.

SELF-TEST OF CHAPTER OBJECTIVES

Fill-In Questions
1. The division of the skeletal system that consists of skull, vertebral column, and rib cage is the _____ skeleton.
2. The _____ girdle serves as the connecting link between the leg and the vertebral column.
3. The _____ girdle consists of four separate bones, whereas the _____ girdle consists of only two bones.
4. The _____ girdle serves as the connecting link between the upper extremity and the vertebral column.
5. The _____ is a cranial bone in the skull possessing three pairs of winglike extensions.
6. The _____ is a cranial bone that articulates with the vertebral column.
7. The cribiform plate is a part of one of the cranial bones called _____.
8. The _____ suture separates the two parietal bones from one another.
9. A bone that forms in a tendon and does not articulate solidly with other bones is a _____ bone.
10. Most vertebrae possess a large opening called the _____.
11. The first and second cervical vertebrae are called the _____ and the _____, respectively.
12. The most superior section of the sternum is called the _____.
13. The costal cartilage connects true ribs to the _____ and false ribs to the _____.
14. The constriction called the neck on a rib lies between the _____ and the _____.
15. In the foot, the _____ longitudinal arch extends from the calcaneus through the talus and navicular, whereas the _____ longitudinal arch extends from the calcaneus through the cuboid.

Matching Questions
Match the anatomical landmarks with the skull bones on the right:

16. sella turcica
17. coronoid process
18. external auditory meatus
19. foramen magnum
20. mastoid process

a. occipital
b. sphenoid
c. temporal
d. mandible

Match the anatomical feature with the specific vertebra on the right:

21. facets for rib articulations
22. transverse foramina
23. odontoid process
24. auricular process
25. promontory

a. axis
b. sacrum
c. thoracic vertebra
d. lumbar vertebra

Match the anatomical landmarks with the bones on the right:

26. olecranon fossa
27. semilunar notch
28. intertubercular groove
29. radial tuberosity
30. lesser tubercle

a. humerus
b. radius
c. ulna

Match the anatomical landmarks with the bones on the right:

31. lateral malleolus
32. greater trochanter
33. linea aspera
34. medial malleolus
35. gluteal tuberosity

a. femur
b. tibia
c. fibula

Match the anatomical landmarks with the bones on the right:

36. acetabulum
37. conoid tubercle
38. infraspinous fossa
39. iliac crest
40. glenoid fossa

a. clavicle
b. scapula
c. os coxae

Multiple-Choice Questions
41. Which of the following bones is *not* a part of the appendicular skeleton?
    a. clavicle   c. hyoid
    b. femur   d. radius
42. Which of the following bones is *not* a part of the axial skeleton?
    a. coccyx   c. sternum
    b. scapula   d. mandible
43. Which of the following is *not* a part of the facial group of bones?
    a. lacrimal   c. vomer
    b. frontal   d. zygomatic
44. The inner ear is found in which of the following skull bones?
    a. temporal   c. occipital
    b. sphenoid   d. temporal
45. A group of cavities located within the bones that surround the nasal area are known as
    a. sinuses   c. fossae
    b. foramina   d. fissures
46. The upper extremity articulates with the pectoral girdle through the
    a. acetabulum   c. intertubercula fossa
    b. glenoid fossa   d. coronoid fossa
47. The lower extremity articulates with the pelvic girdle through the
    a. acetabulum   c. intertubercular fossa
    b. glenoid fossa   d. gluteal fossa
48. Which of the following is an unpaired bone?
    a. vomer   c. nasal
    b. palatine   d. maxilla
49. These bones are found in the sutures of the skull:
    a. sesamoid bones   c. fossorial bones
    b. wormian bones   d. floating bones
50. How many bones make up the carpus?
    a. five   c. seven
    b. six   d. eight

Essay Questions
51. List the bones that comprise the cranial group of skull bones and describe the major function of each.
52. List the bones that comprise the facial group of skull bones and describe the major function of each.

53. Describe a typical vertebra.
54. List the characteristics that can be used to distinguish between (a) cervical and thoracic vertebrae, (b) cervical and lumbar vertebrae, and (c) lumbar and thoracic vertebrae.
55. Describe the features used to distinguish among true ribs, false ribs, and floating ribs.
56. Describe the arches found in the foot.
57. Discuss the importance of the talus in distributing body weight to the arches of the foot.
58. List the foramina that pierce the sphenoid bone and the structures that pass through them.
59. Describe the curves of the vertebral column.
60. Discuss the major functional differences between the axial and appendicular skeletons.

Clinical Application Questions
61. Describe the condition known as craniosynostosis.
62. Discuss fractures of the cribriform plate, nasal bones, and zygomatic bones in terms of effects on the senses.
63. Compare and contrast congenital metatarsus varus (pigeon toe) with congenital inversion of the foot.
64. Discuss the effects of scoliosis and hyperlordosis on the curvature of the spine.
65. Describe some of the effects that may result from a herniated intervertebral disk.

8
Articulations: The Joints Between Bones

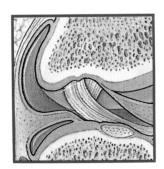

In the preceding chapter we concentrated on the bones as individual structures that fit into a framework that provides body support. But in addition to providing structural support, most of the bones of the skeleton are connected to one another by **articulations** or movable **joints.** Contraction of the muscles connected to these bones produces movement at their junctions. This action is how the body moves from one place to another and manipulates things in the environment. In this chapter the joints that make such movement possible will be studied.

Two general approaches to the study of joints can be taken. One is based on function and the other on structure. Both approaches are included in this chapter. First, articulations between skeletal elements such as cartilages and bones are discussed in terms of their general ability to allow the movement of body parts. Next, the structural nature of the junctions between bones and cartilages is discussed.

In this chapter we will examine

1. the classification of the joints of the skeleton based on both structural and functional differences,
2. the differences among synarthroses, amphiarthroses, and diarthroses with respect to the amount of movement these joints allow,
3. the structure of fibrous, cartilaginous, and synovial joints and examples of each,
4. the typical movements of the human skeleton as they relate to the nature of the joints that link the bones together, and
5. some of the common diseases and disorders of the joints.

synarthroses: (Gr.) *syn*, together + *arthron*, joint
amphiarthroses: (Gr.) *amphi*, on both sides + *arthron*, joint
diarthroses: (Gr.) *dis*, two + *arthron*, joint

synostosis: (Gr.) *syn*, together + *osteon*, bone + *osis*, condition
syndesmosis: (Gr.) *syn*, together + *desmos*, ligament + *osis*, condition

Articulations

After completing this section, you should be able to:

1. **Distinguish between functional and structural classifications of joints.**
2. **Describe types of joints based on the amount of movement they allow.**
3. **List the types of materials present in the joint cavity that allow classification on a structural basis.**
4. **Describe the structure of fibrous, cartilaginous, and synovial joints.**

Functional Classification

The functional classification of articulations is based on the amount of movement that can occur between the bones involved. In some articulations the skeletal elements are held so tightly together that virtually no movement is possible. These joints are called **synarthroses** (sĭn″ăr·thrō′sēs). The sutures between the skull bones (discussed in Chapter 7) are examples of synarthroses. **Amphiarthroses** (ăm″fē·ăr·thrō′sēs) are a second type of joint that permits a small amount of movement between the bones. The pubic symphysis and the joint between the bodies of two vertebrae and the intervening intervertebral cartilage are examples of amphiarthroses. While only a slight amount of movement is possible between two adjacent vertebrae, considerable bending and twisting of the vertebral column is possible when these movements are added up along the entire length of the spine.

Most of the joints in the body are **diarthroses** (dī″ăr·thrō′sēs), a third type of articulation that allows a considerable amount of movement. Freely movable joints such as the knee, elbow, wrist, and shoulder are good examples of diarthroses.

Although all joints can be placed in one of these functional categories, complications arise when we consider the developmental histories of some joints. For example, the joint between the auricular surface of the sacrum and the ilium (the sacroiliac joint) may pass through all three stages of movement during an individual's lifetime. This joint develops as a diarthrosis, with a narrow cavity between the two bones. In elderly people movement in the joint is restricted as connective tissue fills the joint cavity; thus it becomes an amphiarthrosis. In extreme cases sacroiliac movement may be prevented altogether by the arthritic fusion of the two bones (synarthrosis).

Structural Classification

While a functional classification is useful in understanding the *amount* of movement possible between two bones, it does not explain the reason for this movement associated with each type of joint. To understand the structural differences that make this movement possible, it is necessary to study the histology of the joints.

During the development of the skeleton, a space is present between adjacent embryonic bones. This space is the **joint cavity.** In some cases this cavity becomes filled with fibrous connective tissue, and that joint eventually develops into a **fibrous joint.** In other cases the joint cavity fills with cartilage, becoming a **cartilaginous** (kăr″tĭ·lăj′ĭ·nŭs) **joint.** The majority of joints, however, have cavities that persist as fluid-filled spaces called **synovial** (sĭn·ō′vē·ăl) **cavities,** and these joints develop into **synovial joints.**

Fibrous Joints

In fibrous joints, embryonic mesenchyme tissue between developing bones produces a dense network of short, sturdy, collagenous fibers. These fibers stretch across the joint cavity and hold the bones securely together, thus restricting their movement. In some instances the two bones grow close to one another in an interlocking pattern, and the cavity is filled with very short fibers. Such fibrous joints are the **sutures** found between the skull bones (Figure 8-1). Sutures normally assume their adult form when an individual reaches the age of 25 years. At this time they are considered to be true synarthroses, or immovable joints. After about age 50, the fibrous connective tissue in the sutures is slowly replaced by bone, and the skull bones slowly fuse in a condition known as **synostosis** (sĭn″ŏs·tō′sĭs), resulting in an obliteration of the suture in elderly people. This ossification process also occurs in certain cartilaginous joints. Epiphyseal plates of long bones and the three bones that make up the os coxa are examples of synostoses.

The joint between the tibia and fibula at their distal ends also contains dense, fibrous connective tissue, but the fibers here are longer and denser than those found in a suture. This type of joint is an example of a **syndesmosis** (sĭn″dĕs·mō′sĭs) (Figure 8-2). The syndesmosis of the tibia and fibula is essentially immovable and is functionally a synarthrosis. Other syndesmoses are found along the borders of the tibia and fibula and along the radius and ulna. Fibers here are long, and the amount of movement is sufficient for these two joints to be classified functionally as amphiarthroses.

The articulation of the teeth within sockets in the maxilla and mandible constitute a special type of fibrous joint called

gomphosis: (Gr.) *gomphosis*, to bolt together
periodontal: (Gr.) *peri*, around + *odous*, tooth
synchondrosis: (Gr.) *syn*, together + *chondros*, cartilage + *osis*, condition

a **gomphosis** (gŏm·fō´sĭs). In this joint one skeletal element fits into another like a peg in a pegboard. The fibrous material in the joint cavity forms the **periodontal ligament**.

Cartilaginous Joints

During development of cartilaginous joints, the cavity becomes filled with cartilage. Such a joint that has its cavity

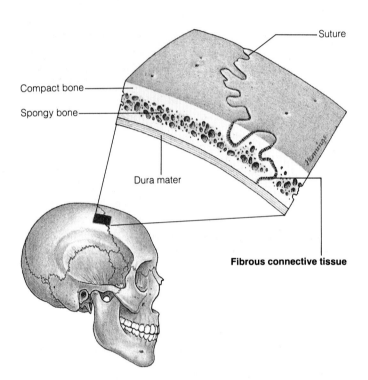

Figure 8-1 Sutures are narrow articulations found between the bones of the cranium. Fibrous connective tissue within the suture serves to firmly anchor the two bones together.

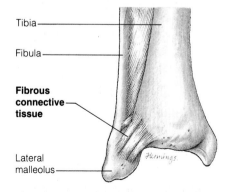

Figure 8-2 The joint between the tibia and fibula at their distal ends is a type of fibrous joint called a syndesmosis. The connective tissue fibers here are longer than in a suture.

CLINICAL NOTE

GOUT AND PSEUDOGOUT

Gout is a type of acute arthritis caused by the accumulation of uric acid crystals in the synovial fluid of the joints. Uric acid is a waste product formed by the metabolism of proteins. Increased blood levels of uric acid are present in about 10% of the population, although only 1 in 20 of these patients go on to develop gout. The disease is usually seen in obese males in their 40s.

Precipitation of uric acid crystals in the soft tissues of joints causes inflammation and severe, disabling pain. Typically, the big toe is affected and often the instep, ankles, heels, and knees as well. An attack may be triggered by trauma, alcohol ingestion, or eating rich foods. Pain usually lasts from one day to one week and is followed by an asymptomatic phase. Recurrent attacks are common and usually more severe.

Deposits of crystalline urate may occur in the cartilage, tendons, synovial membranes, and soft tissues of untreated patients, usually along the ear or ulnar surface of the forearm. Current treatment includes antiinflammatory and prophylactic drugs to decrease uric acid levels in the blood.

Pseudogout (or **chondrocalcinosis**) is also a type of acute (or subacute) arthritis, but in contrast to gout, it is caused by deposition of calcium crystals in the joint. Microscopic examination of the synovial fluid is the only way to distinguish it from gout. The reason calcium is deposited in the joints is uncertain. Pain, swelling, redness, and warmth can be associated with the onset of the disease, which usually involves the knee and other large joints. It most commonly affects the elderly.

filled with hyaline cartilage is known as a **synchondrosis** (sĭn˝kŏn·drō´sĭs). One example of a synchondrosis in the human skeleton is the epiphyseal plate. During growth of long bones through endochondral ossification, the diaphysis is held to the epiphyses by the epiphyseal plate, a flattened area of hyaline cartilage (see Chapter 6). This connection constitutes an immovable synchondrosis. This joint persists for many years during development of the bone, but eventually it is eliminated when the epiphyseal plate ossifies. Another example of synchondroses are the costal cartilages that connect the ribs to the sternum; these persist throughout an individual's life.

A **symphysis** (Figure 8-3) is a cartilaginous joint in which the joint cavity has become filled with fibrocartilage. The intervertebral disks found in the vertebral column maintain this kind of joint between the bodies of adjacent vertebrae. Another example is the pubic bones, which are joined by the pubic symphysis. Both of these symphysis joints have slight movement and thus are functionally classified as

> **CLINICAL NOTE**
>
> ### SYNOVITIS
>
> Inflammation of the synovial membrane, the lining of synovial joints, is called **synovitis** (sĭn″ō·vī′tĭs). This membrane is the most active tissue in the joints and is responsible for most of the pathologic changes that occur there. Injury to a joint readily results in an outpouring of synovial fluid and swelling of the joint, called **effusion**. Inflammation of the joint, as in many types of arthritis, can cause thickening of the synovial membrane, effusion, and in some cases destruction of the cartilaginous surfaces by invasion and erosion. Rheumatiod and psoriatic arthritis, for example, are characterized by inflammation and thickening of the synovial membrane.
>
> **Villous pigmented synovitis** is an uncommon cause of painful swollen joints, usually in young men. The synovial membrane is greatly enlarged with branching formations of brownish nodules which may cause the joint to lock. The cause is unknown and the treatment consists of surgically removing the damaged synovial membrane.

beyond the anatomical position, as in bending the head backward) of a joint which could result in injury to the synovial membrane. The ends of the ligaments are also continuous with the periosteum of the bones in the joint. In some joints, such as the mandibular and knee joints, pads of fibrocartilage present between the bones serve as additional

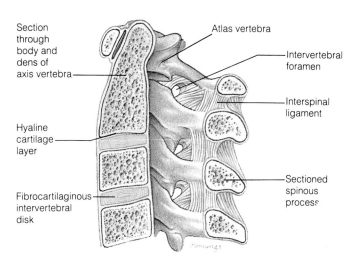

Figure 8-3 The fibrocartilage connection between two adjacent vertebrae is referred to as a symphysis.

amphiarthroses. Generally speaking, a symphysis is more flexible than a synchondrosis.

Synovial Joints

In a synovial joint the fluid-filled joint cavity persists. Figure 8-4 shows a diagrammatic view of a typical synovial joint. It is more complex than either fibrous or cartilaginous joints. The ends of two bones connected by a synovial joint are covered by a thin layer of hyaline cartilage (called the **articular cartilage**), whose free surface lacks a perichondrium. This articular cartilage protects the ends of the two participating bones from the trauma of continual contact. The joint cavity is enclosed in a capsule composed of dense, fibrous connective tissue. The capsular fibers penetrate the periosteum of the bones, creating a strong bridge between them. The inner surface of the capsule is modified into a **synovial membrane** which is provided with an abundant blood supply (see Clinical Note: Synovitis). The membrane secretes **synovial fluid** composed of lymph (see Chapter 24) and **hyaluronic acid,** a major component of the matrix of loose connective tissue. Synovial fluid serves as a lubricant reducing friction and general wear and tear on the articular cartilages of bones.

Synovial joints usually allow free and sometimes extensive movement between the bones involved and are thus functionally classified as diarthroses. **Ligaments** are elongate straps of fibrous connective tissue that reinforce the joint capsule. They help to prevent **hyperextension** (extension

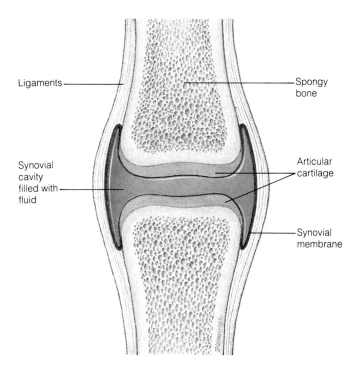

Figure 8-4 Synovial joints, such as the idealized one shown here, are characterized by the presence of a fluid-filled sac between the two skeletal elements. The fluid contained within the sac serves as a friction-reducing lubricant.

168 Chapter 8 Articulations: The Joints Between Bones

bursa:	(Gr.) *bursa*, a leather sac	
ginglymus:	(Gr.) *ginglymos*, hinge	
condyloid:	(Gr.) *kondulos*, knuckle + *eidos*, form, shape	

shock absorbers and may also increase the mobility of the joint.

The knee joint is a complex articulation (Figure 8-5) and contains all the structures usually associated with a synovial joint (see section on The Knee: A Special Joint). In addition to the synovial cavity, several sacs called **bursae** are present. They are usually continuous with the synovial cavity and are filled with synovial fluid. Bursae are located near the joints between muscles, muscles and ligaments, muscles and bones, ligaments and bones, and skin and bone. They increase the freedom of movement of body parts by reducing friction (see Clinical Note: Bursitis). **Tendon sheaths** are special bursae that surround elongate **tendons,** connective tissue that attaches a muscle to a bone (see Chapter 10). The internal fibrocartilage pads in the knee are called **articular disks** or **menisci** (mĕn·ĭs´kē). The muscles and tendons surrounding a knee joint aid in maintaining the proper position of the bones.

CLINICAL NOTE

BURSITIS

A **bursa** is a soft sac filled with lubricating fluid that eases the friction of muscles and tendons over bony prominences. There are about 150 separate bursae in the body. Occasionally, overuse or pressure cause inflammation of these sacs leading to **bursitis** (bŭr·sī´tĭs). The patient experiences aching and pain localized in a specific area over a joint which is aggravated by pressure and movement. Joggers may develop bursitis involving the feet or hips. Treatment consists of rest, heat, and aspirin to decrease pain and inflammation.

Movements of Synovial Joints

After completing this section, you should be able to:
1. List and describe six groups of synovial joints.
2. Describe the types of movements allowed by each of the six types of synovial joints.

Synovial joints exhibit several recognizable types of movement. They have been classified into six groups, characterized either by the type of movement they permit or by their appearance: gliding, hinge, ellipsoidal, pivot, ball-and-socket, and saddle joints (Figure 8-6).

Gliding Joints

Gliding joints, or **arthrodial** (ăr·thrō´dē·ăl) **joints,** allow a simple type of back-and-forth and side-to-side sliding movement between two bones. Articular surfaces involved are usually small and flat. Gliding joints are found between the articulating processes of vertebrae and between carpals and tarsals.

Hinge Joints

Hinge joints, or **ginglymus** (jĭng´lĭ·mŭs) **joints,** allow the motion typical of a hinge on a door, specifically, movement through an arc in one plane. The articular surface of one bone in a hinge joint is usually concave, while that of the other is convex. This produces a complementary surface that facilitates the hingelike movement of the joint. Typical hinge joints include the knee, elbow, ankle, and the interphalangeal joints of the fingers and toes.

Ellipsoidal Joints

Ellipsoidal joints, or **condyloid** (kŏn´dĭ·loyd) **joints** involve contact between an oval-shaped condyle on one bone

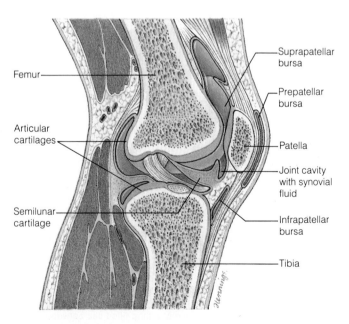

Figure 8-5 The knee joint. Note how the positions of the synovial cavities promote free movement of the bones.

prosthesis: (Gr.) *pros*, in addition + *tithenai*, to put
trochoid: (Gr.) *trokos*, wheel + *eidos*, form, shape

CLINICAL NOTE

HIP PROSTHESIS

The hip joint is a source of many problems in medicine. It is a common site for fractures in the elderly, particularly those with osteoporosis. It is subject to deformity in elderly patients with arthritis and in young people who have had congenital dislocation of the hips. Fortunately, surgeons are now able to replace a patient's hip with a functioning artificial hip joint called a **prosthesis** (prŏs′thē·sĭs) (see figure). A metallic ball and socket are inserted that resemble the ball-and-socket joint of the hip. This restores the function and comfort of even a severely damaged hip joint.

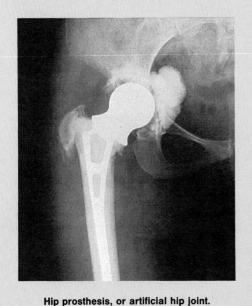

Hip prosthesis, or artificial hip joint.

articulation between the atlas and the axis in the neck and the proximal articulation between the head of the radius and the radial notch of the ulna.

Ball-and-Socket Joints

Ball-and-socket joints consist of a spherical knob or ball on one bone inserted into a concave spherical socket on

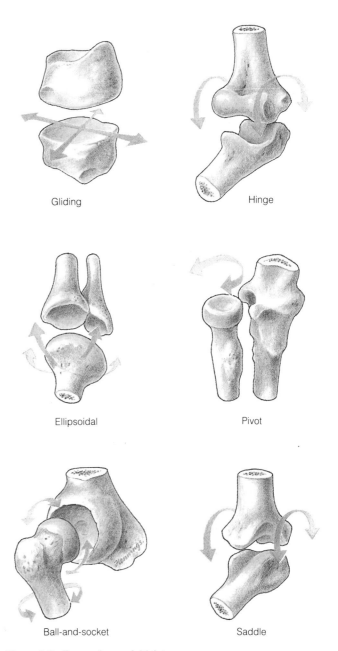

Figure 8-6 Types of synovial joints.

and an elliptical depression on another. Condyloid joints allow movement sometimes referred to as **biaxial movement**, which means through two arcs in two planes. Examples of condyloid joints include the major wrist joint between the radius/ulna and the carpals and between the metacarpals and the proximal phalanges.

Pivot Joints

Pivot joints, or **trochoid** (trō′koyd) **joints**, allow rotation of one bone relative to another. The articulating surfaces vary in shape, but they usually consist of a roughly cylindrical bone or process that rotates within a concave depression in the other bone. Examples of pivot joints are the

CLINICAL NOTE

RHEUMATOID ARTHRITIS

Rheumatoid arthritis is a chronic disease that most commonly affects women between the ages of 20 and 50. The usual manifestation of rheumatoid arthritis is bilaterally symmetrical inflammation of the small, non-weight-bearing joints of the hands and feet. However, the knees, hips, and jaw may eventually be affected. The disease begins in an unremarkable fashion with nonspecific complaints of tiredness, loss of appetite, loss of weight, and fever. The small joints of the hands may be sore and stiff. Sometimes, however, the disease begins abruptly with many painful, swollen joints.

Although this is considered to be an immune system disorder, both hereditary and environmental factors probably play a part in its cause. Most people who have rheumatoid arthritis have the rheumatoid factor present in their serum. This factor is an antibody to one of our own immunoglobulins (see Chapter 24). It is also found in the synovial fluid and synovial membranes of persons who have rheumatoid arthritis. The antibody combines with the immunoglobulin to form immune complexes which activate the complement system; this in turn begins the inflammatory process. This inflammatory process in the joint causes the synovial cells to enlarge. Eventually, the inflammatory process erodes the articular cartilage extending into the joint capsule. Damage to the ordinarily smooth joint surfaces causes instability of the joint. The pull of ligaments and tendons surrounding this unstable joint causes deviation and displacement of bones (see figure). Adjacent bones may also fuse.

Rheumatoid arthritis may also develop in children where it involves only one of a few large joints, usually the knee or ankle. This type of arthritis is usually seen in children between the ages of two and four, and it has a good prognosis, only rarely leading to joint destruction and disability. Initially, development of a limp should alert the clinician to the possibility of arthritis in the absence of injury.

About half of the children with arthritis experience symptoms unrelated to joints, including fever and skin rash or, more seriously, inflammation of the eyes, which may lead to loss of vision.

Treatment for both adult and juvenile rheumatoid arthritis centers around adequate rest, exercise, and anti-inflammatory drugs to reduce inflammation and pain.

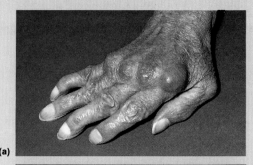

(a)

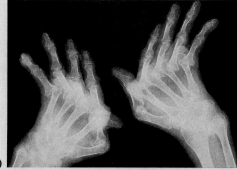

(b)

Rheumatoid arthritis. (a) Rheumatoid hand showing extreme deformities of the fingers and thumb. (b) X-ray view that reveals the dislocation of all metacarpophalangeal joints.

---

another bone. Ball-and-socket joints allow the maximum movement in a joint. The shoulder and hip joints are ball-and-socket types (see Clinical Note: Hip Prosthesis).

Saddle Joints

In **saddle joints**, also called **sellaris** (sĕl·ăr′ĭs) or **biaxial joints**, the articulating ends of the bones are saddle shaped, and this type of joint allows movement in two planes. The articulation between the metacarpal of the thumb and the trapezium of the carpus is a saddle joint, and it permits opposing the thumb against other fingers on the hand in a motion such as picking up a dime from a desk top.

Table 8-1 is a summary of the functional and structural classifications of the major joints in the human body.

Movements of the Skeleton

After completing this section, you should be able to:

1. Describe some of the anatomical factors that limit the extent to which joints can be opened.
2. Name and describe the major types of movement that have been designated with descriptive names.

Synovial joints contain a fluid-filled cavity between two bones that facilitates movement by reducing friction. However, the exact nature and extent of the movement of each joint depends on several factors. Ligaments and muscles

Table 8-1 A SURVEY OF IMPORTANT JOINTS IN THE HUMAN BODY

Joint	Structural Classification	Functional Classification	Movements
Skull			
Suture	Fibrous	Synarthrosis	None
Temporomandibular	Synovial (two synovial cavities separated by a fibrocartilage disk)	Hinge diarthrosis	Flexion, extension, abduction, and adduction
Vertebral Column			
Atlantooccipital	Synovial	Gliding diarthrosis	Slight lateral, and forward nodding
Atlantoodontoid process	Synovial	Pivot diarthrosis	Rotation
Articular process	Synovial	Gliding diarthrosis	Gliding
Vertebrae-intervertebral disk	Cartilage	Amphiarthrosis	Very slight
Pectoral Girdle			
Both clavicular	Synovial	Gliding diarthrosis	Gliding
Shoulder	Synovial	Ball-and-socket diarthrosis	Flexion, extension, abduction, adduction, rotation, and circumduction
Thoracic			
Rib-vertebra	Synovial	Gliding diarthrosis	Gliding
Sternocostal			
First rib	Synchondrosis	Synarthrosis	None
Remaining ribs	Synovial (cavity becomes filled with cartilage in older people)	Gliding diarthrosis	Gliding
Manubrium-sternum	Cartilage (synostosis fuses this joint in older people)	Amphiarthrosis	Slight
Sternum-xiphoid process	Cartilage (synostosis usually fuses this joint in older people)	Amphiarthrosis	Slight
Upper Extremity			
Elbow	Synovial	Hinge diarthrosis	Flexion and extension
Proximal radioulnar	Synovial	Pivot diarthrosis	Rotation
Distal radioulnar	Synovial	Pivot diarthrosis	Rotation
Wrist	Synovial	Saddle diarthrosis	Biaxial
Carpal	Synovial	Gliding diarthrosis	Gliding
Metacarpal 1–trapezium	Synovial	Saddle diarthrosis	Biaxial
Carpometacarpal	Synovial	Gliding diarthrosis	Gliding (slight)
Metacarpophalangeal	Synovial	Ellipsoidal diarthrosis	Biaxial
Interphalangeal	Synovial	Hinge diarthrosis	Flexion and extension
Pelvic Girdle			
Sacroiliac	Part fibrous and part synovial	Part amphiarthrosis and part diarthrosis	Very slight
Pubic symphysis	Cartilage	Amphiarthrosis	Very slight
Femur-acetabulum	Synovial	Ball-and-socket diarthrosis	Flexion, extension, abduction, adduction, rotation, and circumduction
Lower Extremity			
Knee	Synovial	Hinge diarthrosis	Flexion and extension
Proximal tibiofibular	Synovial	Gliding diarthrosis	Slight gliding
Distal tibiofibular	Fibrous	Amphiarthrosis	Very slight
Ankle	Synovial	Hinge diarthrosis	Dorsiflexion and plantar flexion
Tarsal	Synovial	Gliding diarthrosis	Slight
Tarsometatarsal	Synovial	Gliding diarthrosis	Slight
Metatarsophalangeal	Synovial	Ellipsoidal diarthrosis	Flexion, extension, abduction, and adduction
Interphalangeal	Synovial	Hinge diarthrosis	Flexion and extension

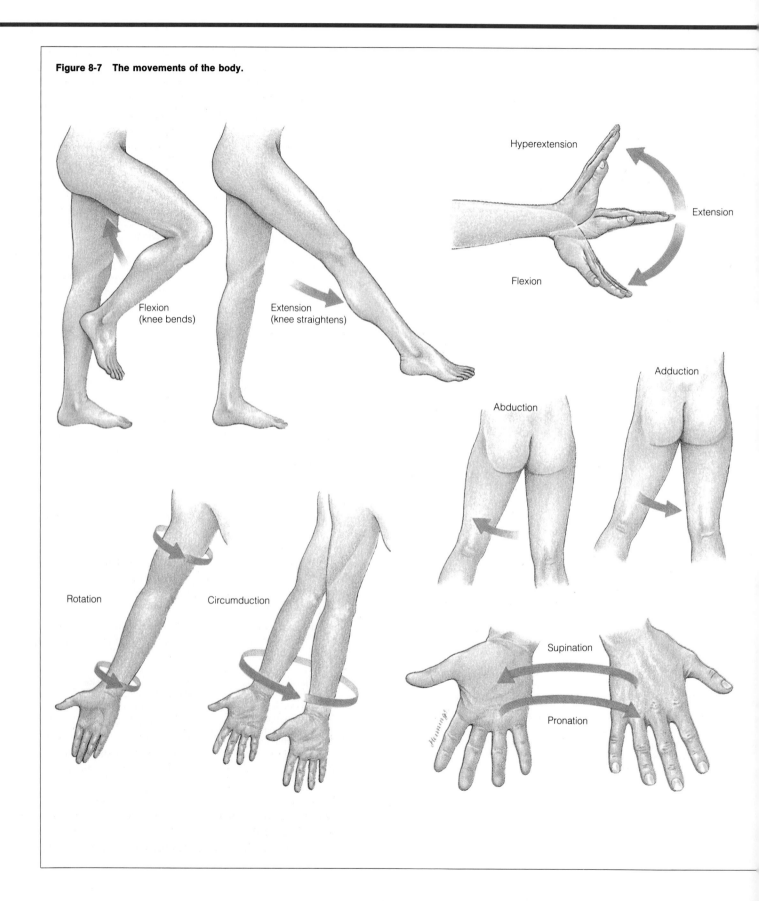

Figure 8-7 The movements of the body.

plantar: (L.) *planta*, sole of foot
dorsiflexion: (L.) *dorsum*, back + *flectionem*, bending
abduction: (L.) *abducere*, to lead away
adduction: (L.) *adducere*, to bring toward

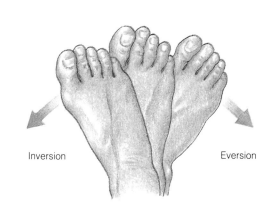

Inversion / Eversion

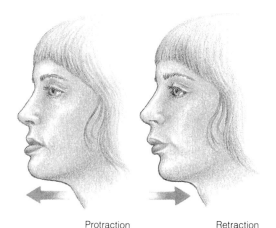

Protraction / Retraction

restrict and direct movement at a joint. For example, raising the arm, as in reaching for a book at the top of a bookshelf, is restricted by muscles from the chest and back that attach to the humerus and limit the ability to reach. Various forms of arthritis, or inflammation of the joints, may inhibit movement (see Clinical Note: Rheumatoid Arthritis). Finally, the mere presence of soft tissues often limits the degree of body movement. For example, muscles of the calf and thigh press against one another when the knee bends, thereby limiting the extent to which a knee joint can be bent. "Musclebound" athletes are an example of individuals who may suffer from restricted movement due to the excessive development of certain muscles.

Several specific joint movements are easily recognized and have been designated with descriptive names. These are shown in Figure 8-7 and described here.

Flexion is a movement that *decreases* the angle between two bones. The most common examples of flexion are bending the elbow and knee joints and closing the fingers in a clenched fist.

Extension, the opposite of flexion, *increases* the angle between two bones. Straightening out the arm and leg at the elbow and knee joints are examples of extensions. Flexion and extension at the ankle joint have special names. Movement of the foot in a downward direction is **plantar flexion** and movement of the foot upward is **dorsiflexion.**

Hyperextension is overextension of two bones at a joint beyond a 180° angle, as in bending the head backward. Due to sexual differences in the amount of mineral matrix deposited in the skeleton, females can usually hyperextend their elbows, while males usually cannot.

Abduction is movement of a body part *away* from the midline. Lifting the arm at the shoulder joint out from the body is abduction of the arm. In the case of the fingers and toes, the midline is in reference to the arm and leg, respectively. Consequently, abduction of the fingers and toes is spreading them.

Adduction is the opposite of abduction, that is, movement *toward* the midline. Lowering an outstretched arm is adduction. Adduction of fingers and toes involves bringing the outstretched digits back together.

Rotation is movement of a bone around its own axis or around another bone. Turning the head from side to side is rotation of the head and atlas around the odontoid process of the axis. One can rotate the humerus in the glenoid cavity by twisting the entire arm. Rotation of the head of the femur is also possible in the acetabulum. The most common type of rotation involves the forearm where the radius rotates freely around itself

arthroscopy: (Gr.) *arthron*, joint + *skopein*, to examine
supination: (L.) *supinatio*, to bend backward
circumduction: (L.) *circum*, around + *ducere*, to lead

CLINICAL NOTE

SPORTS INJURIES

Sports injuries have increased dramatically along with the recent renewed awareness of fitness and increased participation in sports. Joint dislocations, fractures, cartilage tears, sprains, and strains are very common.

Joint dislocations repeatedly involve the shoulder or fingers. When the humeral head is separated from the glenoid cavity, shoulder dislocation occurs. Stability of the shoulder joint is formed by a cuff of four muscles and their tendons, which together make up the rotator cuff. Baseball pitchers and football quarterbacks are prone to injuries involving the rotator cuff. Dislocation may occur when the arm, raised and ready to throw, is forcefully bent backward.

Torn cartilage is especially common in the knee. A sudden pop or snap usually signals the injury, followed by a locking of the joint or decreased motility and pain. Repair is likely to help restore function and decrease the chance of arthritis. **Arthroscopy** (ăr·thrŏs′kō·pē) is a surgical procedure designed to determine the extent of an injury and to repair it with a minimum of trauma. Prior to its development, a major operation with prolonged convalescence was required to expose the joint and repair the damage (for more on arthroscopy, see Chapter 6 Clinical Note: Torn Cartilage).

Fractures near the joints may occur if someone falls forward on outstretched hands. These are caused by the tremendous force imparted to the radius bone of the forearm and the clavicular bones. The radius most commonly fractures transversely about 3 cm from the wrist, which displaces the hand upward and backward. The clavicle, which helps to stabilize the shoulder to the chest, will fracture if there is sufficient posteriorly directed force against the shoulder joint. (In fact, the clavicle is the most commonly fractured bone in the body.)

Sprains are caused by exaggerated twisting or bending movements of a joint that stretch or tear the joint capsule and the supporting ligaments and tendons. Injury to blood vessels and nerves at the joint may also occur. Acute painful swelling limits the mobility and weight-bearing function of the joint.

Muscle strains are related to the stretching of a muscle but without tearing ligaments or tendons and without disrupting the continuity of the joint. Usually these are healed fairly quickly with application of local heat along with rest and analgesics.

---

and the ulna to turn the hand and wrist. This movement has special names: **supination** is the rotation of the forearm so that the palm faces forward, and **pronation** is turning the palm to the rear.

Circumduction is a complex movement. It involves abduction, adduction, flexion, extension, and some rotation. The movement is usually described as circumscribing a circle in midair. Two children turning a long jump rope for another child jumping in the middle execute this movement with their shoulder joints.

Inversion and **eversion** apply to movements of the foot at the ankle. Inversion turns the foot so the sole faces inward (medially). Eversion is turning the foot so that the sole faces outward (laterally). A person's first experience on ice skates is usually a long and painful series of inversions and eversions of the ankle.

Protraction is moving a body part straight out and away from the midline parallel to the ground. Two examples of protraction are jutting the chin out and sliding the scapulae out away from the vertebral column, as in "rounding" the shoulders forward.

Retraction is the opposite of protraction and involves return of the body part straight back toward the midline.

The Knee: A Special Joint

After completing this section, you should be able to:

1. Describe the anatomical features of the knee joint.
2. Explain the functions of these features.

Now that we have studied anatomical features and characteristic movements associated with joints, we will take a closer look at one of the most complex joints in the body: the knee.

The knee is the largest joint in the body and is actually composed of three joints that work together: an anterior **patellofemoral joint** between the patella and the patellar surface of the femur, a medial **tibiofemoral joint** between the medial condyles of the femur and tibia, and a lateral tibiofemoral joint between their corresponding lateral condyles (Figure 8-8a).

The patella is surrounded by the **patellar ligament,** which is a portion of the lower end of the tendon of the quadriceps femoris muscles of the thigh. Three bursae separate the patella from surrounding tissues (Figure 8-8a). These

The Knee: A Special Joint 175

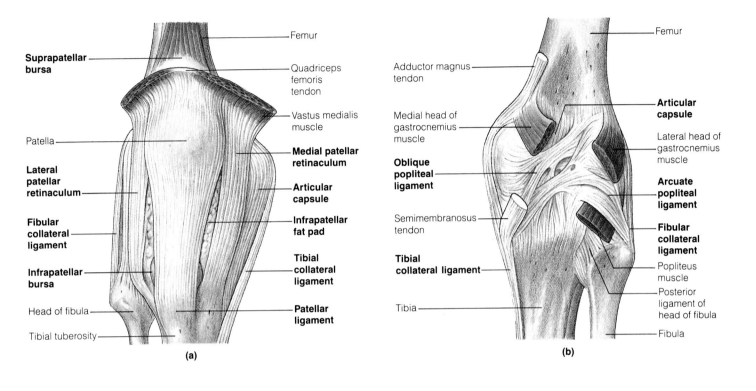
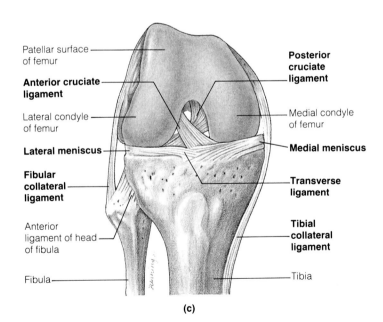
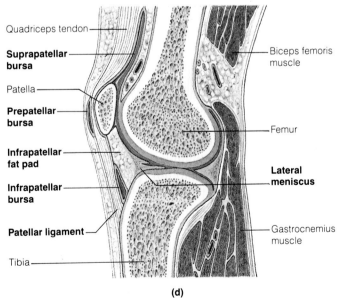

Figure 8-8 A detailed look at the knee. (a) Anterior, (b) posterior, and (c) flexed anterior views of the knee joint. (d) Sagittal section of the knee joint.

popliteal: (L.) *popliteus,* ham

| CLINICAL NOTE |

STABILITY OF JOINTS

The articulations of the body are to be admired for their marvelous complexity. Although we classify articulations according to their function and structure, many of them are unique. Often, the special character of an articulation is the very cause of its instability or dislocation. The shoulder joint, for example, has the greatest and most complex range of motion and is also the most common one to dislocate.

The stability of articulations is very important to the proper functioning of the joint. An injured joint may not fully dislocate, but it may be functionally unstable if it partially gives way. There are two types of stability: static and dynamic. **Static stability** is due to the intrinsic construction of the joint, the shape of the bones involved, the integrity of the cartilaginous surfaces, and the tension of the supporting ligaments and joint capsule. The **dynamic stability** of a joint depends on the strength and coordination of the muscles and tendons that produce the movements in the joint. Very often the dynamic stabilizing mechanisms can overcome a serious defect in the static stability of a joint. This is often the basis for **physical therapy** treatment. For example, in a knee injury that tears the anterior cruciate ligament, strengthening of the quadriceps muscle of the thigh (which is at the front of the knee near the ligament) can compensate for the torn ligament and result in a normally functioning knee.

---

fluid-filled sacs contribute to the relatively free gliding movements of the patella by reducing friction between the bone and adjacent ligaments, tendons, and skin.

Several other ligaments also connect the femur, tibia, and fibula. The **tibial collateral ligament** lies on the medial side of the joint and is attached to the medial meniscus. The **fibular collateral ligament** lies on the lateral side, and unlike its tibial counterpart, is external to the capsule. These two ligaments prevent side-to-side movements of the joint. The **oblique popliteal ligament** and the **arcuate popliteal ligament** are located on the posterior side of the knee (Figure 8-8b). They strengthen the lower posterior portion of the joint. The **anterior cruciate** and **posterior cruciate ligaments** are called **intraarticular ligaments** because they are located within the joint (Figure 8-8c). They connect the tibia and fibula and prevent twisting movements of the knee.

In addition to ligaments, the tendons of the quadriceps muscles hold the bones together at their anterior surfaces. These tendons are the **medial** and **lateral patellar retinacula.** Posteriorly, two fibrocartilage disks, the **medial** and **lateral menisci,** lie between the medial and lateral condyles (Figures 8-8c and d). They prevent the bones from rubbing together when moving, and they also act as shock absorbers. Menisci are connected to each other by a **transverse ligament** and to the head of the tibia by **coronary ligaments.**

Chapters 6, 7, and 8 have concentrated on the skeletal system. We have studied how the skeleton develops, how the bones fit together to support the body, and the nature of the articulations and movements that are possible between adjacent bones. Support, however, is not the only function of the skeletal system. In the next two chapters, 9 and 10, we will study how muscles are able to contract, shorten, and move the bones, allowing the body to move around and manipulate the environment.

STUDY OUTLINE

Articulations (pp. 165–168)
Articulations or joints are the connections between bones. They can be classified either according to function or structure.

Functional Classification *Synarthroses* are immovable joints, such as sutures. *Amphiarthroses* are joints that permit a small amount of movement between the bones, such as between the vertebrae. *Diarthroses* are joints that allow a considerable amount of movement, such as the knee and shoulder.

Structural Classification A structural classification is based on what develops in the space or *joint cavity* between adjacent bones.

In *fibrous joints,* embryonic mesenchyme tissue between developing bones produces a dense network of sturdy collagenous fibers. Sutures and syndesmoses are examples of fibrous joints.

In *cartilaginous joints,* the joint cavity becomes filled with cartilage. Joints called *synchondroses* (filled with hyaline cartilage) and *symphyses* (filled with fibrocartilage) are examples.

In *synovial joints,* the fluid-filled joint cavity persists and is lined with a secretory membrane called the *synovial membrane.* These usually allow free and extensive movement.

Movements of Synovial Joints (pp. 168–170)
Synovial joints exhibit several types of movement, which have been classified into six groups.

Gliding Joints These joints, such as between carpals, allow a simple sliding movement between two bones.

Hinge Joints These joints allow the motion typical of a hinge on a door. The knee, elbow, and ankle are examples of hinge joints.

arthritis: (Gr.) *arthron*, joint + *itis*, inflammation

THE CLINIC

DISORDERS OF THE JOINTS

The articulations of bones, like other specialized organs, are subject to many pathological conditions. In spite of the limited types of tissues found in the joints, they are frequently involved in many different disease processes, both localized and systemic.

Inherited or Congenital Diseases

A disfiguring congenital disease is **arthrogryposis multiplex congenita,** a generalized fixation of joints present at birth. It may be caused by a variety of changes in the spinal cord, muscles, or connective tissues.

Another common inherited defect seen most frequently in breech (buttocks first) deliveries and in female infants is congenital dislocation of the hips. The cause is uncertain, but it appears to be related to the looseness of the ligaments around the hip joint. If the condition is discovered early, it can easily be treated by maintaining the hips in **abduction** (movement of similar parts away from the median plane) and by external rotation with the use of padded diapers (pillow pants). If the condition remains neglected, it may cause a painful limp and early arthritis in adulthood.

There are also many less disfiguring inherited conditions that cause only minor functional disability. Most of them are rare and of interest only in their recognition. **Turner syndrome** is a sex-linked abnormality in females that may be suspected because of the increased **valgus** (turning away from the body) or "carrying angle" of the elbows. **Madelung's deformity** is an inherited shortening and angulation at the wrists.

Infectious Diseases

Infections in a joint can be caused by any of the microbial agents: bacteria, viruses, or fungi. Bacterial infections in a joint cavity are particularly serious and can result in total destruction of the joint unless they are treated vigorously with surgical drainage and massive, prolonged antibiotic therapy. The joint is swollen, hot, and extremely painful. Infections with **staphylococcus** (stăf″ĭl·ō·kŏk´ŭs) and **gonococcus** (gŏn″ō·kŏk´ŭs) (the bacterium causing the sexually transmitted disease gonorrhea) are most common.

Trauma

Injuries to the joints are very common to all of us. The ankle joint is the most often injured, followed closely by the small joints of the fingers. The most common injury is a sprain (see Clinical Note: Sports Injuries). Fractures may also occur within the joint and may involve the cartilage alone **(chondral fracture),** just the bone **(osseous fracture),** or both **(osteochondral fracture).** Fractures that involve the joint surface **(intraarticular fractures)** must be meticulously aligned to ensure proper functioning of the joint.

Immunological or Allergic Diseases

The most common type of immunological disease of the joints is **arthritis,** which is inflammation of the joints. There are many forms of arthritis, including rheumatoid arthritis, gout, and pseudogout (see previous clinical notes in this chapter). **Ankylosing spondylitis** (ăng″kĭ·lō´sĭng spŏn″dĭ·lī´tĭs) is another type of arthritis that affects the spine and sometimes other joints and adjacent soft tissues. It is much more common in men than women and occurs usually between the ages of 15 and 40. Initial complaints consist of low back pain and stiffness. The advanced disease may cause the patient to have a bent-over posture, a rigid spine, and a waddling gait. Bony erosions destroy the normal anterior concavity of the vertebrae. Ossification sometimes occurs in the outer layer of the anulus fibrosis causing the characteristic "bamboo spine" that can be seen on an X ray.

Another less common immune disease is called **serum sickness,** which is a delayed allergic reaction frequently caused by drugs, that results in skin rash, swollen painful joints, and fever. Another is **lupus erythematosus** (lū´pŭs ĕr″ĭ·thē·mă·tō´sĭs), which is a puzzling disease frequently resembling arthritis in women that may involve many systems of the body. It exhibits a peculiar skin rash in the central area of the face (called a butterfly rash) that is exacerbated by exposure to the sun. It is associated with several unusual serum antigens.

Tumors

All of the tumors associated with bone and cartilage can involve a joint. In addition, the synovial lining of the joint can produce both benign and malignant tumors.

Aging

The ravages of time and the multiple traumas associated with living produce joint changes known as **osteoarthritis or degenerative arthritis.** This degeneration of the joints is related to excess wear and tear. Those at high risk are obese persons over 50 years of age. The initial degenerative process causes loss of compressibility of the joint cartilage. The surface of the cartilage begins to loosen, erode, and crack, leading to a narrowing of the joint space. New bone formation may develop at the margin of the articular cartilage.

The arthritis is sometimes related to trauma, infection, or an anatomical abnormality of the joint, although occasionally there is no predisposing abnormality. Symptoms include aching, pain, and stiffness of affected joints when lifting or moving. This type of arthritis usually occurs in the large weight-bearing joints, such as the hips and knees as well as the small joints of the fingers. Severe or repeated injury to the joints may produce degenerative changes at an earlier age. The normal aging processes also affect the intervertebral disks of the spine causing shrinkage of the disk and loss of stature in the elderly.

Recent developments in surgical techniques and the production of new metals and plastics have made possible the replacement of severely damaged joints in the hips, knees, fingers, and toes, relieving the pain and disability of many types of arthritis.

Ellipsoidal Joints These joints allow movement through two arcs in two planes and include the wrist joint.

Pivot Joints The articulation between the atlas and the axis is an example of a pivot joint, which allows rotation of one bone relative to another.

Ball-and-Socket Joints These joints allow the maximum freedom of movement in a joint and include the shoulder and hip.

Saddle Joints These joints allow movement in two planes, such as between the thumb and the hand.

Movements of the Skeleton (pp. 170–174)

Several specific movements of the synovial joints are recognized, specifically: *flexion* (decrease in angle), *extension* (increase in angle), *hyperextension* (increase in angle beyond 180°), *abduction* (movement away from body midline), *adduction* (movement toward body midline), *rotation* (twisting around long axis), *circumduction* (circumscribing a circle), *inversion* (turning the sole inward), *eversion* (turning the sole outward), *protraction* (movement straight away), and *retraction* (movement straight toward).

The Knee: A Special Joint (pp. 174–176)

The knee, the largest joint in the body, is composed of three joints: an anterior *patellofemoral joint*, a medial *tibiofemoral joint*, and a lateral tibiofemoral joint. The joint is maintained by several ligaments and the tendons of the quadriceps muscles. The fibrocartilaginous *medial* and *lateral menisci* in the joint prevent the bones from rubbing together and act as shock absorbers.

SELF-TEST OF CHAPTER OBJECTIVES

Fill-In Questions

1. A _____ classification of skeletal joints indicates the type of connective tissue present in the joint cavity.
2. A _____ classification of skeletal joints indicates the amount of movement possible in the joint, but not how the joint cavity develops.
3. _____ is movement of a body part away from the midline of the body.
4. _____ is movement that reduces or decreases the angle between two bones.
5. In the movement of the foot, _____ is the opposite of dorsiflexion.
6. The combination of complex movements of the humerus involved in throwing a softball underhanded is called _____.
7. The movement opposite to retraction is _____.
8. Functionally, the elbow joint is classified as a _____.
9. Structurally, the temporomandibular joint is classified as a _____.
10. The cartilaginous pads found in the knee joint are called _____.

Matching Questions

Match the joints on the left with their type on the right:

11. interphalangeal joints
12. elbow joint
13. coronal suture   a. synarthrosis
14. knee joint   b. amphiarthrosis
15. distal tibiofibular joint   c. diarthrosis
16. intervertebral disk joint   d. none of these
17. wrist joint
18. ankle joint

Match the joints on the left with terms describing their structure on the right:

19. maxillotemporal joint
20. knee joint
21. intervertebral disk joint
22. interphalangeal joints
23. elbow   a. suture
24. epiphyseal plate-epiphysis joint   b. syndesmosis
25. interosseous borders of radius and ulna   c. synchondrosis
26. frontonasal suture   d. symphysis
27. distal tibiofibular joint   e. synovial
28. occipital condyle-atlas joint
29. femur-acetabular joint
30. humerus-glenoid fossa joint

Multiple-Choice Questions

31. Which of the following joints does not have a fluid-filled joint cavity?
    - a. elbow joint
    - b. ankle joint
    - c. intervertebral disk joint
    - d. interphalangeal joint
    - e. none of the above
32. Sometimes two bones will fuse. Such a condition is known as
    - a. suture
    - b. syndesmosis
    - c. gout
    - d. synostosis
    - e. rheumatism
33. The joint between the metacarpal of the thumb and the trapezium of the carpus is
    - a. a saddle joint
    - b. a hinge joint
    - c. a pivot joint
    - d. an ellipsoidal joint
    - e. none of the above
34. The occipital condyle-atlas joint is a
    - a. saddle joint
    - b. hinge joint
    - c. pivot joint
    - d. gliding joint
    - e. none of the above
35. The knee joint is a
    - a. saddle joint
    - b. hinge joint
    - c. pivot joint
    - d. gliding joint
    - e. none of the above

Essay Questions

36. Compare the knee and shoulder joints in terms of the structure of the joint.
37. Circumduction is a complex movement. Describe the separate components of circumduction in terms of other designated body movements.

38. Describe the differences between the articular cartilage and menisci found in the knee.
39. How does a synchondrosis differ from a syndesmosis?
40. List the joints of the hand, and classify each in terms of structure and function.
41. List the ligaments of the knee, and describe how they help hold the bones of the knee together.
42. Describe the differences between the movements in gliding and saddle joints.
43. How are bursae and synovial cavities similar? How are they different?
44. How does a symphysis differ from a synchondrosis?
45. How does a syndesmosis differ from a synostosis?

Clinical Application Questions
46. Compare and contrast gout and pseudogout.
47. How does bursitis differ from synovitis?
48. Explain how a joint can be statically unstable yet dynamically stable.
49. Describe the causes and manifestations of rheumatoid arthritis.
50. How does osteoarthritis differ from serum sickness?

9

Muscles: Cellular Organization and Physiology

Much of human physiology involves movement. Individual organs may show movement, such as the contractions of the heart that propel blood through the circulatory system or the movements of the stomach and intestines that push food through the digestive tract. Movement of the limbs of the body itself also occurs, as one moves from place to place. Both types of movement are accomplished through the action of **muscles,** which are organs and tissues that have the ability to **contract,** that is, to shorten in length. Three different categories of muscle are found in the body, each distinctive in appearance, structure, role, and location: skeletal muscle, cardiac muscle, and smooth muscle. In spite of their differences, these three types probably function in similar ways. In this chapter we will look at all three types and examine the structure of muscle tissue and the mechanisms involved in contraction, considering

1. different types of muscle tissue,
2. differences among muscles with respect to function and location,
3. the mechanism of muscle contraction,
4. the structure of muscle tissue at the microscopic level, and
5. effects of exercise on muscle structure and physiology.

tendon: (L.) *tendere*, to stretch

Organization of Skeletal Muscle

After studying this section, you should be able to:

1. Diagram a typical skeletal muscle in cross section and identify the structural components that make up the muscle.
2. Define a motor unit and describe the relationship between nerve and muscle tissue present in a motor unit.
3. Diagram a neuromuscular junction showing how nerve and muscle fibers are associated.
4. Diagram the microscopic appearance of a skeletal muscle fiber and explain the molecular basis for its appearance.

Of the three muscle types, **skeletal muscle** is the only **voluntary muscle,** called this because it is generally under conscious control. Cardiac muscle (which forms the heart) and smooth muscle (which lines many internal organs) are both involuntary muscles, meaning that their control is not conscious. (These two muscle types are discussed in separate sections later.)

Skeletal muscle is responsible for the movement of bones (Figure 9-1). It is also called **striated** (strī′ā·tĕd) **muscle** because it has a striated or banded appearance when viewed through a microscope. Skeletal muscles are attached to bones by **tendons** (tĕn′dŭns) at each end. Tendons are bundles of connective tissue that pull on the bones when a muscle contracts, causing one of the bones to pivot at its joint with another bone (see Clinical Note: Tendonitis).

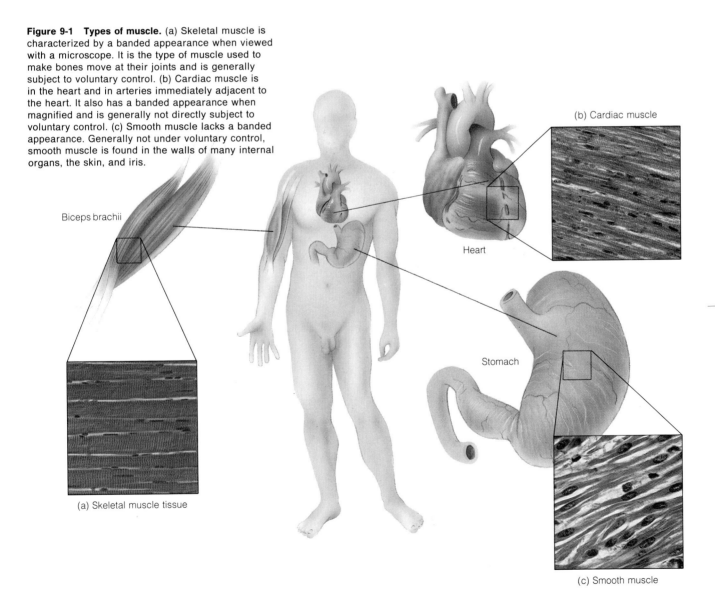

Figure 9-1   Types of muscle. (a) Skeletal muscle is characterized by a banded appearance when viewed with a microscope. It is the type of muscle used to make bones move at their joints and is generally subject to voluntary control. (b) Cardiac muscle is in the heart and in arteries immediately adjacent to the heart. It also has a banded appearance when magnified and is generally not directly subject to voluntary control. (c) Smooth muscle lacks a banded appearance. Generally not under voluntary control, smooth muscle is found in the walls of many internal organs, the skin, and iris.

somite: (Gr.) *soma*, body
myotome: (Gr.) *mys*, muscle + *tome*, incision
fascia: (L.) *fascia*, band
epimysium: (Gr.) *epi*, upon + *mys*, muscle
perimysium: (Gr.) *peri*, around + *mys*, muscle

Embryonic Development of Muscles

During the early development of the embryo, three tissue layers differentiate: **ectoderm, mesoderm,** and **endoderm.** Ectoderm, the outermost layer, gives rise to the skin, nervous system, and sense organs. Endoderm, the most internal layer, develops into the digestive and respiratory tracts and several large internal organs. Mesoderm lies sandwiched between and develops into the bulk of body tissues including bone, muscle, and the circulatory system.

The muscular system is derived from paired, repetitious blocks of mesoderm called **somites.** The first somite appears at the end of the third week of embryonic life, and during the next two weeks 40 to 43 more develop alongside the neural tube (Figure 9-A). Each somite subdivides into three regions: a **dermatome**, which contributes to the dermis of the skin; a **sclerotome** (sklē´rō·tōm), which contributes to the vertebrae; and a **myotome,** which develops into muscle. Myotomes in the central portion of the embryo spread ventrally from the somites and meet along the ventral midline to form the muscles of the trunk (Figure 9-B). Limb muscles develop from somites adjacent to the limb buds and from general mesoderm that differentiates within the buds. Segmental nerves associated with the somites grow into the muscles, providing innervation for voluntary control. Somites do not develop in the head, therefore, muscles in that region differentiate from generalized head mesoderm.

Smooth muscle differentiates from mesodermal cells surrounding the embryonic digestive tract, arteries and veins, and numerous ducts. As these tubular organs develop, muscle cells multiply and become organized into concentric sheets in the walls. The young cells become oriented in different directions

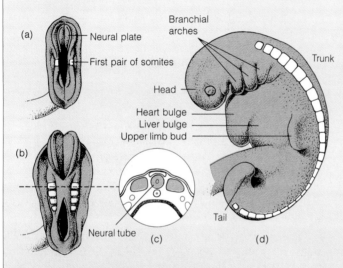

Figure 9-A Somite formation. During the fourth week of development, the spinal area of the embryo becomes divided into a segmented chain of mesodermal somites. (a) At the beginning of the first week, the first pair appears near the middle of the embryo. (b) By about the middle of the fourth week, several large somites are apparent. (c) In this transverse section, the neural tube can be seen developing between the somites. (d) By the end of four weeks, the chain of somites extends the length of the embryo.

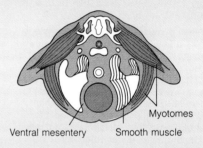

Figure 9-B Embryonic formation of muscles. Skeletal muscles form from myotomes, portions of the somites that spread into the limb buds and around the trunk. Smooth muscle differentiates from other mesodermal cells in the intestinal wall in the region called the ventral mesentery.

in the intestinal wall, producing the characteristic longitudinal, circular, and oblique bands of muscle. Arteries and veins only have circular muscles in their walls.

Cardiac muscle differentiates from mesodermal cells that migrate to the embryonic heart. This migration occurs while the heart is still a tubular structure and before nerve connections have become established. Cardiac muscle cells are inherently contractile; they begin rhythmic pulsations even before they connect to each other in a functional tissue and before they are connected to nerve cells. These contractions eventually come under the control of the **pacemaker** of the heart, which organizes the individual cellular contractions into a single coordinated heartbeat (see Chapter 22).

There are about 400 skeletal muscles in the body where they control operations as delicate as winking an eye or as complex as fielding a baseball. A skeletal muscle consists of several different tissues that collectively comprise the **muscle trunk.** In addition to muscle tissue, a skeletal muscle contains connective tissue, circulatory vessels, blood, and nervous tissue. The muscle trunk is surrounded by a layer of dense connective tissue called the **deep fascia** (fāsh´ē·ă). The deep fascia blends with an underlying layer of more delicate connective tissue called the **epimysium** (ĕp˝ĭ·mĭz´ĭ·ŭm) (Figure 9-2). The muscle tissue itself is organized into subunits called **fascicles** (făs´ĭ·kls), each of which is surrounded by a sheath of connective tissue called the **perimysium** (pĕr´ĭ·mĭs´ĭ·ŭm). Each fascicle, in turn, consists of subunits called **muscle fibers,** which are separated from one another by sheaths of connective tissue

endomysium:	(Gr.) *endo*, within + *mys*, muscle
neuromuscular:	(Gr.) *neuron*, nerve + (L.) *musculus*, muscle
myoneural:	(Gr.) *mys*, muscle + *neuron*, sinew
sarcoplasm:	(Gr.) *sarkos*, flesh + *plasma*, form

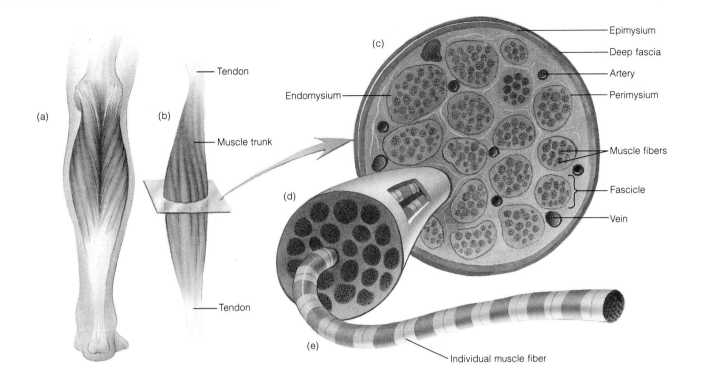

Figure 9-2 Organization of a skeletal muscle. (a) Skeletal muscles are connected through tendons to the bones that they move. (b) Each muscle consists of several tissues organized into a muscle trunk. (c) A muscle trunk is organized into bundles called fascicles, each of which is surrounded by connective tissue. Blood vessels and nerves that supply the muscle fibers pass through the connective tissues of the trunk. (d) Each fascicle is a bundle of long, slender cells called muscle fibers. (e) Individual fibers have a banded appearance that results from the organization of their subunits.

called **endomysium** (ĕn″dō·mĭs′ĭ·ŭm). The endomysium, perimysium, and epimysium are all continuous and together provide a framework that holds the muscle together (Figure 9-2). At each end of the muscle these tissues come together to form the tendons that connect the muscle to a bone. Arteries, veins, capillaries, and nerves pass through these layers of connective tissue, delivering nutrients and the nerve impulses that control contraction.

Motor Units

Each muscle is controlled by a nerve, a bundle of nerve cells that delivers signals that cause the muscle to contract. Upon entering the connective tissue of the muscle, the nerve branches into finer and finer subdivisions, with the branches ending at the surface of individual muscle fibers.

All the muscle fibers reached by the branches of a single nerve cell are controlled by that nerve cell. Together, these muscle fibers constitute a **motor unit.** The number of fibers within a motor unit varies depending on the amount of branching of the nerve cell. In large, powerful muscles, a motor unit may include hundreds of individual muscle fibers. In muscles used for very fine operations, such as those that move the eyeball, a motor unit may consist of as few as five or ten muscle fibers.

Each nerve cell branch terminates in an enlarged end called a **motor end plate** (Figure 9-3). The motor end plate lies very close to, but does not touch, the muscle fiber, their membranes lying only 50 to 100 nm apart. This region is called a **neuromuscular** (nū″rō·mŭs′kū·lăr) **junction,** or **myoneural** (mī″ō·nū′răl) **junction,** and the space or gap between the membranes is called a **neuromuscular cleft** (or **myoneural cleft**). The motor end plate is highly irregular in shape, corresponding to indentations in the muscle fiber membrane. The complex folding in this neuromuscular junction increases the surface area of the nerve cell and muscle fiber membranes of the junction.

Structure of the Skeletal Muscle Fiber

The outer membrane of a muscle fiber, called a **sarcolemma** (sar″kō·lĕm′ă), is similar in appearance to the plasmalemma of other kinds of cells (Figure 9-4). It encloses the **sarcoplasm** (sar′ko·plăzm), the muscle fiber equivalent of the cytoplasm of other cells. An elaborate system of membranes, vesicles,

184 Chapter 9 Muscles: Cellular Organization and Physiology

sarcomere: (Gr.) *sarkos*, flesh + *mere*, part

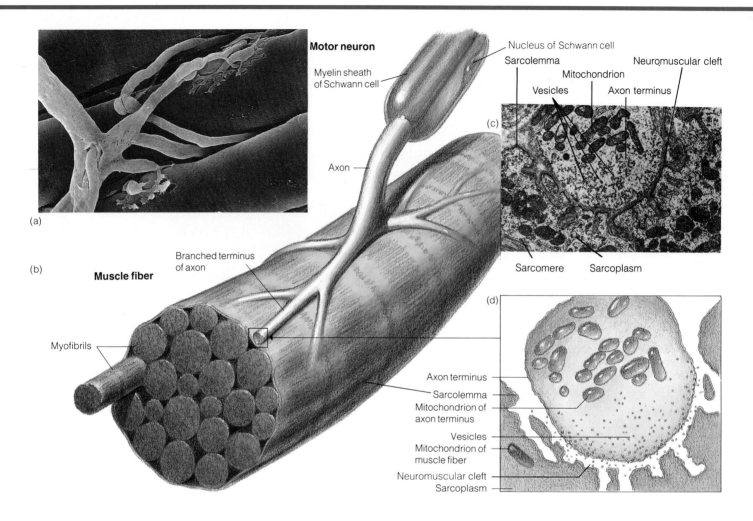

Figure 9-3 **A motor end plate.** (a) Scanning electron micrograph showing motor nerve and two end plates on adjacent muscle fibers (mag. 94,149×). (b) This drawing shows how the tip of each motor neuron branch is closely associated with the surface of a muscle fiber. (c) Electron micrograph of a neuromuscular cleft (mag. 10,863×). (d) Interpretative drawing of electron micrograph in part (c). The membranes of a nerve cell ending and a muscle fiber lie close to one another, separated by a narrow space, the neuromuscular cleft.

and tubules is present in the fiber, collectively comprising the **sarcoplasmic reticulum** (sar″kō·plaz′mĭk rĕ·tĭk′ū·lŭm). Folds of the membranes of the sarcoplasmic reticulum divide the fiber into many long, cylindrical subunits called **myofibrils** (mĭ·ō·fĭ′brĭls) (Figure 9-4).

Sarcomeres

The myofibril is divided into units of equal length called **sarcomeres** (sar′kō·mērs) (Figure 9-5). Sarcomeres are particularly important elements in the structure of a muscle fiber because reactions occur in them that are responsible for the contraction of the muscle as a whole. Consequently, sarcomeres are considered to be the functional units of muscle contraction. In an electron micrograph, a sarcomere consists of a series of dark and light bands (Figure 9-5a). The myofibrils in a muscle fiber are all aligned with their sarcomeres in phase with one another, giving the muscle fiber its characteristic striated appearance. It can be seen in a micrograph that each end of a sarcomere is marked by transverse lines called **Z lines** (Figure 9-5b). The Z line is the edge of a circular disk called a **Z disk.** Just inside the Z lines at each end of the sarcomere are lighter regions called **I bands,** and between the I bands there lies a darker middle region called the **A band.** The A band itself contains a lighter center called the **H zone,** and there is a dense line in the middle of the H zone termed the **M line.**

These light and dark bands in the sarcomere are produced by the overlapping of hundreds or thousands of long, narrow filaments that lie oriented lengthwise throughout the interior of the sarcomere. There are two kinds of filaments, one

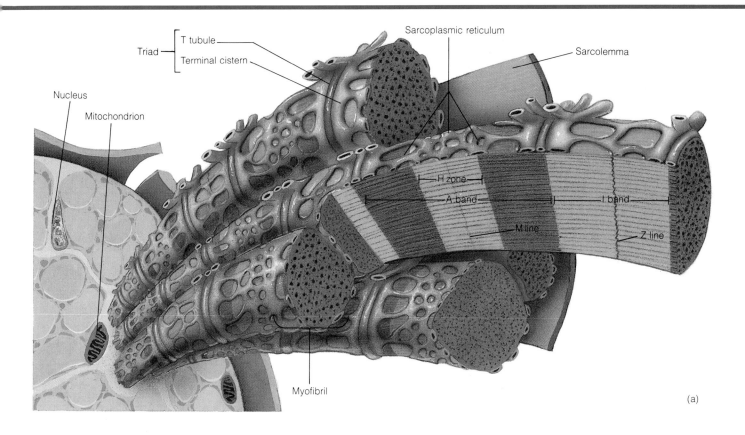

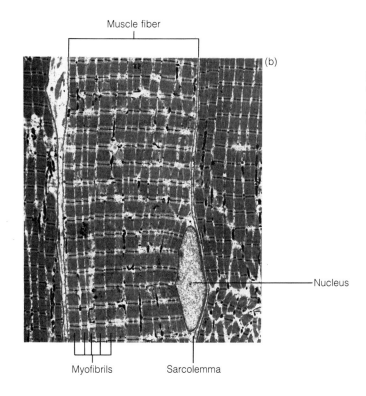

Figure 9-4 Organization of a striated muscle fiber. (a) The interior of the fiber consists of a large number of myofibrils, each of which is surrounded by a portion of the network of membranes called the sarcoplasmic reticulum. Each myofibril, in turn, contains numerous thick and thin filaments whose arrangement is responsible for the striated appearance of the muscle. (b) Transmission electron micrograph of skeletal muscle.

tenosynovitis: (Gr.) *tenon*, tendon + *syn*, with + (L.) *ovum*, egg + (Gr.) *itis*, inflammation
tropomyosin: (Gr.) *trope*, a turn + *mys*, muscle

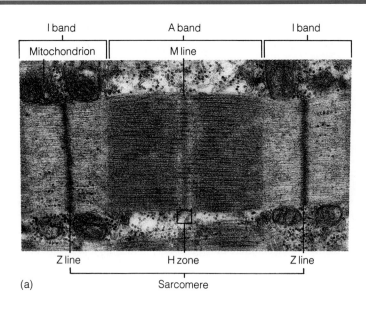

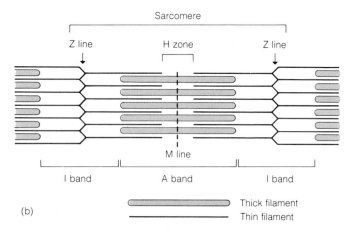

Figure 9-5 Arrangement of filaments in a myofibril. (a) This electron micrograph shows how overlapping thin and thick filaments produce light and dark bands in a sarcomere. Thin filaments extend into the sarcomere from the Z lines and are overlapped by thick filaments suspended in the middle region of the sarcomere (mag. 46,560×). (b) Diagram of a sarcomere showing the relationship of the overlapping filaments to the bands of the sarcomere.

relatively thin and the other relatively thick in diameter. **Thin filaments** extend into the sarcomere from each Z disk, whereas **thick filaments** are suspended in the central part of the sarcomere, overlapping the thin filaments at each end. Thus, the I zone contains thin filaments only, the H zone contains thick filaments only, and the darker regions of the A band are where the filaments overlap.

> **CLINICAL NOTE**
>
> ### TENDONITIS
>
> **Tendonitis** (tĕn·dŏn·ī′tĭs) is an inflammation of the tendons; **tenosynovitis** (tĕn″ō·sĭn″ō·vī′tĭs) is an inflammation of the tendon sheaths, usually of the hands or wrists. Both cause severe pain and, because inflammation is present, the area may be swollen, reddened, and tender. Movement of the joint is usually limited. Tenosynovitis may indicate systemic disease such as rheumatoid arthritis. Prolonged exercise such as the activities of typists and concert pianists may also be responsible since tendonitis is caused by excessive joint movement and/or trauma.
>
> The extensor tendon of the thumb is subject to a particularly painful type of tendonitis known as **de Quervain's disease.** It occurs in patients who repeatedly grasp with the thumb and twist the wrist, such as carpenters who use hammers and screwdrivers all day and nursing mothers who manually express milk.
>
> The useful drugs for these conditions are massage and anti-inflammatory corticosteroids given orally or injected directly into the painful joint. Analgesics, such as codeine or acetaminophen, are also prescribed. The joint has limited mobility during the acute phase of both tendonitis and tenosynovitis, but exercise should be planned as soon as motion becomes painless to promote joint mobility.

The filaments form a regular crystal-like array when viewed in transverse section (Figure 9-6a). In the region of the A band where the thin and thick filaments overlap, each thick filament is surrounded by six thin filaments, and each thin filament is surrounded by three thick filaments (Figure 9-6b).

Molecular Organization of Filaments

Thin filaments consist of three kinds of proteins, called **actin** (ăk′tĭn), **tropomyosin** (trō″pō·mī′ō·sĭn), and **troponin** (trō′pō·nĭn). Actin molecules are joined together to form two long, narrow filaments wrapped around one another to form a helical double filament (Figure 9-7a). Tropomyosin molecules lie end to end in each of the grooves formed by the twisted actin filaments, and a troponin molecule is bound to each tropomyosin molecule.

Each thick filament contains hundreds of **myosin** (mī′ō·sĭn) subunits, which are long, slender filaments with two globular heads attached at one end. The filamentous portions of the myosin subunits lie alongside one another to form the backbone of the filament, leaving the head sections projecting outward in a spiral path around the surface of the thick filament (Figure 9-7b).

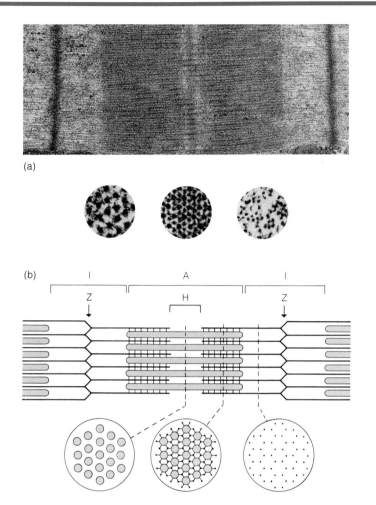

Figure 9-6 Arrangement of thick and thin filaments. (a) Electron micrograph of longitudinal section through a sarcomere (mag. 53,760×). Below this are three electron micrographs showing cross sections through an H band, an A band, and an I band. (b) Diagram showing relationship of thin and thick filaments in a longitudinal section along with the same three cross sections as above.

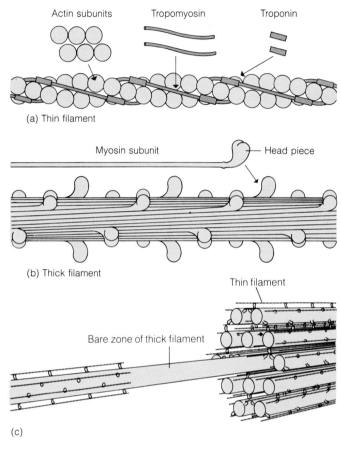

Figure 9-7 Molecular organization of thin and thick filaments. (a) Assembly of a thin filament from actin, troponin, and tropomyosin subunits. (b) Assembly of a thick filament from myosin subunits. (c) Perspective view. Every third cross bridge is in contact with the same thick filament. The remaining cross bridges are in contact with neighboring thick filaments. This arrangement causes the filaments to form a tightly cross-linked network.

Myosin subunits are oriented in opposite directions in the two halves of the filament, forming a central section that lacks the projecting heads. The resulting smooth portion of the filaments (the bare zone) forms the M line that lies in the middle of the H zone in an electron micrograph.

Between the thick and thin filaments lie **cross bridges,** formed where the heads of the myosin subunits combine with actin subunits of the thin filament. The cross bridges hold the thin and thick filaments in a tightly bound and stable matrix of hundreds of overlapping filaments (Figure 9-7c).

Membranes of the Sarcomere

Membranes of the sarcoplasmic reticulum extend throughout the muscle fiber, encircling each myofibril with a network of channels (Figure 9-8) through which nutrients are delivered to the inner portions of the fiber. Mitochondria suspended in these membranes provide ATP used in muscle contraction.

Thin tubules of the sarcoplasmic reticulum extend from the I band at each end of the sarcomere to the center where they fuse in the region of the H band. The membranes also fuse where the I and A bands meet, producing a saclike

cistern: (L.) *cisterna*, a closed cavity

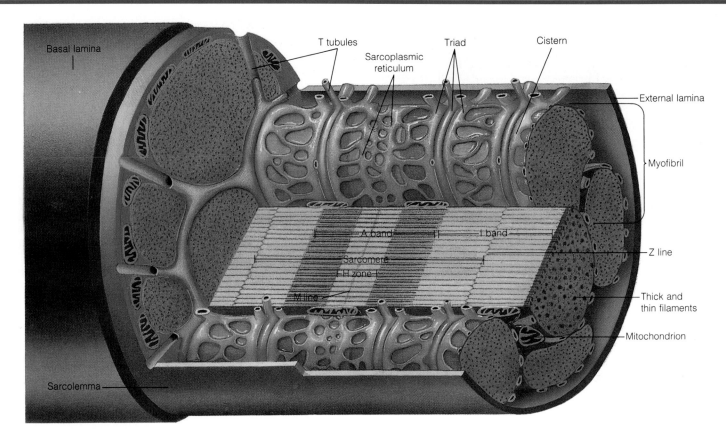

Figure 9-8 Organization of a sarcomere. This diagram shows a cutaway view of a muscle fiber to reveal the sarcomere and its relationship to other structures. The arrangement of the T tubules, cisterns, and sarcoplasmic reticulum allow contraction to occur nearly simultaneously throughout the sarcomere as well as in all the sarcomeres of the fiber. The close association of mitochondria with the sarcoplasmic reticulum provides a ready source of energy for contraction.

cavity called a **cistern** (sĭs′tĕrn) that encircles the filaments in the myofibril (Figure 9-8). A similar cistern passes around the circumference of the myofibril in the region of the Z disk, forming a chamber that is continuous with a cistern on the opposite side of the Z disk. Cisterns are believed to be involved in the storage and release of calcium ions used to induce muscle contraction.

At each end of the sarcomere, the sarcolemma forms tubules that branch into the fiber to form a system of **T tubules.** The T tubules branch and encircle the sarcomeres, lying sandwiched between neighboring cisterns at each end of the sarcomere. A T tubule and its two neighboring cisterns are referred to as a **triad** (trī′ăd). The branching arrangement of the T tubules enables them to carry stimuli to all the sarcomeres of a muscle fiber. This allows all the sarcomeres in the fiber to contract almost simultaneously and the fiber to act as a single contracting unit.

Muscle Contraction

After studying this section, you should be able to:

1. Explain how a nerve impulse can stimulate an impulse in a muscle cell.
2. Describe how an impulse in a muscle cell can induce a contraction in the muscle cell.
3. Describe the source, role, and fate of Ca^{2+} involved in muscle contraction.
4. Describe and diagram the interaction of thin and thick filaments that leads to muscle contraction.
5. Describe the events following a contraction that lead to relaxation of the muscle.

Initiation of Contraction

Muscle contraction is normally initiated by a nerve impulse delivered to the muscle at the motor end plate. The nerve impulse (the nature of which is examined in Chapter 11) reaches the motor end plate of the nerve and causes the release of a substance called **acetylcholine** (as″ĕ·tĭl·kō′lēn) into the neuromuscular cleft. Acetylcholine belongs to a general class of chemical compounds called **transmitter substances,** chemicals that carry impulses across neuromuscular clefts. Acetylcholine is manufactured in the terminal region of the nerve cell where it is stored in vesicles (Figure 9-9). When a nerve impulse reaches the end of a nerve cell, these vesicles migrate to the plasmalemma and fuse with it, releasing the acetylcholine into the cleft. Acetylcholine diffuses across the neuromuscular cleft and initiates an impulse in the muscle fiber on the opposite side. In this way, the nerve impulse is transmitted across the cleft to the muscle fiber. Once induced in the muscle fiber, the impulse proceeds over the surface of the fiber and passes down the T tubules into the interior of the fiber, where the sarcolemma comes close to the membranes of the sarcoplasmic reticulum.

The passage of the impulse along the membrane of the T tubules has an effect on the membrane of the sarcoplasmic reticulum, making it suddenly permeable to calcium ions stored in the sarcoplasmic reticulum. The calcium ions then diffuse into the fluid that surrounds the thin and thick filaments of the sarcomere and trigger the mechanism that contracts the muscle.

Removal of Acetylcholine

Acetylcholine that has been released into the neuromuscular cleft is rapidly removed by the action of an enzyme called **choline esterase** (kō′lĭn ĕs′tĕr·ās). This enzyme is important

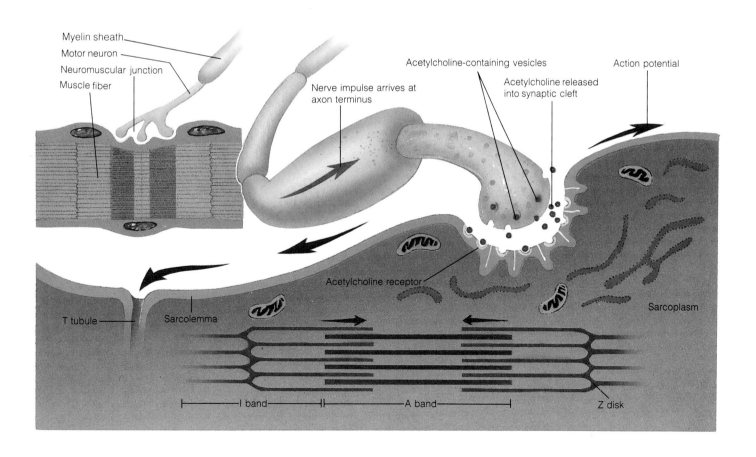

Figure 9-9   Release of acetylcholine into a neuromuscular cleft. Arrival of a nerve impulse at a neuromuscular junction causes acetylcholine-containing vesicles to fuse with the nerve cell membrane. Fusion of a vesicle with the membrane releases the acetylcholine into the cleft, and the acetylcholine diffuses to the muscle fiber membrane opposite. Combination with receptors in the membrane induces an impulse that travels over the surface of the fiber to the T tubules and down into the fiber where contraction is stimulated.

> **CLINICAL NOTE**
>
> **DRUG-INDUCED NEUROMUSCULAR JUNCTION BLOCK**
>
> Organophosphate insecticides, such as parathion, and most nerve gases are cholinesterase inhibitors, which means that they prolong the action of acetylcholine and thus cause persistent depolarization at the neuromuscular cleft (see Chapter 11 for explanation of depolarization). Exposure to these chemicals can cause symptoms of **miosis** (constriction of the pupils), tightness in the chest, excessive bronchial secretions, gastrointestinal hyperactivity, and prolonged muscular weakness and twitching. These symptoms are sometimes seen in farmers who spray their crops with organophosphates.

in the process of muscle contraction because it removes acetycholine released in response to one nerve impulse and readies the neuromuscular cleft for the transmission of the next impulse. Choline esterase also plays a regulatory role, requiring the nerve to release sufficient acetylcholine to overcome the effect of the enzyme if a nerve impulse is to stimulate the muscle fiber.

All-or-None Effect

The amount of acetylcholine required to initiate a response varies, depending on the state of the fiber at the time. The cell will only respond if a certain threshold value is reached, and anything less than the amount required to reach the cell's threshold will have no effect. Once a given fiber is stimulated, however, the strength of the fiber's contraction is independent of the amount of acetylcholine released. This is called an **all-or-none effect** (Figure 9-10).

The strength of contraction of a muscle depends on the number of muscle fibers in it that have been induced to contract. The smooth, graded response of a muscle to a stimulus is the sum of the responses of the muscle fibers in the muscle, each with its own particular threshold value that must be reached in order to induce it to contract.

Mechanism of Contraction

Passage of an impulse along the membrane of a T tubule stimulates the neighboring membrane of the sarcoplasmic reticulum to become permeable to Ca^{2+} stored in the sarcoplasmic reticulum. The Ca^{2+} diffuses into the fluid that surrounds the thin and thick filaments and reacts with the troponin-tropomyosin complexes of the thin filaments (see Clinical Note: Calcium Channel Blockers). Recall that thick filaments are made up of many individual myosin molecules, with the head portion of each one sticking out at regular intervals along the thick filament. These heads form cross bridges by combining with actin units in the neighboring thin filaments.

In a resting muscle (that is, one that has not been stimulated to contract), actin sites are covered by tropomyosin molecules, making it impossible for the myosin heads to form cross bridges. In an activated muscle, Ca^{2+} released from the sarcoplasmic reticulum reacts with the troponin-tropomyosin complexes, which causes the tropomyosin molecules to shift position, exposing actin sites and allowing cross bridges to form.

Formation of a cross bridge involves ATP (adenosine triphosphate). When it is not attached to an actin site, a myosin head will combine with a molecule of ATP and cleave it into ADP (adenosine triphosphate) and inorganic phosphate. These products remain attached to the myosin head until an actin binding site becomes available (that is, when a tropomyosin molecule has been moved out of the way by calcium ions), at which time the head binds to the new site and simultaneously releases the ADP and inorganic phosphate produced in the earlier cleavage of ATP. These reactions cause a change in the angle of attachment of the myosin head to the thick filament. The change in configuration of many thousands of myosin heads pulls the thin fila-

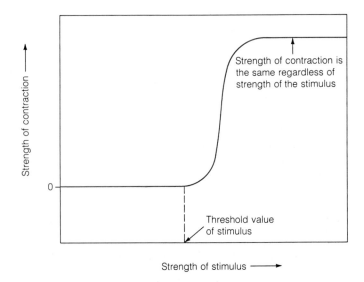

Figure 9-10 The "all-or-none" effect in a muscle fiber. Initiation of contraction in an individual muscle fiber requires that sufficient acetylcholine be released by the nerve cell to contract it. The magnitude of the signal produced and the contraction that results is independent of the strength of the stimulus as long as the threshold is surpassed.

CLINICAL NOTE

CALCIUM CHANNEL BLOCKERS

Drugs known as **calcium channel blockers** interfere with the movement of calcium across the cell membrane. In cardiac muscle cells, the drugs cause a reduction in the amount of calcium that is available for reaction with the troponin-tropomyosin complex. These drugs also inhibit calcium movement in smooth muscles of the vascular system, causing blood vessels to be more relaxed. The clinical effect of these drugs varies with the exact drug used, the method of its administration, and the underlying patient pathology. In general, calcium channel blockers increase blood flow through relaxed vessels of the heart and its periphery and decrease the rate of heart muscle contraction. The most serious side effect of these drugs is a dangerously low blood pressure and a slowed heart rate.

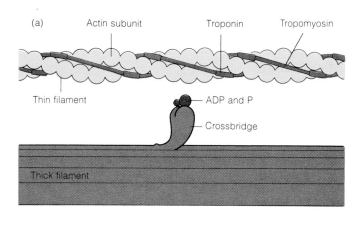

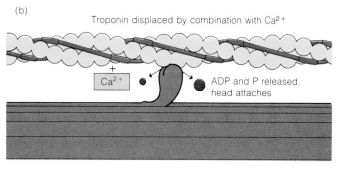

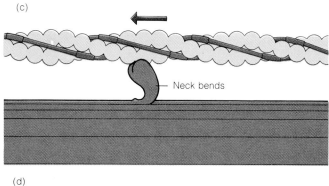

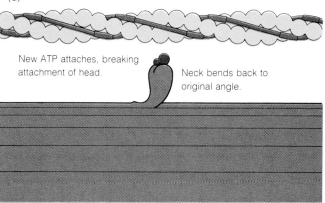

ments along the thick filaments and the sarcomere shortens (Figure 9-11).

The head, now free of ADP and phosphate and binding to an actin site, binds with a new molecule of ATP. This causes the head to release its attachment to the actin site, immediately after which the ATP is cleaved to ADP and inorganic phosphate. The ADP and phosphate remain attached to the now free myosin head, which bends back to its original angle. If actin sites continue to be available, the cycle is repeated over and over and the thin filament is pulled step by step along the thick filament.

Each thick filament contains as many as several hundred myosin heads that jut out to the side and react with neighboring thin filaments. The combined stepwise movement of the heads along the neighboring thin filaments results in a smooth sliding motion of the filaments over one another. As the thin filaments slide along the thick filament toward the center of the sarcomere, they pull the Z disk along with them and, consequently, the sarcomere shortens. This mechanism, the **sliding filament model** for muscle contraction, is illustrated in Figure 9-11 and summarized in Table 9-1.

Figure 9-11 Sliding filament model of muscle contraction. (a) In a resting muscle fiber, actin sites are covered by tropomyosin molecules, preventing the attachment of cross bridges between the thick and thin filaments. ATP combines with the free head and is split into ADP and inorganic phosphate. (b) In response to Ca^{2+}, actin sites are exposed to the free heads. Attachment of a head releases the ADP and inorganic phosphate. (c) Attachment is followed by a bending of the neck of the head, pulling the filaments past each other. (d) A new ATP molecule attaches to the head, causing release of the attachment to the actin site. The ATP is split and the head bends back to its original angle from which it can attach to the next available actin site.

rigor mortis: (L.) *rigor*, stiffness + *mors*, death

The simultaneous contraction of the sarcomeres shortens the entire muscle fiber. Since all the fibers in a motor unit receive the same nerve-transmitted stimulus, the entire motor unit contracts at once. The degree to which the muscle contracts depends on the number of fibers involved and the degree to which they have contracted.

Relaxation

Calcium ions used in muscle contraction are removed from the fluid around the filaments by a calcium pump in the membranes of the sarcoplasmic reticulum. As these ions are removed, tropomyosin molecules move back into position over the binding sites on the thin filament. As myosin heads are released from actin sites, the tension that caused the sarcomere to contract is relieved. The sarcomere remains shortened until it is stretched back to its original length when the muscle is stretched. In the absence of tension, the muscle is said to be *relaxed*.

It should be noted that ATP is used to cause the myosin head to be released from the thin filament. The formation of new cross bridges and the change in the angle of the head that causes the filaments to slide occurs *spontaneously*, as long as new sites of attachment are available and as long as ADP and inorganic phosphate are attached to the head. After death, the tissue loses its ability to produce the ATP. This is why muscles become rigid after death in a phenomenon called **rigor mortis** (rĭg´ŏr mōr´tĭs) (see Clinical Note: Rigor Mortis). The myosin heads remain attached to the thin filaments, and the muscle remains contracted until it starts to deteriorate.

> **CLINICAL NOTE**
>
> ### RIGOR MORTIS
>
> **Rigor mortis** occurs upon death when ATP ceases to be supplied. The muscles remain contracted and the joints become stiff and immovable. Although its appearance varies with the individual, rigor mortis usually begins in the involuntary muscles and then moves to the voluntary muscles two to four hours after death. Actually, the time of the onset of rigor mortis depends on many factors, such as the previous state of health, the cause of death, and the ambient temperature. It ends after about 48 hours when degradation of muscle protein occurs and the muscles become flaccid.
>
> The onset of rigor mortis is often significant in **forensic medicine** (legal medicine). Readers of detective novels will appreciate the time-honored importance of the appearance and disappearance of rigor mortis to the solution of the mystery.

Table 9-1 SUMMARY OF STEPS IN MUSCLE CONTRACTION AND RELAXATION IN SKELETAL MUSCLE

1. A nerve impulse reaches the motor end plate of a nerve cell and causes the release of acetylcholine into the synaptic cleft.
2. Acetylcholine diffuses across the synaptic cleft and combines with the muscle membrane, where it may cause an impulse to be initiated in the muscle fiber. The acetylcholine is destroyed by choline esterase.
3. The impulse passes over the sarcolemma and down T tubules into the cell, where it stimulates the release of calcium ions stored in the sarcoplasmic reticulum. The ions pass into the fluid that bathes the thin and thick filaments.
4. Calcium ions combine with the troponin-tropomyosin molecules present in the thin filaments, causing the tropomyosin portions to move, exposing sites to which myosin heads from the thick filaments can attach.
5. ATP molecules attach to myosin heads and are split into ADP and inorganic phosphate (P_i). The myosin head releases the ADP and P_i as it attaches to the next available actin site. Attachment causes the head to pivot on its neck, pulling the thin filament in the process. This action, duplicated at hundreds or thousands of myosin heads, causes the thin filaments to slide in from either end of the sarcomere, pulling the end plates to which they are attached along with them. The sarcomere shortens.
6. Calcium ions are actively transported back into the sarcoplasmic reticulum. In the absence of calcium ions, tropomyosin covers the binding sites on the thin filaments. Uptake of ATP by the myosin heads causes them to detach from the actin sites, and the filaments slide apart as the muscle relaxes.

Physiology of Muscle Contractions

After studying this section, you should be able to:

1. Define and describe the contraction phenomena known as twitch, latent period, summation, refractory period, treppe, and tetanus.
2. Explain the chemical and cellular events responsible for each of the phenomena listed above.
3. List the sources of energy for muscle contraction.
4. Describe why oxygen often continues to be used at an accelerated rate after strenuous exercise has ceased.
5. List the chemical waste products of muscle contraction.

Certain aspects of contraction can be studied using skeletal muscles that have been surgically removed from animals (see Clinical Note: Abnormal Muscle Contractions). Various chemicals or electric shock can induce these muscles to contract, and the characteristics of the contraction can then be studied by firmly holding one end of the muscle in a

fibrillation: (L.) *fibrilla*, little fibers

stationary clamp and attaching the other end to a stylus that enscribes a path on a moving strip of chart paper. Such an apparatus is shown in Figure 9-12. The tracing produced is called a **myogram** (mī′ō·grăm).

Twitches

When a muscle is stimulated with an electric shock of sufficient magnitude, either by application of the shock directly to the muscle itself or to the nerve that enters the muscle, the muscle contracts. When the stimulus is a single brief shock, the contraction is referred to as a **twitch.** Figure 9-13 shows a myogram produced by a twitch. Notice that there is a short delay after the stimulus is given, after which the muscle contracts rapidly and then returns to its original length. The delay that precedes contraction is called the **latent period;** it results from the time it takes for the action potential to travel over the surface of the fiber, pass down the T tubules, and induce the filaments to slide over one another.

Individual twitches vary in their duration depending on the type of muscle involved. Muscles that are called upon to react rapidly, such as muscles that are responsible for the blinking movement of the eye, typically show twitch responses as fast as 10 msec (0.01 sec) or less. Muscles that are required to react more slowly, such as muscles involved in arm movements, typically display twitches that last 40 to 50 msec. Muscles that are called upon to provide slower

> **CLINICAL NOTE**
>
> ### ABNORMAL MUSCLE CONTRACTIONS
> Although most muscle contractions are a normal part of a healthy, functioning body, some types of muscle contractions may be symptomatic of minor (or even major) problems.
>
> **Fibrillation** (fĭ″brĭl·ā′shŭn), the contraction of single muscle fibers, is manifested as small, involuntary, and local contractions. Fibrillation may occur in the heart in either the atria or the ventricles. If many single fibers are involved, the rhythm of the heart is affected, leading to poor cardiac performance. Life-threatening ventricular fibrillation can be countered by electric shock administered to the heart to interrupt the fibrillation, a process known as **defibrillation.**
>
> A **twitch** is the normal contraction of muscle after a single brief stimulus. However, twitches occurring spontaneously in muscles at rest may signal disease of the neuron supplying the muscle.
>
> **Fasciculations** (fă·sĭk″ū·lā′shŭns) are small, local muscular contractions, visible on the skin, that stop and start spontaneously for no apparent reason. They represent the spontaneous discharge of a number of muscle fibers receiving innervation from a motor neuron.
>
> A **tic** is a spasmodic, rhythmic twitch, usually of the muscles of the face, primarily of the eyelid. A tic of the diaphragm is a hiccup.
>
> A **spasm** is a spontaneous and involuntary contraction of a single muscle or a group of muscles. It can occur in involuntary muscles and cause constriction of hollow passages, such as the esophagus or bladder. A spasm in the smooth muscle of a blood vessel is called a **vasospasm,** which is associated with heart attacks and strokes.
>
> Spasms are sometimes painful; these are thought to occur when the nerve supply to the muscle is irritated. There are three main types. A **clonic spasm** is followed by immediate relaxation of the muscle. In a **tonic spasm,** unlike the clonic type, the muscles continue to contract. Tetany, as well as certain diseases of the central nervous system, may cause tonic spasms. In a **convulsive spasm,** all the muscles of the body are involved and movement is jerky and violent. Epilepsy is marked by convulsive spasms.

and more sustained contractions, such as muscles of the vertebral column that are used in maintaining posture, are the slowest of all skeletal muscles in their twitch response, taking as long as 1 sec from the time they are stimulated until relaxation is complete (Figure 9-13).

When a second stimulus is given before the period of relaxation has ended, the second contraction builds upon the first and results in a contraction greater than either of the individual ones would have been by themselves. This phenomenon is known as **summation** (Figure 9-14).

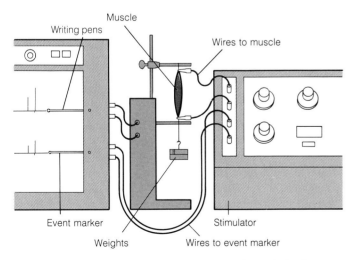

Figure 9-12 Obtaining a myogram. A suspended muscle is given an electric shock, which makes it contract. When it contracts, a writing pen on a moving strip of paper is caused to move, and the resulting figure reflects the timing, duration, and magnitude of the contraction.

treppe: (Ger.) *treppe*, staircase
tetanus: (Gr.) *tetanos*, stretched

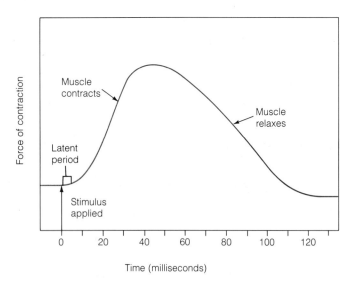

Figure 9-13 Typical myogram of a muscle twitch.

Summation occurs when an additional impulse travels over a fiber and induces a new contraction before relaxation from the previous contraction is complete. Additional acetylcholine released into neuromuscular junctions before it has been completely removed enables the threshold level of more muscle fibers to be reached. As a result, the contraction from the second stimulus involves more fibers than did the contraction following the first stimulus, and the second contraction is thus stronger than the first.

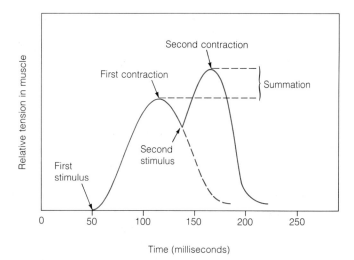

Figure 9-14 Summation. Although the stimuli given to the muscle are the same magnitude, the contraction induced by the second stimulus is greater than the contraction induced by the first stimulus.

However, the second stimulus will not induce a contraction if it is given too quickly after the first one, regardless of its magnitude. This is because a short time is required for the membrane to become sensitive again. This period of insensitivity to additional stimuli is called the **refractory period**.

If an isolated muscle is stimulated repeatedly, but slowly enough for the individual responses to be observed, we would find that the strength of each contraction gradually increases until a maximum is reached. This phenomenon is called the **staircase effect**, or **treppe** (trĕp´ĕ) (Figure 9-15a). Treppe is one reason that warm-up exercises are helpful to athletes who are preparing to call upon their muscles for maximum output. As a result of the warm-up, their muscles are able to contract more strongly.

A possible explanation for treppe is that repeated passage of impulses into the T tubules causes more and more Ca^{2+} to pass into the fluid around the fibers. With the consequent increase in the concentration of Ca^{2+} in the fibers, more and more actin sites are exposed and contraction becomes progressively stronger.

If successive stimuli are given slowly enough, the rise to maximum contraction occurs in a series of steps. However, if the stimuli are given more rapidly, the individual contractions will not be discernible because the muscle will not have an opportunity to relax between each stimulus. The myogram will appear as a smooth curve, rising to a sustained maximum, and then declining slowly as the muscle **fatigues** (Figure 9-15b). A muscle held in such a continuously contracted state is said to be in **tetanus** (tĕt´ă·nŭs).

Fatigue occurs as a result of the depletion of acetylcholine reserves in the nerve endings at the motor end plates, the depletion of oxygen and nutrients, and the accumulation of metabolic waste products. Some muscles fatigue more readily than others, reflecting differences in the efficiency of nutrient and oxygen supply, the efficiency of the removal of waste products, and the density of contractile apparatus within the muscles. The pain that is felt when muscles are fatigued is thought to be caused by a peptide produced by nerve cells called **substance P**. During rigorous exercise, substance P is produced faster than the blood can carry it away, and it accumulates to levels that stimulate nerves that carry signals to pain centers in the brain (see Chapter 15). Less rigorous or intermittent exercise is not painful because the circulatory system is able to remove the substance P as rapidly as it is produced. Muscle fatigue, however, is probably not responsible for the overall sense of "tiredness" that one feels after intensive exercise. The basis for that kind of tiredness is not well understood, but it may be due to responses of the brain to changes in blood pH that result from extensive and strenuous exercise.

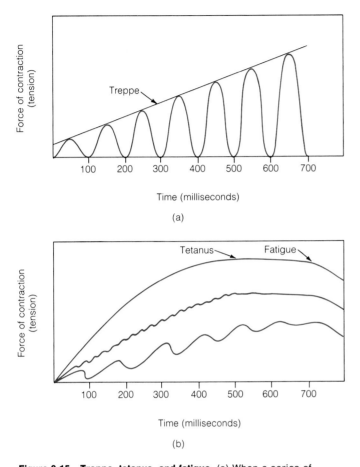

Figure 9-15 Treppe, tetanus, and fatigue. (a) When a series of stimuli is given at a frequency that allows the muscle to relax between stimuli, each successive contraction is stronger than the preceding one, producing a staircase effect, treppe, in a myogram. (b) As the frequency of stimuli increases, the muscle has insufficient time to relax completely. If the frequency is rapid enough, a smooth curve results. Tetanus occurs when a muscle is held in a constant state of contraction. After a time strength of the contraction begins to decline as the muscle fatigues.

Energy Supply for Contraction

Because of the role of ATP in contraction, active muscle generally requires much larger amounts of ATP than do other tissues. The increased utilization of ATP in a muscle when it is exercised compared to when it is resting can be dramatic, some muscles using as much as 200 times as much ATP when active than when at rest. Nevertheless, the amount of ATP in muscle tissue is only sufficient to sustain less than a second of contraction and is about the same whether the muscle is resting or exercising. This is because the muscle produces ATP at the same rate as it is used, and the amount present at any instant represents ATP that has been produced but not yet used.

There are two mechanisms used to produce large amounts of ATP quickly. As ATP is used, the ADP produced from it is rapidly rephosphorylated by transfer of phosphate from **creatine phosphate**, a low molecular weight nitrogen-containing acidic compound stored in the muscle. Creatine phosphate readily transfers its phosphate group (P) to ADP in the reaction:

Creatine phosphate ⇄ ADP → P
Creatine ⇄ ATP → Energy for movement of filaments

As a result, as fast as ATP is used, it is replenished at the expense of creatine phosphate.

As the supply of creatine phosphate diminishes, however, ADP begins to accumulate and a second mechanism for producing ATP gains in importance. In this reaction, two ADP molecules interact, and a phosphate from one is transferred to the other. This results in a molecule of adenosine *mono*phosphate (AMP) and a molecule of ATP:

ADP → AMP
ADP → ATP

This reaction is catalyzed by **adenylate kinase,** an enzyme that is activated by Ca^{2+}.

After exercise is over, the resting muscle replenishes its supply of creatine phosphate and ATP by catabolizing glycogen stored in the muscle. Glycogen breakdown releases **glucose phosphate,** which when metabolized releases energy for the production of ATP, and the ATP is used to rephosphorylate creatine and AMP. In active muscles, glycogen stores can be depleted rapidly, and the muscle must then rely on nutrients delivered in the blood. Fatty acids become the primary source of energy at this time, although glucose and other energy-rich compounds may also be used.

Fast Twitch Fibers

There are two categories of muscle fibers that differ in their efficiency at recovering energy from glycogen. **Fast twitch** muscles fibers contract relatively quickly and because of this, they are responsible for more precise and delicate types of activity. This type of fiber might be especially prevalent, for example, in the muscles of the fingers and hands. Fast twitch muscle fibers are not supplied with blood as efficiently as are slower muscle fibers, so oxygen is depleted more quickly during exercise, and the fiber must resort relatively quickly to anaerobic respiration for the production of ATP. This is a much less efficient way to produce ATP than respiration in the presence of oxygen, so carbohydrate reserves must

myoglobin: (Gr.) *mys*, muscle + (L.) *globus*, globe
myoglobinuria: (Gr.) *mys*, muscle + (L.) *globus*, globe + (Gr.) *ourin*, urine

---

> **CLINICAL NOTE**
>
> **FAST AND SLOW TWITCH FIBERS**
>
> Recently, exercise physiologists have been able to identify neuromuscular units according to their speed of contraction. Two types of motor units have been identified: **slow twitch** (Type I) and **fast twitch** (Type II). There is considerable variation in the percentage of these fibers within individual muscles and also within different people. In general, endurance-type athletes, such as long distance runners, are thought to have a higher percentage of Type I fibers and speed-oriented athletes, such as sprinters, are thought to have more Type II fibers. These fiber types are identified by taking a biopsy of the muscle. The fiber type appears to be genetically determined rather than influenced by training techniques.

> **CLINICAL NOTE**
>
> **MYOGLOBINURIA**
>
> **Myoglobin** is a protein found in muscle similar to the hemoglobin found in the red blood cells. At times when there is damage to the skeletal muscles due to crush injuries, cold injuries, or even following excessive muscular activity, large amounts of myoglobin will be released from the damaged muscles and will filter out through the kidneys, causing the urine to turn a reddish-brown color. This condition of having myoglobin in the urine is called **myoglobinuria** (mī′ō·glō·bĭn·ū′rē·ă). Myoglobin in the urine can be easily differentiated from hemoglobin in the urine, which occurs with bleeding into the urinary tract. Excessive amounts of myoglobin released from the injured muscle can damage the kidney and result in renal failure.

be used more rapidly to produce adequate amounts of ATP. As a result, glycogen stores are used up more rapidly, and there is a correspondingly rapid accumulation of lactic acid (see next section) and fatiguing of the muscle. To counteract this, these fibers are supplied with **myoglobin** (mī″ō·glō′bĭn), an oxygen-binding pigment used to store oxygen. This oxygen is used for the catabolism of glucose, enabling the muscles to function longer before fatiguing.

Slow Twitch Fibers

Slow twitch muscle fibers, in contrast, are relatively well supplied with oxygen and can manufacture ATP for longer periods while using smaller amounts of glycogen in the process. Examples of muscles rich in this type of fiber include, for example, muscles in the lumbar region of the back that undergo relatively sustained contractions. Slow twitch muscle fibers contain relatively more mitochondria than do fast twitch fibers, enabling them to produce ATP more efficiently from the carbohydrates used as energy sources.

Because of the presence of myoglobin in fast twitch muscles, they have a deeper red color than do slow twitch muscles. Dark meat and white meat in chickens and turkeys are examples of fast and slow twitch muscles, respectively, their coloration reflecting the amounts of myoglobin present (see Clinical Note: Fast and Slow Twitch Fibers).

Oxygen Debt

When a skeletal muscle is exercised strenuously, its demand for oxygen may be greater than the rate at which oxygen can be supplied by the blood. Individuals who suddenly begin exercising vigorously show a rise in oxygen consumption almost immediately (Figure 9-16). When the exercise stops, oxygen consumption continues at a relatively high rate for a short time before it begins to decline. This additional oxygen is used to pay an **oxygen debt** that built up during the intensive exercise.

Oxygen consumed after exercise stops is used to replenish the supply of oxygen stored in myoglobin and to replenish supplies of ATP and creatine phosphate. Continued oxygen consumption is also due in part to the generally elevated metabolic rate that results from the continued presence of adrenalin released into the blood during the exercise.

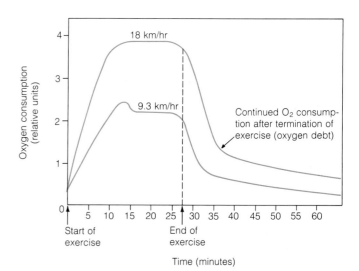

Figure 9-16 Oxygen debt. This figure shows the relative amount of oxygen consumed during exercise on a treadmill at two different speeds. Note that the rate of oxygen consumption remains elevated for a time after exercise has stopped. This continued elevation in uptake of oxygen is replacing an oxygen debt. (Adapted figure from R. Margaria, ed., *Exercise at Altitude* [Amsterdam: Exerpta Medica, 1967])

Heat

Exercising muscles produce heat. The heat is carried away from the muscle by the blood and distributed to the rest of the body where it contributes to body temperature. Muscles are also partially contracted even when resting, in a phenomenon referred to as **muscle tone,** and produce heat then, too. Additional contractions produce even more heat, making exercise an effective way to increase body temperature. **Shivering** is an involuntary and synchronized rhythm of muscle contractions that increases body temperature in response to cold.

Chemical Waste Products

In addition to heat, contracting muscles produce chemical wastes in the form of **lactic acid** and **creatinine** (krē·ăt´in·in). Lactic acid is a waste product of anaerobic metabolism produced when muscles use oxygen faster than it can be supplied. Although lactic acid is a waste product as far as the muscle is concerned, it is not a waste product of the body since it is transported to the liver, where it is converted back to usable products such as glycogen and pyruvic acid. Creatinine is produced from creatine in the muscle and is excreted by the kidney at an approximately constant rate (see Clinical Note: Serum Creatinine). Its production increases as a result of muscle damage, fever, menstruation, and strenuous muscular activity.

Cardiac Muscle

After studying this section, you should be able to:

1. Diagram a typical cardiac muscle cell, showing how it is associated with neighboring cells.
2. Compare an impulse that causes contraction in a cardiac muscle cell with one in a skeletal muscle fiber.
3. Explain the advantage of a slower traveling impulse and longer refractory period in cardiac muscle as compared with skeletal muscle.

Cardiac muscle is found in the heart and in the walls of the large arteries immediately adjacent to the heart (Figure 9-1). Like skeletal muscle, it has a banded appearance, although the striations are not as pronounced and regular. Contraction of cardiac muscle is involuntary and controls the rhythmic pumping of blood through the heart and into major arteries.

The heart consists of four chambers, two of which are called **atria** and two of which are called **ventricles** (see

> **CLINICAL NOTE**
>
> **SERUM CREATININE**
>
> **Creatinine** is a degradation product of muscle metabolism. The levels of creatinine present in the blood and urine are used as quick indices of kidney function. It is excreted solely by glomerular filtration in the kidney of normal persons. Blood creatinine level and the rate that creatinine is removed from the blood by the kidneys are sometimes used to monitor renal function in patients with kidney disease. If the kidneys are functioning properly, the level of creatinine present in an individual's blood is proportional to the muscle mass. Blood creatinine levels remain fairly constant unless there is unusual muscular exercise or the kidneys are damaged and unable to excrete creatinine from the body.

Chapter 22). Atria collect blood as it arrives in major veins, one atrium receiving blood from the body and the other receiving blood from the lungs. Contraction of the muscular walls of the atria forces this blood into the ventricles, where contraction of ventricular walls forces the blood out of the heart and into the major arteries that lead from the heart. Unlike skeletal muscle, cardiac muscle is not induced to contract by nerve impulses, but by stimuli that originate in the heart itself.

Individual cells in heart muscle are branched and each one contains a single nucleus (Figure 9-17a). The cells lie beside one another and meet other muscle cells at their ends in highly convoluted pairs of membranes called **intercalated disks.** In cross section an intercalated disk appears as a dense, wavy line with structures that resemble desmosomes at various points in the interface. Gap junctions are found on the longitudinal surfaces of neighboring cells. The two kinds of junctions provide points through which impulses can pass directly from cell to cell, rather than through a transmitter substance that transmits the impulse from cell to cell, as in skeletal muscle.

Cardiac muscle is striated somewhat like skeletal muscle. Overlapping thick and thin filaments create a banded appearance, although the filaments are not organized as uniformly as they are in skeletal muscle. In addition, the filaments are not organized into myofibrils within the cardiac muscle fiber, but are less evenly distributed throughout the cell (Figure 9-17c).

Impulses travel through cardiac muscle at about one-tenth the speed of an impulse in skeletal muscle. The contraction of cardiac muscle is also of longer duration, and as a result, cardiac muscle contracts and relaxes relatively slowly. Cardiac muscle cells also have a longer refractory period than that of skeletal muscle fibers (about 60 times as long). This prevents a second contraction from occurring

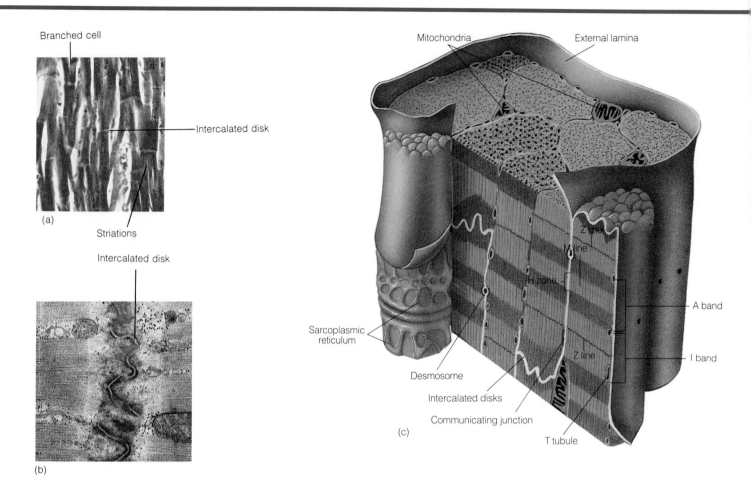

Figure 9-17 Cardiac muscle. (a) Photomicrograph of cardiac muscle showing the branched structure of many of the cells (mag. 190×). (b) Electron micrograph (at a higher magnification) of one intercalated disk. (c) This perspective diagram of cardiac muscle shows its striated appearance and the convoluted junctions that lie between the ends of successive cells. Note the numerous mitochondria and elaborate sarcoplasmic reticulum.

until after the muscle has relaxed from a previous contraction, and thus the heart "beats" in a series of rhythmic contractions. This in turn prevents summation and tetany from occurring in cardiac muscle, both of which would interfere with heart function and lead to death.

Smooth Muscle

After studying this section, you should be able to:

1. List areas of the body in which smooth muscle is found.
2. Describe the functional and structural differences between multiunit and visceral smooth muscle.
3. Compare a smooth muscle contraction with a skeletal muscle contraction.

Smooth muscle is another type of involuntary muscle, and it lacks the striated appearance of skeletal muscle. It is found in the walls of many internal organs, including those of the digestive system, the reproductive tract, and the respiratory and urinary systems. It is also found in the skin, blood vessels, and iris of the eye.

There are two kinds of smooth muscle: **multiunit smooth muscle** and **visceral smooth muscle** (Figure 9-18). Multiunit smooth muscle consists of individual fibers or bundles of fibers that are individually controlled by nerve endings. It is found around sweat glands in the skin, in the arrector pili muscles that cause goose bumps and hairs to stand erectly, and in the iris of the eye where it controls the diameter of the pupil. Visceral smooth muscle forms sheets or layers of tissue in hollow organs such as the stomach and other organs of the digestive tract, where it

atrophy: (Gr.) *a*, absence of + *trophe*, nourishment

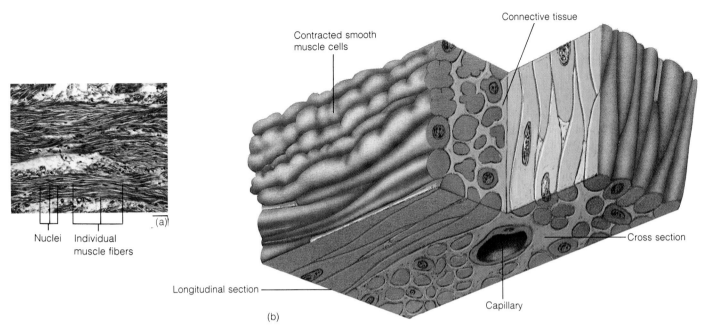

Figure 9-18 Smooth muscle. Smooth muscle consists of individual spindle-shaped cells that lack the striations characteristic of skeletal and cardiac muscle. (a) Photomicrograph of a longitudinal section from a human bladder (mag. 64,940×). (b) Diagram showing two layers of smooth muscle.

controls the diameter of the organ, and by contracting, forces the contents through the organ (Figure 9-1). It is also found in the urinary bladder and the uterus.

Both kinds of tissue consist of individual spindle-shaped fibers that contain a single nucleus. The fibers lack the striations seen in skeletal or cardiac muscle (Figure 9-18), but are nevertheless thought to contract by a mechanism similar to that of skeletal and cardiac muscle. Thick filaments are found in smooth muscle cells, but they are not found in the orderly array characteristic of other types of muscle. Although actin can be extracted from smooth muscle in significant amounts, thin filaments are not found. It is thought that thin filaments may be transitory, forming in response to the stimulus that causes a contraction.

The chemical mechanism involved in contraction differs from that of skeletal and cardiac muscle in that myosin molecules of smooth muscle must be phosphorylated in order to interact with actin subunits, and dephosphorylation is necessary for relaxation. Phosphorylation and dephosphorylation depend on particular enzymes, and the enzymes, in turn, require calcium ions bound to a protein called **calmodulin** (kăl″mŏd′ū·lĭn) to be able to function.

Smooth muscle can be stimulated to contract by nerve impulses, by chemicals carried to the muscle by the blood, or by mechanical stimuli, such as stretching. The nerves that control smooth muscles are part of the nervous system called the **autonomic** (ŏ·tō·nŏm′ĭk) **nervous system** (see Chapter 14). A nerve impulse arriving at the myoneural junction results in the release of acetylcholine or another type of transmitter substance called **norepinephrine** (nōr·ĕp″ĭ·nĕf′rĭn) (see Chapter 11).

As in cardiac muscle, stimulation of smooth muscle is followed by a relatively long refractory period. As a result, smooth muscles contract relatively slowly compared with skeletal muscle, causing organs that contain smooth muscle to carry out long, sustained contractions, such as the slow churning motions of the large intestine (see Chapter 18 on the digestive system).

Effects of Training on Muscles

After studying this section, you should be able to:

1. Describe the effects of exercise on muscle structure.
2. Distinguish between the effects of isometric and isotonic exercise on muscle structure and function.
3. List examples of exercises that are primarily isometric and examples of other exercises that are primarily isotonic.

The strength of a muscle depends on how much it is used. A muscle that is not used diminishes in cross-sectional area and loses portions of its contractile apparatus, a condition called **atrophy** (a′trō·fē). **Muscular atrophies** result

dystrophy: (Gr.) *dys*, bad + *trophe*, nourishment
isometric: (Gr.) *isos*, equal + *metron*, measurement
isotonic: (Gr.) *isos*, equal + *tonos*, tone

> **CLINICAL NOTE**
>
> **CHARLEY HORSE**
>
> **Charley horse** is a generic term used by athletes to describe any minor injury to a muscle. The term may be used to describe a muscle tear, strain (an overstretching of a muscle), bruise, or cramp. Athletic physicians usually reserve the term to describe a contusion (bruise) with bleeding into the muscle belly.

from disorders of the nervous system that prevent muscle function and that result in a wasting away of the affected muscle. **Muscular dystrophies** (dĭs′trō·fēs) are a group of inherited diseases in which there is a progressive deterioration of the muscle fiber itself, without any nerve disorder as a cause (see Chapter 10, The Clinic: Muscular Diseases). In **myasthenia gravis** (mī′ăs·thē′nē·ă grăv′ĭs), muscle function is impaired at the neuromuscular junction, possibly through interference with the ability of acetylcholine to combine with receptors on the muscle fiber membrane. In all these cases, lack of use results in muscle deterioration. Muscular atrophy also occurs in people who are beddriden or otherwise limited in physical activity.

Strength and Endurance

Just as lack of activity can result in a reduction of muscle function, exercise can increase the size, strength, and endurance of a muscle. Muscular size and strength depend largely on the number and size of individual fibers in the muscle and the number of thick and thin filaments in the fibers. **Endurance,** which should be distinguished from strength, is the ability of a muscle to continue contracting against a resistance.

As a muscle becomes stronger, there is an increase in diameter. This increase is not brought about by an increase in the number of muscle fibers, which remains the same, but by an increase in the number of thick and thin filaments in the fibers. Since the strength of contraction is largely a function of the size of the muscle and the density of the filaments, the larger and denser the muscle, the greater the strength of contraction it can develop. In general, the maximum strength of contraction that a muscle can develop is about 3 kg/cm^2 (43 lb/in.^{2}) of cross-sectional area. A guard for the Chicago Bears, for example, might have a quadriceps muscle (the large anterior muscle of the thigh, see Figure 10-23) with a cross-sectional area of 30 in.^{2}, enabling him to develop 1300 lb or more of force in each leg from this one muscle alone!

Isometric and Isotonic Contractions

There are two kinds of contractions to which a muscle can be subjected. When a muscle pulls against a load but is unable to shorten because the load is too great, the muscle maintains a constant length and the energy of the contraction is dissipated as heat rather than through the performance of work. This kind of contraction is called an **isometric** (ī′′sō·mĕt′rĭk) contraction (Figure 9-19a). In contrast, when a muscle contracts against a load and is able to shorten in length while maintaining a constant tension, the contraction is an **isotonic** (ī′′sō·tŏn′ĭk) contraction (Figure 9-19b).

Isometric and isotonic exercises have different effects on muscle size and ability. In general, isometric exercises increase muscle size and strength with relatively little effect on muscle endurance. They subject the muscle to anaerobic conditions as the enlarged (but not shortened) muscle compresses the blood vessels around it and inhibit the delivery of oxygen.

Body building exercises such as weight lifting rely on isometric contraction to increase the strength of the muscle. Before the weights can be lifted, the muscles must develop sufficient force of contraction to move the weights, and during the time the force is developing, the muscles are held at a constant length. It is during that time that weight lifting is an isometric type of exercise. Exercise prescribed for lower back pain is also generally isometric and is intended primarily to strengthen the muscles necessary to maintain correct posture.

Isotonic exercises, in contrast, place stress on the circulatory system and result in increased oxygen delivery to the muscles. With time, this can stimulate increased cardiac output and oxygen exchange in the lungs. At the muscle level there is increased capillary flow, increased oxygen storage in the muscle fiber, and increased efficiency of ATP production. None of these effects by itself increases the strength or size of the muscles, but together they enable the muscle to continue contracting for longer periods before fatiguing.

Running, swimming, and bicycling are examples of exercises that are primarily isotonic. There is relatively little increase in muscle size or strength as a result of this kind of exercise, at least when compared to the results experienced by a weight lifter, but endurance can be increased dramatically through the use of isotonic exercises.

Figure 9-19 Isometric versus isotonic contraction. (a) In isometric contractions, the muscle is prevented from shortening and the energy used is liberated as heat. (b) In isotonic contractions, the muscle shortens and work is performed, liberating less energy as heat.

steroid: (Gr.) *steros*, solid + *eidos*, form

> ### THE CLINIC
>
> **TRENDS IN BODY BUILDING**
>
> A great deal of interest has recently surfaced in the study of muscle response to exercise. In the United States there has been a floodtide of interest in fitness, and riding the crest has been the art of muscle building or **body building.** Most athletes and many fitness enthusiasts spend a great deal of time "pumping iron" (lifting weights) to increase their muscle mass and strength. There are many theories and techniques proposed to help increase muscle mass. Most of the techniques are based on the theory that to build muscle, one must subject the muscle to repeated stress of submaximal loads. This technique causes the muscle fibers to increase in size but not in number. American industry has joined the trend by producing a great number of machines used to exercise specific muscle groups. One must continue the program indefinitely to maintain the size and strength desired.
>
> Along with body building has come an interest in **steroids.** Steroids are a group of naturally occurring hormones and membrane components, but in body building parlance "steroids" denotes a group of natural or synthetic substances related to the male sex hormone testosterone. They are also called **anabolic steroids** because of their ability to induce weight gain and muscle mass. They were originally developed to help cancer patients and victims of starvation to gain weight and strength. Steroids are sometimes used by athletes in those sports that require weight, strength, and aggressiveness such as weight lifting and football. These drugs are now considered illegal by most athletic associations because they result in unfair competition and can have serious side effects, such as testicular atrophy, induction of malignant tumors of the liver, and antisocial, aggressive behavior.
>
> Those people who have an interest in body building should be aware that so-called fitness is not the same as health. Therefore, great caution must be used to avoid drugs and the overuse of muscles to prevent injury and harmful side effects.

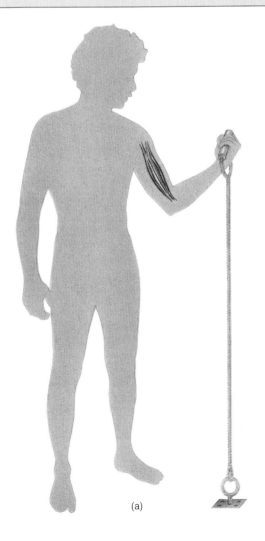
(a)

(b)

In this chapter, we studied the cellular organization of muscle tissue, concentrating on the elongate fibers of skeletal muscle. You learned the chemical reactions and energy exchanges that establish cross-links between intracellular proteins, resulting in contraction of a muscle fiber. We also studied how such factors as the nutrient supply, temperature, blood flow, and the concentration of waste products affect the strength of contractions and fatigue in a whole muscle. In the following chapter, we will study the anatomical arrangement of muscles in the body and the details of their attachments to the skeleton.

STUDY OUTLINE

Organization of Skeletal Muscle (pp. 181–188)

Skeletal muscle, also called *striated* or *voluntary* muscle, is responsible for the movement of bones. Skeletal muscles are attached to bones by *tendons*.

A *muscle trunk* contains a collection of tissues surrounded by the *deep fascia* and *epimysium*. It is subdivided into *fascicles* surrounded by *perimysium*. Muscle fibers are in fascicles and are surrounded by *endomysium*.

Motor Units Muscle fibers controlled by a single nerve cell comprise a *motor unit*. The terminal portion of a nerve cell branch, the *motor end plate*, forms a *neuromuscular (myoneural) junction* with the muscle fiber. The gap between the nerve cell and muscle fiber membranes is the *neuromuscular cleft*.

Structure of the Skeletal Muscle Fiber A muscle fiber is surrounded by a *sarcolemma*, which encloses the cell's *sarcoplasm*. Membranes in the fiber form the *sarcoplasmic reticulum*, which partitions the fiber into *myofibrils*.

Sarcomeres The basic units of muscle contraction are *sarcomeres*, which are bordered at each end by *Z lines* and contain a series of light and dark bands called the *I bands*, *A bands*, *H zone*, and *M line*.

The bands are produced by overlapping *thin* and *thick filaments*. Thin filaments consists of *actin*, *tropomyosin*, and *troponin* subunits. Thick filaments consist of *myosin* subunits. The filaments are held in a three-dimensional matrix by *cross bridges* extending from myosin subunits that attach to actin subunits.

The sarcoplasmic reticulum forms *cisterns* in close association with *T tubules*, regions of the sarcolemma that extend into the fiber. A T tubule and associated cisterns form a *triad*.

Muscle Contraction (pp. 188–192)

Initiation of Contraction A nerve impulse causes *acetylcholine* to be released into the neuromuscular cleft. Acetylcholine induces an impulse in the muscle fiber that passes into the fiber and induces Ca^{2+} release, which causes contraction to occur. Acetylcholine is degraded by *choline esterase*.

A *threshold* must be reached to induce a contraction. The strength of the contraction is independent of the strength of the stimulus, referred to as an *all-or-none* effect. Graded muscle responses result from the presence of many fibers in a muscle with varying thresholds.

Mechanism of Contraction Ca^{2+} released in response to a stimulus causes tropomyosin to move and expose actin sites. Cross bridges form between the sites and myosin heads. ADP and inorganic phosphate is released. Attachment of the myosin head causes the angle of the head to change and exert a force on the thin filament.

Combination of a new ATP molecule with the myosin head causes the head to release its attachment to the actin and to bend back to its original angle. The cycle is repeated as long as actin sites are exposed. Movement of thousands of myosin heads causes the thick filaments to be pulled along the thin filaments, and the sarcomere shortens. This is the *sliding filament model* of muscle contraction.

Relaxation Ca^{2+} is pumped back into the sarcoplasmic reticulum. Tropomyosin molecules revert to positions that block attachment of myosin heads to actin sites. In absence of attachment, the sarcomere is stretched back to its original length. The muscle is *relaxed*.

Physiology of Muscle Contractions (pp. 192–197)

Twitches A muscle contraction induced by an electric shock is a *twitch*. The twitch is preceded by a *latent period*. When electrical shocks are given faster than the muscle can return to normal, the contractions build on one another and *summation* occurs. A *refractory period* follows an electric shock during which the muscle is insensitive to additional shocks.

Treppe is the increased strength of response of a muscle to repeated stimulations. If stimuli are given quickly enough, individual contractions will not be discernible and the muscle will reach a state of maximum contraction, *tetanus*. If held in tetanus, the muscle contraction will slowly weaken as the muscle *fatigues*.

Energy Supply for Contraction ATP supplies the energy for muscle contraction. ATP is converted to ADP rapidly as filaments slide past one another. The ADP is rapidly rephosphorylated by creatine phosphate, which is stored in the muscle. As the creatine phosphate supply is used up, accumulating ADP molecules interact, and phosphate groups are transferred from ADP to ADP, producing ATP and AMP in a reaction catalyzed by adenylate cyclase. ATP is also produced by the metabolism of glucose phosphate, glycogen breakdown, and the metabolism of fatty acids and other nutrients delivered in the blood.

Fast twitch fibers use their glycogen reserves quickly because they produce ATP for rapid muscle movements. They contain *myoglobin*, an oxygen-storing molecule that prolongs muscle function. *Slow twitch* fibers use their glycogen more slowly and metabolize it more efficiently.

Oxygen Debt Exercise is accompanied by increased oxygen uptake by muscles. The rate of oxygen use remains elevated for a time after exercise ends, as the muscle repays an *oxygen debt*. Continued oxygen use replenishes the supply of ATP and the amount of stored oxygen.

Heat One product of muscle activity is heat. Muscles are partially contracted even when resting, displaying *muscle tone*. *Shivering* is an involuntary rhythmic contraction of muscles and has the effect of producing heat in response to cold.

Chemical Waste Products *Lactic acid* and *creatinine* are two chemical products of muscle contraction.

Cardiac Muscle (pp. 197–198)

Cardiac muscle is in the heart and walls of major arteries adjacent to the heart, and it is striated and involuntary. Cardiac muscle cells are branched and form *intercalated disks* where they meet other cardiac muscle cells. They are not organized into myofibrils. Impulses travel through cardiac muscle more slowly than they do through skeletal muscle. Their longer refractory period prevents tetanus from occurring in cardiac muscle.

Smooth Muscle (pp. 198–199)

Smooth muscle is located in the walls of many internal organs, skin, iris of the eyes, and arteries. It is not striated and is involuntary. *Multiunit* smooth muscle consists of individually controlled cells or groups of cells; *visceral* smooth muscle consists of cylinders or layers of cells that are controlled as a unit.

Contraction involves phosphorylation of myosin directly, involving particular enzymes, Ca^{2+}, and *calmodulin*. Smooth muscle contraction is induced by autonomic nerve impulses, mechanical stimuli, or chemicals in the blood. Smooth muscle fibers contract relatively slowly and have a long refractory period compared with skeletal muscle.

Effects of Training on Muscles (pp. 199–202)

The strength of a muscle depends on how much it is used. Unused muscles *atrophy* as contractile elements are lost. *Muscular atrophies* may result from neurological disorders that interfere with delivery of stimuli to muscles. *Muscular dystrophies* result from disorders of the muscle itself.

Strength and Endurance Muscle size, strength, and endurance depend primarily on the number and size of muscle fibers and the number of filaments in the fibers. *Endurance* is the ability of a muscle to continue contracting against a resistance. Exercise causes the number of thick and thin filaments in a muscle to increase, but not the number of individual fibers. The greater the size of a muscle and the density of fibers in it, the stronger the muscle.

Isometric and Isotonic Contractions Contraction against a force that prevents the muscle from shortening is *isometric* contraction. Contraction that allows the muscle to shorten while maintaining a constant tension is *isotonic* contraction. Isometric exercise primarily increases muscle size and strength with relatively little effect on endurance. Isotonic exercise primarily increases endurance with relatively less effect on muscle size or strength.

SELF-TEST OF CHAPTER OBJECTIVES

True-False Questions
1. The outermost covering of the muscle trunk is the perimysium.
2. The neuromuscular cleft is a region where nerve and muscle membranes fuse together.
3. The thin filaments in a sarcomere are anchored in Z disks.
4. The T tubules, sarcolemma, and sarcoplasmic reticulum are all different parts of the same membrane.
5. The muscles that rely most heavily on stored fuels are generally the fast twitch muscles.
6. The oxygen uptake that often continues after exercise is due to the continued metabolism of lactic acid that was produced by the muscle while contracting in the absence of oxygen.
7. Muscle contraction against an unmovable resistance is called isometric contraction.
8. In a muscle twitch, the short lag period that follows a stimulus before the contraction starts is called the latent period.
9. A summation of contractions takes place when stimulations are occurring faster than the fiber can relax, but slowly enough to allow a refractory period to pass.
10. Progressive deterioration of muscle due to lack of use is referred to as muscular atrophy.

Multiple-Choice Questions
11. The type of muscle you use to move your forearm is
    a. cardiac muscle   c. involuntary muscle
    b. striated muscle   d. smooth muscle
12. There is a layer of muscle in the wall of your stomach. It is made of
    a. cardiac muscle   c. somatic muscle
    b. striated muscle   d. smooth muscle
13. In skeletal muscle, individual muscle fibers are surrounded by a layer of connective tissue called the
    a. fascia   c. perimysium
    b. epimysium   d. endomysium
14. The fundamental unit of contraction in a skeletal muscle is the
    a. sarcomere   c. muscle fiber
    b. perimysium   d. motor unit
15. Acetylcholine must be removed from the neuromuscular cleft if the muscle fiber is to be able to react to subsequent stimuli. The acetylcholine is removed by
    a. diffusion   c. enzymatic breakdown
    b. reabsorption   d. osmosis
16. Which of the following is the earliest effect to follow the stimulation of a muscle fiber?
    a. myosin cross bridges change their angles
    b. ATP is cleaved
    c. an actin site is exposed
    d. Ca^{2+} is released from the sarcoplasmic reticulum
17. If continuous stimulation is provided to a muscle, the contraction reaches a maximum and the muscle is in a state of
    a. summation   c. fatigue
    b. tetanus   d. treppe

18. After reaching a maximum of contraction during continuous stimulation, the degree of contraction slowly diminishes. This is called
    a. summation   c. fatigue
    b. tetanus   d. treppe
19. A muscle held in continuous contraction is said to be in
    a. treppe   c. tetanus
    b. fatigue   d. twitch
20. A muscle about 2 cm² in circular cross-sectional area should be able to develop a force of contraction of about
    a. 2 kg   c. 6 kg
    b. 4 kg   d. 8 kg
21. Which of the following lists muscle subunits in the order of increasing size?
    a. myofibril, thin filament, sarcomere, troponin
    b. thin filament, sarcomere, myofibril, troponin
    c. troponin, thin filament, sarcomere, myofibril
    d. sarcomere, myofibril, thin filament, troponin
22. Which of the following lists the regions of a sarcomere in sequence, beginning from either end?
    a. A band, H zone, Z line, I band
    b. Z line, I band, A band, H zone
    c. H zone, Z line, A band, I band
    d. I band, A line, H zone, Z line
23. In which of the following regions of a sarcomere are both thick and thin filaments found?
    a. I band
    b. A band outside the H zone
    c. A band inside the H zone
    d. M line
24. Which one of the following is not found in thin filaments?
    a. myosin   c. troponin
    b. actin   d. tropomyosin
25. Which of the following regions of a sarcomere will stay the same length as the sarcomere contracts?
    a. I band
    b. A band
    c. H zone
26. What happens to the Ca^{2+} that initiates a muscle contraction when the contraction is over?
    a. It is carried back into the sarcoplasmic reticulum by active transport.
    b. It diffuses from the fibers and is carried off by the blood.
    c. It accumulates and remains between the thick and thin filaments.
    d. It remains attached to the thin filaments.
27. Rigor mortis, the strongly contracted state of muscles following death, is due to which of the following?
    a. lack of nervous stimuli to induce relaxation
    b. lack of ATP required to break bonds between thin and thick filaments
    c. a breakdown in the structure of the membrane that prevents it from conducting an impulse
    d. a lack of choline esterase in the neuromuscular cleft to break down the acetylcholine that remains from stimuli that occurred before death
28. The latent period probably corresponds to the period when
    a. the action potential is passing over the surface of the muscle fiber
    b. ATP is being broken down to ADP and inorganic phosphate
    c. acetylcholine is diffusing across the neuromuscular cleft
    d. sufficient change is occurring in the membrane potential to pass the membrane's threshold
29. What effect does a particularly long refractory period have on the ability of a muscle cell to contract?
    a. It cannot be stimulated as rapidly as muscle cells with a shorter refractory period.
    b. It requires a stronger stimulus to cause a contraction.
    c. Tetanus can be reached more quickly.
    d. Less ATP is used in a contraction.
30. The specialized junction between the ends of two cardiac muscle cells is called a(n)
    a. motor end plate   c. cardiomyal junction
    b. intercalated disk   d. myoneural junction
31. Which of the following is used as a transmitter substance for a cardiac muscle impulse as it passes over and through the heart?
    a. acetylcholine   c. epinephrine
    b. norepinephrine   d. none of these
32. Which type of exercise has the greatest positive effect on the heart and circulatory system?
    a. isotonic
    b. isometric
    c. weight lifting
33. When a skeletal muscle is exercised strenuously and then exercise ceases, which of the following will probably occur?
    a. Lactic acid will continue to be made.
    b. Glucose delivery to the muscle will stop.
    c. Increased levels of oxygen uptake will continue.
    d. ATP levels in the muscle will decline.
34. What is the role of creatine phosphate in muscle contraction?
    a. to provide energy directly to the thin and thick filaments
    b. to serve as a reservoir for storing wastes produced by muscle contraction
    c. to provide phosphates for the rapid resynthesis of ATP used during muscle contraction
    d. to accept phosphate from ATP during muscle contraction.
35. Which of the following activities, performed regularly and strenuously, would have the greatest relative effect on the strength of the muscles involved?
    a. push-ups   c. swimming
    b. jumping rope   d. playing tennis

Essay Questions
36. Prepare a table comparing skeletal, cardiac, and smooth muscle with respect to (a) location, (b) cell structure, and (c) appearance.
37. Draw a typical skeletal muscle in cross section, illustrating and identifying the muscle, circulatory, nervous, and connective tissues present.

38. List at least ten events involved in the contraction and relaxation of a skeletal muscle from the time an impulse reaches the neuromuscular junction until the relaxed muscle is stretched to its original length.
39. Explain why the absence of ATP will cause muscles to remain contracted in rigor mortis.
40. Explain why a relatively long refractory period present in cardiac muscle is important in the functioning of the heart.
41. List and cite the source of products of muscle contraction.
42. Would you expect to find relatively more or fewer muscle fibers in motor units of your hand compared with the number of muscle fibers you would find in the motor units of your leg? Explain your answer.
43. Many kinds of exercises involve a combination of isometric and isotonic effects. Explain why this is so.
44. Draw the pattern of thick and thin filaments that would exist in a cross-sectional view of the I band, the A band, and the H zone of a sarcomere.
45. Skeletal muscles can contract so strongly that the H zone disappears. Using a diagram, explain why this is so.
46. Smooth muscle does not show the highly ordered arrangement of thin and thick filaments organized into sarcomeres as is found in skeletal muscle. Yet the mechanism of contraction is thought to be similar in the two types of muscle. Describe the basis for this assumption, and suggest how smooth muscle might contract in spite of the absence of highly ordered sarcomeres.
47. Suggest what the effect would be if transmitter substance were not removed from the neuromuscular cleft following a stimulus.
48. Describe the chemical events probably responsible for the phenomena of (a) the latent period, (b) summation, (c) treppe, (d) tetanus, and (e) fatigue.
49. Explain why having a supply of creatine phosphate increases a muscle's capacity to contract.

Clinical Application Questions
50. Distinguish between tendonitis and tenosynovitis and describe the medical treatment generally used for their control.
51. Define and distinguish among fibrillation, twitches, fasciculations, tics, and spasms as abnormal forms of muscle contractions.
52. Describe the effect that the intake of anabolic steroids has on muscle mass and why their use to stimulate muscle buildup in athletes is illegal.
53. Explain how a drug that blocks calcium channels in smooth muscle can increase blood flow in coronary arteries.

10
Musculature of the Human Body

A human body includes three different types of muscle tissue: skeletal, smooth, and cardiac. In Chapter 9, we discussed the histology and chemical mechanisms that allow muscles (particularly skeletal muscles) to contract and hence shorten in length. In this chapter selected representatives of more than 600 muscles that move the skeleton and some soft external tissues, such as the lips and eyelids, will be studied. Specifically, the topics to be discussed will explain how movements, such as walking, running, eating, smiling, reading, and writing are accomplished when skeletal muscles shorten and produce movements of joints or soft tissues. Our major goals in this chapter are to examine

1. the different shapes of skeletal muscles,
2. the difference between the origin and insertion of a muscle, based on the skeletal action upon contraction,
3. how the positioning of muscles around a joint is related to their function as antagonists, and
4. how the muscular movement of the skeleton can be compared to simple levers.

pennate: (L.) *penna*, feather
fusiform: (L.) *fusus*, spindle + *forma*, shape

Muscles and Their Movements

After completing this section, you should be able to:

1. List and discuss the bases for naming muscles.
2. Describe the major shapes of muscles.
3. Define the terms origin, insertion, and action with reference to muscles.
4. Describe the major role of muscles as agonists and antagonists.
5. Discuss the differences between first, second, and third class levers and give examples of muscles that fall into each category.

Naming the Muscles

As the muscular system is presented, be aware that the names of most muscles refer to some prominent feature of the muscle. For example, many muscles are named with reference to their size, such as *pectoralis major* (large chest muscle) and *pectoralis minor* (small chest muscle). The terms *major* and *minor* are used to distinguish between two similar muscles of different sizes. Some muscles are named for the position they occupy in the body, such as *tibialis anterior* (anterior tibial surface) and *tibialis posterior* (posterior tibial surface). Muscles may be named for movements they accomplish, such as *flexor digitorum* (flexing of the fingers) and *levator scapulae* (elevation of the scapula). Muscles are also named for their shape; for example, *trapezius* (trapezoidal shape) and *orbicularis oculi* (circular muscle that surrounds the eye). Some muscles have been named for bones to which they attach; for example, *zygomaticus* (attaches to the zygomatic bone) and *temporalis* (attaches to the temporal bone). Muscles have been named for the number of divisions in the muscle, such as *biceps brachii* (a two-headed muscle of the arm) and *quadriceps femoris* (a four-headed muscle on the femur). Finally, keep in mind that a muscle's name may contain two or more references to prominent features; for example, the name *flexor carpi ulnaris* not only indicates that the muscle is a flexor of the wrist (carpus) but also indicates its position adjacent to the ulna.

The preceding chapter described the function of a nerve in initiating muscular contraction. As muscles are presented in this chapter, the specific nerves that supply each muscle are indicated in the summary table. This is done because the nerves that exit the spinal cord develop in a segmental pattern similar to that seen in muscular development. These segmental muscles are innervated by complementary segmental spinal nerves. Thus, related muscles (those that develop from the same embryonic region) are innervated by the same spinal or cranial nerves. This additional knowledge may help to make sense out of a large mass of anatomic information.

Shapes of Skeletal Muscles

Chapter 9 described how muscle fibers are grouped into bundles of fascicles (or fasciculi). The arrangement of fascicles within the whole muscle reveals four distinct patterns: parallel, convergent, pennate, and circular. **Parallel** fascicles lie parallel to one another along the longitudinal axis of the muscle and form wide flat muscles. A **convergent** pattern is found in triangular-shaped muscles, where fascicles attach along a broad tendon at one end of the muscle and converge to a narrow tendon at the other end. Fascicles in a **pennate** pattern are attached like plumes of a feather to an elongate tendon that is nearly as along as the entire muscle. The fascicles may be attached only to one side of the tendon **(unipennate)** or to both sides **(bipennate)**. A **circular** pattern of fascicles is characteristic of sphincter muscles that surround openings, such as the mouth or anus.

Because of the different ways in which fascicles attach to tendons, skeletal muscles exist in a variety of sizes and shapes. Some common ones are illustrated in Figure 10-1. **Fusiform** (fū′zĭ·form) muscles are columnar in shape because the muscle fibers tend to run in the same direction. The tendons that attach fusiform muscles to bones are restricted to the ends of the muscle. The thickest part of the muscle, its **belly**, the fleshy portion in a fusiform muscle, is usually near the middle of the muscle. **Penniform** muscles are flattened, and one or both of the tendons will extend for some distance along the length of the belly. The tendon in penniform muscles may extend along one side of the belly **(unipenniform)**, or it may course along the center of the belly **(bipenniform)**. **Radiate** muscles are roughly triangular in shape, with the muscle fibers converging from a wide tendinous attachment at one end into a narrow tendinous connection at the opposite end (Figure 10-1). In addition to these generalized shapes, muscles can also be **rectangular, triangular, rhomboid,** and **circular** in shape; other shapes defy simple description.

Origin, Insertion, and Action

A muscle produces skeletal movements because its fibers or tendons are attached across a movable joint connecting two different bones. Skeletal muscles may also move soft tissues that are not strictly skeletal parts, such as the eyelids, eyeballs, and lips. In most cases of movement, the joint connecting the two bones is a synovial joint, which permits

Chapter 10 Musculature of the Human Body

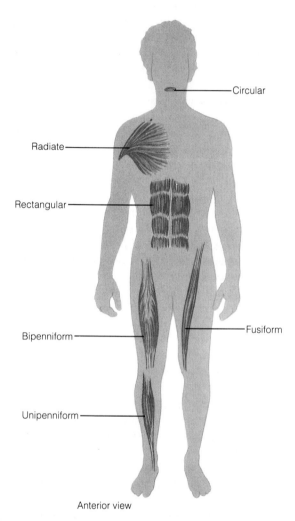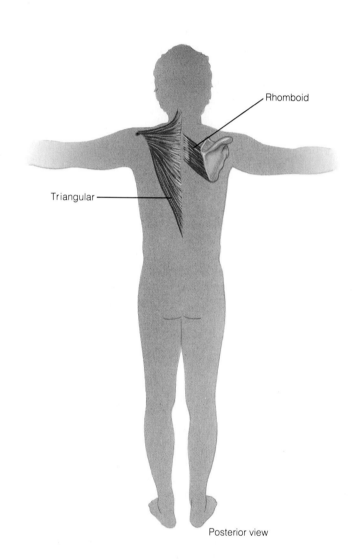

Figure 10-1 Shapes of muscles. Skeletal muscles exist in a variety of sizes and shapes directly related to their function. Common examples of each shape are illustrated here.

agonist:	(Gr.) *agon*, contest
antagonist:	(Gr.) *anti*, against + *agon*, contest
synergist:	(Gr.) *syn*, together + *ergon*, to work

significant movement between the two bones involved. When a muscle contracts and shortens, it moves one of the two bones attached to it. Usually only one of the two bones will move. As will be seen shortly, this is accomplished by the stabilizing contractions of other muscles attached to the nonmoving bone. The site of attachment that does not move, or moves least, when the muscle contracts is the **origin** of the muscle. The place of attachment that moves is the **insertion**. For example, flexing the fingers to form a clenched fist is accomplished by muscles located on the front of the forearm that are attached to the phalanges of the fingers at one end and to bones of the arm near the elbow at the other end. When the muscle contracts to flex the fingers, only the fingers move, not the entire hand. In this case the origin of the muscle is near the elbow and the insertion is on the phalanges of the fingers. The specific movement caused by a muscle is its **action.** In the example just given, the muscle's action is flexion.

As muscles are described throughout this chapter, we will consider their origins, insertions, and actions. These will also be summarized in reference tables for review and study.

Muscles are capable of only one activity, namely, contraction, and only muscles contract. This limits the action of any given muscle to one specific movement or sometimes a few closely related movements. Thus, to move a bone in a variety of directions, differently oriented muscles are required. For example, in the fist-clenching action, a muscle on the anterior surface of the forearm is used to flex the fingers. To unclench the fist, muscles on the back of the forearm are used.

A muscle of primary importance in any specific action is a **prime mover** or **agonist** (ăg´ŏn·ist). In some cases only one muscle is used in certain actions, as in abducting the arm away from the body at the shoulder joint. (Recall from Chapter 8 that *abduction* is the movement of the arm away from the body.) At other times a group of prime movers or agonists is used, as in adducting the arm toward the body at the shoulder joint. (*Adduction* is drawing the arm in toward the body.) Most muscles of the body are arranged in opposing pairs on opposite sides of a joint. Each muscle produces opposing movements. When a prime mover contracts, its opposing muscle relaxes. For example, when the fingers are flexed in a clenched fist, the opposing muscles—in this case the extensor on the other side of the forearm—relaxes. Similarly, when the fingers are extended in unclenching the fist, the opposing flexors relax. Any muscle that performs an opposing action to an agonist, such as in flexing and extending or abducting and adducting, is referred to as an **antagonist** (ăn·tăg´ŏ·nĭst).

In addition to producing significant movements, muscles can also function as **synergists** (sĭn´ĕr·jĭsts), which are stabilizing muscles that help another prime mover accomplish some desired action. For example, two relatively powerful muscles positioned on the back of the forearm act as prime movers in extending the wrist joint, as in hitting a backhand shot in ping-pong. The same muscles act as synergists to the flexors of the fingers during the clenching of the fist by contracting just enough to keep the wrist joint from flexing simultaneously with the fingers. Most muscles act as prime movers at certain times, as antagonists at other times, and as synergists at still other times. Two muscles at a joint can be either agonistic or antagonistic to each other; they cannot act in both ways simultaneously.

When a muscle is acting to immobilize a joint it is called a **fixator.** A fixator can act as a synergist, such as the wrist muscles described above. Fixators may also function in stabilizing a body part without specifically helping a prime mover. For example, several groups of muscles lie in the posterior region of the trunk attached to the vertebrae, pelvic girdle, and ribs. The chief function of these muscles is to keep the vertebral column erect. Whenever a person stands or sits upright on a backless stool or bench, these muscles remain contracted, maintaining proper postural position. The cooperation of several kinds and groups of muscles is necessary to maintain proper posture and to execute coordinated movements of body parts.

Levers

Movements of the skeleton can be compared to movements of simple levers. A **lever** is a rigid rod that moves around a pivot point, the **fulcrum.** A force, or **effort,** is required to move the lever, and the weight it must move is the **resistance.** In skeletal movements a bone acts as a lever moving around a joint that acts as the fulcrum. The strength required to move the bone is provided by the contraction of a muscle attached to the bone. The resistance in this case is the weight of the bone.

In abduction of the arm (Figure 10-2), the lever is the humerus, the fulcrum is the shoulder joint, the effort is provided by contraction of the deltoid muscle, and the resistance is the weight of the whole arm. To increase the resistance, the person could perform the same action while holding an exercise dumbbell. An increased resistance requires a greater effort; this provides the basis for physical fitness programs involving weight lifting.

Three classes of levers are recognized. In a **first class (class I) lever,** the fulcrum is located between the point where the force is applied (**lever arm**) and the point of resistance (**resistance arm**) (Figure 10-3a). Common examples of first class levers include a seesaw and a pair of scissors. A first class lever is used in the body to nod the head (as shown in the figure). It is also used to extend the forearm at the elbow joint. In this case, the effort is applied by the triceps muscle, the fulcrum is the elbow joint, and the resistance is the hand or something held in the hand. The actions of throwing a baseball, throwing a forward pass in football, or throwing a javelin are all examples of activities accomplished through the use of a first class lever.

In a **second class (class II) lever,** the resistance or weight is located between the fulcrum and the point where the effort is applied (Figure 10-3b). Examples of second class levers include a wheelbarrow and a door hinge. You use a second class lever to stand on your tip-toes. In this case the fulcrum is the ball of your foot, the resistance is your body weight, and the effort is applied by your calf muscles pulling on your heel. Any kind of running, hopping, or jumping

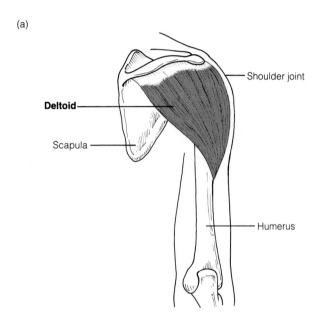

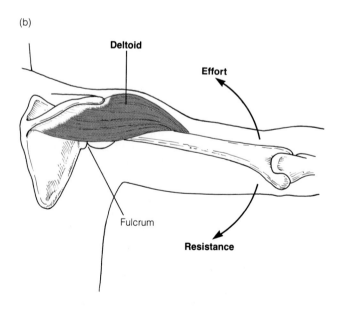

FIGURE 10-2 A lever. (a) The deltoid muscle at rest. (b) When the deltoid muscle contracts, it exerts a force (effort) to move the humerus (the resistance) away from the body at the shoulder joint (the fulcrum).

activity involves use of leverage gained through a second class lever.

In a **third class (class III) lever,** effort is applied between the fulcrum and the resistance (Figure 10-3c). A tweezers (forceps) is a common example of a third class lever. Most of the muscles act as part of third class levers. For example, flexing the forearm at the elbow joint involves a third class lever. The elbow joint is the fulcrum, the resistance is the hand, and the effort is applied by the biceps muscle. Performing chin-ups on a horizontal bar and squeezing a hand exerciser also involve a third class lever.

Before proceeding to the next section, it will be useful to review the section in Chapter 8 that describes specific movements of synovial joints (pp. 168–170). The terms describing these movements are the same as the terms used in describing actions produced by muscles. Figures 10-4 and 10-5 provide superficial anterior and posterior views of the skeletal muscles. Use these figures as a review of the body's musculature and as a reference for your study of the remainder of the chapter.

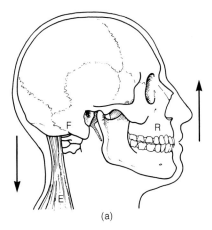

(a)

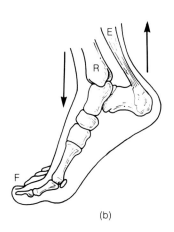

(b)

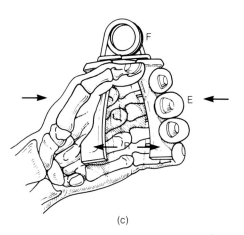

(c)

Figure 10-3 Examples of how levers operate in moving parts of the body. (a) first class lever, (b) second class lever, and (c) third class lever. The characteristics of each kind of lever are defined in the text: E = effort, R = resistance, F = fulcrum.

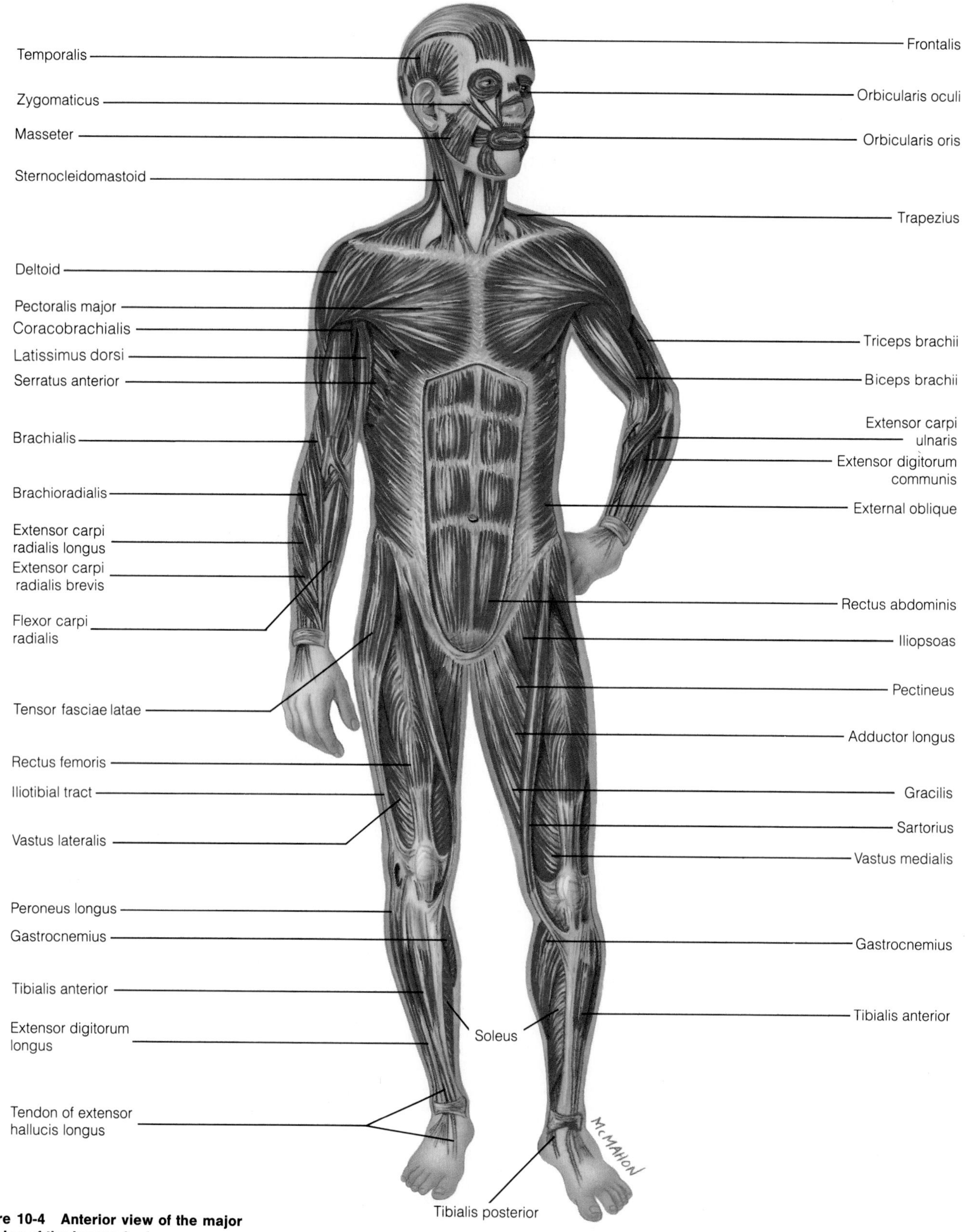

212   Chapter 10   Musculature of the Human Body

- Temporalis
- Zygomaticus
- Masseter
- Sternocleidomastoid
- Deltoid
- Pectoralis major
- Coracobrachialis
- Latissimus dorsi
- Serratus anterior
- Brachialis
- Brachioradialis
- Extensor carpi radialis longus
- Extensor carpi radialis brevis
- Flexor carpi radialis
- Tensor fasciae latae
- Rectus femoris
- Iliotibial tract
- Vastus lateralis
- Peroneus longus
- Gastrocnemius
- Tibialis anterior
- Extensor digitorum longus
- Tendon of extensor hallucis longus

- Frontalis
- Orbicularis oculi
- Orbicularis oris
- Trapezius
- Triceps brachii
- Biceps brachii
- Extensor carpi ulnaris
- Extensor digitorum communis
- External oblique
- Rectus abdominis
- Iliopsoas
- Pectineus
- Adductor longus
- Gracilis
- Sartorius
- Vastus medialis
- Gastrocnemius
- Tibialis anterior

Soleus

Tibialis posterior

Figure 10-4   Anterior view of the major muscles of the human body.

Muscles and Their Movements 213

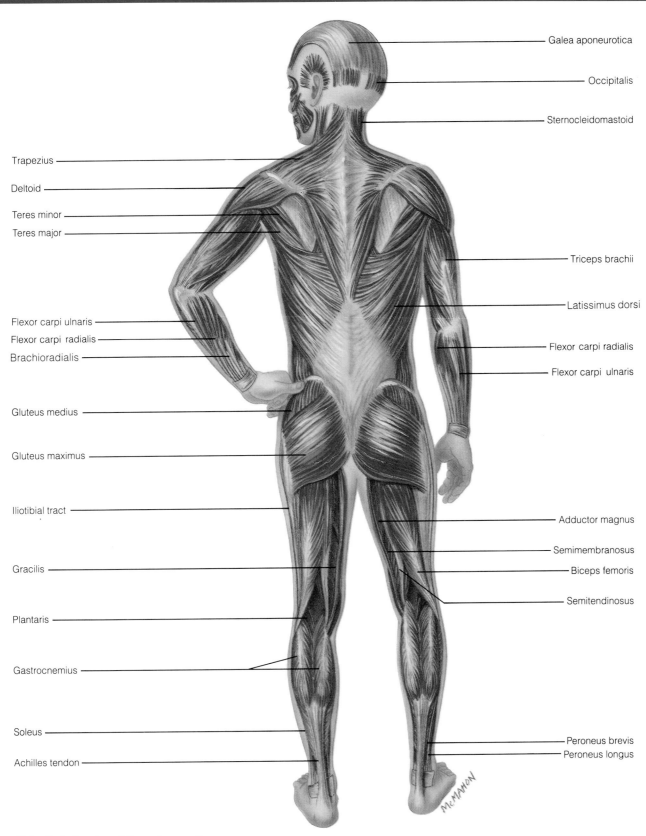

Figure 10-5 Posterior view of the major muscles.

epicranius:	(Gr.) *epi*, upon + *kranion*, cranium
galea:	(L.) *galea*, helmet
aponeurotic:	(Gr.) *apo*, from + *neuron*, nerve, tendon + *ic*, pertaining to
orbicularis:	(L.) *orbis*, orb, sphere
corrugator:	(L.) *corrugare*, to wrinkle

Muscles of the Head and Neck

After completing this section, you should be able to:

1. Describe the muscles of the head and neck.
2. Discuss how the facial expressions of a smile and a frown are caused and list the muscles associated with each.
3. Describe the muscles that accomplish chewing motions.
4. Locate and describe the anterior and posterior triangles on the neck.

Muscles of the head can be divided conveniently into four major groups: facial expression, extrinsic muscles of the eye, mastication, and muscles of the tongue.

Muscles of Facial Expression
(Table 10-1 and Figure 10-6)

Larger muscles of the facial region move the soft, fleshy parts of the face and produce the familiar expressions of surprise, joy, sadness, and other nuances of body language. The **epicranius** (ĕp″ĭ·krā′nē·ŭs) fits over the cranium like a skull cap. The part that lies under the scalp is the **galea aponeurotica** (gā′lē·ă ăp″ō·nū·rŏt′ĭk·ă). Its fleshy part is divided into two regions: the **frontalis,** which covers much of the forehead, and the **occipitalis** (ŏk·sĭp″ĭ·tā′lĭs), at the rear of the skull (see Clinical Note: Musculotension Headaches). Contraction of these muscles either moves the scalp or elevates the eyebrows, producing, for example, the horizontal ridges in the forehead associated with a surprised look. The **orbicularis oculi** (ŏr·bĭk″ū·lā′rĭs ŏk′ū·lī) muscles surround the eye and flesh out the eyelids. Contraction of these muscles may involve only superficial fibers that close the eyes in a blink or while sleeping. A forceful contraction of the deep fibers of this muscle produces squinting expressions and winks. The vertical furrows in the forehead between the eyebrows associated with frowning, expressing sorrow, and crying are produced by the small **corrugator supercilii** (kŏr″ū·gā″tŏr sū″pĕr·sĭl′ē·ī) muscle.

The **orbicularis oris** (ōr′ĭs) brings the lips together and purses or puckers them. A smile is produced by several different muscles acting together. The **zygomaticus** (zī″gō·măt′ĭk·ŭs) **major** and **minor**, the **levator labii superioris** (lē·vā′tŏr lā′bē·ī sū·pĕ″rē·ōr′ĭs), and the **buccinator** elevate and retract the corners of the mouth in the facial expression that is universally recognized as approval or happiness, namely, a smile. The buccinator is used by musicians who play brass instruments to force air out the mouth. It is commonly known as the "trumpeter's muscle." The **platysma** (plă·tĭz′mă) and **depressor labii inferioris** retract and depress the skin of the neck and lower jaw in a snarl and bare the lower teeth.

CLINICAL NOTE

MUSCULOTENSION HEADACHES

The **musculotension headache** is one of the most common kinds of headaches. It is due to abnormal or prolonged contraction of the muscles of the epicranius, usually the occipitalis. This type of headache may be caused by fatigue, emotional stress, or poor posture. It is characterized by a dull, aching pain at the base of the skull and over the forehead. Associated symptoms may be blurring of vision, **photophobia** (sensitivity to light), and mild nausea. A musculotension headache may last for hours or even days, but is usually relieved by simple analgesics and rest. Chronic recurring headaches of this type may result from faulty posture, such as allowing the head to slump forward, which keeps the occipital muscles under chronic tension. Relief may be provided by correcting the faulty posture and by daily exercises for the muscles of the neck.

supercilium: (L.) *super*, above + *cilium*, eyelashes
levator: (L.) *levare*, to raise
labia: (L.) plural of *labium*, lip
buccinator: (L.) *buccinator*, trumpeter
platysma: (Gr.) *platysma*, flat plate

Table 10-1 MUSCLES OF FACIAL EXPRESSION

Muscle	Origin	Insertion	Action	Innervation
Epicranius				
frontalis	Galea aponeurotica	Skin above the eyebrows	Elevates eyebrows, wrinkles brow horizontally	Facial (cranial nerve VII)
occipitalis	Occipital bone and mastoid process	Galea aponeurotica	Pulls scalp posteriorly	Facial
Orbicularis oculi	Frontal and maxillary bones	Skin of the eyelids	Closes the eye	Facial
Corrugator supercilii	Frontal bone; between eyelids	Skin of the eyebrows	Retracts eyebrows, wrinkles brow vertically	Facial
Procerus	Skin over nasal bone	Skin over glabella	Depresses skin between eyebrows	Facial
Orbicularis oris	Fascia of muscle fibers surrounding the mouth	Skin at base of lips and corners of mouth	Closes, purses, and puckers lips	Facial
Zygomaticus major and minor	Zygomatic bone	Skin at the corners of the mouth	Elevates corners of the mouth	Facial
Levator labii superioris	Maxillary and zygomatic bones, below orbit	Skin above upper lip	Elevates upper lip	Facial
Buccinator	Alveolar processes of mandible and maxilla	Fascia and skin at the corners of the mouth	Retracts corners of the mouth, compresses cheeks	Facial
Risorius	Fascia over masseter muscle	Fascia at corners of mouth	Retracts corners of mouth	Facial
Depressor labii inferioris	Body of mandible	Skin of lip	Depresses lower lip	Facial
Platysma	Fascia overlying pectoralis major and deltoid muscles	Skin overlying mandible	Depresses mandible, draws lower lip and skin over mandible downward	Facial
Mentalis	Mandible	Skin overlying chin	Elevates lower lip and skin overlying chin	Facial

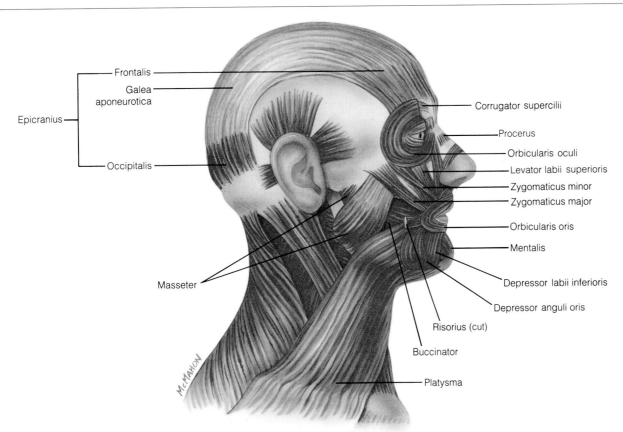

Figure 10-6 Muscles of the head and neck.

Muscles That Move the Eyeball
(Table 10-2 and Figure 10-7)

Each eyeball has six extrinsic (meaning attached to the outer surface) muscles that produce coordinated movements necessary for proper vision. (Intrinsic muscles within the eyeball are described later.) Four **rectus muscles** are concerned primarily with moving each eyeball up and down and from side to side, whereas two **oblique muscles** rotate the eyeball.

Table 10-2 MUSCLES THAT MOVE THE EYEBALL

Muscle	Origin	Insertion	Action	Innervation
Superior rectus	Rear of orbit, near optic foramen	Superior surface of eyeball	Rotates eyeball upward	Occulomotor (cranial nerve III)
Inferior rectus	Rear of orbit, near optic foramen	Inferior surface of eyeball	Rotates eyeball downward	Occulomotor
Lateral rectus	Rear of orbit, near optic foramen	Lateral surface of eyeball	Rotates eyeball laterally	Abducens (cranial nerve VI)
Medial rectus	Rear of orbit, near optic foramen	Medial surface of eyeball	Rotates eyeball medially	Occulomotor
Superior oblique	Rear of orbit, near optic foramen	Superior surface of eyeball, underneath superior rectus	Turns eyeball downward and rotates it laterally	Trochlear (cranial nerve iV)
Inferior oblique	Anterior medial surface of orbit	Inferior, lateral surface of eyeball	Turns eyeball upward and rotates it laterally	Occulomotor

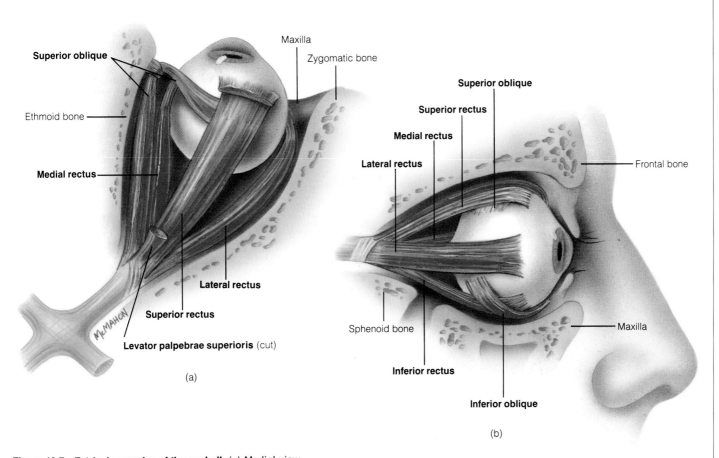

Figure 10-7 Extrinsic muscles of the eyeball. (a) Medial view. (b) Lateral view.

masseter:	(Gr.) *masseter*, chewer
pterygoid:	(Gr.) *pteryx*, wing + *eidos*, form, shape

Muscles of Mastication (Table 10-3 and Figure 10-8)

Muscles of mastication are involved in biting and chewing. The large **masseter** (măs·sē´tĕr) and **temporalis** (tĕm˝pō·rā´lĭs) are powerful elevators of the mandible assisted by the **medial pterygoid** (tĕr´ĭ·goyd). The **lateral pterygoid**, assisted by the digastric muscle (discussed under Muscles of the Neck), opens the mouth by depressing and protracting the mandible. The two pairs of pterygoid muscles may also be contracted one side at a time, producing the side-to-side rocking motion of the mandible associated with chewing and grinding food.

Table 10-3 MUSCLES OF MASTICATION (CHEWING)

Muscle	Origin	Insertion	Action	Innervation
Masseter	Zygomatic arch	Ramus of mandible	Elevates and protracts mandible	Trigeminal (cranial nerve V)
Temporalis	Temporal fossa	Coronoid process of mandible	Elevates and retracts mandible	Trigeminal
Medial pterygoid	Medial surface of lateral pterygoid process of sphenoid and maxilla	Medial surface of angle and ramus of mandible	Elevates and protracts mandible and rocks from side to side	Trigeminal
Lateral pterygoid	Greater wing and lateral surface of lateral pterygoid process of sphenoid	Condyloid process of mandible	Depresses and protracts mandible and rocks from side to side	Trigeminal

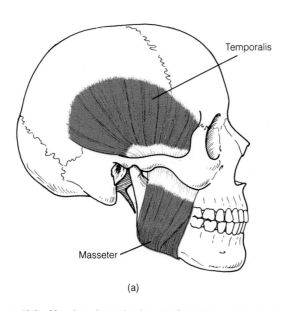

(a)

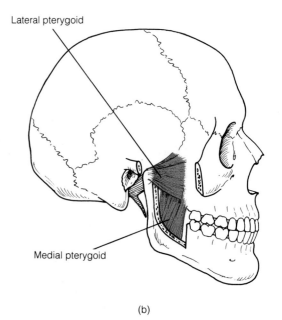

(b)

Figure 10-8 Muscles of mastication. (a) Superficial view showing the masseter and temporalis. (b) Deep view showing the pterygoid muscles, which lie inside the zygomatic arch and mandible.

styloglossus: (Gr.) *stylos*, pillar + *glossa*, tongue
genioglossus: (Gr.) *geneion*, chin + *glossa*, tongue

Muscles of the Tongue (Table 10-4 and Figure 10-9)

The tongue is a very important muscular organ used in manipulating food, in speaking, and in swallowing. It is composed of both intrinsic and extrinsic muscles. **Intrinsic muscles** within the tongue consist of three groups of muscle fibers oriented in longitudinal, transverse, and vertical bundles. Most people possess all three groups of muscles and are able to twist and turn the tongue in a variety of actions. However, some people lack one or more of the bundles. In a genetically determined variation, the transverse bundles may be missing, making it impossible for the individual who lacks them to roll the tongue. An even rarer occurrence is the individual who lacks the longitudinal group, making it impossible to poke the tongue out and lift it up at the tip.

Extrinsic muscles of the tongue, namely, the **styloglossus** (stī″lō·glŏs′ŭs), **hyoglossus** (hī″ō·glŏs′ŭs) and **genioglossus** (jē″nē·ō·glŏs′ŭs) are at the rear and along the sides of the tongue, and the latter extends into the tongue. They move the tongue up and down in the mouth and protract and retract the tongue.

Table 10-4 EXTRINSIC MUSCLES THAT MOVE THE TONGUE

Muscle	Origin	Insertion	Action	Innervation
Styloglossus	Styloid processes of temporal bone	Lateral and inferior sides of tongue	Elevates and retracts tongue	Hypoglossal (cranial nerve XII)
Hyoglossus	Hyoid bone	Lateral sides of tongue	Depresses and retracts tongue	Hypoglossal
Genioglossus	Medial surface of mandible	Inferior surface of tongue, hyoid bone	Protracts, retracts, and depresses tongue	Hypoglossal

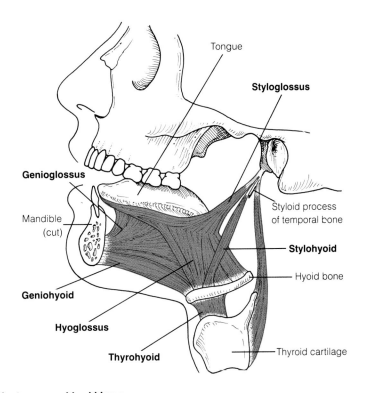

Figure 10-9 Extrinsic muscles of the tongue and hyoid bone.

sternocleido-mastoid:	(Gr.) *sternon*, sternum + *kleidos*, clavicle + *mastos*, breast + *eidos*, form, shape
digastric:	(Gr.) *di*, two + *gaster*, belly
infrahyoid:	(L.) *infra*, below + *hyoid*, hyoid bone

Muscles of the Neck (Table 10-5 and Figure 10-10)

If one looks at the side of the neck, the most prominent muscle is the **sternocleidomastoid** (stĕr″nō·klī″dō·măs′toyd). This muscle passes from the sternum and clavicle to the mastoid process of the temporal bone. Acting together, this pair of muscles flexes the head and neck forward; acting along, each rotates the head in the opposite direction.

The diagonally positioned sternocleidomastoid also divides the neck area into two well-defined areas: the **anterior triangle** and the **posterior triangle**. Anatomists use these triangles to locate the positions of various structures. Muscles in the anterior triangle are usually divided into two groups based on their positions relative to the hyoid bone. The **suprahyoid** muscles elevate the hyoid bone when the mandible is fixed in place and lower the mandible when the hyoid is held firm by muscles beneath it. One of the suprahyoid muscles, the **digastric** muscle (see Muscles of Mastication), is divided lengthwise by its central tendon into an **anterior belly** that extends forward to the mandible and a **posterior belly** that extends backward to the mastoid process of the temporal bone. The **infrahyoid** muscles act antagonistically to the suprahyoid muscles, depressing the hyoid bone. One can easily feel these muscles moving by placing the fingers on the larynx ("Adam's apple") and then swallowing.

Several muscles located on the posterior side of the neck are involved in additional movements of the head and neck (see Figure 10-11a). The **semispinalis capitis** (sĕm″ē·spī·nal′ĭs kăp′ĭ·tĭs), **splenius capitis** (splē′nē·ŭs kăp′ĭ·tĭs), and **longissimus capitis** (lŏn·jĭs′ĭ·mŭs kăp′ĭ·tĭs) act as antagonists to the sternocleidomastoid by extending the head and neck backward when acting together or by rotating the head and neck when only the muscles on one side of the neck are used.

> **CLINICAL NOTE**
>
> **TORTICOLLIS**
>
> **Torticollis** (tōr″tĭ·kŏl′ĭs), or "stiff neck," is a common, painful muscle spasm of the large neck and upper back muscles, specifically, the trapezius and sternocleidomastoid muscles. It causes severe discomfort, tilting of the head to one side, and an inability to move the head to the opposite side of the tight muscles. In most cases there is no history of injury and the patient awakens with the problem in the morning. It is often caused by an abnormal sleeping posture and is readily relieved by application of ice and gentle stretching exercises.

Table 10-5 MUSCLES OF THE NECK

Muscle	Origin	Insertion	Action	Innervation
Sternocleidomastoid	Manubrium and clavicle	Mastoid process of temporal bone	Acting together, both muscles flex head and neck; acting separately, each muscle rotates the head to the opposite side	Spinal accessory (cranial nerve XI)
Suprahyoid Muscles				
Digastric				
Anterior belly	Medial surface of mandible	Hyoid bone	Both bellies elevate the hyoid, and depress the mandible	Trigeminal
Posterior belly	Mastoid process of temporal bone	Hyoid bone		Facial
Mylohyoid	Medial surface of mandible	Hyoid bone	Elevates hyoid	Trigeminal
Stylohyoid	Styloid process of temporal bone	Hyoid bone	Elevates and retracts hyoid	Facial
Geniohyoid	Medial surface of mandible	Hyoid bone	Protracts hyoid	Hypoglossal
Infrahyoid Muscles				
Sternohyoid	Manubrium and sternal end of clavicle	Hyoid bone	Depresses hyoid	Hypoglossal
Omohyoid	Superior border of scapula	Hyoid bone	Depresses hyoid	Hypoglossal
Sternothyroid	Manubrium	Thyroid cartilage of larynx	Depresses larynx	Hypoglossal
Thyrohyoid	Thyroid cartilage of larynx	Hyoid bone	Depresses hyoid	Hypoglossal

semispinalis: (L.) *semi*, half + *spinalis*, relating to spine
capitis: (L.) *caput*, head
torticollis: (L.) *tortus*, twisted + *collum*, neck

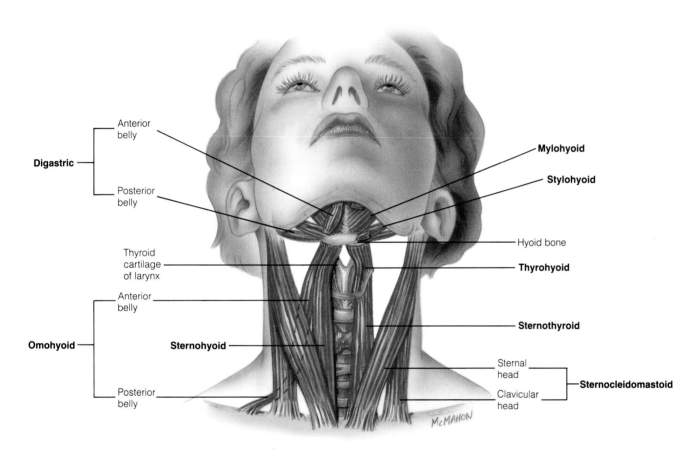

Figure 10-10 Muscles of the neck.

spina:	(L.) *spina*, thorn
collum:	(L.) *collum*, neck
cervicis:	(L.) *cervix*, neck

Muscles of the Trunk

After completing this section, you should be able to:

1. Describe the muscles of the trunk.
2. Discuss the importance of the erector spinae group of muscles.
3. Describe the muscles associated with breathing.

Trunk muscles are classified into four functional groups: those of the vertebral column, thoracic wall, abdominal wall, and pelvic floor.

Muscles of the Vertebral Column
(Table 10-6 and Figure 10-11)

Muscles of the vertebral column, often referred to as the **erector spinae** (spī′nā) group, are separated into bundles that stabilize the spine and extend it (arch the back) when used

Table 10-6 MUSCLES OF THE VERTEBRAL COLUMN

Muscle	Origin	Insertion	Action	Innervation

Semispinalis
  capitis
  cervicis
  thoracis
	Transverse processes of seventh cervical and upper ten thoracic vertebrae	Occipital bone and spinous processes of upper thoracic and cervical vertebrae	Extend and rotate head and head and cervical vertebrae	Spinal

Interspinales	Superior surface of each vertebra	Inferior surface of next vertebra above	Extend the vertebral column	Spinal
Intertransversales	Transverse process of each vertebra	Transverse process of next vertebra above	Abduct vertebral column from side to side	Spinal
Rotatores	Transverse process of each vertebra	Spinous process of next vertebra above	Extend and rotate the vertebral column	Spinal
Multifidus	Posterior surface of ilium and sacrum, and transverse processes of all lumbar, thoracic, and lower four cervical vertebrae	Spinous process of next vertebra above	Extend and rotate the vertebral column	Spinal
Scalenes	Transverse processes of third to seventh cervical vertebrae	First and second ribs	Flex and rotate cervical vertebrae, elevate ribs	Cervical nerves 5 to 8

Splenius
  capitis
  cervicis
	Ligamentum nuchae and spinous processes of lower cervical and upper thoracic vertebrae	Superior nuchal line, mastoid process of temporal bone, transverse processes of upper three cervical vertebrae	Extend, abduct, and rotate head and neck	Spinal

Rectus capitis	Transverse and spinous processes of atlas and axis	Occipital bone	Flexes, extends, and rotates head	Cervical nerves 1 and 2
Longus colli	Lower cervical and upper thoracic vertebrae	Anterior surfaces of upper six cervical vertebrae	Flex and rotate cervical vertebrae	Cervical nerves 2 to 7
Erector Spinae (also called the sacrospinalis)				

Iliocostalis
  cervicis
  thoracis
  lumborum
	Angles of ribs 3 to 12, spinous processes of lumbar vertebrae, crests of sacrum and ilium	Transverse processes of cervical vertebrae, angles of ribs	Extend and abduct vertebral column from side to side	Spinal

Longissimus
  capitis
  cervicis
  thoracis
	Transverse processes of lower cervical, thoracic, and lumbar vertebrae	Mastoid process of temporal bone, transverse processes of cervical and thoracic vertebrae, and ribs	Extend, abduct head and vertebral column from side to side	Spinal

Spinalis
  cervicis
  thoracis
	Spinous processes of seventh cervical, last two thoracic, and first two lumbar vertebrae	Spinous processes of axis and upper thoracic vertebrae	Extend vertebral column	Spinal

quadratus: (L.) *quadratus,* squared
lumborum: (L.) *lumbus,* loin
psoas: (Gr.) *psoa,* loin

together or move the spine from side to side when used singly. Some smaller muscles rotate the spine (see Table 10-6). Several are relatively short, extending over only one to three vertebrae. Others are long, covering from the sacrum and ilium to the rib cage or from the thoracic vertebrae to the occipital bone.

In the cervical or neck region, several small muscles (the **rectus capitis, longus colli, splenius capitis,** and **splenius cervicis** (sĕr′vĭ·sĭs)) flex, extend, and rotate the vertebrae when a person turns his or her head (Figure 10-11b). Three muscles in the lumbar region—the **quadratus lumborum** (kwŏd·rāt′ŭs lŭm·bō′rŭm), **psoas** (sō′ăs) **major,** and **psoas minor** (see Figure 10-21a)—rotate and bend the vertebral column laterally.

Figure 10-11 Muscles that move the vertebral column.
(a) Posterior view; superficial muscles have been removed on the right side. (b) Anterior view of the cervical region. (c) Posterior view of intervertebral muscles.

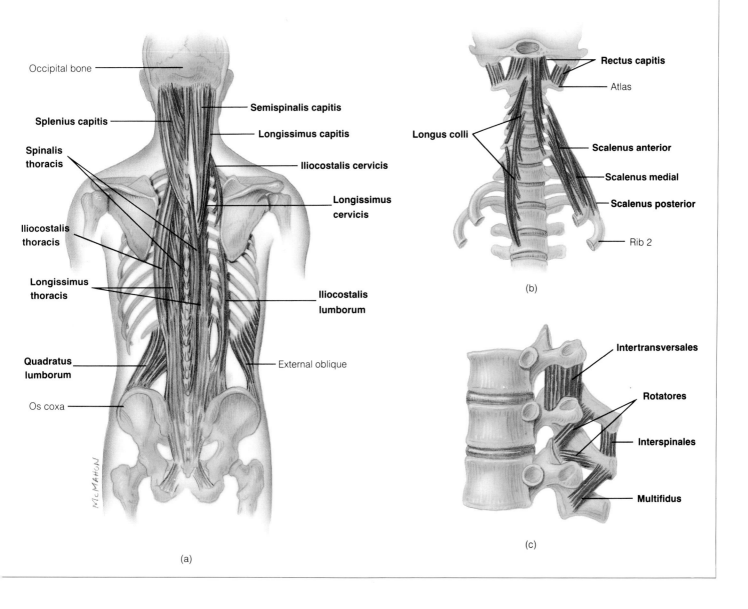

serratus:	(L.) *serra*, saw
hernia:	(L.) *hernia*, rupture

Muscles of the Thoracic Wall
(Table 10-7 and Figure 10-12)

The thoracic wall includes muscles involved primarily with breathing movements. The most important of these muscles are the **external intercostals, internal intercostals, serratus** (sĕr·ā′tŭs) **posterior, transverse thoracis,** and **diaphragm.** The intercostals fill spaces between the ribs, and their fibers are oriented at right angles to the ribs. The external intercostals are positioned closer to the spine than the internal intercostals. They extend from the vertebrae to the costal cartilages, while the internals extend from the angles of the ribs to the sternum. The external intercostals overlie the internals. The latter move the ribs during breathing.

The diaphragm separates the thoracic and abdominal cavities. The fibers of this dome-shaped muscle pass radially from the lower ribs, costal cartilages, and vertebrae and insert on a thick **central tendon** (Figure 10-12d). Contraction of the muscle fibers of the diaphragm pulls on the central tendon, which lowers the dome. This action expands the thoracic cavity and draws air into the lungs. When the diaphragm relaxes, the dome rises and elastic recoil of lung tissue forces air out. Several large structures, including the esophagus, aorta, and vena cava, pierce the diaphragm (see Clinical Note: Diaphragmatic Hernias). In all other respects, the diaphragm is a typical skeletal muscle.

CLINICAL NOTE

DIAPHRAGMATIC HERNIAS

The opening in the diaphragm through which the esophagus, aorta, and vena cava traverse is called the **hiatus.** At times the hiatus may be enlarged either congenitally or following trauma so that part of the abdominal contents may protrude or **herniate** into the chest cavity. This condition is known as a **hiatus hernia.** Usually, the upper portion of the stomach herniates, causing symptoms of chest pain, **dysphagia** (pain with swallowing), and sometimes bleeding from the ulceration of the herniated portion of the stomach. These symptoms frequently occur at night when the weight of the abdominal contents push up on the diaphragm during sleep. The symptoms can also appear during heavy weight lifting.

The symptoms of a hiatus hernia can usually be treated by taking medications that reduce the gastric acid and by elevating the head of the bed at night so that gravity pulls the stomach away from the hiatus. Rarely, surgical repair of the defect is required to relieve patients with persistent symptoms. The vast majority of individuals with hiatal hernias have no symptoms.

Table 10-7 MUSCLES OF THE THORACIC WALL

Muscle	Origin	Insertion	Action	Innervation
External intercostals	Inferior borders of ribs, extending from vertebrae to costal cartilages	Superior borders of adjacent ribs	Elevates ribs during inspiration	Intercostals
Internal intercostals	Inferior borders of ribs, from angle of rib to sternum	Superior borders of adjacent ribs	Assists external intercostals during quiet breathing; depresses ribs during forced expiration	Intercostals
Serratus posterior superior	Spinous processes of seventh cervical upper three thoracic vertebrae	Superior borders of ribs 2 to 5	Elevates ribs during inspiration	Thoracics
Serratus posterior inferior	Spinous processes of last two thoracic and first three lumbar vertebrae	Inferior borders of lower four ribs	Depresses ribs during expiration	Thoracics
Transverse thoracis	Medial surface of sternum	Medial surface of costal cartilages of ribs 2 to 6	Depresses ribs during expiration	Intercostals
Diaphragm	Lumbar vertebrae, inferior surfaces of lower ribs, lower costal cartilages, and xiphoid process	Central tendon of the diaphragm	Depresses central tendon	Phrenic

Muscles of the Trunk

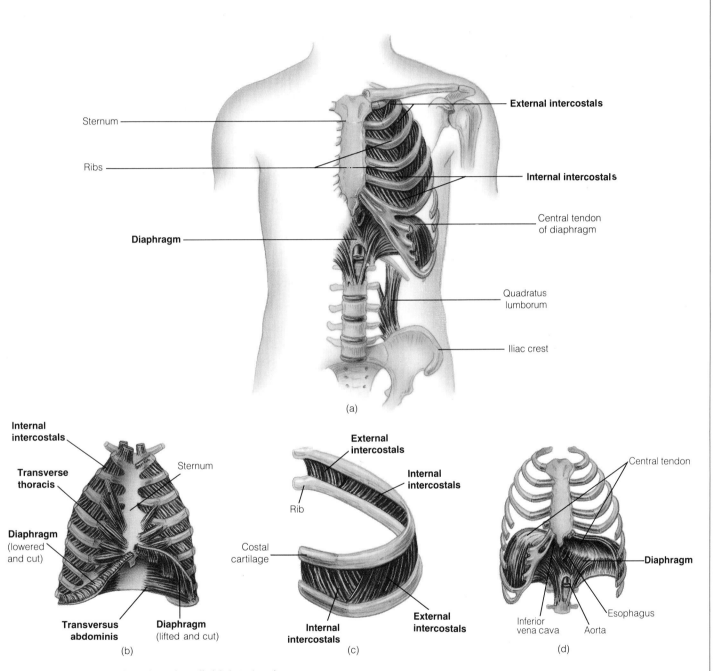

Figure 10-12 Muscles of the thoracic wall. (a) Anterior view. (b) Posterior view from inside the thoracic cavity. (c) Detail of intercostal muscles. (d) Anterior view of diaphragm.

Muscles of the Abdominal Wall
(Table 10-8 and Figure 10-13)

The chief function of abdominal wall muscles is maintenance of physical support for organs in the abdomen. This act is accomplished by maintaining muscle tone, not by forcible contractions. The **external abdominal oblique, internal abdominal oblique,** and **transversus abdominis** stretch across the lateral abdominal wall. Fibers of these muscles course in different directions, resulting in three flat layers of muscular tissue that reinforce one another. The **rectus abdominis** extends from the sternal xiphoid process, lower ribs, and costal cartilages inferiorly to the pubes. In addition to forming the abdominal wall, these four muscles may be voluntarily contracted to compress the abdomen during urination, defecation, exercise, and childbirth or in anticipation of a trauma to the abdomen. The rectus abdominis, external oblique, and internal oblique also help in flexing and rotating the trunk and are used in forceful exhalation. The quadratus lumborum (see section on Muscles of the Vertebral Column) lies on either side of the lumbar vertebrae just anterior to the transversus abdominis. It helps form the posterior abdominal wall and also rotates the vertebral column or depresses the lower ribs during forced breathing.

Table 10-8 MUSCLES OF THE ABDOMINAL WALL

Muscle	Origin	Insertion	Action	Innervation
External abdominal oblique	Inferior borders of lower eight ribs	Linea alba and from iliac crest to pubic symphysis	Compresses abdomen, flexes and rotates trunk	Intercostals (8 to 12), iliohypogastric and ilioinguinal
Internal abdominal oblique	Lumbodorsal fascia, iliac crest, inguinal ligament	Lower four ribs, linea alba, xiphoid process	Compresses abdomen, flexes and rotates trunk	Intercostals (8 to 12), iliohypogastric, and ilioinguinal
Transversus abdominis	Lower ribs and costal cartilages, lumbodorsal fascia, iliac crest, inguinal ligament	Xiphoid process, linea alba, pubis, inguinal ligament	Compresses abdomen, flexes trunk	Intercostals (7 to 12), iliohypogastric, ilioinguinal
Rectus abdominis	Pubic bone	Xiphoid process, costal cartilages of lower ribs	Compresses abdomen, flexes trunk	Intercostals (7 to 12)
Quadratus lumborum	Iliac crest, iliolumbar ligament	Upper four lumbar vertebrae and last rib	Depresses ribs, flexes vertebral column	Twelfth thoracic first lumbar

Muscles of the Trunk 227

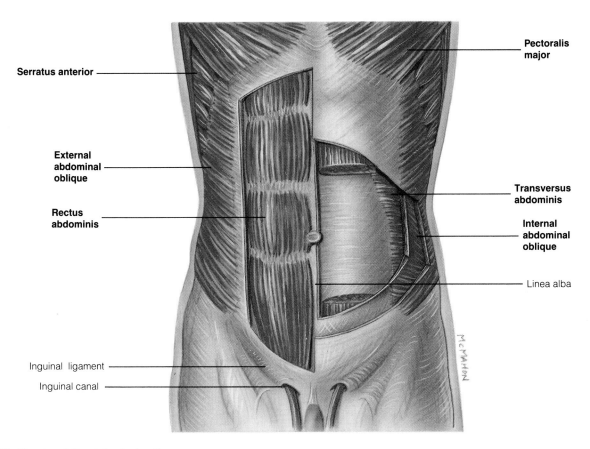

Figure 10-13 Muscles of the abdominal wall.

sphincter: (Gr.) *sphingein*, to bind tightly

Muscles of the Pelvic Floor
(Table 10-9 and Figure 10-14)

Muscles of the pelvic floor support the organs that project into the pelvis from the abdominal cavity. Two muscles are involved in this task: the **levator ani,** which surrounds the rectum, urethra, and vagina (in females), and the **coccygeus** (kŏk·sĭj´ē·ŭs). In addition to providing support, these muscles can be voluntarily contracted during defecation, urination, and sexual intercourse. Several smaller muscles are associated specifically with the reproductive organs, and two **anal sphincters,** one under voluntary control, surround the lower (distal) end of the intestinal tract.

Table 10-9 MUSCLES OF THE PELVIC FLOOR

Muscle	Origin	Insertion	Action	Innervation
Levator ani	Medial surface of pubis, ischial spine	Anterior surface of coccyx	Supports pelvic organs, elevates rectum, and constricts vagina	Sacrals 3 to 5
Coccygeus	From ischial spine to sacrospinal ligament	Lateral surfaces of sacrum and coccyx	Supports pelvic organs	Sacrals 4 to 5
Anal sphincters	Connective tissue surrounding the anal opening (not connected to skeletal)	Connective tissue surrounding the anal opening	Close the anal opening	Pudendal

Muscles of the Trunk 229

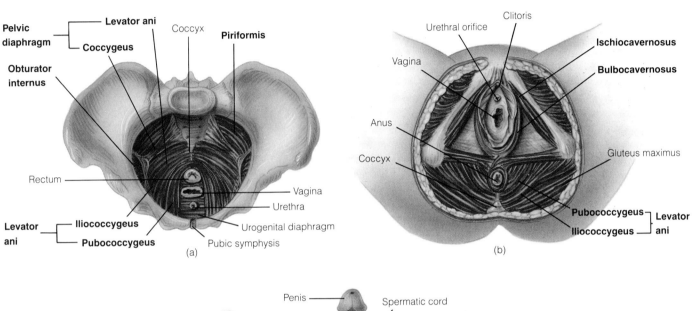

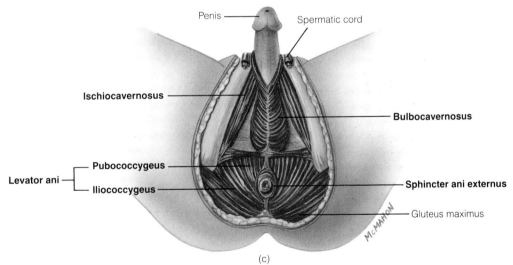

Figure 10-14 Muscles of the pelvic floor. (a) Superior view of female pelvis. (b) Inferior view of female perineum. (c) Inferior view of the male perineum.

trapezius: (Gr.) *trapezion*, small table
pectoralis: (L.) *pectus*, breast

Muscles of the Pectoral Girdle and Upper Extremity

After completing this section, you should be able to:

1. Describe the muscles that move the humerus.
2. Describe the antagonistic muscles involved in pronation and supination of the forearm.
3. Discuss the movements produced by the trapezius.

For convenience, we divide the muscles of the pectoral girdle (shoulder area) and upper extremity into six groups. These include muscles that move the scapula, humerus, forearm, hand, and fingers. Each of these groups will be considered in turn.

Muscles That Move the Scapula
(Table 10-10 and Figures 10-15 and 10-16)

The scapula, a large triangular bone, articulates solidly only with the small clavicle. Because the scapula is not held firmly in place, it is free to move in several directions. The **trapezius** (tră·pē′zē·ŭs) is a large, flat muscle that covers much of the upper back region. Its fibers extend in several directions, and it rotates the scapula quite freely. The **rhomboideus** (rŏm·bō·ĭd′ē·ŭs) muscles and the **levator scapulae** retract and elevate the scapula. Acting agonistically to these muscles, the **pectoralis** (pĕk″tō·rā′lĭs) **minor** and **serratus anterior** protract the scapula forward, as when a person reaches for something. The **subclavius** (sŭb·klā′vē·ŭs) muscle in this region attaches to and acts indirectly through the clavicle (which firmly articulates with the scapula) to pull the scapula toward the sternum.

Table 10-10 MUSCLES THAT MOVE THE SCAPULA

Muscle	Origin	Insertion	Action	Innervation
Trapezius	Occipital bone, ligamentum nuchae, spinous processes of seventh cervical and all thoracic vertebrae	Acromial end of clavicle, spine and acromion process of scapula	Retracts, elevates, depresses, and rotates scapula; extends head	Spinal accessory (cranial nerve XI), cervical nerves 3 to 4
Levator scapulae	Upper four cervical vertebrae	Medial border of scapula, above the spine	Elevates scapula, abducts neck	Dorsal scapular (which is cervical nerves 3 to 5)
Rhomboideus major	Spinous processes of upper five thoracic vertebrae	Medial border of scapula, below spine	Retracts and rotates scapula	Dorsal scapular
Rhomboideus minor	Spinous processes of last cervical and first thoracic vertebrae	Medial border of scapula, below spine	Retracts and rotates scapula	Dorsal scapular
Pectoralis minor	Lateral surfaces of ribs 3 to 5	Coracoid process of scapula	Protracts scapula, elevates ribs during forced breathing	Medial pectoral (from eighth cervical and first thoracic spinal nerves)
Serratus anterior	Lateral surfaces of ribs 1 to 9	Medial border of scapula (from underneath)	Protracts and rotates scapula	Long thoracic (from cervical spinal nerves 5 to 7)
Subclavius	Lateral surface of first rib	Acromial end of clavicle	Depresses acromial end of clavicle	Cervical spinal nerves 5 to 6

Figure 10-15 Muscles of the trunk (opposite). (a) Posterior view. (b) Anterior view of trunk muscles that move the scapula. (c) Lateral view.

Muscles of the Pectoral Girdle and Upper Extremity 231

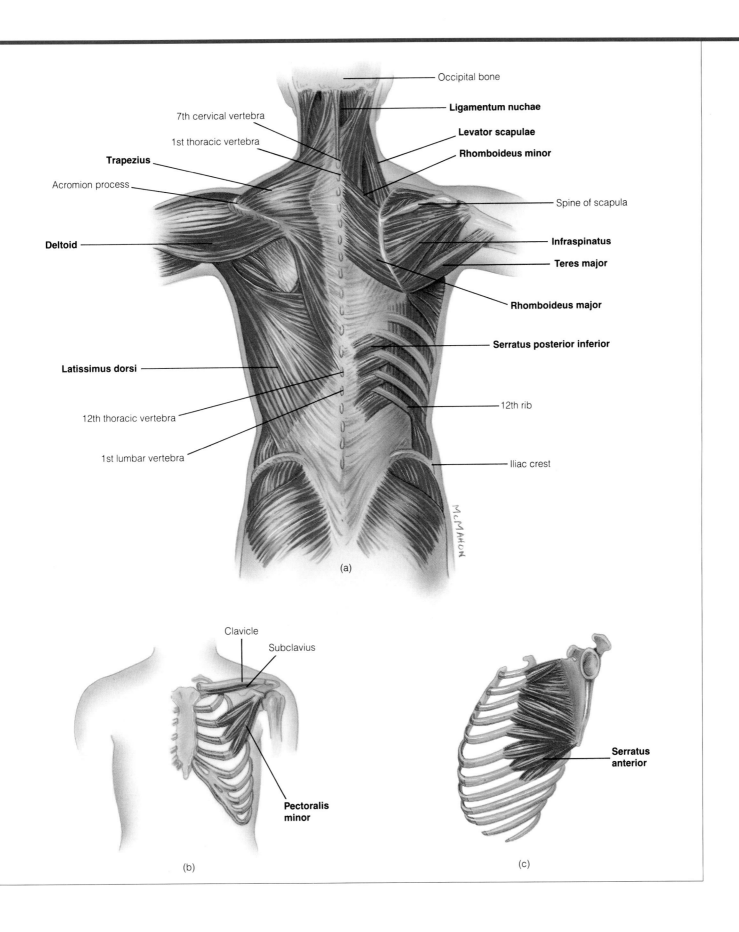

teres:	(L.) *teres*, rounded
coracobrachialis:	(Gr.) *korax*, raven + (L.) *brachiolis*, arm

Muscles That Move the Humerus
(Table 10-11 and Figures 10-15 and 10-16)

Two groups of muscles move the humerus. One group inserts close to the shoulder joint, produces small movements, and is used primarily in maintaining the articulation of the humerus with the glenoid fossa and in rotating the humerus. This group includes the **supraspinatus** (sū″-prăˑspĭˑnăˊtŭs), **infraspinatus, subscapularis** (sŭbˑskāpˊūˑlăˮrĭs), and **teres minor.** The term **rotator cuff** is often applied to an area where tendons of these muscles fuse with tissues of the shoulder joint and rotate the humerus. This area is subject to trauma and pain in athletes who make extensive use of their arm, such as baseball pitchers and tennis players (see Unit II Case Study: Overuse Syndromes).

The second group of humeral muscles includes the **deltoid, pectoralis major, latissimus dorsi** (lăˑtĭsˊĭˑmŭs dŏrˊsī), **teres major,** and **coracobrachialis** (korˮăˑkōˑbrāˮkēˑălˊĭs). The deltoid raises the arm and contributes to movements of the humerus associated with such activities as climbing. The pectoralis major, latissimus dorsi, teres major, and coracobrachialis oppose the deltoid by lowering the humerus.

Table 10-11 MUSCLES THAT MOVE THE HUMERUS

Muscle	Origin	Insertion	Action	Innervation
Articular Muscles of the Shoulder				
Supraspinatus	Supraspinous fossa of scapula	Greater tubercle of humerus	Abducts humerus	Suprascapular
Infraspinatus	Infraspinous fossa of scapula	Greater tubercle of humerus	Laterally rotates and adducts humerus	Suprascapular
Teres minor	Lateral border of scapula	Greater tubercle of humerus	Laterally rotates and adducts humerus	Axillary
Subscapularis	Subscapular fossa of scapula	Lesser tubercle of humerus	Rotates humerus medially	Subscapular
Prime Movers of Humerus				
Deltoid	Acromial end of clavicle, spine and acromion process of scapula	Deltoid tuberosity of humerus	Abducts, flexes, extends and rotates humerus	Axillary
Pectoralis major	Clavicle, sternum, costal cartilages of first seven ribs	Shaft of humerus below greater tubercle	Adducts, flexes, and medially rotates humerus	Medial and lateral pectoral
Latissimus dorsi	Spines of lower six thoracic vertebrae, all lumbar vertebrae, sacrum, iliac crest	Shaft of humerus below lesser tubercle	Adducts, extends, and medially rotates humerus	Thoracodorsal
Teres major	Posterior surface of scapula, near inferior angle	Shaft of humerus, below lesser tubercle	Adducts, extends and medially rotates humerus	Subscapular
Coracobrachialis	Coracoid process of scapula	Shaft of humerus	Adducts and flexes humerus	Musculocutaneous

Muscles of the Pectoral Girdle and Upper Extremity 233

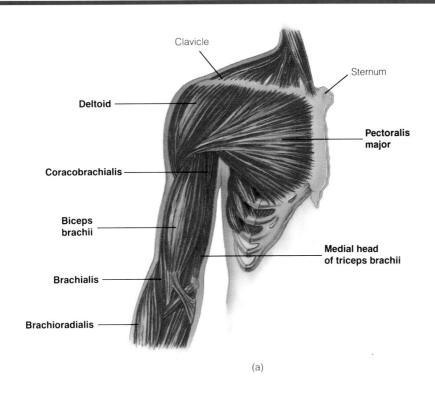

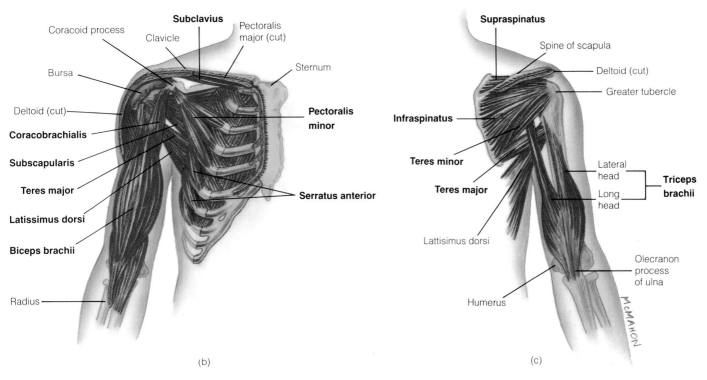

Figure 10-16 Muscles that move the humerus. (a) Superficial, anterior view. (b) Deep anterior view. (c) Deep posterior view.

biceps: (L.) *bis*, twice + *caput*, head
anconeus: (Gr.) *ankon*, elbow

Muscles That Move the Forearm
(Table 10-12 and Figure 10-17)

The forearm includes two large bones, the ulna and radius. Two opposing sets of muscles move the forearm in two different directions at the elbow. The **biceps brachii, brachialis** (brā″kē·ăl′ĭs), and **brachioradialis** (brā″kē·ō·rā″dē·ă′lĭs) flex the forearm. The biceps brachii is also a powerful supinator, which means it is used to rotate the arm and palm forward. Extension of the forearm at the elbow is accomplished by the large **triceps brachii** and the smaller **anconeus** (ăn·kō′nē·ŭs).

Table 10-12 MUSCLES THAT MOVE THE FOREARM

Muscle	Origin	Insertion	Action	Innervation
Biceps brachii (two heads)	Long head from supraglenoid tuberosity of scapula, short head from coracoid process of scapula	Radial tuberosity and fascia of proximal forearm muscle	Flexes forearm; rotates radius laterally, which supinates hand	Musculocutaneous
Brachialis	Shaft of humerus; distal, anterior region	Coronoid process of ulna	Flexes forearm	Musculocutaneous, radial, medial
Brachioradialis	Lateral supracondyloid ridge of humerus	Styloid process of radius	Flexes forearm	Radial
Triceps brachii (three heads)	Long head from infraglenoid tuberosity of scapula, lateral head from shaft of humerus, medial head from shaft of humerus	Olecranon process of ulna	Extends forearm, long head adducts arm	Radial
Anconeus	Lateral epicondyle of humerus	Olecranon process of ulna	Extends forearm	Radial

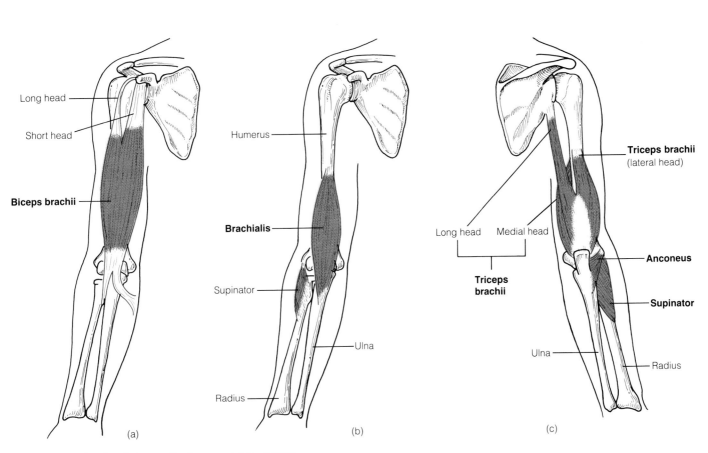

Figure 10-17 Muscles that move the forearm. (a) Superficial anterior view. (b) Deep anterior view. (c) Posterior view.

profundus:	(L.) *profundus*, deep
pollicis:	(L.) *pollex*, thumb

Forearm Flexors of the Wrist and Hand
(Table 10-13 and Figures 10-18 and 10-19)

The anterior surface of the forearm contains three layers of muscles, most of which are involved in flexing the wrist and fingers. The superficial layer includes the **pronator teres** (which rotates the radius, turning the palm down) and three flexors of the wrist: the **flexor carpi ulnaris** (ŭl·nā′rĭs), **flexor carpi radialis** (rā″dē·ā′lĭs), and the **palmaris longus** (păl·mā′rĭs lŏng′gŭs). The palmaris longus is a genetically variable muscle. If you possess the palmaris longus muscle you will see its prominent tendon protruding vertically along your wrist and along the fleshy bulge at the base of the thumb by pinching your fingers together and flexing your wrist. If you lack the palmaris longus, you will notice a definite depression in this region. Whether or not the palmaris longus is present, the wrist is adequately flexed by the other two muscles.

The **flexor digitorum superficialis** (sū″pĕr·fĭsh·ē·ā′lĭs) muscle is the middle muscle on the anterior side of the forearm. Its tendon of insertion divides at the wrist and attaches to the middle phalanges of fingers two through five (Figure 10-18b). Along with the flexor digitorum profundus, which lies deep to it, the flexor digitorum superficialis flexes fingers two through five, as in making a fist.

The deep layer of these forearm muscles includes the **flexor digitorum profundus** (prō·fūn′dŭs), **pronator quadratus**, and **flexor pollicis** (pŏl′ĭs·ĭs) **longus.** The flexor digitorum profundus inserts on the terminal phalanges of fingers two through five and assists its superficial counterpart in flexing fingers. The pronator quadratus pronates the hand, which means it turns the hand so the palm faces backward or downward. The thumb is special and has its own flexors, mainly the flexor pollicis longus, which flexes the thumb into the palm.

Table 10-13 FOREARM FLEXORS OF THE WRIST AND HAND

Muscle	Origin	Insertion	Action	Innervation
Pronator teres	Medial epicondyle of humerus, coronoid process of ulna	Shaft of radius	Rotates radius medially, thereby pronating hand	Median
Flexor carpi radialis	Medial epicondyle of humerus	Metacarpals 2 to 3	Flexes wrist and forearm, abducts hand	Median
Flexor carpi ulnaris	Medial epicondyle of humerus and proximal, posterior surface of ulna	Hamate, pisiform, fifth metacarpal	Flexes wrist, adducts hand	Ulnar
Palmaris longus	Medial epicondyle of humerus	Palmar aponeurosis	Flexes wrist	Median
Flexor digitorum superficialis	Medial epicondyle of humerus, coronoid process of ulna, shaft of radius	Second phalanx of fingers 2 to 5	Flexes fingers, flexes wrist	Median
Flexor digitorum profundus	Shaft of ulna, interosseous membrane	Distal phalanges of fingers 2 to 5	Flexes fingers, flexes wrist	Median and ulnar
Flexor pollicis longus	Shaft of radius, coronoid process of ulna	Distal phalanx of thumb	Flexes thumb, flexes wrist	Median
Pronator quadratus	Distal shaft of radius	Distal shaft of radius	Pronates hand	Median

Muscles of the Pectoral Girdle and Upper Extremity 237

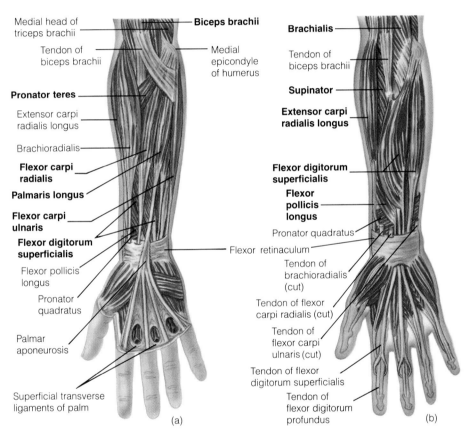
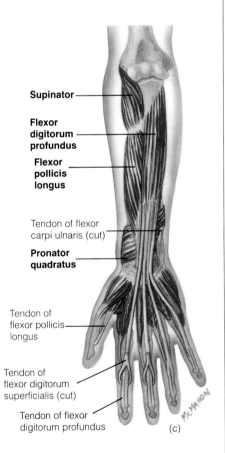

Figure 10-18 Anterior muscles of the forearm. (a) Superficial layer. (b) Middle layer. (c) Deep layer.

brevis:	(L.) *brevis*, short
indicis:	(L.) plural of *index*, indicator, pointer

Forearm Extensors of the Wrist and Hand
(Table 10-14 and Figure 10-19)

The posterior aspect of the forearm has two layers of muscles involved in extending the wrist and fingers. An outer layer consists of five muscles. Three of the these are wrist extensors: the **extensor carpi radialis longus, extensor carpi radialis brevis** (brĕv′ĭs), and **extensor carpi ulnaris** (ŭl·nā′rĭs). The other two are finger extensors: the **extensor digitorum communis** (kŏ·mū′nĭs), which has tendons that can be seen fanning over the back of the hand and the knuckles, and the **extensor digiti minimi** (mĭn′ĭ·mī). The deep layer moves the fingers and includes the **abductor pollicis longus, extensor pollicis longus, extensor pollicis brevis,** and **extensor indicis** (ĭn′dĭ·cĭs).

Table 10-14 FOREARM EXTENSORS OF THE WRIST AND HAND

Muscle	Origin	Insertion	Action	Innervation
Extensor carpi radialis longus	Lateral supracondyloid ridge of humerus	Metacarpal 2	Extends wrist, abducts hand	Radial
Extensor carpi radialis brevis	Lateral epicondyle of humerus	Metacarpal 3	Extends wrist, abducts hand	Radial
Extensor carpi ulnaris	Lateral epicondyle of humerus	Metacarpal 5	Extends wrist, adducts hand	Deep radial
Extensor digitorum communis	Lateral epicondyle of humerus	Phalanges of fingers 2 to 5	Extends fingers and wrist	Deep radial
Extensor digiti minimi	Lateral epicondyle of humerus	Phalanx of finger 5	Extends fifth finger	Deep radial
Supinator	Lateral epicondyle of humerus, shaft of ulna	Shaft of radius	Supinates hand	Deep radial
Abductor pollicis longus	Shaft of ulna and radius, interosseous membrane	Metacarpal 1	Abducts thumb, abducts hand	Deep radial
Extensor pollicis longus	Shaft of ulna, and interosseous membrane	Distal phalanx of thumb	Extends thumb, abducts hand	Deep radial
Extensor pollicis brevis	Shaft of radius, interosseous membrane	First phalanx of thumb	Extends thumb, abducts hand	Deep radial
Extensor indicis	Shaft of ulna, interosseous membrane	Fuses with tendon of extensor digitorum communis on index finger	Extends index finger	Deep radial

Muscles of the Pectoral Girdle and Upper Extremity 239

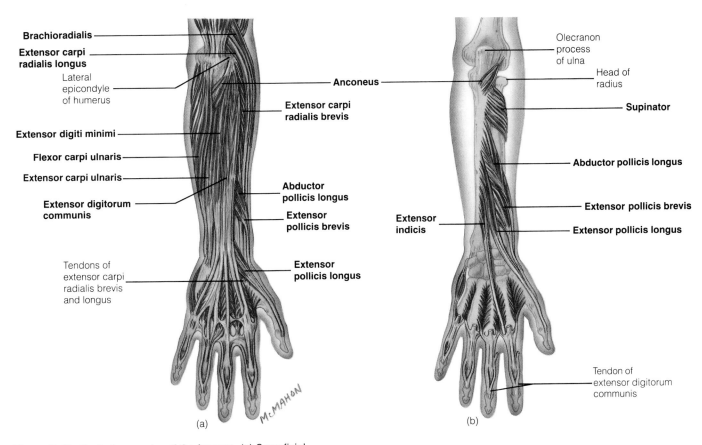

Figure 10-19 Posterior muscles of the forearm. (a) Superficial layers. (b) Deep layer.

hypothenar: (Gr.) *hypo*, under + *thenar*, palm

Intrinsic Muscles of the Hand
(Table 10-15 and Figure 10-20)

In addition to tendons from forearm muscles, the hand is operated by three groups of muscles on the palmar surface itself. They permit a versatile range of movements of the thumb and fingers. The first group of muscles, which moves the thumb, is called the **thenar** (thē′năr) **eminence,** and it forms a fleshy pad at the base of the thumb. On the opposite side of the hand is the narrower **hypothenar eminence** that extends from the base of the little finger. It consists of four muscles, three of which move the little finger (Figure 10-20a). In between these two eminences is a V-shaped depression with the three sets of **midpalmar muscles** that move fingers two through five.

Table 10-15 INTRINSIC MUSCLES OF THE HAND

Muscle	Origin	Insertion	Action	Innervation
Muscles of the Thenar Eminence				
Abductor pollicis brevis	Navicular and trapezium carpal bones	Proximal phalanx of thumb	Abducts thumb	Median
Opponens pollicis	Trapezium carpal bone	Metacarpal 1	Effects opposition of thumb to other fingers	Median
Flexor pollicis brevis	Transverse carpal ligament, trapezium	Proximal phalanx of thumb	Flexes and adducts thumb	Median
Adductor pollicis	Capitate carpal bone, metacarpals 2 to 3	Proximal phalanx of thumb	Adducts thumb	Ulnar
Midpalmar Muscles				
Lumbricales	Tendons of flexor digitorum profundus	Tendons of extensor digitorum communis	Flexes bases of fingers 2 to 5, extends tips of fingers 2 to 5	Median and ulnar
Dorsal interossei	Adjacent sides of metacarpals 2 to 5	Proximal phalanges, fingers 2 to 5	Abducts and flexes fingers 2 to 5	Ulnar
Palmar interossei	Metacarpals 2, 4 to 5	Proximal phalanges, fingers 2, 4 to 5	Adducts fingers	Ulnar
Muscles of the Hypothenar Eminence				
Palmaris brevis	Transverse carpal ligament	Skin along medial border of palm	Adducts skin over hypothenar eminence	Ulnar
Abductor digiti minimi	Pisiform carpal bone, tendon of flexor carpi ulnaris	Proximal phalanx of fifth finger	Abducts fifth finger	Ulnar
Flexor digiti minimi brevis	Hamate, transverse carpal tendon	Proximal phalanx of fifth finger	Flexes fifth finger	Ulnar
Opponens digiti minimi	Hamate carpal bone, transverse carpal tendon	Metacarpal 5	Effects opposition of fifth finger to thumb	Ulnar

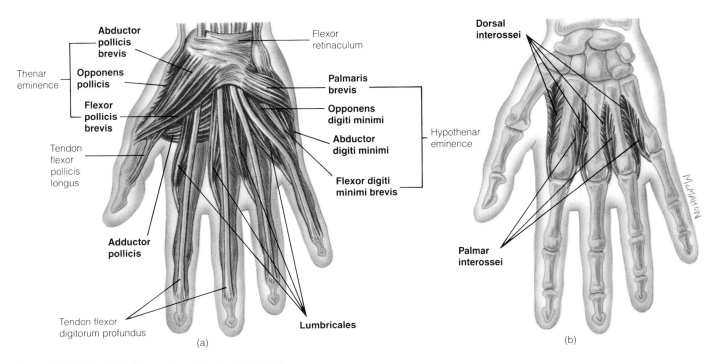

Figure 10-20 The intrinsic muscles of the hand as seen in palmar view. (a) Superficial muscles. (b) Deep muscles.

gluteus: (Gr.) *gloutos*, buttock
gracilis: (L.) *gracilis*, slender
pectineus: (L.) *pecten*, comb
piriformis: (L.) *pirum*, pear + *forma*, shape
obturator: (L.) *obturare*, to close
gemellus: (L.) *gemellus*, twin

Muscles of the Lower Extremity

After completing this section, you should be able to:

1. Locate the five groups of muscles that move the thigh.
2. Identify the longest muscle in the body.
3. Identify the muscles in the hamstring group.
4. Discuss the major differences between the intrinsic muscles of the hand and foot.

Unlike many other vertebrates, the lower extremity of the human body has a different function than the upper extremity. While our arms are used to manipulate objects, our legs act in locomotion. This difference in function is reflected in the organizational complexity of the muscles in the two appendages. Muscles that move the lower ones are more involved with strength and stability rather than complex manipulation. We divide the muscles of the lower extremity into functional groups based on their actions.

Muscles That Move the Thigh
(Table 10-16 and Figure 10-21)

Because the femur articulates with the pelvic bone through a ball-and-socket joint, a variety of movements are possible. Flexion of the thigh is accomplished by three muscles—the **iliacus** (il″ē·āk″ŭs), **psoas major**, and **psoas minor**—which lie anterior to the vertebral column and pelvic bone. The psoas minor is another genetically variable muscle, being absent in about 40% of the population. Because the tendons of insertion of these three muscles fuse on the femur's lesser trochanter, the three muscles are often referred to as the **iliopsoas** (il″ē·ō·sō″ăs) muscle group. In addition to flexing the thigh, they also help in lateral rotation of the thigh. Extension of the thigh is accomplished chiefly by the large **gluteus** (glū″tē·ŭs) **maximus** muscle that forms most of the fleshy buttocks.

Abduction of the femur is accomplished by three muscles: the **tensor fasciae latae** (lā′tā), **gluteus medius**, and **gluteus minimus.** These three muscles also rotate the femur medially. The tensor fascia latae is attached to the **iliotibial band** or **tract** (or **fascia lata**), a broad superficial sheet of connective tissue (Figure 10-21b). This band also attaches to the gluteus maximus and several other muscles on the anterior and lateral surface of the femur. The iliotibial band extends to the lateral condyle of the tibia. Contraction of the several muscles that attach to it serves to strengthen the extended knee joint during walking and running (see Clinical Note: Iliotibial Band Friction Syndrome).

Five major adductors of the femur are the **gracilis** (grăs″ĭ·lĭs), **pectineus** (pĕk·tĭn·ē″ŭs), **adductor longus**, **adductor brevis**, and **adductor magnus.** The gracilis is longer than the other four, extending from the pubic bone to the tibia, just below the medial condyle. Consequently, it also flexes the leg. The other four muscles in this group assist in lateral rotation of the thigh. A group of six smaller muscles close to the hip joint rotate the femur laterally. They are the **piriformis** (pĭr″ĭ·fŏrm″ĭs), **obturator internus**, **obturator externus**, **gemellus** (jĕm·ĕl″ŭs) **superior**, **gemellus inferior**, and **quadratus femoris** (Figure 10-21e).

CLINICAL NOTE

ILIOTIBIAL BAND FRICTION SYNDROME

The iliotibial band (or tract) is of particular interest to the athlete because excessive tightness of the band can cause rubbing of its distal portion over the lateral femoral condyle. It is a common problem for runners. It causes pain in the lateral aspect of the knee after running only a short distance, which creates emotional distress in runners since they can only run short distances. A physical examination of the runner may be entirely normal except for the pain located on the lateral side of the knee. Special tests will demonstrate the tight band. Treatment consists of application of ice and special exercises to stretch the band.

CLINICAL NOTE

PIRIFORMIS SYNDROME

The piriformis muscle is just one of a group of muscles that rotates the leg within the hip joint. The great sciatic nerve exits from the pelvis beneath the piriformis. Thus, abnormality of the piriformis muscle can produce pressure on the sciatic nerve to cause **sciatica** (pain in the area supplied by the sciatic nerve). The **piriformis syndrome** can mimic the symptoms of a herniated intervertebral disk except that there is usually no associated motor weakness or negative tests. Piriformis syndrome is usually due to prolonged sitting on a hard surface. The symptoms may be relieved with anti-inflammatory drugs, physical therapy to encourage stretching of the muscle, and proper sitting posture.

Table 10-16 MUSCLES THAT MOVE THE THIGH

Muscle	Origin	Insertion	Action	Innervation
Iliacus	Iliac crest and iliac fossa	Lesser trochanter of femur	Flexes and rotates femur laterally	Femoral
Psoas major	Anterior surfaces of last thoracic and all lumbar vertebrae	Lesser trochanter of femur	Flexes and rotates femur laterally	Lumbars 2 to 3
Psoas minor	Anterior surfaces of last thoracic and first lumbar vertebrae	Tendon of psoas major	Flexes and rotates femur laterally	Lumbar 1
Gluteus maximus	Posterior gluteal line of ilium, and posterior surface of sacrum and coccyx	Gluteal tuberosity of femur, fascia lata, shaft of femur	Extends and rotates femur laterally	Inferior gluteal
Gluteus medius	Lateral surface of ilium	Greater trochanter of femur	Abducts and rotates femur medially	Superior gluteal
Gluteus minimus	Lateral surface of ilium	Greater trochanter of femur	Abducts and rotates femur medially	Superior gluteal
Tensor fasciae latae	Anterior superior iliac spine	Iliotibial tract	Abducts, flexes, and rotates femur medially	Superior gluteal
Gracilis	Pubic symphysis	Proximal, medial surface of tibia	Adducts femur, flexes leg	Obturator
Pectineus	Superior ramus of pubis	Shaft of femur, below, lesser trochanter	Adducts, flexes, and rotates femur laterally	Femoral
Adductor longus	Superior ramus of pubis	Linea aspera of femur	Adducts, flexes, and rotates femur laterally	Obturator
Adductor brevis	Inferior ramus of pubis	Linea aspera of femur	Adducts and rotates femur laterally	Obturator
Adductor magnus	Ischial tuberosity, inferior ramus of pubis	Linea aspera and adductor tubercle of femur	Adducts, flexes, and rotates femur laterally	Obturator
Piriformis	Anterior surface of sacrum	Greater trochanter of femur	Rotates femur laterally	Sacral nerves 1 to 2
Obturator internus	Margin of obturator foramen	Greater trochanter of femur	Abducts and rotates femur laterally	Last lumbar and first two sacral spinal nerves
Obturator externus	Margin of obturator foramen	Trochanteric fossa of femur	Rotates femur laterally	Obturator
Gemellus superior	Spine of ischium	Greater trochanter of femur	Rotates femur laterally	Last lumbar and first two sacral spinal nerves
Gemellus inferior	Ischial tuberosity	Greater trochanter of femur	Rotates femur laterally	Last two lumbar spinal nerves
Quadratus femoris	Ischial tuberosity	Shaft of femur	Rotates femur laterally	Last two lumbar spinal nerves

244 Chapter 10 Musculature of the Human Body

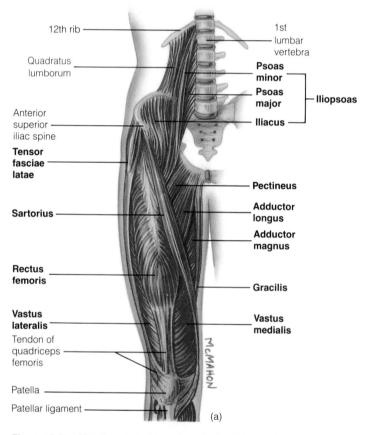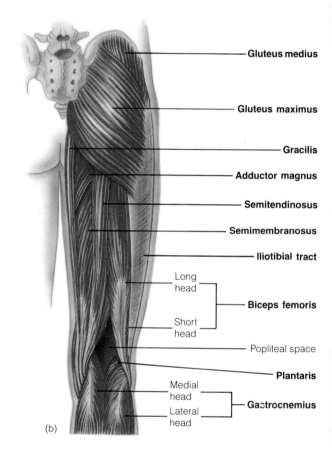

Figure 10-21 Muscles of the hip and thigh. (a) Anterior view. (b) Posterior view. (c) and (d) Deep anterior views. (e) Deep posterior view.

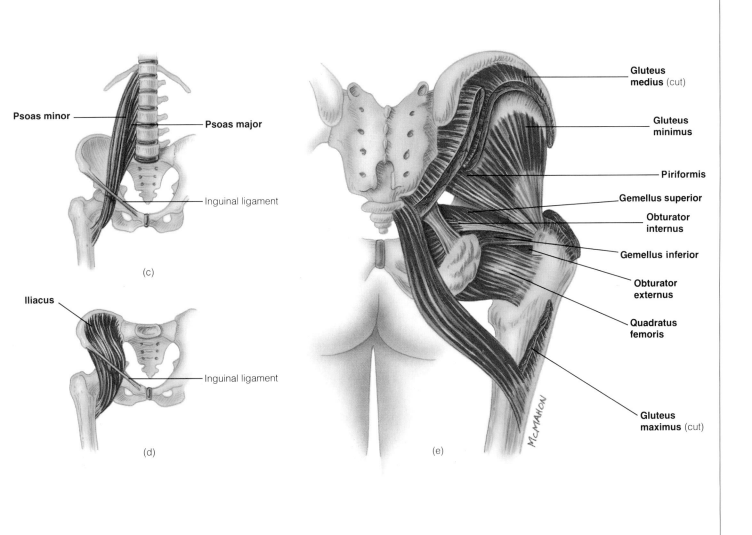

sartorius:	(L.) *sartor,* tailor	
rectus:	(L.) *rectus,* straight	
semitendinosus:	(L.) *semi,* half + *tendo,* tendon	
semimembranosus:	(L.) *semi,* half + *membranosus,* pertains to membrane	

Muscles That Move the Leg
(Table 10-17 and Figures 10-21 and 10-22)

Leg muscles are organized into anterior and posterior functional groups of the thigh. (Some anatomists also recognize a third group of muscles consisting of the five adductors previously discussed.) The anterior group includes the **sartorius** (săr·tō´rē·ŭs), the longest muscle in the body, and the **quadriceps femoris** group, which consists of four separate muscles: the **rectus femoris, vastus lateralis, vastus medialis,** and **vastus intermedius.** The sartorius extends from the ilium diagonally over the anterior surface of the thigh, eventually passing behind the medial side of the knee. Its relationship to these two joints allows it to flex both the thigh and leg. The massive quadriceps muscles fuse into a common tendon of insertion, the **patellar tendon,** which surrounds the patella and continues with the **patellar ligament** to insert on the tibial tuberosity (Figures 10-21a and 10-22). The quadriceps group, a powerful extensor of the leg, is used in walking, running, and specifically kicking. The rectus femoris originates on the ilium and also assists in flexing the thigh.

Antagonists to these muscles form the posterior group of thigh muscles. They are usually referred to as the "hamstring" group because tendons of these muscles are used by butchers to attach curing hams to meat hooks. The hamstring consists of three muscles: the **biceps femoris, semitendinosus** (sĕm˝ē·tĕn´dĭ·nō´sŭs), and **semimembranosus** (sĕm˝ē·mĕm´bră·nō´sŭs) (Figure 10-21b). All three muscles originate on the ischial tuberosity. Their tendons of insertion are easily felt posterior to the knee joint, the biceps laterally and the other two medially. All three hamstring muscles extend the thigh and flex the leg. The popliteus also flexes the leg, but it is described in the next section.

> **CLINICAL NOTE**
>
> **MUSCLE TEARS OR PULLS**
>
> Sudden overexertion, especially when a muscle is not "warmed up," can result in tearing of the muscle fibers and hemorrhaging into the muscle bundle. These injuries are usually not serious and will heal with adequate rest. Repeated tearing of a muscle, however, can result in scar tissue replacement of the muscle fibers with resultant weakness and inability to extend the muscle. The most common sites for muscle tears are in the hamstring group, the sartorius, and the quadriceps mucles. Tears high up near the origin of the sartorius are called "groin pulls."
>
> The initial treatment for muscle tears is RICE—rest, ice, compression, and elevation. When the acute stage is over and the bleeding stops, ice massage and gentle stretching of the muscle will help restore the length and flexibility of the muscle. Great care must be taken to prevent reinjury until complete healing has taken place.

Table 10-17 MUSCLES THAT MOVE THE LEG

Muscle	Origin	Insertion	Action	Innervation
Sartorius	Anterior superior iliac spine	Proximal, medial shaft of tibia	Flexes femur and lower leg, rotates femur laterally	Femoral
Quadriceps Femoris				
Rectus femoris	Anterior inferior iliac spine and superior margin of acetabulum	Patella, tibial tuberosity via the patellar ligament	Flexes femur, extends leg	Femoral
Vastus lateralis	Greater trochanter and linea aspera of femur	Same as above	Extends leg	Femoral
Vastus medialis	Proximal shaft of femur	Same as above	Extends leg	Femoral
Vastus intermedius	Shaft of femur	Same as above	Extends leg	Femoral
Hamstrings				
Biceps femoris	Ischial tuberosity, linea aspera of femur	Lateral to lateral condyle of tibia, head of fibula	Extends femur, flexes leg	Tibial
Semitendinosus	Ischial tuberosity	Proximal, medial shaft of tibia	Extends femur, flexes leg	Tibial
Semimembranosus	Ischial tuberosity	Proximal, medial shaft of tibia	Extends femur, flexes leg	Tibial

Muscles of the Lower Extremity

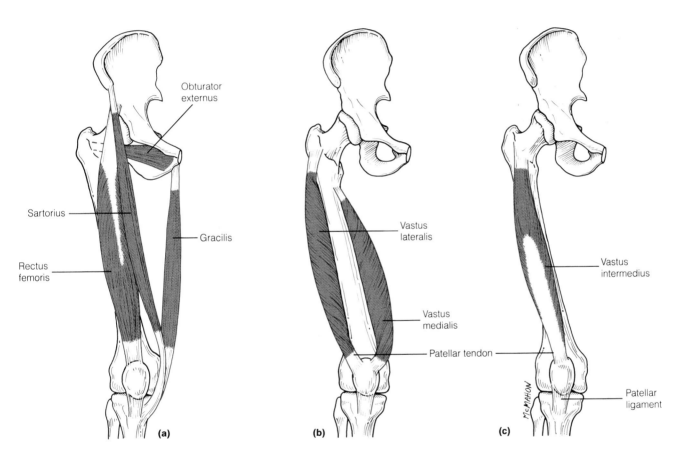

Figure 10-22 Individual muscles of the thigh shown separately.

peroneus: (Gr.) *perone*, pin
tertius: (L.) *tertiarius*, third in sequence
hallucis: (L.) *hallux*, great toe
gastrocnemius: (Gr.) *gaster*, stomach + *kneme*, lower leg

Muscles That Move the Foot and Toes
(Table 10-18 and Figure 10-23)

The foot attaches to the leg at the ankle joint, a diarthrosis that allows movement in four directions: up, down, and side to side. Because of their positions, the muscles that extend the toes also help lift the foot, and the muscles that flex the toes help lower the foot.

A group of four muscles lies on the anterior surface of the leg anterior to the interosseous membrane between the tibia and fibula. The **tibialis** (tib″ē·ā′lĭs) **anterior** lifts and inverts the foot, while the smaller **peroneus tertius** (pĕr″ō·nē′ŭs tĕr′shē·ŭs) lifts and everts it. The **extensor hallucis** (hă·lū′sĭs) **longus** and **extensor digitorum longus** extend the toes and dorsiflex the foot.

A lateral pair of leg muscles, the **peroneus longus** and **peroneus brevis,** extend along the lateral surface of the fibula. Both muscles lower and evert the foot.

A posterior group of muscles in the "calf" of the leg is composed of seven muscles divided into superficial and deep groups. The thick, fleshy **gastrocnemius** (găs″trŏk·nē′mē·ŭs), **soleus** (sō′lē·ŭs), and **plantaris** (plăn·tăr′ĭs) belong to the superficial group (see Clinical Note: Torn Plantaris Muscle). The tendons of these three muscles fuse at their lower ends to form the large **calcaneal tendon** or **Achille's tendon** which inserts on the calcaneus. All three muscles lower the foot. The gastrocnemius and plantaris also help in flexing the knee joint. The deep group includes four muscles. The relatively short **popliteus** (pŏp·lĭt′ē·ŭs) flexes the leg (Figure 10-23d). The **tibialis posterior** lowers and inverts the foot. The **flexor hallucis longus** and **flexor digitorum longus** flex the toes and also help lower the foot.

CLINICAL NOTE

TORN PLANTARIS MUSCLE

The plantaris muscle has a very small muscular portion situated high up behind the knee. It has a long tendon that extends all the way to its insertion on the calcaneus. The muscle has little function in humans except to serve as a good source for a tendon graft. The muscle is, however, subject to tear, usually in an older athlete. A torn plantaris can result in a prolonged, painful disability. The usual victim of a torn plantaris muscle is a golfer or tennis player who feels a sudden, painful, loud pop in the calf area in the middle of a stroke, almost as if he or she had been shot in the calf. There is a marked limp, swelling, and tenderness high up in the calf. It is painful to walk and impossible to run. Treatment includes analgesics, rest, and crutches, and recovery may take as long as four to six weeks.

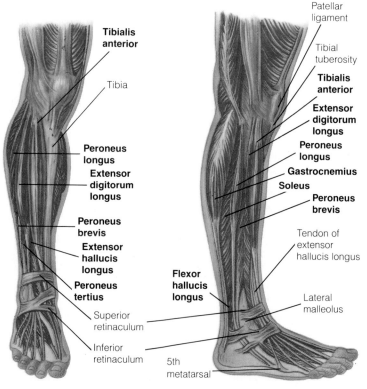

Figure 10-23 Muscles of the leg. (a) Anterior view. (b) Lateral view. (c) Posterior superficial view. (d) Posterior view, middle layer. (e) Deep posterior view.

soleus: (L.) *solea*, sole of the foot
plantaris: (L.) *planta*, sole of the foot
calcaneal: (L.) *calcaneus*, heel bone

Table 10-18 MUSCLES THAT MOVE THE FOOT AND TOES

Muscle	Origin	Insertion	Action	Innervation
Tibialis anterior	Proximal, lateral shaft of tibia, interosseous membrane	First cuneiform and metatarsal 1	Dorsiflexes and inverts foot	Deep peroneal
Peroneus tertius	Distal, anterior shaft of fibula, interosseous membrane	Metatarsal 5	Dorsiflexes and everts foot	Deep peroneal
Extensor hallucis longus	Anterior shaft of fibula, interosseous membrane	Distal phalanx of great toe	Extends great toe, dorsiflexes and inverts foot	Deep peroneal
Extensor digitorum longus	Anterior shaft of fibula, interosseous membrane	Phalanges of toes 2 to 5	Extends toes, dorsiflexes and everts foot	Deep peroneal
Peroneus longus	Lateral shaft of fibula	First cuneiform and metatarsal 1	Plantar flexes and everts foot	Superficial peroneal
Peroneus brevis	Lateral shaft of fibula	Metatarsal 5	Plantar flexes and everts foot	Superficial peroneal
Gastrocnemius	Lateral and medial condyles of femur	Calcaneus	Flexes leg, plantar flexes foot	Tibial
Soleus	Proximal ends of tibia and fibula	Calcaneus	Plantar flexes foot	Tibial
Plantaris	Distal posterior shaft of femur	Calcaneus	Flexes leg, plantar flexes foot	Tibial
Popliteus	Lateral condyle of femur	Proximal, posterior shaft of tibia	Flexes and rotates leg medially	Tibial
Tibialis posterior	Posterior shaft of tibia and fibula, interosseous membrane	Navicular, cuneiforms cuboid, metatarsals 2 to 4	Plantar flexes and inverts foot	Tibial
Flexor hallucis longus	Distal shaft of fibula	Distal phalanx of great toe	Flexes great toe, plantar flexes and inverts foot	Tibial
Flexor digitorum longus	Distal shaft of tibia	Phalanges of toes 2 to 5	Flexes toes, plantar flexes and inverts foot	Tibial

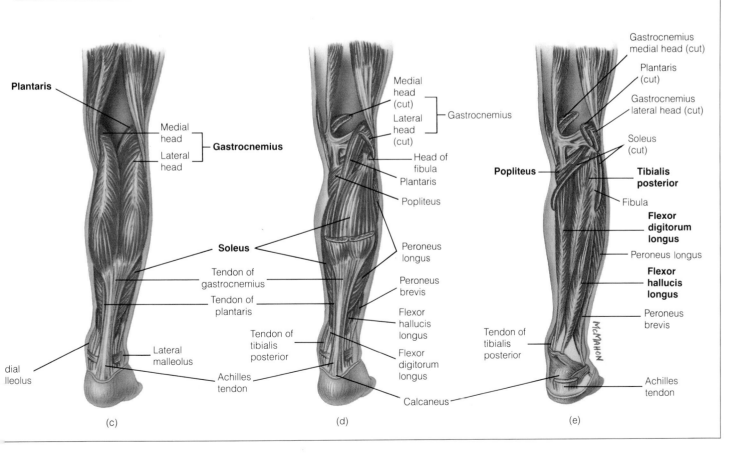

Intrinsic Muscles of the Foot
(Table 10-19 and Figure 10-24)

The hand and foot are composed of three similar groups of bones and possess similar sets of muscles. Because of the very different functions of the two structures, however, significant differences exist in the muscles. The foot lacks the flexibility of the hand, but possesses more strength for support and locomotion.

The foot possesses a dorsal muscle, the **extensor digitorum brevis,** which is lacking in the hand. The remaining ten muscles lie in four overlapping layers on the bottom surface of the foot. They consist of three functional groups; one group moves only the great toe, another moves only the fifth toe, and a third group moves the second through the fifth toes. These have the same names as their counterparts in the hand (compare Figure 10-20 with Figure 10-24).

Table 10-19 INTRINSIC MUSCLES OF THE FOOT

Muscle	Origin	Insertion	Action	Innervation
Extensor digitorum brevis	Superior surface of calcaneus	Tendons of extensor digitorum longus	Extends toes 2 to 5	Deep peroneal
First Layer				
Abductor hallucis	Calcaneus	Proximal phalanx of great toe	Abducts great toe	Medial plantar
Flexor digitorum brevis	Calcaneus	Phalanges of toes 2 to 5	Flexes toes 2 to 5	Medial plantar
Abductor digiti minimi	Calcaneus	Proximal phalanx of toe 5	Abducts toe 5	Lateral plantar
Second Layer				
Quadratus plantae	Calcaneus	Tendons of flexor digitorum longus	Flexes toes 2 to 5	Lateral plantar
Lumbricales	Tendons of flexor digitorum longus	Tendons of extensor digitorum longus	Flexes toes 2 to 5	Medial plantar
Third Layer				
Flexor hallucis brevis	Cuboid, third cuneiform	Proximal phalanx of great toe	Flexes great toe	Medial plantar
Adductor hallucis	Metatarsals 2 to 4, ligaments of metatarsophalangeal joints	Proximal phalanx of great toe	Adducts great toe	Lateral plantar
Flexor digiti minimi brevis	Metatarsal 5	Proximal phalanx of toe 5	Flexes toe 5	Lateral plantar
Fourth Layer				
Plantar interossei	Metatarsals 3 to 5	Proximal phalanges of toes 3 to 5	Adducts toes 3 to 5	Lateral plantar
Dorsal interossei	Adjacent sides of metatarsals 1 to 4	Proximal phalanges of toes 2 to 4	Abducts toes 2 to 4	Lateral plantar

Muscles of the Lower Extremity

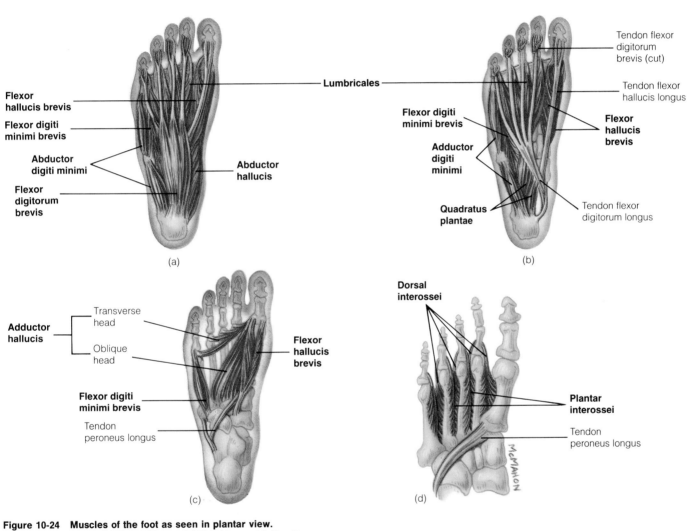

Figure 10-24 Muscles of the foot as seen in plantar view.
(a) Superficial layer. (b) Second layer. (c) Third layer. (d) Fourth (deepest) layer.

myalgia:	(Gr.) *mys*, muscle + *algos*, pain	
arthrogryposis:	(Gr.) *arthron*, joint + *grypos*, curved + *osis*, condition	
myopathy:	(Gr.) *mys*, muscle + *pathos*, disease	
myotonia:	(Gr.) *mys*, muscle + *tonos*, tension	

THE CLINIC

MUSCULAR DISEASES

Research in muscle physiology has progressed at an accelerated rate with the recent interest in fitness. The discovery of fast and slow twitch muscle fibers has stimulated investigation into the relative numbers of these fibers in world class athletes of various sports (see Chapter 9 Clinical Note: Fast and Slow Twitch Fibers). Scientists are attempting to answer these questions: Is there a difference in the number of fiber types in runners versus weight lifters? Is athletic ability genetically determined? Can training change slow twitch fibers into fast twitch fibers? Physiologists have also looked into the cause of postexercise pain: Is it really due to excessive lactic acid production during anaerobic metabolism?

Physicians have many more questions concerning muscle physiology: What is the cause of the muscle aching **(myalgia)** during an infectious process? What really is nonarticular rheumatism? Why do we see so many patients with recurrent pain in the area of the large flat muscles around the shoulder girdle but rarely in the large bulky muscles of the lower extremity? On a daily basis physicians see patients with chronic pain syndromes that are called myositis or fibrositis; however, physicians have little knowledge of the true pathology of these ailments. Although a great deal is known about the anatomy and physiology of muscles, much remains unknown about the pathophysiology of the musculoskeletal system.

The following are descriptions of some of the diseases of the muscular system.

Hereditary and Developmental Diseases
Congenital Absence of Muscle Some muscles may be completely absent at birth. The upper portion of the pectoralis major is frequently absent on one or both sides, as is the palmaris longus. These deletions are largely cosmetic and cause no serious problems. Rarely, there is a congenital absence of various layers of the abdominal musculature. This may be associated with the condition of genital and urinary malformations and undescended testicles (cryptorchidism) in males known as **prune belly syndrome.**

Arthrogryposis Arthrogryposis (ăr″thrō·grĭ·pō′sĭs) is a rare congenital disorder resulting in permanent contractions of the limbs. The cause is variable but is thought to be related to reduced fetal movement, which appears to be necessary for proper joint and muscular development. In arthrogryposis, the joints are fixed in flexion, and the muscular development is deficient and often shows the persistence of fetal muscle fibers. Extensive surgery and physical therapy is sometimes helpful for children with this condition.

Muscular Dystrophies Muscular dystrophies are a group of inherited disorders of the muscles that produce progressive weakness and degeneration of the muscle fibers without neural involvement. **Pseudohypertrophic muscular dystrophy** (Duchenne's type) is the most common type of the disease. Men are usually affected more than women. It normally becomes evident in three- to seven-year-old boys who demonstrate a waddling gait and great difficulty in standing and climbing stairs. It is a progressive ailment. Most of its victims are confined to a wheelchair by the age of 10. About 50% of these unfortunate boys have a lower than normal IQ.

An increase of muscle enzymes such as creatine phosphokinase (CPK) and lactic dehydrogenase (LDH) are found in the blood of most persons affected by muscular dystrophy. Measuring these enzymes can be used as screening tools, but there is no specific treatment. Mobility is maintained as long as possible through muscle strengthening exercise and with the use of appropriate bracing and support.

Myotonic Myopathies The myotonic myopathies are an inherited group of disorders characterized by an abnormally slow relaxation time after the contraction of voluntary muscles. **Myotonia atrophica** (ăt·rō′fĭ·că) (Steinert's disease) is the most common type. It may develop at any age with variable severity. Muscle spasms and rigidity of the hands and drooping of the eyelids (ptosis) are the usual symptoms. **Myotonia congenita** (mī″ō·tō′nē·ă cŏn·jĕn′ĭ·tă) (Thomsen's disease) is a rarer form and begins earlier in life; muscle stiffness is the main difficulty.

Glycogen Storage Diseases Glycogen storage diseases are a rare group of diseases characterized by abnormal accumulation of glycogen in skeletal muscles due to the absence of various muscle enzymes. McArdle's disease, Tarier's disease, and Pompe's disease are examples of this disease, each due to the absence of a different enzyme. They are all characterized by symptoms of muscle pains, cramping, and exercise intolerance.

Familial Periodic Paralysis Familial periodic paralysis is a group of inherited diseases of muscles that is characterized by sudden attacks of flaccid paralysis without loss of consciousness. This group of diseases is associated with abnormalities in potassium metabolism. It is interesting to note that these diseases also occur in goats. This group of rare hereditary diseases is also of historical

myasthenia:	(Gr.) *mys*, muscle + *astheneia*, weakness		
leiomyoma:	(Gr.) *leios*, smooth + *mys*, muscle + *oma*, tumor	fibromyositis:	(L.) *fibra*, fiber + (Gr.) *mys*. muscle + *itis*, inflammation
sarcoma:	(Gr.) *sarx*, flesh + *oma*, tumor	trichonosis:	(Gr.) *trichos*, hair + *osis*, condition

interest in that the diseases were well documented long before modern molecular techniques discovered the basic defects.

Trauma
There are a great number of injuries that involve the muscular system. Some specific injuries have been discussed in the previous clinical notes and others appear in the Unit II Case Study entitled Overuse Syndromes.

Immunology
Myasthenia gravis Myasthenia gravis is a disease of the motor end plate that produces sporadic muscular fatigue and weakness, usually in the muscles that are supplied by cranial nerves, such as the eye and eyelid muscles. Diplopia (double vision) and ptosis (dropping eyelids) often herald the disease. Weakness of the jaw muscles and the proximal muscles of the arms, legs, and respiratory system may also occur. It is sometimes accompanied by a **thymoma** (tumor of the thymus gland) in adults over 40. Slurred words and a nasal tone are noticeable when the speech muscles are weakened. These symptoms worsen during the day when the muscles are used, but after a night's rest, strength is improved. If left untreated, muscle atrophy will occur.

Myasthenia gravis is probably an autoimmune disease in which the receptor sites for acetylcholine, the transmitter substance, have been damaged by autoantibodies. Thus, the muscle membrane does not respond sufficiently to the nerve impulses to produce muscle contractions. Since the transmitter substance seems to be present in sufficient quantities, the disorder appears to be isolated to the receptor molecules.

Drugs that have an anticholinesterase action are used for therapy. These drugs delay the removal of acetylcholine from the neuromuscular cleft and thus prolong its availability to receptor molecules. Excessive use of anticholinesterase, however, may cause weakness of voluntary muscles. Removal of a thymoma, if present, is also a possible cure.

A drug-induced neuromuscular blockade can be produced by exposure to organophosphate insecticides (parathion) and military nerve gases; this will produce severe symptoms similar to myasthenia gravis.

Dermatomyositis Dermatomyositis is another example of an immunologically related disease of the muscular system (see Chapter 4, The Clinic: Collagen Vascular Diseases).

Tumors of Muscles
Benign tumors of muscle are called **myomas**. Those of smooth muscle are called **leiomyomas** (lī″ō·mī·ō′măs); the most common is the fibroid tumor of the uterus. Leiomyomas can occur in the cardiac muscle and are a very rare cause of congestive heart failure.

Malignant tumors of muscles are called **sarcomas** (săr·kō′măs). These deadly tumors can occur anywhere in the muscles and have been known to occur after simple trauma. In the past, surgeons used massive, disabling amputations in an attempt to eradicate these cancers. More recently, the combination of more limited surgery plus radio- and chemotherapy have been shown to be equally effective.

Other Muscle Abnormalities
The term **myositis** is used to describe a wide variety of unexplained muscular aches and pains. **Fibromyositis** (fī″brō·mī″ō·sī′tĭs) is a term ascribed to a group of nonspecific symptoms of pain, stiffness, and tenderness in muscle that may be associated with trigger areas (localized tender nodules in muscles). One of these conditions, **fibromyalgia,** has symptoms of chronic sore and stiff muscles with multiple symmetrical trigger areas and is associated with symptoms of depression and sleep disturbances. Another name for these common muscle problems is **nonarticular rheumatism.**

Trichinosis is a parasitic disease of muscles caused by the invasion of muscle by the pork roundworm, *Trichinella spiralis*. It causes muscular pain, swelling around the eyes, and an increased number of eosinophil cells in the blood (which are leukocyte cells that stain with eosin). The diagnosis is established by a muscle biopsy.

Aging
All of us have observed the processes of aging on the muscular system. Muscular mass, strength, and reflex reactions gradually dissipate with age. The muscle cells appear to have little capacity to regenerate and are gradually replaced with fat and fibrous tissues. Fitness experts attempt to tell us that these changes are due to lack of exercise, improper diet, and the sedentary lifestyles of the elderly, and that with increased activity one can maintain vigor and muscular power to an advanced age. This theory is only partially true; perhaps these efforts delay the inevitable decline for a few years.

In the last two chapters we studied the muscular system at two distinct levels. We started by examining molecular events that cause contraction of microscopic muscle fibers. These contractions result in the shortening of entire muscles, which move the body parts to which they are attached. It is obvious that these movements must be controlled so that muscular activity results in the coordinated body movements described in this chapter. In Unit III we will study the nervous system, that part of the body that coordinates and controls muscular activities and body movements.

STUDY OUTLINE

Muscles and Their Movements (pp. 207–213)
Muscle tissue contracts, moving body parts and materials within the body.

Naming the Muscles Muscles are named for their prominent features, including size, shape, position, the bones to which they attach, and the movements they produce.

Shapes of Skeletal Muscles Muscles are grouped in bundles (fascicles) arranged in four distinct patterns: *parallel, convergent, pennate,* and *circular. Fusiform* muscles are columnar in shape, while *penniform* muscles are flattened, and *radiate* muscles are triangular. Muscles can also be *rectangular, triangular, rhomboid,* and *circular.*

Origin, Insertion, and Action The site of muscular attachment that does not move when a muscle contracts is the *origin* of the muscle. The place of attachment that moves is the *insertion.* The specific movement caused by a muscle is its *action. Prime movers,* or *agonists,* are muscles of primary importance in an action. Muscles that perform opposing actions are *antagonists. Synergists* are muscles that help prime movers accomplish actions. *Fixators* immobilize a joint.

Levers Muscles move the skeleton similar to movements of simple levers. In a *first class lever,* the joint (fulcrum) is located between the point where the force is applied and the point of *resistance.* In a *second class lever,* the resistance is located between the fulcrum and the force. In a *third class lever,* the force is applied between the resistance and fulcrum.

Muscles of the Head and Neck (pp. 214–221)
Muscles of the head and neck can be divided into five major groups. One group moves the soft, fleshy parts of the face and produces familiar facial expressions. Another group moves the eyeballs. A third group is responsible for the chewing motions of the jaw (mastication). The fourth group moves the tongue. The fifth group (muscles in the neck) moves the head and neck.

Muscles of the Trunk (pp. 222–229)
Trunk muscles are grouped into four functional groups. The bones of the vertebral column are stabilized and moved by numerous small muscles that attach to the spine. Muscles of the thoracic wall are active in breathing movements. Muscles of the abdominal wall maintain physical support for organs in the abdomen. Muscles of the pelvic floor support organs that project into the pelvis from the abdominal cavity.

Muscles of the Pectoral Girdle and Upper Extremity (pp. 230–241)
We divide the muscles of the pectoral girdle and upper extremity into six functional groups. One group is responsible for the movements of the scapula, while another group moves the humerus. Five muscles organized into two opposing sets accomplish the movements of the forearm. Three layers of muscles on the anterior surface of the forearm flex the wrist and hand, while two layers on the posterior surface extend the wrist and hand. Each hand contains 20 intrinsic muscles that produce the complex movements of the fingers.

Muscles of the Lower Extremity (pp. 242–254)
Muscles of the lower extremity are divided into four functional groups. One group produces a complex set of movements of the ball-and-socket joint connecting the thigh to the pelvic bone. A second set of muscles moves the leg at the knee joint. Another group of muscles in the shin and calf region move the foot at the ankle joint. Each foot has four layers of intrinsic muscles that are responsible for moving the toes.

SELF-TEST OF CHAPTER OBJECTIVES

True-False Questions
1. In a first class lever, the fulcrum is located between the point of resistance and the point of force application.
2. Antagonistic muscles are often positioned on the same side of a joint.
3. The pectoralis minor is smaller than the pectoralis major.
4. The adductor longus is included in the functional group of muscles that rotates the tibia around the fibula.
5. The origin of a muscle is the end that moves least when the muscle contracts.
6. The deltoid is a prime mover of the humerus.
7. The gastrocnemius muscle inserts on the calcaneus.
8. Synergists are muscles responsible for moving the skeleton in locomotion.
9. The deltoid muscle is antagonistic to the pectoralis major.
10. Flexion of the forearm involves the use of a second class lever.
11. More muscles are contracted to produce a frown than a smile.
12. The trapezius is named because of its shape.
13. The action of a muscle is always described in terms of its place of insertion.
14. Antagonistic muscles perform opposing actions.
15. Most muscles in the body act as part of a third class lever.

Matching Questions

Match the muscles on the left with the specific actions on the right:

16. orbicularis oris
17. buccinator
18. masseter
19. sternocleidomastoid
20. orbicularis oculi

a. elevates mandible
b. purses lips
c. rotates head and neck
d. closes eyes
e. flattens cheeks

Match the muscles on the left with their actions on the right:

21. supraspinatus
22. deltoid
23. teres major
24. subscapularis
25. biceps brachii

a. supinates hand
b. abducts humerus
c. rotates humerus laterally
d. rotates humerus medially

Match the muscles on the left with their agonists on the right:

26. biceps brachii
27. triceps brachii
28. rectus femoris
29. gastrocnemius
30. brachioradialis

a. vastus medialis
b. anconeus
c. brachialis
d. plantaris

Match the muscles on the left with their antagonists on the right:

31. flexor digitorum superficialis
32. gastrocnemius
33. trapezius
34. biceps brachii
35. triceps brachii

a. tibialis anterior
b. brachialis
c. triceps brachii
d. extensor digitorum communis
e. pectoralis minor

Essay Questions

36. Muscles are named for their size, shape, and position, among other things. List as many bases for naming muscles as you can and give examples for each.
37. Describe the muscles that move the scapula. Identify the agonistic and antagonistic groups.
38. The biceps brachii flexes the forearm. Discuss this action in terms of levers.
39. Discuss the fact that, at different times, the same muscle may act as a prime mover or a synergist.
40. Locate and describe the anterior and posterior triangles on the neck. Which muscle divides the two areas?
41. Describe the major movements of the thigh, and identify the muscles responsible for these movements.
42. Describe the antagonistic muscles involved in pronation and supination of the forearm.
43. Discuss the muscles of facial expression and indicate the common expressions they produce.
44. How does a fusiform muscle differ from a penniform muscle?
45. Identify the erector spinae group of muscles and discuss their actions.

Clinical Application Questions

46. Describe causes of the musculotension headache.
47. What do musculotension headaches and torticollis have in common?
48. How does iliotibial band friction syndrome differ from piriformis syndrome?
49. The initial treatment for trauma and tears of muscles is abbreviated "RICE." What do these letters indicate?
50. Compare and contrast muscular dystrophy and myasthenia gravis.
51. List some of the overuse syndrome problems that might afflict someone who exercises strenuously but infrequently.

periostitis: (Gr.) *peri*, around + *osteon*, bone + *itis*, inflammation

UNIT II CASE STUDY

OVERUSE SYNDROMES

Doctor James is a general physician who has had a long interest in sports medicine and office orthopedics. This morning he has been working with a fourth year medical student from a local medical school. They are taking a break from a busy morning to discuss the cases they have seen in the outpatient clinic.

The first patient is an 18-year-old member of the university track team who has been plagued with recurrent attacks of pain in his anterior shin areas after only about 2 miles of cross-country running. The pain is so bad that he has to stop and rest for 15 to 20 minutes before continuing and is then able to run again for only a few miles until the pain returns. When he persists in running and tries to "run through it," the pain will persist for the rest of the day. He is very anxious about these symptoms because this is his first year with the track team and he wants to do well. Examination showed diffuse tenderness all along the medial borders of his tibias and marked hyperpronation of his subtalar joints. X-ray examination of his lower legs showed diffuse calcification all along the medial edge of his tibias but no evidence of stress fractures. His running shoes are well worn and showed wear patterns along the outer heel and the medial aspect of the sole. Doctor James remarked that this was a common problem in young runners called **periostitis** (pĕr″ē·ŏs·tī′tĭs) caused by the constant traction on the periosteum of the bone where the tibialis muscle originates. (Periostitis is one type of shin splint, a term that can also mean a stress fracture of the tibia or fibula.) Correcting his hyperpronation with orthotics (molded insoles) and new shoes should eliminate the problem.

The second patient is a music student who has developed a persistent pain in the medial aspects of her feet, which seems to be related to her job as organist for a local church. Examination of her feet showed a tender, warm, swollen area along the tendons of her anterior tibialis muscles. Plantar flexion of her feet produces a grating sensation along the tendon. Further history revealed that the organ she plays is an old mechanical type with very heavy pedal action. Doctor James remarked that the findings were typical of **tenosynovitis** (tĕn″ō·sĭn″ō·vĭ′tĭs) (inflammation of a tendon sheath). Treatment with anti-inflammatory medications should help relieve some of her symptoms, and the use of a good ankle support would perhaps prevent recurrence.

The next patient is a member of the varsity swim team who has had pain in his left shoulder for the past week. The team is preparing for its active season, holding three practice sessions per day and concentrating on distance swimming. The patient is very concerned that the pain would limit his participation should it progress. He really is a sprinter and swimming at this distance is new to him. Examination showed tenderness over the supraspinatus tendon in the front of the shoulder; his pain was precipitated by extension and abduction of the shoulder. This condition is called **shoulder impingement syndrome**, commonly known as swimmer's shoulder. There was no crepitation (rubbing feeling) with shoulder motion, and his strength was excellent. They discussed reducing his

Overuse syndromes can affect both athletes and nonathletes alike, whenever tired muscles are repeatedly overworked.

mileage, using ice massage after workouts, and exercising his shoulders in a position at less than 90° abduction to prevent impingement of the tendon between the corocoid process and the head of the humerus.

The final person they discussed is not really a patient but rather the office secretary. She has been experiencing chronic neck and shoulder discomfort for several months. Her neurological examination was entirely normal, as was the examination of her shoulder girdle. The only positive finding was a forward posture of her neck and tightness and tenderness over her trapezius muscles. She works long hours at her word processor, and the computer screen was positioned so that she constantly worked with her chin forward and neck and shoulders slumped anteriorly. Adjustments were made to her chair and the angle of her computer screen, and she was instructed in exercises to strengthen her neck muscles and improve her posture, which should provide relief of her symptoms.

Doctor James remarked about the diversity of these problems, but added that they were all really examples of the same problem: **overuse syndromes.** Overuse syndromes are a varied group of pain symptoms and signs due to repeated, unaccustomed use of a muscle, tendon, or bone. Overuse syndromes are often related to intrinsic factors involving abnormalities of the person's body structure, such as hyperpronation of the feet or poor posture. Extrinsic factors such as improper techniques and poor equipment can also be involved. Overuse syndromes most commonly affect novice athletes, but as our examples have shown, they often occur in nonathletes as well. Also, overuse syndromes are not only limited to physical problems, but may be manifested as an emotional problem called "burnout." Perhaps the best advice to fitness buffs (and to all of us) is "Use it or lose it, but don't abuse it."

Unit III

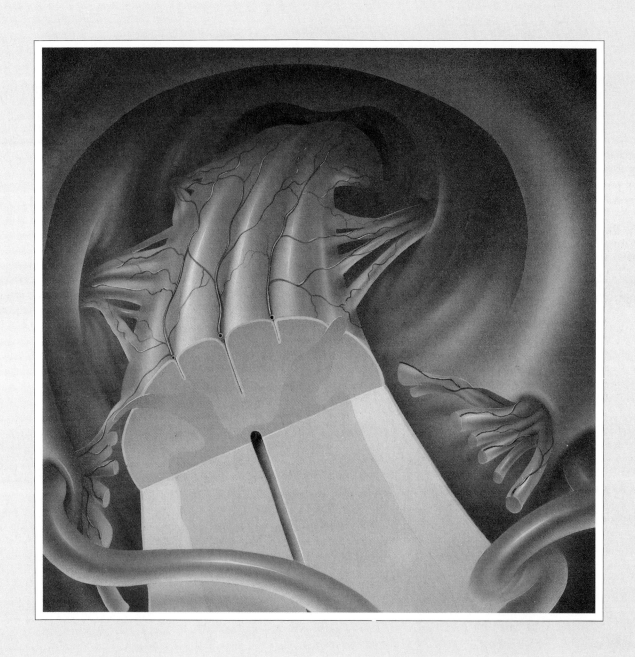

Nervous and Chemical Integration and Control Systems

In this unit we study the two systems of the body that are responsible for controlling metabolic activities. Cells of the nervous system connect to all body organs and provide precise control over specific regions. Endocrine glands, on the other hand, influence all cells in the body by secreting regulatory chemicals into the blood. Changes in the external and internal environment stimulate sensory nerve cells that relay information to control centers in the brain and spinal cord. Information from different sources is integrated, and signals from the central nervous system provide coordinated responses to different stimuli. The nervous system is also responsible for the complex processes of abstract and creative thinking.

11
The Nervous System: Neurons and Impulses

12
The Central Nervous System

13
The Peripheral Nervous System

14
The Autonomic Nervous System

15
Functional Aspects of the Nervous System

16
Special Sense Organs

17
The Endocrine System

11
The Nervous System: Neurons and Impulses

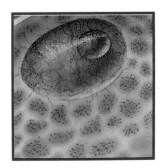

The **nervous system** is an elaborate system of cells that extends throughout the body receiving information about the body's internal and external environment, assessing that information, and then sending signals to organs that cause an appropriate response. As such, it is one of two major systems used to regulate body processes. The other is the *endocrine system*, which acts through the use of chemical regulators called *hormones* (see Chapter 17). In the nervous system, stimuli received by specialized **sensory cells,** such as heat-sensitive cells in the skin or light-sensitive cells in the eye, are transmitted as electrochemical impulses to the spinal column and brain. When a response is called for, signals are sent to **effector organs,** which are glands and muscles that respond by secreting a hormone or by contracting.

The nervous system is thus an elaborate communication system involving receptors that perceive stimuli and transmit signals to a central processing region, which may then send signals to effectors where they cause a response. In this chapter we will examine the structure of nervous tissue and the mechanisms by which signals are transmitted, considering

1. the organization of a typical nerve,
2. the various cell types that comprise the nervous system,
3. how a membrane potential is established in a nerve cell and used for the transmission of an impulse,
4. how impulses are passed from cell to cell, and
5. how the signals are controlled to provide an integrated system of impulses.

afferent: (L.) *afferre*, to bring toward
efferent: (L.) *efferre*, to carry away
epineurium: (Gr.) *epi*, upon + *neuron*, nerve
perineurium: (Gr.) *peri*, around + *neuron*, nerve
endoneurium: (Gr.) *endon*, within + *neuron*, nerve

Organization of the Nervous System

After studying this section, you should be able to:

1. Describe the organization of the nervous system, distinguishing between the central and peripheral nervous systems.
2. Distinguish between sensory, motor, and mixed nerves.
3. Draw a cross-sectional view of a typical nerve that shows the organization of connective tissue coverings and other types of tissue in the nerve.

The CNS and the PNS

The nervous system consists of two major portions: the **central nervous system** (abbreviated **CNS**) and the peripheral nervous system (abbreviated **PNS**). The CNS includes the brain and spinal cord, and the PNS consists of all the nervous tissue outside (or "peripheral" to) the CNS (Figure 11-1).

The **brain** consists of the nervous tissue contained within the skull. It is in the brain that evaluation of impulses from sensory organs is performed, where consciousness originates, where personality and emotion originate, and where the overall general control of other systems occurs.

The **spinal cord** extends from the brain through the column of vertebrae that comprise the backbone, reaching from the lower portion of the brain to the lower spine. Within the spinal cord nervous tissue is arranged in bundles of cells that provide circuits for the transmission of signals. Such bundles are referred to as **tracts**.

The PNS is organized into **nerves**, which are bundles of nervous tissue that emanate like cables from the CNS throughout the body where they provide pathways for signals traveling to and from the CNS. Functionally, the PNS consists of two portions called the afferent and efferent divisions. The **afferent** division carries signals to the CNS from sensory organs, and the **efferent** division carries signals from the CNS to effector organs. The efferent division is further subdivided into the **somatic nervous system**, which carries signals specifically to skeletal muscles, and the **autonomic nervous system**, which carries signals to glands, smooth muscle, and the heart. Signals carried by the somatic nervous system are often produced by voluntary, conscious activity of the brain, whereas signals produced by involuntary, subconscious activity of the brain are usually carried by the autonomic nervous system.

Organization of a Nerve

Specific nerves either carry signals toward the central nervous system, away from it, or both. Those that carry signals exclusively to the CNS are therefore totally afferent in function; they are called **sensory nerves**. Nerves that carry signals exclusively away from the CNS, and are therefore completely efferent in function, are referred to as **motor nerves**. Some nerves carry signals in both directions, and are both efferent and afferent. They are called **mixed nerves**.

A typical nerve is shown in cross section in Figure 11-2. The nerve contains not only cells involved in the transmission of impulses, but also circulatory and lymphatic vessels that pass through layers of connective tissue in the nerve. The **epineurium** (ĕp″ĭ·nū′rē·ŭm) is a layer of connective tissue that surrounds the nerve and extends into it to form the **perineurium** (pĕr″ĭ·nū′rē·ŭm). The perineurium divides the nerve into separate bundles of nerve cells called **fascicles**. Within each fascicle individual nerve cells are surrounded by a thin layer of connective tissue called the **endoneurium** (ĕn″dō·nū′rē·ŭm). The epineurium, perineurium, and endoneurium are continuous and together form a framework that holds the nerve together as a cablelike unit, much like the components of a muscle are held together by layers of connective tissue.

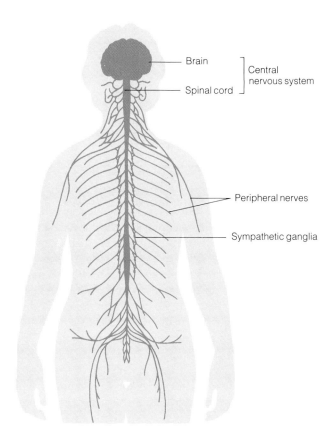

Figure 11-1 Organization of the nervous system. The nervous system consists of the central nervous system (the brain and spinal cord) and the peripheral nervous system (all the nervous tissue outside the central nervous system).

Chapter 11 The Nervous System: Neurons and Impulses

neuroglia:	(Gr.) *neuron*, nerve + *glia*, glue
soma:	(Gr.) *soma*, body
dendrite:	(Gr.) *dendron*, tree
perikaryon:	(Gr.) *peri*, around + *karyon*, nucleus

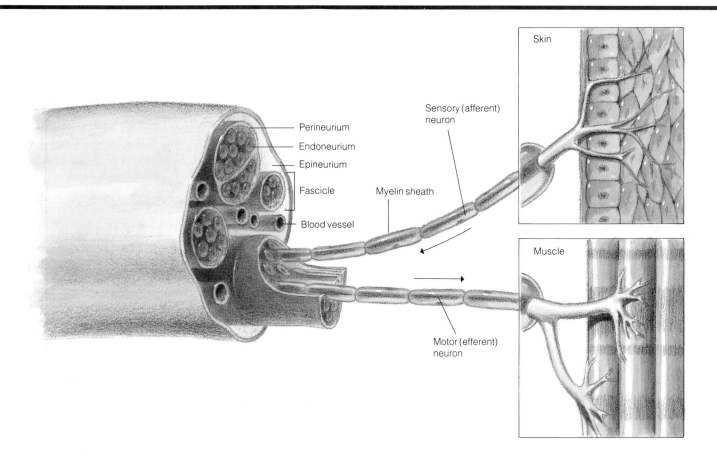

Figure 11-2 Cross section through a peripheral nerve. Nerve fibers and blood vessels are held together by connective tissue sheaths that surround the individual fibers and vessels. Arrows along the neurons indicate the direction of travel of nerve impulses.

Cells of the Nervous System

After studying this section, you should be able to:

1. Describe in general terms the function and organization of a neuron, distinguishing among dendrites, cell body, axon, and axon terminals.
2. Describe the structure of unipolar, bipolar, and multipolar neurons.
3. Explain the functional difference between neurons and neuroglia.
4. List the different functional roles and locations within the nervous system of astrocytes, ependyma, microglia, oligodendrocytes, satellite cells, and Schwann cells.
5. Describe the organization of a myelin sheath formed around an axon by a Schwann cell in the peripheral nervous system and by oligodendrocytes in the central nervous system.

There are two general categories of cells in the nervous system: cells that are responsible for carrying impulses, called **neurons,** and cells that support, insulate, and protect the neurons, called **neuroglia** (nū·rŏg´lē·ă) or **glia.**

Neurons

Neurons usually have four distinct regions: the **cell body,** also called the **soma** (sō´mă); long, tapering extensions of the cell body called **dendrites** (děn´drīts) that receive signals from neighboring cells; an extension that carries signals away from the cell body called an **axon** (ăk´sŏn); and fine subdivisions of the tip of the axon called **axon terminals.** Collectively, extensions that emanate from the cell body are referred to as **processes** or **fibers** (Figure 11-3).

The cell body of a neuron contains the nucleus and a surrounding film of cytoplasm called the **perikaryon** (pĕr˝ĭ·kăr´ē·ŏn). The perikaryon contains an extensive rough endoplasmic reticulum, indicative of the large amount of protein produced in the cell body. Ribosomes are often present in such high density in the perikaryon that they form masses that can be stained and seen with a light microscope. Such masses of ribosomes are called **Nissl bodies** (Figure 11-3). The perikaryon also contains a well-developed Golgi apparatus and an especially large number of mitochondria, indicative of the high level of metabolic activity that usually occurs in a neuron. Virtually all protein synthesis occurs in the cell body, and many products are enclosed within vesicles while in the perikaryon. They are then transported in vesicles to other parts of the cell.

Dendrites often contain Nissl bodies also, as well as rough endoplasmic reticulum and a Golgi apparatus. Dendrites are characterized by large amounts of microtubules and filaments that are arranged throughout the length of each dendrite and involved in the transport of materials from the cell body to the tip of the dendrite.

Although some neurons lack an extension that serves specifically as an axon, most neurons possess one. Axons lack Nissl bodies and an extensive endoplasmic reticulum but possess a large number of microtubules and filaments called **neurofilaments** that run the length of the axon. Neurofilaments transport materials made in the perikaryon to the axon terminals and move other materials from the terminals to the cell body.

In some neurons the axon is about the same length as the dendrites and is hard to identify because of its similarity to dendrites. In other neurons, however, the axon is significantly longer than the dendrites and may extend a considerable distance from the cell body (such as shown in Figure 11-3). Some neurons in humans have axons that are

Figure 11-3 A typical myelinated neuron (a somatic motor neuron). (a) A schematic diagram of a motor neuron showing the major types of processes. Note that the length of the processes relative to the cell body size is in actuality much greater than shown here. (b) Scanning electron micrograph of the cell body of a motor neuron with the axon shown extending downward at the bottom (mag. 11,160×).

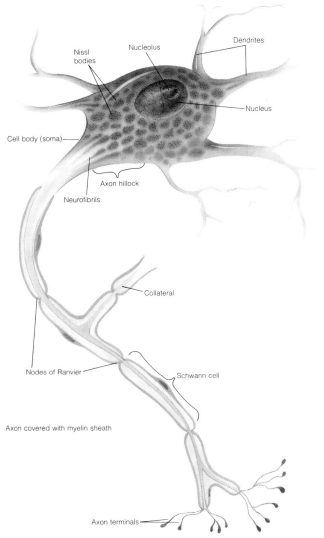
(a)

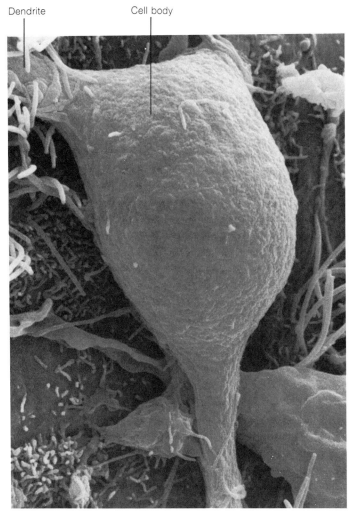
(b)

hillock: (M.E.) *hilloc*, little hill

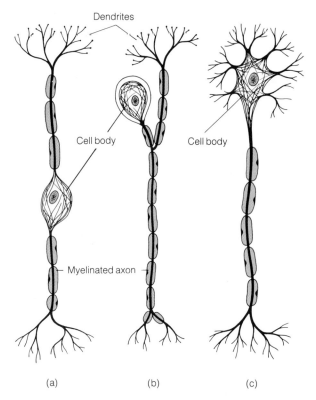

FIGURE 11-4 **Types of neurons.** (a) Bipolar. (b) Unipolar. (c) Multipolar.

a meter or more in length, reaching from the spinal cord to muscles in the farthest portions of the lower extremities.

Axons usually begin as an enlarged shoulder of the cell body called a **hillock** (hĭl ′ŏk). Some axons are branched. The branches, called **collateral branches,** often occur successively and diminish in diameter with each division. Ultimately, each branch ends in many, perhaps hundreds, of finely divided **axon terminals,** all communicating with the dendrites, cell bodies, and axons of other neurons. In the case of neurons carrying signals outward from the CNS, communication is often with the surface of a muscle or gland cell.

There are three kinds of neurons: bipolar, unipolar, and multipolar (Figure 11-4). **Bipolar neurons** have a single dendrite and a single axon emanating directly from the cell body. Bipolar neurons are rare in humans but are found, for example, in the nervous tissue of the ear. **Unipolar neurons** possess a single process that emanates from the cell body and then branches to form a single axon and dendrite. Unipolar neurons are typically found in sensory nerves, where they transport signals from sensory organs to the CNS. **Multipolar neurons** possess many dendrites emanating from the cell body and a single axon that emanates either from the cell body or the base of one of the dendrites. This type of neuron is found in many regions of the brain and in motor nerves.

Neuroglia

In spite of their important role in carrying signals throughout the nervous system, neurons represent a minority of the cells present in nervous tissue. In the brain, for example, neuroglia outnumber neurons by as much as 50 to 1 and may account for more than one-half the weight of the brain. Four different kinds of neuroglia are present in the CNS: *astrocytes, ependyma, microglia,* and *oligodendrocytes* (Figure 11-5). There are two types of neuroglia in the PNS. Neuroglia associated with cell bodies in particular portions of the peripheral nervous system are referred to as *satellite cells.* Others associated specifically with axons are referred to as *Schwann cells.*

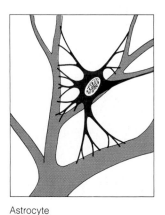
Astrocyte

Ependyma

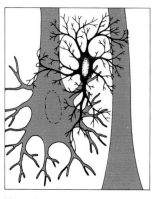
Microglia

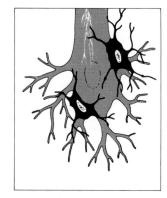
Oligodendrocytes

Figure 11-5 Types of neuroglia include astrocytes, ependyma, microglia, oligodendrocytes.

sclerosis: (Gr.) *skleros*, hard + *osis*, condition
astrocyte: (Gr.) *astron*, star + *kytos*, cell
ependyma: (Gr.) *ependyma*, outer garment
microglia: (Gr.) *mikros*, small + *glia*, glue
oligodendrocyte: (Gr.) *oligos*, few + *dendron*, tree + *kytos*, cell
myelin: (Gr.) *myelos*, marrow

CLINICAL NOTE

MULTIPLE SCLEROSIS

The myelin covering the axons of nerves produces the means for fast transmission of nerve impulses. Any loss of myelin will cause the cessation or slowing of the action potential transmission. In the disease of **multiple sclerosis** (sklĕ·rō´sĭs), portions of the white matter in the spinal cord and brain are affected by **sclerotic plaques** that occur randomly. The cause of these sclerotic plaques is unknown, but a viral origin in conjunction with altered immune response is suspected. The plaques destroy myelin and alter transmission along the neuron. The disorder resulting from the altered nerve transmission varies depending on the neurons affected and the severity of the attack. Multiple sclerosis is manifested by periods of active destruction of myelin interspersed with periods of no destruction. Thus, the patient suffers varied loss of function, such as weakness, lack of coordination, areas of diminished sensation, and speech and vision problems. The disease is experienced as a long, relapsing one. The National Multiple Sclerosis Society is a source of information and encouragement for people with the disorder, their families, and health practitioners.

Neuroglia of the CNS

Astrocytes (ăs´trō·sīt) are so named because they are roughly star shaped, with many processes that radiate outward from the cell body. These neuroglia are primarily responsible for holding neurons in place, and they prevent sudden changes in the composition of the fluid that surrounds the neuron by regulating the exchange of substances between the fluid and blood (see Blood Supply in Chapter 12).

Ependyma (ĕp·ĕn´dĭ·mă) are epithelial cells that line ventricles, fluid-filled chambers of the brain. They also line the inner wall of the central canal that runs through the spinal cord. Ependymal cells range in shape from cuboidal to columnar and are sometimes ciliated. Movement of the cilia may assist in exchanging nutrients between the fluid in the chambers and the fluid that bathes neurons and other neuroglia.

Microglia (mī·krŏg´lē·ă) are the smallest of the neuroglia. It is unclear whether they are truly nervous tissue or a portion of the circulatory system. They appear to play an important role in removing dead neurons and foreign matter from the CNS, and their numbers increase in inflamed nervous tissue, indicating that they may be components of the body's defense system.

Oligodendrocytes (ŏl˝ĭ·gō·dĕn´drō·sīts) are characterized by having relatively few and delicate extensions that radiate from the cell body. They are found in groups surrounding blood vessels or in close association with neurons. In many parts of the CNS, oligodendrocytes form a sheath of fatty material called **myelin** (mī´ă·lĭn) that is wrapped around the axon of neighboring neurons. A single oligodendrocyte often forms sheaths around the axons of several neurons simultaneously, forming a scaffolding that supports the neurons (Figure 11-6). In addition, the sheaths provide an insulating covering necessary for proper axon functioning (see Clinical Note: Multiple Sclerosis).

Neuroglia of the PNS

Satellite cells are small cells that encapsulate the cell bodies of certain neurons clustered together in groupings called **ganglia** (Figure 11-7). **Schwann cells**, in contrast, are associated with the axons of certain peripheral neurons (see Figure 11-3), wrapping the axons with a myelin sheath similar to that of the oligodendrocytes in the CNS. Unlike oligodendrocytes, however, each Schwann cell is associated with a single neuron. During development of the nervous system, a Schwann cell wraps its membrane repeatedly around the axon, forming, in cross section, a spiral around the axon.

The myelin wrappings are separated from one another at regular intervals along the axon, leaving gaps in the sheath called **nodes of Ranvier** (rŏn·vē·ā´) (see Figure 11-3). Neurons whose axons possess a Schwann cell covering are termed **myelinated neurons;** those that lack such a covering are called **nonmyelinated neurons.**

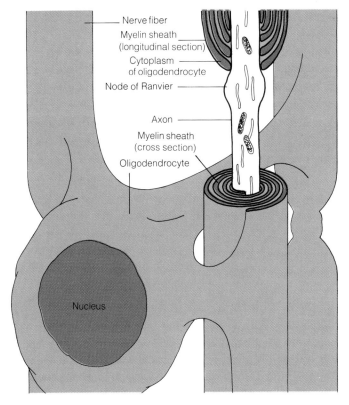

Figure 11-6 Diagrammatic view of an oligodendrocyte showing the formation of myelin sheaths around neighboring axons.

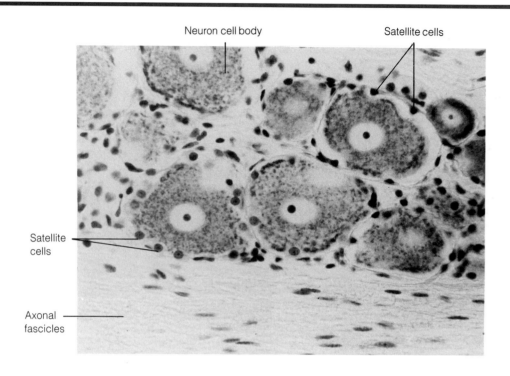

FIGURE 11-7 Photomicrograph of satellite cells (mag. 530×).

Membrane Potential

After studying this section, you should be able to:

1. Explain the concept of membrane potential.
2. Describe the mechanism for establishing a membrane potential.
3. Explain the basis for leakage of ions through the membrane at different rates.

Like other mammalian cells, a neuron maintains significantly different concentrations of certain ions on opposite sides of its cell membrane. Table 11-1 shows the difference with respect to Na^+, K^+, and Cl^-, the main ions involved in neuron function. (Not shown in the table are the concentrations of negatively charged proteins and certain other ions. See Figure 26-2.) Although the overall number of positively and negatively charged ions is equal, there is a slight excess of positive ions outside and a corresponding excess of negative ions inside, which results in a charge difference across the membrane. If the membrane did not act as a barrier to the movement of the ions, they would spontaneously diffuse through the membrane until their concentrations on the two sides were equal. Charges carried by them would also be equally distributed, and the difference in charge on the two sides of the membrane would disappear.

Electric Potentials

Whenever conditions exist such that electrically charged particles would migrate from one point to another were it not for some barrier, there is said to be an **electric potential** between the points. In the case of a cell, the electrically charged particles are ions and the barrier to their migration is the cell membrane. Electric potentials that exist across cell membranes are called **membrane potentials.** Generally, electric potentials are measured in **volts** (V), but in the case

Table 11-1 DISTRIBUTION OF MAJOR IONS INVOLVED IN NEURON FUNCTION

Ion	Intracellular ion concentration (mmol/L water)	Extracellular ion concentration (mmol/L water)
Na^+	15	150
K^+	150	5
Cl^-	10	110

Adapted from D. S. Luciano, A. J. Vander, and J. H. Sherman, *Human Anatomy and Physiology, Structure and Function*, 2nd ed. (McGraw Hill Publishing Company, 1983), 86.

of a membrane potential, the voltage is so small that it is more common to express the potential in **millivolts** (mV), which is 1/1000 of a volt.

Electric (and membrane) potentials can be either *positive* or *negative*, depending on the direction in which a charged particle would travel if given the opportunity. By convention, a membrane potential is classified on the basis of the net charge of the ions in the *interior* of the cell. Since there is usually an excess of negative ions on the inside of a cell, its interior is usually negative and the membrane potential is expressed as a negative value. In the case of a resting neuron, one that is not carrying an impulse, the membrane potential is often about −70 mV. Membranes across which there is a negative membrane potential are said to be **polarized.**

A membrane potential is maintained by a **sodium-potassium-ATPase pump,** the name given to carrier proteins in the membrane that actively transport Na^+ and K^+ in opposite directions across the membrane. These carrier proteins combine with Na^+ on the inside of the cell and carry it to the outside; conversely, they combine with K^+ on the outside and carry it to the inside. ATP provides the energy for this transport. Transport of each ion is linked to the transport of the other in a sodium-potassium-ATPase pump, so the ions are transported in an approximate 3:2 ratio. The precise mechanism responsible for this ratio is unknown, but it may reflect the presence of three Na^+ binding sites on the inner surface and two K^+ binding sites on the outer surface of a carrier protein that extends through the membrane. Perhaps attachment of ions to these sites induces an ATP-dependent configurational change that carries the ions through the membrane and releases them on the opposite side (Figure 11-8). As a result of the action of the pump, a concentration gradient is established across the membrane for each of the ions, with Na^+ in excess on the outside and K^+ in excess on the inside (Table 11-1).

Ions leak back through the membrane at specific "leak channels" through which some ions pass more readily than others. The membrane is about 100 times more permeable to K^+ than it is to Na^+, so the ions leak back through the membrane at different rates. The net result is a difference in ion concentration on the two sides of the membrane that produces the membrane potential.

Resting Potentials

To understand how a membrane potential is established, consider a container of pure water in which the water is divided into two portions by a membrane permeable to Na^+ but not to Cl^- (Figure 11-9a). Suppose NaCl is added to the water in the left half but not the right. As the NaCl dissolves, the ions become distributed throughout the left half. Since the membrane is permeable to Na^+, some Na^+ will diffuse

Figure 11-8 Mechanism of action of a sodium-potassium pump. The sodium-potassium pump is an active transport system that carries Na^+ and K^+ ions in opposite directions across the neuronal membrane. The result is an accumulation of Na^+ on the outside of the membrane and K^+ on the inside.

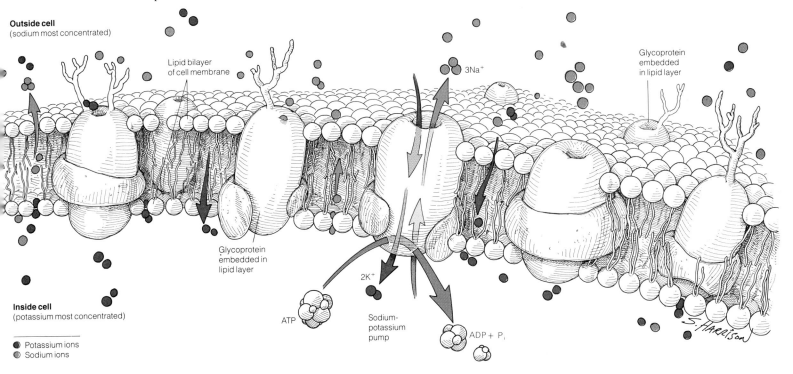

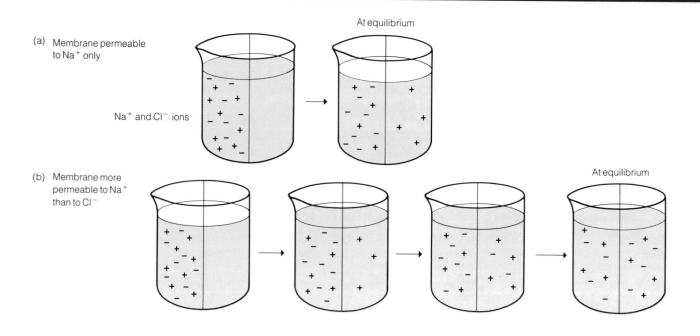

through to the right side, but Cl⁻ will not. With the movement of each Na⁺ ion through the membrane, there will be an increase of one positive charge on the right and a decrease of one positive charge on the left. As the number of positive charges accumulates, they exert a repulsive effect that inhibits diffusion of additional Na⁺. A corresponding excess of negative charges also develops on the left side A that exerts an attractive force on Na⁺ which also inhibits its diffusion. In time the tendency of Na⁺ to diffuse across the membrane because of the difference in its concentration between the two sides is offset by the inhibiting effects of Na⁺ and Cl⁻. There will then be no further net flow of Na⁺ across the membrane, but there will be a potential difference across the membrane because of the difference in electric charge caused by the unequal distribution of the ions.

Now suppose the demonstration is repeated, but instead of using a membrane that is permeable to Na⁺ and totally impermeable to Cl⁻, one is used that is highly permeable to Na⁺ and only slightly permeable to Cl⁻ (Figure 11-9b). In this case Cl⁻ will also diffuse through the membrane, although more slowly than Na⁺ diffuses. Initially, the greater rate of Na⁺ diffusion than of Cl⁻ diffusion will establish a potential difference across the membrane, but in time the amount of Cl⁻ that has diffused across will catch up with Na⁺ diffusion, and there will be equal concentrations of both ions on either side of the membrane. From that point on there will be no further net movement of ions across the membrane and the membrane potential will be zero.

Figure 11-9 Membrane potentials. (a) If NaCl is added to the water in the left half and only Na⁺ can diffuse through the membrane, an electric potential will develop across the membrane as a result of the difference in ion concentration on the two sides. At equilibrium the tendency for Na⁺ to diffuse from the left to the right is balanced by the repulsion of additional Na⁺ caused by the excess of positive charges in the solution on the right. The positive ions remain close to the membrane. (b) If Cl⁻ also diffuses through the membrane but more slowly than Na⁺, an electric potential will be established initially but gradually diminish as the ions become equally distributed.

A cell membrane is like the membrane in the second example; that is, it shows different degrees of permeability depending on the ions involved. Initially, equal concentration gradients are established for Na⁺ and K⁺ on the two sides of the membrane by the sodium-potassium pump. However, K⁺ leaks back out through the membrane much more rapidly than Na⁺ diffuses back in because the membrane is much more permeable to K⁺ than it is to Na⁺. This causes the outside of the membrane to become more positive than the inside. This charge difference attracts Cl⁻ to the outside, and since the membrane is about 50 times more permeable to Cl⁻ than it is to Na⁺, the Cl⁻ also moves out more rapidly than Na⁺ diffuses back in. At any instant there is a potential difference across the membrane that is the net sum of the potential difference caused by each ion. But since the membrane is permeable to all three ions, in time all will become equally distributed, and there will no longer be a net membrane potential.

In a resting neuron, maintenance of a membrane potential in spite of continued ion leakage is made possible by the continued activity of the sodium-potassium pump. As the pump continues to move Na^+ to the outside and K^+ to the inside, it offsets the diffusion of the ions that is simultaneously occurring. The result is an equilibrium between ion diffusion and ion pumping. The ion concentration difference that exists while that equilibrium is maintained produces a membrane potential called the **resting potential.**

An Action Potential

After studying this section, you should be able to:

1. Define an action potential and diagram the changes in membrane potential that constitute an action potential.
2. Explain the processes of depolarization and repolarization in an action potential.
3. Distinguish between voltage-dependent and voltage-independent ion channels in the neuronal membrane.
4. Compare the properties and roles of voltage-dependent Na^+ channels with those of voltage-dependent K^+ channels.

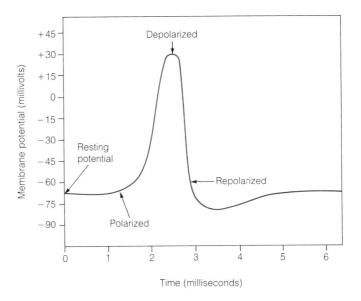

Figure 11-10 An action potential. An action potential is a temporary reversal of the membrane potential. This figure shows the potential at a point on the membrane as the action potential occurs.

The feature that makes a neuron unique is not the presence of a membrane potential, since membrane potentials are probably common to most, if not all, cells. The feature that does make it unique is its ability to respond to stimuli by changing the magnitude of the membrane potential. This characteristic is termed **excitability,** and the change in the membrane potential induced by a stimulus is called an **action potential.** Action potentials typically begin at one end of an axon and travel to the other end where they cause events to occur that stimulate the next cell in the sequence. The movement of an action potential along the neuron is called a **nerve impulse.**

Depolarization

Figure 11-10 shows the changes in membrane potential that comprise an action potential. In this case the membrane potential rises from a resting value of -70 mV to $+30$ mV, a span of 100 mV. After reaching this maximum, the membrane potential declines, diminishing more slowly than it rose, until it becomes even more negative than the original membrane potential. The potential then slowly returns to its original value. During the interval when the potential is changing, the membrane is said to be **depolarized.** Once the original potential is reestablished, the membrane is said to be **repolarized.**

The rapid rise of the membrane potential that begins an action potential is the result of a sudden change in the permeability of the membrane to Na^+. This increase is caused by the opening of large numbers of Na^+-specific **pores,** or **channels,** in the membrane that open in response to a stimulus. Na^+ flows through the pores, driven into the cell by the electrical and concentration gradients that exist across the membrane for that ion. Na^+ movement through the channels occurs for only a very brief time, because depolarization of the membrane causes the channels to close and block the passage of additional Na^+. In the time during which the channels are open, only a few hundredths of a percent of the excess Na^+ on the outside of the membrane is able to pass through, but this is sufficient to depolarize the membrane. Once closed, the channels cannot be reopened until the membrane is repolarized (Figure 11-11).

Repolarization

Repolarization of the membrane occurs as a result of an increase in the membrane permeability to K^+. The positive membrane potential that exists after Na^+ has passed into the cell causes K^+-specific channels to open. Now both an electrical gradient and a concentration gradient exist which

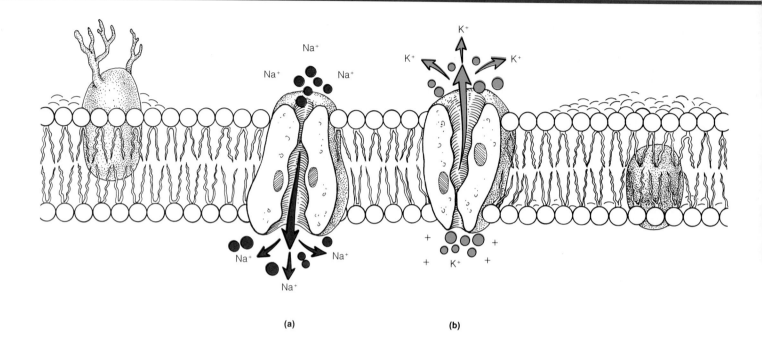

Figure 11-11 Voltage dependent ion channels. Separate channels exist for Na$^+$ and K$^+$. (a) Na$^+$ channels open in response to depolarization but close spontaneously, halting the rise in the membrane potential. (b) K$^+$ channels open in response to depolarization and remain open until the membrane resting potential is reestablished.

drive K$^+$ through the membrane. More than enough K$^+$ passes out of the cell to repolarize the membrane, and the membrane potential becomes even more negative than it had been at the beginning of the action potential. Return to a negative membrane potential causes the K$^+$ channels to close, and the resting membrane potential is reestablished.

The Na$^+$- and K$^+$-specific channels through which these ions pass during an action potential are different from the Na$^+$ and K$^+$ leak channels that account for the permeability of the membrane in a resting state. Leak channels do not open or close in response to membrane potential as do the ones involved in an action potential and are thus **voltage independent**. Channels through which Na$^+$ and K$^+$ pass during an action potential are induced to open in response to the membrane potential and are thus **voltage dependent**. Voltage-dependent Na$^+$ channels open in response to depolarization and close spontaneously. Their closing causes the rise in membrane potential to stop. Voltage-dependent K$^+$ channels, in contrast, open in response to a positive membrane potential and remain open until the membrane potential returns to a negative value.

Propagating an Action Potential

After studying this section, you should be able to:

1. Describe how an action potential that occurs at one point in a neuronal membrane induces an action potential in neighboring regions of the membrane.
2. Describe the principle of passive depolarization.
3. Explain why an action potential usually occurs first in the axon hillock.
4. Explain the concepts of threshold value and all-or-none response.
5. Explain why increased axon diameter and the presence of a myelin sheath increase the speed with which an action potential passes along an axon.
6. Describe the basis for the transmission of impulses in volleys in a peripheral nerve.

One way to induce an action potential is to stimulate it electrically; that is, to give it an "electric shock." If the shock depolarizes the membrane sufficiently, Na$^+$ channels will open and Na$^+$ will flow in. An action potential follows spontaneously. If one were to stimulate an axon by giving it an electric shock at a particular point on its surface, the action potential induced by the shock would spread rapidly from the point of stimulation over the surface of the axon.

The progressive movement of an action potential over the surface of an axon is the nerve impulse.

To understand the mechanism of this propagation, we must refer back to our discussion of the movement of Na^+ and K^+ through voltage-dependent channels that are specific for each ion. Initially, Na^+ movement into the axon occurs precisely at the point where the membrane is depolarized by the electric shock. As Na^+ ions pass through the channels, they are replaced by Na^+ ions from the neighboring region. As they emerge from the channel in the interior of the cell, they diffuse away from the channel opening. As a result, the membrane potential in the region surrounding the point of stimulation is reduced as well.

Reduction of the membrane potential in these neighboring regions has exactly the same effect as the electric shock had in the first place, and the result is the same. Neighboring Na^+ channels open and Na^+ begins to flow through these channels as well. As a result, there is a successive opening of Na^+ channels beginning at the point of stimulation and spreading progressively outward over the surface of the membrane. Once sufficient Na^+ has passed through to depolarize the membrane in each new region, K^+ channels open and K^+ begins to pass outward, continuing until the membrane is repolarized in each new region. The result is a traveling action potential that sweeps, wavelike, over the surface of the membrane (Figure 11-12).

Initiating an Impulse

Although an impulse can travel in both directions simultaneously when the axon is stimulated at some point along its length, impulses normally travel only in one direction in an axon. Neurons usually receive stimuli at the dendrites and cell body and induce an action potential at the end of the axon.

Stimulation of a dendrite or cell body opens Na^+ channels in the membrane of those parts of the neuron. However, an action potential is not usually induced directly. Instead, there is a localized migration of positively charged ions, mostly K^+, away from the point of entry of the Na^+, which depolarizes the surrounding membrane. Since an action potential does not occur, the depolarization is not self-propagating and may die out within a short distance as ions leak back through the membrane. If enough Na^+ enters, however, sufficient depolarization can develop to offset leakage and spread to the hillock. The spread of a membrane depolarization in the absence of an action potential is called **passive depolarization.**

Whether or not an action potential is induced in the neuron depends on two things: the magnitude of the depolarization that reaches the hillock, and the membrane potential in the hillock itself. If the membrane of the hillock is depolarized sufficiently to induce an action potential, an impulse will be triggered. If not, an impulse will not be triggered. Thus, a particular magnitude of depolarization must be reached in the hillock for the neuron to initiate an impulse. The value that must be reached is referred to as the **threshold value** of the neuron, and the hillock generally has a lower threshold than other portions of the axon. Unless its threshold is reached, the neuron will not fire. If it is reached, the neuron will fire, and the magnitude of the action potential that is propagated will be the same regardless of the degree of depolarization that induced it. This is referred to as an **all-or-none** effect. Recall from Chapter 9 that this also occurs in muscles (see Figure 9-10).

Sometimes neurons are stimulated directly by changes in the environment that surround their dendrites. For example, free dendrites located in hair follicles initiate nerve impulses in response to very slight movement of the hair, such as might be caused by a light touch or a gentle breeze blowing across the surface of the skin. Many sensory neurons have dendrites that respond to movement, heat, cold, light, heavy pressure, or muscle contraction. In most cases the dendrites are encased within connective tissue or are closely associated with specialized groups of muscle fibers. Such **sensory receptors** are described in Chapter 15.

Speed of an Impulse

Speed of an impulse along an axon depends primarily on two factors: axon diameter and the presence or absence of a myelin sheath. Generally, the greater the axon diameter, the more rapid the impulse, and in axons of similar diameter, impulses travel faster in myelinated neurons than they do in nonmyelinated ones.

The reason for the greater speed in axons of greater diameter is that they have a higher density of Na^+ channels than do smaller axons. For example, the density of Na^+ channels in certain small axons of fish may be only about 3 channels per square micrometer, and in larger axons of rabbits, about 25 channels per square micrometer. The effect of the greater density of channels is to allow Na^+ into the cell more rapidly and thus bring about a more rapid depolarization of the neighboring membrane.

In myelinated neurons the myelin sheath of the axon is interrupted periodically along its length by nodes of Ranvier. Na^+ channels are present at these nodes in very high density and are the only places along the axon where Na^+ is able to pass through the membrane in response to depolarization. When an axonal membrane is depolarized, the Na^+ enters the cytoplasm at the node and diffuses away from the nodal region. The high concentration of Na^+ that

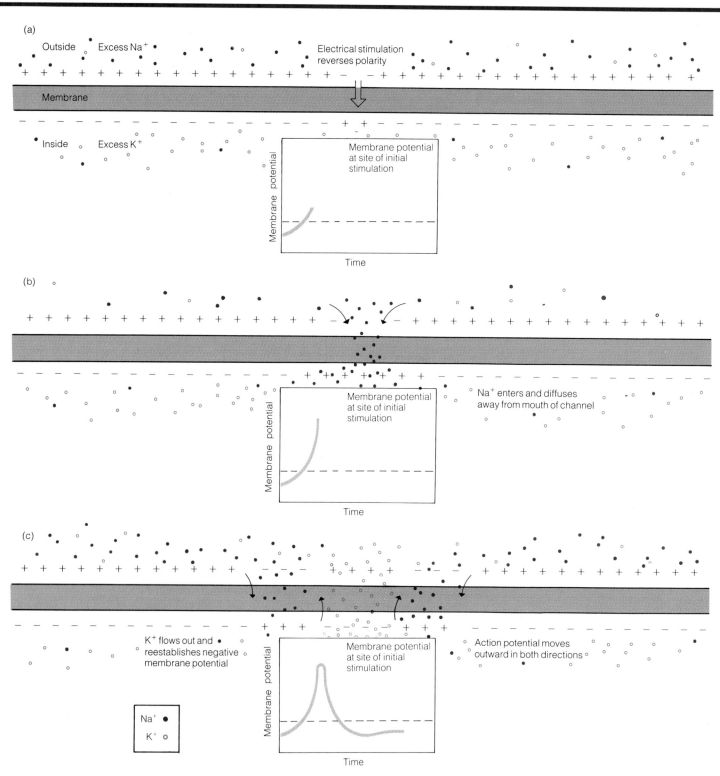

Figure 11-12 Propagation of an action potential. (a) Sodium channels are opened by depolarization of the membrane. When the membrane potential becomes more depolarized than the threshold (dashed line in the graph), the potential change becomes regenerative. (b) Sodium flows in and spreads outward, depolarizing the neighboring membrane. (c) Potassium flows out through channels opened by the depolarization, reestablishing polarity of the membrane. The process spreads outward from the point of initial stimulation and travels as a wave along the axon.

develops in the vicinity of the node depolarizes the neighboring membrane. Since the myelin sheath blocks the flow of ions in adjacent regions, action potentials cannot be induced. As the cytoplasmic concentration of Na^+ becomes higher, its depolarizing effect extends farther from the node until it reaches as far as the next node. When depolarization is sufficient to reach the threshold at that point, an action potential is induced. Consequently, the impulse travels along the axon by occurring at successive nodes in a series of rapid "jumps." This jumping or discontinuous pattern is called **saltatory** (săl′tă·tō″rē) **conduction** (Figure 11-13).

The insulating effect of the myelin sheath also provides an important advantage to the neuron because the axon has to expend less energy to achieve and maintain a resting potential. The active transport necessary to concentrate Na^+ outside the membrane occurs only at the nodes rather than

Figure 11-13 Saltatory conduction. (a) In a myelinated neuron, Na^+ channels are present in high density at the nodes of Ranvier. Large amounts of Na^+ flow through the membrane at a node. (b) Depolarization of the membrane at the next node initiates an action potential in it. (c) The process is repeated at each successive node.

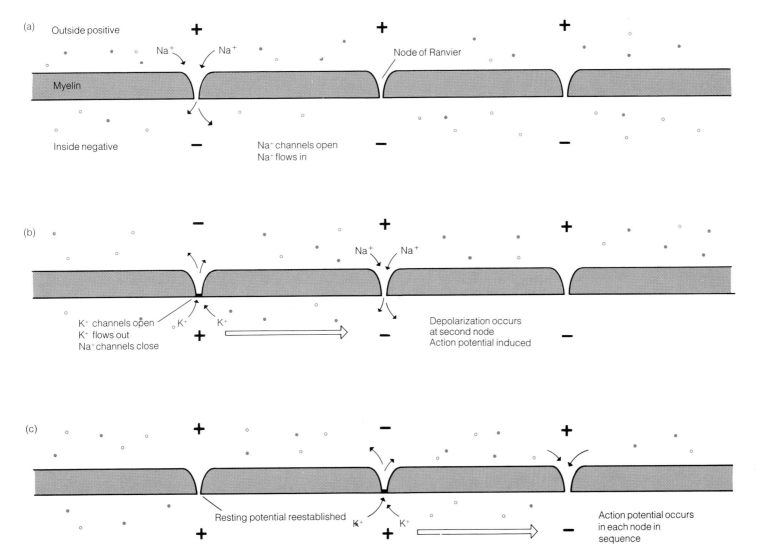

over the entire surface of the axon, as it does in nonmyelinated neurons. Thus, less of the cell's energy reserves need to be used. In addition, the sheath prevents diffusion of ions through the membrane, and less energy has to be expended to compensate for leakage.

Volleys

Nerves often contain neurons of different diameter, some of which are myelinated and some of which are not. Consequently, impulses that are initiated simultaneously in several neurons of a nerve travel along the nerve at different rates. This causes them to arrive at a point some distance from the point of stimulation at slightly different times. Such a cluster of impulses arriving at a point is called a **volley**, or a **compound action potential.** Figure 11-14 shows a typical volley consisting of three major groups or waves of action potentials that are designated A, B, and C. Group A or A wave is further subdivided into α, β, and γ action potentials.

The neurons responsible for these action potentials are correspondingly designated A-, B-, and C-type neurons, with the A-type neurons subdivided into Aα, Aβ, Aγ, and Aδ. A-type neurons are myelinated and relatively large in diameter. They are associated with nervous reactions that require the greatest speed: the sensing of heat, cold, touch, and pressure and the rapid response of skeletal muscle often required in reaction to such sensations. B- and C-type neurons are probably unmyelinated and carry impulses to smooth muscles and glands. C-type neurons, the slowest of the neurons, are relatively small in diameter. They serve both sensory and motor functions and carry impulses from touch, pressure, heat, and cold sensors in the skin to the CNS or impulses to smooth muscles and glands from the CNS. B- and C-type neurons are members of the autonomic part of the peripheral nervous system.

Transmission from Neuron to Neuron

After studying this section, you should be able to:

1. **Draw and identify the structures in a synapse.**
2. **Distinguish among axodendritic, axosomatic, and axoaxonic synapses.**
3. **Explain the difference between presynaptic and postsynaptic neurons.**
4. **Define a neurotransmitter and describe how neurotransmitters are used to transmit signals across a synapse.**
5. **Explain why transmission of a signal across a synapse is slower than transmission of an action potential along an axon.**

Nerve cells communicate with one another through close associations that exist between the finely branched axon terminals of one neuron and the membrane of the cell body or dendrites of another neuron. Points where neurons associate with other cells in this way are called **synapses** (sĭn·ăp'sēz"). The axon terminals are not in direct contact with the dendritic and cell body membrane in a synapse, but are separated by a very narrow space of only about 20 to 25 nm called a **synaptic cleft** (Figure 11-15).

A neuron with axon terminals that form a synapse with another neuron is called a **presynaptic neuron** (*pre*- meaning before the synapse), and the neuron on the opposite side of the synapse is called a **postsynaptic neuron** (*post*- meaning after the synapse). In a synapse the membrane of the presynaptic neuron is the **presynaptic membrane,** and the membrane of the postsynaptic neuron is the **postsynaptic membrane.** Synapses formed by the axon of a presynaptic neuron and the *dendrite* of a postsynaptic neuron are termed **axodendritic** (ăk"sō·děn·drĭt'·ĭk) synapses. If the synapse is formed with the *body* of a postsynaptic neuron, it is called an **axosomatic** (ăk"sō·sō·măt'ĭk) synapse, and if the synapse is formed with the *axon* of a postsynaptic neuron, it is called an **axoaxonic** (ăk"sō·ăk·sŏn'ĭk) synapse (Figure 11-16).

When a neuron forms a synapse with a muscle cell, the synapse is called a **neuromuscular** or **myoneural junction.** When the synapse is formed with a cell of a gland, it is called a **neuroglandular junction.**

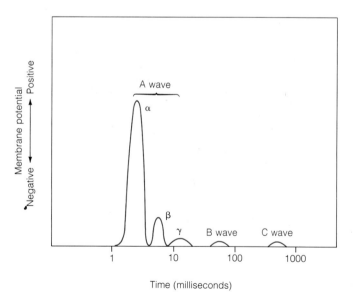

Figure 11-14 A volley. When a nerve consists of fibers that conduct impulses at different rates, impulses that begin at the same time arrive at a point further along the nerve at different times. Modified from Bell *et al.*, *Textbook of Physiology and Biochemistry*, 9th ed. (Churchill Livingstone, 1976).

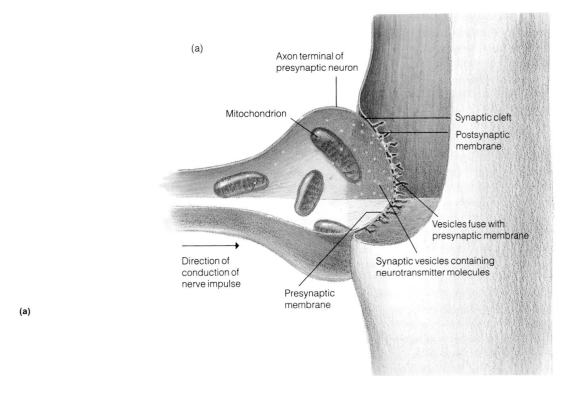
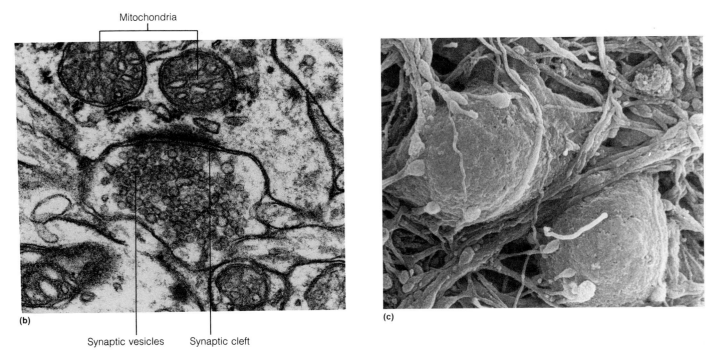

Figure 11-15 A synapse. (a) A synapse is the region where nerve cells communicate with other neurons or with muscle or gland cells. The narrow gap between the membranes of the two cells is called a synaptic cleft and is a region through which chemical transmitter substances carry an impulse from one cell to the other. (b) Electron micrograph of a synapse (mag. 51,528×). (c) Scanning electron micrograph of two neurons that receive axosomatic synapses (mag. 4522×).

bouton: (Fr.) *bouton*, button

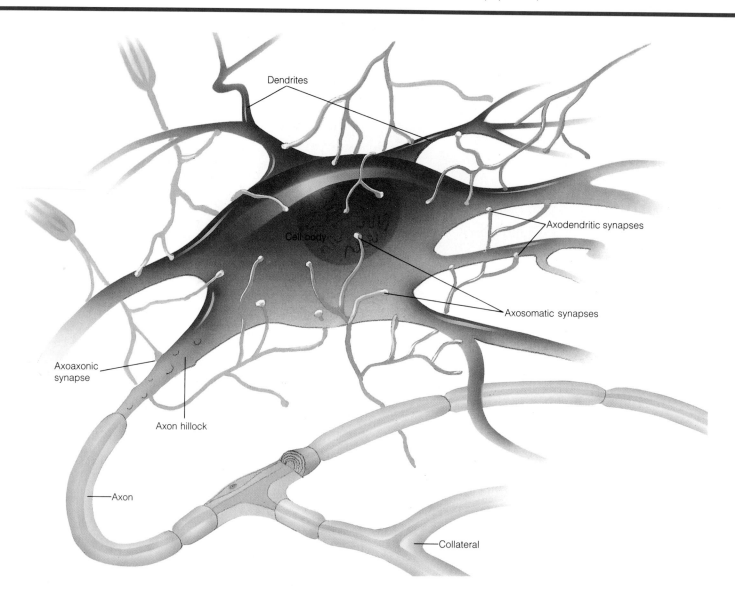

Figure 11-16 Synapse types. A neuron usually receives signals at many different synapses. The terminals of the presynaptic neurons form synapses with the dendrites, cell body, and axon of the postsynaptic neuron.

Communication Across a Synapse

Impulses are transmitted across a synaptic cleft by chemical substances called **neurotransmitters** (nū″rō·trăns′mĭt·ĕrs). Neurotransmitters are made in an axon and stored in vesicles in an enlarged region at the tip of the axon terminal called a **synaptic knob,** or **bouton** (bū·tŏn′) (Figure 11-17). Each vesicle contains a single type of neurotransmitter, but there may be vesicles that carry different types of neurotransmitters in the same axon terminus.

Each vesicle typically contains about 10,000 molecules of a particular neurotransmitter. When an action potential arrives at an axon terminus, a few hundred of these vesicles, which may number in the thousands in the synaptic knob, fuse with the presynaptic membrane and release their contents into the synaptic cleft. The neurotransmitter molecules then diffuse across the cleft and combine with specific **receptors** in the postsynaptic membrane. The combination of neurotransmitters with receptors causes a change in the permeability of the postsynaptic membrane to some of the ions on either side of it. As a result, ions flow through the membrane and cause a change in its potential that may result in an action potential in the postsynaptic neuron.

Transmission of an impulse across a synapse is quite slow compared with the rate of an action potential along an axon because of the time it takes to release the packets of neurotransmitters and for a sufficient number of molecules to diffuse across the synapse. As a result, passage of a signal through the synapse introduces a delay in the travel of the impulse. This delay is called a **synaptic delay.** In larger myelinated axons, where the speed of an impulse along the axon may be as high as 100 m/sec, the rate of travel across the synapse is only about 0.00005 m/sec—a difference by a factor of 2 million.

Release of Neurotransmitters

The best understood mechanism for the release of neurotransmitters is in the case of acetylcholine, a common neurotransmitter in motor nerves of the somatic nervous system. Acetylcholine is synthesized in the axon and stored in vesicles just inside the presynaptic membrane. Arrival of an action potential at the axon terminus causes Ca^{2+}-specific channels in the presynaptic membrane to open, and Ca^{2+} diffuses into the terminus from the synaptic cleft. Through mechanisms that are as yet unclear, increased Ca^{2+} in the cytoplasm causes the acetylcholine-containing vesicles to fuse with the presynaptic membrane and to release their contents into the synaptic cleft. The Ca^{2+} is quickly pumped back into the synaptic cleft where it can be used again in response to another action potential.

Synaptic fatigue occurs when a neuron is stimulated so rapidly that neurotransmitter is released faster than the presynaptic neuron can replenish it. In time there is too little transmitter left to stimulate the postsynaptic neuron, and transmission of impulses across the synapse stops.

Mode of Action of Neurotransmitters

There are two means by which neurotransmitters make it possible for ions to diffuse through a postsynaptic membrane. In some cases, as with acetylcholine, the receptor itself is potentially an ion channel. Combination of the neurotransmitter with the receptor changes the conformation of the receptor and opens the channel. In most cases, however, combination of a neurotransmitter with its receptor initiates a sequence of chemical reactions that end up opening ion channels that are distinct from the receptor. This kind of mechanism often results in a delay of several milliseconds in the induction of a response because of the additional steps that must be accomplished before the ion channels open.

Removal of Neurotransmitters

Once it has done its job, it is important that a neurotransmitter be removed rapidly from a synaptic cleft so that the synapse will be able to conduct another impulse. Depending on the neurotransmitter, removal is accomplished by enzymatic degradation, reabsorption into the presynaptic neuron, or diffusion from the synapse. Acetylcholine is removed by diffusion or is degraded to acetic acid and choline by an enzyme called cholinesterase. Norepinephrine, a

Figure 11-17 Release of neurotransmitter into a synaptic cleft. The neurotransmitter is carried to the presynaptic membrane in vesicles, which fuse with the membrane and release their contents into the synaptic cleft. The neurotransmitters diffuse across the cleft and combine with receptors in the postsynaptic membrane.

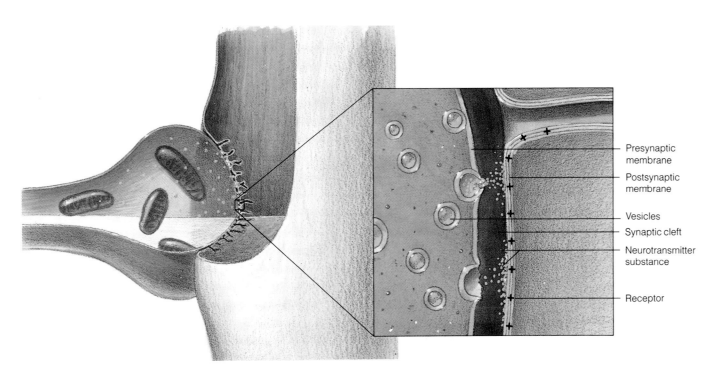

neurotransmitter of the autonomic nervous system (discussed later in this chapter and in Chapter 14), is either degraded in the synaptic cleft by an enzyme called catechol-O-methyl transferase or is taken back into the presynaptic neuron. When it is reabsorbed, some of the norepinephrine is repackaged in vesicles and used over, and some is degraded by an enzyme called monamine oxidase.

Electrical Synapses

Although the vast majority of synapses transmit impulses chemically, mention should be made of electrical synaptic transmission through gap junctions that lie between certain neurons in crayfish, leeches, frogs, turtles, and most likely, mammals. As we saw in Chapter 3, a gap junction is a region where the membranes of adjoining cells lie very close to one another, separated by a distance of only about 2 nm compared with the usual synaptic cleft of about 20 to 25 nm. Gap junctions in synapses probably provide a low resistance path by which an impulse can travel directly from one cell to the next. This is undoubtedly faster than chemical transmission, and indeed, such synapses have been found in motor neurons of goldfish where they are responsible for the especially rapid reflex action of the flip of the fish's tail. Electrical synapses, however, probably do not provide a point where the transmission of impulses can be controlled the way that chemical synapses can.

Postsynaptic Membrane Potentials

After studying this section, you should be able to:

1. Explain the difference between excitatory and inhibitory postsynaptic membrane potentials and describe the mechanisms for producing them.
2. Describe how the effects of many action potentials in presynaptic neurons can be combined to induce a single action potential in a postsynaptic neuron.
3. Explain how a neurotransmitter can inhibit the firing of a postsynaptic neuron.

Stimulating a neuron to fire requires that the membrane potential of the neuron be reduced to a point where an action potential will occur. Sometimes, however, a neurotransmitter has just the opposite effect and causes the membrane to become even more negative. When this happens, the membrane is said to be **hyperpolarized** (hī″pĕr·pō′lăr·īzd). The effect of hyperpolarization is to inhibit the neuron from firing. A postsynaptic membrane potential that *causes* an impulse is called an **excitatory postsynaptic potential (EPSP)**, and the hyperpolarized potential that *inhibits* an impulse is called an **inhibitory postsynaptic potential (IPSP)**.

Excitatory Postsynaptic Potential

An **excitatory neurotransmitter** causes an excitatory postsynaptic potential by causing the postsynaptic membrane to become more permeable to Na^+. Driven by the difference in concentration on the two sides of the membrane and by the potential difference across the membrane, Na^+ rushes from the synaptic cleft into the cytoplasm of the postsynaptic cell. Although the flow is brief, lasting only about 1 msec, enough Na^+ can rush in to raise the resting potential in the membrane to a new level. This is the excitatory postsynaptic potential (Figure 11-18a).

If the EPSP is only slightly greater than the resting potential, there may be no further effect, and the membrane potential will return to its original resting value after a few milliseconds. However, if the EPSP is raised high enough, an action potential will be triggered in the postsynaptic neuron.

Spatial Summation

The amount of excitatory neurotransmitter released into a synaptic cleft in response to a single action potential is usually not enough by itself to raise the EPSP to the cell's threshold value. Instead, the threshold value of the postsynaptic neuron is usually reached through the combined effects of many action potentials. One way this is accomplished is by release of neurotransmitters at several synapses at about the same time. If enough neurotransmitters are released, their combined effect can raise the membrane potential past the threshold and an impulse will be initiated. This is called **spatial summation.**

Temporal Summation

Another way to induce an action potential is by accumulating neurotransmitters in a relatively few synaptic clefts. When many impulses arrive in rapid succession, neurotransmitters are released faster than they can be removed and they build up, stepwise, in the synaptic cleft. Each increment of neurotransmitters causes additional Na^+ to enter the postsynaptic cell, and the membrane becomes more and more depolarized. If the threshold is reached, the neuron will fire. This mechanism is called **temporal summation.**

Inhibitory Postsynaptic Potential

Neurotransmitters that cause an inhibitory postsynaptic potential are called **inhibitory neurotransmitters.** Inhibitory neurotransmitters act either by opening K^+- or Cl^--specific

---

hyperpolarize: (Gr.) *hyper*, above + (L.) *polus*, pole

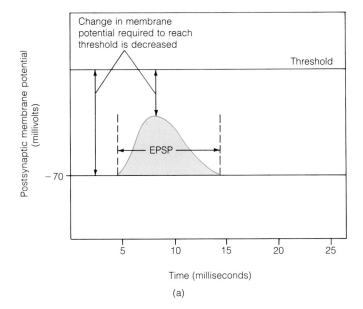

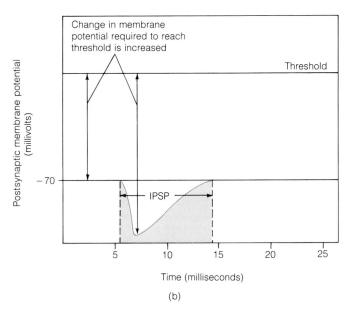

Figure 11-18 (a) Excitatory postsynaptic membrane potential (EPSP). (b) Inhibitory postsynaptic membrane potential (IPSP).

channels, depending on the neurotransmitter, without simultaneously opening Na^+-specific channels. Even though they must diffuse against an electrical gradient, the concentration gradient is great enough for each of these ions that some are able to diffuse through. In the case of K^+, diffusion is from the inside to the outside, and in the case of Cl^-, diffusion is from the outside to the inside. In both cases the membrane potential becomes hyperpolarized and an inhibitory postsynaptic membrane potential results (Figure 11-18b). To overcome the hyperpolarization of the membrane and trigger an action potential, more Na^+ must diffuse through the postsynaptic membrane than would be required if the membrane were not hyperpolarized. This requires that more excitatory neurotransmitters be released by the presynaptic neuron so that enough Na^+-specific channels are opened to supply the additional amount of Na^+ needed to reach the cell's threshold.

In **presynaptic inhibition,** impulses from neurons whose endings lie close to the ends of excitatory neurons reduce the amount of neurotransmitter released into the cleft. As a result, the impulses that reach the end of the presynaptic neuron are not as effective in inducing an action potential in the postsynaptic membrane.

Integrating the Stimuli

After studying this section, you should be able to:

1. Distinguish between divergence and convergence as means of distributing or focusing the distribution of nerve impulses within the nervous system.
2. Explain why a negative membrane potential has to be reestablished in an axon hillock after an action potential has occurred before another action potential can be initiated.
3. Explain how maintaining an excitatory potential above the neuron's threshold value causes the neuron to initiate a series of action potentials.

The number of synapses in the nervous system is exceedingly large, and it has been estimated that there are on the order of 10^{15} synapses in the brain alone. Proper functioning of the nervous system requires a carefully integrated pattern of impulse transmission of a complexity that is difficult to comprehend. It is clear, however, that transmission of signals throughout the nervous system involves far more than the simple propagation of an action potential along a neuron to a synapse and then the chemical transmission of the signal to the next cell in a line of cells.

Divergence and Convergence

The axon terminals of most neurons form synapses with many different postsynaptic neurons. This structural organization, referred to as **divergence,** allows a signal passing through a neuron to be delivered to many other neurons at the same time. Alternatively, the axon terminals of many different neurons may form synapses with a single postsynaptic neuron, an organizational pattern referred to as **convergence** (Figure 11-19). Indeed, a neuron may receive signals through literally thousands of synapses formed with many hundreds of different presynaptic cells.

Transmitters released into many of these synapses are excitatory and cause a decrease in the potential of the postsynaptic membrane. In other synapses, however, inhibitory transmitters are released, perhaps from cells that originate in different parts of the nervous system than those delivering the excitatory signals. Any given cell may have a variety of types of receptors at one end and may release different types of transmitters into different synapses at the other end. The effects of signals arriving from several different sources at the same time may be summed and will either increase the magnitude of the membrane potential and inhibit the cell from firing, or decrease it and make it easier to initiate an impulse. Likewise, if signals arrive in rapid succession, their effects may be combined to excite or inhibit the postsynaptic cell. All these factors combine to produce some net potential in the dendrites and body of the postsynaptic cell. Only if that effect is to lower the membrane potential enough to reach the cell's threshold will the cell fire.

Encoding

If a cell does fire, the voltage-dependent Na^+ channels in the hillock can only become operable after a negative potential is reestablished, and a period of time must pass for this to occur. However, if the neuron continues to receive sufficient excitatory signals to keep the hillock in a depolarized state, the Na^+ channels will remain inoperable and the neuron will be incapable of transmitting another impulse unless there is a mechanism to compensate for the continuous stimulation. This is provided by several K^+ and Ca^{2+} channels of a type unique to the hillock. Ion movement through these channels restores a negative membrane potential even in the presence of a sustained net excitatory membrane potential and thereby enables the Na^+ channels to return to an operable state. Once they are operable, another action potential can be initiated.

The result is a series of action potentials of a specific magnitude (because of the all-or-none effect) that continue as long as the net excitatory potential is maintained above the threshold. Moreover, the higher the excitatory potential, the more frequent the action potentials occur (Figure 11-20). Consequently, the neuron responds to changes in the magnitude and frequency of the stimuli it receives by changing

Figure 11-19 (a) Convergence. Axon termini of several neurons form synapses with a single neuron. (b) Divergence. The axon termini of a single neuron form synapses with several different neurons.

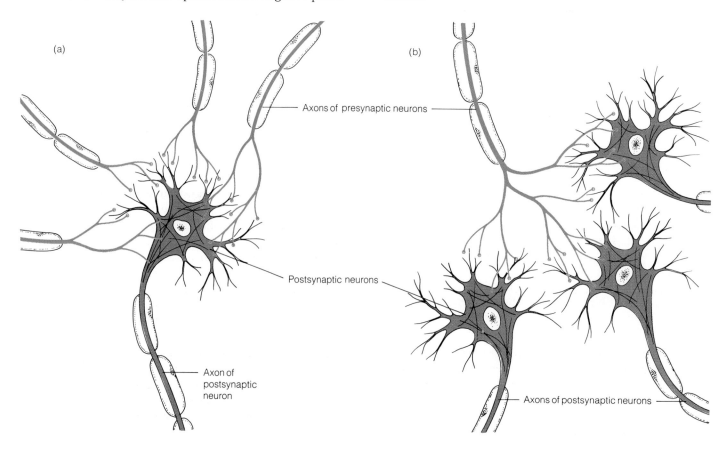

endorphin: (Gr.) *endo*, within + (m)*orphine*
enkephalin: (Gr.) *enkephalos*, brain
epinephrine: (Gr.) *epi*, upon + *nephros*, kidney + *ine*, chemical suffix

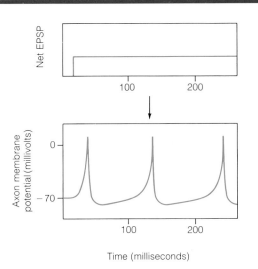
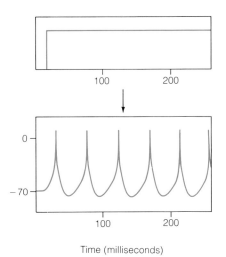
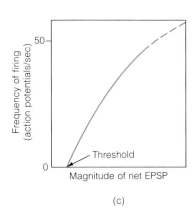

(a) 1 action potential per 100 milliseconds

(b) 2.4 action potentials per 100 milliseconds

Figure 11-20 Encoding an excitatory membrane potential.
(a) Dendritic and cell body membranes may be held in a sustained depolarized state if the neuron continues to receive signals. When this happens, the hillock fires repeatedly, producing a sequence of action potentials of equal magnitude. (b) The frequency, but not the magnitude, of the action potentials depend on the magnitude of the depolarization. (c) The frequency of action potentials increases as a function of the EPSP magnitude.

the frequency with which it initiates its own impulses, halting those impulses only when the net excitatory potential drops below the cell's threshold. This phenomenon, called **encoding**, allows the cell to respond to the input of varying amounts of stimulation in a graded, proportionate manner.

Neurotransmitters

After studying this section, you should be able to:

1. List the characteristics of a neurotransmitter.
2. List several neurotransmitters of the CNS and PNS.
3. Distinguish between adrenergic and cholinergic synapses.
4. Explain how alpha and beta cholinergic receptors function.

For a compound to be identified as a neurotransmitter, it must be shown to have the following characteristics: it must be present in presynaptic vesicles, it must be released from those vesicles in response to a nerve impulse, when added to a synapse experimentally it must induce the same response as does stimulation of the presynaptic neuron, and

CLINICAL NOTE

ENDORPHINS AND ENKEPHALINS

Opiates are substances extracted from the opium plant and include such refined substances as morphine and heroin. In 1974 an internal opiate called an **opioid** was found in the brain of mammals. Since there are chemical receptors for opiates in the brain, it really was not surprising that there existed an internal chemical to occupy those sites. **Endorphins** (ĕn·dor′fĭns) are these internal pain-blocking substances and neurotransmitters that appear to aid in the body's management of pain and emotions. **Enkephalins** (ĕn·kĕf′ă·lĭns) are two endorphin-like compounds that are thought to be responsible for the "high" that many runners and other athletes experience during exercise.

it must be removed rapidly from the synapse. Because neurotransmitters are present in such small quantities, it is generally very difficult to demonstrate that a particular compound meets all these criteria. As a result, it has been difficult to identify with certainty many compounds that are suspected to be neurotransmitters. At present, three categories of compounds have been shown to meet these criteria in the CNS, two of which are also active in the PNS. In the CNS, the compounds are acetylcholine (already discussed), **catecholamines** (kăt″ĕ·kōl′ă·mēns), and certain amino acids. Catecholamines include the compounds **epinephrine** (ep″ĭ·nĕf′rĭn), **norepinephrine** (nōr·ep″ĭ·nef′rĭn), and **dopamine** (dō′pă·mēn). In addition to these compounds, several peptides ranging in length from as short as three amino acids to as long as about thirty amino acids, are suspected to be neurotransmitters of the CNS (Table 11-2).

Table 11-2 REPRESENTATIVE NEUROTRANSMITTERS AND PRESUMED NEUROTRANSMITTERS

Amines, Amino Acids, and Amino Acid Derivatives

Acetylcholine

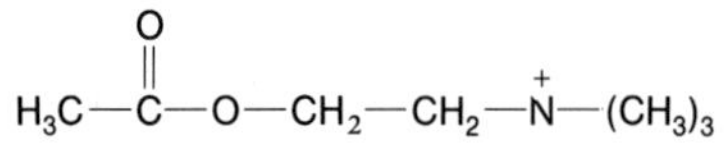

$$H_3C-\overset{\overset{O}{\|}}{C}-O-CH_2-CH_2-\overset{+}{N}-(CH_3)_3$$

Dopamine

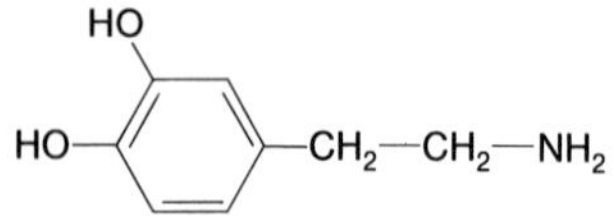

Norepinephrine

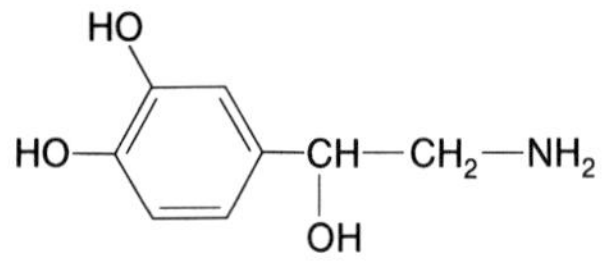

Epinephrine

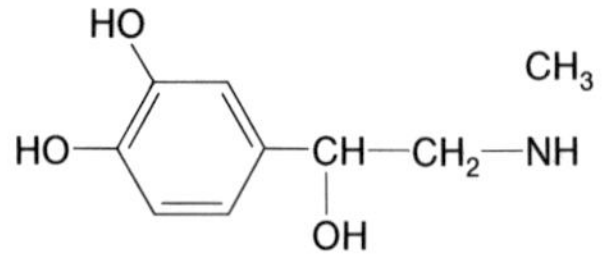

Octopamine

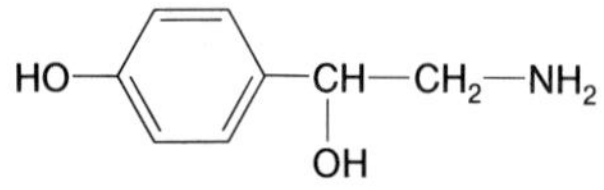

Serotonin

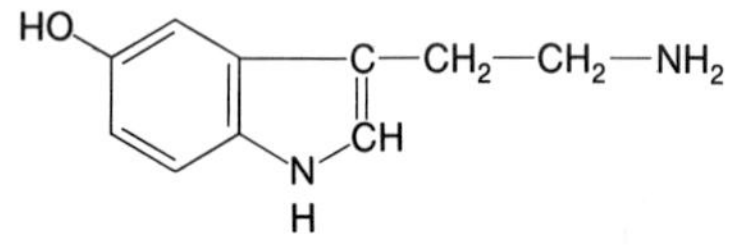

Glycine

$$H_2N-CH_2-COOH$$

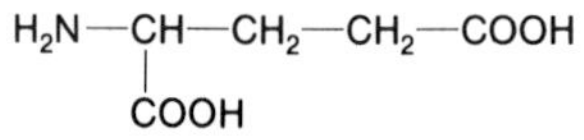

Selected Peptides

Enkephalin

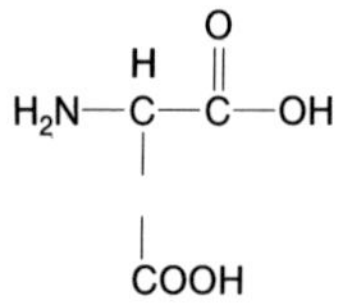

(Tyr)(Gly)(Gly)(Phe)(Met)

Substance P

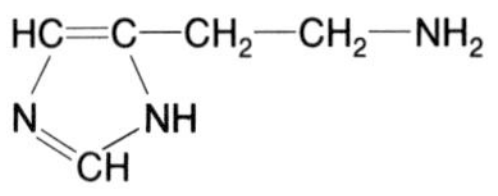

(Arg)(Pro)(Lys)(Pro)(Gln)(Gln)(Phe)(Phe)(Gly)(Leu)(Met)

Aspartic Acid

Beta-alanine

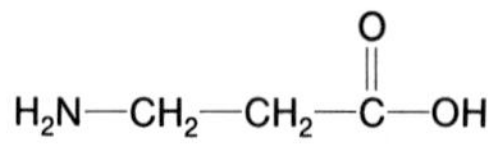

$$H_2N-CH_2-CH_2-\overset{\overset{O}{\|}}{C}-OH$$

Histamine

Taurine

$$H_2N-CH_2-CH_2-\overset{\overset{O}{\|}}{\underset{\underset{O}{\|}}{S}}-OH$$

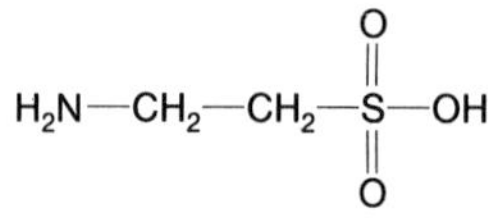

Gamma-aminobutyric acid (GABA)

$$H_2N-CH_2-CH_2-CH_2-COOH$$

Glutamic acid

$$H_2N-\underset{\underset{COOH}{|}}{CH}-CH_2-CH_2-COOH$$

Thyrotrophin-releasing hormone

(Glu)(His)(Pro)

Neurotensin

(Glu)(Leu)(Tyr)(Glu)(Asn)(Lys)(Pro)(Arg)(Arg)(Pro)(Tyr)(Ile)(Leu)

chorea: (Gr.) *choreia*, dance

Table 11-2 REPRESENTATIVE NEUROTRANSMITTERS AND PRESUMED NEUROTRANSMITTERS (*continued*)

Angiotensin I

Angiotensin II

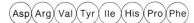

Vasoactive intestinal peptide

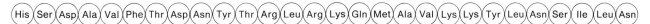

Somatostatin

Beta-endorphin

Note: Amino acids are abbreviated according to a three-letter convention that, with a few exceptions, corresponds to the first three letters in the amino acid name. Sequences begin with the amino acid at the—COOH terminus.

In addition, at least four amino acids themselves have been identified as neurotransmitters in the CNS. Glutamic acid and aspartic acid are excitatory transmitters in many parts of the brain and may be the most common excitatory transmitters of the CNS. Glycine and gamma-aminobutyric acid (GABA) are inhibitory transmitters of the CNS. Glycine, the simplest of all amino acids, acts as an inhibitory transmitter in certain parts of the spinal cord, and GABA is found in both the brain and the spinal cord. Although GABA is an amino acid, it is not used in the synthesis of proteins, and it appears to function only as an inhibitory neurotransmitter. Synthesized from glutamic acid, GABA is the most common of all neurotransmitters in the CNS.

Acetylcholine and catecholamines have also been identified as PNS neurotransmitters. Nerves in which acetylcholine is the neurotransmitter are referred to as **cholinergic** (kō″lĭn·ĕr′jĭk), and nerves that use norepinephrine or epinephrine are referred to as **adrenergic** (ăd·rĕn·ĕr′jĭk). This name comes from the pioneering research in this field involving epinephrine, which was referred to by its British name **adrenaline.**

Catecholamines can act either to excite or inhibit postsynaptic membranes, depending on the kind of receptors in the postsynaptic membrane. Two classes of catecholamine

---

CLINICAL NOTE

HUNTINGTON'S DISEASE

Huntington's disease (also called **Huntington's chorea**) is an inherited neurological disorder manifested by bizarre, ceaseless, rapid, jerking body movements. The person also experiences progressive deterioration of mental function and death occurring within 10 to 15 years from onset. Each child of a person with this disorder has a 50% chance of inheriting the disease. The disease manifestations do not occur until the age of 30 or later, usually after the afflicted person has had children. A test has recently been developed that allows a person to find out if he or she is likely to have the disease before its symptoms appear. Ironically, the decision of whether or not to be tested is a very difficult one, and some potential victims are opting not to be tested because they would rather not know that they have the disease.

A person with Huntington's disease is deficient in the neurotransmitters **gamma-aminobutyric acid (GABA)** and acetylcholine. The chorea occurs in Huntington's disease probably because of an excess of the neurotransmitter dopamine relative to the levels of GABA and acetylcholine. If the balance among these neurotransmitters can be restored, the bizarre movements abate, but there is no treatment for the loss of mental capacity.

> **CLINICAL NOTE**
>
> **BETA-BLOCKERS**
>
> **Beta-blockers** are drugs that block the effect of the neurotransmitter epinephrine at beta-adrenergic receptors located on the cells of effector organs. The outcome of such a receptor blockade is to decrease the work of the heart; thus, these drugs are useful after a heart attack. Beta-blockers are also used for the control of hypertension and migraine headaches.

receptors are known, called **alpha**- and **beta**-receptors. Combination of a catecholamine with an alpha-receptor causes Na^+-specific channels to open in the postsynaptic membrane and, if enough Na^+ passes through the membrane, induces an action potential. When a catecholamine combines with a beta-receptor, however, K^+ channels are opened, the membrane becomes hyperpolarized, and an action potential is inhibited. Norepinephrine only combines with alpha-receptors and is therefore only stimulatory in its action. Epinephrine, in contrast, combines with both types of receptors, so it serves as an excitatory neurotransmitter in synapses where alpha-receptors are present and as an inhibitory neurotransmitter in synapses where beta-receptors are present. (For more discussion of neurotransmitters, see Chapter 14.)

In this chapter we studied the histology of the nervous system, concentrating on the neuron. You learned that a difference in electrical potential is maintained between the inside and outside of a neuron which can be manipulated to produce an impulse that travels from one end of the cell to the other. We also studied the synapse and the mechanisms responsible for transmitting an impulse across this junction from one neuron to another. In the next chapter we will examine how neurons are arranged in the brain and spinal cord, an integral functional unit called the central nervous system.

STUDY OUTLINE

Organization of the Nervous System (pp. 261–262)
The nervous and endocrine systems are responsible for communication within the body. The nervous system is cell-based, the endocrine system uses circulating chemicals.

The CNS and the PNS The *central nervous system* consists of the brain and spinal cord. The *peripheral nervous system* consists of nervous tissue outside the brain and spinal cord. The *afferent* division of the PNS carries signals to the CNS; the *efferent* division carries signals from the CNS. The *somatic* portion of the efferent division carries signals to skeletal muscles; the *autonomic* portion carries signals to glands, smooth muscle, and the heart.

Organization of a Nerve Nerves are *sensory, motor*, or *mixed*, depending on the direction in which signals travel in them. A nerve is surrounded by an *epineurium*, which is continuous with the *perineurium*. The perineurium divides the nerve into *fascicles*. Individual nerve cells are surrounded by *endoneurium*.

Cells of the Nervous System (pp. 262–265)
Cells of the nervous system include *neurons* and *neuroglia*.

Neurons The four regions of most neurons are the *cell body (soma), dendrites, axon,* and *axon terminals*. Extensions from the cell body are collectively called *processes* or *fibers*. The cell body contains the cell *nucleus* and *perikaryon*. *Nissl bodies* are clumps of ribosomes in the perikaryon.

Dendrites and axons contain numerous microtubules and *neurofilaments*. Axons usually begin as an enlarged *hillock*, may possess *collateral branches,* and end in very fine branches called *axon terminals*.

Unipolar neurons have a single process emanating from the cell body that branches into a single axon and dendrite. *Bipolar* neurons have a single dendrite and axon emanating from the cell body. *Multipolar* neurons have several dendrites and a single axon emanating from the cell body.

Neuroglia *Astrocytes* are star shaped and hold CNS neurons in place; they regulate fluid composition surrounding neurons. *Ependyma* form the epithelial lining of CNS chambers. *Microglia* play a defensive and protective role in the CNS. *Oligodendrocytes* form a scaffolding for CNS neurons and provide a *myelin sheath* about axons.

PNS neuroglia include *satellite* cells, which are in ganglia, and *Schwann cells*, which form myelin sheaths around axons of *myelinated* PNS neurons. Gaps in the covering are *nodes of Ranvier*.

Membrane Potential (pp. 266–269)
A membrane potential exists across the membrane of all living cells.

Electric Potentials *Electric potentials* exist whenever there is a tendency for charged particles to move from one region to another. In cells, those charged particles are ions, and the membrane is a barrier to free flow of the ions. In cells, the electric potential across the membrane is measured in *millivolts*. Membrane potentials are measured on the basis of net charge within the cell and are negative. The charge difference that creates the potential is established by a sodium-potassium-ATPase pump. Differences in permeability of the membrane allow some ions to diffuse across the membrane more readily than others.

Resting Potentials The sodium-potassium pump moves Na^+ to the outside and K^+ to the inside. K^+ leaks back. This creates a charge difference that attracts Cl^- outside. While this is going on, the sodium-potassium pump continues to work. An equilibrium is reached at which there is a net excess of positively charged ions outside and negatively charged ions inside. This creates the *resting potential*.

An Action Potential (pp. 269–270)

An *action potential* is a transient reversal of the membrane potential.

Depolarization *Depolarization* of the membrane occurs as a result of the inflow of Na^+ following an opening of Na^+-specific channels. Sufficient Na^+ flows in to reverse the polarity of the membrane.

Repolarization Reversal of the membrane potential closes the Na^+-specific channels and opens K^+-specific channels. Sufficient K^+ flows out to reestablish a negative membrane potential. The membrane is then *repolarized*. The channels through which Na^+ and K^+ pass during an action potential are *voltage dependent*.

Propagating an Action Potential (pp. 270–274)

A nerve impulse consists of an action potential that is propagated along the axon. The occurrence of an action potential at one point on the axon initiates an action potential in neighboring regions, which initiates an action potential in the next neighboring region, and so forth, and the action potential travels along the axon.

Initiating an Impulse Action potentials are usually triggered in the axon hillock as a result of a *passive depolarization* that spreads to the hillock following stimulation of the dendrites or cell body by impulses from other neurons. If the *threshold value* of the hillock is reached, an action potential will occur in the hillock and travel along the axon. Sometimes, as in *sensory receptors*, dendrites are stimulated directly by environmental conditions.

Speed of an Impulse The speed of an impulse depends on axon diameter and the presence or absence of myelination. Larger diameter axons have higher densities of voltage-dependent Na^+-specific channels, allowing more rapid depolarization.

The sheath in myelinated neurons is interrupted at nodes of Ranvier. Action potentials jump from node to node in *saltatory conduction*, which is more rapid than progression along a nonmyelinated neuron. The myelin sheath also insulates the axon, minimizing ion leakage.

Volleys Nerves contain neurons that transmit action potentials at different speeds. If several neurons are stimulated simultaneously, action potentials travel at different rates in the nerve, arriving at a point in a *volley*. Neurons are classified as A-, B-, and C-types depending on how rapidly impulses travel along them. A-type neurons are fastest; C-type are slowest.

Transmission from Neuron to Neuron (pp. 274–278)

Nerve cells communicate with other cells through *synapses*. A *presynaptic* neuron delivers an impulse to a synapse; a *postsynaptic* neuron receives an impulse at a synapse. *Axodendritic, axosomatic,* and *axoaxonic* synapses are formed between axon terminals and dendrites, soma, and axons, respectively. *Neuromuscular* and *neuroglandular* junctions form between neurons and muscles and glands, respectively.

Communication Across a Synapse Neurons communicate across synapses with *neurotransmitters*. Neurotransmitters combine with *receptors* in the postsynaptic membrane. Transmission across a synapse is slower than transmission along an axon and results in *synaptic delay*.

Release of Neurotransmitters Neurotransmitters are released from vesicles in response to an action potential. *Synaptic fatigue* results when a transmitter is used faster than it can be replenished.

Mode of Action of Neurotransmitters Combination of a neurotransmitter with a receptor either opens an ion channel directly or results in chemical reactions that open ion channels.

Removal of Neurotransmitters Neurotransmitters are removed by degradation, reabsorption, or diffusion.

Electrical Synapses Electrical synapses, while rare, provide points for direct transmission of an action potential from one cell to another.

Postsynaptic Membrane Potentials (pp. 278–279)

Excitatory Postsynaptic Potential *Excitatory neurotransmitters* cause postsynaptic membranes to become more permeable to Na^+. This changes the membrane potential to a new level that increases the cell's potential for firing. This can be accomplished through the combined effects of neurotransmitters operating at several synapses simultaneously, *spatial summation*, or the accumulative effect of action potentials arriving over a period of time, *temporal summation*.

Inhibitory Postsynaptic Potential *Inhibitory neurotransmitters* open K^+- or Cl^--specific channels. Diffusion of either of these ions hyperpolarizes the membrane and makes it more difficult to reach a threshold. Neurons can also be inhibited by *presynaptic inhibition*, in which impulses from nearby neurons reduce the amount of neurotransmitter released into the cleft.

Integrating the Stimuli (pp. 279–281)

Divergence and Convergence *Divergence* results when the axon branches from one or a few neurons form synapses with many more postsynaptic neurons. *Convergence* results when axon branches from many different presynaptic neurons form synapses with relatively few postsynaptic neurons.

Encoding If the membrane of a cell body is kept in a depolarized state, a sequence of action potentials will be initiated in the axon hillock. The frequency of firing depends on the degree to which the cell body is depolarized.

Neurotransmitters (pp. 281–284)

Neurotransmitters of the CNS include *acetylcholine, catecholamines,* and at least four amino acids—glutamic acid, aspartic acid, glycine, and GABA. Glycine and GABA are inhibitory neurotransmitters. Catecholamines are excitatory or inhibitory depending on the nature of the receptors in the postsynaptic membrane.

anencephaly: (Gr.) *an.* not + *enkephalos,* brain
spina bifida: (L.) *spina,* spine + *bis,* twice + *findere,* to cleave
encephalitis: (Gr.) *enkephalos,* brain + *itis,* inflammation

THE CLINIC

DISEASES OF THE NERVOUS SYSTEM

The nervous system is so closely related to all of the organs and functions of the body that almost every disease produces manifestations of neurologic disease. There are also myriads of diseases that originate primarily in the nervous system and produce dysfunction in other organs and systems. In many instances it can be difficult to ascribe the disease state primarily to a disorder of the nervous system alone. Here we discuss some examples of diseases whose origins are primarily in the nervous system.

Note that the broad coverage of this clinic makes it applicable to all of the chapters on the nervous system (Chapters 11 to 16), thus, it is put here in the first of these chapters as a preview and a framework of reference for the chapters to come.

Hereditary and Developmental Conditions

There are many bizarre, and fortunately rare, hereditary conditions that affect the nervous system. **Huntington's disease,** for example, is an autosomal dominant trait that manifests itself in adult life with ceaseless jerking movements and mental deterioration (see Clinical Note: Huntington's Disease). There are many hereditary diseases associated with mental retardation as well as other defects, such as **Down's syndrome** (see Chapter 28), **Tay-Sachs disease,** and a few **trisomies** (genetic conditions that are due to an extra chromosome).

Developmental abnormalities also produce neurologic complications. The most common are **neural tube defects (NTDs),** which result from various degrees of failure of the neural tube to close during early fetal development. There is a genetic predisposition toward neural tube defects, and there are several types. **Anencephaly** (ăn·ĕn·sĕf′ă·lē) is an absence of the cerebral hemispheres with or without covering of bone or skin; it is a lethal condition. Another type is **hydrocephalus** (hī·dro·sĕf′ă·lŭs), which is a swelling of the ventricals of the brain due to local obstruction (see Chapter 12 Clinical Note: Hydrocephalus). Premature closure of the cranial sutures can produce a small head and thus compromise brain development. It has been treated successfully by surgically fragmenting the skull and allowing the brain to develop normally.

Neutral tube defects may also occur at the tail end of the neural tube and produce **spina bifida** (spīn′ă bī′fĭ·dă). Spina bifida is a failure in the development of the spinal cord's bony covering that results in the defective closure of the vertebral column. In **spina bifida occulta,** the bony spinal canal is incomplete but the meninges covering the cord do not protrude. The afflicted person may have a dimple on the back or a patch of hair on the skin covering the bony defect. Approximately 5% of the population is affected by this condition. Treatment is not usually required. However, in **spina bifida cystica,** a protrusion of matter through the bony defect results in a cystic swelling or sac. This sac may contain meninges, spinal cord, or both. Therapy depends on the amount of neurological damage sustained and the extent of the defect. Infection of the spinal cord is a great risk. Treatment is usually surgical.

Today, these defects can be tested during early pregnancy so a decision can be made whether or not to continue the pregnancy.

Infectious Diseases

Infections of the nervous system are often tragic illnesses. Inflammation of the meninges of the brain and spinal cord is called **meningitis.** It can be caused by anything that is irritating to the meninges and gains access to them, such as bacteria, viruses, lead poisoning, or malignant cells. Bacterial infection of the brain proper is known as **cerebritis** (sĕr″ĕ·brī′tĭs). Acute bacterial meningitis can be caused by a number of different bacteria (such as *Neisseria meningitidis* and *Haemophilus influenzae*) that invade the nervous system from an external source (for example, from throat, ear, sinus, or skin infections). Symptoms begin with respiratory infection, fever, headache, stiff neck, and vomiting. Suspicion of meningitis should be followed by an immediate lumbar puncture test. An increase in white blood cells (called **pleocytosis**), a decrease in glucose concentration, and identification of the organism will confirm the diagnosis. Treatment includes high doses of appropriate antibiotics.

Inflammation of the central nervous system is known as **encephalitis** (ĕn·sĕf·ă·lī′tĭs) if only the brain is involved; it is an **encephalomyelitis** (ĕn·sĕf′ă·lō·mī·ĕl·ī′tĭs) if the spinal cord is also involved. These two diseases are simply called **aseptic meningitis** when there is no evidence of bacterial infection in the spinal fluid. Encephalitis and encephalomyelitis are usually caused directly by viral infection or are secondary to a viral disease, such as influenza or measles.

On rare occasions, isolated **brain abscesses** can develop. These usually develop from an infection in the skull, sinuses, or mastoid air cells. Abscesses usually have symptoms similar to those of brain tumors, such as headache, vomiting, localized paralysis, and convulsions. Treatment includes surgical drainage and antibiotics.

Some organisms cause unusual infections of the nervous system. The *herpes simplex* virus frequently invades local nerve ganglia where it may remain dormant for long periods of time and later become activated to produce recurrent **canker sores** on the lips (Herpes I) or genitalia (Herpes II). The **rabies** virus and **tetanus** bacteria invade local nerves and travel along them to the central nervous system. Such organisms as the **botulism** and **diphtheria** bacilli produce neurotoxins that affect the nervous system causing respiratory paralysis and death.

hematoma: (Gr.) *haimatos*, blood + *oma*, tumor
polyneuropathy: (Gr.) *polys*, many + *neuron*, nerve + *pathos*, disease

Trauma of the Head
Injury to the head causes more deaths and disability in people under 50 years of age than any other neurologic condition. Head injury may be caused by penetrating wounds or, more frequently, by rapidly accelerating or decelerating forces. These injuries may lead to a variety of other injuries to the skull, meninges, brain, or cranial nerves. **Concussion** is perhaps the most common injury from a blow to the head and results in a temporary disruption of the senses and often loss of consciousness without structural damage to the brain. Temporary **retrograde amnesia** (loss of memory), headache, blurring of vision, and unstable gait are symptoms of concussion. Recovery occurs spontaneously but may require several days or weeks. More serious injuries to the head include cerebral contusions, lacerations, and edema. **Cerebral edema** is a swelling of the brain tissue with fluids. It may cause severe damage by compressing the brain structures inside the skull or even forcing the brain stem into the narrow foramen magnum located at the base of the skull (see Chapter 12). Fractures of the skull itself require no special treatment unless they are depressed into the brain tissue; they then need surgical elevation.

Injury to the coverings of the brain may result in the tearing of blood vessels and accumulation of blood, which can compress the brain tissues. **Epidural hematoma** can occur with injuries to the temporal or occipital areas of the head and cause increasing headache, gradual loss of consciousness, and neurologic deterioration 24 to 96 hr after the initial injury. Surgical drainage of the growing blood clot relieves the damaging pressure. **Chronic subdural hematoma** is a late complication of head injury and may not become symptomatic for several weeks. It results in late neurologic deterioration manifested by increasing daily headaches, fluctuating drowsiness, and mental changes. Treatment is by surgical drainage. (Trauma of the spinal cord is discussed in Chapter 12 Clinical Note: Spinal Cord Injury.)

Immunologic Diseases
There are several diseases of the nervous system that are thought to be due to immunologic problems. Multiple sclerosis (see Clinical Note in this chapter) is thought by some to be an immune reaction to a viral infection, perhaps the rubella (measles) virus. Myasthenia gravis (see Chapter 10, The Clinic: Muscular Diseases) is a neuromuscular disease in which the neurotransmitter acetylcholine is made unavailable due to an immunologic defect. The **Guillain-Barré syndrome** is an acute progressive **polyneuropathy** characterized by muscular weakness and mild sensory loss following a mild infectious illness. The illness usually begins with distal muscle weakness that progresses centrally and may involve the cranial nerves and sympathetic system, resulting in paralysis of the muscles of the extremities, face, and respiratory system. Spinal fluid examination reveals an absence of cells and an increase in protein. Recovery over a prolonged period is usually complete.

Tumors
Tumors of the central nervous system occur in great variety; they cause difficulty because they occupy space in a cavity confined by a bony shell. As tumors grow, they either directly invade and destroy nervous tissue or indirectly destroy nervous tissue by compressing it. It is of interest that tumors arise in supportive tissue and not in nerve cells themselves.

Tumors of the central nervous system are divided into two groups: **primary** growths arising from the tissues of the central nervous system, and **secondary** or **metastatic** growths arising from outside the central nervous system.

Primary intracranial neoplasms are classified into six groups: (1) tumors of the skull, (2) tumors of the meninges, (3) tumors of the cranial nerves, (4) tumors of the supportive tissues (gliomas), (5) tumors of the pituitary, and (6) congenital tumors (usually from developmental problems). Secondary tumors are frequently due to cancer of the lung or breast. Brain tumors are found in about 2% of routine autopsies. Tumors of the central nervous system usually cause symptoms of increased intracranial pressure, such as headache, vomiting, and personality changes. They are frequently difficult to diagnose in the early stages.

Spinal cord tumors are more rare than intracranial neoplasms. Only about 10% of spinal cord tumors occur within the cord itself (intramedullary). Early symptoms usually involve compression of the nerve roots that result in pain and numbness followed by sensory loss.

Aging
There are many physiological and disease-related changes to the central nervous system with age. The number of axons within the large nerves declines to about two-thirds of those in a young adult by the end of the seventh decade. The brain itself loses about 40% of its weight in the same time span. The aging brain also shows accumulations of a lipid-derived pigment called **lipofuscin** (lĭp·ō·fŭs′sĭn) that appears in the **brown atrophy** of the aging brain.

Diseases of old age also cause deterioration of the nervous system. One of the most common problems of the aging brain is **cerebrovascular disease** such as strokes, emboli, and blockage of the blood supply due to **atherosclerotic** (ăth″ĕr·ō″sklĕ·rō′tĭk) **plaques.** In recent years we have seen a marked increase in **Alzheimer's disease** (see Chapter 12 Clinical Note: Alzheimer's Disease: The False Epidemic?), a devastating degeneration of the central nervous system that results in memory loss and character deterioration in the aged.

SELF-TEST OF CHAPTER OBJECTIVES

True-False Questions
1. The gaps in the myelin sheath of myelinated peripheral nerves are called nodes of Ranvier.
2. Some axons possess collateral branches.
3. Myelin is located only in the peripheral nervous system.
4. The neuron's membrane potential is negative, which, by convention, means that the outside is negative with respect to the inside.
5. The membrane potential of a neuron at rest results primarily from an excess of Na^+ on the outside and an excess of K^+ on the inside of the membrane.
6. If an electric potential is applied across the membrane such that the potential is made more negative, the neuron would be closer to its threshold and thus be more readily stimulated.
7. The only place a neuron can be stimulated is at its dendrites.
8. Impulses travel faster in a nonmyelinated axon than they do in a myelinated axon of the same diameter.
9. Transmission across a chemical synapse is faster than the speed of the impulse itself along the axon.
10. A transmitter substance may be excitatory in some synapses and inhibitory in others.

Matching Question
Match each of the functions on the left with the cell type that is probably responsible for it.

11. lining ventricles of the brain
12. phagocytosis
13. provide a scaffolding that supports neurons
14. provide myelin sheaths for neurons in the CNS
15. regulate exchange between blood capillaries and fluid around neurons

    a. astrocytes
    b. ependyma
    c. microglia
    d. oligodendrocytes

Multiple Choice Questions
16. Which of the following lists structures in order of increasing numbers of neurons enclosed?
    a. perineurium, epineurium, endoneurium
    b. endoneurium, epineurium, perineurium
    c. perineurium, endoneurium, epineurium
    d. endoneurium, perineurium, epineurium
17. Which of these types of neurons are typically found in sensory nerves?
    a. multipolar
    b. unipolar
    c. bipolar
18. The single long process that emanates from a neuron and that carries the impulse away from the cell body is called a(n)
    a. dendrite   c. soma
    b. hillock   d. axon
19. Which of the following might be an example of a typical membrane potential?
    a. -70 mV   c. $+50$ mV
    b. -10 mV   d. $+90$ mV
20. When an action potential occurs, the membrane potential changes from
    a. negative to positive
    b. negative to positive and back to negative
    c. positive to negative
    d. positive to negative and back to positive
21. The first change in potential during an action potential is due to the rush of
    a. incoming sodium ions   c. incoming potassium ions
    b. outgoing sodium ions   d. outgoing potassium ions
22. A synaptic cleft is typically how wide?
    a. 5 nm   c. 50 nm
    b. 20 nm   d. 100 nm
23. Transmitter substance is released through the _____ membrane and combines with _____ in the _____ membrane. Which of the following provides the appropriate terms in the correct order?
    a. postsynaptic, receptors, presynaptic
    b. presynaptic, receptors, postsynaptic
    c. vesicle, sodium pores, dendritic
    d. presynaptic, cholinesterase, postsynaptic
24. Sometimes more than one impulse is required to stimulate the next neuron in a sequence. If the impulses reach several synapses at about the same time, it is an example of
    a. temporal summation   c. additive summation
    b. spatial summation   d. synaptic summation
25. An example of a common inhibitory transmitter substance is
    a. acetylcholine   c. gamma-aminobutyric acid
    b. seratonin   d. cholinesterase
26. In which regions of a neuron does Na^+ diffuse during saltatory conduction?
    a. through layers of myelin surrounding an axon
    b. through the nodes of Ranvier
    c. through the membrane of the dendrites and cell body
    d. through the membrane at the axon terminals
27. Repolarization of a neuronal membrane during an action potential occurs most directly from which of the following?
    a. leakage of K^+ out through the membrane
    b. active transport of K^+ out of the cell
    c. opening of voltage-dependent K^+ channels and diffusion of K^+ out of the cell
    d. opening of voltage independent K^+ channels and diffusion of K^+ out of the cell
28. Release of neurotransmitters into a synaptic cleft involves which of the following?
    a. active transport
    b. diffusion
    c. exocytosis
    d. synthesis of neurotransmitter at the presynaptic membrane
29. An action potential is normally initiated in which part of a neuron?
    a. dendrite   d. axon
    b. cell body   e. axon terminal
    c. hillock

30. What would be the likely effect of a neurotransmitter that opened only K^+ channels in a postsynaptic membrane?
    a. hyperpolarization of the membrane and inhibition of an action potential
    b. depolarization of the membrane and stimulation of an action potential
    c. depolarization of the membrane and inhibition of an action potential
    d. opening of Na^+ channels
31. If an aqueous solution of NaCl is placed on one side of a membrane that is permeable to Na^+ but not to Cl^-, and pure water is placed on the other side, what will happen?
    a. Na^+ will be unable to diffuse through the membrane because of the attraction of Cl^-.
    b. Na^+ will diffuse through the membrane until its concentration is the same on both sides of the membrane.
    c. Some Na^+ will diffuse through, but not enough to reach equal concentrations on both sides of the membrane.
    d. Na^+ will diffuse through and carry Cl^- along with it by electrical attraction.
32. Removal of neurotransmitters from a synaptic cleft is accomplished by which of the following mechanisms?
    a. degradation
    b. diffusion
    c. endocytosis
    d. any of the above, depending on the neurotransmitter and the synapse
33. What do oligodendrocytes and Schwann cells have in common?
    a. Both are involved in the conduction of impulses.
    b. Both are present in the peripheral nervous system.
    c. Both are responsible for forming myelin sheaths around axons.
    d. Both are present in the central nervous system.
34. Which of the following accounts most directly for the increased speed of an impulse along a myelinated axon?
    a. The myelin sheath allows for the passive depolarization of the membrane over greater distances than would occur in the absence of the sheath.
    b. Myelinated neurons have a greater density of voltage-dependent Na^+ and K^+ channels.
    c. The myelin sheath conducts the impulse more rapidly than the unmyelinated axon.
    d. Nothing; myelinated axons carry impulses more slowly than do nonmyelinated ones.
35. What is the likely effect of continuous excitation of the body and dendrites of a neuron if the level of depolarization exceeds the cell's threshold?
    a. The cell will fire once and then be dormant until excitation is reduced.
    b. The cell will be unable to initiate an action potential at all.
    c. A series of action potentials will occur with a frequency determined by the magnitude of the excitation.
    d. The cell will respond with an action potential of a magnitude that depends on the strength of the excitation.

Essay Questions
36. Outline the organization of the nervous system, distinguishing between the central and peripheral nervous systems and the subdivisions of each.
37. List the types of cells found in nervous tissue and describe the role of each.
38. Draw a typical nerve in cross section, identifying regions of connective and nervous tissue present in the nerve.
39. Distinguish between voltage-dependent and voltage-independent ion channels in the neuronal membrane and the roles they play in the transmission of a nerve impulse.
40. Na^+ and K^+ are transported across the neuronal membrane in a ratio of 1:1. Nevertheless, differences in ion concentration develop that result in a membrane potential. Describe how the membrane potential is established following active transport of Na^+ and K^+ through the membrane.
41. An action potential begins if sufficient voltage-dependent Na^+ channels open. Describe the mechanism responsible for opening and closing the Na^+ channels involved in an action potential.
42. Describe how diffusion of Na^+ through voltage-dependent channels in one region of the neuronal membrane leads to the opening of similar channels in neighboring regions of the membrane.
43. List three ways in which a transmitter substance might be removed from a synaptic cleft.
44. Receptors for a transmitter substance are present in the postsynaptic membrane of a synapse where they may serve to induce or inhibit the transmission of an impulse across the synapse. Describe the difference between a mechanism that leads to induction and one that inhibits the transmission of an impulse.
45. Distinguish among axoaxonic, axosomatic, and axodendritic synapses.
46. Distinguish between cholinergic and adrenergic synapses.
47. Whether or not an impulse occurs in a postsynaptic neuron depends upon the net result of stimuli from presynaptic neurons acting to stimulate or inhibit the neuron. List and describe several factors that determine whether or not an action potential will be initiated in a postsynaptic neuron.
48. Draw a graph that shows in a general way the relationship between the magnitude of a net excitatory postsynaptic potential and the frequency of action potentials in a postsynaptic cell.
49. List four characteristics of a neurotransmitter.

Clinical Application Questions
50. Explain the basis for multiple sclerosis.
51. List and compare neurologic conditions that result from incomplete development of the brain and spinal column.
52. Distinguish among meningitis, encephalitis, and encephalomyelitis.
53. Describe several types of damage that can occur to the nervous tissue of the head as a result of trauma.
54. List and describe the origin of tumors of the CNS.
55. Briefly describe changes in the CNS often associated with aging.

12
The Central Nervous System

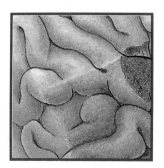

In the preceding chapter we described how nervous tissue receives and transforms sensory stimuli into nerve impulses and how these impulses are transmitted to other nerve cells, glands, and muscle cells. In this chapter we will study the **central nervous system (CNS)**, which controls and coordinates body activities through electrical impulses. The CNS consists only of the **brain** and the **spinal cord.**

The adult brain is organized into four major areas (Figure 12-1). The most anterior portion consists of the *cerebrum* (telencephalon), subdivided into two *cerebral hemispheres*, which fill most of the skull's cranial cavity. Posterior to these are the *cerebellar hemispheres* tucked under the rear of the cerebrum. Between the cerebellum and cerebrum lies the *diencephalon*, composed of the *thalamus* and *hypothalamus*. The remainder of the brain is organized into a compact *brain stem*, which sits at the top of the spinal cord.

In this chapter we will discuss

1. the embryological development of the nervous system,
2. the major anatomical structures of the brain,
3. the major functional areas of the brain,
4. the three meningeal layers and their relationship to cerebrospinal fluid,
5. the anatomical and functional makeup of the spinal cord, and
6. spinal reflex arcs.

sulcus: (L.) *sulcus*, furrow
gyrus   (L.) *gyrus*, circle

Telencephalon (Cerebrum)

After completing this section, you should be able to:

1. Identify the four major regions of the adult brain.
2. Describe the anatomy of the cerebral hemispheres.
3. Locate and discuss the functional areas of the cerebral cortex.
4. Identify the basal ganglia of the cerebrum.
5. Describe the organization of the cerebral white matter into fiber tracts.

The **cerebrum**, which develops from the embryonic telencephalon, consists primarily of two large hemispheres called the **cerebral hemispheres**, which are composed of several subdivisions (Figure 12-2). The *cerebral cortex* is composed of nonmyelinated neurons (neurons that lack the myelin sheath produced by Schwann's cells) and appears dark grayish in color; hence, it is frequently referred to as the **gray matter**. The interior of the cerebral hemispheres is composed mostly of myelinated nerve fibers (neurons with fibers that are wrapped with a myelin sheath), organized into discrete bundles or tracts. These tracts connect areas of the cortex that perform similar functions on opposite sides of the brain. These myelinated fibers form the **white matter** because they appear white. Embedded in the white matter are additional isolated masses of gray matter, the *basal ganglia* or *basal nuclei*. A horn-shaped cavity occupies a small space within each cerebral hemisphere forming two *lateral ventricles*. These cavities, along with other cavities of the brain and the *central canal* of the spinal cord, contain *cerebrospinal fluid*, the special extracellular fluid of the central nervous system.

Cerebral Cortex

The **cerebral cortex**, the relatively thin (2 to 4 mm) outer layer of the cerebral hemispheres (see Figure 12-4a), consists of densely packed, nonmyelinated neurons that make numerous interconnections with each other and with fibers in the underlying white matter. The cerebral surface is highly convoluted, consisting of numerous depressions called **sulci** (sŭl′sē) separated by equally numerous ridges called **gyri** (jī′rī). Several sulci that are noticeably deeper than others

Figure 12-1 Four major regions are identified in the adult brain: the cerebral and cerebellar hemispheres, the diencephalon (which lies between these two regions), and the brain stem at the base of the brain.

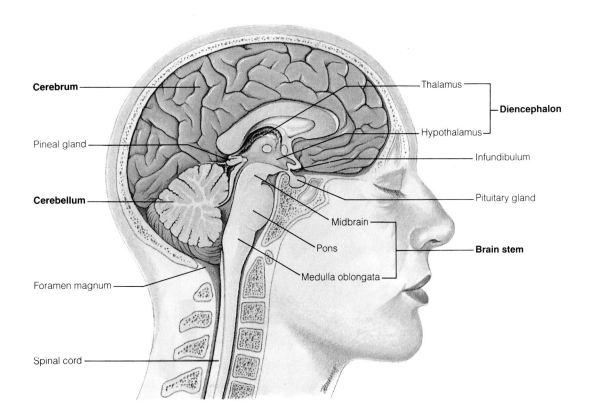

Embryonic Development of the Nervous System

By the third week of embryonic development, the embryo has differentiated into an elongate flattened structure composed of three tissue layers: **ectoderm, mesoderm,** and **endoderm.** The entire nervous system forms from a strip of ectoderm called the **neural plate** that lies along the longitudinal axis of the embryo. The plate thickens and a median **neural groove** marks the initiated of nervous system development. **Neurulation** (nū″rū·lā′shŭn), the formation of a hollow **neural tube,** begins simultaneous with the development of somites (Figure 12-A). The neural groove pushes down into the area destined to become the spine, while the margins of the plate enlarge and become **neural folds.** As the folds meet along the midline, they push the newly formed neural tube beneath the surface ectoderm. The initial fusion occurs near the middle of the embryo and proceeds in both directions like a double zipper closing a jacket.

As the neural tube is forming, a group of ectodermal cells pinches off from the neural folds and is sandwiched between the surface and the neural tube. These **neural crest** cells become organized along the length of the neural tube, resembling a chain whose links are in tandem with the lengthwise series of mesodermal somites. Neural crest cells develop into several structures, including sensory and motor neurons in cranial and spinal nerves, support cells that surround neurons, cells of the adrenal medulla, and other structures not directly connected with the nervous system.

By the fourth week, the neural tube is essentially closed. At its anterior end, a slight bulge marks the primitive

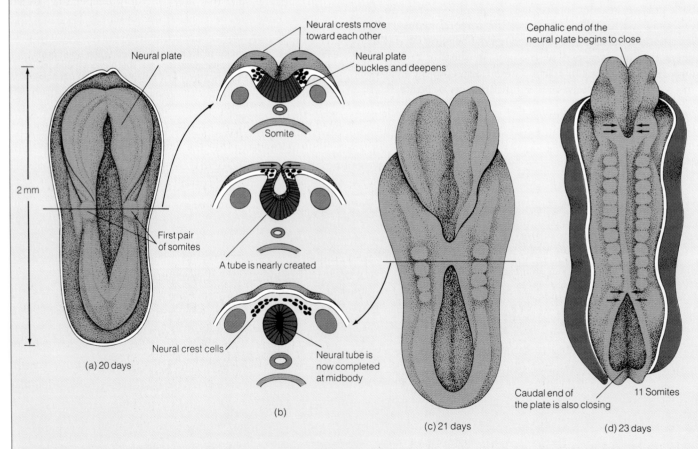

Figure 12-A Embryonic development of the nervous system. (a) Near the end of the third week of development, the nervous system begins to differentiate. (b) Cross sections of the developing neural tube at successively later stages (top to bottom), showing formation of the neural tube. (c) The neural tube begins forming in the middle of the embryo, and (d) closing advances simultaneously toward both ends.

prosencephalon: (Gr.) *pros*, before + *enkephalos*, brain
rhombencephalon: (Gr.) *rhombos*, rhomb + *enkephalos*, brain
telencephalon: (Gr.) *tele*, far + *enkephalos*, brain
myelencephalon: (Gr.) *myelos*, marrow + *enkephalos*, brain

brain, which at this stage of development consists of three small, fluid-filled vesicles: the **forebrain** or **prosencephalon** (prŏs″ĕn·sĕf′ă·lŏn), **midbrain** or **mesencephalon**, and **hindbrain** or **rhombencephalon** (rŏm″bĕn·sĕf′ă·lŏn) (Figure 12-B). During the next two weeks, the prosencephalon divides into two distinct regions, the **telencephalon** and **diencephalon**, and the rhombencephalon subdivides into the **metencephalon** and **myelencephalon**. Together these five vesicles will produce the adult brain (Table 12-1). During the next few weeks, strong flexures develop between the regions. Enlargement of the telencephalon and metencephalon produces a globular brain that gradually assumes the adult form. Posterior to the brain, the neural tube develops into the spinal cord.

Table 12-1 RELATIONSHIP OF EMBRYONIC AND ADULT BRAIN REGIONS

Primary Embryonic Divisions	Secondary Subdivisions	Adult Region
Prosencephalon (Forebrain)	Telencephalon	Cerebrum
	Diencephalon	Thalamus
		Hypothalamus
Mesencephalon (Midbrain)	Mesencephalon	Corpora Quadrigemina
		Cerebral Peduncles
Rhombencephalon (Hindbrain)	Metencephalon	Cerebellum
		Pons
	Myelencephalon	Medulla Oblongata

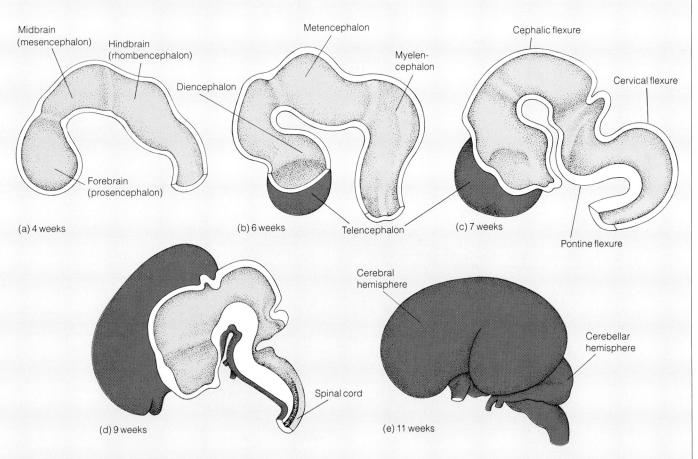

Figure 12-B Formation of brain vesicles. (a) Three vesicles develop early in the differentiation of the brain. (b) Forebrain and hindbrain subdivide. (c) Several flexures cause the vesicles to fold over one another. (d) and (e) The cerebral hemispheres enlarge greatly and grow over most of the other areas. (Times are approximate.)

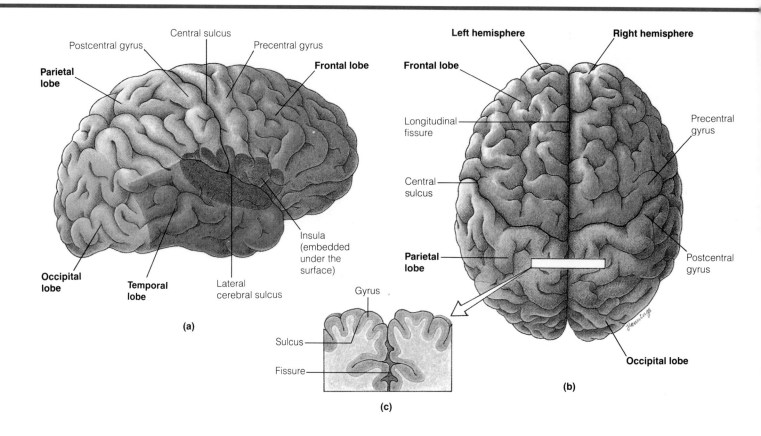
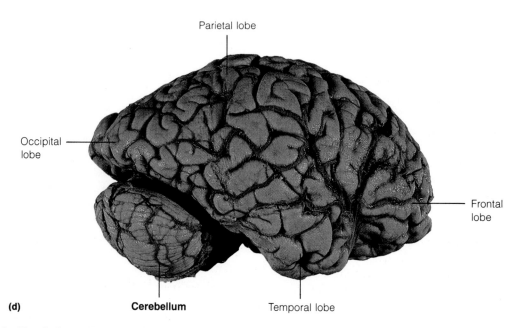

Figure 12-2 Cerebral hemispheres. Each hemisphere has several large lobes separated by sulci and shallower gyri. (a) Lateral view. (b) Superior view. (c) Transverse section through cerebral cortex. Note the distinction between the more superficial gray matter (shown in dark pink) and the deeper white matter (shown in light pink). (d) Photograph of human brain.

insula: (L.) *insula*, island
proprioception: (L.) *proprius*, one's own + *capio*, to take

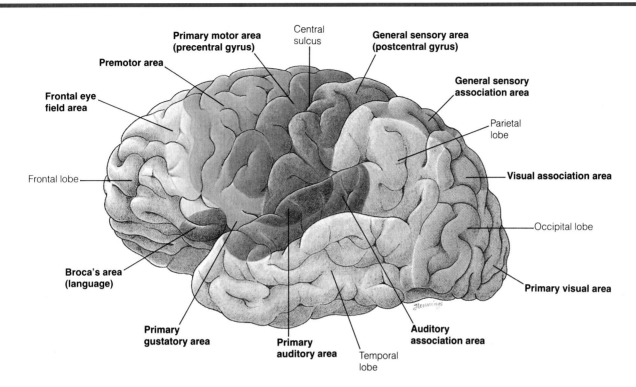

Figure 12-3 Functional areas of the cerebral cortex. In this view of the left cerebral hemisphere, the surface appears relatively uniform in structure, yet specific areas function quite differently. Some areas receive sensory impulses, others initiated motor impulses, while others perform association functions.

are referred to as **fissures**. Figure 12-2 illustrates the fissures of the cerebrum and the major areas or **lobes** they define. Each hemisphere is composed of a **frontal lobe, parietal lobe, occipital lobe,** and **temporal lobe** named after the cranial bones they underlie. In addition to these externally visible lobes, a large internal fold of cortex called the **insula** lies deep to the **lateral fissure.**

The anatomical significance of the gyri and sulci is that they increase the surface area (and thus the volume) of the cerebral cortex. Functionally, this expanded volume increases the volume of cells, synapses, and pathways available in the cerebral cortex, thereby increasing the efficiency of the brain in analyzing incoming information and generating a complex variety of motor impulses in response to sensory input.

Functional Areas of the Cerebral Cortex

Although the cerebral cortex appears relatively uniform, it consists of numerous specific centers divided into three major functional areas: *sensory, motor,* and *association.*

Sensory Areas

The **primary sensory area** (or **general sensory area**) is located in the postcentral gyrus (Figure 12-3). This region receives sensory impulses carrying information relating to touch, pressure, pain, temperature, and **proprioception** (prō″prē·ō·sĕp′shŭn). Proprioception involves the reception of internal stimuli, such as the amount of tension in muscles and tendons. This sensory area receives stimuli from all parts of the body, but because the sensory fibers from each body part terminate in a specific area of the primary sensory area, the cortex is able to sort out the precise body part involved. The size of a specific body part is not directly proportional to the size of the primary sensory area devoted to that part. Rather, it is the *number of sensory neurons* in the body part that determines the size of the sensory area. Consequently, much more of the primary sensory area is concerned with interpreting sensory impulses from the hands and face than, for example, from the shoulders, arms, or trunk.

In addition to the primary sensory area, other sensory areas have been identified in the cerebral cortex (Figure 12-3). The **primary visual area** located in the occipital lobe receives and interprets impulses coming from sensory cells in the eyes. Both the **primary auditory area** and the **primary olfactory area** are found in the temporal lobe. The latter is concerned with interpreting the numerous odors

lobotomy: (Gr.) *lobos*, lobe + *tomos*, cutting

and aromas one encounters daily. Neural impulses of taste are received and interpreted in the **primary gustatory** (gŭs′tă·tō·rē) **area**, in the parietal lobe near the lateral fissure.

Motor Areas

The **primary motor area** located in the precentral gyrus (anterior to the primary sensory area) controls the conscious, voluntary movements of the skeletal muscles (Figure 12-3). Like the primary sensory area, it contains specific groups of neurons that control movement of specific body parts. The size of the motor area involved with a specific body part is proportional to the *number of muscles* located in that body part and the relative complexity of the movements possible. For example, movement of the hand, which involves the contraction of many muscles, is controlled by more than five times the volume of cerebral cortex as that which controls the relatively few muscles moving the entire trunk.

In addition to the primary motor area, several special motor areas have been identified (Figure 12-3). Learned activities, such as writing, typing, and playing a musical instrument are controlled by the **premotor area** located in the frontal lobe just anterior to the primary motor area. The special **motor area for speech** is found in the frontal lobe just above the lateral fissure. This area is also called **Broca's area** for Paul Broca, a French anthropologist and neurophysiologist of the nineteenth century. Broca's area is often referred to as a **language area** because it involves the production of spoken words from conscious thoughts. This process requires the coordination of different muscles in the mouth, pharynx, and larynx and involves a considerable amount of learning over months or years.

Also located within the primary motor area are pyramid-shaped motor neurons with axons that descend through the spinal cord as **pyramidal tracts** or **pathways** (see Figure 15-10d). These tracts involve the innervation of voluntary muscles. All additional tracts of motor fibers are **extrapyramidal tracts** or **pathways.**

Association Areas

Much, perhaps most, of the cerebral cortex is composed of **association areas** (Figure 12-3). These relatively large areas connect sensory and motor areas with one another and also relate different sense modalities with one another to form a composite sensory awareness. Although it is not possible to portray accurately the intricacy and complexity of the interconnections of the association areas, a simple example can

> **CLINICAL NOTE**
>
> **LOBOTOMY**
> The frontal lobes of the brain anterior to the precentral motor areas have been associated with higher intellectual and psychic functions. Prior to the discovery of psychoactive drugs, an uncontrollable mental patient was often subjected to a **prefrontal lobotomy,** an operation that severed the connections between the frontal lobes and the rest of the brain. This operation produced changes in moral and social attributes, disinterest in the environment, intellectual deterioration, and distractibility. Patients with chronic violent and antisocial behavior were rendered docile automatons. Today, surgical removal of portions of the cerebral cortex or even one hemisphere may be used in rare instances to control severe, uncontrollable seizure disorders. Fortunately, psychoactive drugs have proved to be much more effective in treatment of these unfortunate patients.

serve to illustrate the magnitude of this complexity. Imagine you are reading a book and that you have just come to the end of the page. You become consciously aware that you have just read the last word on the page. Motor impulses are sent out that cause your finger to grasp the page and turn it over. Your eyes, and probably your head also, move up and to the left to continue reading at the upper left corner of the new page, while your hand goes back to its original resting position. Several brain centers would be involved in this very simple activity. The primary visual area responds to the visual images of reading. The primary motor area directs the movements of the eyes and hands. Other parts of the brain yet to be discussed control and coordinate the movements involved so that you can turn only one page and not ten. Coordination of these activities is made possible by the association areas and related tracts in the underlying white matter.

White Matter of the Cerebrum

Underneath the nonmyelinated neurons of the cerebral cortex are numerous bundles (tracts) of myelinated nerve fibers that form the white matter (Figure 12-4). These bundles are usually classified into three functional tracts: projection, association, and commissural. **Projection tracts,** formed from **projection fibers,** transmit either sensory impulses from the spinal cord to the sensory areas of the cortex or motor impulses from the motor areas of the cortex to the spinal

ipsilateral:	(L.) *ipse*, same + *latus*, side	callosum:	(L.) *callosus*, hard
commissural:	(L.) *commissura*, seam	caudate:	(L.) *cauda*, tail
corpus:	(L.) *corpus*, body	amygdaloid:	(Gr.) *amygdale*, almond

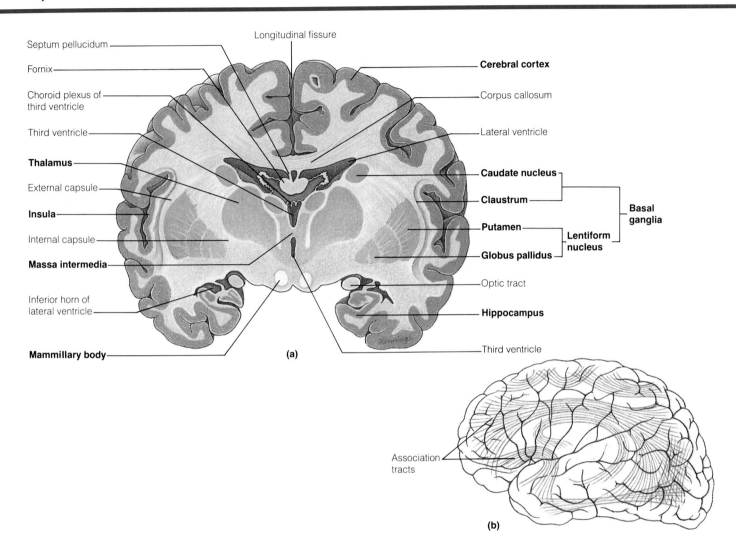

Figure 12-4 White matter of the cerebrum. (a) Frontal section of the brain. Gray matter is not restricted to the cerebral cortex but is also found in interconnected masses deep within the white matter. (b) Lateral view showing the positions of association tracts. These tracts of white fibers lie deep to the cerebral cortex and connect different regions of the cortex to other areas of the CNS.

cord. **Association tracts,** formed by **association fibers,** transmit impulses from one area of the cortex to another within the same hemisphere (**ipsilateral**) (Figure 12-4b). **Commissural** (kŏm·mĭs′ū·răl) **tracts,** formed from **commissural fibers,** transmit impulses from one hemisphere to the other (**contralateral**). Examples are the large **corpus callosum** (kŏr′pŭs kă·lō′sŭm) (Figure 12-4) and the smaller **anterior commissure** (see Figure 12-7).

Gray Matter of the Cerebrum

Paired masses of isolated gray matter called **basal ganglia** (**basal nuclei** or **cerebral nuclei**) are buried deep within the white matter of the cerebrum (Figure 12-5). Each hemisphere contains four basal ganglia: the caudate nucleus, amygdaloid nucleus, lentiform nucleus, and claustrum (see also Figure 12-4). The **caudate nucleus** is a long, curved mass of cells (shaped like the horn of a ram) that forms the most medial portion of the basal ganglia. Attached to the tail of the caudate nucleus is the **amygdaloid** (ă·mĭg′dă·loyd) **nucleus,**

dementia: (L.) *dementare*, to make insane
pallidus: (L.) *pallidus*, pale
putamen: (L.) *putamen*, shell
claustrum: (L.) *claustrum*, bar, barrier

CLINICAL NOTE

ALZHEIMER'S DISEASE: THE FALSE EPIDEMIC?

During the past few years there has been a great deal of concern in the medical and public press about the dramatic increase in the number of people diagnosed as having **Alzheimer's** (ălts´hī·mĕrz) **disease**. Recent reports suggest that there are 2.5 million Americans suffering from this debilitating disease and that it is the fourth leading cause of death in this country.

Alzheimer's disease was first described by a German physician, Alois Alzheimer, in 1906 as a cause of memory loss and mental deterioration in adults during the fourth and fifth decades of life. It was quite rare and was described as **presenile dementia**. Dementia is a deterioration of the intellectual faculties, reasoning power, memory, and will and is characterized by varying degrees of confusion, disorientation, apathy, and stupor. In contrast, **senile dementia** was ascribed to the aging process of the brain and was very common in the very elderly. In 1976 both of these conditions were grouped together as Alzheimer's disease since they were both shown to produce the same microscopic changes in the brain: atrophy of the anterior portion of the frontal and temporal lobes with tangles of decaying nerve cells and deposition of lipoprotein plaques. This discovery enormously increased the prevalence of the diagnosis of Alzheimer's disease. In addition, the number of people living beyond 75 years of age has increased tenfold since 1900. About 5% of individuals over 65 years of age suffer from severe dementia. All of these factors have resulted in a greater concern about the supposed "increase" in Alzheimer's disease. Neurologists point out, however, that there is no evidence that the disease is affecting younger people or that the percentage of older people acquiring the condition is increasing.

Alzheimer's disease is a particularly serious problem since the patients gradually lose their personalities and their ability to care for themselves. They become totally dependent on families or society for years before their death.

PET scans of brain showing comparison of (a) normal brain to (b) brain affected by Alzheimer's disease.

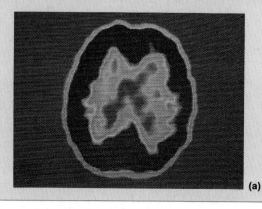
(a)

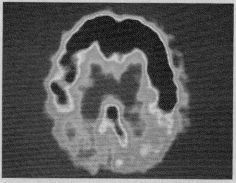
(b)

---

an oval body of cells. The **lentiform nucleus** lies lateral and slightly inferior to the head of the caudate nucleus It is composed of two connected masses of cells, the medially positioned **globus pallidus** (glō´bŭs păl´ĭ·dŭs) and the more lateral **putamen** (pū·tā´mĕn). The **claustrum** (klŏs´trŭm) is a thin sheet of cell bodies lying between the lentiform nucleus and the cortex of the insula.

The **internal capsule** is white matter that separates the basal ganglia from the thalamus. The myelinated fibers of the internal capsule carry sensory and motor impulses to and from the cerebral cortex. Because of their striped appearance, the caudate nucleus, lentiform nucleus, and internal capsule together are often referred to as the **corpus striatum** (kŏr´pŭs strī·ā´tŭm).

It is now thought that three additional areas, the *substantia nigra* (see p. 303) of the midbrain, the *subthalamic nucleus* (which lies next to the internal capsule), and the *red nucleus* (see p. 303) may also be a functional part of the basal ganglia.

These ganglia are associated with the transmission of motor impulses. They are important in the control and coordination of voluntary muscular movements.

diencephalon: (Gr.) *dia*, between + *enkephalos*, brain
thalamus: (Gr.) *thalamos*, chamber
cephalalgia: (Gr.) *kephale*, head + *algos*, pain

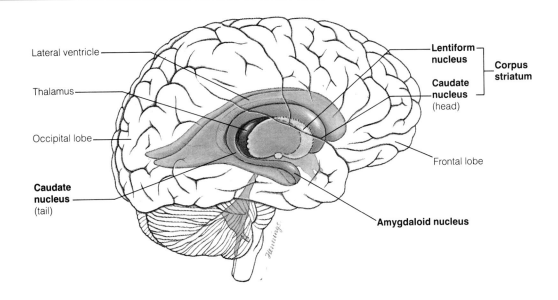

Figure 12-5 The basal ganglia consist of paired masses of gray matter buried within the white matter of the cerebral hemispheres. (The claustrum is shown in Figure 12-4.)

Diencephalon

After completing this section, you should be able to:

1. Describe the positions of the three major thalamic regions in the diencephalon.
2. Describe the function and structures of the thalamus.
3. Discuss the functions of the hypothalamus relative to the nervous and endocrine systems.
4. Discuss the functions of the limbic system.

The **diencephalon** (dī″ĕn·sĕf′ă·lŏn) is the adult development of the second subdivision of the fetal brain. It is almost totally surrounded by the enlarged cerebral hemispheres, and it is composed of three major thalamic regions: the thalamus, hypothalamus, and epithalamus.

Thalamus

The **thalamus** (thăl′ă·mŭs) is shaped roughly like a dumbbell in cross section (Figure 12-4). Two oval masses of nonmyelinated cell bodies form the major portions of the thalamus.

CLINICAL NOTE

HEADACHES

A headache or **cephalalgia** (sĕf·ă·lăl′jē·ă), is a very common problem. It may result from simple causes, such as fatigue, fever, or alcohol ingestion, or more serious causes, such as intracranial infections, intracranial tumors, head injury, severe hypertension, cerebral anoxia, and many diseases of the eyes, ears, nose, and throat.

Chronic and recurrent headaches can be diagnosed by the distinctive location and character of the headaches. **Migraine headaches** usually affect only one side of the head, are frequently preceded by an **aura** (visual or sensory symptom), and are associated with nausea and vomiting. Migraine attacks may last for several days and can be incapacitating. There is often a strong family history of migraine attacks. The cause is not known, but the mechanism appears to be a dilation of the blood vessels in the brain. Treatment with vasoconstrictor drugs can abort or prevent the attacks.

Muscle tension headaches are usually moderate in intensity, involve the frontal-occipital area of the head, and are accompanied by a feeling of stiffness or tightness of the neck and scalp. These headaches are usually relieved with rest and analgesics such as aspirin. A much rarer type called **psychogenic headaches** affect the entire head, can be constant, and are made worse by emotional disturbances. Another type called **cluster headaches** (or histamine headaches) are abrupt, severe, and associated with symptoms of vasodilation resulting in tearing, running nose, or swelling below the eye. Treatment is by use of corticosteroids or vasoconstrictor drugs.

geniculate:	(L.) *geniculum*, little knee
hypothalamus:	(Gr.) *hypo*, under + *thalamos*, chamber
chiasma:	(Gr.) *chiasma*, cross
infundibulum:	(L.) *infundibulum*, funnel
tuber:	(L.) *tuber*, swollen knob

These two masses of tissue are joined medially by a short rod-shaped **massa intermedia.** The thalamus functions as a relay center for both types of impulses between the cerebral cortex and other neural areas.

Although they are difficult to define anatomically, studies show that each mass of thalamic tissue consists of numerous paired **thalamic nuclei** (Figure 12-6). The more prominent nuclei can be divided into those that relay sensory impulses and those that relay motor impulses. The sensory nuclei include the **lateral geniculate** (jĕn·ĭk´ū·lāt) **nucleus,** which relays visual impulses; the **medial geniculate nucleus,** which relays auditory impulses; and the **inferior posterior nucleus,** which relays sensations of taste. The prominent motor nuclei include the **inferior anterior nucleus** and the **inferior lateral nucleus,** both of which relay impulses along voluntary motor pathways. Along with these nuclei, other thalamic areas relay general and specific sensory impulses (except olfaction) to other areas of the brain, specifically, the sensory areas of the cerebral cortex and the hypothalamus.

Hypothalamus

The **hypothalamus** (hī˝pō·thăl´ă·mŭs), a small mass of nerve tissue, is continuous with the inferior end of the thalamus (Figures 12-1 and 12-7). It consists of numerous **hypothalamic nuclei,** which regulate a variety of homeostatic activities, including control of the autonomic nervous system, control of the endocrine system through the pituitary gland (see Chapter 17), hunger and thirst, body temperature, wakefulness and sleepiness, and emotions, such as anger, rage, and aggression.

Prominent external landmarks associated with the hypothalamus are the **optic chiasma** (kī·ăz´mă), **infundibulum** (ĭn˝fŭn·dĭb´ū·lŭm), **tuber cinereum** (tū˝bĕr sĭn·ē´rē·ŭm), and the **mammillary bodies** (Figure 12-7). The optic chiasma is an **X**-shaped area formed by the crossed-over fibers of the optic nerves. Posterior to this chiasma is the infundibulum, a stalk of neurons that connects several hypothalamic nuclei with the posterior lobe of the **pituitary gland** or **hypophysis** (hī·pŏf´ĭ·sĭs). The tuber cinereum and mammillary bodies (see Figure 12-9) project from the posterior hypothalamic surface. The tuber cinereum transports neural hormones from the hypothalamus to the infundibulum, and the mammillary bodies are relay stations involved in olfactory reflexes.

Perhaps the most important integrative function of the hypothalamus is control of the autonomic nervous system.

Figure 12-6 The thalamic nuclei. The arrows in the figure indicate directions of interchange of information between the thalamus and the cerebral cortex. The medial geniculate, lateral geniculate, and inferior posterior nuclei receive and relay sensory impulses to conscious areas in the cortex, while the inferior anterior and inferior lateral nuclei relay impulses along voluntary motor pathways.

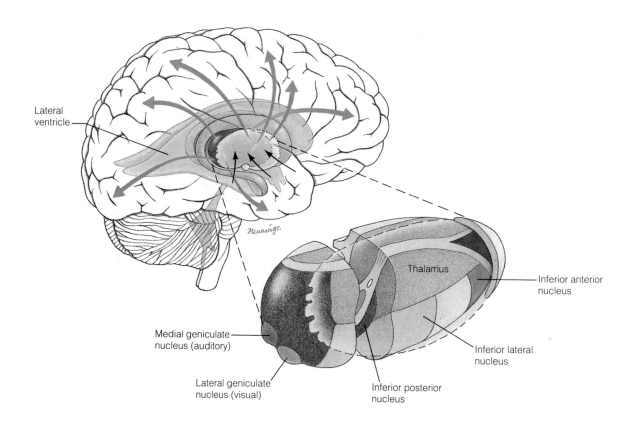

cinereum:	(L.) *cinereus,* ashen-gray
mammillary:	(L.) *mammilla,* nipple
hypophysis:	(Gr.) *hypo,* under + *physis,* growth
pineal:	(Fr.) *pineal,* pine cone
epithalamus:	(Gr.) *epi,* upon + *thalamos,* chamber
limbic:	(L.) *limbus,* border

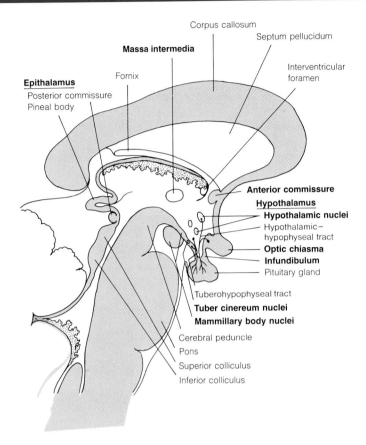

Figure 12-7 Sagittal section through the diencephalon. Note the epithalamus, massa intermedia of the thalamus, and the hypothalamus.

Such actions include shivering, inhibition of sweat glands, and constriction of the blood vessels in the skin, which diverts blood back to the interior of the body. If the blood temperature rises above normal, the hypothalamic thermostat causes the autonomic nervous system to stimulate sweat gland activity and to increase the flow of blood to the skin so heat can be lost.

Epithalamus

The most posterior portion of the diencephalon consists of two discrete structures, the **pineal body** (or **gland**) and the **posterior commissure**, collectively referred to as the **epithalamus** (ĕp″ĭ·thăl′ă·mŭs) (Figure 12-7). The pineal body is an endocrine gland, while the posterior commissure contains fibers that connect the cerebrum with the midbrain.

Limbic System

Certain parts of both the cerebral hemispheres and the diencephalon function together in the control of emotional behavior. These structures are known as the **limbic system** (Figure 12-8). This system is composed of the olfactory

Connections with several organs influence control of many internal activities such as breathing rate, heart rate, and **peristalsis** (pĕr·ĭ·stăl′sĭs), which is the muscular churning associated with the digestive tract. In addition to regulating movements of this tract, the hypothalamus also receives stimuli of sensations of hunger and thirst and the feeling of satiety.

The hypothalamus also provides communication between the nervous and endocrine systems through the anterior pituitary gland. The endocrine system, with its several glands that secrete hormones, regulates chemicals that in turn control body activities (see Chapter 17). Stimulation or inhibition from the hypothalamus increases or decreases the output of these hormones.

Certain cells in the hypothalamus are extremely sensitive to temperature. When the temperature of the blood flowing around these cells varies significantly from normal, these cells act like a thermostat and return the body temperature to normal. For example, if the blood's temperature drops below normal (a common occurrence on a cold day), the hypothalamus stimulates the autonomic nervous system to increase activities that generate and conserve body heat.

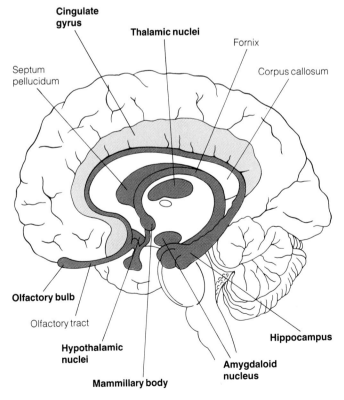

Figure 12-8 Limbic system of the brain. This complex interconnection of brain areas regulates aspects of behavior associated with memory and emotions.

cingulate:	(L.) *cingulum*, girdle
hippocampal:	(Gr.) *hippokampos*, sea horse
mesencephalon:	(Gr.) *mesos*, middle + *enkephalos*, brain

bulbs, the **cingulate** (sĭn´gū·lāt) **gyrus** and **hippocampal** (hĭ´pō·kăm˝păl) **gyrus** of the cerebrum, the amygdaloid nucleus of the basal ganglia, the mammillary bodies and other nuclei of the hypothalamus, and the anterior nuclei of the thalamus.

Much of what is known about the limbic system is based on observations of individuals who suffered damage to that part of the system and on results of animal studies. These studies indicate that the limbic system regulates aspects of behavior associated with memory and emotions. Experiments with animals have shown that stimulation of certain parts of the limbic system result in behavior associated with extreme pleasure. Stimulation of other parts elicit aggressive behavior. Some patterns of stimulation cause the animal to assume a defensive posture, while others elicit docile behavior in the same animal. These experimental results strongly suggest that the limbic system is essential to the experience of emotions.

Mesencephalon

After completing this section, you should be able to:
1. Describe the anatomy of the mesencephalon.
2. Discuss the function of the corpora quadrigemina relative to visual and auditory reflexes.
3. Describe the function of the fiber tracts in the cerebral peduncles.

The **mesencephalon** (mĕs·ĕn·sĕf´ă·lŏn), or "midbrain," is a relatively small structure lying approximately in the middle of the brain (Figures 12-1 and 12-9). A narrow canal called the **cerebral aqueduct** passes longitudinally through the midbrain. This aqueduct contains cerebrospinal fluid, and it connects two other fluid-filled cavities, the third ventricle above it, and the fourth ventricle below.

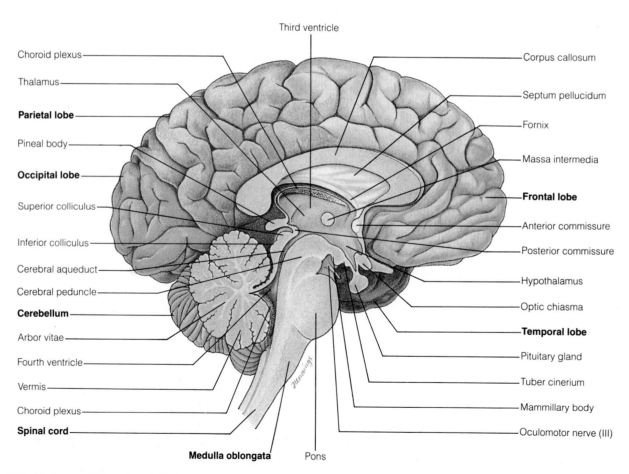

Figure 12-9 Median sagittal section of the brain showing details of the mesencephalon.

tectum:	(L.) *tectum*, roof		lemniscus:	(Gr.) *lemniskos*, ribbon
quadrigemina:	(L.) *quattuor*, four + *geminus*, twin		metencephalon:	(Gr.) *meta*, after + *enkephalos*, brain
colliculus:	(L.) *colliculus*, small mound		vermis:	(L.) *vermis*, worm
peduncle:	(L.) *pedunculus*, little foot		arbor vitae:	(L.) *arbor*, tree + *vita*, life
rubrospinal:	(L.) *ruber*, red + *spinal*, thorn		dentate:	(L.) *dentatus*, toothed

A sagittal section through the midbrain (Figure 12-9) reveals the cerebral aqueduct dividing the midbrain into posterior and anterior regions. The posterior region (**tectum**) contains four rounded masses of nerve tissue (the corpora quadrigemina), and the anterior region (cerebral peduncles) includes many fibers and nuclei.

Corpora Quadrigemina

The **corpora quadrigemina** (kŏr′pŏ·ră kwŏd″rĭ·jĕm′ĭn·ă) are organized into two pairs of rounded protrusions (Figure 12-7). The upper one is the **superior colliculus** (kŏl·lĭk′ū·lŭs). It receives sensory impulses from several sources and functions primarily as a reflex center for visual, auditory, and tactile stimuli. For example, suppose you are sitting in a dimly lit room and a flash of light occurs to your right side (perhaps someone has taken a snapshot). Your immediate response would be to turn your head in the direction of the flash of light. This type of reflex is controlled by the superior colliculus. The lower bulge, the **inferior colliculus**, serves as a reflex center for auditory stimuli. Imagine your response now to a loud noise instead of a bright flash of light. You would also give this type of stimulus your immediate attention, a reflex controlled by the inferior colliculus.

The **substantia nigra** (sŭb·stăn″shē·ă nī′gră) is a nucleus of cells located near the colliculi that seems to function in regulating involuntary muscular movements.

Cerebral Peduncles

The **cerebral peduncles** (pē·dŭn′kls) are paired bulges of nerve fibers in the midbrain that project anteriorly from the brain stem. They are composed primarily of fiber tracts that connect the cerebrum with other parts of the nervous system. Tracts located deep within the peduncles carry sensory impulses up to reflex centers in the diencephalon. More superficial tracts of the peduncles consist of motor fibers that carry impulses from the cerebrum down to the spinal cord. The **red nucleus** is a small mass of gray matter located deep within the peduncles. It functions as a relay station for impulses originating in the cerebrum and cerebellum. Cells located within the red nucleus also give rise to long axons that pass down the spinal cord as the **rubrospinal** (rū′brō·spī′năl) **tract** (see p. 366 for more discussion of this tract).

A tract of white fibers called the **medial lemniscus** (lĕm·nĭs′kŭs) passes from the brain stem through the midbrain to the thalamus. It transmits sensory impulses involved with fine touch and proprioception from the medulla oblongata, located at the base of the brain stem, to the thalamus.

Metencephalon

After completing this section, you should be able to:

1. Describe the anatomy of the metencephalon.
2. Describe the connections between the cerebellum and other parts of the brain.
3. Discuss the reflex centers in the pons.

The **metencephalon** (mĕt″ĕn·sĕf′ă·lŏn) is the second largest region of the brain and is located beneath the cerebral hemispheres. It consists of a large posterior cerebellum, and a smaller pons (Figure 12-9).

Cerebellum

The **cerebellum** (sĕr·ĕ·bĕl′ŭm) bears a superficial resemblance to the cerebrum, being divided into two **cerebellar hemispheres.** They are shaped like the shells of an open clam and are joined in the middle by the narrow **vermis.** The nerve tissue of the cerebellum is organized much like that of the cerebrum. A thin cortex composed of gray matter surrounds an interior composed mostly of white matter. The cerebellar cortex is folded into numerous horizontally oriented ridges called **folia cerebelli** (fō′lē·ă sĕrĕ·bĕl′lī). Several prominent fissures subdivide the folia into a number of **lobules.** A sagittal section through the vermis reveals a unique branching of the white matter called the **arbor vitae** ("tree of life") (Figure 12-9). Several gray matter nuclei are buried deep within the cerebellar white matter, the most prominent being the **dentate nucleus.**

The cerebellum is connected to other parts of the brain through three paired bundles (peduncles) of fibers. The **superior cerebellar peduncles** connect with the mesencephalon, the **middle cerebellar peduncles** with the pons, and the **inferior cerebellar peduncles** with the medulla oblongata (see p. 305).

The major function of the cerebellum is reflex control and coordination of skeletal muscles. It also maintains the partial contraction of skeletal muscles responsible for muscle tone by reflexively causing individual fibers to contract alternately.

The cerebellum functions on several levels of muscular activities. At a basic level, for example, the cerebellum controls muscles that maintain our balance. It receives sensory input from the inner ear that detects posture and body equilibrium. These impulses inform the cerebellum of the position of the body in space. That is, they tell reflex centers in the cerebellum whether the body is sitting, standing, reclining, or moving forward or backward. The inner ear also sends

epilepsy: (Gr.) *epilambanein*, to take hold of
etiology: (Gr.) *aitia*, cause + *logos*, study
idiopathy: (Gr.) *idios*, distinct + *pathos*, disease

CLINICAL NOTE

EPILEPSY

Epilepsy is defined as a recurrent seizure disorder of cerebral function characterized by sudden, brief attacks of altered consciousness, motor activity, sensory phenomena, or inappropriate behavior. It affects about 2% of the population. Epilepsy can be classified by its **etiology,** or probable cause, as **symptomatic,** suggesting that the cause is known, or **idiopathic,** meaning that the cause has not been determined. A seizure disorder can also be classified according to the pattern of onset: **generalized seizures** affect both consciousness and motor function and **focal seizures** are characterized by specific localized motor or sensory phenomena such as twitching, chewing, or numbness. Focal manifestations that immediately precede a generalized seizure (aura) include flashing lights, strange smells, or feelings of dizziness or unreality.

There are many types of seizures described, but they are usually consistent in a particular patient. **Grand mal seizures** (or primary generalized tonic-clonic seizures) usually begin with an aura and proceed to a loss of consciousness, uncontrollable contractions of the muscles, and loss of bowel or bladder control lasting two to five minutes. The seizures are followed by a **postictal state** characterized by a deep sleep, headache, muscle soreness, and eventual recovery. These attacks may occur at any age. **Jacksonian seizures** are focal seizures that start in a hand or foot and then "march" up the extremity. **Petit mal attacks** (or absence seizures) are brief, generalized seizures with loss of consciousness for 10 to 30 seconds. The patient suddenly stops in the middle of an activity and then resumes it after the attack. These attacks are often seen in children and can occur many times during a day. They rarely indicate gross brain damage; in fact, many affected children are highly intelligent. **Akinetic seizures** are brief generalized seizures in children that pitch the child to the ground and carry the risk of serious injury. **Psychomotor attacks** are focal seizures that last 1 to 2 minutes accompanied by loss of contact with the surroundings manifested by staring, staggering, or performing automatic, purposeless movements. They are usually associated with lesions in the temporal lobes. In **status epilepticus,** seizures follow one after another for hours or days and may be fatal.

Some of the causes of epileptic seizures include high fever, meningitis, metabolic disturbances, toxic substances, cerebral anoxia (lack of oxygen to the brain), expanding brain lesions, and trauma. The diagnosis of a particular seizure disorder is usually made based on the patient's history or by observing an attack. It can often be confirmed by obtaining an electroencephalogram of the patient's brain waves (see figure). The treatment consists of correcting the cause (if possible) and administering specific medication to suppress the attacks. Recently, encouraging results have been obtained in relieving the severity and occurrence of epileptic seizures by partially separating the left and right cerebral hemispheres by surgical incision through the corpus callosum.

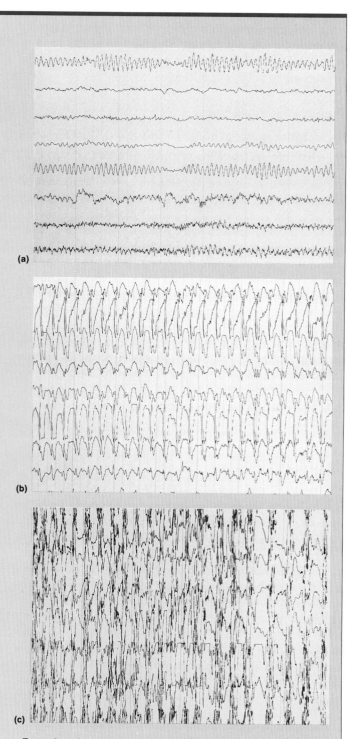

Examples of normal and epileptic EEGs. Each of these is a partial EEG showing several representative tracings from the many electrodes placed on the head. (a) Normal awake state. (b) Petit mal seizure. (c) Grand mal seizure. Note the erratic and intense brain wave activity in the latter.

pons: (L.) *pons*, bridge
apneustic: (Gr.) *a*, not + *pnoe*, breathing
medulla: (L.) *medulla*, marrow
oblongata: (L.) *oblongus*, long

CLINICAL NOTE

BRAIN DEATH

The problem of defining irreversible death has become more difficult with the advent of modern medical technology. Prior to the 1950s, when modern emergency resuscitation and intensive care techniques were developed, it was agreed that cardiac arrest at normal body temperature for longer than 4 to 5 minutes could not be reversed to allow normal brain function. The development of **cardiopulmonary resuscitation (CPR)** and **advanced life support (ALS)** techniques in the 1960s resulted in an increasing number of comatose CPR patients being admitted to the intensive care units (ICUs) of hospitals. Many of these patients developed secondary brain death or cardiac failure within days. Some survived for long periods with permanent brain damage. The social and economic cost of such patients is enormous. Since it is now possible to maintain cardiopulmonary function for prolonged periods, it has become necessary to change our definition of death.

Clinical death is defined as loss of breathing movements and total cardiac arrest with suspension of cerebral activity. It is this very state that may be reversed by effective CPR techniques. **Biologic death** invariably follows clinical death if CPR measures are not instituted. It is an autolytic (self-destructive) process that affects the neurons in about 1 hour and the heart, kidneys, lungs, and liver after 2 hours. **Social death** is a persistent vegetative state in which the patient remains unconscious and unresponsive but with some intact reflexes and brain wave activity. **Cerebral death** (cortical death) is irreversible destruction of the cerebrum with coma but spontaneous respirations.

Brain death is death and necrosis of the entire brain with no hope of recovery of neurological function and a straight line electroencephalogram (EEG). Determination of brain death is a meticulous procedure involving repeated clinical and EEG examinations demonstrating no electrical activity of the brain and is attested to by two physicians. It is accepted as the legal definition of death in the United States, and all life support measures may be removed following the determination.

---

impulses from different sides of the body that indicate whether one is leaning to one side or the other. Using this information, the cerebellum, acting through reflex centers in the brain stem, influences muscles on both sides of the body to contract in such a way as to maintain equilibrium.

A more complex example of cerebellar coordination would be the activity of playing the guitar. In this example, arm muscles on either side of the body are doing different things, and they are doing them simultaneously. The left hand is depressing combinations of strings at different positions on the fingerboard, while the fingers of the right hand are plucking or strumming the strings. The cerebellum receives impulses generated in the cerebral cortex relayed through the midbrain and pons. It also receives information from the arms regarding which muscles are contracting in each hand. These impulses differ depending on the song being played by the guitarist and must be translated into coordinated contractions of groups of muscles for the song to be played correctly. The cerebellum responds by sending impulses to the motor portion of the cerebral cortex that controls hand and finger movements, thereby stimulating and inhibiting the proper muscles to produce the desired sounds.

Pons

The **pons** lies just anterior to the cerebellum. It is approximately the same length as the midbrain (2.5 cm) and is separated from the cerebellum by a triangular, fluid-filled cavity called the **fourth ventricle** (Figure 12-9). The pons ("bridge") serves as a link between parts of the brain and between the brain and spinal cord.

The pons is composed of white fibers with several nuclei scattered within. The most prominent nuclei are the **pneumotaxic** (nū″mō·tăk′sĭk) and **apneustic** (ăp·nū′stĭk) areas. Both centers are important in regulating the breathing cycle. Other nuclei in the pons are associated with paired cranial nerves. The *trigeminal nerves* (V) relay sensory and motor (chewing) impulses from the head; the *abducens nerves* (VI) control certain eyeball movements; the *facial nerves* (VII) conduct sensory and motor impulses related to taste, salivation, and facial expressions; and branches of the *vestibulocochlear nerves* (VIII) function in balance. (These nerves are all discussed in detail in Chapter 13.)

Myelencephalon

After completing this section, you should be able to:

1. Describe the anatomy of the medulla oblongata.
2. Discuss the reflex centers located in the medulla oblongata.
3. Describe the reticular formation and discuss its function.

Medulla Oblongata

The myelencephalon, which is usually called the **medulla oblongata** (mĕ·dūl′lă ŏb″lŏng·gah′tă) or just **medulla**, lies

decussation: (L.) *decussare*, to form an X
cuneatus: (L.) *cuneus*, wedge

at the base of the brain stem (Figure 12-9). It is about 3.5 cm in length. It forms the connection between the spinal cord and brain, thus all sensory and motor tracts connecting the brain and spinal cord must cross it. Each lateral surface of the medulla contains a prominent swelling called the **olive**, which connect the medulla to the cerebellum. The posterior surface of the medulla forms the floor of the fourth ventricle, which continues into the spinal cord as the **central canal**.

The white matter in the anterior portion of the medulla oblongata consists of both sensory and motor tracts. The most prominent of these tracts (called **pyramids**) are triangular in transverse section. They transmit motor impulses from the cerebral cortex to voluntary muscles. An interesting thing happens to these fibers as they pass from the brain into the spinal cord: those originating on the left side of the cerebral cortex descend the right side of the spinal cord and those originating on the right side of the cerebral cortex descend the left side of the spinal cord. This crossing over of fibers is known as the **decussation** (dĕk·ŭ·sā′shŭn) **of pyramids.** It explains why an injury to the right side of the brain may result in paralysis of areas on the body's left side and vice versa.

In addition to these pyramidal tracts, the interior of the medulla oblongata contains several nuclei that act as reflex centers. The **nucleus gracilis** (grăs′ĭ·lĭs) and **nucleus cuneatus** (kū·nē·ā′·tŭs) are located in the posterior region of the medulla. They receive sensory impulses from related spinal tracts and relay them to higher regions in the brain. Like motor fibers of the pyramidal tracts, these sensory tracts also cross over in the medulla so that certain sensations affecting the left side of the body (touch, pressure, and conscious muscle sense) are perceived on the right side of the brain, and vice versa. In addition to these two prominent nuclei, other reflex centers in the medulla help regulate several essential body activities. The **cardiac center** regulates heart rate, the **vasomotor center** regulates blood pressure, and the **medullary rhythmicity** area regulates breathing rate. Other centers in the medulla regulate reflexes of swallowing, coughing, sneezing, and vomiting.

Finally, a number of cranial nerves arise from nuclei located within the medulla. The vestibulocochlear nerves (VIII) carry sensory impulses related to hearing and equilibrium and arise from nuclei located within the medulla. The glossopharyngeal nerves (IX) carry impulses involved with taste, salivation, and swallowing. The vagus nerves (X) carry impulses to and from a variety of thoracic and abdominal organs. The spinal accessory nerves (XI) carry impulses that control movements of the head and shoulders. The hypoglossal nerves (XII) carry impulses that are concerned with movements of the tongue. (These nerves are discussed in greater detail in Chapter 13.)

Reticular Formation

In addition to reflex centers and tracts of fibers, a fine network of fibers and gray matter called the **reticular formation** (or **reticular activating system, RAS**) is also present within the medulla. This network extends from the upper spinal cord, through the medulla, pons, and midbrain, and into the diencephalon (Figure 12-10). Its fibers make extensive connections with centers in the diencephalon, cerebellum, and cerebrum, as well as with all important sensory and motor tracts within the brain stem.

The reticular formation seems to have several essential functions. First, it apparently controls general wakefulness or alertness of the brain. This function is based on evidence showing that injury to the formation often results in unconsciousness or deep coma. Second, it seems to filter sensory impulses entering the brain—accepting certain impulses and enhancing their relay to the cerebrum, while rejecting others. In this way it protects the brain from a constant barrage of sensory "noise" that would certainly become overwhelming. The reticular formation also influences certain motor activities, particularly the coordination of muscular movement. Some experts believe the reticular formation influences the processes of thinking and abstraction in the cerebrum.

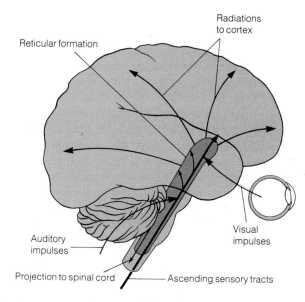

Figure 12-10 Reticular formation of the brain. This system of neurons appears to control conscious alertness of the brain and filters out unimportant sensory impulses before they reach the cerebrum.

choroid:	(Gr.) *chorion*, skin + *eidos*, form, shape	hydrocephalus:	(Gr.) *hydor*, water + *kephale*, head
plexus:	(L.) *plexus*, interwoven, braided	septum:	(L.) *septum*, partition
arachnoid:	(Gr.) *arachne*, spider + *eidos*, form, shape	pellucidum:	(L.) *pellucidus*, transparent
villus:	(L.) *villus*, tuft of hair		

Ventricles and Meninges of the Central Nervous System

After completing this section, you should be able to:

1. Describe the anatomy of the ventricles of the brain.
2. Locate the ventricles of the brain in terms of the major structures they lie within.
3. Describe the three meningeal layers that surround the brain and spinal cord.
4. Discuss the flow of cerebrospinal fluid through the ventricles and meninges.
5. Describe the blood-brain barrier and discuss how it regulates the flow of material from the blood to brain cells.

> **CLINICAL NOTE**
>
> **HYDROCEPHALUS**
>
> **Hydrocephalus** (hī·drō·sĕf´ă·lŭs) is a condition created by an unusual accumulation of cerebrospinal fluid (CSF) in the brain and an increase in intracranial pressure. This results in a rapid, grotesque enlargement of the soft skull in the developing child. In the type of hydrocephalus known as **communicating hydrocephalus,** an abnormal amount of CSF collects because of a decrease in the absorption of fluid from the subarachnoid space on the brain's surface. Tumors, infections, trauma, and abnormal accumulations of blood can lead to communicating hydrocephalus. In **noncommunicating hydrocephalus,** however, an obstruction is present somewhere in the system; therefore the fluid does not circulate freely to reach an area where the fluid can be reabsorbed.
>
> Hydrocephalus is usually a congenital defect. If left untreated, the mortality rate is about 50%. Three-fourths of those who survive have severe mental and physical handicaps. The usual treatment for hydrocephalus is the insertion of a tube with valves (a shunt or catheter) into the ventricles, which provides for a one-way flow of fluid and allows drainage into the abdominal cavity and the right atrium of the heart. This process relieves the cranial pressure. If the procedure is performed on a child, periodic lengthening of the tube will be necessary as the child grows.

In the embryo, the development of the brain involves the differentiation of a hollow neural tube that enlarges anteriorly into the adult structures (see box Embryonic Development of the Nervous System). The internal cavity of the tube enlarges simultaneously with the surrounding structures and persists in the adult brain as a series of cavities filled with cerebrospinal fluid.

Cerebrospinal fluid (CSF) is similar in composition and consistency to the fluid surrounding most tissues of the body (see Chapter 26 for a detailed description of tissue fluid). It is formed through a filtration process from three clumps of highly branched blood vessels, each called a **choroid plexus** (kō´royd plĕks´ŭs), and the neuroglial cells that cover these vessels. Each choroid plexus is associated with a specific part of the ventricular system (Figure 12-11). The constant addition of cerebrospinal fluid from the choroid plexuses to the ventricular system produces a regular circulation of fluid through the ventricles and spinal cord (see Clinical Note: Hydrocephalus). Eventually, the fluid makes its way to the exterior surface of the brain where it is reabsorbed by the tiny fingerlike projections called **arachnoid villi** that cover the brain (Figure 12-11). Once reabsorbed, the cerebrospinal fluid returns to the blood from which it was formed. Cerebrospinal fluid functions as a physical shock absorber, diminishing the effects of any sudden violent blow or movement to delicate tissues of the brain. In spite of the presence of this fluid, the brain is still susceptible to concussion.

Ventricles of the Brain

The series of cavities of the brain form four distinct regions called **ventricles:** two lateral ventricles (first and second ventricles); the third ventricle, a narrow, interconnecting cerebral aqueduct; and the fourth ventricle (Figure 12-12). (They are numbered based on the direction of flow of CSF.) The **lateral ventricles** are U-shaped cavities that occupy the medial portions of the two cerebral hemispheres. The open portion of the U is directed anteriorly. Three specific extensions are recognized: the **anterior horn** in the frontal lobe, the **inferior horn** in the parietal lobe, and the **posterior horn**, which projects from the base of the U into the occipital lobe. The **septum pellucidum** (sĕp´tŭm pĕl·lū´sĭd·ŭm) is a vertical membrane that separates the two lateral ventricles from each other. A small opening called the **interventricular foramen** (or **foramen of Monro**) connects each lateral ventricle and the adjoining third ventricle, allowing cerebrospinal fluid to circulate among these ventricles.

The **third ventricle,** a very thin cavity, separates the thalamic halves. It connects the lateral ventricles via the interventricular foramen and surrounds the massa intermedia of the thalamus. Cerebrospinal fluid enters the third ventricle from above and passes into the next ventricular region, the **cerebral aqueduct.** This narrow canal (also called the **aqueduct of Sylvius**) conducts cerebrospinal fluid from the third ventricle to the pons, cerebellum, medulla oblongata, and spinal cord.

The **fourth ventricle** is triangular and located between the cerebellum and pons and the medulla oblongata (Figure

308 Chapter 12 The Central Nervous System

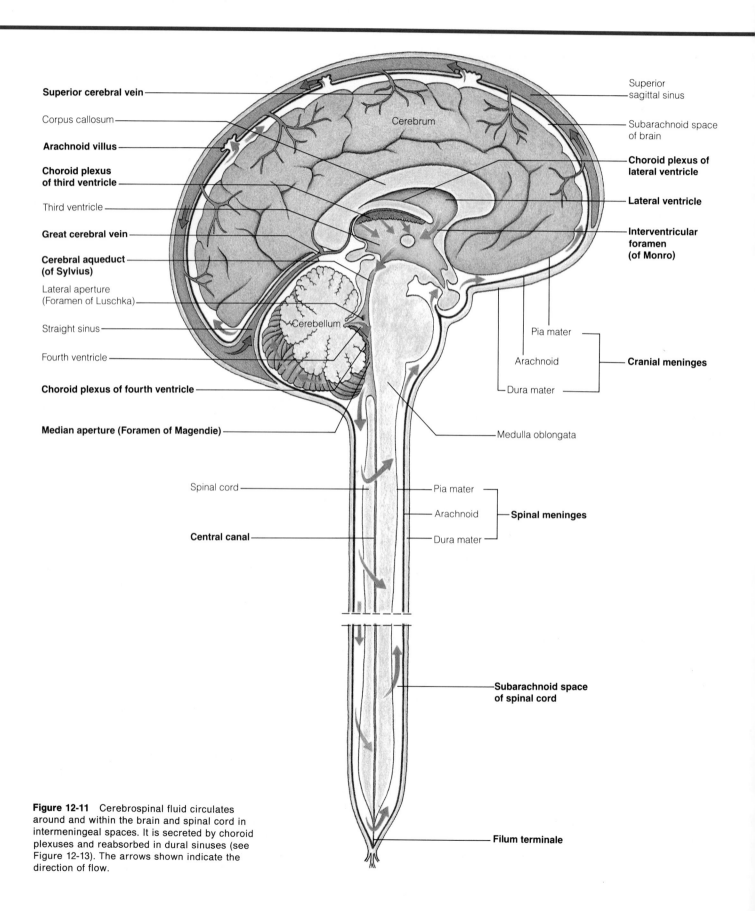

Figure 12-11 Cerebrospinal fluid circulates around and within the brain and spinal cord in intermeningeal spaces. It is secreted by choroid plexuses and reabsorbed in dural sinuses (see Figure 12-13). The arrows shown indicate the direction of flow.

dura:	(L.) *durus*, hard		
mater:	(L.) *mater*, mother	tentorium:	(L.) *tentorium*, tent
falx:	(L.) *falx*, sickle	pia:	(L.) *pia*, soft

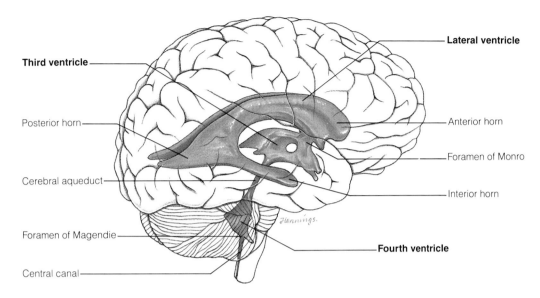

Figure 12-12 Ventricles of the brain.

12-12). Located in its roof, inferior to the cerebellum, is one choroid plexus responsible for production of the cerebrospinal fluid that passes through the narrow subarachnoid space that surrounds the brain and spinal cord as part of the meninges (see the next section). The fourth ventricle has three prominent foramina. Two of these (called the **foramina of Luschka**) lie in the superior lateral walls of the fourth ventricle. The third foramen (called the **foramen of Magendie**) lies in its posterior, inferior wall. The fourth ventricle is continuous inferiorly with the central canal of the spinal cord.

Meninges of the Brain and Spinal Cord

The brain and spinal cord are covered by three membranous tissue layers called **meninges** (měn·ĭn′jēz) that channel the flow of cerebrospinal fluid around the brain and spinal cord and provide additional physical protection to soft nerve tissues. The three meninges are the dura mater, pia mater, and arachnoid (Figure 12-13).

Dura Mater

The **dura mater** is the outermost of the three meninges (Figure 12-13). It consists of a double layer of tough ("dura"), fibrous connective tissue that covers the brain and attaches to the cranial bones. The external sublayer terminates at the foramen magnum (the large opening at the base of the skull), but the internal sublayer extends through this opening and continues onto the spinal cord. The two dural layers are fused over most of the brain forming a single tough membrane whose outer surface is the inner lining of the skull bones. In several places these two layers separate forming **dural sinuses** that collect venous blood and cerebrospinal fluid from the brain and return it to large veins that drain blood from this area.

The internal layer of dura mater extends into three fissures of the brain forming three prominent partitions: the **falx** (fălks) **cerebri, falx cerebelli,** and **tentorium cerebelli** (Figure 12-14). The first one extends from the crista galli of the ethmoid bone (see Figure 7-11), where it attaches posteriorly through the longitudinal fissure, which separates the cerebral hemispheres. The falx cerebelli extends between the cerebellar hemispheres, and the tentorium cerebelli extends horizontally through the fissure that separates the cerebrum and cerebellum. Another small partition called the **diaphragma sellae** (dī′ă·frăm·ă sěl′ě) forms a roof over the pituitary gland. A small circular hole in this partition contains the infundibulum (see Figure 12-1), which connects the hypothalamus to the pituitary gland.

Pia Mater

The **pia mater**, the innermost meninx (singular of meninges), is a thin, vascularized, connective tissue membrane that adheres closely to the outer surface of the brain, folding into sulci and fissures (Figure 12-13). This mater and part of the arachnoid form the choroid plexuses, which produce cerebrospinal fluid. The pia mater extends through the foramen magnum and covers the entire spinal cord medial to the dura mater. At the inferior end of the spinal cord (at about the level of the second lumbar vertebra), the pia mater continues

filum: (L.) *filum*, thread

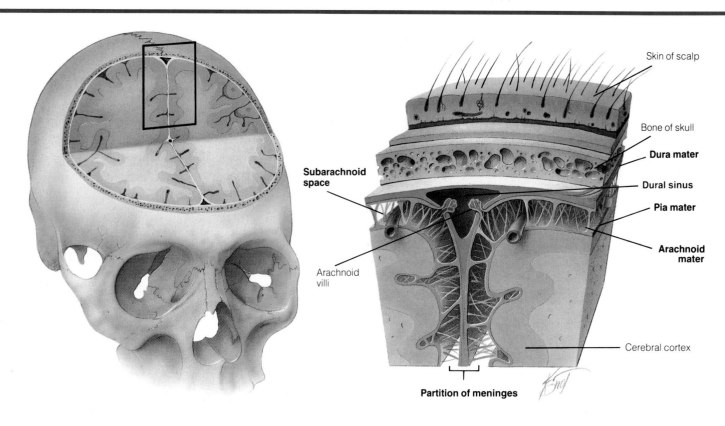

Figure 12-13 Meninges of the brain. This section shows the meninges and the fluid-filled spaces associated with the surface of the brain.

as a narrow, threadlike extension called the **filum terminale**, which extends caudally and attaches to the second coccygeal segment.

Arachnoid

The **arachnoid** (ă·răk'noyd) occupies the space between the other two meninges, although it does not attach equally to both. Externally, the arachnoid adheres closely to the dura mater, except for the presence of a narrow subdural space between them containing a thin film of cerebrospinal fluid. Internally, the arachnoid is loosely attached to the pia mater by delicate strands of fibrous connective tissue called **trabeculae** (tră·běk'ū·lē) that span a larger **subarachnoid space** filled with cerebrospinal fluid (Figure 12-13). The arachnoid layer extends into the large longitudinal and transverse cerebral fissures, and it bridges the sulci and other brain depressions.

From the arachnoid, the fingerlike arachnoid villi project into the dural sinuses. Here, cerebrospinal fluid returns from the subarachnoid space to the blood in the dural sinus, completing the circulatory route of this important fluid (Figure 12-11).

Blood Supply

At rest, four large arteries carry approximately 20% of the blood leaving the heart to the brain (see Figure 23-9), bringing a constant supply of nutrients and oxygen and carrying away wastes. Brain tissue is one of the more metabolically active tissues in the body, yet unlike highly active muscle cells, brain cells do not store large quantities of energy-rich carbohydrates. Their need for a constant blood supply is so critical that an interruption of blood flow for as short a time as 15 sec can result in a loss of consciousness, and 4 min without fresh blood can result in death of irreplaceable brain cells.

The capillaries that enter the brain penetrate all regions; however, unlike the situation in other organs, the capillaries do not come into intimate contact with the functional cells, the neurons. Instead, the capillaries are encased by exten-

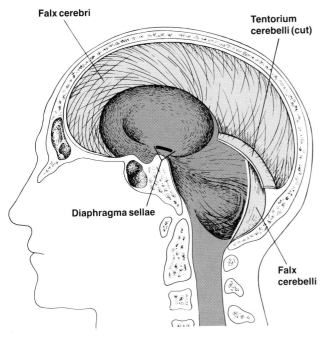

Figure 12-14 The falx cerebri and falx cerebelli are vertical extensions of dura mater that partition the left and right hemispheres of the cerebrum and cerebellum, respectively. The tentorium cerebelli lies between the cerebrum and cerebellum.

sions from neuroglial astrocytes (see Chapter 11) that impose a **blood-brain barrier (BBB)** between the blood and the neurons. This barrier restricts the movement of larger molecules from the blood and thus controls which materials can enter the extracellular brain fluid. Glucose, oxygen, carbon dioxide, water, and certain fat-soluble materials such as alcohol cross the BBB quite readily. Certain other substances, however, such as sucrose, chloride ions, creatinine, and inulin, pass through very slowly. Proteins and other large molecules are effectively prevented from crossing the barrier.

The BBB is beneficial because it prevents the entrance of toxins and other harmful materials into the highly sensitive brain tissue. However, it must be taken into consideration in certain drug therapy situations. Antibiotics do not cross the BBB, therefore the blood cannot deliver them to infections in the brain. It has been shown that an increase in the concentration of dopamine (a neurotransmitter) in the brain is an effective treatment for Parkinson's disease. However, because dopamine cannot cross the blood-brain barrier, it cannot be used directly to relieve symptoms in a patient. Instead, treatment involves administration of L-dopa, a precursor molecule that can cross the barrier and enter brain tissue where it is transformed into dopamine.

> **CLINICAL NOTE**
>
> **SPINAL TAP**
>
> A **spinal tap** or lumbar puncture is the insertion of a needle between the lumbar vertebrae below the end of the spinal cord. The needle is usually introduced between the L4 and L5 vertebrae into the subarachnoid space. The fluid obtained in this manner is cerebrospinal fluid. CSF can be examined microscopically for the presence of blood cells to determine if there is bleeding into this space, as in the case of a subarachnoid hemorrhage. The fluid can also be examined for sugar, protein, and other constituents that could give the clinician more information about possible diseases of the brain. The pressure of the fluid can also be measured to determine if the intracranial pressure has increased. Radiopaque dye can be inserted into the subarachnoid space to allow constrictions in the spinal canal to be seen by X ray.

Spinal Cord

After completing this section, you should be able to:

1. Describe the anatomy of the spinal cord.
2. Discuss the functions of the fiber tracts within the spinal cord.
3. Compare the functions of gray and white matter within the cord.
4. Describe typical spinal reflexes and discuss their importance.
5. Distinguish between monosynaptic and polysynaptic reflexes and give examples of each.

Anatomy of the Spinal Cord

At the lower end of the medulla oblongata where the brain stem terminates, columns of gray and white matter in the medulla become organized into the **spinal cord.** It serves two major functions. First, it contains neurons that connect sensory and motor areas of the brain with other parts of the body. These neurons provide pathways for conducting impulses in either direction—from sensory receptors to the brain then back along motor neurons to effectors, muscles, and glands. Second, the spinal cord directly connects sensory neurons with appropriate motor neurons that produce responses independent of brain influences **(spinal reflexes).**

The spinal cord is a cylinder of nerve tissue somewhat flattened anteriorly and posteriorly. It is contained within the **verебral canal** of the vertebrae and begins at the foramen magnum, passing through the vertebral foramen. In

conus: (Gr.) *konos*, cone
equina: (L.) *equus*, horse
internuncial: (L.) *inter*, between + *nuncius*, messenger

an embryo the spinal cord and vertebral column are approximately the same length. However, during late embryonic and postnatal growth, the spinal cord elongates at a slower rate than the vertebral column, so that the adult spinal cord does not extend the full length of the vertebral canal. Instead, it terminates in a cone-shaped **conus medullaris** (kō′nŭs mĕd′ū·lār″ĭs) between the first and second lumbar vertebrae (Figure 12-15). The spinal nerves that exit the lower lumbar and sacral regions are very long and they angle inferiorly until they reach their point of exit from the vertebral canal. Early anatomists thought this group of spinal nerves resembled a horse's tail, and they gave it the name **cauda equina** (kaw′dă ē·kwĭn′ă) (Figure 12-15).

The anterior surface of the spinal cord has a deep vertical groove called the **anterior median fissure** (Figure 12-16), and the posterior side bears a narrower, yet deeper slitlike groove, the **posterior median sulcus.** The lateral sides of the cord show two shallow sulci: the **anterior lateral sulcus** and **posterior lateral sulcus.** An additional groove, the **posterior intermediate sulcus,** lies between the posterior median sulcus and posterior lateral sulcus in the cervical and upper thoracic regions.

The spinal cord is divided into three major regions that correspond to the regions of the vertebral column it passes through: cervical, thoracic, and lumbar. In the cervical and lumbar regions are two swellings called the **cervical** and **lumbar enlargements,** respectively (Figure 12-15). These bulges mark places where spinal nerves leave the cord and innervate the upper and lower extremities.

The same meninges that cover and protect the brain also cover and shield the spinal cord. The dura mater, pia mater, and arachnoid extend through the foramen magnum, forming a continuous wrapping of the spinal cord. At its lower end the dura mater and arachnoid terminate. Only the pia mater continues inferiorly to form the filum terminale (fī′lŭm tĕr·mĭ·nă lē), which reaches the coccyx at the lower end of the vertebral column (Figure 12-15).

Gray and White Matter of the Spinal Cord

As in the brain, nerve tissue in the spinal cord is divided into gray and white matter. Note that in the spinal cord, however, the locations of the two are reversed.

Gray matter of the spinal cord consists mainly of neurons, unmyelinated axons, dendrites, and neuroglia. In a cross section of the spinal cord, gray matter forms an **H** shape in the center of the cord (Figure 12-16). The central canal is a remnant of the channel formed during the embryonic development of the neural tube. This canal is continuous with the fourth ventricle of the brain.

The horizontal "bar" of spinal gray matter is called the **gray commissure,** further subdivided by the central canal

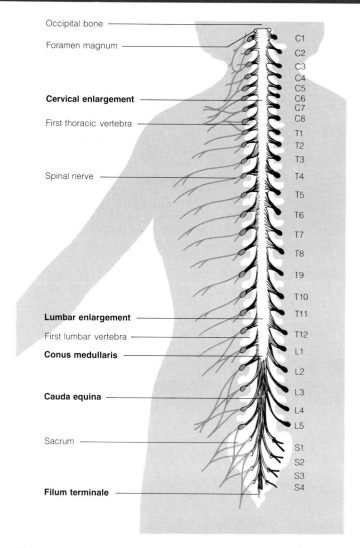

Figure 12-15 Posterior view of the spinal cord and vertebral column. Note that the spinal nerves are named after adjacent vertebrae and that the cord itself terminates near the first lumbar vertebra. The cervical, thoracic, lumbar, and sacral spinal nerves are abbreviated C, T, L, and S.

into anterior and posterior portions. Two **anterior columns** of gray matter extend anteriorly from the horizontal bar. These anterior columns are composed of the cell bodies of motor neurons that transmit effector impulses through spinal nerves. Two narrower **posterior columns** of gray matter extend from the gray commissure. They approach the surface of the cord more closely than the anterior columns. These posterior columns consist of sensory nerves, which send impulses into the cord from spinal nerves, and cell bodies and axons of **internuncial** (ĭn″tĕr·nŭn′shĭ·ăl) (or **association**) **neurons** that transmit sensory information, either to other neurons at the same level in the cord or toward higher centers in the cord or brain.

funiculi: (L.) *funiculus*, little cord
fasciculus: (L.) *fasciculus*, little bundle

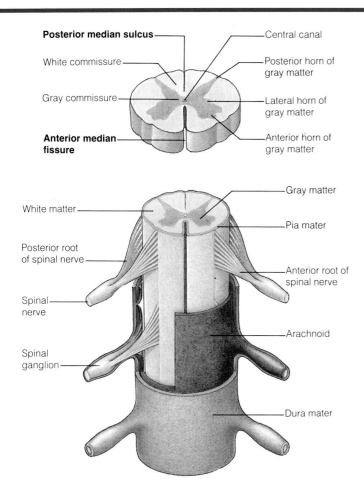

Figure 12-16 The organization of the spinal cord and related spinal nerves.

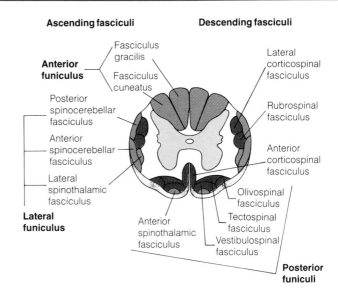

Figure 12-17 Spinal cord white matter is organized into both ascending and descending fasciculi. Ascending tracts are labeled on the left side of the diagram and descending tracts on the right side, although both types occur on both sides.

The spinal white matter is mainly axons of myelinated nerve fibers embedded in neuroglia. Small numbers of nonmyelinated fibers are also present but are greatly outnumbered by myelinated ones. Practically all the fibers are organized into tracts that run lengthwise, except for the **white commissure** that lies anterior to the gray commissure (Figure 12-16). It contains horizontal tracts that cross the spinal cord. The white matter of the cord conducts impulses up and down the cord.

Extensions of the anterior and posterior columns of gray matter divide the white matter into the **anterior funiculi** (fū·nĭk′ū·lī), **lateral funiculi,** and **posterior funiculi** on each side (Figure 12-17). Within these loosely defined areas are specific tracts (bundles) or **fasciculi** (fă·sĭk′ū·lī). Each fasciculus contains axons functionally related to one another; that is, the axons either originate or terminate in the same general region. The fasciculi are functionally either ascending or descending tracts, depending on the direction in which its fibers conduct impulses (Table 12-2).

Ascending and Descending Fasciculi

The **ascending fasciculi** (Figure 12-17) conduct impulses from sensory neurons in the peripheral nervous system through the spinal cord to higher brain centers. The **anterior spinothalamic** (spī″nō·thăl·ăm′ĭk) **fasciculus** conducts impulses of touch and pressure up the anterior part of the cord to thalamic centers, which in turn relay the impulses to appropriate areas in the cerebral cortex. The **lateral spinothalamic fasciculus** conducts impulses that register temperature and pain up the lateral side of the cord to centers in the thalamus, which relay the information to cortical portions of the cerebrum. Both **anterior** and **posterior spinocerebellar** (spī″nō·sĕr·ĕ·bĕl′ăr) **fasciculi** conduct impulses carrying information to the cerebellum regarding the amount of tension or stretch in muscles and tendons. The **fasciculus cuneatus** and **fasciculus gracilis** transmit impulses to centers in the medulla oblongata, namely, the nucleus cuneatus and nucleus gracilis. The fasciculus cuneatus conducts impulses from the upper part of the body and upper extremity, while the fasciculus gracilis transmits information from the lower part of the body. These latter two spinal tracts transmit sensory stimuli that inform a person of the location of a body part without the necessity of looking at it. They also transmit discriminatory touch stimuli informing a person of the exact place on the skin where an object touches the body, and they provide information about the size, shape, and texture of that object.

tectospinal: (L.) *tectum*, roof + *spina*, thorn
hemiplegia: (Gr.) *hemi*, half + *plege*, stroke
aphasia: (Gr.) *a*, not + *phasis*, speaking

Table 12-2 SUMMARY OF MAJOR ASCENDING AND DESCENDING TRACTS IN THE SPINAL CORD

Tract	Function
Ascending Fasciculi	
Anterior spinothalamic	Conducts impulses off touch and pressure to thalamus which relays them to cerebral cortex
Lateral spinothalamic	Conducts impulses of temperature and pain to thalamus which relays them to cerebral cortex
Anterior and posterior spinocerebellar	Conduct impulses of muscle tone and stretching of tendons to cerebellum
Fasciculus cuneatus and fasciculus gracilis	Conduct impulses of discriminatory touch and body awareness to medulla oblongata
Descending Fasciculi	
Anterior and lateral corticospinal	Conduct impulses from primary motor area of cerebral cortex to body muscles
Vestibulospinal	Conducts impulses controlling equilibrium from medulla to skeletal muscles
Tectospinal and rubrospinal	Conduct impulses controlling posture from midbrain to skeletal muscles

Descending fasciculi conduct motor impulses from brain centers down the spinal cord and to skeletal muscles. The **anterior** and **lateral corticospinal** (kŏr″tĭ·kōspī′năl) **fasciculi** (Figure 12-17) transmit impulses from the primary motor area of the cerebral cortex to muscles of the body. Skeletal muscles receive impulses from the **vestibulospinal** (vĕs·tĭb″ū·lō·spī′năl) **fasciculus** that are involved with equilibrium and posture. The **rubrospinal** (rū″brō·spī′năl) and **tectospinal** (tĕk″tō·spī′năl) **fasciculi** transmit impulses from nuclei in the midbrain to various skeletal muscles.

Thus far, we have described tracts within the ascending and descending fasciculi in terms of whether they conduct sensory or motor impulses up or down the cord. However, some impulses enter the right side of the cord and are transmitted through association neurons to the proper fasciculus on the opposite side of the cord. Similarly, motor impulses that originate on the left side of the brain may ultimately stimulate muscles on the right side of the body, and vice versa. This crossover of fibers, either in the spinal cord or brain, causes the right side of the brain to receive sensory impulses from and send motor impulses to the left side of the body, whereas the left side of the brain sends and receives impulses to and from the right side of the body.

Spinal Cord Reflexes

One of the functions of the spinal cord is to accommodate **spinal reflexes.** Such a reflex is a mechanism for producing an automatic response to a stimulus and usually involves only neurons in the spinal cord and not in the brain. Because reflexes in general are the basic units of behavior, this function is of prime importance.

Monosynaptic Reflexes

As the term implies, a **monosynaptic** (mŏn′ō·sĭn·ăp″tĭk) **reflex** involves only one synapse between a sensory and a

---

CLINICAL NOTE

STROKE

Strokes, or **cerebrovascular accidents (CVAs),** are the most common cause of neurologic disability in Western society. Most CVAs are the results of **atherosclerotic** (ăth″ĕr·ō″sklĕ·rō′tĭk) disease, which is hardening of the arteries or formations of cholesterol and lipid plaques in the arteries, or **hypertension** (high blood pressure). A CVA is a devastating occurrence, causing a previously capable person to become quite ill, dependent, and often handicapped.

One cause of a CVA is a traveling blood clot, or **embolus** (ĕm′bō·lŭs), that blocks a major cerebral blood vessel, leading to a decreased blood supply and eventual death of tissue. A CVA can also be caused by a break in a vessel creating a hemorrhage in the brain tissue (cerebral hemorrhage). A CVA is only one of many types of cerebrovascular diseases. Others include **cerebral insufficiency** (transient disruptions of blood flow); **cerebral infarction** (the death of brain cells due to a lack of blood flow); and **thrombosis** (the formation of a plug of plaque or fibrin clot in a blood vessel) of an intra- or extracranial artery.

Strokes are associated with increased blood lipids, smoking, diabetes, high blood pressure, and advanced age. Manifestations of CVAs include headaches, vomiting, and **hemiplegia** (paralysis of half of the body) on the side of the body opposite the affected side of the brain. This opposite effect is due to the crossover of motor and sensory fibers in the central nervous system. If the stroke occurs on the dominant side (left for right-handed people and right for left-handed people) then aphasia, due to involvement of the speech area, is likely to accompany the CVA. **Aphasia** (ă·fā′zē·ă) is a deficit related to language and can be the loss of verbal expression, the loss of the ability to comprehend language, or both.

Related to cerebrovascular accidents are **transient ischemic** (ĭs·kē′mĭk) **attacks (TIAS),** which are short episodes of disordered cerebral function. The episodes last minutes or hours, sometimes up to 24 hours, after which the symptoms of the disordered neuronal function disappear without any residual effects. Their symptoms include loss of vision in one eye, loss of sensation in one-half of the body, difficulty in performing familiar acts, inability to recognize familiar objects or faces, double vision, and staggering. Nausea, vomiting, and dizziness may also occur. TIAs are thought to be caused by tiny clots or local alterations in cerebral blood flow. While the effects of the TIA do not remain, the occurrence is considered to be a forewarning of a stroke.

CLINICAL NOTE

SPINAL CORD INJURY

Accidents and trauma may cause damage to the spinal cord resulting in **transection** (cut across) or **partial section** of the cord. Such injuries cause partial or complete cessation of the transmission of impulses up and down the spinal cord and have a variety of effects depending on the location and extent of the damage to the cord. Spinal cord injuries are among the most tragic of accidents since they cause disability and dependence in otherwise active, healthy individuals.

There are basically two mechanisms by which the damage to the cord occurs. The first mechanism is the initial insult to the nervous pathways that go through the spinal cord. The second mechanism is the progressive damage caused by decreased nourishment to the neurons due to resulting inflammation, local edema, or decreased blood supply to the cord that may accompany severe accidents or trauma.

Spinal shock describes the period of about three weeks after the initial injury. During this period the person may have little control over bowels and bladder, blood may pool in the lower extremities (because there is no muscular activity to return blood to the heart), all sensation is lost, and adequate circulation cannot be maintained. All these are effects of the impedance of impulses from the higher brain control center through the damaged area of the cord. As spinal shock ends, some sensation, reflexes, and motion return depending on the location and extent of injury.

If the cord injury is not a complete transection but a **hemisection** (half cut), then motor losses will occur on the same side as the injury (motor neurons cross in the brain, not in the cord). The loss of sensation due to hemisection of the cord vary; the pattern and type of loss give indications of the location of the lesion. In compression injuries, the posterior columns of the white matter at the level of the injury are involved. Sensory fibers carrying information about proprioception (the sense of the position of the body in space) may be injured.

If the cord is completely transected, all areas of skin innervated by nerves below the level of the injury will lose all sensation and motion; all sensation and motion above the transection will be preserved. If the injury is between vertebrae T1 and L2, **paraplegia** will result which is paralysis of the lower part of the body and both legs. If the injury is between C4 and T1, all four extremities and usually the trunk will be involved and **quadriplegia** will result. Injuries above C4 are often fatal. (See Unit III Case Study: Paralysis.)

In paraplegia the person may lose bowel and bladder control and may not have control over abdominal or lower back muscles. However, the reflex arcs are still intact so the person may experience mass reflex after stimulation. Upper motor neuron control is lost in both paraplegia and quadriplegia; this may lead to a dangerous situation called **autonomic hyperreflexia** (hī″pĕr·rē·flĕk′sē·ă). Because spinal reflexes are intact but not controlled by the upper motor neuron, reflex vasoconstriction may occur below the lesion. Ordinarily, the higher autonomic centers would regulate the degree of vasomotor constriction and dilation, but because of the spinal lesion, impulses from the control centers cannot be transmitted downward. Widespread and unremitting vasoconstrictions can occur below the spinal lesion and drive the blood pressure higher and higher. This rise is sensed appropriately, but body mechanisms to decrease the blood pressure are ineffective. The person may appear extremely flushed as blood vessels above the lesion dilate; massive sweating may occur and the severe hypertension may lead to a cerebrovascular accident unless the stimulus is located and removed. Stimuli include a full bladder, a wrinkle in the sheet or chair cushion, unusual pressure somewhere below the spinal lesion, or even an attendant's touch.

Treatment of spinal cord injuries may include surgery if a fracture of a vertebra is involved. Immobilization to limit further damage is usually employed. Sometimes ice or corticosteroids are used to limit local swelling. Extensive and skilled nursing care is necessary to see the person through the course of injury. Rehabilitation teams made up of many health professionals assist the individual to reach full potential. If function can return, it may take as long as 2 years.

Quadriplegic being aided by a specially trained monkey.

motor neuron in a spinal reflex arc (Figure 12-18). A **spinal reflex arc** is the pathway whereby nerve impulses from sensory neurons reach motor neurons without traveling to the brain. Since the brain is not involved in the pathway, a response to a stimulus can be elicited without conscious input. One example of a monosynaptic reflex is the **stretch reflex.** It is one of the least complicated reflexes, yet it illustrates how the nervous system operates when responding to stimuli. The receptor in this reflex is a **muscle spindle.** When a muscle is stretched, its muscle spindles respond, each one fires and relays impulses along its axon to the spinal cord. The axon synapses with a motor neuron that sends impulses to other fibers in the same muscle, causing the muscle to contract, thereby counteracting the stretching. This action is the stretch reflex, and it is very important in maintaining muscle tone. One detrimental effect of being confined to bed is that muscles are not stretched to the same degree as they are in the movements of everyday living. This lack of stretching causes the muscles of a bedridden individual to lose tone and become flabby. Some physical therapy programs are designed to overcome this condition and to maintain muscle tone in nonambulatory patients.

Impulses of monosynaptic reflexes, such as the stretch reflex, enter and leave the spinal cord on the same side. Any reflex confined to one side of the spinal cord is called an **ipsilateral reflex.** Ipsilateral reflex arcs are often used in medical examinations to determine the condition of the pa-

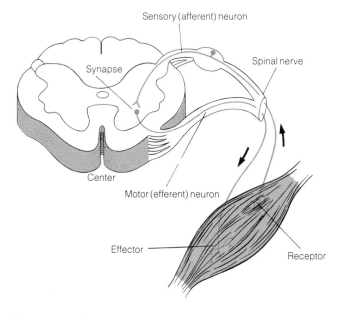

Figure 12-18 **The stretch reflex diagrammed here is an example of a monosynaptic, ipsilateral, spinal reflex arc.** Note that only one synapse is involved between two neurons.

CLINICAL NOTE

REFLEXES

A variety of reflexes can be elicited by an examiner seeking to discover the intactness of the nervous system, including the location of lesions, if they exist.

Reflexes are elicited by presenting a stimulus—some sensory input—and then observing the response to this sensory stimulation. The response can occur in smooth muscle or skeletal muscle. Since "automatic" circuits are involved, the person need not be conscious and, by definition, cannot control the reflex response.

The **muscle stretch reflex** or **deep tendon reflex** is elicited by tapping briskly on the tendons of a muscle group. This is the familiar one from the doctor's office, which is sometimes called the "knee jerk response." This reflex is the only monosynaptic reflex arc in the body; most reflex pathways involve numerous synaptic connections. The response is contraction of the muscle group stimulated by the tap. If there is no response, then some part of the reflex arc is not operating. This information added to other data can help locate neurological damage. The muscle stretch reflexes are rated on a scale of 0 to 4, with 0 for absent reflex and 4 for markedly hyperactive. These scalar judgments are then recorded on a body reflex map. The major sites of the muscle stretch reflex are the interior elbow for the biceps, the exterior elbow for the triceps, the wrist for the brachioradialis, the knee just below the patella for the patellar, and the back of the ankle for the achilles.

The **Babinski's sign** or **reflex** is present in children, but its presence after the age of 3 is considered a sign of pathological processes involving the pyramidal tract. The response is generated when the side of the sole of the foot is stroked firmly. If present, the great toe extends upward and the other toes fan outward. If absent, all the toes bend downward.

Normal reflexes in children are tested in the newborn nursery. **Moro's reflex,** or the **startle reflex,** is bilateral abduction and extension of the baby's arms and legs in response to a loud noise. After the legs and arms are extended, the baby pulls the arms and legs close to the body. This reflex appears at birth and disappears at about four months. The **tonic neck reflex** is extension of the leg and arm on the same side to which the examiner turns the head. The opposite arm and leg flexes. The baby looks like a fencer as it assumes this fencing position. This reflex appears soon after birth and disappears sometime between 4 and 6 months. The **palmar grasp** is elicited by placing an object, such as a finger, in the child's hand so that the baby tightly grasps the object. This appears at birth and disappears by about 4 months. The **rooting reflex** is elicited by stroking the infant's cheek. The baby turns the head to that side and opens its mouth slightly. This reflex is present at birth and disappears by 4 months. The important **sucking reflex** is an actual sucking motion of the mouth and tongue when the mouth is touched gently. This appears at birth and disappears by about 10 months. In most of these infant reflexes, their absence immediately after birth or their persistence beyond the usual age for disappearance signals pathology requiring further investigation.

contralateral: (L.) *contra*, against + *latus*, side

tient's nervous system. Perhaps the most familiar is the **knee jerk** (or **patellar**) **reflex.** When an examiner taps the patellar ligament just below the patient's knee cap, parts of the quadriceps femoris muscles are *stretched*. The reflex response is *contraction* of other parts of the quadriceps femoris, causing the knee to jerk. People suffering from certain illnesses, such as advanced syphilis and advanced diabetes, do not exhibit this reflex.

Polysynaptic Reflexes

Reflexes involving three or more neurons and more than one synapse are called **polysynaptic reflexes.** Many such reflexes exist in the repertoire of human response, but we will examine only two as examples of the pathways involved.

A typical and easily understood example of a polysynaptic reflex is the **flexor** (or **withdrawal**) **reflex.** This one occurs when part of an extremity is injured, for example, when a bee stings the sole of your foot. Pain receptors in the sole respond to the sting by sending impulses to the spinal cord (Figure 12-19), where the afferent neuron synapses with one or more association neurons in the gray matter of the cord. Association neurons, in turn, synapse with motor neurons leading to the large leg flexors. Once begun, the reflex withdraws your leg from the source of the stimulus (the bee). Because the pathway enters and leaves the spinal cord on the same side, this action/reaction is an ipsilateral reflex.

An impulse in a **contralateral reflex** crosses from one side of the spinal cord to the other (Figure 12-19). Such a reflex occurs in the bee sting described above. The nerve impulse generated by the sting travels along branches of the neuron, enters the spinal cord, and is transferred to association neurons in the gray commissure. These association neurons (often called **commissural interneurons** because they are located within the gray commissure) cross from one side

Figure 12-19 Polysynaptic reflex arcs involve several synapses among different neurons. (a) A sensory neuron transmits sensory impulses into the spinal cord. (b) Contralateral muscles respond to these impulses by stabilizing the opposite side of the body while the injured leg withdraws. The extensor muscle is stimulated to contract while contraction of the flexor muscle is inhibited.

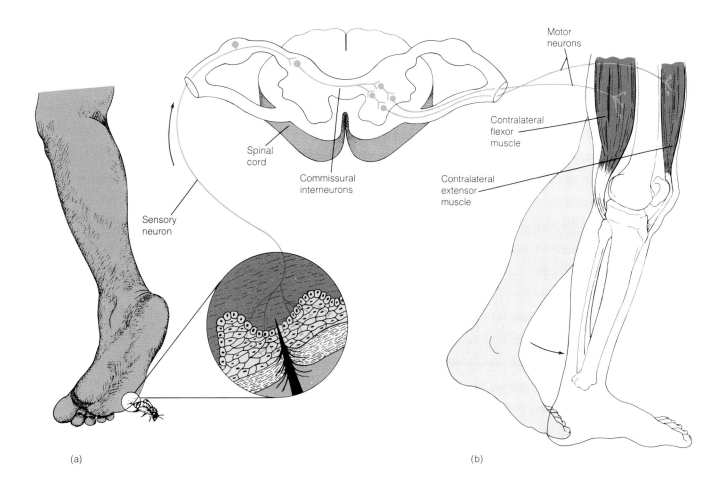

(a) (b)

of the cord to the other where they synapse with motor neurons. These motor neurons innervate the extensor muscles in your opposite leg, which are stimulated to contract simultaneously with the contraction of the flexors in the leg that was stung. The reflex initiated by the bee sting causes you to flex your stung leg and extend your other leg. This is a reflex called a **crossed extensor reflex.** Because this arc involves impulses that enter the spinal cord on one side and leave on the opposite side, it is *contralateral.*

The functional significance of the crossed extensor reflex is that it helps you to automatically maintain your balance. When the ipsilateral withdrawal reflex causes you to raise your right leg, all your body weight is transferred to your left leg. The crossed extensor reflex extends your left leg at the knee joint, thereby enabling the weight shift to be handled smoothly without a loss of balance.

Polysynaptic reflexes need not always be contralateral. For example, in the leg suffering the bee sting, two groups of muscles govern the hingelike movements of the knee. The hamstring group produces flexion, while the quadriceps femoris group extends the leg (see Chapter 10). Contraction of one group requires relaxation of the other group; simultaneous contraction of both muscle groups would result in functional paralysis of the knee. Coordinated motion of the leg would be lost. The ipsilateral flexor reflex cited here also involves synapses with spinal association neurons that have an inhibitory influence on motor neurons that innervate the extensors on the same side. Thus, while the flexors are stimulated, the ipsilateral extensors are inhibited, thereby permitting efficient movement of the leg. This simultaneous stimulation and inhibition of antagonistic groups of muscles is called **reciprocal innervation,** and it occurs in similar reflexes involving other joints.

In this discussion of polysynaptic reflexes, only the aspects involving the spinal cord have been mentioned. Obviously, when a person is stung, the pain sensation eventually reaches consciousness and the person is able to localize the injury site. In addition, the experience will most likely be remembered so that later, the mere sight of a bee will probably initiate actions to avoid a recurrence of the sting. These activities require transmission of the impulse in the cord to higher brain centers.

In this chapter we studied the central nervous system, consisting of the brain and spinal cord. You learned the locations of the sensory, association, and motor areas of the cerebral cortex, the major reflex centers in the brain stem, and the conduction tracts in the spinal cord. We also studied how spinal reflexes can provide automatic responses without the necessity of involving the brain. In the following chapter we will examine the peripheral nerves that extend from the brain and spinal cord and study how they receive stimuli and elicit muscular contractions.

THE CLINIC

EXAMINATION OF THE NERVOUS SYSTEM

The complete examination of the nervous system is probably the most difficult and time consuming of all medical examinations. The nervous system is involved in the function and integration of all of the activities of the body, and each must be evaluated in the complete neurologic examination. The major areas covered in the neurologic examination include (1) mental status, (2) cranial nerves, (3) motor system, (4) sensory system, and (5) reflexes.

As in all complete examinations, the patient's medical history is of utmost importance. Attention should be given to the patient's chief complaint, the mode of onset, and the course of the disease symptom by symptom. Social factors, history of recent travel, work, and home environment may all give clues to possible exposure to toxins or infectious agents. A thorough review of systems and family history may give clues to generalized or hereditary diseases. The past medical history may also produce information about previous trauma.

The actual neurologic examination begins when the patient enters the office as the physician observes the patient's gait, posture, and balance. The physician will also observe the patient's speech, facial expressions, and the vigor of the handshake. The formal examination begins as the patient explains his medical history and the doctor observes speech patterns, attention span, memory, orientation to time and place, and clarity of thought patterns. A complete evaluation of the cranial nerves and assessment of the special senses (hearing, vision, taste, and smell) follows. Examination of the motor functions includes the speed and strength of motion of all extremities and facial muscles including muscle bulk, tone, and coordination. Coordination is evaluated by asking the patient to perform tasks such as touching the finger to the nose, walking on heels and toes, and performing rapid alternating movements. The **deep tendon reflexes** are evaluated with the familiar percussion hammer, with the doctor noting the presence or absence of the reflexes and whether they are affected on one or both sides of the body. A search is made for **pathologic reflexes,** such as **Babinski's reflex** (extension of the great toe while stroking the bottom of the foot) or **Hoffman's reflex** (flexion of the thumb when snapping the fingers). There are many pathologic reflexes that may be present when damage has occurred to the nervous system. **Clonus** is a rapid jerking alternation of contraction and relaxation of a muscle that may be detected by sudden passive stretching of the muscle (as in dorsiflexing the foot).

A final assessment is made of the sensory function. Tests are conducted for sharp (pin prick) and dull (a whisp of cotton) sensation of the face, trunk, and extremities searching for areas of **hypoesthesia** (decreased sensation) or **anesthesia** (complete loss of sensation). The physician may also test for sense of hot or cold and for vibratory senses. Standard diagrams may be consulted that delineate the known distribution of cutaneous nerves. The physician may perform other tests if certain diseases are suspected. After completion of the detailed neurologic examination, specific laboratory procedures may be needed to confirm a suspected diagnosis.

STUDY OUTLINE

Telencephalon (Cerebrum) (pp. 291–298)

The *cerebrum* consists primarily of two large hemispheres, each composed of several major subdivisions: cerebral cortex, gray matter, white matter, and basal ganglia. Cavities within the brain are filled with *cerebrospinal fluid*.

Cerebral Cortex The *cerebral cortex* (referred to as gray matter) is the outer layer of tissue of the *cerebral hemispheres*. Its surface is convoluted into *sulci* and *gyri*. Each hemisphere is composed of *frontal, parietal, occipital* and *temporal lobes*.

Functional Areas of the Cerebral Cortex The cerebral cortex contains numerous specific centers divided into three major functional types: sensory, motor, and association. The *primary sensory area*, which receives impulses for touch, pressure, pain, temperature, and *proprioception* is located in the postcentral gyrus. The *primary visual area* is located in the occipital lobe. Both *primary auditory* and *olfactory areas* are in the temporal lobe. The *primary gustatory area* is in the parietal lobe.

The *primary motor area*, which controls movement of specific body parts, is in the precentral gyrus. The *premotor area*, which controls learned activities such as writing, is in the frontal lobe, as is the *motor area for speech*.

Most of the cerebral cortex is composed of *association areas* that relate different sense modalities with each other to form a composite sensory awareness, which then may elicit a co-ordinated motor response to a stimulus or set of stimuli.

White Matter of the Cerebrum *White matter* of the cerebrum consists of fibers organized into tracts that connect different parts of the brain with each other and with the spinal cord.

Gray Matter of the Cerebrum The *basal ganglia* are masses of *gray matter* buried within the cerebrum. They are important in transmitting motor impulses used to control and coordinate voluntary muscular movements.

Diencephalon (pp. 299–302)

The *diencephalon* is composed of three major regions: the thalamus, hypothalamus, and epithalamus.

Thalamus The *thalamus* consists of several nuclei that act as relay centers for both sensory and motor impulses.

Hypothalamus The *hypothalamus* regulates several homeostatic activities, including control of the autonomic nervous system, control of the endocrine system through the pituitary gland, hunger and thirst, body temperature, wakefulness and sleepiness, and emotions such as anger, rage, and aggression.

Epithalamus The *epithalamus* consists of the *pineal body* (an endocrine gland) and the *posterior commissure*, which contains fibers that connect the cerebrum and midbrain.

Limbic System The *limbic system* consists of parts of the cerebral hemispheres and the diencephalon, and it plays an active part in controlling memory and emotional behavior.

Mesencephalon (pp. 302–303)

The *mesencephalon* (midbrain) is composed of the corpora quadrigemina, cerebral peduncles, and *cerebral aqueduct*.

Corpora Quadrigemina The *corpora quadrigemina* (including the *superior* and *inferior colliculi*) receive sensory impulses and function primarily as reflex centers for visual, auditory, and tactile stimuli.

Cerebral Peduncles The *cerebral peduncles* contain fiber tracts that connect the cerebrum with other parts of the nervous system.

Metencephalon (pp. 303–305)

The *metencephalon*, the second largest region of the brain, consists of the cerebellum and pons.

Cerebellum The *cerebellum* consists of two hemispheres joined by a narrow *vermis* in the center. It is connected to other parts of the brain through *cerebellar peduncles*, and it functions as a reflex center for control and coordination of skeletal muscles.

Pons The *pons* is a bridge of white fibers that connects parts of the brain with one another and with the spinal cord.

Myelencephalon (pp. 305–307)

Medulla Oblongata The *myelencephalon* (*medulla oblongata*) connects the brain to the spinal cord. In addition to white fibers, it also contains nuclei that are reflex centers for cutaneous sensations, heart rate, blood pressure, breathing rate, swallowing, coughing, sneezing, and vomiting. Cranial nerves VIII, IX, X, XI, and XII also arise from nuclei in the myelencephalon.

Reticular Formation The *reticular formation* (or RAS) is a fine network of fibers and gray matter within the myelencephalon that controls wakefulness and alertness in the brain. It also acts as a filter for sensory impulses impinging on the brain, accepting important ones and rejecting others.

Ventricles and Meninges of the Central Nervous System (pp. 307–311)

Ventricles of the Brain The brain contains a series of cavities: *lateral ventricles, third ventricle, cerebral aqueduct,* and *fourth ventricle*. Cerebrospinal fluid fills the cavities and acts as a shock absorber.

Meninges of the Brain and Spinal Cord The brain and spinal cord are covered by three membranous tissue layers called *meninges* that channel the flow of cerebrospinal fluid around the brain and spinal cord and provide additional physical protection to soft nerve tissues.

The *dura mater* is the outermost of the three meninges and consists of a double layer of tough, fibrous connective tissue.

The *pia mater*, the innermost layer, is a thin vascularized connective tissue membrane that adheres closely to the outer surface of the brain, folding into sulci and fissures.

The *arachnoid* lies between and bridges the other two meninges.

Blood Supply Four large arteries supply the brain with

blood. A *blood-brain barrier* of neurological astrocytes restricts movement of large molecules from the blood into the extracellular brain fluid.

Spinal Cord (pp. 311–318)

Anatomy of the Spinal Cord The *spinal cord* connects sensory and motor areas of the brain with other parts of the body and directly connects sensory neurons with appropriate motor neurons that produce *spinal reflexes*.

Externally, the spinal cord is divided into three major regions: cervical, thoracic, and lumbar. Internally, it consists of a central core of gray matter surrounded by a cylinder of white matter. The gray matter is composed primarily of neurons. The white matter consists mainly of myelinated nerve fibers and neuroglia organized into *ascending* and *descending fasciculi* (tracts). These tracts conduct sensory impulses up the cord to the brain, and motor impulses down to muscles and glands.

Spinal Cord Reflexes *Spinal reflexes* are automatic responses to stimuli and usually involve only neurons in the spinal cord. *Monosynaptic reflexes* involve only one synapse between a sensory and a motor neuron. Polysynaptic reflexes involve three or more neurons. Both types of reflexes may occur on the same side of the body (*ipsilateral*) or on different sides of the body (*contralateral*).

SELF-TEST OF CHAPTER OBJECTIVES

True-False Questions
1. Commissural tracts in the white matter of the cerebrum transmit impulses from one hemisphere to the other.
2. The cerebellum, pons, and medulla form the brain stem.
3. The orientation of gray matter and white matter is the same in both the brain and spinal cord.
4. The nervous system develops from embryonic ectoderm tissue.
5. Cranial nerves are classified as part of the central nervous system.
6. Generally speaking, cerebrospinal fluid circulates in a superior to inferior direction in the brain.
7. The arachnoid is the most interior layer of the three meningeal layers.
8. The cerebral hemispheres contain only sensory areas, whereas the cerebellum contains only motor areas.
9. Spinal reflexes are responses to stimuli that do not require the involvement of neurons in the brain.
10. The hypothalamus is the principal area of the nervous system that controls emotional behavior.
11. The cerebral aqueduct connects the two lateral ventricles.
12. Neural crest cells along the lateral sides of the embryonic neural tube may give rise to peripheral neurons.
13. The basal ganglia are composed mainly of myelinated nerve fibers.
14. Injury to the reticular formation of the brain often results in unconsciousness or deep coma.
15. The lateral ventricles are located within the cerebral hemispheres.

Matching Questions
Match the specific anatomical structures with the five major regions of the brain:
16. pons
17. basal ganglia
18. fourth ventricle
19. hypothalamus
20. medulla oblongata
21. superior colliculi
22. infundibulum
23. insula
24. cerebellum
25. corpus callosum

a. telencephalon
b. diencephalon
c. mesencephalon
d. metencephalon
e. myelencephalon

Multiple-Choice Questions
26. Of the following, which lists the correct anterior to posterior positions of the regions of the brain?
    a. diencephalon, mesencephalon, prosencephalon, metencephalon, myelencephalon
    b. telencephalon, mesencephalon, diencephalon, metencephalon, myelencephalon
    c. telencephalon, diencephalon, mesencephalon, metencephalon, myelencephalon
    d. diencephalon, telencephalon, mesencephalon, metencephalon, myelencephalon
27. The filum terminale at the inferior end of the spinal cord is formed primarily by
    a. dura mater
    b. pia mater
    c. arachnoid
    d. spinal falx
    e. none of these
28. Learned activities such as writing, typing, or piano playing are controlled by
    a. the premotor area
    b. Broca's area
    c. the caudate nucleus
    d. the nucleus gracilis
    e. the globus pallidus
29. The primary sensory area of the brain is located in the
    a. cerebellum
    b. basal ganglia
    c. medulla oblongata
    d. cerebral cortex
    e. none of these
30. Which of the following is *not* one of the basal ganglia?
    a. caudate nucleus
    b. globus pallidus
    c. corpus callosum
    d. lentiform nucleus
    e. amygdaloid nucleus
31. The inferior and superior colliculi of the brain are referred to as the
    a. cerebral peduncles
    b. rubrospinal tract
    c. limbic system
    d. corpora quadrigemina
    e. pyramids
32. The specific part of the brain that controls the endocrine system through the pituitary gland is the
    a. fornix
    b. hypothalamus
    c. epithalamus
    d. cerebral peduncle
    e. pineal body

33. The triangular middle portion of the cerebellum is the
    a. vermis
    b. rubrospinal tract
    c. dentate nucleus
    d. middle cerebellar hemisphere
    e. folia cerebelli
34. Portions of the spinal cord that conduct sensory (afferent) impulses to the brain are
    a. descending fasciculi
    b. ascending fasciculi
    c. monosynaptic reflex arc
    d. polysynaptic arc
    e. anterior columns
35. The diaphragma sella is a part of the
    a. pia mater
    b. lateral ventricle
    c. arachnoid mater
    d. fourth ventricle
    e. dura mater

Essay Questions
36. Diagram the anatomical elements that make up a spinal reflex arc and give a simple explanation of what happens.
37. Describe the relationship of the five embryonic brain regions to the four major adult regions.
38. Trace the flow of cerebrospinal fluid from the lateral ventricles of the brain, through the ventricular system, and back to its starting point.
39. Compare and contrast the composition of gray and white matter in the brain and spinal cord.
40. Describe the functional differences between the ascending and descending fascicles in the spinal cord white matter, and name the most important ones.
41. Describe the cerebral cortex and discuss the functional significance of the gyri and sulci.
42. Discuss the primary sensory areas of the cerebral cortex. Locate them on the brain and discuss the differences in sizes they exhibit.
43. In general, the right side of the brain controls movements of the left side of the body, and vice versa. Describe the anatomical specializations that allow this to happen.
44. Describe how the right side of the brain knows what the left side is doing, and vice versa.
45. Discuss the functions of the meninges that cover the brain and spinal cord.

Clinical Application Questions
46. Why are prefrontal lobotomies no longer used to control chronic violence in psychiatric patients?
47. Compare and contrast presenile dementia and senile dementia.
48. Discuss the various causes of headaches.
49. Describe the characteristics of a grand mal epileptic seizure
50. How does clinical death differ from brain death?
51. Discuss the types of paralysis that may result from a stroke. Be sure to consider how paraplegia differs from quadriplegia.
52. List the reflexes normally found in a newborn infant, and discuss the survival value of each.

13
The Peripheral Nervous System

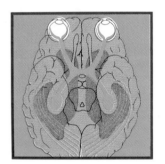

In the preceding chapter we examined one of the two major subdivisions of the human nervous system, namely, the central nervous system (CNS), consisting of the brain and spinal cord. The other major subdivision is the **peripheral nervous system (PNS),** consisting of the cranial nerves, the spinal nerves, and the autonomic nervous system. In this chapter we will look at the cranial nerves and spinal nerves.

Throughout this discussion of peripheral nerves, we make reference to the presence of autonomic (either sympathetic or parasympathetic) neurons and fibers without a detailed explanation of their positions or functions, which is reserved for the next chapter. In general, these neurons control automatic unconscious actions, primarily the contraction of involuntary smooth muscles and the secretory activities of glands. In the next chapter we discuss the relationship of the autonomic nervous system to the voluntary neurons discussed here. In our study of this chapter we will describe

1. the 12 cranial nerves, their places of origin, their composition, and the organs they innervate;
2. the anatomical relationships of the 31 pairs of spinal nerves to the spinal cord and vertebral column;
3. the major areas innervated by the anterior versus the posterior branches of the spinal nerves; and
4. the three major plexuses that innervate the upper and lower extremities.

peripheral: (Gr.) *peri*, around + *pherein*, to carry

Cranial Nerves

After completing this section, you should be able to:

1. Name the 12 pairs of cranial nerves in sequence and indicate the general region of the brain from which each emerges.
2. Distinguish among sensory, motor, and mixed nerves.
3. Identify the cranial nerve with the widest distribution within the head and indicate the regions it innervates.
4. Describe the vagus nerve in terms of its wide distribution.

I—Olfactory
II—Optic
III—Oculomotor
IV—Trochlear
V—Trigeminal
VI—Abducens
VII—Facial
VIII—Vestibulocochlear
IX—Glossopharyngeal
X—Vagus
XI—Accessory
XII—Hypoglossal

A total of 12 pairs of cranial nerves extend from the base of the brain. All but the first pair arise from the brain stem (Figure 13-1). They are numbered as they emerge from the anterior to the posterior brain, and each pair has the following name and number:

Some cranial nerves include primarily sensory or motor neuron fibers and are called **sensory** or **motor nerves**, respectively. Others are known as **mixed nerves** because they include both sensory and motor fibers. Only the axons and dendrites form the nerves, while the cell bodies of the motor neurons form nuclei within the brain tissue and those of sensory neurons are grouped in ganglia outside but adjacent to the CNS.

For the most part, cranial nerves are associated with voluntary activities, although the presence of certain sensory

Figure 13-1 The 12 pairs of cranial nerves are visible in this inferior view of the brain. Except for the olfactory nerves (I), all originate from the brain stem, and except for the trochlear nerve (IV), all originate from the inferior surface.

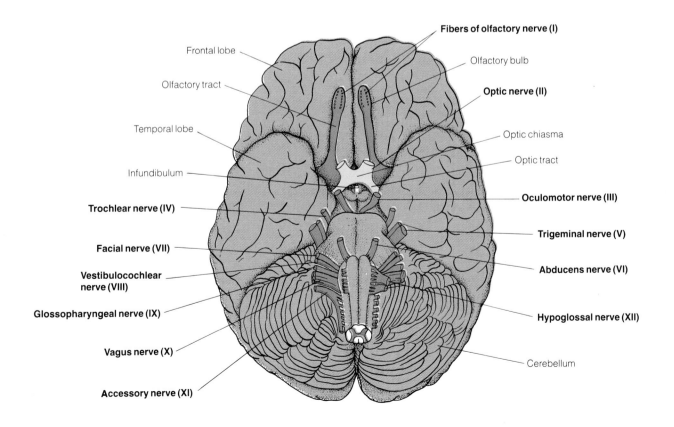

Embryonic Development of Peripheral Nerves

As part of the discussion of the development of the central nervous system in the preceding chapter, we described the relationship of neural crest tissue to the embryonic spinal cord. It consists of a chain of cells along the lateral sides of the neural tube (see box on Embryonic Development of the Nervous System in Chapter 12). Peripheral nerves, both cranial and spinal, are derived from this neural crest tissue and from cells within the embryonic brain stem and nerve cord.

Neural crest cells differentiate along two anatomical and functional lines (Figure 13-A). Some remain adjacent to the brain and spinal cord, developing into sensory neurons. Their cell bodies remain outside the brain and spinal cord, with the exception of proprioceptive neurons that apparently migrate into the brain. Fibers of sensory neurons grow in two directions: into the central nervous system and out to sensory organs throughout the body. Other neural crest cells migrate away from the cord and mature into motor cells of the autonomic nervous system.

Axons from certain **neuroblasts** (embryonic nerve cells) within the basal region of the developing brain and spinal cord grow to all parts of the body. These cells develop into voluntary and involuntary motor neurons. Some synapse with neurons of the autonomic nervous system, while others extend to all skeletal muscles.

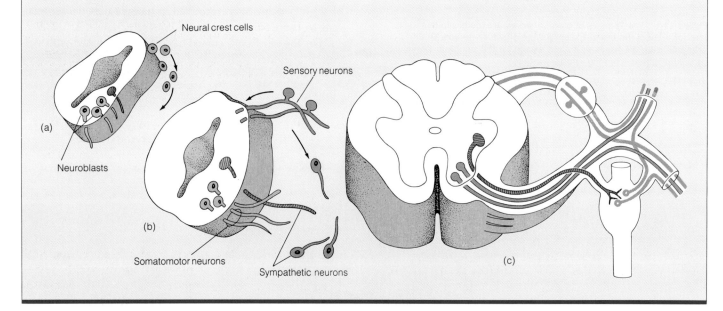

Figure 13-A Embryonic development of peripheral nerves.
(a) Some neural crest cells migrate away from the spinal cord while others remain close by. Neuroblasts within the cord send out axons. (b) Neural crest cells become sensory neurons near the cord and sympathetic neurons away from the cord. Neuroblasts differentiate into somatomotor or sympathetic neurons. (c) Distribution of spinal neurons in the adult.

---

fibers demonstrates their role in regulating some autonomic functions. In addition to neurons associated with the special senses (vision, hearing, taste, balance, and smell), several autonomic and proprioceptive neurons transmit sensory information regarding muscle tension through cranial nerves to reflex centers in the brain. This information influences these brain centers to send impulses to muscles in the head and neck regulating muscle tone and subconscious movements.

Generally speaking, cranial nerves innervate structures in the head and neck region. However, one exception is the vagus nerve, which is a mixed nerve that innervates the palate, neck, thorax, and abdominal cavity.

The 12 cranial nerves, their sites of origin, and mode of action are summarized in Table 13-1.

Olfactory Nerves (I)

The **olfactory** (ŏl·făk′tō·rē) **nerves** (Figure 13-2) are the first pair of cranial nerves. They are purely sensory nerves and are involved in smell, or olfaction. Neurons of each

Table 13-1 THE CRANIAL NERVES

Nerve	Origin	Location of Exit from Skull	Functions
Olfactory (I)	Cerebral hemispheres	Cribriform plate of ethmoid bone	Sensory nerve; olfaction, smell
Optic (II)	Diencephalon	Optic canal	Sensory nerve; vision
Oculomotor (III)	Midbrain	Superior orbital fissure	Mixed nerve; proprioceptive and motor fibers to extrinsic eye muscles; parasympathetic fibers to ciliary muscle of lens and circular muscle of iris
Trochlear (IV)	Midbrain	Superior orbital fissure	Mixed nerve; proprioceptive and motor fibers to superior oblique muscle of eye
Trigeminal (V)	Pons		
Ophthalmic nerve		Superior orbital fissure	Sensory nerve; from eye, face, and head
Maxillary nerve		Foramen rotundum	Sensory nerve; from nasal cavity, mouth and face
Mandibular nerve		Foramen ovale	Mixed nerve; sensory fibers from mouth and face; proprioceptive and motor fibers to and from chewing muscles
Abducens (VI)	Pons	Superior orbital fissure	Mixed nerve; motor and proprioceptive fibers to and from lateral rectus muscle
Facial (VII)	Pons	Stylomastoid foramen	Mixed nerve; sensory fibers; taste from tongue; motor and proprioceptive fibers from facial muscles; parasympathetic fibers to salivary and lacrimal glands
Vestibulocochlear (VIII)	Medulla oblongata	Internal acoustic meatus	Sensory nerve; hearing and equilibrium
Glossopharyngeal (IX)	Medulla oblongata	Jugular foramen	Mixed nerve; sensory fibers; taste and touch from mouth and pharynx; motor fibers to pharyngeal muscles and parotid salivary gland
Vagus (X)	Medulla oblongata	Jugular foramen	Mixed nerve; sensory fibers from external ear, taste from epiglottis, and visceral sensations from larynx and numerous thoracic and abdominal organs; motor and autonomic fibers to palate, pharyngeal, and laryngeal muscles and numerous thoracic and abdominal organs
Accessory (XI)	Medulla oblongata	Jugular foramen	Motor and proprioceptive fibers to and from sternocleidomastoid and trapezius muscles
Hypoglossal (XII)	Medulla oblongata	Hypoglossal canal	Motor nerve; motor fibers to and from muscles of the tongue

olfactory nerve comprise a diffuse network of cells in which dendrites and cell bodies are located in the mucous membrane lining the most superior recesses of the nasal cavity. Axons from these neurons penetrate the holes in the cribriform plate and terminate in the **olfactory bulb,** an extension of the forebrain above the cribriform plate. Occurring within the bulb are synapses with additional olfactory neurons in the olfactory tract underneath the frontal lobes. These fibers in the tract terminate in the primary olfactory area of the cerebral cortex.

Optic Nerves (II)

The **optic nerves,** the second pair of cranial nerves, are sensory nerves involved in vision. Although listed among the cranial nerves, they are actually specialized parts of the brain. (The anatomy and physiology of the eyes are discussed in detail in Chapter 16.) Axons of neurons in the retina of the eye extend inward as the optic nerve, which passes through the optic canal into the cranial cavity. Optic nerves from the eyes fuse into the X-shaped **optic chiasma** (Figure 13-3). An interesting distribution of neurons occurs within the optic chiasma. Neurons from the medial (inside) half of each eye cross to the opposite side, while those from the lateral (outside) half of each eye remain on the same side. This mixture of axons extends from the optic chiasma as the **optic tracts,** which enter the brain under the thalamus. Each optic tract contains fibers from both eyes, and this crossover of fibers partially accounts for three-dimensional vision in humans.

Once within the brain, the optic tract fibers proceed to two areas. Most go to the lateral geniculate nucleus of the thalamus where synapses are made with neurons leading to the primary visual area in the cerebral cortex of the occipital

hemianopia: (Gr.) *hemi*, half + *an*, not + *ops*, eye
oculomotor: (L.) *oculus*, eye + *motus*, moving

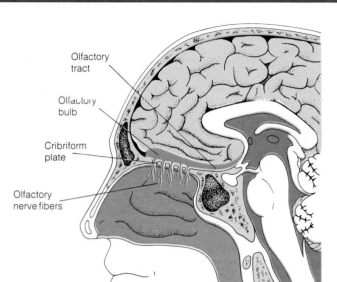

Figure 13-2 The olfactory nerve is a sensory nerve that functions in the sense of smell.

CLINICAL NOTE

OPTIC CHIASMA AND PITUITARY TUMORS

Tumors of the pituitary gland, which lies beneath the optic chiasma, may result in compression of the medial tracts. This compression may cause a bilateral defect in the temporal visual fields known as **bitemporal hemianopia** (hĕm″ē·ă·nō′pē·ă). Noting this type of field defect in a patient with severe headaches is a good clue to the diagnosis of a pituitary tumor.

lobes. In these lobes, visual stimuli are interpreted. The remaining fibers of the optic tracts proceed to the superior colliculi. Here, they synapse with cranial neurons that connect to nuclei controlling movements of the eyes. Through these connections, reflexes control eye movements.

Oculomotor Nerves (III)

The **oculomotor** (ŏk″ū·lō·mō′tŏr) **nerves** are mixed cranial nerves because they contain both sensory and motor fibers. Efferent fibers of the parasympathetic division of the auto-

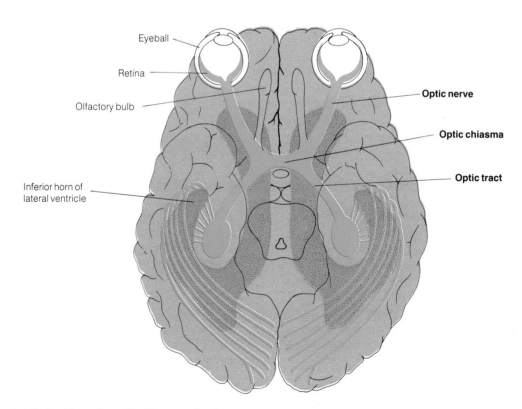

Figure 13-3 This ventral view shows the optic chiasma and optic nerves, which transmit impulses initiated by visual stimuli from the eyes to the brain.

papilledema: (L.) *papilla*, nipple + (Gr.) *oidema*, swelling
neuritis: (Gr.) *neuron*, nerve + *itis*, inflammation
trochlear: (Gr.) *trochilia*, pulley

CLINICAL NOTE

"EYE SIGNS" AND DIAGNOSIS

Careful examination of the eyes can provide many clues to the condition of the central nervous system. Abnormalities of the **extraocular motions** (movements of the entire eyeball) may make one suspect abnormalities in the tracts of the oculomotor, trochlear, or the abducens nerves. Examination of the pupils for equality, shape, and reactions to light are important in the diagnosis of brain tumors and evaluation of head injuries.

Direct visualization of the optic nerve with an opthalmoscope may reveal swelling of the nerve, or **papilledema** (păp″ĭl·ĕ·dē′mă), in the presence of increased intracranial pressure or depression of the nerve head if there is increased intraocular pressure. Pallor of the optic nerve head may be a clue to an **optic neuritis** (inflammation of the optic nerve) seen in several obscure diseases.

nomic nervous system also course through these nerves (see Chapter 14).

Oculomotor nerves originate in nuclei of the lower portion of the midbrain (Figure 13-4). From there, fibers extend anteriorly through the superior orbital fissure where they branch and enter the orbit as superior and inferior subdivisions. The **superior branch** innervates the superior rectus muscle, an extrinsic eyeball muscle, and the levator palpebrae superioris, a muscle that raises the upper eyelid. The **inferior branch** innervates the inferior oblique muscle and the medial and inferior rectus muscles of the eyeball. Another small extension innervates the **ciliary ganglion,** a functional part of the autonomic nervous system responsible for regulating intrinsic eyeball muscles.

Impulses in the oculomotor nerve serve three functions. First, somatic motor impulses control contractions of the extrinsic eye muscles, thereby helping to control eye movements. Second, stretch receptors send impulses through this nerve informing reflex centers in the brain of the amount of muscle tone in each eye muscle. Third, autonomic fibers within the inferior branch of the oculomotor nerve control intrinsic eye muscles. The autonomic motor fibers (discussed further in Chapter 14) of the oculomotor nerve course from the brain to the ciliary ganglion located behind the eyeball (Figure 13-4). Within this ganglion the fibers synapse with neurons whose axons leave the ganglion and connect with the ciliary muscle of the eye and the sphincter muscles of the iris. Based on the amount of light reaching the retina, the autonomic neurons of the oculomotor nerve regulate the pupil's diameter, hence the intensity of light reaching the retina. These neurons also regulate the curvature of the lens, hence the ability of the eye to focus on objects at various distances (see Chapter 14).

Trochlear Nerves (IV)

The **trochlear** (trŏk′lē·ăr) **nerves** are the smallest of the cranial nerves (Figure 13-5). They are also the only ones that arise on the dorsal surface on the midbrain. They are motor nerves that begin in small nuclei below the inferior colliculi and pass laterally and forward to the superior orbital fissures, where they enter the orbit along with the oculomotor

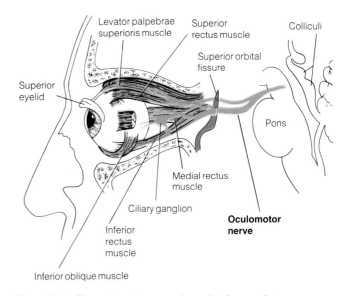

Figure 13-4 The oculomotor nerve is a mixed nerve that innervates both internal and external muscles of the eye and a muscle of the upper eyelid.

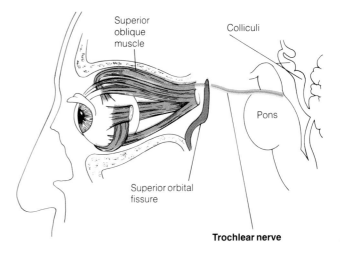

Figure 13-5 The trochlear nerve is the smallest of the cranial nerves and innervates only the superior oblique muscle of the eyeball.

trigeminal: (L.) *tri*, three + *geminus*, twin
semilunar: (L.) *semi*, half + *luna*, moon
neuralgia: (Gr.) *neuron*, nerve + *algos*, pain

nerves. There, they innervate a single extrinsic eye muscle, the superior oblique. Autonomic fibers within the trochlear nerve also help control movements of the eyeball.

Trigeminal Nerves (V)

The largest of the cranial nerves are the **trigeminal** (trī·jĕm′ĭn·ăl) **nerves,** which, as their name suggests, consist of three major subdivisions each: the **ophthalmic** (ŏf·thăl′mĭk) **nerve, maxillary nerve,** and **mandibular nerve** (Figure 13-6). The base of each trigemial arises from nuclei in the pons. Just beyond where the trigeminal nerve exits the pons is a large **semilunar ganglion** (or gasserian ganglion), which contains the cell bodies of the sensory neurons found in the three branches of the nerve. The three branches diverge distal to the semilunar ganglion. The ophthalmic nerve passes through the superior orbital fissure, the maxillary nerve enters the foramen rotundum, and the mandibular nerve goes into the foramen ovale. Dendrites of these sensory neurons conduct impulses of touch, heat, cold, and pain to the semilunar ganglion. From there these signals go to nuclei in the pons.

Ophthalmic nerve dendrites receive stimuli from the eye, the skin of the forehead and upper eyelid, the skin at the sides of the nose, the lacrimal gland, and the lining of the nasal cavities. Impulses from the maxillary nerve travel from the maxillary teeth and gingivae (jĭn′jĭ·vē) (or gums), the palate, the mucosal lining of the nose and upper pharynx, and the skin of the upper lip and lower eyelid. Dendrites of the mandibular branch are concentrated in the lower jaw. They conduct impulses of touch and pain from the mandibular teeth, gingivae, and tongue; the mucous membrane lining the floor of the mouth; and the skin of the cheek and temporal region of the head. Because all mandibular

> **CLINICAL NOTE**
>
> **TRIGEMINAL NEURALGIA**
>
> **Trigeminal neuralgia** is a condition of recurrent bouts of severe, piercing pain in one or more divisions of the trigeminal nerve. Attacks may be triggered by a light touch on the cheek or by activities such as chewing or brushing the teeth. The pain is intense and may incapacitate the patient. The cause of this condition is unknown. A similar condition may affect the glossopharyngeal nerve, producing bouts of pain in the back of the pharynx and tongue or in the middle ear.

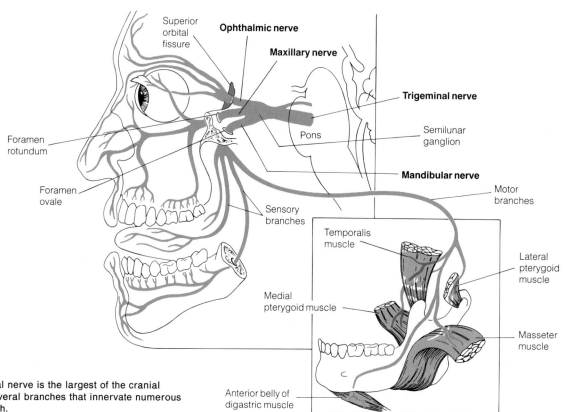

Figure 13-6 The trigeminal nerve is the largest of the cranial nerves and divides into several branches that innervate numerous areas of the face and mouth.

anesthesia: (Gr.) *an*, not + *aisthesis*, sensation
abducens: (L.) *abducere*, to lead away
geniculate: (L.) *geniculum*, little knee
vestibulocochlear: (L.) *vestibulum*, vestibule + *cochlea*, snail shell

CLINICAL NOTE

ANESTHESIA

The word **anesthesia** (ăn″ĕs·thē′zē·ă) denotes the loss of sensation, usually involving pain, either through injury to the nervous system or induced as a temporary method for relieving pain during a surgical procedure. There are many methods of inducing anesthesia, ranging from the simple method of applying cold to a localized area to the complicated science of inducing surgical anesthesia by putting the patient to sleep using a combination of drugs. Anesthetic drugs are usually lipid soluble substances that can dissolve in the lipid substances that surround the nervous system.

The three basic techniques of anesthesia are **infiltration anesthesia,** which is injection of a local anesthetic into the margins of a wound to be sutured; **regional anesthesia,** which is blockage of a nerve to the area of surgery by injecting the anesthetic around the nerve; and **general anesthesia,** which is rendering the patient unconscious by drugs that are inhaled or taken intravenously. Physicians who specialize in administering anesthetics are called **anesthesiologists.**

A specific type of regional anesthesia is induced by injecting drugs (usually procaine) into the spinal fluid space of the lower spinal column. This technique, called a spinal block, can produce a loss of sensation in the lower abdomen and legs without rendering the patient unconscious. It is used for cesarean sections and surgery on the lower abdomen and legs that will take less than two hours.

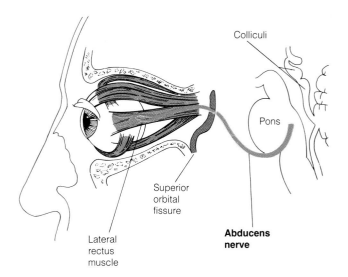

Figure 13-7 The abducens nerve is a mixed nerve that innervates only the lateral rectus muscle of the eye.

teeth on one side of a jaw are innervated by one subdivision of this branch, it is relatively easy for a dentist to anesthetize the lower jaw with one injection (see Clinical Note: Anesthesia).

The motor branch of the trigeminal nerve leaves the pons alongside the mandibular nerve and passes through the foramen ovale. It conducts impulses to the anterior belly of the digastric, masseter, mylohyoid, temporalis, and medial and lateral pterygoids. It also includes sensory proprioceptive fibers that transmit information to the pons concerning the state of contraction of the chewing muscles.

Abducens Nerves (VI)

The **abducens** (ăb·dū′sĕnz) are the sixth pair of cranial nerves. They arise from small nuclei near the posterior end of the pons (Figure 13-7). Each is a mixed nerve, carrying both somatic motor neurons and sensory neurons. It is one of the smaller of the cranial nerves and enters the orbit through the superior orbital fissure where it innervates the lateral rectus muscle of the eye. The abducens controls voluntary movements of the eyeball and provides reflex centers in the pons with information regarding muscle tone in the lateral rectus.

Facial Nerves (VII)

The **facial nerves** are the seventh pair of cranial nerves. They are mixed nerves, containing somatic motor neurons, somatic sensory neurons, and proprioceptors. Originating in nuclei near the lower end of the pons, each facial nerve enters the petrous portion of the temporal bone through the internal acoustic meatus and exits through the stylomastoid foramen. The somatic motor fibers of the facial nerve branch into numerous fibers that innervate the voluntary muscles of the face and scalp, lacrimal glands, and sublingual and submandibular salivary glands (Figure 13-8). Proprioceptors located in these muscles also send information concerning their state of contraction back to reflex centers in the pons.

Sensory fibers in the facial nerve accept stimuli from taste buds in the anterior two thirds of the tongue. Cell bodies of these neurons lie in the **geniculate** (jĕn·ĭk′ū·lāt) **ganglion** located in the temporal bone. Impulses are transmitted along these fibers to reflex centers in the medulla oblongata. From there they are relayed through the thalamus to the primary gustatory area in the parietal lobe of the cerebral cortex.

Vestibulocochlear Nerves (VIII)

The **vestibulocochlear** (vĕs·tĭb″ū·lō·kŏk′lē·ăr) **nerves** are the eighth pair of cranial nerves. They have also been called the **auditory** (or **acoustic**) **nerves** because their function was originally assumed to be concerned exclusively with hearing. Each nerve, however, consists of two branches: the

glossopharyngeal: (Gr.) *glossa*, tongue + *pharynx*, pharynx

> **CLINICAL NOTE**
>
> **BELL'S PALSY**
>
> **Bell's palsy** is the name given to a unilateral paralysis of the facial nerve. The cause is unknown but thought to be due to swelling of the nerve, caused by immune or viral diseases, as it courses through the temporal bone. Complete recovery is expected in 90% of cases.

drites of the vestibular nerve originate in the semicircular canals, sacculus, and utriculus of the inner ear (Figure 13-9). They respond to impulses concerning equilibrium and balance transmitted over one or more of several routes to reflex centers in the medulla oblongata and cerebellum and to conscious centers in the cerebral cortex. Association tracts in the cerebral hemispheres and in reflex centers of the cerebellum and pons use this information in coordinating muscular movement to maintain balance.

Glossopharyngeal Nerves (IX)

The **glossopharyngeal** (glŏs″ō·fă·rĭn′jē·ăl) **nerves** are the ninth pair of cranial nerves. They originate in nuclei along the lateral aspect of the medulla oblongata, exit the skull through the jugular foramen, and innervate regions in the tongue and throat (Figure 13-10). Glossopharyngeal nerves are mixed nerves and include visceral sensory as well as voluntary and parasympathetic motor fibers.

cochlear (kŏk′lē·ăr) **nerve,** which transmits auditory stimuli, and the **vestibular** (vĕs·tĭb′ū·lăr) **nerve,** which transmits stimuli related to equilibrium. Thus, the word *vestibulocochlear* better reflects the function of cranial nerve VIII. Cell bodies of the cochlear nerve neurons lie in the **spiral ganglion** beside the spiral organ of Corti in the ear (discussed in Chapter 16). Those of the **vestibular ganglion** are in the petrous portion of each temporal bone (Figure 13-9).

The cochlear nerve receives auditory stimuli in the organ of Corti. Impulses transmitted along these fibers pass to nuclei in the medulla oblongata, through the thalamus, and to the primary auditory area of the cerebral cortex. Den-

The visceral sensory neurons carry impulses from taste buds in the posterior third of the tongue. The nerve also carries pressure stimuli from the epithelium of the tongue, pharynx, and tonsils. A separate branch innervates chemo-

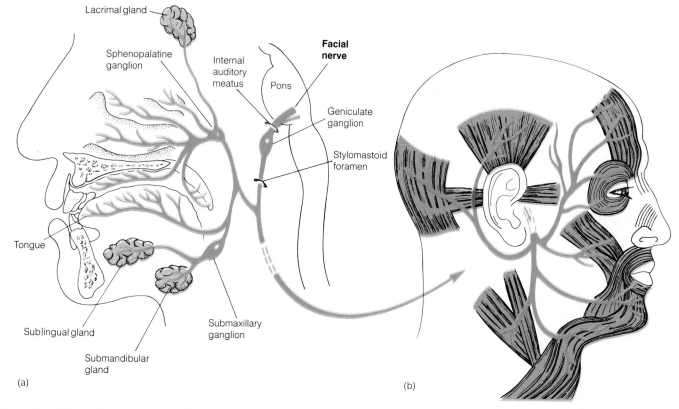

Figure 13-8 The facial nerve is a mixed nerve that conducts impulses to and from several areas in the face and neck. (a) Distribution of sensory and parasympathetic motor fibers. (b) Distribution of voluntary motor fibers.

otic: (Gr.) *otos*, ear
vagus: (L.) *vagus*, wandering
jugular: (L.) *jugulum*, collarbone
nodose: (L.) *nodosus*, knotted

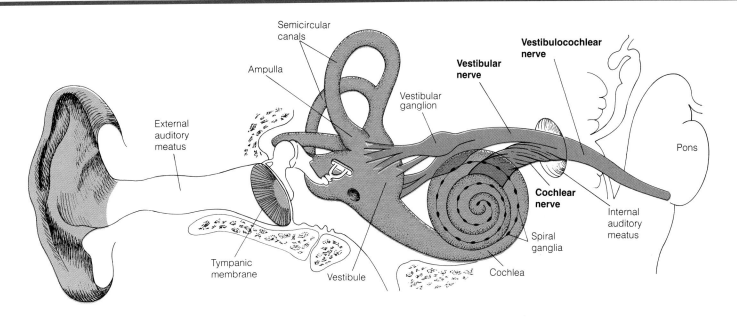

Figure 13-9 The vestibulocochlear nerve is a sensory nerve that transmits impulses generated by auditory stimuli and stimuli related to equilibrium, balance, and movement.

receptors sensitive to the concentration of carbon dioxide in the blood and pressure-sensitive cells in the carotid sinus and carotid body (both located in the carotid arteries in the neck). The cell bodies of these sensory neurons are located in two ganglia—the **superior** and **inferior ganglia**—just outside the jugular foramen. These sensory impulses are used in the control of swallowing, blood pressure, and breathing rate.

The somatic motor fibers of the glossopharyngeal nerves innervate the stylopharyngeus muscle. Parasympathetic motor fibers of the glossopharyngeal nerves originate in the medulla and synapse with neurons in the **otic (ō′tĭk) ganglion** near the foramen ovale. These fibers stimulate the parotid salivary gland to produce saliva.

Vagus Nerves (X)

The **vagus nerves**, the tenth pair of cranial nerves, innervate organs and tissues in the head, neck, thorax, and abdomen. They are mixed nerves, carrying visceral sensory and visceral motor fibers.

Each vagus has several roots in nuclei along the lateral aspect of the medulla oblongata (Figure 13-11). These roots fuse into a single nerve before leaving the cranial cavity through the jugular foramen. Two ganglia, the **superior jugular ganglion** and the **inferior nodose (nō′dōs) ganglion** lie outside this foramen. Somatic sensory and motor

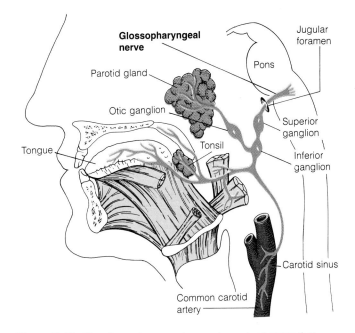

Figure 13-10 The glossopharyngeal nerve is a mixed nerve that innervates structures in the mouth and throat.

and parasympathetic fibers carry impulses to and from the many organs illustrated in Figure 13-11. These fibers aid in regulating conscious movements of muscles in the pharynx and larynx and unconscious reflex activity of glands and muscles in the abdominal cavity (viscera). The vagus also functions in the regulation of heart rate, breathing rate, and blood pressure.

hypoglossal: (Gr.) *hypo*, under + *glossa*, tongue

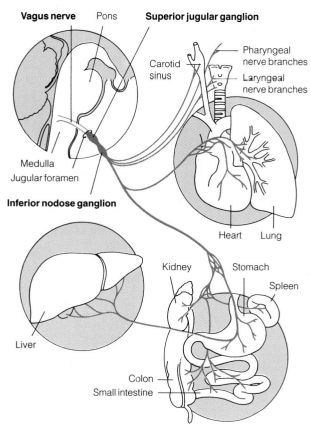

Figure 13-11 The vagus nerve is not like the rest of the cranial nerves, in that it sends branches to numerous organs in the thorax and abdomen, as well as the neck.

Accessory Nerves (XI)

The **accessory nerves** (formerly called the **spinal accessory nerves**) have two origins: a cranial root arising in the medulla and a spinal root (spinal accessory) arising from the posterior medulla and from the first few segments of the spinal cord (Figure 13-12). The spinal root passes through the foramen magnum into the cranial cavity where it fuses with the cranial root. The accessory nerve exits through the jugular foramen.

The accessory nerve is essentially a motor nerve. The spinal root sends axons to the trapezius and sternocleidomastoid muscles, while fibers from the cranial root mix with the vagus nerve and pass to muscles in the pharynx and larynx. Sensory fibers detect muscle tone in the innervated structures.

Hypoglossal Nerves (XII)

The **hypoglossal** (hī″pō·glŏs′ăl) **nerves** are associated with tissues under the tongue. Each arises from a nucleus within the medulla as a series of small rootlets that fuse into a nerve on each side of the brain stem (Figure 13-13). This

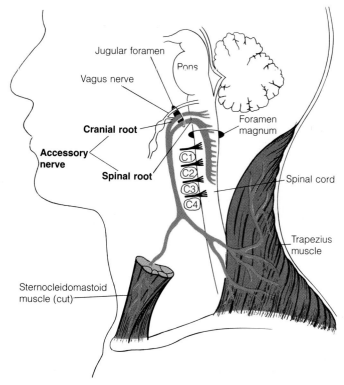

Figure 13-12 The accessory nerve functions primarily as a motor nerve, innervating muscles in the neck.

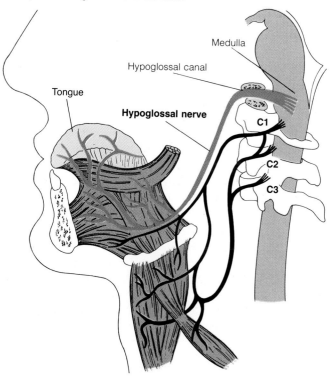

Figure 13-13 The hypoglossal nerve innervates structures within and below the tongue. Additional innervation to these muscles is provided by cervical nerves C1–C3 as shown.

nerve exits the cranial cavity through the hypoglossal canal and passes to the extrinsic and intrinsic muscles of the tongue. It includes somatic motor neurons that regulate movement of the tongue as well as proprioceptive sensory fibers that register tension in these muscles.

Spinal Nerves

After completing this section, you should be able to:

1. Describe the anatomy of a typical spinal nerve.
2. Describe the cervical, brachial, and lumbosacral plexuses and indicate the general regions each innervates.
3. Compare the differences in the pattern of innervation of the skin and skeletal muscles by spinal nerves.
4. Discuss the differences between anterior and posterior roots of a spinal nerve.

Spinal nerves originate in the spinal cord. These nerves leave through foramina of the vertebral column and go to the skin, muscles, bones, and joints of the posterior head region, the trunk, and the appendages (Figure 13-14). In all, 31 pairs of spinal nerves are normally present: 8 cervical, 12 thoracic, 5 lumbar, 5 sacral, and 1 (occasionally more) coccygeal. The first cervical nerve passes between the occipital bone and the atlas. Cervical nerves 2 to 7 exit the spinal region between cervical vertebrae. The eighth cervical nerve exits between the seventh cervical nerve and the first thoracic vertebra. The rest of the spinal nerves exit the spinal cord inferior to the vertebra for which they are named.

Anatomy of Adult Spinal Nerves

Each spinal nerve originates as two extensions from the spinal cord: an **anterior root** and a **posterior root** (Figure 13-15). The first includes both voluntary and involuntary motor neurons. The second includes only sensory neurons.

Voluntary motor neurons in the anterior root are called **somatic motor (or somatomotor) neurons.** Each one extends from the spinal cord gray matter to the effector, in this case, a skeletal muscle. Some neurons, such as those innervating the foot, have very long axons. These neurons are responsible for all the voluntary movements produced by contractions of skeletal muscles attached to bones and to certain superficial skin structures in the head and neck.

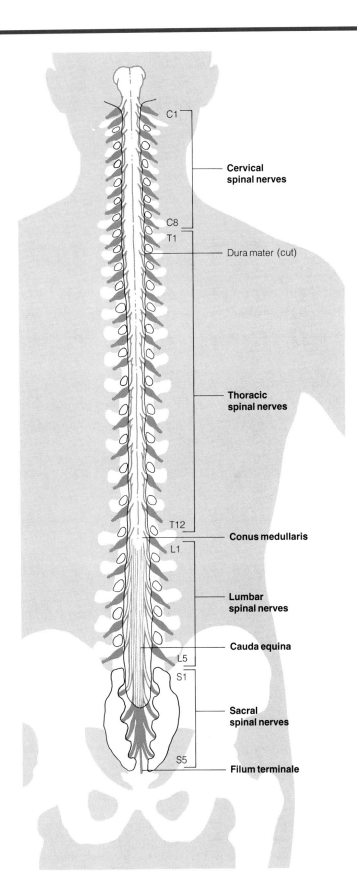

Figure 13-14 A total of 31 pairs of spinal nerves originate in the spinal cord and pass through the spinal foramina of the vertebral column to innervate the peripheral regions of the body.

ramus:	(L.) *ramus*, branch
cutaneous:	(L.) *cutis*, skin
intercostal:	(L.) *inter*, between + *costa*, rib
phrenic:	(L.) *phrenicus*, diaphragm

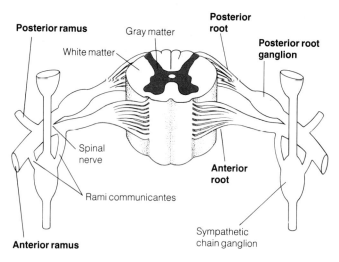

Figure 13-15 The spinal cord and a pair of spinal nerves. Each spinal nerve is formed from extensions from both the anterior and posterior horns of the spinal cord.

In addition to these voluntary nerve cells, the anterior roots contain fibers of the autonomic nervous system that control unconscious involuntary activities. The exact positions and functions of these autonomic neurons will be discussed in the following chapter.

Posterior roots of spinal nerves have only sensory neurons. Dendrites of these neurons are situated in tissues and organs. Some receive stimuli relayed to conscious centers in the brain and are classified as **general sensory neurons.** They carry impulses of touch, pressure, and pain. Other sensory neurons are receptive to information of proprioception in muscles, tendons, and blood vessels. These impulses course to reflex centers in the CNS, which usually respond through the autonomic system. In this way, numerous reflexes are regulated. Sensory neuron cell bodies in each posterior root of each spinal nerve are grouped in a **posterior root ganglion** (Figure 13-15).

Just distal to the posterior root ganglion, the posterior and anterior roots fuse and form a spinal nerve. Because each spinal nerve thus includes both sensory and motor fibers, each one is a mixed nerve. Cervical, thoracic, and lumbar spinal nerves extend laterally and exit the neural canal through intervertebral foramina. Immediately distal to these foramina, each spinal nerve divides into two branches called the **anterior ramus** and **posterior ramus** (Figure 13-15). In this respect, sacral nerves are slightly different from other spinal nerves because they split within the sacral canal, and the anterior and posterior rami exit separately through corresponding sacral foramina.

Compared to its anterior counterpart, the posterior ramus of each spinal nerve is relatively short. Posterior rami innervate only restricted areas of skin and muscles along the back of the head and the sides of the vertebral column. Anterior rami of the spinal nerves innervate the remainder of the trunk and limbs.

Peripheral Regions Innervated by Spinal Nerves

The pattern of spinal nerve innervation of the skin can be mapped as shown in Figure 13-16. A regular sequential pattern is most prominent on the trunk and is slightly modified in the limbs. The nerves that innervate the skin, called **cutaneous** (kū·tā′nē·ŭs) **nerves,** contain only sensory and autonomic motor neurons and lack somatic motor neurons.

Spinal nerve innervation of skeletal muscles follows a similar sequential pattern, that is, spinal nerves in the cervical and thoracic region innervate skeletal muscles in the neck, arm, and chest, while nerves from the lumbar and sacral regions innervate abdominal and leg muscles.

Many areas of both the skin and muscles are innervated by nerves of mixed spinal origin. Such nerves are produced by a fusion and exchange of neurons from the anterior rami of certain neighboring spinal nerves. The resulting mixed nerves form a **plexus** (discussed in the next section).

Nerves supplying the skeletal muscles follow one of two pathways. The first pathway involves the thoracic nerves. These 12 nerves leave the intervertebral foramina and pass between the ribs as **intercostal nerves.** In general, intercostal nerves T1 to T6 innervate the skin and skeletal muscles in the thoracic region. Fibers from their posterior rami are confined to a narrow area bordering the neural spines of the vertebrae, while fibers of the anterior ramus run to the remaining anterior and lateral regions. Intercostal nerves T7 through T12 stimulate corresponding intercostal muscles as well as skin and voluntary abdominal muscles. Nerves T2 also extend into the arms and supply the skin of the axillary (underarm) region and the surface of the back of the arm.

The second pathway involves anterior rami of the remaining spinal nerves in collections of mixed nerves, or plexuses.

Cervical Plexus

The **cervical plexus** is formed by a mixing of cervical nerves 1 to 4, with contributions from C5. The pattern of fusion and branching is illustrated in Figure 13-17. The area innervated by each peripheral nerve is summarized in Table 13-2. The cervical plexus consists of two subdivisions: a superficial **cutaneous branch** and a **deep branch.** The cutaneous branch innervates the skin of the posterior part of the head, neck, and shoulder region, while the deep branch supplies skeletal muscles in the neck. An elongate **phrenic** (frĕn′ĭk)

Spinal Nerves 335

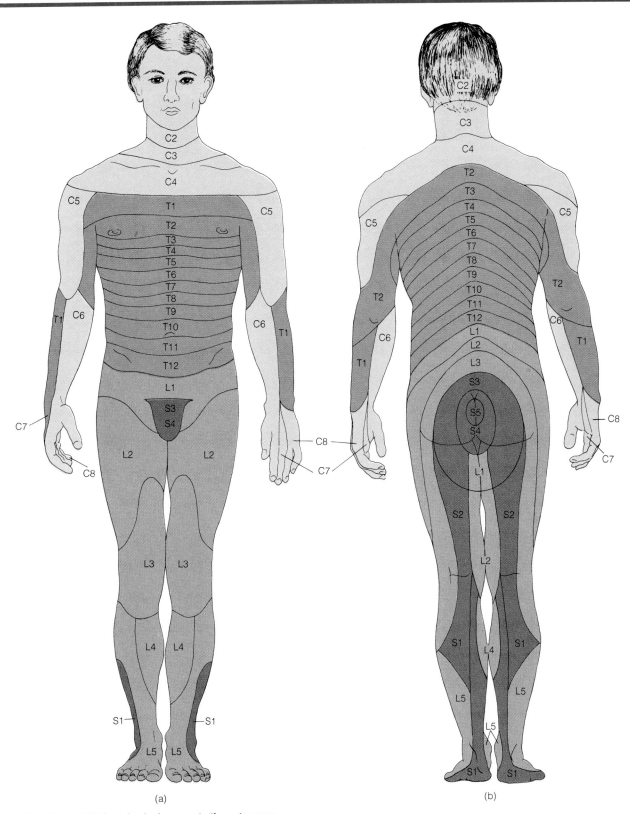

Figure 13-16 The distribution of spinal nerves to the cutaneous regions of the body. (a) Anterior view. (b) Posterior view. C, T, L, and S stand for cervical, thoracic, lumbar, and sacral, respectively.

ansa: (L.) *ansa*, handle
electromyography: (Gr.) *elektron*, amber + *mys*, muscle + *graphein*, to write

> **CLINICAL NOTE**
>
> ### ELECTRODIAGNOSIS
>
> Clinical medicine makes good use of the action potentials of nerve conduction and the electrical responses of the motor unit in the diagnosis of injuries to peripheral nerves and of diseases of the neuromuscular unit.
>
> **Electromyography (EMG)** is the study of motor nerves performed by inserting tiny needles along the course of a motor nerve that record the conduction velocity of the nerve impulse at intervals along the nerve pathway. The evoked muscle potential is recorded by a surface electrode over the innervated muscle (Figure 1a). EMG studies can help determine the site of injury in a motor nerve or a disease in the muscle unit itself.
>
> **Electroneurography (ENG)** is the study of the sensory nerve conduction velocity and is determined by applying a stimulus to a sensory nerve distally and recording the conduction velocity at various proximal sites (Figure 1b). Evaluation of these studies can be very helpful in determining the site and extent of injury to a peripheral nerve.
>
>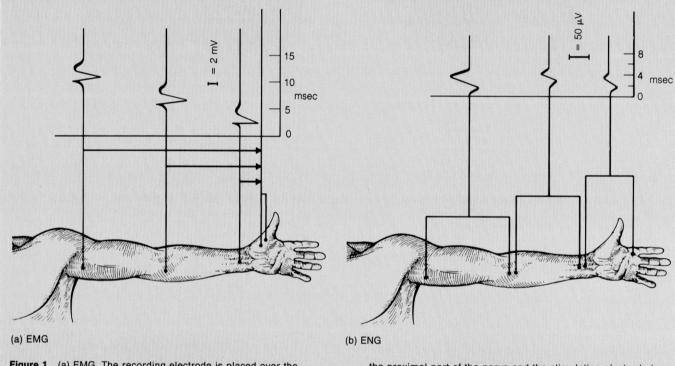
>
> (a) EMG
> (b) ENG
>
> **Figure 1** (a) EMG. The recording electrode is placed over the belly of the opponens pollicis. The median nerve is stimulated at the wrist, elbow, and axilla, and the evoked muscle potential is recorded. (b) ENG. The recording electrode is placed over the proximal part of the nerve and the stimulating electrode is placed over the distal part of the nerve. The distal part is stimulated and the evoked nerve potential is recorded.

nerve extends from the deep branch to the diaphragm, where it performs the important function of stimulating the muscular contractions of breathing.

Fibers from C1 to C3 of the cervical plexus form a loop called the **ansa cervicalis** (ăn′să sĕr·vĭ·kă′lĭs) (Figure 13-17). C1 joins the hypoglossal nerve for a short distance, then branches away as the cranial limb of the ansa.

Brachial Plexus

The **brachial** (bra′ke·ăl) **plexus** is the peripheral nerve supply for the upper extremities. It is formed by the exchange of fibers between the anterior rami of cervical nerves 5 to 8 and the first thoracic nerve (Figure 13-18). Table 13-3 summarizes the destinations of the major nerves of the brachial plexus.

Rami of the brachial plexus unite to form three **trunks**. Spinal nerves C5 and C6 form the **superior** trunk, root C7 is alone and forms the **middle trunk**, while roots C8 and T1 form the **inferior trunk**. Each trunk divides into anterior and posterior divisions. Neurons of the first division innervate the anterior skin and muscles of the forearm and hand, while the posterior division innervates the posterior portion of the entire extremity.

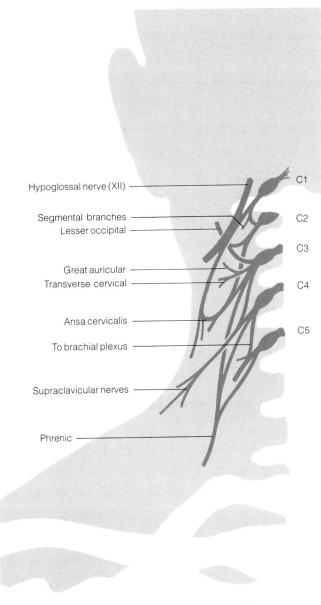

Figure 13-17 The cervical plexus. The distribution of the nerves is summarized in Table 13-2.

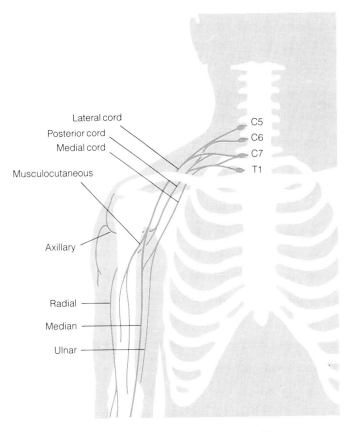

Figure 13-18 The brachial plexus. The distribution of the nerves is summarized in Table 13-3.

Table 13-2 THE CERVICAL PLEXUS

Nerve	Spinal Nerve Origin	Distribution
Lesser occipital	C2–C3	Scalp behind ear
Great auricular	C2–C3	Skin surrounding ear
Transverse cervical	C2–C3	Anterior skin of neck
Supraclaviculars	C3–C4	Skin of upper thorax
Ansa cervicalis	C1–C3	Pharyngeal muscles
Phrenic	C3–C5	Diaphragm

Table 13-3 THE BRACHIAL PLEXUS

Nerve	Spinal Nerve Origin	Distribution
Posterior cord		
Axillary	C5–C6	Skin and muscles of shoulder region
Radial	C5–C8, T1	Skin and muscles of posterolateral region of arm, forearm, and hand
Lateral cord		
Musculocutaneous	C5–C7	Skin and muscles of lateral region of forearm
Medial cord		
Ulnar	C8–T1	Skin of medial portion of hand, and muscles of front of forearm and hand
Median	C5–T1	Skin of lateral portion of hand, and muscles of lateral region of forearm and hand

lumbar: (L.) *lumbus*, loin
pudendal: (L.) *pudeda*, external genitalia
sciatic: (Gr.) *ischion*, hip

> **CLINICAL NOTE**
>
> ### SCIATICA AND RUPTURED DISKS
>
> The term **sciatica** (sī·ăt′ĭ·kă) literally means having a pain in the hips or in the back of the thigh along the course of the sciatic nerve. This pain is due to sciatic nerve injury which may be either primary or secondary. Primary injury to the sciatic nerve can result from a fall, stabbing, gunshot, or other direct trauma. Secondary injury, which is more common, results from pressure on the nerve from a rupture of the intervertebral nucleus pulposus, which is called **a ruptured disk.** The offending disk is usually at L5 to S1 and occasionally at L4 to L5 intervertebral space.
>
> A herniated or ruptured nucleus pulposus occurs when the annulus cartilage degenerates because of age and subsequent loss of elasticity and increased rigidity. As the cartilage becomes increasingly less plastic, the disk flattens and bulges; any sudden twist or additional pressure from bending or lifting may contribute to the annulus tearing. The person may not actually be lifting tremendous weights or pushing heavy objects for the tear to occur, although these are the usual strains which cause the rupture. A sneeze or cough while the body is bent forward can cause the break in the annulus. The nucleus pulposus then extrudes through the tear. It can rupture anteriorly toward the spinal cord, laterally, or posteriorly. The mass presses on the dura mater surrounding the cord or on the nerve root; this causes pain in the back and pain radiating down the nerve (see Chapter 7, The Clinic: Disorders of the Spine).
>
> The incident may be resolved in two ways. If the entire contents have not ruptured, the nucleus may move back inside the disk. The person is usually prescribed bed rest with a firm mattress, anti-inflammatory medication, physical therapy, and muscle relaxants until the acute phase has passed. If the mass remains outside the disk, it can adhere to the nerve roots and add to inflammation and pain. This distortion of the intervertebral space can lead to further degeneration of the joint. If traction over the joint and the measures enumerated above do not allow the mass to return, then surgery may be necessary to remove the pressure on the nerve roots. The extruded nucleus and the damaged annulus is removed or the joint is fused or both.
>
> Sciatica may also occur during the later stages of pregnancy. As the fetus grows within the amniotic sac, abdominal and pelvic organs are displaced, and sufficient pressure may be placed on the sciatic nerve to cause chronic pain in the legs. This type of sciatica disappears shortly after the baby is delivered and the abdominal organs have returned to their normal positions.
>
> Good body mechanics throughout life can lessen the chance of sciatica. Care when bending, lifting, or pushing heavy or clumsy objects is imperative. Children should be taught good body mechanics and adults should practice them. The back is vulnerable to painful injuries, and the recovery period is long and expensive.

Distally, an additional series of fusions occurs to form three **cords.** Fusion of the posterior divisions of all three trunks forms the **posterior cord,** the anterior divisions of the superior and middle trunks form the **lateral cord,** and continuation of the anterior division of the inferior trunk forms the **medial cord.** Additional branchings and even more fusions of the three cords form the major peripheral nerves that innervate the arm.

In addition to the major cords, smaller nerves that emanate from the brachial plexus are not detailed here. They arise along the length of the plexus from the roots through the cords.

Lumbosacral Plexus

The **lumbosacral** (lŭm″bō·sā′krăl) **plexus** combines three groups of spinal nerves: the **lumbar plexus,** nerves L1 to L4; the **sacral plexus,** nerves L4 to L5 and S1 to S3; and the **pudendal** (pū·děn′dăl) **plexus,** nerves S2 to S4. In a few cases the most inferior thoracic nerve (T12) contributes to the lumbar plexus through nerve L1.

Rami of this plexus follow a pattern different from those in the brachial plexus. Instead of forming trunks as in the brachial plexus, extensions from the spinal roots divide into anterior and posterior divisions. These produce the large nerves shown in Figure 13-19. As in the brachial plexus, smaller nerves arise directly from the anterior rami of the nerves involved. The largest nerve in the body, the **sciatic** (sī·ăt′ĭk) **nerve,** arises in the lumbosacral plexus (see Clinical Note: Sciatica and Ruptured Disks). Although it is common to refer to the sciatic nerve as a single structure, it consists of two branches, the common peroneal and tibial nerves, both enclosed within the same sheath of fibrous connective tissue.

The origin and distribution of nerves in the lumbosacral plexus are summarized in Table 13-4.

In this chapter we have studied the anatomy and general functions of the voluntary portion of the peripheral nervous system. In the next chapter we will examine the structural and functional aspects of the involuntary peripheral nerves, the autonomic nervous system.

neuropathy: (Gr.) *neuron*, nerve + *pathos*, disease

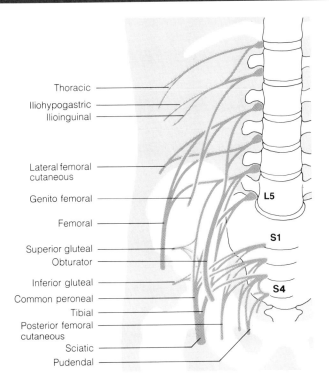

Figure 13-19 The lumbosacral plexus. The distribution of the individual nerves is summarized in Table 13-4.

Table 13-4 THE LUMBOSACRAL PLEXUS

Nerve	Spinal Nerve Origin	Distribution
Lumbar plexus		
Iliohypogastric	T12–L1	Skin and muscles of lower trunk
Ilioinguinal	L1	Skin of upper thigh and external genitals in males, and muscles of lower trunk
Genitofemoral	L1–L2	Skin of external genitals in both sexes and internal genitals in females, and front of thigh
Lateral femoral cutaneous	L2–L3	Skin covering the lateral region of the thigh
Femoral	L2–L4	Skin of the medial thigh, leg, and foot, and certain muscles of the thigh
Obturator	L2–L4	Skin of medial thigh and certain muscles of the thigh
Sacral plexus		
Superior gluteal	L4–S1	Certain muscles of the thigh
Inferior gluteal	L5–S2	Gluteus maximus muscle
Posterior femoral cutaneous	S1–S3	Skin of external genitals in both sexes and internal genitals in females, and posterior surface of thigh
Sciatic	L4–S3	Composed of tibial and common peroneal nerves
Tibial	L4–S3	Skin of posterior leg and foot, and several muscles of thigh, leg, and foot
Common peroneal	L4–S2	Skin of anterior leg and foot, and several muscles of the leg
Pudendal plexus	S2–S4	Skin of the perineum and genitals

THE CLINIC

PERIPHERAL NEUROPATHY

The term **peripheral neuropathy** (nū·rŏp′ă·thē) or **neuritis** is used to describe a syndrome of sensory, motor, reflex, and vasomotor symptoms produced by injury to peripheral nerves or nerve roots. These conditions should be differentiated from **neuralgias** (nū·ral′jē·ăs) which are recurrent paroxysms of acute pain along the distribution of a nerve (see Clinical Note: Trigeminal Neuralgia). The main diagnostic sign of peripheral neuropathy is **Tinel's sign**—eliciting **paresthesia** (prickling or tingling) by tapping over the affected nerve.

There are many causes of peripheral neuropathies, the major ones of which are listed here.

Mechanical Stress Direct injury to the peripheral nerves may be caused by compression, direct blows, penetrating injuries, contusions, and detachments. They usually result in a **mononeuritis,** or injury to a single nerve. Neuropathies frequently occur in association with fractures. **Pressure palsies** occur from prolonged external pressure on a peripheral nerve from leaning on crutches, being bound in a tight cast, or prolonged sitting in a cramped posture. For example, falling asleep with an arm over the back of a chair may compress the radial nerve against the humerus, causing a radial nerve palsy frequently seen in alcoholics. Superficial nerves may be compressed by internal structures and are known as **entrapment neuropathies.** One common entrapment neuropathy is called **carpal tunnel syndrome,** which is compression of the median nerve at the wrist that is common during pregnancy and in people who perform repeated tasks with their wrists and hands, such as typists.

Vascular or Collagen Vascular Conditions Many systemic conditions, such as atherosclerosis and rheumatoid arthritis, may cause multiple peripheral neuropathies. **Volkmann's ischemic paralysis** may occur with occlusion of a major artery of an extremity and is feared when swelling occurs beneath a cast for treatment of fractures.

Infections Direct infection of peripheral nerves occurs in leprosy, malaria, and tuberculosis. **Polyneuritis** (inflammation of many nerves) can occur from bacterial toxins, such as diphtheria or botulism, or autoimmune reactions as seen in Guillain-Barré syndrome (see Chapter 11, The Clinic: Diseases of the Nervous System).

Toxic Agents Many chemicals can cause mono- or polyneuropathies. Some useful medications, such as sulfa drugs, streptomycin, and Dilantin, can cause neuropathies and limit their usefulness.

Metabolic Many nutritional deficiencies result in polyneuropathies. Deficiency of the vitamin B complex can cause polyneuropathy, such as in alcoholism, beriberi, pernicious anemia, malabsorption syndromes, and starvation. Diabetes mellitus is a common cause for many neuropathies (see Chapter 17, The Clinic: Diabetes Mellitus).

Malignancy Many forms of cancer cause neuropathies by direct invasion of nerves, compression, and other unknown mechanisms.

STUDY OUTLINE

Cranial Nerves (pp. 323–333)
Twelve pairs of cranial nerves extend from the base of the brain.

Olfactory Nerves (I) The *olfactory nerves* are purely *sensory nerves*, involved in smell, or olfaction.

Optic Nerves (II) The *optic nerves* are sensory nerves involved in vision. Optic nerves from the eyes fuse at the X-shaped *optic chiasma*, where neurons from the medial half of each eye cross to the opposite side, while those from the lateral half of each eye remain on the same side.

Oculomotor Nerves (III) The *oculomotor nerves* are *mixed nerves* that contain both sensory and motor fibers. They have both *superior* and *inferior branches*. They innervate extrinsic and intrinsic muscles of the eyeball and control certain eye movements and regulate adjustments of the pupil and lens.

Trochlear Nerves (IV) The *trochlear nerves* are *motor nerves* that innervate the superior oblique eye muscle.

Trigeminal Nerves (V) Sensory fibers of the *trigeminal nerves*, the largest cranial nerves, conduct impulses from several areas in the head. Motor fibers stimulate contractions of muscles involved in chewing. Each is divided into the *ophthalmic, maxillary,* and *mandibular nerves*.

Abducens Nerves (VI) The *abducens* nerves are mixed nerves that innervate the lateral rectus muscle.

Facial Nerves (VII) The facial nerves are mixed nerves. Sensory fibers carry taste sensations from the tongue, while motor fibers innervate voluntary muscles of the face and scalp, lacrimal glands, and sublingual and submandibular salivary glands.

Vestibulocochlear Nerves (VIII) The *vestibulocochlear nerves* are sensory nerves that carry impulses related to hearing and balance. Each has two branches: the *cochlear* and *vestibular nerves*.

Glossopharyngeal Nerves (IX) The *glossopharyngeal nerves* are mixed nerves. Sensory fibers carry taste impulses from the tongue and touch stimuli from the mouth. Specialized receptors carry information concerning carbon dioxide concentration in the blood and blood pressure. Motor fibers innervate the stylopharyngeus muscle and the parotid salivary gland.

Vagus Nerves (X) The mixed *vagus nerves* innervate many organs in the thorax and abdomen. Sensory fibers carry many kinds of sensations, while motor fibers stimulate reflex activities of glands and muscles and regulate heart rate, breathing rate, and blood pressure.

Accessory Nerves (XI) Essentially motor nerves, the *accessory nerves* innervate several muscles in the neck.

Hypoglossal Nerves (XII) The *hypoglossal nerves* are mixed nerves that innervate intrinsic and extrinsic muscles of the tongue.

Spinal Nerves (pp. 333–339)
A total of 31 pairs of spinal nerves originate in the spinal cord and leave through foramina of the vertebral column.

Anatomy of Adult Spinal Nerves Each spinal nerve has two extensions from the spinal cord: an *anterior root* containing motor neurons and a *posterior root* containing sensory neurons. Just distal to the *posterior root ganglion*, the roots fuse and form a spinal nerve. Just distal to the intervertebral foramen, each spinal nerve splits into a *posterior ramus*, which innervates only restricted areas of the skin and muscles along the back of the head and the sides of the vertebral column, and an *anterior ramus*, which innervates the remainder of the trunk and limbs.

Peripheral Regions Innervated by Spinal Nerves A regular sequential pattern of *cutaneous nerve* innervation occurs on the trunk and is modified in the limbs. Several spinal nerves fuse in collections of mixed nerves called *plexuses*.

Cervical Plexus The *cervical plexus* is formed by a mixing of cervical nerves 1 to 4, with contributions from C5. This plexus innervates skin and muscles in the posterior part of the head, neck, and shoulder regions. The *phrenic nerve* stimulates muscles used in breathing.

Brachial Plexus The *brachial plexus* is formed from cervical nerves 5 to 8 and the first thoracic nerve. The brachial plexus innervates the arm. It branches to form *trunks* and fuses to form *cords*.

Lumbosacral Plexus The *lumbosacral plexus* is formed from the lumbar and sacral nerves. It contains the *sciatic nerve*, the largest nerve in the body. This plexus innervates organs in the lower trunk and the leg.

SELF-TEST OF CHAPTER OBJECTIVES

True-False Questions
1. Optic nerves are usually studied with cranial nerves, even though they are really specialized extensions of the diencephalon.
2. The ciliary ganglion is a mass of nerve cell bodies associated with the trochlear nerve.
3. The peripheral nervous system contains only sensory fibers of cranial and spinal nerves.
4. As a general rule, cranial nerves do not contain autonomic fibers.
5. The trigeminal nerves are the largest cranial nerves.
6. The trochlear nerves provide the major innervation for the muscles used in chewing.
7. As a general rule, body appendages are innervated by the posterior rami of the spinal nerves in their area.
8. The abducens is a mixed nerve, innervating only the lateral rectus muscle.
9. Spinal nerves innervate both the skin and skeletal muscles.

10. The facial nerve is a large mixed cranial nerve that innervates, among other structures, salivary glands under the tongue.
11. The posterior root of a spinal nerve contains only sensory neurons.
12. The vagus nerve arises as a cranial nerve, but unlike the other cranial nerves, its fibers extend throughout the thorax and abdomen.
13. Cervical nerves contribute to the formation of the brachial plexus, but they do not innervate neck regions.
14. The olfactory nerves are pure sensory nerves that transmit impulses of the sense of smell.

Matching Questions
Match the specific nerve with the correct descriptive statement.

15. trigeminal nerve
16. sciatic nerve
17. phrenic nerve
18. accessory nerve
19. vestibulocochlear nerve

a. stimulates contractions of the diaphragm used in breathing
b. the largest of the cranial nerves
c. innervates the sternocleidomastoid muscle
d. carries impulses related to equilibrium and balance
e. the largest nerve in the body

Multiple-Choice Questions
20. One pair of cranial nerves exchanges fibers from the left side of the body to the right member of the nerve pair and vice versa. This nerve is the
    a. olfactory   d. glossopharyngeal
    b. optic   e. accessory
    c. trigeminal
21. A mass of nerve cell bodies located adjacent to the spinal cord is called
    a. a plexus   c. a nerve ganglion
    b. the ansa cervicalis   d. a lateral cord
22. Which of the following nerves does *not* arise in the pons?
    a. trigeminal   c. abducens
    b. facial   d. vagus
23. Which of the following is *not* a portion of the brachial plexus?
    a. sciatic nerve   c. medial cord
    b. superior trunk   d. inferior trunk
24. Which of the following cranial nerves sends branches throughout the thorax and abdomen?
    a. oculomotor   c. trigeminal
    b. vagus   d. glossopharyngeal
25. The posterior root ganglion of a spinal nerve contains cell bodies of
    a. sensory neurons only
    b. both voluntary and involuntary neurons
    c. voluntary motor neurons only
    d. involuntary motor neurons only
26. The anterior root of a spinal nerve contains
    a. sensory neurons only
    b. both voluntary and involuntary motor neurons
    c. only voluntary motor neurons
    d. only involuntary motor neurons
27. The pudendal plexus contributes to the formation of the
    a. brachial plexus   c. cervical plexus
    b. phrenic nerve   d. lumbodorsal plexus
28. Teeth in both the upper and lower jaw are innervated by the
    a. trigeminal nerve   c. abducens nerve
    b. glossopharyngeal nerve   d. facial nerve
29. Spinal nerves are considered to be
    a. sensory nerves   c. mixed nerves
    b. motor nerves   d. autonomic nerves

Essay Questions
30. List the 12 cranial nerves in order of their emergence from the brain (anterior to posterior).
31. Describe the cervical plexus in general terms and indicate the major regions innervated by its nerves.
32. Distinguish between the major areas innervated by the anterior and posterior branches of a spinal nerve.
33. Compare and contrast the optic and oculomotor nerves in terms of their composition and the structures they innervate.
34. Describe the brachial plexus in general terms and indicate the major regions innervated by its nerves.
35. Identify the route traveled by the vagus nerve. Discuss the organs it innervates and some of its effects on those organs.
36. Describe the anatomy of a typical spinal nerve and discuss the differences between the anterior and posterior roots that emerge from the cord to form the nerve.
37. Discuss the pattern of innervation of the skin by cutaneous nerves.
38. Describe the lumbosacral plexus in general terms and indicate the majors regions innervated by its nerves.
39. Describe the differences among sensory, motor, and mixed nerves.

Clinical Application Questions
40. Explain how a pituitary tumor may adversely affect the optic chiasma.
41. Describe tests used to detect abnormalities in the oculomotor, trochlear, and abducens cranial nerves.
42. Compare and contrast regional anesthesia versus general anesthesia.
43. How does electromyography differ from electroneurography?
44. Discuss primary and secondary sciatica.
45. List some of the major causes of peripheral neuropathy.

14

The Autonomic Nervous System

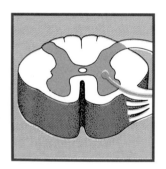

The nervous system is divided *anatomically* into two major segments: the central and peripheral nervous systems. Historically, the nervous system has also been divided into two major *functional* units based on whether or not conscious control was exerted over a particular activity. For example, the control of muscle movements required to ride a bicycle involve different parts of the nervous system than those that function in the control of heart beat, movements of the digestive system, and glandular activities. In the past physiologists thought those activities that were clearly controlled at a subconscious level were regulated by a functionally separate, hence, *autonomous* part of the nervous system. They called this portion of the nervous system the **autonomic nervous system (ANS).** We now know the autonomic system is not truly autonomous. Indeed, it is regulated by other parts of the central nervous system, specifically, centers in the cerebral cortex, hypothalamus, and medulla oblongata. During study of the ANS in this chapter, we will discuss

1. preganglionic versus postganglionic neurons and how they are organized into functional units,
2. the anatomic positions of the sympathetic versus parasympathetic divisions,
3. the variety of routes traveled by sympathetic preganglionic and postganglionic fibers, and
4. the physiology of the autonomic nervous system in regulating metabolic activities.

autonomic: (Gr.) *autos*, self + *nomos*, law
visceral: (L.) *viscera*, internal organs
thoracolumbar: (Gr.) *thorakos*, chest + (L.) *lumbus*, loin

Distribution of Neurons

After completing this section, you should be able to:

1. Identify visceral efferent neurons and discuss their relationship to the autonomic nervous system.
2. Describe the functional units of the ANS that consist of two units each.

The details of the autonomic nervous system can be simplified by an overview of the distribution and pattern of its neurons. Although some experts still believe that the ANS is entirely motor in function, some neurophysiologists now include in this system visceral sensory neurons in the cranial and spinal nerves. Nevertheless, the major part of the autonomic nervous system consists of **visceral efferent neurons.** They carry motor impulses to cardiac muscle, certain glands, and visceral muscles in blood vessels and organs of the thoracic and abdominal cavities. In other words, they affect heart rate; they regulate the amount of constriction of arteries, and hence, blood pressure; and they stimulate or inhibit glandular secretions.

In general, motor neurons of the autonomic nervous system are organized into functional units of two neurons each. Each unit includes a **preganglionic** (prē″găng·lē·on´ĭk) or **presynaptic neuron** with a cell body that lies within the CNS (Figure 14-1). The axon of this cell synapses with a second or **postganglionic** (pōst″găng·lē·ŏn´ĭk) or **postsynaptic neuron.** These nerves synapse either within a ganglion that lies close to the spinal cord or within one of several ganglia located in the thoracic and abdominal cavities. Axons from postganglionic neurons terminate in a specific, predetermined organ or tissue. Impulses carried along these neurons either stimulate or inhibit metabolic activities of these organs.

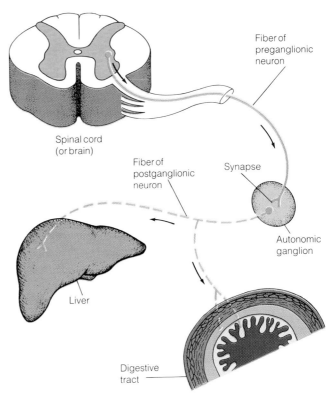

Figure 14-1 Preganglionic and postganglionic neurons are so named for their roles in carrying information to and from an autonomic ganglion.

Embryonic Development of the Autonomic Nervous System

Preganglionic neurons of the ANS develop from embryonic spinal cord cells. These produce long axons that grow from the gray matter and terminate in a chain of ganglia adjacent to the spinal cord or in a ganglion attached to an organ. Postganglionic neurons, however, develop from neural crest cells located outside the embryonic spinal cord. Bodies of these cells accumulate in ganglia either adjacent to the spinal cord or near a visceral organ. Axons grow from these ganglia and terminate in thoracic or abdominal organs. (See Figures 12-A and 12-B in Embryonic Development of the Nervous System in Chapter 12.)

Anatomy of the Autonomic Nervous System

After completing this section, you should be able to:

1. Describe the anatomical layout of the autonomic nervous system.
2. Compare and contrast the effects of the sympathetic and parasympathetic divisions of the ANS.
3. Define the relationship of gray and white communicating rami to the sympathetic chain.
4. Discuss the relationship between sympathetic chain ganglia and collateral ganglia.
5. Compare the sympathetic chain with the abdominal and pelvic plexuses of the parasympathetic system relative to the distribution of impulses to target organs.

In general, the autonomic nervous system has two distinct anatomical and physiological subdivisions: the sympathetic and parasympathetic systems. The **sympathetic system**, also called the **thoracolumbar** (thō·ră″kō·lŭm´băr)

parasympathetic:	(Gr.)	*para,* beside + *sympathetikos,* sympathy
craniosacral:	(Gr.)	*kranion,* skull + (L.) *sacrum,* sacred
ganglion:	(Gr.)	*ganglion,* knot
celiac:	(Gr.)	*koilia,* belly

system because its neurons emerge from thoracic and lumbar regions of the spine, innervates smooth muscles of the arteries. Just as arteries penetrate all parts of the body, so do postganglionic sympathetic fibers. Sympathetic fibers also innervate several abdominal organs (Figure 14-2). The general effect of the sympathetic nervous system is to prepare the body for action in stressful situations such as the fight-or-flight response.

The other division, the **parasympathetic system,** or **craniosacral** (krā″nē·ō·sā′krăl) **system,** functions as the principal nerve supply to certain structures in the head, digestive organs, and other viscera, such as the lungs and heart. Its neurons emerge with cranial and sacral spinal nerves. Effects of the parasympathetic system are in some ways opposite those of the sympathetic. The parasympathetic system stimulates activities of the digestive organs and glands and slows the heart beat and respiratory rate. It tends to calm the body after a stress-producing experience, and it promotes activities that maintain life-support systems.

In many instances, fibers from both divisions innervate the same organs and the interplay between the two determines the net activities of an organ or the reactions of the whole body.

Sympathetic System

Preganglionic neurons of the sympathetic division are anchored in the lateral columns of the spinal cord, and their axons emerge from the cord through the anterior roots of all thoracic spinal nerves and through the first two lumbar spinal nerves (Figure 14-2). These axons remain within a spinal nerve for only a short distance, exiting through one of two **communicating rami** that connect with a chain of ganglia paralleling the vertebral column. These ganglia are the **paravertebral ganglia,** and they comprise the **sympathetic trunk** (or **chain**). Since preganglionic sympathetic fibers are myelinated, the rami they run through to enter the sympathetic trunk are called **white communicating rami.** Figure 14-3 shows pathways available to these motor neurons entering the sympathetic chain.

An axon may synapse within the ganglion it enters or may pass through and synapse in a ganglion at a different level in the chain (Figure 14-3). Other axons run through the sympathetic ganglia and do not synapse until they join neurons in ganglia within or near visceral organs. The nerves formed by these latter preganglionic neurons are called **splanchnic** (splăngk′nĭk) **nerves.**

Impulses in sympathetic neurons travel two different pathways to get to target organs depending on which synaptic pattern occurs. In the first pattern, impulses are generated in postganglionic fibers that leave the sympathetic trunk ganglia and pass out through spinal nerves. Postganglionic sympathetic fibers are nonmyelinated and pass through **gray communicating rami** (Figure 14-3). Because this happens along the entire length of the cord, all spinal nerves contain at least some postsynaptic sympathetic fibers. Some of these postganglionic fibers leave the sympathetic trunk ganglia in the cervical region and accompany major arteries in the head until they reach their target organs. Within the head they follow the external carotid, internal carotid, and vertebral arteries. The postganglionic fibers are called **plexuses** and are named according to the artery involved. Thus, these fibers are distributed into three plexuses: the **external carotid plexus, internal carotid plexus,** and **vertebral plexus.**

A second pattern is formed by preganglionic neurons that run through splanchnic nerves and synapse with their postganglionic partners within **abdominal** (or **collateral**) **ganglia.** Collateral ganglia form masses of cell bodies of postganglionic neurons. Here, synapses occur with preganglionic axons from the spinal cord. Three collateral ganglia—the **celiac** (sē′lē·ăk) (Figure 14-2), **superior mesenteric,** and **inferior mesenteric**—lie near the large arteries for which they are named. Smaller collateral ganglia lie in the pelvic cavity.

Postganglionic axons exit the collateral ganglia to form a series of **extrinsic autonomic plexuses.** After reaching target organs, these axons penetrate the organ and are distributed as **intrinsic autonomic plexuses** within the organ.

One exception to the two-neuron plan in the sympathetic system is known. It involves the adrenal gland, attached to the top of the kidney. Preganglionic sympathetic nerves that innervate cells of the inner portion (medulla) of the gland do not synapse with postganglionic neurons. The significance of this exception is discussed in the section on Membrane Receptors.

Parasympathetic System

The second part of the ANS is the parasympathetic or craniosacral system. Its neurons arise from opposite ends (cranial and sacral ends) of the spinal cord (Figure 14-2). It functions as the principal nerve supply for certain structures in the eye, glands in the head, heart, reproductive organs, and smooth muscles and glands of the digestive system. The fibers of this system accompany four of the cranial nerves and the anterior roots of spinal nerves S2, S3, and S4. Spinal nerves proper do not include parasympathetic motor fibers.

Figure 14-2 (opposite) Efferent fibers of the autonomic nervous system. In this diagram the sympathetic division is indicated in blue and the parasympathetic is in red. Solid lines indicate preganglionic fibers and dashed lines indicate postganglionic fibers.

Anatomy of the Autonomic Nervous System 345

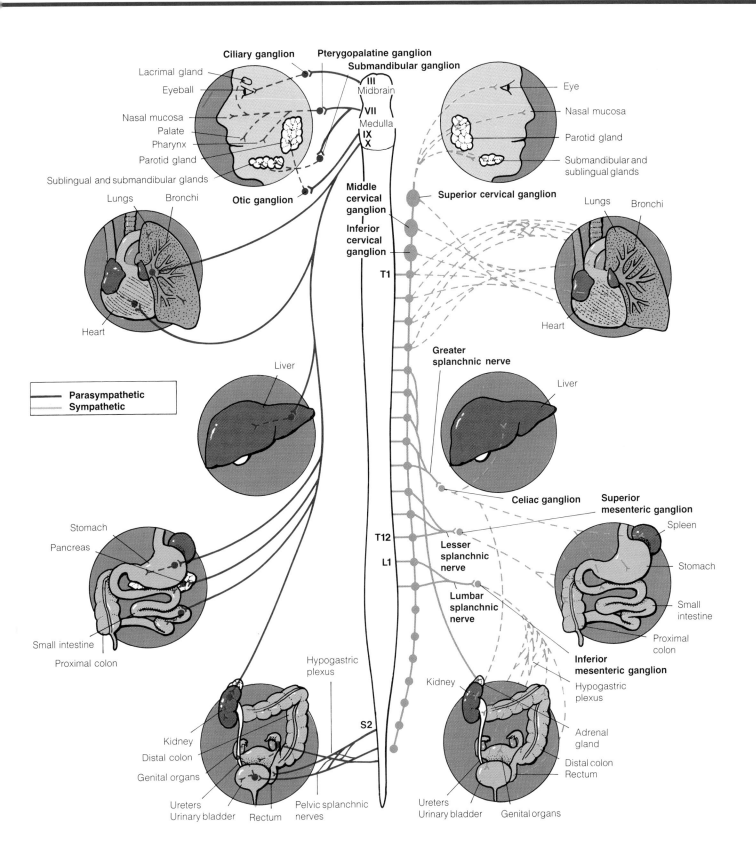

pterygopalatine: (Gr.) *pterygodes*, wing shaped + (L.) *palatinus*, palate

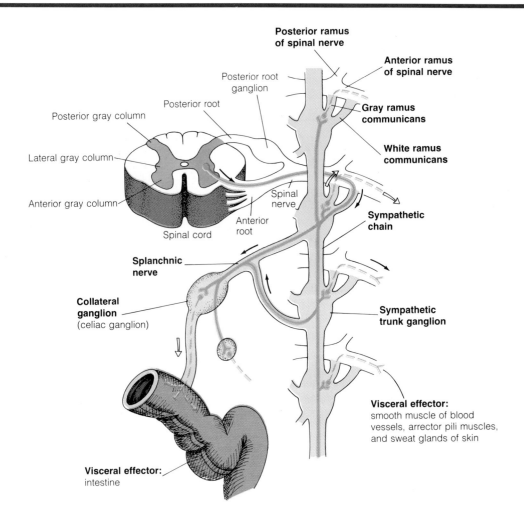

Figure 14-3 **Motor routes of the sympathetic division of the autonomic nervous system.** Solid lines indicate preganglionic fibers and dashed lines indicate postganglionic fibers.

The preganglionic parasympathetic fibers that emerge from the head have their cell bodies in nuclei of the brain stem. The positions of the nuclei and the cranial nerves they use in leaving the brain have been discussed in Chapters 12 and 13. We will only summarize the preganglionic situation here.

All preganglionic parasympathetic fibers that exit in the brain do so in four cranial nerves (III, VII, IX, and X), yet once outside the brain, the axons do not remain associated with their nerve. Instead, fibers that exit with CN III, VII, and IX become associated with parts of the large, diversified trigeminal nerve. This shift in pathways apparently occurs because the trigeminal nerve provides the easiest, and sometimes only, nerve route leading to target organs.

Preganglionic fibers emerging with cranial nerve III arise from cell bodies in the mesencephalon (midbrain) (Figure 14-2). These axons leave with CN III and terminate in the **ciliary ganglion** within the orbit. There they synapse with postganglionic neurons from which axons leave the ganglion and course with trigeminal branches to intrinsic eyeball muscles.

Cranial nerve VII is accompanied by preganglionic fibers that originate in nuclei in the pons (Figure 14-2). These axons leave with CN VII, but some soon branch and join parts of the trigeminal nerve, terminating in the **pterygopalatine** (tĕr″ĭ·gō·păl′ă·tĭn) **ganglia** and **submandibular ganglia.** Postganglionic fibers go from here to the lacrimal gland, the sublingual and submandibular salivary glands, and glands in the lining of the nasal cavity.

Preganglionic fibers originate in the medulla oblongata and leave with CN IX (Figure 14-2). These axons go to the **otic ganglion,** a swelling on one branch of the trigeminal nerve near the foramen ovale. They synapse with postganglionic neurons that leave through another branch of the trigeminal nerve and run to the parotid gland.

Other preganglionic fibers arise in nuclei located in the medulla oblongata and leave the central nervous system with the vagus nerve. As the name *vagus* (wandering) implies, this nerve reaches many parts of the body. It innervates the heart, respiratory system, intestinal tract and their associated glands.

Axons of these neurons branch into several thoracic and abdominal extrinsic autonomic plexuses leading to the heart, lungs, liver, pancreas, spleen, and organs of the digestive, reproductive, and excretory tracts. The principal extrinsic plexuses in the thoracic cavity are the **esophageal, cardiac,** and **pulmonary,** each associated with the organ for which it was named. Within the abdominal cavity are the **celiac** and **superior mesenteric plexuses,** which send fibers to several abdominal organs. Preganglionic fibers exit through these plexuses to synapse with postganglionic neurons that form intrinsic plexuses within the abdominal organs.

Sacral parasympathetic neurons have preganglionic fibers that arise in the lateral columns of the gray matter in the sacral spinal cord. These axons leave with the anterior roots of sacral nerves 2, 3, and 4. They form the **pelvic splanchnic nerves** and travel to the **inferior mesenteric plexus** and other extrinsic plexuses in the lower abdominal and pelvic regions. From these plexuses, preganglionic axons extend to intrinsic plexuses within the target organs. The primary organs innervated by the sacral parasympathetic fibers are the descending and sigmoid colon, rectum and anus, urinary bladder, and the external genitals.

Physiology of the Autonomic Nervous System

After completing this section, you should be able to:

1. Describe the relationship of neurotransmitters to the functions of the ANS.
2. Distinguish between adrenergic and cholinergic fibers.
3. Discuss the role of membrane receptors in mediating the effects of neurotransmitters.
4. Describe how dual innervation by sympathetic and parasympathetic fibers maintains homeostatic control of autonomic functions.
5. Discuss the manner in which sensory information is transmitted to autonomic centers in the central nervous system.
6. Describe the role of biofeedback and meditation in conscious control of autonomic functions.

Neurotransmitters

The autonomic nervous system controls the automatic activities of muscles, glands, and organs through the secretion of the neurotransmitters acetylcholine, norepinephrine, and epinephrine (these were discussed in detail in Chapter 11). Acetylcholine is secreted by all somatic neurons in the central and peripheral nervous system and by all presynaptic axons of the ANS. All parasympathetic postganglionic axons also secrete acetylcholine. In contrast, postganglionic axons of sympathetic neurons secrete norepinephrine. Another part of the sympathetic system, the adrenal medulla, secretes epinephrine directly into the blood (see discussion next section). Norepinephrine and epinephrine are very similar chemically and functionally.

Membrane Receptors

The cells that are targets of these neural secretions have specialized **membrane receptors** that determine the specific effect of each neurotransmitter. Cells that respond to norepinephrine and epinephrine have either **alpha-** or **beta-receptors** (see Chapter 11). In general, cells that have alpha-receptors are stimulated when one of these neurotransmitters combines with the receptor. Conversely, when norepinephrine or epinephrine combine with a beta-receptor, the activities of the effector cell are inhibited (Table 14-1).

Most effectors have either alpha- or beta-receptors. However, even though it seems self-defeating for a cell to be simultaneously stimulated and inhibited by the same chemical, both alpha- and beta-receptors are present on some cells. Perhaps the net effect on such a cell is determined by the relative number of receptors. If this is the case, a cell with more alpha-receptors would be stimulated and one with more beta-receptors would be inhibited by the same amount of norepinephrine or epinephrine.

Responses to all neurotransmitters seem to be mediated by membrane receptors, which have been classified based on their responses to other chemicals. Acetylcholine receptors on all postsynaptic autonomic neurons also respond to the drug *nicotine* and are thus classified as **nicotinic** (nik″ō·tĭn′ĭk) **receptors.** But acetylcholine receptors on smooth muscles and glands do not respond to nicotine. Instead, they are stimulated by *muscarine*, a poison produced by some mushrooms. Such membrane receptors are found in visceral organs and are classified as **muscarinic** (mŭs″kă·rĭn′ĭk) **receptors.** The physiological significance of this sensitivity to chemicals not usually encountered is unknown.

Epinephrine and norepinephrine bring about cellular responses that prepare the body for emergency situations. The specific actions are listed in Table 14-1. The bronchial

Table 14-1 PHYSIOLOGICAL EFFECTS OF THE AUTONOMIC NERVOUS SYSTEM

Organ	Receptor	Effects of Sympathetic Stimulation	Effects of Parasympathetic Stimulation
Eye			
Iris	Alpha	Dilates pupil	Constricts pupil
Ciliary muscle	Beta	Relaxation accommodates lens for far vision	Contraction accommodates lens for near vision
Heart	Beta	Increases rate and strength of contraction	Decreases rate and strength of contraction
Lungs (bronchioles)	Beta	Dilation	Constriction
Stomach	Alpha and beta	Decreases activity	Increases activity
Intestine	Alpha and beta	Decreases activity	Increases activity
Liver	Beta	Increases release of glucose, decreases bile production	Decreases release of glucose, increases bile production
Pancreas	Alpha and beta	Decreases secretion	Increases secretion
Spleen	—	Discharges stored blood into blood vessels	No innervation
Kidney	Beta	Decreases urine production	No known effect
Urinary bladder	Alpha and beta	Relaxes muscles in wall	Constricts muscles in wall
Reproductive organs	Alpha and beta	Vasoconstriction, ejaculation	Vasodilation produces swelling and erection
Arrector pili muscles	Alpha	Contractions produce "goose bumps"	No innervation
Blood vessels of			
Skin	Alpha	Vasoconstriction	No innervation
Skeletal muscles	Alpha and beta	Vasodilation	No innervation
Visceral organs	Alpha and beta	Vasoconstriction	Vasodilation
Glands			
Lacrimal	—	No innervation	Stimulates tear secretion
Salivary	Alpha	Decreases secretion	Stimulates secretion
Digestive tract	—	Inhibits secretion	No innervation
Sweat	Alpha	Stimulates secretion	No innervation

passages are dilated, which increases the amount of oxygen–carbon dioxide exchange in the lungs. Cardiac output is increased as the heart rate is stimulated. The flow of blood to skeletal muscles is increased because arteries supplying them are dilated. At the same time, constriction of arteries leading to visceral organs and glands inhibits their activity. Thus, the sympathetic system prepares the body for action while simultaneously inhibiting visceral activities such as digestion and absorption, thereby providing more blood to the muscular system.

The **adrenal medulla** is a functional part of the sympathetic system that never grows long postganglionic axons. It develops from embryonic postganglionic sympathetic cells that migrate away from the sides of the nerve tube and become incorporated into the adrenal gland. They become secretory cells that, upon maturity, manufacture a product that is approximately 80% epinephrine and 20% norepinephrine. It is not quite accurate to call this secretion a neurotransmitter, because instead of being released into a synaptic or neuromuscular junction, it is released directly into the blood. Unlike synaptic secretions, adrenal medulla secretions have long-lasting effects over a wide part of the body. These secretions function as hormones, and the adrenal medulla is often classified as part of the endocrine system (see Chapter 17).

The parasympathetic division counteracts the sympathetic, returns the active body systems to a maintenance level, and keeps them there (Table 14-1). Acetylcholine secreted from these fibers maintains the activities of the visceral organs at a normal level. The heart and breathing rates are controlled at resting rates, and the activities of the digestive and reproductive systems are inhibited.

exteroceptive: (L.) *exterus*, outside + *receptus*, to receive
enteroceptive: (Gr.) *enteron*, gut + (L.) *receptus*, to receive

Control Through Dual Innervation

The heart, smooth muscles, and many glands receive **dual innervation,** that is, they are innervated by both sympathetic and parasympathetic neurons. Sometimes the combined effects of the two divisions are antagonistic. Heart rate, for example, is increased by sympathetic secretions and decreased by parasympathetic secretions. Sometimes the combined effects are complementary, as in control of the male sexual response. Parasympathetic secretions cause vasodilation in the penis and result in erection. Sexual stimulation causes increased sympathetic secretions, which eventually results in ejaculation and orgasm. Thus, both divisions cooperate to initiate, maintain, and complete the male sexual response. These two examples illustrate how the interplay of the two autonomic systems can maintain homeostatic control of autonomic body processes.

To summarize, the parasympathetic system keeps the body functioning at normal rates under normal conditions, whereas the sympathetic system takes over at times of stress to prepare the body for quick, forceful actions.

Sensory Input into the Autonomic Nervous System

Thus far, we have discussed activities of the autonomic nervous system only in terms of efferent motor neurons. The activities of these motor neurons are dependent on information concerning the body in general and the specific organs innervated. Such sensory information enters autonomic centers through either somatic sensory neurons or visceral sensory neurons. These neurons do not follow the two-neuron chain pattern of autonomic motor neurons, but are arranged in the same way as conscious sensory neurons.

Somatic or **exteroceptive** (ĕks″tĕr·ō·sĕp′tĭv) **sensory neurons** receive information about the external environment. They transmit data concerning temperature, touch, pain, vision, smell, hearing, and similar stimuli from receptors in the skin and sense organs that are keyed to occurrences outside the body, and they relay information to autonomic reflex centers within the central nervous system. They are mainly concerned with stimuli received at a conscious level of experience. **Visceral** or **enteroceptive** (ĕn″tĕr·ō·sĕp′tĭv) **sensory neurons** transmit information from internal organs to autonomic reflex centers. These data include sensations from physical and chemical stimuli in the visceral organs. Enteroceptive sensory neurons transmit information at the subconscious level.

No matter what the source of the sensory information, the impulses arrive at autonomic reflex centers in the brain where they trigger appropriate responses. An example of such a reflex involves the unconscious response of a newborn to the filling of its urinary bladder. As urine fills the bladder, its walls distend like a balloon. Visceral sensory neurons, in this case *pressoreceptors*, transmit impulses concerning the amount of stretch in the muscular walls of the bladder to a parasympathetic center in the sacral spinal cord. The parasympathetic motor fibers, in turn, transmit impulses to the bladder muscles and the internal urethral sphincter. These impulses start contraction of the muscles and relaxation of the sphincter. The result: the bladder is emptied and the child needs a diaper change!

Obviously, the sequence of events just outlined becomes more complicated with toilet training of a toddler, a process involving voluntary control of an additional sphincter muscle (see Chapter 25). In other words, it is possible to exert some voluntary control over the so-called involuntary processes governed by the autonomic nervous system.

Conscious Control of the Autonomic Nervous System

The bulk of material in this chapter has been devoted to demonstrating that the autonomic nervous system controls various aspects of metabolic activity without conscious control. However, such control of the system is possible under certain special circumstances.

Voluntary control of the ANS is possible because this division of the nervous system is connected to the CNS at several junctions, primarily in the brain stem and spinal cord. Motor neurons of the sympathetic division are carried with spinal nerves that include somatic motor nerves. These even participate in several spinal plexuses. Of more importance is the presence of several autonomic control centers within the brain stem. From here neuronal connections are made with several other regions of the brain, particularly the cerebral cortex. Through these connections conscious events, real or imagined, may stimulate or repress either autonomic division.

For example, an individual in an empty house on a dark night may interpret every strange noise as a potential intruder and remain in a state of anxiety, a sympathetic reaction, throughout the night. On the other hand, if an individual who has not eaten all day thinks of delicious food, the parasympathetic system will stimulate salivary glands to secrete saliva, even though nothing has been eaten. These are examples of conscious influence of responses regulated by activities of the autonomic nervous system.

In recent years researchers have learned that it is possible to control many functions of the autonomic nervous system through training. Applications of this knowledge are being used by people who practice biofeedback training or forms of mediation.

CLINICAL NOTE

SUBSTANCE ABUSE

All drugs, whether legal or illegal, can be abused. Other substances that are often abused are food and alcohol. Alcohol is probably the most abused substance in the world; alcohol abuse ranks as the third leading cause of death in the United States. **Substance abuse** is defined as the use of a substance that causes legal, emotional, social, and health problems for the individual and those around him. Two problems associated with substance abuse are dependency and tolerance. **Dependency** means that a person needs a substance to function, and two types exist. Psychological dependency occurs when a person needs the substance to experience positive feelings and a sense of self-esteem. Physical dependency means that the person needs the substance to avoid physical withdrawal symptoms. **Tolerance** occurs when a person needs to have increasing amounts of the substance to achieve the same effects as a smaller amount achieved previously. Increasing amounts or increasing frequency of use of the substance is known as **progression** of the disease.

There are many competing theories attempting to explain substance abuse that incorporate genetic, social, and psychological roots of abuse. No one is entirely satisfactory because abuse is probably multifactorial in its cause. Many clinicians have focused on treatment rather than the puzzle of cause. Self-help groups such as Alcoholics Anonymous, Overeaters Anonymous, and Synanon have proved exceedingly helpful either alone or in combination with other therapies in halting the progression of the disorders. Self-help groups such as Alanon and Alateen are available to assist spouses and children of alcoholics. Professional treatment includes **detoxification** (a period of time spent in a protected environment while the substance removes itself from the body), accompanied by individual, family, or group counseling. Sometimes Antabuse, a drug which causes severe nausea and vomiting when alcohol is taken, is used in conjunction with other therapies.

In addition to alcohol and food, there are many drugs that are frequently abused. The **narcotics,** including the **opiates** (opium, morphine, heroin, and codeine) and the **synthetic nonopiates** (Methadone and Demerol), are often abused. Their street names include snow, stuff, H, junk, smack, and scag. These are all central nervous system depressants with effects lasting from 3 to 24 hours. They may be injected into a vein, under the skin, or into a muscle or smoked or sniffed. They impart the user with a temporary feeling of well-being or euphoria; all care disappears and the person feels transported to a state of bliss. These drugs also eliminate the physical and mental pain associated with withdrawal.

The **barbiturates** (Nembutal, Seconal, and Amytal), the **sedatives** (Equanil, chloral hydrate, and Quaalude), and the **minor tranquilizers** (Valium and Librium) are all also central nervous system depressants that are legal through prescription but are frequently abused. They are known on the street as sleepers, goofballs, redbirds, yellow jackets, red devils, barbs, and downers. They are used predominantly as oral drugs, but they can be prepared for injection into the vein or muscle. The user experiences freedom from care, dissipation of anxiety, and relaxation of tense muscles. Unfortunately, these drugs are often abused in

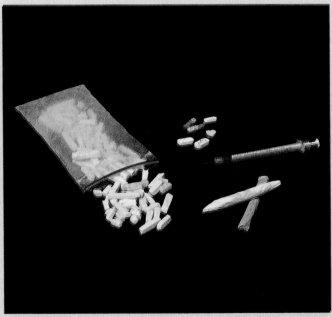

conjunction with alcohol abuse. This combination is highly dangerous and sometimes lethal. Men and women who suffer stress often have the drug prescribed for them and then begin taking them in an abusive way rather than seeking some other way of decreasing the stress in their lives.

The class of drugs called **stimulants,** which includes amphetamines and cocaine, is another group of drugs frequently abused. Amphetamines are known as speed balls, uppers, speed, crystal, or dexies, and cocaine is commonly known as snow and is usually sniffed. However, a particularly potent form of smokable cocaine called crack is now in vogue. These drugs stimulate the central nervous system and make the person feel and be at peak alertness. Calculation and cognitive abilities are heightened. The person has increased initiative, is creative, and feels and acts excited. Cocaine is known as a recreational drug, and this nomenclature tends to hide the problems it can cause in the abuser's life. Psychological dependence is very strong with abuse of this drug, but true physical dependence is debated.

The **natural** and **synthetic hallucinogens** comprise another category of substances that are frequently abused. The best known of the synthetic hallucinogens is LSD, also known as acid. The drug is usually taken by mouth and is occasionally injected. Its effect, or "trip," lasts between 10 and 12 hours. A "mind-expanding" drug, LSD confers feelings of amazing insight on the personal and human condition. The senses are distorted and one feels exhilarated and energetic but detached. Feelings existing before taking the drug may continue into the trip and be heightened. Negative, aggressive, hostile, or self-destructive feelings will lead to a "bad trip" with possibly dangerous outcomes.

The natural hallucinogens, mescaline from cactus and psilocybin from mushrooms, have been used by native people in their ceremonial events. They may also be abused. A person may develop a tolerance to these drugs, but physical dependence does not occur. Psychological dependence, however, is possible.

Marijuana and hashish are difficult to classify since they both stimulate and depress the central nervous system. Marijuana is also known as joints, reefers, pot, or grass, and hashish is often referred to as hash. Both are usually inhaled but may be added to food and eaten. Users of these drugs feel relaxed and free from tension. There is an idea that perception about oneself and the world around one are increased. Day-to-day problems are forgotten. The person may get very giggly, and the appetite is increased while taste is enhanced. Tolerance to these drugs seems not to develop and neither does physical dependence, although there are possibly some long-term physical side effects. The major risk, however, lies in the probable psychological dependence on these drugs: the user escapes problems rather than dealing with them.

One must not overlook nicotine and caffeine as very commonly abused substances. An individual can become physically and psychologically dependent on both of these substances and also tolerant to them.

Biofeedback

The technique of controlling body activities is called **biofeedback.** It is currently being developed by researchers working with individuals whose ailments do not respond to traditional methods of treatment, such as drugs. A biofeedback program involves training in conscious control of body processes normally under subconscious control. This requires a mechanical device that registers an internal condition such as blood pressure, heart rate, or skin temperature. The subject is trained to practice certain exercises, some physical and some mental, that modify the level of the internal condition being monitored. Depending on the body function under study, the individual learns to use the acquired techniques to regulate it.

For example, a person with hypertension (high blood pressure) can be taught to lower blood pressure with biofeedback techniques. The person is attached to a device that registers blood pressure and signals when the pressure rises above the desired level. This signal can be accomplished through a buzzer, light, or some other signal easily perceived and understood. The subject is instructed how to relax the body through reciting certain phrases or visualizing certain scenes. Immediate "biofeedback" is provided when the recitation or visualization technique stops the signal. The subject quickly learns that certain behaviors and practices lower blood pressure and others raise it. It does not take long before a person can learn to use these techniques, even without the reinforcement of the biofeedback machine.

Forms of Meditation

Yoga, meditation, and **breathing therapy** are other techniques practiced to relax and slow down body processes. This modification is accomplished by inhibiting the sympathetic nervous system through a repetition and practice of physical, verbal, and mental techniques. These procedures minimize the activities of that part of the autonomic nervous system that normally stimulates body activities. At the present time, no one has explained how these meditation techniques work. The interesting thing is they do work for many people. Obviously, this area of human activity offers a promising avenue for researchers interested in studying the relationship between the autonomic nervous system and its control of body activities.

In this chapter we examined the anatomy and physiology of the autonomic nervous system. You learned that it is divided into a sympathetic division that readies the body for action and a parasympathetic division that slows down and relaxes the body. In the next chapter we will study the manner in which sensory, association, and motor functions of the brain, spinal cord, and peripheral nerves are integrated into a coordinated nervous system.

352 Chapter 14 The Autonomic Nervous System

sympathomimetic:	(Gr.) *sympathetikos*, sympathy + *mimetikos*, imitating
anaphylactic:	(Gr.) *ana*, again + *phylaxis*, protection
sympatholytic:	(Gr.) *sympathetikos*, sympathy + *lytikos*, dissolving

THE CLINIC

INFLUENCING THE ANS WITH DRUGS

One gains an appreciation of the importance of the autonomic nervous system by reviewing the number of diseases and abnormal conditions that are related to imbalance between the sympathetic and parasympathetic systems. For many years medical science has used drugs to change the balance within the system to alleviate conditions causing illness. Most of the drugs used to influence the autonomic system either stimulate or depress neurotransmitters in the system. Drugs that stimulate neurotransmission have the suffix *-mimetics* because they mimic the effects of the particular system. Drugs that depress the action of a system have the suffix *-lytics* because they cut or reduce the system's effect. This is a very simplified classification because the two systems tend to have opposing actions. At times it is difficult to distinguish between a medication that stimulates the sympathetic system and one that depresses the parasympathetic system.

Sympathomimetic Drugs

There are three main groups of **sympathomimetic** (sĭm″pă·thō·mĭm·ĕt′ĭk) or adrenergic drugs that stimulate the sympathetic nervous system. They are all either naturally occurring neurotransmitters or synthetic (man-made) analogues. These three are epinephrine, norepinephrine, and isoproterenol. Although these drugs are similar, they produce different effects when introduced into the body because there are different receptor sites in the ANS for each drug (see text).

Epinephrine is a naturally occurring neurotransmitter and hormone in the body. It has both alpha- and beta-receptor activity. It is a very important drug in the emergency treatment of **anaphylactic** (ăn″ă·fĭ·lăk′tĭk) **shock** or collapse from an acute allergic reaction. Epinephrine is also the drug of choice in the emergency treatment of asthma and bronchospasm because it causes dilation of the bronchial tubes. It also causes fast heart rate, elevation of blood pressure, and vasoconstriction.

Norepinephrine is an important naturally occurring neurotransmitter. It also has both alpha- and beta-receptor activity. Its chief medical use is to elevate blood pressure in hypotensive states. It is given intravenously. Its side effects of vasoconstriction in the brain, kidneys, and extremities limit its usefulness.

Isoproterenol (ī″sō·prō″tĕ·rē′nŏl) has limited clinical use since it is poorly absorbed from the gastrointestinal tract. It can be given sublingually (absorbed beneath the tongue) for treatment of certain cardiac arrhythmias, but more often it is administered as an inhaled spray to treat asthma.

Another adrenergic drug is **dopamine,** which is the immediate precursor of norepinephrine. Dopamine has the advantage that at low dosages it has almost pure beta-receptor effects. At increased dosages the alpha-reception predominates. Dopamine is used to treat hypotensive states and resistant congestive heart failure.

There are many other adrenergic agents of limited or specialized use. **Ephedrine** is similar to epinephrine but slower in onset of action and longer in effect. **Terbutaline** is a pure beta-receptor stimulant and is very useful in the treatment of asthma since it causes little cardiac stimulation. **Phenylephrine** is often used in cold remedies to produce nasal vasoconstriction.

Sympatholytic Drugs

There are two types of **sympatholytic** (sĭm″pă·thō·lĭt′ĭk) **drugs,** or adrenergic blocking agents, which cut or block the effects of the sympathetic nervous system. The **alpha-blocking agents,** which are of little medical use, are the first type of sympatholytic drugs. They mainly cause vasodilation and are sometimes used to treat such diseases as Raynaud's disease (constriction of the small arterioles of the digits on exposure to cold), postfrostbite syndrome, and arteriosclerosis obliterans (narrowing of small arteries in the extremities due to deposition of cholesterol plaques in

STUDY OUTLINE

Distribution of Neurons (p. 343)

Visceral efferent neurons comprise the major part of the ANS. They innervate cardiac muscle, certain glands and visceral muscles in blood vessels and organs of the thoracic and abdominal cavities. These neurons are organized into functional units of two neurons each: *preganglionic* and *postganglionic neurons.*

Anatomy of the Autonomic Nervous System (pp. 343–347)

The autonomic nervous system has two distinct anatomical and physiological subdivisions: *sympathetic* (or *thoracolumbar*) and *parasympathetic* (or *craniosacral*) *nervous systems.*

Sympathetic System Preganglionic neurons of the sympathetic system emerge through anterior roots of all thoracic spinal nerves and through the first two lumbar spinal nerves. These neurons exit through *communicating rami* and enter *paravertebral ganglia,* where they may synapse with postganglionic cells, while others (*splanchnic nerves*) pass through the ganglia and synapse with postganglionic cells near their target organs where they form *plexuses.* Preganglionic neurons that innervate cells of the adrenal medulla pass directly to that organ without synapsing with postganglionic cells.

Parasympathetic System Preganglionic neurons of the parasympathetic system emerge through cranial nerves III, VII, IX, and X and through the anterior roots of sacrospinal

the arterial walls). Examples of these drugs include tolazoline, phenoxybenzamine, and phentolamine. Phentolamine (Ritalen) is of interest because it is used as a diagnostic test in hypertension caused by a rare tumor that produces adrenergic compounds. Ritalen is also used to treat children with attention deficit disorder, a common cause of learning disability.

The **beta-blocking agents** are the second type of sympatholytic drugs, which produce prominent effects on the cardiovascular system. They cause reduced heart rate, myocardial contraction, and reduced stroke volume resulting in reduced cardiac output and oxygen consumption. They are very useful in the treatment of angina pectoris (chest pain due to reduced circulation to the myocardium). Over a prolonged period of administration, the beta-blockers are effective in lowering blood pressure. They are helpful in any condition that would benefit from reduced sympathetic activity. The most common drug of this type is propranolol. Caution must be exercised in the use of this type of medication in patients who have asthma since they may cause increased bronchospasm.

Parasympathomimetic Drugs
The **parasympathomimetic** (păr″ă·sĭm·pă·thō·mĭm·ĕt′ĭk) **drugs** are also known as **cholinergics** (kō″lĭn·ĕr′jĭks) since they act at sites where acetylcholine is the neurotransmitter. These drugs act directly on the peripheral blood vessels causing vasodilation, and they stimulate the autonomic ganglia and neuromuscular junctions. The cholinergic drugs are classified into three groups. **Choline esters,** in the form of bethanechol chloride, work largely on the bladder and bowel, increasing motility. They are used to relieve urinary retention. **Cholinergic alkaloids,** represented by Pilocarpine, are used clinically to cause constriction of the pupil in the treatment of glaucoma (increased pressure in the anterior chamber of the eye). **Anticholinesterases** (ăn″tĭ·kō·lĭn·ĕs′tĕr·ās·ĕs) are drugs capable of blocking enzymes that terminate the action of acetylcholine. They act as cholinergics since they allow for prolonged action of the acetylcholine. Physostigmine is a plant alkaloid used as a miotic in ophthalmology and to reverse the effects of poisoning due to atropine (see below) and tricyclic antidepressant poisoning. Edrophonium chloride is a very short-acting drug of this type used to diagnose myasthenia gravis (see Chapter 10, The Clinic: Muscular Diseases).

Parasympatholytic Drugs
The **parasympatholytic** (păr″ă·sĭm″pă·thō·lĭt′ĭk) **drugs** are also called **anticholinergics** because they oppose the action of acetylcholine. They do not block the neuromuscular junction. They cause pupil dilation, fast heart rate, reduction of secretions of the exocrine glands, and relaxation of the smooth muscles of the bronchi, gastrointestinal tract, and urinary bladder. **Atropine** is a prototype of this group of drugs. Atropine is a plant alkaloid found in many native plants, such as the deadly nightshade. Atropine has many uses in medicine. It is a long acting mydriatic; it reduces smooth muscle spasm in the gastrointestinal tract, gallbladder, urinary bladder, and ureters; and it is used to reverse bradycardia and to dry up secretions prior to surgery. The classic toxic symptoms of atropine poisoning suggest its actions:

"Mad as a hatter":	cerebral stimulation
"Hot as a pistol":	decreased sweating and vasodilation
"Dry as a bone":	decreased secretions
"Blind as a bat":	pupil dilation and paralysis of ciliary muscle

nerves 2, 3, and 4. The neurons in CN III, VII, and IX synapse with postganglionic cells in cranial ganglia. Those in CN X and the sacral region pass directly to target organs.

Physiology of the Autonomic Nervous System (pp. 347–351)

Neurotransmitters The ANS controls the automatic activities of muscles, glands, and organs through the secretion of chemical neurotransmitters. The sympathetic division secretes norepinephrine and epinephrine, which tend to speed up certain body activities and prepare the body for emergency situations. The parasympathetic division secretes acetylcholine, which counteracts the effects of the other two chemicals. The adrenal medulla secretes epinephrine and norepinephrine directly into the blood.

Membrane Receptors The cells that are targets of neurotransmitters have specialized *membrane receptors* for these chemicals. Cells that respond to norepinephrine and epinephrine have either *alpha-* or *beta-receptors.* Cells with alpha-receptors are stimulated by the chemicals, while those with beta-receptors are inhibited. Cells responsive to acetylcholine have either *nicotinic* or *muscarinic receptors.*

Control Through Dual Innervation *Dual innervation* of the heart, smooth muscles, and many organs and glands by both sympathetic and parasympathetic fibers allows for both antagonistic and complementary effects.

Sensory Input into the Autonomic Nervous System Sensory information enters autonomic centers through either *somatic sensory (exteroceptive)* or *visceral sensory (enteroceptive) neurons*. These data involve stimuli from external sources, such as temperature, touch, pain, vision, smell, and hearing, and internal sensations from physical and chemical stimuli in the visceral organs.

Conscious Control of the Autonomic Nervous System Connections between the ANS and CNS allow some conscious control of autonomic activities. *Biofeedback* is a form of control using mechanical devices to record the results of control techniques. *Yoga* and *meditation* are forms of control that inhibit the sympathetic division.

SELF-TEST OF CHAPTER OBJECTIVES

True-False Questions
1. Neurons that comprise the autonomic nervous system are primarily sensory neurons.
2. Efferent neurons of the autonomic nervous system that are anchored in the spinal cord are called preganglionic neurons.
3. All spinal nerves contain some autonomic nerve fibers.
4. Preganglionic neurons of the autonomic nervous system arise from cells in the lateral columns of the spinal cord.
5. The sympathetic division of the autonomic nervous system is also called the craniosacral division because of the anatomical position of its preganglionic fibers.
6. The sympathetic chain ganglia are connected to the anterior roots of the spinal nerves through short communicating rami.
7. The sympathetic nervous system is often characterized as the body's primary nerve supply for the arteries.
8. In the head and neck, the sympathetic nerves are distributed in plexuses that follow the major arteries in the region.
9. The parasympathetic nervous system readies the body for strong, fast actions as in the fight-or-flight response.
10. Adrenergic fibers in the autonomic nervous system release acetylcholine from the ends of their axons.

Matching Questions
Match the structures on the left with the appropriate statement on the right.

11. preganglionic neurons
12. postganglionic neurons
13. epinephrine
14. acetylcholine
15. collateral ganglia

a. secreted by all preganglionic axons
b. cell bodies lie in the lateral horns of the spinal cord
c. secreted by sympathetic postganglionic axons
d. cell bodies lie in paravertebral ganglia
e. celiac ganglia

Multiple-Choice Questions
16. Cholinergic fibers release which of the following substances from the ends of their axons?
    a. epinephrine   c. norepinephrine
    b. pseudoepinephrine d. acetylcholine
17. Sensory input into the autonomic nervous system is accomplished by
    a. epinephrine   c. norepinephrine
    b. somatic sensory neurons d. acetylcholine
18. Postganglionic cells of the autonomic nervous system arise from
    a. neuroblasts in the lateral horns of the spinal cord
    b. neural crest cells
    c. preganglionic cells that migrate out of the spinal cord
    d. spinal nerve cells that leave the nerve and become autonomic cells
19. Sympathetic ganglia found in the abdominal cavity are called
    a. plexuses   c. collateral ganglia
    b. spinal ganglia d. extrinsic autonomic ganglia
20. Postganglionic sympathetic fibers that permeate tissues of an abdominal organ are
    a. intrinsic autonomic plexuses c. collateral ganglia
    b. extrinsic autonomic plexuses d. splanchnic ganglia
21. All of the preganglionic parasympathetic fibers that exit the brain do so in only four cranial nerves:
    a. III, V, VII, X   c. II, III, VII, IX
    b. III, VII, IX, X   d. III, IV, VII, X
22. Which of the following substances slows down body activities such as heart and breathing rate?
    a. acetylcholine   c. pseudoepinephrine
    b. epinephrine   d. norepinephrine
23. Visceral efferent neurons are organized into functional units composed of how many cells?
    a. one   c. three
    b. two   d. four
24. Which of the following is *not* a collateral ganglion?
    a. superior mesenteric ganglion
    b. celiac ganglion
    c. inferior mesenteric ganglion
    d. otic ganglion
25. Which of the following plexuses is a functional part of the sympathetic system?
    a. external carotid plexus c. pulmonary plexus
    b. aortic plexus   d. superior plexus

Essay Questions
26. Discuss the differences in function between the somatic efferent nervous system and the autonomic nervous system.
27. Describe the anatomical differences between the preganglionic and postganglionic neurons of the autonomic nervous system.
28. How is it possible that the action of the same neurotransmitter from a postsynaptic neuron can be excitatory to some tissues and inhibitory to others?
29. Compare and contrast the general functions of the sympathetic and parasympathetic systems.

30. Discuss the concept of biofeedback. How can it be used to control a chronic internal situation that might otherwise require the use of drug therapy?
31. Discuss the role of dual innervation in maintaining homeostatic control of autonomic functions.
32. Distinguish between adrenergic and cholinergic neurotransmitters, and discuss the distribution of each in the two divisions of the ANS.
33. Describe membrane receptors relative to neurotransmitters in the ANS.
34. Describe the differences between sympathetic chain ganglia and collateral ganglia.
35. Discuss the relationship of the adrenal medulla to the ANS. What neurotransmitters does it secrete, and how does their action differ from that of the same neurotransmitters secreted by other sympathetic neurons?

Clinical Application Questions
36. Describe the difference between dependence and tolerance in drug abuse.
37. Compare the effects of barbiturates versus those of cocaine.
38. Define the term sympathomimetic.
39. How do sympathomimetic drugs differ from sympatholytic drugs?

15

Functional Aspects of the Nervous System

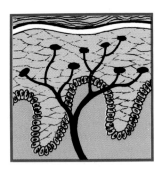

Functional aspects of the nervous system fall into three categories: sensory, motor, and integrative. First, sensory stimuli from external and internal sources are received, and impulses are transmitted into the central nervous system. (The special sense organs are covered in the next chapter.) Second, voluntary and involuntary motor impulses that arise in the CNS are sent to muscles and glands. Third, an intermediate network of association neurons integrates all body activities, either as reflexes or as perceived and controlled acts. Integrative activities in the brain are responsible for such phenomena as consciousness (wakefulness), sleep, abstraction, memory, learning, and emotions. As we study the functional aspects of the nervous system in this chapter, we will discuss

1. the nature of sensory reception and perception,
2. the differences between special and general senses and the nature of the receptors associated with the
3. the functions of sensory and motor pathways in reflexes and voluntary activities,
4. the sensory and motor functions of the brain, and
5. the integrative functions of the brain, including consciousness, alertness, sleep, memory, and learning.

visceroreceptor: (L.) *viscera*, body organs +
photoreceptor: (Gr.) *photos*, light +
mechanoreceptor: (Gr.) *mechane*, machine +
thermoreceptor: (Gr.) *therme*, heat +
chemoreceptor: (Gr.) *chemeia*, chemistry +
baroreceptor: (Gr.) *baros*, weight +
nociceptor: (L.) *nocere*, to injure +

(L.) *receptor*, a receiver

Sensory Reception

After completing this section, you should be able to:

1. List the four major steps leading to perception of a sensation.
2. Compare and contrast exteroceptors, visceroceptors, and proprioceptors.
3. Classify the sensory receptors of the body based on the type of stimuli they detect.
4. Define "adequate stimulus."
5. Describe the relationship between a receptor potential and an action potential in a sensory neuron.
6. Discuss the phenomena of adaptation and afterimage.

An Overview of Sensory Reception and Perception

The major functions of the central nervous system include monitoring the external and internal environments and responding to significant changes by stimulating or inhibiting cellular activities. Sensations are the raw material upon which responses are based. A **sensation** is a sensory impulse that stimulates the CNS. **Perception** is the conscious awareness of a sensation. In order to perceive a sensation, four things must happen:

1. A *stimulus* sufficient to evoke a response must be present.
2. A *receptor*, a specialized cell or tissue associated with the nervous system, must respond to the stimulus by generating an impulse.
3. The impulse must be *conducted along a pathway* to the brain.
4. A specific part of the brain must *interpret* the impulse as a perception.

An impulse that terminates in a region of the CNS below the cerebral cortex is not consciously perceived, although it may elicit a reflex action.

Classification of Senses

Senses can be classified as general or special. **General senses** involve sensations received over a wide area of the body. These senses include perception of touch, pressure, pain, heat, and cold. Receptors of these sensations are relatively simple. They are either naked dendrites or the encapsulated ends of neurons (Figure 15-1). The pathways traveled by general sense impulses are uncomplicated. **Special senses**, on the other hand, involve specific sensations received only by highly specialized sense organs that are concentrated in the head. The nerve pathways associated with these special senses are more complex than those in the general senses. Special senses include vision, smell, taste, equilibrium, and hearing; they are the subject of Chapter 16.

Classification of Receptors

Receptors can be classified by location. **Exteroceptors** are located near the surface of the body and respond to stimuli in the external environment. These receptors relay sensations of vision, taste, smell, hearing, touch, pressure, temperature, and pain. **Visceroreceptors** (vĭs″ĕr·ō·rē·sĕp′tŏrz) (or **enteroceptors**) are located within the blood vessels and visceral organs and respond to internal sensations such as hunger, thirst, fatigue, pressure, pain, and nausea. **Proprioceptors** are located within the inner joints, muscles, and tendons and transmit information regarding equilibrium and body position.

A more functional classification of receptors recognizes the stimuli they detect. **Photoreceptors** are sensitive to light stimuli, while **mechanoreceptors** respond to touch, pressure and vibrations. **Thermoreceptors** are stimulated by temperature changes. **Chemoreceptors** respond to chemicals and function in taste, smell, and changes in concentration of certain chemicals in body fluids. **Baroreceptors** are sensitive to changes in blood pressure, and **nociceptors** (nō″sī·sĕp′tŏrz) respond to pain.

Stimulation of Receptors

Adequate Stimulus

The stimulus that normally elicits a response is called the **adequate stimulus.** Just as certain chemicals are the adequate stimuli for taste and olfactory receptors, light is the adequate stimulus for the photoreceptors of the eye. If one closes the eyelid and presses on the eyeball, a series of light flashes are perceived even though the sensory cells are being stimulated only by pressure and not by light. A sensory neuron normally is so specialized that it always produces the same sensation no matter what the stimulus.

Receptor Potentials

The electrical response of sensory receptors to stimuli is similar to that of other neurons, as outlined in Chapter 11. The stimulus initiates local depolarizations in the dendrites proportional to the strength of the stimulus. Each depolarization is a **receptor potential** (or **generator potential**) and is analogous to the excitatory postsynaptic potential (EPSP) in other neurons. If the receptor potential is of sufficient intensity, an action potential is generated in the sensory fiber.

pseudounipolar: (Gr.) *pseudes*, false + (L.) *unus*, one + *polaris*, pole

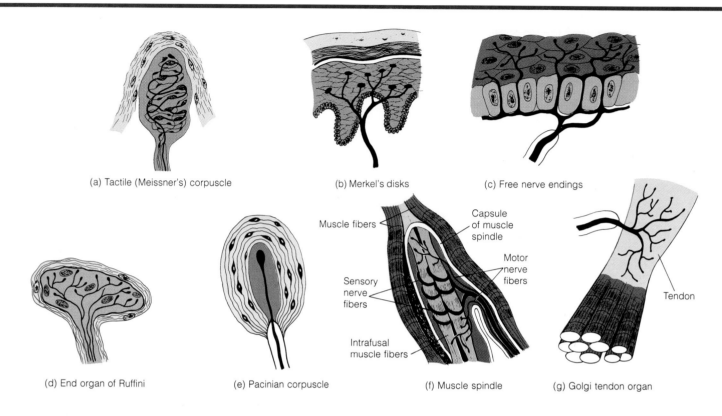

Figure 15-1 The nerve endings depicted here detect general stimuli such as touch, pressure, pain, and stretch in muscles, tendons, and ligaments.

The impulse resulting from this action potential travels from the dendrite toward the cell body in the posterior root ganglion (Figure 15-2). Although the fiber consists of afferent and efferent portions relative to the cell body, the short, right-angle connection of the cell body allows both portions to extend as a single fiber from the dendrite to the terminus of the axon. The sensory neuron is said to be **pseudounipolar** (sū′dō·ū·nĭ·pō′lăr), and it conducts an impulse in the same manner as a truly unipolar neuron does over a single axon.

Adaptation and Afterimage

The pattern of processing stimuli in the nervous system allows the body to differentiate between those that have been stimulating a sense organ at a low level for an extended period of time and those that have just been initiated. This phenomenon of **adaptation** involves the reduction of impulse production in certain sensory neurons even though the stimulus is still being applied to the receptor. Thus, the senses respond vigorously to beginnings and endings of events and less to constant stimulation. For example, among the articles of clothing you put on in the morning is probably one that is elastic, such as a watchband or a belt that binds clothing close to your body. The phenomenon of adaptation elimi-

nates your constant awareness of this tight clothing all day. Olfactory adaptation accounts for you becoming quickly accustomed to food aromas that are so strong when you first enter a kitchen where someone is cooking. Sensory neurons that adapt rapidly are called **phasic receptors;** those that do not are **tonic receptors.**

If one looks at a bright light for a short time and then glances away, an image of the light remains. This is called an **afterimage.** Such afterimages seem to be the opposite of adaptation. During the afterimage formation, one is consciously aware of a sensation even though the stimulus that produced it has been removed.

The phenomena of adaptation and afterimage formation indicate that sensory neurons are capable of more than merely responding or not responding to a stimulus.

Projection

If a sensory organ detects a sufficiently intense stimulus, the impulse initiated will be transmitted along one or more axons

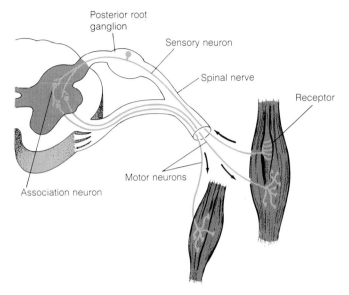

Figure 15-2 Sensory and motor neurons. The dendrites and axon fibers of the sensory neuron are connected such that an impulse is conducted along the fiber as it would be in a unipolar neuron.

in a sensory pathway to the CNS. If the impulse travels only to the spinal cord, the individual does not become consciously aware of the stimulus. Only when the impulse follows a pathway to the cerebral cortex will a person become aware of the sensation. The cerebral cortex is able to localize a sensation to a specific part of the body because impulses travel definitive pathways from the point of stimulation. This process is called **projection,** and it allows you to determine, for example, that your left index finger is hurting and not your left thumb.

General Sense Receptors

After completing this section, you should be able to:

1. Describe the two types of mechanoreceptors and differentiate between their locations.
2. Describe the two-point discrimination test.
3. Differentiate between somatic and visceral pain.
4. Indicate how proprioceptors contribute to an awareness of body position.

Mechanoreceptors

Two types of mechanoreceptors exist: **touch receptors** and **pressure receptors.**

Touch Receptors

The sense of **touch** involves the mechanical stimulation of receptors in the skin, mouth, nasal cavity, eyes, and several areas of the digestive tract. They are not evenly distributed throughout the body and are most concentrated in the tongue, lips, fingertips, certain areas of the facial skin, and the skin of the genitals.

A common demonstration of this difference in distribution of touch receptors involves a subject being tested with a pair of calipers. In the test, the points of the calipers are set very close together and touched to a subject's skin. When the points nearly touch one another, the subject is unable to distinguish them as two discrete points and will report only one point touching the skin. If the points are separated progressively and the test repeated, subjects can eventually distinguish two points of contact. This test is the **two-point discrimination test.** It is used to measure the distance between touch receptors in different parts of the body. On the tongue most people can distinguish two points separated by only 1.5 mm. The discriminatory distance increases from the lips, fingertips, palms, and chest, until, finally, on the back of the neck the subject cannot distinguish two points until they are almost 37 mm apart!

Touch receptors are divided into two general classes: those of **crude touch** sensations and those that register **fine touch.** The former receptors are usually enteroceptors and merely register the mechanical pressure of being moved or stretched. Anyone who has swallowed a whole peanut or a piece of ice has been able to follow the progress of the object through the esophagus and into the stomach through the progressive stimulation of consecutive regions of crude touch receptors in the esophagus.

Fine touch receptors, however, can pinpoint specific sites of stimulation. Three types of neurons are involved in the receptions of fine touch stimuli (Figures 15-1a–c). **Tactile** (or **Meissner's**) **corpuscles** consist of the dendritic ends of sensory neurons surrounded by oval-shaped capsules of connective tissue. They are in dermal papillae. They are distributed throughout the skin, but are most numerous in the fingertips, palms, soles, tip of the tongue, lips, eyelids, nipples, glans penis, and clitoris. **Merkel's disks** are receptors with terminal dendritic endings that are disk shaped. They are dermally situated and register fine touch stimuli. **Free nerve endings** (also called **root hair plexuses**) are also found in the dermis. Each surrounds the root of a hair. Whenever a hair is even slightly bent, the dendrite is stimulated, alerting the nervous system that a specific hair has been moved. In this way we can follow an insect crawling over our forearm, although it may never actually touch our skin. A third touch receptor is called the **end organ of Ruffini** (Figure 15-1d). Located deeper in the dermis than other such receptors, it seems to require greater pressure to be stimulated.

Pressure Receptors

Pressure receptors register somewhat deeper pressure than touch receptors. Specific cells responsible for detecting pressure stimuli are **pacinian corpuscles** (Figure 15-1e). These cells consist of a single dendrite surrounded by several concentric layers of connective tissue, resembling a section of onion. These connective tissue layers most likely shield the dendrite from light touch stimuli but cannot do so against heavier pressure. Pacinian corpuscles are distributed over the surface of the body in the deep subcutaneous connective tissues, within tissues surrounding joints and muscles, and within the mammary glands, external genitals, and intestinal tract.

Thermoreceptors

Thermoreceptors respond to stimuli of heat and cold and are widely distributed over the skin. The specific receptors for heat and cold have not been identified, but they are probably free nerve endings similar to pain receptors.

Nociceptors

Receptors of pain, or nociceptors, are the least specialized of cutaneous receptors. These cells have dendritic terminal endings that are branched, free nerve endings without either a myelin sheath or neurilemma. The skin is not the only location of pain receptors. They are found in almost all tissues of the body. Pain receptors respond to chemicals released by injured cells. Tissue damage can result from overstimulation by many different stimuli, including heat, cold, pressure, stretching, and physical injury. Pain receptors are tonic receptors and exhibit little adaptation, if any at all. If they did to any great extent, it could be disadvantageous to the survival of an individual. If the sensation of pain were to disappear while serious harm were happening to a body part, irreparable damage could occur.

The sensation of pain occurs in two forms: somatic and visceral. **Somatic pain** originates in the skin and superficial tissues, whereas **visceral pain** arises in the viscera.

In most circumstances, conscious areas of the cerebral cortex are able to associate a specific area of the body with a pain sensation. If the thumb on your left hand gets caught while closing a door, you can localize the pain to your thumb. If you get hit on the right knee, that knee hurts. However, the brain is unable to locate accurately the source of some types of visceral pain. For example, a gallbladder attack may be felt as pain in the diaphragm or in the skin on the right side of the neck. This phenomenon is known as **referred pain** (Figure 15-3). In these examples, the pain is felt in an area other than the actual location (see Clinical Note: Referred Pain). The most well-known example of referred pain often occurs during a heart attack. Instead of the pain being localized in the interior of the chest, it is often felt in the left arm.

Referred pain occurs when different areas of the body send sensory impulses to the cerebral cortex along the same major spinal cord routes. Under such circumstances, the brain is sometimes unable to sort out the two signals. The exact origin of the sensory signals is undeterminable. Hence, the brain incorrectly interprets the source of the pain.

In the example of a heart attack, impulses of pain associated with a heart attack enter the spinal cord through the same pathways as those from the medial side of the left arm, specifically the area between the first and fourth thoracic vertebrae. For some unexplained reason, the brain is unable to differentiate between impulses coming from the arm and those coming from the heart. Consequently, the pain associated with the attack is perceived as coming from the left arm and shoulder.

Proprioceptors

Proprioceptors are neurons that register the amount of tension created by muscle contractions, particularly those of the arms and legs. These are receptors that allow us to walk or run, for example, without consciously thinking about every step. Finely tuned proprioceptors would even allow such a complex task as a blind person playing a guitar. Proprioceptors cause the brain to be aware of spatial relationships of body parts without the necessity of looking at them.

Two major types of proprioceptors have been identified: **muscle spindles** and **Golgi tendon organs** (Figures 15-1f

---

CLINICAL NOTE

REFERRED PAIN

The concept of **referred pain** is an important one in the practice of medicine. Many pain syndromes are ill defined and difficult to diagnose if the physician is not alert to the common pain referral patterns. Obscure knee pain often results from diseases in the hip joint; irritation of the diaphragm from blood or infection in the abdomen often presents itself as pain in the left shoulder; the everpresent low back pain may originate from a wide variety of sources from the foot to the renal or pelvic areas.

Another cause of referred pain originates during embryologic development when organs migrate from their site of origin and carry their nerve supply with them. The testes are a good example of this phenomenon. They originate in the embryo near the kidney and later migrate to the scrotum taking their nerves and blood supply along with them. Testicular pain is frequently described as deep in the upper abdomen and can confuse the unwary physician.

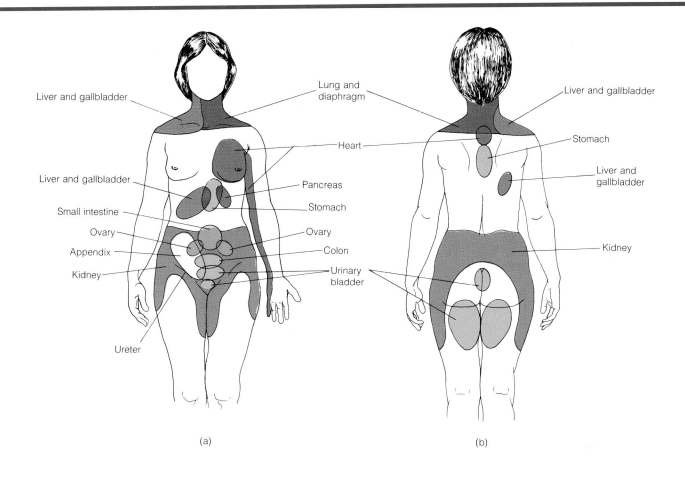

Figure 15-3 Referred pain. The shaded area in the diagram indicates the area in the skin where pain from internal organs is actually felt. (a) Anterior view. (b) Posterior view.

and g). Muscle spindles consist of branched dendrites wrapped around certain specialized skeletal muscle cells called **intrafusal fibers.** When a muscle is stretched, the intrafusal fibers are also stretched. When stretched, the spindles send impulses to the spinal cord, the number of impulses increasing with increased stretching. Thus, the central nervous system is kept aware of the activities within a muscle. Using this information, reflexes may counteract the stretching by contracting other fibers in the muscle.

Golgi tendon organs are highly branched sensory neurons found in the tendons of skeletal muscles. These receptors are sensitive to tendon stretching and relay impulses indicating the amount of stretching to the CNS.

Sensory Pathways

After completing this section, you should be able to:

1. Identify the three orders of neurons involved in afferent conduction pathways and describe their function.
2. List the conduction pathways used for sensations of touch, pressure, and proprioception.
3. Discuss which senses the spinocerebellar sensory tracts transmit.

In general terms, an impulse travels one of two major routes to the CNS. Some impulses enter the spinal cord and initiate spinal reflexes without involving higher nerve centers. Others are transmitted on specific pathways through the spinal cord to centers in the brain stem and on through projection fibers to the cerebral cortex. These pathways, called **sensory** or **afferent conduction pathways,** are localized in ascending tracts (fasciculi) of spinal white matter (see Chapter 12). If an impulse travels only to the spinal cord, a person does not become consciously aware of the stimulus. Only when the nerve impulse follows a pathway that leads to the cerebral cortex can the individual become aware of the sensation.

Afferent pathways that transmit impulses of general senses from peripheral receptors through the spinal cord to the brain differ in the number of neurons involved and in the specific routes traveled. Three classes of neurons together form a pathway (Figure 15-4). **First-order neurons** are sensory neurons that receive the stimulus and transmit the resulting impulse to the spinal cord. **Second-order neurons** are association neurons that carry impulses in ascending spinal tracts which synapse in the brain, primarily the thalamus. **Third-order neurons,** also association neurons, transmit impulses primarily from the thalamus to the cerebral cortex where the sensations intrude into the individual's consciousness.

Touch and Pressure

The conduction pathway for the general senses of fine touch and pressure is illustrated in Figure 15-5. The two spinal tracts are the **fasciculus gracilis** and **fasciculus cuneatus.** Crossover occurs within this pathway, within the **medial lemniscus** of the brain stem. Recall from Chapter 12 that the impulse travels from the side of the cord where it entered to the opposite side of the brain. As a result of crossover, the right side of the cerebral cortex eventually receives impulses from the left side of the body and vice versa.

First-order neurons terminate in the nucleus gracilis and nucleus cuneatus of the medulla oblongata. Here they synapse with second-order neurons which, in turn, synapse in the thalamus with third-order neurons. These project into the general sensory area of the brain where memory helps to interpret the perceived sensations.

This sensory pathway can be demonstrated in a popular child's game called "grab bag." At the beginning the host places unknown yet familiar objects in a bag. Each child is blindfolded and given a chance to reach into the bag and describe the objects using only the sense of touch. In this instance, sensations of fine touch and pressure allow the children to detect size, shape, texture, resiliency, temperature, and other characteristics of the unknown objects. Cutaneous sensations from the objects are transmitted along the afferent tracts illustrated in Figure 15-5 to centers in the brain. Memory and the brain's interpretive abilities (integration) facilitate the child's ability to describe and perhaps name the unseen objects. The child who best describes or identifies the most objects wins a prize.

This example shows how the nervous system allows an individual to interpret sensory stimuli in a way to increase

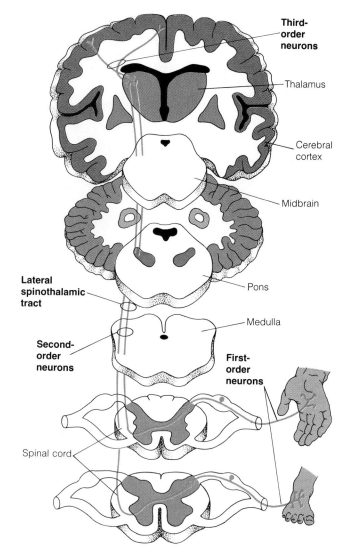

Figure 15-4 The sensory (afferent) pathway for pain and temperature in the lateral spinothalamic tract. The three neurons in the pathway are labeled consecutively in the direction of the transmission of the impulse: first-order neuron, second-order neuron, and third-order neuron.

knowledge about the environment. Our example involved a child's game, but the same principles apply to a physician who is manipulating and feeling the organs in the abdominopelvic cavity during a physical examination to determine their condition.

Proprioception

Impulses of proprioception are carried along spinocerebellar tracts similar to those used in fine touch and pressure, but they do not extend to the cerebral cortex. Instead, these pathways terminate in the cerebellum and midbrain. Con-

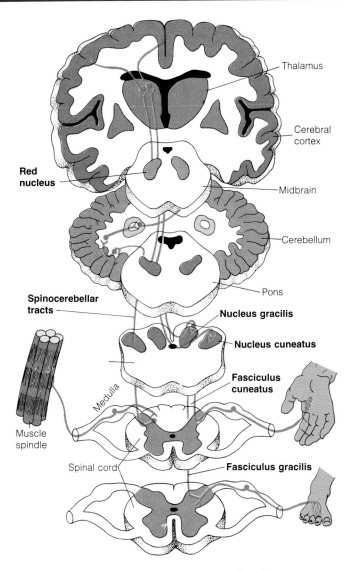

Figure 15-5 **The afferent pathways for fine touch and pressure.** These sensory pathways are located in the fasciculus cuneatus, fasciculus gracilis, and spinocerebellar tracts.

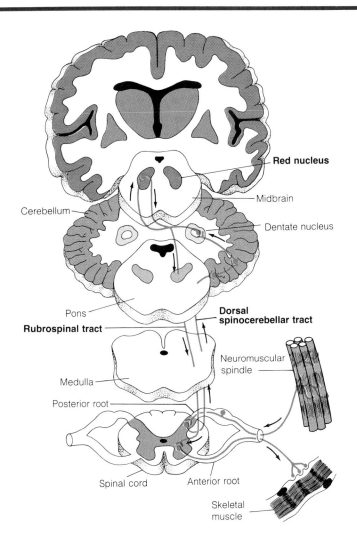

Figure 15-6 **Central nervous system tracts for proprioception.** These tracts contain both afferent (sensory) and efferent (motor) pathways.

sequently, impulses carried along these tracts do not reach consciousness.

Two pairs of spinocerebellar tracts are identifiable (Figure 15-6), one anterior and one pair posterior (see Chapter 12). First-order neurons of these pathways synapse with second-order neurons within the spinal cord. The latter axons pass up the spinal cord through the inferior cerebellar peduncles and synapse with third-order neurons in the gray matter of the cerebellum. During this traverse of the spinal cord, some fibers in the anterior tract cross over, but all in the posterior tract remain on the same side of the cord.

These impulses inform the reflex centers in the cerebellum and midbrain about the tone and position of muscles. They also stimulate these centers to initiate muscle contractions in response to these stimuli, thereby allowing coordinated actions.

Temperature and Pain

Pathways for temperature and pain sensations involve slightly different routes than for touch, pressure, and proprioception. Second-order neurons for temperature and pain cross over at the same level of the spinal cord at which they enter (Figure 15-4). They ascend the spinal cord through the lateral spinothalamic tracts, eventually terminating in the thalamus. Here they synapse with third-order neurons and

> **CLINICAL NOTE**
>
> ### CONTROL OF PAIN
>
> Stimulation of the large afferent fibers running between the peripheral receptors and the central nervous system will block or depress pain signals. These signals may originate in the same area of the body as the origin of the large sensory fibers or from some area located many spinal segments away. Simple therapies such as rubbing the skin in a painful area may stimulate the large fibers and diminish transmission of the pain signals. This is believed to be the logic behind counterirritants, liniments of various kinds, and **acupuncture.** In acupuncture, a method used to relieve or prevent pain, the long needles inserted under the skin in prescribed patterns may stimulate the large fibers and prevent pain transmission by the small fibers (see figure). The perception of pain has a large psychogenic component; this can either diminish or enhance the pain experience. The ritual of acupuncture probably plays some role in initiating excitation of the brain's chemical central analgesia system and thus helps in diminishing pain.
>
> Clinicians dispensing pain medication also take advantage of the psychological component of pain. The **placebo** (plă·sē′bō) **effect** is invoked when the clinician not only administers the medication for pain but also provides the person with the information that this is medication for pain, it will help to diminish the pain, and it will begin to work in such and such a time period. Pain medication is much more effective when given in conjunction with such information.
>
>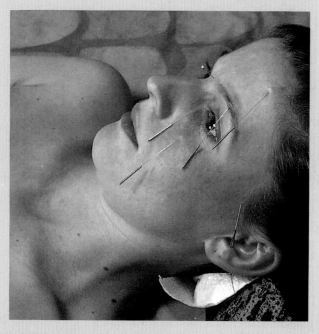
>
> Patient undergoing acupuncture treatment for persistent headaches.

transfer the impulses to the primary sensory area in the postcentral gyrus. In this region the brain interprets the impulses as either temperature or pain and localizes the sensation in a specific part of the body.

Motor Pathways

After completing this section, you should be able to:

1. Discuss reflex motor pathways and indicate how they differ from other spinocerebellar pathways.
2. Compare and contrast pyramidal and extrapyramidal pathways.

Body activity is dependent upon the conduction of motor (efferent) pathways in the nervous system. These pathways conduct motor impulses from reflex centers and from consciously controlled regions of the cerebral cortex to muscles and glands.

Reflex Motor Pathways

We have discussed the sensory spinocerebellar tract and how proprioceptive information is transmitted to the cerebellum and midbrain. Once informed of muscular conditions (involving involuntary and voluntary movements), centers in the cerebellum and midbrain stimulate motor neurons that carry impulses down the spinal cord through the rubrospinal tract. Afferent and efferent neurons cross over in the midbrain, resulting in ipsilateral responses (Figure 15-6). Reflexive responses are not under conscious control, but they help produce coordinated, smooth movements for a whole range of behaviors, including walking, driving, or playing baseball, or playing the piano.

Voluntary Motor Pathways

Voluntary motor pathways conduct impulses to skeletal muscles that produce the whole spectrum of voluntary movement we associate with human behavior. These impulses originate in primary and special motor areas of the cerebral cortex (Chapter 12) and are carried down the spinal cord through fiber tracts in the white matter (Figure 15-7). As

corticospinal: (L.) *cortex*, rind + *spina*, thorn
corticobulbar: (L.) *cortex*, rind + *bulbus*, bulb

with sensory pathways, crossover occurs in transit, hence, the right side of the cerebrum controls movements on the left side of the body and vice versa.

Impulses associated with voluntary movements arise in two types of cells in the cerebral cortex. Some arise in the large pyramidal cells of the precentral gyrus and travel through the pyramids of the medulla oblongata in **pyramidal pathways.** The remainder originate in other cells of the cerebral cortex and in subcortical regions. Because they do not originate in pyramidal cells, these routes are called **extrapyramidal** (ĕks″tră·pĭ·răm′ĭ·dăl) **pathways.**

Pyramidal Pathways

Pyramidal pathways include two neurons. The first in the sequence is an **upper motor neuron** (or **pyramidal fiber**). The cell bodies of upper motor neurons lie in the cerebral cortex, mostly the precentral gyrus. Axons of these motor neurons extend from the cortex through fiber tracts in the white matter of the brain. At specific levels for each neuron in the brain and spinal cord, synapses occur with the second nerve cell in the sequence, a **lower motor neuron** (or **peripheral fiber**). Cell bodies of lower motor neurons are located in nuclei of cranial nerves that innervate voluntary muscles in the head and neck and in the anterior columns of gray matter of the spinal cord. These lower motor neurons exit with a cranial nerve or with anterior roots of the cord as spinal nerves and terminate in various skeletal muscles.

Pyramidal pathways are also called **corticospinal** (kor″tĭ·kō·spī′năl) pathways because they connect the cortex of the brain with the spinal cord. Several of these pathways exist; here we study three as representatives.

More than three-fourths of the upper motor neurons in the pyramidal pathway cross over in the medulla oblongata, forming the decussation of pyramids described in Chapter 12. These upper motor neurons pass down the spinal cord in the **lateral corticospinal tract** that extends its length. Most of the remaining upper motor neurons proceed from the cerebral cortex down the cord and do not cross over until they reach the level at which they synapse with lower motor neurons (Figure 15-7). This latter group of upper motor neurons forms the **anterior corticospinal tract,** which usually does not extend past the thoracic level of the spinal cord. Upper motor neurons from both corticospinal tracts synapse with lower motor neurons in the anterior columns of the spinal cord gray matter. The axons of these lower motor neurons extend out through the anterior roots of the spinal nerves. Impulses traversing the corticospinal tracts stimulate skeletal muscles in the trunk and limbs.

A third pyramidal pathway, the **corticobulbar** (kor″tĭ·kō·bŭl′băr) **tract,** serves the voluntary muscles of the head

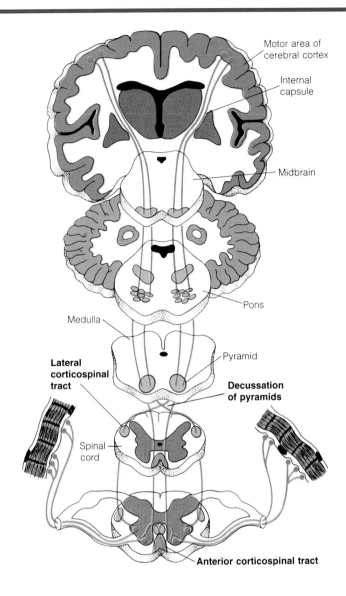

Figure 15-7 The central nervous system tracts involved in the transmission of voluntary motor impulses to the skeletal muscles include both pyramidal and extrapyramidal pathways.

and neck and does not enter the spinal cord. Cell bodies of the upper motor neurons of this tract lie in the cerebral cortex, and their axons extend through the internal capsule of the cerebrum along with those of the other pyramidal tracts. Crossover occurs in the brain stem, and the axons terminate in the nuclei of cranial nerves V, VI, VII, IX, X, XI, and XII. There, they synapse with lower motor fibers that pass out with the cranial nerves to voluntary muscles in the head and neck.

rubrospinal: (L.) *ruber*, red + *spina*, thorn

Extrapyramidal Pathways

Extrapyramidal pathways include motor pathways that do not originate in the cerebral cortex. Instead, the cell bodies of these upper motor neurons are located in the brain stem, primarily in the basal ganglia and reticular formation.

The major extrapyramidal pathways are the **rubrospinal tracts, vestibulospinal** (vĕs·tĭb″ū·lō·spī′năl) **tracts,** and **tectospinal tracts.** Upper motor neurons in the rubrospinal tract are situated in the red nucleus of the midbrain. The tract extends the length of the spinal cord. Upper motor neurons of the vestibulospinal tract lie in the vestibular nucleus of the midbrain. The tract extends through the anterior column of white matter in the spinal cord. Upper motor neurons of the tectospinal tract lie in the superior colliculus of the midbrain, and the tract extends through the anterior column of white matter only through the cord's cervical region.

Motor impulses are initiated in these pathways by association neurons carrying proprioceptive information from skeletal muscles or by impulses from the cerebellum. Extrapyramidal pathway impulses control muscle tone, posture, balance and equilibrium, and unconscious movements of the head. In addition to stimulating muscle contractions, some signals in the extrapyramidal tracts inhibit muscular contractions.

The Diver

The sensory and motor functions we have been discussing do not operate in a vacuum. Instead, they work together to allow the complex activities that constitute human behavior.

Picture a swimmer stepping onto a diving board. The diver brings into play all elements of the nervous system. His senses inform him of the length and height of the board, as well as the water depth in the pool. He walks to the end of the board and lightly bounces it to test its resiliency. All this sensory information is recorded by the central nervous system. Proprioceptors in the arm and leg muscles transmit information concerning the contraction of all muscles used in the action; the central nervous system responds by increasing the readiness of the muscles. The sympathetic nervous system increases its output of epinephrine, which readies the body for any anticipated action. Finally, the diver makes a conscious decision to dive. He returns to the base of the diving board, turns around, takes four steps forward, flexes his knees, bounces once on the end of the board and executes a perfect swan dive into the pool (Figure 15-8). This action has required the cooperation of sensory, motor, and association neurons in a functional sequence that brings about the desired action. The cerebral cortex initiated nerve impulses along motor neurons to cause muscular contrac-

> **CLINICAL NOTE**
>
> ### EXTRAPYRAMIDAL DISORDERS
>
> Disorders of the extrapyramidal system result in involuntary movements (tremors and tics), impairment of voluntary movements, and changes in muscle tone and posture. **Parkinsonism** (păr′kĭn·sŏn·ĭzm″) is an example of an extrapyramidal disease. It is a chronic disorder that results in slowness, lack of purposeful movements, muscular rigidity, and tremors. The lesions are usually in the area of the caudate nucleus and putamen, and it is accompanied by a decrease in the dopamine content of the axon terminals. The cause is usually unknown.
>
> Drug-induced parkinsonian symptoms can occur in some people who are given **phenothiazine** (fē″nō·thī′ă·zēn) drugs, including tranquilizers and antiemetics (drugs to stop vomiting). The patient will experience uncontrollable oral and facial grimaces, tongue movements, and twisting of the neck (torticollis). These symptoms are very frightening to the patient. Fortunately, they can be quickly reversed by an injection of the antihistamine Benadryl.

Figure 15-8 An athlete executing a dive into a swimming pool brings into play all of the elements of the nervous system to coordinate the senses and skeletal muscles in an integrated dynamic activity.

circadian:	(L.) *circa*, about + *dies*, day
electroencephalogram:	(Gr.) *elektron*, amber + *enkephalos*, brain + *gramma*, a writing

tions, and the cerebellum coordinated these muscular movements. In this manner, the nervous system functions to control everyday body activities. Do not forget that the swimmer could have decided not to dive if the board had seemed too high or the water too shallow, and then the same elements of the nervous system would have functioned to prevent the dive rather than to produce it.

Integrative Functions of the Brain

After completing this section, you should be able to:

1. Relate consciousness and unconsciousness to the integrative functions of the brain.
2. Describe the role of the reticular formation in sleep and wakefulness.
3. Define the stages of sleep and indicate when REM and NREM sleep occur.
4. Compare short-term and long-term memory.

> **CLINICAL NOTE**
>
> **JET LAG**
>
> The normal circadian rhythm of the body may be disturbed by long distance air flights, particularly if several time zones are crossed. This disorder is called **circadian dysrhythmia** or "jet lag." People suffering from this disorder complain of insomnia, anorexia, nausea, and mental confusion. They usually recover spontaneously after several days, or if the symptoms are severe, they may require sedation or antiemetics for relief.

Because of the lack of experimental laboratory work on humans, much of our knowledge of brain function is based on experiments in which certain regions of an animal's brain have been severed. Humans who have suffered injuries to specific parts of the central nervous system have provided specific information to supplement animal research. By noting the effects on body functions that result from lesions, injuries, or strokes, inferences can be made about the normal control functions of a given area of the brain in an uninjured person. Additionally, computer studies have become invaluable to physiologists and behavioral scientists because they allow the design and study of complex models that test the possible relationships between interconnected association neurons of the brain. Such studies have led to the hypothesis that the processes of thinking and abstraction are based on the activation of complex patterns of interconnections among association neurons, primarily those located in the cerebrum. According to this hypothesis, each separate thought that rises to consciousness involves a different pattern of interconnections. Such hypotheses are interesting and may be close to the truth. However, many currently held ideas concerning the manner of brain functioning are speculative and will certainly be modified as additional information emerges from continued laboratory and theoretical studies.

Consciousness Versus Unconsciousness

One amazing feature of the brain is its ability to operate at several different levels of awareness and efficiency. The levels of awareness range all the way from alert consciousness to **coma**, a state of unconsciousness from which a person cannot be awakened. (A person under general anesthesia in the operating room of a hospital is in a temporary coma.) The human brain can function at many points along the spectrum of consciousness, and it usually undergoes a rhythmic cycle of alternation between consciousness and sleep. This type of regular cyclic behavioral activity is a **circadian** (sĭr″kā·dē′ăn) **rhythm.** Humans have a circadian cycle of approximately 24 hours which is tied to the hourly cycle of one day (see Clinical Note: Jet Lag).

Consciousness is a state of alertness that allows an individual to become aware of external and internal stimuli and to make decisions about whether or not to respond to the stimuli. This condition involves activity in all parts of the brain, particularly the cerebral cortex and the reticular formation.

Recordings can be made of the electrical activity of the brain. Such a recording is called an **electroencephalogram** (ē·lĕk″trō·ĕn·sĕf′ă·lō·grăm) **(EEG).** Figure 15-9 shows typical EEG tracings under a variety of normal brain conditions (see also Chapter 12 Clinical Note: Epilepsy). These tracings reflect changes that occur in the activities of cerebral cortex cells when one is awake, alert, drowsy, and asleep. Changes in EEG wave patterns involve variations in their frequency and amplitude (height of the waves). During consciousness the EEG is characterized by **alpha waves** (higher frequency but lower amplitude).

During consciousness continuous communication appears to be maintained between the association neurons of the cerebral cortex and other areas of the brain, particularly the thalamus and the reticular formation. Impulses travel into and out of the cortex and other brain centers bringing sensory information to the cortex. This information is integrated with various sensations so that an individual gains a perception of the whole body structure and not merely a series of disjointed parts. Similarly, motor impulses do not arise

368 Chapter 15 Functional Aspects of the Nervous System

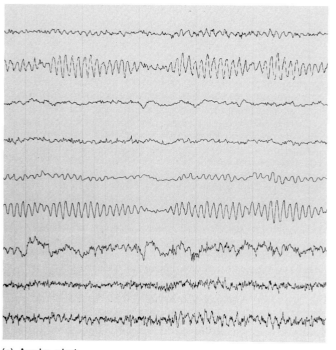
(a) Awake, alert

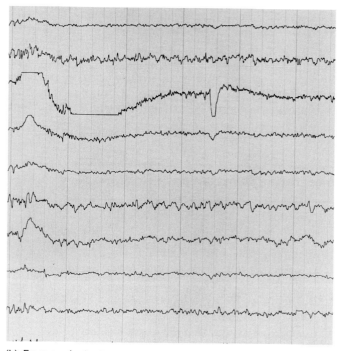
(b) From awake to drowsy

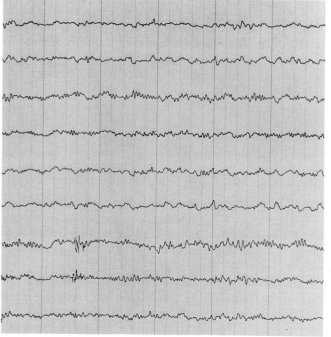
(c) NREM sleep

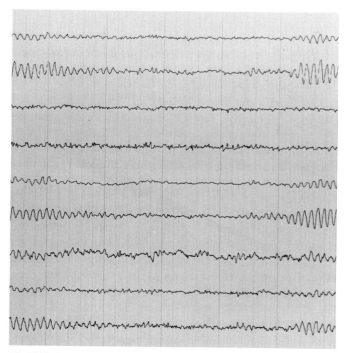
(d) REM sleep

Figure 15-9 The electroencephalogram tracings shown here indicate the range of variability of the electrical activity of brain cells. Note the similarity of the EEG tracings taken during an awake state and during REM sleep.

narcolepsy: (Gr.) *narke*, numbness + *lepsis*, seizure

> ### CLINICAL NOTE
>
> #### SLEEP DISORDERS
>
> Sleep is a state of unconsciousness from which the person can be easily awakened. Normal sleep is a rhythmic and usually refreshing occurrence. **Narcolepsy,** however, is uncontrolled sleepiness. Everyone occasionally has difficulty staying awake; wakefulness may be especially difficult after a heavy meal, after staying up late, in a warm room, or in a particularly boring but comfortable situation. Narcolepsy, on the other hand, is a consistent difficulty in retaining wakefulness at any time during the day. Many times throughout the day, the person is virtually attacked by an overwhelming and irresistible desire to sleep. The eyes close, breathing slows, muscles relax, and the person drops into sleep. The individual is easily awakened and is refreshed by the nap. The period of sleep may not last longer than 15 minutes, but this sleep may be necessary many times throughout the day. The person may not be able to stay awake during classes, meetings, or even while driving. Narcolepsy usually begins during adolescence or young adulthood, and males seem to be more frequently affected than females.
>
> **Insomnia** is a subjective feeling that one is not getting enough sleep. Once again this is a common experience for many of us at certain times during our lives. Actually, insomnia means difficulty in falling asleep either initially or after awakening after a period of sleep. Insomnia usually has some direct psychological cause, such as anxiety, depression, sexual arousal, or nightmares. There is some evidence that not getting sufficient physical exercise can predispose an individual to insomnia. Individuals can sometimes be taught to manage this sleep disturbance. Drinking milk, doing relaxation exercises, meditating, listening to soft music, and reading are often helpful. Worrying about not sleeping tends to increase insomnia; letting go seems to help.
>
> Many people suffer from **sleep interruption.** This is common among night workers who try to sleep in bright, noisy daytime environments, or for anyone sleeping in an unusual place, such as a hospital or when away from home. Internal causes of sleep interruption include pain, a full bladder, hunger, alcohol, caffeine, or coughing. Parents experience sleep interruption because of the nighttime crying of infants. The resulting fatigue from frequent and consistent sleep interruption can lead to emotional instability and depression.

in a vacuum of the cerebral cortex. Association neurons connecting different parts of the cortex and centers in the subcortical region inform widespread areas of the brain of muscular activities, so that the body is moved as a unit, not just as a conglomeration of parts.

Sleep

The reticular formation, or the reticular activating system (RAS) (see Chapter 12) appears to function in activating the body to consciousness after a period of sleep. According to current theory, the RAS is sensitive to many stimuli, both external and internal. External stimuli may include an alarm clock, a barking dog, a bright light, or anything that elicits one's attention. Some people, however, do not need an external stimulus to awaken them because their circadian rhythm seems to be strong enough to activate the RAS. They seem to have an "internal alarm clock" that awakens them. In either case, whether the wake-up stimulus is external or internal, the RAS perceives the stimulus, evaluates it, and passes the stimuli along to the cerebral cortex. The process seems to involve two anatomically distinct subdivisions of the RAS. First, the fibers of the RAS in the midbrain receive sensory impulses generated by the stimulus causing the person to awaken. If the sensations are of sufficient magnitude, the thalamic component of the RAS causes complete arousal. This process is only beginning to be understood, and additional research will help us to understand why consciousness involves many different levels, ranging from just barely awake to the alertness of a runner at the starting line awaiting the starter's gun. We still do not understand why a mother instantly awakens at the first cry of her infant, yet manages to sleep soundly through many other nighttime disturbances.

Sleep is a phenomenon similar to, but not the same as, unconsciousness. Most people spend from one-fourth to one-third of their lives asleep. Sleep occurs on a cyclical basis in normal individuals as part of the circadian rhythm. The stimuli that initiate sleep are not known, although several factors are involved. A reduction in the number of external stimuli impinging on the sensory receptors usually promotes sleep. In other words, it's usually easier to go to sleep in a darkened, quiet room than in a brightly lit one in which someone or something is making loud noises. It is also easier to go to sleep if you are tired or fatigued, although it is not clear how fatigue promotes the sleep process. Finally, centers in the brain stem promote sleep when stimulated, and the reticular formation is somehow inhibited so that normal alertness is suppressed, allowing sleep to overtake the body.

All these factors undoubtedly affect one's ability to sleep, yet there must be more to the initiation of sleep. Even under ideal conditions, it is sometimes difficult to go to sleep, and conversely, most people have had the experience of nodding off to sleep when they thought they were alert and paying attention. There are individuals who can supposedly function on less than one hour of regular sleep per day, and others who can substitute several short naps for one long sleep period. In any event, sleep seems to be an essential part of the body's physiology (see Clinical Note: Sleep Disorders).

When sleep occurs, it does so in a regular sequence of physiological stages. As one stretches out in bed, the eyes close and a feeling of drowsiness sets in. The EEG pattern of a relaxed person entering the first stages of sleep is similar to that of an alert person being dominated by alpha waves. Gradually, the person passes from this relaxed stage of light sleep into progressively deeper levels. Several physiological changes accompany the onset of deep sleep. Breathing becomes slower and deeper (snoring may even occur), blood pressure and body temperature drop, and overall muscle tone decreases. Usually about one-half hour elapses before this stage of deep sleep is reached and a series of changes in the brain waves occurs. The high frequency, low amplitude alpha waves are gradually replaced by low frequency, high amplitude delta waves (Figure 15-9c). This type of sleep is called **slow wave sleep** or **nonrapid eye movement (NREM) sleep** in reference to the low frequency brain waves and the slow movements of the eyeballs from side to side under the closed eyelids.

Throughout the night (or sleep period), NREM sleep alternates with a second major type of sleep called **paradoxical sleep** or **rapid eye movement (REM) sleep** because the eyes move rapidly under the closed eyelids. The physiological changes associated with NREM sleep are reversed: breathing rate increases, blood pressure and pulse rate increase, and the brain waves (EEG) revert to the high frequency, low amplitude pattern characteristic of an alert person (Figure 15-9d). Most dreaming occurs during REM sleep. Interestingly, if you awaken during or shortly after a period of REM sleep you will remember a dream quite vividly and will often think you have actually experienced the dream event while awake. A person who is awakened during a frightening dream usually awakens quite alert, and the physiological changes associated with REM sleep may be exaggerated. The person may awaken soaked in sweat, with a rapid pulse and breathing rate. REM and NREM sleep alternate with each other throughout the sleep period, with REM periods increasing in length toward the end of the sleep period. An adult typically spends about 25% of the sleep period in REM sleep. People deprived of REM sleep for a long period of time, a brainwashing technique, may exhibit symptoms of physiological and emotional stress, and they may be susceptible to suggestions and propagandizing.

Memory

The ability to recall sensations and experiences is called **memory.** Two basic types of memory have been identified: short-term and long-term memory. Neither is well understood, but research has indicated that different mechanisms may be involved in each type.

Short-Term Memory

As its name implies, **short-term memory** is distinguished by the recall of information from sensations or experiences that occurred a short time (a few minutes to an hour or so) before. For example, it is used when you transfer information from one source to another, such as taking notes during a lecture or jotting down someone's phone number. The ability to remember what you have read long enough to turn to a notebook page and write it down is an example of short-term memory. This type of memory is involved in other situations. For example, imagine you are at a party and the host introduces you to a new arrival. A short time later, your host asks you to introduce the new guest to others in the room. Short-term memory enables you to remember the new guest's name long enough to make the introductions. However, if you memorize the person's name in short-term memory only, you may be unable to remember his or her name when you meet a week later.

One of the most widely held hypotheses to explain short-term memory involves **reverberating neuron circuits** in the cerebral cortex. According to this hypothesis, these circuits work in the following way. In the simplest case, a single association neuron relays sensory information into the cerebral cortex. The impulse is relayed to one of the cortical neurons which, in addition to having synapses with other neurons, has branches that circle back and restimulate the original association neuron (Figure 15-10). In this way the cortical neuron is restimulated many times by a single sensory input, and the phenomenon produces memory. Reverberating neuron circuits may also involve sequences of neurons interconnected in more complex arrangements, involving both stimulation and inhibition in the circuit.

Short-term memory can persist for relatively long periods of time, particularly if one reinforces the items to be remembered by numerous repetitions. Repetition of the items to be remembered reinforces a reverberation circuit over and over again. This repetition lowers the threshold of the cortical synapses, thereby facilitating establishment of the circuit. In this way memory persists for a longer period of time and is easier to recall. This hypothesis could explain what happens when a student "crams" for an exam the night before. Numerous repetitions of the material to be memo-

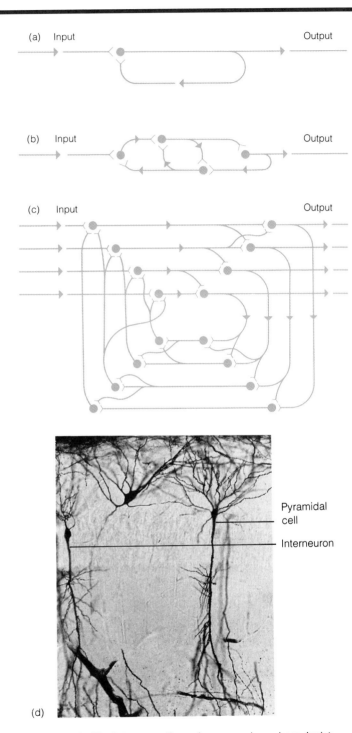

Figure 15-10 The interconnections of neurons shown here depict the reverberating circuits hypothesis used to explain how memory acts in the cerebral cortex. (a) and (b) The pathways may range from very simple two-neuron circuits to (c) highly complicated multineuron circuits involving both stimulation and inhibition of different neurons in the circuit. (d) The interneuron shown in this photomicrograph of the hippocampal region is the type of cell that presumably is involved in reverberating neuron circuits in the brain (mag. 130×).

rized, along with little organizational tricks, allow the student to remember the material for several hours, perhaps only long enough to get through the exam. Anyone who has ever studied like this knows the memory working here is definitely short-term. If, without further study, the same person is given an exam on the same material a few months later, he or she invariably performs at a much lower level. Also, if the individual is asked about this material several years later, he or she may not even remember having taken the exam.

Long-Term Memory

In contrast, **long-term memory** allows a person to recall experiences ranging from fleeting sensations to complex events years after the sensations or events were experienced. It is not unusual for some stimulus, such as a smell, sound, or sight, to stimulate the memory in an adult of an event that occurred in childhood. Most people have little trouble remembering certain incidents that occurred in kindergarten, such as the name of the teacher, a favorite friend, or some picture or project made. Interestingly, the emotional importance or meaning of an event affects the persistence of memory.

Even less is known about long-term than short-term memory. Some authorities believe long-term memory is an extension of short-term memory, involving the same basic processes of reverberating neuron circuits and facilitated synapses. Persistence of memory for such a long period of time may involve many, perhaps unconscious, repetitions of the neuron circuit.

Other authorities have proposed another possible explanation for long-term memory. They base this postulate on the results of experimental tests on laboratory animals. Evidence has shown that brains of animals forced to learn laboratory routines undergo both histological and biochemical changes. Cortical neurons in these animals increase in size and in the complexity of terminal branches of dendrites and axons. Additionally, an increase in the amount of protein in the cortical neurons has been demonstrated. Although no direct evidence proves this hypothesis, some workers postulate that a concomitant synthesis of protein accompanies long-term memory. This idea suggests that a special protein is synthesized for each memory by RNA. If this hypothesis is correct, persistence of the RNA is instrumental in the recall of a memory because it can allow the protein associated with the memory to be resynthesized when properly activated. But solid experimental evidence showing the existence of such RNA or of any special molecules associated with memory is lacking.

Learning

Learning is the ability to acquire new sensations and memories and to use them to modify thinking and behavior. It involves short-term and long-term memory and requires integration of sensory, association, and motor functions of the brain. In its simplest form, learning is little more than **conditioning,** the ability of humans and animals to change an associated action or response from one stimulus (or no stimulus at all) to a new one. The classic experiments in conditioning were performed at the end of the nineteenth century by Ivan Pavlov, a Russian physician and physiologist. In his experiments he conditioned dogs to salivate at the sound of a bell by first providing them with food while simultaneously ringing a bell. The dogs became conditioned to respond to the secondary stimulus (the bell) with the same response they normally gave to the primary stimulus (food). Afterward, they could be stimulated to salivate merely by a ringing of the bell without any food.

Although Pavlov's experiments were performed on animals and not humans, many instances of conditioning in people have been reported. Two common examples involve the desirable conditioning that occurs in young children when they are taught to differentiate between red and green traffic signals and the meaning of each. Another example is their response to the bell ringing at the beginning and end of class periods in school. No innate, unlearned reflex in humans dictates that red means stop and green means go, or that when a bell rings one is supposed to stand and go to another classroom. However, after a week or two of this bell ringing in school, the conditioned response of the children becomes so well entrenched that teachers must modify the response so that the class does not disintegrate into chaos when the bell rings.

More complex types of learning are possible in animals and humans, but present knowledge of the actual physiological processes and learning mechanisms is still primitive. For example, an individual can be taught to perform four basic operations used in mathematics: addition, subtraction, multiplication, and division. Without instruction and challenge, the person may never progress beyond these basic skills. But with additional training and encouragement, the same person can learn to use these basic skills in solving mathematic operations and in applying them to problems of everyday living. This type of learning integrates information from many sensory and association sources and directs motor sequences in modified or completely new patterns to accomplish a learned task.

In this chapter, we studied how the sensory, association, and motor functions of nerve cells are coordinated into an integrated nervous system. You learned how sensory and motor pathways occupy discrete tracts in the spinal cord. You also learned some of the characteristics of conscious and unconscious states and some of the latest theories regarding learning. In the following chapter, we will examine the special sense organs and study how they inform the brain of external and internal conditions.

STUDY OUTLINE

Sensory Reception (pp. 357–359)

An Overview of Sensory Reception and Perception To perceive a *sensation* (a sensory impulse that stimulates the CNS), a *receptor*, responding to a sufficiently intense stimulus, must generate an impulse that is conducted along a pathway to the brain, where it is interpreted as a *perception*. Impulses that terminate below the cerebral cortex are not consciously perceived.

Classification of Senses Senses are classified either as *general* (touch, pressure, pain, heat, and cold), or *special* (vision, smell, taste, equilibrium, and hearing).

Classification of Receptors Receptors are classified either by location (*exteroceptors, visceroreceptors,* and *proprioceptors*) or by the stimuli they detect (*photoreceptors, mechanoreceptors, thermoreceptors, chemoreceptors, baroreceptors,* and *nociceptors*).

Stimulation of Receptors The stimulus that normally elicits a response is called the *adequate stimulus*. Stimulation of a receptor initiates a local depolarization, the *receptor potential*, which may initiate an action potential. Unlike *tonic receptors* that respond to all adequate stimuli, *phasic receptors* adapt to prolonged low level stimulation by reducing the number of impulses they generate. *Afterimage* formation is the opposite of *adaptation*. *Projection* is the ability to localize a sensation to a specific part of the body because impulses travel definitive pathways from the point of stimulation to specific parts of the cerebral cortex.

General Sense Receptors (pp. 359–361)

Mechanoreceptors Mechanoreceptors respond to stimuli associated with touch and pressure. Enteroceptors respond to crude touch, while *tactile (Meissner's) corpuscles, Merkel's disks,* and *free nerve endings* respond to fine touch. *End organs of Ruffini* and *pacinian corpuscles* respond to pressure stimuli.

Thermoreceptors Thermoreceptors have not been specifically identified, but they respond to heat and cold stimuli.

Nociceptors Nociceptors are branched free nerve endings that respond to pain. *Somatic pain* originates in the skin, while *visceral pain* arises in the viscera. *Referred pain*, the inability of the brain to accurately locate the source of some types of visceral pain, occurs when different areas of the body send sensory impulses to the cerebral cortex along the same major spinal cord routes.

dyslexia: (Gr.) *dys*, bad + *lexis*, diction

> ### THE CLINIC
>
> ## DEVELOPMENTAL DYSLEXIA
>
> It is estimated that 10 to 15% of the U.S. population suffers from a poorly understood condition involving particular difficulty in learning language and reading in spite of normal or even superior intelligence. It goes by several different names, including minimal brain dysfunction (MBD), dyslexia, and specific learning disability (SLD), but perhaps the most accurate term is **developmental dyslexia** (dĭs·lĕk'sē·ă). It is defined as a constitutional and often genetically determined disparate reduction in the rate and quality of written language skills, such as reading, writing, and spelling. Victims may exhibit problems with some or all of the following:
>
> 1. Mirror writing, for example, the inability to distinguish *b* from *d*, *p* from *q*, or *saw* from *was*.
> 2. Difficulty with manipulation of symbols, such as +, −, ÷, and ×.
> 3. Time and counting concepts, such as time, dates, and monetary amounts.
> 4. Disordered spacial concepts, for example, right from left and order from disorder.
>
> Although this condition has been known since 1900, it is still poorly recognized today in our school systems. The problem goes unrecognized in many people well into their teens and early 20s. Often it is suspected first by parents, sometimes by the dyslexics themselves, and only occasionally by teachers.
>
> Students with developmental dyslexia experience a very difficult time in our school systems and continually fall behind as the educational program progresses. Some are able to compensate somewhat by spending inordinate amounts of time doing homework; others may be able to use "tricks" to hide their disability, such as memorization of assigned oral reading tasks; others rely heavily on the more workable audiovisual senses. Most, however, persistently fall behind and become more and more frustrated with continual failure, which often leads to psychological and behavioral problems.
>
> Those with genetically determined dyslexia more often tend to be males who are left handed and who have several male relatives who are dyslexic. Some may have a **crossed dominance** in which the dominant eye is the right, but the dominant hand is the left. To complicate the problem, some also have difficulty with mathematics (dysmetria), problems with spacial orientation (spacial dyslexia), or difficulty with writing (dysgraphia). Some also demonstrate hyperactivity and a short attention span, which is called **attention deficit disorder (ADD)**. Students with ADD may be helped by being given a stimulant drug (Ritalin) on school days. Why a stimulant drug helps in ADD is not entirely known.
>
> The pathology of developmental dyslexia is largely unknown. Some researchers postulate that differential activity between the right and left brain is responsible for some aspects of this disorder. Using this hypothesis, one could propose that decreased activity in the left cerebrum of a dyslexic, where the language center would be affected, might be masked by other normal motor activities directed by a dominant right brain. Recent research using PET scans has given some credence to this theory.
>
> There is no specific cure for the dyslexic, but a few special teaching techniques seem to help, such as the **Orton-Gillingham system** of repetitively breaking words up into their parts and learning the parts. Most people with developmental dyslexia eventually function quite well in society, and there are many examples of dyslexics who have had brilliant, productive careers. Information on all aspects of developmental dyslexia may be obtained from the Orton Dyslexia Society, 724 York Road, Baltimore, Maryland 21204.

Proprioceptors *Muscle spindles* and *Golgi tendon organs* are proprioceptors that register the amount of tension created by muscle contractions.

Sensory Pathways (pp. 361–364)

Sensory (afferent) conduction pathways are ascending tracts of fibers in the spinal white matter. Three classes of neurons form a pathway. *First-order neurons* are sensory neurons. *Second-order neurons* are association neurons in ascending spinal tracts. *Third-order neurons* are association neurons that transmit impulses from the thalamus to the cerebral cortex.

Touch and Pressure The conduction pathway for fine touch and pressure passes through the *fasciculus gracilis* and *fasciculus cuneatus*. Crossover occurs within the *medial lemniscus* of the brain stem.

Proprioception Impulses of proprioception are carried along spinocerebellar tracts similar to those for fine touch and pressure. However, they terminate in the cerebellum and midbrain and do not reach consciousness.

Temperature and Pain Pathways for temperature and pain ascend the spinal cord through the lateral spinothalamic tracts. These impulses are relayed from the thalamus to the primary sensory area in the postcentral gyrus.

Motor Pathways (pp. 364–367)

Motor impulses are conducted along efferent pathways down the spinal cord.

Reflex Motor Pathways Reflex motor pathways pass down the rubrospinal tract and help produce coordinated, smooth body movements.

Voluntary Motor Pathways Voluntary motor impulses pass down the spinal cord in two groups of tracts. Some impulses arise in large pyramidal cells of the precentral gyrus, and pass along *pyramidal pathways* in the corticospinal tracts. The remainder arise in other cortex and subcortical cells and pass down *extrapyramidal pathways* in the *rubrospinal, vestibulospinal,* and *tectospinal tracts.*

Integrative Functions of the Brain (pp. 367–372)

Consciousness Versus Unconsciousness The brain is capable of operating at several levels of *consciousness,* and these levels alternate in a *circadian rhythm* of approximately 24 hours in humans. *Electroencephalograms (EEGs)* are tracing of electrical activity in the brain and can indicate different stages of consciousness.

Sleep Sleep is a cyclical phenomenon similar to unconsciousness that allows the body to rest. A typical sleep cycle involves four recognizable stages and includes *nonrapid eye movement (NREM) sleep* and *rapid eye movement (REM) sleep.* Dreaming occurs during REM sleep.

Memory Memory is the ability to recall both *short-term* and *long-term* sensations and experiences. Although poorly understood, memory may involve *reverberating neuron circuits* in the cerebral cortex.

Learning Learning is the ability to acquire new sensations and memories and to use them to modify thinking and behavior. *Conditioning* is the simplest form of learning.

SELF-TEST OF CHAPTER OBJECTIVES

True-False Questions
1. Intrafusal fibers are specialized cells that detect fine touch stimuli in the skin.
2. The rubrospinal tracts of the spinal cord are extrapyramidal motor pathways.
3. An adult typically spends about 90% of the sleep period in REM sleep.
4. REM sleep is characterized by slow breathing, decreased blood pressure, and alpha waves on an EEG tracing.
5. Adaptation is the ability of sensory receptors to cease transporting impulses of a redundant nature.
6. Third-order neurons in a sensory conduction pathway relay impulses from the thalamus to conscious centers in the cerebral cortex.
7. Both sensory and motor pathways in the nervous system actively contribute to the human ability to respond to complex stimuli.
8. Visceroreceptors are also called exteroceptors.
9. Dreaming is most common during paradoxical sleep.
10. Pacinian corpuscles and Ruffini's corpuscles are both heat receptors.
11. Golgi tendon organs are proprioceptors that detect stretching in a tendon.
12. Extrapyramidal motor pathways differ from pyramidal motor pathways in that the former do not originate in the cerebral cortex whereas the latter do.
13. Dreaming commonly occurs during NREM sleep.
14. Spinocerebellar sensory pathways differ from other sensory tracts in that the former do not extend all the way up to the cerebral cortex.
15. Ipsilateral spinal reflexes enter and leave the spinal cord on the same side.
16. Nociceptors are widespread throughout the body.
17. The adequate stimulus for an end bulb of Ruffini is chemical.
18. Nociceptors are "tonic" in terms of their ability to adapt to a continuous stimulus.
19. Pacinian corpuscles are thermoreceptors.
20. Second-order neurons are parts of reflex motor pathways.

Multiple-Choice Questions
21. Which of the following is *not* an extrapyramidal pathway?
    a. rubrospinal tract   c. vestibulospinal tract
    b. tectospinal tract   d. corticobulbar tract
22. Merkel's disks are general sensory receptors that detect
    a. heat   c. cold
    b. fine touch   d. pain
23. The sensory receptor that registers blood pressure within vessels is a/an
    a. baroreceptor   c. proprioceptor
    b. exteroceptor   d. nociceptor
24. Reverberating neuron circuits have been postulated as the possible mechanism for
    a. paradoxical sleep   c. insomnia
    b. short-term memory   d. dreams
25. Which of the following structures relays sensory impulses from the thalamus to conscious centers in the cerebral cortex?
    a. first-order neurons   c. third-order neurons
    b. second-order neurons   d. fourth-order neurons
26. Which of the following neurons can generate an action potential but *not* a receptor potential?
    a. first-order neurons   c. pacinian corpuscle
    b. second-order neurons   d. Merkel's disk
27. Which of the following motor pathways does not pass through the spinal cord?
    a. anterior corticospinal tract   c. rubrospinal tract
    b. tectospinal tract   d. corticobulbar tract
28. Free nerve endings in the skin are sense receptors for
    a. pain   c. fine touch
    b. pressure   d. deep pressure
29. The sensory pathway in the spinal cord used to relay impulses of the sense of touch is the
    a. spinocerebellar tract   c. fasciculus gracilis
    b. spinothalamic tract   d. corticobulbar tract
30. The sensory pathway in the spinal cord used to carry impulses of the sense of pain is the
    a. spinothalamic tract   c. fasciculus cuneatus
    b. spinocerebellar tract   d. corticobulbar tract

Essay Questions

31. List the four necessary steps that lead to the perception of a sensation.
32. Discuss the advantages and disadvantages of classifying receptors by location versus the type of stimulus they detect. How do the two classifications overlap?
33. Describe the relationship between a receptor potential and an action potential in a sensory neuron.
34. Differentiate between somatic and visceral pain. Are the two types of pain detected by the same receptors?
35. Describe the sensory pathways used for sensations of touch and pressure stimuli.
36. Discuss the types of motor pathways in the corticospinal tracts of the spinal cord.
37. How do the spinocerebellar pathways that conduct proprioception sensations differ from those that conduct touch sensations?
38. Compare and contrast the motor pathways involved in reflex activities with those involved in voluntary movement.
39. Describe the role of the reticular formation in sleep and wakefulness.
40. Describe the stages of sleep and indicate when REM and NREM sleep occur.

Clinical Application Questions

41. Define "referred pain." Why is this problem dangerous?
42. Describe how acupuncture can relieve pain.
43. How does narcolepsy differ from circadian dysrhythmia?
44. Discuss some of the more common causes of insomnia.
45. Describe the more common manifestations of developmental dyslexia.

16
Special Sense Organs

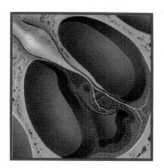

The preceding chapter described the manner in which sense organs respond to internal and external stimuli and the routes traveled by impulses generated by those stimuli. These are the generalized senses, including touch, pressure, pain, heat, and cold, which can be detected in almost all tissues of the body. We will now study the more specialized sense organs, those that are involved in smelling, tasting, vision, hearing, and balance. As we study the special sense organs, we will describe

1. the structure of embryonic and adult special sense organs,
2. the physiology of the special sense organs, and
3. the major pathways traveled by impulses generated in these organs.

sustentacular: (L.) *sustentaculum*, prop, support

Olfaction: The Sense of Smell

After completing this section, you should be able to:

1. Distinguish olfactory from sustentacular cells.
2. Outline the pathway traveled by olfactory sensations to the brain.
3. Discuss the current hypothesis used to explain how cells are able to distinguish among different odors.
4. Describe olfactory adaptation.

The mucous membrane lining the upper reaches of the adult nasal cavity includes three major types of cells (Figure 16-1). Columnar epithelial cells serve as **supporting cells**, or **sustentacular** (sŭs″tĕn·tăk´ū·lăr) **cells.** The bases of these cells are anchored to the underlying connective tissue, the **lamina propria,** and their distal ends extend out to the surface of the olfactory epithelium. Scattered among the supporting cells are numerous bipolar neurons called **olfactory cells** (or **chemoreceptors**) and columnar mucous cells. Sustentacular cells provide physical support for the olfactory cells, while mucous cells secrete mucus. This secretion moistens the free surface of the olfactory epithelium.

Each olfactory cell has an oval-shaped cell body and tapered proximal and distal ends. The distal (free) end is composed of nonmyelinated dendritic branches or cilia called **olfactory hairs.** These fine fibers extend into and are surrounded by mucus. The inner end of the cell narrows into a thin axon that combines with other axons to form **olfactory nerves.**

Stimulation of Olfactory Receptors

The manner in which olfactory cells function is still unknown. The most persuasive hypothesis states that they are stimulated by substances dissolved in the mucus surrounding them. This hypothesis seems logical, yet it does not explain how seemingly identical cells—those that generate the same kind of impulse from a variety of chemical stimulants—can differentiate among the smells of onions, perfume, rotten eggs, and thousands of other odors.

Two similar explanations have been offered by researchers in the field. A **chemical theory** states that the cell membrane and dendrites of olfactory receptors contain numerous special receptor sites that combine with specific chemicals capable of stimulating the cell. Once established, the chemical combination depolarizes the membrane and initiates a generator potential followed by a nerve impulse. A **physical theory** specifies that physical receptor sites (different from chemical receptor sites) on a cell membrane combine with olfactory stimuli to produce a nerve impulse in the olfactory cell. Neither hypothesis is an adequate explanation of what we experience as the sense of smell. However, it appears that odors can be classified into groups based on the sensations they evoke. Thus, we are able to characterize odors as acrid, pungent, putrid, floral, or musky, among

Figure 16-1 Olfactory receptors. (a) Detail of receptors interspersed among numerous supporting cells. (b) The placement of the olfactory receptors in the nasal cavity.

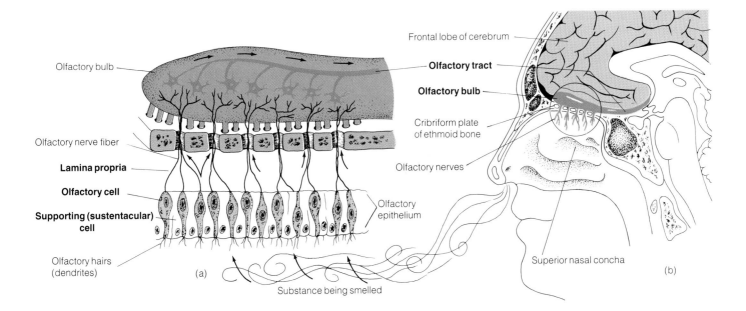

other classes. It is likely that molecules in the same odor class share similar shapes and stimulate similar receptor molecules in the membrane of the olfactory cell. In any event, olfactory cells are very sensitive receptors, and a single molecule of an odorant may be sufficient to generate an impulse (see Clinical Note: Anosmia).

Impulses initiated in the olfactory receptors are transmitted to neurons within the olfactory nerves and tracts and are passed on to the general area of olfaction in the cerebral cortex (see Chapter 12). Presumably, stimulation of different patterns of cerebral cortex cells produces sensations of specific smells. Future stimulation of the same or a very similar pattern of receptor sites activates the same circuit of cerebral cortex cells, resulting in the same sensation of smell. Short-term and long-term memory are strongly activated by olfactory sensations, and individuals usually have no trouble identifying smells recently experienced as well as ones they have smelled in the past. In fact, most people have had the experience of smelling an odor that reminded them of some occurrence in the distant past, perhaps even in early childhood.

Olfactory Adaptation

The sense of smell is subject to relatively rapid adaptation, at least with most odors. For example, if you are in the kitchen cooking with garlic, you quickly become adapted to the smell and are surprised when your newly arrived guest reacts strongly to the smell. But if you leave the kitchen, upon returning your olfactory sense is fully stimulated again and you can empathize with your guest.

Such rapid adaptation can have dangerous, even fatal, consequences under certain circumstances. For example, in the past, coal miners working in deep excavations faced a constant danger of being overcome by an accumulation of poisonous gases in underground caverns. They were subject to physiological and psychological adaptation of their olfactory sense because their concentration on the job at hand tended to downplay any changes detected by the sense of smell. Rapid adaptation of their sense of smell allowed a concentration of a gas to reach fatal levels before the miners became aware of it. Today, miners use sensitive electronic devices that sound an alarm when the concentration of any one of a number of gases reaches dangerous levels.

Olfactory Pathway

The olfactory nerves formed by axons from the olfactory cells extend through the holes in the cribriform plate and join the **olfactory bulbs.** These bulbs are extensions of the brain tissue and lie on both sides of the ethmoid bone's crista galli. Synapses occur within the olfactory bulbs between

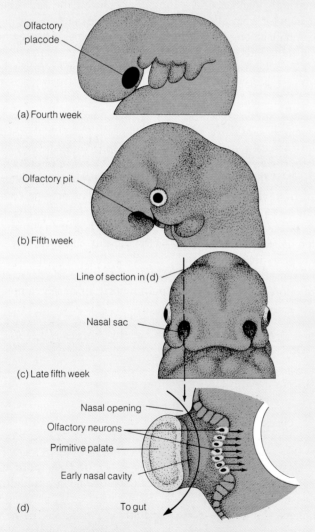

Embryonic Development of the Olfactory Organs

The fourth week of embryonic development marks the appearance of the **olfactory** (or **nasal**) **organs** as rounded **olfactory placodes** at the front of the head. During the fifth week they sink into the head as depressions that deepen into **nasal sacs** (Figure 16-A). **Olfactory neurons** differentiate within the sacs, and their long axons grow toward the olfactory bulbs of the forebrain. Synapses occur between these axons and the neurons within the bulbs. The cribriform plate of the ethmoid bone then ossifies around the olfactory axons.

Figure 16-A Development of the olfactory organs. (a) Olfactory placodes appear during the fourth week of development. (b) The placodes deepen and sink into the head. (c) Nasal sacs are evident by the fifth week. (d) Section through a nasal sac showing connection between exterior and opening into the foregut (curved arrow). Parallel arrows on the right indicate where axons will grow from the olfactory neurons

anosmia: (Gr.) *an*, not + *osme*, smell

> **CLINICAL NOTE**
>
> **ANOSMIA**
>
> Testing the sense of smell in the physician's office is a simple matter of having the patient sniff several well-known odors with each nostril separately. Several small bottles of pungent substances, including turpentine, coffee, oil of wintergreen, and oil of cloves, are usually available for this use. In most cases, however, this testing is unnecessary because the patient will just relate his or her inability to smell. The most common cause of temporary **anosmia** (ăn·ŏz′mē·ă) is the common cold. Permanent anosmia is a rare condition that may follow a fracture of the cribriform plate (see Chapter 7 Clinical Note: Fracture of the Cribriform Plate). A very rare genetic condition known as **Kallmann's syndrome** also causes anosmia. In it a defect in the hypothalamus causes very low production of gonadotropin-releasing hormone, which is the hormone that stimulates the sex organs. It also prevents the olfactory bulb from developing. Thus, these individuals lack secondary sex characteristics as well as a sense of smell. Treatment with male or female sex hormone will help the development of the secondary sex characteristics but not the anosmia.

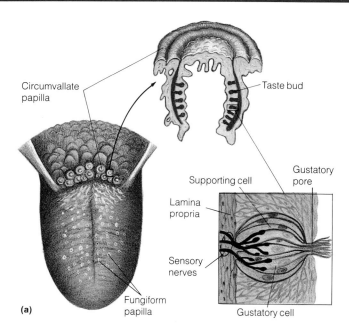

Figure 16-2 (a) Histology of a taste bud. (b) Scanning electron micrograph of taste buds.

olfactory cells and neurons in the bulbs. The axons of the olfactory bulb cells extend into the brain as the **olfactory tracts,** which pass under the frontal lobes of the brain (Figure 16-1a). These axons run to the olfactory centers in the temporal lobes. There, the sensations of smell are brought to conscious awareness.

Gustation: The Sense of Taste

After completing this section, you should be able to:

1. Compare and contrast the senses of taste and smell.
2. Describe a taste bud and distinguish between gustatory and sustentacular cells.
3. Outline the nerve pathway traveled by gustatory sensations to the brain.

Gustatory Cells

The sense of **taste,** or **gustation,** is accomplished by **gustatory** (gŭs′tă·tō·rē) **cells,** which are modified epithelial cells in the mouth and throat. They are found in the roof of the mouth and palate, but their greatest concentration is on the upper surface of the tongue where they are grouped within concentrations called **taste buds** (Figure 16-2).

A taste bud is an oval-shaped group of cells attached to the connective tissue (tunica propria) below the tongue's epithelium. The bud opens through a **gustatory pore** on the surface of the tongue. Two kinds of cells are found in a taste bud. Sausage-shaped sustentacular cells form a peripheral capsular layer that supports and encloses the bud. Within it are more supporting cells interspersed with **gustatory cells,** which are the receptors of taste stimuli originating from substances dissolved in saliva. Gustatory cells are narrower than the supporting ones and are more numerous near the center of the bud. The tapered distal ends of gustatory cells bear microvilli that extend into the mucus covering the tongue.

conjunctiva: (L.) *conjungere,* to join together

> **CLINICAL NOTE**
>
> **TESTING GUSTATION**
>
> Evaluation of the ability to taste is done by placing small amounts of salt and sugar solution on the anterior portion of the protruded tongue. When one is sure that the subject has had sufficient time to taste the test material, the mouth is rinsed with water and another test substance applied. Each side of the tongue is tested separately. Loss of taste sensation indicates a defect in the sensory portion of the facial nerve (VII).

Although most people claim the ability to identify many different tastes, such as garlic, fish, steak, cabbage, and pistachio ice cream, only four definite "tastes" have been identified: sweet, salty, sour, and bitter. Although four different complementary types of gustatory cells have not been identified, different regions of the tongue respond differently to these four taste groups (Figure 16-3). Most sensations identified as specific tastes seem to be combinations of taste and smell. One demonstration of this phenomenon is the inability of many people who have a stuffy nose to "taste" their food. What they are really experiencing is an inability to "smell" their food.

Gustatory Pathway

Gustatory receptors are innervated by three different cranial nerves. The facial nerve (VII) supplies the anterior two thirds of the tongue, the glossopharyngeal (IX) innervates the posterior third of the tongue, and the vagus (X) innervates the pharyngeal epithelium. Gustatory cells do not act as the first-order neurons in the sensory pathway. Instead, myelinated cranial neurons serve as first-order neurons, extending to the gustatory cells. Stimulation of gustatory cells by substances dissolved in the saliva initiates impulses in the first-order neurons. These impulses are transmitted through the medulla to the thalamus and from there to the somatic sensory region in the parietal lobe of the cerebral cortex.

Anatomy of the Eye

After completing this section, you should be able to:

1. Describe the anatomy of the eye.
2. Discuss the functions of the accessory structures of the eye.
3. Describe the cellular layers of the retina.
4. Trace the path of light from the cornea to the retina.

An adult eye consists of two major regions, the **eyeball** and the **accessory structures** surrounding it.

The Eyeball

The paired eyeballs are spherical structures about 2.5 cm in diameter. Each lies in an **orbit**, a deep recess in the facial region of the skull. Each eyeball has three layers of tissue: the fibrous tunic, vascular tunic, and retina.

Fibrous Tunic

The outermost layer of the eyeball is called the **fibrous tunic** (Figure 16-4). Except for its most anterior portion, it is composed of tough connective tissue commonly known as the "white of the eye." The anterior portion is clear and bulges out as the **cornea**. The outer corneal surface is covered with a thin, clear epithelium called the **bulbar conjunctiva** (bŭl′băr kŏn″jŭnk·tī′vă) that is continuous with the inner layer of the eyelids. The white portion of the fibrous tunic is the **sclera;** its chief functions are to maintain the shape of the eyeball and to provide protection for the soft inner structures. The clear cornea allows light to penetrate the eyeball, providing visual stimulation of the retina.

Vascular Tunic

Sandwiched between the inner and outer layers of the eye is the **vascular tunic.** As its name implies, the vascular tunic contains most of the vessels that provide blood to the eyeball.

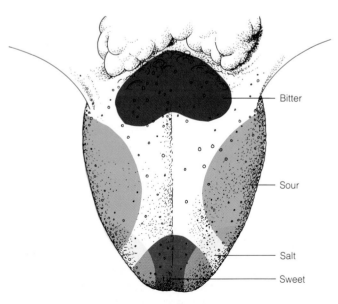

Figure 16-3 The tongue is sensitive to different tastes in different regions. Sweet, salty, bitter, and sour are detected in these areas of the tongue.

iris: (L.) *iris*, rainbow

Embryonic Development of the Eye

The eyes develop from ectoderm and mesoderm during several months of embryonic and fetal gestation. By four weeks, paired lateral outgrowths extend from the forebrain, nearly reaching the surface ectoderm (Figure 16-B). During the fifth week a concavity develops on each outgrowth which is called the **optic cup.** The optic cup eventually develops into the retina of the eye. Surface ectoderm adjacent to the depression thickens and differentiates into the **lens placode.** Mesodermal mesenchyme separates the lens from the cup and surrounds the latter. During the next two weeks the optic cup depression deepens as the future lens enlarges and pushes inward, forming a hollow vesicle that fills the depression. The base of the cup narrows and elongates as the **optic stalk.** The **choroid fissure,** a groove on the ventral surface of the stalk that is continuous with the cup's concavity, provides a pathway for blood vessels supplying the interior of the eyeball. Nerve fibers leaving the eye also pass through the fissure and contribute to the optic nerve. The choroid fissure eventually closes around these vessels.

With additional development the outer layer of the cup remains relatively thin, while the inner layer thickens. The thin outer sheet of cells will become the pigmented layer of the retina, while the thick inner layer differentiates into the sensory neurons of vision.

As development proceeds, the lens vesicle detaches from the surface and enlarges. The optic cup develops into the eyeball, and the hollow vesicle of the lens fills with cells. Upper and lower **eyelid folds** grow toward one another and finally meet and fuse in front of the lens, leaving only a narrow space between them. The eyelids remain fused until shortly before birth. A thin layer of mesoderm between the lens and eyelids gives rise to the cornea. Additional mesenchyme surrounding the eye produces muscles, connective tissue, and protective tunics.

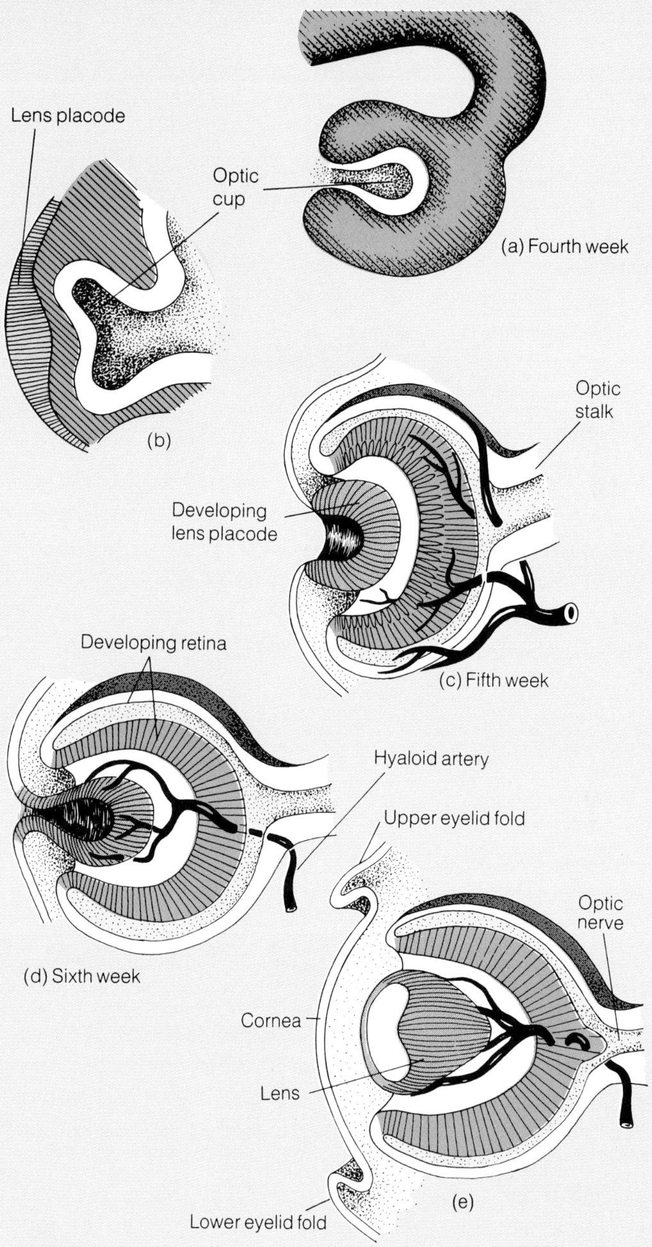

Figure 16-B Embryonic development of the eye. (a) Optic cup in four-week-old embryo. (b) Section through optic cup. (c) The developing lens placode buckles inward. (d) By the sixth week, the eye has begun to take shape. (e) By the seventh week, the lens has separated from the surface.

The vascular tunic is divided into three major regions: the choroid, ciliary body, and iris.

The **choroid** provides a connective tissue route for small blood vessels that carry blood to the eyeball. The **ciliary body** contains muscles that change the shape of the lens, which fine focuses images within the eyeball. This change in shape is accomplished with the help of the **suspensory ligament** (Figure 16-4), which holds the lens in place. The **iris,** an anterior extension of the ciliary body in front of the lens, contains circular and radially positioned muscle fibers. Contraction of the circular fibers reduces the size of the **pupil,** the circular opening in the middle of the iris, thereby reducing the amount of light entering the eye. Contraction of the radial fibers increases the size of the pupil.

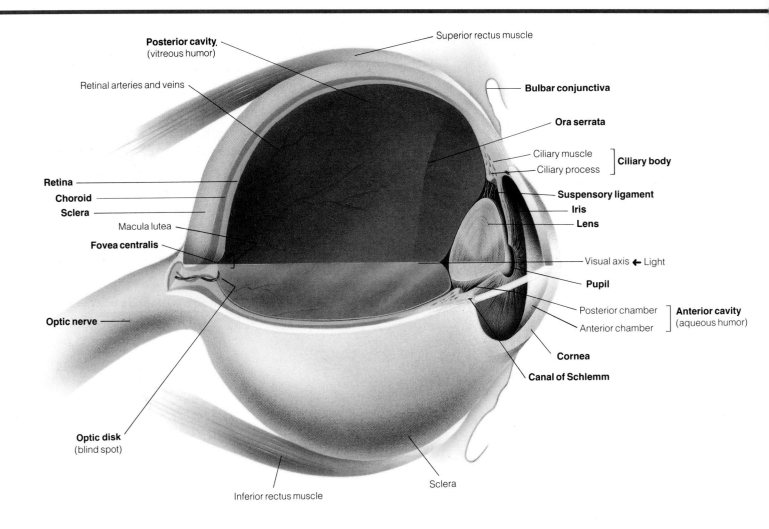

Figure 16-4 Internal view of the eyeball.

Retina

The **retina** (also called the **nervous tunic**) is the inner layer of the eye. It includes two layers, an outer pigmented one and an inner nervous layer (Figure 16-5). The pigmented layer, only one cell thick, extends from the posterior end of the eyeball to an area underneath the ciliary body and iris. This retinal layer is not light sensitive. The thicker nervous layer of the retina coats the interior of the pigmented portion and extends forward to the peripheral edge of the iris. The anterior circular edge of the visual portion is called the **ora serrata** (ō'ră sĕr·ā'tă).

The nervous part of the retina is light sensitive. It consists of three layers of neurons. The outer layer of these neurons contains two types of **photoreceptive cells: rods** for black and white vision and **cones** for color vision (Figure 16-5). Rods greatly outnumber cones and are sensitive to low intensities of light. In contrast, cones are sensitive only to high intensities of light and are the primary cells

> **CLINICAL NOTE**
>
> ### CORNEAL TRANSPLANTS
>
> **Corneal transplantation** is one of the most common and successful organ transplants in use today. The corneas from deceased individuals are surgically removed and used to replace the diseased corneas of patients whose vision is otherwise normal except for the damaged cornea. The technique does not require the close tissue typing or the immunosuppression that other organ transplants require. Corneal transplantation is highly successful in restoring useful vision in patients who have suffered damage to the cornea through injury, infection, or birth defects.
>
> The donation of one's corneas after death is a great humanitarian gift that can restore vision to a grateful recipient.

fovea: (L.) *fovea,* depression
macula: (L.) *macula,* spot

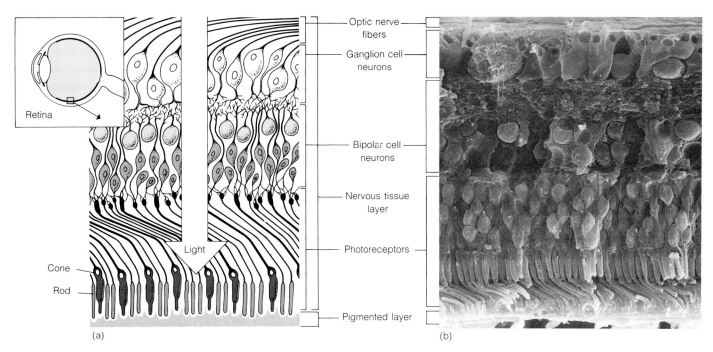

Figure 16-5 A detail view of the retina. (a) Schematic diagram. (b) Scanning electron micrograph (mag. 1246 ×).

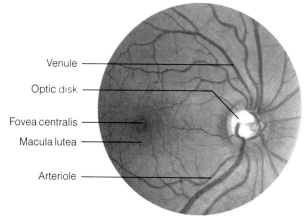

Figure 16-6 The surface of the retina showing the fovea centralis.

for receiving visual stimuli of different colors. Cones tend to be concentrated near the rear of the retina where light rays coming into the eye are most sharply focused. In fact, a small depression called the **fovea centralis** (fō′vē·ă sĕn·tră′lĭs) at the center of this area of greatest visual acuity contains only cones (Figure 16-6). It is surrounded by a yellowish area called the **macula lutea** (măk′ū·lă lŭ′tē·ă). From the macula outward, progressively fewer and fewer cones occur, until at the anterior edges of the visual retina only rods are present.

The second layer of cells in the visual retina are **bipolar neurons** that relay impulses generated in the rods and cones to the inner layer of cells. Several rods make contact with one bipolar cell, but only a few cones synapse with each bipolar cell. In the fovea centralis, individual cones synapse with individual bipolar cells, resulting in the sharpest detection of visual images here.

The third and inner layer of the retina is composed of **ganglion neurons.** Ganglion neurons are stimulated by bipolar neurons, usually several bipolar cells synapsing with each ganglion cell. The axons of ganglion neurons converge at the rear of the eyeball where they exit the eye and form the **optic nerve.** The point where the optic nerve exits the eyeball contains no photoreceptive cells and is thus called the **blind spot** (or **optic disk**) (Figure 16-6).

Lens

The **lens** is a solid but pliable, proteinaceous structure located near the anterior end of the eyeball. It is enclosed within the eye (Figure 16-4). The lens is oval-shaped when viewed from its edge and completely transparent. It is composed of concentric layers of protein arranged like the layers of an onion. It is held in place by thin fibers of the suspensory ligament, which in turn attaches to the ciliary body. Contraction and relaxation of the ciliary body causes the flexible lens to change shape and allows a person to focus visually on objects at different distances from the eye. Although the

ophthalmoscope: (Gr.) *ophthalmos*, eye + *scopium*, to see

CLINICAL NOTE

THE EYE: WINDOW TO THE BODY

A complete examination of the eyes can provide important clues to the presence of innumerable disorders and diseases. Not only can one look outward with the eye, but the physician can look into the eye using an instrument called an **ophthalmoscope** (ŏf·thăl′mō·skōp). This instrument allows a direct view of the internal structures of the eye, and more importantly, a direct observation of the functioning blood vessels and nerve endings on the retina. Figure 1 shows an ophthalmoscopic view of the eye of a person with diabetes. The walls of the retinal arterioles are rendered transparent by the light from the examiner's instrument so what is actually seen is a column of blood as it moves through the blood vessel. **Ophthalmologists** (eye physicians) will often observe retinal changes that denote the presence of systemic disease and will refer the patient for diagnosis and evaluation. This is particularly true for such diseases as diabetes mellitus, hypertension, and systemic vascular diseases.

cholesterol in their walls, displacement of veins at crossings ("A-V nicking"), and hemorrhaging into the retina (Figure 2). Systemic vascular diseases may cause characteristic changes in the retina, such as hemorrhages, exudates (fluid accumulations), degeneration of the macula, and changes in the nerve endings.

Many other diseases also produce retinal changes that may make a physician suspect the presence of a particular disease. A few examples are listed here:

Disease	Changes
Retinoblastoma	Tumor
Retinitis pigmentosa	Color changes of retina
Leukemia	Hemorrhage
Renal disease	Exudates and edema
Retinal detachments	Folds in the retina
Macular degeneration	Atrophy of macula
Optic neuritis	Pallor of optic disk
Glaucoma	"Cupping" of disk (nerve head)
Brain tumors	Swelling of the nerve head

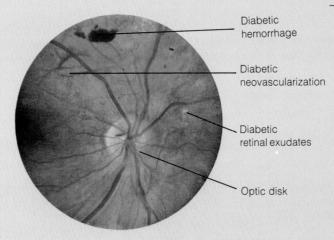

Figure 1 View of the retina through the ophthalmoscope showing characteristic signs of diabetes.

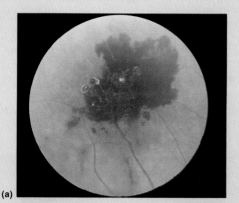

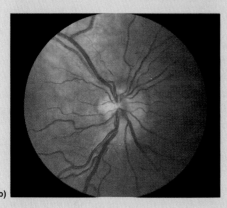

Figure 2 Changes in retina resulting from hypertension. (a) Hemorrhage (red area). (b) Papilledema (swelling of the optic nerve) causing an indistinct nerve edge.

The changes in blood sugar level caused by diabetes mellitus affect the sugar content of the lens and aqueous humor and thus their refractive characteristics (see Chapter 17, The Clinic: Diabetes Mellitus). One clinical effect of this is frequent changes in prescription glasses for affected individuals. An ophthalmologist may see changes in the blood vessels of the retina that are characteristic of diabetes, including microaneurysms (tiny bubbles on the vessels) and microhemorrhages. Diabetes is also a major cause of blindness, especially in elderly diabetics.

Hypertension (high blood pressure) will cause narrowing of the tiny arteries of the retina, changes in the color of these arteries ("copper wiring") due to deposition of

CLINICAL NOTE

CONTACT LENSES

The use of contact lenses, instead of eyeglasses, to correct refractive errors in vision has become very popular. Contact lenses often provide better visual acuity and better peripheral vision than eyeglasses do. There are two types of contact lenses in common use. **Hard corneal contact lenses** are thin, curved disks of polymethyl methacrylate, 7 to 10 mm in diameter. They float on a layer of tears and cover only part of the cornea. Hard contact lenses require an adaptation period of a week or longer and frequently cause temporary blurring of vision ("spectacle blur") when changing to eyeglasses. They may also cause abrasions of the cornea if not handled properly, resulting in painful sensitivity to light and tearing. **Soft hydrophilic contact lenses,** in contrast, are made of 2-hydroxyethyl methacrylate, are larger (13 to 15 mm in diameter), and cover the entire cornea. They mold to the eye and require very little adjustment time for comfortable wear. A recent development is **extended wear contact lenses,** which may be worn for days or weeks. All contact lenses require meticulous fitting and care for successful use.

---

refraction of light through the cornea provides almost 75% of the focusing of light within the eye, the lens is the chief structure used in focusing light rays on the retina for sharp vision.

Chambers of the Eyeball

The lens and iris near the anterior end of the eye divide its interior into a small **anterior cavity** and a larger **posterior cavity.** The anterior cavity lies in front of the lens and has two chambers. The **anterior chamber** is between the iris and cornea. It is shaped like a thick saucer, convex in front and concave behind. The **posterior chamber** of the anterior cavity is a narrow, doughnut-shaped space between the iris and the suspensory ligament and lens (Figure 16-4).

Both anterior cavity chambers contain the **aqueous humor,** a watery fluid produced by a plexus of capillaries in the ciliary body. It empties into the posterior chamber at a rate of about 5 to 6 mL per day. From this posterior chamber, the aqueous humor is forced through the pupil into the anterior chamber. From there it goes to a small venous sinus, the **canal of Schlemm,** then into the venous circulation. Thus, a continuous production and circulation of the aqueous humor passes through the anterior cavity sufficient to maintain an **intraocular pressure** of about 20 mm Hg. This pressure pushes the interior of the eyeball to the periphery, keeping the retina pressed evenly against the choroid. This pressure is measured during a simple, quick glaucoma test (see Clinical Note: Diseases Affecting the Eye).

The aqueous humor acts in much the same way as cerebrospinal fluid does, that is, it acts as a circulatory fluid bringing oxygen and nutrients to the structures in the anterior part of the eyeball.

The eye's **posterior cavity** is much larger than the anterior one, and it contains the **vitreous** (vĭt′rē·ŭs) **humor,** a more viscous, jellylike substance. The vitreous humor is formed during fetal development and remains with a person for life; it is not constantly replaced as is the aqueous humor. The vitreous humor maintains the intraocular pressure established by the aqueous humor. It also helps to maintain the shape of the eyeball and to keep the retina pressed against the choroid.

Accessory Structures of the Eye

The eyeball lies within the orbit, a deep recess in the facial bones. This region helps protect against physical damage. In addition to this secure location, several accessory structures contribute to protection of the eye. These are the eyebrows, eyelids, eyelashes, and the lacrimal gland.

Eyebrows

The **eyebrows** are transverse bands of skin located on top the superciliary ridges of the frontal bone. In addition to being amply supplied with short, coarse hairs, the eyebrow skin contains many sebaceous glands that secrete oil. Eyebrows shade the eyes from sunlight and protect against physical injury from above.

Eyelids

The upper and lower **eyelids,** or **palpebrae** (păl′pĕ·brā), are folds of skin that extend from the respective edges of the orbit. A thin line of short, curved hairs, **eyelashes,** projects from the free borders of the eyelids; these are physical barriers to small particles entering the eye. The superior eyelid is more mobile than the lower because it has a muscle, the **levator palpebrae superioris,** not present in the lower lid. This muscle assists in raising the upper eyelid. Other movements of the eyelids are under conscious and unconscious control and involve contraction of the orbicularis oculi muscles (see Chapter 10). Under normal circumstances the eyelids blink between 20 and 30 times every minute, sweeping tears and minute dust particles from the surface of the cornea into the medial corner of the eye while lubricating the surface between the cornea and the eyelid.

The eyelids are composed of several layers of tissue. The free border contains tough connective tissue, the **tarsal**

fornix:	(L.) *fornix*, arch, vault
canthus:	(Gr.) *kanthos*, corner of the eye
caruncle:	(L.) *caruncula*, a small piece of flesh

plate, which maintains structural support of the eyelids. The tarsal plates have sebaceous glands called **tarsal** or **meibomian** (mī·bō′mē·ăn) **glands** associated with them. These glands secrete oil onto the inner surfaces of the eyelids, helping to maintain a moist environment for the corneas. Modified sweat glands, **ciliary glands,** are located at the base of the eyelashes. The interior lining of the eyelids is called the **palpebral conjunctiva** and consists of a thin layer of mucous membrane. This epithelial layer extends from the inner surface of the eyelids over the sclera, where it becomes the bulbar conjunctiva. The junctions of the palpebral and bulbar conjunctivae have special designations: the **superior conjunctival fornix** lies above and the **inferior conjunctival fornix** lies below. Inflammation of this epithelial layer is called **conjunctivitis** (see Clinical Note: Diseases Affecting the Eye).

The opening between the eyelids is the **palpebral fissure.** The superior and inferior eyelids meet at the **lateral canthus** at the lateral corner of the eye and at the **medial canthus** at the medial corner of the eye. A small papillary projection called the **caruncle** can be seen in the medial canthus (Figure 16-7). This caruncle includes sebaceous and sudoriferous glands that produce the characteristic white material that accumulates in the medial canthus.

Lacrimal Gland

The eyes possess a **lacrimal apparatus** with special functions to moisten and clean the eyes. The **lacrimal gland** is elongate and located in the upper, lateral corner of the eye (Figure 16-7). It produces **tears** which drain through several (usually 6 to 12) **lacrimal ducts** and empty onto the bulbar conjunctiva. Movement of the tears across the surface of the cornea is facilitated by blinking. In addition to water, tears contain salt, mucin, and the enzyme **lysozyme.** The water in tears prevents the conjunctival cells from drying out, the mucin lubricates the conjunctiva and facilitates movements of the eyelids, and the lysozyme destroys numerous bacteria that enter the eye.

During the course of a normal day, lacrimal glands produce approximately 1 mL (20 drops) of fluid. Any irritation will usually cause an increase in tear production, as will certain emotional conditions. The small amount of tears produced under normal conditions usually evaporates; any excess tends to gather in the area of the medial canthus. At

Figure 16-7 (a) Anterior view and (b) sagittal section through the eye, showing the accessory structures.

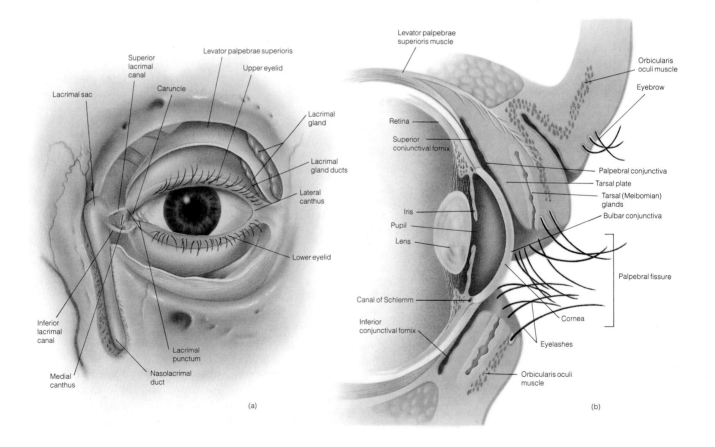

conjunctivitis: (L.) *conjungere*, to join together + *itis*, inflammation
punctum: (L.) *punctum*, point

> **CLINICAL NOTE**
>
> ## DISEASES AFFECTING THE EYE
>
> **Conjunctivitis** (pink eye) is an inflammation of the superficial lining that covers the inner eyelids and eyeball. It can be due to viruses, bacteria, chemicals, or allergies. The conjunctiva becomes inflamed and red and produces watery or puslike drainage. Like the common cold, it is highly contagious, and proper hygiene is recommended.
>
> **Macular degeneration** is an age-related deterioration of the retina and choroid in the area of central vision. Although it does not affect peripheral sight, it is a leading cause of blindness in the United States.
>
> Damage to blood vessels in the eye, such as hemorrhages and exudates, can be caused by diabetes and are evidence of a deficiency of oxygen supply to the retina (see Clinical Note: The Eye: Window to the Body). New blood vessels that form to replace the damaged ones often bleed and grow into the vitreous humor. This process is called **neovascularization** and can be slowed down by laser treatment. Fibrous bands can then form within the vitreous and can contract, causing **retinal detachment**. Retinal detachment can also occur from trauma to the eye, deterioration, and diseases other than diabetes. It can lead to permanent loss of vision if not repaired.
>
> **Cataracts** are opacities within the lens. There are many different types, but not all of them cause visual distortion. If the natural lens does not permit clear vision, it can be removed surgically and a plastic lens can be implanted within the eye. There are several advantages of implants over conventional cataract glasses: they give a patient better side vision and less distortion than do the thick cataract glasses.
>
> **Chronic glaucoma** refers to visual loss resulting from damage to the optic nerve by high intraocular pressure. Aqueous humor is produced in the ciliary body in the posterior chamber and flows into the anterior chamber. Drainage occurs through the trabecular meshwork into the canal of Schlemm and out to the bloodstream. Poor outflow of aqueous humor may produce painless elevation of pressure within the eye. This results in gradual damage to the optic nerve. The central vision remains good until very late in the disease making it difficult to detect. A test involving a tiny burst of air on the eye is now done routinely by ophthalmologists to test for high intraocular pressure, which forewarns glaucoma. Two drugs commonly used in glaucoma are pilocarpine, which increases aqueous outflow, and timolol, which decreases the production of aqueous humor.
>
> **Acute glaucoma** is caused by a total block of aqueous humor outflow. This occurs when the anterior chamber is narrow and the pupil dilates, permitting the iris to block the outflow of fluid from the eye at the trabecular meshwork. This painful condition requires immediate treatment. Pilocarpine is used to constrict the iris in an attempt to pull it away from the trabecular meshwork and break the block. Usually a surgical hole is created in the iris (iridectomy) to prevent pressure buildup within the posterior chamber and forward bulging of the iris.

the corner of each eyelid, the medial canthus bears a small **lacrimal papilla** punctured by a tiny hole called the **lacrimal punctum** (Figure 16-7). Each punctum leads into a short, medially directed **lacrimal canal** that empties into an enlarged **lacrimal sac** that fills the lacrimal canal in the lacrimal bone. The lower end of the lacrimal sac narrows into a **nasolacrimal duct**, which opens into the nasal cavity under the inferior nasal concha. Excess tears enter the lacrimal puncta and eventually drain into the nasal cavity. This explains why a person's nose runs when crying.

Physiology of Vision

After completing this section, you should be able to:

1. Differentiate between the far and near points of vision.
2. Describe the different methods of visual accommodation.
3. Discuss the reversible reaction linking opsin with retinaldehyde.
4. Describe light and dark adaptation.
5. Trace the pathway traveled by visual impulses to the brain.

Formation of the Image

The eye translates light rays into visual sensations. At least six coordinated steps have been identified in the process of "seeing" an object:

1. Light penetrates the cornea and lens.
2. The light rays are focused onto the retina.
3. The photoreceptive cells of the retina are stimulated to produce nerve impulses.
4. Nerve impulses are relayed through the bipolar cells to the ganglion cells.
5. Ganglion cells carry the impulses through the optic nerve into the thalamus.
6. The impulses are carried along projection fibers to the visual cerebral cortex where they are interpreted.

The seemingly effortless process of vision involves the cooperation of several coordinated parts of the nervous system.

Visual Accommodation

The process of **visual accommodation** involves several adjustments of the eye resulting in the production of a focused image on the retina. To form an image light rays must pass from the atmosphere through the cornea, aqueous humor,

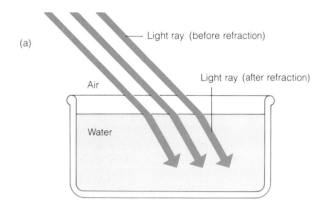
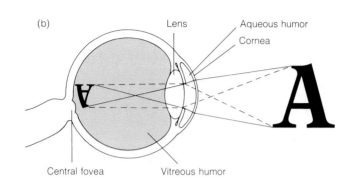

Figure 16-8 Refraction of light. (a) Light rays are bent (refracted) as they pass from one medium into another. (b) Refraction of light rays as they pass through the structures of the eye.

lens, and vitreous humor before they strike the photoreceptive cells of the retina. When light rays pass from one transparent medium to another of a different density, the rays are bent at the interface of the two media. This bending, called **refraction** (Figure 16-8), occurs as the light rays pass through the boundaries of each transparent region of the eye. In a normal eye this bending produces a focused image on the retina. Light rays coming from a distant object (at least 6 m away) are traveling parallel to one another when they strike the outer surface of the cornea, and the refraction produced by a relaxed eye causes the image to fall on the fovea centralis, the area of sharpest vision on the retina. This distance (6 m) is often called the **far point of vision.** But light rays coming from objects closer than 6 m from the eye are not parallel when they strike the cornea. Instead, they are diverging and must undergo additional bending (refraction) if they are to fall into focus on the fovea. The principal structure used to accomplish this additional refraction is the lens (Figure 16-9).

The pliable lens is held in place by the suspensory ligament and the ciliary body, the latter of which can contract. When it does, it reduces the stretching effect on the lens of the suspensory ligament. Relieved of this pressure, the lens "relaxes" and decreases its diameter by thickening. This increase in the thickness of the lens increases the length of the path that light must travel through the lens and, consequently, the amount of refraction of the light. Focusing of light rays from very close objects requires greater contractions of the ciliary muscle, and it is more difficult for the lens to accommodate and focus these divergent light rays onto the fovea centralis. There is a limit to the ability of the lens to change shape and, hence, a limit to the nearness at which an object can be focused by the eye. For most young adults this distance is about 10 cm in front of the eye. This distance is called the **near point of vision.** As one matures, the lens loses its elasticity and ability to focus, and by age 40, the near point has increased to about 20 cm and by age 60 to about 80 cm.

Figure 16-9 Visual accommodation. (a) Object farther than 6 m from the eye (b) Object closer than 6 m. Note that an inverted image is formed on the retina.

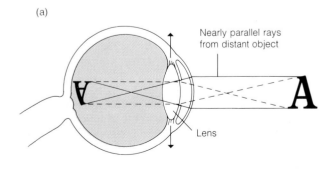
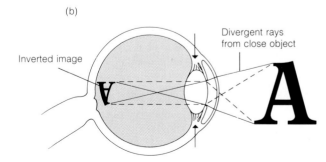

emmetropia: (Gr.) *emmetros*, in measure + *opsis*, sight
myopia: (Gr.) *myein*, to close + *ops*, eye
hyperopia: (Gr.) *hyper*, above + *opsis*, sight
astigmatism: (Gr.) *a*, not + *stigma*, point + *ismos*, condition

CLINICAL NOTE

VISUAL DISORDERS

Emmetropia (ĕm″ĕ·trō′pē·ă) is the term used to describe normal vision. The eye of an emmetrope can produce a clear image of a distant object without focusing or accommodating (Figure 1a). These people do not require glasses for seeing at a distance, but they often require reading glasses as they lose the ability to accommodate, which usually occurs about age 40.

Myopia, or nearsightedness, occurs when the eye focuses light from a distant object in front of the retina instead of directly on it, and the resulting image is blurred (Figure 1b). The eyes of myopes tend to be more elongated than those of emmetropes, and glasses are necessary to move the focal point back onto the retina. Moving the object closer to the eye also moves the focal point back, which is why myopes require glasses for distance but can see close objects without them.

Hyperopia (or **hypermetropia**) is farsightedness, which occurs in an eye that focuses distant objects behind the retina (Figure 1c). These eyes tend to be shorter from front to back than emmetropic eyes. These people can see clearly by accommodating: making the lens shorter and fatter to focus the image upon the retina. The closer an object, the more accommodation is necessary. Young hyperopes have very elastic lenses and can accomplish this easily; however, with age they lose this ability and require glasses.

Astigmatism results when the amount of curvature of the cornea along a vertical plane is different from that along a horizontal plane. A cornea with a perfectly spherical surface produces no astigmatism (Figure 2a). However, a cornea with uneven curves (similar to the surface of a football) produces noticeable astigmatism (Figure 2b). An object can be focused along two separate lines 90° apart with blurred images between. A person with astigmatism might see the vertical lines of the pattern in Figure 2c sharply while seeing the horizontal lines as blurred. Hard contact lenses worn over an astigmatic cornea correct this error, but soft lenses take the shape of the cornea they overlie and do not correct it.

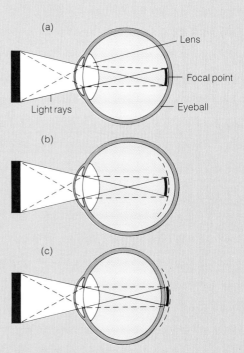

Figure 1 Optics of the eyeball. (a) Emmetropia (normal vision). (b) Myopia. Light is focused in front of the retina. (c) Hyperopia. Light is focused behind the retina.

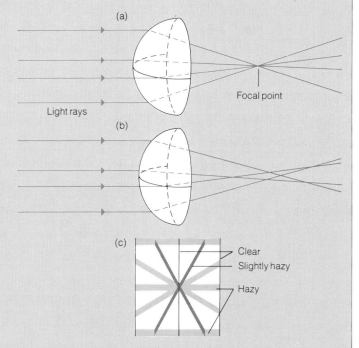

Figure 2 Astigmatism. (a) A normal eye is spherical and has a single focal point. (b) An astigmatic eye demonstrates separate vertical and horizontal focal points. (c) A person with astigmatism might see a star pattern this way (with the vertical lines clear and the horizontal lines blurred) if the vertical lines are focused closer to the retina than the horizontal lines.

esotropia: (Gr.) *eso*, inward + *tropos*, turning
exotropia: (Gr.) *exo*, outward + *tropos*, turning

In addition to adjustments of the lens, the eye is also capable of accommodating in other ways. For example, whenever you leave a brightly lit place and enter a darker place, you are aware of the darkness in the new place. It is almost impossible to see at first. However, within a few seconds, the radially arranged dilator muscles of the iris contract, increasing the diameter of the pupil, thereby allowing more light to enter the eye. (Biochemical changes also occur that are discussed in the section on Biochemistry of Vision.) Within a few minutes, the pupil has opened sufficiently to allow you to see within the darkened room. The iris of the eye is composed of involuntary muscles innervated by sympathetic and parasympathetic nerves. Both radial (dilator) and circular (constrictor) muscles are present. The dilator muscles are innervated by sympathetic fibers, the constrictors by parasympathetic fibers. Contraction of the dilator muscles increases the diameter of the pupil and allows more light to penetrate the eye. This action of the iris is another type of accommodation, since the eye accommodates itself to the changing intensity of light striking the retina. Conversely, bright light causes constriction of the pupil and a reduction in the intensity of light entering the eye. Thus, when you leave a theater and go back outdoors, the pupillary accommodation reflex reduces the bright light coming into the eye.

In addition to regulating the amount of light entering the eye, **pupillary accommodation** also reduces the area of the lens involved in focusing light rays onto the retina. This reduces the amount of the periphery of the lens that is active in focusing light on the retina and reduces the possibility of blurring on the retina. Pupillary accommodation also increases the "depth of field" (see p. 391) and therefore increases the focusing ability of the eyes.

The discussion thus far has centered on adaptations of our eyes that manipulate the rays of light that enter. Eventually, those rays must fall upon rods and cones. No matter whether the object is near or far, the image clear or blurred, large or small, it falls upon the retina in an inverted position. Refraction through the cornea, aqueous humor, lens, and vitreous humor bends the light rays in such a way that they converge within the vitreous humor and focus in an inverted position on the retina (Figure 16-9). The human brain learns to interpret these upside-down images so that one is not conscious of an inverted image.

Convergence and Binocular Vision

Accommodation by the lens and iris is accompanied by convergence of the eyeballs on near objects (those closer than 6 m). For example, if a person views an object from less than 6 m, the right eye sees more of one side of the object and the left eye sees more of its other side. The reader can test this by holding an object at arm's length and alternately closing one eye and then the other. Not only is the difference in viewing angle pronounced, but also the difference in background visible to each eye accentuates the difference in viewing angle. The extrinsic eye muscles, particularly the medial rectus muscles, are responsible for this convergence of vision on a near object and produce single binocular vision.

Single binocular vision is peculiar to those animals whose eyes are located at the front of a relatively flat face (Figure 16-10). This condition is pronounced in such animals as owls, primates, and humans. Autonomic centers in the brain monitor the amount of convergence and translate this information into distance readings, thereby allowing **depth perception,** the ability to perceive the three-dimensional

> **CLINICAL NOTE**
>
> ## STRABISMUS
>
> **Strabismus** (stră·bĭz′mŭs) is a general term referring to an abnormality in eye position, and it includes both esotropia and exotropia. **Esotropia** (crossed eyes) occurs when one eye is turned inward with respect to the other. **Exotropia** describes the condition of one eye being turned outward. Adults with either of these conditions may have double vision, but when these occur at an early age the brain may suppress one image and **amblyopia** (ăm″blē·ŏ′pē·ă), or lazy eye, develops.

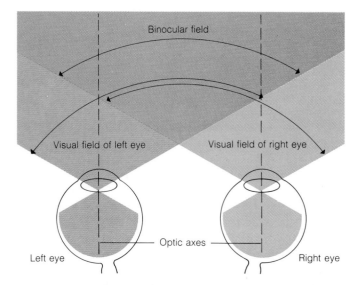

Figure 16-10 Binocular vision. The positions of the left and right eyes provide different visual fields of an object, and this allows vision in three dimensions.

rhodopsin: (Gr.) *rhodon*, rose + *opsis*, sight
chromophore: (Gr.) *chroma*, color + *pherein*, to bear

nature of the environment. Thus, binocular vision in humans allows us to determine an object is close enough to touch or to be in the range of danger.

Depth of Field

An additional complication to the proper interpretation of visual stimuli is that the left and right eyes do not view objects from the same perspective (binocular vision). Thus, objects at different distances from the eye cannot be focused on the retina simultaneously. The individual must choose which specific object will be in sharp focus, leaving the images of those objects both in front of and behind the plane of focus fuzzy in appearance (Figure 16-11). As a result, the visual cortex receives different information from each eye. With experience, the brain learns to accept and deal with such out-of-focus images, and they contribute to one's ability to see the environment in three dimensions. We perceive a "depth of field" and can only focus on a certain interval of space in much the same way that a camera does. This interpretation is another of the functions of the visual cortex of the cerebrum.

Biochemistry of Vision

Once an image has been formed on the retina, photoreceptive cells are stimulated to initiate impulses transmitted through the bipolar cells to the ganglion cells and out through the optic nerve. Stimulation of the photoreceptive cells involves several chemical reactions which have been studied for some time, yet are not fully understood. It appears that two sets of reactions occur, one in the rods and the other in the cones. Both sets of reactions involve photosensitive pigments found within the cells.

Visual Pigments in Rods

Rods contain the pigment **rhodopsin** (rō·dŏp'sĭn), **visual purple.** A rhodopsin molecule has two subunits: **retinal** (also referred to as **retinene** and **retinaldehyde**), which is a derivative of a vitamin A molecule, and a protein called **opsin** or **scotopsin** (skō·tŏp'sĭn). Retinal is a **chromophore**, a light-absorbing molecule. It exists in two shapes: a stable, relatively straight form known as all-*trans* retinal and a less stable, curved 11-*cis* form. The 11-*cis* shape is the form of retinal that combines with opsin to form rhodopsin. The all-*trans* form does not fit the receptive site on the opsin molecule and is found dissociated from opsin.

Rhodopsin appears purple because it reflects red and blue light and absorbs yellowish-green light with a wavelength of about 500 nm (Figure 16-12). Absorption of light energy by a rhodopsin molecule causes the 11-*cis* retinal to

Figure 16-11 Depth perception. When the eye focuses on near objects, distant objects are not in focus (top), and vice versa (bottom).

break away from the opsin portion and become free as all-*trans* retinal (Figure 16-13). This dissociation reaction initiates a generator potential in the rod, which results in an action potential in the bipolar and ganglion neurons.

The reaction that splits retinal and opsin molecules is reversible. Rhodopsin can be resynthesized in rods from all-*trans* retinal in a reaction that requires energy and enzymes. The degradation portion of this reversible reaction in the

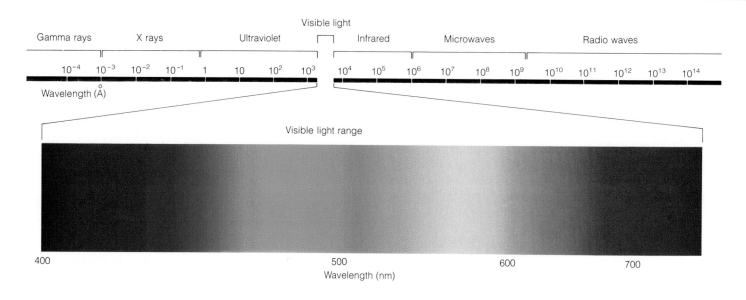

Figure 16-12 **The electromagnetic spectrum.** Only a small part of the spectrum is visible to humans. ($1 \text{ Å} = 10^{-10}$ m; $1 \text{ nm} = 10^{-9}$ m)

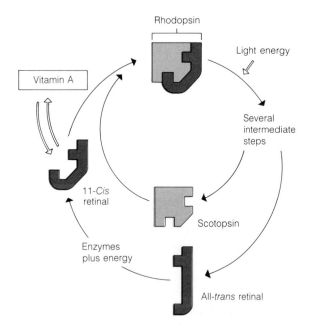

Figure 16-13 **The retinal-rhodopsin cycle.** When light strikes the rhodopsin molecule, it splits into scotopsin and all-*trans* retinal, triggering a visual impulse in a rod cell. An enzymatically controlled reaction transforms all-*trans* retinal into 11-*cis* retinal, which recombines with scotopsin to form rhodopsin. A reservoir of vitamin A in the pigmented layer of the retina is used to manufacture additional retinal.

rods requires only small amounts of light energy, a characteristic that has two consequences. First, rods can be stimulated in very dim light. This allows an individual to see at night or in other poorly lit situations. Under these low light conditions, the rods "see" in terms of brightness or shades of gray; color cannot be detected. Second, because rhodopsin reacts so easily, the cell's supply can be depleted quickly in brightly lit situations, causing a delay in the ability to see in a dark situation. This is what happens when you go from brightly lit daylight into a darkened room. The rods require a few minutes to resynthesize enough rhodopsin to respond to the low light level, and during that time you have trouble seeing. Only after sufficient rhodopsin has been made can you begin to see objects in the room. Because retinal is a vitamin A derivative, sufficient quantities of this vitamin must be in the diet to produce enough rhodopsin for normal vision in dim light. A deficiency of vitamin A can result in less retinal and rhodopsin. Affected individuals have difficulty seeing under dim light conditions, and they suffer from **night blindness.** Inability to see at night is also due to lack of stimulation of the cones which are unable to respond to low light intensity.

This chemical process of adaptation to changing intensities of light is augmented by pupillary accommodation, which allows changes in the size of the pupil, thereby regulating the amount of light striking the retina. Together, these processes constitute **light** and **dark adaptation.**

Visual Pigments in Cones

Cones have visual pigments that are similar to those in rods. Retinal is present in both rods and cones, but the protein portion of the molecule is different. Some cones contain a pigment that is sensitive to red light, while other cones are sensitive to green light or blue light. Just as the screen of a color television set uses combinations of tiny dots of only three colors (red, green, and blue) to produce all colors, the retina uses combinations of the three photosensitive pig-

tympanic: (Gr.) *tympanon*, drum

> **CLINICAL NOTE**
>
> ## COLOR BLINDNESS
>
> Cone receptors for red, green, and blue are located within the retina. All three are required for normal color vision. Persons born with only one type of cone are called **monochromats,** and everything they see has the same color. If only two types of cones are present, a person will confuse pairs of colors, such as red-green or blue-yellow. This is commonly known as **color blindness.** About 8% of males but only 0.4% of females are affected by some disorder of color vision.

ments to respond to the variety of colors in the environment, thus allowing the perception of many colors. In contrast to rhodopsin, pigments of the cones do not respond to low light levels, and so cones function only in relatively bright light.

Visual Pathways to the Brain

Nervous impulses that begin in the retina in response to visual stimuli are transmitted to ganglion cells that leave the eyeball through the optic nerve (see Chapter 13). This nerve passes through the optic foramen in the lesser wing of the sphenoid bone and enters the cranial cavity anterior to the pituitary gland. At this point, the fibers merge in the X-shaped optic chiasma (Figure 16-14). A partial crossover of fibers occurs from each eye to the path on the opposite side. The pathway from the optic chiasma consists of two optic tracts leading to the brain. Most optic tract fibers terminate in the lateral geniculate bodies of the thalamus where they synapse with third-order neurons that terminate in the occipital lobes. A lesser number of optic tract fibers are diverted to the superior colliculi in the mid-brain. Here they provide information used in reflex actions involving coordination of body movements during activities such as playing tennis, dancing, or playing the piano.

Anatomy of the Ear

After completing this section, you should be able to:

1. Describe the anatomic and functional relationships of the outer, middle, and inner ears.
2. Discuss the function of the ear ossicles in transmitting vibrational stimuli from the tympanum to the cochlea.
3. Distinguish between endolymph and perilymph.
4. Describe the semicircular canals and discuss their function.

External Ear

The adult ear has three portions: the external, middle, and inner ears. The **external** (or **outer**) **ear** consists of the cartilage-filled **auricle** (or **pinna**) and the canal that extends into the skull. The auricle protrudes from the side of the temporal region of the skull and is supported by an internal framework of elastic cartilage covered by skin. Individual portions of the auricle have been given special names detailed in Figure 16-15.

At the anterior border of the auricle is the **external auditory meatus** (or **canal**) that pierces the petrous ("hard") portion of the temporal bone. The entrance to the meatus is surrounded by numerous hairs that tend to prevent the accidental entrance of foreign objects. The skin in this region contains **ceruminous** (sĕ·rū′mĭ·nŭs) **glands,** which are specialized sebaceous glands that secrete a thick, waxy material called **cerumen** (commonly known as **earwax**). The auditory meatus ends at the **tympanic membrane** or **tympanum** (tĭm′păn·ŭm) **(eardrum),** a thin vascular membrane stretched across the terminus of the canal. The tympanum is cone-shaped and forms the boundary between the external and middle ears. Its external surface is covered with a thin layer of skin, while the internal surface is covered with a thin mucous membrane.

Middle Ear

The **middle ear** is a short tube connecting the external and internal ears (Figure 16-16). It is bounded by the tympanic membrane externally and the inner ear internally. It is lined

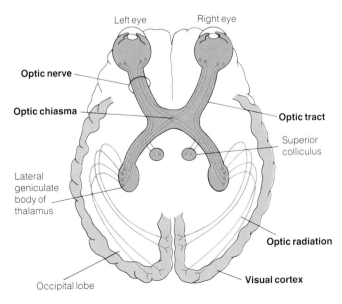

Figure 16-14 **Nerve pathways for vision.**

Embryonic Development of the Ear

The first indication of ear development is the appearance early in the fourth week of two round, thickened ectodermal regions called **otic placodes** on each side of the embryonic head (Figure 16-C). Each placode enlarges and pushes into the underlying mesodermal mesenchyme of the head, forming a deepening depression that becomes a hollow ball, the **otocyst** (or **otic vesicle**). During week five the otocyst loses all connection with the surface ectoderm and sinks into the surrounding mesenchyme. Innervation of the otocyst occurs as nerve fibers from the vestibulocochlear nerve (cranial nerve VIII) make contact. Further development of the otocyst involves differentiation of a complex structure consisting of two major components, the **cochlear** (hearing) **organ** and **vestibular** (balance) **organ**.

Simultaneous with the development of the otocyst, the lateral walls of the pharynx develop bulges that expand into a series of **pharyngeal pouches** (Figure 16-C). The first two pouches fuse and expand toward the surface ectoderm. At the same time a series of depressions (**branchial grooves**) develop in the surface ectoderm. One of these grooves deepens and approaches the pharyngeal pouch. Further development of the groove produces the external auditory meatus, while the pouch becomes the auditory tube. The thin barrier of tissue remaining between them develops into the tympanic membrane. The middle ear ossicles ossify from mesenchyme between the branchial groove and otocyst.

The external ear differentiates from swellings surrounding the external auditory meatus. These swellings enlarge into tubercles that fuse and develop into the auricle.

Figure 16-C Embryonic development of the ear. (a) Otic placodes appear during the fourth week. (b) The placodes deepen into pits and (c) become detached otocysts during the fifth week. (d) The otocysts enlarge into the ducts of the inner ear, while the middle ear ossicles ossify from adjacent mesenchyme.

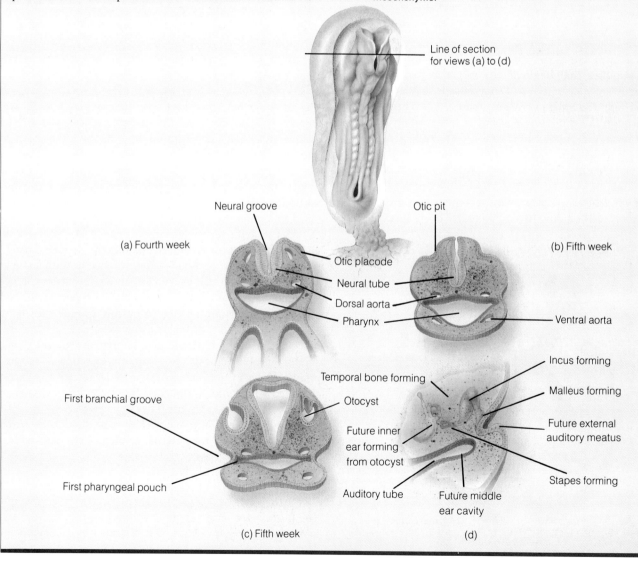

fenestra: (L.) *fenestra*, window
antrum: (L.) *antrum*, cavity

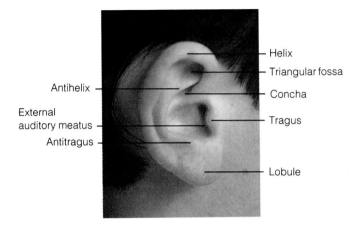

Figure 16-15 Lateral view of the external ear.

with mucous membrane and is filled with air. Three small bones called **ossicles** occupy much of the middle ear. The **malleus** (or **hammer**) is the most lateral of the three; its "handle" rests against the interior surface of the tympanum. The **incus** (or **anvil**) is the intermediate of the three, and the **stapes** (or **stirrup**) is the innermost. (The function of these bones is described in the next section.) The base of the stapes inserts into the **oval window** (or **fenestra ovalis**), an oval opening into the inner ear. The **round window** (or **fenestra rotundum**) forms another opening from the inner ear into the middle ear. Both windows are covered by thin membranes that prevent fluid from leaking into the middle ear. The middle ear is also connected to the mastoid sinus of the temporal bone via the **tympanic antrum**.

An opening is maintained between the middle ear and the nasopharynx through the **auditory tube** (or **pharyngeal tympanic tube**). In a sense, this tube is a safety device. Its major function is to allow the air pressure to equalize on both sides of the tympanum. Whenever a change occurs in the atmospheric pressure, a danger exists that the fragile tympanum could be stretched to the point of rupture, similar to a balloon being inflated too much. This does not normally happen unless a significant change in pressure pushes on the eardrum. For example, when an individual climbs a mountain or travels in an airplane, the increase in altitude causes a corresponding decrease in the atmospheric pressure. Conversely, if a person dives under water, the pressure of the

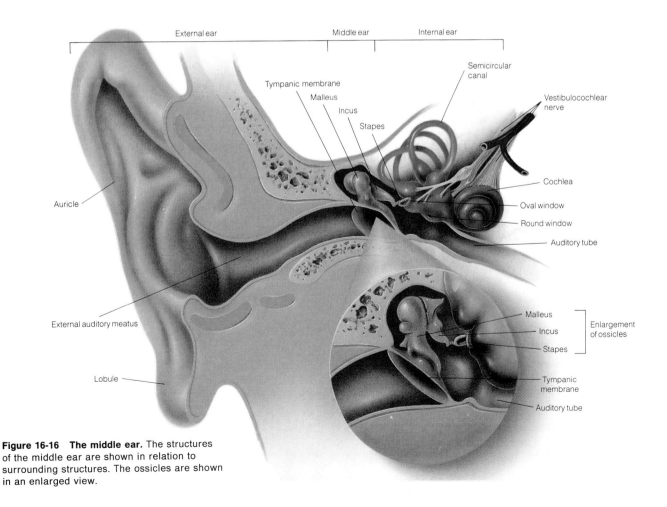

Figure 16-16 The middle ear. The structures of the middle ear are shown in relation to surrounding structures. The ossicles are shown in an enlarged view.

mastoiditis:	(Gr.) *mastos*, breast + *eidos*, form + *itis*, inflammation	
endolymph:	(Gr.) *endon*, within + (L.) *lympha*, clear fluid	
perilymph:	(Gr.) *peri*, around + (L.) *lympha*, clear fluid	
cochlea:	(Gr.) *kochlias*, snail	

heavier water is transferred through the air in the auditory tube and causes increased pressure on the membrane. These conditions cause the tympanic membrane to bow outward or inward, respectively, usually accompanied by discomfort and perhaps even pain. When this happens, the pressure on both sides of the tympanum can be equalized by consciously opening the auditory tube, thereby allowing the tympanum to resume its normal shape. Yawning, swallowing, and blowing the nose all have an almost immediate effect of opening the pharyngeal entrance to the auditory tube, allowing atmospheric pressure to press equally on both sides of the tympanum.

Unfortunately, the open connection between the middle ear and pharynx provides access to infectious agents (viruses and bacteria) from the mouth, nose, and throat. Therefore, it is not unusual for infections of the respiratory tract to move into the middle ear. In fact, these agents may also enter the tiny openings in the tympanic antrum of the temporal bone's mastoid process. Inflammation of the membrane in the mastoid process is called **mastoiditis**.

Internal Ear

The **internal** (or **inner**) **ear** is embedded within the petrous portion of the temporal bone. It is composed of two fluid-filled sacs, one almost surrounded by the other. The inner sac, the **membranous labyrinth**, includes ducts and chambers containing the fluid **endolymph**. A **perilymphatic** (pĕr′ĭ·lĭmf·at′ĭk) **labyrinth** encloses the membranous labyrinth and is composed of bony canals and chambers filled with the fluid **perilymph** (Figure 16-17).

The perilymphatic labyrinth (often called the **bony** or **osseous labyrinth**) occupies a hollow space within the petrous portion of the temporal bone. This space is bounded by a thin, membranous sheath that lies in close association with the endosteum of the bony cavity where it is found. The perilymphatic labyrinth is divided into three discrete compartments: the semicircular canals, vestibule, and cochlea.

Semicircular Canals and Vestibule

The **semicircular canals** occupy the outer section of the perilymphatic labyrinth. They extend from the vestibule in three planes at approximately right angles to one another. The **vestibule** is the central portion of the perilymphatic labyrinth that connects the semicircular canals with the cochlear canals. These canals (**anterior, lateral,** and **posterior canals**) and the membranous ducts they contain are located within the perilymphatic labyrinth.

Three **semicircular ducts** (**anterior, lateral,** and **posterior ducts**) lie within the semicircular canals surrounded by the perilymphatic labyrinth (Figure 16-17). Each one has

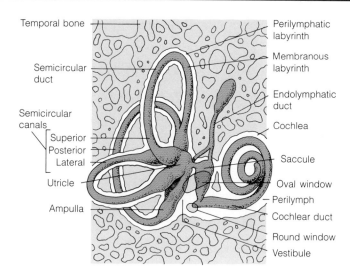

Figure 16-17 Anatomy of the internal ear. The endolymphatic ducts are completely enclosed and surrounded by the perilymphatic space.

a small swelling, or **ampulla**, near its base. The ampullae are characterized by specialized cells attached to low ridges called the **ampullary crests**. These cells detect stimuli that indicate the body's position in space during movement (see section on Physiology of Equilibrium).

The vestibule of the membranous labyrinth has two regions: the **utricle** (ū′trĭk·l) and the other small sac that connects to it called the **saccule** (săk′ūl). The utricle also connects to the semicircular ducts at their base (Figure 16-17). A region of cells similar to those in the ampullae of the semicircular ducts but within the utricle is called the **macula**. The saccule contains an area of sensory cells. The utricle and saccule are joined through a short, narrow duct called the **endolymphatic duct**, a narrow extension of the membranous labyrinth.

Cochlea

The **cochlea** (kŏk′lē·ă) is a snail-shaped system of tubes (**cochlear canals**) that are modifications of the perilymphatic labyrinth. The cochlear canals are coiled in two and a half tight turns producing the snailshell appearance of the cochlea, but if they were straightened out (as shown in Figure 16-18), they could be likened to a wiener enclosed by a hot dog bun. Like the bun, the perilymphatic canals are divided into two segments, with the membranous **cochlear duct** (or **scala media**) sandwiched in between like the wiener. In our comparison, the upper canal, the **scala vestibuli** (skā′lă vĕs·tĭb′ū·lē), corresponds to the top of the bun, and the lower canal, the **scala tympani** (skā′lă tĭm′pă·nē), corresponds to the bottom of the bun. The scalas merge at the **helicotrema** (hĕl″ĭ·kō·trē′mă) (Figure 16-18).

scala: (L.) *scala,* ladder
helicotrema: (Gr.) *helix,* coil + *trema,* hole
modiolus: (L.) *modiolus,* small measure

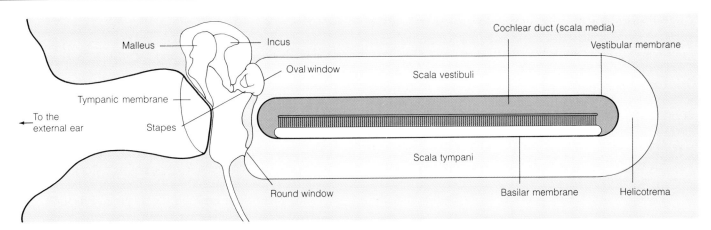

Figure 16-18 Schematic diagram of the cochlear canal in an uncoiled position.

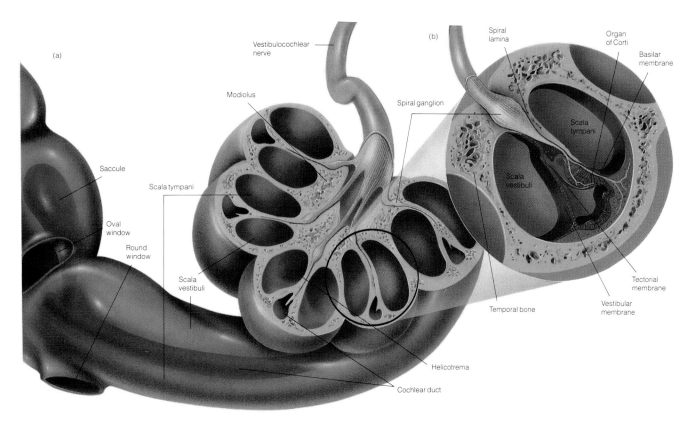

Figure 16-19 (a) A section through the cochlea showing its internal anatomy. (b) An enlarged view of the organ of Corti.

The third cochlear canal, the cochlear duct, is most intimately associated with hearing. It is a narrow extension from the base of the saccule between the scala vestibuli and scala tympani. The snail-shaped cavity in the temporal bone housing the cochlea shelters a rod-shaped, bony core called the **modiolus** (mō·dī´ō·lŭs). The modiolus resembles a screw because, along its length, a thin shelf called the **spiral lamina** spirals from the modiolus like threads of a screw following the coils of the cochlea (Figure 16-19). The spiral lamina separates the two sections of the perilymphatic labyrinth, and its free edge attaches to the interior of the cochlear duct. This duct is triangular in cross section with the apex resting on the spiral lamina. The narrow **basilar membrane,** also attached to this aspect of the lamina, extends across the

tectorial: (L.) *tectorium*, covering

cochlear space. It forms the base of the interior wall of the cochlear duct and separates the scala vestibuli from the scala tympani.

The most important specialization of the basilar membrane is the **organ of Corti** (Figures 16-19 and 16-20). It is a ridge of sensory cells that runs almost the entire length of the basilar membrane. Associated with the organ of Corti is another membranous extension from the spiral, the **tectorial** (těk·tō´rē·ăl) **membrane.** The tectorial membrane parallels the basilar membrane and contacts hairlike extensions from sensory cells in the basilar membrane. These cellular processes are bathed with the endolymph within the cochlear duct.

Innervation

The external ear is innervated by branches from cranial nerves V, VII, X, and XI. The tympanic membrane is innervated by branches from CN V, X, and XI. Two very small muscles associated with the ear ossicles receive impulses along branches of CN V and VII. The vestibular branch of CN VIII innervates the vestibule and semicircular canals and ducts, and the cochlear branch of the same nerve innervates the cochlea.

Physiology of Hearing

After completing this section, you should be able to:

1. Describe the current theory used to explain how the cochlea can differentiate among sound frequencies.
2. Describe how the cochlea can differentiate among sound intensities.
3. Trace the nerve pathway traveled by sound sensations from the inner ear to the brain.

Sound

Sound is a form of energy transmitted by an alternating series of compressions and decompressions of molecules produced in air by a vibrating object, such as a tuning fork, the plucked string on a guitar, or a vocal cord. Sound waves have two principal characteristics: amplitude and frequency.

Amplitude is the intensity of the vibration and is directly related to loudness. Intensity is measured in **decibels** (dB). Under the correct circumstances, humans can hear sounds as low as 1 dB. Normal conversation occurs in the range of 45 dB. Above 115 dB, sound is painful.

Frequency is the number of vibrations made by a vibrating object in a given period of time. It is measured in hertz (Hz) or cycles per second (cps). The frequency of a sound is directly related to pitch or tone. Humans are able to hear sounds between 20 and 20,000 Hz. The musical note "A" used by concert musicians to standardize the tuning of their instruments has a frequency of 440 Hz.

Transmission of Sound to the Inner Ear

One of the two functions performed by the ear is to collect sound waves and transform these physical stimuli into nerve impulses to be interpreted by the brain as perceptions.

The external ear executes the simplest of the many functions involved in converting sound into nerve impulses. The ear's auricle directs the waves into the external auditory meatus which leads them to the tympanic membrane. The successive compressions and rarefactions in the air of the ear canal forms waves that vibrate the tympanic membrane in unison with the sound waves. This movement sets the three ear ossicles (the malleus, incus, and stapes) in motion, which is amplified by the leverage arrangements of the bones. The base of the stapes, anchored in the oval window of the perilymphatic labyrinth, transfers this vibration to the perilymph within the scala vestibuli. The surface of the tympanic membrane is approximately 20 times that of the oval window and thus is capable of collecting 20 times the amount of energy generated by the sound waves as could an unassisted oval window.

Functions of the Cochlea

The following sequence of events seems to result in the transfer of vibrational movements from the perilymphatic fluid to the sensory cells of the organ of Corti. Movement of the stapes within the oval window causes the perilymphatic fluid in the scala vestibuli to vibrate at the same frequency as the sound waves that originally initiated the movement. An examination of Figure 16-20 will show that the continuity between the scala tympani and the scala vestibuli allows pressure waves to be transmitted throughout the perilymphatic fluid. Because this fluid is not compressible, the movement of the perilymphatic fluid in the scala vestibuli is transferred to the endolymph in the cochlear duct, which in turn causes the hairlike cellular extensions and probably the entire sensory cells of the basilar membrane to vibrate.

The exact means by which these cells discriminate among different frequencies of sound waves (which translates to tone or pitch) is unknown. The following theory is advanced by some experts.

The basilar membrane extends the length of the scala media, but its width and flexibility increase steadily from the base near the oval window to the apex near the helicotrema. Pressure waves of different frequencies travel differently through the endolymph. Presumably, the greater the amplitude of a pressure wave at a specific point on the organ of Corti, the greater the stimulation of the sensory cells at that point. The pressure wave propagated by a high frequency sound reaches its greatest amplitude close to the oval window, whereas one of low frequency reaches its greatest amplitude toward the opposite end of the organ of Corti (Figure 16-20c). Intermediate frequency waves reach their greatest amplitude in the middle of the organ of Corti, stimulating these sensory cells more than at either end. Thus differential stimulation of sensory cells along the length of the organ of Corti results from sound waves of different frequency having different points of maximum force.

This process explains how the ear can discriminate among different frequencies. The ability to differentiate sound intensities is easier to explain. As the intensity of a sound increases, movement of the tympanum, ossicles, and endolymphatic fluid increase proportionately. Greater stimulation of the sensory cells causes proportionately more impulses (per unit time) to be initiated in the spiral ganglion cells, resulting in more impulses traveling to the brain. Therefore, while pitch of a sound may be determined by *which* cells are stimulated on the basilar membrane, loudness is determined by the *rate* at which impulses are initiated in those cells.

Sensory Pathways for Hearing

Axons from the spiral ganglion form the cochlear nerve, which leaves the petrous portion of the temporal bone as a branch of the vestibulocochlear cranial nerve (VIII). The auditory pathways enter the medulla and travel through the midbrain to the inferior colliculus and the medial geniculate body. Pathways involved in the medulla and midbrain are complex and not fully understood. Crossover, even double crossover, of fiber tracts occurs (Figure 16-21). Fibers from both ears eventually reach both sides of the brain. Connections made in the inferior colliculus appear to be important in governing the reflex of turning one's head in response to sounds such as a loud noise that attracts one's attention. Fiber tracts that pass through the thalamus eventually terminate in the primary auditory area in the temporal lobe of the cerebral cortex.

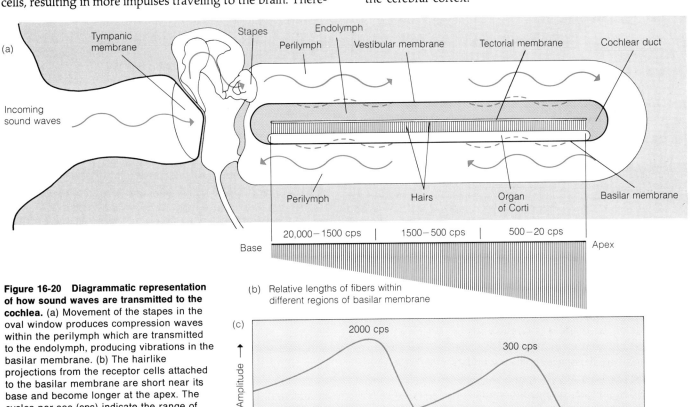

Figure 16-20 Diagrammatic representation of how sound waves are transmitted to the cochlea. (a) Movement of the stapes in the oval window produces compression waves within the perilymph which are transmitted to the endolymph, producing vibrations in the basilar membrane. (b) The hairlike projections from the receptor cells attached to the basilar membrane are short near its base and become longer at the apex. The cycles per sec (cps) indicate the range of frequencies that must stimulate the cells at that part of the basilar membrane. (c) This graph indicates the relative amount of response (amplitude) of different positions of the membrane to two specific sound frequencies, 2000 and 300 cps.

otolith: (Gr.) *otos*, ear + *lithos*, stone

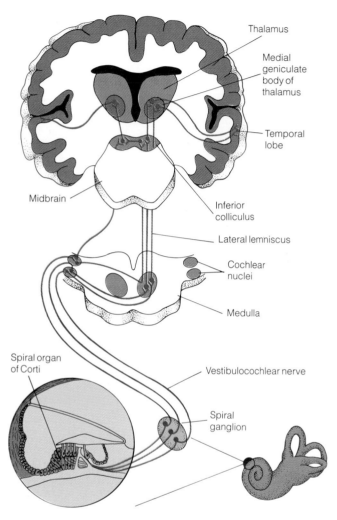

Figure 16-21 Transmission pathway for auditory stimuli from the inner ear to the cerebral cortex.

Pathways from specific portions of the basilar membrane terminate in specific areas of the cortex on both sides of the brain. High frequency sounds are projected onto anterior portions of the auditory area, while impulses generated by low frequency sounds arrive at more posterior regions. Because fibers carrying information from both ears reach the same areas of the brain, it is possible for the brain to interpret certain differences in information from the two ears in a meaningful way. For example, a sound coming from one side of the head reaches the nearer ear sooner than the one on the other side of the head; consequently, sensory cells in the nearer ear are stimulated before those in the other ear. The brain can detect this difference in arrival time of impulses from the two ears and uses this information to determine the general direction from which the sound came. This determination is referred to as **localization of sound.**

Physiology of Equilibrium

After completing this section, you should be able to:

1. Compare and contrast static and dynamic equilibrium.
2. Discuss the function of otoliths in sensing body position.
3. Trace the nerve pathway traveled by sensations of equilibrium in reaching the brain.

Two problems faced by the body are the need to locate its position in space and the need to accurately detect its movement. Both abilities are necessary to maintain **equilibrium,** or balance. Balance can be accomplished partially by relying on information obtained from visual stimuli, but detailed awareness of position takes into account sensations from the semicircular canals and the utricle and saccule. Balance involving posture, position of the head, and general body position is **static equilibrium.** Balance involving movement of the head is **dynamic equilibrium.**

Static Equilibrium

The capacity for sensing static equilibrium (balance while stationary) is centered in the utricle and saccule. Recall the two areas of specialized sensory cells called **maculae** within these structures. Each macula is a swelling of unique cells anchored in the wall of the membranous labyrinth. Each cell has a hairlike projection that extends out into the endolymphatic fluid. The "hair" actually consists of a type of cilium (called a **kinocilium**) that is surrounded by numerous microvilli (or **stereocilia**). (See Chapter 3 for a complete description of cilia and microvilli.) Surrounding these hairs is a gelatinous material containing small deposits of calcium carbonate called **otoliths** (Figure 16-22). The mass of the otoliths (also called **otoconia** or **statoconia**) increases the inertia of the gelatinous material, making it more sensitive to outside forces, including gravity. Any change in the position of the head results in a change in the direction of the force of gravity relative to the sensory cells of the maculae. Different positions of the head pull the gelatinous mass, otoliths, and embedded kinocilia in corresponding directions, stimulating the sensory cells differentially. Nerve impulses generated in neurons of the vestibular branch of CN VIII (connected to the sensory cells) are relayed to conscious and reflex brain centers. Analysis of the impulses allows an individual to be consciously aware of the position of the head. Such treatment also allows reflex centers in the cerebellum and medulla to stimulate skeletal muscles that keep one's balance in walking, running, and other activities.

cupula: (L.) *cupula*, small tub

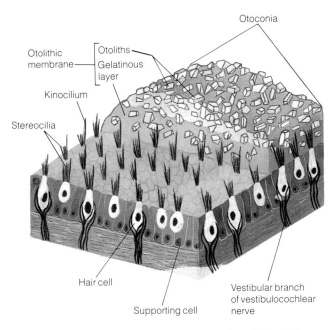

Figure 16-22 Sensory cells located in the maculae of the utricle and saccule are used to detect the position of the head.

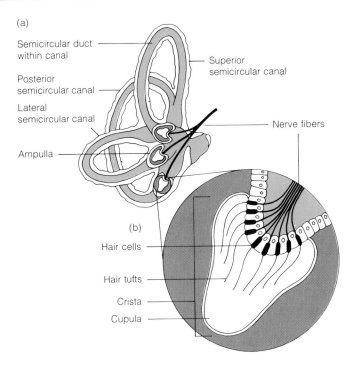

Figure 16-23 Ampulla. (a) Location of ampullae at the bases of the semicircular ducts. (b) Detailed structure of an ampulla.

Dynamic Equilibrium

The sense of dynamic equilibrium (maintaining balance while moving) centers in the three semicircular membranous canals. Because the three are positioned simultaneously at right angles to each other (frontal, sagittal, and transverse), they can detect movement in three-dimensional space. Specifically, the ampulla within each semicircular duct detects movement.

Each ampulla contains a group of sensory cellular extensions called a **crista,** which forms a small papillary elevation. The projections in the crista are arranged in a row that extends transversely across the lumen of the duct (Figure 16-23), a position that allows easy detection of the endolymphatic fluid movements within the duct. The projections of the sensory cells are surrounded by the **cupula,** a small mass of gelatinous material. The difference in density between the cupula and endolymph allows detection of certain types of movement.

When the head is moved, the endolymph in the appropriate duct resists the change of inertia, hence, it does not move as rapidly in the same direction. The slower moving endolymph pushes the cupula, moving the gelatinous material, which stimulates the sensory cells; the more forceful the movement, the more intense the stimulation. If the movement of the body is at a constant velocity, the endolymph catches up with the duct and the stimulation of the crista is ended. This phenomenon explains why one can travel 50 mph in an automobile or 500 mph in an airplane and be unconscious of the movement until either acceleration or deceleration occurs. When the movement ends, the processes are reversed, and the brain interprets the new barrage of impulses in terms of deceleration. Thus, mechanisms that operate in the inner ear only allow the brain an awareness of *changes* in motion. Detection of *constant* movement requires visual stimuli.

Detection of rotational movement is slightly more complicated. The same mechanics are involved, but since the ears are located on opposite sides of the median axis of the body, endolymph in the opposite ducts is moved in opposite directions. This means that the brain receives different signals from opposite sides of the head and, as a result, can interpret the phenomenon as rotational movement. A skater beginning a spin or a blindfolded person being spun on a rotating stool does not need visual information to know the direction of the spin.

Some of the most unpleasant feelings a person can experience are associated with the sense of balance or, more accurately, disruptions of the sense of balance. **Motion sickness** is one of the most common results of experiencing continual small movements of the body. An example is the feeling of sea sickness that some people experience while

nystagmus: (Gr.) *nystagmos*, to nod
vertigo: (L.) *vertigo*, whirling about

CLINICAL NOTE

DISTURBANCES TO EQUILIBRIUM

The most common symptom of disturbance of the vestibular apparatus is **vertigo**. Evaluation of a patient with symptoms of vertigo in the office consists of several tests. **Romberg's sign** is performed by asking the patient to stand up straight with the feet together and the eyes closed; wavering or falling is a positive sign of vestibular malfunction. Eliciting **nystagmus** (nĭs·tăg′mŭs), an oscillatory movement of the eyeballs, by quickly placing the patient in various positions is suggestive of malfunction of the vestibular apparatus. **Caloric testing** is a more specific way of checking each vestibular apparatus by syringing cold water into the ear canal; an active vestibular apparatus will produce severe vertigo and nystagmus. This test is very uncomfortable to the patient and may even induce vomiting. Well-equipped audiology laboratories can test vestibular function using sophisticated electrostimulators.

The most common problem of the vestibular apparatus is **motion sickness** (nausea and vomiting caused by repetitive angular, linear, or vertical motion). Treatment consists of avoidance of the extremes of motion and the occasional use of vestibular sedatives, such as Dramamine, prior to travel.

A more serious problem is **Meniere's disease** characterized by recurrent prostrating attacks of vertigo, hearing loss, and **tinnitus** (a subjective ringing, roaring, or hissing sound in the ears). The cause is poorly understood, but is thought to be due to generalized dilation of the membranous labyrinth. Treatment may be the use of vestibular sedatives or various surgical procedures.

Some medications can also cause disturbances of the vestibula and the cochlea. **Streptomycin** is an antibiotic used to treat tuberculosis and other severe infections, but it must be used with care because of its detrimental effect on the organs of the inner ear.

---

riding in a boat, especially through rough water. **Vertigo** is a feeling of disorientation and dizziness that occurs after a person has experienced a series of rapid rotational movements, such as one undergoes in many rides at amusement parks. Even after the movement stops, one still has the sensation that his or her surroundings are spinning around. A person emerging from one of these rides often staggers around for a few minutes until the feeling subsides (see Clinical Note: Disturbances to Equilibrium).

Nerve Pathways in Equilibrium

Sensory impulses are transmitted along the vestibular branch of CN VIII into the brain. Most of the sensory fibers terminate in vestibular nuclei of the medulla. The remaining ones pass through the inferior cerebellar peduncles and terminate in the cerebellum. Additional cerebral-cerebellar pathways integrate sensory information from the inner ear with visual stimuli to keep the body consciously and unconsciously aware of balance and equilibrium.

Motor impulses originate in several areas, and all maintain balance in the body, whether sitting, standing, or moving. Reflex impulses leave through nuclei of CN III, IV, and VI. These help coordinate eye movements with balance and equilibrium. Motor impulses leave through CN XI and help coordinate movements of the head and neck. In addition, motor pathways from the cerebrum, cerebellum, and reflex centers in the medulla course to skeletal muscles that maintain balance, equilibrium, and muscle tone.

Thus far in Unit III, we have studied the nervous system and how it controls body activities. Sensory receptors detect external and internal stimuli, and this information is then used in generating reflex responses to the stimuli or it is transmitted to the brain where it is integrated with other stimuli being received simultaneously or compared with what is stored in memory. The brain integrates both conscious and unconscious activities and coordinates body movements and the secretions of glands. In general terms, the nervous system provides *immediate* responses to stimuli. In the following chapter we will describe the endocrine system, that part of the body's control systems that produces *long-term* effects through secretion of cellular products into the blood.

STUDY OUTLINE

Olfaction: The Sense of Smell (pp. 377–379)
  Olfaction, the sense of smell, is centered in *olfactory cells* located in the mucous membrane of the nasal cavity.

  Stimulation of Olfactory Receptors   The manner in which olfactory cells function is unknown. The most persuasive hypothesis suggests the existence of receptor sites on the membranes of these cells that are stimulated by different chemicals.

  Olfactory Adaptation   The sense of smell is subject to relatively rapid physiological and psychological adaptation.

  Olfactory Pathway   Axons from olfactory cells join the *olfactory bulbs*, extensions of brain tissue. *Olfactory tracts* extend from the bulbs into the olfactory centers in the temporal lobes of the brain.

Gustation: The Sense of Taste (pp. 379–380)
  Gustatory Cells   *Gustatory cells* are concentrated in taste buds on the tongue and the roof of the mouth and palate. Taste buds are sensitive to four tastes: sweet, salty, sour, and bitter.

  Gustatory Pathway   Gustatory receptors are innervated by cranial nerves VII, IX, and X. The gustatory pathway extends through the medulla to the thalamus and from there to the somatic sensory region in the parietal lobe of the cerebral cortex.

otosclerosis: (Gr.) *otos*, ear + *skleros*, hard

THE CLINIC

HEARING TESTS AND DISORDERS

Hearing Tests

The office evaluation of hearing problems requires only minimal time and equipment. An idea of the patient's gross level of hearing may be obtained by the physician speaking in a normal voice at a distance of 15 feet, testing each ear separately. Gross hearing is recorded as the distance in feet that the patient first hears the normal speaking voice. It is recorded as 15/15 for the normal subject and 12/15 for a patient with slightly reduced hearing acuity. Another common method is to record the distance in inches that a watch tick is heard in each ear. Also used are pure tone **audiometers** that allow the physician to obtain an estimation of the level of hearing at various sound frequencies. If significant hearing loss is determined, one may use a tuning fork to discern if the loss is conductive or perceptive. **Conductive loss** implies that something has interfered with the conduction of sound waves in the external or middle ear. **Perceptive loss** suggests a more serious problem with the sensorineural mechanism of the inner ear or brain. In **Weber's test** the tuning fork is struck and applied to the forehead. The normal patient will hear the tone equally in both ears; in nerve deafness it will be heard better in the good ear; and it will be heard best in the poorer ear if conductive deafness is present. To perform the **Rinne test** the tuning fork is struck and first placed on the mastoid process. When the patient can no longer hear the tone, the fork is moved in front of the external auditory meatus. The Rinne test differentiates air conduction from bone conduction. A person with normal hearing hears better by air conduction. If bone conduction is better than air conduction, one would suspect a conductive hearing loss. Frequently, conductive loss is due to infection or fluid behind the tympanic membrane and is readily treated. Hearing loss due to sensory nerve damage is usually more difficult to treat and may require the use of a hearing aid. Patients with serious hearing problems may be referred for **audiology analysis** in which more sophisticated equipment will clarify the problem. The **conventional audiogram** records the amount of hearing loss at specific frequencies in both air and bone conduction (see figure).

Hearing Disorders

One of the most important causes of acquired deafness is **serous otitis media**. This insidious condition occurs when fluid collects behind the eardrum following respiratory infection or chronic allergy. It is secondary to a blockage of the auditory tube and replacement of the normal air in the inner ear with fluid. The fluid prevents normal motion of the eardrum and thus a conductive hearing loss. Treatment consists of restoring the open passageway of the auditory tube by the use of nose drops or oral decongestants, removing obstructing adenoid tissue, and avoiding respiratory infections. If the condition persists or the fluid reaccumulates, a tiny plastic tube may be inserted through the eardrum and left in place for several months to ventilate the middle ear. This is commonly done in small children. Failure to recognize or alleviate chronic serous otitis can result in permanent hearing loss.

An important cause of conductive hearing loss in the adult is **otosclerosis** (ō″tō·sklē·rō′sĭs). Otosclerosis affects about 4% of the total population. This is a disease that affects the bones of the middle ear. The process begins by resorption of bone around the otic capsule and progresses with deposition of new bone around the oval window and the footplate of the stapes, eventually preventing the transmission of sound waves to the inner ear. Treatment may involve the use of a hearing aid or restoring the mechanism of the stapes using microsurgery.

Perceptive hearing loss may be congenital or may occur following an infection of measles, mumps, or encephalitis. Some of these patients may be helped with hearing aids. Recent research is progressing in the development of **cochlear implants,** which are multiple tiny electrodes placed within the cochlea and attached to small transmitters placed behind the external ear. Cochlear implants have restored some hearing, but the sound they produce has not been perfected. Many new surgical techniques using the **operative microscope** have helped to restore hearing to the hearing-impaired patient.

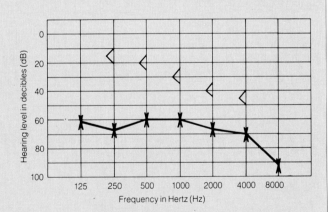

Example of an audiogram. This one shows moderately severe hearing loss. The top graph (<) shows bone conduction and the bottom graph shows air conduction.

Anatomy of the Eye (pp. 380–387)

The adult eye consists of the eyeball and accessory structures surrounding it.

The Eyeball The *eyeball* is a spherical structure about 2.5 cm in diameter. The *fibrous tunic* is the outermost layer of the eyeball and consists of the transparent *cornea* and the opaque *sclera*. The *vascular tunic* is sandwiched between the inner and outer layers of the eye and consists of the *choroid, ciliary body,* and *iris*.

The *retina* (*nervous tunic*) is the inner layer of the eyeball. It contains two types of photoreceptive cells: *rods* (for black and white vision) and *cones* (for color vision). The centrally located *fovea centralis* contains only cones and is the area of greatest visual acuity. The *optic nerve* leaves the retina at the *blind spot* and forms the visual pathway to the brain.

The *lens* is a solid proteinaceous structure that focuses light rays onto the retina. The *anterior* and *posterior chambers* of the *anterior cavity* lie in front of the lens. Both chambers of this cavity contain a watery fluid, the *aqueous humor*. The posterior cavity contains the more viscous *vitreous humor*.

Accessory Structures of the Eye *Eyebrows* are transverse bands of skin above the eye containing short, coarse hairs and sebaceous glands. *Eyelids* are folds of skin composed of several tissue layers that extend from the edges of the orbit. The *lacrimal gland* produces tears that moisten and clean the eye. In addition to water, tears contain salt, mucin, and the enzyme *lysozyme* that destroys bacteria.

Physiology of Vision (pp. 387–393)

Formation of the Image During formation of a visual image, the lens focuses light rays onto the retina. Once stimulated, the rods and cones initiate impulses that are relayed to the visual cerebral cortex, where they are interpreted.

Visual Accommodation *Visual accommodation* involves several adjustments in the eye resulting in a focused image. The thickness of the lens changes and the diameter of the pupil changes, resulting in light rays being *refracted* through a limited part of the lens onto the retina. Adjustment of the iris also allows different amounts of light to enter the eye.

Convergence and Binocular Vision The extrinsic eye muscles cause the eyes to *converge* on a near object, thereby producing *single binocular vision*. Autonomic centers in the brain monitor the amount of convergence and translate this information into distance readings, allowing *depth perception*.

Depth of Field Due to binocular vision, one must choose which specific objects will be in sharp focus, leaving the images of those objects both in front of and behind the plane of focus fuzzy in appearance. This produces *depth of field*

Biochemistry of Vision Formation of an image on the retina stimulates reactions involving several *visual pigments*. *Rhodopsin*, found in rods, splits into *retinal* and *opsin* when light strikes the molecule. This initiates an impulse in the cell that is relayed to the brain. Insufficient amounts of rhodopsin result in *night blindness*. Cones contain visual pigments that are sensitive to red, green, or blue light.

Visual Pathways to the Brain Visual impulses travel along the optic nerves and optic tracts to the lateral geniculate bodies of the thalamus. From here they are relayed to the visual cortex in the occipital lobes of the brain.

Anatomy of the Ear (pp. 393–398)

External Ear The *external ear* consists of the *auricle* and *external auditory meatus*. It terminates at the *tympanic membrane*.

Middle Ear The *middle ear*, a short tube connecting the external and internal ears, contains three small *ossicles*: the *malleus, incus,* and *stapes*. These bones transmit vibrations from the tympanic membrane to the internal ear.

Internal Ear The *internal ear* consists of the *membranous labyrinth* and *perilymphatic labyrinth*. The *semicircular canals* are a part of the perilymphatic labyrinth and are involved in detecting stimuli that indicate the body's position in space during movement. The *cochlea* is a snail-shaped system of three tubes—the *scala vestibuli, scala tympani,* and *cochlear duct*—which are modifications of the perilymphatic labyrinth. The *basilar membrane* extends through the cochlear duct. The *organ of Corti* on the basilar membrane bears the cells specialized for interpreting acoustic stimuli.

Innervation The ear is innervated by CN V, VII, X and XI (external ear); CN V, X and XI (tympanic membrane); CN V and VII (ear ossicles); and CN VIII (internal ear).

Physiology of Hearing (pp. 398–400)

Sound Sound is a form of energy transmitted by an alternating series of compression and decompression waves of air molecules. Sound waves vary in *amplitude* (intensity) and *frequency* (pitch).

Transmission of Sound to the Inner Ear Vibrations of the tympanic membrane caused by sound waves are amplified and transmitted through movements of the ear ossicles to the perilymphatic labyrinth within the cochlea.

Functions of the Cochlea Sensory cells on the organ of Corti are stimulated by movements of the endolymphatic fluid within the cochlear duct. It is theorized that one can discriminate among different pitches because different cells are stimulated by waves generated by different frequencies of sound. Higher intensity sounds produce greater stimulation of the sensory cells.

Sensory Pathways for Hearing The vestibulocochlear nerve carries impulses through the midbrain to the inferior colliculus and the medial geniculate body. Fiber tracts that pass through the thalamus terminate in the primary auditory area of the temporal lobe of the cerebral cortex.

Physiology of Equilibrium (pp. 400–402)

Static Equilibrium The sense of *static equilibrium* (balance while stationary) is controlled by *cilia* (kinocila) and *microvilli* (stereocilia) projecting from cells in the *macula* of each saccule and utricle. These hairs are stimulated by movements of a gelatinous fluid containing small deposits of calcium carbonate (*otoliths*). Different positions of the head pull the gelatinous mass, otoliths, and cellular extensions in corresponding directions, stimulating the hairs differently.

Dynamic Equilibrium The sense of *dynamic equilibrium* (maintaining balance while moving) centers in the three semicircular membranous canals. A group of sensory cells (*crista*) located within the canals is stimulated by movement of the endolymph that occurs when the head moves.

Nerve Pathways in Equilibrium Sensory impulses are carried along the vestibular branch of CN VIII to the medulla and cerebellum. Motor impulses issue along several nerves, including CN III, IV, VI, and XI as well as many spinal nerves, to skeletal muscles that maintain balance, equilibrium, and muscle tone.

SELF-TEST OF CHAPTER OBJECTIVES

True-False Questions
1. The "far point of vision" is an important distance from the eye because light rays that strike a normal eye from this distance or greater require effort on the part of the lens to focus them on the retina.
2. The auditory tube connects the external and middle ear.
3. The lens contains internal muscles that allow it to adjust its own thickness.
4. The lacrimal gland lies in the lacrimal sac that fills the lacrimal canal.
5. Gustatory cells in the taste buds act as first-order neurons in the taste nerve pathway.
6. The bulbar conjunctiva is a thin layer of mucous membrane that covers the cornea.
7. The pigmented layer of the retina is not sensitive to light.
8. The sense of taste, like the sense of smell, differentiates thousands of taste sensations.
9. The membranous labyrinth is surrounded by the perilymphatic labyrinth.
10. The vestibular branch of cranial nerve VIII carries impulses generated by sound waves.
11. The three ossicles of the middle ear lie between the cochlea and the semicircular canals.
12. The vibrations created by sound waves are detected by cells in the basilar membrane of the endolymphatic perilymph.
13. Cones of the eye utilize the light-sensitive pigment rhodopsin to interpret images in color.
14. The stapes rests within the oval window of the inner ear.
15. Pigments of the cones do not respond to low light levels.
16. Olfactory stimuli produce impulses that are carried to the brain through cranial nerve VII.
17. The retina is also known as the vascular tunic.
18. Vision is sharpest at the fovea centralis.
19. Absorption of light energy by a rhodopsin molecule causes it to split into opsin and retinaldehyde subunits.
20. Rhodopsin is a derivative of vitamiin A.

Matching Questions
Match the specific anatomic structure with its associated sense on the right.

21. sclera
22. olfactory tract
23. semicircular duct
24. cochlear duct
25. gustatory cell
26. malleus
27. cranial nerve VII
28. suspensory ligament
29. vestibule
30. cochlear nerve

a. taste
b. smell
c. vision
d. hearing
e. balance

Multiple-Choice Questions
31. Light rays enter the eye by passing through the
    a. cornea   c. vitreous humor
    b. iris   d. none of these
32. Which of the following is an example of a chemoreceptor?
    a. cone cell   c. ganglion neuron
    b. roll cell   d. gustatory cell
33. The rear of the tongue is especially sensitive to substances that taste
    a. sour   c. sweet
    b. bitter   d. salty
34. The opening between the eyelids that exposes the eyeball is called the
    a. palpebral conjunctiva   c. palpebral fissure
    b. lateral canthus   d. medial canthus
35. The area of greatest visual acuity on the retina is the
    a. macula lutea   c. optic disc
    b. fovea centralis d. ora serrata
36. The cellular structure in the inner ear that detects vibrational movements of the endolymph and initiates nerve impulses is the
    a. spinal lamina   c. tectorial membrane
    b. modiolus   d. basilar membrane
37. Which of the special senses is carried by multiple cranial nerves?
    a. taste   c. vision
    b. smell   d. hearing
38. The region in the semicircular ducts that detects changes in acceleration and other movements of the head is the
    a. crista   c. macula
    b. cupula   d. otolith
39. Which of the following visual disorders results from excess pressure within the eyeball?
    a. myopia   c. glaucoma
    b. hyperopia   d. cataracts

Essay Questions
40. Discuss olfactory adaptation. Are olfactory cells phasic or tonic in terms of adaptation?
41. How are the senses of taste and smell similar? How do they differ?
42. Compare how the nerve pathways for smell and taste sensations travel to reach the brain.
43. Compare the fibrous and nervous tunics of the eye.
44. Discuss the phenomenon of depth of field. What is its cause and how do we deal with it?
45. Trace the pathway that light rays travel from the cornea to the retina.
46. Discuss the reaction that occurs when light strikes a rhodopsin molecule.
47. Describe the processes responsible for color vision.
48. How does the cochlea differentiate among sound frequencies?
49. Trace the pathway that sound waves travel from the pinna to the cochlea.

Clinical Application Questions
50. Define anosmia and describe some possible causes.
51. Discuss the effects that diabetes mellitus has on vision.
52. How does chronic glaucoma differ from acute glaucoma?
53. Compare and contrast myopia and hyperopia.
54. Which of the senses would you be testing if you were using the Romberg sign and caloric testing?
55. Compare and contrast serous otitis media and otosclerosis.

17
The Endocrine System

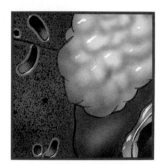

Complementing the nervous system, the endocrine system is the second of the two major systems responsible for controlling the body's response to internal and external stimuli. Unlike the nervous system, however, which communicates via action potentials sent through neurons, the endocrine system acts through chemicals that are distributed throughout the body by the circulatory system. Action potentials are fundamentally the same throughout the nervous system, and precision of control is achieved by sending impulses along precise pathways to specific destinations. In contrast, each endocrine gland releases specific chemicals that affect particular **target** tissues or organs. The fundamental difference in the way the two systems work enables the nervous system to respond rapidly to stimuli, whereas the endocrine system causes slower, generally longer lasting changes in affected tissues. In this chapter we will discuss

1. the chemical nature of these substances,
2. the organs and tissues that produce them,
3. their mechanisms of action, and
4. the mechanisms used to control their production.

endocrine: (Gr.) *endon*, within + *krinein*, to secrete
exocrine: (Gr.) *exo*, outside + *krinein*, to secrete

Hormones

After studying this section, you should be able to:

1. Describe the chemical nature of hormones, distinguishing water soluble from lipid soluble ones.
2. List the organs of the endocrine system.
3. Define a second messenger and describe how the combination of a hormone with a membrane receptor enables a second messenger to cause a change in a cell's chemical activity.
4. Describe how steroid hormones change a cell's chemical activity.
5. Explain what a negative feedback mechanism is and how it is used to control hormone production.

The specialized chemicals produced by the endocrine system are called **hormones.** Altogether, about 40 to 50 hormones have been identified in humans, with effects as diverse as causing an immediate increase in heart and breathing rates to causing the growth and development that occur over many years.

Chemical Composition of Hormones

Collectively, hormones are a diverse group of molecules, many of which are amino acid derivatives, short chains of amino acids (peptides), proteins, or glycoproteins (Table 17-1). *Epinephrine* and *thyroxine*, for example, are derived from the amino acid tyrosine. They are involved in preparing the body to cope with stress and in the control of metabolic rate. Peptide hormones may contain as few as nine or as many as hundreds of amino acids and control many aspects of growth, development, and metabolism.

Protein hormones, one major category of hormones, are produced by the same mechanisms used for the production of other proteins. That is, RNA is synthesized in the cell nucleus using a portion of DNA as a template. The RNA is then modified into a messenger RNA molecule, which moves to the cytoplasm where it specifies the amino acid sequence of the hormone. Processing of the protein may also be necessary to produce the functional hormone.

A second major category of hormones consists of steroids, molecules that contain 18 to 21 carbon atoms and associated oxygen and hydrogen atoms. There are five major classes of steroid hormones, all of which are derived from cholesterol. These include *progestogens*, which prepare the uterus for the growth of a developing embryo; *glucocorticoids*, which regulate the metabolism of carbohydrates and the degradation of proteins and fats; *mineral corticoids*, which regulate sodium, chloride, and bicarbonate removal by the kidneys; and *androgens* and *estrogens*, which control the development of male and female secondary sexual characteristics.

In general, water-soluble hormones are responsible for reactions that occur relatively rapidly, perhaps within a few minutes of stimulation. The hormones are broken down rapidly by enzymes within serum, so they must perform their function quickly. This provides a way to control the concentration of these hormones, since they can be removed quickly after stimuli that cause their release have passed. In contrast, steroid hormones are lipid-soluble hormones that must be carried in the serum by specialized carrier molecules. They are not degraded quickly and may last for days following a stimulus. Steroid hormones are responsible for reactions that occur over longer periods of time, perhaps days or weeks. For example, hormones that most directly regulate growth and development are steroids.

Concentration of Hormones

A characteristic feature of a hormone is that an extremely small amount is required to stimulate a response. **Thyroxine** (thī·rŏks′ĭn), a hormone that increases the rate of many metabolic reactions, is usually present in a concentration of 0.004 to 0.011 mg (4 to 11 millionths of a gram) per 100 mL of blood. Of this, all but about 0.000001 mg (1 billionth of a gram) per 100 mL is bound to special carrier protein. The level of the sex hormone testosterone is only about 0.0006 mg per 100 mL of blood in adult males; gastrin, a hormone involved in digestion, is present in blood at only 0.000003 mg per 100 mL. Even in such small amounts, the level of any of these hormones is quite critical, and very slight changes in concentration can lead to dramatic changes in physiology.

Source of Hormones

Hormones are synthesized in organs called **endocrine** (ĕn′dō·krĭn) **glands** and are released from these glands directly into the blood. A second category of glands, **exocrine** (eks′ō·krĭn) **glands** do not release their products directly into the blood. Instead, their products are released onto the surface of the gland itself or into tubular **ducts** that lead a short distance to the area where the secretions perform their function. For this reason, exocrine glands are often called **ducted glands.** (Endocrine glands are referred to as **ductless glands** because they lack such ducts.) Substances released by exocrine glands are not hormones; saliva, sweat, and mother's milk are examples of exocrine gland products. A third category of glands perform both endocrine and exocrine functions, possessing some tissues that secrete

heterocrine: (Gr.) *heteros*, other + *krinein*, to secrete

Table 17-1 CHEMICAL CLASSIFICATION OF HORMONES

Chemical Class	Hormone	Major Source
Amines	Norepinephrine	CNS, adrenal medulla
	Epinephrine	Adrenal medulla
	Melatonin	Pineal gland
Iodinated thyronines	Thyroxine (T_4)	Thyroid
	Triiodothyronine (T_3)	Peripheral tissues (from T_4)
Small peptides	Antidiuretic hormone (ADH)	Posterior pituitary
	Oxytocin	Posterior pituitary
	Melanocyte-stimulating hormone (MSH)	Anterior pituitary
	Thyrotrophin-releasing hormone (TRH)	Hypothalamus, CNS
	Gonadotrophin-releasing hormone (GnRH)	Hypothalamus, CNS
	Somatostatin	Hypothalamus, pancreas
	Angiotensin	Blood
Proteins	Insulin	Pancreas
	Glucagon	Pancreas
	Human growth hormone (HGH)	Anterior pituitary
	Placental lactogenic hormone	Placenta
	Prolactin	Anterior pituitary
	Parathyroid hormone (PTH)	Parathyroid glands
	Calcitonin	Thyroid (C cells)
	Adrenocorticotrophic hormone (ACTH)	Anterior pituitary
	Secretin	Small intestine
	Cholecystokinin	Small intestine
	Gastrin	Stomach
	Gastric-inhibitory peptide	Small intestine
Glycoproteins	Follicle-stimulating hormone (FSH)	Anterior pituitary
	Luteinizing hormone (LH)	Anterior pituitary
	Chorionic gonadotrophin	Placenta
	Thyroid-stimulating hormone (TSH)	Anterior pituitary
Steroids	Estrogen	Ovary, placenta
	Progesterone	Corpus luteum, placenta
	Testosterone	Testis
	Dihydrotestosterone	Tissues sensitive to testosterone
	Glucocorticoids	Adrenal cortex
	Aldosterone	Adrenal cortex
	Vitamin D products	Liver, kidney

Source: Adapted from J. Tepperman, *Metabolic and Endocrine Physiology*, 4th ed. (Chicago: Year Book Medical Publishers, Inc., 1980), 7.

hormones to the blood and other tissues that release their products to specific neighboring regions. Glands in this category are called **heterocrine** (hĕt´ĕr·ō·krīn) **glands.** Figure 17-1 shows the location in the body of the major endocrine glands.

Mechanism of Action

There are two primary methods by which hormones cause target organs to react. Water soluble hormones, that is, those other than steroid or thyroid hormones, are unable to diffuse through the lipid portion of a cell membrane. Instead, they attach to receptor molecules in the membrane, which initiates reactions that activate cytoplasmic enzymes. Steroid hormones, in contrast, activate (or inhibit) specific genes. All cells do not respond to the same hormones in the same way.

This is because every cell contains specific receptors that determine the precise response of the cell to a hormone. When a cell does not respond to a hormone, it is because the cell lacks receptors for that hormone.

Adenylate Cyclase

In most cases, water-soluble hormones act by combining with a receptor to form a complex that catalyzes the addition of guanosine triphosphate (GTP) to a membrane protein called **G protein.** G protein has a molecule of guanosine diphosphate (GDP) bound to it which is displaced in the reaction. The resulting G protein–GTP complex is an enzyme that activates another protein called **adenylate cyclase** (ă·dĕn·ĭ·lāt´ sī´klās). The G protein–GTP complex also catalyzes the removal of a phosphate group from itself, causing

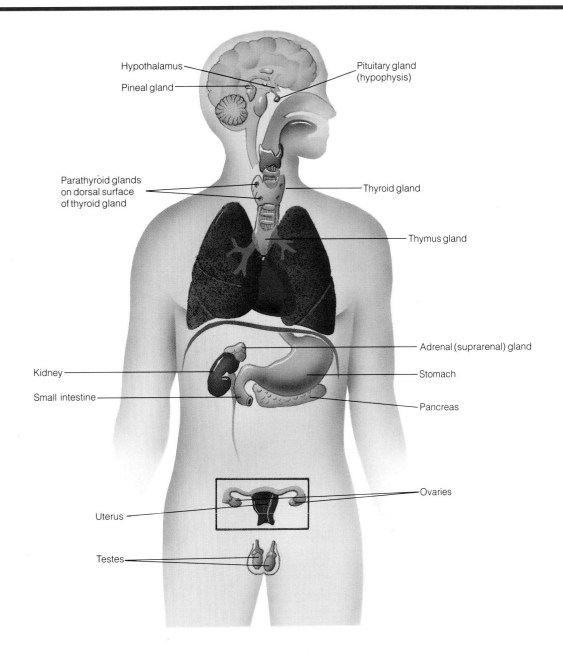

Figure 17-1 The endocrine glands and associated structures.

it to become a G protein–GDP complex that immediately loses the ability to activate adenylate cyclase.

When adenylate cyclase is activated, it catalyzes the removal of two phosphate groups from adenosine triphosphate and rearranges the resulting adenosine monophosphate into a cyclic (or ringed) molecule called **cyclic AMP (cAMP)**. Cyclic AMP diffuses into the cytoplasm and combines with proteins called **protein kinases**. Once a cAMP molecule combines with a protein kinase molecule, the protein kinase acquires the ability to catalyze the addition of phosphate groups from ATP to still other inactive enzymes in the cytoplasm. This is perhaps the key step in the reaction sequence, because the addition of phosphate groups to these proteins causes them to become enzymes that catalyze specific chemical reactions (Figure 17-2).

More than a dozen hormones are known that induce their effect through a cAMP mechanism, and each hormone induces characteristic reactions in particular kinds of cells, bypassing all other cells with which it comes into contact. Whether or not a cell responds to the hormone depends on

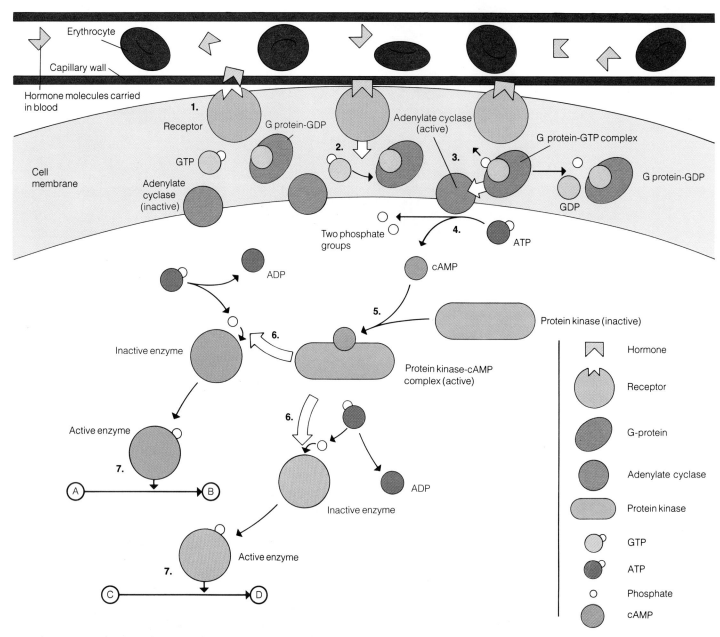

Figure 17-2 The cAMP-induced mechanism of hormone action. (1) Combination of a hormone with a receptor activates the receptor. **(2)** This enables the receptor to catalyze the addition of GTP to G-protein. **(3)** The G protein–GTP complex then catalyzes the activation of adenylate cyclase. **(4)** Adenylate cyclase next catalyzes the conversion of ATP to cAMP. **(5)** cAMP combines with inactive protein kinases and activates them. **(6)** Activated protein kinase molecules catalyze the addition of phosphate to inactive enzymes, activating them. **(7)** Activated enzymes catalyze new reactions in the cell that lead to the hormonal response.

the receptor molecules in its membrane. If appropriate receptors are present, the response that follows depends on which proteins are present in the cell's cytoplasm that can be activated. Water-soluble hormones are thus only indirectly involved in the chemical changes they induce in a cell, cAMP being more directly involved in inducing the response than the hormone itself. For this reason, cAMP has been referred to as a **second messenger** in those cells that respond to a hormone through a cAMP-mediated mechanism.

Steroid Hormones

Steroid hormones are like water-soluble hormones in that they combine with specific receptor molecules. But in the

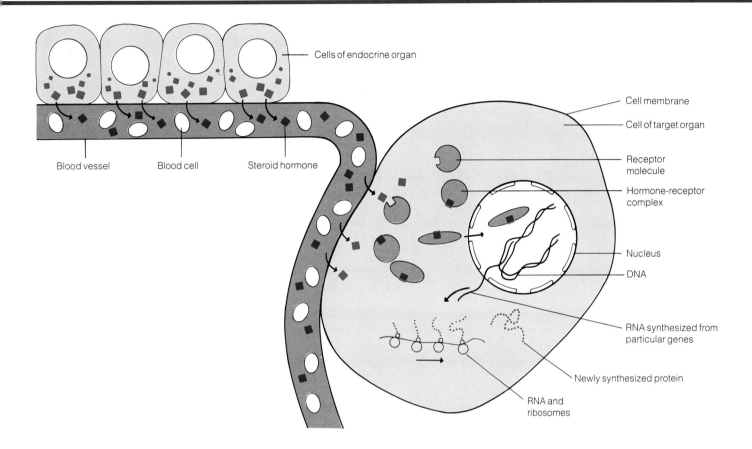

Figure 17-3 Activation of genes by steroid hormones. Steroid hormones pass through the cell membrane and combine with receptor molecules in the cytoplasm. The hormone-receptor complex migrates into the nucleus where it induces particular genes to become active.

case of steroid hormones, the receptor molecules are in the cytoplasm rather than in the membrane. Steroids diffuse easily through the lipid portion of the membrane and enter the cytoplasm. Combination with a receptor protein causes the protein to change its shape, a phenomenon referred to as **receptor activation** (Figure 17-3). The hormone-receptor complex passes from the cytoplasm into the nucleus through pores in the nuclear membrane. Once in the nucleus, the complex combines with particular portions of the DNA, or with proteins in the chromatin, and activates specific genes.

Regulation of Hormone Activity

Implicit in the role of hormones as regulators of metabolism is the need to control the hormones themselves. That is, there must be mechanisms to regulate the concentration or activity of those hormones that are produced in response to a stimulus. This control is accomplished with nerve impulses, changes in concentration of nutrients and ions in the fluid that bathes an endocrine gland, or by other hormones for which the endocrine gland is the target (Table 17-2).

Synthesis or secretion of a hormone is usually controlled by the effect that the hormone has on its target tissue. For example, when the concentration of Ca^{2+} in blood serum falls, the parathyroid glands increase their production of parathyroid hormone, a hormone that causes the level of serum Ca^{2+} to rise. As the concentration of Ca^{2+} rises, the glands are increasingly inhibited in their production of parathyroid hormone. Consequently, the more Ca^{2+} in the serum, the less parathyroid hormone is produced. With decreased production, parathyroid hormone in the serum declines, which causes Ca^{2+} concentration to decline. The concentration of parathyroid hormone and Ca^{2+} alternately rise and fall, each dependent on the other for control. This type of regulatory mechanism, where the product of the hormone's action inhibits the production of more hormone, is called **negative feedback** (Figure 17-4).

pituitary: (L.) *pituita*, phlegm
adenohypophysis: (Gr.) *aden*, gland + *hypo*, under + *physis*, growth

Table 17-2 CONTROLLING FACTORS FOR HORMONE SECRETION

Primary Controlling Factor	Hormones
Metabolites, nerve impulses, or hormones	Oxytocin ADH Hypothalamic-releasing hormones
Hypothalamic-releasing hormones	Growth hormone Thyroid-stimulating hormone ACTH Gonadotrophins Prolactin
Anterior pituitary hormones	Thyroid hormone Cortisol Estrogen Progesterone Testosterone
Impulses from autonomic neurons	Catecholamines Renin Insulin Glucagon Gastrointestinal hormones
Ion or nutrient concentration in blood Ca^{2+} Glucose Glucose Na Ca^{2+}	 Parathyroid hormone Insulin Glucagon Aldosterone Calcitonin

Source: Adapted from D. S. Luciano, A. J. Vander, and J. H. Sherman, *Human Anatomy and Physiology*, 2nd ed. (New York: McGraw-Hill, 1983), 391.

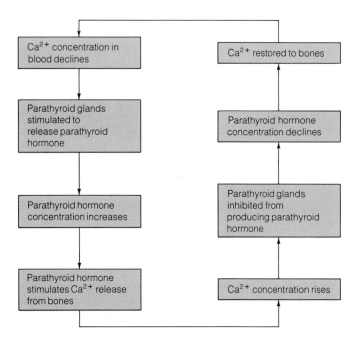

Figure 17-4 A negative feedback mechanism for the control of hormone release. This example demonstrates how the concentrations of Ca^{2+} and parathyroid hormone in the blood depend on one another.

Hypothalamus and Pituitary Gland

After studying this section, you should be able to:

1. Describe the anatomical relationship of the hypothalamus to the pituitary gland.
2. Diagram the pituitary gland, identifying the anterior and posterior lobes.
3. Name two hormones released by the posterior lobe of the pituitary gland and give their sources and functions.
4. Describe the mechanism by which the hypothalamus controls the release of hormones by the anterior pituitary gland.
5. List four anterior pituitary lobe hormones that control the release of hormones by other endocrine glands.
6. List three anterior pituitary lobe hormones that control nonendocrine tissue.

One of the primary regions of the brain involved in the control of hormone production is the **hypothalamus** (hī″pō·thăl′ă·mŭs), which lies just below the thalamus where it forms the floor and lower walls of the third ventricle (Figure 17-5). Hanging from the underside of the hypothalamus is another structure important to hormone production called the **pituitary** (pĭ·tū′ĭ·tār″ē) **gland** or **hypophysis** (hī·pŏf′ĭ·sĭs). The pituitary gland is connected to the hypothalamus by the **hypophyseal stalk**, or **infundibulum** (ĭn″fŭn·dĭb′ū·lŭm), and rests in the hypophyseal fossa (sella turcica), a depression of the sphenoid bone.

The pituitary gland is about 1 cm in diameter and consists of two major portions: a **posterior lobe** (also called the **posterior pituitary, neurohypophysis,** or **pars nervosa**), which is continuous with the hypophyseal stalk, and an **anterior lobe** (also called the **anterior pituitary** or **pars distalis,** părz dĭs·tā′lĭs). The superior end of the anterior lobe becomes the **pars tuberalis** (parz tū″bĕr·ā′lĭs). In some animals there is a smaller third lobe called the **pars intermedia** between the posterior and anterior lobes, but in humans it is reduced to a thin layer of epithelial cells. The anterior lobe, pars tuberalis, and pars intermedia collectively comprise the **adenohypophysis.**

The hypothalamus responds to information regarding many aspects of the state of the body. This information is received either directly, by hypothalamic cells sensitive to the solute concentration of the fluid that surrounds them, or indirectly, as signals sent through sensory nerves from other parts of the body. In response to these stimuli, the hypothalamus sends nervous signals that control such diverse and general phenomena as mental alertness, heart rate, blood pressure, hunger, stomach activity, body temperature, and

oxytocin: (Gr.) *oxys*, sharp + *tokos*, childbirth
neurophysin: (Gr.) *neuron*, nerve + *physis*, nature

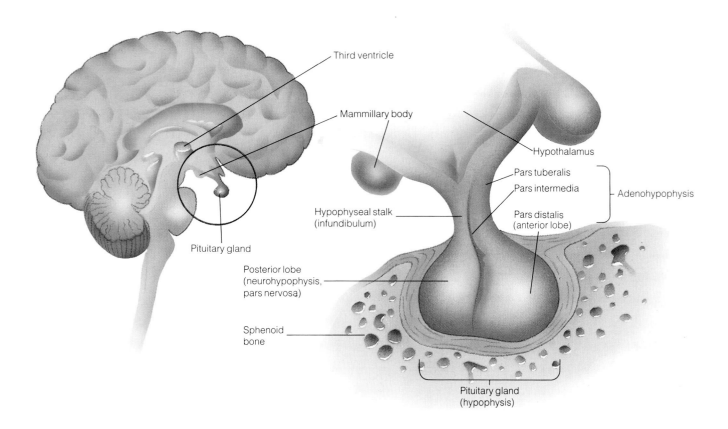

Figure 17-5 Structure of the hypothalamus and pituitary gland. The drawing of the brain on the left shows the location of the detailed drawing.

water and mineral balance. Among the many regulatory functions performed by the hypothalamus is control of hormone release by the pituitary gland.

Hormones of the Posterior Pituitary Gland

Passing through the hypophyseal stalk are close to 100,000 individual nerve fibers, each with its cell body located in the hypothalamus and its axon terminating in the posterior lobe of the pituitary gland. Two hormones, **antidiuretic** (ăn″tĭ·dī′ū·rĕt′ĭk) **hormone (ADH)** and **oxytocin** (ŏk″sē·tō′sĭn), are synthesized in these neurons and flow along the axons from the hypothalamus, through the hypophyseal stalk, and into the posterior lobe. They travel bound to specific binding proteins called **neurophysins** (nū″rō·fī′sĭns) and appear as tiny granules in the neurons. As the granules reach an axon terminal, they accumulate and form larger masses of granules called **Herring bodies.** Herring bodies are stored in the terminal bulbs of the axon until an action potential causes the hormone to be released into neighboring capillaries.

Neurons of two distinct types synthesize and carry the hormones through the hypophyseal stalk. The cell bodies of these neurons are located in regions of the hypothalamus called **paraventricular** (păr″ă·věn·trĭk′ū·lăr) **nuclei** and **supraoptic** (sū″pră·ŏp′tĭk) **nuclei;** oxytocin is produced primarily in cell bodies of paraventricular nuclei and ADH is produced primarily in cell bodies of supraoptic nuclei (Figure 17-6). Neurophysins associated with each of the hormones also appear to be of two different types, each specific for the hormone with which it forms a granule.

Antidiuretic Hormone

Diuresis (dī″ū·rē′sĭs) means excessive discharge of urine. As its name suggests, antidiuretic hormone is a hormone that opposes production of excessive urine. ADH causes the kidney to recover water from urine, returning water to the blood that would otherwise be excreted. (See Chapter 25 for more on diuresis.)

ADH is a small peptide, containing only nine amino acids, that operates through a cyclic AMP mechanism. It is released continuously by the posterior lobe of the pituitary

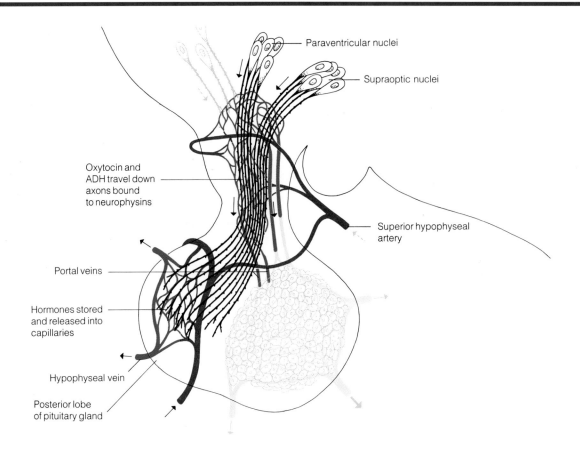

Figure 17-6 Posterior lobe production of oxytocin and antidiuretic hormone. Oxytocin and antidiuretic hormone (ADH) are produced in neurons that pass through the hypophyseal stalk. The cell bodies of the neurons are located in paraventricular and supraoptic nuclei located in the hypothalamus. Oxytocin is primarily made in paraventricular neurons, and antidiuretic hormone is primarily produced by the supraoptic neurons.

gland and is continuously destroyed by the kidneys and liver. As a result, an equilibrium exists and ADH is maintained in a more or less constant concentration.

Changes in ADH concentration result from a change in the volume of body fluids, as might occur when drinking large amounts of water or from dehydration or hemorrhage. Changes in fluid volume cause changes in osmotic properties of the fluid that bathes the supraoptic neurons of the hypothalamus. For example, when the solute concentration increases, the supraoptic neurons send signals more rapidly to the posterior pituitary gland and stimulate the gland to release more ADH. When water is taken in, the solute concentration decreases, and the supraoptic neurons send signals at a slower rate. When this happens, the posterior pituitary reduces its release of ADH, thereby reducing the stimulus for recovering water. ADH release is thus controlled by a feedback mechanism in which increased water volume reduces the production of ADH and allows increased water excretion, and vice versa. A rare form of diabetes, diabetes insipidus, results from insufficient ADH (see Clinical Note: Diabetes Insipidus).

Oxytocin

Oxytocin is similar to ADH in structure, also consisting of a peptide of only nine amino acids. All but two of the amino acids are identical to those of ADH. Considering the similarity in structure between the two hormones, it is not surprising that oxytocin has ADH-like activity and can cause water retention. Its main functions, however, are to stimulate the ejection of milk from mammary glands and to stimulate contraction of the uterus during labor (see Chapter 28).

Oxytocin has the unusual property of being controlled by a positive feedback mechanism. When a baby suckles, the suckling and the visual and auditory stimuli associated

diabetes: (Gr.) *diabetes*, a passing through
trophic: (Gr.) *trophe*, nourishment

> **CLINICAL NOTE**
>
> **DIABETES INSIPIDUS**
>
> **Diabetes insipidus** is an example of a relatively rare disease that results from a deficiency of antidiuretic hormone (ADH). It is characterized by the excretion of large quantities of dilute urine, accompanied by continuous and pronounced thirst. Two forms of the disease are known that differ in their sensitivity to ADH. In **ADH-sensitive diabetes insipidus,** there are insufficient neurons in the hypothalamus, perhaps as a result of damage by tumors, injury, or infection. The posterior pituitary gland's ability to produce ADH is reduced in this form of the disease, but the symptoms can be treated by administering ADH to the patient. In **ADH-resistant diabetes insipidus,** the kidneys have lost their ability to respond to the hormone, often as a result of infection or due to a genetic disorder. In this form of the disease, added ADH has no effect because the cells that contain receptors for the hormone have been lost or damaged. In the case of a genetically based disorder, symptoms usually appear soon after birth and can result in permanent brain damage due to dehydration and electrolyte imbalance. Treatment consists of drinking sufficient water to compensate for the excessive amounts of urine that are excreted.

with the nursing infant cause neural signals to be sent to the mother's brain. Some of these signals are carried to the hypothalamus where they induce the release of oxytocin from the posterior lobe. The hormone is then carried by the blood to the mammary glands, where it stimulates the smooth muscles in the glands to contract and force out milk that has already been produced and stored by the gland. As milk begins to flow and the baby continues to suckle, additional nervous stimuli are sent from the nipple to the hypothalamus, where they stimulate the release of more hormone. The additional hormone stimulates the gland further and causes release of even more milk. Thus, the more a mother nurses, the more milk she releases.

Oxytocin also operates by positive feedback in the induction of labor. As pregnancy progesses, the number of oxytocin receptors in cells of the uterine wall increases, making the uterus increasingly sensitive to oxytocin in the blood. With time there are enough receptors for oxytocin to stimulate the smooth muscle of the uterus to contract. During labor the contractions become so frequent and so strong that the fetus is forced downward, toward the cervix of the uterus. This stretches the cervix, which in turn causes nervous impulses to be sent to the hypothalamus, where they stimulate the release of still more oxytocin. The additional oxytocin stimulates the uterus even more, and the contractions become stronger and stronger until the fetus is forced through the cervix and the baby is born (see Chapter 28).

Hormones of the Anterior Pituitary Gland

Anterior pituitary gland hormones are stored in cytoplasmic granules in many of the epithelial cells of the gland. The granules stain brightly, and cells that contain them are referred to as **chromophils.** Five different kinds of chromophils can be discerned based on their staining properties and distribution in the anterior pituitary tissue. Each type of cell produces one or two distinct hormones; thus, **somatotrophs** produce *growth hormone* (*GH*), **mammotrophs** produce *prolactin,* **thyrotrophs** produce *thyroid-stimulating hormone* (*TSH*), **gonadotrophs** produce *follicle-stimulating hormone* (*FSH*) and *luteinizing hormone* (*LH*), and **corticotrophs** produce *adrenocorticotrophic hormone* (*ACTH*) and *melanocyte-stimulating hormone* (*MSH*). (All of these hormones are discussed in depth in the following sections.) ACTH and MSH are derived from a large precursor protein called **pro-opiomelanocortin (POMC),** from which certain other peptides are also produced. Among these are β-endorphin and β-lipotrophin. β-Endorphin is a naturally occurring opiate of the CNS (see Chapter 15). The function of β-lipotrophin has yet to be determined.

Anterior pituitary gland hormones fall into two functional categories: those that stimulate endocrine glands to produce hormones and those that have a direct effect on non-endocrine tissues. Anterior pituitary hormones that stimulate endocrine glands are collectively called **trophic hormones.** These include TSH, ACTH, LH, and FSH. Anterior pituitary gland hormones that stimulate nonendocrine tissues include GH, prolactin, and MSH.

The hypothalamus controls the release of anterior pituitary hormones by a different mechanism than the strictly neurological one used to control the release of posterior pituitary hormones. In this case, control is accomplished by **releasing** and **release-inhibiting hormones** secreted by the hypothalamus and carried in the blood the short distance to the anterior pituitary lobe.

The blood that carries these hormones flows from the internal carotid artery to a capillary network in the hypophyseal stalk called the **primary capillary plexus.** Neurons with cell bodies that lie in the hypothalamus terminate in close proximity to this capillary network. These neurons synthesize the hypothalamic hormones and release them into the capillaries. After passing through the capillary network, the blood, now carrying the hypothalamic hormones, passes through **portal vessels** to a second capillary network (the **secondary capillary plexus**) in the anterior pituitary gland. (Portal vessels are blood vessels that lie between

somatostatin:	(Gr.) *soma*, body + *stasis*, a standing
prolactin:	(L.) *pro*, before + *lac*, milk
lactogenic:	(L.) *lac*, milk + (Gr.) *gennan*, to produce

separate capillary networks.) The hormones then diffuse from this second capillary bed into the anterior pituitary gland, where they control the synthesis and release of anterior pituitary hormones. Blood leaves the second capillary network carrying the hormones (Figure 17-7).

Several releasing hormones or release-inhibiting hormones are produced by the hypothalamus and are named for the specific hormones they control. Thus, **thyrotrophin-releasing hormone (TRH)** causes the release of thyroid-stimulating hormone, and **corticotrophin-releasing factor (CRF)** stimulates the release of adrenocorticotrophic hormone. **Growth hormone release inhibitory hormone (GHRIH)** (also called **somatostatin**) inhibits the release of growth hormone. Gonadotrophin release is regulated by a hypothalamic hormone called **gonadotrophin-releasing hormone (GnRH)**, and prolactin release is regulated by **prolactin-releasing factors (PRF)**.

Prolactin

The hormone **prolactin** stimulates the synthesis of milk (rather than its secretion, which is controlled by oxytocin) and also promotes breast development. Because it stimulates milk production, it is also called **lactogenic** (lăk″tō·jĕn´ĭk) **hormone.** It has a variety of effects on other tissues, including development and activity of the uterus, growth in general, lipid metabolism, and blood production.

Prolactin production is controlled by two prolactin-releasing factors, one called **prolactin-releasing hormone (PRH)** and the other **prolactin-inhibiting hormone (PIH)**, which act by way of a feedback mechanism. Thus, suckling causes nerve impulses to be sent to the hypothalamus, which, in turn, releases PRH to the portal system. The anterior pituitary gland responds to PRH by releasing prolactin, which stimulates the production and storage of more milk by the breasts. Unlike control of other pituitary hormones, however, prolactin also appears to be under inhibitory control. That is, PIH inhibits prolactin release, and stimulation of

Figure 17-7 Hypothalamic control of anterior pituitary hormone production. Hypothalamic hormones that control the release of anterior pituitary gland hormones are released to a capillary network in the hypophyseal stalk, from which they are carried by portal vessels to capillaries in the anterior pituitary gland. Once delivered to the anterior lobe, they regulate release of hormones by the lobe.

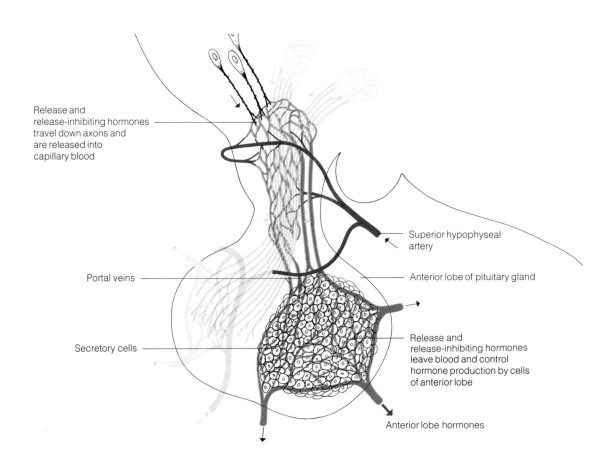

somatomedin: (Gr.) *soma*, body + (L.) *mediari*, to mediate
melanocyte: (Gr.) *melanos*, black + *kytos*, cell
acromegaly: (Gr.) *akron*, extremity + *megas*, big

prolactin production probably also involves reduction of the inhibitory effect of PIH.

Growth Hormone

Human growth hormone (HGH), like prolactin, has numerous effects. It is important in regulating the growth of the skeleton, connective tissue, muscle, and internal organs. These effects are probably caused by the stimulating effect the hormone has on the synthesis of proteins and nucleic acids in general. When injected into the body, growth hormone reverses the effect of insulin, thereby interfering with the ability of the body to use glucose. At least some of the effects of the hormone are indirect, in that they are caused by a hormone called **somatomedin** (sō″mă·tō·mēd′ĭn). Somatomedin is made in the liver and kidneys, either directly from HGH or as a result of the action of HGH.

HGH is very similar in structure to prolactin. Both are proteins with nearly 200 amino acids, close to 50% of which are identical. Many of the biological effects of the two hormones overlap, probably due to their similarity in structure.

Irregularities in the production of HGH can result in growth abnormalities such as dwarfism and gigantism (see Clinical Note: Growth Abnormalities).

> **CLINICAL NOTE**
>
> **GROWTH ABNORMALITIES**
>
> Excess or insufficient production of human growth hormone (HGH) in the pituitary gland can lead to abnormal growth patterns. Deficient production of HGH is one cause of **pituitary dwarfism**, a condition resulting from lack of bone deposition and lengthwise bone growth at the epiphyseal plates. These people have a very short stature, but normal body proportions are maintained.
>
> In contrast, excess production of HGH leads to gigantism and acromegaly. Usually due to a tumor of the pituitary gland, **gigantism** is the result of excessive bone growth prior to puberty. Bones of the arms and legs are particularly affected and can cause an afflicted person to grow to 8 ft or more in height. Internal organs are also larger than normal in these people. After the onset of puberty, the epiphyses have fused and additional growth in height is impossible, but the periosteum surrounding the bones continues to thicken, as does much of the skin and subcutaneous tissues. The result is massive enlargement of the bones of the extremities (particularly of the hands and feet) and also of the mandible and forehead, giving the person a characteristic squared-off masculine appearance. This condition is referred to as **acromegaly** (ăk″rō·mĕg′ă·lē).

Melanocyte-Stimulating Hormone

Melanocyte- (mĕl′ăn·ō·sīt) **stimulating hormone (MSH)** exists in two forms: α-MSH and β-MSH. Both are peptides, α-MSH containing 13 amino acids and β-MSH containing 22 amino acids. The amino acid sequence of α-MSH is identical to the first 13 amino acids in adrenocorticotrophic hormone (ACTH), another anterior pituitary hormone. β-MSH is not so similar in structure, but nevertheless has a sequence of 7 of its 22 amino acids in common with α-MSH and ACTH. The function of MSH in humans is unknown, but based on its ability to darken the skin of frogs and other amphibia, it is possible that it plays a role in causing the development of pigmentation in humans.

Adrenocorticotrophic Hormone (ACTH)

Adrenocorticotrophic (ă·drē″nō·kŏr″tĭ·cō·trŏf′ik) **hormone** (also called **adrenocorticotrophin**), a 39-amino acid protein, controls the production of corticosteroid hormones by the adrenal glands. Cortisteroids control the level of various mineral ions in body fluid and regulate the metabolism of fat, protein, and carbohydrate (see section on Adrenal Glands).

Blood concentrations of ACTH are controlled by CRF, the production of which depends partly on the levels of steroids in the blood. This control is accomplished through a negative feedback mechanism: elevated levels of the steroids inhibit CRF production and decrease the stimulus for ACTH production. When the concentration of ACTH in the blood declines, there is a resulting decrease in steroid concentration and CRF production rises. Increased CRF causes more ACTH to be released, which stimulates the adrenal cortex to release more corticosteroids.

Thyroid-Stimulating Hormone (TSH)

Thyroid-stimulating hormone (TSH), also called **thyrotrophin**, is a glycoprotein. This hormone is responsible for regulating the activity of the thyroid gland, stimulating it to release thyroid hormones. In the absence of TSH, the thyroid gland atrophies and a decrease in metabolic activity of the body occurs due to decreased thyroid function. TSH enables the thyroid gland to absorb iodine from the blood more effectively and to secrete *thyroxines*, hormones that help control metabolic rate and the use of carbohydrates, lipids, and protein (see section on Thyroid and Parathyroid Glands).

Gonadotrophins

Hormones that stimulate the production of sex hormones by ovaries and testes are called **gonadotrophins** (gō·năd·ō·trō′fĭns). They include **follicle-stimulating hormone (FSH)** and **luteinizing hormone (LH),** both of which,

thyroxine: (Gr.) *thyreos*, shield

like TSH, are glycoproteins. The protein part of each hormone contains two distinct polypeptides designated α and β. The α subunits are identical in FSH and LH and the β subunits are different. Consequently, differences in the biological properties of the two hormones must be due to differences in the β subunits. FSH and LH are named for their effects in the female, but both are present and active in the male as well.

In females FSH causes the ovaries to develop at puberty and subsequently to initiate periodic development of **follicles,** saclike structures in the ovary that contain the eggs. In males FSH promotes the production of sperm.

LH causes the continued development of a follicle, which leads eventually to follicle rupture and release of an egg. This process is called **ovulation.** After an egg is released, LH causes a large number of cells to be produced and to accumulate within the cavity of the follicle. The result is an endocrine structure called the **corpus luteum** (kōr'pŭs lū'tē·ŭm). During their development the follicle and corpus luteum release the sex hormones estrogen and progesterone. In males LH stimulates the proliferation of certain cells in the testes that secrete the male sex hormone testosterone (see section on Gonads).

Production of gonadotrophins is regulated by the hypothalamic hormone GnRH. GnRH is a peptide that contains 10 amino acids. Its production is regulated in a negative feedback mechanism by the levels of estrogen and progesterone in women and by the level of testosterone in men.

Thyroid and Parathyroid Glands

After studying this section, you should be able to:

1. Describe the organization and structure of the thyroid gland.
2. Describe how thyroid hormones are produced in the colloid of a thyroid gland.
3. List several functions performed by thyroid hormones.
4. Describe the effects of under- and overproduction of thyroid hormone.
5. Cite the source of production of calcitonin and explain the function of that hormone.
6. Cite the source of parathyroid hormone and compare the function of parathyroid hormone with that of calcitonin.

Anatomy of the Thyroid Gland

The thyroid gland is a mass of spongy tissue that consists of two lobes lying on either side of the trachea and larynx (Figure 17-8a). The lobes are connected by a region called the **isthmus,** which lies on the anterior side of the trachea. The gland is surrounded by a fibrous capsule, extensions of which reach down into the gland separating it into a number of regions called **lobules.** Within the lobules are found the functional units of the thyroid gland, spherical sacs of tissue called **thyroid follicles.**

Each follicle is a single spherical layer of epithelial cells surrounding a lumen filled with a gelatinous mass of protein called **colloid.** The outer surface of the follicle is in contact with a basement membrane that surrounds the cells of the follicle like a tightly fitting bag (Figure 17-8b). Other cells are also occasionally found within the basement membrane but not in contact with the colloid; these are **parafollicular** (păr"ă·fōl·lĭk'ū·lăr) **cells** or **C cells.**

The thyroid gland receives its blood through the thyroid arteries, which branch from the carotid and subclavian arteries. The arteries continue to branch until they form a capillary bed that permeates the connective tissue that separates the lobules and follicles from one another. Because of the elaborate branching of the blood vessels, each follicle is in close proximity to an abundant blood supply.

The thyroid gland is also supplied with numerous small nerve bundles that pass through the connective tissue, and many nerve fibers form neuromuscular junctions with smooth muscle in the arterioles of the thyroid gland. Lymphatic vessels also enter the thyroid gland and drain it of lymph.

Thyroid Hormones

The epithelial cells surrounding the colloid synthesize a large glycoprotein, **thyroglobulin** (thī"rō·glŏb'ū·lĭn), and secrete it into the colloid. At the same time, ionic iodine (I^-, also called **iodide**), is oxidized to atomic iodine and secreted into the colloid. In the colloid, iodine is added to tyrosine, an amino acid present in thyroglobulin. Tyrosine is then modified further to form two compounds called **triiodothyronine** (trī·ī"ō·dō·thī'rō·nēn), or T_3, and **thyroxine** (thī·rŏks'ĭn), or T_4, both of which remain as part of the thyroglobulin molecule. T_3 and T_4 are named for the number of iodine atoms in the subunit, T_3 containing three atoms and T_4 containing four.

Under the influence of TSH, iodinated thyroglobulin molecules are reabsorbed by cells that surround the colloid. The molecules are degraded within the cells, causing the release of T_3 and T_4 into the bloodstream, where they adsorb to proteins in the blood. They are carried in this bound form to other parts of the body. Of the two, T_3 is more active as a thyroid hormone. T_4 is more concentrated in the blood, however, and a significant portion of T_4 is converted to T_3 in target tissues by removal of an iodine atom.

cretinism: (Fr.) *cretin*, idiot + (Gr.) *ismos*, denoting condition
myxedema: (Gr.) *myxa*, mucus + *oidema*, swelling
exophthalmos: (Gr.) *ex*, out + *ophthalmos*, eye
endemic: (Gr.) *en*, in + *demos*, people

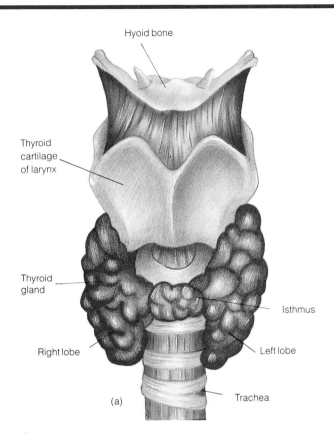
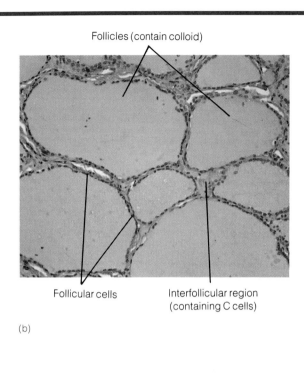

Figure 17-8 The thyroid gland. (a) Anatomy of the gland. The thyroid gland consists of two lobes connected by the isthmus. (b) Histology of the thyroid gland (mag. 650×). The lumen of the follicles contains colloid, in which thyroid hormones are formed. Calcitonin is produced by C cells located in the interfollicular space and in the epithelial layer.

Effects of Thyroid Hormones

Thyroid hormones are responsible for a wide range of effects in both the developing and adult body. These effects include increased O_2 utilization and increased ATP and heat production by tissues in general. The liver, kidneys, muscles, and gastric mucosa respire more rapidly in response to thyroid hormones. Protein synthesis is enhanced, mineral balance is affected, and fat metabolism is increased, causing a decrease in lipids stored in fat and carried in the plasma.

Because of their profound effect on growth and metabolism, inadequate or excessive production of thyroid hormones causes serious problems. Insufficient production, called **hypothyroidism** (hī·pō·thī′royd·ĭzm), decreases metabolic rate; when this occurs in children, there is reduced growth and development. The result is a condition called **cretinism** (krē′tĭn·ĭzm).

When thyroid hormone production is lost or decreases in the adult, that is, after growth and development is complete, the results are different from those in the developing child. In adults the skin thickens causing a characteristic puffiness in the face and hands. The tongue thickens, various nervous reflexes are delayed, and mental and physical apathy may occur. This condition is referred to as **myxedema** (mĭks·ĕ·dē′mă) (Figure 17-9).

Occasionally, the thyroid gland secretes excessive amounts of hormone, a condition called **hyperthyroidism**, which has results that are generally opposite those of deficient thyroid hormone production. In **Graves' disease**, for example, one form of hyperthyroidism, metabolic rate increases and causes a loss of weight (in spite of an increased appetite). The patient may be nervous, hypertensive, and have heart problems. Because of a retraction of the eyelids and a forward projection of the eyeballs, there is a characteristic protrusion of the eyes from their sockets, a condition known as **exophthalmos** (ĕks″ŏf·thăl′mōs).

Whenever the thyroid gland becomes enlarged, the condition is referred to as **goiter**, and it is sometimes associated with inadequate hormone production and disease. In **endemic goiter**, however, the enlargement of the gland (Figure 17-10) is due to insufficient iodine in the diet, a condition

oxyphil: (Gr.) *oxys*, sharp + *philein*, to love
parathormone: (Gr.) *para*, beside + *thyreos*, shield + *hormaein*, to excite

---

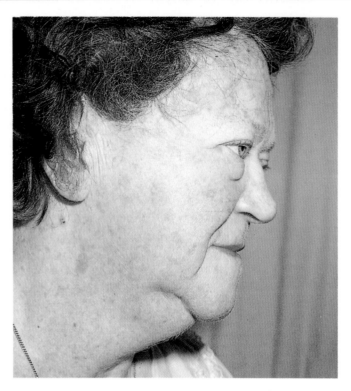

Figure 17-9 A patient with myxedema, a condition that results from a deficiency of thyroid hormone.

Figure 17-10 A woman with an endemic goiter caused by lack of iodine in the diet. Iodine deficiency prevents the production of adequate amounts of thyroid hormone.

that can be alleviated by adding iodine to the food. Iodized salt was first devised to counteract the effects of diets deficient in iodine that were leading to endemic goiter.

Calcitonin

Calcitonin (kăl″sĭ·tō′nĭn) is a hormone produced by C cells that lie between follicles in the thyroid gland or in the epithelium of the follicles. Calcitonin is a peptide with 32 amino acids. It is responsible for regulating the level of Ca^{2+} in the blood. Calcitonin lowers the level of Ca^{2+} by inhibiting removal of Ca^{2+} and by stimulating its uptake by bones when the plasma level of Ca^{2+} is too high. Calcitonin also causes a lowering of phosphate ion in the blood serum by causing the phosphate to enter bone.

Calcitonin secretion is stimulated when Ca^{2+} is taken orally through a mechanism that involves a hormone produced by cells in the intestinal wall. These cells are sensitive to Ca^{2+} and release their hormone, called **gastrin,** in response to elevated Ca^{2+} concentrations. Gastrin is then carried to C-cells and induces them to release calcitonin. The concentrations of calcitonin and gastrin are interdependent, gastrin stimulating calcitonin production and calcitonin inhibiting gastrin secretion.

Parathyroid Glands

Most people have four parathyroid glands, each about 6 mm long. Two are located on the posterior surface of each of the two lobes of the thyroid gland (Figure 17-11a). The parathyroid glands change their cellular composition as one ages. Initially, and up to about the age of 10, the glands consist mainly of **chief cells,** small cells with relatively little cytoplasm. As one ages, additional cells appear called **oxyphil** (ŏk′sē·fĭl) **cells** (Figure 17-11b). They are larger than the chief cells, possess more mitochondria, and stain differently. Oxyphil cells are generally found in clumps close to the surrounding capsules, and their number continues to increase with age. A third type of cell develops later in life that produces large fat-containing vacuoles.

Of the three cell types, the function of only the chief cells is known. These cells produce a protein called **parathyroid hormone (PTH)** or simply **parathormone.** Parathyroid hormone helps regulate the level of Ca^{2+} in the blood, causing it to be released from the bones by stimulating osteoclast activity when blood levels of Ca^{2+} become too low. It also stimulates the intestines to absorb Ca^{2+} more effectively and inhibits Ca^{2+} excretion by the kidneys, both functions serving to increase the supply of Ca^{2+} in the blood. PTH also increases the rate at which phosphate ion is removed from the blood by the kidneys.

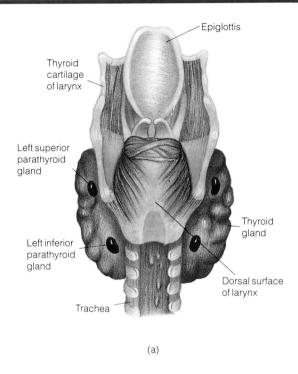
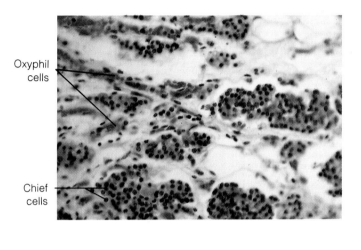

Figure 17-11 The parathyroid glands. (a) Location of the parathyroid glands on the dorsal surface of the thyroid gland. (b) Histology of the parathyroid glands (mag. 650×). The chief cells are the source of parathyroid hormone, a hormone that regulates the level of Ca^{2+} in the blood. The function of oxyphil cells is not known.

Since parathyroid hormone and calcitonin are both involved in the control of Ca^{2+} concentration in the blood, it is not surprising that their regulatory mechanisms are interrelated. As noted earlier, elevated Ca^{2+} levels cause the thyroid gland to release additional calcitonin. At the same time, however, these elevated levels cause the parathyroid glands to release less parathyroid hormone (Figure 17-12). Increased calcitonin and reduced parathyroid hormone have the same effect: to cause more Ca^{2+} to be absorbed by bones, thus causing its concentration in the blood to diminish. Lowered blood levels of Ca^{2+} stimulate the parathyroid glands to release more parathyroid hormone, reducing the stimulus to the thyroid gland to release more calcitonin. Parathyroid hormone also causes Ca^{2+} to be absorbed more effectively from the gastrointestinal tract and from urine in the kidneys and causes it to be released from bone. Concomitant with reduced calcitonin release from the thyroid gland, the concentration of Ca^{2+} in the blood rises.

As a result of disease or injury, parathyroid hormone production can be altered and can lead to increased or decreased secretion of hormone. **Hypoparathyroidism** (hī·pō·păr·ă·thī·royd·izm), in which too little hormone is secreted, can result from injury to the glands or their surgical removal. The lowered blood Ca^{2+} levels that result have a profound effect on nerve and muscle activity, since both nerve and muscle tissues require Ca^{2+} for proper functioning. Uncontrolled muscular spasm and tetany can occur and be fatal. **Hyperparathyroidism**, in which too much hormone is secreted, can result in bones weakened from excessive removal of calcium salts. Afflicted bones become deformed and can break easily. The excessive Ca^{2+} levels in the blood associated with hyperparathyroidism can lead to the production of kidney stones formed as calcium phosphate precipitates in the kidneys.

Production of PTH is controlled through a negative feedback mechanism in which increased concentrations of PTH are associated with decreased concentrations of calcitonin, and vice versa. The primary stimulus to PTH secretion is a low concentration of Ca^{2+} in the extracellular fluid that bathes the cells of the parathyroid gland. Increased concentrations of phosphate also stimulate PTH release.

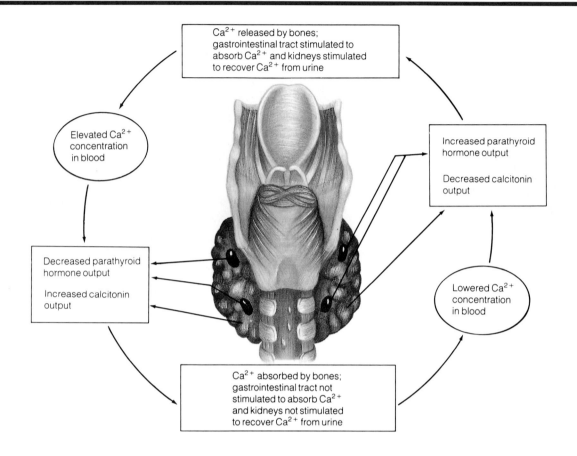

Figure 17-12 Feedback control of parathyroid hormone and calcitonin release.

Adrenal Glands

After studying this section, you should be able to:

1. Describe the organization of the adrenal glands.
2. Identify three regions within the adrenal cortex and cite the corticosteroid hormones produced in each region.
3. Explain the role of mineral corticoids, glucocorticoids, and gonadocorticoids.
4. Describe the structure of the adrenal medulla.
5. Cite the function of the catecholamines produced by the adrenal medulla.

There are two adrenal glands in the body, one located at the superior end of each of the kidneys. They are enclosed within the surrounding membrane and embedded in a layer of fat that surrounds the membrane (Figure 17-13a). Each gland consists of two distinct regions that differ from one another in appearance, function, and embryologic origin. The adrenal gland is surrounded by a fibrous **capsule**, immediately within which is located a yellowish layer of tissue called the **cortex**. The cortex, in turn, surrounds an interior mass of reddish tissue called the **medulla** (Figures 17-13b–d).

Adrenal Cortex

The region of the adrenal cortex immediately within the capsule consists of undifferentiated cells that are actively dividing. The cells produced in this region differentiate while simultaneously migrating inward through the cortex, forming three structurally and functionally different zones. The outermost region, just within the capsule, is the **zona glomerulosa** (zō′nă glō·mĕr″ū·lō′să). The cells in this region form irregularly shaped clumps permeated by passageways called **sinuses** and separated from one another by septa that extend from the capsule.

Extending inward from the zona glomerulosa, the cell clumps undergo a gradual transition to become arranged in

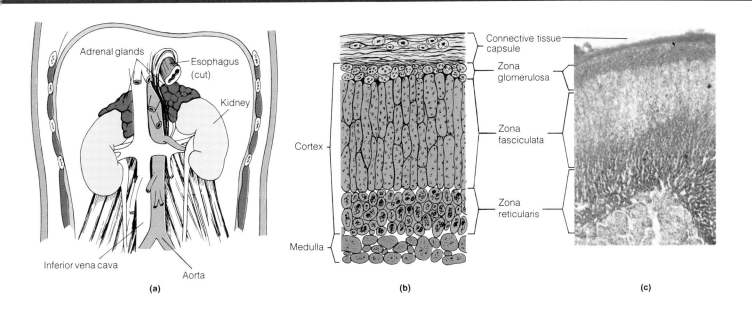

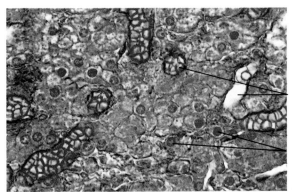

Figure 17-13 The adrenal glands. (a) Anatomy in relation to the kidneys. Each adrenal gland is divided into a cortex and medulla and surrounded by a fibrous capsule. (b) Drawing and (c) photomicrograph of the zones of the cortex (mag. 20×). The cortex appears as three different zones as a result of the progressive differentiation of cells that are produced in the outer portion of the cortex and migrate inward. (d) Photomicrograph showing the cellular organization of the adrenal medulla (mag. 348×).

roughly parallel cords. This region is the **zona fasciculata** (fă·sĭk″ū·lă′tă) and comprises about 80% of the cortex. The innermost region of the cortex is the **zona reticularis** (rĕ·tĭk·ū·lăr′ŭs), a region of smaller cords than are found in the zona fasciculata (Figures 17-13b and c). These cords are no longer in parallel bundles but branch repeatedly to form a network of cords. After the cells have reached this zone, they die and disintegrate and their debris is carried off in the capillary blood that passes through the gland.

Under the control of ACTH from the anterior pituitary, the adrenal cortex secretes nearly 50 steroid compounds collectively called **corticosteroids**. Many of these compounds are intermediates in the production of active hormones by the adrenal cortex. All corticosteroids are produced from cholesterol.

Hormones of the Zona Glomerulosa

The hormones produced by each of the three zones of the cortex differ in their primary functions. The zona glomerulosa produces a class of corticosteroids called **mineral corticoids**. These hormones, of which **aldosterone** (ăl·dŏs′tĕr·ōn) is the primary example, regulate the retention of sodium by the kidney and help maintain the balance of potassium and sodium in the body by regulating mineral loss from the kidney. Aldosterone synthesis and release,

angiotensin:	(Gr.)	*angeion*, vessel + *teinen*, to stretch
muralium:	(L.)	*murus*, wall
chromaffin:	(Gr.)	*chroma*, color + (L.) *affinis*, to have affinity for

CLINICAL NOTE

ALDOSTERONISM

By definition, **aldosteronism** (ăl·dō·stĕr´ōn·ĭzm˝) refers to a state characterized by an inappropriately increased production of aldosterone by the adrenal gland. About two-thirds of people with aldosteronism have an adrenal tumor, which causes more aldosterone to be secreted than the feedback control system can regulate. Excessive aldosterone results in excessive reabsorption of sodium, which means that the whole body has too much sodium, and as a result, an excess amount of water is retained. This excess extracellular fluid leads to increased blood pressure. Thus, if a patient shows increased blood pressure, aldosteronism, though not a common disease, should be considered.

Because sodium is retained in the body, potassium is preferentially lost, leading to a low concentration of potassium in the extracellular and intracellular fluids. A low serum potassium level, in turn, leads to muscle weakness, fatigue, and alterations in the conductivity of the heart muscle. Hydrogen ion is also preferentially lost as sodium is retained. This decrease of hydrogen ion concentration leads to **metabolic alkalosis,** or excessive alkalinity of body fluids.

The treatment of aldosteronism depends on the cause. It may require surgical removal of the adrenal gland. It may require blocking the renal receptor for aldosterone. Spironolactone, a diuretic that blocks the effect of aldosterone, may be administered. Sodium is then not retained to excess and therefore fluid is not retained, potassium is spared, and hydrogen ion is not lost in an uncontrolled manner.

unlike that of the other corticosteroids, is not controlled by ACTH from the anterior pituitary. Instead, its level is controlled by the sodium and potassium concentration in the blood, and a decline in sodium or a rise in potassium is a direct stimulus for aldosterone secretion from the zona glomerulosa. Aldosterone is also produced in response to a blood pressure-regulating hormone called **angiotensin II** that circulates in the blood.

Hormones of the Zona Fasciculata

The zona fasciculata is the source of a group of steroids called **glucocorticoids.** These hormones are primarily responsible for regulating carbohydrate and protein metabolism; for example, they stimulate the synthesis of glycogen in the liver and the breakdown and conversion of protein into carbohydrate. Many glucocorticoids suppress inflammation and are used medically for that purpose.

Glucocorticoid secretion is stimulated by stress, and the effect of the hormone is to produce an increase in available carbohydrate. The carbohydrate is used as an energy source for combating trauma, shock, excessive blood loss, and other effects that may result from injury.

Glucocorticoid secretion is controlled by a negative feedback mechanism in which elevated levels of glucocorticoids inhibit CRF production by the hypothalamus (Figure 17-14). CRF released by the hypothalamus in response to stress or low concentrations of glucocorticoids in the blood stimulates the anterior pituitary gland to release ACTH. ACTH stimulates glucocorticoid release by the adrenal cortex. The elevated levels of glucocorticoids that result provide increased glucose that is used to combat the stress that induced production in the first place, reducing the production of additional CRF by the hypothalamus.

The main glucocorticoid is **cortisol** (kŏr´tĭ·sŏl), which is released to the blood and transported loosely bound to a protein called **transcortin** (trăns·kŏr´tĭn). Cortisol is not active as a hormone when it is bound to transcortin, but becomes active when released at the surface of a target tissue.

Hormones of the Zona Reticularis

The zona reticularis is the source of small quantities of sex hormones called **gonadocorticoids,** primarily male sex hormones referred to as **adrenal androgens.** The female sex hormones estrogen and progesterone are also produced, but in even smaller amounts. These hormones are produced in such small quantities that they generally have little effect. Only when the adrenal cortex is diseased or when the ovaries or testes are removed do the quantities reach high enough levels to play an important role. (For a discussion of adrenal cortex problems, see Clinical Note: Diseases of the Adrenal Cortex.)

Adrenal Medulla

The dividing line between the cortex and medulla is marked by a transition in cell type at the border between the zona reticularis and the medulla. The medulla consists of a system of interconnecting passages called a **muralium** (mū·răl´ē·ŭm) (Figure 17-13d). Hormones are delivered from cells that line these walls into the blood that flows through the muralium. These cells, called **chromaffin cells,** contain many small granules that are believed to contain the hormones (or substances from which they are produced) secreted by the medulla.

Hormones of the adrenal medulla are collectively called **catecholamines** (kăt˝ĕ·kōl´a·mēns), which consist of about 80% epinephrine and 20% norepinephrine. Norepinephrine is also a neurotransmitter.

In general, epinephrine causes a rapid increase in the delivery of blood to skeletal muscles. It also increases the

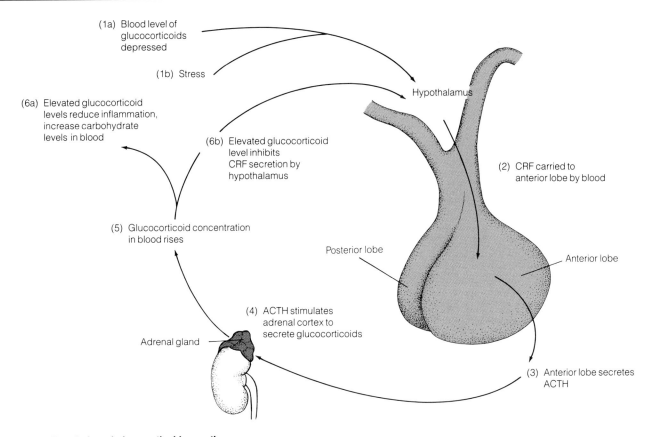

Figure 17-14 Regulation of glucocorticoid secretion.

CLINICAL NOTE

DISEASES OF THE ADRENAL CORTEX

The adrenal cortex is subject to diseases that modify its rate of hormone production. In **Addison's disease,** for example, the cortex produces insufficient quantities of aldosterone and cortisol. This hypofunction of the gland prevents regulation of sodium and potassium levels, and too much sodium is lost and too much potassium is retained. The mineral imbalances that result interfere with water retention, leading to dehydration. At the same time, inadequate cortisol secretion leads to altered metabolism, with insufficient amounts of protein being converted to carbohydrate. The metabolic effects that result can be lethal if not counteracted with adrenal hormone.

Cushing's syndrome and adrenal virilism are conditions that result from an overactive adrenal cortex. In **Cushing's syndrome,** excessive production of cortisol leads to an accumulation of fat in the neck, face, and trunk. Muscles are generally weak and diminished in size, the skin is thin, and bruises appear easily. Affected women may have menstrual difficulties, and children may have their growth stunted. When caused by a tumor of the pituitary gland that causes excess production of ACTH, treatment may require surgical removal of the tumor. In other cases, the problem lies in the adrenal glands themselves, and it may be necessary to remove the glands. Since adrenal secretions are necessary for survival, patients who have had their adrenal glands removed must receive steroid medication from that point on.

Adrenal virilism is a syndrome that results from excessive production of adrenal androgens. The effect is considerably more pronounced in women than in men since it causes masculinization in the form of facial hair, baldness, and increased muscle development. At the same time, normal female characteristics may be suppressed, menstruation may stop, the uterus may atrophy, and the breasts may decrease in size. Treatment of this condition, like Cushing's syndrome, often requires removal of the adrenal glands.

pheochromocytoma: (Gr.) *phaios*, dusky + *chroma*, color + *kytos*, cell + *oma*, tumor

---

> **CLINICAL NOTE**
>
> ## PHEOCHROMOCYTOMA
>
> **Pheochromocytoma** (fē·ō·krō″mō·sī·tō′mă) is a tumor of the adrenal medulla that causes secretion of epinephrine and norepinephrine in an uncontrolled manner. Excess amounts of these hormones cause vasoconstriction, increased resistance to blood flow within the narrowed vessels, and ultimately increased blood pressure. This sustained hypertension is the most outstanding clinical manifestation of pheochromocytoma; blood pressure measurements can be in the range of 200/150 to 300/175. During the time that the tumor is secreting the excess catecholamines, the person will experience severe headaches, a feeling that the heart is beating irregularly or too fast, and sweating. The person also feels apprehensive and may be emotionally unstable, having periods of elation or depression. Dilated pupils may be seen and the individual will either be very pale or flushed. Very severe hypertension may result in a life-threatening stroke, kidney disease, or eye damage. Excision of the tumor will remove the source of uncontrolled secretion of epinephrine and norepinephrine.

a dilation of blood vessels in skeletal muscles but norepinephrine causes these same vessels to constrict. Norepinephrine also causes a rise in both the systolic and diastolic blood pressure, whereas epinephrine causes a rise only in systolic blood pressure, and then only when administered in high concentrations (see Clinical Note: Pheochromocytoma).

Pancreas

After studying this section, you should be able to:

1. Describe the structure of the pancreas.
2. Cite the role of the islets of Langerhans in hormone production.
3. List the hormones produced by the pancreas.
4. Describe the effects of insulin.
5. Describe the effects of glucagon.
6. Describe the effects of somatostatin.

The pancreas is a heterocrine gland that secretes four hormones: *glucagon* and *insulin*, hormones that regulate the body's metabolism of carbohydrates and fats; *somatostatin*, the inhibitory releasing hormone also produced by the hypothalamus; and *pancreatic polypeptide*, a hormone of uncertain function associated with digestion. Cells that secrete these hormones are localized in groups scattered throughout the pancreas called **islets of Langerhans** (Figure 17-15). There are about one million islets in the pancreas, each containing about 3000 cells embedded in the nonendocrine tissue. There are four different kinds of hormone-producing cells in the islets, designated **alpha, beta, delta,** and **F cells**, which are responsible for producing glucagon, insulin, somatostatin, and pancreatic polypeptide, respectively. Islets are well supplied with capillaries, permitting the release of hormones directly to the blood. The nonendocrine tissue of the pancreas secretes enzymes and bicarbonate ions into the small intestine where they are used for digestion.

Insulin

Insulin is a protein that consists of two separate polypeptides, designated as **A** and **B chains**, held together by covalent bonds. The chains are derived from a single protein produced by the beta cells, which is then cleaved to produce the functional hormone. Once insulin is formed, it is released to the blood, where most of it becomes bound to protein carriers in the plasma. Since blood is drained from the pancreas by the hepatic portal vein, the bound insulin is carried to the liver where much of it is trapped and degraded before

Table 17-3 COMPARISON OF EFFECTS OF EPINEPHRINE AND NOREPINEPHRINE

Effect	Epinephrine	Norepinephrine
Cardiac output	Increase	Decrease
Blood pressure	Decrease	Increase
Blood glucose concentration	Decrease	Decrease
Heat production	Decrease	Increase
Fatty acid release to blood	Decrease	Increase
CNS stimulation	Increase	Decrease

Source: Adapted from W. F. Ganong, *Review of Medical Physiology*, 7th ed. (Los Altos, CA: Lange Medical Publications, 1975), 269.

supply of glucose to those muscles and increases the cell's ability to metabolize it. In general, these are reactions that prepare the body to cope with sudden stress, and for that reason, they are often referred to as "fight-or-flight" reactions. Norepinephrine, on the other hand, plays a more restricted role, primarily controlling blood pressure. Because epinephrine and norepinephrine induce responses similar to those induced by the sympathetic nervous system, they are sometimes described as being **sympathomimetic** in function.

In spite of a structural similarity that exists between epinephrine and norepinephrine, each of the two compounds has its own distinctive role to play, and in some cases the effects of the two compounds are exactly opposite (Table 17-3). For example, epinephrine causes an increase in heart rate but norepinephrine causes a decrease; epinephrine causes

hyperinsulinism: (Gr.) *hyper*, above + (L.) *insula*, island + (Gr.) *ismos*, denoting condition

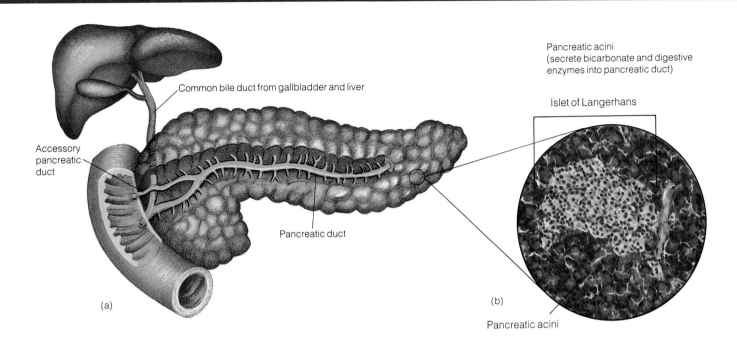

Figure 17-15 The pancreas. (a) Longitudinal section through the pancreas, a mixed gland that produces both endocrine and nonendocrine secretions. (The liver is not drawn to scale.) (b) Photomicrograph showing a single islet of Langerhans embedded in pancreatic tissue.

it can perform any function. Controlling the rate of this destruction provides a means for regulating the amount of insulin that reaches the general circulation.

Insulin production is affected by several substances. Glucagon, secreted by alpha cells, stimulates insulin secretion by beta cells. Amino acids, cAMP, acetylcholine and certain drugs also stimulate beta cells to release insulin. Release of digestive hormones into the small intestine is accompanied by the release of insulin into the blood. Beta cells are probably stimulated directly by these hormones as they are carried in the blood from the intestines. This coordinates digestive hormone release and insulin release and also provides a stimulus for metabolism of sugars absorbed in response to digestive hormones. Insulin secretion is also stimulated and inhibited by autonomic nerves that terminate in the islets of Langerhans, providing an additional mechanism for controlling insulin secretion.

The main effect of insulin is to make glucose available to most of the cells of the body. Insulin combines with specific receptors in the cell membrane, forming a complex that permits more rapid diffusion of glucose into the cell. Insulin-induced glucose transport across cell membranes is especially effective in muscle and fat tissue, but insulin does not significantly enhance glucose transport into brain cells. Insulin also stimulates the enzymatic incorporation of glucose into glycogen in the liver and muscle, permitting the storage of glucose in this form.

Transport of glucose into cells has secondary effects on fat and protein metabolism, since metabolism of fat and protein is influenced by the availability of glucose. In adipose tissue, for example, glucose can be metabolized to glycerol

CLINICAL NOTE

HYPERINSULINISM

Increased insulin production results in the rare disorder of **hyperinsulinism** caused frequently by a benign tumor of the islets of Langerhans and only rarely by a malignant tumor. Insulin is produced in a wild, uncontrolled, and inappropriate way. Sometimes, in the case of malignant tumors, the metastatic sites of the tumors can secrete insulin, resulting in massive amounts of insulin that require the administration of large amounts of glucose.

Although the central nervous system does not require insulin to assist in the utilization of glucose by neuronal cells, high levels of insulin with the consequent low levels of blood glucose leads to decreased metabolism of the central nervous system. The person feels anxious and nervous, sweats, and trembles. As the blood sugar continues to fall, convulsions occur. Finally, the person may fall into a coma. The treatment of this extreme hyperinsulinism is to give vast quantities of glucose in an attempt to control the blood sugar. The administration of glucagon will stimulate glycogenolysis by the liver and will also increase blood sugar.

which can then be used in the synthesis of fat. The presence of glucose also inhibits the breakdown of proteins in many cells, and so these proteins are said to be **spared** from degradation. Amino acid transport into the cell is also stimulated by insulin, and increased amino acids in the cell permit additional protein synthesis.

The major illness associated with insulin production is **diabetes mellitus** (dī″ă·bē′tēz měl·lī′tŭs), a condition that is due to a deficiency in the production of insulin. Because of the important role that insulin plays, its deficiency has profound effects (see The Clinic: Diabetes Mellitus.)

Glucagon

Like insulin, **glucagon** (glū′kă·gŏn) is derived from a precursor protein. This precursor, termed **proglucagon**, is modified to produce glucagon, a peptide of 29 amino acids. Proglucagon is produced by alpha cells of the pancreas.

The main role of glucagon is to increase the level of glucose in the blood. It does this by stimulating liver cells to convert stored glycogen into glucose, which is then released into the blood. Glucagon also stimulates the conversion of amino acids to glucose, the breakdown of lipids within the liver and fatty tissues, and the production of insulin by pancreatic beta cells.

Glucagon and insulin production are interrelated, the effects of one influencing the secretion of the other. Alpha cells increase their secretion of glucagon in response to decreased concentrations of glucose in the blood caused by insulin. Increased sugar in the blood stimulates beta cells to secrete insulin at the same time that it removes the stimulus for glucagon secretion. As a result, glucagon secretion declines as insulin secretion rises. With increased insulin in the blood, glucose is removed as it is transported into cells. Glycogen utilization and fat and protein breakdown are also inhibited. With the decline of blood glucose, alpha cells are again stimulated to make glucagon as beta cells reduce their production of insulin, and the cycle is repeated (Figure 17-16).

Glucagon is a hormone that acts through a cAMP mechanism. It combines with the membrane of receptive cells in the liver causing a sequence of reactions that ends with the cleavage of glucose from glycogen. Each step in the sequence is accompanied by an amplification of the number of molecules that act as activators for the next step, and one molecule of hormone can lead to the release of hundreds of millions of glucose molecules.

Occasionally, following an insulin injection, glucose is depleted from the blood of diabetics more rapidly than it can be replenished. As a consequence, the level of glucose drops below the level required for normal metabolism, a condition known as **hypoglycemia** (hī″pō·glī·sē′mē·ă). If the drop is severe enough, coma or death can occur. Injection of glucagon into the patient greatly accelerates the release of glucose from the liver, raises the sugar level in the blood quickly, and enables the patient to recover. Therapy is then

hypoglycemia: (Gr.) *hypo*, under + *glykos*, sweet + *haima*, blood

Figure 17-16 Regulation of glucagon and insulin secretion.

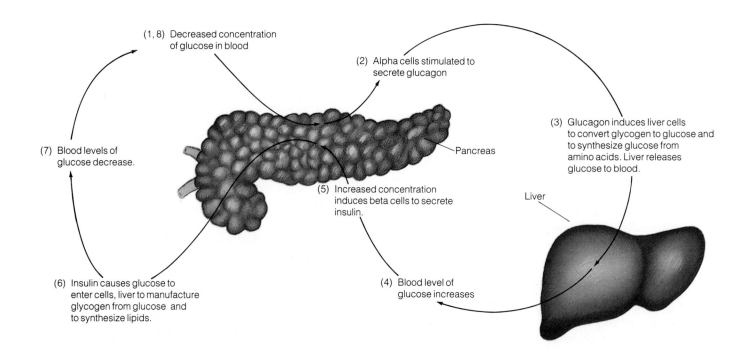

gonad: (Gr.) *gonos*, offspring

continued by eating glucose or having it administered directly to the blood by intravenous injection.

Somatostatin

Somatostatin (sō·măt′ō·stăt″ĭn) was originally discovered because of its inhibitory effects on the release of growth hormone in the anterior pituitary gland (which is the basis for its name). It has since been identified in other tissues as well, including the gastric and intestinal epithelium and parafollicular cells (C cells) of the thyroid gland.

Somatostatin is a peptide that contains 14 amino acids. It inhibits the secretion of several different hormones, including insulin, glucagon, several produced by the stomach and intestines, calcitonin from the thyroid gland, parathyroid hormone, immunoglobulins (proteins involved in the immune response), and renin (a product of the kidney involved in regulating blood pressure). Somatostatin also inhibits glucose absorption in the small intestine, secretion of hydrochloric acid by the stomach, and bicarbonate and enzyme secretion into the small intestine from the pancreas.

Somatostatin is not detected as readily in blood as are other hormones, suggesting that it exercises a local influence on neighboring cells in the organ in which it is produced. If this is the case, somatostatin does not conform to the traditional definition of a hormone. Instead, it should be considered a "local" hormone, a substance that influences cells the way hormones do but is not transported away from the organ in which it is produced.

Other Endocrine Glands

After studying this section, you should be able to:

1. Cite the hormones produced by the ovaries and the testes and describe the function of each.
2. Describe the structure of the thymus gland and its function in the immune system and as an endocrine organ.
3. Describe the structure and function of the pineal gland.
4. List several sources of additional hormones and the functions of the hormones they produce.
5. Explain how prostaglandins differ from hormones in general and describe their function.

Gonads

Gonads (gō′năds) are glands that are responsible for controlling sexual development and behavior. In females the gonads are the **ovaries**, a pair of oval organs about the size of large almonds located on each side of the pelvic cavity (see Figure 27-13). In addition to producing and releasing eggs during the reproductive years, the ovaries are a source of the sex hormones **estrogen** (ĕs′trō·jĕn) and **progesterone** (prō·jĕs′tĕr·ōn). These hormones accelerate growth of the uterus, vagina, external genitalia, pelvis, breasts, and pubic and axillary hair in the developing female. In the adult nonpregnant female, estrogen stimulates growth of the epithelial lining of the uterus and growth of the smooth muscle within its wall. Estrogen and progesterone also stimulate breast development in pregnant females.

Production of estrogen and progesterone is induced by gonadotrophins from the anterior lobe of the pituitary gland, themselves under the control of the releasing hormones described earlier. Control is exercised via a negative feedback mechanism in which increased levels of estrogen and progesterone cause a decrease in releasing hormones and a consequent reduction in gonadotrophin secretion.

In the male the gonads are the **testes**, two oval-shaped glands located in the scrotum (see Figure 27-1). Certain cells in each testis, called **interstitial cells of Leydig**, are stimulated by luteinizing hormone to produce the male sex hormone **testosterone** (tĕs·tŏs′tĕr·ōn). Testosterone is liberated from the testes and carried in the blood bound to a protein produced in the liver known as **steroid hormone-binding globulin (SHBG)**. Testosterone causes the development and maintenance of sexual characteristics in the male. These include development of genitalia, production of body hair, development of masculine voice quality, and many other aspects of sexual behavior.

Other cells in the testis, called **Sertoli's** (sĕr·tō′lēz) **cells**, respond to follicle-stimulating hormone (FSH) and testosterone, both of which enable the Sertoli's cells to assist in the production of sperm by the testis. FSH production is controlled in males by a feedback mechanism involving a hormone released by the Sertoli's cells called **inhibin**. A more detailed description of the function of these hormones is presented in Chapter 27.

Thymus Gland

The **thymus gland** is located in the superior portion of the thoracic region, posterior to the sternum (Figure 17-17a). It is in reality two glands, or lobes, joined to one another by connective tissue. The thymus is relatively large at birth, growing only slowly until puberty is reached. After puberty, the thymus usually begins to decrease in size as its tissues are replaced by fat and loose connective tissue. In middle-aged people, the thymus gland may be so reduced in size that it is difficult to recognize.

Also part of the body's system for defense against disease, the thymus gland is classified as a **lymphoepithelial**

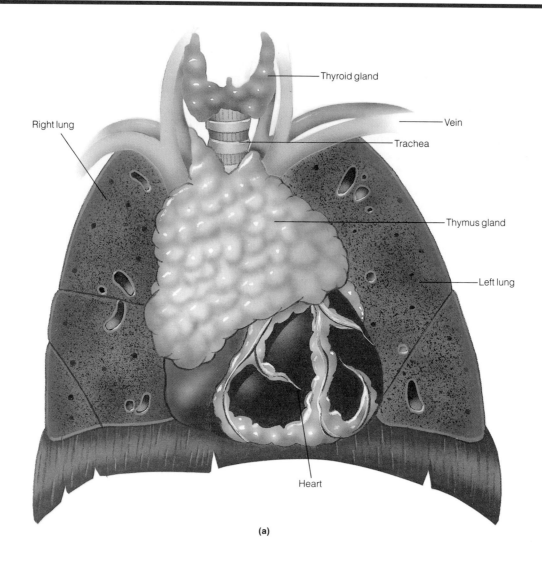

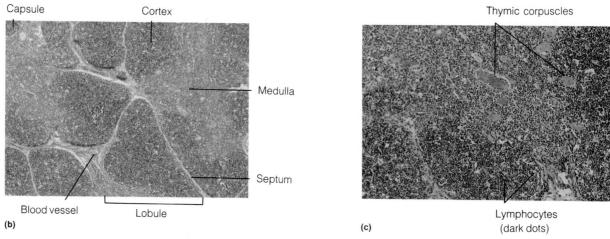

Figure 17-17 The thymus gland. (a) The thymus gland is a relatively large gland in a young child, as shown here, but usually decreases in size with age. (b) The thymus is organized into lobules separated from one another by septa formed from connective tissue (mag. 202×). (c) Photomicrograph at higher magnification showing cellular organization within the lobule (mag. 808×).

corpora arenacea:	(L.) *corpus*, body + *arenaceus*, sandy
somatomammotrophin:	(Gr.) *somatos*, body + (L.) *mamma*, breast + (Gr.) *trophe*, nourishment
prostaglandin:	(Gr.) *prostates*, prostate + (L.) *glans*, acorn

organ. The hormones it produces cause cells to differentiate into forms capable of producing chemicals used to destroy invading organisms. Each lobe of the thymus gland is surrounded by a fibrous **capsule**, invaginations of which divide the lobe into a number of **lobules**. Each lobule contains a **cortex** of cells called **lymphocytes** (see Chapter 24) held in a fibrous network made up of tissue called a **reticular epithelium**. The cortex surrounds an inner **medulla**, which consists of a looser mixture of reticular epithelial cells and lymphocytes. Also included within the medulla are aggregates of cells surrounding a dense core of protein, the **thymic corpuscles** (Figure 17-17c).

The hormones produced by the thymus are probably produced by the reticular epithelium, and they constitute a family of compounds collectively referred to as **thymosins** (thī′mō·sĭns). Thymosins are thought to promote the maturation of lymphocytes, enabling them to recognize and attack foreign proteins and other substances, such as those found in infecting bacteria and viruses or in the membranes of transplanted tissues.

Pineal Gland

The **pineal** (pĭn′ē·ăl) **gland** (also called the **epiphysis**) is a structure within the diencephalon (the posterior subdivision of the forebrain). In adults it is an encapsulated, oval, stalked structure that measures only 5 to 8 mm in length and 3 to 5 mm in width. The stalk is hollow for a portion of its length, with the hollow portion an extension of the third ventricle. Cells of the pineal gland are of two primary types: **pinealocytes** (pĭn′ē·ă·lō·sīts), which may be secretory in function, and neuroglia, which support the pinealocytes. The organ is divided into numerous septa by invaginations of the fibrous capsule, giving the gland the appearance of a small pine cone (which is the basis for its name).

From the time of birth, calcium and magnesium phosphate and carbonate salts may form deposits in the pineal gland called **corpora arenacea** (kōr′pō·ră ăr″ē·nā′sē), or "brain sand." These grains provide a useful marker that shows the location of the pineal gland in X rays of the brain. The function of these particles, if any, is unknown.

The endocrine function of the pineal gland in humans is not well understood, but the gland secretes a hormone called **melatonin** (mĕl″ă·tō′nĭn) in response to norepinephrine from sympathetic fibers that terminate in the gland. In humans approximately 70% of the melatonin is secreted at night between the hours of 11:00 P.M. and 7:00 A.M. Its production during these hours is inhibited by bright light. Melatonin affects pigmentation in lower animals, such as frogs and other amphibians, but in humans it inhibits the release of gonadotrophin by the pituitary gland and gonadotrophin-releasing hormones by the hypothalamus.

Sources of Additional Hormones

In addition to the hormones described so far, several hormones are secreted by other tissues and organs that are not considered to be primarily endocrine organs. For example, the placenta produces a hormone called *human chorionic gonadotrophin* (*HCG*), which helps stimulate the ovary to produce progesterone during pregnancy. As pregnancy continues, the placenta makes increasing amounts of estrogen and progesterone, while simultaneously decreasing its HCG output. In addition, the placenta produces *somatomammotrophin*, *placental corticotrophin*, and *placental thyrotrophin*, all of which help maintain the mother's body during pregnancy. There are several hormones secreted by various portions of the digestive tract that aid in digestion, such as *secretin*, *cholecystokinin*, *gastrin*, *gastric-inhibitory peptide*, and *somatostatin*. The kidney, blood, and liver are sources of still other hormones, the functions of which are discussed in chapters that deal specifically with those organs.

Prostaglandins

Prostaglandins (prŏs′tă·glănd·ĭns) are chemicals that alter the response of a cell to a hormone. There are over a dozen different prostaglandins known, all of which consist of a modified fatty acid. Prostaglandins are classified into four categories on the basis of differences in structure (Figure 17-18). In addition to the prostaglandins, there are several other compounds with similar chemical structure and biologic activity called **prostacyclins** and **thromboxanes**.

Prostaglandins were originally discovered in the 1930s in human semen and in extracts obtained from sheep prostate

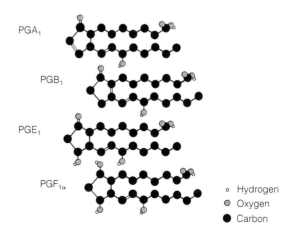

Figure 17-18 Some representative prostaglandins. These compounds are produced by many, if not all, cells and facilitate the action of hormones on cells in which they are produced. Each molecule is a fatty acid with a five-membered ring in the middle that causes the chain of carbon atoms to fold back on itself.

| mellitus: | (L.) *mellitus*, honey, sweet |
| microangiopathy: | (Gr.) *mikros*, small + *angeion*, vessel + *pathos*, disease |

THE CLINIC

DIABETES MELLITUS

Diabetes mellitus is a disorder associated with a total or partial lack of the pancreatic hormone insulin. This lack affects the metabolism of carbohydrates, fat, and protein and thus has varied and widespread effects on individuals who have this disorder. The two main types of diabetes mellitus are **insulin-dependent diabetes mellitus (IDDM)** and **non-insulin-dependent diabetes mellitus (NIDDM)**. Although these two disorders share the characteristic lack of insulin, they are distinctly different.

Insulin-dependent diabetes mellitus usually has its onset before the age of 30 years, often in an adolescent of normal body weight. Approximately 5 people in 1000 are affected by this type of diabetes. While heredity does seem to play a part in its cause, it does not seem to have as big a role as it does in NIDDM. Insulin-dependent diabetes mellitus has recently been identified as an autoimmune disease because 70% of affected people have circulating antibodies to their own pancreatic islet cells. Coxsackievirus, mumps, and rubella viruses may either directly destroy the beta cells or may induce changes in the islet cells so that they are recognized as foreign by one's own immune system and are attacked.

Early in the disease, the secretion of insulin is decreased, but eventually insulin production may stop completely. The lack of circulating insulin leads to **hyperglycemia** (increased level of sugar in the blood) because sugar is not being taken up by peripheral fat and muscle cells. Additionally, the liver begins to overproduce sugar through glycogenolysis and gluconeogenesis (see Chapter 19). Synthesis of protein, ATP, DNA, RNA, and fat is impaired. The sparing effect on protein and fat degradation is also lost, and these compounds begin to be broken down and used in place of glucose. Fatty acids accumulate in the blood, leading to diabetic ketosis, which produces excess acidity in the body fluids. Acids spill into the urine along with excess sugar where they can both be detected and measured for easy diagnosis. Excess water is eliminated due to osmotic effects of sugar in the urine resulting in dehydration.

In contrast, persons with non-insulin-dependent diabetes mellitus are usually over 30 when the disorder appears, are generally overweight, exercise very little, and show no evidence of a recent viral infection. Heredity does seem to play some part since there is 95% concordance in twins. The disorder affects between 1% and 5% of the population. NIDDM is a disorder of insulin reception by peripheral target cells. It has been proposed that a decrease in insulin receptors exists in the obese individual with this type of diabetes mellitus. Diabetes-induced ketosis is rare, possibly because enough insulin activity occurs to prevent the breakdown of fats, but not enough occurs to prevent hyperglycemia.

Both IDDM and NIDDM cause a great number of secondary disorders. Hyperglycemia damages the Schwann cells that produce and maintain the myelin sheaths around axons of the peripheral nervous system. Damage to these nerve coverings leads to neuropathies. Also, hyperglycemia allows the attachment of glucose to proteins of the capillary basement membranes, leading to damage of these small vessels known as **microangiopathy**. Microangiopathy is the precise cause of damage in the kidneys of diabetics **(nephropathy)** and in the eyes of diabetics **(retinopathy)** (see Chapter 16 Clinical Note: The Eye: Window to the Body). Damage to large vessels, or **macroangiopathy**, is another complication of both types of diabetes mellitus. In addition, **atherosclerosis** (ăth″ĕr·ō·sklĕr′ō·sĭs) is greatly accelerated in the heart and large vessels of diabetics. Complications of atherosclerosis include heart attack and stroke. In general, persons with IDDM develop microangiopathies, such as kidney failure, as a result of nephropathy, and persons with NIDDM develop macroangiopathies, such as heart disease, as a cause of death.

Patients with IDDM are generally treated with a special controlled diet having a restricted amount of fats and carbohydrates specified for their age and activity. They are supplied with carefully regulated amounts of exogenous insulin. Persons with NIDDM are generally required to lose weight and increase exercise and are occasionally given insulin.

Diabetes mellitus is a chronic disease that requires the patient to participate actively in the treatment. Self-care is a big part of diabetic therapy and requires self-testing of urine or blood, self-injection of insulin, careful control of diet, and the participation in a specified activity level or exercise program. This requires motivation, will power, and support from the health provider and family.

---

glands. They have since been found in most, if not all, organs that have been examined, including the lungs, brain, spinal cord, placenta, kidneys, iris of the eye, and thymus gland. Unlike hormones, prostaglandins act on the cells that produce them rather than on cells to which they are carried. However, because of the similarity in response of many cells to both hormones and prostaglandins, the latter have sometimes been referred to as local hormones.

Prostaglandins are involved in many aspects of physiology, including reproduction, cardiovascular and renal functions, digestion, and respiration. They stimulate inflammation of tissue, they inhibit glucagon and other hormones in fat breakdown and antigen-induced histamine release in the immune system, they are involved in platelet aggregation in the clotting of blood, and they influence the central nervous system.

In any one reaction, opposite effects may be caused by closely related prostaglandins. For example, one prostaglandin called prostaglandin F (PGF) stimulates the circular smooth muscle in the intestine to contract. Another, called prostaglandin E (PGE), inhibits contraction (see Chapter 18). Because of these effects, PGE and PGF are thought to help control the contractions that force food through the intestine. PGE also causes the smooth muscle of bronchi (the tubes through which air passes on its way to the lungs) to relax, whereas PGF causes it to contract. This suggests that these prostaglandins are involved in regulating air flow into the lungs (see Chapter 20).

Because some prostaglandins increase and others decrease the rate of cAMP production, it has been suggested that prostaglandins regulate the activation of adenylate cyclase by other hormones. One hypothesis is that when a hormone interacts with its receptor on a target cell membrane, the interaction stimulates the production of prostaglandins. Depending on which prostaglandins are produced, activation of adenylate cyclase may be stimulated or inhibited.

In this chapter we studied the endocrine system, which consists of glands that secrete hormones into the blood. You learned the characteristics of several hormones and the mechanisms they use in exerting long term control over many metabolic processes. In the next unit we will study several of these metabolic processes in detail.

STUDY OUTLINE

Hormones (pp. 407–412)

Hormones are regulatory chemicals that affect *target* organs and tissues.

Chemical Composition of Hormones Hormones consist of amino acid derivatives, peptides, proteins, glycoproteins, or steroids. Five major classes of steroid hormones include *progestogens, glucocorticoids, mineral corticoids, androgens,* and *estrogens.* Water-soluble hormones work quickly and are degraded rapidly. Steroid hormones are bound to carrier molecules. They are responsible for long-term responses.

Concentration of Hormones Hormones function in low concentration, sometimes measured in a billionth of a gram per 100 mL of blood.

Source of Hormones Hormones are produced in *endocrine* (or *ductless*) glands, which release the hormones directly to the bloodstream. *Exocrine* glands are *ducted* and do not produce hormones. Some hormone-producing glands are *heterocrine,* having both endocrine and exocrine function.

Mechanism of Action Nonsteroid hormones attach to receptors in the cell membrane, which initiates a series of reactions that activate *adenylate cyclase.* Adenylate cyclase catalyzes the production of *cyclic AMP* (cAMP), which diffuses into the cell and activates *protein kinases.* These enzymes activate proteins, converting them to enzymes that catalyze reactions that characterize the hormone. cAMP is referred to as a *second messenger.*

Steroid hormones combine with receptors in the cytoplasm. The hormone-receptor complex activates specific genes, which produces the reaction that characterizes the hormone.

Regulation of Hormone Activity Hormone production is induced by nerve impulses, chemical environment, or other hormones. The result of a hormone's action is usually the stimulus to reduce production or release of the hormone, referred to as *negative feedback.*

Hypothalamus and Pituitary Gland (pp. 412–418)

The *hypothalamus* forms the floor and lower walls of the third ventricle of the brain. The *pituitary* gland is suspended from the hypothalamus and consists of a *posterior lobe* and an *anterior lobe.* The hypothalamus controls many phenomena, including hormone release by the pituitary gland.

Hormones of the Posterior Pituitary Gland *Antidiuretic hormone* (ADH) and *oxytocin* are produced by the hypothalamus and travel bound to *neurophysins* into the posterior pituitary gland. *Herring bodies* are masses of granules of neurophysin-hormone complexes. Oxytocin is produced in *paraventricular nuclei.* ADH is produced in *supraoptic nuclei.*

ADH is a small peptide that increases water recovery from urine. If supraoptic nuclei are bathed in fluid in which solute concentration is elevated, more ADH is released. When the solute concentration diminishes, less ADH is released.

Oxytocin is similar in structure to ADH. It works through positive feedback to stimulate ejection of milk from mammary glands and contraction of the uterus during labor.

Hormones of the Anterior Pituitary Gland Anterior gland hormones are produced by distinct cells in the gland: *somatotrophs, mammotrophs, thyrotrophs, gonadotrophs,* and *corticotrophs.* TSH, ACTH, LH, and FSH are *trophic hormones,* which stimulate endocrine glands.

The hypothalamus produces *releasing hormones* and *release-inhibiting hormones* carried in portal vessels to the anterior pituitary gland where they induce or inhibit the synthesis and release of anterior pituitary hormones. Hypothalamic hormones include *thyrotrophin-releasing hormone, corticotrophin-releasing factor, growth hormone release inhibitory hormone, gonadotrophin-releasing hormone,* and *prolactin-releasing factors.*

Prolactin is a protein of 200 amino acids that stimulates milk synthesis and breast development. There are two prolactin-releasing factors, *prolactin-releasing hormone* and *prolactin-inhibiting hormone.*

Human growth hormone (HGH) is a protein of 200 amino acids that primarily regulates growth of the skeleton. It also reverses the effect of insulin. Some effects are mediated by *somatomedin.*

Melanocyte-stimulating hormone has two forms: α-MSH and β-MSH. Both forms are peptides. The function of MSH is unclear, but by analogy to its function in amphibia, it may play a role in pigment development.

Adrenocorticotrophic hormone (ACTH) controls corticosteroid production by adrenal glands. ACTH and CRF levels are coregulated through a negative feedback mechanism.

Thyroid-stimulating hormone (TSH) regulates the activity of the thyroid gland.

Gonadotrophins stimulate the production of sex hormones by ovaries and testes. They include *follicle-stimulating hormone* (FSH) and *luteinizing hormone* (LH). Both are glycoproteins. FSH causes ovaries to develop in females and sperm in males. LH causes continued development of a follicle and production of a corpus luteum after ovulation. In males, LH stimulates proliferation of cells that produce testosterone.

Thyroid and Parathyroid Glands (pp. 418–421)

Anatomy of the Thyroid Gland The thyroid gland consists of two lobes lying on either side of the trachea and larynx. Its *lobules* contain *thyroid follicles*. Each follicle is a spherical layer of epithelia surrounding a lumen filled with *colloid*. *Parafollicular* cells (*C cells*) are also present.

Thyroid Hormones Thyroid hormones include *triiodothyronine* (T_3) and *thyroxine* (T_4), both of which are formed by iodination of tyrosine. They are included in a glycoprotein, *thyroglobulin*, which is absorbed into the epithelial cells around the colloid in response to TSH.

Thyroid hormones stimulate metabolism, respiration, protein synthesis, fat metabolism, and mineral balance. ATP and heat production are also increased. *Hypothyroidism* can lead to *cretinism* and *myxedema*. *Hyperthyroidism* can result in *exophthalmos* and *goiter*.

Calcitonin is produced by the parafollicular cells. It is a peptide hormone that stimulates uptake of Ca^{2+} and phosphate by bones. Calcitonin production is stimulated by *gastrin*.

Parathyroid Glands There are four *parathyroid glands*, two on the posterior surface of each lobe of the thyroid gland. They contain *chief cells* in children; *oxyphil cells* also appear in an adult. Chief cells produce *parathyroid hormone*, which causes Ca^{2+} to be released from bones. It also stimulates Ca^{2+} absorption in the intestines and inhibits Ca^{2+} excretion by the kidneys. Production of parathyroid hormone is controlled by a negative feedback mechanism.

Adrenal Glands (pp. 422–426)

Adrenal glands are located at the superior end of each kidney surrounded by fatty tissue and a fibrous *capsule*.

Adrenal Cortex The adrenal cortex lies within the capsule. Its three subdivisions are the *zona glomerulosa*, *zona fasciculata*, and *zona reticularis*, which are formed by cells produced just beneath the capsule that migrate inward. These regions produce *mineral corticoids* (including *aldosterone*), *glucocorticoids*, and *gonadocorticoids*, respectively. Mineral corticoids regulate sodium and potassium levels by controlling their release from the kidneys. Glucocorticoids regulate carbohydrate metabolism. Gonadocorticoids are sex hormones.

Adrenal Medulla The adrenal medulla has a system of passages called a *muralium*. *Chromaffin cells* in the walls of these passages release *catecholamines*, primarily *epinephrine* and *norepinephrine*. These hormones are *sympathomimetic* in function. Epinephrine stimulates metabolism and the delivery of blood to the tissues; norepinephrine controls blood pressure. The two hormones often have opposite effects.

Pancreas (pp. 426–429)

The *pancreas* is a heterocrine gland. Its hormone-secreting cells (*alpha*, *beta*, and *delta*) are localized in *islets of Langerhans*.

Insulin Insulin is derived from a protein synthesized in beta cells. Its main effect is to make glucose available to cells by stimulating glucose transport across the membrane. It also stimulates incorporation of glucose into glycogen in the liver. Insulin production is regulated by levels of glucagon, amino acids, cAMP, digestive hormones, and acetylcholine, as well as by neural mechanisms.

Glucagon Glucagon is derived from *proglucagon*. It is a peptide produced by the alpha cells. It primarily increases the level of glucose in the blood by inducing glycogen breakdown and conversion of amino acids to carbohydrates. It acts through a cAMP mechanism.

Somatostatin Somatostatin is a small peptide that inhibits secretion of several other hormones, glucose absorption in the small intestine, HCl secretion in the stomach, and bicarbonate and enzyme secretion from the pancreas. It is produced by delta cells. Somatostatin may be a local hormone, performing its activities in the region in which it is synthesized.

Other Endocrine Glands (pp. 429–433)

Gonads *Gonads* include *ovaries* and *testes*. Ovaries produce *estrogen* and *progesterone*, which stimulate development of sexual characteristics. Testes produce *testosterone*, which causes the development of male sexual characteristics. Testosterone is produced by *interstitial cells of Leydig*. *Sertoli's cells* assist in the production of sperm.

Thymus Gland The *thymus gland* produces hormones called *thymosins* that stimulate lymphocytes to differentiate. The gland is located in the superior portion of the thoracic region.

Pineal Gland The pineal gland is in the diencephalon. It contains *pinealocytes*, which may be sources of hormones, and neuroglia. Calcium and magnesium phosphate and carbonate salts accumulate in some of the cells forming *corpora arenacea* (brain sand). This gland produces *melatonin*, which may influence pigment development.

Sources of Additional Hormones Several additional hormones are produced by organs that are not primarily endocrine organs. Hormone-secreting organs include the placenta, digestive tract, kidneys, blood, and liver.

Prostaglandins Prostaglandins alter the response of a cell to a hormone. Over a dozen are known, all of which are derived from fatty acids. They are generally local hormones and have a variety of functions. Different prostaglandins may have opposing effects.

SELF-TEST OF CHAPTER OBJECTIVES

True-False Questions
1. The hormone system works independently of the nervous system in regulating the body and in maintaining homeostasis.
2. All hormones are secreted by endocrine glands.
3. The specificity of a certain cell's response to specific peptide hormones probably results from the presence of particular receptors in the cell membrane.
4. Glucagon is an example of a hormone that acts through a cAMP mechanism.
5. Neurophysins are involved in the transport of posterior pituitary hormones.
6. The pituitary gland exercises both nervous control and hormonal control of other processes.
7. Some hypothalamic hormones are carried to the pituitary gland by the blood and others are delivered directly by nerves that connect the two glands.
8. POMC is a peptide precursor of ACTH in the anterior pituitary gland.
9. Graves' disease is associated with decreased metabolic rate because of a diminished production of thyroid hormone.
10. Somatomedin is made primarily in the hypothalamus.

Matching Questions
Match the organ in the right column with the hormones that it produces.

11. oxytocin
12. somatostatin
13. thyroid-stimulating hormone
14. melatonin
15. gonadotrophins
16. antidiuretic hormone
17. thyrotrophin-releasing hormone

a. anterior pituitary gland
b. hypothalamus
c. posterior pituitary gland
d. pineal gland

Indicate which of the hormones in the left column are produced directly in response to changes in body conditions and which are produced in response to a pituitary hormone.

18. glucocorticoids
19. thyroxine
20. calcitonin
21. parathyroid hormone
22. insulin
23. estrogen

a. change in condition
b. pituitary hormone induced

Match each of the hormones on the left with the gland that produces it.

24. aldosterone
25. HGH
26. glucagon
27. thyroxine
28. PTH
29. epinephrine
30. insulin

a. thyroid gland
b. anterior pituitary
c. pancreatic beta cells
d. adrenal cortex
e. adrenal medulla
f. parathyroid glands
g. pancreatic alpha cells

Multiple Choice Questions
31. Activation of which of the following occurs latest when a nonsteroid hormone affects a target cell?
    a. ATP
    b. protein kinase
    c. adenylate cyclase
    d. prostaglandin synthetase
32. Which gland depends on one of the others for control of its activity?
    a. pancreas
    b. hypothalamus
    c. pituitary
    d. pineal
33. Iodine is added to the amino acid tyrosine in the part of the thyroid gland called the
    a. parafollicular cells
    b. colloid
    c. isthmus of the thyroid
    d. corpus luteum
34. All of the following but one probably use ATP as a second messenger. Which one doesn't?
    a. thyroid-stimulating hormone
    b. ACTH
    c. ADH
    d. glucocorticoid hormones
35. Which of the following hormones is involved in regulating skeletal growth?
    a. insulin
    b. glucagon
    c. HGH
    d. ADH
36. Which of the following hormones enables cells to use glucose?
    a. ADH
    b. glucagon
    c. insulin
    d. thymosin
37. Prostaglandins are unlike other hormones in all of the following ways but one. Which one is it?
    a. They are active in very small amounts.
    b. They are produced by most, if not all, cells.
    c. They often exert their effect on nearby cells.
    d. They are derivatives of fatty acids.
38. Which of the following hormones probably combines with its receptor inside the cell?
    a. estrogen
    b. prolactin
    c. oxytocin
    d. human growth hormone
39. Which of the following hormones is synthesized in a neuron and travels to the end of the axon from which it is released into the blood?
    a. oxytocin
    b. thyroid-stimulating hormone
    c. glucocorticoid
    d. thyroxine
40. Which of the following hormones influence the maturation of lymphocytes?
    a. gastrin
    b. thymosin
    c. thyroxine
    d. epinephrine

41. Which of the following can be expected to occur as a result of an increase in Ca^{2+} in the blood?
    a. Parathyroid hormone release will be inhibited.
    b. More Ca^{2+} will be released from the bones.
    c. Calcitonin release will be inhibited.
    d. Phosphate levels in the blood will increase.
42. Alpha, beta, delta, and F endocrine cells are characteristic of which structure?
    a. colloid
    b. adrenal medulla
    c. adrenal cortex
    d. islet of Langerhans
43. To which of the following hormones is glucagon most closely related in a negative feedback mechanism?
    a. human growth hormone
    b. thyroid hormone
    c. insulin
    d. thyroxine

Essay Questions

44. Describe the anatomical relationship between the hypothalamus and the pituitary gland.
45. Compare the mechanism of action of hormones whose receptors are in the target cell membrane with those whose receptors are in the cytoplasm.
46. Compare the ways in which hormones are delivered to the two lobes of the pituitary gland.
47. Describe a negative feedback mechanism involving two different hormones.
48. Name a heterocrine organ and list its endocrine products.
49. List three hormones that are also found in nervous tissue.
50. Describe the organization of an adrenal gland and list the hormones that are produced in each region of the gland.
51. Explain the role of alpha, beta, delta, and F cells in the islets of Langerhans.
52. Protein hormones are degraded relatively rapidly compared with steroid hormones. Explain why this is consistent with their role in controlling relatively rapid responses.
53. List several hormones that control the synthesis or release of other hormones.

Clinical Application Questions

54. Describe the symptoms associated with excess insulin production, the treatment employed, and its desired effects.
55. Explain the different bases for ADH-sensitive and ADH-resistant diabetes insipidus and explain why the treatment used for one form is inappropriate for the other.
56. Describe the source and effects of excessive aldosterone production.
57. Compare the effects of a hypoactive adrenal cortex with those of a hyperactive adrenal cortex and name the conditions that result from each.
58. Define pheochromocytoma and describe its effects.
59. Compare the physiological bases of insulin-dependent and noninsulin-dependent diabetes mellitus.
60. Prepare a table listing endocrine disorders, the hormones involved, and the symptoms associated with the disorders.
61. Patients receiving steroid therapy following an organ transplant sometimes develop symptoms characteristic of Cushing's syndrome. Suggest a reason for the similarities.
62. Define acromegaly and describe a cause due to an endocrine disorder.
63. The effects of adrenal virilism are generally more pronounced in women than in men. Suggest a reason.

UNIT III CASE STUDY

PARALYSIS

As an overview of our unit on nervous and chemical integration and control of the body, we look at the case of a young girl affected by one of the most severe problems of the nervous system: paralysis. **Paralysis** is the loss of motor function in the affected area of the body, and it results from severe injury to the brain or spinal cord. There are two types: paraplegia, if the injury is below vertebra C7, and quadriplegia, if between C2 and C7 (see Chapter 12 Clinical Note: Spinal Cord Injury). Injury to the spinal cord above the C2 level is often lethal since it can involve the nerves to the heart and respiratory muscles.

Our patient is a 13-year-old girl who jumped from a dock into shallow water while swimming with her friends. She was seen to jump feet first into the water, appeared to hyperextend her neck as her feet struck the bottom, and collapsed unconscious into the water. She was quickly removed from the water by her friends, and she slowly regained consciousness but was unable to move her arms or legs and complained of pain in her neck. When the emergency rescue team arrived, they quickly examined her, placed a protective neck splint on her, strapped her to a spine board, and transported her to a nearby hospital. She was given oxygen by mask during the transport. Her vital signs recorded her blood pressure as 90/60, pulse rate at 100 per minute, and respirations at 12 per minute. The emergency medical technician started an IV line in her right arm and noted that the arm was flaccid (limp) and that she did not complain of pain or attempt to withdraw the arm when the needle was inserted.

The emergency room doctor quickly noted that she was able to speak, was frightened, pale, sweating, and immobile. Repeat vital signs confirmed hypotension (low blood pressure), rapid pulse, and slow respirations, all suggesting surgical shock. The intravenous fluid rate was increased, her head lowered, oxygen continued, and cortisone administered through the intravenous line to help combat the shock. A complete neurologic examination confirmed a flaccid paralysis of all four extremities, absence of the deep tendon reflexes, and absence of sensation to pin prick and light touch below her collar bones. X rays of her cervical spine, which were obtained without removing the neck collar, showed a fracture dislocation of cervical vertebrae 5 and 6. She was returned to the emergency room where a neurosurgeon inserted a traction apparatus into her skull under local anesthesia to stabilize her neck. She was then admitted to the intensive care unit where her hypotension was stabilized and preparations were made for surgery.

In the operating room, she was anesthetized with intravenous Demerol, an endotracheal tube was inserted into her trachea to control her respirations, and she was carefully positioned face down while maintaining the alignment of her neck at all times. Surgical exploration showed her spinal cord to be intact but bruised by bone fragments from the fracture. The bone fragments were carefully removed, the dislocation reduced, and finally, the C5 and C6 vertebrae were fused together to provide stability. At the completion of the surgery, a brace was installed to hold her neck rigid, and she was placed on a special frame bed that would allow her to be turned from her back to her stomach easily.

In the intensive care unit following surgery, her spinal cord shock slowly subsided over a period of two weeks and she gradually developed more sensation in her arms and then her legs. During the next several weeks, she first regained use of her arms and then her legs. At 8 weeks the neck brace was removed. Months of intense physical therapy gradually restored her muscle strength, and she progressed from a wheel chair to walking with a crutch. A year following her injury, she was walking, but still required a leg brace to correct a residual foot drop.

This patient was very fortunate in making an excellent recovery from a very serious injury. Many patients with similar injuries sustained in motor vehicle accidents, athletic activities, or severe falls are left with permanent paraplegia or quadriplegia and require constant assistance for the rest of their lives.

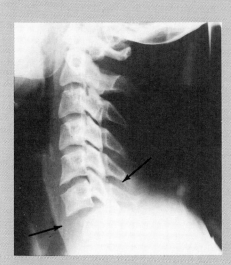

X ray of fractured spine showing displacement of vertebrae (arrows).

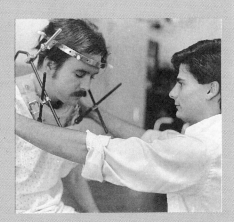

Common neck brace used in neck and back injuries.

Unit IV

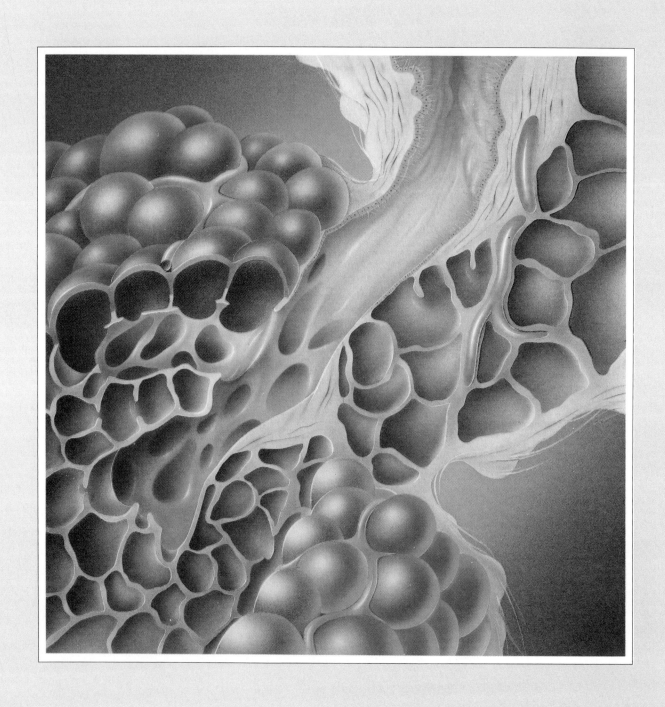

Metabolic Processes: Digestion and Respiration

Unit IV deals with the metabolic processes involved in the exchange of matter and energy between the body and its environment. We begin with the digestive system, which is specialized to procure and process food. Within it, large complex molecules are broken down into smaller molecules that are absorbed into the blood and carried to the cells. Some molecules are used to produce new body tissues, while others are split to yield heat and energy to power chemical reactions within the cells. The rates of these reactions are regulated by homeostatic mechanisms so that a constant body temperature is maintained. The unit ends with an examination of the respiratory system, which provides oxygen needed for energy transfer reactions and rids the body of carbon dioxide produced in those processes.

18
The Digestive System

19
Metabolic Processes

20
The Respiratory System

18
The Digestive System

The food we eat serves two purposes: it provides raw material for building tissue, and equally important, it provides the energy needed to drive the chemical reactions of life. Most food, however, consists of molecules that cannot be used directly because of their size and complexity, and it must be converted to forms that can be absorbed and used. The digestive system performs this function, breaking down large complex molecules and producing molecules that are absorbed and delivered to the blood for distribution.

The digestive system includes the organs of the **alimentary canal** (also called the **gastrointestinal tract**, or **GI tract**) and accessory organs (Figure 18-1). The alimentary canal includes the *mouth, esophagus, stomach, small intestine, large intestine,* and *anus*. Accessory organs include the *teeth, tongue, salivary glands, liver, gallbladder,* and *pancreas*.

In this chapter we will examine

1. the structure of the organs of the digestive system,
2. the mechanisms used for the physical breakdown of food,
3. the chemical processes used in digestion,
4. how and where in the digestive system nutrients are absorbed, and
5. mechanisms employed to regulate the activity of the digestive system.

alimentary: (L.) *alimentum*, nourishment

Histology

After studying this section, you should be able to:

1. Describe in general the histological organization of the alimentary canal, listing the tunics that make up the wall.
2. List the subdivisions present, if any, in each of the tunics.
3. List the tissues present in each of the subdivisions.

The wall of the alimentary canal is composed of four concentric layers of tissue, or **tunics** (Figure 18-2): the tunica mucosa, tunica submucosa, tunica muscularis, and tunica serosa. Modifications and specializations of each layer enable them to perform specialized functions in different parts of the canal.

The **tunica mucosa** (tū′nĭ·kă mū·kō′să) is the innermost tunic. It is a composite tissue consisting of three subdivisions: **epithelium, lamina propria** (prō′prē·ă), and **muscularis mucosa**. The epithelium, the innermost layer of the tunic, makes direct contact with the contents of the canal. It is protected by its own secreted mucus and that of the lamina propria. The lamina propria contains loose connective tissue, blood vessels, nerves, and lymph vessels. The outermost division of the tunica mucosa, the muscularis mucosa, is a thin cylinder of muscular tissue lying medial to the lamina propria.

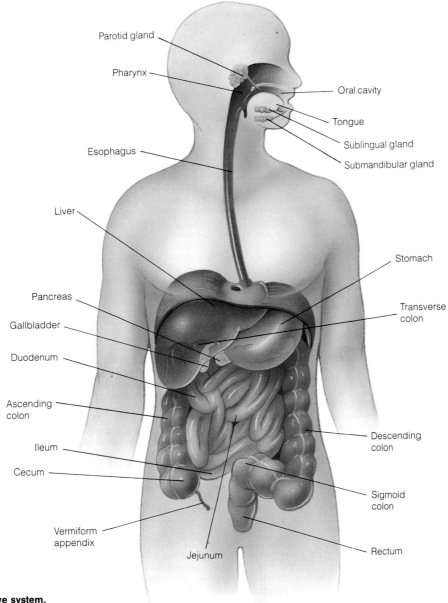

Figure 18-1 The digestive system.

Embryonic Development of the Digestive System

Examination of a 19-day-old embryo reveals a three-layered disk perched atop a fluid-filled **vitelline** (vī·tĕl′ēn) **sac** (Figure 18-A). Ectoderm forms the embryo's outer layer and mesoderm the middle layer; the inner endoderm is continuous with the sac; the sac's roof is the presumptive gut. The endoderm contacts the ectoderm of the embryonic disk at opposite ends, forming the oropharyngeal and **cloacal membranes.**

Differential growth of the head and tail ends of the disk result in the development of a distinctly tubular **foregut** and **hindgut,** with the **midgut** in between. Continued elongation of the embryo isolates the gut from the sac. The vitelline sac soon becomes incorporated into the umbilical cord and makes no further contribution to the gastrointestinal tract.

By the third week of development, depressions have formed on the ectodermal side of the oropharyngeal and cloacal membranes. These depressions, the **stomodeum** (stō″mō·dē′ŭm) and **proctodeum** (prŏk·tō·dē′ŭm), soon break through, forming the openings into the mouth and anus, respectively. Continued development sees the appearance of outgrowths from the tube that produce the thyroid and parathyroid glands, salivary glands, liver, gallbladder, and pancreas. The thyroid and parathyroids lose their connection to the gut and become part of the endocrine system. The salivary glands, liver, gallbladder, and pancreas remain connected to the gut through ducts that conduct their secretions into the mouth and intestine.

Figure 18-A Embryonic development of the digestive system. (a) The beginnings of the digestive system appear late in the third week of development. (b) Elongation of the embryo occurs during the fourth week and (c) turns the foregut and hindgut into pouches, separate from the vitelline sac. (d) By the end of the fourth week, the tubular nature of the gut is established and several organ buds have appeared.

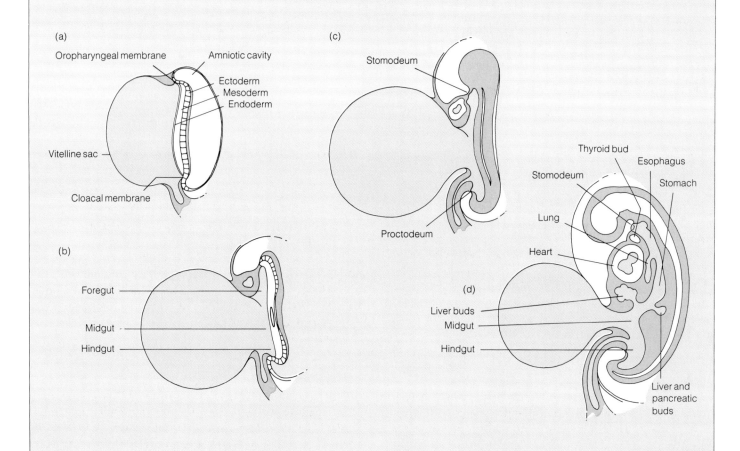

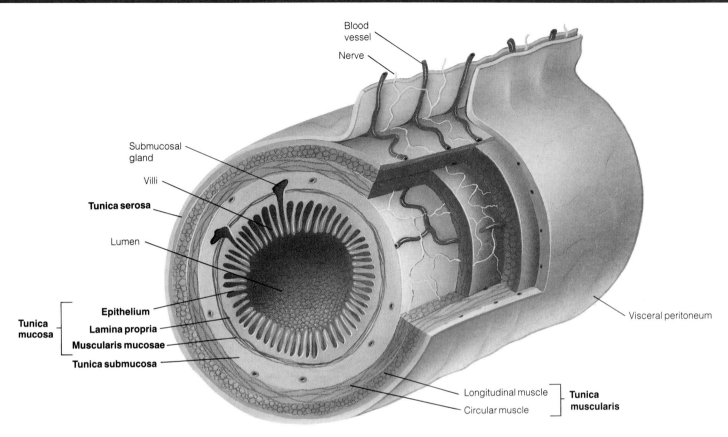

Figure 18-2 Histological organization of the digestive tract. The tract is composed of four concentric tunics. From the inside out they are the mucosa, submucosa, muscularis, and serosa.

The **tunica submucosa**, the second tunic, is a thick layer of loose connective tissue that contains blood and lymph vessels, autonomic nerves, and the bottom portions of many glands that extend through the tunica mucosa. The tunica submucosa also binds the mucosal tunic to the tunica muscularis.

In most areas of the alimentary canal, the **tunica muscularis** consists of an inner layer of muscle in which muscle fibers oriented in a circular direction surround the tunica submucosa and an outer layer in which fibers are oriented longitudinally. Muscle in the upper end of the esophagus undergoes a transition from voluntary striated fibers to involuntary smooth fibers. The remaining muscle throughout the canal, except for the anal opening, is entirely smooth muscle.

The **tunica serosa** (sē·rō′să), the outermost tunic, is absent for most of the length of the esophagus. It first appears at the base of the esophagus, where it forms the **tunica adventitia** (ăd″věn·tĭsh′ē·ă) and extends through the remainder of the alimentary canal.

Mouth

After studying this section, you should be able to:

1. Identify and describe the structures associated with the mouth.
2. Describe the structure and organization of the tongue.
3. List the salivary glands, describe their location, and describe how their secretions are delivered to the mouth.
4. Describe the composition and function of saliva.
5. Describe the muscular contractions involved in swallowing.

The **mouth** is the entrance to the alimentary canal (Figure 18-3). It is formed by the lips anteriorly, the cheeks laterally, the palate above, and the floor of the mouth and tongue below. The volume enclosed by these organs is the **oral** or **buccal** (bŭk′ăl) **cavity**. The tongue extends from the floor of the mouth into the oral cavity.

frenulum: (L.) *frenulum*, little bridle
papilla: (L.) *papilla*, nipple
dentin: (L.) *dens*, tooth

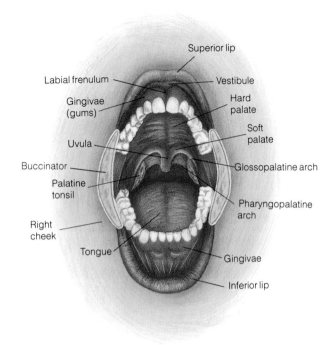

Figure 18-3 The oral cavity.

The **lips** or **labia** (lā´bē·ă) and **cheeks** are muscular structures. Both are lined internally with stratified squamous epithelium that contains mucous glands. Externally, the cheeks are covered by skin. The lips lack the keratinized outer layer characteristic of skin and are instead covered by a thin, nonkeratinized epithelium that is kept moist by licking the lips. Blood in underlying tissue shows through the transparent surface layers, giving lips their characteristic red color. Large numbers of sensory receptors make the lips especially sensitive to touch and temperature. The space between the lips and cheeks and the teeth is the **vestibule.** The lips are attached to the upper and lower gums by an **upper** and **lower labial frenulum** (frĕn´ū·lŭm).

The **palate** consists of two parts: an anterior **hard palate** and a posterior **soft palate.** The hard palate, formed by the palatal process of the maxillae and the palatine bones, is covered by a mucous membrane. The soft palate lies between the oral cavity and the upper portion of the pharynx, a space in the throat region shared by the digestive and respiratory systems (see p. 449). A small fingerlike structure, the **uvula** (ū´vū·lă), hangs from the posterior border of the soft palate. The soft palate is attached laterally to the tongue by the **glossopalatine** (glŏs″ō·păl´ă·tīn) **arch** or **anterior pillar,** a muscular fold of tissue, and posteriorly to the walls of the pharynx by another muscular fold, the **palatopharyngeal** (păl″ă·tō·făr·ĭn´jē·ăl) **arch** or **posterior pillar.** **Palatine tonsils** are masses of lymphatic tissue located in the depression between the two arches. Because they are prone to infection, it is often necessary to remove the palatine tonsils in a tonsillectomy.

Tongue

The **tongue** covers most of the floor of the oral cavity (Figure 18-4a). It consists of connective tissue (Figure 18-4b) and three sets of muscles oriented longitudinally, horizontally, and vertically. The tongue is covered by a mucous membrane and attaches to the floor of the mouth by the **lingual** (lĭng´gwăl) **frenulum.** Several muscles attach the tongue to the mandible, the hyoid bone, and the styloid process of the temporal bone. These muscles move the tongue.

The upper and lateral surfaces of the tongue contain three kinds of specialized projections called **papillae** (pă·pĭl´ē). Small cone-shaped **filiform** (fĭl´ĭ·form) **papillae** are distributed over most of the upper surface of the tongue (Figure 18-4). Larger mushroom-shaped **fungiform** (fŭn´jĭ·form) **papillae** are interspersed among the filiform papillae, being most numerous near the tip of the tongue. About a dozen **vallate** (văl´āt) **papillae** form an inverted V-shaped cluster near the rear of the tongue. Vallate papillae are the largest of the three papillae (Figure 18-4c). Rounded masses of lymphatic tissue, collectively called the **lingual tonsil,** lie just behind the vallate papillae covering the posterior third of the tongue.

Taste buds are groups of specialized nerve cells sensitive to sweet, salty, sour, and bitter tastes. They are present in the fungiform and vallate papillae. (See Chapter 16, and specifically Figures 16-2 and 16-3, for more on the taste buds.)

Gingivae and Teeth

The **gingivae** (jĭn·jī´vē) or **gums** are a layer of nonkeratinized stratified squamous epithelium and dense fibrous connective tissue that cover the alveolar processes of the maxillae and mandible (Figure 18-3). The epithelium also attaches to the enamel of the teeth, and the connective tissue is continuous with connective tissue that surrounds the roots of the teeth.

Teeth protrude from **sockets (alveoli)** in the alveolar borders of the maxillae and mandible. Figure 18-5 depicts a longitudinal section of a tooth. A tooth has three principal regions: a **crown,** which is an exposed portion covered by an extremely hard crystalline **enamel;** one to three **roots,** which are elongate extensions that anchor the tooth in the alveolus; and a **neck,** which is a narrow constriction between the crown and roots. Much of the tooth interior is **dentin** (dĕn´tĭn), a substance similar to bone. Dentin surrounds a **pulp cavity** that contains blood vessels, nerves, and a soft connective tissue called the **pulp.** The pulp cavity narrows into one or more **root canals** that extend into the roots. Blood vessels and nerves pass through a small **apical foramen** at the end of each root canal.

periodontal: (Gr.) *peri*, around + *odous*, tooth
cuspid: (L.) *cuspis*, point

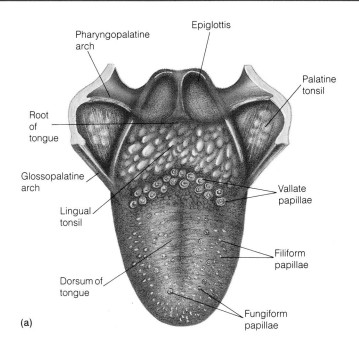

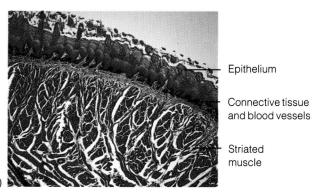

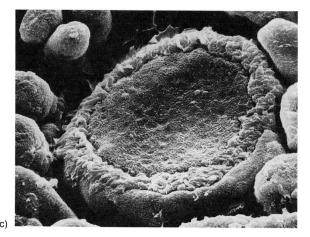

Figure 18-4 The tongue. (a) Drawing of the dorsal surface. (b) Cross-sectional photomicrograph showing tissues of the tongue (mag. 32×). (c) Scanning electron micrograph of a vallate papilla.

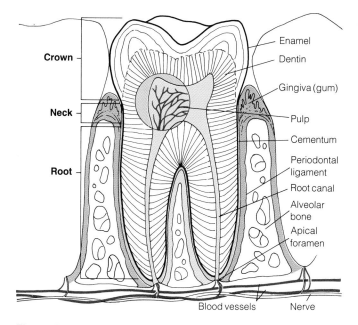

Figure 18-5 Cross-sectional view of a molar tooth.

Roots are anchored in the alveolus by **cementum** (sē·měn′tŭm), another bonelike material that surrounds the roots. The **periodontal** (pĕr″ē·ō·dŏn′tăl) **ligament** lining the alveolus is a modification of the periosteum of the bone that forms the tooth socket. In addition to helping anchor the tooth, the periodontal ligament contains blood vessels that deliver nutrients to the tooth.

Adult humans possess four types of teeth (Figure 18-6). **Incisors** are flat, chisel-shaped teeth located at the front of the mouth. They are used for cutting food. On either side of the incisors are cone-shaped **cuspids** (kŭs′pĭds) (or **canines**) used in tearing food. Posterior to the cuspids are **bicuspids** (or **premolars**) and **molars**, both of which possess relatively broad crowns equipped with rounded **cusps** (or **tubercles**) used to grind and crush food.

A person normally grows two sets of teeth: a deciduous set (sometimes called milk or baby teeth) and a permanent set. **Deciduous teeth** begin erupting through the gums at about 6 months of age with the appearance of incisors. Eruption is usually complete by about 2 years of age. **Permanent teeth** begin to replace deciduous ones in 6-year-old children. Time of completion of eruption varies, usually occurring between 17 and 21 years. The most posterior molars (**wisdom teeth**) are present only in the permanent set and sometimes never erupt. If the jaws are not long enough to accommodate them, wisdom teeth may be unable to grow through the angular space between the molar in front and the coronoid process of the mandible to the rear (Figure 18-7).

CLINICAL NOTE

TEETH AND GUM DISORDERS

The vast majority of dental problems are either disorders of the teeth or of the periodontium. The **periodontium** refers to the supporting structures of the teeth, that is, the gums and underlying bone.

Tooth Decay

Tooth decay, or **dental caries** (commonly called cavities), is a familiar malady for most of us and the principal cause of tooth loss up to age 35. Cavities in teeth result from the action of certain bacteria, principally *Streptococcus mutans,* on sugars or other fermentable carbohydrates in the mouth. The sugars are converted by the bacteria to acid, which begins at once to demineralize the surface of the teeth. Along with the demineralization comes a loss of the organic matrix and thus the formation of a cavity (Figure 1). As the decay process proceeds through the very hard enamel, it reaches the inner bulk of the tooth, the dentin, which decays much more quickly because it is less dense. Allowed to go unchecked, the decay will eventually reach the central pulpal chamber causing the death of the pulp and resulting in an abscess in the alveolar bone surrounding the root tip. Once the process reaches this stage, root canal therapy in addition to restorative procedures is necessary to save the tooth.

Researchers are discovering that saliva plays a very important role in the prevention of tooth decay. It has a limited ability to neutralize and rinse the caries-producing acids off the surface of the teeth. Saliva also supplies the tooth area with calcium and phosphorus. These minerals fill in tiny cracks and scratches in the tooth enamel, and some remineralization can occur in the early stages of the decay process.

Tooth structure lost to decay, however, does not replace itself. Fortunately, artificial materials such as silver-amalgam, gold, and the newer composite resins can be used by the dentist to restore teeth and help protect them from further decay. Although the silver and gold alloys have been improved over the years, their use dates back to the earliest days of dentistry. Most recent advances in dental restorative materials have been in the area of composite resins, first introduced in the 1950s. At this stage of their development they are not as durable as the alloy restorations, but they are tooth colored and thus have more aesthetic appeal. Additionally, because they may be bonded to tooth structure, they allow more conservative restorative procedures by the dentist and are often used to close gaps between front teeth and as an alternative to crowns.

Periodontal Disease

The most common cause of tooth loss in the adult is **periodontal disease,** or "gum" disease. Teeth may be lost to gum disease that have never had any decay. Again, bacteria are the principal cause, especially the bacteria that reside in the dental plaque. **Dental plaque** is an almost invisible film that forms on the teeth. Unless it is constantly removed by scrupulous hygiene, it becomes a protected refuge for not only decay-causing bacteria, but bacteria that cause periodontal disease as well. As the organisms become organized and colonized in the plaque, their bacterial by-products are a source of irritation to the gum tissue. This first causes **gingivitis,** the red, tender, bleeding gums that are the early manifestation of periodontal disease (Figure 2a). At this stage the process is easily reversed by proper cleaning procedures, but if allowed to proceed, destruction of the attachment between tooth and gum tissue occurs, resulting in pocket formation (Figure 2b). Because the dental pockets provide an even more protected environment for the destructive bacteria, professional intervention is generally required to prevent further loss of gum tissue, underlying bone, and ultimately the tooth. Usually, mineral salts are also deposited in some areas of the plaque causing it to form a hard substance commonly called **calculus** or "tartar." This is also irritating to the gum tissue and must be periodically removed by the dentist or dental hygienist.

Although other factors such as genetics, diet, and general health all play a role, the primary preventive and therapeutic focus in dental disease must be on the continued, lifelong battle to remove the constantly reforming plaque. This continues to be a major research area in dentistry, and new chemical agents to remove plaque or prevent its formation are currently under investigation.

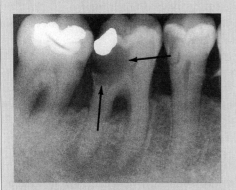

Figure 1 X ray of a tooth with dental caries.

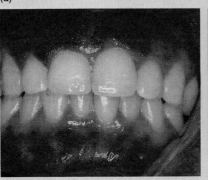 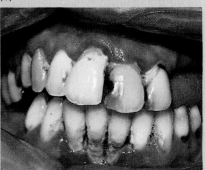

(a)   (b)

Figure 2 Gum disorders. (a) Gingivitis. (b) Advanced periodontal (gum) disease.

parotid: (Gr.) *para*, beside + *ous*, ear

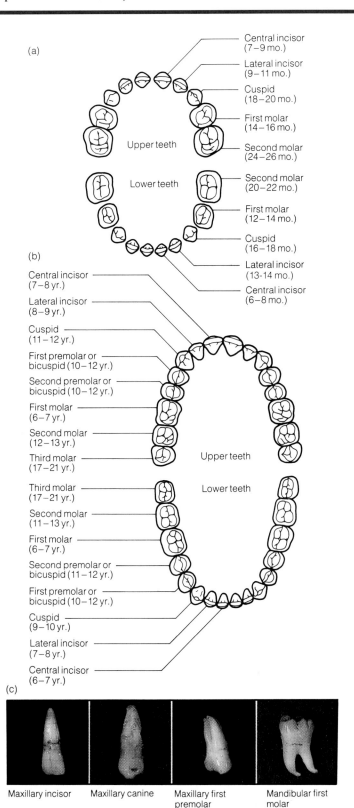

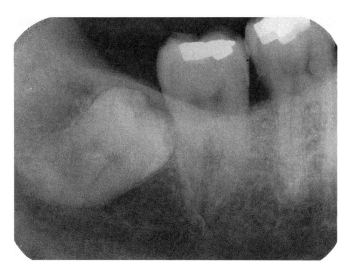

Figure 18-7 X ray of an impacted wisdom tooth. Continued growth of the tooth on the left will cause the teeth in front of it to be pushed out of alignment.

Figure 18-6 (a) Deciduous teeth in the upper and lower jaws. (b) Permanent teeth. Times of eruption are shown in parentheses. (c) Photographs of four types of teeth.

Such a condition is referred to as **impaction** of the wisdom teeth. Impacted wisdom teeth are usually removed surgically to prevent their pushing anterior teeth out of alignment.

Salivary Glands

There are three pairs of salivary glands: the parotid, submandibular, and sublingual glands (Figure 18-8). The **parotid** (pă·rŏt′ĭd) **glands** lie just below and in front of the ears. They are the largest salivary glands. Each empties into the side of the mouth through the **parotid** (or **Stensen's**) **duct**, which passes through the buccinator muscle. The duct opens in the inner surface of the cheek near the upper second molar.

The **submandibular** (or **submaxillary**) **glands** lie along the rear interior surface of the body of the mandible. Because these glands contain serous as well as mucous cells, saliva from them tends to be more viscous than that from the other two glands. Submandibular glands empty into the floor of the mouth near the frenulum of the tongue through **submandibular** (or **Wharton's**) **ducts.**

The **sublingual glands** lie in the floor of the mouth under the tongue. They are the smallest salivary glands and empty into the floor of the mouth through several small ducts.

In addition to the three large pairs of salivary glands, numerous small **buccal glands** are in the mucous membrane that lines the mouth. These glands also secrete saliva but in smaller quantities than the salivary glands.

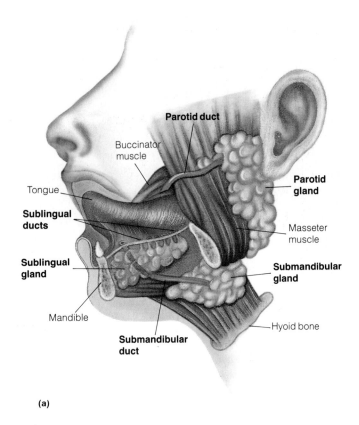
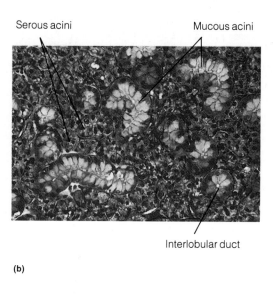

Figure 18-8 (a) The salivary glands. In addition to these glands, there are many smaller glands in the submucosa that also produce saliva. (b) Photomicrograph of a submandibular gland.

Digestion in the Mouth

Mechanical breakdown of solid foods usually begins with **mastication,** or chewing, as food is manipulated by the tongue and mouth and ground between the teeth. At the same time, saliva is secreted and mixed with the food to create a semifluid mass that is more easily swallowed. Grinding the food into smaller particles also increases its surface area and makes subsequent chemical digestion more efficient.

Altogether, approximately 1000 mL of saliva are secreted daily, the submandibular glands contributing about 700 mL, the parotid glands about 200 mL, and the sublingual glands about 50 mL. The remaining 50 mL is secreted by the buccal glands.

Saliva consists of about 99.5% water and about 0.5% proteins and ions of sodium, potassium, calcium, bicarbonate, chloride, and phosphate. Its pH ranges between 6.0 and 7.0. Most salivary protein is **mucin** (mū´sĭn), a glycoprotein that lubricates the food, as well as salivary amylase, an enzyme involved in starch digestion. Also present in smaller amounts are **albumin** and **globulin** (proteins present in blood serum), various enzymes, urea, and dissolved oxygen and carbon dioxide. In addition to dissolved substances, saliva contains squamous cells from the lining of the mouth, gland cells that have broken loose from salivary glands, leucocytes, and bacteria.

Saliva has many functions. In addition to making swallowing easier, its lubricating effect makes speech easier by decreasing friction between the tongue and lips. It is a solvent for components of food that stimulate taste buds. The enzymatic action of amylase helps dislodge food particles from the teeth, and another enzyme, **lysozyme** (lī´sō·zīm), helps control the bacterial population and minimizes the effect of microorganisms on the teeth and gums.

The only component of food subjected to chemical digestion in the mouth is starch, which is cleaved into maltose by salivary amylase. The extent to which starch is digested depends on the amount and time it is exposed to amylase. Normally, food is swallowed so quickly that amylase has time to digest only a small amount before the mixture enters the stomach. Once in the stomach, the acidic conditions quickly denature the amylase and stop its action.

deglutition: (L.) *de*, down + *glutire*, to swallow
peristaltic: (Gr.) *peri*, around + *stalsis*, contraction

Pharynx and Esophagus

After studying this section, you should be able to:

1. Describe the structure of the pharynx and esophagus.
2. List three stages of swallowing.
3. Distinguish between the voluntary and involuntary components of swallowing.
4. Describe the neurological control and mechanism of a swallowing reflex.
5. Explain how food is prevented from entering the larynx and is forced into the stomach during swallowing.

Pharynx

The **pharynx** (făr'ĭnks) lies between the nasal and oral cavities and the larynx and esophagus, where it serves as a common passageway for food going to the stomach and air going to the lungs (Figure 18-1). The major portion of the pharynx, the **oropharynx** (ōr″ō·făr'ĭnks), is located posterior to the soft palate, beginning at a narrow region formed by the base of the tongue, the soft palate, and tissues that surround the palatine tonsils. The oropharynx leads into the **laryngopharynx,** a short tube that leads to the esophagus.

Esophagus

The **esophagus** (ē·sŏf'ă·gŭs) is a flattened tube about 25 cm long that passes from the laryngopharynx to the superior end of the stomach (Figure 18-1). The first few centimeters of the esophagus contain skeletal muscle, which undergoes a transition to two layers of smooth muscle: an outer longitudinal layer and an inner circular layer. Mucous-secreting glands lie in the lamina propria of the upper and lower portions of the esophageal wall, and their secretions lubricate the wall. The wall of the lower 2 to 6 cm is a region of thick muscle that forms the **lower esophageal sphincter,** a constriction that normally prevents backflow of stomach contents.

Swallowing

Swallowing, or **deglutition** (dē″glū·tĭsh'ŭn), is a complex sequence of muscular contractions that forces the mixture of food and saliva into the stomach. The process involves three stages.

Buccal Stage

The first stage of swallowing, the **buccal stage,** is voluntary and involves pushing the liquid (if one is drinking) or the wad of chewed food called a **bolus** (bō'lŭs) into the posterior region of the oral cavity. During this stage the tongue presses upward against the hard palate and forces the bolus past the tonsillar pillars into the pharynx (Figure 18-9a). At this point swallowing enters the pharyngeal stage and becomes involuntary.

Pharyngeal Stage

The **pharyngeal stage** is initiated by nerve impulses from **swallowing receptor areas** located around the opening to the pharynx. Impulses pass from these receptors through the trigeminal and glossopharyngeal nerves to a **swallowing center** in the medulla oblongata. The swallowing center responds by sending signals through cranial and cervical nerves to muscles in the pharyngeal wall.

During the pharyngeal stage the posterior wall of the pharynx moves forward, causing the soft palate to rise (Figure 18-9b). Meeting of the palate and pharyngeal wall causes the nasopharynx to close and the pharynx to narrow to a slit-shaped opening that prevents the passage of larger, unchewed particles. At the same time the larynx moves upward and forward, raising the opening of the larynx above the level of the bolus. As the bolus is forced through the pharynx, the epiglottis covers the opening to the larynx and stretches the opening to the esophagus. Simultaneously, folds of tissue in the larynx tighten and close the passageway into the larynx. At this point the **upper esophageal** (or **hypopharyngeal**) **sphincter,** a muscular ring at the superior end of the esophagus, relaxes and allows the bolus to be forced into the esophagus. This is the end of the pharyngeal stage.

Esophageal Stage

The bolus moves through the esophagus in the third stage of swallowing called the **esophageal stage** (Figure 18-9c). With relaxation of the hypopharyngeal sphincter, the bolus is forced into the esophagus by the muscular contraction of the pharyngeal walls.

Entry of the bolus stimulates the esophageal wall to contract, forcing the bolus still further into the esophagus. The bolus continues to be forced downward by a **peristaltic** (pĕr″ĭ·stăl'tĭk) **wave,** a coordinated involuntary contraction of circular muscle just above the bolus and of longitudinal muscle just below it (Figure 18-10). **Primary peristaltic waves** originate in the pharynx and pass into the esophagus. **Secondary peristaltic waves** originate in the esophagus itself, where they are stimulated by distension of the esophagus caused by a bolus that has been passed over by a primary wave. The combined effect of primary and secondary waves is to force the bolus through the esophagus and

450 Chapter 18 The Digestive System

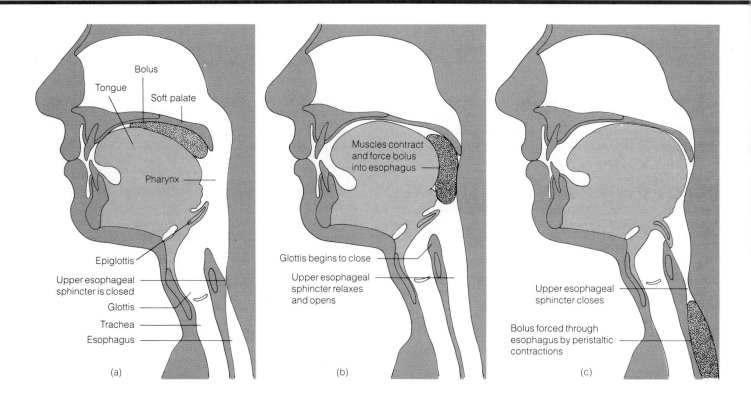

Figure 18-9 **The swallowing reflex.** (a) Buccal stage. (b) Pharyngeal stage. (c) Esophageal stage.

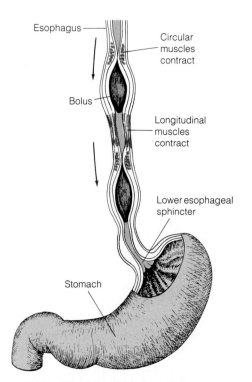

Figure 18-10 **Peristalsis.** Coordinated contractions of the longitudinal and circular muscles of the esophagus squeeze the bolus through the esophagus and into the stomach.

into the stomach, a transit that normally takes about 4 to 8 seconds for solid and semisolid food.

As a peristaltic wave reaches the lower esophageal sphincter, the sphincter relaxes and allows passage of the bolus into the stomach. Sometimes the resting pressure of the lower sphincter is not enough to close the lower sphincter, and stomach contents flow back into the esophagus. When the stomach is irritated by spicy foods, infection, or even nervousness, pressure can develop that overcomes the lower esophageal sphincter and forces stomach contents into the esophagus. The acidic stomach contents can irritate the esophageal wall and lead to **pyrosis** or "heartburn." Sometimes the sphincter pressure is chronically low and stomach contents enter the lower portion of the esophagus often enough to cause esophageal ulcers (see The Clinic: Disorders of the Gastrointestinal Tract).

Aside from contributing a path from the pharynx to the stomach, the esophagus contributes little to digestion. No digestive enzymes are known to be secreted and no absorption is known to occur.

pylorus: (Gr.) *pyloros,* gatekeeper

Stomach

After studying this section, you should be able to:

1. Describe the anatomy of the stomach.
2. List the glands in the wall of the stomach, describe their cellular structure, and cite their products.
3. Explain how protein-digesting enzymes are activated.
4. Describe how HCl is produced and released into the stomach and discuss its role in digestion.
5. Describe the mechanism used to control stomach emptying.

At the lower end of the esophagus, the gastrointestinal tract enlarges into a pouchlike organ called the **stomach.** It is usually illustrated as a J-shaped organ (Figure 18-11), but it is variable in size and shape depending on how full it is and the extent to which the muscle in its wall is contracting. The opening into the stomach at the base of the lower esophageal sphincter is the **cardiac orifice.** The exit at the other end of the stomach is the **pyloric** (pī·lōr′ĭk) **orifice,** which is guarded by a muscular **pyloric sphincter.** The right side of the stomach forms the **lesser curvature** and the left side the **greater curvature.** The lumen (internal space) of the stomach is divided into three regions: the **fundus** (fŭn′dŭs), which bulges to the left and above the cardiac orifice, the central **body,** and a terminal region that narrows to form the **pylorus** (or **pyloric antrum**).

The four tunics of the stomach wall are highly modified. The tunica muscularis is composed of three layers of fibers, in contrast to two layers in other portions of the alimentary canal. In addition to the circular and longitudinal muscles, there is an internal layer of fibers oriented in an oblique direction. Contraction of these muscle layers when the stomach is empty produces thick longitudinal folds in the two inner tunics called **rugae** (rū′gē). When the stomach fills, its walls stretch, and the rugae flatten. The stratified squamous epithelium on the surface of the esophageal tunica mucosa changes to simple columnar epithelium at the cardiac orifice. This epithelium lines the remainder of the digestive tract.

Digestion in the Stomach

When empty, the stomach is a flaccid and extensible bag. There is little tone in the muscular wall, which allows the bag to stretch as food is delivered to it. Only as the stomach reaches its limit of elasticity does the wall begin to tighten and the pressure within it begin to increase. An adult stomach can accept about 1 L of food before this limit is reached.

Figure 18-11 **The stomach.** This drawing shows the muscle layers and rugae.

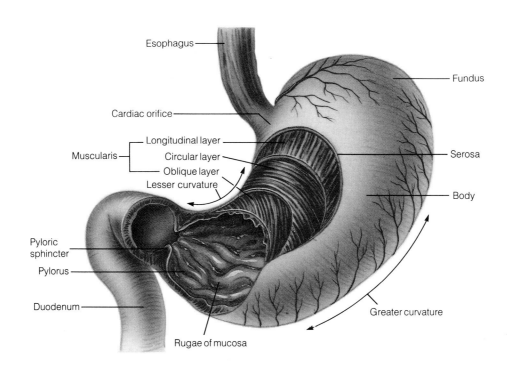

gastric: (Gr.) *gaster*, stomach
chyme: (Gr.) *chymos*, juice

CLINICAL NOTE

PEPTIC ULCER DISEASE

Pure gastric juice is capable of digesting and destroying all living tissue. The mucosa of the stomach is protected from the destructive effects of the gastric juice by compactly arranged epithelial cells with tight junctions between them, a thick layer of mucus, rapidly regenerating mucosal cells, alkaline secretions of the pancreas and intestine, and a good blood supply. If these mechanisms fail and adequate repair of mucosal injury does not occur, craters or pits form in the mucosa. The craters are produced by the loss of inflamed tissue. This condition is called **peptic ulcer disease (PUD)** (see figure).

Peptic ulcer disease can occur in any region of the gastrointestinal tract that is exposed to gastric juices: the lower esophagus, the stomach, or the duodenum. PUD most commonly affects the lesser curvature of stomach (**gastric ulcers**) or the first 3 to 4 inches of the duodenum (**duodenal ulcers**). Decreased tissue resistance plays a part in both types but appears especially important in persons with gastric ulcers. Excessive secretion of acid, due to neurogenic or humoral influences, seems to play a bigger part in duodenal ulcers. A susceptibility to PUD also appears to be hereditary. Gastric ulcers are one-tenth as common as duodenal, the age range is usually higher, and they affect more women than men. Women have a lower incidence and lesser severity of duodenal ulcers. People with blood group O are 35% more likely to have duodenal ulcers, and their ulcers are more likely to bleed and perforate the wall of the organ involved.

A chronological parallelism exists between emotional pressure and the onset of PUD. Emotional tension, associated with a prolonged inability to deal with problems and internal conflicts in susceptible individuals, seems to be associated with ulcers. Prolonged emotional turmoil tends to produce vascular engorgement, increased secretory and muscular activity in the stomach, and increased susceptibility of the mucosa to injury.

PUD pain occurs in the abdomen above the stomach and is described as being gnawing, boring, or burning. The pain is usually absent before breakfast, but occurs 1 to 4 hours after breakfast, lasting about a half hour or until lunch. It then occurs 1 to 4 hours after lunch and is more severe. It occurs again in the evening (but is usually less severe than in the afternoon) and may occur again after midnight and wake the person from sleep. These symptoms may persist for weeks then disappear only to return again several months later. Nausea is not unusual. Vomiting may occur, especially if the ulceration compromises the pyloric sphincter. Affected persons may either gain or lose weight depending on whether they are too distressed to eat or if they overeat to control the discomfort.

Peptic ulcer disease is treated with medications, avoidance of aggravating substances (such as cigarettes, coffee, aspirin, and alcohol), and stress-reducing lifestyle changes. Several classes of drugs are used in treatment, the most effective being the H-2 receptor antagonists cimetidine and rautidine. They cause marked decrease in acid production by the stomach.

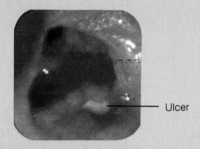

Endoscopic photograph of a peptic ulcer in the stomach.

As the stomach stretches upon the entry of a bolus, its wall reacts by initiating a series of contractions called **mixing waves** that occur about three times per minute. Mixing waves combine the bolus with secretions of the stomach wall called **gastric juice** and convert the semisolid bolus to a milky fluid called **chyme** (kīm). Mixing waves also force the digesting bolus toward the pylorus.

As it approaches the pyloric sphincter, the mixture of chyme and gastric juice is subjected to stronger peristalsis that occurs about every 20 seconds. These contractions comprise the **pyloric pump** and propel about 1 to 3 mL of chyme into the duodenum with each contraction.

Source and Composition of Gastric Juice

The inner wall of the stomach is studded with countless tiny depressions called **gastric pits** that secrete gastric juice (see Clinical Note: Peptic Ulcer Disease). These pits lead into the mucosa for varying depths (Figure 18-12) and generally branch to form about three secondary pits, which in turn branch into three tubular glands. Three functionally distinct kinds of glands are present in these pits, depending on the location of the pit and the product it secretes. They are the cardiac, gastric, and pyloric glands.

Cardiac and Gastric Glands

Cardiac glands are confined to pits in the cardiac region, where they secrete mucus. **Gastric glands** are found in long, thin pits of the fundus and body. Several different cell types line the wall of these pits. The uppermost region of each pit contains **mucous neck cells** that secrete an acidic glycoprotein. Many of these cells are actively dividing, and as new cells are formed some migrate to the stomach interior where

pepsinogen: (Gr.) *pepsis,* digestion

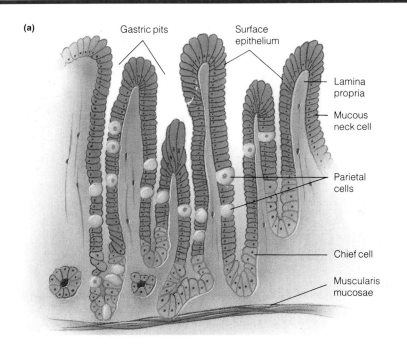
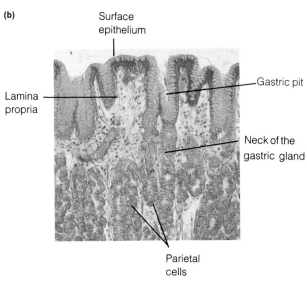

Figure 18-12 A gastric gland. (a) Drawing showing internal structure of gland. (b) Photomicrograph of mucosa of the fundic wall of the stomach (mag. 218 ×).

they replace mucus-secreting cells in the gastric epithelium. Other cells migrate deeper into the pit and differentiate into parietal and chief cells.

Parietal (pă·rī′ĕ·tăl) **cells** produce hydrochloric acid that is pumped into the channel of the gastric gland that leads to the stomach interior. H^+ is produced in a reaction catalyzed by carbonic anhydrase in which water and carbon dioxide combine to form carbonic acid:

$$CO_2 + H_2O \xrightarrow{\text{carbonic anhydrase}} H_2CO_3$$

Ionization of carbonic acid then produces H^+ and bicarbonate:

$$H_2CO_3 \longrightarrow H^+ + HCO_3^-$$

H^+ is pumped from the cells in a reaction that uses ATP. Cl^- follows passively, attracted by the positively charged hydrogen ions. HCO_3^- is transported to the blood and carried away, eventually to be exhaled as CO_2.

Parietal cells are the source of **intrinsic factor**, a glycoprotein that combines with vitamin B_{12}. It is only in this form that the vitamin can be absorbed later in the small intestine. Individuals who lack parietal cells, perhaps because of surgical removal of large portions of the stomach or the death of parietal cells, may develop pernicious anemia. Pernicious anemia is a disease that results from a deficiency in vitamin B_{12} and is characterized by insufficient production of red blood cells and numerous neurologic and hormonal abnormalities (see Chapter 21 Clinical Note: Anemia).

Chief cells, or **zymogenic** (zī″mō·jĕn′ĭk) **cells,** are smaller than parietal cells and comprise the lining of fundic and body glands. They produce **pepsinogen,** an inactive enzyme that is converted to active **pepsin** by stomach acid. Pepsins are a family of enzymes that cleave proteins into polypeptide chains, primarily by attacking peptide bonds adjacent to the amino acids phenylalanine and tyrosine. Because of this specificity in point of attack, few, if any, free amino acids are released in the stomach from the digestion of proteins. Instead, pepsins split proteins into fragments that consist of approximately four to ten amino acids each.

Pyloric Glands

The third category of glands in gastric pits are **pyloric glands.** Located in the pyloric antrum and pyloric canal, these glands also secrete mucus. Some of their cells also secrete **gastrin** (găs′trĭn), a polypeptide hormone that has numerous effects on the digestive tract. These effects include the following: (1) increased secretion of HCl, intrinsic factor, and pepsinogen by gastric glands; (2) increased secretion of digestive chemicals by similar glands in the small intestine; (3) increased secretion of insulin by the pancreas and of calcitonin by the thyroid gland; and (4) increased muscle tone in the lower esophageal sphincter and walls of the stomach and intestine. All these effects increase the efficiency of digestion.

Control of Gastric Gland Secretion

Digestive gland activity is regulated by a combination of hormonal and nervous mechanisms that operate in three phases of digestion: cephalic, gastric, and intestinal phases.

Cephalic Phase

The **cephalic** (sĕ·făl′ik) **phase** is characterized by neural activity that results from the sight, taste, and smell of food. Nervous impulses delivered to the cerebral cortex from the eyes, tongue, or nose are relayed to the medulla oblongata, which then transmits signals through parasympathetic fibers in the vagus nerve to the **myenteric** (mī″ĕn·tĕr′ĭk) **plexus,** a nerve complex in the stomach wall. From the myenteric plexus, impulses stimulate glands in the cardiac, gastric, and pyloric pits causing them to secrete their respective products. This prepares the stomach for digestion before food reaches it.

Gastric Phase

The **gastric phase** begins as food enters the stomach. Stomach distension causes a reflex arc in which nerve impulses leave the stomach in the vagus nerve, travel to the medulla oblongata, and return to the stomach. Some impulses go directly to gastric glands and cause them to increase their rate of secretion. Other signals go to gastrin-producing cells inducing them to release gastrin into the blood. The gastrin is then transported to other portions of the stomach.

Gastrin-producing cells are also stimulated by distension of the pyloric antrum and by partially digested proteins and certain other substances in the pyloric antrum. There is also evidence for a direct neural mechanism in which nervous impulses originate in receptor cells on the surface of the pyloric antrum and travel directly to the gastrin-secreting cells. In both cases gastrin is carried by the blood to gastric glands and induces the glands to increase their activity.

Intestinal Phase

Gastrin is also secreted by the mucosal cells of the upper portion of the duodenum. This form of the hormone, called **intestinal gastrin,** is released in response to peptides in chyme passing from the stomach to the small intestine. Like gastrin from the pyloric antrum, intestinal gastrin is carried by the blood to gastric glands, which respond by increasing their secretion of HCl and pepsinogen. This is the **intestinal phase** of digestive gland activity.

gastritis: (Gr.) *gaster*, belly + *itis*, inflammation

> **CLINICAL NOTE**
>
> ### GASTRITIS
>
> Inflammation of the stomach, or **gastritis,** occurs in two forms: acute and chronic. The most common cause of **acute gastritis** is the chronic use of aspirin. Approximately 70% of people with rheumatoid arthritis and other disorders requiring daily doses of aspirin have gastritis. Aspirin, or acetylsalicylic acid (ASA), in its un-ionized form is fat soluble and able to penetrate the lipid cell membrane of the mucosal cells. Once inside the cells, it ionizes and injures the cells causing the cells to shed. This leads to the back diffusion of hydrogen ion. The increase in acidity worsens the cellular injury and acute gastritis ensues. Heavy smoking, excess coffee, and alcohol can also contribute to gastritis. An evening of drinking and smoking followed the next day by copious consumption of coffee and aspirin can easily lead to acute gastritis. Symptoms include nausea, vomiting (sometimes of blood), and pain.
>
> **Chronic gastritis** can be viewed as the end stage of a chronic inflammatory process that begins with gastritis. As the inflammatory process continues, the cells become more and more abnormal. Eventually the cells that secrete pepsin and acid may become so depleted that **hypochlorhydria** (low acid content of stomach secretions) can result. This lack of acid permits growth of bacteria in the stomach and may lead to an increase of nitrate concentration, which may lead in turn to the production of nitrosamines and possible carcinogenesis.

Control of Stomach Emptying

Once a meal has been eaten, the food is stored in the stomach while protein is digested and the mass is converted to chyme. Chyme is "squirted" by the pyloric pump into the duodenum, a few milliliters at a time, until the stomach is emptied. Emptying the stomach usually takes about 1 to 2 hours, depending on volume, composition, and fluidity of the chyme reaching the pyloric pump (Figure 18-13).

Neural Control

The rates of emptying shown in Figure 18-13 are influenced by pressure receptors in the stomach wall that are stimulated by distension and that control gastric emptying by a neural mechanism. The greatest control of stomach emptying, however, comes from the duodenum. Increased pressure in the duodenum resulting from a large volume, increased acidity, or osmotic effects that reduce the flow of water from the chyme initiate an **enterogastric** (ĕn″tĕr·ō·găs′trĭk) **reflex.** This reflex is a complex one and involves localized nerve pathways in the stomach and duodenal walls as well

duodenum: (L.) *duodeni*, twelve (fingerbreadths in length)
jejunum: (L.) *jejunum*, empty
ileum: (L.) *ileum*, groin

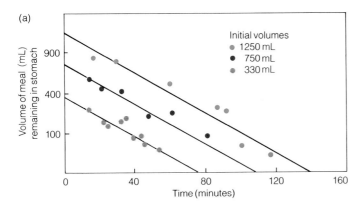

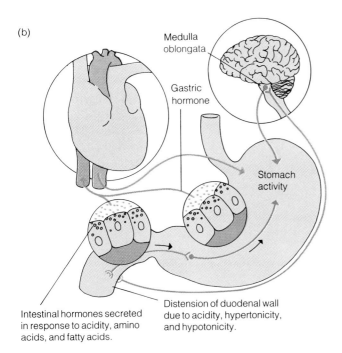

Figure 18-13 (a) Relationship between time and the volume of a liquid meal in one individual. (b) Hormonal and neural mechanisms involved in regulating rate of stomach emptying.

Small Intestine

After studying this section, you should be able to:

1. Describe the anatomy of the small intestine, identifying the duodenum, jejunum, and ileum.
2. Describe the structure of the intestinal wall.
3. Explain the function of secretions of the gallbladder and pancreas in digestion and the mechanisms used to induce their secretion.
4. List the enzymes secreted in the small intestine, their source, their substrates, and the products of their activity.
5. Explain how and where carbohydrates, proteins, and lipids are absorbed.

The **small intestine** extends from the pyloric sphincter to the cecum of the large intestine and consists of three regions: the **duodenum** (dū″ō·dē′nŭm), **jejunum** (jē·jū′nŭm), and **ileum** (ĭl′ē·ŭm) (Figure 18-14). It measures about 2.5 cm in diameter and about 7 m in length. The duodenum is the first 25 to 30 cm of the small intestine; the jejunum is the middle region, measuring about 2.5 m in length; and the ileum is the third section, measuring about 4 m. These measurements vary at any given instant because of contraction of longitudinal muscles in the tunica muscularis.

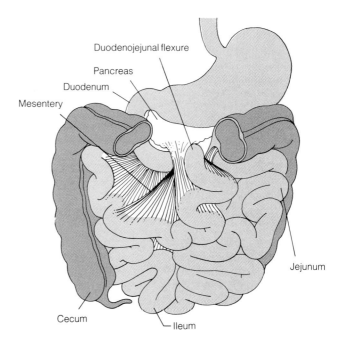

Figure 18-14 The small intestine.

as a reflex arc that travels from duodenal receptors through the vagus nerve to the medulla oblongata and back to the pyloric wall. The reflex inhibits the pyloric pump and reduces the amount of chyme delivered to the duodenum.

Hormonal Control

The duodenum also regulates stomach emptying by a hormonal mechanism. Certain substances, emulsified fat in particular, extract **gastric inhibitory protein (GIP)** from the intestinal mucosa. GIP is reabsorbed by the blood and carried to the stomach where it inhibits the pyloric pump.

endoscopy: (Gr.) *endon*, within + *skopein*, to examine

> **CLINICAL NOTE**
>
> ## GASTROINTESTINAL DIAGNOSTIC TECHNIQUES
>
> Examination of the gastrointestinal system by the use of X rays has been the standard practice for many years. The esophagus, stomach, duodenum, and small intestine can be visualized by having the patient drink a radiopaque substance, usually a barium salt, and examining its progress with a fluoroscope. Standard X-ray films are taken to provide a permanent record of all areas and of specific findings.
>
> The colon can be examined by giving the barium solution as an enema. A more refined technique called an **air contrast enema** is performed by injecting air into the colon after the barium. This provides a thinner coating to the surfaces, and the contrast between the radiopaque barium and the radiolucent air produces much finer detail in the final films.
>
> Special examinations of the bile ducts and gallbladder can be performed by giving the patient a radiopaque dye that is excreted by the liver into the bile ducts. This type of examination is called a **cholecystogram** (kō″lē·sĭs′tō·grăm). The liver and spleen can be visualized by giving specially prepared radioactive materials intravenously and scanning the abdomen with instruments that detect the radiation and produce a picture of the organs.
>
> New developments in instrumentation have made gastrointestinal diagnosis more accurate and more acceptable to the patient. The use of **ultrasound scans,** based on the technology of sonar, has given us a noninvasive, repeatable method of diagnosing tumors and gallstones with little discomfort to the patient. The development of **fiberoptics** has led to the improvement of **endoscopy** (ĕn·dŏs′kō·pē), a technique using a device consisting of a tube and an optical system that is inserted into the gastrointestinal tract (see figure). Using an endoscope, the entire length of the lower digestive tract and the important structures of the upper tract can be examined by direct visualization and biopsy samples can be obtained.
>
> **Endoscopy.** (a) Endoscope used in examination of upper gastrointestinal tract. (b) Endoscopic photograph of vocal cords. (c) Endoscopic photograph of transverse colon. This latter photograph was taken with a colonoscope, a type of endoscope used for the lower gastrointestinal tract.
>
>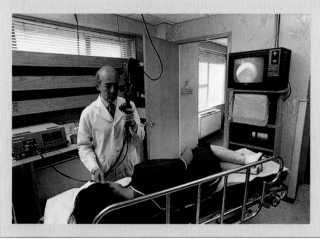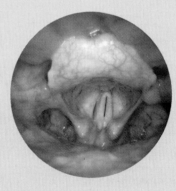

All four tunics previously described are present in the small intestine, but are modified in several ways not found elsewhere. The tunica mucosa forms fingerlike projections called **villi** (vĭl′ī) that are so numerous that they give the surface the appearance of a plush carpet (Figure 18-15a). Additionally, the mucosa and submucosa are modified into **circular folds,** or **plicae circulares,** that extend into the intestinal lumen. The effect of these modifications is to greatly increase the surface area of the small intestine, facilitating absorption of nutrients.

The surface of each villus is covered by a layer of epithelial cells that encloses a core of loose connective tissue (Figure 18-15b). Within the core lies a capillary network and a small lymphatic vessel, a **lacteal** (lăk′tē·ăl), surrounded by smooth muscle fibers. The epithelial cells are attached to a delicate basement membrane and consist mostly of two distinct types of cells: a surface layer of **columnar cells** interspersed by occasional **goblet cells.** As their names suggest, columnar cells are long and narrow and goblet cells are enlarged at one end, in the shape of a wine goblet. The

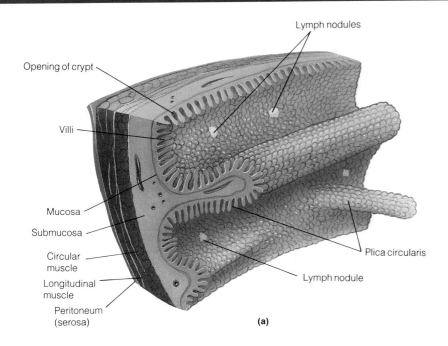
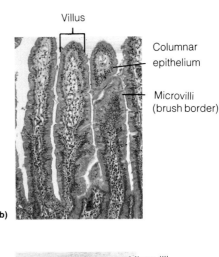

Figure 18-15 Villi and microvilli. (a) The inner surface of the small intestine is covered with small fingerlike villi. (b) Photomicrograph of several villi (mag. 59 ×). (c) Electron micrograph of the epithelial surface of a villus. Note the numerous microvilli. The dark lines extending into the cytoplasm from each microvillus are filaments formed by actin.

enlarged end faces the intestinal lumen. Columnar cells are responsible for absorbing nutrients from the intestinal lumen; goblet cells secrete mucus that lubricates the internal intestinal surface. A third type of cell, **enteroendocrine cells,** are less common. These cells are thought to secrete hormones that regulate digestive activity in the stomach and small intestine and the release of bile and pancreatic secretions.

The border of the columnar cells that faces the intestinal lumen is covered with fine extensions called **microvilli** that collectively give the surface of the cell a striated appearance when examined with an electron microscope (Figure 18-15c). These microvilli form a **brush border** on the surface of the cell. The final stages of the enzymatic degradation of macromolecules are catalyzed by enzymes located in the plasmalemmae of the microvilli.

Simple tubular **intestinal glands** (also called **crypts of Lieberkühn**) secrete digestive enzymes into the spaces between villi. Relatively large compound tubuloalveolar glands **(Brunner's glands)** are in the duodenal submucosa. These glands empty into the lumen and secrete protective mucus. The duodenum receives secretions from the liver and pancreas through the common bile duct and the pancreatic duct (see section on Accessory Structures). These ducts join and empty into the duodenum through a common opening called the **ampulla of Vater** (fah′tĕr) located at the end of a small projection called the **duodenal papilla** (Figure 18-16). In some people an accessory pancreatic duct **(duct of Santorini)** empties separately near the duodenal papilla.

There are no clear demarcations separating the jejunum and ileum, rather a gradual transition occurs in diameter and wall thickness, the intestine becoming narrower and its wall thinner as it approaches the distal end. There are fewer villi and circular folds in these regions than in the duodenum, and the ileum wall is marked by large clumps of lymph nodes **(Peyer's patches)**. Lymph from lacteals is filtered through these nodes. The ileum ends in a slitlike opening into the large intestine that is controlled by the **ileocecal** (ĭl″e·ō·se′kăl) **valve.**

Digestion in the Small Intestine

The small intestine completes the digestion of macromolecules in the chyme, aided by pancreatic and gallbladder secretions. Efficiency is enhanced by **segmentation,** contractions

cholecystokinin: (Gr.) *chole*, bile + *kystis*, cyst + *kinein*, to move
pancreozymin: (Gr.) *pan*, all + *kreas*, flesh + *zyme*, leaven

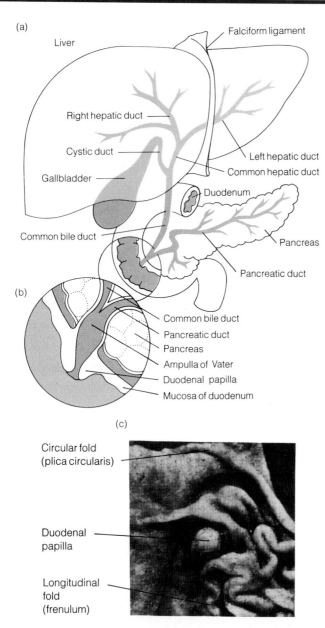

Figure 18-16 Duodenal ducts and duodenal papilla. (a) Anatomy of the ducts leading into the duodenum from the liver and pancreas. (b) Detailed anatomy of the duodenal papilla. Note that the common bile duct and pancreatic duct join to form the ampulla of Vater. (c) Electron micrograph showing the duodenal papilla from the small intestine.

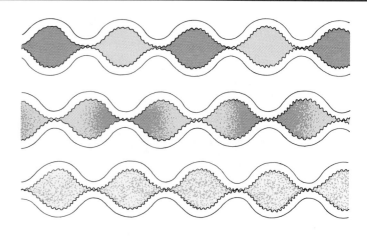

Figure 18-17 Segmentation in the small intestine. Chyme mixes with enzymatic secretions, increasing the efficiency of breakdown of the macromolecules in the chyme. The yellow and red shows how chyme from different areas mixes.

of the intestinal wall that occur intermittently throughout the length of the intestine and cause ringlike constrictions that mix the chyme with intestinal secretions (Figure 18-17). In addition, relatively weak peristaltic contractions begin at various points in the intestine, travel a short distance, and then subside. This results in a slow movement of chyme through the intestine, taking as long as 10 hours.

Neutralization of Chyme

Chyme enters the duodenum as a highly acidic milky fluid rich in protein-digesting enzymes. The presence of chyme in the duodenum, and particularly of HCl in the chyme, causes release of **secretin** (sē·krē′tĭn) from the duodenal mucosa. Secretin is a polypeptide hormone that is stored in an inactive form. It is carrried to the pancreas where it stimulates pancreatic duct cells to release bicarbonate ions. The bicarbonate ions are carried to the duodenum, where they react with hydrogen ions in chyme to produce carbonic acid:

$$H^+ + HCO_3^- \longrightarrow H_2CO_3$$

Absorption of H^+ causes the pH to rise to about 7 or 8, and the chyme is neutralized.

Pancreatic Enzymes

The pancreas releases several digestive enzymes that continue the digestion of macromolecules (Table 18-1). Release of these enzymes is triggered hormonally. The hormone involved also causes contraction of the gallbladder and is named in reference to both functions: **cholecystokinin-pancreozymin** (kō″lē·sĭs″tō·kī′nĭn-păn′krē·ō·zī″mĭn) or **CCK-PZ.** Chyme in the duodenum, particularly polypeptides, causes the duodenal mucosa to release CCK-PZ into

trypsinogen: (Gr.) *tryein*, to rub down + *pepsis*, digesting + *genes*, producing

Table 18-1 PANCREATIC ENZYMES

Enzyme (Inactive Form)	Activating Agent	Substrate	Product (or Function)
Trypsin (trypsinogen)	Enterokinase	Proteins, polypeptides	Cleaves peptide bonds next to arginine or lysine
Chymotrypsins (chymotrypsinogens)	Trypsin	Proteins, polypeptides	Cleaves peptide bonds adjacent to certain classes of amino acids
Elastase (proelastase)	Trypsin	Elastin*	Cleaves peptide bonds adjacent to alanine, glycine, and serine
Carboxypeptidases (procarboxypeptidases)	Trypsin	Proteins, polypeptides	Cleaves terminal amino acid at —COOH end of polypeptide
Pancreatic lipase	Emulsifying agents	Triglycerides	Fatty acids, di- and monoglycerides, glycerol
Pancreatic amylase	Cl$^-$	Starch	Maltose, maltotriose
Ribonuclease	—	RNA	Ribonucleotides
Deoxyribonuclease	—	DNA	Deoxyribonucleotides

*Elastin is a protein found in fibers of elastic connective tissue.
Source: Modified from W. F. Ganong, *Review of Medical Physiology*, 7th ed. (Los Altos, CA: Lange Medical Publications, 1975), 348.

the blood. Carried by blood to the pancreas, it causes pancreatic acinar cells to release enzymes (see section on Pancreas).

Protein Digestion

At the time of their release, the enzymes are in an inactive form, just as pepsin is before it is released into the stomach. The first enzyme to be produced is **trypsin** (trĭp′sĭn), which is converted from an inactive form, **trypsinogen** (trĭp·sĭn′ō·jĕn). Trypsinogen is activated by **enterokinase** (ĕn″tĕr·ō·kī′nās), an enzyme in the intestinal mucosa that catalyzes activation as trypsinogen enters the duodenum.

Trypsin activates other trypsinogen molecules as well as two other proteins, **chymotrypsinogen** (kī″mō·trĭp·sĭn′ō·jĕn) and **procarboxypeptidase** (prō·kăr·bŏk″sē·pĕp′tĭ·dās), which are converted into **chymotrypsin** and **carboxypeptidase**, respectively. Trypsin, chymotrypsin, and carboxypeptidase continue to digest peptides in the chyme.

Even these enzymes, however, generally do not cause the release of free amino acids because they are specific in the kind of peptide bonds that they cleave. Trypsin, for example, only cleaves bonds between the carboxyl (—COOH) group of arginine or lysine and the amino (—NH$_3$) group of the neighboring amino acid. Chymotrypsin cleaves peptide bonds between the carboxyl group of tyrosine and similar amino acids and the amino group of the adjacent amino acid, and carboxypeptidase cleaves only the terminal amino acid at the COOH end of a protein. Consequently, polypeptides in the stomach are shortened further, with only an occasional release of a free amino acid.

Carbohydrate Digestion

Pancreatic acinar cells also release **pancreatic amylase,** an enzyme identical to the salivary amylase produced in the mouth. Most starch remains undigested when it reaches the small intestine because salivary amylase is inactivated in the stomach. Starch digestion is resumed in the less acidic environment of the duodenum. Hydrolysis of starch by amylase produces maltose, **maltotriose** (a carbohydrate fragment consisting of three glucose subunits), and **dextrins** (dĕks′trĭns), short chains of simple sugars.

Fat and Nucleic Acid Digestion

Pancreatic lipase, also released in response to CCK-PZ, attacks neutral fats (triglycerides), converting them to free fatty acids and glycerol and to glycerides (molecules of glycerol to which one or two fatty acids are attached). This action is made more efficient by bile salts released into the duodenum from the gallbladder (see next section).

The pancreas also releases **ribonucleases** (rī″bō·nū′klē·ās·ĕs) and **deoxyribonucleases** (dē·ŏk′sē·rī″bō·nū′klē·ās·ĕs). These enzymes break down RNA and DNA into individual nucleotides. Since nucleotides are complex molecules, they require additional breakdown before they can be absorbed.

Role of Bile

Bile is a yellowish fluid produced by the liver that aids in lipid digestion. It is an aqueous mixture of bile salts, bile pigments such as **bilirubin** (a breakdown product of hemoglobin), lecithin (a phospholipid), cholesterol, sodium chloride, and bicarbonate ions. About 1 L of bile is produced each day, much of which is stored in the gallbladder. Bile is concentrated by absorption of electrolytes from the gallbladder. As minerals are removed, water flows out by osmosis, reducing the volume in the gallbladder by as much as 90%.

cholelithiasis: (Gr.) *chole*, bile + *lithos*, stone + *iasis*, condition
glycocalyx: (Gr.) *glykys*, sweet + *kalyx*, cup

> **CLINICAL NOTE**
>
> ### GALLSTONES
>
> The presence of **gallstones** in the gallbladder is called **cholelithiasis** (kō″lē·lĭ·thī′ă·sĭs). The precise cause of gallstones is unknown, but they seem associated with obesity. They do not necessarily cause any problems, but they can be associated with indigestion, nausea, and pain in the abdomen. All symptoms worsen after eating a meal containing a large amount of fat.
>
> A complication of cholelithiasis will occur if a stone escapes from the gallbladder, travels along one of the bile ducts, and becomes stuck. The blocking of a duct leads to obstruction of the flow of bile to the small intestine, which causes jaundice. Severe pain accompanies the obstruction of the duct and is known as **biliary colic**. An obstructed duct requires surgery; usually the entire gallbladder is removed to prevent a recurrence.

> **CLINICAL NOTE**
>
> ### LACTOSE INTOLERANCE
>
> **Lactose intolerance** is one of the **malabsorption syndromes**. It is a disorder resulting from the lack of one enzyme, lactase, normally found in the brush border of the intestinal villi. Lactase is needed for the breakdown of lactose, a disaccharide found in milk and milk products. In the United States about one-tenth of the white population has this disorder, but it is more common in blacks, native Americans, Asians, and Hispanics.
>
> Because of the missing enzyme, lactose is not broken down into the simple sugars glucose and galactose for absorption in the small intestine. Instead, lactose is acted on by the lactase produced by intestinal bacteria, which results in the production of hydrogen gas, carbon dioxide, and organic acids. Additionally, the increased concentration of undigested lactose in the small intestine leads to increased hypertonicity of the chyme. This in turn leads to increased fluid secretion into the large intestine and the resultant diarrhea. Thus, the symptoms of lactose intolerance are colonic irritability, bloating, and diarrhea.
>
> An individual with lactose intolerance must follow a low lactose or lactose-free diet. If the disorder occurs in postmenopausal women, calcium supplements are necessary.

As chyme enters the small intestine, it triggers the release of CCK-PZ and secretin. These hormones cause muscles in the gallbladder to contract, forcing the bile through the cystic and bile ducts into the duodenum.

Bile salts are emulsifying agents that disperse fat droplets in the chyme into much smaller droplets. Fats are hydrolyzed more efficiently by lipase in this dispersed state because lipase is water soluble, whereas fat droplets are not. Dispersing the droplets increases their total surface area, which makes the enzymes more efficient at hydrolyzing the fat. Bile salts also cause the release of additional CCK-PZ and help activate lipase within the intestine.

Sometimes bile becomes so concentrated in the gallbladder that cholesterol precipitates and forms **gallstones**, which may have to be surgically removed (see Clinical Note: Gallstones).

Intestinal Enzymes

Final breakdown of nutrients is accomplished by **intestinal enzymes** located in the brush border of the villi. Each microvillus is surrounded by a **glycocalyx** (glī″kō·kăl′ĭks), a complex assembly of protein and polysaccharide. Embedded in this glycocalyx and attached to the cell membrane are numerous spherical particles, each of which contains a package of intestinal enzymes.

Carbohydrate Digestion

Included in the intestinal enzymes are **maltase, lactase, sucrase,** and **dextrinase**. These enzymes hydrolyze maltose, lactose, sucrose, and dextrins, respectively. Although a small amount of maltose may be ingested directly, most results from the action of amylase on starch. Lactose is milk sugar, and sucrose is common table sugar. Cleavage of each of these molecules yields two monosaccharide molecules, both of which can be absorbed (but see Clinical Note: Lactose Intolerance). Dextrinase cleaves dextrins directly to glucose.

Peptide Digestion

The short chains of amino acids that remain are broken down further in the brush border by **amino peptidase** and **dipeptidase**. Amino peptidase cleaves polypeptides into still shorter lengths, even to the point of producing free amino acids. Dipeptidases cleave the peptide bond remaining between pairs of amino acids, thereby completing the catabolism of the original proteins to free amino acids.

Fat and Nucleic Acid Digestion

Other enzymes of the brush border include a small amount of **intestinal lipase**, which hydrolyses fats that have escaped pancreatic lipase digestion, and various nucleases and phosphatases that complete the hydrolysis of DNA and RNA. Table 18-2 shows enzymes of the brush border and the functions they perform.

Absorption

Nearly all nutrients contained in chyme are absorbed in the small intestine. Depending on the diet, virtually 100% of

Table 18-2 ENZYMES OF THE BRUSH BORDER OF THE SMALL INTESTINE

Enzyme	Substrate	Product (or Function)
Enterokinase	Trypsinogen	Trypsin
Aminopeptidases	Peptides	Cleaves the amino acid at NH_3^- end of the polypeptide
Dipeptidases	Dipeptides	Amino acids
Maltase	Maltose	Glucose
Lactase	Lactose	Glucose and galactose
Sucrase	Sucrose	Glucose and fructose
Isomaltase	Dextrins	Glucose
Nucleases	DNA and RNA	Nucleotides
Phosphatases	ATP, nucleic acids	Nucleotides, ribose, deoxyribose, bases
Intestinal lipase	Monoglycerides	Glycerol, fatty acids

Source: Modified from W. F. Ganong, *Review of Medical Physiology,* 7th ed. (Los Altos, CA: Lange Medical Publications, 1975) 348.

nutrient molecule from chyme into an intestinal cell is accompanied by transport of a sodium ion from the cell into the intestinal lumen.

Most organic nutrients are absorbed in the proximal portion of the intestine, so that by the time chyme reaches the ileum, very little organic nutrient remains. Bile salts are absorbed in the ileum by specialized carrier molecules. Absorption is highly efficient, and almost all bile salts are returned to the liver and stored in the gallbladder. Bile salts are recycled as often as six times per day with only about 5% loss. Vitamin B_{12}, associated with intrinsic factor, is also absorbed in the ileum by specific carrier molecules.

Nutrients absorbed through the villus wall are delivered to blood in the capillaries or to lymph in the lacteal (Figure 18-18). Blood enters a villus through peripherally located arterioles, passes through the capillary bed, and exits through a centrally located venule. The branched lacteal leads outward to the lymphatic vessels.

Absorption of Carbohydrate

Monosaccharides are absorbed intact, and at least some of them are absorbed by facilitated diffusion. Glucose, however, appears to be absorbed by active transport and by facilitated diffusion. Sugars such as fructose appear to be absorbed by diffusion.

Once monosaccharides enter the epithelial cells of a villus, they continue through to the capillary bed. This movement is passive, and the monosaccharides diffuse from the epithelial cells, where they are in high concentration, to the blood, where they are in low concentration.

usable carbohydrates, 90% of protein, and 85% of fat are absorbed, along with much of the water and many of the dissolved ions.

Absorption occurs either passively as simple or facilitated diffusion or by active transport across the epithelial cell membranes of the villi. When active transport occurs, it is coupled to a sodium pump such that movement of a

Figure 18-18 Absorption of nutrients into a villus. Amino acids, nucleotides, and monosaccharides enter the capillary blood and are carried off. Most lipids enter the lacteal and are transported in lymph.

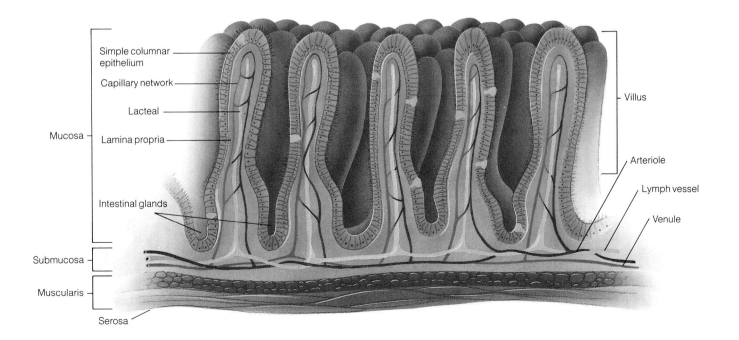

Some people lose their ability to produce certain disaccharidases and consequently lose their ability to use the corresponding sugar. Lactase deficiency causes people to be unable to use lactose. When these people drink dairy products that contain lactose, the sugar is not broken down and absorbed. Consequently, the sugar causes a hypertonic chyme that interferes with the osmotic flow of water into the intestinal wall, and the chyme remains more fluid than normal. The lactose enters the large intestine where it serves as a nutrient for the organisms that live there (see section on Intestinal Flora). Bathed in such a rich fluid, bacteria and other organisms proliferate and carry on a great deal of metabolic activity. Gases from their activity produce flatulence, and lactic acid and other products derived from the lactose irritate the intestine, often causing diarrhea and other discomfort.

Absorption of Amino Acids

Amino acids are absorbed by active transport. Several different carrier systems are involved, each specific for a particular class of amino acids. For example, the basic amino acids lysine, ornithine, arginine, and cysteine are absorbed by specific carriers; glutamic acid and aspartic acid employ a second carrier; while proline, hydroxyproline, and glycine share a third class of carrier molecule.

When dipeptides are released by the action of pancreatic enzymes, they diffuse into the glycocalyx of the brush border. There, dipeptidase and aminopeptidase hydrolyze the dipeptides to individual amino acids, which are immediately transported across the membrane. The amino acids pass through the cells into the interior of the villus and eventually into the blood in the capillary bed.

Protein digestion and absorption of the resulting amino acids is a rapid process, the amino acids appearing within the blood about 1 hour after ingestion of the protein.

Not all amino acids absorbed in the small intestine are from food, however. A sizable amount of intestinal protein consists of enzymes secreted into the gastrointestinal tract and of proteins in epithelial cells that are constantly sloughing off the wall of the tract. Together, these two sources account for over 150 g of protein that is digested and reabsorbed daily (compared to about 20 g of protein in a hamburger patty). Since recovery is not 100% efficient, some protein is lost in the feces. Consequently, digestion would result in a net loss of protein if protein were not included in the diet.

Absorption of Fat

Most of the fat eaten in the course of a day is in the form of triglyceride, with smaller amounts of other lipids. Not all the fat in chyme is of dietary origin, however, since some

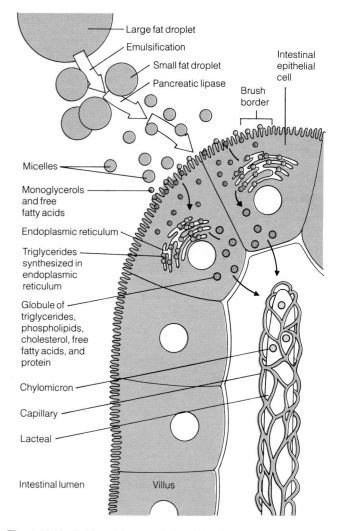

Figure 18-19 Fat breakdown and absorption into chylomicrons. Monoglycerides and free fatty acids are absorbed by the epithelial cells of the villus. They are then incorporated into chylomicrons, which pass into the lacteal.

fat is also present in bile and in the cells that slough off the intestinal wall. The action of lipase on triglycerides yields free fatty acids and monoglycerides, which are single fatty acids with an attached glycerol. Bile salts and the monoglycerides spontaneously form **micelles** (mī·sĕls′), spherical structures about 3 to 10 nm in diameter in which the glycerol-containing heads of the monoglycerides lie on the surface and the fatty acids extend into the interior (Figure 18-19). Free fatty acids, cholesterol, other lipids, and fat-soluble vitamins dissolve in the interior of the micelle. The micelle is then referred to as a **mixed micelle.**

Because the surface of the micelle contains the water-soluble portion of the molecules, the micelles remains soluble. Dissolved in the chyme, they either fuse with the membranes of the cells of the brush border or are taken in to the cell by pinocytosis. Once within the microvillus cell, the micelle dissociates. The smaller fatty acids, those that contain perhaps a dozen or fewer carbon atoms, pass from the cell directly into the blood and are carried off. Fatty acids containing more than about a dozen carbon atoms are reattached to monoglycerides to produce triglycerides again.

The cell now synthesizes a protein coat around groups of triglycerides, forming **chylomicrons** (kī″lō·mī′krŏns), which migrate to the interior of the villus and pass into the lacteal. Once in the lacteal, wavelike contractions of smooth muscle in the villus pump the fluid from the lacteal into lymphatic ducts that carry the chylomicrons in lymphatic fluid, eventually to be delivered to the blood.

Absorption of Water and Minerals

Organic nutrients are not the only substances absorbed through the wall of the small intestine. Several liters of water enter the small intestine each day, partly consumed as fluids and present in food and partly from secretions produced by the gastrointestinal tract itself. In spite of this, only about 1 L passes into the large intestine, the remainder having been absorbed by the small intestine.

Most water absorption goes on in the proximal portion of the small intestine where the bulk of organic nutrients are also absorbed. As the nutrients are actively transported through the intestinal wall, the intestinal fluids become hypotonic with respect to the surrounding tissue and water diffuses out, following the nutrients that have been absorbed.

The small intestine also absorbs several ions from chyme, including sodium, potassium, chloride, hydrogen, iron, calcium, magnesium, bicarbonate, and sulfate. Most are absorbed by active transport, although some also appear to move passively.

The functions of the small intestine are summarized in Figure 18-20.

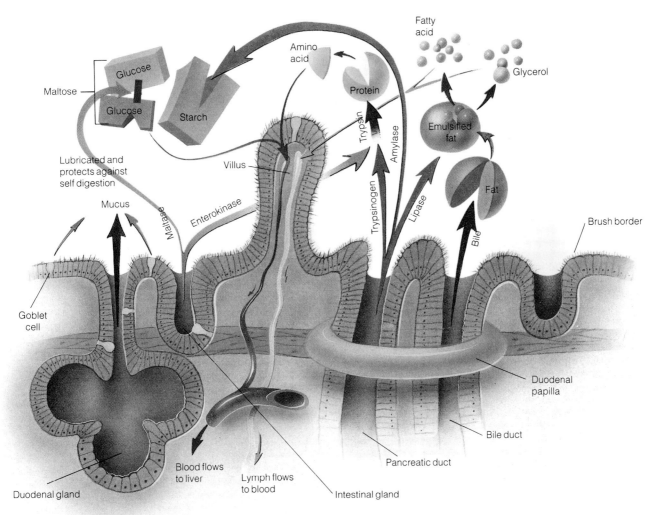

Figure 18-20 Summary of functions of the small intestine.

haustra: (L.) *haustor*, drawer
cecum: (L.) *caecum*, blindness

Large Intestine

After studying this section, you should be able to:

1. Describe the anatomy of the large intestine.
2. Compare the function of the large intestine with that of the small intestine.
3. Describe the movements of the large intestine that mix and move intestinal contents.
4. Explain the role and effect of intestinal organisms.
5. Describe how feces are formed and where they are stored.
6. Describe the defecation reflex.

The **large intestine** is so named because its diameter is greater than that of the small intestine (Figure 18-21). It measures about 1.5 m in length and about 7 cm in diameter. A characteristic feature of the large intestine are saclike pouches, or **haustra** (haws′tră), along its length. Instead of stretching around the circumference of the large intestine in a continuous sheet, the layer of longitudinal muscle in the tunica muscularis consists of three elongate bands of fibers called the **taenia coli** (tē′nē·ă kō′lī). These bands constantly contract, producing transverse internal folds (**plicae semilunares**) between adjacent haustra.

The circular folds so prominent in the small intestine are absent in the large intestine. Solitary lymph nodes are present in the mucosa. Externally, small **epiploic** (ĕp″ĭ·plō·ĭk) **appendages** emanate from the large intestine. These appendages are fat-filled folds of the tunica serosa.

There are three anatomical regions in the large intestine: the cecum, colon, and rectum. The ileocecal valve opens into the large intestine approximately 7 cm from the end, producing a blind sac, the **cecum** (sē′kŭm). Attached to the lower end of the cecum is the **vermiform appendix** (Figure 18-22). This organ has the same histological organization as the rest of the intestine and contains lymphatic tissue that helps combat infection. The appendix performs no digestive function, but it may be a vestigial remnant of an organ that was functional in human ancestors (see Clinical Note: Appendicitis).

The second part of the large intestine, the **colon,** has four segments. The **ascending colon** rises vertically along the right side of the abdomen. At the level of the liver, it makes an abrupt left turn at the **hepatic flexure** and continues across the abdominal cavity as the **transverse colon.**

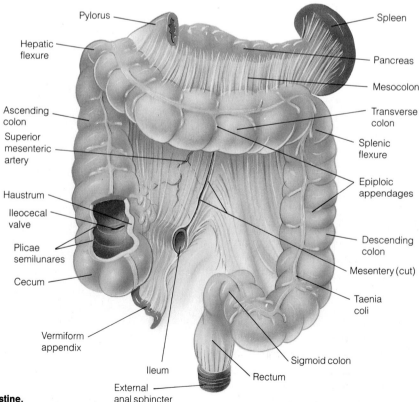

Figure 18-21 The large intestine.

CLINICAL NOTE

APPENDICITIS

The vermiform appendix may become inflamed, causing localized abdominal pain, low grade fever, loss of appetite, occasional vomiting, and an elevated white blood cell count. The abdominal pain of the resulting **appendicitis** may be localized at a point halfway between the navel and the hip bone (McBurney's point) on the right side. If the tender place on the right lower abdominal quadrant is prodded by the examiner's finger, the person will feel more pain as the finger is removed (rebound tenderness); any coughing or moving of the abdominal wall will exacerbate the pain. The afflicted person may bend the knees in an attempt to reduce the tension on the abdominal muscles.

Appendicitis is caused by the long, thin tube of the appendix being obstructed. This obstruction may result from a small piece of dried and hardened stool, kinking of the appendix, or edema of the lymphoid tissue. The obstruction of the appendix leads to hypoxia (deficiency of oxygen). This in turn will cause the mucosa to ulcerate. Finally, bacteria invade the wall, and the infection leads to even more edema which enhances the impediment to blood flow. Gangrene and perforation of the appendix may follow. Perforation, or a "burst appendix," leads to very serious consequences because the contents of the colon are released into the peritoneal cavity.

Appendicitis may be difficult to diagnose, especially in the very young and the elderly. There seems to be a higher incidence in adolescents and young adults and a slightly higher incidence in males than in females. Treatment is surgical removal of the inflamed and infected appendix.

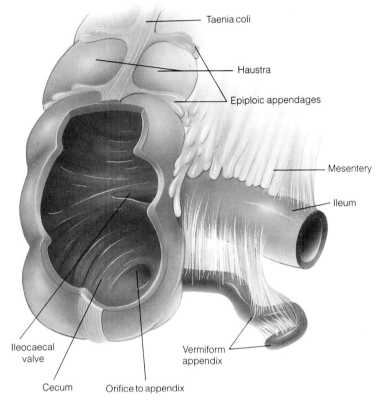

Figure 18-22 Connection between the large and small intestines. The ileocecal valve controls the flow of chyme from the ileum into the large intestine. Note the location of the cecum and appendix.

As it approaches the spleen, the transverse colon makes an abrupt downward turn at the **splenic** (splĕn´ĭk) **flexure** and continues along the left abdominal region as the **descending colon** (Figure 18-21). In the vicinity of the pelvic rim, the descending colon begins an S-shaped curve, marking the beginning of the **sigmoid** (sĭg´moyd) **colon.**

Near the sacrum, the sigmoid colon turns downward and becomes the **rectum** (Figure 18-23). The rectum lacks haustra because the taenia coli terminate in the sigmoid colon. The rectum leads into the **anal canal** and terminates at the **anus.** The simple columnar surface epithelium of the rectum changes to stratified squamous epithelium in the anal canal. Thick circular muscles in the tunica muscularis of the lower rectum and anal canal partially constrict the internal mucosal and submucosal layers into longitudinally oriented **anal columns** separated by **anal sinuses.** The submucosal layers of the rectum and anus have an abundant blood supply, and some of the veins in this region are oriented longitudinally within the anal columns. Two sphincters are associated with the anus: an **internal anal sphincter,** which

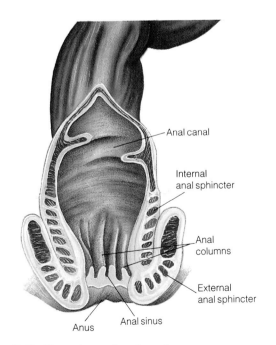

Figure 18-23 The rectum and anal canal.

is a continuation of the smooth, involuntary muscle layer of the intestine, and an **external anal sphincter,** which is composed of skeletal muscle.

Digestion in the Large Intestine

Approximately 1 L of fluid passes from the small intestine into the colon daily, about 900 mL of which is absorbed. The remainder is stored in the sigmoid colon and rectum as a semisolid mass of fecal matter or **feces.** The colon absorbs minerals and other substances from the fluid.

The wall of the large intestine is studded with numerous pits that extend down into the mucosa (Figure 18-24a). These pits are the openings of simple tubular glands that are richly lined with goblet cells (Figure 18-24b). Goblet cells are also common in the surface epithelium. No villi are present. The abundance of goblet cells reflects the fact that the wall of the large intestine secretes copious amounts of mucus. This mucus lubricates the wall of the large intestine and protects it from abrasion by the roughage present in the colon. The mucus also gives the feces the form characteristic of the stool of a healthy person.

Movement through the large intestine is accomplished much like it is in the small intestine, except that it is considerably slower. Whereas chyme may pass through the 3 to 5 m of small intestine in 3 to 10 hours, it normally takes 40 to 70 hours for material to pass through the 1.5 m of the colon.

There are two kinds of muscular motions in the colon. One, strictly a mixing motion, occurs as the circular muscle in the tunica muscularis contracts in different regions of the colon. As the constrictions squeeze the lumen of the colon, the muscles of the adjacent portions contract. This causes the uncontracted regions of muscle between the constrictions to bulge outward to form haustra. The alternating contractions and relaxations slowly mix the contents of the colon as weak peristaltic contractions push the contents through the colon. These peristaltic contractions are unlike those in other portions of the digestive tract in that they only travel a short distance before they die out, to be reinitiated elsewhere in the colon. Such peristaltic waves occur only a few times each day and serve to push the fecal matter into the sigmoid colon and rectum. The feces are stored in the rectum until defecation (see section on Defecation).

Intestinal Flora

The large intestine and, to a smaller extent, the small intestine contain large numbers of microorganisms important to digestion. When a baby is born, its digestive tract is largely free of bacteria, but bacteria immediately begin to enter the alimentary canal as they are ingested with the baby's food. Although most are killed in the stomach, some are able to pass into the intestines where they take up residence. There may be as many as 10^5 microorganisms per

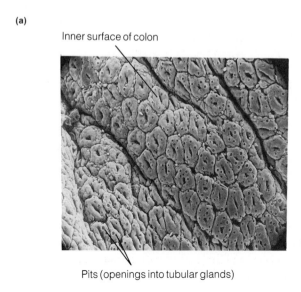
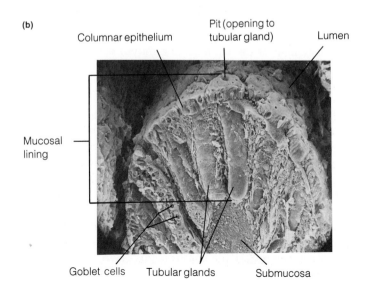

Figure 18-24 Histology of the large intestine. (a) The inner surface of the colon lacks the villi present in the small intestine and therefore appears relatively smooth (mag. 122×). (b) This photomicrograph shows the intestinal glands rich in goblet cells, reflecting their role in producing mucus (mag. 147×).

dysentery: (Gr.) *dys*, bad + *enteron*, gut
tenesmus: (Gr.) *teinesmos*, a straining

> **CLINICAL NOTE**
>
> ## ULCERATIVE COLITIS
>
> **Ulcerative colitis** (kŏ·lī′tĭs), a type of inflammatory bowel disease, is a serious and not uncommon disorder involving diffuse superficial ulcerations of the colon. Other signs and symptoms may occur outside of the gastrointestinal tract, such as arthritis, uveitis, and skin lesions. The disorder is more prevalent in young adulthood and seems to affect slightly more females than males. Its cause is unclear, but it seems to involve immunologic derangements that may cause or perpetuate the disorder. Autoantibodies against one of the polysaccharides that form the mucous cells of the colon are present in people with the disorder and in asymptomatic relatives. Lymphocytes from patients with ulcerative colitis destroy human colon cells.
>
> Ulcerative colitis is a disorder of the left colon but may spread to the entire colon. In about one-third of the cases, the ileum is also involved. Tiny abscesses begin in the crypts of the colon, then enlarge and begin to underline the mucosa. The mucosa may slough off leaving small ulcerations. The ulcers enlarge and coalesce and may erode into the muscle layer and involve the entire colonic wall. The mucosa may appear grossly normal under examination until the excavation is quite extensive, and then virtually the entire mucosa may slough off at once.
>
> The disorder may be insidious and may manifest itself by diarrhea with as many as 30 defecations a day. The diarrhea contains blood, mucus, and flecks of feces. The person may experience ineffectual and painful straining during defecation and lower left abdominal pain. Fever and weight loss may accompany the disorder.
>
> The course of the disorder is relapsing. Flare-ups of the disorder seem associated with emotional or physical stress. Massive hemorrhage, perforation of the colon from peritonitis, and obstruction of the colon may result from it. Treatment includes medication or surgical removal of the colon. Death from chronic ulcerative colitis may occur early in its course from massive hemorrhage, but the incidence of death decreases with time. However, the danger of malignancy increases as the disease becomes more chronic.

> **CLINICAL NOTE**
>
> ## DYSENTERY
>
> **Dysentery** is not one disorder but many, all of which involve inflammation of the mucous membrane of the colon. This results in abdominal pain, a feeling of having to move one's bowels **(tenesmus)**, and frequent bloody and mucus stools. Dysentery may result from chemicals, parasitic worms, or microorganisms in the water or food one consumes. It is especially dangerous to infants and the elderly because of the loss of fluid and electrolytes. It is not as prevalent today in industrialized nations because of the separation of sewage from drinking water.

Defecation

The amount of water in the feces depends on diet, health, and elimination behavior. At one extreme, as a result of irritation or infection of the mucosa, feces can be propelled through the large intestine so rapidly that water cannot be withdrawn and they are eliminated as a liquid. This is called **diarrhea**, and it can lead to fluid volume deficit and loss of electrolytes (see also Clinical Note: Dysentery). In one sense, diarrhea can be thought of as a protective mechanism since the rapid evacuation of the bowel also eliminates the infectious or chemical agents that are irritating the intestinal tract. At the other extreme, if defecation is delayed, more than the usual amount of water may be absorbed and result in hard, dry stool. This condition, called **constipation**, can result from lack of dietary fiber, insufficient intake of fluids, insufficient exercise, and certain drugs.

Normally, however, enough water is retained to give feces a semisolid consistency, normally containing about 75% water. The remaining 25% consists of bacteria, cellular debris from the walls of the gastrointestinal tract, lipids and inorganic matter, and a small amount of protein. Only a small amount of fecal matter comes directly from food.

As feces are pushed into the rectum, nerve receptors in the wall of the rectum are stimulated to initiate a **defecation reflex**. Impulses are sent from the rectum to the spinal cord and back to the colon where they intensify peristaltic contractions. Simultaneously, involuntary muscles of the internal anal sphincter relax. If the voluntary muscles of the external anal sphincter also relax, the peristaltic contractions force the feces out through the anus. If the defecation reflex is occurring at an inopportune time, one can maintain the external anal sphincter in a closed condition and prevent extrusion of the feces.

When defecation is prevented voluntarily, the defecation reflex diminishes and the urge to defecate subsides until feces again stimulate the reflex.

milliliter of chyme and even more in the mucus lining. The role of these microorganisms is unclear, but in experimental animals in which an intestinal flora is prevented from developing, the intestinal wall develops a thick mucous lining and the composition of the feces is altered. There is some evidence that certain vitamins and other nutrients are produced by intestinal bacteria, but the amounts produced are small. Bacterial metabolism produces a mixture of gases, including carbon dioxide, methane, hydrogen sulfide, and hydrogen gas, although much of the gas in the colon is swallowed air.

omentum: (L.) *omentum*, a covering

Accessory Structures

After studying this section, you should be able to:

1. Describe the organization and structure of the liver and its anatomical relationship to the gallbladder.
2. Describe the organization and structure of the pancreas, citing the tissues responsible for products used in digestion.
3. List the membranes of the abdominal cavity and describe their anatomical relationship to digestive organs.

The gastrointestinal tract utilizes several accessory organs to carry out its digestive functions, including the salivary glands, liver, gallbladder, and pancreas. With the exception of the salivary glands, these organs play significant roles in other systems as well, and those functions are described in the relevant chapters. This section deals with the anatomy of the liver, gallbladder, and pancreas.

Liver

The liver is the largest gland in the body, weighing almost 2 kg (4.4 lbs) in adults. It is divided into two major portions, the **right** and **left lobes** (Figure 18-25). Smaller regions, the **quadrate** and **caudate lobes,** are attached to the right lobe. The liver is connected to the inferior side of the diaphragm by the **falciform** (făl′sĭ·form) **ligament,** which also joins the right and left lobes. The **ligamentum teres** (lĭg″ă·měn′tum tē′rēz), or **round ligament,** is a remnant of the fetal umbilical vein that attaches the anterior surface of the liver to the umbilical cord by way of the falciform ligament. Attachment of the liver to the diaphragm is also secured by the **coronary** and **triangular ligaments.** Inferiorly, the liver is joined to the stomach by the **lesser omentum** (ō·měn′tŭm).

The surface of the liver is a thin **capsule** of fibrous connective tissue, which extends into the interior of the liver and divides it into columnar segments known as **lobules.** Each lobule consists of rows of cuboidal cells arranged radially around a central vein that serves as a connection between the hepatic portal vein, which brings blood from the digestive organs into the liver, and the hepatic veins, which drain the liver (Figure 18-26). (See Chapter 23 for more on these veins.)

Liver lobules receive blood from the hepatic artery (oxygenated blood) and the hepatic portal vein (deoxygenated blood). These two sources of blood join as **radial branches** (or **sinusoids**) within the lobules and fuse at the lobule periphery.

Cells in liver lobules have several functions, including production of bile, a substance formed from the breakdown of blood that is used in fat digestion. The liver also removes old red blood cells, regulates blood glucose levels, stores vitamin A, detoxifies harmful chemicals, converts excess amino acids to carbohydrates, and regulates nitrogen levels in the blood.

Bile collects in small radially positioned **bile canaliculi,** which empty into **bile ducts** located between the lobules.

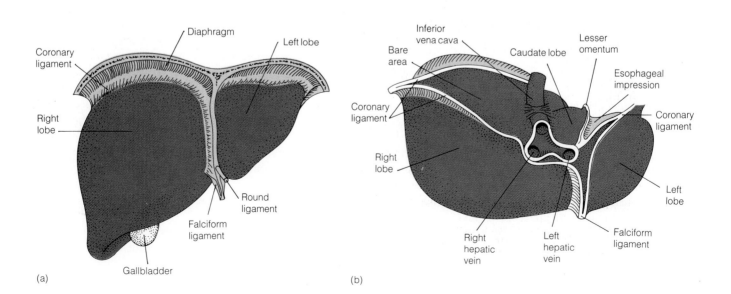

Figure 18-25 External anatomy of the liver. (a) Anterior view. (b) Superior view.

Accessory Structures 469

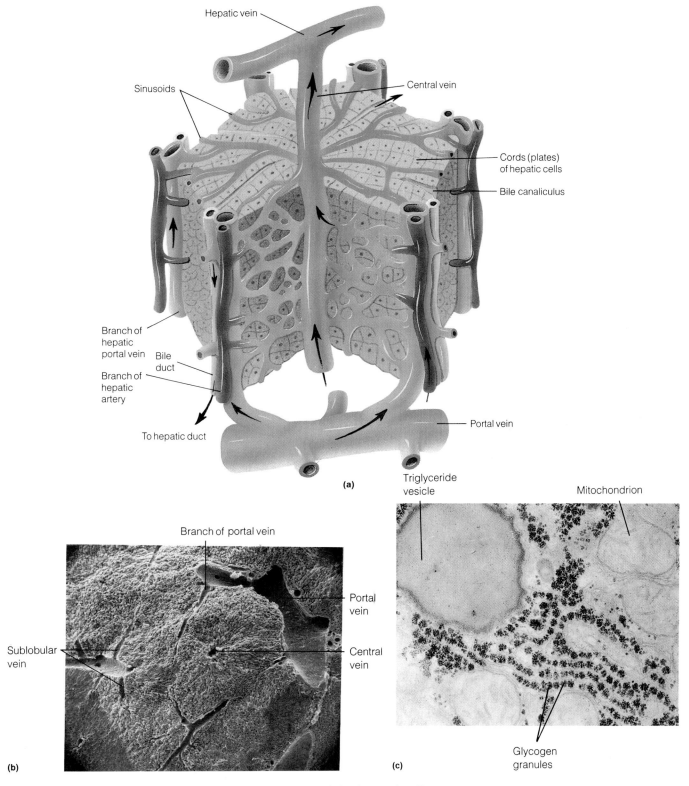

Figure 18-26 Histology of the liver. (a) Internal organization of a liver lobule. Arrows show the direction of flow of blood and bile in the vessels. (b) Scanning electron micrograph of a section of a lobule (mag. 44×). (c) Transmission electron micrograph of liver tissue showing numerous glycogen granules (mag. 2540×).

cirrhosis: (Gr.) *kirrhos*, orange-yellow color + *osis*, denoting a condition
cystic: (Gr.) *kystis*, sac, bladder

CLINICAL NOTE

CIRRHOSIS OF THE LIVER

Cirrhosis (sĭ·rō′sĭs) is a major cause of death in men between the ages of 25 and 64 years. It is a chronic disease of the liver in which diffuse destruction and regeneration of hepatic cells has occurred. This leads to a diffuse increase in connective tissue in the liver and results in a disorganization of the lobular structure. Problems with the liver that are consistent in all cases of cirrhosis are necrosis, regeneration, and scarring.

Cirrhosis has been associated with excessive alcohol consumption, viral hepatitis, drug-induced injury, prolonged obstruction of the bile ducts outside of the liver, and even dietary deficiency. Not every heavy user of alcohol develops cirrhosis, and it is not yet clear whether the amount of alcohol abuse or the duration of alcohol abuse plays a greater part in cirrhosis. It is clear, however, that alcohol abuse and cirrhosis are associated. The manifestations of cirrhosis vary but generally include deterioration of health, weight loss due to muscle wasting, fatigue, and an inability to metabolize bilirubin, leading to jaundice. Early in the disorder, the liver is fatty and enlarged; as scarring occurs, the liver becomes fibrotic and smaller.

The fibrosis that accompanies cirrhosis leads to an increase in fluid pressure within the portal vein system. This leads to vascular changes as the blood draining from the splanchnic area must find a new route to the bloodstream past the liver vascular obstruction. Collateral channels are opened leading to increase in pressure in the abdominal, periumbilical, esophageal, and hemorrhoidal veins (see Chapter 23). These veins become enlarged and tortuous. The veins in the lower esophagus may rupture leading to severe and life-threatening loss of blood.

The liver's function in extracting and modifying portal blood is bypassed because of the blocked hepatic drainage of the portal system. Substances normally removed from the portal blood increase in the blood and can be dangerous. Notably, ammonia formed in the colon and usually removed by the liver is diverted into the circulation and may produce neurological changes. Addition of blood to the gastrointestinal tract by bleeding veins in the esophagus will increase ammonia production and therefore hasten neurological changes.

Estrogen, which is normally metabolized in the liver, is not metabolized and reaches high levels in the blood. This is thought to be associated with enlarged breasts and a decrease in body hair seen in males with liver disease. Some putative vasodilating compounds are not metabolized, and these compounds, reaching high levels in the blood, are believed to account for the red palms and dilated vessels on the face and abdomen seen in the cirrhotic individual. It is thought that excess bile salts gain access to the blood, are deposited on the skin, and lead to itching.

Aldosterone may also not be detoxified in the liver (see Chapter 17). Excess levels of this adrenal hormone lead to sodium and water reabsorption by the kidney, thus causing excess circulating fluid volume. In cirrhosis, the flow of blood and the flow of lymph is disturbed, leading to an accumulation of fluid in the abdominal cavity. This is known as **ascites**. Because of the liver damage, the serum albumin concentration is also low, making it difficult for fluid lost to the tissue to return to the vascular space. Due to poor synthesis of clotting factors by the disturbed liver and lack of vitamin K, persons with cirrhosis have a tendency to bleed.

Cirrhosis is a life-threatening disorder that is perhaps more easily prevented than treated. If liver damage is discovered early and the cause of the damage eliminated, then the liver will heal.

The ducts fuse to form two **hepatic ducts**, one coming from each lobe, which in turn fuse to form a **common hepatic duct**. The common hepatic duct then fuses with the **cystic duct** to form the **common bile duct**.

Gallbladder

The gallbladder is a small sac located under the right lobe of the liver (Figures 18-16 and 18-25). Bile stored in it drains through the **cystic duct**. The common bile duct is joined by the pancreatic duct and both enter the small intestine through the duodenal papilla. The opening of the common bile duct into the small intestine is surrounded by the circular **sphincter of Oddi** (ŏd′ē). This sphincter is constricted when the gastrointestinal tract is inactive, allowing bile to collect in the gallbladder.

Pancreas

In adults the pancreas has an elongate, triangular shape and is about 15 cm long (see Figure 17-15). It is located just under the stomach, within the first loop of the duodenum. Four regions of the pancreas are, from right to left, the head, neck, body, and tail. The **pancreatic duct** (or **duct of Wirsung**) drains exocrine secretions used in digestion through the ampulla of Vater and into the duodenum. The pancreas is surrounded by a capsule of fibrous connective tissue, which extends into the gland and divides it into numerous lobules.

There are two distinctive types of tissue within the pancreas, reflecting its heterocrine function. The major part of each lobule consists of **alveolar clusters** of serous cells called **pancreatic acini** (ăs′ĭ·nī) (Figure 18-27). Secretions from the acini pass through small ducts that lead into

peritoneum: (Gr.) *peri*, around + *teinein*, to stretch

> ## CLINICAL NOTE
>
> ### HEPATITIS
>
> **Hepatitis** is an acute, systemic, viral infection that predominantly affects the liver. There are three distinct clinical forms, each caused by a different hepatitis virus. Although treatment is similar, the viruses differ in their transmission, immunology, and clinical features. The three causative agents are hepatitis A virus (HAV), hepatitis B virus (HBV), and unknown agents for non-A/non-B hepatitis.
>
> The pathology occurring in the liver is largely due to inflammation and mononuclear cell infiltration. The result is hepatic cell necrosis, proliferation of the immunologically active Kupffer's cells, and accumulation of cell and necrotic debris in the liver lobules and bile ducts. The end result is disturbance in bilirubin excretion. While cellular damage is occurring, cellular regeneration is taking place concurrently. Complete regeneration usually occurs within 2 to 3 months. If the liver fails to regenerate in this time, the result is **severe, fulminating hepatitis,** which is often fatal. Continuation of the inflammatory process and necrosis results in **chronic active hepatitis.** Fulminating hepatitis and chronic active hepatitis are more likely to occur with hepatitis B and non-A/non-B hepatitis. Chronic active hepatitis does not occur following hepatitis A.
>
> There is a great deal of variability in the manifestations of hepatitis regardless of type, ranging from lassitude and malaise to profound jaundice, vomiting, prostration, and pain in the upper right abdominal quadrant.
>
> The virus of **hepatitis A** is acquired through the eating of uncooked shellfish obtained from waters contaminated with sewage, through the consumption of food or water contaminated with sewage, or directly through the fecal-oral route. Once having gained entrance to the body, the virus enters the liver where it replicates and causes the inflammation; it eventually enters the intestine via the bile and is lost from the body in the feces.
>
> **Hepatitis B,** in contrast, is transmitted by body fluids: blood, secretions containing serum (oozing wounds or burns), saliva, vaginal secretions, and semen. Persons can thus become infected by being injected with contaminated blood or serum or by using injection equipment contaminated with infected blood or serum. Intravenous drug abusers who share contaminated needles are likely to share the virus also. The virus can also enter from contaminated fluid through a cut or scrape in the recipient's skin. The contaminated body fluids may also enter mucosal surfaces of the mouth or eye through absorption on other mucosal surfaces during sexual contact. An infant delivered from an infected mother might also be infected. It is of interest that HIV, the virus of AIDS, is transmitted in exactly the same way as hepatitis B virus (see Chapter 24 Clinical Note: AIDS). Hepatitis B is not transmitted in the feces. A vaccine is available for administration to individuals who are at high risk for getting the disorder. Health care workers, individuals in hemodialysis units, homosexual males, injectable drugs users, and prison inmates are among those who are at high risk for infection.
>
> **Non-A/non-B hepatitis** is clinically similar to hepatitis B, but immunological differences make it fairly clear that a different causative agent is involved. Although milder than hepatitis B, this disorder has a greater chance of becoming chronic.
>
> Rest, preferably bed rest, is a must in the treatment of these disorders. The infected individual will also need supportive care for nausea and pain. Because the liver is inflamed, all unnecessary medications should be avoided.

interlobular ducts. These in turn empty into the pancreatic duct. Scattered among the pancreatic acini are the islets of Langerhans described in Chapter 17.

Membranes of the Abdominal Cavity

The stomach and intestines, as well as the abdominal cavity in which they lie, are covered by a thin serous membrane called the **peritoneum** (pĕr″ĭ·tō·nē′ŭm) (Figure 18-28). The **parietal peritoneum** lines the abdominal cavity, and the **visceral peritoneum** (which is continuous with the tunica serosa of many digestive organs) wraps around and covers many of the internal organs. The space between the parietal and visceral layers is called the **peritoneal cavity.**

In the upper part of the abdominal cavity, the falciform and coronary ligaments are formed by folds of the peritoneum. The lesser omentum is a double layer of serous membrane that attaches the stomach and duodenum to the

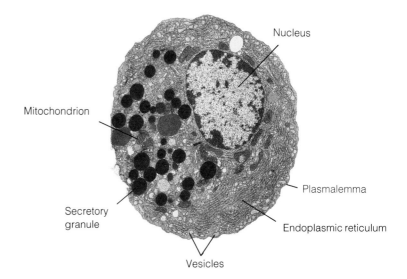

Figure 18-27 Transmission electron micrograph of an isolated pancreatic acinar cell. Dark structures are secretory granules containing the pancreatic product secreted by this cell.

mesentery: (Gr.) *mesos*, middle + *enteron*, gut
retroperitoneal: (L.) *retro*, backward + *peri*, around + *teinein*, to stretch
dysphagia: (Gr.) *dys*, bad, abnormal + *phagein*, to eat
achalasia: (Gr.) *a*, not + *chalasis*, relaxation

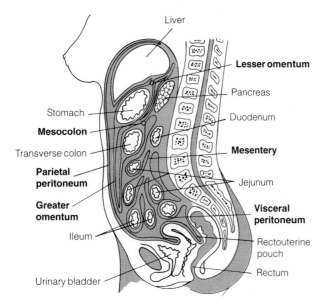

Figure 18-28 This saggital section through the abdominopelvic cavity shows the peritoneum and its modifications.

underside of the liver. The **greater omentum** is an apron-like flap of serous membrane that descends from the greater curvature of the stomach and attaches to the transverse colon and the parietal peritoneum. The greater omentum usually contains large areas of adipose tissue as well as numerous lymph nodes.

The **mesentery** (měs'ĕn·tĕr"ē) is a fan-shaped fold of peritoneum that attaches the jejunum and ileum to the posterior abdominal wall, and the **mesocolon** (měs"ō·kō'lŏn) attaches the large intestine to the same wall. The mesentery and mesocolon are segments of the visceral peritoneum that provide a pathway for the passage of blood and lymph vessels and nerves of the digestive organs.

Because of the manner in which it folds, several organs of the abdominal cavity are not completely enclosed within the peritoneum. The duodenum, pancreas, rectum, kidneys, urinary bladder, and female reproductive organs are located outside of the peritoneal cavity and are thus referred to as **retroperitoneal** (rē"trō·pĕr"ĭ·tō·ne'ăl) **organs.** In males the peritoneum also expands down into the scrotum and envelops the testes.

In this chapter we have examined the structures and processes involved in converting complex chemicals to forms that can be absorbed. Once those chemicals enter the blood, they are carried to all parts of the body where they provide the raw material for the synthesis of new compounds and cell structures as well as the energy required to perform the activities of life. In Chapter 19 we will examine many of those processes.

THE CLINIC

DISORDERS OF THE GASTROINTESTINAL TRACT

Disorders of the gastrointestinal system are related to the embryologic origin, the anatomy of the particular organ, and its physiological function.

Mouth
The embryologic development of the mouth is complex, involving tissues from all three primary layers of ectoderm, mesoderm, and endoderm. Two developmental deformities that may result are **cleft lip** and **cleft palate,** which can interfere with the ability to suck and later to chew and swallow. Neurologic diseases such as multiple sclerosis can interfere with the process of swallowing and result in starvation. The most common tumor of the mouth is **squamous cell carcinoma,** which seems to be related to poor dental hygiene, use of tobacco, and alcohol abuse. These tumors tend to spread to the neck and require wide excision or radiation for treatment.

The absence of teeth or poor dentition resulting from disease or an accident can lead to severe malnutrition since it is difficult to provide adequate sustenance with a liquid diet for prolonged periods of time. Patients who have their jaws wired together because of fractures of the mandible almost always lose considerable amounts of weight in spite of frequent feedings of high caloric liquid diets. Reduced or absent production of saliva occurs at times following irradiation of the mouth for treatment of cancer. This causes a very uncomfortable dry mouth and makes swallowing very difficult, which can lead to severe malnutrition.

The most common disorders of the mouth are tooth decay and gum disease, which are discussed in the Clinical Note: Teeth and Gum Disorders.

Esophagus
Disorders of the esophagus usually result in **dysphagia,** or difficulty in swallowing because of impaired transport of matter from the pharynx to the stomach. Dysphagia may be due to obstruction, as in tumors of the esophagus, or to external pressure from abnormal blood vessels or tumors outside the esophagus. Dysphagia may also result from motor disorders that impede the normal progression of peristalsis; this is called **achalasia** (ăk"ă·lā'zē·ă). This is a lower esophageal sphincter disorder that results from increased sphincter pressure and its failure to relax in response to a peristaltic wave. As a result, peristaltic contractions are unable to force a bolus through the sphincter and food accumulates in the esophagus. This food may be regurgitated, or it may accumulate and remain in place long enough for bacteria to grow and infect the esophageal mucosa.

Reflux of gastric acid into the lower esophagus may result in inflammation and even ulceration of the mucosal lining. Reflux of gastric acid when the patient is recumbent may actually spill over into the lungs and result in asthma-like symptoms (see Unit IV Case Study: The Strange Asthmatic). Protrusion of the stomach above the diaphragm is called a **hiatus hernia** and may be the result of trauma to the upper abdomen or a developmentally short esophagus. Systemic diseases such as **scleroderma** can cause diffuse

ileitis: (L.) *ileum*, groin + (Gr.) *itis*, inflammation

thickening of the wall of the esophagus. Laceration of the gastroesophageal junction may occur with protracted vomiting causing massive hemorrhage, a complication of the eating disorder bulimia. This is known as the **Mallory-Weiss syndrome.**

Stomach

The stomach is spared most developmental disorders except for hiatus hernias, as discussed above, and **pyloric stenosis,** which is a congenital enlargement of the pyloric muscle resulting in obstruction of the emptying of the stomach contents into the small intestine. Pyloric stenosis appears to occur mainly in firstborn males and causes persistent projectile vomiting in newborns. Careful examination allows the physician to feel the enlarged muscle (the size of an olive) through the abdominal wall and confirm the diagnosis. Treatment consists of surgically cutting part of the muscular fibers, thus relieving the obstruction. Tumors of the stomach are mostly carcinomas resulting in ulcerations that fail to heal with treatment. A rare stomach cancer is **linitis plastica,** a slow-growing tumor that infiltrates the wall of the stomach and results in a thickened, stiff wall.

The stomach reacts to stress by developing **stress ulcers,** which are usually multiple superficial ulcerations that occur following intense stress such as CNS lesions, trauma, burns, surgery, shock, or infection. These ulcerations usually produce severe hemorrhage.

An interesting condition that is rare today is **pernicious anemia.** This results from the inability of the terminal ileum to absorb vitamin B_{12} from animal protein because the parietal cells of the stomach fail to produce the substance called intrinsic factor necessary for the absorption of vitamin B_{12}. It is caused by achlorhydria (absence of hydrochloric acid in the stomach), atrophic gastritis, or surgical removal (gastrectomy) of large portions of the stomach. Pernicious anemia appears as a severe, slowly progressive anemia (see Chapter 21 Clinical Note: Anemia). A Nobel Prize was awarded to the physicians who provided a cure by injecting liver extracts that contained large amounts of vitamin B_{12}.

Small Intestines

The small intestine is subject to a large group of conditions called **malabsorption syndromes.** These conditions result from a wide variety of diseases that alter the ability of the small intestine to digest and absorb nutrients. Surgical removal of large portions of the stomach or the intestine itself will result in malabsorption syndromes; lack of pancreatic enzymes, bile salts, and disaccharidases also cause them. A chronic intolerance to gluten in the diet is a cause for these syndromes in children. Whatever the cause, they result in chronic diarrhea, weight loss, weakness, pallor, protuberant abdomen, and bleeding.

A common problem specific to the duodenum is the **duodenal ulcer** caused by repeated subjection to highly acid stomach contents and the action of digestive enzymes from the stomach (see Clinical Note: Peptic Ulcer Disease).

Duodenitis or inflammation of the duodenal bulb is even more common than ulcer disease and results in moderate discomfort in the right upper quadrant of the abdomen.

The ileum is subject to a single common disease called **regional ileitis** (ĭl″ē·ī′tĭs) or **Crohn's disease.** This is a nonspecific inflammatory disease involving all layers of the ileum and occasionally the colon. It usually begins before age 40, with a peak incidence in the 20s. The cause is unclear but seems to involve autoimmune factors, hereditary factors, and infection. It causes pain, diarrhea, fever, and weight loss. It is frequently mistaken for acute appendicitis. This disease is often complicated by intestinal obstruction and perforation. Patients may also have symptoms of arthritis and inflammation of the eyes and skin. It often runs a chronic, recurring course.

Colon

The **irritable bowel syndrome** is one of the most common problems of the digestive system. It is a motility disorder involving the small intestine and large bowel and has symptoms of chronic, recurrent abdominal pain, constipation, and diarrhea. It is thought to be a reaction to stress and frequently is associated with episodes of depression. A diagnosis of irritable bowel syndrome is made from a history of typical symptoms in an otherwise healthy person and by ruling out other causes of the symptoms. Treatment consists of encouraging a healthy lifestyle, prescribing a high fiber diet, and alleviating the stress.

Also affecting the colon is **ulcerative colitis,** a chronic, nonspecific, inflammatory and ulcerative disease of the colon causing bloody diarrhea (see Clinical Note: Ulcerative Colitis). Another type of colitis, **infectious colitis,** may occur as a result of bacterial, viral, or parasitic organisms in the colon. **Diverticulosis** is the presence of small, saccular, mucosal herniations through the muscular wall of the colon. They are usually asymptomatic. **Cancer of the colon** is a common disease of the elderly. It usually begins as malignant degeneration of a polyp, which is a neoplastic nodule from the mucosa. Early removal of polyps can prevent the disease; therefore, diligent search for colon polyps is an important preventive health measure.

A high fiber diet is a relatively new concept in treating many bowel disorders. In the past it was customary to put the bowel at rest with a soft or liquid diet. At this time physicians have found that providing a high fiber, bulk diet is more helpful in treating many disorders of the digestive system and may even prevent cancer of the bowel.

Rectum and Anal Canal

The rectum and anal canal are subject to both mucosal and skin diseases. **Hemorrhoids** are a mass of dilated, tortuous veins of the perianal plexus. **Anal fissures** are linear, ulcerated lacerations of the lower anal canal. **Proctitis** is an inflammation of the rectal mucosa. **Anorectal abscesses** are abscesses that start in the anal and rectal glands and erode into the surrounding soft tissues. Drainage through the perianal skin is called **fistula in áno.**

STUDY OUTLINE

Histology (pp. 441–443)

The *tunica mucosa* is the innermost tunic of the *alimentary canal*. It consists of an *epithelium*, the *lamina propria*, and the *muscularis mucosa*. The epithelium secretes mucus that protects the wall. The lamina propria contains blood vessels, nerves, and connective tissue, and the muscularis mucosa contains muscle tissue.

The *tunica submucosa* contains blood and lymph vessels, nerves, and glands that secrete materials into the lumen of the alimentary canal.

The *tunica muscularis* contains layers of muscle oriented circumferentially and longitudinally. Except for a region at the superior end of the esophagus, the muscle tissue is smooth and involuntary.

The *tunica serosa* is the outermost tunic and begins at the base of the esophagus.

Mouth (pp. 443–448)

The *mouth* is the opening to the alimentary canal and is formed by the *lips, cheeks, palate, tongue,* and *floor*. The volume formed by these structures is the *oral* or *buccal cavity*.

Tongue The *tongue* is a muscular organ that contains *papillae* on its upper and lateral surfaces. *Taste buds* sensitive to sweet, salty, sour, and bitter tastes are present in certain of these papillae.

Gingivae and Teeth *Gingivae* (*gums*) cover the maxillae and mandible and attach to the enamel of the teeth. The *teeth* are embedded in sockets in the maxillae and mandible. Each tooth consists of a *crown, roots,* and a *neck*. Enamel covers the exposed portion; the interior contains *dentin* and *pulp*.

The four types of teeth are *incisors, cuspids, bicuspids,* and *molars*. *Cusps* on the surface of the bicuspids and molars help grind food.

Deciduous teeth are replaced by *permanent teeth* in childhood.

Salivary Glands The three pairs of salivary glands are the *parotid, submandibular,* and *sublingual glands*, as well as numerous *buccal glands*.

Digestion in the Mouth Mechanical and chemical breakdown of food begins with *mastication* (chewing) and the mixing of food with saliva.

Saliva is a complex fluid mixture that lubricates food, contributes to starch digestion, and helps control bacteria.

Pharynx and Esophagus (pp. 449–450)

Pharynx The *oropharynx* lies posterior to the oral cavity and leads through the *laryngopharynx* to the esophagus.

Esophagus The *esophagus* is a tube that leads from the laryngopharynx to the stomach. Its wall contains muscle tissue that helps push food to the stomach. Its lower end is constricted at the lower *esophageal sphincter*, which usually prevents backflow of stomach contents.

Swallowing Swallowing (or *deglutition*) consists of three stages, controlled by voluntary actions and by involuntary signals from the swallowing center in the medulla oblongata. The *buccal stage* is a voluntary stage during which a *bolus* (chewed food) is forced to the rear of the oral cavity. The *pharyngeal stage* is an involuntary stage that closes off the opening to the larynx and forces the bolus to the esophagus. The *esophageal stage* is an involuntary stage during which the bolus is forced through the esophagus by peristaltic contractions.

Stomach (pp. 451–455)

The opening to the stomach from the esophagus is the *cardiac orifice*. The opening at the other end is the *pyloric orifice*. Regions in the stomach are the *fundus, body,* and *pyloric antrum*. The right side forms the *lesser curvature*, and the left side forms the *greater curvature*.

Digestion in the Stomach Solid food matter is converted to liquid *chyme* in the stomach.

Source and Composition of Gastric Juice Gastric juice is formed and secreted by glands at the base of gastric pits in the stomach wall.

Cardiac and *gastric glands* are located in the cardiac region and in the fundus and body, respectively. Cardiac glands secrete mucus. Gastric glands secrete HCl, *intrinsic factor,* and *pepsinogen*. Pepsinogen is converted to *pepsin* by stomach acid.

Pyloric glands secrete mucus and *gastrin*, a hormone that helps regulate stomach activity.

Control of Gastric Gland Secretion Activity of gastric glands is controlled by a combination of hormonal and neural mechanisms.

The *cephalic phase* consists of neural activity generated by the sight and sound of food and stimulates glands in all parts of the stomach to increase secretion of their products.

The *gastric phase* begins as food enters the stomach. Neural and hormonal activity further increase gland secretion.

The *intestinal phase* is initiated as chyme enters the duodenum, where it stimulates the release of gastrin to the blood, which stimulates additional release of HCl and pepsinogen into the stomach.

Control of Stomach Emptying The stomach empties into the small intestine in response to *neural* and *hormonal controls* that inhibit the pyloric pump.

Small Intestine (pp. 455–463)

The *small intestine* begins with the *duodenum*, continues as the *jejunum*, and ends with the *ileum*. *Villi* and *microvilli* on the surface increase the surface area through which nutrients are absorbed. Each villus contains capillaries and a central *lacteal*.

Glands in the space between villi secrete intestinal enzymes. Secretions from the gallbladder and pancreas enter the duodenum and contribute to digestion.

Digestion in the Small Intestine The acid in chyme is neutralized by bicarbonate from the pancreas, secretion of which is controlled by *secretin*. Enzymes from the pancreas continue the digestion of protein, carbohydrates, fats, and nucleic acids. *Bile* from the gallbladder emulsifies fat and increases the efficiency of fat digestion.

Intestinal Enzymes Final breakdown of nutrients is performed by enzymes in the *brush border* of the villi.

Carbohydrate digestion is completed by *maltase, lactase, sucrase,* and *dextrinase.* Peptide breakdown is completed by *amino peptidase* and *dipeptidase.* Additional enzymes include *intestinal lipase,* nucleases, and phosphatases.

Absorption Most nutrients are absorbed by the time chyme reaches the ileum. Bile salts are reabsorbed in the ileum, as are vitamin B_{12} and intrinsic factor. Organic nutrients are digested to their basic subunits before absorption. Fatty acids are absorbed in mixed *micelles,* reassembled into triglycerides, and carried away in lymphatic vessels.

Large Intestine (pp. 464–467)

The three anatomical regions of the large intestine are the *cecum, colon,* and *rectum*. The *vermiform appendix* is attached to the cecum. The colon consists of *ascending, transverse,* and *descending* portions. The rectum terminates at the *anus*.

Digestion in the Large Intestine Movement through the large intestine is caused by slow and sustained peristaltic contractions. Absorption of minerals and fluids reduces the liquid contents to semisolid *feces*.

Intestinal Flora Bacteria and other microorganisms live in the large intestine in large numbers and influence the development and activity of the large intestine.

Defecation Fecal matter is stored in the rectum and sigmoid colon. Irregularities in feces water content can cause *diarrhea* or *constipation*. A *defecation reflex* is initiated in response to distension of the rectum and causes peristaltic contractions to force feces through the anus. Involuntary muscles in the internal anal sphincter relax during the reflex. Voluntary muscles in the external anal sphincter can inhibit the reflex.

Accessory Structures (pp. 468–472)

Liver The *liver* is organized into *lobes*. It connects to the inferior surface of the diaphragm through the *falciform, coronary,* and *triangular ligaments*. It is joined to the stomach by the *lesser omentum*. Blood is delivered to the liver by the hepatic portal vein. The liver forms bile from breakdown products of blood, and it also removes waste and toxic products from the blood.

Gallbladder The *gallbladder* is located under the right lobe of the liver, where it is drained by the *cystic duct*. The cystic duct joins with the *common hepatic duct* to form the *common bile duct*. The common bile duct fuses with the *pancreatic duct* and both empty into the duodenum through pores in the duodenal papilla. The *sphincter of Oddi* regulates the opening into the duodenum.

Pancreas The pancreas consists of the head, neck, body, and tail regions. It is composed of endocrine and exocrine tissue. *Pancreatic acini* secrete the products that are delivered to the duodenum. Islets of Langerhans distributed among the acini produce hormones.

Membranes of the Abdominal Cavity The stomach and intestines are covered by the *peritoneum*. The regions of the peritoneum are the *parietal* and *visceral peritonea* and the *greater omentum*. The *mesentary* and *mesocolon* are portions of the visceral peritoneum through which nerves, blood vessels, and lymph vessels pass. Several organs of the alimentary system are not located within the peritoneum and are referred to as *retroperitoneal organs*.

SELF-TEST OF CHAPTER OBJECTIVES

True-False Questions
1. Palatine tonsils are masses of lymphatic tissue located on both sides of the pharynx.
2. The space between the cheeks and teeth is called the labial frenulum.
3. The organ that absorbs the greatest amount of water in the digestive tract is the colon.
4. Fungiform papillae on the tongue do not contain taste buds.
5. Peristaltic contractions generally begin at one end of the colon and then travel uninterrupted along its entire length.
6. The tunica muscularis in the stomach consists of three muscle layers, whereas in the small intestine it consists of only two layers.
7. The upper region of the esophagus is composed of only three tunics of tissue.
8. The parotid salivary gland is located below the tongue.
9. The exposed portion of a tooth is covered externally with a bony material called dentin.
10. The mucus that coats the inner wall of the stomach comes from special secretory cells found in the wall.
11. The common bile duct and pancreatic duct empty into the jejunum region of the small intestine.
12. CCK-PZ stimulates emptying of the gallbladder.
13. The cecum is a part of the small intestine.
14. The brush border is found on the inner wall of the large intestine.
15. Goblet cells are found in the pits and epithelium of the large intestine.
16. Movement of chyme through the large intestine is generally faster than through the small intestine.
17. Feces consist mainly of indigestible foodstuff.

Matching Questions

Match the macromolecule on the left with the organ(s) on the right in which it is digested:

18. protein
19. starch
20. lipid
21. nucleic acid

a. mouth
b. esophagus
c. stomach
d. small intestine
e. large intestine

Match each enzyme in the left column with its substrate on the right:

22. pepsin
23. carboxypeptidase
24. pancreatic amylase
25. sucrase
26. lipase

a. fats
b. sucrose
c. starch
d. polypeptides
e. proteins

Match the structure on the left with the digestive organ on the right in which it is found:

27. uvula
28. haustra
29. Peyer's patches
30. villi
31. pylorus
32. gingivae
33. taenia coli
34. Brunner's glands
35. gastric pits
36. hepatic flexure

a. small intestine
b. esophagus
c. stomach
d. mouth
e. large intestine

Multiple Choice Questions

37. The largest of the papillae found on the tongue are the
    a. filiform papillae
    b. fungiform papillae
    c. vallate papillae
    d. lingual papillae
38. The cutting teeth are classified as
    a. incisors
    b. cuspids
    c. canines
    d. bicuspids
39. By the time the intestinal contents reach the rectum, they form a semisolid mass. Water has been removed by
    a. active transport
    b. facilitated diffusion
    c. osmosis
    d. pinocytosis
40. Monoglycerides and bile salts join together to form
    a. fatty acids
    b. lipids
    c. micelles
    d. cholesterol
41. The hormone that induces the release of bicarbonate from the pancreas is
    a. secretin
    b. prosecretin
    c. cholecystokinin
    d. trypsinogen
42. The pyloric pump is inhibited by the
    a. volume of the meal one has eaten
    b. volume of chyme in the duodenum
    c. acidity of the stomach contents
    d. protein content of the meal
43. The enterogastric reflex controls the rate of
    a. food entering the stomach
    b. chyme leaving the stomach
    c. peristalsis along the small intestine
    d. passage through the ileocecal valve
44. The bulged regions of the colon located between muscular constrictions are called the
    a. lumen
    b. tunica muscularis
    c. haustra
    d. rectum
45. The main ingredient of bile is
    a. cholesterol derivatives
    b. fat-digesting enzymes
    c. cholecystokinin
    d. secretin
46. Gallstones are formed largely from
    a. condensed mucus
    b. bile acids
    c. precipitated cholesterol
    d. calcium bicarbonate
47. The villus is a characteristic structure of the inner wall of the
    a. stomach
    b. small intestine
    c. large intestine
    d. both intestines
48. Virtually all nutrients are absorbed by the time they pass through the
    a. stomach
    b. duodenum
    c. jejunum
    d. ileum
49. Bile salts are reabsorbed by the
    a. gallbladder
    b. duodenum
    c. ileum
    d. colon
50. When triglycerides are reformed with a protein coat in the cells of the small intestine they form a structure called a
    a. micelle
    b. mixed micelle
    c. chylomicron
    d. free fatty acids
51. "Intrinsic factor" makes possible the absorption of
    a. amino acids
    b. sugar
    c. fatty acid
    d. vitamin B_{12}
52. The lamina propria is part of which intestinal tunic?
    a. tunica mucosa
    b. tunica submucosa
    c. tunica muscularis
    d. tunica serosa
53. The part of the peritoneum that connects the liver to the diaphragm is the
    a. lesser omentum
    b. greater omentum
    c. mesentery
    d. falciform ligament
54. Hydrochloric acid is produced in the stomach by
    a. chief cells
    b. parietal cells
    c. goblet cells
    d. Brunner's glands

Essay Questions

55. Trace the path of a glucose unit present in a starch molecule from the time it is ingested until it enters the blood.
56. Describe the glandular structures of the digestive system, indicating their location, their products, and the use of those products in digestion.
57. Describe the tunics of the alimentary canal.
58. Describe the muscular contractions that occur in different portions of the alimentary canal and explain the role they play in digestion.

59. Draw a molar and identify its parts.
60. Describe the hormonal and neural mechanisms used to regulate digestion.
61. Explain how dispersion of fat droplets by bile salts increases efficiency of fat digestion.
62. Describe the steps, organs, and enzymes involved in protein digestion.
63. List and describe the three stages of swallowing.
64. Diagram and identify the ducts from the liver, gallbladder, and pancreas that drain into the duodenum.
65. Describe the embryological origin of the digestive system.

Clinical Application Questions

66. Define anorexia nervosa and describe its effects.
67. Peptic ulcers usually occur in the stomach, esophagus, or duodenum, but not in the lower portions of the small intestine or in the large intestine. Explain why.
68. One of the possible outcomes of long term, sustained gastritis is the growth of bacteria in the stomach. Bacteria are not normally able to grow in the stomach. Why are they able to as a result of chronic gastritis?
69. Describe the devices and techniques a physician might use to examine the gastrointestinal system.
70. Suggest what digestive functions might be lost or reduced as a result of surgery for cholelithiasis.
71. Many people, as they grow older, find that they no longer can drink milk or eat ice cream without incurring gastrointestinal discomfort. Explain what their problem might be, and the measures they can take to avoid the problem.
72. Describe what happens when one has an "attack" of appendicitis.
73. Cirrhosis is an extremely dangerous illness involving progressive destruction of the liver. Describe the causes and effects of the disease on the structure and function of the liver.
74. Using a table format, list the organs of the digestive tract, disorders that can occur in each, and possible effects of each disorder.

19
Metabolic Processes

Nutrients absorbed during digestion are used for two purposes. First, they serve as a fuel, providing energy for such diverse processes as muscle contraction, establishment of membrane potentials necessary for nerve impulses, and synthesis of macromolecules. When used as an energy source, the chemicals are broken down, and energy released in the process is used to produce ATP from ADP and inorganic phosphate. The ATP then serves as the direct source of energy in those reactions that require an input of energy, being cleaved back into ADP and inorganic phosphate in the reaction. Second, nutrients provide the matter for the synthesis of new compounds, which in turn may be incorporated into macromolecules, including the nucleic acids in which the cell's genetic information is stored, the enzymes that catalyze the tens of thousands of reactions that go on in a cell, and the lipids that provide the structural basis of cell membranes.

In this chapter we will examine

1. the reactions used to remove energy from carbohydrates for the production of ATP,
2. the mechanisms used for the synthesis and degradation of fats and proteins,
3. the role of vitamins and minerals in these processes, and
4. the way that heat produced by the reactions is used and regulated.

Overview of Metabolism

Metabolism refers to all those chemical processes involved in the formation and degradation of cellular materials, including the atomic and molecular rearrangements that occur in chemical reactions and the energy transfers that accompany them. Metabolic processes in which simple compounds are used to form larger and more complex compounds are referred to as **anabolism.** The opposite of anabolism,

gluconeogenesis: (Chem.) glucose + (Gr.) *neos*, new + *genesis*, origin

catabolism, involves metabolic reactions that change complex chemical substances into smaller and simpler compounds. Anabolic reactions require energy to produce complex substances and to store some of the energy in the bonds of those substances. Catabolism releases that energy, often making it available for other cellular processes, while at the same time producing heat used to maintain body temperature. Metabolism performs a profound role in maintaining a relatively constant internal environment, a homeostatic role that enables the body to function efficiently.

The body alternates between two different metabolic states, depending on whether the chemicals being metabolized have just been absorbed from the digestive tract or are being obtained from stored reserves. The **absorptive state** exists during the time when carbohydrates, lipids, amino acids, and other compounds absorbed from the digestive tract are being distributed throughout the body by the blood or lymph. Reactions involving the storage or interconversion of these compounds occur during this period. It usually continues for a few hours following a meal, then gradually subsides as the body enters the **postabsorptive state.** During this state, substances that had been stored during the absorptive state are broken down and used in metabolism. Glucose breakdown is repressed in favor of fat breakdown as a source of energy in all tissues other than nervous tissue. The glucose is said to be **spared** during this period. With extended fasting or starving, however, fat and carbohydrate reserves are used up and energy begins to be obtained through the breakdown of proteins.

Carbohydrate Metabolism

After studying this section, you should be able to:

1. Describe how glucose is obtained, stored, and released.
2. Explain how the glucose level in the blood is controlled.
3. Describe the reactions and the products of glycolysis, the Krebs cycle, and the respiratory chain.
4. Explain where in the reactions of cellular respiration ATP is produced and why more ATP is produced when oxygen is available.
5. Describe how glucose can be used as a source of other molecules when it is not degraded entirely in cellular respiration.

Carbohydrates are obtained from the diet and by synthesis from other compounds. Many different forms of carbohydrate are usually present in food, ranging in complexity from such simple sugars as glucose and fructose to the complex polysaccharides starch and glycogen. However, regardless of their form, all usable carbohydrates are converted to monosaccharides in the digestive tract and absorbed in that form. Most monosaccharides obtained in the diet or produced by digestion are glucose molecules, although smaller amounts of fructose and galactose may also be present depending on how much fruit, sugar, or milk is consumed. Even these monosaccharides are converted to glucose in the liver, thus the metabolism of dietary carbohydrate is essentially the metabolism of glucose.

Synthesis of carbohydrates from noncarbohydrates is **gluconeogenesis** (glū″kō·nē″ō·jĕn′ĕ·sĭs), which occurs principally in the liver and, to a lesser extent, the kidneys. The primary noncarbohydrates used in gluconeogenesis are lactic acid, amino acids, and glycerol. **Lactic acid** is produced by skeletal muscles when they are using oxygen faster than it can be supplied, such as during periods of vigorous exercise. Amino acids used in gluconeogenesis are derived from digested proteins and from catabolism of muscle proteins during periods of starvation. Glycerol is produced by the catabolism of triglycerides, a reaction that releases one glycerol molecule and three fatty acid molecules from each triglyceride (see Chapter 2). Fatty acids produced in this reaction continue to be degraded and are converted to CO_2, which is eliminated (Figure 19-1).

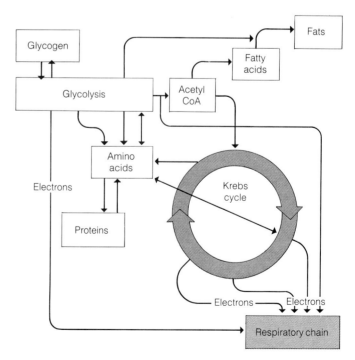

Figure 19-1 An overview of the metabolism of carbohydrates amino acids, and fats. This diagram shows how the reactions involving these classes of compounds are interrelated.

glycogenolysis: (Chem.) glycogen + (Gr.) *gennan*, to produce + *lysis*, dissolving

Metabolism of Glucose

There are three potential ways glucose can be used. It may be stored, catabolized to carbon dioxide and water, or used to produce new substances. Sometimes, depending on how much is consumed, glucose may also be excreted.

Glucose Storage

When greater amounts of glucose are consumed than used, glucose molecules are joined to glycogen molecules in the liver and skeletal muscles. In this process, called **glycogenesis** (glī″kō·jĕn′ĕ·sĭs), glucose receives a phosphate group from ATP, which converts the glucose to **glucose phosphate** (see Chapter 2). Glucose phosphate reacts with a substance produced by the cell called **uridine triphosphate (UTP)** and becomes **UDP-glucose** (Figure 19-2a). Only in this form can glucose be added to a growing glycogen chain. Upon addition of the glucose portion of UDP-glucose to glycogen, the uridine is released and resynthesized to UTP.

Although many glucose molecules absorbed in the small intestine are stored in the liver as glycogen, others pass through the liver and enter the general circulation. Once in the blood, they are carried to tissues where they are metabolized or, in the case of skeletal muscle, stored as glycogen. About 500 g of glycogen may be stored in the liver and skeletal muscles of a 70-kg man, representing about 15 times as much energy as in the free glucose in body fluids.

Once glycogen reaches about 5% of the mass of the liver, incorporation of glucose into glycogen declines. At this point, glucose phosphate in the liver is diverted to reactions that synthesize fats. Most of these fats are released and carried by blood to **adipose tissue** where they are stored.

Glucose Release

Glucose is recovered from glycogen by **glycogenolysis** (glī″kō·jĕn·ŏl′ĭ·sĭs), a process in which glucose molecules are sequentially removed from the ends of glycogen molecules. Inorganic phosphate is added to each glucose molecule in this reaction, producing glucose phosphate (Figure 19b). The reaction that releases glucose from glycogen is not simply the reverse of the reaction that adds glucose to glycogen. As a result, different mechanisms can be used to control glycogen synthesis and degradation, allowing for more precise control of glucose levels.

Control of Glycogen Synthesis and Degradation

Glycogen synthesis and degradation are primarily controlled by epinephrine (Figure 19-3). Epinephrine is released during periods when additional glucose is required, such as when glucose levels have been lowered by cellular respiration or when the body is reacting to stress. Epinephrine operates through a cyclic-AMP (cAMP) mechanism, activating adenylate cyclase at the cell membrane. Activated adenylate cyclase produces cAMP. cAMP then stimulates the activation of protein kinase, an enzyme that simultaneously activates enzymes that cleave glucose from glycogen and deactivates enzymes that catalyze the addition of glucose to glycogen. In this way epinephrine induces the release of glucose at the same time that it stops the incorporation of glucose into glycogen. As the glucose level rises, epinephrine release diminishes and its effects are reversed (see Chapter 17).

Catabolism of Glucose

Catabolism of glucose to carbon dioxide and water is the primary means by which cells obtain energy. Collectively, the reactions the cell uses in this process comprise **cellular respiration.** Cellular respiration should not be confused with respiration involving gas exchange by the lungs, although that process provides oxygen used by cells and

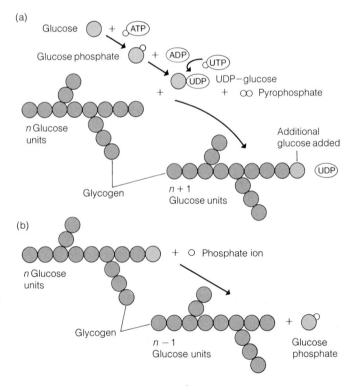

Figure 19-2 (a) Glycogenesis and (b) glycogenolysis. Although these reactions have opposite effects, one is not simply the reverse of the other. These two different pathways go in opposite directions, which provides a more efficient mechanism for regulating glycogen synthesis and breakdown.

glycolysis: (Gr.) *glykys*, sweet + *lysis*, dissolving

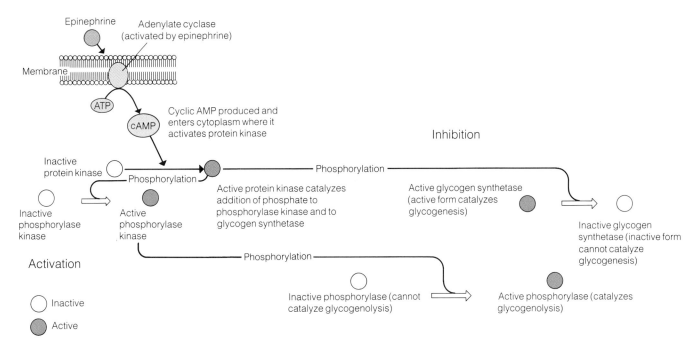

Figure 19-3 Epinephrine control of glycogen synthesis and breakdown. Epinephrine acts through a membrane receptor, as shown in Figure 17-2.

eliminates carbon dioxide produced by them as they respire. In cellular respiration there is a stepwise degradation of glucose that is accompanied by synthesis of ATP. As a result, much of the energy in bonds of glucose is retained in ATP, where it is available for use by the cell. Since this process involves the release of electrons, cellular respiration is also referred to as the **oxidation** of glucose.

Cellular respiration involves three major phases: glycolysis, the Krebs cycle, and the respiratory chain. Glycolysis occurs in the cytosol, whereas the Krebs cycle and respiratory chain occur in mitochondria.

Glycolysis

In the reactions of **glycolysis,** the first phase of cellular respiration, glucose is catabolized in a stepwise sequence of reactions until it has been converted to two molecules of a three-carbon compound called **pyruvic** (pī·rū′vĭk) **acid.** The process begins when the glucose molecule is carried into the cell by facilitated or active transport or is produced in the cell from glycogen. Glucose obtained from glycogen has a phosphate added to it as it is cleaved, but glucose molecules that are transported in must have phosphate added before they can be used. This is accomplished by the transfer of a phosphate group from ATP to the glucose molecule (Figure 19-4). Glucose phosphate produced by this reaction is then rearranged to fructose phosphate, and a second phosphate is added from another ATP molecule. At this point in the sequence, two ATP molecules have each supplied a phosphate group and the glucose molecule has been converted to **fructose diphosphate.**

In the next reaction fructose diphosphate reacts with water and splits into two halves, each of which contains three carbon atoms, a phosphate group, and associated oxygen and hydrogen atoms. These two halves are called **phosphoglyceraldehyde** (fŏs″fō·glĭs″ĕr·ăl′dĕ·hīd) **(PGAl)** and **dihydroxyacetone phosphate (DHAP).** DHAP is rapidly converted to PGAl, so in effect, two molecules of PGAl are produced by the cleavage of fructose diphosphate. None of these reactions has resulted in the synthesis of ATP, but instead have occurred at the expense of two ATP molecules.

The next step in the sequence occurs as PGAl is oxidized by a molecule of **nicotine adenine dinucleotide (NAD$^+$)** (Figure 19-4). Two electrons and a hydrogen ion are transferred to NAD$^+$ in this reaction, converting it to **NADH.** Oxidized PGAl combines with phosphate available in the surrounding fluid and becomes diphosphoglyceric acid. This phosphorylation is different from the phosphorylation of

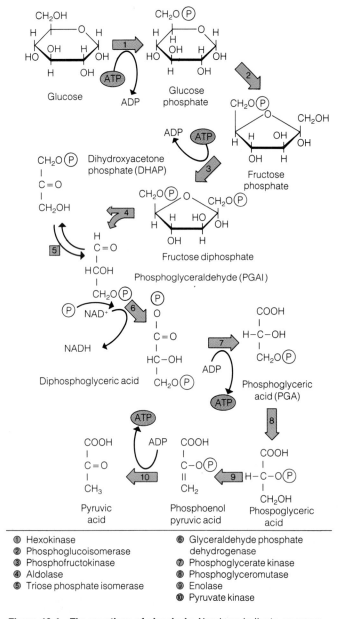

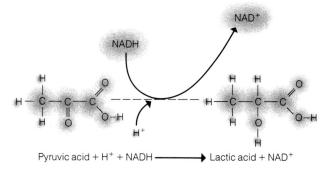

Figure 19-5 Formation of lactic acid. Lactic acid is produced by the reduction of pyruvic acid. The source of electrons for this reduction is NADH, which is oxidized to NAD^+ in the process. This reaction resupplies NAD^+ that is used in the oxidation of PGAl to PGA.

Figure 19-4 The reactions of glycolysis. Numbers indicate enzymes that are used to catalyze the reactions (see list at bottom). (P) represents a phosphate group, $H_2PO_4^-$. The curved arrows in reactions 1 and 3 indicate the transfer of a phosphate group from ATP to the sugar molecule.

① Hexokinase
② Phosphoglucoisomerase
③ Phosphofructokinase
④ Aldolase
⑤ Triose phosphate isomerase
⑥ Glyceraldehyde phosphate dehydrogenase
⑦ Phosphoglycerate kinase
⑧ Phosphoglyceromutase
⑨ Enolase
⑩ Pyruvate kinase

glucose and fructose that occurred earlier because ATP does not supply the phosphate group. Phosphate obtained directly from the surrounding fluid is referred to as **inorganic phosphate (P_i)**.

The step that occurs next is the first one in glycolysis that produces ATP. In this reaction the phosphate group just added is transferred to ADP, producing a molecule of ATP and a molecule of **phosphoglyceric acid (PGA)**. Since this reaction occurs with both phosphoglyceric acid molecules derived from a single glucose molecule, two ATP molecules are produced. At this point, the amount of ATP produced is the same as the amount used, and any additional ATP production represents a "profit" of ATP.

The next two steps involve the rearrangement of PGA to produce first a molecule of phosphoenol pyruvic acid and then a molecule of pyruvic acid. Pyruvic acid is produced from phosphoenol pyruvic acid by the removal of the remaining phosphate group, which is transferred to ADP. At this point in the catabolism of a glucose molecule, the cell has used two and produced four ATP molecules, and glycolysis is complete.

Fate of Pyruvic Acid

Two possible fates await pyruvic acid. If oxygen is not available, pyruvic acid is reduced to **lactic acid** using electrons provided by NADH produced when PGAl was oxidized. In this way conversion of pyruvic acid to lactic acid resupplies NAD^+ needed for the oxidation of additional PGAl and enables glycolysis to continue (Figure 19-5).

If oxygen is available, however, pyruvic acid enters the Krebs cycle where it is broken down further. First, however, pyruvic acid loses a carbon atom as CO_2, leaving a two-carbon fragment known as an **acetyl group**. The acetyl group is attached to a molecule of **coenzyme A (CoA)**, which "carries" the acetyl group into the reactions of the Krebs cycle. Two electrons are also transferred to NAD^+ in this step.

Krebs Cycle

The next phase of cellular respiration is called the **Krebs cycle**, which is also known as the **citric acid** or **tricarboxylic acid cycle**. The name Krebs comes from Hans Krebs, a

flavin: (L.) *flavus*, yellow
fumaric: (L.) *Fumaria*, a plant related to a poppy
malic: (L.) *Malus*, the genus of apple

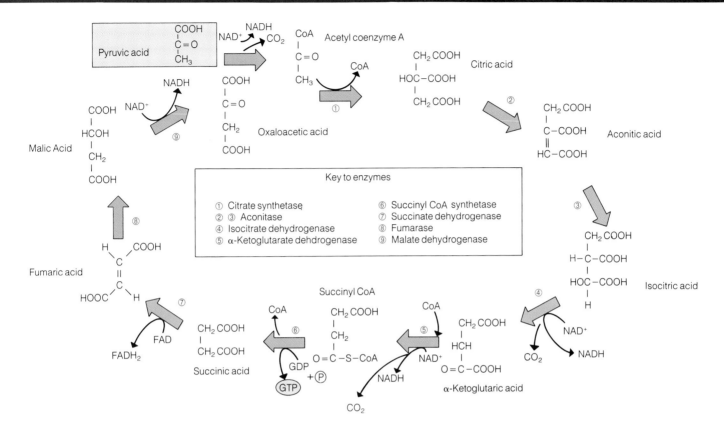

Figure 19-6 Reactions of the Krebs cycle. Numbers indicate enzymes that are used to catalyze the reactions.

biochemist who made most of the fundamental discoveries of this phase of cellular respiration. The acetyl group mentioned above enters the Krebs cycle by combining with **oxaloacetic acid,** a compound that contains four carbon atoms. The CoA molecule that carried the acetyl group is released and reused to carry another acetyl group to the Krebs cycle (Figure 19-6). Addition of the acetyl group to oxaloacetic acid creates a six-carbon product called **citric acid.** Citric acid undergoes two rearrangements to produce **isocitric acid,** a compound that still contains six carbon atoms. In the next step, however, a carbon atom is released as CO_2 when citric acid is converted to **α-ketoglutaric acid.** This reaction also involves the transfer of two electrons to NAD^+.

A similar reaction happens to α-ketoglutaric acid, and another CO_2 molecule is released and another molecule of NAD^+ is reduced. The four-carbon product of this reaction combines with coenzyme A to produce **succinyl CoA.** In the next step, energy in succinyl CoA is used to add a molecule of inorganic phosphate to an ADP-like compound called **guanosine diphosphate (GDP).** The product, **guanosine triphosphate (GTP),** readily gives up the phosphate group to ADP, so this is an ATP-producing step in the Krebs cycle. Coenzyme A is also released at this point, leaving **succinic** (sŭk·sĭn'ĭk) **acid** as a product.

The remaining reactions of the Krebs cycle regenerate oxaloacetic acid from succinic acid. This is accomplished in three steps. The first step consists of an oxidation-reduction reaction in which electrons are released to another electron-receiving compound called **flavin adenine dinucleotide (FAD).** The product of this step is **fumaric acid,** which is rearranged to **malic acid.** Malic acid then releases electrons to NAD^+ and is oxidized to oxaloacetic acid. The oxaloacetic acid is then available to begin another cycle.

In summary, the Krebs cycle releases the remaining two carbon atoms from pyruvic acid as CO_2 and transfers electrons to NAD^+ and FAD. In addition, one molecule of GTP is produced by the direct phosphorylation of GDP, and the phosphate group is then transferred to ADP, producing a molecule of ATP (Figure 19-6).

Respiratory Chain

The **respiratory chain,** the final phase of cellular respiration, is a series of oxidation-reduction reactions in which electrons are transferred from NADH and $FADH_2$ to a series of electron-carrying molecules in the inner membrane of the mitochondrion (Figure 19-7). What makes this an especially important part of cellular respiration is that much of the energy present in these electrons is released as they pass

cytochrome: (Gr.) *kytos*, cell + *chroma*, color
chemiosmosis: (Gr.) *chemeia*, chemistry + *osmos*, a thrusting

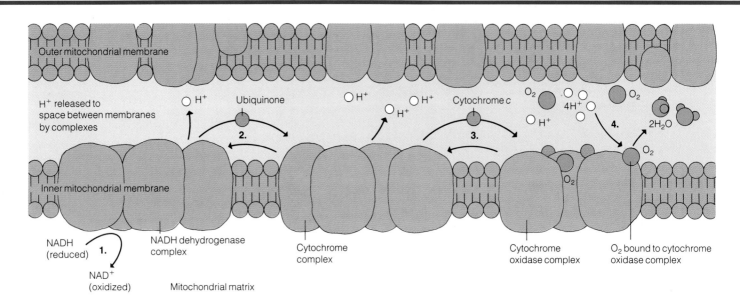

Figure 19-7 Steps in the respiratory chain. (1) NADH delivers electrons to NADH dehydrogenase complex. (2) After passing through NADH dehydrogenase complex, electrons are carried one at a time to cytochrome complex by ubiquinone. (3) Electrons pass through cytochrome complex and are carried by cytochrome *c* to cytochrome oxidase complex. (4) Bound O_2 molecules receive four electrons, attracting four hydrogen ions as a result. The oxygen molecule dissociates and produces two water molecules.

along the chain and is used for the synthesis of ATP. Since there are so many steps in the oxidation of glucose where electrons are released, the electron transport that occurs in the respiratory chain is a source of large amounts of ATP.

The first few reactions in the chain involve the passage of electrons through a complex of molecules that includes **cytochrome *b***, a protein containing iron and sulfur called **Fe—S protein**, and a small lipid molecule called **ubiquinone** (ū·bĭk'wĭ·nōn). Once through these molecules, the electron moves through a sequence of iron-containing proteins called **cytochrome c_1, cytochrome *c*, cytochrome *a*,** and **cytochrome a_3**. From cytochrome a_3, the electron passes to an atom of oxygen. Electrons are passed along individually, and four electrons are required to reduce a molecule of O_2. Each oxygen atom receives two electrons and attracts two hydrogen ions (H^+) from the surrounding fluid, so two H_2O molecules are produced for every four electrons that pass along the respiratory chain. The disassembly of the atoms of the original glucose molecule is now complete (Figure 19-7).

Oxidative Phosphorylation

The production of ATP from ADP and inorganic phosphate (P_i) that occurs as an electron passes through the respiratory chain is called **oxidative phosphorylation.** The energy required to attach P_i to ADP is provided by a mechanism that establishes a concentration gradient of H^+ across the inner membrane of the mitochondrion. According to the **chemiosmotic** (kĕm"ē·ŏz·mŏt'ĭk) hypothesis, transfer of electrons at three points in the respiratory chain is accompanied by the transport of a hydrogen ion from the inner matrix of the mitochondrion into the space between the two mitochondrial membranes. This occurs when electrons are passed from NADH to FAD, from cytochrome c_1 to cytochrome *c*, and from cytochrome *a* to cytochrome a_3. The accumulation of H^+ in the space between the membranes causes a membrane potential to be established, with the fluid between the membranes being electrically positive compared to the fluid in the matrix.

The membrane is impermeable to the diffusion of H^+ except at certain points where **hydrogen transport protein complexes** are located, which are groups of proteins that extend through the inner membrane. These complexes provide channels through which the hydrogen ions can pass back into the matrix. The complexes also include an inactive form of **ATPase,** an enzyme that catalyzes the addition of P_i to ADP.

Driven by the concentration and electrical gradients that exist across the membrane, H^+ flow through the channels into the matrix. As they do, they activate the ATPase in

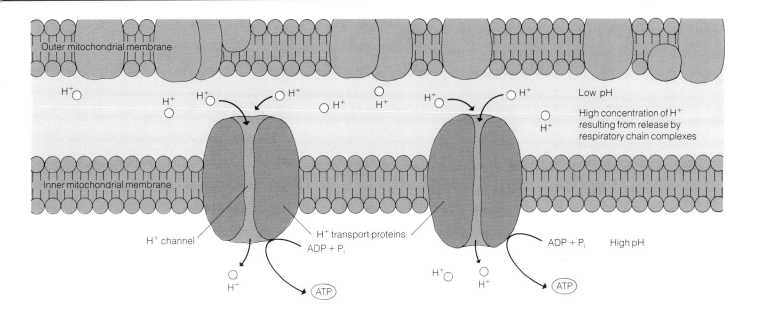

Figure 19-8 Production of ATP by the respiratory chain. According to the chemiosmotic hypothesis, movement of electrons at three points in the respiratory chain results in the transport of H$^+$ from the matrix to the space between the two mitochondrial membranes. The resulting concentration and electrical gradients drive the H$^+$ back through the membrane through special carrier molecules in a reaction coupled with the addition of a phosphate group to ADP.

the complex, and the activated ATPase catalyzes the synthesis of ATP (Figure 19-8). As a result, an ATP molecule is produced for every H$^+$ that flows back into the matrix.

An Overview of Cellular Respiration

Electrons are released at several steps in glucose oxidation, usually to NAD$^+$, but at one point in the Krebs cycle to FAD. In each case, the electrons pass through the respiratory chain to oxygen, reducing the oxygen to water. Three ATP molecules are produced for each molecule of NAD$^+$ reduced when pyruvic acid is converted to acetyl CoA and for each molecule of NAD$^+$ reduced in the Krebs cycle. Production of NADH when PGAl is converted to PGA and when FAD is reduced in the Krebs cycle results in the production of only two ATP molecules, because the initial reaction in which ATP is produced is bypassed in these cases. In total, the respiratory chain accounts for 32 ATP molecules for every glucose molecule. In addition, there are two reactions in glycolysis in which ATP is produced directly and one in the Krebs cycle in which ATP is produced from GTP, producing six more ATP molecules for each glucose molecule.

This is a total of 38 ATP molecules per glucose molecule. Two ATP molecules were used to phosphorylate glucose when glycolysis began, so the net production of ATP is 36 molecules per molecule of glucose consumed.

The overall reactions of cellular respiration can be summarized in a single chemical equation:

$$2\,ATP + glucose + 38\,ADP + 38\,P_i + 36\,H^+ + 6\,O_2 \longrightarrow$$
$$6\,CO_2 + 38\,ATP + 42\,H_2O + 2\,ADP$$

The amount of energy represented by the 36 newly produced ATP molecules represents 38% of the energy released by the oxidation of glucose. That is, cellular respiration is 38% efficient in storing the energy of glucose in ATP (Table 19-1). The remainder is released as heat.

Glucose as a Source of Carbon

Although serving as an energy source in cellular respiration is an important role of glucose, it is not the only one. Equally important is its role as a source of new compounds. Many glucose molecules never make it through all the reactions of cellular respiration because they are diverted at various points in the process into reactions that lead to the production of other compounds. Indeed, if one were to trace the paths followed by carbon atoms of glucose as the glucose is metabolized by the cell, those paths would lead not only to the carbon dioxide released by cellular respiration, but also to proteins, lipids, nucleic acids, and every other carbon-containing substance made by the cell (Figure 19-9).

Table 19-1 ATP PRODUCED BY THE COMPLETE OXIDATION OF GLUCOSE

Reactions	ATP Molecules Produced per Glucose	Relative ATP Energy Yield (%)
Glycolysis (Conversion of Glucose to Pyruvic Acid)		
Phosphorylation of glucose to fructose diphosphate	−2	−2.1
Conversion of two molecules of diphosphoglyceric acid to PGA	2	2.1
Conversion of two molecules of phosphoenolpyruvic acid	2	2.1
Net yield of glycolysis	2 ATPs	2.1%
Conversion of Pyruvic Acid to Acetyl CoA	0	0
Krebs Cycle		
Conversion of succinyl CoA to succinic acid produces 2 GTP; phosphate are then transferred to ADP	2	2.1
Net yield of Krebs cycle	2 ATPs	2.1%
Respiratory Chain		
Transport of electrons released from 2 NADH formed in glycolysis (2 ATP/NADH)	4	4.3
Transport of electrons released from 2 NADH formed in the conversion of pyruvic acid to acetyl CoA (3 ATP/NADH)	6	6.4
Transport of electrons released from 2 $FADH_2$ formed in Krebs cycle (2 ATP/$FADH_2$)	4	4.3
Transport of electrons released from 6 NADH formed in Krebs cycle (3 ATP/NADH)	18	19.2
Net yield of respiratory chain	32 ATPs	34.2%
Total Net ATP Produced per Glucose Molecule	36 ATPs	38.3%

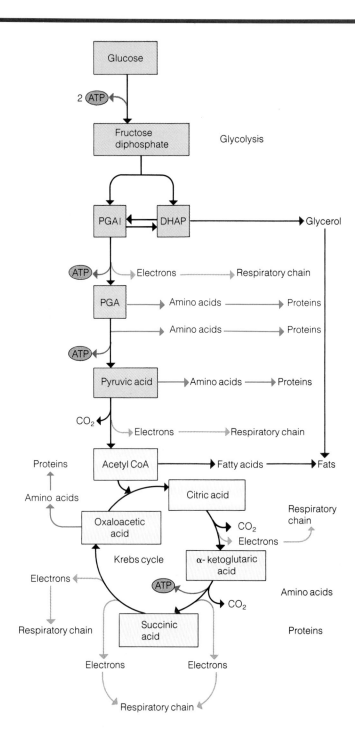

Figure 19-9 Overview of carbohydrate metabolism. Glucose is either oxidized to carbon dioxide and water in the combined reactions of glycolysis, Krebs cycle, and the respiratory chain, or its products are used as a source of amino acids, fatty acids, and other compounds made by the cell.

For example, DHAP resulting from the cleavage of fructose diphosphate may be converted to glycerol and incorporated into fat. Phosphoglyceric acid, phosphoenolpyruvic acid, pyruvic acid, α-ketoglutaric acid, and oxaloacetic acid can all be used to produce amino acids. Likewise, the acetyl group of acetyl CoA is the starting material for the synthesis of fatty acids, steroids, and nucleotides.

It should be noted, however, that the cell can also use many of these compounds directly if it is supplied with them. Some amino acids cannot be made by human cells at all, or at least in quantities sufficient to meet the cell's needs, and must be supplied in food.

adipocyte: (L.) *adiposis*, fatty + (Gr.) *kytos*, cell

Lipid Metabolism

After studying this section, you should be able to:

1. Describe how lipids are transported in the blood, and distinguish between high density, low density, and very low density lipoproteins.
2. Explain how lipids are stored, and distinguish between white and brown adipose tissue.
3. Compare the reactions involved and the energy released in fat oxidation with the reactions and energy release of glucose oxidation.
4. Describe how fats are formed.

Lipids consist of a large number of different kinds of compounds, including phospholipids, glycolipids, steroids, triglycerides, and free fatty acids (see Chapter 2). Phospholipids, glycolipids, and steroids are the major components of cell membranes. Certain steroids act as hormones, and others, particularly cholesterol, play an important digestive role by emulsifying fats in the small intestine. The bulk of neutral fats are stored in fatty tissue where they serve as a source of energy and as an insulator against hot and cold temperatures and physical shock.

Transport of Lipids

Lipids are transported in the blood in association with micelles, which are protein aggregates. Chylomicrons formed in the intestine from fat digestion and absorption are the largest micelles (see Chapter 18). Three other kinds of micelles in plasma are **high density lipoprotein (HDL), low density lipoprotein (LDL),** and **very low density lipoprotein (VLDL)** micelles (Table 19-2). HDL micelles contains a protein component known as **α-apoprotein** and a lipid component that consists largely of phospholipids. These micelles are probably the form in which phospholipids and cholesterol are transported from the liver to other parts of the body. The level of HDL in the plasma is influenced by one's physical activity, and the more active a person is, the higher the level of HDL in the plasma. There is also a correlation between high levels of HDL and a decreased likelihood of coronary heart disease, causing some authorities to suggest that HDL helps protect against heart attack.

LDL micelles contain a protein component designated **β-apoprotein** and a lipid component that consists mostly of cholesterol, with smaller amounts of phospholipids and triglycerides. LDL micelles transport cholesterol to the liver from the tissues that synthesize it. LDL may be involved in the production of the atherosclerotic plaques responsible for atherosclerosis, possibly by delivering cholesterol to damaged arterial walls. There is a correlation between increased levels of LDL and arterial disease (see Chapter 23).

VLDL micelles are richest in triglycerides, and they are probably the primary transport vehicle for serum triglycerides. Most of the protein of VLDL is β-apoprotein with a smaller amount of α-apoprotein.

Storage of Fats

When the diet contains more fat than is necessary for energy production, the excess is stored primarily in adipose tissue, with smaller amounts stored in certain other tissues. Adipose tissue contains **adipocytes** (ăd′ĭ·pō·sīts), which are cells specialized in storing neutral fats (Figure 19-10). When lipoprotein micelles that carry neutral fats reach these cells, enzymes produced by the cells catalyze the release of the fats from the particles and their conversion to free fatty acids and glycerol. Fatty acid and glycerol enter the adipocytes where they are reassembled into neutral fats. A rapid turnover of

Table 19-2 CONTENTS OF LIPOPROTEIN MICELLES

	High Density (HDL)	Low Density (LDL)	Very Low Density (VLDL)	Chylomicrons
Lipid-protein component	α-Lipoprotein	β-Lipoprotein	pre-β-Lipoprotein	—
Size (nm)	7–10	20–30	30–80	70–100
Composition (%)				
Cholesterol	18	43	13	5
Triglycerides	2	10	65	90
Phospholipid	30	22	12	4
Protein	50	25	10	1

Source: (Adapted from G. H. Bell, D. Emslie-Smith, and C. R. Paterson, *Textbook of Physiology and Biochemistry*, 9th ed. (Edinburgh, London, and New York: Churchill Livingston, 1976), 189.

Figure 19-10 Photomicrograph of adipocytes (mag. 305×). Adipocytes are specialized for the storage of fat in a large vacuole. Note the thick collagenous fibers and thinner elastic fibers that provide a lattice for the globular fat-filled cells.

this fat occurs, with most fat being replaced in only two to three weeks.

White adipose tissue and **brown adipose tissue** are two types of adipose tissue that differ in color due to the amount of blood carried in each tissue. Brown adipose tissue, found in fetuses and newborn babies, is more highly vascularized than white adipose tissue and thus carries more blood. This type of fat is used for heat production and is gradually replaced by white adipose tissue as a baby gets older. In adults, white adipose tissue makes up at least 15% of body weight in males and 22% in females. An excess of this fat-storage tissue causes a person to be overweight or obese (see Clinical Note: Obesity).

Metabolism of Fats

Fats represent a greater store of energy than do carbohydrates because they are present in greater quantity and because their catabolism produces more ATP than does the catabolism of an equal weight of carbohydrate. Fat stored in adipose tissue is released to the blood after carbohydrate reserves are depleted, such as following a fast or during a low calorie diet. Fat breakdown is normally greatest in the morning because carbohydrate reserves have been depleted during the night.

Catabolism of Fats

Fat catabolization begins with cleavage of fatty acids from glycerol. The glycerol is then converted to DHAP, which in

---

CLINICAL NOTE

OBESITY

Obesity is defined as the condition of weighing more than 20% above the ideal body weight for frame and height. It can also be considered as that point at which excess body fat leads to a health risk.

Being overweight or obese greatly increases the risk factors for many disorders, including heart disease, hypertension, type II diabetes mellitus, infection, and cardiovascular disorders. In addition, the overweight or obese person often suffers psychological pain due to altered appearance and low self-esteem. The extra weight cannot be hidden and makes up part of one's self-presentation. Women particularly suffer from the psychological effects of obesity, partially because societal expectations seem to be different for women than for men.

Treatment regimens for obesity are varied and largely unsuccessful. They include appetite-depressing drugs, surgery (intestinal bypass, jaw wiring, and stomach stapling), hospitalization, and fasting. Less drastic intervention is a program of moderate exercise and modified calorie intake (see The Clinic: Diets). Several self-help groups are available and are perhaps the most effective means for managing obesity. These include Overeaters Anonymous (modeled after Alcoholics Anonymous) and Weight Watchers (a proprietary self-help group). These groups work well because they provide guidance and support for the person. Interestingly, Overeaters Anonymous does not focus on food or weight loss but on modifying compulsive behavior.

---

turn is converted to PGAl and catabolized in the remaining reactions of glycolysis and the Krebs cycle. Fatty acids are transported into mitochondria and degraded by the sequential removal of two carbon atoms at a time; this process is called **β-oxidation.** The two-carbon fragments are oxidized by NAD^+ and attached to CoA to produce acetyl CoA. Provided that sufficient oxaloacetic acid and oxygen are available, the fragments are converted to carbon dioxide and water through the reactions of the Krebs cycle and the respiratory chain, and ATP is produced (Figure 19-11a).

On a carbon-to-carbon basis, oxidation of a fatty acid yields more ATP than does glucose oxidation. For every glucose molecule that undergoes glycolysis and is oxidized in the Krebs cycle, there is a net yield of 36 ATP molecules, corresponding to a yield of 6 ATP molecules per carbon atom. For a fatty acid such as palmitic acid, which has 16 carbon atoms, complete oxidation yields 129 ATP molecules. This corresponds to a ratio of 8.06 ATP molecules produced per carbon atom. Fatty acid oxidation is thus about 34% more effective than glucose oxidation in producing ATP.

anorexia: (Gr.) *an*, not + *orexis*, appetite
bulima: (Gr.) *bous*, ox, cow + *limos*, hunger
ketosis: (Chem.) keto + (Gr.) *osis*, a condition

CLINICAL NOTE

ANOREXIA NERVOSA

A serious eating disorder known as **anorexia nervosa** affects an alarming number of young women in the United States. The victim of this disorder will severely limit caloric intake in an effort to lose weight or in a frantic attempt not to gain weight. This will occur despite the fact that she is already underweight to the point of emaciation. Most individuals suffering from this disorder are adolescent girls, with an overall incidence of about 1 in 250 girls. It has been estimated, however, that 1 in 4 women college students is or has been a victim of this disorder. Since the group most at risk is white girls from upper and middle class backgrounds with overachieving, compliant personalities, it is not surprising that a college population of women should have a high incidence of sufferers. The American emphasis on thinness for women has probably contributed to the prevalence of this tragic disorder and its recent increase. Expected normal weights for women have been decreasing steadily for about 20 years as the ideal body type for women has moved more and more toward thinness. It is not surprising, then, that a postpubescent girl, already concerned about body image, would develop anxiety about being overweight and a fear of overeating. Since it is clear that not every girl goes through this anxiety, many theories—psychological, sociological, and biological—have been advanced to explain the occurrence of anorexia. None is entirely satisfactory.

Because the girl or young woman is severely malnourished, various effects of the dietary disturbance occur. The individual appears emaciated, yet may deny that hunger, fatigue, or even thinness exists. They may be physically active and participate in an almost ritualistic exercise program. Unfortunately, if social or business situations do not allow the individuals to avoid eating, they may eat and then take an opportunity to find a bathroom and vomit the food before it can be digested and absorbed. Loss or absence of menses frequently accompanies anorexia nervosa. Often there is an increase of very fine body hair, but sometimes the person can have obvious hairiness. The pulse is often slow, respiration is slow, blood pressure is low, the skin is often dry, and body temperature is often hypothermic. If **bulimia,** or purging, is common, then there may be sores around the mouth or discoloration of the tooth enamel because of these structures coming in contact with the stomach acid.

Since cause often dictates control, the fact that the etiology of this disorder is unknown means that there is no specific cure. Some women have benefited from such self-help organizations as Overeaters Anonymous (OA) because the underlying problem is associated with obsessive-compulsive behavior surrounding the issue of food. Certainly the girl or woman needs an understanding and supportive environment in which she is encouraged to discuss her eating disorder in a safe setting. Mental health specialists who have an interest in eating disorders are available in large communities; they are extremely helpful and, in some cases, life saving.

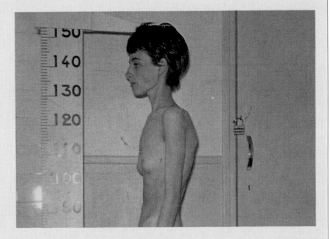

Young woman suffering from anorexia nervosa.

Since most fatty acids have an even number of carbon atoms, most of the carbon is released as acetyl CoA fragments. A few fatty acids have an odd number of carbon atoms, however, and acetyl groups are cleaved from these until all that is left is a three-carbon fragment called **propionic** (prō″pē·ŏn´ĭk) **acid**. The propionic acid residue then combines with CoA and produces **propionyl CoA**. A carbon atom is then added to the propionyl CoA to make a four-carbon CoA derivative that is rearranged to succinyl CoA and used in the Krebs cycle or converted to another compound.

When more acetyl CoA molecules are produced than there are oxaloacetic acid molecules to accept them, the excess acetyl CoA molecules react with one another and produce **acetoacetic acid**. In this reaction two acetyl CoA molecules combine to produce one acetoacetic acid molecule and two free CoA molecules. This reaction provides a means of releasing CoA that would otherwise be tied up in acetyl CoA molecules.

Much of the acetoacetic acid is reduced to **β-hydroxybutyric acid** or decarboxylated to yield **acetone**. All three compounds are called **ketone bodies** because they contain a "keto" group, which is a double-bonded oxygen attached to a carbon atom. If ketone bodies accumulate in the blood, they create a condition called **ketosis** (kē·tō´sĭs). Ketosis is dangerous because the accumulation of acetoacetic acid

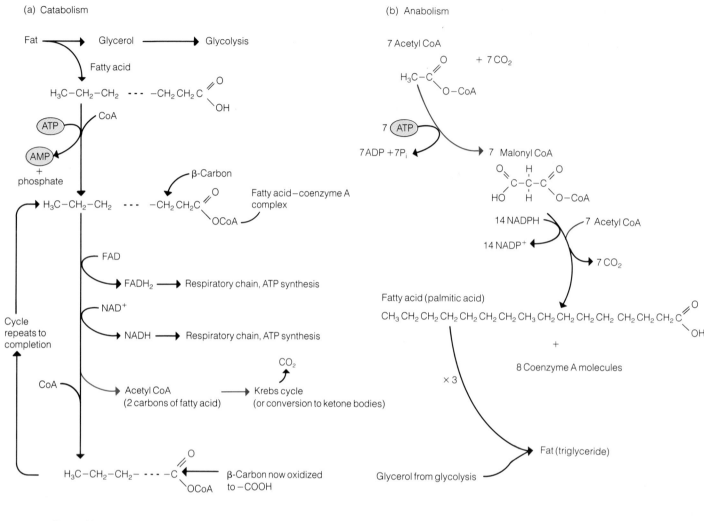

Figure 19-11 Fat catabolism and anabolism. (a) Glycerol released from a fat enters the glycolytic pathway. The fatty acids are cleaved by β-oxidation to produce acetyl CoA groups that can enter the Krebs cycle. (b) Fatty acids are synthesized from acetyl CoA molecules, which combine with glycerol derived from three-carbon intermediates of glycolysis.

and β-hydroxybutyric acid raises the acidity of the blood. Maintenance of proper H^+ is critical in the tissues, and serious problems or even death can result if the H^+ concentration rises even slightly. Normally, however, the ketone bodies are carried to tissues where they are rapidly absorbed, converted again to acetyl CoA and oxidized. Ketosis occurs when unusually large amounts of fats are being oxidized, such as following a fat-rich meal or when fasting or starvation cause stored fats to be oxidized rapidly.

Anabolism of Fats

Not all triglycerides that the body uses are obtained from the diet, although certain **essential fatty acids** must be obtained from food. Fatty acids can also be synthesized from excess acetyl CoA produced from carbohydrate metabolism. Most fatty acid synthesis occurs in the liver through the addition of a CO_2 molecule to acetyl CoA, a reaction that produces a product called **malonyl CoA** (Figure 19-11b). A fatty acid such as stearic acid, which has 18 carbon atoms, is produced by the combination of eight malonyl CoA molecules and one acetyl CoA molecule. The eight carbon dioxide molecules used to produce the malonyl CoA molecules in the first place are released in the process. These reactions require energy, which is provided by **NADPH**, a compound that is similar to NADH but differs by having an additional phosphate group. A substantial amount of energy is conserved in the bonds of the fatty acid molecules and is released when the fatty acids are later oxidized.

The final step in the synthesis of a triglyceride is the attachment of fatty acids to glycerol. When large quantities of glucose are undergoing glycolysis, DHAP becomes available for conversion to glycerol and incorporation into fat. Conversely, when there is only a limited supply of glucose, DHAP is used preferentially for glycolysis and the reactions of the Krebs cycle. As a consequence, most fat synthesis occurs when glucose is in excess.

Protein and Amino Acid Metabolism

After studying this section, you should be able to:

1. **Describe how excess amino acids are converted to carbohydrates.**
2. **Explain how the amino groups of excess amino acids are converted to forms that are excreted.**
3. **Explain how amino acids can be used to produce components of the Krebs cycle.**

Unlike glycogen and fat, excess amino acids are neither stored nor excreted. If they are not used for the synthesis of proteins, they are converted to components of glycolysis or the Krebs cycle. Consequently, any discussion of the metabolism of proteins and amino acids deals primarily with their supply, the removal of excess quantities, and the chemical reactions by which they are converted to other compounds.

During digestion proteins are gradually broken down to their constituent amino acids, which are then absorbed and carried to the liver. Many are absorbed in the liver and converted to carbohydrates by removal of the nitrogen portion of the molecules. Amino acids not absorbed in the liver are carried by the blood to cells in the rest of the body where they are used for the synthesis of proteins.

Metabolism of amino acids by the liver is accomplished in different ways depending on the amino acid, but the product is always a **keto acid**, a compound similar to an amino acid except that the amino group is replaced by an oxygen atom. The process by which the amino group is removed from an amino acid is called **deamination** (Figure 19-12).

The most important deamination mechanism involves the exchange of the amino group on one amino acid for an

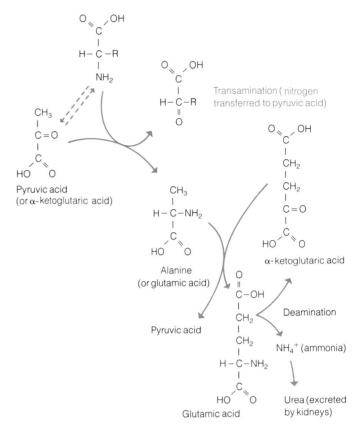

Figure 19-12 Transamination and deamination of amino acids. These processes enable the body to rid itself of excess nitrogen. Glutamic acid and alanine derived from the breakdown of proteins are a source of pyruvic acid and α-ketoglutaric acid. (The R represents the variable side group that characterizes each of the amino acids.)

oxygen atom on some other molecule. Such transfer of the amino acid to another compound is called **transamination**. Most amino acids are deaminated this way, exchanging their nitrogen for the oxygen of pyruvic acid or α-ketoglutaric acid. While the process converts the amino acid to a keto acid, it does not by itself reduce the number of amino acids in the cell because the pyruvic and α-ketoglutaric acids are changed to amino acids when they receive the amino group.

When pyruvic acid receives an amino group through transamination, it is converted to alanine. When α-ketoglutaric acid receives an amino acid group, it is converted to **glutamic acid** (Figure 19-12). Alanine produced from pyruvic acid is also transaminated, however, and its amino group is transferred to α-ketoglutaric acid. This restores the pyruvic acid molecule and produces another molecule of glutamic acid. Consequently, the amino groups of most of the amino acids deaminated in the liver end up in glutamic acid molecules.

If this were to occur extensively, the cell's supply of α-ketoglutaric acid would be used up rapidly, and the cell would not be able to carry on the reactions of the Krebs cycle. Exhaustion of α-ketoglutaric acid is avoided by removal of the amino group of glutamic acid and its conversion to **ammonium ion** (NH_4^+). Removal of the amino group converts the glutamic acid back to α-ketoglutaric acid. Most of the ammonia produced is converted to urea and eliminated by the kidneys (see Chapter 25).

Cofactors in Metabolism

After studying this section, you should be able to:

1. Define cofactors and explain their role in metabolism.
2. List several minerals and vitamins that serve as cofactors.
3. Distinguish between water-soluble and fat-soluble vitamins.
4. Describe the specific role of several minerals and vitamins.

Cofactors are substances that must be present in small amounts for many metabolic reactions to occur. Although not consumed in these reactions, they are often needed to combine with an enzyme in order for the enzyme to be functional. There are two general categories of cofactors: minerals and vitamins.

Minerals

Minerals are inorganic elements necessary for many aspects of cellular metabolism. They are present in the soil, from which they are absorbed by plants. We obtain them when we eat the plants or when we eat animals that have eaten the plants, and our bodies use them as cofactors in many chemical reactions. More than a dozen elements are essential to metabolism, and these are categorized on the basis of the amounts needed as major elements or trace elements. **Major elements** comprise about 4% of body weight and include calcium, phosphorus, potassium, sulfur, sodium, chlorine, magnesium, and iron. Trace elements together comprise only a few hundredths of a percent of body weight and include manganese, copper, iodine, cobalt, zinc, fluorine, selenium, molybdenum, tin, silicon, and vanadium. Although care must be taken to ensure that sufficient amounts of certain of the major elements are included in the diet, some of them, such as sodium and potassium, are so common that it is difficult to avoid acquiring enough. Trace elements are generally required in such small quantities that enough occurs naturally in normal diets to meet the body's needs.

Table 19-3 MINERALS REQUIRED BY THE BODY

Mineral	Examples of Sources	Examples of Function
Major Elements		
Sodium	Beef, cheese, fast foods, salty foods	Maintaining fluid balance, pH; nerve and muscle function; membrane potential
Potassium	Bananas, parsley, prunes, raisins	Nerve transmission; muscle contraction; fluid balance, pH; membrane potential
Calcium	Milk and dairy products, fish, eggs, cereals, fruits and vegetables	Formation of bones and teeth; muscle contraction; blood clotting
Sulfur	Protein-rich foods	Component of certain amino acids and proteins
Chlorine	Sodium chloride (table salt)	Primary negative ion of intercellular fluids; nerve function; membrane potential
Phosphorus	Milk and dairy products, legumes, fish, oatmeal, liver	Formation of bones and teeth; energy metabolism; synthesis of DNA and RNA
Magnesium	Green leafy vegetables, seafood, cereals	Enzyme cofactor; formation of bones and teeth; nerve transmission
Trace Elements		
Iron	Beef, spinach, peas, liver, kidney	Formation of hemoglobin, myoglobin, cytochromes; enzyme cofactor
Iodine	Seafood, iodized salt, dairy products	Thyroid hormones; regulation of metabolism
Fluorine	Drinking water, seafood, tea	Formation of teeth and bones
Zinc	Vegetables, many other sources	Enzyme cofactors, vitamin A metabolism
Copper	Kidneys, liver, whole grain cereals	Enzyme cofactor; oxygen utilization
Cobalt	Green leafy vegetables	Component of Vitamin B_{12}
Manganese	Cereals and tea	Enzyme cofactors and component
Chromium	Brewer's yeast, wine, beer	Glucose utilization

The minerals and the functions they serve are listed in Table 19-3.

Vitamins

Vitamins are naturally occurring substances obtained from various plant and animal sources that are required in small amounts for the maintenance of metabolic processes and the control of growth and development. There are two categories of vitamins classified on the basis of solubility in water or fat and identified by the letters A, B, C, D, E, and K. **Water-soluble vitamins** include several compounds grouped together as the **vitamin B complex** and a single compound called **vitamin C. Fat-soluble vitamins** include **vitamins A, D, E, and K** (see Chapter 2, The Clinic: Vitamins).

Table 19-4 WATER-SOLUBLE VITAMINS AND THEIR EFFECTS

Vitamin	Examples of Sources	Examples of Function
Vitamin C (Ascorbic acid)	Citrus, tomatoes, leafy green vegetables	Collagen synthesis; wound healing; bone formation
Vitamin B complex		
Vitamin B_1 (thiamine)	Whole grain, nuts, legumes, yeast, green vegetables	Coenzyme to several enzymes; carbohydrate metabolism; nerve function; muscle function
Vitamin B_2 (riboflavin)	Milk, cheese, meat, liver, eggs	Energy and carbohydrate metabolism (FAD production)
Niacin (nicotinic acid)	Liver, meat, fish, legumes, whole grain cereals, conversion from tryptophan by intestinal flora	Carbohydrate metabolism (NAD and NADP production)
Vitamin B_6 (pyridoxine)	Liver, whole wheat, cereals, fish, legumes	Coenzyme in amino and fatty acid synthesis; important in nervous function, maintenance of blood vessels
Pantothenic acid	Most foods, especially eggs, yeast, liver	Component of coenzyme A
Folic acid	Dark green leafy vegetables, liver, organ meats, fruit, intestinal flora	Coenzyme in red blood cell maturation; synthesis of purines and pyrimidines
Biotin	Intestinal flora, egg yolk, liver, tomatos	Coenzyme in reactions that involve incorporation of CO_2, such as the production of oxaloacetic acid
Vitamin B_{12} (cobalamin)	Liver, beef, pork, organ meats, milk, eggs	Coenzyme involved in succinyl CoA production; folic acid metabolism

Table 19-5 FAT-SOLUBLE VITAMINS AND THEIR EFFECTS

Vitamin	Examples of Sources	Examples of Function
Vitamin A	Liver, eggs, milk, butter	Necessary for normal vision; body growth; bone and teeth growth
Vitamin D	Fish liver oil, butter, egg yolk, enriched milk	Increased absorption of calcium and phosphate in intestine
Vitamin E	Egg yolk, wheat germ, leafy vegetables, vegetable oil	Protects vitamin A and unsaturated fatty acids from oxidation; may protect certain respiratory enzymes from oxidation
Vitamin K	Dark green leafy vegetables, intestinal flora	Normal blood clotting

Water-Soluble Vitamins

The vitamin B complex consists of a group of vitamins that are particularly active as cofactors in reactions in which carbohydrates, lipids, and proteins are oxidized. The complex includes **thiamine (vitamin B_1)**, **riboflavin (vitamin B_2)**, **niacin**, **pyridoxine** (pĭ·rĭ·dŏks´ēn) **(vitamin B_6)**, **pantothenic** (păn·tō·thĕn´ĭk) **acid**, **cobalamin** (kō·băl´ă·mĭn) **(vitamin B_{12})**, **folic acid**, and **biotin** (bī´ō·tĭn). The specific chemical roles played by vitamin C are not known, although it is known to be important in the production of collagen (a protein found in connective tissues), in the metabolism of certain amino acids, and in wound healing. Water-soluble vitamins are excreted in urine rather than being stored, so they must be provided on a regular basis. These vitamins and their functions are summarized in Table 19-4.

Fat-Soluble Vitamins

Fat-soluble vitamins are obtained in fatty foods and are carried in the blood in lipid micelles. Each of these vitamins is actually a group of structurally similar compounds. The specific molecular roles of fat-soluble vitamins are not as well known as for water-soluble vitamins, although the effects of a deficiency of each are known. Two of these vitamins are not obtained from the diet. One member of the vitamin D group, designated **vitamin D_3**, is synthesized in the skin in response to sunlight. A member of the K group, designated **vitamin K_2**, is produced by bacteria that grow in the large intestine. The fat-soluble vitamins and their effects are summarized in Table 19-5.

Heat as a By-Product of Metabolic Reactions

After studying this section, you should be able to:

1. Define a calorie and distinguish between small calories and large calories.
2. Explain how heat is produced and distributed in the body.
3. Describe the role of the hypothalamus in regulating body temperature.
4. Describe the mechanisms used to warm and cool the body.
5. Explain the cause of fever.
6. Define basal metabolic rate, describe how it is measured, and explain its significance.

When new compounds are formed by metabolic reactions, the amount of energy in the reaction products is less than the energy in the molecules consumed in the reaction. The difference is given off by the reaction as heat. Although this heat may be regarded as a "waste" product of a specific chemical reaction, it is in no sense a waste product of metabolism as a whole because it is used to maintain a temperature in the body at which enzymes function most efficiently.

Temperature

The amount of heat present in a system is measured as degrees of **temperature.** In the Celsius system of measuring temperature, 0°C is defined as the temperature at which water freezes, and 100°C is defined as the temperature at which water boils (at sea level). Using the Fahrenheit system, these temperatures are 32°F and 212°F, respectively. The usual temperature of the body, when measured with a thermometer in the mouth, is 37°C (98.6°F).

Although measured in degrees of temperature for convenience, the amount of heat present in a system can sometimes be more usefully indicated by units called **calories (cal).** A calorie is useful because it expresses the amount of energy present and can be used to relate one form of energy to another. For example, the amount of energy potentially available from the oxidation of 1 g of glucose is about 3800 cal, and the ATP produced when it is oxidized by a cell contains about 1450 cal (assuming that each glucose molecule is oxidized completely to CO_2 and water). This corresponds to an efficiency of about 38%. Most of the remaining 2350 cal are released as heat, which can be measured with a thermometer.

A calorie is defined as the amount of heat required to raise the temperature of 1 g of water from 14.5°C to 15.5°C. Sometimes it is more convenient to use a unit called a **kilogram calorie** or **kilocalorie (kcal),** which is the amount of heat required to raise the temperature of 1 kg of water 1°C. A kilocalorie, also called a large calorie (abbreviated Cal), is equal to 1000 small calories. This unit is used to describe the amount of energy in food. To avoid confusion, we will use the term kilocalorie when referring to the larger unit and not "calorie" as is commonly used by dieters.

Production and Distribution of Heat

The amount of heat produced is not the same in all parts of the body because some organs are metabolically more active than others and because some parts of the body lose heat at a faster rate than others. If organs such as the liver or active muscles are generating heat, for example, blood is warmed as it passes through them and the heat is carried to other cooler areas of the body. If the skin is cold because of the outside temperature, the blood passing through the layers of skin loses heat to the surrounding air. The body has a number of mechanisms that it can use to regulate this heat loss and even to increase it when the surroundings are warm.

Regulation of Body Temperature

Normally the interior of the body is protected from heat loss by the insulating effect of the skin and underlying layers of fat. It is common to speak of an inner body **core temperature** as compared with the temperature in the peripheral regions, called the **shell temperature.** While the shell temperature may fluctuate several degrees, the core temperature remains relatively constant (Figure 19-13).

Control by the Hypothalamus

Control of core temperature is maintained primarily by the hypothalamus, which regulates both the rate of heat production and the rate of heat loss. There are two functionally distinct regions in the hypothalamus for temperature control: a region in the anterior hypothalamus responsible for causing increased heat loss and a region in the posterior hypothalamus responsible for heat retention. For the hypothalamus to control temperature, it must receive signals that tell it whether the core temperature is rising or falling. These

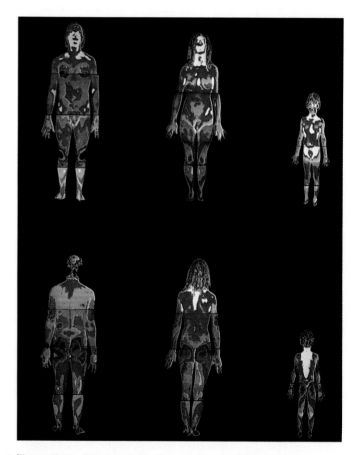

Figure 19-13 Thermographs. These infrared photos show the distribution of heat in the bodies of an adult male and female and a child. The colors indicate the degree of heat: white (warmest), red orange, blue, purple, and pink (coolest).

signals are generated by sensory cells that communicate directly with the hypothalamus and stimulate one of the regulatory regions.

One sensory area is in the hypothalamus itself, in an anterior region called the **thermostatic control center** (Figure 19-14). Neurons in this center are sensitive to the temperature of blood passing through the hypothalamus, firing more frequently as the temperature rises and more slowly as it falls. Increased frequency causes stimulatory signals to be sent to the hypothalamic control area that causes cooling responses and inhibitory signals to the hypothalamic control area that causes warming responses. Decreased frequency of firing has the opposite effect, inhibiting the region responsible for cooling reactions and stimulating the region responsible for warming reactions.

Mechanisms for Warming the Body

The hypothalamus uses three methods for increasing body temperature. These include constriction of blood vessels in the skin, increasing muscle activity to generate additional heat, and stimulating the release of increased amounts of hormones that cause increased metabolic activity.

Figure 19-14 Regulation of body temperature. The hypothalamus contains a thermostatic control center that provides primary control of response to changes in body temperature. Dashed lines indicate inhibition.

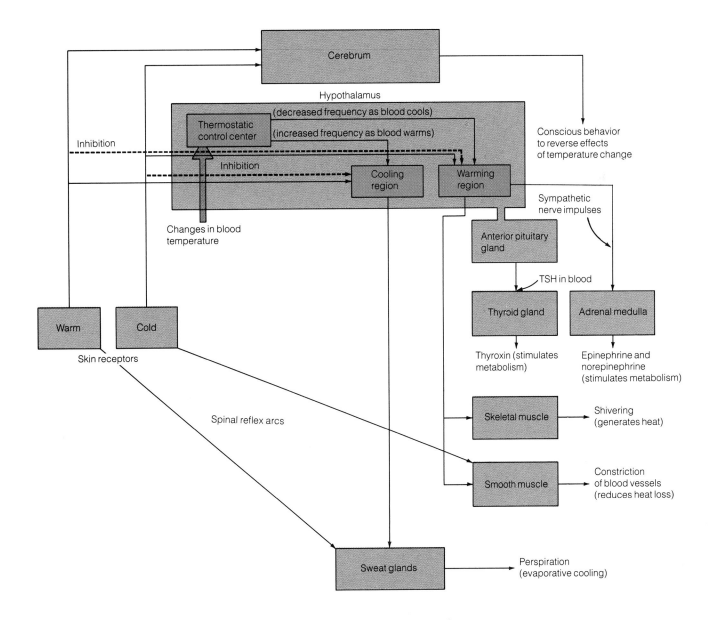

In response to signals from sensory regions, particularly in the skin and gut, the warming region or the cooling region of the hypothalamus responds by sending signals that counteract changes in temperature sensed by the sensory regions. When the core temperature drops below about 37°C, signals are sent to blood vessels in the skin that cause them to constrict. This reduces blood flow to the skin and reduces the amount of heat carried to the skin where it is easily lost. Constriction of the blood vessels is also induced by a spinal reflex arc that emanates directly from the cold-sensitive cells in the skin. If the temperature drops enough, the hypothalamus sends additional signals to skeletal muscles and causes them to contract and relax rapidly. This is **shivering.** The chemical activity of shivering muscles produces heat that is carried by the blood to the rest of the body.

Hormones used to counteract lowered temperature include thyroxine and catecholamines. Lowered blood temperature causes thyroid-releasing factor (TRF) to be released from the hypothalamus. TRF travels to the anterior pituitary gland where it causes the release of thyroid-stimulating hormone (TSH). TSH is carried by the blood to the thyroid gland where it causes the gland to release thyroxine, which has a general effect of increasing the body's metabolic rate. Increased metabolism increases the rate of heat production and helps to offset the drop in temperature that initiated the response in the first place.

The adrenal medulla is also stimulated by sympathetic impulses from the hypothalamus when the body is cooled and releases epinephrine and norepinephrine in response. Like thyroxine, these hormones cause a general increase in metabolic rate and a resulting increase in heat production.

Mechanisms for Cooling the Body

If body temperature rises too much, mechanisms are brought into play that reverse the action of the heating mechanisms (see Clinical Note: Heat Stroke). The shivering response is inhibited, as are sympathetic impulses sent to arteries in the skin. As the arteries lose tone, blood pressure causes them to dilate, which allows more blood to pass through and more heat to be delivered to the skin. The heat is lost as it radiates from the skin. But perhaps the most effective response to increased body temperature is that of the sweat glands, which are stimulated to release sweat by signals from the hypothalamus and by reflex arcs emanating from warm-sensitive skin receptors. The evaporation of sweat cools the body (see next section).

Heat Loss

The body loses heat by several different mechanisms: radiation, conduction, evaporation, and excretion. **Radiative heat loss** takes the form of infrared electromagnetic waves that radiate from the body. Whether or not there is a net loss of heat by radiation depends on the surroundings, and the heat gained from a radiating object nearby, such as a warm stove, may be greater than the heat lost through radiation. **Conductive heat loss** is the heat that diffuses into matter at a lower temperature that is in contact with the body. Lying on a cold stone bench after vigorous physical activity can transfer so much heat to the bench that it is noticeably warmed. Most conductive heat loss, however, occurs as a result of transfer to the surrounding air, and the greater the temperature difference between the body's surface and the air, the more rapid the heat loss. Transfer is most efficient if the air is moving over the surface of the body, because when the air is stationary, it heats up and acts as an insulator against further loss. This is why wind is so important in cold weather and can increase the cooling effect of cold air dramatically.

Evaporative heat loss is based on the physical principle that energy is required to transform liquid to gas. For water,

---

> **CLINICAL NOTE**
>
> ### HEAT STROKE
>
> The amount of heat the human body can withstand depends to a great extent on whether the atmosphere around the body is wet or dry. In a very dry atmosphere with some air movement, the person can withstand temperatures up to 150°F for several hours without any long-term harmful effects. However, if the air is extremely humid or if the person is in water, the body temperature begins to rise after the ambient temperature reaches 94°F. If the person is working hard doing physical labor, the body temperature may begin to rise when the ambient temperature is only 84° to 89°F. Even with maximal sweating, there is a limit to the heat-losing ability of the body. Also, if the hypothalamus becomes heated, its ability to promote heat loss will be compromised and sweating will decrease, leading to further heat gain.
>
> When the body temperature reaches 106°F, **heat stroke** will occur. This is manifested by dizziness, pains in the abdomen, delirium, and loss of consciousness. Because of the severe loss of fluid that accompanies sweating in such high ambient temperatures, the body is probably in some degree of circulatory shock; this may account for the symptoms. The high body temperature is dangerous because the extreme heat may damage neurological tissue. Tissues of the kidney, liver, and other organs might be damaged by the extreme heat also. Since neurological tissue cannot be repaired, the loss of such tissue by heat damage will be devastating to the person, thus treatment must begin immediately. Treatment consists of cooling the body as rapidly as possible. Care should be taken not to make the person shiver during this cooling process, as that would increase heat production. Cooling sprays or sponges are recommended rather than ice water baths.

pyrogen: (Gr.) *pyr*, fire + *gennan*, to produce

600 cal are required to transform 1 mL of liquid water to water vapor. Even when not exercising, the adult body can lose 30 mL or more of water per hour as a result of simple diffusion through the skin. Water lost in this way is referred to as **insensible perspiration** and can carry away about 25% of the heat produced by metabolism.

Sensible perspiration, in contrast, is sweat. Sweat is produced by sympathetic stimulation of sweat glands in the skin. Large amounts of water and dissolved salt can be secreted from the sweat glands as a result of strenuous exercise or warm surroundings. For the sweat to cool the body, it must evaporate, and for it to evaporate, the surrounding air must not already be saturated with water vapor. If it is, the sweat cannot evaporate and the cooling effect is lost. This is why our skin remains so damp on hot, humid days and why we often feel more uncomfortable on humid days at 90°F than at 100°F on dry days.

Fever represents a situation in which heat gain exceeds heat loss. It results from an alteration in the body's normal temperature set point and usually accompanies infections (see Clinical Note: Fever).

Basal Metabolic Rate

It is often useful for medical purposes to know the rate at which metabolism is occurring. Some diseases are associated with elevated or depressed metabolic rates, and a patient's metabolic rate can be a useful measure of the nature of one's illness. For example, certain diseases of the thyroid gland alter thyroxine production, and thyroxine influences metabolic rate in general. In hypothyroidism, thyroxine production is reduced and there is a resulting depression of metabolic rate. Hyperthyroidism causes the metabolic rate to increase as a result of excess production of thyroxine. Fever also causes an increase in the metabolic rate because chemical reactions often occur more rapidly at higher temperatures. Stress increases metabolic rate, too, because stress causes increased levels of catecholamines to be produced and they cause a generalized increase in metabolic rate.

Because heat is a product of biochemical reactions, the amount of heat released in a given time can be used as an overall measure of metabolism. It is impractical to measure the total amount of heat produced by a person, however, and alternative methods have been devised for estimating metabolic rate from the amount of oxygen consumed. Based on laboratory measurements of the amount of heat released when carbohydrates, fats, and proteins are burned, it is known that an average of 4.825 kcal (kilocalories) of heat are released for every 1 L of oxygen used. Using this figure, the amount of heat released can be estimated on the basis of the amount of oxygen used. Results of such tests are usually expressed in kilocalories per unit time even though measurements are made of the volume of oxygen consumed.

> **CLINICAL NOTE**
>
> **FEVER**
>
> **Fever** is an elevated body temperature that results from a "resetting" of body temperature by the hypothalamus in response to chemicals called **pyrogens.** Pyrogens are substances produced by leucocytes that have attacked invading bacteria or other microorganisms. Pyrogens may also be released by the bacteria themselves during an infection. These chemicals stimulate the thermostatic control center in the hypothalamus to raise body temperature. This is done by stimulating the mechanisms that produce and conserve heat that would normally be employed if the body's temperature had fallen. Aspirin is an effective agent at counteracting the responses of the hypothalamus to a pyrogen. It is possible that pyrogens indirectly affect the hypothalamus by stimulating the production of a prostaglandin, which is the actual stimulant, and that aspirin reduces fever by inhibiting the synthesis of the prostaglandin.
>
> It is not really understood clearly if fever is beneficial in humans or not. Certainly it occurs in many species and its universal presence would argue an adaptive function rather than a destructive one. Some experiments have attempted to elucidate this problem and have shown that fever may be helpful in fighting off infection. Very high fevers may bring their own hazards and should probably be lowered, but moderate fevers may offer some benefit. It is believed that moderate fevers inhibit infection by affecting the microorganism or by enhancing body defenses.

One's metabolic rate can vary considerably due to temporary conditions of stress, exercise, or even the presence of food in the intestinal tract. Thus, standard conditions have to be established to compare the measured value with established norms. A measurement made under such standardized conditions is called a **basal metabolic rate (BMR).** To determine BMR, oxygen consumption is measured early in the day, after a night's rest, and after fasting for 12 to 14 hours. The test itself is conducted with a **metabolator,** a bellows-like device filled with oxygen that the subject breathes through a mouthpiece. It allows the amount of oxygen used in a given time to be determined. Once this value has been adjusted mathematically to standard temperature and pressure, it is related to body surface area, age, weight, and sex of the individual and compared with established norms.

The BMR of the average adult male is about 2000 kcal per day and that of a female about 1800 kcal per day. There is a great deal of variation among healthy people even when these standardized conditions are used in the measurement, so care must be used in interpreting results. Results are expressed as a percentage over or under the average BMR for individuals of the same sex and age. Thus, a BMR of +10 indicates that the patient's BMR is 110% of the

Table 19-6 THE MAYO FOUNDATION NORMAL STANDARDS OF BASAL METABOLIC RATE

Males		Females	
Age	BMR (kcal/m²/hr)	Age	BMR (kcal/m²/hr)
6	53.0	6	50.6
7	52.5	$6\frac{1}{2}$	50.2
8	51.8	7	49.1
$8\frac{1}{2}$	51.2	$7\frac{1}{2}$	47.8
9	50.5	8	47.0
$9\frac{1}{2}$	49.4	$8\frac{1}{2}$	46.5
10	48.5	9–10	45.9
$10\frac{1}{2}$	47.7	11	45.3
11	47.2	$11\frac{1}{2}$	44.8
12	46.7	12	44.3
13–15	46.3	$12\frac{1}{2}$	43.6
16	45.7	13	42.9
$16\frac{1}{2}$	45.3	$13\frac{1}{2}$	42.1
17	44.8	14	41.5
$17\frac{1}{2}$	44.0	$14\frac{1}{2}$	40.7
18	43.3	15	40.1
$18\frac{1}{2}$	42.7	$15\frac{1}{2}$	39.4
19	42.3	16	38.9
$19\frac{1}{2}$	42.0	$16\frac{1}{2}$	38.3
20–21	41.4	17	37.8
22–23	40.8	$17\frac{1}{2}$	37.4
24–27	40.2	18–19	36.7
28–29	39.8	20–24	36.2
30–34	39.3	25–44	35.7
35–39	38.7	45–49	34.9
40–44	38.0	50–54	34.0
45–49	37.4	55–59	33.2
50–54	36.7	60–64	32.6
55–59	36.1	65–69	32.3
60–64	35.5		
65–69	34.8		

Source: Modified from W. M. Boothby, J. Berkson, and H. L. Dunn, *American Journal of Physiology*, 116:468 (1936) as printed in E. E. Selkurt, ed., *Basic Physiology for the Health Sciences*, 2nd ed. (Boston: Little, Brown, 1982), 522.

average BMR of people of the same sex and age. Patients whose BMR is 85% of average for their age and sex are said to have a BMR of −15. Similar tests carried out in conjunction with exercise can also be used as a measure of physical fitness. Table 19-6 shows the normal basal metabolic rates for males and females from childhood to adulthood.

In this chapter we studied the mechanisms in the body that metabolize carbohydrates, lipids, and proteins. In addition to studying the fates of the products of digestion, we also learned about energy exchanges associated with anabolic and catabolic reactions and how the presence of oxygen increases the efficiency of energy release. Finally, we studied the mechanisms that regulate heat production and their importance in maintaining a constant core body temperature. In the following chapter we will look at the respiratory system which brings the oxygen into the body used in metabolic reactions and removes the carbon dioxide produced in those reactions.

THE CLINIC

DIETS

Discussions of diets are almost as popular as discussing the weather. All of us at one time or another have altered our diets in response to illness, the weather, how we perceive ourselves (thin or fat), or even how busy we are. Newspapers and magazines deluge us with the latest popular or fad diets to lose weight, cure disease, promote well being, and remedy all manner of minor afflictions from dry skin to falling hair.

In addition to the basic normal diet, consisting of the four major food groups (milk, meat, vegetables and fruit, and breads), there are two other important types of diets that have specific functions: low calorie (or reducing) diets and therapeutic diets.

Low Calorie Diet*

A **low calorie diet** that is to be followed for many months for the purpose of losing weight must be planned carefully and the patient instructed in its preparation and use. It should contain all essential nutrients and should include standard staple foods rather than special foods that may not be generally available. This will allow a more normal social life and reassures the obese person that he or she is not different from other people. Commercially prepared formula diets, regardless of their nutritional merit, may be useful for several weeks but are usually unsatisfactory for continuous use.

Many fad or crash diets have been widely publicized, such as high protein diets, high fat diets, grapefruit diets, no carbohydrate diets, and macrobiotic diets. Many lack any reasonable or scientific basis and may be harmful. All are of limited value for long maintained weight reduction. Theoretically, with a diet that provides a sustained caloric intake that is 500 kcal per day lower than output, 1 lb per week should be lost and with a 1000 kcal deficit, 2 lb per week. Unfortunately, this simple arithmetic does not always hold true because of fluid retention and because poorly understood internal metabolic adjustments reduce caloric needs.

The best type of diet for weight reduction has only a moderate caloric deficit (providing about 1200 kcal per day) and consists of limited servings of the widest variety of foods that meet all vitamin and mineral allowances, with 20 to 25% of the calories from protein, 30 to 35% from fat (with a 1:1 ratio of saturated to polyunsaturated fatty acids), and 45 to 50% from carbohydrates. The accompanying table provides a typical day's menu. The diet calls for five or six daily feedings, each containing some protein, so that the constant ingestion and metabolism of protein suppresses appetite and eliminates the need for anorectic agents. Multiple feedings also reduce the stimulus for insulin production and release.

To reach a steady state of calorie balance that will maintain improvement once the desired weight is achieved re-

* This section adapted from *The Merck Manual*, 13th ed. (Rahway, N.J.: Merck Sharp & Dohme Research Laboratories, 1977) 1182.

quires a transition period in which the basic diet is modified by small stepwise additions of previously forbidden foods. For example, a small amount of baked potato might be added to dinner for 1 week, while body weight is monitored. If no gain has occurred, a half slice of bread may be added at the noon meal. Rich sweet desserts with large amounts of refined carbohydrate should be the last food to be added; most obese people must permanently avoid them. It will soon be evident how much freedom can be tolerated in food selection. After several weeks or months of adaptation, during which the new body weight is maintained and new food habits become established, some people will be able to control food intake in a correct dynamic balance with energy output. Regular exercise is an important part of the process as well.

Therapeutic Diets

Therapeutic diets include a variety of diets used to treat a specific disease or condition. **Routine hospital diets** usually vary in texture and consistency depending upon the patient's condition. They include regular house diets, soft diets for patients with digestive disorders, liquid diets for those patients awaiting surgery, and **special tests** prescribed as part of the treatment program. For example, all patients with diabetes mellitus should have a specific dietary prescription to normalize their weight and reduce the amount of sugar to help reduce the glucose level in the blood and urine. Some patients require a **restricted diet** to reduce the intake of specific substances that affect their illness. Patients who have fluid retention problems due to heart failure or liver or kidney disease are frequently placed on **low sodium diets** of variable levels. Low sodium diets are also used to help reduce blood pressure in patients suffering from persistent hypertension.

At times diets are prescribed to eliminate or markedly reduce nutrients that cause toxic reactions in the body due to abnormalities of metabolism. Gluten-free diets are used to treat children with malabsorption syndromes. Low purine diets are helpful in patients suffering from gout (an arthritic condition caused by deposition of uric acid around joints). Low protein diets are used to treat patients with advanced kidney disease. **Elimination diets** are sometimes used to eliminate or incriminate certain foods that may act as allergens. This type of dietary prescription may be time consuming in its search for the food substance that is causing the patient's symptoms.

Physicians in the industrialized world are very concerned about the incidence of coronary artery atherosclerosis and strokes in our population and have promoted diets low in cholesterol to help reduce the amount of this fatty material in the bloodstream before it can be deposited in arterial walls and cause obstruction to blood flow. Low cholesterol diets discourage the excessive intake of animal (unsaturated) fats and eggs that are high in cholesterol.

SAMPLE 1200-KCAL REDUCING DIET

	Percentage	Grams	Kilocalories
Protein	20	60	240
Fat	35	47	420
Carbohydrate	45	135	540
			1200

Breakfast
4 oz ($\frac{1}{2}$ cup) orange juice
1 egg, poached or boiled
or
1 oz dry cereal (not presweetened)
1 slice bread
1 tsp margarine
4 oz ($\frac{1}{2}$ cup) skim milk
black coffee or tea if desired

Lunch
4 oz lean meat, fish, poultry or cottage cheese
1 slice bread or 1 roll
1 tsp margarine
1 tomato and lettuce salad
1 tsp mayonnaise
4 oz ($\frac{1}{2}$ cup) skim milk
black coffee or tea if desired

Midafternoon
6 oz tomato juice or bouillon
1 apple or pear
4 oz ($\frac{1}{2}$ cup) skim milk

Dinner
6 oz lean meat, fish, poultry
4 oz carrots, peas, or squash
4 oz broccoli, cabbage, or spinach
$\frac{1}{2}$ cup berries or 2-in. wedge of melon
black coffee or tea if desired

Evening
6 oz tomato juice, bouillon, or mixed vegetable juice
4 oz ($\frac{1}{2}$ cup) skim milk

Source: Adapted from *The Merck Manual*, 13th ed. (Rahway, N.J.: Merck Sharp & Dohme Research Laboratories, 1977) 1182.

STUDY OUTLINE

Overview of Metabolism (p. 478)

Metabolism consists of the chemical processes involved in life. Formation of larger compounds from smaller ones is *anabolism*. Breakdown of compounds to smaller parts is *catabolism*. *Absorptive state* is the time when nutrients absorbed during digestion are distributed throughout the body. *Postabsorptive state* is a period of catabolism following the absorptive state when chemicals are used for energy.

Carbohydrate Metabolism (pp. 479–486)

Carbohydrates serve both as a source of energy and as a source of material for the synthesis of new compounds. The most important carbohydrate in metabolism is glucose.

Carbohydrates are obtained in the diet and synthesized from other compounds. *Lactic acid*, amino acids, and glycerol are the primary compounds used in *gluconeogenesis*, the synthesis of carbohydrate from noncarbohydrate sources.

Metabolism of Glucose Glucose is stored in glycogen molecules in the liver and skeletal muscle through the process of *glycogenesis*, which involves ATP and *uridine triphosphate*. When glycogen reaches about 5% of the mass of the liver, *glucose phosphate* is used to produce fat, which is stored in *adipose tissue*.

The recovery of glucose from glycogen is called *glycogenolysis*. ATP is used in the process. Glycogen synthesis and degradation are controlled primarily by epinephrine. Epinephrine acts through a cyclic-AMP mechanism that stimulates the release of glucose from glycogen and simultaneously inhibits the addition of glucose to glycogen.

Cellular respiration is the primary process involved in glucose catabolism. Overall, it is an oxidative process that converts glucose to carbon dioxide and water. Energy in glucose is used to synthesize ATP. The three components of cellular respiration are glycolysis, Krebs cycle, and the respiratory chain.

Glycolysis Glycolysis begins with the addition of phosphate to glucose from ATP and leads to the production of two molecules of *pyruvic acid*. Energy released in the process is used to synthesize ATP, and there is a net gain of two ATP molecules and two reduced *NAD* molecules for every glucose molecule used.

Fate of Pyruvic Acid When oxygen is available, pyruvic acid is catabolized further in the Krebs cycle through use of *coenzyme A*. When oxygen is not available, those reactions cannot go on, and pyruvic acid is reduced to lactic acid. Reduced NAD from glycolysis is a source of electrons for the reaction.

Krebs Cycle When oxygen is available, pyruvic acid loses a carbon atom as carbon dioxide, leaving a two-carbon intermediate that enters the reactions of the *Krebs cycle*. In the Krebs cycle the remaining two carbon atoms are released as carbon dioxide, and electrons are transferred to the respiratory chain by NAD (or *FAD* in one reaction).

Respiratory Chain Electrons pass through the *respiratory chain* in a series of oxidation-reduction reactions, ultimately to combine with oxygen to produce water. *Cytochromes* are major components of the respiratory chain.

Oxidative Phosphorylation Energy released in the process is used to produce a concentration gradient of hydrogen ions. Diffusion of the hydrogen ions through the mitochondrial membrane in response to the concentration gradient is coupled to the addition of phosphate to *ADP*.

An Overview of Cellular Respiration The complete oxidation of glucose uses six molecules of oxygen, yields six molecules of carbon dioxide, and produces a net of 36 molecules of ATP. The ATP molecules contain 38% of the energy of the glucose molecule released by cellular respiration.

Glucose as a Source of Carbon Glucose is also a source of carbon for synthesizing other compounds. Several compounds in glycolysis and the Krebs cycle can be converted directly to amino acids. Some can be used for the synthesis of fat, other lipids, and nucleic acids.

Lipid Metabolism (pp. 487–491)

Lipids are a diverse group of compounds that includes phospholipids, *glycolipids*, *steroids*, triglycerides, and free fatty acids.

Transport of Lipids Lipids are transported in blood in association with micelles. *Chylomicrons* are the largest micelles. *High density*, *low density*, and *very low density lipoprotein* micelles contain α- or β-apoprotein and transport phospholipids, cholesterol, and triglycerides in plasma.

Storage of Fats Lipoprotein micelles carry neutral fats to *adipocytes* in adipose tissue. The fat molecules are disassembled, their components enter the adipose cells, and neutral fats are reassembled. The most common type of fat tissue is *white adipose tissue*. *Brown adipose tissue*, colored by its higher blood content, is found in fetuses and newborn babies.

Metabolism of Fats Fats are used as an energy source after carbohydrate reserves are depleted. In neutral fat oxidation, glycerol and fatty acids are separated and glycerol is converted to DHAP and catabolized. Fatty acids are disassembled two carbon atoms at a time, and the acetyl units are catabolized in the Krebs cycle. On a carbon-to-carbon basis, more ATP is made from fatty acid oxidation than from glucose oxidation. Excess acetyl groups produce *ketone bodies* (acetoacetic acid, acetone, and β-hydroxybutyric acid), which are normally converted back to *acetyl CoA* and oxidized.

Most fatty acid synthesis occurs in the liver from the addition of acetyl groups to the growing fatty acid. The process requires energy, which is supplied by reduced *NADPH*.

Completed fatty acids are added to glycerol, derived from DHAP, to produce triglycerides.

Protein and Amino Acid Metabolism (pp. 491–492)
Excess amino acids are *deaminated* to produce keto acids. Amino groups are transferred to α-*ketoglutaric acid* and pyruvic acid, converting them to the amino acids *glutamic acid* and *alanine*, respectively. Alanine may also lose its amino group to α-ketoglutaric acid. Glutamic acid may lose its amino group as *ammonium ion*, which is used to form urea.

Cofactors in Metabolism (pp. 492–493)
Cofactors are substances that must be present for many enzymes to function.

Minerals *Major elements* comprise about 4% of body mass. *Trace elements* comprise only a few hundredths of a percent of body mass.

Vitamins Vitamins are water soluble or fat soluble. Water-soluble vitamins include *vitamin B complex* (thiamine, riboflavin, niacin, pyridoxine, pantothenic acid, cobalamin, folic acid, and biotin), and *vitamin C*. B complex vitamins are involved in oxidation reactions. Fat-soluble vitamins include *vitamins A, D, E, and K*.

Heat as a By-Product of Metabolic Reactions (pp. 493–498)
Temperature *Temperature* is a means of measuring the heat of a system. Calories are units of energy in a system and can be used to relate different forms of energy. A *calorie* (cal) is the amount of heat required to raise the temperature of 1 g of water from 14.5°C to 15.5°C. A *kilogram calorie* or *kilocalorie* (kcal) is equal to 1000 small calories.

Production and Distribution of Heat Heat is a by-product of metabolic reactions. It is produced and lost in different degrees depending on the tissue, metabolic activity, and environment.

Regulation of Body Temperature The inner body temperature is the *core temperature*, and the temperature of peripheral body regions is the *shell temperature*. Core temperature is normally more constant than shell temperature.

There are two regions in the hypothalamus involved in controlling temperature: one in the anterior hypothalamus that increases heat loss and one in the posterior hypothalamus that increases heat retention. These regions receive signals from sensory regions that indicate core temperature. Body temperature is increased by constricting peripheral blood vessels, increasing muscle activity by *shivering*, and stimulating the release of thyroxine and catecholamines, which increase metabolic activity. When the core temperature is too warm, the mechanisms for generating heat are inhibited, blood vessels dilate, and sweat is produced.

Heat Loss Heat is lost by radiation, conduction, evaporation, and excretion. *Insensible perspiration* occurs even when not exercising. Sweat produced by exercise is *sensible perspiration*. Fever occurs when heat gain exceeds heat loss.

Basal Metabolic Rate The *basal metabolic rate* (BMR) is an indirect measure of the heat produced by the body during periods of minimal stress and activity. Variations from normal values are associated with various metabolic diseases.

SELF-TEST OF CHAPTER OBJECTIVES

True-False Questions
1. The formation of protein from amino acids is an example of an anabolic process.
2. PGA results from the oxidation of PGAl.
3. When glucose is oxidized, most ATP is synthesized during glycolysis.
4. The respiratory chain is responsible for producing CO_2 and using O_2 during respiration.
5. Fats and steroids are generally transported through the blood in solution.
6. Ketone bodies are the products of the breakdown of carbohydrates.
7. The immediate source of ammonia in the liver is an amino acid called alanine.
8. The thermostat that "sets" the body's temperature is located in the hypothalamus.
9. Nervous impulses cause the loss of water through insensible perspiration.
10. Fever is induced by chemical substances called pyrogens.

Multiple Choice Questions
11. When all three are available, which of the following is normally used first by the body as an energy source?
    a. glycogen   c. stored fat
    b. starch   d. protein
12. Which of the following is least likely to happen to a molecule of glucose as it is circulating in the blood?
    a. absorbed by a cell and metabolized to CO_2 and H_2O
    b. absorbed by a cell and stored as fat
    c. absorbed by the liver and stored as glycogen
    d. excreted by the urine
13. Which of the following hormones increases the level of glucose in the blood most directly?
    a. insulin   c. growth hormone
    b. glucagon   d. estrogen
14. Which of the following substances can be produced directly from pyruvic acid?
    a. lactic acid   c. ketone bodies
    b. oxaloacetic acid d. fatty acid
15. Which of the following terms refers to the production of glucose from noncarbohydrate sources?
    a. glycolysis   c. glycogenesis
    b. glycogenolysis   d. gluconeogenesis

16. Which of the following compounds probably has the greatest amount of available energy in its bonds?
    a. a molecule of PGA
    b. a molecule of PGAl
    c. a molecule of pyruvic acid
    d. a molecule of ATP
17. Production of ATP by the respiratory chain probably involves which of the following?
    a. reduction of oxygen
    b. establishment of a pH gradient within the mitochondrion
    c. release of carbon dioxide
    d. production of water
18. In which type of micelle are phospholipids and cholesterol transported?
    a. high density lipoprotein micelles
    b. low density lipoprotein micelles
    c. very high density lipoprotein micelles
    d. very low density lipoprotein micelles
19. Which component of the lipids is found in highest concentration in the low density lipoprotein micelles?
    a. phospholipids
    b. cholesterol
    c. fatty acids
    d. glucose
20. Which lipoprotein category may be inversely correlated with the likelihood of coronary disease?
    a. high density lipoprotein
    b. low density lipoprotein
    c. very high density lipoprotein
    d. very low density lipoprotein
21. Once the liver's store of glycogen is used, which of the following becomes the body's primary source of energy?
    a. fatty acids
    b. steroids
    c. proteins
    d. amino acids
22. The carbon atoms of a glucose molecule can be used to form
    a. carbon dioxide
    b. amino acids
    c. fatty acids
    d. all of the above
23. Which of the following compounds used in cellular respiration can be produced from the catabolism of either glucose or fatty acids?
    a. pyruvic acid
    b. PGAl
    c. acetyl coenzyme A
    d. propionic acid
24. Which of the following is a product of fatty acid catabolism when oxaloacetic acid is in short supply?
    a. propionic acid
    b. acetoacetic acid
    c. succinic acid
    d. pyruvic acid
25. β-Oxidation is a term that refers to the way that carbon atoms are cleaved from a fatty acid molecule
    a. one at a time
    b. two at a time
    c. three at a time
    d. four at a time
26. Which of the following effects can result from excess fat degradation?
    a. lowered blood pH
    b. elevated blood pH
    c. reduced blood production
    d. excessive blood production
27. Exchange of the —NH_2 group of one amino acid for an oxygen atom of α-ketoglutaric acid produces
    a. alanine and a carbohydrate
    b. alanine and a lipid
    c. glutamic acid and urea
    d. glutamic acid and a carbohydrate
28. When too much α-ketoglutaric acid has been used in transamination, it is replenished by
    a. deamination to produce N_2
    b. transamination to produce alanine
    c. transamination to produce urea
    d. deamination to produce ammonia
29. Which of the following lists consists solely of fat-soluble vitamins?
    a. vitamins A, C, D, and K
    b. vitamins B, D, E, and K
    c. vitamins A, D, E, and K
    d. vitamins A, B, C, and D
30. Which of the following vitamins is synthesized by bacteria growing in the large intestine?
    a. vitamin A
    b. vitamin D
    c. vitamin E
    d. vitamin K
31. The thermostatic control center of the hypothalamus
    a. increases its rate of firing as a result of increased blood temperature
    b. decreases its rate of firing as a result of increased blood temperature
    c. increases its rate of firing as a result of increased osmotic pressure of the blood
    d. decreases its rate of firing as a result of increased osmotic pressure of the blood
32. Which of the following is likely to be happening when one is shivering?
    a. Increased amounts of thyroxine are released into the blood.
    b. Sympathetic nerve impulses are sent to the sweat glands.
    c. The BMR is repressed.
    d. The blood vessels of the skin become dilated.
33. When you become chilled while swimming in cold water, your body is losing heat primarily by
    a. radiation
    b. evaporation
    c. conduction
    d. convection
34. A person has her BMR measured and a value of +100 is determined. This person could be suffering from
    a. hyperthyroidism
    b. hypothyroidism
    c. hypothermia
    d. hyperlipoproteinemia

Essay Questions
35. Define metabolism and distinguish between anabolic and catabolic processes.
36. Describe the reactions involved in the storage and retrieval of glucose.
37. Diagram the metabolic reactions by which glucose is converted to carbon dioxide and water.
38. Cite the steps in cellular respiration in which energy from glucose is used to make ATP.
39. Diagram the metabolic pathways for the interconversion of amino acids, sugars, and fats.
40. Describe how lipids are transported, stored, and metabolized.
41. Describe the reactions involved in metabolizing excess amino acids.
42. List the sources of heat and heat loss.

43. Explain how body temperature is regulated.
44. Define basal metabolic rate and cite several factors that influence it.

Clinical Application Questions
45. Describe the dangers of obesity and describe the type of diet that helps in weight reduction.
46. Explain what heatstroke is, the conditions that cause it, and the treatment for it.
47. List and describe three therapeutic diets and the conditions for which they might be prescribed.
48. Discuss the reasons why anorexia nervosa is so widespread, and describe its effects on the body.
49. Describe how the effect of a pyrogen or other agent acting on the thermostatic control center might cause a fever.

20
The Respiratory System

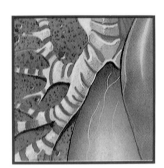

The acquisition of oxygen and the elimination of carbon dioxide is accomplished by the **respiratory tract,** a system of passageways that leads from the nose and mouth to tiny chambers in the lungs. Oxygen diffuses into blood from the gas in these chambers and is carried to the rest of the body as the blood is pumped through the circulatory system. As oxygen diffuses into the blood, carbon dioxide carried by the blood from respiring tissues diffuses into the chambers. Respiration, as studied in this chapter, should not be confused with cellular respiration studied in Chapter 19. Cellular respiration is the process by which cells oxidize and extract energy from nutrients. Respiration considered here, in contrast, involves three processes: **pulmonary ventilation** or **breathing**, the mechanisms used to move air into and out of the lungs; **external respiration,** the exchange of gases between the lungs and the blood; and **internal respiration,** the exchange of gases between the blood and body tissues. Pulmonary ventilation is further divided into *inspiration*, the mechanism that draws air into the lungs, and *expiration*, the mechanism that forces air out of the lungs.

The respiratory system also serves an important role in maintaining the pH of blood and other body fluids, a process that is controlled mainly by regulation of the level of carbon dioxide in the blood. In addition, the upper portion of the respiratory tract is used to produce words and other sounds and contains the organs responsible for the sense of smell.

In this chapter we will examine

1. the organization and anatomy of the respiratory system,
2. the mechanics of breathing,
3. the properties of gases on which the functioning of the respiratory system is based,
4. the mechanisms involved in the exchange of gases between the atmosphere and blood, and
5. the processes involved in regulating respiration.

Organization of the Respiratory Tract

After studying this section, you should be able to:

1. Distinguish the organs of the upper respiratory tract from those of the lower tract.
2. Describe the anatomy of the nasal cavity and associated sinuses.
3. Diagram the pharynx, showing the organization of the nasopharynx, oropharynx, and laryngopharynx.
4. Describe the structure of the larynx and explain the role of the tissue folds within it.
5. Explain how the structure of the trachea contributes to the distribution of air to the lungs.
6. Describe the organization of the lungs, including the manner in which they are covered by pleural membranes.
7. Identify the subdivisions of the pulmonary tree and explain where and how gas exchange occurs in it.

The respiratory tract consists of an **upper respiratory tract** and a **lower respiratory tract** (Figure 20-1). The upper respiratory tract includes the external *nose*, a *nasal cavity* that lies posterior to it, a chamber behind the nasal cavity called

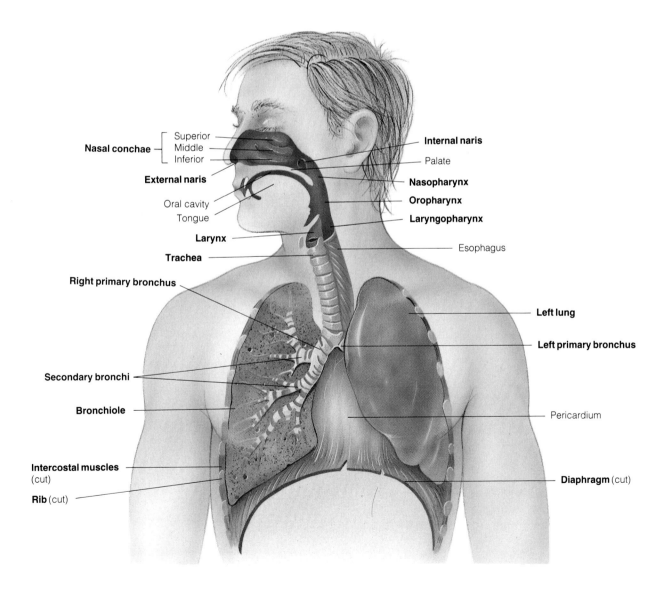

Figure 20-1 The respiratory tract. A portion of the anterior wall and lung has been removed to reveal the pulmonary tree.

nares: (L.) *nares* (pl. of *naris*), the nose

Embryonic Development of the Respiratory System

The respiratory system first appears during the fourth week of embryonic development as a ventral pouch from the foregut (Figure 20-A). Composed of endoderm, this **laryngotracheal** (lăr·in″go·trā′kē′ăl) **bud** divides, and the branches thus produced elongate. The proximal portion of the bud develops into the trachea, while the distal branches become bronchi. Early in development, the slitlike glottis differentiates as the connection between the trachea and foregut.

During ensuing weeks, the embryonic bronchi continue to branch and rebranch producing a mass of bronchioles. In the latter part of development, the blind ends of the bronchioles expand into the alveoli. The respiratory epithelium of the trachea, bronchial tubes, and alveoli is derived from endoderm, while the tracheal and bronchial cartilages, muscles, blood vessels, and connective tissues develop from mesodermal mesenchyme.

Figure 20-A Embryonic development of the respiratory system. (a) The appearance of the laryngotracheal bud during the fourth week of development marks the initiation of growth of the respiratory system. (b) Further development of the system involves extensive branching and elongation of the bronchial tubes.

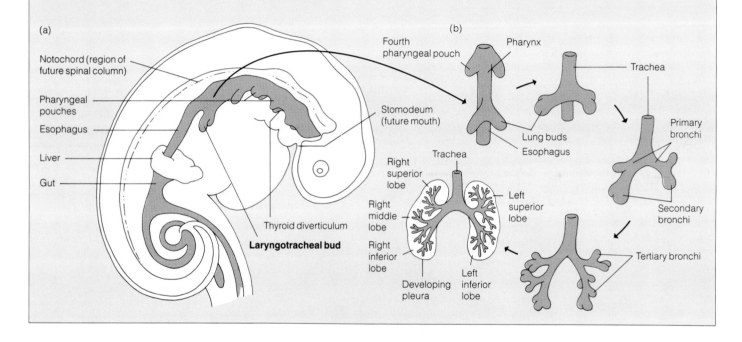

the *nasopharynx*, and the *larynx*. The mouth can also be involved in inhalation and exhalation, but it is not normally considered to be part of the respiratory tract. The pharynx is a region common to both the respiratory and digestive tracts and is traversed both by food on its way to the stomach (see Chapter 18) and by air on its way to the lungs.

The lower respiratory tract consists of a network of branching tubes and tubules that lead to a system of sacs where gas exchange occurs. The first portion of the tubular system consists of the *trachea*, a tube that descends downward from the larynx. The trachea branches into *bronchi*, which in turn branch into successively smaller subdivisions called *bronchioles*. The bronchioles ultimately terminate as blind sacs called *alveoli*. Collectively, the network of tubules and alveoli comprise the *lungs*.

Upper Respiratory Tract

The **nose** provides the opening to the respiratory system. It is formed by cartilage and bone covered by integument (Figure 20-2). The two openings at its base are the **nostrils**, or **external nares** (nă′rēz). Each nostril leads into a **vestibule** (věs·tĭb′ūl), an enlarged region immediately behind the nostrils. The vestibule leads into the **nasal cavity**, which extends from the internal portion of the nose into the skull and terminates at the **choanae** (kō′ă·nē), or **internal nares**,

choanae: (Gr.) *choane*, funnel

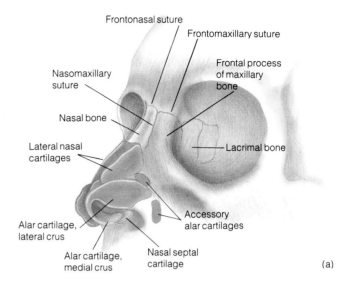

> **CLINICAL NOTE**
>
> ### UPPER RESPIRATORY INFECTION
>
> **Upper respiratory infections** account for 15% of all visits to physicians. Symptoms include sneezing with nasal obstruction and discharge and a sore throat. The two major clinical syndromes are **coryza** (kō·rī′ză) (the **common cold**) and **influenza** (the **flu**). Coryza is basically an acute viral inflammation of the mucous membranes associated with discomfort but usually no fever. There are more than 95 different rhinoviruses known to cause it; this huge number of involved viruses makes it almost impossible to produce an effective vaccine to prevent the common cold. The viruses are spread directly by coughing, sneezing, and more importantly, contaminated hands. They can also spread through the conjunctiva as well as the respiratory passages. Symptoms usually end spontaneously in 5 to 10 days unless secondary bacterial infections result in otitis media, sinusitis, or bronchitis.
>
> Influenza is manifested by the sudden onset of dry cough, sore throat, aching joints, runny nose, sweating, and often severe headaches. Although these infections are usually mild, they can cause necrosis of the epithelium in the nose, throat, trachea, or bronchi. The influenza virus is very frustrating to treat not only because of the annual variation in the predominate group causing the disease, but also because of the ability of the virus to mutate rapidly. These characteristics make it necessary to constantly change the content of the vaccines used every year in the attempt to predict which organism will predominate in the upcoming year's epidemic. It usually runs its course in about 10 days, but may progress to bronchitis or pneumonia. It is associated with high mortality in the elderly and in people with chronic illnesses, such as heart disease, diabetes, or chronic respiratory diseases.

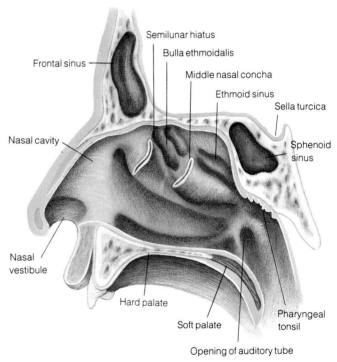

Figure 20-2 The nose. (a) The external nose. (b) The nasal cavity.

a pair of openings in the posterior portion of the nasal cavity. The external nares are rimmed by large numbers of relatively coarse hairs than filter particles from inhaled air.

Nasal Cavity

The roof of the nasal cavity is formed by the cribriform plate of the ethmoid bone. This plate is perforated by olfactory sensory cells. The floor of the nasal cavity is formed by the superior surface of three structures: the maxilla, the palatine bone of the hard palate, and posteriorly, the soft palate. The cavity is divided laterally into two **nasal fossae** by the vertical **nasal septum** that extends from the floor of the cavity to the roof. The nasal septum consists of an anterior cartilaginous portion that attaches to the flat, perpendicular plate of the ethmoid bone suspended from the roof of the cavity. The lower and posterior portion of the septum consists of another flat, perpendicular bone that rests on the floor of the cavity formed by the vomer. Each fossa is also divided into passageways **(meatuses)** by three bony projections of the lateral walls of the cavity called **nasal conchae** or **nasal turbinates** (see Chapter 7).

The wall of the nasal cavity is a mucous membrane consisting of epithelial tissue richly supplied with blood

sinusitis: (L.) *sinus*, a hollow region + (Gr.) *itis*, inflammation

CLINICAL NOTE

SINUSITIS

Sinusitis (sī·nŭs·ī′tĭs) is an infectious disease of the paranasal sinuses. **Acute sinusitis** usually follows an upper respiratory infection (or "cold") that creates an obstruction to the openings of the sinus cavities in the nose. When the outlet of a sinus is obstructed, the oxygen is absorbed from the cavity by the vascular system leaving a negative pressure in the sinus **(vacuum sinusitis)** that causes initial pain. Vigorous blowing of the nose can cause sufficient pressure to force the infection into the sinuses. If the negative pressure persists, it causes a fluid from the mucous lining of the sinus to form. This fluid makes an ideal medium for growth of bacteria from the nasal cavity and finally results in an infection with pus formation and increased pressure in the sinus cavity, resulting in an increase in pain. Fever usually accompanies the symptoms. The typical infecting organisms in acute sinusitis are streptococci, pneumococci, and staphylococci. If sinusitis persists, **chronic sinusitis** may result as the mucous membranes thicken and continue to impede the drainage of the sinuses. Chronic sinusitis is frequently caused by bacilli, rod-shaped microorganisms. About 25% of chronic maxillary sinusitis is caused by dental infections. Air pollution and underwater swimming can also lead to chronic sinusitis.

Decongestants that decrease the inflammatory response will promote drainage of the sinuses. If an actual infection is present, antibiotics and other measures to restore adequate drainage may be necessary to treat the condition. Prevention consists of humidifying the air, avoiding pollutants, and ensuring adequate fluid intake.

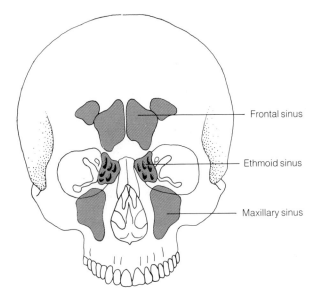

Figure 20-3 The sinuses. The sphenoid sinus is not visible in this view because it lies posterior to the nasal cavity (see Figure 20-2b).

vessels and mucus-secreting glands. Blood carried in the vessels delivers heat to the inspired air, warming the air as it passes over the tissue. At the same time, air is moistened by evaporation of water from the mucus-covered surface. Thus, air passing through the nasal cavity is warmed (or cooled, depending on the temperature outside), cleaned, and moistened. Mucus is sticky so it traps much of the material that has passed through the hairs at the nostrils. As the air is moistened, the mucous surface becomes drier and forms a semisolid material in which the fine particulate matter is trapped. This material is carried toward the posterior portion of the nasal cavity by a current of fluid established by ciliated epithelial cells in the tissue lining the cavity. Rhythmic and coordinated motion of these cilia moves the mucus out the posterior exit of the nasal cavity and into the oropharynx. Once in the oropharynx, the material is either swallowed or spat out.

During development, outcroppings from the nasal membrane form and are surrounded by bony tissue of the developing skull. After development is complete, the cavities remain as the **paranasal sinuses. Maxillary sinuses,** the largest of the paranasal sinuses, are located in the body of the maxillary bones on either side of the nasal fossae. **Frontal sinuses** are located in the frontal bone above the eyes, and **ethmoid sinuses** are found in the ethmoid bone in the lateral wall of the nasal cavity (Figure 20-3). Unlike other sinuses, ethmoid sinuses consist of numerous individual air cells clustered in groups which collectively comprise the sinuses. The **sphenoid sinus** is a single chamber in the sphenoid bone just above and behind the junction of the ethmoid bone and vomer in the posterior portion of the nasal septum (Figure 20-2b).

Each sinus is lined with epithelial tissue similar to that of the nasal cavity itself. The tissues also produce a mucus that drains through ducts that pass from the sinus into the nasal cavity. These ducts are relatively narrow, and when they swell as a result of infection, drainage of the mucus is blocked. When this happens, the mucus and other fluids collect in the sinuses, causing a build-up of pressure that can be very painful (see Clinical Note: Sinusitis).

Pharynx

The **pharynx** is a chamber about 12 cm long that connects the oral and nasal cavities with the esophagus and larynx. As such, it provides a common pathway for food on its way to the digestive tract (see Chapter 18) and air on its way to the respiratory tract (Figure 20-4). Air passing through the nasal fossae exits the nasal cavity through the choanae and enters the upper part of the pharynx in a region above the soft palate called the **nasopharynx** (nā″zō·făr′ĭnks).

adenoid: (Gr.) *adenoides*, glandular
epiglottis: (Gr.) *epi*, upon + *glottis*, glottis
cricoid: (Gr.) *krikos*, ring + *eidos*, form
cricoarytenoid: (Gr.) *krikos*, ring + *arytaina*, pitcher + *eidos*, form

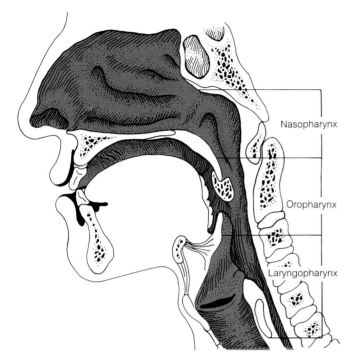

Figure 20-4 The pharynx. The pharynx consists of three sections: the nasopharynx, oropharynx, and laryngopharynx.

The roof of the nasopharynx is formed by the sphenoid bone at the base of the skull and its floor by the soft palate. The lateral walls of the nasopharynx contain the openings of the auditory tubes called the **pharyngeal** (făr·ĭn′jē·ăl) **apertures.** The posterior portion of the nasopharynx contains a broad, flat mass of lymphoid tissue called the **pharyngeal tonsils,** or **adenoids** (ăd′ĕ·noyds) (see Chapter 24). Infection can cause the adenoids to swell and interfere with the passage of air and mucus through the nasopharynx.

At the posterior edge of the soft palate, the nasopharynx opens into the **oropharynx.** The oropharynx communicates anteriorly with the oral cavity, making it directly visible through the opened mouth. The oral cavity communicates with the oropharynx through the **isthmus of the fauces** (fŏ′sēz), a narrow passageway formed by the base of the tongue, the soft palate, and two curved folds of tissue that lie anterior and posterior to the palatine tonsils. The anterior folds form the **palatoglossal** (păl″ă·tō·glŏs′al) **arch** and mark the boundary between the oral cavity and the oropharynx. The oropharynx leads downward into the **laryngopharynx** (lăr·ĭn″gō·făr′ĭnks), a short tube that lies posterior to the larynx.

Larynx

After passing from the nasal cavity into the nasopharynx, inhaled air passes through the oropharynx and enters an opening in the anterior wall just below the base of the tongue called the **laryngeal aperture.** Separating the aperture from the base of the tongue is a leaflike cartilaginous flap of mucous membrane-covered tissue, the **epiglottis** (ĕp″ĭ·glŏt′ĭs). During swallowing the larynx is elevated, forcing the epiglottis down over the opening into the larynx and preventing food or liquid from passing into the respiratory tract.

Air passing by the epiglottis and through the laryngeal aperture enters the **larynx** (lăr′ĭnks), which acts as a valve that controls access to the tubular system that lies below it (Figure 20-5).

The larynx consists of bony and cartilaginous structures held together by ligaments, muscles, and other tissues. It lies below the horseshoe-shaped hyoid bone. Suspended from the hyoid bone by ligaments and membrane is the **thyroid cartilage,** the largest unit of the larynx. The thyroid cartilage produces a bulge in the front of the neck called the **laryngeal prominance;** this bulge is better known as the **Adam's apple.**

Suspended from the thyroid cartilage by ligaments and muscle is the **cricoid** (krī′koyd) **cartilage,** a ringlike structure that nests at the base of the thyroid cartilage. The cartilages give the larynx its form and also support the tissues used by the larynx in acting as a valve and sound-producing organ.

The interior of the larynx is lined with mucous membrane that forms folds of tissue that extend into the passageway (Figure 20-6). Just within the laryngeal aperture is an expanded region called the **vestibule,** which is limited inferiorly by two folds called the **vestibular folds,** or "false vocal cords." Just below these folds there is another widening of the channel, followed by two more folds of the membrane. These folds are the **vocal folds,** or "true vocal cords," and are responsible for producing sound as air passing over them from the lower respiratory tract causes them to vibrate. The space between the two vocal folds is the **glottis** (glŏt′ĭs). Inferior to the vocal folds, the laryngeal channel widens and continues into the trachea, the next organ in the tract.

The larynx contains several muscles that control the diameter of the passageway through it. With the exception of the posterior **cricoarytenoid** (krī″kō·ă·rĭt′ĕn·oyd) muscles, which lie on the anterior surface of the base of the larynx and dilate the passage, spasm of the laryngeal muscles can close the folds and prevent airflow into the lungs.

Voice and Singing

The vestibular folds are controlled by muscle tissue in the wall that can be contracted to open and close the glottis. These muscles help close the larynx to the expulsion of air from the lungs when you hold your breath, lift a weight, or are otherwise straining.

510 Chapter 20 The Respiratory System

trachea: (Gr.) *tracheia*, rough

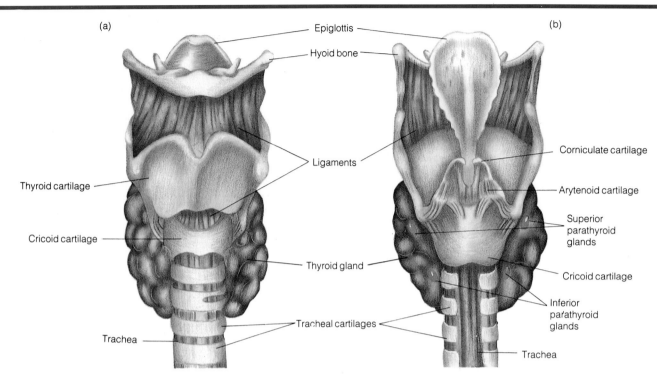

Figure 20-5 The larynx. (a) External anterior view. The isthmus of the thyroid gland has been omitted to show the cricoid cartilage. (b) External posterior view.

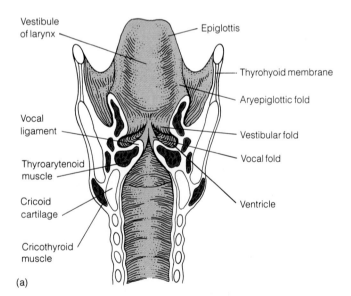

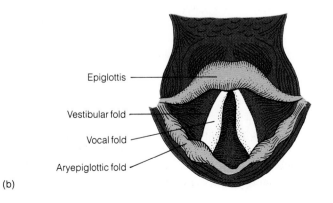

The vocal folds are rimmed by **vocal ligaments** that border the glottis. Within the folds themselves are muscles that control the tension on the ligaments and control the size of the opening between the folds. Several other laryngeal muscles are also involved in determining the shape and orientation of the vocal folds by adjusting the orientation of the various cartilages. The quality of sound produced by the vibrating folds is determined in this way. The muscles primarily involved in establishing tension in the vocal ligaments are the paired **cricothyroid** (krī·kō·thī′royd) and **thyroarytenoid** (thī″rō·ă·rĭt′ĕn·oyd) muscles.

Lower Respiratory Tract

The **trachea** (trā′kē·ă), more commonly called the "windpipe," is a tube that extends downward from the base of the larynx and branches into two **primary bronchi** (brŏng′kī) (Figure 20-7). The trachea is about 12 cm long by about 2.5 cm in diameter in an adult and lies anterior to the esophagus.

Figure 20-6 Interior organization of the larynx. (a) Posterior view. (b) View from the pharynx, shown with glottis open. The vestibular folds (or false vocal cords) delimit the lower portion of the vestibule of the larynx. The vocal folds and the muscles involved in sound production are located below the vestibular folds.

mediastinum: (L.) *mediastinum*, in the middle of

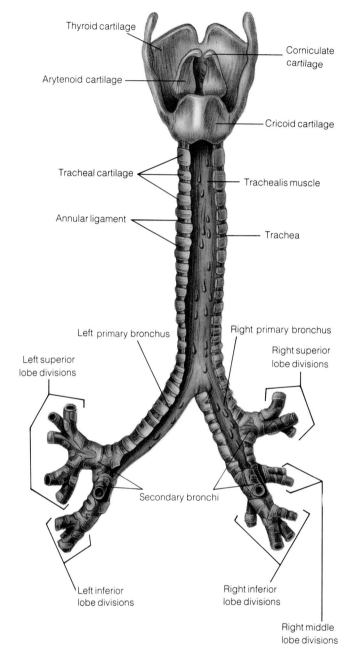

Figure 20-7 The trachea and bronchi in a posterior view. Note the trachealis muscle that spans the opening of the C-shaped cartilaginous bands. The trachea maintains its shape because of these bands.

> **CLINICAL NOTE**
>
> **LUNG CANCER**
>
> **Lung cancer**, or **bronchogenic carcinoma**, most commonly affects male smokers at or past middle age. The incidence in women is increasing, and by the end of the 1980s, lung cancer is expected to overtake breast cancer as the most common malignancy in women in the United States. The cancerous growth may be an indurating, ulcerating, or fungating mass lying within the bronchial lumen or wall (see The Clinic, p. 531, for a photograph of lungs destroyed by lung cancer). These tumors are associated with inflammatory changes and collapse of distal lung tissue. Symptoms include coughing, lung and throat irritation, and occasionally blood-tinged **sputum** (substance from the respiratory tract expelled by coughing) from ulceration of the bronchial wall. The four main types are squamous, adenocarcinoma, oat cell, and anaplastic. The prognosis for the first two is reasonably good if diagnosed early. The prognosis for the latter two is quite poor due to their rapid growth and early metastatic spread.

Unlike the esophagus, which is soft and pressed flat when empty, the trachea is kept permanently open. This is accomplished by 16 to 20 horizontally oriented C-shaped bands of cartilage in the tracheal wall. The open portion of the C is spanned by the **trachealis** (trā″kē·ă′lĭs) muscle. The bands are connected to one another by intervening annular ligaments and are oriented so that their open portions are on the posterior side of the trachea next to the esophagus.

The inner surface of the trachea is lined with ciliated mucous membrane laid on a supporting submucosa. This lining is similar in composition to the mucosal lining of the lower portion of the larynx, the nasal cavity, and the nasopharynx. It also helps to filter particulate matter from air passing through the trachea. Externally, the trachea is surrounded by a sheath of connective tissue.

Gross Morphology of the Lungs

The **lungs** are located in the chamber of the chest formed by the ribs laterally, the vertebral column posteriorly, the sternum anteriorly, and the diaphragm inferiorly. This region, the thoracic cavity (see Figure 1-4 in Chapter 1), also contains several other structures, including the heart, esophagus, thoracic lymphatic duct, major nerves, and major arteries and veins that carry blood to and away from the lungs. These organs are located in the **mediastinum** (mē″dĭ·ăs·tī′nŭm), the medial portion of the thoracic cavity. The lungs themselves lie on either side of the mediastinum, enclosed in **pleurae** (plū′rē), a pair of double-walled sacs of serous membrane (Figures 20-8a and 20-9). The pleura on either side of the thoracic cavity is a single bag that folds back on itself, much like a partially inflated balloon would look if pushed with a fist until the opposing inner surfaces were in contact.

Each pleura consists of a thin serous membrane that lies tightly pressed to the inner wall of the rib cage, folding back on itself in the region of the mediastinum, and continuing back out and around the outer surface of the lung, to which

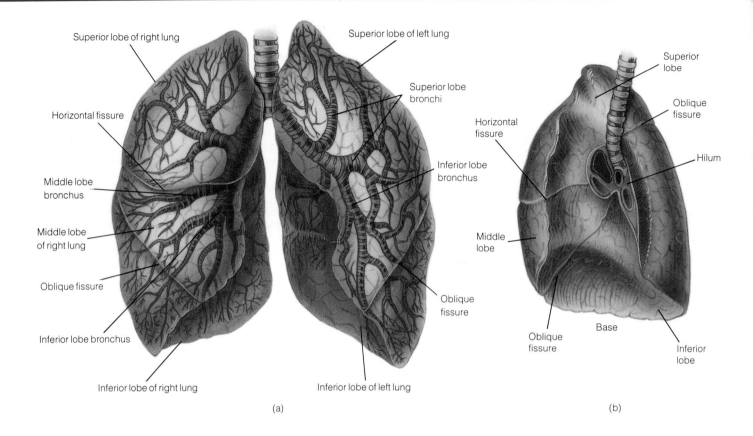

Figure 20-8 The lungs. (a) Anterior view of both lungs. Note the fissures that divide the lungs into lobes. (b) Medial view of right lung showing the hilum.

it is also closely pressed. The **parietal pleura** consists of that portion of the serous bag that is not closely adherent to the lungs. The portion that adheres to the lung surfaces is the **pulmonary** or **visceral pleura.** The portion of the parietal pleura that adheres to the thoracic wall, the **costal pleura,** goes on to adhere to a portion of the diaphragm, where it forms the **diaphragmatic** (dī″ă·frăg·măt′ĭk) **pleura.** The diaphragmatic pleura folds upward from the diaphragm and outlines the mediastinum. This portion is the **mediastinal pleura.**

Returning to the analogy of the partially filled balloon with a fist in it, the fist corresponds to the lungs and the inner surfaces of the balloon in contact with one another are analogous to the surfaces of the parietal and pulmonary pleurae. The portion of the balloon where the fist has entered and which would surround the wrist is equivalent to a region known as the **pulmonary hilum** (hī′lŭm). The pulmonary hilum is a roughly triangular hole through which the primary bronchus, a pulmonary artery and vein, bronchial arteries and veins, nerves, and lymphatic vessels pass into the lung (Figure 20-8b). Collectively, these structures comprise the **root** of the lung.

The two lungs are not mirror images of one another, the left lung being slightly smaller than the right. Each lung has deep fissures in its surface that divide it into **lobes.** The right primary bronchus branches from the trachea and gives rise to three secondary bronchi, each of which enters a lobe. In contrast, the left primary bronchus gives rise to only two secondary bronchi, each of which enters a lobe. Consequently, the right lung has three lobes and the left lung has two. The pulmonary pleura follows the contours of the lobes, descending to the base of each fissure and out again, adhering closely to the surface of the lungs at all points except at the root (Figure 20-9).

Pulmonary Tree

Inspired air passes into the lungs through the primary bronchi and is distributed throughout the lungs by an elaborate system of tubules that comprises the **pulmonary tree** (Figure 20-10). The pulmonary tree consists of two functionally distinct regions, one in which air passes and another in which

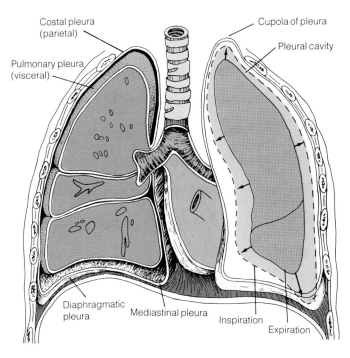

Figure 20-9 Organization of the pleurae.

CLINICAL NOTE

PNEUMONIA

Pneumonia is caused by viruses, bacteria, or fungi that reach the alveoli through the inspired air. New pathogens are constantly appearing as causes, such as the mycoplasma and the Legionella organisms, which challenge physicians. Pneumonia usually occurs only if a person inhales a high concentration of pathogen or has an improperly functioning immune system. Initially there is an outpouring of edema fluid into the alveolar spaces with a rapid spread of fluid to adjacent lung tissue. This is followed by replication of microorganisms in the edema fluid and invasion by white blood cells in response to the infection. The process can spread rapidly, leading to consolidation of the segment or entire lobe of the lung. Oxygen exchange with the blood may be diminished. The treatment of pneumonia depends on isolating the specific organism involved, then employing the most specific antibiotic available. In bacterial pneumonia, prompt treatment with appropriate antibiotics promotes resolution of a potentially fatal disease.

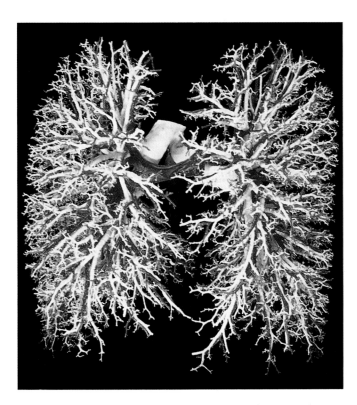

Figure 20-10 **The pulmonary tree.** The successive branches of the bronchi and bronchioles produce the passageways that deliver oxygen to the terminal bronchioles. This resin cast of the pulmonary tree of a human lung shows the bronchial passages in white and the main arteries in red.

gas exchange occurs. The first of these is the **conducting division** and the second is the **respiratory division** of the pulmonary tree.

Conducting Division

The two primary bronchi descend downward and laterally into the chest cavity for a few centimeters before passing through the hilum. The right branch extends a shorter distance than the left one and also has a slightly larger diameter and is oriented more vertically. As a result, if a foreign body succeeds in passing through the larynx, it tends to fall through and lodge in the right bronchus more readily than the left.

The bronchial wall in this region has a cartilaginous skeleton similar to that of the trachea, although the bands are not as regular. In the region where the bronchus passes into the lung, the bands become even more irregular and are gradually replaced by cartilaginous plates of variable shape and size. Continued branching produces tubes with reduced cartilaginous support, and after a few branchings, the cartilage is missing entirely. At that point the tubules are referred to as **bronchioles** (brŏng′kĭ·ōls). Bronchioles continue to branch smaller and smaller until they become **terminal bronchioles.** Normally 16 branch points occur from the trachea to the terminal bronchioles, producing nearly 66,000 terminal bronchioles. These are distributed about equally between the right and left lungs.

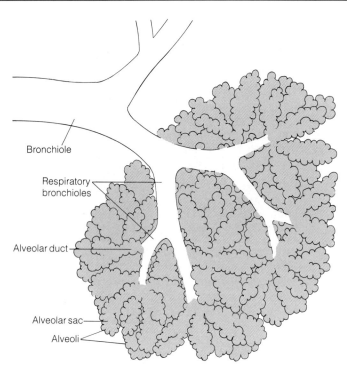

Figure 20-11 A primary lobule. Lobules consist of the branches of respiratory bronchioles that branch from a terminal bronchiole. Each one consists of successive branches of respiratory bronchioles and alveolar ducts, with the ducts terminating in alveolar sacs.

Lacking a cartilaginous skeleton, bronchioles are softer and more flexible than bronchi. They also contain smooth muscle, constriction of which reduces their diameter and the amount of air that can pass through them.

Respiratory Division

Terminal bronchioles branch still further to produce a network of **respiratory bronchioles.** Typically, three levels of branching of respiratory bronchioles occur, with the final respiratory bronchiole in the sequence branching to produce a pair of **alveolar** (ăl·vē·ō′lăr) **ducts.** Each alveolar duct may branch further and terminate in an **alveolar sac.** The walls of the passageways from the respiratory bronchioles to the alveolar ducts are perforated by openings that lead into small chambers called **alveoli.** Alveoli are the functional units of the lung where gas exchange occurs. A respiratory bronchiole and its subsequent divisions collectively comprise a unit of the lung called a **primary lobule** (Figure 20-11).

There are typically 23 levels of branching between the trachea and the alveoli, giving rise to about 150 million alveoli in each lung. The average diameter of an alveolus is about 300 μm (micrometers), so the entire functional surface of the lungs through which gas exchange occurs amounts to about 75 m², equal to a little over 800 sq ft, or a square measuring more than 28 ft per side.

Structure of the Alveolar Wall

The alveoli are supported by a network of fine bundles of smooth muscle fibers and fibrous protein that lie in the walls formed where alveoli come into contact with one another. These walls are called **interalveolar septa.** The septa also have the important function of carrying the capillaries in which blood flows as it absorbs oxygen from the alveolar air and releases carbon dioxide to it.

Figure 20-12 shows the organization of one of these extremely delicate septa. The surfaces on either side of the septum that face the interior of an alveolus are covered

---

tuberculosis: (L.) *tuberculum*, small swelling

> **CLINICAL NOTE**
>
> ## TUBERCULOSIS
>
> **Tuberculosis** is an infectious disease caused by tubercle bacilli. Infection usually spreads by inhalation of the bacteria through contact with an infected person. Patients often do not have symptoms, but they can experience cough, fever, weight loss, night sweats, and blood-tinged sputum. The bacteria cause a nonspecific foreign body reaction within the lungs. Some of the bacteria are destroyed, some multiply, and some spread to the lymph nodes and then throughout the body. Usually the primary disease is not detectable clinically, but reactivation tends to be progressive, forming small nodules (or **tubercles**). During this process the lung tissue can be destroyed, resulting in scar formation. Diagnostic procedures include isolation of the tubercle bacilli in the sputum and tuberculin skin testing. Therapy consists of prolonged treatment with one or more antituberculous drugs, the main one being **isoniazid.** Its drawbacks include early bacterial resistance and liver damage.
>
>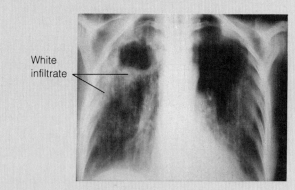
>
> The white infiltrate in the lung on the left is a result of tuberculosis.

atelectasis: (Gr.) *ateles*, imperfect + *ektasis*, expansion

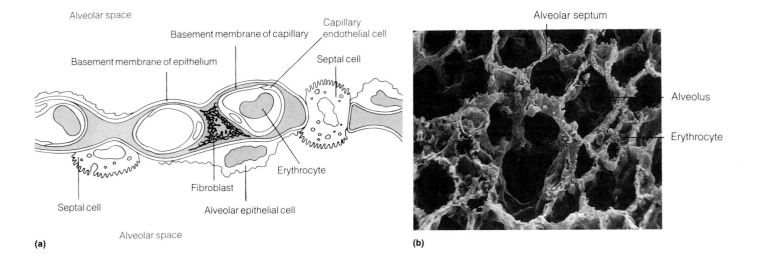

Figure 20-12 An interalveolar septum. (a) Drawing showing how the septum forms the wall of the alveolus. It contains the capillaries through which the blood flows as it gains oxygen and releases carbon dioxide. (b) Scanning electron micrograph showing histology of interalveolar septum (mag. 119×).

primarily by an epithelium that is so thin that much of it can only be visualized by the electron microscope. The only parts that are thick enough to see are the regions where the nuclei are located. The surface toward the septum rests on a basement membrane of mucopolysaccharides secreted by the epithelial cell.

Sandwiched between the epithelial cells on either side of the septum is a region in which the capillaries are found. Capillaries are fine tubular vessels formed by flat endothelial cells wrapped in cylindrical form. Blood passes through these cylinders. Like the epithelial cell of the alveolus, the endothelial cell of a capillary produces a thin basement membrane which at some points merges with that of the epithelium.

Also included in the septum are occasional **septal cells** that secrete a phospholipid material that adheres to the alveolar surface of the epithelial cells. This substance acts as a **surfactant**, a substance that reduces the surface tension of the moisture on the cell surface. This in turn facilitates the diffusion of oxygen into the epithelium from the alveolus. Septal cells are roughly cuboidal in shape and may be present on only one surface of a septum or extend all the way through, providing surfactant for both sides of the wall (see Clinical Note: Atelectasis).

An alveolus contains macrophages that remove microorganisms and particulate matter that has been inhaled. These

CLINICAL NOTE

ATELECTASIS

The collapse of an area of normal lung or the incomplete expansion of a portion of the lung, which is capable of expansion, is referred to as **atelectasis** (ăt″ĕ·lĕk′tă·sĭs). It is caused by compression of lung tissue from fluid or air accumulation, tumors, or the pressure of other organs. Atelectasis may occur after abdominal surgery at the bottoms of the lungs in the area immediately above the diaphragm. People with abdominal wounds who are in pain and a little groggy from anesthesia may not breathe deeply or move around enough to allow the compressed portion of the lungs to expand. To prevent atelectasis, patients are taught to practice breathing exercises before surgery and are assisted in doing these exercises after surgery.

Any person confined to bed has some degree of atelectasis and compression of lung tissue, which may lead to infection. Turning the person frequently and encouraging coughing and deep breathing allows reexpansion of the lung.

When atelectasis occurs in newborns it is called **respiratory distress syndrome.** This syndrome occurs most frequently in premature infants, and is mainly caused by immaturity of the surfactant system. Surfactant is secreted into the alveoli by special alveolar cells where it acts as a detergent, reducing the surface tension of the fluids in the alveoli. When the alveoli empty during respiration, they grow smaller, and without surfactant, the surface tension of the fluids in the smaller sphere tend to collapse the alveoli. Infants with respiratory distress syndrome are maintained in an intensive care nursery where hypoxia and hypotension can be controlled. When well cared for, these infants have good chance of survival and normal development.

macrophages are unusual in that they spend considerable time in the alveolus, foraging over the epithelial surface where they ingest dust and soot particles as well as microbes. Once macrophages have ingested this foreign matter, they are carried out through the bronchi to the pharynx and swallowed.

Breathing Mechanics

After studying this section, you should be able to:

1. Distinguish between inspiration and expiration, describing the mechanisms involved in each.
2. Explain why enlargement of the thoracic cavity causes the lungs to fill.
3. Identify the components of lung volume.
4. Describe how certain diseases can decrease lung function.

Inspiration and Expiration

Breathing consists of **inspiration** (or **inhalation**), the drawing of air into the lungs, and **expiration** (or **exhalation**), the subsequent expulsion of that air. At rest, an average of 12 to 15 inspiration-expiration cycles occur per minute. Each breath draws in about 500 mL of air, for a total volume of 6 to 7.5 L per minute. In the short time that the alveoli are inflated, sufficient diffusion goes on between the alveolar air and the blood to supply oxygen to the tissues and to remove some of the carbon dioxide carried by the blood.

Inspiration is an active process primarily involving the contraction of the diaphragm, and secondarily the contraction of the external intercostal muscles. Contraction of these muscles enlarges the thoracic volume, reducing intrathoracic pressure, and essentially sucking air into the lungs. Expiration, in contrast, is largely passive (when at rest), resulting from the elastic properties of the tissues that cause the thoracic volume to decrease when the diaphragm and external intercostal muscles relax (Figure 20-13).

When one is breathing harder, for example, following strenuous exercise, inspiration also involves the sternocleidomastoid, the scalenes, and the pectoralis minor muscles in the neck and upper portion of the chest. Contraction of these muscles increases the thoracic volume even more. Expiration is partly an active process during heavy breathing since internal intercostal muscles pull the ribs and sternum back down. At the same time, the diaphragm is forced back up by the abdominal viscera as the muscles of the abdominal wall contract and create pressure on the abdominal cavity.

The diaphragm consists of radially oriented muscle fibers that extend outward from a central fibrous plate, the **central**

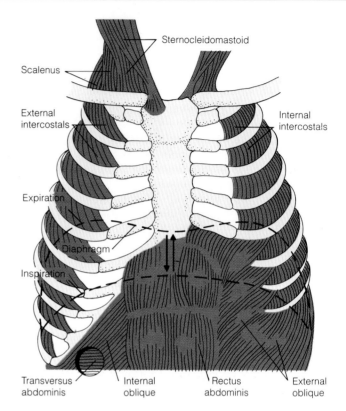

Figure 20-13 Muscles involved in inspiration and expiration. Dashed lines show positions of top of diaphragm at expiration and inspiration.

tendon of the diaphragm, to the surrounding thoracic wall. Posteriorly they insert on the lower thoracic and upper lumbar vertebrae, laterally and anteriorly on the lower costal cartilages and ribs, and anteriorly on the lowest part of the sternum, the xiphoid process (see Chapter 7). The diaphragm covers the superior surface of the liver and stomach and forms a continuous domelike floor to the thoracic cavity (and a ceiling to the abdominal cavity). Contraction of the diaphragm pulls the dome downward only about 1.5 cm under resting conditions but as much as 7.5 cm during strenuous exercise. This increases the volume of the thoracic cavity and causes a reduction of the atmospheric pressure in the lungs. The difference between the pressure in the lungs and external atmospheric pressure then forces air through the respiratory passageways to the alveoli.

As the diaphragm moves downward during inspiration, the external intercostal muscles contract and cause the ribs to pivot upward where they articulate with the vertebral column. This also moves the sternum upward and outward. As the ribs and sternum rise, the anterior-posterior dimension of the thoracic cavity is increased, although the lateral dimension stays the same. The resulting increase in thoracic

spirometer: (L.) *spirare*, to breathe + (Gr.) *metron*, to measure
embolism: (Gr.) *embole*, a throwing in + *ismos*, a condition

volume also contributes to the reduction of air pressure in the lungs and helps to draw air in.

As the lungs fill, they slide over the inner wall of the thoracic cavity. The sliding occurs between the outer surfaces of the visceral pleura and the inner surface of the costal pleura. These two surfaces are separated by a thin film of fluid that lubricates the surfaces as they slip over one another. As long as the pleural surfaces are in contact, changes in lung shape and volume follow changes in the shape and volume of the chest.

If the thoracic cavity consisted only of the chest wall, the mechanism for filling the lungs would be relatively easy to understand: muscle activity would cause the chest volume to expand. Increased volume would reduce the pressure of the gas in the interior to less than atmospheric pressure, and air would be drawn in. The structure of the chest is not so simple, however, because the surface of the lungs is separated from the inner surface of the chest wall. Increased thoracic volume causes the lungs to fill because pressure at the interface between the opposing pleurae is less than the pressure inside the lungs. As the thoracic volume increases, a chain of events is initiated: the interpleural pressure decreases, increasing the difference between interpleural and lung pressures. The lungs expand in response to the drop in interpleural pressure, which increases lung volume. Increased lung volume decreases the gas pressure in the lungs to less than atmospheric pressure. The difference in pressure between the air outside and the gas inside causes air to flow in as they expand. Thus, expansion of the lungs causes air to flow in.

During expiration, essentially the opposite happens. As the thoracic volume decreases, interpleural pressure increases to greater than the pressure in the lungs. This causes the lungs to compress, which, in turn, increases the gas pressure in them to greater than atmospheric pressure and the gas is forced out.

The importance of interpleural pressure in maintaining lung volume can be demonstrated by allowing air to enter the interpleural space between one of the lungs and the thoracic wall. Since pressure in the space is less than atmospheric pressure, anything that punctures the chest will allow air to enter. When this happens, the interpleural pressure becomes equal to atmospheric pressure. In the absence of a positive pressure difference between the lung and the interpleural space, elasticity of the lung is sufficient to cause it to force gas out and the lung collapses.

Respiratory Volumes

The amount of inspired or expired air can be measured with an instrument called a **spirometer** (or **respirometer**). A spirometer is essentially a hose and mouthpiece attached to a container of air that increases or decreases in volume,

> **CLINICAL NOTE**
>
> **PULMONARY EMBOLISM**
>
> **Pulmonary embolism** is blockage of blood flow through a pulmonary artery by a blood clot **(embolus)**. It is common in bedridden patients, especially those who have had pelvic surgery or those immobilized with leg fractures. The embolus forms in the leg and breaks off, flowing downstream. The blood vessels continually widen on the way to the heart and the clot passes unimpeded. Once it enters the pulmonary circulation, the vessels begin to narrow and the embolus becomes lodged in a pulmonary artery. The accompanying symptoms depend on the amount of blood flow obstructed. Sudden unexplained shortness of breath is a hallmark of the disease. The portion of the lung not perfused by blood becomes "physiological dead space." This may lead to decreased oxygenation of blood and tissue death. Approximately 10% of pulmonary emboli are fatal. Prompt treatment with oxygen and anticoagulants help to prevent further clot formation.

depending on the amount of air in the container. The container in turn is attached to a pen that writes on a moving strip of chart paper. A subject breathes through the mouthpiece, causing the volume of the container to increase and decrease alternately. The volume inspired or expired can be calculated from the movement of the pen. Using this instrument, characteristics of a person's lungs can be measured and used to determine the condition of the lungs with respect to many diseases.

When a person breathes normally, the volume of air that is exchanged is the **tidal volume** of the respiratory tract (Figure 20-14). In a healthy adult male this is normally about 500 mL. More air can be inspired with additional effort, however, and the amount in excess of the tidal volume that can be inspired with maximum effort is the **inspiratory reserve volume.** To determine the inspiratory reserve volume, the subject breathes in the normal way and then, at the end of an inspiration, inhales again as deeply as possible. Conversely, if one exhales as completely as possible, the additional volume exhaled over normal exhalation is the **expiratory reserve volume.** The total of the inspiratory reserve volume and the tidal volume corresponds to the total volume that can be inspired at the end of a normal expiration. This volume is the **inspiratory capacity.** The total of the expiratory and inspiratory reserve volumes plus the tidal volume is referred to as the **vital capacity.** In an adult male the vital capacity is typically about 5 L and in a female about 3 L.

Not all the gas in the respiratory tract can be exhaled because the passageways and alveoli remain open even after

emphysema: (Gr.) *emphysan*, to inflate
asthma: (Gr.) *asthma*, panting

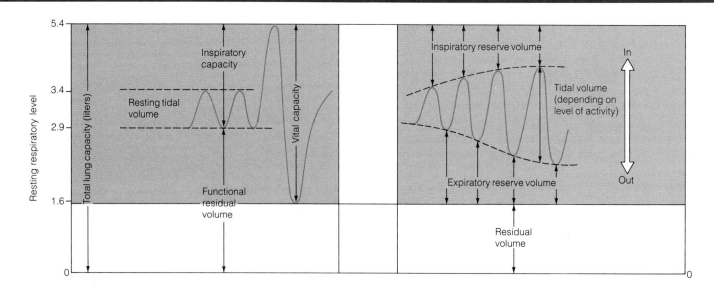

Figure 20-14 Various lung volumes and capacities in a typical adult male as determined with a spirometer. The graph on the right shows the changes in these volumes that result from increased physical activity.

maximum expiration. The volume that remains after the expiratory reserve volume has been expired is the **residual volume.** The sum of the residual volume and the vital capacity is the **total lung capacity** (TLC), which is typically about 6 L in an adult male and 4 L in an adult female.

About 150 mL of the tidal volume represents space in the passageways, where gas is unavailable for exchange with blood, rather than in alveoli. This portion of the tract is referred to as **anatomical dead space.** Even in the alveoli, some gas may not be exchanged because of damage to the alveolar walls, perhaps due to disease or injury. This volume plus the anatomical dead space is the **physiological dead space** (see Clinical Note: Pulmonary Embolism).

In addition to simple volumes, another useful measure of the condition of the lungs is the rate at which they can be filled. This can be measured in two ways: the maximum amount that can be expired in 1 sec is the **timed vital capacity,** and the maximum amount of gas that can be breathed in and out in 1 min is the **maximal voluntary ventilation.** Many of these values depend in part on the elasticity of the lungs. The more elastic the lungs, the more they can stretch in response to incoming air. To measure this property, a subject inhales or exhales a measured volume of air through a tube with a valve in it. The air is inspired 50 to 100 mL at a time and the valve is closed. The subject then relaxes his or her respiratory muscles and the air pressure in the mouth is measured. Since the glottis is open and since there is no air movement, the air pressure throughout the respiratory tract is equalized at this point (the pressure in the mouth equals that in the lungs). The difference at that moment between the respiratory tract pressure and the barometric pressure of the surrounding air is called the **relaxation pressure.** The ability of the lungs to stretch in response to different pressures is determined by calculating the ratio between the change in volume of the lungs that occurs and the change in pressure that caused the volume change, a value called the **compliance** of the lungs. Because different people have different lung values, however, a more useful measure is **specific compliance,** the ratio of compliance to total lung volume.

Specific compliance is a useful parameter in measuring the extent of certain lung diseases. **Pulmonary emphysema** (ĕm″fĭ·sē·mă), for example, is a disease in which many of the alveolar walls disintegrate and combine to form large air sacs. These sacs add to the physiological dead space as surfaces involved in gas exchange are lost. In more advanced stages of the disease, there is an accumulation of fibrous tissue in the damaged walls that reduces the lungs' resistance to expansion. Consequently, for a given change in internal pressure, there is a greater than normal change in lung volume. In other words, specific compliance of the lungs increases.

Bronchial asthma is a disease that causes a temporary narrowing of the bronchiolar airways (see The Clinic: Diseases of the Respiratory System). This is brought on by spasmotic contractions of the smooth muscles lining these passages and by their swelling and inflammation, all of which cause a narrowing of the passages. As a result, ventilation of the lungs is made more difficult, breathing may be labored, and there can be wheezing and violent and uncontrolled coughing.

The rate at which the lungs can be filled or emptied progressively decreases as an asthma attack proceeds. Air drawn in cannot be expelled efficiently because the increased pressure causes the passageways to narrow still more. Pressure buildup in the alveoli and in the small chambers leading to them can cause them to enlarge to a point where they may rupture. The total lung capacity and the residual volume are increased because of the inflation of the lungs and the resulting impediment to air movement. Reduced ventilation efficiency deprives the tissues of oxygen and causes a buildup of carbon dioxide in the tissues. As the carbon dioxide reacts with water, it is converted to carbonic acid which releases H^+, and tissue acidity increases.

Properties of Gases

After studying this section, you should be able to:

1. **Define partial gas pressure.**
2. **Describe how the partial pressure of a gas influences its concentration in a fluid.**
3. **Explain how pressure, volume, and temperature influence the properties of a gas.**

The primary function of the lungs is to provide oxygen for respiring tissues and to eliminate carbon dioxide produced by those tissues. Exchange of gases with the atmosphere occurs at the alveolar membrane where oxygen passes successively through several layers: the surfactant that covers the wall, a thin epithelial cell, the basement membrane of the epithelial cell, the basement membrane of the endothelial cell of the capillary, plasma, and finally, the erythrocyte membrane. Once in an erythrocyte, most of the oxygen combines with hemoglobin and is carried to the systemic capillaries and released. Carbon dioxide, produced mainly by metabolic reactions in tissues throughout the body, moves in the opposite direction, eventually to diffuse from the blood into the alveoli.

Gas Pressure

To understand the gas exchange that goes on in the lungs and body tissues, it is necessary to understand the properties of gases and the factors that govern their behavior. Dry air is a mixture consisting of oxygen (20.98%), nitrogen (78.06%), carbon dioxide (0.04%), and various inert substances (0.92%), including helium, argon, and neon. Air at sea level pressure can also contain as much as 6% water vapor, which lowers the percentage of the other gases proportionately.

Gas in a container exerts a force on the walls of the container. The force results from individual gas molecules striking the container wall. The amount of force exerted in a given unit of area of the wall is called **gas pressure**, and it is proportional to the number of molecules of the gas in the container. Gas pressure can be described in two ways: as the force exerted by the gas on a unit of area, such as grams per square centimeter or pounds per square inch (psi), or as the distance the pressure would force a fluid up an evacuated cylinder. For example, at sea level the force exerted by the atmosphere is usually about 14.7 psi. If an evacuated cylinder were placed in a container of water, air pressure would force the water up the cylinder about 10 m. If the cylinder were placed in a container of mercury (Hg) exposed to air at this pressure, air pressure would force the mercury up the cylinder a distance of 760 mm. At 760 mm, the weight of the mercury in the cylinder creates a force equal to the force exerted by the air. Thus, a gas pressure of 14.7 psi can also be expressed as 760 mm of mercury.

Partial Pressure

In a mixture of gases the pressure of each gas contributes proportionately to the total gas pressure, which is the sum of the pressures contributed by each component. The pressure contributed by each gas in such a mixture is termed its **partial pressure (P)**, and it is determined by multiplying the proportion of the gas in the mixture by the total gas pressure. For example, at an air pressure of 760 mm Hg, oxygen, which comprises 20.98% of dry air, has a partial pressure (P_{O_2}) of 159 mm Hg (0.2098 × 760). In the same air the partial pressure of CO_2 (P_{CO_2}) (0.04% of the total volume) is 0.30 mm Hg (0.0004 × 760), and nitrogen and other inert gases, which collectively comprise about 79% of the air, contribute a partial pressure of 600 mm Hg (Figure 20-15).

Ideal Gas Laws

Three factors determine the pressure exerted by a gas (or gas mixture): the amount of gas (in weight or number of molecules), the volume it occupies, and its temperature. These factors have certain constant physical relationships referred to as **ideal gas laws**. For example, **Boyle's law** states that at constant temperature, the volume of a gas is inversely proportional to its pressure. In other words, as gas pressure increases, the volume occupied by the gas decreases. This law is expressed by the equation

$$p_1 v_1 = p_2 v_2$$

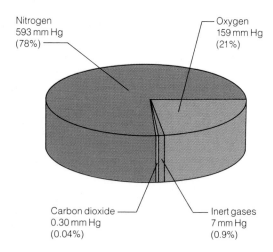

Figure 20-15 The composition of air. The various components are given here as percentages and as partial pressures (in mm Hg).

in which p_1 and v_1 are the initial pressure and volume of the gas and p_2 and v_2 are the pressure and volume after a change has occurred.

A second gas law, **Charles' law,** states that at constant pressure, the volume of a given mass of gas is directly proportional to its absolute temperature. In other words, if a gas is heated, it will expand; if it is cooled, it will contract, provided that the pressure is kept constant. This law is expressed by the equation

$$\frac{v_1}{t_1} = \frac{v_2}{t_2}$$

in which v_1 and t_1 are the initial volume and temperature of a gas and v_2 and t_2 are the volume and temperature after a change has occurred.

Boyle's and Charles' laws can be combined into a single equation called the **combined gas law** that shows the relationship between pressure, volume, and temperature:

$$\frac{p_1 v_1}{t_1} = \frac{p_2 v_2}{t_2}$$

Boyle's law combined with the principle that gases flow from regions of higher pressure to regions of lower pressure explains why air moves into and out of the lungs when a person breathes. During inspiration, movement of the diaphragm and rib cage increases the volume of the thoracic cavity. According to Boyle's law, as the volume of the thoracic cavity increases, air already in the lungs expands to fill the increased volume, with a resulting decrease in its pressure. This creates a pressure difference between the air inside and outside the lungs, and the air flows in. During expiration, the volume of the thoracic cavity decreases as the diaphragm and intercostal muscles relax. This causes the pressure of the gas in the lungs to become greater than the external atmospheric pressure, and the air flows out.

Solubility of Gases in Liquids

When a substance is present in a gas and also dissolved in a liquid with which the gas is in contact, molecules of the substance will pass back and forth between the two phases. (For example, imagine a container of water in contact with the air, and O_2 molecules passing back and forth between them.) If the concentration of the substance in the liquid is initially low, most movement will be from the gas into the liquid. As more and more molecules accumulate in the liquid (that is, the concentration of the substance increases), the rate at which they move back into the gas increases. In time, molecules will move from the liquid to the gas as fast as they are moving from the gas to the liquid, and the system will be in equilibrium. The precise concentration of the substance in the fluid once equilibrium is established depends on temperature, solubility of the substance in the fluid, and its pressure in the gas over the liquid.

Although concentration of most substances is expressed in moles per liter or grams per liter, the concentration of dissolved gases is often expressed as the gas pressure that would be required to produce that concentration in the liquid. Thus, if a container of water is placed in a chamber that contains O_2 at a pressure of 760 mm Hg, the equilibrium concentration of O_2 will be "760 mm Hg." If the gas is a mixture of components, the concentration of each component in the water can be expressed as the partial pressure of the component in the gas. For example, when air is in equilibrium with water, partial pressures of the gases in the air and in the water are 149 mm Hg for O_2, 0.3 mm Hg for CO_2, and 564 mm Hg for N_2 and other inert substances. These values are less than the values for dry air because some of the total pressure of the gas is due to water vapor. In air that is saturated with water and at 37°C and 760 mm Hg, water vapor has a partial pressure of 47 mm Hg.

It should be noted that components at the same partial pressure are not necessarily present in the same chemical concentration in the water because their solubilities in water may differ. The "true" concentration of a substance in water when it is at equilibrium with its gaseous phase can be determined by multiplying the partial pressure for the substance by a value called the **solubility coefficient** of the substance. Thus, the solubility coefficient for O_2 in water is 3×10^{-5} mL/mL H_2O/mm Hg partial pressure. This means that 0.00003 mL of O_2 will dissolve in each 1 mL of water for each 1 mm Hg of O_2 pressure. The solubility coefficient for CO_2 is 7×10^{-4} mL/mL H_2O/mm Hg partial pressure

hemoglobin: (Gr.) *haima*, blood + (L.) *globus*, globe

and for molecular nitrogen (N_2) 1.6×10^{-5} mL/mL H_2O/mm Hg partial pressure. Thus, CO_2 is about 20 times more soluble in water than O_2, and N_2 is about half as soluble in water as O_2 is.

Gas Exchange in Alveoli

After studying this section, you should be able to:

1. Describe the factors influencing oxygen and carbon dioxide exchange between alveolar air and blood.
2. Define respiratory quotient and explain why it depends on substances being catabolized in respiration.
3. Explain how O_2 and CO_2 are transported in blood.
4. Describe the structure of hemoglobin and explain what factors affect the ability of hemoglobin to carry O_2.
5. Define chloride shift.

Lung volume at the end of inspiration in someone at rest is typically about 3.5 L. Since the amount of air breathed in is typically about 0.5 L, each breath is only one-seventh of the total volume of gas in the lungs at the end of the inspiration. Because the amount of air inspired is only a small proportion of the amount already there, the concentration of oxygen in the lungs does not change dramatically with each breath. Fluctuations that do occur become less and less down the branching of the pulmonary tree, until at the level of the alveolus, there is no fluctuation at all. At that level oxygen diffuses into the alveolus at a steady rate governed by the rate of its diffusion into the blood.

Blood is delivered to the lungs by the pulmonary arteries, which branch within the lungs roughly following the branching pattern of the bronchi. By the time the vessels reach the alveolar walls, they have been reduced to capillaries. It takes less than 1 sec for a cell to pass through a capillary in an alveolar septum, during which time it collects oxygen and loses carbon dioxide. Because there is a nearly constant concentration of oxygen in the alveolar gas and a steady flow of blood in the capillaries, oxygenation of the blood occurs at a steady state. After passing through the capillaries, the blood enters an ascending series of venules and veins, finally leaving the lungs in four pulmonary veins that carry the blood to the left atrium (see Chapter 22).

Blood is then carried from the lungs to the heart where it is pumped into the systemic circulatory system. As blood passes through systemic capillaries, it loses oxygen and gains carbon dioxide produced by the respiring tissues. Leaving these tissues, the blood is carried back to the heart, from which it is pumped again to the lungs, completing one circuit of the circulatory system.

After blood leaves the lungs it travels to the left side of the heart, from which it is pumped into the systemic circulation. Blood that delivers oxygen to the lung tissues themselves is returned directly to the pulmonary veins, which dilutes the blood in those veins by about 2%. The bypassing of the alveoli by this oxygen-poor and carbon dioxide-enriched blood is referred to as a **physiologic shunt.**

Factors Affecting Gas Exchange

The amount of oxygen gained and carbon dioxide lost by the blood as it passes through alveolar capillaries depends on many factors: the composition of the alveolar gas, the amount of carbon dioxide and oxygen already present in the blood, the rate at which the blood flows through the capillaries, the number of red blood cells in the blood and the amount of hemoglobin in them, and the physical condition of the alveolar wall.

Figure 20-16 shows the partial pressures of different gases present in inspired and expired air and typical concentrations of these substances in the blood of a person who is breathing quietly. Several points can be drawn from this information: expired air still carries a considerable amount of O_2 (15% of expired air versus 21% of inspired air), and the percentage of CO_2 is higher in expired air (0.04% of inspired air versus 3.5% of expired air). When blood arrives in the lungs, it still has nearly half as much O_2 as it had when it left (40 mm Hg when entering the lungs versus 95 mm Hg when leaving). Conversely, when it leaves the lungs, it still has nearly 90% of the CO_2 it had when it entered (40 mm Hg versus 46 mm Hg). The relatively small percentage change in O_2 and the relatively large percentage change in CO_2 in inspired versus expired air reflects the large difference in concentration of these two gases in air. The actual amount of oxygen consumed and carbon dioxide released are usually about the same.

The ratio of the volume of CO_2 released to the volume of O_2 consumed in respiration is the **respiratory quotient (RQ),** and it depends primarily on the mix of substances catabolized in cellular respiration. When carbohydrate is used exclusively, the RQ is 1.00, meaning that the volume of CO_2 liberated is the same as the volume of O_2 consumed. When fat is metabolized, less CO_2 is released than O_2 consumed, producing an RQ of about 0.70. The actual volumes of these gases exchanged may change dramatically, for example, during strenuous exercise, but the RQ remains the same for a given mix of substances catabolized.

Oxygen Transport

Oxygen is carried by blood two ways. Normally about 3% is in solution, dissolved in plasma and in cytoplasm of cells and platelets. The remaining 97% is bound to **hemoglobin**

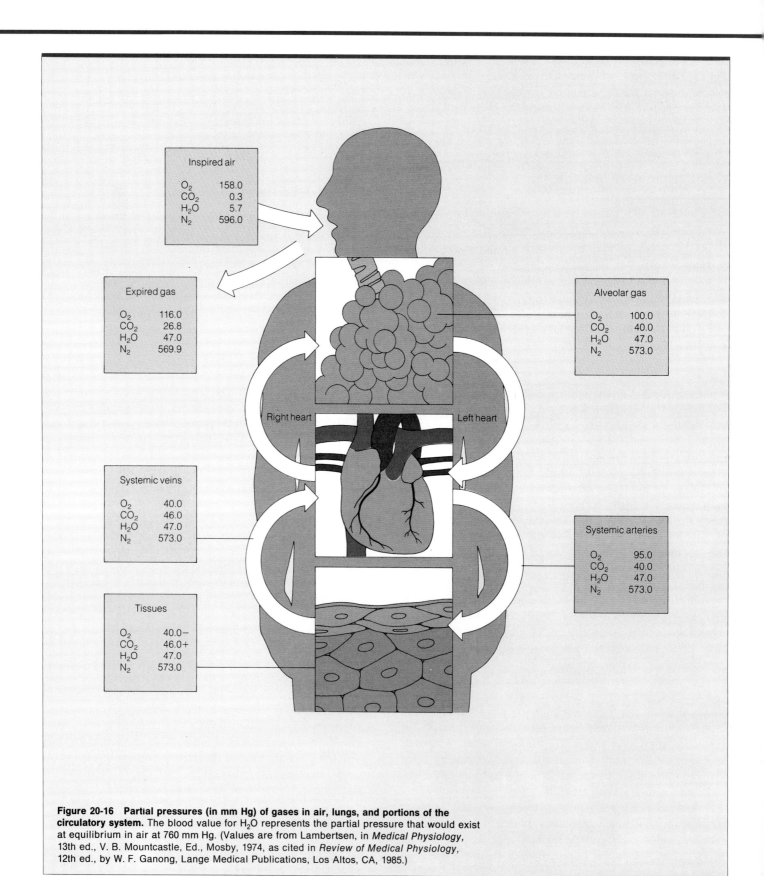

Figure 20-16 Partial pressures (in mm Hg) of gases in air, lungs, and portions of the circulatory system. The blood value for H_2O represents the partial pressure that would exist at equilibrium in air at 760 mm Hg. (Values are from Lambertsen, in *Medical Physiology*, 13th ed., V. B. Mountcastle, Ed., Mosby, 1974, as cited in *Review of Medical Physiology*, 12th ed., by W. F. Ganong, Lange Medical Publications, Los Altos, CA, 1985.)

(hē′mō·glō″bĭn), the principal protein in erythrocytes (see Chapter 21). Hemoglobin molecules are such effective carriers of oxygen that they increase the oxygen-carrying capacity of blood by nearly 300-fold, compared with the amount of oxygen that could be carried in solution by the same volume of plasma. It has been estimated that if oxygen were simply carried in solution, rather than bound to hemoglobin, the heart would have to pump nearly 1500 L of blood per minute to meet the needs of the body instead of its usual 5 L per minute.

Structure of Hemoglobin

A hemoglobin molecule consists of four protein subunits, each of which is associated with a **heme** (hēm) group (Figure 20-17). In adults with normal blood, 98% of the hemoglobin molecules consist of two molecules each of **α-globin** and **β-globin**, combined in a form called **hemoglobin A.** The remaining 2% consists of molecules that contain two molecules of α-globin and two of **δ-globin**. This type of hemoglobin is called **hemoglobin A_2.** The β and δ subunits of these hemoglobins are similar to one another, differing only in 2 of the 146 amino acids present in each. Hemoglobin of embryos, fetuses, and newborn babies contains slightly different globin molecules.

The heme unit associated with each globin molecule consists of an organic molecule called **protoporphyrin** and an iron atom. The iron atom, which is bound by the protoporphyrin unit, is the part of the hemoglobin molecule that binds oxygen. Since there are four heme groups in each hemoglobin molecule, each molecule can carry up to four oxygen molecules.

Oxygen Binding by Hemoglobin

When oxygen-poor blood passes through pulmonary capillaries, oxygen that has diffused through the alveolar wall and into the blood enters the erythrocyte and combines with hemoglobin to form **oxyhemoglobin.** With the binding of each oxygen molecule, the hemoglobin changes to a shape that binds the next oxygen molecule even more readily. That is, hemoglobin's *affinity* for oxygen increases with the absorption of each molecule of oxygen until all four binding sites are filled. Once all four sites contain an oxygen molecule, the oxyhemoglobin molecule is said to be **saturated.** This process is represented by the following reaction (where Hb stands for hemoglobin):

$$Hb + 4 O_2 \longrightarrow HbO_2 + 3 O_2 \longrightarrow$$
$$Hb(O_2)_2 + 2 O_2 \longrightarrow Hb(O_2)_3 + O_2 \longrightarrow Hb(O_2)_4$$

Figure 20-18 shows an oxygen dissociation curve for human blood at 37°C, pH 7.0, and 40 mm Hg of carbon dioxide. It is important to specify the temperature, pH, and carbon dioxide concentration because each affects the degree to which oxygen is absorbed by blood.

Any one hemoglobin molecule can only exist at 0, 25, 50, 75, or 100% saturation because it has exactly four binding sites, and the number that are filled depends on how much oxygen is available. In blood there are typically about 2.5×10^8 molecules of hemoglobin in each erythrocyte and 5×10^9 erythrocytes in each cubic centimeter

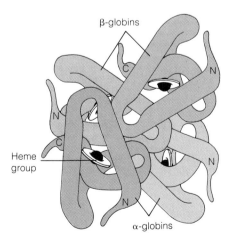

Figure 20-17 A model of a hemoglobin molecule. Note the presence of two α- and two β-globin chains, each of which is associated with a heme group. The iron atom in each heme group combines with an oxygen molecule, giving the hemoglobin molecule the capability of carrying four oxygen molecules.

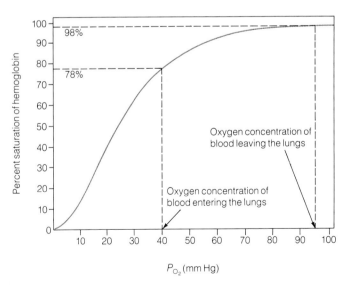

Figure 20-18 Oxygen dissociation curve of hemoglobin. The partial pressure of oxygen P_{O_2} in blood leaving the lungs is 95 mm Hg, corresponding to about 98% saturation. When blood reenters the lungs, the P_{O_2} is about 40 mm Hg, corresponding to about 78% saturation of the hemoglobin.

of blood. Oxygen concentration in the blood is typically about 95 mm Hg as it leaves the lungs and has decreased to about 40 mm Hg by the time it leaves the systemic capillaries.

When blood travels through capillaries in the alveolar septa, it is exposed to the relatively high concentration of oxygen (high P_{O_2}) diffusing in from the alveolus. As a result, most hemoglobin molecules become saturated, and the blood leaves the alveolus with a nearly full load of oxygen. When the blood reaches the systemic capillaries, however, it enters an environment made poor in oxygen by cellular respiration. Since the concentration of oxygen is lower in these tissues (low P_{O_2}), oxygen leaves the hemoglobin and diffuses into them.

Increased cellular activity results in an increased rate of cellular respiration, which means increased oxygen consumption. Consequently, tissues that are respiring most rapidly are surrounded by fluids that contain the least oxygen. Since the rate of diffusion depends partly on the difference in concentration of a substance in two areas, oxygen diffuses more rapidly from the blood into metabolically active tissues than it does into tissues that are relatively nonactive.

Factors Affecting Oxygen Release

Figure 20-19 shows how P_{CO_2}, temperature, and hydrogen ion concentration (pH) affect O_2 release from oxyhemoglobin. Increases in any one result in a reduction in the oxygen-carrying capacity of hemoglobin, regardless of the P_{O_2} of the surrounding fluid. The figure shows that O_2 will be released from hemoglobin most readily in precisely those tissues where it is being used most, that is, in tissues that are respiring most rapidly. In contrast, blood passing through the alveolar capillaries is in an environment where the P_{O_2} is high and where heat and CO_2 are dissipated. In that environment the affinity of hemoglobin for oxygen is high, and hemoglobin effectively absorbs a fresh supply of oxygen.

Carbon Dioxide and Hydrogen Ion Transport

Carbon dioxide is carried to the lungs as dissolved CO_2 or as bicarbonate ion (HCO_3^-) in plasma and in the cytoplasm of erythrocytes, or it is bound directly to proteins in the plasma and erythrocytes. Hydrogen ion from carbonic acid is carried primarily in the erythrocytes, where it is bound to hemoglobin. Most carbonic acid is produced in erythrocytes, and about 70% diffuses into the plasma. Diffusion of the negatively charged HCO_3^- from the erythrocyte is balanced by diffusion of chloride ions (Cl^-) from the plasma into the cell. As a result, erythrocytes leave the systemic capillaries carrying about 1.5% more chloride than they had when they entered. This exchange of Cl^- and HCO_3^- is referred to as a **chloride shift.**

Carbon dioxide combines with certain amino acids in hemoglobin to form a loosely bound complex called **carbaminohemoglobin** (kăr·băm″ĭ·nō·hē″mō·glō′bĭn). Formation of this compound occurs most readily with deoxygenated hemoglobin, making the blood in the venous system especially effective in transporting carbon dioxide.

Figure 20-19 Effect of (a) P_{CO_2}, (b) temperature, and (c) pH on oxygen binding of hemoglobin.

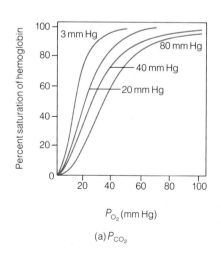

(a) P_{CO_2}

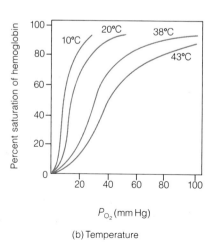

(b) Temperature

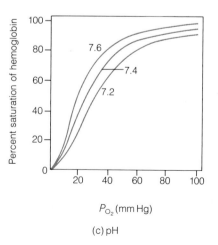

(c) pH

> **CLINICAL NOTE**
>
> **DIVING ILLNESS**
>
> There are several medical conditions associated with exposure to increased pressures in the environment. These conditions were previously limited to deep sea divers and tunnel and caisson workers, but now they are seen more frequently in scuba enthusiasts. In scuba diving the pressure increases rapidly. In seawater, every 33 ft of descent adds 1 atmosphere (14.7 psi) of pressure on the body.
>
> **Barotrauma** is the effect of the increased pressure on rigid, closed cavities of the body, such as the sinuses or the middle ear. If the diver cannot clear his or her airway because of a cold or stuffed nose, bleeding into the cavity or even eardrum rupture can result. Severe vertigo (dizziness) and dysequilibrium may result if cold water enters the middle ear.
>
> The effects of increased pressure on the gases dissolved in the blood and body fluids can also cause problems. Oxygen can produce toxic effects to the lungs and convulsions with prolonged exposure at increased partial pressure. Increased partial pressure of dissolved nitrogen can result in **nitrogen narcosis,** which is similar to alcoholic intoxication and causes hallucinations and disorientation. It may be incapacitating at a depth of 300 ft.
>
> Breath-holding divers (those who do not use scuba) who hyperventilate prior to a dive will blow off carbon dioxide and may lose consciousness from hypoxia before the CO_2 increases sufficiently to cause them to surface and breathe. This is called **shallow water blackout.**
>
> Rapid ascent from depth may also be hazardous. It may cause overinflation of the lungs, forcing gas bubbles into the capillaries and arteries of the brain and causing abrupt loss of consciousness. This is called **gas embolism.** The classic result of a too rapid ascent is **decompression sickness** (or the "**bends**"), which results from gas bubbles being released into the tissues and blood vessels causing pain and occasional neurologic symptoms. Both of these conditions require prompt and adequate decompression treatment in a special decompression chamber under the care of a diving expert.

Regulation of Respiration

After studying this section, you should be able to:

1. Describe the location and function of the respiratory center in regulating breathing rate.
2. Compare the action of the apneustic and pneumotaxic centers in regulating the respiratory center.
3. Define the Hering-Breuer reflex.
4. Describe how carotid and aortic bodies influence breathing rate in response to changes in blood chemistry.
5. Describe how chemoreceptors in the medulla influence breathing rate.

Vigorous exercise can have a dramatic effect on the body. After only a few minutes, one begins to perspire, the heart beats more rapidly and stronger, and breathing becomes deeper and more rapid. In addition, extensive hormonal and metabolic changes provide nutrients to meet the increased demands of the exercising muscles. These responses are coordinated to make the delivery of nutrients and oxygen and the removal of heat and waste chemicals from the active muscles as efficient and effective as possible. Fundamental to these changes is an increase in breathing rate that delivers the greater amounts of O_2 needed by the respiring tissues and removes the increased amounts of CO_2 produced by them.

The regulatory mechanisms that control breathing rate are both voluntary and involuntary. Unlike other systems, one can control the respiratory system and use it to perform other tasks in addition to breathing. Speaking, singing, whistling, and other such activities are ways we can control the respiratory system.

Voluntary activity is controlled by the cerebral cortex, from which neurons carry signals directly to the thoracic muscles, diaphragm, and other muscles used to control breathing. Using these nerves it is possible to suck air in and hold one's breath just before diving to the bottom of a swimming pool, or to blow into a trumpet with the force necessary to make music. During these kinds of activities, the involuntary control systems that regulate respiration are essentially overridden.

There is a limit, however, to the extent to which voluntary control can override involuntary control. You can hold your breath only so long, and then it becomes impossible to hold it any longer as you forcefully expel the gas in your lungs and gulp a fresh breath. As time passes without breathing, signals from the suppressed involuntary system become stronger and stronger, until they overwhelm the voluntary signals that have been operating.

apneusis: (Gr.) *a*, not + *pnoe*, breathing

CLINICAL NOTE

SUDDEN INFANT DEATH SYNDROME

Sudden infant death syndrome (SIDS), more commonly known as **crib death,** is a completely unexpected and unexplained death of an apparently well infant. It is the most common cause of death in infants from 2 weeks to 1 year of age and occurs most commonly in premature infants. In the United States over 10,000 infants succumb to this syndrome each year. The cause is usually unknown.

Almost all of the infants simply stop breathing and die in their sleep, which has suggested to some researchers that abnormal sleep patterns are responsible. Alarm devices have been invented for use on high risk infants that summon the parents if depression of respiration or heart rate occurs. The problem of identifying those infants who are at risk has not been solved, and the anxiety induced in the parents of a child selected for monitoring is tremendous.

In years past, SIDS was thought to be caused by an oversized thymus gland and increased lymphoid tissue found in some of these infants. Many infants found to have a large thymus on chest X ray were subjected to irradiation to the thymus area. This treatment resulted in an increased incidence of cancer of the thyroid gland in later years, and the treatment was discontinued.

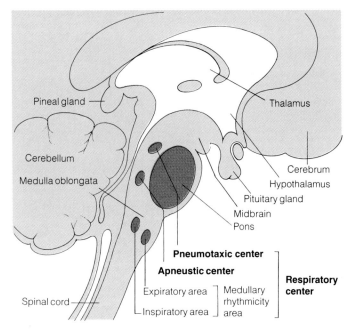

Figure 20-20 The respiratory center, located in the medulla oblongata, consists of regions of neurons that control inspiration and expiration.

Neural Control of Breathing

Involuntary activities of the respiratory tract are largely controlled by regions in the brain stem, which act in response to neural signals sent by peripheral receptors or to chemical changes occurring in the extracellular fluid that bathes them. Nerve impulses that stimulate respiratory muscles originate in diffuse regions of the medulla oblongata collectively called the **respiratory center** (Figure 20-20). This center consists of two regions situated bilaterally in the medulla, one comprising the **dorsal respiratory group (DRG)** and the other the **ventral respiratory group (VRG).** The DRG primarily contains cells responsible for stimulating inspiration, sending signals directly to the muscles that cause inspiration. The DRG also sends neural signals to the VRG, over which it exercises control. The VRG contains neurons involved both in controlling inspiration and expiration; it sends neural signals to various muscles involved in both aspects of respiration. During heavy breathing, impulses emerging from the center alternate between those traveling to inspiratory muscles (mainly the diaphragm and external intercostal muscles) and those traveling to expiratory muscles (mainly the internal intercostal muscles). Consequently, the muscles are induced to contract alternately, creating the sequence of inspirations and expirations that make up respiration.

The respiratory center is influenced by impulses from several other regions that affect rate, depth of breath, and coordination of inspiration and expiration. Two such regions, the **apneustic** (ăp·nū′stĭk) **center** and the **pneumotaxic** (nū·mō·tăk′sĭk) **center,** are also in the brain stem. The apneustic center, located in the lower portion of the pons, stimulates inspiration and, if unchecked, can induce **apneusis** (ap·nū′sĭs), a sustained contraction of inspiratory muscles.

The pneumotaxic center controls the apneustic center, preventing apneusis by inhibiting apneustic signals at the conclusion of each inspiration. The pneumotaxic center also regulates the output of the medullary centers, ensuring that breathing is accomplished as a series of smoothly alternating periods of inspiration and expiration. Under normal resting conditions, inspiration and expiration cycles occur about 10 to 18 times per minute. Inspirations are slightly longer in duration than expirations. At a rate of 15 breaths per minute, each inspiration lasts about 3 sec and each expiration about 1 sec.

Peripheral Receptors

The respiratory center, especially the DRG, receives signals from peripheral sensory cells located in several regions of the body. Two major categories of peripheral receptors

apnea: (Gr.) *a*, not + *pnoe*, breathing

> **CLINICAL NOTE**
>
> ## HYPERVENTILATION AND HYPOVENTILATION
>
> **Hyperventilation** (or overbreathing) can be caused by a central nervous system disease, chemical imbalances within the blood, or anxiety. Strokes associated with coma can be accompanied by deep regular breathing. **Cheyne-Stokes respiration** (periodic respiration) is a pattern of hyperventilation followed by periods of **apnea** (suspension of respiration) and is usually associated with a reduction of cerebral circulation. The slow, deep respiratory pattern seen in diabetics is called **Kussmaul breathing**. **Biot's respiration** is an intense hyperventilation sometimes seen in patients with brain hemorrhages.
>
> Extreme anxiety or fear can lead to rapid, shallow breathing, and hyperventilation can occur in women left unsupervised during labor and advised to "pant." Hyperventilation causes the CO_2 concentration in the blood to fall with resulting increase in blood pH. Tingling of the fingers, chest tightness, or shortness of breath are all common symptoms. Relief is obtained by holding one's breath or by breathing into a paper sack, which serve to increase the CO_2 concentration in the blood.
>
> In contrast, **hypoventilation** is a reduced rate and depth of breathing. Like hyperventilation, it occurs in persons with normal lungs but with disease elsewhere in the body. An interesting syndrome called **pickwickian syndrome** affected the character Joe in the Dicken's novel *The Pickwick Papers*. This syndrome involves hypoventilation associated with extreme obesity. Affected patients often weigh over 300 pounds, and they tend to fall asleep very easily. Their shallow breathing is especially evident during sleep and is thought to be due to excessive weight of the chest wall. Weight loss helps to relieve the problem.

located in the respiratory system are **pulmonary stretch receptors** and **pulmonary irritant receptors**. Stretch receptors are stimulated during deep breathing. Impulses from these receptors are sent through the vagus nerve (the tenth cranial nerve) to the respiratory center where they inhibit the inspiratory neurons. When inhibited by such signals, the inspiratory neurons cease transmission to the inspiratory muscles, preventing overinflation of the lungs. Except in newborn infants, this response occurs only following deep inspiration and only as long as the vagus nerves are intact. This reflex, called the **Hering-Breuer** (hĕr'ĭng broy'ĕr) **reflex**, is initiated by stretch receptors in smooth muscle of the trachea, bronchi, and bronchioles.

Irritant receptors lie between epithelial cells that line passageways in the lungs. These receptors are stimulated by lung distension as well as various kinds of irritants. They may be important factors in asthma attacks, responding to histamine by stimulating rapid breathing and vasoconstriction.

A third category of pulmonary receptors are the **juxtacapillary receptors**, so named because they are located in the walls of pulmonary capillaries. These receptors are sensitive to increased amounts of extracellular fluid associated with edema and congestion.

Sensory receptors are also located in muscles of the chest wall where they are sensitive to muscle strain when breathing is made difficult by narrowing of the passageways or constriction of the chest. Swallowing temporarily inhibits breathing, probably due to sensory elements in the pharynx that transmit inhibitory impulses to the respiratory center. Coughing, sneezing, and yawning are other reflex actions that temporarily interrupt respiration, presumably by inhibiting the respiratory center.

Chemical Regulation of Breathing

The respiratory centers must be notified to speed up or slow down the rate of ventilation when respiratory requirements of the body change. Notification is provided by two categories of receptors sensitive to chemical changes. These are **peripheral chemoreceptors** and **medullary chemoreceptors**. The primary peripheral chemoreceptors include **carotid** (kă·rŏt'ĭd) **chemoreceptors** and **aortic** (ā·ōr'tĭk) **receptors**. Carotid chemoreceptors are located in **carotid bodies**, small masses of tissue located on the surface of each of the common carotid arteries where they divide to form the internal and external carotid arteries. Aortic chemoreceptors are located in **aortic bodies**, small masses of tissue similar in appearance to carotid bodies located on the aortic arch and between it and the pulmonary artery (see Chapter 22). Medullary chemoreceptors are located bilaterally on the ventrolateral surface of the medulla oblongata, just below the junction of the medulla and the pons.

Peripheral Chemoreceptors

Carotid and aortic bodies are flattened, somewhat ovoid in shape, and about 3 to 6 mm long. The two types of structures contain free unmyelinated nerve endings of glossopharyngeal and vagal nerve fibers, respectively, and small groups of cells separated from one another by fine passageways called **sinusoids** (sī'nŭs·oyds). Blood flows through the sinusoids where it is in intimate contact with sensory cells that detect changes in blood chemistry and stimulate the nerve endings in response to those changes.

Carotid bodies are particularly sensitive to changes in the O_2 concentration in blood, and send neural impulses to the medulla with increasing frequency when O_2 concentration begins to fall. These signals stimulate the respiratory centers to increase breathing rate. The response of the carotid bodies to decreases in O_2 concentration is graded, with the frequency of impulses increasing rapidly as O_2 concentration continues to fall. Near maximum impulse frequency is reached at O_2 levels about half the normal concentration. Although carotid bodies begin to transmit impulses in response to only slight decreases in oxygen concentration, the effect on breathing rate itself is not so rapid. This is because decreased blood O_2 inhibits cellular respiration, and consequently, CO_2 production by respiring tissue cells is also inhibited. The decline in blood CO_2 that results has a greater depressing effect on the medullary chemoreceptors than decreased O_2 has on the stimulatory effect of the carotid bodies. If CO_2 is maintained at a normal level while O_2 levels decline, breathing rate increases almost immediately. Carotid bodies are also slightly sensitive to pH and CO_2 changes in the blood, increasing their impulse rate as pH declines and CO_2 concentration rises. However, they are not nearly as sensitive to these substances as they are to changes in O_2.

Aortic bodies also respond to decreased blood O_2 levels by increasing impulse transmission, but their effect is far less important than that of carotid bodies. If aortic bodies are removed and the carotid bodies are intact, decreased O_2 levels have no perceptible effect on breathing rate. Like carotid bodies, they are slightly sensitive to CO_2 levels but appear to be insensitive to pH. Although aortic bodies can influence breathing rate, they appear to be primarily involved with regulating arterial blood pressure.

Medullary Chemoreceptors

If one breathes an atmosphere only slightly enriched in CO_2, there is a rapid increase in respiratory rate (Figure 20-21). Medullary chemoreceptors are responsible for much of this reaction, sending impulses to the nearby respiratory centers following an increase in CO_2 concentration. The effect of CO_2 on medullary chemoreceptors is indirect, however, and results not from the CO_2 itself, but from increased H^+ concentration produced by the reaction of CO_2 and H_2O. CO_2 and H_2O, you will recall, combine to form H_2CO_3, which ionizes to H^+ and HCO_3^-. CO_2-enriched blood does not bathe the medullary chemoreceptors directly; instead, CO_2 readily diffuses into the interstitial fluids of the medulla, where it reacts with H_2O and lowers the pH. (H^+ and HCO_3^- do not diffuse readily into the medulla.)

Because of the time it takes for CO_2 to diffuse into the medulla, the neural response to increased CO_2 of medullary chemoreceptors is relatively slow compared with that of carotid bodies to changes in O_2 or CO_2. Carotid bodies are in direct contact with relatively large volumes of blood so the effect of chemical changes are virtually immediate.

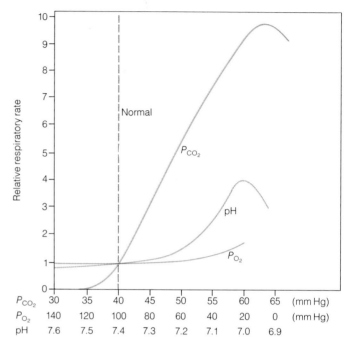

Figure 20-21 Respiratory rate as a function of P_{CO_2}, pH, and P_{O_2} in arterial blood. (Adapted from A. C. Guyton, *Basic Human Physiology*, W. B. Saunders Company, Philadelphia, 1971.)

Interaction Between Medullary and Peripheral Chemoreceptors

Chemoreceptors can have an effect on one another since the activity of one can cause changes in the conditions to which the other is sensitive. For example, when one goes rapidly from low to high altitudes, such as backpackers might do as they traverse a mountain pass, resting respiratory rate increases only slightly in spite of a relatively large decrease in atmospheric O_2 (about 84 mm Hg at 14,000 ft compared with about 100 mm Hg at sea level). Although decreased O_2 stimulates the carotid bodies, the increased respiration that results removes CO_2 more effectively, thereby reducing the stimulus to the medullary chemoreceptors. Medullary receptors are sensitive to very slight changes in CO_2 concentration, so their decreased stimulation of breathing offsets the increase that would have been caused by lowered O_2. Within a few hours, however, changes develop in the chemical composition of the fluid surrounding the medullary chemoreceptors that decrease their activity. Consequently, there is a gradual increase in effectiveness of respiration that causes CO_2 concentration to stabilize. When this has happened one is said to have "acclimated" to the higher altitude.

hypoxemia: (Gr.) *hypo*, under + *oxys*, acid + *haima*, blood

> **CLINICAL NOTE**
>
> **ALTITUDE SICKNESS**
>
> The recent interest in skiing, hiking, and mountain climbing above 8000 ft has resulted in an increased incidence of **altitude sickness.** Altitude sickness is a result of **hypoxemia,** or insufficient oxygenation of the blood. Three types are recognized based on severity. **Acute mountain sickness (AMS)** is characterized by morning headache, insomnia, limb swelling, nausea, and weakness. It is usually self-limited and relieved by resting at the same or a lower altitude. A more serious condition is **high-altitude pulmonary edema (HAPE),** an effusion of fluid into the pulmonary alveoli. Prevention of HAPE depends on a slow ascent, increased fluid intake, and decrease in dietary sodium. An even more serious complication is **high-altitude cerebral edema (HACE),** which results in swelling of the brain and may lead to paralysis and unconsciousness. Treatment for HACE is an immediate and rapid descent to lower altitudes. Bleeding into the retina of the eye can also occur, resulting in loss of vision. The medication **acetazolamide** (ăs″ĕt·ă·zŏl′ă·mīd) has been used to guard against altitude sickness, but a slow ascent of not more than 1200 ft per day with a rest day approximately every 3000 ft is a useful guide to prevention.

In this chapter we studied the anatomy and physiology of the respiratory system. We began with an examination of the gross anatomy and histology of the lungs and proceeded to a study of the mechanics of gas exchange during breathing. Next we learned how gases are carried in the blood and the importance of hemoglobin in this task. Our study of the respiratory system ended with an examination of the mechanisms used to control the respiratory rate. In the next unit we will examine the circulatory system, and it will become evident that the blood transports many materials in addition to oxygen and carbon dioxide.

STUDY OUTLINE

Organization of the Respiratory Tract (pp. 505–516)
  The respiratory tract consists of an upper portion, which includes the nose, nasal cavity, nasopharynx, and larynx, and a lower portion, which includes the trachea, bronchi, bronchioles, and alveoli.

  Upper Respiratory Tract The *upper respiratory tract* begins at the *nostrils*, or *external nares*. Each nostril opens into a *vestibule* that leads into the *nasal cavity*. The nasal cavity terminates at the *choanae*, or *internal nares*.

  The nasal cavity is formed by two bones and the hard and soft palates. It is divided laterally into two *nasal fossae* by the *nasal septum. Nasal conchae* projecting from the septum divide each fossa into *meatuses*, passageways through which air flows. The cavity wall is covered by mucus membrane that cleanses the air as it passes through the cavity. *Paranasal sinuses* are located about the cavity and open into it.

  The *pharynx* connects the oral and nasal cavities with the esophagus and larynx. It consists of the *nasopharynx*, superior to the soft palate, the *oropharynx*, posterior to the oral cavity, and the *laryngopharynx*, lying posterior to the larynx. The *pharyngeal tonsils* lie in the posterior portion of the nasopharynx.

  The *larynx* controls passage of air into the lower respiratory tract. Its opening is controlled by the *epiglottis*. The larynx contains *vocal folds* (true vocal cords) that produce sound as air is forced past them.

  Lower Respiratory Tract The *lower respiratory tract* begins with the *trachea*, a tube that descends from the larynx and branches to form two *bronchi*.

  Gross Morphology of the Lungs The lungs lie in the thoracic cavity, on either side of the *mediastinum* and enclosed within the *pleurae*. The *root* of the lung includes the *primary bronchus*, pulmonary artery and vein, bronchial arteries, nerves, and lymphatic vessels, all of which pass through the *pulmonary hilum*. The right lung has three lobes and the left lung has two.

  Pulmonary Tree The *pulmonary tree* consists of a *conducting division*, through which gas is conducted, and a *respiratory division*, in which gas is exchanged. The conducting division consists of *bronchi* and successively branching *bronchioles*. The ultimate branches in the conducting division are *terminal bronchioles*. The respiratory division consists of *respiratory bronchioles, alveolar ducts*, and *alveolar sacs*.

  Structure of the Alveolar Wall The alveoli are separated from one another by *interalveolar septa*, which consist of epithelia, capillaries, and septal cells. *Septal cells* secrete *surfactant*, which enhances diffusion of gases into and out of the septum.

Breathing Mechanics (pp. 516–519)
  Inspiration and Expiration Breathing consists of *inspiration* and *expiration*. Inspiration normally results from contraction of the diaphragm and external intercostal muscles, but the sternocleidomastoid and scalene muscles are also involved during exercise. Expiration is passive at rest because of elastic properties of the lungs. During exercise, the internal intercostal and abdominal wall muscles help decrease thoracic volume. Increased thoracic volume reduces interpleural pressure to less than lung pressure. The lungs expand, and air flows in. Reduction of thoracic volume increases interpleural pressure to greater than lung pressure. This reduces lung volume and causes gas to flow out.

  Respiratory Volumes The amount of gas exchanged with each normal breath is the *tidal volume*. Forced breathing

bronchitis: (Gr.) *bronchos*, windpipe + *itis*, inflammation

THE CLINIC

DISEASES OF THE RESPIRATORY SYSTEM

Because the respiratory system is vital to the normal functioning of the entire organism, any serious defect in its proper functioning is often fatal. There are a great many diseases that affect or involve this system.

Hereditary and Congenital Diseases
Some of the congenital respiratory diseases are of interest because they demonstrate the results of failure on a microcellular level. **Respiratory distress syndrome (RDS)** is a disease of newborns causing respiratory distress and **atelectasis**, which is a failure of the air cells to expand (see Clinical Note: Atelectasis). This syndrome is due to a failure of the premature lung to manufacture pulmonary surfactant (mainly lecithin), which stabilizes the alveoli. The air sacs collapse on expiration, which leads to atelectasis. RDS is the leading cause of death in premature infants (25,000 deaths per year in the United States). It is now treated with a technique called **continuous positive airway pressure** (CPAP) with good results.

Cystic fibrosis is an inherited disorder of the exocrine glands which is diagnosed by excessive amounts of sodium and chloride in the sweat. It is manifested by the secretion of very thick and sticky mucus and an unusual secretion of sweat and saliva. In the lungs, thick and sticky mucus blocks the smaller respiratory channels, making the movement of air into and out of the alveoli difficult. Air becomes trapped behind the mucus obstruction, leading either to overexpansion of the alveoli or collapse (atelectasis). Respiratory infections often lead to progressive lung damage. In the gastrointestinal system, thick mucus also prevents pancreatic secretions from reaching the small intestine. This leads to poor digestion and absorption of nutrients from the intestine. The mucus obstruction also affects the sweat glands, resulting in heat sensitivity. Recent discovery of the gene locus for this disease may allow for detection of the disease in families. Treatment with antibiotics and respiratory therapy has allowed many children with this disease to live into adulthood.

Infections of the Respiratory System
Respiratory infections are some of the most common diseases affecting humans. Only the skin is in more direct contact with our environment. With each breath, we inhale a large volume of air that is often contaminated with a wide variety of infectious organisms. **Upper respiratory infections**, including the common cold and the flu, affect more people worldwide than any other illness (see Clinical Note: Upper Respiratory Infection).

Acute tonsillitis and **pharyngitis** represent another group of common infections of the upper respiratory system. Both are usually viral in nature and self-limited. Infection may also be bacterial, however, usually involving the streptococcus bacteria. Sometimes it is a manifestation of **infec-**

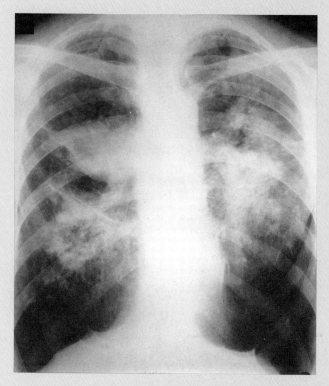

Figure 1 Silicosis. This lung disease is caused by inhalation of silica dust, which stimulates lung tissue fibrosis (the white patches on the X ray).

tious mononucleosis caused by the **Epstein-Barr** virus. Streptococcal tonsillitis and pharyngitis have been implicated as the cause of the autoimmune disease **rheumatic fever** and can also spread beyond the tonsils to produce a peritonsillar abscess called **Quinsy** (kwĭn′zē), a severely painful infection that can cause an obstruction of the pharynx due to the massively swollen tonsils. Streptococcal infections are treated with the antibiotic penicillin.

Infections may invade the lower respiratory system causing **laryngitis** if the infection is localized to the larynx, **tracheitis** if the trachea is involved, and **bronchitis** when the smaller bronchi are affected. **Acute epiglottitis** is a particularly severe, rapidly progressive infection of the epiglottis that may be quickly fatal due to airway obstruction by the inflamed epiglottis. When an infection reaches the air sacs and terminal bronchioles, it is known as **pneumonia** (see Clinical Note: Pneumonia).

Allergic Reactions and Hypersensitivity
The respiratory system is particularly prone to hypersensitivity reactions, probably due to its close contact with the

external environment. One of the most common allergic respiratory diseases is **bronchial asthma.** Bronchial asthma is characterized by repeated attacks of reversible spasms of the bronchial smooth muscles causing airway obstruction due to narrowing of the large and small bronchi. It is accompanied by edema and inflammation. Symptoms of asthma include expiratory wheezing and **dyspnea** (shortness of breath). Asthma can be subdivided into two types: **extrinsic asthma,** which has attacks precipitated by allergenic exposure to pollens, dust, or molds, and **intrinsic asthma,** which is triggered by nonallergenic factors, such as infections, irritants, or emotions and stress. Asthmatic attacks are treated by removing the precipitating cause and by giving the patient bronchodilator drugs, such as adrenaline and theophylline.

Anaphylaxis (ăn″ă·fĭ·lăk′sĭs) is an acute, life-threatening allergic reaction in highly allergic people that is manifested by hives, laryngeal edema, and vascular collapse. It may be rapidly fatal if not treated promptly with adrenaline, anti-histamines, and life-support measures.

There are many other allergic pulmonary diseases that have pneumonia-like symptoms. For example, **farmer's lung** is the accumulation of fluid in the alveoli due to inhalation of dusts or fungi. Occupational exposure to specific toxic dusts, such as coal (**pneumoconiosis** or **black lung**), asbestos fibers (**asbestosis**), silicon (**silicosis**), and beryllium (**berylliosis**) can result in chronic lung disease (see Figure 1).

Tumors of the Lungs

The lung may be the site of both primary and malignant tumors and is a common site of metastatic cancers from other organs. **Lung cancer** is the most common fatal cancer in the United States. There is a strong association between bronchogenic carcinoma and cigarette smoking (see Clinical Note: Lung Cancer). Benign tumors of the lung may be asymptomatic if situated in the periphery of the lung or may cause bronchial obstruction with resultant atelectasis and infection if higher up in the bronchi (see Figure 2).

Aging

Aging does not usually cause specific changes to the lung itself. However, the vital capacity of a person decreases with age. This is probably due to loss of muscle strength, particularly in the intercostal muscles. The whole chest cage becomes less elastic as the cartilage of the chest becomes stiffer and less flexible. If the older person is obese, the work of breathing is that much greater for the weakened muscles and stiffened joints. Arthritis of the dorsal spine and rib articulations may also reduce the vital capacity and ability to clear the lungs of secretions, making the lung more susceptible to pneumonia.

Pulmonary emphysema is a chronic disease of the pulmonary alveoli and alveolar ducts with destruction of their walls resulting in decreased surface area for gas exchange, decreased elastic recoil of the lung, and if severe, pulmonary hypertension. With the elastic forces of the lungs reduced, the muscular forces that expand the thoracic wall are more effective and pull the muscles of the chest into a barrel shape, which is a characteristic of advanced emphysema. The breakdown of an alveolar wall is thought to be due to proteolytic enzyme action associated with the inflammation of the lungs. Anything that can induce a chronic state of inflammation can increase the likelihood of emphysema. This includes smoking and the chronic inhalation of dust and soot.

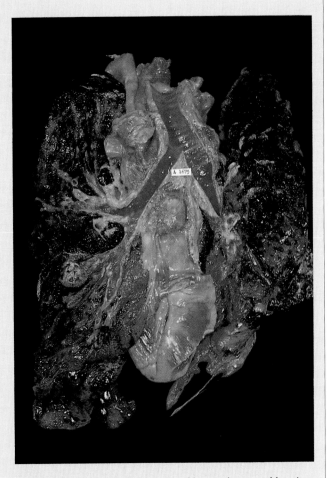

Figure 2 Lung cancer. Photograph of human lungs and heart showing effects of cigarette smoking.

allows more gas to be inhaled or exhaled, representing *inspiratory* and *expiratory reserve volumes*, respectively. Inspiratory reserve volume plus tidal volume equals *inspiratory capacity*. Tidal volume plus inspiratory and expiratory reserve volumes equals *vital capacity*. The volume of gas that remains after maximum exhalation is the *residual volume*. *Total lung volume* is equal to residual volume plus vital capacity.

Those parts of the respiratory system where gas exchange does not occur constitute *anatomical dead space*. Anatomical dead space plus space where gas exchange capability has been lost is *physiological dead space*.

Timed vital capacity is the maximum volume of air that can be exhaled in 1 sec. *Maximal voluntary ventilation* is the amount that can be inhaled and exhaled in 1 min. *Relaxation pressure* is the difference between the pressure in the lungs and the surrounding air. *Compliance* is a measure of the ability of the lungs to stretch in response to internal pressure. *Specific compliance* is the ratio of compliance to total lung volume.

Properties of Gases (pp. 519–521)

Gas Pressure *Gas pressure* is the force a gas exerts on a given area of the wall of a container. It is usually expressed as psi, g/cc^2, or mm Hg. *Partial pressure* is the portion of total pressure exerted by each of the gases in a mixture. Air is a mixture of O_2 (20.98%), CO_2 (0.04%), and N_2 and other inert gases (79%).

One *ideal gas law* is *Boyle's law*, which states that the product of volume and pressure of a gas are a constant, that is, increase one and the other decreases. Another is *Charles' law*, which states that the ratio of volume to temperature is constant. The two laws together illustrate the relationship of pressure, volume, and temperature in the *combined gas law*.

Solubility of Gases in Liquids The number of molecules of a substance in a liquid depends on temperature, solubility in the liquid, and the partial pressure of the substance in the gas over the liquid. When the gas is a mixture, each component will dissolve in the liquid in proportion to its solubility and partial pressure.

Gas Exchange in Alveoli (pp. 521–524)

Blood that delivers oxygen to lung tissues returns directly to pulmonary veins where it mixes with blood coming from the alveolar capillaries; this is called a *physiologic shunt*. Consequently, this blood does not have its oxygen supply replenished or its carbon dioxide removed before returning to the heart.

Factors Affecting Gas Exchange The amount of each gas exchanged depends on the partial pressures of O_2 and CO_2 in the alveolar gas, the amount of each already present in the blood, the rate of blood flow through the capillaries, the number of red blood cells, the amount of hemoglobin in the red blood cells, and the condition of the alveolar wall.

The ratio of the volume of CO_2 released to the volume of O_2 consumed is the *respiratory quotient* (*RQ*), which depends on the mix of fats and carbohydrates being respired. Pure carbohydrate catabolism produces an RQ of 1.00. Pure fat catabolism produces an RQ of about 0.70.

Oxygen Transport Most O_2 (about 97%) is carried bound to *hemoglobin*, which increases the O_2-carrying capacity 300-fold over plasma alone. The remaining 3% is dissolved in plasma and the cytoplasm of blood cells and platelets.

Hemoglobin consists of four globin subunits, each of which is associated with a *heme* group. Most hemoglobin (98% in adults) consists of two *α-globin* and two *β-globin* chains, designated *hemoglobin A*. In the remaining 2% the β-globin chains are replaced by *δ-globin* chains. The heme unit associated with each chain consists of *protoporphyrin* and an iron atom.

Hemoglobin can carry up to four O_2 molecules. The affinity for oxygen increases with the addition of each O_2 molecule until all four binding sites are filled. Hemoglobin releases its O_2 most readily to tissue poor in O_2, as in respiring tissues, and binds it most readily in O_2-rich environments, as in alveolar capillaries. Binding of oxygen by hemoglobin is affected by heat, CO_2 concentration, and pH.

Carbon Dioxide and Hydrogen Ion Transport CO_2 is carried in solution or as HCO_3^- in plasma and cytoplasm. Some binds to deoxygenated hemoglobin and is carried as *carbaminohemoglobin*. Excess H^+ is carried bound to hemoglobin. HCO_3^- that diffuses from erythrocytes is replaced by Cl^-, so the chloride concentration of erythrocytes increases in the systemic capillaries (*chloride shift*).

Regulation of Respiration (pp. 525–529)

Regulatory mechanisms in respiration are neurological, based on organs sensitive to carbon dioxide and oxygen concentration in the blood.

Neural Control of Breathing The *respiratory center* consists of the *dorsal respiratory group* (*DRG*) and the *ventral respiratory group* (*VRG*). The DRG contains inspiratory neurons, and the VRG contains both inspiratory and expiratory neurons that send signals alternately to inspiratory and expiratory muscles. The respiration center is influenced by impulses from other regions, including the *apneustic* and *pneumotaxic centers* in the brain stem and higher centers in the brain.

Peripheral Receptors Lung overinflation is prevented by stretch receptors that initiate a *Hering-Breuer reflex* that inhibits inspiratory neurons in the respiratory center. *Pulmonary irritant receptors* are stimulated by lung distension and irritants.

Chemical Regulation of Breathing Chemoreceptors located in the periphery include *carotid bodies* and *aortic bodies*, which respond to O_2 and CO_2 levels in the blood. Carotid bodies also respond to blood pH. Medullary chemoreceptors respond to changes in pH. Decreased pH stimulates respiration in both kinds of chemoreceptors; carotid bodies also stimulate respiration when O_2 levels decline.

SELF-TEST OF CHAPTER OBJECTIVES

True-False Questions
1. Chambers that drain into the nasal cavity are called nares.
2. The bronchi are held open permanently by circular bones in their walls.
3. The largest element in the larynx is the thyroid cartilage.
4. The surface area of the nasal cavity is increased by the nasal conchae.
5. The portion of the gas in the lungs that remains after maximum effort at expiration is called the inspiratory reserve volume.
6. The maximum volume of the lungs that can be inspired after maximum expiration is the vital capacity.
7. The right lung is larger than the left.
8. The right lung has two lobes and the left has three.
9. The pulmonary pleura around each lung consists of two separate membranous bags.
10. All of the CO_2 in the blood is released to the alveolus as the blood passes through the pulmonary capillaries.

Multiple Choice Questions
11. The vomer is found in which of the following parts of the respiratory tract?
    a. nasal cavity   c. larynx
    b. pharynx   d. bronchial tree
12. Which of the following is a passageway through which food passes on its way to the stomach and air passes on its way to the lungs?
    a. nasopharynx   c. larynx
    b. oropharynx   d. esophagus
13. The ultimate subdivision of the pulmonary tree is the
    a. trachea   c. bronchiole
    b. primary bronchus d. alveolus
14. During swallowing, the laryngeal aperture is covered and closed by the
    a. hyoid bone   c. epiglottis
    b. glottis   d. laryngeal valve
15. Which of the following is more superiorly located in the larynx?
    a. vestibular folds c. cricoid cartilage
    b. vocal folds   d. thyroarytenoid muscle
16. Annular ligaments are located in the
    a. larynx   c. bronchioles
    b. trachea   d. alveoli
17. Which of the following is not supplied with a cartilaginous skeleton?
    a. trachea   c. secondary bronchi
    b. primary bronchi   d. bronchioles
18. Each of the lobes of a lung ultimately receives air through a single
    a. trachea   c. secondary bronchus
    b. primary bronchus d. bronchiole
19. Through which of the following must an O_2 molecule first pass on its way to blood in an aveolar septum?
    a. epithelial basement membrane
    b. erythrocyte plasmalemma
    c. endothelial plasmalemma
    d. epithelial plasmalemma
20. The function performed by surfactant is to
    a. filter debris from inspired air
    b. provide lubrication between layers of pleurae
    c. increase efficiency of gas exchange through the alveolar wall
    d. lubricate the trachea
21. The role of the septal cell in the alveolar septum is to
    a. remove dust and microbes
    b. secrete surfactant
    c. maintain the pH of the alveolus
    d. regulate the flow of oxygen into the blood
22. Which of the following muscles contributes the most to enlarging the thoracic cavity during inspiration?
    a. internal intercostals   c. diaphragm
    b. external intercostals   d. pectoralis minor
23. Inspiratory capacity is the total of which of the following?
    a. tidal volume plus inspiratory reserve volume
    b. expiratory and inspiratory reserve volumes
    c. inspiratory reserve volume plus residual volume
    d. residual volume plus vital capacity
24. Vital capacity is the total of which of the following?
    a. tidal volume plus inspiratory reserve volume
    b. tidal volume plus inspiratory and expiratory reserve volumes
    c. inspiratory and expiratory reserve volumes
    d. residual volume plus inspiratory and expiratory reserve volumes
25. The ratio between an increase in lung volume and the pressure required to cause the increase is called
    a. respiratory quotient   c. compliance
    b. elasticity   d. specific compliance
26. When lung tissue loses its elasticity so that it has less resistance to expansion, its compliance will
    a. increase
    b. decrease
    c. remain the same
27. In which of the following tissue environments would hemoglobin probably have the greatest affinity for oxygen?
    a. low temperature, low pH, low P_{CO_2}
    b. low temperature, high pH, low P_{CO_2}
    c. high temperature, high pH, low P_{CO_2}
    d. high temperature, high pH, high P_{CO_2}
28. If there are 5×10^9 erythrocytes per cubic centimeter (cm^3) and 2.5×10^8 hemoglobin molecules per cell, what is the maximum number of O_2 molecules that can be carried in 1 cm^3 of blood?
    a. 1×10^9 c. 1.25×10^{18}
    b. 4×10^9 d. 5×10^{18}

29. If the partial pressure of oxygen is 95 mm Hg in blood as it enters a part of the circulatory system and 40 mm Hg when it emerges, what part of the circulatory system is the blood probably in?
   a. systemic capillaries
   b. pulmonary capillaries
   c. pulmonary arteries
   d. a major vein leading to the heart from the systemic circulation
30. The following are typical CO_2 concentrations in various parts of the circulatory system. Which one is probably in blood leaving the lungs?
   a. 40 mm Hg
   b. 42 mm Hg
   c. 45 mm Hg
31. CO_2 is carried in the blood
   a. in erythrocytes as bicarbonate ion
   b. in erythrocytes in combination with hemoglobin
   c. in the plasma as dissolved CO_2
   d. in the plasma combined with plasma proteins
   e. all of the above
32. The effect of signals from the apneustic center is to
   a. increase heart rate
   b. decrease heart rate
   c. increase inspiratory rate
   d. decrease inspiratory rate
33. When one breathes into a plastic bag, breathing rate increases quickly. This is due most directly to stimulation of the
   a. carotid bodies
   b. aortic bodies
   c. respiratory center
   d. chemosensitive areas in the medulla oblongata
34. If the O_2 level of the blood falls dramatically, signals from which of the following are sent at an increased rate to the respiratory center?
   a. carotid bodies
   b. aortic bodies
   c. apneustic center
   d. chemosensitive areas in the medulla oblongata
35. The Hering-Breuer reflex
   a. involves inhibition of inspiratory neurons in the respiratory center
   b. involves inhibition of expiratory neurons in the respiratory center
   c. is initiated by signals from the cerebrum
   d. is initiated by signals from the apneustic center

Essay Questions

36. Describe the organization of the respiratory tract, listing the organs through which air passes as it is inspired and travels from the nose to the alveoli.
37. List three functions performed by the mucus-covered lining of the nasal cavity.
38. Diagram the structure of the larynx and explain its functions.
39. List the barriers through which oxygen must travel as it passes from the alveolus to hemoglobin.
40. Describe the mechanisms involved in inspiration and expiration that cause air to be drawn in and forced out of the lungs.
41. Diagram the gross morphology of the lungs, showing how they are covered by pleurae and how they are entered by the bronchi and the major pulmonary arteries and veins.
42. Diagram a pulmonary tree, identifying its regions and explaining the structure and function characteristic of each region.
43. Describe the structure of hemoglobin and explain how pH, temperature, and CO_2 concentration influence the O_2-binding properties of hemoglobin.
44. Describe the chemistry of CO_2 transport in the blood.
45. List the organs and tissues involved in regulating respiration and explain the function performed by each.

Clinical Application Questions

46. Describe sinusitis and explain the treatments used in its control.
47. What is the derivation of the term *tuberculosis*?
48. Explain why a pulmonary embolism is life threatening.
49. Compare the conditions of chronic bronchitis and emphysema with respect to cause, effect, and treatment.
50. Describe the causes and effects on the pulmonary system of bronchial asthma.
51. List several medical conditions associated with exposure to increased environmental pressure. Explain the physiological effect of each condition and the treatment used to control or prevent it.
52. Hyperventilation and hypoventilation can occur in people who have healthy respiratory systems. Suggest how emotional, chemical, or physical factors might cause improper ventilation.
53. Describe the condition known as sudden infant death syndrome.

UNIT IV CASE STUDY

THE STRANGE ASTHMATIC

Jack W. was an 18-year-old college freshman who was in general good health, even playing on the ice hockey team. He had been plagued, however, for the past several years with repeated episodes of sudden attacks of severe shortness of breath and wheezing, usually in the middle of the night. Jack had had a complete allergic workup prior to his entering college, which failed to turn up any significant findings. A review of his family history failed to reveal other members of his family with significant allergic problems. The only other complaint that he seemed to have was an occasional "acid stomach" after overeating pizza or other rich foods.

One night Jack had a particularly bad asthma attack and was taken by his roommate to the emergency room of a nearby hospital. The physician was called in to see him at 2:00 A.M. and found him sitting up on the examination table leaning forward and struggling to get his breath. He was using all of his accessory muscles to breathe and appeared pale and frightened. His physical exam was normal except for his appearance, and listening to his chest revealed coarse wheezes in all lung fields in both inspiration and expiration. His vital signs showed a temperature of 97.4°F, blood pressure at 130/80, pulse at 100, and a respiratory rate of 22 per minute. Measurement of his peak respiratory flow rate (PF) was only 250 (the estimated norm for his height and weight is 600). The nurse had already given him a dose of adrenaline subcutaneously without much relief of his symptoms. The physician ordered terbutylene to be administered intravenously and gave him an oral dose of cortisone. He improved a little following this treatment, but

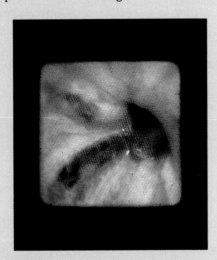

Endoscopic photograph of esophagus showing inflamed area (dark patches) caused by esophagitis.

his peak flow was still only about 350. Two further doses of adrenaline were required to produce significant relief. He was then admitted to the hospital, and the next morning he was feeling fine and had a peak respiratory flow of 550. His chest X ray was normal, as was his complete blood count, urinalysis, and pulmonary function studies.

The puzzling thing about this case was that the attacks occurred in the middle of the night with no apparent cause. A collaborating doctor was reminded that "all that wheezes is not asthma." He suggested that Jack's acid stomach problem be examined because on rare occasions, asthma-like attacks can be produced by refluxing gastric acid into the lungs. To test this hypothesis, gastrointestinal X rays were made, and they did indeed show evidence of acid reflux into the lower esophagus. An endoscopic examination confirmed evidence of inflammation in the esophagus (esophagitis). Jack was informed of the findings and instructed to change his diet to several small meals throughout the day and to avoid "pigging out" in the evenings. He was also advised to elevate the head of his bed about 6 in. to reduce the chance of acid refluxing up his esophagus while he sleeps. He was given a prescription for cimetidine, a drug that blocks certain histamine receptors (see Chapter 21) and reduces acid production in the stomach. These measures resolved the problem of the strange nighttime asthma.

Unit V

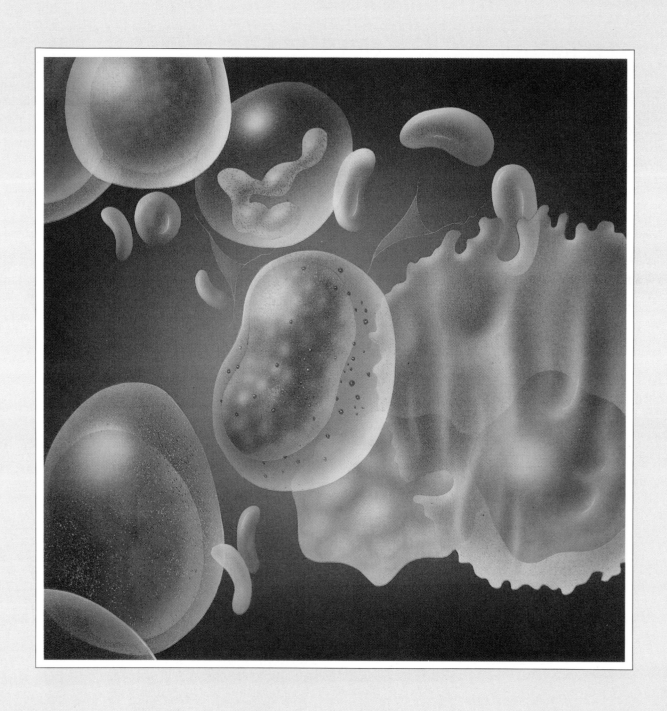

Transport and Circulatory Systems

Unit V describes the cardiovascular system, which transports both solid and dissolved materials throughout the body. We begin with a study of the blood, the circulatory medium. Next, we examine the heart, a specialized muscular organ that contracts to move the blood to almost all the tissues of the body through a vast system of circulatory vessels. The unit ends with an examination of the lymphatic system and the specialized cells and chemicals it releases into the blood. These defensive agents are vital in our constant battle against infection and disease.

21
The Blood

22
The Cardiovascular System: The Heart

23
The Cardiovascular System: Blood Vessels

24
The Lymphatic System and Immunity

21
The Blood

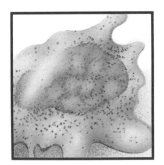

The human body is a complex assemblage of multicellular tissues and organs that work together to produce many functions associated with human life. There are over a trillion cells in a typical adult human, and each one has certain needs. Cells require nutrients that provide raw materials and energy needed to perform specific tasks. Toxic waste materials produced by the cells must be removed so they do not poison themselves. Chemicals used to control processes in some other part of the body must be transported to the cells where they will be used. In addition, cells require a watery environment so that dissolved materials can move into and out of the cell. All these needs are provided for by the **blood.** As we study the blood in this chapter, we will

1. outline the composition of the blood,
2. list the major categories of plasma proteins and identify their functions,
3. describe the formed elements and study the major functions associated with each, and
4. describe the manner in which a blood clot forms.

Functions, Composition, and Production of Blood

After completing this section, you should be able to:

1. List the six major functions of blood.
2. Describe the composition of blood with regard to plasma and formed elements.
3. Discuss the importance of proteins and solutes in plasma.
4. Distinguish among erythrocytes, leukocytes, and thrombocytes.
5. List the five major hematocytoblasts that develop from mesenchyme cells during the production of the formed elements of blood.

Functions of Blood

Blood has six major functions:

1. Blood transports materials from one place to another.
2. Blood provides the watery environment for individual cells.
3. Blood helps to maintain a constant internal temperature.
4. Blood plays a major role in combating disease and infection.
5. Blood helps to regulate the body's pH.
6. Blood prevents the loss of body fluids through a clotting process.

The primary function of the blood is transport. Substances carried in the blood include oxygen, carbon dioxide, other gases, nutrients (such as glucose and amino acids), and hormones that exert an effect on cellular processes in target tissues. Blood also carries cellular wastes such as carbon dioxide and lactic acid. Urea and other nitrogen-containing wastes are carried from the tissues to the kidneys where they are removed.

Another important function of blood is maintenance of a fluid environment for cells. Water moves from the blood into the spaces between the cells where it forms **interstitial fluid** (also called **tissue fluid, extracellular fluid,** or **ECF**). Interstitial fluid normally has a concentration of solutes that maintains the tissues at an appropriate internal pressure by regulating the osmotic flow of water through the cells. For interstitial fluid to function properly, the concentration of dissolved substances must be regulated. This is accomplished by the blood, which regulates the amount of water and dissolved substances present in the interstitial fluid. For example, if too much dissolved salt is present in the interstitial fluid, water will leave the cells, resulting in cell dehydration.

Too little dissolved salt in the interstitial fluid has the opposite effect. The blood helps to prevent this.

The fluid component of blood is capable of absorbing large quantities of heat from the tissues that are producing it, such as active muscles, and moving this heat to another part of the body. This helps to maintain a more or less constant internal temperature.

Blood also plays a major role in combating disease and ridding the body of toxic substances produced by microorganisms that invade the body. This is accomplished through the combined action of certain proteins called *antibodies* (see Chapter 24) and specialized cells that recognize invading microorganisms and destroy them.

Special chemicals called *buffers* contained in the blood prevent large shifts in the pH of the blood and interstitial fluid by absorbing excess acids or bases. (Buffers were discussed in detail in Chapter 2.)

Finally, blood has components that form *clots,* or barriers that can block the loss of blood from an injured vessel.

Composition of Blood

The amount of blood in an individual usually amounts to about 10% of his or her body weight. In adult males this is about 6 L, and in adult females it is from 4 to 5 L. At any one moment, about 10% of the blood is in the lungs, about 5% is in the heart, and about 85% is in the remainder of body circulation. Of the blood outside the heart, only 20% is being moved through arteries away from the heart, 10% is in the capillaries, and the remaining 70% is found in the veins.

Blood is a thick fluid; its viscosity is 4 to 5 times that of water. It is slightly alkaline, with a pH of 7.3 to 7.5. Its temperature is 38°C (100°F), slightly warmer than overall body temperature.

Two major components of blood are a clear, pale, yellow fluid called **plasma** and a cellular portion called the **formed elements** (Figure 21-1). Normally, the latter comprise about 45% of the blood's volume in men and about 42% in women. These values can vary considerably as a result of disease or injury, or if one routinely engages in strenuous physical activity.

Plasma

If 100 mL of whole blood is placed in a centrifuge and spun, the cells are separated from the fluid component, the plasma (Figure 21-2). Plasma is 90 to 92% water. The remaining percentage consists of dissolved proteins, hormones, nutrients, minerals, and wastes. Table 21-1 lists ranges of values for the various constituents of plasma reflecting the variation among individuals.

Chapter 21 The Blood

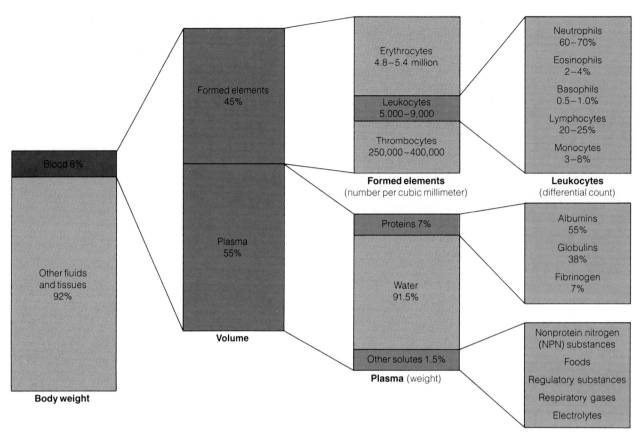

Figure 21-1 The composition of blood.

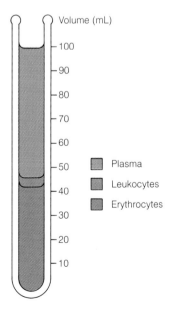

Figure 21-2 If 100 mL of blood is spun in a centrifuge, it separates into two fractions: plasma and the formed elements, which are primarily erythrocytes and leukocytes.

Table 21-1 MAJOR COMPONENTS OF PLASMA

Substance	Approximate Amount
Water	90–92%
Plasma proteins	
Albumins	3.2–5.0 g/100 mL
Fibrinogen	0.20–0.45 g/100 mL
Globulins	
Alpha (α)	0.40–0.98 g/100 mL
Beta (β)	0.56–1.06 g/100 mL
Gamma (γ)	0.44–1.04 g/100 mL
Glucose	61–130 mg/100 mL
Cholesterol	128–347 mg/100 mL
Bilirubin	0–1.1 mg/100 mL
Urea	13.8–39.8 mg/100 mL
Sodium	310–356 mg/100 mL
Potassium	12–21 mg/100 mL
Calcium	8.2–11.6 mg/100 mL
Iron	0.04–0.21 mg/100 mL
Chloride	355–381 mg/100 mL

albumin:	(L.) *albumen*, white of an egg
hemopoiesis:	(Gr.) *haima*, blood + *poiein*, to make
hematocytoblast:	(Gr.) *haima*, blood + *kytos*, cell + *blastos*, germ

Proteins make up the majority of dissolved material in the blood, comprising about 7% of the plasma and nearly 90% of the solutes in the plasma. **Albumins** are the most abundant, being about 60% of plasma proteins. These large molecules prevent excessive movement of water from the blood into the interstitial fluid by osmosis. Because of their size, albumins have difficulty passing from a capillary into the surrounding interstitial space, and consequently, their concentration is greater within the capillary. This greater concentration of albumins and other solutes helps prevent blood from leaving the blood vessel through osmosis. Globulins make up almost 34% of plasma proteins. They are important as transport molecules and for their role as antibodies. Fibrinogen forms almost 5% of plasma proteins. It is important in chemical reactions that result in blood clotting and in preventing excessive loss of blood (see section on Hemostasis).

In addition to proteins, the solutes in plasma include nutrients, such as glucose, amino acids, and lipids, as well as waste materials, such as urea, a product of cellular metabolism. Dissolved ions, particularly sodium, potassium, bicarbonate, and chloride ions; dissolved gases; and numerous other materials, such as hormones, vitamins, enzymes, and waste materials; are all transported as part of the dissolved materials in plasma. Plasma is truly a complex mixture of compounds.

Formed Elements

If we separated plasma from 100 mL of blood in a centrifuge, as mentioned previously, the solid portion that would remain is a group of cells called the formed elements. There are three kinds of formed elements: erythrocytes, leukocytes, and thrombocytes (or platelets) (Table 21-2). Erythrocytes carry oxygen and carbon dioxide, leukocytes are part of the body's defense system, and thrombocytes assist in the formation of blood clots (each is discussed in its own section on the following pages).

Production of Blood Cells

The production of blood cells, which begins early in embryonic and fetal life, is called **hemopoiesis** (hē″mō·poy·ē′sĭs) or **hematopoiesis** (hēm″ă·tō·poy·ē′sĭs). In the early embryo hemopoiesis begins in the yolk sac. As internal organs develop, production of blood cells also occurs in the liver, spleen, lymph nodes, thymus, and the red marrow of the bones of the fetal skeleton. After birth the red bone marrow becomes increasingly more important in this function, and in adults hemopoiesis is restricted to this red marrow in the sternum, ribs, pelvis, vertebrae, and lymphoid tissue.

In adults all blood cells begin their life as undifferentiated mesenchymal cells in red bone marrow. These cells differentiate into **hematocytoblasts** (hēm″ă·tō·sī′tō·blăsts), which are immature cells with the capability of developing along one of five major lines (Figure 21-3). Although all hematocytoblasts possess a full complement of genetic instructions capable of producing all five adult cell types, only one set of instructions can be activated in any one cell. A full explanation of the factors that determine why one line of development will be taken rather than another is beyond the

Table 21-2 FORMED ELEMENTS OF THE BLOOD

Formed Element	Approximate Diameter (μm)	Approximate Abundance (per mm³ of blood)	Function
Erythrocytes	7.5	Males: 5.1–5.8 million Females: 4.3–5.2 million	Transport oxygen and carbon dioxide
Leukocytes			
Granulocytes			
Neutrophils	12–14	3,000–7,000 (60–70% of total leukocytes)	Phagocytic cells
Eosinophils (acidophils)	12	50–400 (2–4% of total leukocytes)	Phagocytic cells
Basophils	9	1–50 (0.5–1% of total leukocytes)	Release histamine and heparin
Agranulocytes			
Monocytes	20–25	100–600 (3–8% of total leukocytes)	Develop into large phagocytic cells called macrophages
Lymphocytes	9 (small) 12–14 (large)	1000–3000 (20–25% of total leukocytes)	Produce antibodies
Thrombocytes (platelets)	2.5	250,000–400,000	Involved in the formation of blood clots

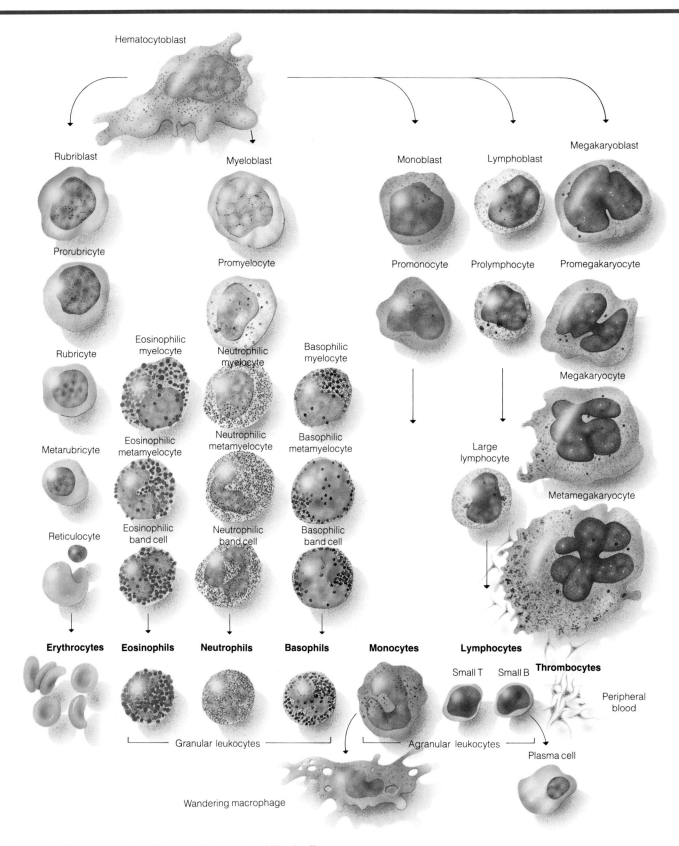

Figure 21-3 Hemopoiesis, the origin and development of blood cells.

rubriblast: (L.) *ruber*, red + (Gr.) *blastos*, germ
erythrocytes: (Gr.) *erythros*, red + *kytos*, cell

scope of this book; however, each hematocytoblast appears to retain this developmental flexibility throughout the life of an individual.

Some hematocytoblasts will develop into **rubriblasts** (rū′brĭ·blăsts), cells that lose their nucleus and accumulate hemoglobin. These cells develop into erythrocytes (see below). Other hematocytoblasts develop into **myeloblasts** (mī′ĕl·ō·blăsts), which accumulate cytoplasmic granules and mature into leukocytes called granulocytes (see p. 546). Other hematocytoblasts develop into **monoblasts,** nongranular cells that mature into monocytes, defensive cells that patrol the body on a perpetual search-and-destroy mission against infective invaders. Certain hematocytoblasts develop into **lymphoblasts** (lĭm′fō·blăsts), which mature into lymphocytes. Lymphocytes are the cells that act in the body's immunological defense system (see Chapter 24). A final group of hematocytoblasts develops into **megakaryoblasts** (mĕg″ă·kăr·ē·ō·blăsts), which fragment into thrombocytes (platelets), cytoplasmic fractions important in the formation of blood clots (see p. 549).

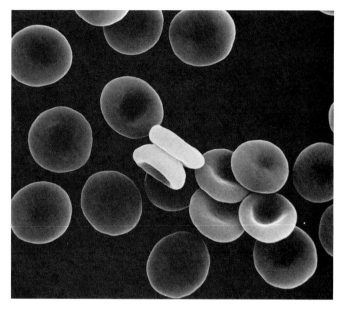

Figure 21-4 Scanning electron micrograph of erythrocytes.

Erythrocytes

After completing this section, you should be able to:
1. Describe a typical erythrocyte.
2. Discuss the functions of erythrocytes.
3. Outline the life history of an erythrocyte.

Structure

Erythrocytes (ĕ·rĭth′rō·sīts), more commonly called **red blood cells,** average 7.5 μm in diameter and have a biconcave surface (Figure 21-4). A mature cell lacks a nucleus, mitochondria, and ribosomes, so it has lost the ability to perform extensive metabolic activities. Erythrocytes depend on glycolysis to meet their energy needs (see Chapter 19). An erythrocyte is surrounded by a plasma membrane, and about 33% of its cytoplasm is **hemoglobin,** the pigment that gives whole blood its crimson red color.

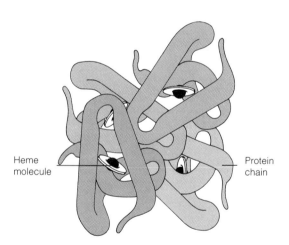

Figure 21-5 **A diagram of the hemoglobin molecule.** Note that four heme molecules are attached to the long protein chains. Each heme molecule can combine with an oxygen molecule, which is how oxygen is carried in the blood. (Also see Figure 20-17).

Function

Erythrocytes are associated with the red pigment hemoglobin, a molecule that consists of four protein (globin) segments and four heme (or iron-containing) molecules (Figure 21-5). As discussed in Chapter 20, each heme molecule can combine with an oxygen molecule; therefore, each hemoglobin molecule is capable of carrying four oxygen molecules.

Oxygen and hemoglobin combine as blood passes through the lungs where oxygen is present in relatively high concentration. The molecule produced is called **oxyhemoglobin.** In the tissues of the body, oxygen leaves the hemoglobin molecule and enters the cells. Once free of the oxygen molecule, the heme group combines with some of the carbon

erythropoiesis: (Gr.) *erythros*, red + *poiein*, to make
rubricyte: (L.) *ruber*, red + (Gr.) *kytos*, cell
reticulocyte: (L.) *reticula*, net + (Gr.) *kytos*, cell

CLINICAL NOTE

MALARIA

Malaria is an acute infectious disease caused by the presence of protozoan parasites in the red blood cells. Next to the common cold, it is the most common infectious disease in the world; 60 million cases occur throughout the world each year. The disease is mainly spread through the bite of any one of four species of mosquito, but can also be spread by transfusion of infected blood and by contaminated needles in drug abusers.

The life cycle of the malarial parasite begins when a female mosquito bites a person with malaria. The ingested blood contains gametocytes, the first stage in their life cycle. These reproduce sexually in the stomach of the mosquito and migrate to the salivary glands where they may be injected into the next victim as sporozoites. The sporozoites reproduce asexually in the liver (preerythrocytic stage) for days to months and are released into the bloodstream as merozoites where they invade the red blood cells initiating the symptomatic disease (erythrocytic stage). The merozoites reproduce asexually in the red blood cells, producing more merozoites and gametocytes.

The erythrocytic stage is characterized by recurrent paroxysms of chills, fever, and sweating and by anemia and enlargement of the spleen. Untreated, the disease runs a chronic, debilitating, relapsing course. One type called **falciparum malaria** is particularly virulent and may cause fatal complications, such as **blackwater fever,** which causes destruction of red blood cells (hemolysis) and renal failure. Another type, **cerebral malaria,** causes coma, convulsions, and neurologic injury.

Malaria may be prevented by avoiding endemic areas, preventing mosquito bites, or taking the drug **chloroquine** weekly before, during, and after visits to tropical endemic areas.

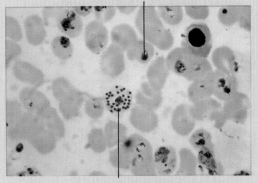

Young parasite (in early stage of development)

Developing parasite (in advanced stage)

Malaria. The parasites (dark spots) are visible in various stages of reproductive development in the red blood cells.

CLINICAL NOTE

RETICULOCYTE COUNTS

Reticulocytes are immature forms of erythrocytes that retain a portion of their cytoplasmic reticulum. They may appear in the peripheral blood when the bone marrow is actively producing large numbers of new red cells. Reticulocytes are readily stained with a special blue dye and counted on a blood smear. They are usually reported as the percentage of reticulocytes per thousand counted red blood cells. An increased number of reticulocytes suggests an attempt by the bone marrow to recover from a loss of red blood cells that has resulted from recent bleeding or from a disease that causes destruction of red blood cells.

---

dioxide molecules produced by the metabolic activities of the cells. This combination is called **carbaminohemoglobin.** About 25% of the carbon dioxide produced in the tissues is carried in this manner. The rest is carried by specialized buffer chemicals in the plasma of the blood.

Production

Erythrocytes are produced by **erythropoiesis** (ĕ·rĭth″rō·poy·ē′sĭs), a specialized process of hemopoiesis (Figure 21-3). Production of erythrocytes in red bone marrow begins with the division of the mesenchymal or stem cells. Hemocytoblasts undergo rapid cell divisions, which result in rubriblasts. Rubriblasts differentiate into **rubricytes,** which begin synthesizing hemoglobin. Once hemoglobin production has begun, a rubricyte undergoes a series of changes that increases its volume and allows a maximum of hemoglobin to be stored within. Soon, the rubricyte develops into a **metarubricyte** (mĕt″ă·rū′brĭ·sīt), then into a **reticulocyte** (rĕ·tĭk′ū·lō·sīt). The reticulocyte extrudes its nucleus and becomes an erythrocyte, which leaves the marrow and enters the bloodstream (see Clinical Note: Reticulocyte Counts).

Once in circulation, erythrocytes live an average of 120 days. This short life span is related to the loss of the nucleus and other cell organelles, with a corresponding loss in adaptability of the cell. Adult males have an average of 5.5 million erythrocytes (ranging from 5.1 to 5.8 million) per cubic millimeter of blood, and adult females average about 4.8 million (ranging from 4.3 to 5.2 million) per 1 mm^3 of blood. Their exact numbers are quite variable and depend on one's general state of health, physical conditioning, and the altitude at which one lives. Given the number of erythrocytes per cubic millimeter and the short life span of each cell, it

macrophage: (Gr.) *makros*, large + *phagein*, to eat
bilirubin: (L.) *bilis*, bile + *ruber*, red
anemia: (Gr.) *a*, not + *haima*, blood

is necessary for the body to produce approximately 2 million new erythrocytes per second every day just to replace those that die.

Replacement

During an erythrocyte's short 120-day life span, it "ages" as chemical changes occur in its membrane lipids and in the enzymes that were present when the cell was first released. Lacking DNA and RNA, the cell is not able to restore enzymes and other proteins that may be lost, nor is it able to carry on the metabolic activity necessary to maintain its membranes. The resulting changes accumulate in time, and eventually the aged cell is consumed by **macrophages** (măk´rō·fāj·ĕs), specialized leukocytes located in the spleen, liver, and bone marrow. The specific changes in the erythrocyte that identify it to macrophages are unknown. However, as the aged cell passes through one of the organs where macrophages are located, it is withdrawn from circulation and consumed by one. Once engulfed, the hemoglobin in the cell is broken down and recycled through a homeostatic mechanism (see next section) that reuses iron from the old hemoglobin molecules to manufacture new ones.

The iron in an erythrocyte is conserved efficiently (Figure 21-6). Of the 30 mg of iron present in erythrocytes destroyed in one day, only 1 mg is lost in excrement and urine. Most of the iron from the hemoglobin combines with **transferrin** (trăns·fĕr´rĭn), a specialized protein in the blood that "transfers" the iron to the bone marrow and liver where it is stored. Other portions of the hemoglobin molecule are also broken down and lead to the production of **bilirubin** (bĭl·ĭ·rū´bĭn) a waste product that accumulates in the liver and is excreted with the bile.

Control of Erythrocyte Production

The homeostatic control of erythrocyte production is a part of a negative feedback mechanism. This mechanism is associated with the production of the hormone **erythropoietin** (ĕ·rĭth˝rō·poy·ĕ´tĭn) by the kidney and other cells. When subjected to a decrease in oxygen concentration (hypoxia), certain cells in the kidney and liver begin to secrete erythropoietin, a glycoprotein that stimulates the red marrow cells to increase their production of erythrocytes. This increase is accomplished primarily by raising the rate of hemocytoblast production by stimulating stem cell division, although subsequent steps in the process of hemopoiesis are also accelerated. After a few days, the level of erythrocytes climbs, and the oxygen level in the blood increases correspondingly. With this increase in oxygen concentration, the stimulus for the production of erythropoietin declines, and the rate of blood cell production diminishes to a normal level.

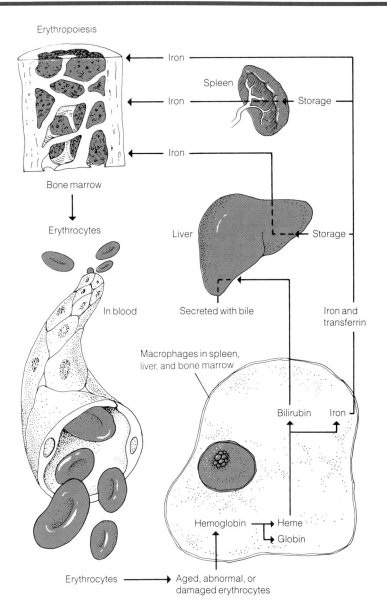

Figure 21-6 Reuse of hemoglobin from erythrocytes. Old red blood cells are destroyed by macrophages. The iron is removed from the hemoglobin molecules and either stored in the spleen and liver or reused immediately to produce new hemoglobin molecules.

Primary causes of hypoxia, and hence increased erythropoiesis, include a reduction in the amount of atmospheric oxygen, which occurs at high altitudes, a reduction in the amount of hemoglobin, or a loss of erythrocytes. The latter two situations are forms of **anemia,** any condition that reduces the ability of the blood to deliver oxygen to the tissues (see Clinical Note: Anemia).

megaloblast:	(Gr.) *megas*, large + *blastos*, germ
hemolytic:	(Gr.) *haima*, blood + *lysis*, dissolve
antigen:	(Gr.) *anti*, against + *gennan*, to produce
granulocyte:	(L.) *granulum*, little grain + (Gr.) *kytos*, cell
neutrophil:	(L.) *neuter*, neither + (Gr.) *philein*, to love

CLINICAL NOTE

ANEMIA

Anemia is any one of a large number of conditions involving a reduction in the number of erythrocytes or the amount of hemoglobin per given volume of blood. The various types are differentiated according to their causes. **Hemorrhagic anemia** is a loss of blood due to profuse bleeding. If the volume of blood lost is not fatal, the reduced quantity of oxygen reaching the tissues stimulates erythropoietin production, and new erythrocytes replace those lost. The body responds similarly to donation of blood and replaces the lost erythrocytes in a matter of weeks.

A deficiency of vitamins B_{12} and folic acid may also lead to anemia as a result of a decline in the rate of erythrocyte production. These vitamins are involved in the synthesis of DNA, and continued cell division is dependent upon sufficient replication of chromosomes. When there is a deficiency of either vitamin, the conversion of hemacytoblasts to metarubricytes is blocked, and immature cells called **megaloblasts** (mĕg´ă·lō·blăsts) are produced instead of erythrocytes. This type of anemia can be reversed by the intake of vitamin B_{12} and folic acid.

Pernicious anemia results from a shortage of vitamin B_{12} or of intrinsic factor, a substance produced by gastric glands of the stomach. Intrinsic factor combines with B_{12} making it absorbable by intestinal cells. Even if sufficient quantities of B_{12} are consumed, a deficiency of intrinsic factor reduces the ability of the intestinal cells to absorb the vitamin and make it available to the red marrow. Thus, damage to the gastric cells that produce intrinsic factor may ultimately affect the blood's ability to deliver sufficient oxygen to the tissues.

Many different kinds of anemia result when erythrocytes are destroyed by macrophages faster than they can be replenished. Such anemias, characterized by shorter than normal life spans for the red blood cells, are **hemolytic** (hē˝mō·lĭt´ĭk) **anemias.** Some are due to genetic deficiencies in the production of erythrocyte enzymes necessary for the metabolism of glucose. Others are caused by inherited defects in the synthesis of the erythrocyte's membrane or due to interactions of the membrane with bacterial toxins or antibodies that circulate in the blood.

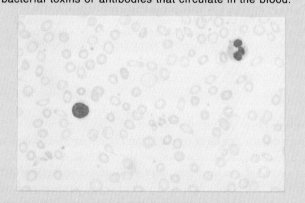

Iron-deficient anemia. Note the central pale area in each red blood cell indicating less hemoglobin than normal cells.

Blood Groups

The surfaces of erythrocytes contain specialized proteins called **antigens,** which are chemicals that stimulate an immune reaction. Such a reaction can occur when blood from certain individuals is mixed, as in a blood transfusion. The antigens found on the erythrocytes, specifically called **agglutinogens** (ă·glū·tĭn´ō·jĕns), in one individual may react with chemicals called **agglutinins** (ă·glū´tĭ·nĭns) in the plasma of the other person to produce agglutination (clumping) of the red cells. The specific types of antigens possessed determines to which blood group a person belongs. Hundreds of different human blood groups have been identified, but most people have heard of only the two major groups: the ABO group, containing blood types A, B, AB, and O, and the Rh group, containing Rh^+ and Rh^- (see Chapter 24 Clinical Note: Hemolytic Disease of the Newborn). Because these blood groups are based on antigen-antibody reactions, we will delay our detailed discussion of them until Chapter 24 where we deal with immunological reactions.

Leukocytes

After completing this section, you should be able to:

1. Distinguish between granulocytes and agranulocytes, and describe the different kinds of each.
2. Describe the phagocytic functions of leukocytes.
3. Discuss nonphagocytic functions of leukocytes in combating disease.

Structure

Leukocytes (lū´kō·sīts) differ from erythrocytes in many ways. Lacking hemoglobin, they are colorless and are thus referred to as **white blood cells.** Unlike erythrocytes, they contain nuclei and other organelles needed to produce proteins and to carry on aerobic respiration.

Two major categories of leukocytes exist, differing in the presence or absence of granules within their cytoplasm (Figure 21-3). Leukocytes with granules are called **granulocytes** (grăn´ū·lō·sīts). These cells develop in the red marrow of bone and possess lobed nuclei. There are three kinds of granulocytes. **Neutrophils** (nū´trō·fĭls) can be stained easily in the laboratory with neutral dyes. They can leave the blood and wander through tissues using ameboid movement. **Eosinophils** (ē˝ō·sĭn´ō·fĭls) stain readily with the acidic dye eosin. Like neutrophils, they also exhibit ameboid

eosinophil:	(Gr.) *eos*, dawn (rose colored) + *philein*, to love	lymphocyte:	(L.) *lympha*, gland + (Gr.) *kytos*, cell
basophil:	(Gr.) *basis*, base + *philein*, to love	monocyte:	(Gr.) *monos*, single + *kytos*, cell
agranulocyte:	(Gr.) *a*, not + (L.) *granulum*, little grain + (Gr.) *kytos*, cell	phagocyte:	(Gr.) *phagein*, to eat + *kytos*, cell
		diapedesis:	(Gr.) *dia*, through + *pedan*, to leap
		heparin:	(Gr.) *hepar*, liver

movement through tissues. **Basophils** (bā´sō·fĭls) stain easily with alkaline dyes. In addition to possessing ameboid movement, they also secrete chemicals active in allergic responses.

Leukocytes that lack cytoplasmic granules are called **agranulocytes.** These cells develop in lymphoid tissue and possess a spheroid nucleus. There are two kinds of agranulocytes. In **lymphocytes** (lĭm´fō·sīts), the smaller of the two, the nucleus occupies practically all of the cell. At any one time about 5% of lymphocytes in the body are circulating in the blood. The remainder are present in lymphatic tissues such as lymph nodes and the spleen where they are active in the immune response (see Chapter 24). **Monocytes,** the other type of agranulocytes, are larger than lymphocytes and have relatively more cytoplasm. Monocytes are capable of ameboid movement and may leave the blood vessels and migrate among tissue cells. Once within a tissue, monocytes enlarge and develop into macrophages, cells that ingest bacteria and dead cells.

Function

Unlike erythrocytes, leukocytes do not carry dissolved gases. Instead they function as part of the body's system of defenses against invasion of microbes and toxic substances. In general, neutrophils and monocytes are active **phagocytes,** which are types of blood cells that ingest bacteria, dead cells, and other foreign materials by means of **phagocytosis** (făg˝ō·sī·tō´sĭs). Once ingested, this material is destroyed through the activity of enzymes within the phagocyte.

Most leukocytes have the ability to move to an area of infection by leaving the bloodstream. This is accomplished by **diapedesis** (dī˝ă·pĕd·ē´sĭs), a process in which leukocytes literally squeeze between the capillary cells (Figure 21-7). The phagocyte is attracted to the area of infection by chemicals released by the microorganism and the damaged cells. Once in contact with the microorganism or damaged cell, the phagocyte engulfs and digests it. As more leukocytes are attracted to the site of infection, they form a thick collection of living and dead cells called **pus.**

Phagocytosis is not the only way a body can combat infection. Some leukocytes produce specialized proteins called **antibodies.** Antibodies are proteins that inactivate antigens, chemicals that stimulate an immune response (see Chapter 24). Although most cells contain or can produce antigens, the majority of antigens that contact blood cells are foreign materials. The surfaces of bacteria, molds, and other chemicals that frequently invade the body are covered with antigens. Certain leukocytes can produce antibodies that attach to a specific antigen and either inactivate or destroy it directly or make it possible for a phagocyte to ingest and destroy the antigen-antibody combination.

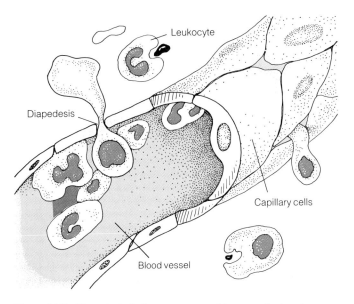

Figure 21-7 Diapedesis is the process used by some leukocytes to leave a blood vessel and wander through tissues.

Neutrophils and monocytes are particularly adapted for phagocytosis. Neutrophils reach an area of infection rapidly. Monocytes travel slower, but arrive in greater numbers than neutrophils. Eosinophils may also be phagocytic. They are known to leave capillaries and move to an area where an antigen has been introduced. They appear to be most active in allergic reactions such as hay fever and asthma, secreting natural antihistamines capable of destroying the antigen-antibody combination.

Basophils are also capable of leaving the capillaries. They squeeze between the cells in the capillary wall and enter connective tissue where they function in a manner similar to that of **mast cells.** Mast cells are large granular cells found in connective tissue. The granules in both basophils and mast cells contain **heparin** and **histamine,** two chemicals active in the fight against bacterial infection. When basophils and mast cells encounter bacteria, they secrete heparin and histamine into the area. Heparin is an **anticoagulant,** a substance that prevents the clotting of blood. Histamine is a **vasodilator,** which means that it dilates or increases the diameter of blood vessels. This in turn increases blood flow and also increases the permeability of blood vessel walls, which facilitates the movement of defense chemicals into the tissue. The increased blood flow brings additional leukocytes to the area where they are used to engulf the invading bacteria. The redness and warmth characteristic of an inflamed area are the result of vasodilation in that locality.

polycythemia: (Gr.) *polys*, many + *kytos*, cell + *haima*, blood
hematocrit: (Gr.) *haima*, blood + *krinein*, to separate
leukocytosis: (Gr.) *leukos*, white + *kytos*, cell + *osis*, condition
leukopenia: (Gr.) *leukos*, white + *penia*, lacking

CLINICAL NOTE

COMPLETE BLOOD COUNT

The **complete blood count (CBC)** is the most widely used and informative of the myriad of laboratory tests available to physicians. It is essentially an analysis of the formed elements of the blood in quantity, quality, and relationship to one another. There are two parts to the CBC (see figure). The first part consists of a count of the total number of white blood cells and red blood cells in 1 dL of blood, an estimation of the amount of hemoglobin, the hematocrit (see below), and the red cell indices (mathematical descriptions of the size and amount of hemoglobin in the red blood cells). The second part of the CBC consists of the **differential count,** a determination of the percentage of each type of white blood cell seen on a thin smear of blood that has been stained with a special dye, as well as an evaluation of the quality of all of the cells seen. The first part of the CBC is now performed by an automatic analyzer, while the second part is done by laboratory technicians if the analyzer shows an abnormality.

An elevated white blood count (or WBC) is called **leukocytosis** and suggests an infection, inflammatory process, or malignancy, such as leukemia. A decrease in the WBC (called **leukopenia**) might occur because of a viral infection, a failure in cell production, or a rapid destruction of the cells as occurs with some drugs.

A significant decrease in the red blood count (or RBC) denotes anemia due to blood loss, failure of production by the bone marrow, or rapid destruction of the cells. Rarely, an increased RBC occurs, called **polycythemia** (pŏl″ē·sī·thē′mē·ă), due to a reduction in the amount of plasma (dehydration) or in a response to a low oxygen concentration in the blood (from pulmonary disease or high altitude exposure).

The **hemoglobin** (Hgb) is the number of grams of hemoglobin in 1 dL of blood. A decrease from normal suggests anemia.

The **hematocrit** (Hct) is the percentage of packed red cells observed when a sample of blood is centrifuged in a tube for a standard speed and time. It is also called the **packed cell volume.** The hematocrit is a more accurate indicator of anemia than the hemoglobin or RBC tests. In addition, it will give clues to the quality of the plasma and the quantity of white blood cells.

The **red cell indices** are derived from the RBC, Hbg, and Hct. They give clues to the average red cell size and the average amount of hemoglobin per cell according to these equations:

$$\text{MCV (mean corpuscular volume)} = \frac{\text{Hct} \times 10}{\text{RBC}} = 82\text{--}92 \; \mu m^3$$

$$\text{MCH (mean corpuscular hemoglobin)} = \frac{\text{Hgb} \times 10}{\text{RBC}} = 27\text{--}30 \; \mu g$$

$$\text{MCHC (mean corpuscular hemoglobin concentration)} = \frac{\text{Hgb} \times 100}{\text{Hct}} = 35.2\%$$

These indices are useful in classifying anemias into large cell size (macrocytes) or small cell size (microcytes); for example, iron deficiency anemia is a microcytic anemia.

The differential count and analysis of the blood smear can provide information about many diseases. A preponderance of **polymorphonuclear** white blood cells (possessing nuclei composed of several lobes or parts) with a few immature forms, referred to as a left shift, suggests a bacterial infection. A preponderance of lymphocytes and

---

Production

Granular leukocytes are produced in the red bone marrow (myeloid tissue). Agranulocytes are produced in myeloid tissue and in lymphoid tissue found in the thymus gland, lymph nodes, and spleen.

Replacement

Unlike erythrocytes, leukocytes have nuclei and mitochondria and are capable of normal metabolic activities. Although they appear to be capable of a long life, it turns out that most leukocytes live only a few days in a healthy person, and only a few hours if they are actively involved in fighting infection. This short life span is due to the activities they perform. Leukocytes are involved in phagocytizing and destroying dead cells and microbes and their toxins. A leukocyte can tolerate only a small amount of this material in its cytoplasm before it begins to interfere with the cell's normal metabolism. Abnormal metabolic activity quickly leads to death of the leukocyte, and the dead cell is removed from circulation and destroyed by the liver or spleen. Continued production of leukocytes by the red marrow and lymphatic tissues replaces old leukocytes as they are destroyed.

Control of Leukocyte Production

The rate of leukocyte production is controlled by the presence or absence of infection in the body (see Chapter 24). **Leukocytosis** is an increase in the number of leukocytes. **Leukopenia** (lū″kō·pē′nē·ă) is a decrease in the number of leukocytes below 5000 per 1 mm^3 of blood.

thrombocyte: (Gr.) *thrombos*, clot + *kytos*, cell
megakaryocyte: (Gr.) *megas*, large + *karyon*, nucleus + *kytos*, cell

---

monocytes would lead one to suspect a viral cause for the infection. The presence of abnormal or very immature forms would suggest cancer (leukemia). An increase in eosinophils would suggest an allergic state or parasitic disease.

In like manner, the presence of immature or abnormal red blood cells can provide clues to the causes of anemias. Also, for bleeding disorders the number and quality of platelets can be estimated.

A well-performed complete blood count can provide an enormous amount of information about the state of the body at minimal expense.

The typical format of the complete blood count (CBC) form used by a medical laboratory.

Thrombocytes (Platelets)

After completing this section, you should be able to:

1. Describe platelets and indicate how they differ from erythrocytes.
2. Indicate the functions of platelets.

Thrombocytes (thrŏm′bō·sīts), more commonly known as **platelets,** are the smallest of the formed elements, averaging only about 2 to 4 μm across. Unlike erythrocytes and leukocytes, they are not cell lineages but are pieces of pinched-off cytoplasm. Platelets lack nuclei but possess sufficient metabolic apparatus to carry on respiration and other activities (Figures 21-3 and 21-8).

Function

Platelets are important because they initiate a series of chemical reactions (hemostasis) that end in the formation of a **blood clot** (see next section). They may also be used as an aid in the control of inflammation by increasing the permeability of blood vessels necessary for movement of blood fluid and leukocytes to injured tissues.

Production

Platelets are produced in the red marrow of bones. Megakaryoblasts in the red marrow enlarge and transform themselves into **megakaryocytes,** a process that also involves the nucleus undergoing changes in shape. Platelets are pieces of cytoplasm surrounded by a membrane that pinch off from

hemostasis: (Gr.) *haima*, blood + *stasis*, stopping

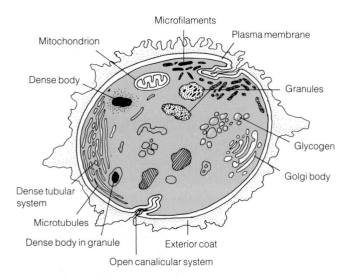

Figure 21-8 The internal structure of a platelet.

the surface of a megakaryocyte. They live an average of 5 to 10 days. In a healthy individual they number between 250,000 and 400,000 per 1 mm^3 of blood.

Hemostasis

After completing this section, you should be able to:

1. Define hemostasis and list its three stages.
2. Describe the importance of platelet plug formation in hemostasis.
3. Discuss the function of plasma coagulation factors in the formation of a blood clot.
4. Distinguish between the intrinsic and extrinsic pathways in the formation of a blood clot.
5. Describe how a clot is removed after a ruptured vessel is repaired.

Thus far, we have discussed the blood and its constituent cells as they function within intact vessels. However, one of the most frequent problems in the body is injury or damage to blood vessels. As one goes through normal daily activities, small bumps and bruises result in ruptures of blood vessels. Because of its limited supply of blood, the body can tolerate hemorrhage (bleeding) for only a short period. **Hemostasis** (hē·mŏs′tă·sĭs) is the complex process of stopping the flow of blood from a broken or damaged blood vessel. This action involves three distinct mechanisms: constriction of the blood vessel, formation of a platelet plug, and coagulation.

Vascular Constriction

When a blood vessel is severed or damaged, contraction of the smooth muscles in the vessel wall constricts the vessel at its severed or damaged end, reducing the size of the opening through which blood can escape. In large vessels this contraction is a result of the transmission of nerve impulses through a reflex arc. The impulses travel through sympathetic nerves to the spinal cord and back to the blood vessel where they initiate the muscle contraction. In addition to the effect of reflexive constriction on slowing blood loss, increased pressure outside of the damaged vessel develops as blood accumulates in the surrounding tissue from the injury. The more the damage, the greater the constriction. Constriction may last for as long as 30 min, providing enough time for two additional mechanisms to operate: formation of platelet plugs and blood clots.

Platelet Plug Formation

Almost immediately after a vessel is damaged, the platelets (thrombocytes) initiate steps to close the hole. Under normal circumstances, platelets do not stick to the smooth interior of an uninjured vessel. However, contact with the rough edges of a damaged vessel or the underlying connective tissues causes the platelets to secrete ADP and enzymes that make the surface of the platelets sticky. They adhere to collagen fibers in the connective tissue and then to one another as more and more platelets enter the region of damage. This mechanism may be sufficient to close a small hole in a vessel and is undoubtedly indispensable in repairing breaks that occur daily in small vessels throughout the body. Additional effects of the third mechanism, coagulation, help reinforce the success of a platelet plug in stopping the loss of blood from an injured vessel, especially a more serious injury to a larger vessel.

Coagulation

Coagulation, or **blood clot formation,** involves a complex series of chemical reactions that produces a solid clot or plug that stops further blood flow. The clot consists of many fibrous molecules of an insoluble protein called **fibrin** that are hooked together to form long fibrous chains. When the chains become connected in the same place, they overlap in random directions and eventually become so numerous that erythrocytes, leukocytes, and other cells are trapped in the growing meshwork. These further block the loss of blood

fibrinogen: (L.) *fibra*, fiber + (Gr.) *gennan*, to produce

> **CLINICAL NOTE**
>
> **SEDIMENTATION RATE**
>
> The **erythrocyte sedimentation rate** (ESR) is a simple test performed by placing anticoagulated blood in a calibrated upright tube and observing how much the red blood cells settle in 1 hour. A settling height above 10 mm in men and 20 mm in women is abnormal and suggests the presence of an infectious, inflammatory, or malignant disease process. The test is dependent on the presence of increased fibrinogen and gamma-globulins in the serum. It is a nonspecific but reliable test often used to separate functional from organic disease or to follow the activity of a disease such as rheumatoid arthritis or lymphoma. Slight elevations are often not significant.

(Figure 21-9). In time this meshwork becomes so dense with fibrin molecules and entrapped cells that even the fluid portion of the blood is unable to flow through. At this point the clot is formed and blood loss is arrested.

Fibrinogen (fī·brĭn′ō·jĕn) is a protein that is dissolved in blood plasma. As a dissolved substance it has no effect on clot formation. However, the events and chemical reactions that lead to formation of a clot cause this soluble protein to be converted into its insoluble form, fibrin. Once formed, the fibrin chains connect together and the clot begins to form. The chemical reactions that convert fibrinogen into fibrin occur like cascading water in a rocky waterfall: the product of one reaction acts as the catalyst for the next. At present, about 13 individual substances have been identified in blood plasma that promote the formation of a clot. They are collectively called **plasma coagulation factors** and have been numbered factors I through XIII (Table 21-3). This numbering system reflects the order in which they were discovered, not the order in which they function. In addition, four **platelet coagulation factors**, which are numbered Pf_1 through Pf_4, have been isolated from platelets.

In spite of the seeming complexity of the clotting process, the formation of a clot may be thought of as occurring in three major phases:

1. The sequence of events is initiated when an enzyme called a **prothrombin activator** is produced in the blood. This enzyme may be synthesized through extrinsic mechanisms that occur when tissues and blood vessels surrounding the fluid blood are damaged, or through intrinsic mechanisms that are triggered when platelets within the blood are damaged.
2. Prothrombin activator acts as a catalyst and promotes the conversion of an inactive plasma protein, **prothrombin,** to its active form, **thrombin.**

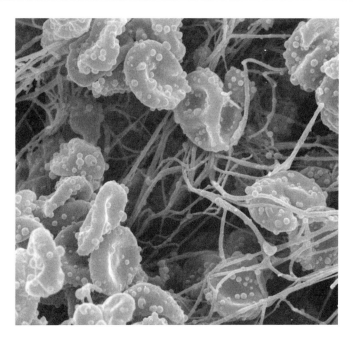

Figure 21-9 During the formation of a blood clot, erythrocytes become trapped in the meshwork of fibrin fibers that forms at the point of injury. This scanning electron micrograph shows erythrocytes, numerous small platelets ("bumps" on erythrocytes). and fibrin fibers (mag. 3848 ×).

Table 21-3 PLASMA COAGULATION FACTORS

Factor	Name	Description
I	Fibrinogen	Plasma protein synthesized in liver
II	Prothrombin	Plasma protein synthesized in liver
III	Tissue thromboplastin	Mixture of phospholipids and lipoproteins produced by damaged tissues
IV	Calcium ions	Ions taken in with food or released from bone
V	Proaccelerin (labile factor or accelerator globulin)	Plasma protein synthesized in liver
VI	—	Not used in coagulation process
VII	Serum prothrombin (stable factor)	Plasma protein synthesized in liver
VIII	Antihemophilic factor (antihemophilic globulin)	Plasma protein synthesized in liver; absence causes hemophilia
IX	Plasma thromboplastin component (Christmas factor)	Plasma protein synthesized in liver
X	Stuart factor (Stuart-Prower factor)	Plasma protein synthesized in liver
XI	Plasma thromboplastin	Plasma protein synthesized in liver
XII	Hageman factor	Plasma protein sythesized in liver
XIII	Fibrin stabilizing factor	Protein found in both plasma and in platelets

3. Thrombin now causes the conversion of soluble fibrinogen into insoluble fibrin. Fibrin initiates the physical formation of the clot.

As soon as a platelet plug begins to form, prothrombin activator is formed. Synthesis of prothrombin activator can occur along either an **intrinsic pathway** or an **extrinsic pathway**. The first process involves only factors already present in the blood. The second uses substances released into the area by the damaged tissue surrounding the ruptured blood vessel. Examine Figure 21-10 as you read the following description of these two pathways.

Intrinsic Pathway

As mentioned before, platelets do not normally adhere to each other or to the endothelium of an undamaged blood vessel. This condition is at least partially because platelets carry a negative charge on their external surface, and they are repelled by a similar negative charge on the blood vessel's endothelial lining. When a vessel is injured, however, its negative charge reverses to positive, and platelets then adhere to the area. The intrinsic pathway begins with the clumping of the sticky platelets at the injury site. The platelets begin to disintegrate and release four platelet coagulation factors into the plasma, which initiates the cascade of reactions that results in the formation of prothrombin activator (Figure 21-10).

The second stage of the clotting sequence involves the conversion of prothrombin into thrombin by prothrombin activator acting with several plasma factors (Figure 21-10). Thrombin now initiates the third stage of the clotting sequence by catalyzing two reactions. First, it causes soluble molecules of fibrinogen to combine, forming the insoluble protein fibrin. This molecule, when first formed, is relatively unstable. Thrombin contributes further to the sequence by activating plasma factor XIII (called **fibrin stabilizing factor**), which helps fibrin molecules form more permanent links among themselves, thereby stabilizing the complex and strengthening the clot.

Figure 21-10 Blood clot formation. Injury to a vessel may trigger formation of a blood clot through an intrinsic mechanism. Chemicals released by damaged tissues surrounding the vessel may also actuate an extrinsic mechanism. Both pathways lead to formation of prothrombin activator, which converts prothrombin into thrombin. Plasma coagulation factors that have been activated are shown by asterisks.

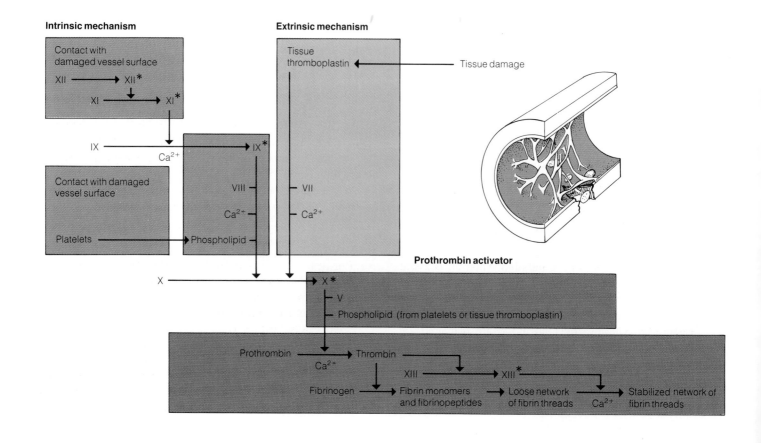

syneresis: (Gr.) *synairesis*, drawing together
fibrinolysis: (L.) *fibra*, fiber + (Gr.) *lysis*, dissolve
thrombus: (Gr.) *thrombos*, clot

Extrinsic Pathway

The extrinsic pathway involves substances released into the tissues surrounding the injured blood vessel. The damaged tissues release a complex of phospholipids and lipoproteins called **tissue thromboplastin,** which combines with several plasma factors to trigger the formation of prothrombin activator. The rest of the sequence is identical to the intrinsic pathway, in which prothrombin activator catalyzes the change of prothrombin to thrombin and the subsequent formation of a clot.

Enhancement of Clot Formation

In addition to catalyzing the change of soluble fibrinogen to insoluble fibrin, thrombin enhances the overall formation of the clot through a positive feedback mechanism. Thrombin stimulates the change of prothrombin to thrombin, which changes more and more fibrinogen to fibrin. In addition, thrombin promotes the ability of platelets to clump together and adhere to the injured endothelial lining. It also speeds up the action of some of the plasma clotting factors through enzymatic activity.

Factors Affecting Clotting Time

Many factors are involved in how quickly the body is able to stop bleeding from a wound. These factors include such things as the size of the wound and the concentration of the platelet and plasma coagulation factors. However, the length of time it takes for three processes to occur are regularly tested in evaluating the ability of a person to stop bleeding: prothrombin time, bleeding time, and clotting time.

A **prothrombin time** test determines the relative amount of prothrombin in the blood. It is measured in a fresh sample of blood from which the calcium ions necessary for coagulation are inactivated. A standard amount of calcium ions, tissue thromboplastin and plasma coagulation factors are added to the sample, and the time needed for the formation of prothrombin is noted. Normal prothrombin time is between 11 and 14 sec.

Bleeding time is the time required for bleeding to cease from a small skin puncture. Normal bleeding time ranges from 1 to 4 min, and it involves all of the stages in hemostasis from vessel constriction to clot formation.

Clotting time is the length of time necessary for blood to coagulate outside of the body. This test can be performed in a number of ways, but they all test the same critical event: the beginning of the formation of fibrin molecules. Clotting time usually ranges from 5 to 15 min and does not take into account constriction of the damaged vessel.

Limiting Growth of the Clot

Success of the clotting mechanism requires that the clot plug the damaged area but not the entire interior of the blood vessel. Therefore, the clotting factors must be removed or inactivated shortly after their job has been completed. Normal circulation removes these factors from the area of the clot and prevents their concentration in the area. Thrombin is adsorbed onto the surface of the fibrin fibers in the clot, thereby reducing its concentration in the area. A specialized plasma protein called **antithrombin III** also aids in deactivating thrombin (as its name implies) by combining with it and removing it from the area.

Retraction and Dissolution of the Clot

Within two or three days after a clot has formed, **clot retraction,** or **syneresis** (sĭn·ĕr′ĕ·sĭs) begins to occur. Platelets trapped in the clot contain protein molecules that have the ability to shorten. As these proteins contract, the clot correspondingly contracts, pulling the damaged ends of the blood vessel together. This reduces the ruptured area and further reduces the likelihood of hemorrhage. As the damaged ends of the vessel close together, it is easier for new cell growth to repair the break. The clot continues to contract, and fluid trapped within its meshwork is squeezed out. This fluid is called **serum,** which is plasma from which clotting factors have been removed.

As the clot slowly dissolves, it begins to decrease in size. Removal of the clot results from the activity of the enzyme **plasmin;** the elimination process itself is called **fibrinolysis.** Normally, one of the inactive enzymes in the plasma is **plasminogen** (plăz·mĭn′ō·jĕn). During the many reactions involved in hemostasis, activated factor XII, thrombin, and other enzymes released by damaged cells cause plasminogen to be converted to the activated form, plasmin. During fibrinolysis plasmin attacks the chemical bonds of the fibrin molecule, eventually causing the weakening and dissolution of the clot. Plasmin is also active in removing small clots that may occur within small blood vessels.

Preventing Abnormal Clotting

Although several mechanisms exist to prevent abnormal and unwanted clotting within blood vessels, clots sometimes occur when they are not needed. One of the most well-studied causes of undesired formation of clots within vessels involves deposition of cholesterol or one of its derivatives. These deposits or **plaques** become rough areas where clots can form. A clot that forms within an unbroken blood vessel is referred to as a **thrombus,** and the process is called **thrombosis.** Such a clot is dangerous if it forms in a vessel that carries blood, nutrients, and oxygen to a vital tissue or

hemoglobinopathy: (Gr.) *haima*, blood + (L.) *globus*, globe + (Gr.) *pathos*, disease
thalassemia: (Gr.) *thalassa*, sea + *haima*, blood
embolus: (Gr.) *embolos*, plug
septicemia: (Gr.) *septikos*, rotting + *haima*, blood

THE CLINIC

DISEASES OF THE BLOOD

We have seen that the blood is composed of many different components both cellular and fluid. Blood may be also considered an organ of the body with many diverse functions. Since it travels to all the cells of the body and is vital to the nutrition of every cell, blood also frequently reflects the overall state of one's health.

Genetic and Hereditary Diseases

There are several inherited diseases of the red blood cells that alter either the hemoglobin or the structure of the red cell itself and result in anemia (see Clinical Note: Anemia). Anemias related to defective hemoglobin synthesis are called **hemoglobinopathies** (hē″mō·glō″bĭ·nŏp′ă·thēz). **Sickle cell anemia** is a chronic hemolytic anemia seen almost exclusively in blacks. It is characterized by sickle-shaped red blood cells caused by the inheritance of a peculiar variation in the hemoglobin molecule (Figure 1a). In the United States about 0.3% of blacks are affected, but 8 to 13% of blacks are carriers of the disease. Sickle cell anemia produces severe anemia, leg ulcers, bone changes, fever, joint and abdominal pain, and local thrombosis. Few patients with sickle cell anemia live beyond age 40. There is no effective treatment.

Thalassemia (thăl·ă·sē′mē·ă) or **Mediterranean anemia** is another hemoglobinopathy occurring in people from countries surrounding the Mediterranean Sea and from Southeast Asia. Thalassemia results from defective production rates of either the α- or β-polypeptide chain in hemoglobin synthesis. It is inherited as a dominant trait. Patients with the most severe form have severe anemia, jaundice, spleen enlargement, and leg ulcers.

There is also a group of hereditary coagulation disorders that result in an abnormal bleeding tendency. The **hemophilias** are typical of the group and result from inherited deficiencies or abnormalities in coagulation factors. It is estimated that there are over 25,000 hemophiliac patients in the United States, 80% with a deficiency of factor VIII (hemophilia A) and the rest with deficiency of factor IX (hemophilia B). Hemophiliacs are treated by supplying them with the missing factor through the infusion of fresh plasma from normal donors or frozen precipitates of plasma. Since these patients require frequent infusions of blood products, they are often infected with the AIDS virus and hepatitis virus.

Infection

Transient infection of the bloodstream is called **bacteremia** and is common following surgical or dental procedures and in patients with venous or urethral catheters. Symptoms of bacteremia are fever, chills, and various skin rashes. Secondary infections of the meninges, joints, and other body cavities may occur. More persistent infection of the bloodstream is called **septicemia** (sĕp·tĭ·sē′mē·ă), which is caused by bacteria. Diagnosis is made by culturing the blood for the specific organism. Treatment is with large doses of specific antibiotics. The bloodstream can also be infected with viruses, fungi, and parasites. For example, malaria is an infection by a parasite that invades the red blood cells (see Clinical Note: Malaria).

Trauma and Injury

The most common injury to the blood system is blood loss. Severe blood loss or **hemorrhage** may occur externally due to injury to the blood vessels or internally with hemorrhage into the gastrointestinal tract, urinary tract, or peritoneal space from a ruptured liver, spleen, any large abdominal blood vessel, or an ectopic pregnancy. Internal hemorrhage may be difficult to detect until large amounts of blood have been lost. Massive hemorrhage results in cardiovascular collapse or shock along with pallor, sweating, low or absent blood pressure, rapid heart rate, and eventually cardiac arrest. Treatment consists of rapidly replacing the blood volume, stopping the bleeding, and restoring the cardiovascular dynamics. Large volumes of blood can also be lost from fractures or blunt trauma to large muscle groups, for example, fractures of the hip.

Injury to the blood may also occur in response to many toxic substances. Many drugs may cause damage to peripheral blood cell lines or to the bone marrow, resulting in characteristic diseases. Sulfa drugs and Butazolidin may cause **agranulocytosis,** a loss of granulocytes, resulting in loss of the ability to fight infection. The antibiotic Chloro-

organ. Deprived of oxygen, the tissue may die. For example, formation of a thrombus in a coronary vessel is a **coronary thrombosis,** which is the major cause of most "heart attacks." Sometimes a thrombus breaks loose and is carried with the circulating blood. Such a circulating clot is an **embolus** (see Chapter 23 Clinical Note: Thrombosis and Embolism). It may become stuck in a small vessel and thus block flow to vital tissues. A **stroke,** or **cerebrovascular accident** (CVA), is caused when an embolus is wedged in a vessel in the brain (see Chapter 12 Clinical Note: Stroke). If clots collect in the lungs, the effect is even more widespread, because oxygenation of the blood is diminished and all the tissues are deprived of oxygen (see Chapter 20 Clinical Note: Pulmonary Embolism).

Several chemicals in the body reduce the possibility of abnormal clotting. In addition to plasmin and antithrombin III, heparin is produced by mast cells and basophils. Heparin accelerates the activity of antithrombin III, which inactivates

leukemia: (Gr.) *leukos,* white + *haima,* blood

mycetin has been known to cause destruction of red blood cells, resulting in **aplastic anemia.** Excessive irradiation with X rays or radioactive particles may completely destroy the bone marrow. The HTLV III virus may invade and destroy a specific line of lymphocytes resulting in the inability to combat infections and cancers, causing AIDS (see Chapter 24 Clinical Notes: AIDS). Massive body burns may also destroy large amounts of peripheral blood and plasma, resulting in shock and a decreased ability to combat infections.

Tumors and Cancer
Cancer of the blood is called **leukemia,** and any of the various types of white blood cells or their precursors may be involved. Most of the cases of leukemia are of the following types: acute lymphoblastic leukemia (ALL), primarily a disease of children; acute myeloblastic leukemia (AML) (Figure 1b) and acute monoblastic leukemia, seen at any age; chronic lymphocytic leukemia (CLL), a disease of later life; and chronic granulocytic leukemia (CGL), seen in adults aged 20 to 50. Leukemias are identified by the abnormal cells observed in the peripheral blood or in the bone marrow. Abnormalities of the other cell types may also occur because the excessive proliferation of the leukemic cells in the bone marrow or lymphatic system may crowd out the normal cells and cause secondary problems such as anemia. Treatment of leukemia has improved remarkably with the advent of antileukemic drugs, total body X radiation, and bone marrow transplants.

Aging
Aging itself does not cause any specific diseases of the blood, probably because the blood cells have a short life span and are continually replaced. Some of the common problems of the elderly, such as poor nutrition, however, may be reflected in the blood by chronic anemias. The immune system does seem to deteriorate in the elderly, making them more susceptible to infectious diseases and perhaps cancer.

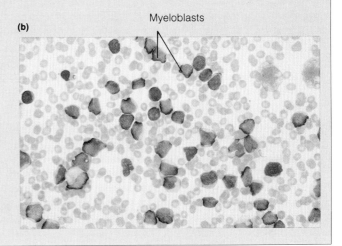

Figure 1 Photomicrographs of diseased blood cells. (a) Sickle cell anemia. Note the sickle-shaped cells. (b) Acute myeloblastic leukemia.

thrombin. Individuals with a history of spontaneous clotting are treated regularly with slower acting anticoagulant chemicals that prevent clotting. These chemicals inhibit the activity of vitamin B_{12}, which suppresses the synthesis of the clotting factors in the liver that are dependent on that vitamin.

Blood is the circulatory fluid of the body, and it is effective in its metabolic functions because it is constantly moving. It is able to distribute materials and regulate concentrations of certain substances by quickly carrying away surpluses that may accumulate in any specific region. The blood is kept moving through the circulatory system by the pumping activities of the heart. In the next chapter we will study this highly specialized muscle that manages to contract regularly throughout a lifetime. Following our look at the heart, we will examine the remaining part of the circulatory system—the blood vessels.

STUDY OUTLINE

Functions, Composition, and Production of Blood (pp. 539–543)

Functions of Blood Blood has six major functions: transport, providing water to cells, temperature regulation, defense against disease and infection, regulation of pH, and clotting to prevent loss of body fluids.

Composition of Blood The 6 L of blood in males and 4 to 5 L in females accounts for 10% of adult body weight. Blood is composed of 55% *plasma* and 45% *formed elements*.

Plasma is 92% water plus plasma proteins and other dissolved solutes. Plasma proteins maintain osmotic balance, transport certain chemicals, act as antibodies, and participate in the clotting reaction. Other plasma solutes include nutrients, waste materials, metabolites, and ions.

Formed elements include *erythrocytes, leukocytes,* and *thrombocytes (platelets)*.

Production of Blood Blood is produced through *hemopoiesis* in the yolk sac, liver, and spleen in a fetus and in red marrow and lymphoid tissue in adults. *Hematocytoblasts* produce *rubriblasts, myeloblasts, monoblasts, lymphoblasts,* and *megakaryoblasts*.

Erythrocytes (pp. 543–546)

Structure *Erythrocytes* average 7.5 μm in diameter, are biconcave, and lack a nucleus, mitochondria, or ribosomes.

Function Erythrocytes carry oxygen to body cells and remove carbon dioxide. *Hemoglobin* in erythrocytes combines with oxygen as *oxyhemoglobin* and with carbon dioxide as *carbaminohemoglobin*.

Production Erythrocytes are produced by *erythropoiesis* in red marrow from mesenchymal stem cells.

Replacement Erythrocytes live about 120 days. As they age, they are consumed by *macrophages*. The iron from hemoglobin in erythrocytes combines with transferrin and is stored, while other fragments are excreted with bile.

Control of Erythrocyte Production Under conditions of hypoxia, cells in the kidney secrete the hormone *erythropoietin*, which stimulates erythropoiesis. *Anemia* is a reduction in the ability of the blood to deliver oxygen to the tissues.

Blood Groups Erythrocytes contain proteins called *antigens* that stimulate an immune reaction, as in a blood transfusion. Antigens determine blood type, of which two major groups exist: the ABO group (blood types A, B, AB, and O) and the Rh group (Rh^+ and Rh^-).

Leukocytes (pp. 546–548)

Structure *Leukocytes* lack hemoglobin and contain normal organelles. *Granulocytes* (*eosinophils, basophils,* and *neutrophils*) have cytoplasmic granules and lobed nuclei, and they develop in the red marrow. *Agranulocytes* (*lymphocytes* and *monocytes*) lack granules, have a spheroid nucleus, and develop in lymphoid tissue.

Function Leukocytes are part of the body's system of defense against infection and disease. Neutrophils and monocytes are active *phagocytes*. Eosinophils may be phagocytic. Basophils fight bacterial infection by secreting *heparin* and *histamine*.

Production Granular leukocytes are made in the red marrow, and agranulocytes in myeloid and lymphoid tissues.

Replacement Leukocytes are replaced every few hours to few days.

Control of Leukocyte Production Leukocytosis is increased by the presence of infectious agents in the body.

Thrombocytes (Platelets) (pp. 549–550)

Thrombocytes or *platelets* are small fragments of cells 2 to 4 μm across. They lack nuclei.

Function Platelets initiate the chemical reactions that lead to formation of a *blood clot*.

Production Platelets are produced in the red marrow by *megakaryocytes*.

Hemostasis (pp. 550–555)

Hemostasis, the process of stopping bleeding, involves vascular constriction, platelet plug formation, and coagulation.

Vascular Constriction A severed or constricted blood vessel will contract at its severed end, reducing bleeding.

Platelet Plug Formation In a damaged blood vessel, platelets stick to the rough edges of the damaged area and accumulate, which begins the formation of a clot to plug the hole.

Coagulation Blood clot formation or *coagulation* involves the conversion by the enzyme thrombin of soluble *fibrinogen* in the plasma to insoluble *fibrin*, which traps cells and platelets in a clot. *Prothrombin activator* formed in the blood converts *prothrombin* into the active form *thrombin*.

In the *intrinsic pathway*, prothrombin activator is released from disintegrating platelets to begin the clotting process.

In the *extrinsic pathway*, prothrombin activator is formed from damaged tissues other than platelets.

Thrombin enhances clot formation through positive feedback by stimulating chemical changes.

Factors Affecting Clotting Time Several factors affect how quickly a clot is formed, including the size of the wound, the concentration of platelets, and coagulation factors. *Prothrombin time, bleeding time,* and *clotting time* are standard times tested to determine one's clotting ability.

Limiting Growth of the Clot Several factors, including circulation of the blood and the activity of *antithrombin III*, act to limit the extent and size of a clot.

Retraction and Dissolution of the Clot *Clot retraction* (or *syneresis*) involves contraction of proteins within the clot and growth of new cells in the vessel wall.

Preventing Abnormal Clotting Several chemicals, notably plasmin, antithrombin III, and heparin, act to prevent the formation of a clot (*thrombus*) within an unbroken vessel. Such clots may cause a coronary *thrombosis* or *stroke*.

SELF-TEST OF CHAPTER OBJECTIVES

True-False Questions
1. The primary function of blood is to transport materials.
2. Leukocytes differ from erythrocytes in part by possessing nuclei, mitochondria, and ribosomes.
3. During formation of a blood clot involving the intrinsic pathway, it is necessary for substances released by the damaged tissue in the surrounding area to initiate the process.
4. Synersis is the retraction of an established blood clot.
5. In 1 mm^3 of blood, platelets are the most abundant of the formed elements.
6. Blood plasma from which all clotting factors have been removed is serum.
7. Fibrinogen is very important in the clotting mechanism because it converts prothrombin to thrombin.
8. Anemia is any condition that reduces the ability of the blood to deliver oxygen to the body tissues.
9. Erythrocytes undergo diapedesis, leaving blood vessels to deliver oxygen directly to body cells.
10. The most abundant substance in blood is water.
11. The primary function of blood is to regulate pH.
12. At any given time approximately 50% of the blood is being oxygenated in the lungs.
13. Albumins are the most abundant plasma proteins.
14. Monoblasts are developmental stages of monocytes.
15. Erythropoiesis is the process whereby blood clots are formed in injured vessels.
16. The average life span of an erythrocyte is approximately four months.
17. Old erythrocytes are removed from circulation by macrophages.
18. Granular leukocytes are produced in the lymph nodes.
19. Thrombocytes are pieces of cytoplasm pinched off from megakaryocytes.

Matching Questions
Match the terms in the left column with their correct descriptions:

20. phagocyte
21. antibody
22. megakaryocyte
23. hemostasis
24. albumin

a. the mechanism that stops bleeding
b. a leukocyte that engulfs other cells
c. produced in response to an antigen
d. the most abundant plasma protein
e. gives rise to thrombocytes

Match the cells on the left with the statement on the right:

25. neutrophil
26. erythrocyte
27. monocyte
28. lymphocyte
29. thrombocyte

a. a wandering macrophage
b. contains the pigment hemoglobin
c. produces antibodies
d. a granular leukocyte
e. important in forming blood clots

Multiple-Choice Questions
30. The enzyme found in blood that is active in dissolving a blood clot is
    a. antithrombin III
    b. plasmin
    c. fibrinogen
    d. fibrin
    e. thromboplastin
31. A loose blood clot circulating in the body is a
    a. thrombus
    b. thrombosis
    c. plaque
    d. embolus
    e. stroke
32. Which of the following is *not* a function of blood?
    a. regulation of the pH (acid-base balance) of the body
    b. production of hormones
    c. transport of oxygen and carbon dioxide
    d. combating disease and infection
    e. maintenance of a constant internal temperature
33. Which of the following is *not* a formed element?
    a. erythrocytes
    b. thrombocytes
    c. eosinophils
    d. monocytes
    e. proteins
34. During the formation of a blood clot, the conversion of fibrinogen to fibrin is catalyzed by
    a. thrombin
    b. platelets
    c. antithrombin III
    d. prothrombin
    e. thromboplastin

Essay Questions
35. List the major categories of plasma proteins and discuss their functions.
36. List the six major functions of blood.
37. Outline the formation of an erythrocyte from a hematocytoblast.
38. Discuss the functions of erythrocytes.
39. Describe thrombocytes and indicate their method of origin and function.
40. Distinguish between granulocytes and agranulocytes and describe the different kinds of each.
41. Define hemostasis and list its three stages.
42. Discuss the process of coagulation and relate the intrinsic and extrinsic pathways of thromboplastin formation to coagulation.
43. Discuss retraction and dissolution of a clot.

Clinical Application Questions
44. In general terms, describe the life cycle of the malaria parasite.
45. Discuss the possible implication of an increase in someone's reticulocyte count.
46. Compare and contrast pernicious anemia and hemorrhagic anemia.
47. Define hematocrit. What information about the condition of the circulatory system is provided by analyzing the hematocrit?
48. Compare and contrast leukopenia and leukocytosis.
49. How many different types of leukemia have been identified? Describe as many as you can recall.

22

The Cardiovascular System: The Heart

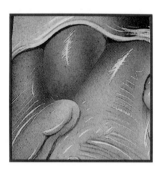

In the preceding chapter we began our study of circulation in the body with a detailed examination of the blood. Blood carries numerous materials to cells and removes their products and wastes. The blood, however, is but one part of the body's circulatory system. The other major division is the **cardiovascular system,** consisting of the heart and the blood vessels. The cardiovascular system contributes to circulation by moving and directing blood through a highly branched series of tubes that brings it very close to nearly every cell in the body.

The efficiency of the transport function of the blood is at least partially due to its constant motion as it flows through the circulatory vessels. Movement of the blood through the vessels of the body is accomplished by the contractions of the heart. The **heart** is a muscular pump that pushes the blood through the circulatory vessels, sending it through every tissue, and returning it to the heart. In this chapter we concentrate on the heart through a study of cardiac anatomy and physiology; we examine the blood vessels in the next chapter. As we study the heart, we will describe

1. the anatomy and histology of the heart,
2. the conduction system of the heart,
3. the mechanical and pressure changes that occur during the cardiac cycle and how they are related to the electrical events recorded in an electrocardiogram, and
4. the physiological factors that control heart rate.

pericardium: (Gr.) *peri*, around + *kardia*, heart
epicardium: (Gr.) *epi*, upon + *kardia*, heart

Anatomy of the Heart

After completing this section, you should be able to:

1. Describe the external and internal anatomy of the heart.
2. Distinguish among the pericardium, epicardium, myocardium, and endocardium.
3. Discuss the relationships of the atria and ventricles to the fibrous skeleton of the heart.
4. Trace the path blood travels through the heart.
5. Distinguish between atrioventricular and semilunar valves.
6. Describe intrinsic cardiac circulation.

The heart lies within the thoracic cavity, just below the sternum. It is roughly cone-shaped, with the apex pointing downward and to the left edge of the diaphragm (Figure 22-1). Large blood vessels join the heart at the base of the cone, which is at the top of the heart just behind the sternum. The position of the heart is slightly off center, so that approximately two-thirds of it lies to the left of the sagittal plane. The adult heart is usually about the size of a clenched fist. Its dimensions average 13 cm in height, 9 cm across at the base, and about 6 cm thick at its thickest anterior-posterior dimension. It weighs about 350 g, or about 0.78 lbs. These figures vary from person to person, but in general, the heart is slightly larger in men than in women. Disease, a prolonged program of physical exercise, and pregnancy can cause an increase in the size of the heart.

Pericardium

The heart is surrounded by a double-layered membrane called the **pericardium** (pĕr″ĭ·kăr′dē·ŭm) or **pericardial sac.** The inner layer of the sac (Figure 22-2), the **visceral pericardium** or **epicardium,** is a thin serous membrane closely attached to the heart's surface. (It is also considered to be the outermost

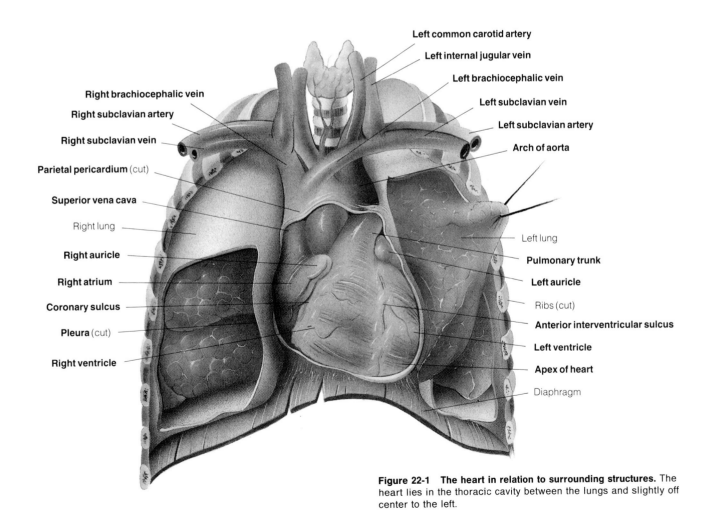

Figure 22-1 The heart in relation to surrounding structures. The heart lies in the thoracic cavity between the lungs and slightly off center to the left.

Embryonic Development of the Heart

Cardiac muscle differentiates very early in the embryo. Late in the second week this tissue becomes organized into a pair of straight **endocardial tubes** lying ventral to the foregut (Figure 22-A). The tubes fuse near their midline, producing a single internal chamber that receives venous blood at its posterior end and pumps it forward to the tissues of the embryo. This embryonic heart first begins to beat after only about 2 weeks of development. At this early stage of heart development, four regions can be identified: the **truncus arteriosus, ventricle, atrium,** and **sinus venosus.**

During continued growth, the heart bends and flexes, assuming an S shape, while a wall of tissue called the **atrial septum** divides the left and right atria. Shortly thereafter, the left and right ventricles become separated by the **ventricular septum.** The anterior truncus arteriosus divides into the pulmonary trunk (conducting blood from the right ventricle) and the aorta (conducting blood from the left ventricle). The vessels of the common sinus venosus shift position to the right atrium and develop into the superior and inferior venae cavae. These complex events transform the simple tubular embryonic heart into the four-chambered adult structure.

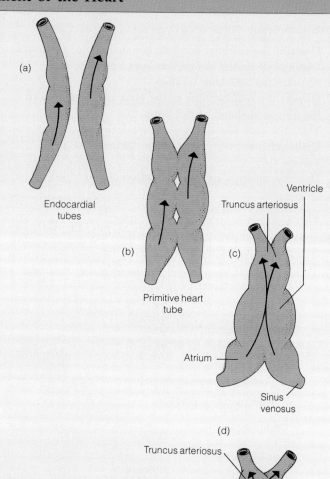

Figure 22-A Embryonic development of the heart. (a) Endocardial tubes differentiate late in the second week of development. (b) and (c) Early in the third week these tubes fuse into a single structure, with four distinct regions. (d) through (f) This structure twists and flexes late in the third week to form the heart. The arrows indicate the direction of blood flow.

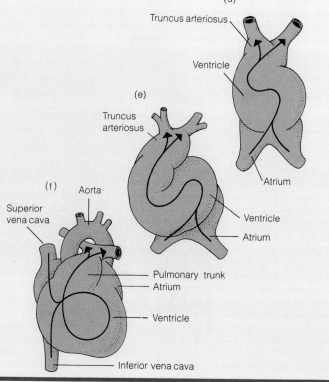

endocardium: (Gr.) *endon*, within + *kardia*, heart
myocardium: (Gr.) *mys*, muscle + *kardia*, heart
trabeculae: (L.) *trabeculae*, little beams
carneae: (L.) *carneus*, fleshy

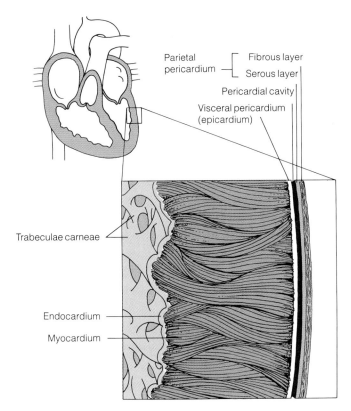

Figure 22-2 The pericardium. The adult pericardium consists of a visceral layer closely attached to the myocardium of the heart and a parietal layer separated from the heart by the pericardial cavity.

of the three tissue layers of the heart, described in the next section.) The outer layer, the **parietal pericardium,** consists of two subdivisions: a thin **serous pericardium** and a thick **fibrous pericardium** (Figure 22-2) that provides strength. The pericardial layers are separated by the **pericardial cavity.** This thin space contains a lubricant called **pericardial fluid** that is secreted by the serous membranes. Thus, the two membranes slide over one another during heartbeats with a minimum of friction. Sometimes, these membranes become infected or irritated and inflamed, resulting in **pericarditis** (pĕr·ĭ·kăr·dī′tĭs) (see The Clinic: Diseases of the Heart). Should the inflammation result in the destruction of their epithelium, the two serous membranes can fuse and prevent normal movements and fillings of the heart. In serious cases, surgery may be required to free them.

Muscular Walls of the Heart

The wall of the heart has three layers of tissue: an internal endocardium, a middle myocardium, and the external epicardium which is also the inner pericardial layer just described (Figure 22-2).

The **endocardium** is a thin layer that lines the interior of the heart, covers all internal structures (such as valves), and extends to the blood vessels that attach to the heart. It consists of two layers, which include an endothelial layer that forms the free surface of the heart's interior and an underlying fibrous layer that supports the endothelium. The fibrous layer contains blood vessels, elastic and collagenous fibers, and specialized muscle cells (Purkinje fibers) that transmit nerve impulses to coordinate contractions of the heart.

The **myocardium,** the thickest of the layers, is composed entirely of muscle tissue. The myocardial layer of the ventricles (Figure 22-2) is much thicker than that of the atria, which allows it to contract with sufficient force to propel blood through the lungs and the body. In the ventricles the muscle fibers are arranged in spiraling bands rather than extending straight from the base to the apex, as in the atria. This arrangement produces contractions that squeeze the blood from the ventricles. It is the myocardial contractions that pump the blood from the ventricles through the circulatory system.

The external surface of the myocardium is relatively smooth, but the internal surface in the ventricles is very irregular. In these areas are many folds, columns, and ridges called **trabeculae carneae** (tră·bĕk′ū·lē kăr′nē·ē) and conical projections called **papillary muscles.** The trabeculae carneae provide additional strength to the myocardium, while adding a minimum of weight. Papillary muscles prevent backward flow of blood through heart valves. (Both are discussed in detail later in this chapter.)

Chambers and Valves of the Heart

The heart has four chambers: right and left **atria** (singular form, **atrium**) specialized for receiving blood from veins and right and left **ventricles** specialized for pumping blood into arteries (Figure 22-3). The atria have thinner walls than the ventricles and lie at the base of the heart, while the ventricles make up the apex and bulk of the heart. The atria and ventricles are separated by muscular walls referred to as **septa.** The **interatrial septum** lies between the left and right atria, while the **interventricular septum** separates the two ventricles. In addition to muscle, the septa also contain the **fibrous skeleton of the heart,** an internal network of fibrous connective tissues (Figure 22-4). This skeleton has two functions: first, it provides a place of attachment for cardiac muscle, and second, it helps support the valves that separate the atria from the ventricles. These valves are the **atrioventricular** (ā″trē·ō·vĕn·trĭk′ū·lăr) valves and are attached to a portion of the fibrous skeleton known as the **coronary trigone** (trī′gōn). The fibrous tissue of this structure provides

562 Chapter 22 The Cardiovascular System: The Heart

vena cava: (L.) *vena*, vein + *cavus*, a hollow

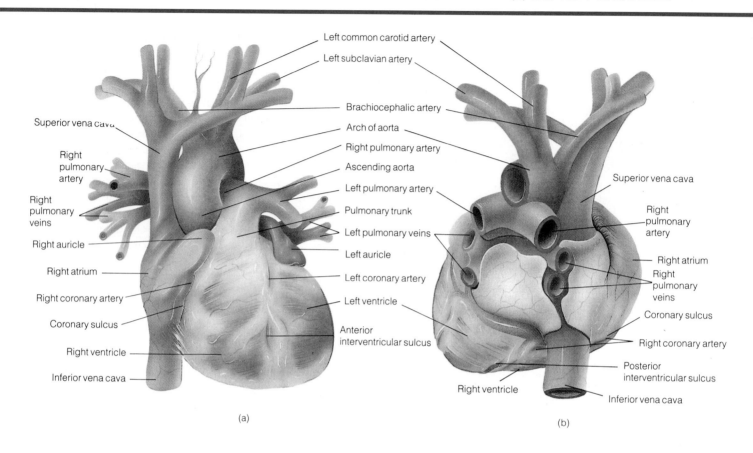

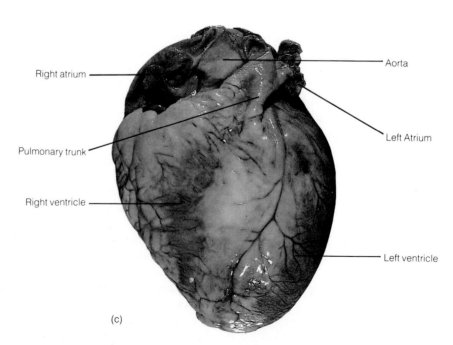

Figure 22-3 The heart in detail. (a) Anterior view. (b) Posterior view. (c) Photograph of heart.

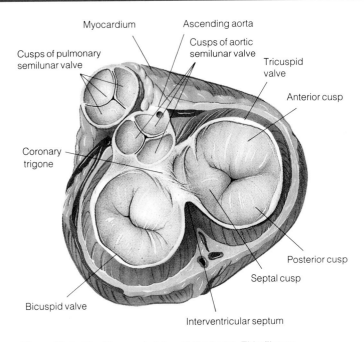

Figure 22-4 The fibrous skeleton of the heart. This fibrous connective tissue serves as a place of attachment for the myocardial tissue of the heart, and it also supports the atrioventricular valves.

support for the valves as they operate during normal cardiac movements while pumping blood.

The chambers in the two sides of the heart form a natural division of the twofold function of the circulatory system. Blood entering the right side of the heart has traveled through the body where its oxygen supply has been lowered and its carbon dioxide level has been raised. This blood is pumped through the heart's **pulmonary circulation** to the lungs. Here the blood exchanges carbon dioxide for oxygen (see Chapter 20). Pulmonary circulation restores a high concentration of blood oxygen and lowers the concentration of carbon dioxide. This blood moves from the lungs and enters the left atrium of the heart. The left side of the heart pumps blood through the **systemic circulation,** that vast system of vessels that carries blood to all the tissues of the body.

Right Atrium

Blood from the body first enters the heart in the **right atrium** from three major veins: the **superior vena cava, inferior vena cava,** and **coronary sinus** (Figure 22-3). The superior vena cava receives blood from the upper part of the body, while the inferior vena cava drains all tissues located below the diaphragm. The coronary sinus receives blood from the muscles of the heart itself. The opening of the coronary sinus is an enlarged vein on the posterior surface of the atrium that collects blood coming from the veins of the cardiac walls.

The **auricles** are ear-shaped extensions of both atria that increase their volume under certain circumstances. During resting conditions an atrium does not open to its full capacity, expanding only enough to hold the moderate amount of blood entering. During times of physical activity and exercise, however, the amount of blood entering the heart may double or triple, and at these times the auricles function as a volume reservoir, expanding to their full extent. They thus increase the volume of both atria and the amount of blood they can accommodate. The interior wall of the right atrium is relatively smooth, whereas the auricle contains numerous ridges of cardiac muscle called the **musculi pectinati** (mŭs′ kū·lē pĕk·tĭ·nă′tē). Musculi pectinati function in much the same way as trabeculae carneae, providing the heart with additional strength with a minimum of weight.

The opening between the superior vena cava and the right atrium is unobstructed, but a flaplike extension from the auricle wall at the base of the inferior vena cava appears to be a remnant of an important functional structure in the fetus. This flaplike structure directs blood through an opening in the fetal heart called the **foramen ovale** (fōr·ā′mĕn ō·vă′lē) (discussed in Chapter 23). This structure has no apparent adult function. In the adult heart the **fossa ovalis,** an oval depression, is visible in the interatrial septum. This is a remnant of the former position of the fetal foramen ovale.

Right Ventricle

Blood then passes from the right atrium into the right ventricle. The **right ventricle** is a triangular chamber occupying the right apex of the heart and resting on the diaphragm. Its walls are thicker than the walls of the atrium. It is separated from the left ventricle by the interventricular septum and from the right atrium by the muscular wall, the fibrous skeleton, and an atrioventricular valve.

The **right atrioventricular** (or **tricuspid**) **valve** is a specialized structure that separates the right atrium and ventricle (Figure 22-5) and consists of two structures: cusps and cords. The three triangular cone-shaped **cusps** extend across the opening between the chambers. These flexible cusps have their peripheral edges attached to the fibrous skeleton (Figure 22-4), and their central edges project into the cavity of the right ventricle. The interior surface of the ventricle is not smooth and flat. Instead, it consists of trabeculae carneae and several fingerlike papillary muscles that project into the cavity of the ventricle (Figures 22-2 and 22-5). The papillary muscles are attached to the free edges of the cusps of the atrioventricular valve by elongate, tendinous cords called **chordae tendineae** (kor′dē tĕn·dĭn′e·ē).

> **CLINICAL NOTE**
>
> ## OPEN HEART SURGERY
>
> In the early days of surgery, the heart was thought to be sacred and its contact to be avoided by surgeons. The first documented case of cardiac surgery was performed by a courageous surgeon who successfully sutured a knife wound in the myocardium of a young man who was dying from the hemorrhage. Once it was demonstrated that the heart could withstand surgical manipulation, progress became dependent on advances in anesthesiology to keep the patient alive while the surgeon performed the operation. The first heart operations were really not on the heart itself but on the vessels around the heart, such as tying the artery to fix patent ductus arteriosus (see The Clinic: Diseases of the Heart). The next advances came in the treatment of congenital heart defects. Many clever procedures were developed to reroute the arterial circulation by anastomosing arteries and veins outside of the heart in children with cyanotic heart disease. The heart was still not penetrated.
>
> One of the first operations on the heart itself was **mitral commissurotomy** (kŏm″ĭ·shūr·ŏt′·ō·mē) in which the mitral valve, fused by scar tissue from the ravages of rheumatic fever, was opened by inserting a finger through a small opening in the auricle and fracturing the scarred valve. Later, a small hooked knife was inserted along the finger to actually cut the valve open. Interauricular septal defects were attacked next by sewing the wall of the auricle around the edge of the defect to close it. A finger was inserted into the auricle to guide a large, hooked needle through the wall of the auricle, through the edge of the septal defect, and out through the auricular wall again. This operation was named "the doughnut" operation because of the appearance of the auricles at the end of the operation.
>
> In the early 1950s the heart was finally opened with the development of an efficient **extracorporeal circulation pump** (heart-lung machine) along with techniques for lowering body temperature and temporarily stopping the heart.
>
> These advancements allowed surgeons to make a direct attack on the interior heart. Surgeons could now replace defective valves with artificial valves or pig valves (see Unit V Case Study: Heart Patient). Septal defects could be closed with plastic patches. Congenital defects could be repaired directly.
>
> Further progress in angiography, cardiac catheterization, and stress testing (see Clinical Note: Special Cardiac Diagnostic Procedures) encouraged surgeons to make a direct attack on the problem of coronary artery disease, the number one cause of death in the United States (see The Clinic). Coronary arteries obstructed with plaques were bypassed with vein grafts from the legs. Over 2 million such **coronary bypass** operations have been performed in the United States to date.
>
> Patients with such severe damage to their myocardiums that death was eminent were given new hope when Christiaan Barnard, a surgeon from South Africa, performed the first **heart transplant operation.** Transplantation surgery is becoming more successful each year. The limiting factors are the availability of donor hearts and the expense. In 1982 there were 45 to 50 transplants performed in the United States. Open heart surgery has advanced to such a degree that recently a patient with severe lung disease had a new heart and lungs transplanted, and his normal heart was transplanted to another patient.
>
> There has been continual search for a suitable **artificial heart** system. The first successful implant of the Jarvik-7 artificial heart was performed in November 1982. Since that time, recurrent problems with strokes and other embolic problems have caused a limitation of this procedure to a temporary measure required to save the patient's life until a donor heart can be transplanted. Research continues for a more effective artificial heart.

As the right atrium fills with blood, it contracts and the blood enters the right ventricle. The force of blood passing through the tricuspid valve forces the cusps down into the right ventricle, providing a clear path into the chamber. After the right ventricle fills with blood, it contracts, squeezing the blood out of the chamber toward the lungs. This action also puts pressure on the undersides of the three atrioventricular cusps and causes them to close by pushing them upward. The simultaneous contraction of the papillary muscles pulls on the chordae tendineae, which prevents the cusps from prolapsing into the right atrium. As a result, the opening is sealed and blood does not leak back into the right atrium.

Blood leaves the right ventricle through the **pulmonary trunk.** The base of this large artery lies in the upper, medial, front region of the right ventricle. Trabeculae carneae are absent here. The entrance to this artery contains a three-part **pulmonary semilunar valve** composed of pocketlike folds (cusps) of endothelium (Figure 22-5). During ventricular contraction, blood forces the cusps open, but they close rapidly when the ventricle relaxes. Closure of the semilunar valve prevents backward blood flow.

Left Atrium

The **left atrium** resembles the right atrium, but its myocardial layer is slightly thicker. Its auricle projects anteriorly and superiorly and partially covers the pulmonary trunk. Blood from four **pulmonary veins** enters the right atrium. Their openings are unobstructed with valves, and relaxation of the atrial walls allows the chamber to fill with oxygenated blood. The musculi pectinati and the left side of the fossa ovalis are also in the left atrium. Contraction of the left atrium sends blood through the atrioventricular opening into the left ventricle (Figure 22-6).

Anatomy of the Heart

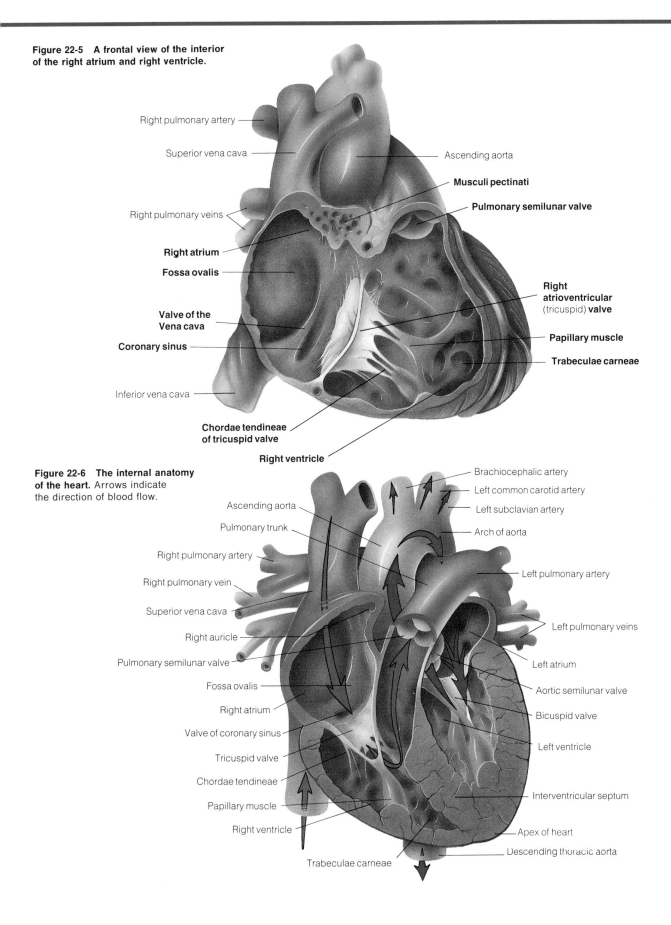

Figure 22-5 A frontal view of the interior of the right atrium and right ventricle.

Figure 22-6 The internal anatomy of the heart. Arrows indicate the direction of blood flow.

Left Ventricle

The **left ventricle** occupies the left apical position of the heart. Its myocardium is approximately three times as thick as that of the right ventricle. This chamber propels blood throughout the entire body when it contracts. Its lining resembles that of the right ventricle with numerous trabeculae carneae and several papillary muscles.

The separation between the left atrium and ventricle is maintained by the **left atrioventricular valve,** also called the **mitral** (mī′trăl) or **bicuspid valve.** It has two instead of three cusps, which connect through numerous chordae tendineae to papillary muscles that project from the wall of the ventricle. The bicuspid valve functions in precisely the same way as the tricuspid valve: both prevent backflow of blood.

Blood leaves the left ventricle through the **aorta,** the largest blood vessel in the body (discussed in more detail in Chapter 23). The base of the aorta lies at the superior, medial edge of the left ventricle. Its orifice is closed by an **aortic semilunar valve** (Figure 22-7) similar in function and structure to the pulmonary semilunar valve. It prevents the backward flow of blood from the aorta.

Cardiac Intrinsic Blood Supply

An adult heart contracts approximately 75 times each minute every day of our lives. This effort results in an extensive amount of work by cardiac muscle cells, and they require a constant supply of oxygen and nutrients. They also produce large amounts of waste materials that must be removed. The myocardium has first call on the blood leaving the left ventricle. Just above the cusps of the aortic semilunar valve, the **right** and **left coronary arteries** originate (Figures 22-3 and 22-7). This is the **coronary circulation,** an intrinsic supply of freshly oxygenated blood to the heart (Figure 22-8). The left and right vessels mainly supply corresponding sides of the heart, but extensive branching along with numerous interconnections called **anastomoses** (ă·năs″tō·mō′sēz) between their branches results in a complex circulatory pattern as the arteries extend from the base of the ventricles to the cardiac apex. One advantage of extensive branching and anastomoses is that blood arriving at any spot in the myocardium may originate in any one of several large branches. Alternative sources of blood for a specific area are called **collateral circulation,** and if blockage occurs in one branch

Figure 22-7 The aortic semilunar valve. (a) The aorta has been cut longitudinally along its anterior surface and opened to show the cusps of the valve. (b) The aorta has been sectioned transversely just above the valve, and the view is looking down on the cusps in their normal closed position. The pulmonary semilunar valve looks virtually identical to this, but lacks the coronary arteries.

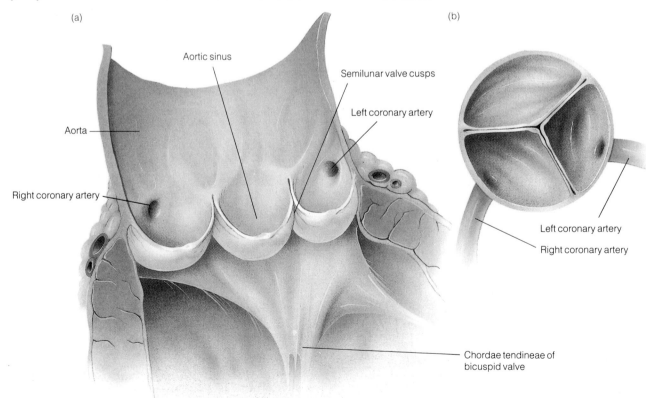

infarction: (L.) *in*, in + *farcire*, to stuff

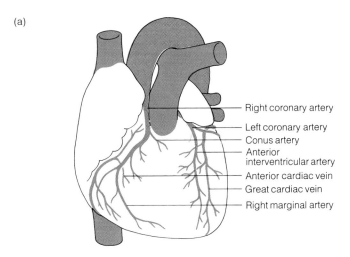

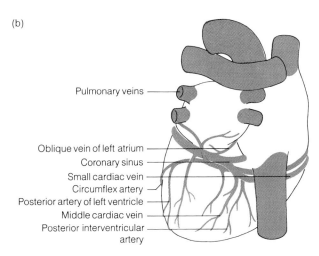

Figure 22-8 The coronary vessels of the heart. (a) Anterior and (b) posterior views.

Venous blood in the myocardium is collected by several **cardiac veins** that anastomose extensively with each other (Figure 22-8). Two of the major cardiac veins are the **great cardiac vein,** which drains blood from the anterior aspect of the heart, and the **middle cardiac vein,** which receives blood from the posterior aspect. Almost all this blood empties into the coronary sinus, an enlarged vein that lacks muscle in its wall and hence cannot change its diameter. The coronary sinus drains into the right atrium. Only a few small veins empty into the right atrium or the ventricles directly.

Coronary circulation is essential to the function of cells in the heart. Blood inside the chambers does not supply heart muscle directly. This function is provided instead by the coronary circulation. Unfortunately, a partial blockage of one coronary artery can result in a decrease in the critical blood supply to the myocardium distal to the blockage. Such a condition may be serious enough to cause severe chest pain, known as **angina pectoris** (ăn·jī'nă pĕk'tō·rĭs). Complete blockage causes a cessation of blood flow and may result in a **myocardial infarction** (mī·ō·kăr'dē·ăl ĭn·fărk'shun) (see The Clinic: Diseases of the Heart). Commonly called "heart attacks," these infarctions contribute to the death of that part of the heart cut-off from nutrients and oxygen, and the dead muscle is replaced with fibrous connective tissue. Because this scar tissue lacks contractility, the heart loses some of its function. Massive infarctions affecting large areas of cardiac tissue can result in death.

leading to a specific area, increased flow through collateral circulation may be sufficient to satisfy the needs of the myocardial cells in that area.

The right coronary artery passes under the right auricle and divides into two branches: the **posterior interventricular artery** and the **marginal artery.** The former supplies blood to the posterior walls of the two ventricles, while the latter carries blood to the right atrium and the anterior wall of the right ventricle. The left coronary artery passes under the left auricle and branches into the **anterior interventricular artery** and the **circumflex artery,** which supply the right atrium and both ventricles. The small arteries in the coronary circulation system eventually divide into networks of microscopic capillaries that provide a link between the incoming blood in the arteries and the outgoing blood in the veins.

Cardiac Physiology

After completing this section, you should be able to:

1. Describe the conduction system of the heart.
2. Discuss how a pacemaker potential relates to an action potential.
3. Discuss the role of the sinoatrial node in regulating heart rate.
4. Discuss Starling's law of the heart relative to differences in intensities of heartbeats.
5. Describe the unique features of the cardiac refractory period.

The heart has but one function—to pump blood through the arteries that lead from it into the pulmonary and systemic systems. This operation has intrigued scientists for centuries, and observations of the heartbeats of many animals are among the oldest physiological recordings. In this section we will study mechanisms by which the heart propels blood through the pulmonary and systemic circuits.

Myocardium

Of fundamental importance to cardiac functioning is the histology of the myocardium. A preliminary look at cardiac muscle was presented in Chapter 4; its structure will now be elaborated.

The myocardium consists of striated, uninucleate cells (fibers) connected at their ends by a thin disklike plasma membrane called an intercalated disk (Figure 22-9). The cells form elongate layers of muscle in the myocardium. Although heart muscle consists of individual cells in the form of fibers, cytoplasmic connections, or gap junctions, exist between them at each intercalated disk. These connections allow impulses to travel rapidly from cell to cell in the myocardium, causing the whole heart to contract as a unit (see Chapter 9).

Recall from Chapter 9 that a system of specialized tubules extends longitudinally and transversely through muscle fibers, particularly the sarcoplasmic reticulum. This tubular system, while present in cardiac muscle, is not extensively developed.

Conduction System of the Heart

During the early formation of the heart, its embryonic cells exhibit the ability to contract independently and spontaneously long before nerve connections are established. Although this ability is beneficial, the heart could not function as a unit if all its cells were contracting at different rates. During development most heart cells become dependent upon the stimulation of specialized muscle cells that form the intrinsic conduction system.

As the heart matures, certain of its cells lose their ability to contract and become specialized to initiate and better conduct electrical impulses that cause contraction in other cardiac muscle cells. These cells form the **conduction system.** Its individual segments include the sinoatrial node, atrioventricular node, the atrioventricular bundle, the bundle branches, and the Purkinje fibers (Figure 22-10). These nodes and fibers (described in the following section) are masses of cells specialized to generate electrical impulses more efficiently, hence more rapidly than other cells.

The heart is innervated by nerves of the autonomic nervous system. These nerves are regulatory devices rather than cells that initiate or stop contractions. Their stimulations only increase or decrease the *rate* of contraction.

Cardiac Contraction

Each contraction (heartbeat) of the heart is initiated by an electrical impulse from the **sinoatrial** (sīn″ō·ā′trē·ăl) or **SA node.** This node spontaneously fires 90 to 100 times per minute, but stimulation from the vagus nerve slows the rate to about 70 times per minute. These impulses set the basic pace at which the heart beats and explains why the SA node is called the **pacemaker** of the heart.

As in neurons, the cells of the SA node have a sodium-potassium pump that can produce a "resting" potential of about −55 mV. However, the SA node fires spontaneously because, instead of returning to a resting state at the end of an impulse, the pacemaker cells become permeable to Na$^+$. This inward leakage of Na$^+$ depolarizes the membrane in a **pacemaker potential.** Upon reaching a depolarization of about +20 mV, an action potential is initiated that stimulates contraction of the muscle cells in the immediate vicinity of the SA node. The stimulus sweeps across the atria, triggering musculature contraction as it goes. This results in the atrial contraction that squeezes blood from the atria into the ventricles. The wave of contraction arrives at the atrioventricular septum, where the fibrous skeleton prevents the

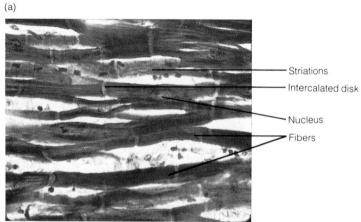

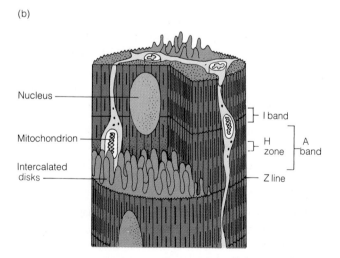

Figure 22-9 Histology of the cardiac muscle tissue. (a) Photomicrograph (mag. 307 ×.) (b) Block diagram through the myocardial tissue. (See Figure 9-17 for a more detailed view of cardiac muscle cells.)

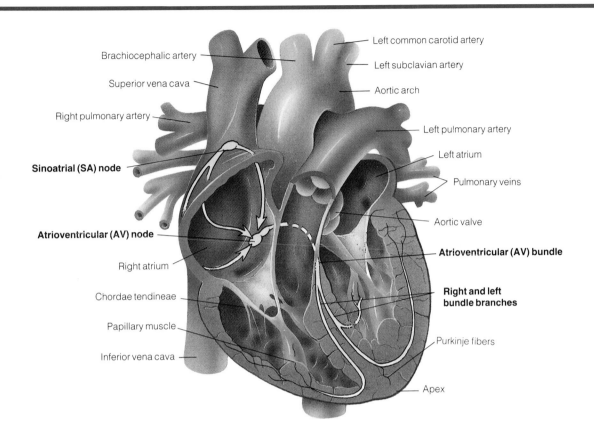

Figure 22-10 The conducting system of the heart. The arrows indicate the direction of the flow of impulses initiating atrial and ventricular contractions.

action potential from spreading directly into the ventricles. Contraction of the ventricles requires the excitation of the **atrioventricular** or **AV node,** another part of the conduction system.

Once stimulated by the action potential from the SA node, the cells in the AV node depolarize and generate another impulse that is transmitted into the ventricular region through the **atrioventricular (AV) bundle** (or **bundle of His**) (Figure 22-10). This network of fibers spreads through the interventricular septum as **bundle branches.** Upon reaching the apex of the heart, the fibers spread out into the external ventricular walls as the **Purkinje** (pŭr·kĭn´jē) **fibers.** Because the atrioventricular bundle leads directly to the apex before extensive branching occurs in the Purkinje fibers, the impulse from the AV node travels to the apex of the heart first and is then spread through the myocardium of the ventricles. This causes the ventricular contraction to begin at the apex and move toward the atrioventricular septum.

This wave of ventricular contraction follows atrial contraction and squeezes blood out of the ventricles and into the pulmonary trunk and aorta. The force of blood moving out of the ventricles simultaneously opens the semilunar valves while closing the atrioventricular valves.

The muscular mass of the heart has two important built-in mechanisms that control the heartbeat. First, when one of the cells of the myocardium contracts, all contract. This characteristic is due to the close connections maintained between the individual cells through the intercalated disks. Like skeletal muscle cells, myocardial cells exhibit an all-or-none response. Myocardial cells differ, however, because the force of their contractions can vary depending on the degree to which they are stretched. Up to a point, increased stretching of cardiac muscle cells results in more forceful contractions, a phenomenon known as Starling's law of the heart (see section on Cardiac Output).

Second, cardiac muscle is unique in having a built-in antifatigue device: a long **refractory period.** Once a skeletal muscle is stimulated and begins to contract, a very short period of time (0.005 sec) exists at the beginning during which the muscle is insensitive to additional stimuli and it will not contract again. Once past this refractory period, a

electrocardiography: (Gr.) *elektron*, amber + *kardia*, heart + *graphein*, to write

> **CLINICAL NOTE**
>
> **FAINTING**
>
> A **syncopal** (sĭn'kō·păl) **attack,** or fainting, is a temporary loss of consciousness associated with an emotional crisis, neurologic event, arrythmia, or heart block. The classic example is the person who "passes out" at the sight of blood, an emotional-mediated response caused by stimuli from the vagus nerve on the blood vessels. The unconsciousness is rarely deep and usually of short duration. The person is usually cool but is perspiring profusely and has a slow pulse. He or she may even have a mild seizure. Recovery is usually spontaneous within a few minutes.
>
> Syncopal attacks related to heart block can be a more serious matter, depending upon the cause of the block. A persistent slow pulse (below 60 beats per minute) or repeated syncopal attacks from temporary cessation of heartbeat are usually indications for insertion of a pacemaker (see The Clinic: Diseases of the Heart).

uses them to diagnose anatomical and physiological conditions of the heart. The device that detects the electrical currents associated with the heartbeat is an **electrocardiograph,** and it produces a tracing called an **electrocardiogram,** commonly referred to as an **EKG** (or **ECG**) (Figure 22-11).

A typical tracing of a normal heartbeat is shown in Figure 22-11. The normal electrocardiogram shows a series of deflections from the horizontal line called "spikes" that represent electrical impulses from contraction of the heart muscles (see Chapter 9). Each deflection is labeled in sequence: P, Q, R, S, and T. The letters merely indicate the order of appearance of the spikes. The **P wave** indicates the first electrical event and marks the depolarization and contraction of the muscles of the atria. The **P-R interval** accounts for the delay preceding the ventricular contraction. It is followed by the **QRS complex,** which marks depolarization and contraction of the ventricles. The **T wave** indicates repolarization of the ventricles. Repolarization of the atria is not visible as a wave because it occurs during the QRS complex, which is strong enough to mask any other event.

Abnormalities in an EKG usually indicate abnormalities in the electrical events associated with the heartbeat, and a trained specialist can learn much about the condition of a subject's heart by analyzing an EKG. For example, the tracing in Figure 22-12a is of a heart undergoing a partial heart block (part of the electrical impulse is blocked). Note that the QRS complex does not follow every P wave; however,

new stimulus will cause the muscle to begin a new contraction, even though it has not completed the first. Chapter 9 described some characteristic happenings under these circumstances, such as summation of twitches, treppe, and tetanus. In cardiac muscle the refractory period is 600 times longer (0.30 sec) than in skeletal muscle. This allows the heart muscle to be over halfway through its contraction period before it becomes sensitive again to another stimulus. This long refractory period minimizes the possibility of a prolonged tetanic contraction and permits the heart to beat continuously for an entire lifetime.

Electrocardiography

After completing this section, you should be able to:
1. Define the science of electrocardiography.
2. Distinguish among P, Q, R, S, and T wave components in a normal electrocardiogram.
3. Describe abnormalities associated with atrial and ventricular fibrillation that may appear on an EKG.

As with other impulses generated in the body, those that arise in and travel through the conduction system of the heart generate electrical currents that can be measured on the skin surface with the proper equipment. A branch of medical science called **electrocardiography** (ē·lĕk″trō·kăr′dē·ŏg′ră·fē) records and analyzes these electrical events and

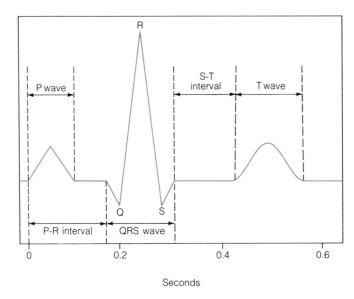

Figure 22-11 A normal electrocardiogram produced by an electrocardiograph machine.

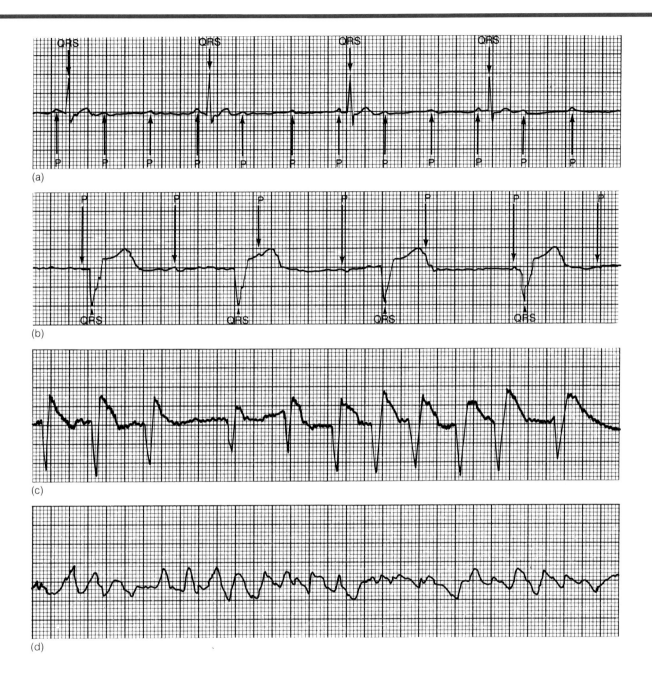

Figure 22-12 Abnormal electrocardiograms. (a) Partial heart block. (b) Complete heart block. (c) Atrial fibrillation. (d) Ventricular fibrillation. See text for an explanation of each tracing.

when the QRS is present, it is in the proper position relative to the P wave on the tracing. The electrocardiogram of a complete heart block (Figure 22-12b) has both P and QRS waves, indicating both atrial and ventricular contractions, but there is no correlation between their positions on the tracing, indicating independent atrial and ventricular rhythms. During atrial fibrillation (rapid, uncoordinated contraction) (Figure 22-12c), there is no regular contraction of the atria, and thus no P wave appears on the tracing. Without the P wave to set the pace for the ventricles, the ventricular contractions are independent and irregular. During ventricular fibrillation (Figure 22-12d), an irregular heartbeat occurs.

diastole: (Gr.) *diastellein*, to expand
systole: (Gr.) *systole*, contraction

Cardiac Cycle

After completing this section, you should be able to:

1. Describe the cardiac cycle.
2. Discuss the relationship of mechanical changes to electrical events during the cardiac cycle.
3. Discuss pressure changes that occur during the cardiac cycle.
4. Discuss causes of the heart sounds.

The function of the heart is to receive venous blood from the body and lungs and to propel that blood away from the heart with sufficient force to send it through both pulmonary and systemic circulations. This process is called the **cardiac cycle,** and it involves two fundamentally different events: relaxation of the muscular walls of the chambers, or **diastole** (dī·ăs′tō·lē), followed by contraction of the walls, or **systole** (sĭs′tō·lē).

This cycle can be best understood if it is broken down into its component parts: the mechanical changes that occur during the cycle and the pressure and volume changes associated with the cycle. Throughout the discussion, the correlation that these changes have with the electrical events elucidated by the electrocardiogram will be discussed.

Mechanical Changes During the Cardiac Cycle

Although the cardiac cycle involves diastolic and systolic changes on both sides of the heart, to simplify the following discussion, we concentrate only on what happens on the left side. Frequent reference to Figure 22-13 will make it easier to understand how these physical events coordinate with changes in pressure and volume within the chambers.

The cardiac cycle begins with the initiation of electrical changes seen as the P wave on the electrocardiogram. This wave indicates that the atria are contracting. Just prior to this, they have been in diastole and oxygenated blood has been entering the left atrium freely through the pulmonary veins. Most of this blood passes straight through the bicuspid valve into the left ventricle. The initiation of the P wave indicates that the contraction of both atria is in progress, and the left atrium is now actively forcing the remainder of the blood from its interior down into the left ventricle.

The amount of time required for the action potential generated in the SA node to reach the apex and initiate a ventricular contraction is approximately 0.1 sec. The stimulation of this impulse causes the left ventricle to contract, sending blood into the aorta. Simultaneously, the atrium begins to relax. Contraction of the ventricle applies pressure

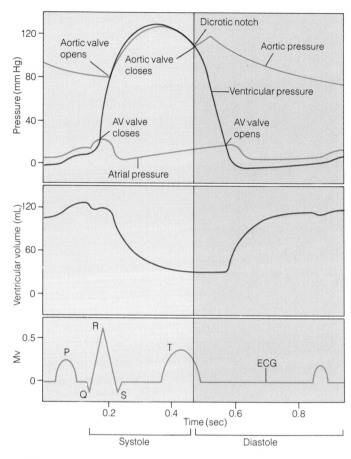

Figure 22-13 The cardiac cycle. See text for an explanation of the pressure changes (top) and volume changes (middle), that occur during the cycle. The corresponding ECG (or EKG) pattern is shown at the bottom.

to the undersides of the cusps of the bicuspid valve, resulting in closure of the valve. This increase in pressure forces the aortic semilunar valve open as blood leaves the ventricle. Contraction of the ventricle, represented by the duration of the QRS wave in Figure 22-13, is approximately 0.3 sec. During this time the relaxed left atrium has been receiving blood from the pulmonary veins. As the ventricle enters diastole, blood accumulating in the left atrium pushes open the bicuspid valve and once more begins to flow into the ventricle. For the next 0.5 sec, the ventricle remains in diastole, passively filling with oxygenated blood. Approximately 0.1 sec before the end of ventricular diastole, the cells of the SA node generate another pacemaker potential, triggering another atrial systole and starting another cardiac cycle. Examination of Figure 22-13 shows some overlap between the systole and diastole of the two chambers. Under resting condition the entire cardiac cycle lasts approximately 0.8 sec.

Pressure Changes During the Cardiac Cycle

The cyclic changes in systole and diastole result in a corresponding cycle of pressure changes in the chambers of the heart and the aorta. The top part of Figure 22-13 shows how these pressures change during a typical cardiac cycle, and the following discussion is based on this graphic presentation. The pressure and volume graphs in Figure 22-13 were derived from measurements taken by inserting sensitive pressure-recording devices through **catheters** (or narrow tubes) inserted into arteries or veins directly into a live, beating heart (Figure 22-14). (See Clinical Note: Special Cardiac Diagnostic Procedures.)

Prior to the beginning of a cardiac cycle, pressure within the left atrium is just slightly above 0 mm Hg. This low pressure level results from the blood entering the left atrium and not remaining there. Instead, it freely flows through into the ventricle and does not push against the atrial walls. At the start of a cardiac cycle, atrial systole squeezes the remaining atrial blood into the ventricle, elevating the pressure in both chambers. The onset of ventricular systole pushes the blood against the bottoms of both the bicuspid valve and the aortic semilunar valve. This backflow of blood causes the bicuspid valve to close, and even causes it to bulge into the atrium, producing a continued increase in atrial pressure. Continued contraction of the ventricle pushes the blood against the aortic semilunar valve, eventually causing it to open and allowing the ventricular blood to flow out of the ventricle into the aorta. The drop in pressure against the bicuspid valve relieves the bulging, causing a sudden drop in atrial pressure. This drop in atrial pressure is followed by a gradual increase in pressure, as blood from the pulmonary veins flows into the atrium. About 0.6 sec into the cardiac cycle, the pressure in the atrium exceeds that of the emptying left ventricle, and the bicuspid valve opens, resulting in another drop in atrial pressure as blood flows through into the ventricle. For the next 0.2 sec, atrial pressure rises gradually, leading to the occurrence of another P wave, which signals the start of another cardiac cycle, with concomitant changes.

Pressure changes within the left ventricle are tied closely to changes in the left atrium. At the beginning of a cardiac cycle, the left ventricle musculature is in a relaxed, diastolic state, and its interior pressure is less than that of the atrium. When the left atrium contracts, it sends additional blood into the ventricle, causing a slight increase in ventricular pressure. With the onset of the QRS wave that indicates ventricular systole, pressure in the left ventricle rises rapidly. This causes the bicuspid valve to close quickly and exerts pressure on the underside of the aortic semilunar valve. Within less than a tenth of a second, the aortic valve opens. Although blood now is moving out of the ventricle into the aorta, the ventricular pressure continues to rise because of the powerful muscular contraction of systole. As blood leaves the ventricle, the pressure begins to abate, and the onset of diastole brings about a rapid decline in ventricular pressure. When ventricular pressure falls below that of the atrium, the bicuspid valve opens and allows blood to enter the ventricle, starting another cardiac cycle.

Aortic pressure changes are not nearly as great as those in the left ventricle. At the beginning of a cardiac cycle,

Figure 22-14 Heart catheter. (a) Arteriogram (an X ray of arteries) showing a catheter that has been inserted into the right coronary artery of the heart. (b) Interpretative drawing of arteriogram in (a) showing catheter in relation to surrounding structures.

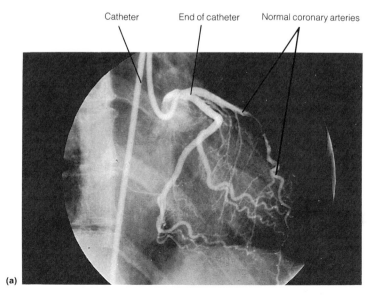
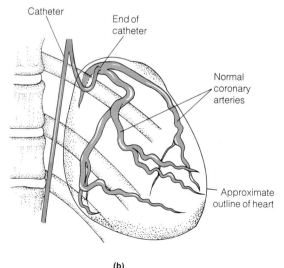

stethoscope: (Gr.) *stethos*, chest + *skopein*, to examine

> **CLINICAL NOTE**
>
> **OFFICE EVALUATION OF THE CARDIAC PATIENT**
>
> A great deal of information concerning the status of a patient's cardiovascular system can be obtained through very simple examination procedures in the physician's office. A thorough history of the patient's complaints of chest pain, palpitation, breathlessness, and fatigue are of prime importance. Careful evaluation of the **vital signs,** which include pulse rate, blood pressure, temperature, and respiratory rate, can provide clues to arrhythmias, hypertension, and low cardiac output. Inspection of the neck veins may provide clues to abnormal pulsations and excessive dilation due to cardiac failure. Visual inspection of the chest wall contours can detect such problems as asymmetry due to cardiac enlargement and abnormal pulsations. Palpation (feeling with fingers) of the apical impulse on the chest wall might suggest cardiac enlargement or arrhythmia. Percussion (tapping the chest wall with the fingers) can elucidate the borders of cardiac dullness and give clues to abnormal chamber enlargement. **Auscultation,** or listening to the heart with the stethoscope, is a much neglected art that can provide enormous information concerning the quality of the heart sounds; this includes the presence of abnormal sounds, such as murmurs, clicks, and arrhythmias. **Heart murmurs** are vibratory sounds that can often reveal a specific valvular heart disease.
>
> A standard electrocardiogram can provide enormous amounts of information about the size of the heart, what chambers may be enlarged, arrhythmias, and the presence of ischemic or injury changes (see text discussion). Cardiac X rays are valuable in demonstrating various chamber enlargements, abnormal calcifications in valves, and the condition of the lungs.
>
> A careful office evaluation will allow a physician to suspect the presence of cardiac disease, make a presumptive diagnosis, and provide a guide in selecting confirmatory tests.

pressure within the aorta reflects that the ventricle is in diastole, and the blood in the aorta is moving away from the heart toward body tissues. The aortic valve has closed, and the pressure that exists within the aorta is caused mainly by the elastic nature of the vessel walls, which recoil during ventricular diastole and squeeze against the diminishing volume of blood. None of the events inside the heart have an effect on the aortic pressure until well into the systole of the left ventricle. Once the ventricular contraction is strong enough to force open the aortic semilunar valve, a sudden rush of blood through the aorta produces a rapid rise in the aortic pressure (Figure 22-13). Shortly thereafter, the ventricle repolarizes and relaxes, resulting in a decrease in the flow of blood into the aorta. When the ventricular pressure falls below that of the aorta, the aortic valve closes. This results in a minor backflow of blood into the cusps of the aortic valve, which raises the aortic pressure momentarily. This sudden increase in pressure is referred to as the **dicrotic notch** (Figure 22-13, top). Aortic pressure now decreases steadily until the next ventricular systole causes a new surge of blood to enter the aorta.

Heart Sounds

Associated with the normal heartbeat are distinctive sounds that can be heard readily if one places an ear against the chest of another or uses a **stethoscope** to amplify the sounds (see Clinical Note: Office Evaluation of the Cardiac Patient). These heart sounds are usually described as a sequence of two closely related thumps that are most often vocalized as "lubb-dupp." The period of time between the dupp and the next lubb is approximately three times longer than between the lubb and the dupp; thus, one hears something like lubb-dupp, pause, lubb-dupp, pause, lubb-dupp, pause, and so on.

The heart sounds are caused primarily by closures of the cardiac valves. First, the atrioventricular valves close, causing the lubb sound, followed closely by the closure of the semilunar valves, the dupp sound. By monitoring the heart sounds, a trained listener using a stethoscope or other sound detectors with amplification can detect many features concerning the heartbeat and the condition of the valves.

Cardiac Output

After completing this section, you should be able to:

1. Compare and contrast end-systolic volume with end-diastolic volume in terms of the amount of blood in the heart at different stages in the cardiac cycle.
2. Define cardiac reserve.
3. Describe specializations in heart muscle that allow the heart to equalize the amount of blood being ejected from both ventricles.
4. Describe Starling's law of the heart.

The mechanical, volumetric, and pressure changes just described in the cardiac cycle are important because they produce an uninterrupted flow of blood from the heart to the tissues of the body. We now consider the details of that output.

ultrasonography: (L.) *ultra*, beyond + *sonus*, sound + (Gr.) *graphein*, to write
angiography: (Gr.) *angeion*, vessel + *graphein*, to write

CLINICAL NOTE

SPECIAL CARDIAC DIAGNOSTIC PROCEDURES

Medical research has led to the development of many sophisticated diagnostic procedures that further delineate and quantify abnormalities of the heart beyond the simple office evaluation. Some of these tests are **noninvasive,** meaning that they do not involve surgical procedures; others are **invasive** and entail all the risks of surgical intervention. The following are a few of the more common and useful procedures.

Ultrasonography (ŭl·tră·sŏn·ŏg′ră·fē), or **echocardiography,** is a noninvasive technique of exploring the heart with high frequency sound waves. A sound transducer is applied to the chest wall, and the waves reflected from structures of the heart are picked up by the same transducer and recorded on an oscilloscope (see figure). Echocardiography is a very useful technique for visualizing the heart valves and for measuring the thickness of the ventricular walls and the size of the various chambers.

Stress testing evaluates the total capacity of the subject to exercise. Subjects are evaluated by continuous electrocardiogram and blood pressure monitoring as they exercise against an increasing load either on a treadmill or a stationary bicycle. Graded stress testing has become a useful test in the diagnosis of coronary artery disease and in the prescription of an exercise program for patients with heart disease.

Angiography (ăn″jē·ŏg′ră·fē) is a special X-ray technique in which radiopaque dyes are injected into an artery or vein and followed through the heart to show the size of the chambers and direction of blood flow.

Cardiac catheterization is a process in which a narrow tube called a catheter is threaded through an artery or vein and into the heart (see Figure 22-14). Pressure measurements, blood samples, and angiograms may be obtained. The coronary arteries can also be catheterized and direct angiograms obtained. Portions of arteries blocked by cholesterol deposits can be dilated with balloon catheters.

Finally, new imaging techniques have appeared with the development of newer equipment such as digital subtraction angiography, which produces high quality images of the heart and coronary arteries.

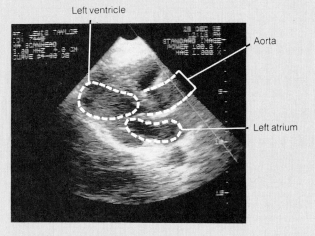

Echocardiogram (oscilloscope image) of a normal heart.

End-Systolic Volume

The net result of all the activities just described during the cardiac cycle is the movement of blood from the left ventricle into systemic circulation. At rest, approximately 70 mL is pumped from the left ventricle into the aorta with each heartbeat. This is the **end-systolic volume** of the heart or is sometimes called the **resting stroke volume** because it represents the volume of blood moved through the left ventricle with each contraction (or stroke) in a person at rest. Multiplying the average stroke volume by the average number of heartbeats per minute (72) reveals that the heart pumps a volume of 5000 mL of blood per minute. This is defined as the **resting cardiac output.** These numbers are average figures that vary among individuals of different ages, weights, and sexes.

The cardiac output is determined by two variables in the stroke volume: the amount of blood that enters the ventricle during diastole and the amount of residual blood left behind after systole.

End-Diastolic Volume

The **end-diastolic volume (EDV)** is the amount of blood in the heart following diastole just before systole. This volume is determined by two factors: the length of diastole and the venous return. The length of diastole is directly proportional to the amount of blood that enters the ventricle. In general, a slow beating heart experiences a longer diastole, and there is more time for blood to enter the chamber than in a rapidly beating heart. Everyone is aware that exercise is one factor that can cause a temporary increase in heart rate, sometimes causing the number of heartbeats per minute to double. When this happens, rapid contraction and relaxation of each heartbeat can significantly reduce the amount of blood entering the ventricle. Recall that most of the blood entering the ventricle does so without the help of atrial systole. It is forced through the atrium by venous pressure and the suction of the expanding ventricle. An increase in venous pressure forces more blood into the ventricle and increases end-diastolic volume. Venous pressure can be influenced by the amount of blood flowing through the veins and by conditions in the peripheral vessels of the circulatory system (see Chapter 23).

Starling's Law of the Heart

Starling's law of the heart states that, within limits, cardiac muscle will increase the force of its contractions as it is stretched. For example, several factors can cause an increase in the end-diastolic volume. This increased volume stretches the ventricular myocardium, and it responds by contracting with greater force, thereby increasing the stroke volume. The volume of blood that can be pumped beyond the resting cardiac output is referred to as **cardiac reserve.** During strenuous physical exercise, this reserve may increase the cardiac output to over 250 mL per stroke (17,000 mL per minute), which is more than three times the normal output.

This ability of the myocardium to respond to stretching with more forceful contractions is a homeostatic device that allows the heart to equalize the amount of blood being ejected from the two ventricles. If for some reason the right ventricle increases its output, additional blood passes through the pulmonary circuit and enters the left ventricle. This larger volume of blood stretches the myocardium, which responds with stronger contractions, increasing its output and accommodating the increased volume of blood. In this way, the output of the left ventricle keeps up with the output of the right ventricle.

> **CLINICAL NOTE**
>
> **CONGESTIVE HEART FAILURE**
>
> **Congestive heart failure (CHF),** also known as cardiac failure or cardiac decompensation, is a consequence of serious heart disease. It is defined as the failure of the heart to pump blood at a rate necessary to meet the requirements of the tissues. Congestive heart failure occurs either because the heart muscle itself is unable to contract efficiently due to disease or damage or because there is an increased demand on the heart. The increased demand on the heart may have several causes: (1) damaged valves, which alter the orderly flow of blood in the heart; (2) increased blood volume; (3) increased metabolic needs of tissues due to such things as fever or increased metabolic rate; and (4) decreased oxygen-carrying capacity of the blood due to anemias.
>
> As the heart begins to fail, various compensatory mechanisms occur. The heart may dilate in an attempt to increase end-diastolic volume and therefore increase stroke volume. The circulating blood volume may expand, and an increase of adrenal hormones will assist compensation. If these mechanisms do not return the system to normal and the cause of the compensation is not rectified (or is not rectifiable), then CHF will occur. The effects of CHF all occur in other organs and are related to a lack of adequate blood circulation. They include increased pressure and fluid leakage in the lungs, liver, and lower legs. People affected may complain of feeling short of breath after only walking a short distance. They may need one or two extra pillows at night. Their legs may swell during the day. Decreased oxygen supply to the brain may lead to irritability and a shortened attention span.
>
> A drug frequently used for the treatment of CHF is **digitalis,** which strengthens the heart contraction and increases the cardiac output. It also causes the kidney to put out more fluid, thereby decreasing fluid load on the heart. The person will also usually be given diuretics to rid the body of excess fluids and decrease circulating fluid volume thereby decreasing the work of the heart.

Physiologic Control of Heart Rate

After completing this section, you should be able to:

1. Describe the activities of the cardioaccelerator and cardioinhibitor centers.
2. Discuss the relationship of the carotid and aortic reflexes to the maintenance of normal circulation.
3. Describe the Bainbridge reflex.
4. Describe the effect of temperature changes on heart rate.

Thus far our discussion has concentrated on the normal activities of the heart without taking into consideration physiologic adaptations of the body that regulate cardiac activities. Such adaptations include neural, chemical, and other physiologic controls over the rate and strength of the heartbeat. These controls allow the heart to change its output in response to changing needs of the body.

Neural Control

There are two main locations in two arteries where cardiac output is monitored through constant measurement of blood pressure. One area is located in the carotid sinus at the base of the internal carotid artery, the large artery that carries blood to the head region (Figure 22-15). The other is in the arch of the aorta. Specialized pressoreceptors (baroreceptors) here are sensitive to the degree of stretching in the tissues of the blood vessel walls where they are located. In addition to these major areas, other minor groups of pressoreceptors also detect pressure changes in the right atrium, pulmonary vessels, left ventricle, and the vena cavae, the large veins that deliver venous blood to the right atrium. Increased pressure caused by increased blood volume in an area stretches the vessel walls, stimulating the pressoreceptors, which generate impulses sent to the brain. A decrease in pressure relieves the tension on the vessel walls, and the pressoreceptors decrease their output of impulses. In either case, these sensory impulses are transmitted to two antagonistic centers in the medulla oblongata; one speeds up the heart rate, while the other slows it down.

The **cardioaccelerator** (kăr″dē·ō·ăk·sĕl′ĕr·ā·tor) **center** is a mass of cells in the medulla oblongata that connects to ganglia in the sympathetic chain (Figure 22-15). Sympathetic neurons pass from these ganglia to the SA node, AV node, and myocardium of the heart as part of the cardiac nerves. Sympathetic neurons release norepinephrine at their target organs. This catecholamine stimulates the heart to contract more rapidly and with greater force, thereby increasing cardiac output.

The **cardioinhibitor** (kăr″dē·ō·ĭn·hĭb′ĭ·tor) **center,** another mass of cells in the medulla, connects with the nucleus that sends parasympathetic fibers through the vagus nerve to both the SA and AV nodes (Figure 22-15). These fibers secrete acetylcholine at their terminals, which inhibits the activities of the pacemaker and the AV node. The strength and frequency of cardiac contractions are decreased by the cardioinhibitor center.

These cardiac centers are involved in the most important control mechanisms for regulating heart rate. These mechanisms operate through three reflex pathways. The first two of these reflexes, the **carotid sinus** and **aortic** reflexes, help maintain normal blood flow to the brain and normal blood pressure in the systemic circulation. The following example will illustrate how they operate.

Any increase in blood pressure stimulates pressoreceptors in the carotid sinus and aorta to send impulses to the medulla. These impulses stimulate the cardioinhibitory center while simultaneously inhibiting the cardioaccelerator center. This stimulation increases the frequency of parasympathetic impulses, while decreasing the number of sympathetic impulses to the heart. The result would be a slowing of the heart rate, which returns the physiological conditions to closer to normal and decreases the blood pressure.

A decrease in blood pressure reduces stimulation of the pressoreceptors and, consequently, reduces stimulation of the cardioinhibitory center. The result is a faster, stronger heartbeat, and the blood pressure rises. The inverse relationship between blood pressure and heart rate is called

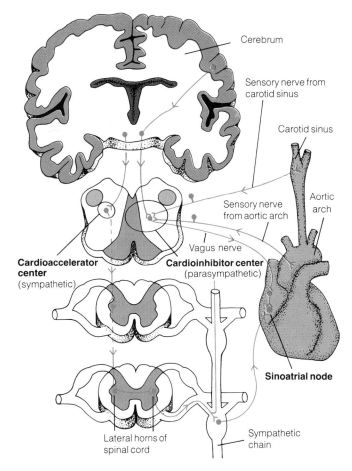

Figure 22-15 Autonomic innervation and nerve pathways that control heart rate.

Marey's law of the heart. (To see how constant high blood pressure may injure the heart, see The Clinic: Diseases of the Heart.)

The third reflex that helps regulate heart rate is the **right atrial** (or **Bainbridge**) **reflex,** which involves the receptors in the vena cavae and the right atrium. These receptors are stimulated by an increase in the flow of blood through these large veins. Stimulation of the pressoreceptors results in impulses that in turn stimulate the cardioaccelerator center, with a resulting increase in heart rate and blood pressure.

In addition to autonomic control, areas in the cerebral cortex also innervate cardioaccelerator and inhibitor centers, hence conscious incidents may also influence heart rate.

Additional Factors Regulating Heart Rule

Although the autonomic nervous system provides the chief control mechanism for heart rate, several other factors also influence its activity. Chief among these are temperature, chemicals, age, and psychological factors that operate via the autonomic nervous system.

Temperature

A rise in body temperature brings on an increased heart rate. Two things are involved in this phenomenon. Warmed blood flowing through the coronary circulation supplying the SA node causes an increase in the rate at which the SA node initiates action potentials, hence increasing heart rate. In addition, an increase in the temperature of blood flowing through the hypothalamus stimulates its reflex centers to influence cardioregulatory centers in the medulla oblongata.

Chemicals

In addition to the acetylcholine and norepinephrine produced by the autonomic fibers innervating the heart, other chemicals in the blood affect the rate and strength of cardiac activity. The most powerful effect is that of epinephrine produced in the adrenal medulla. Under stress the level of this hormone increases in the blood. In the coronary circulation it has an effect similar to that of norepinephrine: it causes an increase in cardiac activity. A rise in the level of blood calcium increases the level of cardiac activity. In contrast, elevated levels of sodium and potassium cause a decrease in heart rate.

Psychological Factors

Because of neural connections between the cardioregulatory centers and the cerebral cortex, psychological factors can influence heart rate. Almost any emotional experience or condition that produces stress, anxiety, anger, or fear also increases heart rate. Conversely, emotional or psychological factors such as depression can cause a decrease in cardiac activity. Heart rate can also be controlled consciously to some degree through practice of meditation and biofeedback techniques.

Age and Sex

Age and sex are important factors in influencing heart rate. Resting heart rate in a newborn infant ranges between 120 to 160 beats per minute when the infant is awake, but it may slow to 90 per minute during sleep. As the infant matures, the heart rate slows, and in adults it averages 70 beats per minute. Heart rate continues to slow, and in elderly individuals it may be as low as 60 beats per minute. If all other factors are equal, the heart rate is usually faster in a female.

In this chapter we studied the anatomy and physiology of the heart. We started with a detailed look at the structure of the chambers and valves and then examined the mechanisms in the cardiac cycle. You also learned about the physiological controls of heart rate and cardiac output. In the next chapter we will study the blood vessels that receive blood from the heart and carry it to all the tissues in the body.

STUDY OUTLINE

Anatomy of the Heart (pp. 559–567)

The heart is a cone-shaped organ in the thorax. Its dimensions average 13 cm in height, 9 cm across at the base, and 6 cm thick. It weighs about 350 g.

Pericardium The *pericardium* is a membrane that surrounds the heart. It consists of an inner *visceral pericardium* and an outer *parietal pericardium*. *Pericardial fluid* is secreted into the space between these two serous membranes.

Muscular Walls of the Heart The wall of the heart has three muscular layers. The internal *endocardium* consists of a thin epithelium and Purkinje fibers. The middle *myocardium* is composed of cardiac muscle. *Trabeculae carneae* and *papillary muscles* are projections of this muscle tissue. The external epicardium is the visceral pericardium.

Chambers and Valves of the Heart The heart has four chambers: two *atria* specialized for receiving blood from veins and two *ventricles* specialized for pumping blood into arteries. The *atrioventricular valves* allow blood to flow from the atria into the ventricles. The chambers allow a functional separation of *pulmonary circulation* from *systemic circulation*.

The *right atrium* receives blood from the *superior* and *inferior vena cavae* and *coronary sinus*. The *auricle*, an ear-shaped extension, increases the volume of the chamber. The *fossa ovalis* is a remnant of a fetal interatrial opening, the *foramen ovale*.

The *right ventricle* receives blood from the right atrium through the *tricuspid atrioventricular* valve. This valve consists of three *cusps* attached to papillary muscles in the wall of the chamber by *chordae tendinae*. Blood leaves the right ventricle through the *pulmonary trunk*. The *pulmonary semilunar valve* prevents blood from flowing backward.

The *left atrium* receives oxygenated blood from the heart. Blood flows from the left atrium into the *left ventricle* through the *bicuspid atrioventricular valve*. Blood leaves the left ventricle through the *aorta*. The *aortic semilunar valve* prevents backward flow of blood.

Cardiac Intrinsic Blood Supply *Coronary arteries* conduct blood to the myocardium. Numerous *anastomoses* produce a complex circulatory pattern with much *collateral circulation*. Blood is collected by several cardiac veins that empty into the coronary sinus. A reduction of blood flow to the myocardium may result in chest pains (*angina pectoris*). Complete blockage of flow results in a *myocardial infarction*.

coarctation: (L.) *coarctare*, to tighten
stenosis: (Gr.) *stenos*, narrow + *osis*, condition
bradycardia: (Gr.) *bradys*, slow + *kardia*, heart
tachycardia: (Gr.) *tachys*, swift + *kardia*, heart

THE CLINIC

DISEASES OF THE HEART

The heart is a miraculous organ. It begins to contract many months before birth, and continues to beat regularly for nearly a century in many people. As you have learned in this chapter, the function of the heart is to send blood through the vessels to every nook and cranny of the body, bringing nutrients, oxygen, and necessary chemicals and removing metabolic wastes and cellular products needed in tissues other than where they were produced. This organ is one of the most vital organs that we possess, and its strength and endurance attest to its importance to life. As strong as it seems to be, the heart still falls prey to a great number of diseases. In fact, heart disease is the number one cause of death in the United States.

Congenital Heart Disease

Any disease present in the human heart or major blood vessels at birth is called a **congenital heart disease.** Such conditions may be caused by genetic mutations (such as incomplete formation of some part of the heart), chromosomal aberrations (such as Down syndrome), or environmental circumstances (such as drugs, infections, exposure to radiation, or disease). Congenital heart defects were among the first cardiac conditions to be solved using surgical techniques.

There are several congenital defects that involve the major blood vessels connected to the heart. **Patent ductus arteriosus** (ăr·tĕ·rē·ō′sŭs) is an abnormal open connection between the pulmonary artery and the aorta. The connection is a normal part of fetal circulation, but it normally closes shortly after birth. The physician is aided in diagnosing this condition by the characteristic heart murmur that sounds like a machine. Patent ductus was the first heart defect to be corrected surgically. The technique involved simply tying off the connection and separating the two vessels. Today, the condition is treated by administering a certain anti-inflammatory drug (indomethacin) to the infant, which causes the open connection to close spontaneously.

Coarctation of the aorta is a narrowing of this large vessel just as it turns and extends down the thorax. This narrowing forces more blood into the arteries leading to the head and arms and reduces blood flow and blood pressure in the legs. It is diagnosed by noting a lower pulse volume and blood pressure in the legs than in the arms. If not corrected, the lower body may be stunted in growth due to the reduction in nutrients. It can be corrected by surgically removing the narrowed portion of the aorta.

Many defects may occur within the heart itself. An opening in the septum between the atria or ventricles will result in mixing of venous and arterial blood within the heart, causing partially deoxygenated blood to enter the arterial circulation. **Atrial septal defects** usually occur when the foramen ovale fails to close during development. **Ventricular septal defects** cause a characteristic heart murmur during systole and may result in heart failure. Both types of defects can be repaired surgically (see Clinical Note: Open Heart Surgery). More serious conditions include partial or total absence of an atrium or ventricle. Failure of a ventricle to develop results in a three-chambered heart, which is very difficult to repair. In addition, **stenosis** or narrowing of one or more valves may occur. **Tetralogy of Fallot** is a rare syndrome of multiple defects. This condition involves four problems: a ventricular septal defect, pulmonary stenosis, displacement of the aorta to the right, and enlargement of the right ventricle. Unique surgical procedures are necessary to relieve symptoms of these difficult conditions.

Cardiac Arrhythmias

Irregularity or loss of rhythm of the heartbeat is called **cardiac arrhythmia** (ă·rĭth′mē·ă). The heart rate may be too slow (less than 60 beats per minute), referred to as **bradycardia,** or too fast (more than 110 heartbeats per minute), called **tachycardia.** Arrhythmias are also classified according to their origin: supraventricular arrhythmias originate in the auricles or the AV node, and ventricular arrhythmias originate in the ventricles. All types of arrhythmias have characteristic patterns on an electrocardiogram, which is used to diagnose which type a patient has.

Atrial arrhythmia is usually not as serious as ventricular, but if sustained for long periods, the former may result in heart failure. There are many types of atrial arrhythmias. **Sinus arrhythmia** is a slightly irregular rhythm produced in the SA node, which is characterized by increased contractions during inspiration and decreased contractions during expiration. **Paroxysmal atrial tachycardia (PAT)** is a condition in which the heart rate suddenly increases to 120–250 beats per minute, and atrioventricular conduction is maintained. It is now thought to be caused by an atrial impulse entering the AV node during its refractory period and being reflected back to the atria, setting up a reentry circuit. It commonly occurs in young women or older people with arteriosclerotic heart disease. It usually lasts only a short time and may be interrupted by massaging the carotid sinus or swallowing against a closed glottis. If the condition persists, various medications may eliminate or control the arrhythmia. **Atrial flutter** is an arrhythmia in which a regular cyclic wave of electrical activity of the atria produces a rate of about 300 beats per minute. **Atrial fibrillation** is a continuous, chaotic reentry of impulses producing a grossly irregular rhythm. **Sinus bradycardia** is a slow rate of less than 60 beats per minute often seen in well-trained athletes. Finally, **atrioventricular block,** which can cause complete heart block, is an arrhythmia in which the atrial impulses are blocked at the AV node, and the artia and ventricles beat at independent rates. Ventricular muscle has an inherent rhythm of its own and will contract at about 50 beats per minute even if no impulses are received from the AV node.

Ventricular arrhythmias disturb the heart in serious ways. The increased contractions reduce nutrients and

(continued)

atheroma: (Gr.) *athera*, porridge + *oma*, tumor
atherosclerosis: (Gr.) *athera*, porridge + *sklerosis*, hardness
angina pectoris: (Gr.) *ankhone*, a strangling + (L.) *pectoralis*, breast

THE CLINIC (continued)

DISEASES OF THE HEART

oxygen in the myocardium and create a need for additional coronary circulation. However, the shortened ventricular filling time associated with increased contractions places less blood into systemic circulation. As a result, both coronary and systemic circulations receive an inadequate supply of blood.

There are many types of ventricular arrhythmias. The most common type is **premature ventricular contractions (PVCs)** in which an extra beat occurs in the ventricle. This happens normally to all of us, and when it does, it feels like your heart "turns over." If the myocardium is injured, PVCs may occur more frequently and trigger a serious arrhythmia. Injury to either branch of the bundle of His may result in blocking of the impulse in this branch called **bundle branch block**. A right bundle branch block (RBBB) may occur normally, but left bundle branch blocks are always associated with disease. **Ventricular tachycardias** are often life threatening. **Ventricular fibrillation** is a rapid chaotic series of heartbeats that is fatal. Ventricular arrhythmias are probably the cause in many cases of sudden death in essentially healthy people, as well as in someone suffering a heart attack.

Cardiac arrhythmias may be treated by drugs or by **cardioversion**, which involves stimulating the heart electrically at a specific time in the cardiac cycle. Persistent arrhythmias may require insertion of a **cardiac pacemaker**, which is an electrical device implanted in the chest and attached to the heart by wires that directly stimulates the heart muscle to contract at a set rate with a small electrical impulse (Figure 1).

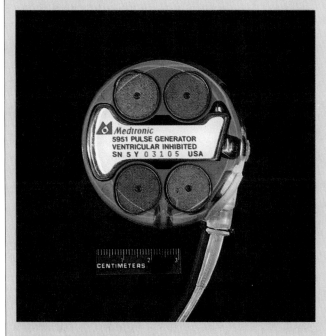

Figure 1 A cardiac pacemaker.

Coronary Artery Disease

Coronary artery disease is the number one cause of death in the United States. In most cases, it is due to deposition of **atheromas** (fatty degeneration or thickening of the walls) in the major arteries serving the heart. The disease is one of the expressions of the more basic disease **atherosclerosis** (ăth″ĕr·ō″sklĕ·rō′sĭs), which clogs the arteries with plaques of fats and cholesterols and causes decreased oxygen delivery to the myocardium. Coronary artery disease produces three complications: angina pectoris, acute myocardial infarction, and sudden cardiac death.

Angina Pectoris **Angina pectoris** is a syndrome (a group of signs and symptoms) caused by myocardial **ischemia**, which is a local and temporary deficiency in the blood supply due to blockage of circulation. It produces discomfort and pressure in the front of the chest and a strangling sensation usually precipitated by exertion and relieved by rest. The diagnosis of angina pectoris is made from the characteristic history of chest pain related to exertion. Between and even during attacks there may be no evidence of organic heart disease. The exercise tolerance test may reveal reduced ability to exercise in the patient, and coronary arteriography may show actual narrowing of the arteries. Immediate treatment involves reducing the atherosclerosis, reducing cardiac work load, and using nitroglycerine to dilate the constricted arteries. Long term treatment of angina involves reducing the risk factors associated with it: smoking, obesity, diabetes, high cholesterol diet, and hypertension. Coronary artery bypass grafts are also performed to permit the flow of blood around the seriously narrowed arteries. The internal mammary artery is frequently used as a graft, or a segment of the saphenous vein is reversed and attached from the ascending coronary artery to another artery past the obstruction.

Myocardial Infarction **Myocardial infarction (MI)**, more commonly called a **heart attack**, is death of the myocardium usually resulting from severe reduction of blood flow to an area of the heart. The MI often results from the sudden closing of an artery already narrowed by atherosclerosis. The occlusion is either the result of a thrombus (clot) or a spasm of the artery. The size and location of the portion of the heart muscle deprived of its blood supply will determine whether the person dies within seconds or hours, or survives. A heart damaged in this manner needs rest and nourishment. The symptoms are similar to angina pectoris but with more prolonged and severe pain, often associated with arrhythmias and loss of function of the left ventricle. Caucasians are at highest risk, and more men are affected than women.

Myocardial infarction differs from angina in that there is actual death of cells in the wall of the heart. Acute myocardial infarction is responsible for 35% of the deaths in men between the ages of 35 and 50. Of all of the deaths

pericarditis: (Gr.) *peri*, around + *kardia*, heart + *itis*, inflammation

from MI, 50% occur within the first 2 hours of the onset of symptoms. An EKG done at this time frequently shows not only the changes associated with ischemia, but also shows inverted T waves, a sign of tissue injury. Should the infarction kill tissue in all three layers of the myocardium (acute transmural infarction), a deep broad Q wave will be evident on the EKG (Figure 2).

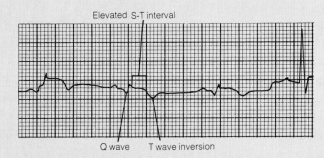

Figure 2 An EKG pattern of a heart attack.

Treatment for MI consists of relieving the pain with injections of morphine, reducing the cardiac load, and increasing the oxygen available to the heart. The treatment of the many complications of MI, such as arrhythmias, cardiac failure, shock, and rupture of the septum or papillary muscles, requires very sophisticated medical techniques. One common treatment is **coronary bypass surgery,** which seems to have improved the quality of life for many patients (see Clinical Note: Open Heart Surgery). Attempts can be made to restore blood flow by stretching the narrowed segment of the coronary artery with intravascular balloons. The intravascular injection of **streptokinase** is a commonly used technique. This is an enzyme that dissolves the clots in the coronary arteries and relieves the ischemia, thus limiting the size of the infarction. Recent release of **tissue plasma activator (TPA),** which has similar action, appears to be more effective than streptokinase.

Sudden Cardiac Death **Sudden cardiac death (SCD)** is death due to cardiac causes in patients previously thought to be free of cardiac disease. In the adult population, over 90% of these deaths are actually due to coronary artery disease. Many of them result from the sudden onset of arrhythmias, such as ventricular fibrillation. Immediate application of cardiopulmonary resuscitation can save some of these otherwise doomed individuals, and it is important that as many people as possible be instructed in the proper techniques of cardiopulmonary resuscitation (CPR). Survivors may be treated with long term antiarrhythmic drugs or fitted with a cardiac pacemaker to prevent recurrence.

Valvular Heart Disease

The human heart may also suffer from **valvular heart disease** characterized by valves that function improperly. Although heart valve disease may be congenital, it is more often acquired as a result of **acute rheumatic fever,** which is an autoimmune reaction following infection with certain streptococcal bacteria. Diseased valves may narrow at their openings (stenosis) or lose their ability to close properly. In rheumatic heart disease, the mitral and aortic valves most commonly suffer damage, but the other valves may also be affected. Valvular heart disease eventually leads to reduced cardiac output, heart failure, and arrhythmias. Diseased valves may require surgical opening or replacement with a prosthetic ball valve or with a valve removed from a pig. Fortunately, early antibiotic treatment of streptococcal infections is reducing the incidence of rheumatic fever.

Hypertensive Cardiovascular Disease

Contraction of the left ventricle with each heartbeat pushes blood under pressure into the aorta and systemic circulation. Normal adult blood pressure in the arteries usually measures less than 140/90 mm Hg. **Hypertension,** or high blood pressure, is a chronic arterial pressure of more than 140/90, and if allowed to persist, it will cause damage to the heart and blood vessels (see Chapter 23 Clinical Note: Hypertension).

Infection and the Heart

On occasion, infectious agents may attack the muscular layers of the heart either by direct invasion or by the toxic effects of their secretions. Infection of the inner lining of the heart and valves is called **bacterial endocarditis** (ĕn″dō·kăr·dī′tĭs). Acute bacterial endocarditis (ABE) is often caused by staphylococcal bacteria and may follow cardiac surgery or self-administration of intravenous drugs. Subacute bacterial endocarditis (SBE) is usually due to an α-hemolytic streptococcus and frequently follows dental or urologic treatment. Endocarditis is characterized by the usual inflammatory signs of infection, and growth of wartlike "vegetations" (actually masses of the protein fibrin) on the heart valves. This infection can also cause production of an embolus (a floating clot) if pieces of the vegetations break off and are carried through the blood to other places, where they may introduce secondary infections.

Myocarditis is an infection of the heart muscle itself and may occur through direct trauma from wounds to the heart or from extension of an infection from elsewhere in the body. Myocarditis may also result from systemic diseases such as rheumatic fever or collagen diseases. **Pericarditis** is an infection of the pericardial sac. It is characterized by a rubbing sound on auscultation, often accompanied by accumulation of fluid within the sac. This fluid can prevent the heart from functioning properly and may result in thickening of the pericardium due to scarring or even heart failure.

Cardiac Physiology (pp. 567–570)

Myocardium The myocardium consists of striated, uninucleate cells connected by intercalated disks and gap junctions.

Conduction System of the Heart The heart has specialized muscle cells that conduct impulses. This *conduction system* is composed of the *sinoatrial* (*SA*) *node*, *atrioventricular* (*AV*) *node*, *atrioventricular* (*AV*) *bundle*, *bundle branches*, and *Purkinje fibers*.

Cardiac Contraction Each contraction of the heart is initiated by a spontaneous electrical impulse from the SA node. Such *pacemaker potentials* result from leakage of Na^+ into the SA node. The stimulus travels over the atria (causing atrial contraction) to the AV node, then through the AV bundle and Purkinje fibers (causing ventricular contraction). Stretching increases the force of contraction of myocardial cells. A long *refractory period* minimizes the possibility of a prolonged tetanic contraction.

Electrocardiography (pp. 570–571)

Electrocardiography is a science that records and analyzes the electrical events associated with heartbeats. A typical tracing on an *electrocardiogram* (*EKG* or *ECG*) includes spikes labeled P, Q, R, S, and T. Abnormalities in a tracing usually indicate abnormalities in the electrical conduction during a heartbeat.

Cardiac Cycle (pp. 572–574)

The cardiac cycle includes contraction (*systole*) and relaxation (*diastole*) of the atria and ventricles.

Mechanical Changes During the Cardiac Cycle During atrial systole, blood is propelled from the atria through the atrioventricular valves into the relaxed ventricles. During ventricular systole, blood moves into the pulmonary trunk and aorta. The atria relax during ventricular systole. The entire cycle takes an average of 0.8 sec.

Pressure Changes During the Cardiac Cycle Pressure within the heart chambers is tied to mechanical changes. As the volume decreases during systole, the pressure rises. Likewise, as the volume increases during diastole, the pressure drops.

Heart Sounds Heart sounds ("lubb-dupp") associated with the normal heartbeat are the result of the closing of the atrioventricular valve (lubb) and semilunar valve (dupp).

Cardiac Output (pp. 574–576)

End-Systolic Volume At rest, the left ventricle pumps 70 mL of blood into the aorta, which is the *end-systolic volume*. This results in a resting cardiac output of 5000 mL per minute.

End-Diastolic Volume *End-diastolic volume* (*EDV*) is the amount of blood in the heart just prior to systole and is determined by the length of diastole and the amount of venous return.

Starling's Law of the Heart This law states that, within limits, cardiac muscle will increase the force of its contractions as it is stretched.

Physiologic Control of Heart Rate (pp. 576–578)

Neural Control Pressoreceptors in the carotid sinus and aortic arch respond to changes in blood pressure in those vessels. Impulses generated here are relayed to *cardioaccelerator* and *cardioinhibitor* centers in the medulla oblongata. The former stimulates the heart to contract more rapidly through sympathetic neurons in a process known as the *carotid sinus reflex*. The latter slows the heart rate through parasympathetic connections in the *aortic reflex*. Pressoreceptors in the vena cavae and right atrium produce a faster heart rate by stimulating the cardioaccelerator center in the *right atrial* (*Bainbridge*) *reflex*.

Additional Factors Regulating Heart Rate A rise in body temperature causes an increased heart rate. Certain chemicals such as epinephrine and calcium cause a faster heart rate, while sodium and potassium slow the heartbeat. Stress, anxiety, fear, and anger increase heart rate, while depression slows it. Heart rate is fastest in infants, and it slows throughout one's lifetime.

SELF-TEST OF CHAPTER OBJECTIVES

True-False Questions

1. Blood returning from the lungs enters the right atrium of the heart.
2. The visceral pericardium is also known as the epicardium.
3. The heartbeat originates in the atrioventricular node.
4. The mitral valve separates the left atrium from the left ventricle.
5. The heart has its own intrinsic blood supply.
6. The chordae tendinae are fibrous cords connected to the pulmonary and aortic semilunar valves.
7. Trabeculae carneae are found in both atria and ventricles.
8. During diastole, the heart relaxes.
9. The coronary trigone is a part of the coronary circulation.
10. The cardioinhibitor center is located in the medulla oblongata.
11. The fibrous pericardium is part of the parietal pericardium.
12. The trabeculae carneae are modifications of the endocardium.
13. Chordae tendineae prevent the atrioventricular valves from prolapsing into the atria.
14. The right atrioventricular valve has three cusps.
15. Intercalated disks maintain a solid separation between adjacent myocardial cells.
16. The atrioventricular node is also known as the cardiac pacemaker.
17. The sinoatrial node is responsible for the pacemaker potential that arises spontaneously in the heart.
18. The cardiac cycle includes only systole and not diastole.
19. The QRS wave in an electrocardiogram is associated with atrial systole.
20. The end-diastolic volume is the amount of blood in the heart just prior to systole.

Matching Questions
Match the terms on the left with their descriptions on the right:

21. auricle
22. Purkinje fibers
23. myocardial infarction
24. trabeculae carneae
25. bundle of His
26. mitral valve
27. foramen ovale
28. sinatrial node

a. part of the heart's conducting system
b. columns in the myocardium
c. reservoir area in the atrium
d. interatrial opening in the fetal heart
e. also called bicuspid
f. heart attack
g. pacemaker of the heart
h. atrioventricular bundle

Multiple-Choice Questions

29. The condition that occurs when none of the impulses originating in the sinoatrial node reach the atrioventricular node is
    a. a partial heart block
    b. a myocardial infarction
    c. a complete heart block
    d. atrial fibrillation
30. During contraction of the left ventricle in a normal cardiac cycle
    a. aortic pressure drops
    b. the bicuspid valve opens
    c. the atrial pressure drops
    d. the aortic semilunar valve opens
31. During contraction of the right ventricle in a normal cardiac cycle
    a. pulmonary trunk pressure drops
    b. the tricuspid valve opens
    c. atrial pressure drops
    d. the pulmonary semilunar valve opens
32. In a normal electrocardiogram tracing, the T wave indicates
    a. contraction of the atria
    b. repolarization of the atria
    c. contraction of the ventricles
    d. repolarization of the ventricles
33. Contraction of the heart is called
    a. diastole
    b. systole
    c. Starling's law of the heart
    d. Marey's law of the heart
34. A drop in blood pressure in the carotid sinus
    a. stimulates the cardioaccelerator center
    b. inhibits the cardioinhibitor center
    c. stimulates release of norepinephrine in the heart
    d. all of the above
35. The muscular columns present in the atrial wall are called
    a. trabeculae carneae
    b. chordae tendineae
    c. musculi pectinati
    d. coronary trigone
36. During a normal cardiac cycle, the P wave is associated with
    a. the dicrotic notch
    b. an increase in atrial pressure
    c. a decrease in ventricular pressure
    d. ventricular contraction
37. During a normal cardiac cycle, the Q part of the QRS wave is associated with
    a. opening of the bicuspid valve
    b. closure of the bicuspid valve
    c. initiation of a pacemaker potential
    d. initiation of an action potential
38. During a normal cardiac cycle, the QRS cycle is associated with
    a. depolarization of the atria
    b. repolarization of the atria
    c. depolarization of the ventricles
    d. repolarization of the ventricles
39. End-systolic volume is also known as
    a. resting stroke volume
    b. resting cardiac output
    c. resting diastolic output
    d. resting systolic ouput
40. Resting stroke volume per minute equals
    a. end-systolic volume
    b. resting cardiac output
    c. resting diastolic output
    d. cardiac reserve

Essay Questions
41. Describe the pericardium and discuss its functional importance to the heart.
42. Describe the anatomy of the four chambers of the heart.
43. Trace the path blood travels through the heart.
44. Discuss the intrinsic cardiac circulation.
45. Compare and contrast a pacemaker potential and an action potential.
46. Discuss the role of the sinoatrial node in regulating heart rate.
47. Describe the electrical events recorded in a typical electrocardiogram.
48. Describe the changes in volume and pressure that occur in the atria and ventricles during a typical cardiac cycle.
49. Discuss the relationship of the carotid and aortic reflexes to the maintenance of normal circulation.
50. Discuss how psychological events can affect heart rate.

Clinical Application Questions
51. Discuss some of the diseases of the heart that may arise as a by-product of streptococcal infections.
52. Fainting (syncope) is a temporary loss of consciousness. Discuss two causes for fainting, one that is not serious and one that is very serious.
53. Describe possible causes for heart murmurs.
54. Describe congestive heart failure. Is it ever a life-threatening condition?
55. Compare and contrast atrial fibrillation and ventricular fibrillation.

23

The Cardiovascular System: Blood Vessels

The cardiovascular system consists of three major components: blood, the heart, and **blood vessels.** The previous chapter discussed the anatomy and physiology of the heart. This chapter presents the blood vessels and the mechanisms involved in the transport of blood through these vessels to the cells. As we study the vessels, we will

1. distinguish anatomically and functionally among arteries, veins, and capillaries;
2. trace the pulmonary and systemic circulatory routes;
3. describe the structural and functional differences between fetal and adult circulations in the cardiac and pulmonary vessels;
4. describe the physiological mechanisms that maintain a relatively constant level of arterial blood pressure; and
5. describe the forces that cause movement of materials through the capillary walls.

tunica adventitia: (L.) *tunica*, a sheath; *adventitius*, coming from abroad
vasa vasorum: (L.) *vas*, a vessel; *vasorum*, of the vessels
vasoconstriction: (L.) *vas*, a vessel + *con*, together + *stringere*, to draw

Blood Vessels

After completing this section, you should be able to:

1. Describe the three tunics in the walls of blood vessels.
2. Distinguish among arteries, veins, and capillaries.
3. Define vasoconstriction and vasodilation.
4. Discuss the general function of vascular valves.

The three types of blood vessels are arteries, veins, and capillaries (Figure 23-1). **Arteries** conduct large quantities of blood from the heart. Distal to the heart, they branch into smaller arteries called **arterioles.** These branch into still smaller vessels that eventually become microscopic **capillaries.** Capillaries connect arteries and veins. They distribute blood to the tissues and bring it into close proximity with the cells. Capillaries form dense networks in almost all body tissues. At the efferent end of a capillary network, they fuse into slightly larger vessels or **venules,** which in turn become **veins.** Veins collect blood from the tissues and return it to the heart. For the most part, blood is confined within the blood vessels, making the human cardiovascular system a "closed" circulatory system. Some of the blood's fluid component does escape from the capillaries and bathes the body tissues in their immediate vicinity. Most of this fluid reenters capillaries in the same network and is returned to the veins. The rest is collected by the lymphatic system before being returned to veins (see Chapter 24).

Arteries

Arteries are blood vessels with thick walls that conduct blood from the heart to tissues. They range in size from about 2.5 cm in diameter to approximately 0.5 mm. The wall of an artery is composed of layers or **tunics** (Figure 23-1). The innermost layer, the **tunica intima,** lines the cavity or lumen of the artery. Also called the **tunica interna,** it is composed of a thin layer of endothelial cells, whose free border is in contact with the blood. The tunica intima is continuous with the cardiac endocardium and extends throughout the cardiovascular system as an endothelium only one cell layer thick.

A thin layer of connective and elastic tissue connects the tunica intima to the **tunica media,** which forms a cylinder external to the tunica intima. The tunica media is a relatively thick layer that consists of smooth muscle and elastic connective tissues.

The **tunica externa** (or **tunica adventitia**) is nearly as thick as the tunica media, but it consists mainly of fibrous

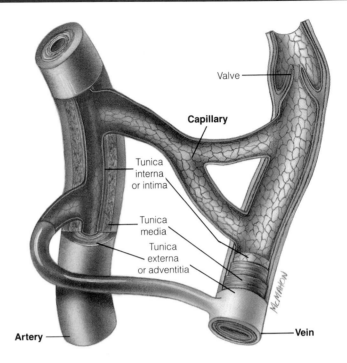

Figure 23-1 Structure of blood vessels. An unbroken connection exists between the arteries, capillaries, and veins. Note the layers of tissue present in the arteries and veins.

connective tissue and only a small number of smooth muscle fibers.

The large aorta and pulmonary trunk and several other equally large arteries contain many layers of elastic tissue sandwiched among the three tunics. These layers allow greater elasticity in these large vessels, which makes it easier for them to absorb the relatively large volume and pressure changes associated with each heartbeat. The presence of these thick layers of elastic tissue has led to these arteries being referred to as **elastic arteries.**

Blood vessels are living tissues requiring all the same things other tissues require. Thus, they have their own set of small blood vessels called the **vasa vasorum** ("vessels of the vessels"), which supply the walls of the large arteries and veins with nutrients and remove wastes. In addition, the muscular layer of the tunica media is also supplied with sympathetic nerve fibers. Sympathetic stimulation of the circular concentric layers of muscles causes **vasoconstriction.** Constriction of a vessel produces an increase in the internal pressure applied by the blood to the walls of the arteries between the ventricle and the area of constriction. This occurs because the heart is trying to push the same amount of blood into a smaller space. When sympathetic stimulation decreases, the muscles relax, and the space within the arteries increases, a process called **vasodilation.** Vasodilation has the opposite effect of vasoconstriction: it produces a decrease in arterial blood pressure between the heart and site of vasodilation. The control of vasoconstriction and vasodilation, and hence blood pressure, is discussed later in this chapter (see section on Control of Blood Pressure).

Arterioles

An arteriole is arbitrarily defined as any artery less than 0.5 mm in diameter. The larger arterioles have well-developed tunica medias composed mainly of smooth muscle with little elastic tissue. Arterioles near the capillaries become progressively smaller in diameter along with a corresponding decrease in smooth muscle thickness. The smallest arterioles meet the capillaries and have an extremely thin tunica media composed of a sparse layer of unconnected muscle cells (Figure 23-2).

The previous chapter described how networks, or anastomoses, of interconnected vessels permitted cardiac tissues to be supplied by more than one source of blood. On a larger scale, anastomoses and collateral circulation allow the systemic arterioles to divert blood from one part of the body to another. Vasodilation in one area combined with vasoconstriction in another results in an accumulation of blood in the dilated area. For example, during strenuous physical activity, arteries and arterioles dilate in the skeletal muscles, while simultaneously constricting in the visceral organs. The result is an increase in the volume of blood flowing through the muscles and a corresponding decrease in blood in the visceral organs. Opposite actions may occur following a large meal when arteries expand in the digestive organs and constrict in the skeletal muscles.

Capillaries

The capillaries are the smallest of blood vessels (Figure 23-1). They are microscopic in size, approximately 0.01 mm in diameter, nearly the same diameter of a red blood cell. Some capillaries are smaller than the red blood cells they carry, forcing the cells to bend into a C shape as they squeeze through the vessel. This increases the surface of the blood cell in contact with the wall of the capillary. Capillaries are usually between 0.5 and 1 mm long. In active tissues, such as skeletal muscles, the liver, the lungs, the nervous system, and the kidneys, capillaries are so numerous that few somatic cells are more than one or two cells away from a capillary. In relatively inactive tissues, such as tendons and ligaments, capillaries are not nearly as numerous. In fact, capillaries are absent from certain tissues including cartilage, the cornea of the eye, the epidermis of the skin, and epithelia.

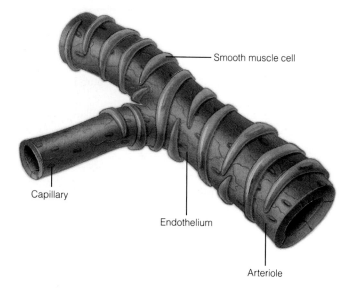

Figure 23-2 Microscopic structure of an arteriole.

The wall of a capillary consists of an endothelial layer only one cell thick. In some very small capillaries, only one elongate endothelial cell wraps itself around the lumen, with its two ends meeting to form the wall. Larger capillaries have several cells cemented together, like tiles on a kitchen floor, to form the capillary wall (Figure 23-2). Capillaries are usually highly branched because they connect the smallest arterioles with the smallest venules.

A circular **precapillary sphincter** of smooth muscle is located at the entrance to each capillary. It can be contracted or relaxed, thereby controlling the amount of blood flowing through a capillary bed.

The capillaries provide the only place where materials can enter and leave an unruptured vessel. Some materials move through the wall of the capillary by means of diffusion or active transport. Serum can leave the capillaries under hydrostatic pressure. Certain white blood cells also possess the ability to squeeze between the endothelial cells, leave the capillary, and wander through body tissues. This ability (discussed in Chapter 21) is called **diapedesis,** and it usually occurs in white blood cells that engulf and destroy dead body cells and microbial invaders.

Venules

Venules are formed where several capillaries join together. They are similar to capillaries in structure, but are slightly larger in diameter. Farther from the capillary bed, a very thin tunica media appears in the venule wall. At this level the

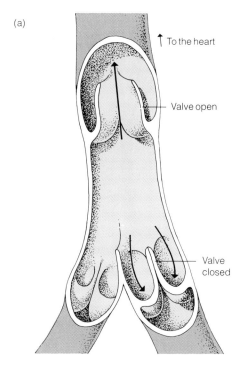
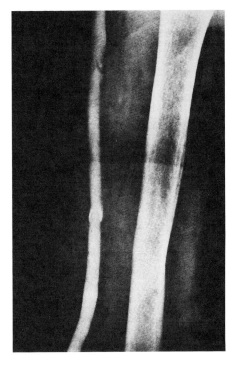
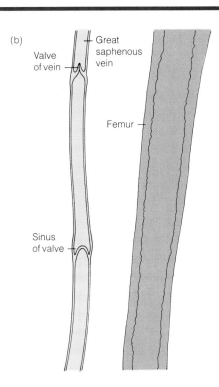

Figure 23-3 Valves in a vein. (a) Constant pressure on the side of the valve away from the heart forces open the cusps and blood flows through. A drop in pressure on the side away from the heart causes the valve to close, preventing blood from flowing backward. (b) An X ray and drawing of valves found in the great saphenous vein of the thigh.

tunica consists of only a few muscle fibers and some fibrous connective tissue. At the point where the tunica externa appears, the venules become veins.

Veins

Veins conduct blood from the venules back to the heart. The smallest veins have thin walls consisting of the same three tunics (interna, media, and externa) that make up the arterial walls (Figure 23-1). The major structural difference between arteries and veins lies in the relative thickness of the tunics. In a vein the tunica media and tunica externa are much thinner than in an artery, and consequently, the walls of veins tend to bulge more easily under pressure than those of arteries.

Another distinctive feature of veins is the presence of **valves,** particularly in the veins of the arms and legs (Figure 23-3). These valves are folds of the tunica intima resembling the semilunar valves of the aorta and the pulmonary artery. As shown in Figure 23-3, the valves allow the blood to flow only in one direction: toward the heart. A drop in pressure in the vein on the side of the valve distal to the heart causes the valve to fill and close (see section on Venous Circulation). On a warm summer day, when the veins just beneath the skin are dilated, their valves may stand out as small lumps on the legs.

Anatomy of the Circulatory Pathways

After completing this section, you should be able to:

1. Describe the pathway blood follows through pulmonary circulation.
2. List the major branches of the aorta and the organs they supply.
3. Describe the arteries that form the circle of Willis.
4. List the major tributaries of the superior and inferior vena cavae.
5. Describe the pathway traveled by a drop of blood flowing from the inferior mesenteric vein to the right atrium.

In this chapter we will study the routes that supply and drain the lungs, forming the pulmonary circulation (see also Chapter 20), and those that supply and drain other tissues,

thrombosis: (Gr.) *thrombos*, clot
embolus: (Gr.) *embolus*, plug

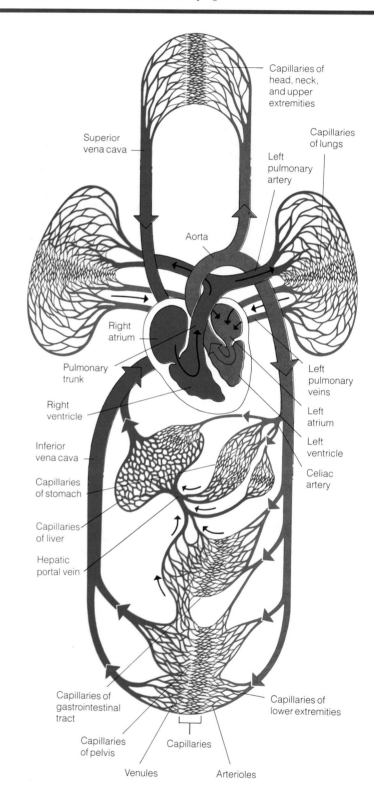

Figure 23-4 Major circulatory routes (shown by arrows).

CLINICAL NOTE

THROMBOSIS AND EMBOLISM

Thrombosis is the development or formation of a blood clot or clots in the vascular system. The clot thus formed, called a **thrombus**, (or thrombi for plural) is made up of platelets, fibrin, and white blood cells. It is therefore paler than a blood clot that forms on a cut, which has red blood cells trapped in the fibrin net.

A thrombus forms when platelets stick to the blood vessel endothelium and gather there. Fibrin forms around the lump of platelets, and more platelets are trapped as well as white blood cells and a few red blood cells. The process repeats itself and the clot grows. It forms a "head," which slows the flow of blood in the vessel, and a "tail," which may extend some distance downstream. Thrombi most commonly occur in the veins, especially of the leg, but they can also form in the heart and in the coronary, cerebral, iliac, and femoral arteries.

Slowing of blood because of bed rest, pressure on vessels (from crossing legs or prolonged sitting), changes in the vessel walls from trauma or inflammation, and changes in the nature of blood itself (increased viscosity or increased red blood cells) are all factors that contribute to thrombus formation.

The danger posed by thrombi is that they can block the vessel in which they form or they can break away from that vessel, travel in the bloodstream, and block another vessel. When a vessel is blocked, blood cannot get by. In an artery this can lead to areas of poor blood supply (ischemia). If a thrombus blocks a small artery in the heart, a **myocardial infarction** (heart attack) may result. The result of a thrombus blocking a cerebral artery is called a **cerebral vascular accident** or stroke (see Chapter 12 Clinical Note: Stroke).

An **embolus** (or emboli for plural) is a mass of undissolved material that is moving in the blood. It can be air, fat, or a thrombus. The most common embolus is a freely floating clot that has become detached from a thrombus. An embolus arising from an artery usually becomes lodged in a smaller arteriole; its effects depend on where it lodges. An embolus arising from a vein flows from a small vessel to a large vessel and then to the heart, where it eventually may become stuck in the arterial system of the lungs **(pulmonary embolism)**. If a large embolus lodges in the main pulmonary artery, death ensues without warning. Small emboli are more common than large ones and become lodged in branches of the pulmonary artery or an arteriole. Common signs and symptoms include difficulty in breathing, chest pain, rapid respiration, abnormally fast heart rate, and fever.

forming the systemic circulation (Figure 23-4). In the preceding chapter we studied coronary circulation, the special systemic branches that supply the heart. In this chapter we will also study the hepatic portal system, a special group of veins that drain blood from the digestive organs into the liver.

Pulmonary Circulation

Blood in the right ventricle of the heart has come from the tissues where it has surrendered much of its oxygen to the surrounding tissues and taken on carbon dioxide produced by the metabolic activities of the cells. Contraction of the right ventricle sends this blood past the pulmonary semilunar valve into the **pulmonary trunk** (Figure 23-5). The short pulmonary trunk branches into right and left **pulmonary arteries** that lead toward the lungs. Shortly before forming the lungs, each pulmonary artery branches into **lobar arteries** leading to lobes of the lungs, two on the left and three on the right. The lobar arteries undergo additional branching into smaller arteries and arterioles, until they finally branch into capillaries closely associated with the alveoli of the lungs (see Chapter 20). Their close proximity to the lumina of the alveoli facilitates diffusion of carbon dioxide from the blood and into the alveoli, and oxygen from the lungs into the capillaries.

After passing through the capillary beds of the lungs, the blood collects in venules, which fuse to form numerous veins. These continue to fuse, eventually forming four large **pulmonary veins,** two from each lung, which lead to the left atrium. Note that the pulmonary arteries carry deoxygenated blood and the pulmonary veins carry oxygenated blood. This situation is the reverse of all the other arteries and veins in the adult body.

The pulmonary circuit is responsible for transporting blood from the heart for the purpose of replenishing its oxygen supply and removing its carbon dioxide. This blood is not used to nourish the tissues of the lungs or respiratory passages, nor is it used as a source of oxygen by these tissues. The lungs have their own set of systemic arteries.

Systemic Circulation: Arteries

Systemic circulation is that part of the vascular system that carries blood from the left ventricle to the tissues and returns it to the right atrium. This blood supplies oxygen and nutrients to the body, while simulatneously removing waste materials and, in some cases, important cellular products such as hormones. It is formed by the systemic arteries and systemic veins, which are examined here in detail.

Figure 23-5 Pulmonary circulation. Blood from the right ventricle enters the pulmonary trunk and is carried to the lungs, where oxygen and carbon dioxide are exchanged. The blood is then returned to the left atrium via the pulmonary vein.

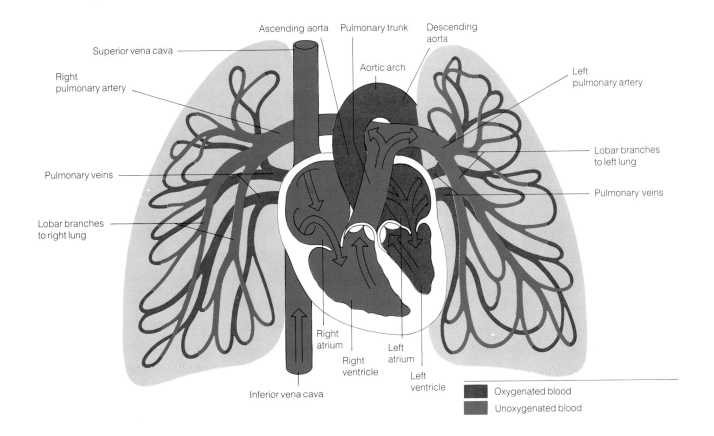

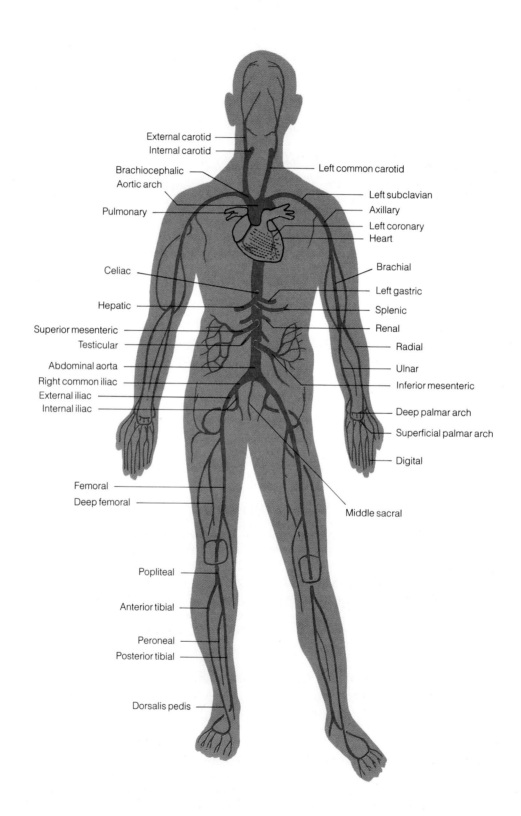

Figure 23-6 Systemic arteries.

brachiocephalic: (L.) *brachium*, arm + *cephalicus*, head
carotid: (Gr.) *karos*, deep sleep
subclavian: (L.) *sub*, under + *clavicula*, little key

The circulatory system is the most variable of all the systems. Because of these variations from individual to individual, the descriptions given here are idealized to represent the most common anatomy. Laboratory study of the blood vessels requires interpretation of the ideal case in terms of the actual vessels present and the organs supplied.

Aorta

The **aorta** is the major artery of the systemic circuit. It begins at the superior medial border of the left ventricle, proceeds upward for a short distance, then executes a sharp U turn (Figure 23-5), and extends down the length of the thorax and abdomen giving off major arteries to all anatomical regions of the body. When it reaches the pelvic girdle, it splits into two divisions (Figure 23-6).

Arteries of the Head and Neck

The arch of the aorta passes superior to the right pulmonary artery (Figure 23-5). Its beginning and terminal portions are often called the **ascending aorta** and **descending aorta** in reference to the pathway of blood carried inside. The superior border of the aortic arch gives rise to three major arteries: the **brachiocephalic** (brā″kē·ō·sě·făl′ĭk), **left common carotid** (kă·rŏt′ĭd), and **left subclavian** (sŭb·klā′vē·ăn) (Figure 23-7). These arteries supply regions of the head, neck, and upper extremities.

To simplify our discussion of arteries, we will only describe the arteries of the right side of the body. The same circulatory pattern exists on the left side of the body, with one exception. Only one brachiocephalic artery exists, which arises on the right side. Thus, the left common carotid and left subclavian arteries arise independently and directly from the aortic arch instead of from the brachiocephalic artery, as do their counterparts on the right side.

The brachiocephalic artery is relatively short and immediately branches into the **right common carotid** and **right subclavian arteries.** Near the base (or origin) of the common carotid artery are two structures, the carotid sinus and carotid body (see Chapter 22). The carotid sinus (see Figure 22-15) contains sensory receptors that detect stretching of the tissues in the vessel. This information is transmitted to brain reflex centers that help regulate blood pressure. The carotid body is a region in the tunica interna of the common carotid artery near its bifurcation. It contains receptors that detect changes in the concentration of blood oxygen and, secondarily, changes in both carbon dioxide concentration and pH.

The right common carotid artery passes upward alongside the neck until it reaches the upper border of the thyroid cartilage. Here it branches into the **internal** and **external**

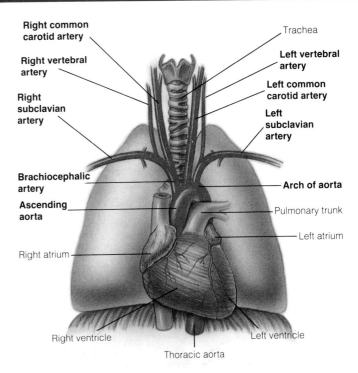

Figure 23-7 Major branches of the aortic arch.

carotid arteries. The internal carotid artery passes through the carotid canal of the skull and conducts blood to the brain. Its contribution is supplemented by an additional source of arterial blood from the vertebral arteries. The internal carotid arteries form a well-known anatomic landmark called the **circle of Willis** (Figure 23-8), which is one prominent example of anastomosis and collateral circulation. The circle of Willis is supplied by both the internal carotid and vertebral arteries, thus it has a dual source of blood. This arrangement provides the brain with collateral circulation, which could be critical to life should a blockage occur in one vessel. Just before terminating in the circle of Willis, the internal carotid gives off the **ophthalmic** (ŏf·thăl′mĭk) **artery.** It conducts blood to the eye, forehead, and part of the nose.

The external carotid artery supplies more superficial structures of the head and neck (Figure 23-9). These include the salivary glands, scalp, teeth, nose, throat, tongue, thyroid gland, and facial muscles.

The second branch of the brachiocephalic artery, the right subclavian artery, passes under the clavicle, as its name implies (Figure 23-7). But before it does, it gives off several smaller arteries. The most important of these is the **vertebral artery.** This artery passes up the neck through the transverse foramina of the six cervical vertebrae (Figure 23-9) and enters the cranial cavity through the foramen magnum. Once within

592 Chapter 23 The Cardiovascular System: Blood Vessels

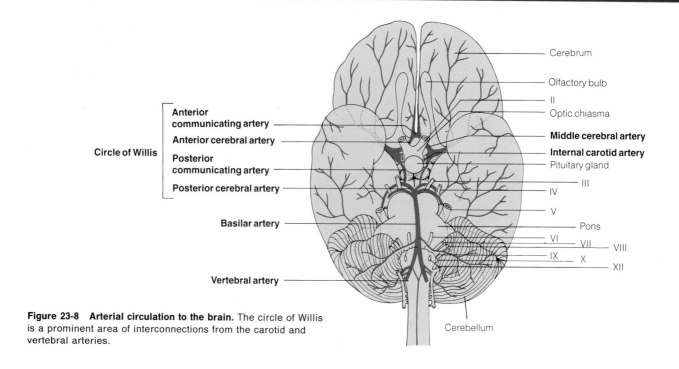

Figure 23-8 **Arterial circulation to the brain.** The circle of Willis is a prominent area of interconnections from the carotid and vertebral arteries.

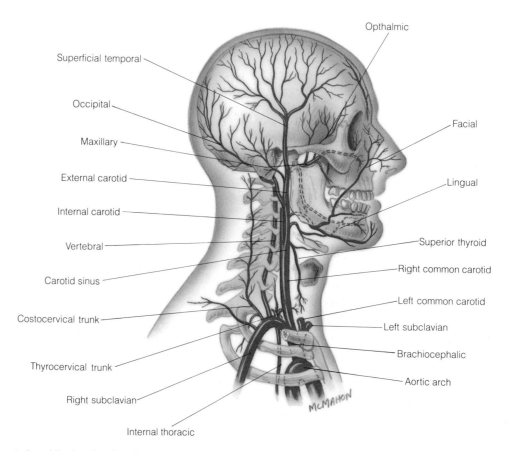

Figure 23-9 Major arteries of the head and neck.

angiography: (Gr.) *angeion*, vessel + *graphos*, written

> **CLINICAL NOTE**
>
> **EVALUATION OF BLOOD VESSEL DISORDERS**
>
> Blood vessel disorders can be evaluated either in the physician's office, if the problem is relatively simple, or through special laboratory testing if more complex.
>
> The office evaluation of peripheral vascular disease is usually straightforward and depends on the patient's history and physical examination. A typical history might reveal pain in the calves accompanying exercise, numbness and coldness of the feet, and skin changes. During the physical examination, the physician would pay particular attention to the peripheral pulses, noting their presence or absence and the difference in volume from one side of the body to the other. The color and temperature of the extremities would be noted. Simply compressing the skin with the tip of a finger and noting the amount of time it takes for the color to return may disclose delayed venous filling time. Close inspection of the skin might show absence of hair on an affected foot, deformed nails, thin parchment-like skin, and perhaps even ulceration of the skin. Listening with the stethoscope over the arteries might reveal a swishing sound over narrowed vessels. Careful abdominal palpation could reveal the pulsating mass of an aneurysm (see Clinical Note: Aneurysm). Calf tenderness and swelling are signs of venous thrombosis (see Clinical Note: Thrombosis and Embolism), which can be confirmed by strongly dorsiflexing the foot thus reproducing the pain. The physician may even feel the thickened cordlike thrombosed vein. These simple office procedures should be sufficient to allow the physician to make a general diagnosis.
>
> When severe peripheral vascular disease is suspected or when surgical intervention is contemplated, a more exact evaluation of the condition of the vessels is necessary. There are many laboratory procedures available for this purpose. **Ultrasonography** is a simple, noninvasive technique that may reveal venous thromboses, and in many cases is the only confirmatory test needed. **Angiography** is more accurate and is performed by injecting a radiopaque dye into the blood vessel and taking high speed X rays to show the interior of the vessel.
>
> **Digital subtraction angiography (DSA)** is an interesting new technique used for visualization of arteries. X-ray images are taken before and after intravenous injection of contrast media. A computer subtracts one image from the other and leaves an unobstructed view of the arterial tree. Unfortunately, this technique is not presently available in all hospitals.
>
> **Angiodynography** (ăn″jē·ō·dĭn·ŏg′rǎ·fē) involves the most advanced technology; it combines the use of ultrasound, doppler, and infrared readings of a suspected blood vessel. A computerized image of the vessel is produced in which the direction of blood flow is shown in contrasting colors along with a measurement of the flow velocity. It is a noninvasive technique that does not require injection of contrast media. It is also available only in a few research hospitals at this time.

this cavity, the right and left vertebral arteries form the **basilar artery.** The basilar artery passes superiorly along the anterior surface of the pons, giving off several paired arteries that carry blood to the more posterior regions of the brain. Just posterior to the pituitary gland, the basilar artery divides into two **posterior cerebral arteries,** each of which gives off a **posterior communicating artery.** These connect to the internal carotid arteries, completing the circle of Willis (Figure 23-8). Other small branches from the subclavian—the **internal thoracic, thyrocervical** (thī″rō·sĕr′vĭ·kăl), and **costocervical** (kŏs″tō·sĕr′vĭ·kăl) **arteries**—carry blood to the thyroid gland and to the muscles of the neck and anterior trunk.

Arteries of the Upper Extremities

The subclavian artery courses under the clavicle and out through the thorax. When it enters the axilla, its name changes from subclavian to the **axillary artery** (Figure 23-10). The practice of changing the name of a vessel when

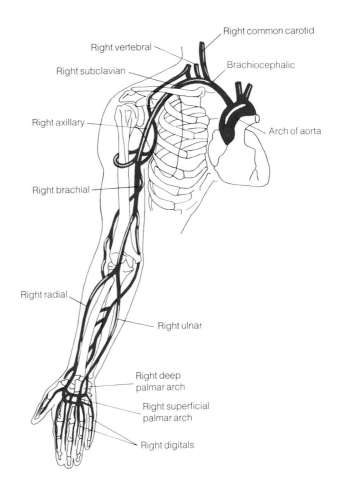

Figure 23-10 Arteries of the right extremity.

it passes into a different region occurs for several prominent vessels. Several small arteries branch from the axillary artery to muscles in this region and in the shoulder.

When the axillary artery enters the arm, it becomes the **brachial artery.** The brachial artery runs the length of the humerus, giving off small branches to the arm. At the anterior surface of the elbow, the brachial artery divides into the **radial** and **ulnar arteries,** amid many mutual and collateral interconnections. These arteries parallel their respective bones supplying the forearm. At the wrist, the two arteries undergo extensive interconnections (Figure 23-10), producing **deep** and **superficial palmar arches** in the palm of the hand. **Digital arteries** extend from the palmar arches into the fingers.

Arteries of the Thorax

The larger branches of the descending aorta in the thorax (where it is called the **thoracic aorta**) and abdomen are often classified into two major categories. Branches that supply organs located within the body cavity are **visceral arteries,** in reference to their connection to visceral organs. Those arteries supplying muscles and organs on the body wall are **parietal arteries.**

The visceral branches of the thoracic aorta include the **bronchial arteries,** which supply nutrients to the lungs and bronchial passages, and the **esophageal artery,** which supplies the esophagus.

Parietal branches include the **posterior intercostal** (Figure 23-11) and **superior phrenic** (frĕn´ĭk) **arteries.** Intercostals supply muscles of the thorax and mammary glands; superior phrenics supply the diaphragm.

Arteries of the Abdomen

The descending aorta penetrates the diaphragm and passes into the abdominal cavity as the **abdominal aorta** (Figure 23-11). This aorta gives off parietal and visceral branches. Parietal branches include the **inferior phrenic arteries,** which supply the diaphragm, the **lumbar arteries,** which supply the vertebral column and muscles of the posterior and lateral abdominal wall, and the **middle sacral artery,** a small vessel that supplies muscles in the sacral region.

Visceral branches of the abdominal aorta supply all organs in the abdominal cavity. Just below the diaphragm an unpaired vessel called the **celiac** (sē´lē·ăk) **artery** projects anteriorly from the abdominal aorta. It usually extends only 2 to 3 cm from the aorta where it immediately divides into three branches: **left gastric, splenic,** and **hepatic** (hĕ·păt´ĭk) **arteries** (Figure 23-12). The left gastric conducts blood to the lower end of the esophagus and the lesser curvature of the stomach. The splenic artery supplies the greater curvature of the stomach, the pancreas, and the spleen. Hepatic arteries transport blood to the liver, gallbladder, pancreas, and duodenum. An examination of Figure 23-12 shows a considerable anastomosing of these three arteries, thereby providing alternate sources of blood to vital visceral organs.

Usually within 2 to 3 cm below the celiac artery, the **superior mesenteric** (mĕs˝ĕn·tĕr´ĭk) **artery** arises from the abdominal aorta (Figure 23-13). It is another unpaired vessel projecting anteriorly from the aorta. The superior mesenteric artery supplies the upper portion of the pancreas, nearly all of the small intestine, and the ascending and transverse colons. The vessels involved lie within the thin mesentery that holds the small intestine to the abdominal wall. Many interconnections occur among the smaller branches of these vessels, as well as with the celiac and those of the other large vessel supplying the intestine, the **inferior mesenteric artery.** It is the third unpaired abdominal artery. It arises from the anterior surface of the aorta and divides into several branches that supply the remainder of the large intestine and rectum. It communicates with branches of the superior mesenteric artery and with the internal iliac artery (see below) supplying the wall of the pelvis and some pelvic viscera.

Paired **adrenal** (or **suprarenal**) **arteries** usually arise from the aorta just below the superior mesenteric artery. They conduct blood to the adrenal glands. The adrenals also receive branches from the inferior phrenics and renals. Paired **renal arteries,** often more than one per kidney, extend between the aorta and kidneys. Their blood is filtered in the kidneys.

The last pair of visceral branches of the abdominal aorta supplies the gonads. In males the **gonadals** are the **testicular** or **spermatic** (spĕr·măt´ĭk) **arteries;** in females they are the **ovarian arteries.** Testicular arteries are longer than ovarian arteries and extend into the scrotum. The ovarian arteries are shorter, leading to the ovaries.

Near the level of the iliac crest, the abdominal aorta divides into two **common iliac arteries.** This bifurcation extends for only 2 to 3 cm, and each common iliac artery divides into **internal** and **external iliacs.** The latter proceed into the legs, but the internal iliacs (also called the **hypogastric arteries**) supply blood to the pelvic cavity and the surrounding area. Specifically, the internal iliac arteries supply the rectum, urinary bladder, and external genitalia in both females and males as well as the uterus and vagina in females. In addition to these organs, the internal iliac arteries also supply muscles in the lower lumbar, gluteal, and hip regions.

Arteries of the Lower Extremities

The external iliac artery passes under the inguinal ligament, a cord of connective tissue that stretches from the anterior

Anatomy of the Circulatory Pathways 595

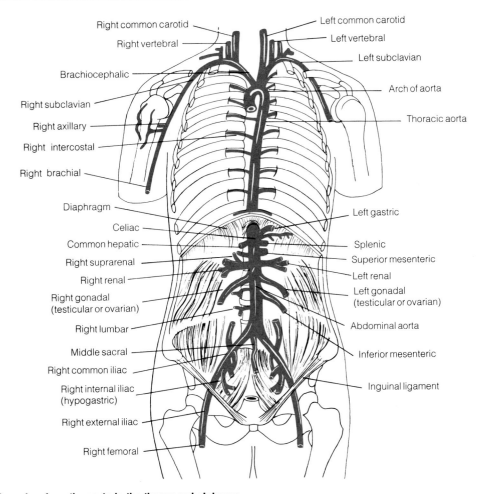

Figure 23-11 Major branches from the aorta in the thorax and abdomen.

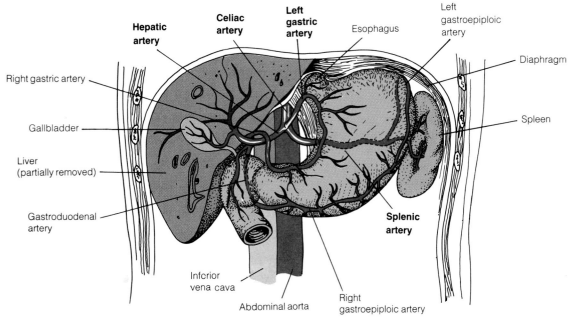

Figure 23-12 The celiac artery and its branches.

popliteal: (L.) *poples*, ham

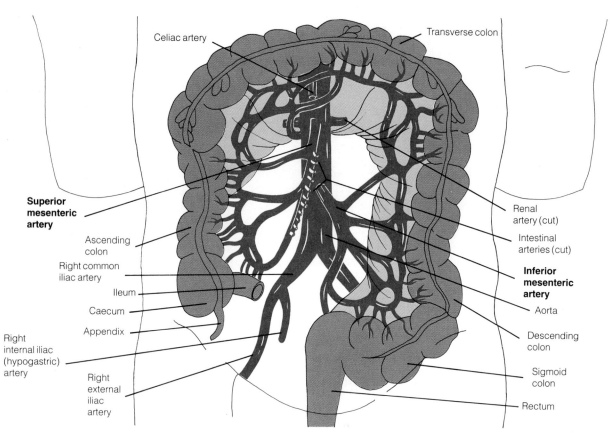

Figure 23-13 Mesenteric arteries.

superior iliac spine to the pubic tubercle. Once in the thigh, the external iliac becomes the **femoral** (fĕm´ōr·ăl) **artery** (Figure 23-14).

The femoral artery runs along the anterior medial region of the thigh. In addition to supplying small arteries in the medial portion of the thigh, the femoral artery produces three relatively large branches. The largest of the three, the **deep femoral**, supplies the posterior portion of the thigh. The **medial femoral circumflex** supplies the proximal portion of the thigh, while the **lateral femoral circumflex** courses along the lateral side of the thigh. Near the distal end of the thigh, the femoral artery passes through a hole (the adductor hiatus) in the tendon of the adductor magnus muscle. The artery continues to the posterior region of the knee. Behind the knee, the femoral artery is called the **popliteal** (pŏp·lĭt´ē·ăl) **artery.** Many interconnections and much collateral branching occurs both in front of and behind the knee.

After passing through the popliteal fossa, a space behind the knee, the popliteal artery divides into the **anterior** and **posterior tibial arteries** that supply the leg. The anterior tibial artery pierces the interosseous membrane between the tibia and fibula and passes along the anterior surface of the membrane to the ankle. Several anastomosing branches are given off at the ankle, and upon entering the foot, the artery is known as the **dorsalis pedis** (dŏr·săl´ĭs pĕ´dĭs) (Figure 23-14b). The dorsalis pedis artery gives off many branches that supply the foot and toes. Some of them join to form the **plantar arch** that supplies blood to the sole and toes.

The other major branch of the popliteal, the **posterior tibial artery**, courses down the posterior aspect of the tibia to the ankle. At its upper end, this artery gives off the **peroneal** (pĕr˝ō·nē´ăl) **artery**, which transports blood to the calf muscles. On the posterior surface of the ankle, the posterior tibial gives off several interconnecting branches, then divides into the **lateral plantar** and **medial plantar arteries.** These plantar arteries anastomose and, along with branches of the dorsalis pedis, form the plantar arch. **Digital arteries** extend from the plantar arch into the toes.

Systemic Circulation: Veins

The systemic veins collect blood from the tissues and return it to the right atrium (Figure 23-15). With few exceptions, capillary blood is collected in venules that aggregate into

jugular: (L.) *jugulares*, throat

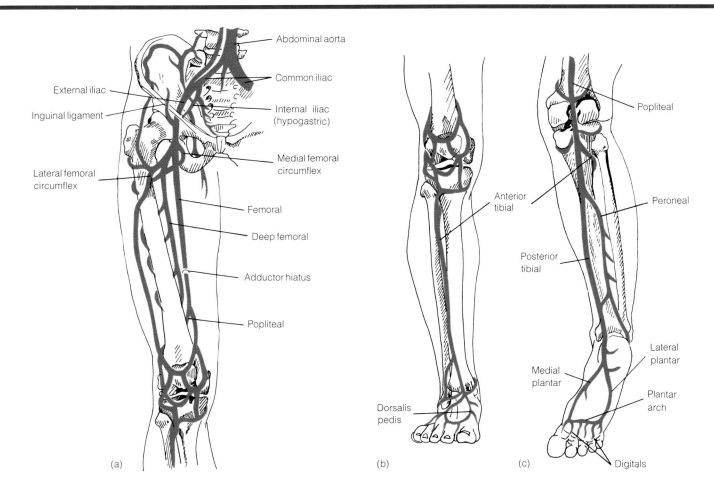

Figure 23-14 Arteries of the lower extremity. (a) Anterior view of the right thigh. (b) Anterior view of the right leg. (c) Posterior view of the right leg.

larger veins. In a few instances, venous blood collects in sinuses, vascular spaces located in organs that accumulate venous blood. Such sinuses are located in the brain **(dural sinuses)**, in the heart **(coronary sinus)**, and in the liver, spleen, and adrenal glands, where they are called **sinusoids** (see Chapter 18). Sinuses and sinusoids are lined with endothelial cells continuous with those of capillaries and veins, but they lack the muscular tunica media of venules or veins. Instead, their walls are chiefly fibrous connective tissue. Eventually, all veins drain into the **superior vena cava, inferior vena cava,** or **coronary sinus** (see Chapter 22), which in turn drain into the right atrium.

Superficial veins (or **cutaneous veins**) lie just beneath the skin in superficial fascia. They receive blood from superficial tissues, yet freely communicate with deeper veins. Superficial veins are usually visible on the surface of the body, especially in the arms and legs. **Deep veins** tend to parallel arteries and usually bear the same names as those arteries.

Veins of the Head and Neck

Most blood from the head and neck regions is collected in three large veins: **internal jugular, external jugular,** and **vertebral veins** (Figure 23-16).

The internal jugular vein forms the major venous drainage of this area. It is a deep vein that parallels the common carotid artery. Its major tributaries include the **transverse sinuses** of the brain (Figure 23-17). These sinuses, in turn, receive blood from several veins and other sinuses, primarily the **sagittal, cavernous, petrosal,** and **straight sinuses.** In addition, several veins from the anterior facial region and the laryngeal and thyroid areas drain into the internal jugular. This major vein begins at the jugular foramen, joins the

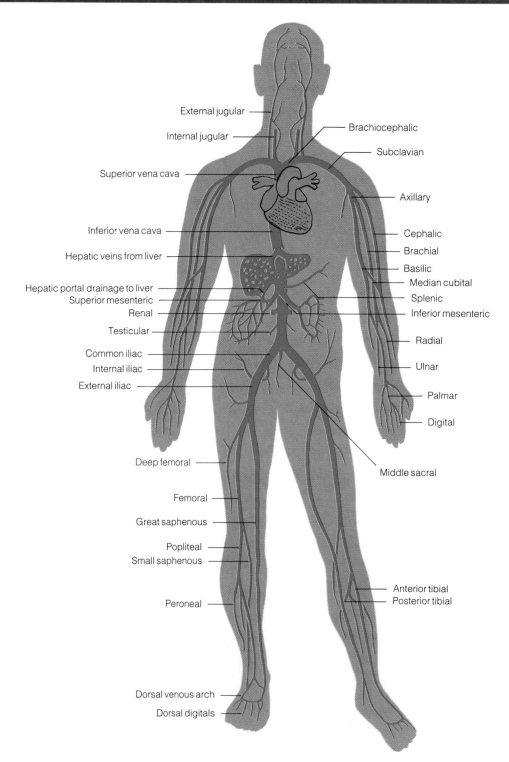

Figure 23-15 Systemic veins.

Anatomy of the Circulatory Pathways 599

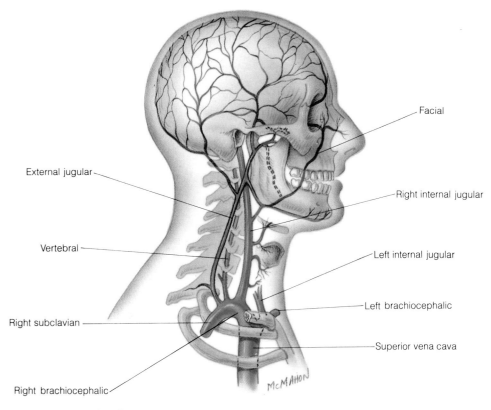

Figure 23-16 Major veins of the head and neck.

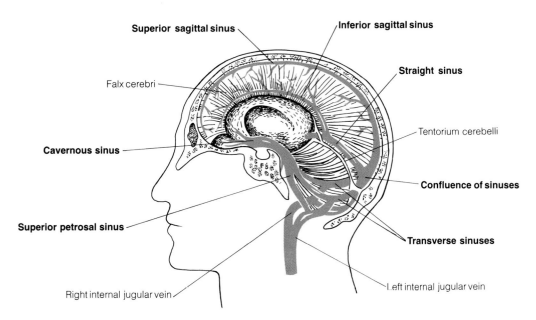

Figure 23-17 Venous sinuses of the brain. Venous blood from the brain collects in these sinuses and empties into the internal jugular vein.

azygos: (Gr.) *a*, not + *zygon*, yoke
hemiazygos: (Gr.) *hemi*, half + *a*, not + *zygon*, yoke

subclavian vein to form the **brachiocephalic vein,** and empties into the superior vena cava.

The external jugular is smaller than the internal jugular and lies superficial to it. It receives blood from superficial regions of the face, scalp, and neck. It connects with the internal jugular and joins the subclavian vein at its base. The latter vein parallels the subclavian artery and passes under the clavicle.

The vertebral vein lies alongside the vertebral artery. Its tributaries transport blood from posterior regions of the brain and superficial regions of the scalp. It is formed by anastomosing interconnections of intracranial sinuses and superficial veins, and it passes through transverse foramina of the cervical vertebrae. Near its base, it fuses with one or more **cervical veins** and joins the subclavian vein.

Veins of the Upper Extremities

The **subclavian vein** is the major drainage of the upper extremity. The subclavian vein parallels the subclavian artery and is known as the **axillary vein** in the axillary region (Figure 23-18). In the axilla it receives several veins from the shoulder, scapular, and upper thoracic regions.

Distal to the axilla lie venous vessels that carry blood from the forearm and arm. Deep veins of the arm lie alongside the arteries in the same region and bear the same names as the adjacent arteries, such as **brachial, radial,** and **ulnar.** However, superficial veins are slightly more complex. They originate as **palmar arches** on the anterior and posterior surfaces of the hand and wrist. These arches are joined by many anastomosing veins from the fingers. Amid many superficial interconnections, large **cephalic** and **basilic veins** ascend the forearm. A prominent connection between them, the **median cubital vein,** occurs at the anterior angle of the elbow (Figure 23-18). It is the most frequently used vein for the extraction of blood for medical purposes. The basilic and brachial veins form the axillary, which is soon joined by the cephalic vein. In addition to these major interconnections, numerous anastomoses exist among the deep and superficial veins of the arm and forearm.

As previously mentioned, the fusion of the internal jugular, external jugular, and subclavian veins forms the brachiocephalic vein. Unlike the asymmetrical condition in the similarly named arteries, two brachiocephalic veins are normal, although the left one is slightly longer. The two brachiocephalic veins become the superior vena cava.

Veins of the Thorax

The superior vena cava is the major drainage of the thorax (Figure 23-19). One large vein and several smaller ones

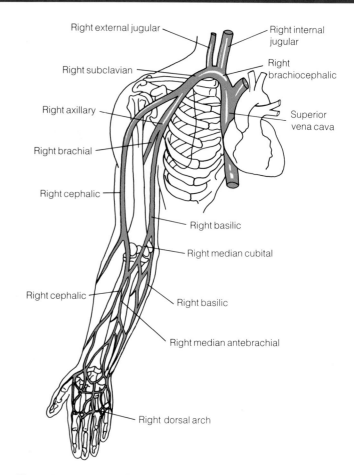

Figure 23-18   Major veins of the upper extremity.

constitute its tributaries from thoracic organs. The major thoracic subsidiary of the superior vena cava is the **azygos** (ăz´ĭ·gŏs) **vein.** It receives blood from the **intercostal veins.** Blood from the intercostal thoracic regions of the right side empties into these veins. Blood from the left side of the thorax indirectly arrives at the **hemiazygos** (hĕm˝ē·ăz´ĭ·gŏs) **vein.** Although these veins are tributaries of the superior vena cava in the thoracic region, they enter the abdominal cavity and make connections with the inferior vena cava.

In addition to intercostal veins, the azygos and hemiazygos veins receive blood from other thoracic areas, primarily via the **esophageal, bronchial,** and **pericardial veins.** Posterior intercostal veins on the left drain into the **accessory hemiazygos** and hemiazygos, which connect across the midline to the azygos.

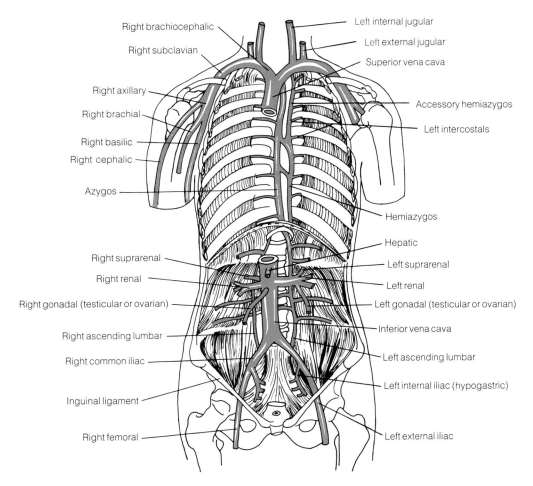

Figure 23-19 Major tributaries of the veins in the thorax and abdomen.

Veins of the Abdominal Cavity

The inferior vena cava begins with the **phrenic veins** of the diaphragm. Beneath these, the **hepatic veins** conduct blood from the liver and gallbladder to the inferior vena cava. Several **lumbar veins** located between the lumbar vertebrae carry blood from the abdominal wall directly to the inferior vena cava. **Suprarenal** and **renal veins** are associated with the adrenal gland and kidneys, while **gonadal veins** (testicular or ovarian) originate in the reproductive glands. On the left side, the suprarenal and gonadal veins empty into the left renal vein instead of directly into the inferior vena cava (Figure 23-19).

The inferior vena cava extends into the pelvic region, where its major branches are the two **common iliac veins.** These form anterior to the lower lumbar vertebrae from the fusion of the **internal** and **external iliac veins.** The internal iliacs drain the urinary and reproductive organs in the pelvic region, while the external iliacs transport blood from the gluteal muscles and other superficial structures in this region before passing out of the abdominal cavity.

Hepatic Portal System

"Portal" systems are circulatory routes that begin and end in capillaries. Digestive organs in the abdominal cavity are supported by special veins that form the **hepatic portal system** (Figure 23-20). It consists of a large **hepatic portal vein** (not to be confused with the hepatic vein), which leads into the liver, and its tributaries. Two major veins contribute to the hepatic portal vein, namely, the **superior mesenteric**, which collects blood from the small intestine, and the **splenic vein,** which drains the spleen and pancreas. The **inferior**

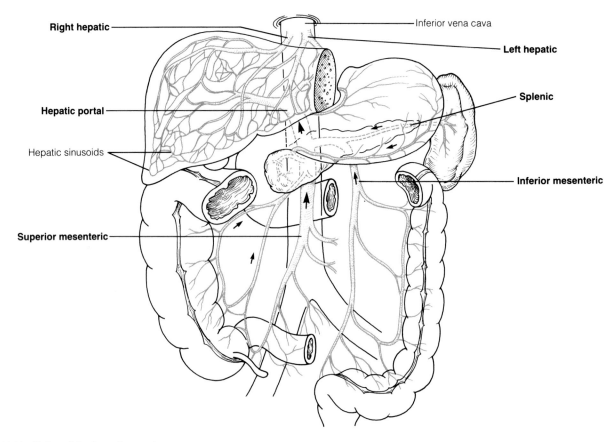

Figure 23-20 Veins of the hepatic portal system.

mesenteric vein empties blood from the large intestine into the splenic vein. Several smaller veins service other digestive organs and their blood empties into the hepatic portal vein. This vein enters the liver and ramifies into smaller veins and venules, eventually terminating in sinusoids. Blood from the sinusoids collects in the hepatic veins and eventually empties into the inferior vena cava.

The liver has two blood sources. Oxygenated blood enters it through the hepatic artery and deoxygenated blood enters through the hepatic portal vein. Although low in oxygen, venous blood from the digestive organs may be rich in nutrients and other substances absorbed in the intestine. As the blood passes through the sinusoids, liver cells may add some materials, remove others, and modify still others. For example, glucose may be removed and stored. Waste products from the breakdown of red blood cells may be modified in the liver. Bacteria are destroyed by liver phagocytosis, and poisons absorbed in the intestine can be detoxified.

Veins of the Lower Extremities

The external iliac vein forms the major drainage of each lower extremity. It receives blood from two sets of veins, one deep and the other superficial. The deep veins parallel the arteries in this area and are named for their arterial counterparts: the **femoral, popliteal, anterior tibial, posterior tibial,** and **peroneal veins.**

The two superficial veins are located on opposite sides of the limb. The **great saphenous** (să·fē′nŭs) **vein** extends the length of the medial surface of the thigh and leg, receiving blood from smaller vessels along its length (Figure 23-21). The **small saphenous** lies along the lateral aspect. Both saphenous veins arise in **plantar arches,** and both extensively anastomose with deep veins of the thigh and leg. The great saphenous is one of the longest unbranched veins in the body. Its superficial position makes it easy to remove surgically. For this reason it is usually the one removed and used when a vein is needed for coronary bypass procedures.

placenta: (L.) *placenta,* a flat cake

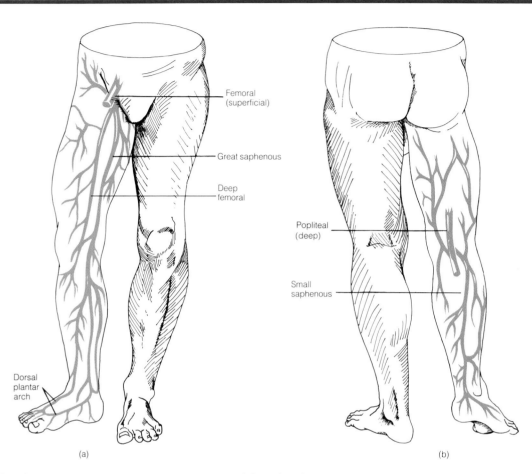

Figure 23-21 Veins of the right lower extremity. (a) Anterior view. (b) Posterior view.

Fetal Circulation

After completing this section, you should be able to:

1. Describe the internal modifications of the heart in a fetus that allow blood to bypass pulmonary circulation.
2. Discuss why it is efficient in a fetus for blood to bypass pulmonary circulation.
3. Describe the fetal arterial shunt, the ductus arteriosus.
4. Discuss the method by which most materials cross the placenta from the mother to the fetus.

Our discussion thus far has emphasized the separation of adult circulatory routes into pulmonary and systemic divisions based on the specialized function of the lungs in adding oxygen and removing carbon dioxide from the blood. Although its lungs are nonfunctional, a fetus has the same requirements as an adult for oxygen for its tissues. In addition, the fetus needs nutrients to supply energy and materials for growth, development, and differentiation of cells and tissues into organs. Embryonic and fetal development is a period of great cellular activity. Waste products produced by this metabolism must be removed so that the fetus does not poison itself. Transportation of all these materials is accomplished by the fetal circulatory system (Figure 23-22).

The fetus is connected to the mother's uterus by a long **umbilical** (ŭm·bĭl'ĭ·kăl) **cord** that terminates in the **placenta**, which is a complex structure of maternal and fetal tissues separated by a thin membrane (see Chapter 28). The placenta and umbilical cord provide the two-way route by which oxygen, nutrients, and other materials pass from the mother to the fetus; waste products and cellular secretions follow this path in reverse.

These materials cross placental membranes mainly by diffusion. As a pregnant woman breathes, eats, drinks, and introduces materials into her body, their concentrations increase. As fetal cells utilize these same materials, their

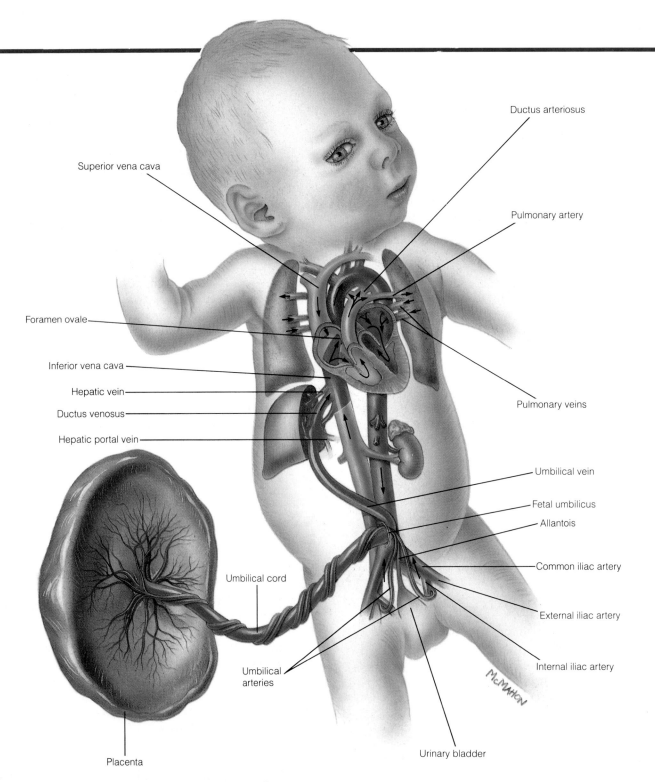

Figure 23-22 Fetal circulation. Modifications in the fetal cardiac and pulmonary circuit greatly reduce the amount of blood that passes through the lungs.

concentrations in its tissues decrease. Conversely, as the fetus produces waste materials and cellular byproducts, their concentrations increase on the fetal side of the placental membrane. Since this membrane is permeable to practically all materials normally present there, movement occurs in both directions across it. The higher concentrations of nutrients and oxygen on the maternal side result in movement (by diffusion) of these molecules into the fetal circulation.

ductus venosus: (L.) *ducere*, to lead; *vena*, vein

The higher concentration of waste materials on the fetal side results in movement of these molecules into the maternal circulation. Blood cells are too large to move through the placental membrane, and unless the placental membrane ruptures, no actual mixing of fetal and maternal blood occurs.

In the fetus, branches from the internal iliac arteries form **umbilical arteries** (Figure 23-22). These arteries pass alongside the fetus's urinary bladder to the anterior wall of its abdomen. They exit the abdominal cavity at the umbilicus (navel) and join the umbilical cord. At the placenta the umbilical arteries split into arterioles, then finally capillaries, which are surrounded by maternal blood. Exchange of materials takes place here by diffusion through the capillaries.

Blood flowing from the placenta to the fetus is returned to the fetal liver through a single **umbilical vein** that branches upon contacting the liver. The largest of these branches, the **ductus venosus** (dŭk'tŭs vē·nō'sŭs), passes through the liver and joins the fetal inferior vena cava. Placental blood entering the inferior vena cava joins blood from other tissues, and all pours into the right atrium of the fetal heart.

In an adult, venous blood entering the right atrium flows to the right ventricle into the pulmonary trunk. In a fetal heart, however, much of the blood in the right ventricle is diverted away from a largely nonfunctional pulmonary circuit. The reason for this is obvious: the fetus is not breathing air, so the blood cannot be oxygenated by the lungs.

Blood entering the fetal right atrium from the superior vena cava is low in oxygen and laden with waste products, while blood from the inferior vena cava contains oxygen and nutrients from the placenta. Blood from these two sources mixes in the right atrium and exits through two openings. Some of the blood, primarily that from the superior vena cava, flows through the tricuspid valve into the right ventricle. The remainder, primarily blood from the inferior vena cava, is diverted by the **valve of the vena cava** through the foramen ovale, an opening between the atria, into the left atrium. This blood flows into the left ventricle, bypassing the pulmonary circulation, and hence the pulmonary tissue. Blood in the right ventricle is pumped into the pulmonary trunk when the heart contracts. Instead of going to the lungs, however, most of it is diverted to the aorta through a connection between the pulmonary trunk and the aorta. This connection, or shunt, is the **ductus arteriosus** (dŭk'tŭs är·tē"rē·ō'sŭs) (Figure 23-22).

The net effect of these modifications in the fetal circulation is to divert blood away from a nonfunctioning pulmonary system into the systemic circulation. Functionally, this increases the amount of blood sent through the placenta, thereby increasing efficiency in the exchange of materials between a fetus and its mother.

At birth, a remarkable physiological event occurs, namely, the respiratory system that was nonfunctional in the fetus suddenly begins to operate. It seems that several stimuli induce the infant to take its first breath. Accumulation of carbon dioxide and a lowering of oxygen concentration in the blood in both the fetus and newborn will result in contractions of the diaphragm and thoracic muscles. The gasps that result bring fluid into a fetus's lungs, but will cause breathing to begin in a newborn. In addition, the physical trauma associated with squeezing through the birth canal causes an elastic rebound in the thorax that sucks in enough air to half fill the newborn's lungs with air, forcing out fluid that has accumulated there.

Changes in intrathoracic pressures associated with filling of the lungs cause blood flow in the ductus arteriosus to reverse within a day or two following birth, and it becomes functionally ineffective. Within a few days, the interior walls of the ductus fuse, and accumulation of connective tissue transforms the vessel into a ligament, the **ligamentum arteriosum.** At the same time, the edges of the foramen ovale fuse, forming the fossa ovalis (see Chapter 22), and the atria become anatomically and functionally separate. If these changes fail to occur, surgical or other medical techniques are needed to achieve the adult configuration.

Arterial Circulation

After completing this section, you should be able to:

1. List the factors that maintain constant blood flow through the body.
2. Describe normal blood pressure.
3. Discuss the relationship between cross-sectional area of vessels and velocity of blood flow.
4. Define "peripheral resistance" and describe its influence on blood pressure.
5. Describe the influence of intrinsic cardiac control on blood pressure.
6. Discuss the influence of the medullary vasomotor center on blood pressure.

The major function of blood is transportation of materials, both dissolved and solid. Nutrients, oxygen, carbon dioxide, wastes, cellular products, and even cells are carried within the vessels of the cardiovascular system. With each heartbeat, a small volume of blood flows into the aorta. The strength of the ventricular contraction starts the blood on its way. Blood flows through the cardiovascular system because of differences in pressure in various regions. The ventricular contraction that forces blood into the aorta establishes an average aortic pressure of about 100 mm Hg. As blood flows through the vessels, the pressure drops. By the time the blood comes back again to the heart and reaches the right atrium, the pressure has decreased to 0 mm Hg.

Several factors maintain an efficient flow of blood through the arteries and capillaries and back through the veins to the heart to complete the circulatory cycle. The most important of these factors are (1) the volume, pressure, and velocity of blood as it flows through the vessels; (2) blood viscosity; and (3) arterial elasticity.

Factors Affecting Blood Pressure

Blood Volume

In a resting individual each contraction of the left ventricle sends approximately 70 mL of blood into the aorta. This volume of blood represents about 1.5% of the total volume of blood in an adult. However, the heart is so efficient, it can pump the entire volume of blood through the body within 1 min.

Normal Blood Pressure

Contraction of the left ventricle squeezes blood into the aorta. Because blood is essentially incompressible, nearly all the pressure applied is passed through the blood to the walls of the aorta. The elastic nature of the aorta allows it to expand, accommodating the additional volume of blood without a dangerous increase in pressure. As the ventricle relaxes, the decrease in force from the ventricle, along with the flow of blood away from the ventricular end of the aorta, allows the aorta to recoil. This decreases its volume, again without involving a drastic change in pressure. If this pressure is measured in young adult males during the peak of systole, the pressure averages 120 mm Hg in the brachial artery, the place where blood pressure is usually measured. During the wane of diastole, the pressure averages 80 mm Hg. Females of the same age and size have slightly lower blood pressure.

As blood flows from the aorta into the arteries, capillaries, and veins, pressure gradually decreases (Figure 23-23). This change is a result of several factors, primarily cardiac output, blood volume, peripheral resistance of the capillaries, and blood viscosity.

Peripheral Resistance

The major influences in reducing pressure and velocity of the blood as it proceeds to and through the capillaries are an increase in cross-sectional surface area and a corresponding increase in **peripheral resistance** (or friction) in the arterioles and capillaries. The progressive branching of the arteries and

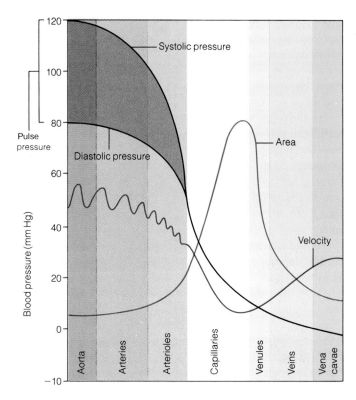

Figure 23-23 This graph shows a comparison of blood pressure, velocity of blood flow, and cross-sectional area in various parts of the vascular system.

arterioles results in an effective systemic capillary bed with a cross-sectional surface area nearly 1000 times that of the aorta and a total surface area of slightly more than 600 m^2. By way of comparison, most single family houses in the United States contain less than 200 m^2 in floor space. This increase in the amount of internal vascular surface area results in a simultaneous increase in contact between the blood and the endothelium of the vessels. The increased surface area and friction are sufficient to slow the velocity of blood from a rate of approximately 30 cm/sec in the aorta to less than 0.04 cm/sec in the capillary beds. The individual effect of either the increase in cross-sectional area or friction would be sufficient to reduce velocity and pressure. Together they decrease pressure markedly and slow the blood to 1/600 of the velocity it had as it entered the aorta. The slower velocity allows more time for exchange of materials between the blood and the tissues.

As blood enters the venules and veins, the cross-sectional area of these individual vessels increases, but a decrease in the number of vessels decreases the total cross-sectional area. This results in an increase in the velocity of the venous blood due to reduced cross-sectional area and less friction, which in turn results in an increase in velocity of the blood in the

aneurysm: (Gr.) *aneurysma*, a widening

> **CLINICAL NOTE**
>
> ## HYPERTENSION
>
> **Hypertension** is the elevation of the systolic or diastolic arterial blood pressure or both. There is no absolute measure of hypertension, but it is agreed that four separate readings of blood pressure above 140 mm Hg for systolic and 90 mm Hg for diastolic is the threshold for hypertension. More than 15% of the U.S. population is estimated to be hypertensive, blacks being affected twice as much as the general population. A strong relationship exists between increased sickness and death and the level of increased blood pressure. The secondary effects of hypertension include left ventricular failure, hypertensive encephalopathy, renal disease, and stroke. These problems can be prevented or modified by adequate pressure control. Unfortunately, the risk of myocardial infarction, another disorder associated with hypertension, does not seem to be easily preventable by blood pressure control.
>
> Hypertension may be secondary to other disorders such as tumors of the adrenal medulla, renal disease, or **aldosteronism**, which is increased aldosterone in the blood. The most common form of hypertension is **primary hypertension**, that is, hypertension for which no other cause can be found. Usually its diagnosis is made by excluding all other probable causes. Primary (or essential) hypertension appears to be partly hereditary. Since it has few symptoms, it is considered to be an insidious silent killer. Risk factors include heredity (as mentioned above), a history of smoking, increased serum lipid levels, and use of excessive salt. Alcoholics are also at high risk for hypertension.
>
> Hypertension is controlled by decreasing the salt content of the diet, not smoking, losing weight if obese, and taking antihypertensive drugs.

> **CLINICAL NOTE**
>
> ## ANEURYSM
>
> An **aneurysm** (ăn'ū·rĭzm) is a sac formed by local stretching or dilation of the wall of an artery, a vein, or the heart. Aneurysms are most frequently associated with atherosclerosis, but any injury to the tunica media (such as syphilis or congenital defects) or prolonged hypertension can lead to an aneurysm.
>
> The most common site for an aneurysm is the aorta. Excessive pressure and excessive stretching of the aortic wall will cause compression of the blood vessels that supply the elastic tissue of the tunica media. As a result, the tunica media becomes ischemic and necrotic. This weakens the wall of the aorta. If pressure elevations persist (that is, if the hypertension is not treated), the tunica media and tunica intima may be torn apart. Blood will then enter the tunica media and cause a longitudinal separation of the wall of the aorta. This type of aneurysm, which is a complication of hypertension, is called a **dissecting aneurysm** and is frequently fatal.

veins on its return to the heart (see section on Venous Circulation). Figure 23-23 shows a reciprocal relationship between pressure and velocity on one hand and the total area of contact on the other hand.

Blood Viscosity

A less important factor that influences blood pressure is its viscosity. Other things being equal, a given quantity of blood that is relatively viscous (syrupy), due to overproduction of red blood cells or a loss of water, will exert greater pressure on vessel walls than will the same quantity of thinner blood flowing at the same velocity.

Elasticity of the Arteries

The elasticity of arteries also affects blood pressure. Arteries stretch with every systolic surge of blood and recoil with each diastole. Any factor, such as atherosclerosis (in which the arterial walls stiffen), that decreases the elasticity of the arteries limits their ability to stretch and respond to the blood volume changes associated with each heartbeat.

Control of Blood Pressure

In a healthy person blood pressure is normally controlled in several ways. The mechanisms involved fall into two categories. Blood pressure is controlled by varying the cardiac output or by changing the peripheral resistance. An increase in cardiac output raises blood pressure, while a decrease in cardiac output lowers it. In addition to intrinsic cardiac control, nervous or chemical mechanisms can change peripheral resistance and thereby cause corresponding changes in blood pressure.

Neural Control of Blood Pressure

In addition to cardiac centers, the medulla contains a **vasomotor center** that controls arterial blood pressure (Figure 23-24). This center may contain separate stimulatory and inhibitory regions. Nerve fibers from the vasomotor center innervate arteries and arterioles throughout the body through sympathetic nerves. Impulses pass along these fibers continuously, providing constant stimulation of the muscles in the vessel walls. The contraction of these muscles, vasoconstriction, maintains arterial blood pressure. A decrease in stimulation of the vascular muscles allows the arteries to relax (vasodilation), which reduces arterial blood pressure. Figure 23-25 illustrates the part the vasomotor center plays in controlling blood pressure.

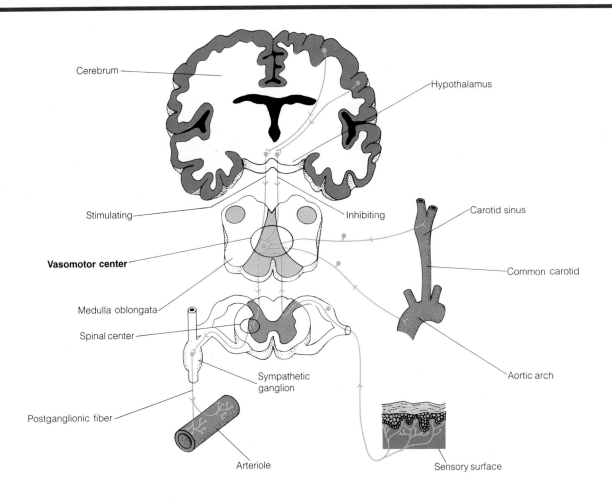

Figure 23-24 **Neural control of arterial blood pressure by the vasomotor center.** Sensory stimuli from several sources shown here can initiate motor impulses to the arteries resulting in an increase in vasoconstriction and an increase in arterial blood pressure.

Pressoreceptors and Chemoreceptors

The vasomotor center receives sensory stimulation from the same sensory nerves that innervate the cardiac control centers, namely, pressoreceptors and chemoreceptors. The pressoreceptors of the carotid sinus and aortic arch have branches that reach the vasomotor center along which information is supplied about the pressure of blood as it leaves the heart (Figure 23-25). In addition, chemoreceptors in the carotid bodies and aortic bodies in the arch of the aorta detect changes in oxygen and carbon dioxide concentrations and blood pH (see Chapter 20). A reduced oxygen level, a higher carbon dioxide level, and a decrease in pH of the blood all stimulate the vasomotor center to produce a rise in blood pressure.

Cerebral Control Centers

In addition to these reflex controls of blood pressure, higher centers in the cerebrum can affect blood pressure. These controls mainly work through the hypothalamus and usually occur during times of emotional stress. For example, anger results in the cerebral cortex stimulating the vasomotor center to increase the number of sympathetic impulses sent to arterioles. This causes vasoconstriction and an increase in blood pressure. Severe depression, on the other hand, results in a decrease of sympathetic impulses, vasodilation, and a drop in blood pressure.

Chemical Control of Blood Pressure

Vasoconstriction, which results in an increase in arterial blood pressure, is caused by several chemicals, particularly hormones discussed in Chapter 17. The most important of these may be epinephrine and norepinephrine secreted by the adrenal medulla. Among other effects, these hormones cause smooth muscle in the walls of cutaneous and abdom-

anaphylactic: (Gr.) *ana*, away from + *phylaxis*, protection

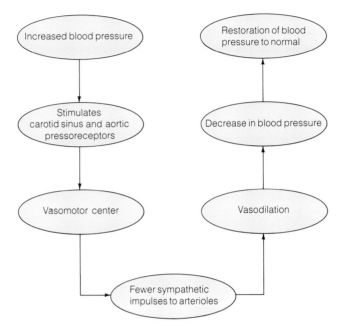

Figure 23-25 Neural control of arterial pressure. An increase in blood pressure has these effects.

inal arteries to constrict, which increases peripheral resistance and hence blood pressure. Antidiuretic hormone (ADH), released from the neurohypophysis, works indirectly to influence blood pressure. In small amounts it promotes retention of water in the blood, increasing its volume and blood pressure. If bleeding involves loss of large quantities of blood, ADH causes additional vasoconstriction and thus an increase in blood pressure.

Vasoconstriction is also stimulated by **angiotensin II,** a chemical produced by a sequence of reactions known as the **renin-angiotensin system** (see Chapter 26 for a more detailed discussion of this process). A drop in blood pressure stimulates certain kidney cells to secrete renin, a hormone that initiates a reaction sequence that produces the protein angiotensin II. This chemical then causes vasoconstriction, which raises blood pressure.

Vasodilation can be caused by two chemicals made in the body. Histamine (produced by mast cells) and kinins (polypeptides in the blood plasma produced by the kidneys and other cells) cause vasodilation, thereby increasing the flow of blood through the affected area. Such activity can be important at certain times, such as during an inflammation response (see Chapter 24).

Local Control of Blood Pressure

In addition to large scale effects of nerves and chemicals that affect the whole system, the body is able to control blood

> **CLINICAL NOTE**
>
> ## SHOCK
>
> **Shock** is circulatory failure; the supply of blood to the tissues is completely inadequate to meet the oxygen and nutritional needs of the cells. Anything that reduces cardiac output predisposes to shock. Shock is classified by the physiologic event that causes it.
>
> **Hypovolemic shock** is brought about by diminished blood volume caused by excessive bleeding, such as in hemorrhage due to trauma, surgery, or coagulation disorders. Hypovolemic shock may also be brought about by loss of plasma such as occurs with severe burns, intestinal obstruction, or dehydration from prolonged vomiting, diarrhea, or diuresis.
>
> **Cardiogenic shock** occurs from a decrease in the cardiac output because cardiac function is compromised. Extensive myocardial damage (as from a myocardial infarction), heart surgery, or prolonged arrhythmia will lead to cardiogenic shock.
>
> **Neurogenic shock** results from decreased vasomotor tone. Although the volume of blood remains the same, the overall size (capacity) of the circulatory system has increased, leading to lower blood pressure and thus diminished cardiac output and poor supply of blood to the tissues. The tone may be diminished because of effects at the vasomotor center, including spinal anesthesia, spinal cord injury, damage to the medulla, or altered function of the vasomotor center because of low blood sugar (**insulin shock**). If the problem is in the blood vessels themselves, this is a type of neurogenic shock called **vasogenic shock.** The most common form is an allergic reaction leading to **anaphylactic** (ăn″ă·fĭ·lăk′tĭk) **shock.** The antibody-antigen reaction leads to a release of histamine, serotonin, and bradykinin. These chemical mediators cause dilation of local blood vessels and increased capillary permeability. All of these effects, if widespread, will increase the capacity of the circulatory system. The volume of blood (which has not changed) is now no longer sufficient to fill the enlarged capacity of the circulatory system. This leads to poor venous return and therefore a cardiac output insufficient to meet the metabolic needs of the cells.
>
> In all forms of shock, treatment is aimed at identifying and treating the underlying cause and restoring tissue blood supply.

flow at a local level, thus increasing blood flow to active organs while decreasing it to less active organs. This is called **autoregulation.** For example, during strenuous exercise, vasodilation increases the flow of blood through active muscles several times over the volume of flow at rest. This is caused by a combination of factors. The increased use of oxygen by the muscle fibers causes a rapid decrease in the oxygen supply locally, with a simultaneous increase in the level of carbon dioxide. The level of cellular waste products, especially lactic acid, increases. The overall effect to the blood is increased carbon dioxide and hydrogen ions, which lower the pH, and decreased oxygen concentration. Any one of these conditions individually would cause vasodilation, and the combined effect causes a marked local vasodilation, thereby increasing the flow of blood to the active tissue.

Capillary Exchange

After completing this section, you should be able to:

1. Describe the difference between blood hydrostatic and blood osmotic pressure.
2. Discuss the effect of blood hydrostatic pressure in moving water through the capillary wall.
3. Discuss the effect of blood osmotic pressure in moving water through the capillary wall.

We have emphasized the structure and functions of arteries and arterioles in delivering blood to the tissues and the role of veins in collecting blood from the tissues and returning it to the right atrium. The walls of these vessels, however, are so thick that the materials (nutrients, oxygen, carbon dioxide, hormones, and other cellular products) carried inside cannot pass through them. Only at the capillary beds does any exchange of materials occur between blood and its surroundings.

One reason capillaries can accomplish such an efficient exchange of material is their extensive surface area and the low velocity of blood that passes through them. As the arteries decrease in diameter distally from the heart through arterioles to capillaries, the decrease in the average diameter of vessels drops from almost 3 cm in the aorta to about 0.01 mm at the capillary level. In addition, most capillaries are only about 1 mm long. The huge surface and relatively slow velocity in the capillaries provides more than enough area and time for adequate diffusion of materials at the tissue level.

Diffusion is a passive movement of molecules from regions of greater concentration to regions of lesser concentration. Most substances in the blood and interstitial fluid are free to diffuse between the plasma and the interstitial fluid following the concentration gradients. However, certain molecules are limited in their ability to penetrate endothelial cells. Chief among these are proteins. Because of their size, protein molecules have difficulty penetrating capillary walls. This fact, along with their greater concentration in the plasma, gives protein molecules great importance in affecting the differences in concentration gradients between the plasma and the interstitial fluid.

Two pressures at the cellular level operate to move materials into and out of the capillary: **hydrostatic** and **osmotic pressure.** Each of these can be subdivided, depending on whether the flow is into or out of the capillary.

Hydrostatic Pressure

Blood hydrostatic pressure (BHP) is the pressure within capillaries that tends to move fluid from the capillary into surrounding tissue. Actual pressures vary among individuals, but BHP averages 30 mm Hg (Figure 23-26) at the arterial

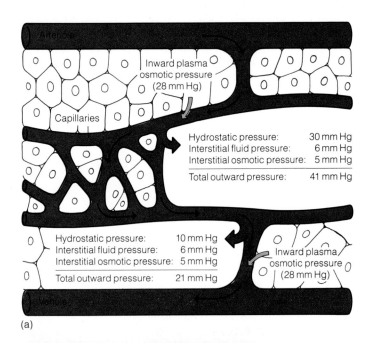

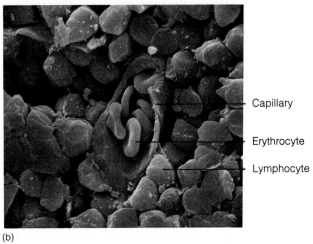

Figure 23-26 Capillary exchange. (a) Factors involved in the exchange of materials between capillaries and cells. (b) Scanning electron micrograph showing a capillary surrounded by cells and interstitial spaces (mag. 2283 ×).

end of a capillary, and about 10 mm Hg at the venous end. **Interstitial fluid hydrostatic pressure (IFHP)** acts on both capillary walls and cells of the surrounding tissue. It is a negative pressure, so it also moves fluid from the capillary. IFHP amounts to about -6 mm Hg along the length of the capillary.

Osmotic Pressure

Blood osmotic pressure (BOP) tends to move water into a capillary. The presence of large amounts of proteins in the plasma causes water and some dissolved substances to enter a capillary. BOP averages about 28 mm Hg for the length of the capillary. Its influence, therefore, is constant throughout the area of capillary exchange. **Interstitial fluid osmotic pressure (IFOP)** is caused by small amounts of proteins in interstitial fluid. Their influence is relatively constant throughout the capillary bed. This pressure averages 5 mm Hg at each end. IFOP pulls fluid into the interstitial spaces.

Movement of water into or out of a capillary depends on the net effect of these forces. At the arterial end, the combined effect of BHP, IFHP, and IFOP is greater than that of BOP, and water leaves the capillaries. At the venous end, however, the combined effect of BHP, IFHP, and IFOP is less than that of BOP, and net flow is into the capillary (Figure 23-26). In a 24-hour period, the amount of fluid that flows out of the capillaries is about 2 to 4 L more than the amount that flows in. This difference is carried from the tissues by lymphatic vessels.

Venous Circulation

After completing this section, you should be able to:

1. **Describe the special adaptation in veins that promote blood flow back to the heart in spite of reduced pressure.**
2. **Describe the changes in blood velocity as it flows through the veins.**
3. **Discuss the function of the veins as blood reservoirs.**

By the time blood passes through the arteries and capillaries, it has encountered enough peripheral resistance to reduce its pressure significantly. At the start of many large veins, the pressure is 20 mm Hg or less, and within the veins of the arms and legs it averages between 5 and 10 mm Hg. The blood from the lower part of the body returns to the heart against the pull of gravity, and its pressure is further reduced to nearly zero at the right atrium. This pressure loss is due not only to the peripheral resistance of arteries, capil-

> **CLINICAL NOTE**
>
> ### VARICOSE VEINS
>
> **Varicose veins,** or **varices** (văr´ĭ·sēz), are veins that are unusually dilated, distended, and tortuous. This condition is commonly produced by venous insufficiency or impairment of the mechanism for returning blood from the legs. The basic disorder in venous insufficiency is valvular incompetence. The structure of venous valves allows blood to flow only in the direction of the heart and prevents backflow. When skeletal muscles in the legs contract, blood flows away from the extremities to the heart; when the muscles relax, valves prevent blood from returning to the extremities. Prolonged pressure from the weight of blood pressing downward, as would happen with prolonged standing without much flexing of muscles, puts a strain on the veins. Eventually, the valves become loose and less efficient; the veins become distended and varicose veins result. Pregnancy is often associated with the development of varicose veins because the weight of the uterus puts pressure on veins coming from the legs. Varicose veins may remain asymptomatic or eventually lead to impaired circulation, ulcers in the lower leg, and discomfort.

laries, and gravity, but also because veins are not as elastic as arteries, hence they tend to expand easily.

In spite of this low pressure, venous return is augmented by several mechanisms, including venous valves, contractions of skeletal muscles, changes in velocity of blood flow, and pressure changes caused by breathing activities.

Venous Valves

Some veins contain valves that permit one-way flow of blood toward the heart (Figure 23-3). Pressure on the side of the valve away from the heart forces it open, permitting blood to flow, while a drop in pressure on the same side causes the valve to close and prevents the flow of blood away from the heart.

Muscular Activity

Many veins are located among skeletal muscles. As Figure 23-27 illustrates, when muscles surrounding a vein contract, they swell, press on the veins located between them, and squeeze the blood through the veins. Because of the valves, the blood can only go in one direction, toward the heart.

Velocity of the Blood

The velocity of the flow of blood is inversely proportional to a vessel's cross-sectional area; the larger this area, the

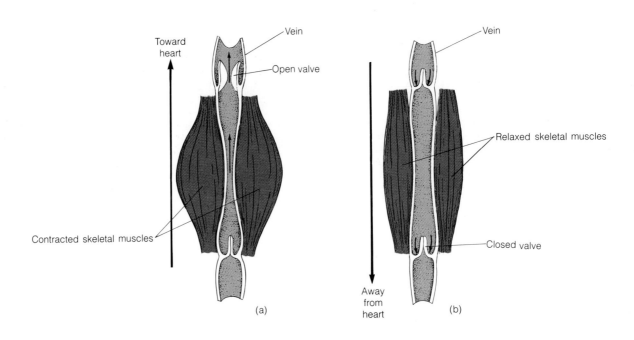

Figure 23-27 The effect of muscular contraction on the movement of venous blood between valves. (a) Contraction of the muscles squeezes the blood in the only direction possible through the open valve. (b) When the muscles relax, the closed valves prevent the backward flow of blood.

slower the movement. Blood entering the aorta (with a cross-sectional area of 2.5 cm^2) has a velocity of approximately 40 cm/sec. By the time it arrives at the capillaries, the total cross-sectional area has expanded by nearly 1000 times to 2500 cm^2, and the blood's velocity has slowed to approximately 0.04 cm/sec. As veins fuse and collect blood, their total cross-sectional area decreases to about 8 cm^2 in the vena cavae, and the velocity increases to nearly 20 cm/sec. This inverse relationship between cross-sectional area and velocity determines the energy necessary to keep blood flowing to the heart.

Breathing Activities

Breathing activities also help blood back to the heart. During each inspiration the thoracic cavity expands. This increase in volume causes a decrease in thoracic pressure. The decrease tends to force blood from the large veins of the abdomen into those of the thorax. During expiration the pressure is reversed, but the venous valves prevent backflow of blood.

Blood Reservoirs

In addition to their role in returning blood to the heart, the veins also serve as **blood reservoirs.** At rest, approximately 60% of the blood is contained in the veins and the venous sinuses, about 14% in the arteries and arterioles, 5% in the capillaries, 10% in the heart, and 11% in the pulmonary circulation. Changing conditions, such as exercise or hemorrhage, stimulate mechanisms of arterial vasoconstriction and vasodilation that can redistribute the blood to meet the changing needs of the body.

In this chapter we studied the anatomy of the blood vessels in both the fetus and adult. You learned about the mechanisms responsible for moving blood through the vessels at a constant rate and about the maintenance of a relatively constant blood pressure. We ended with an examination of the factors that govern exchange of materials between the blood and tissue fluids. In the next chapter we will study a specialized division of the circulatory system, the lymphatic system, and its role in defense against infection and disease.

angioma: (Gr.) *angeion*, vessel + *oma*, tumor
hematoma: (Gr.) *haimatos*, blood + *oma*, tumor
angitis: (Gr.) *angeion*, vessel + *itis*, inflammation
hemangioma: (Gr.) *haimatos*, blood + *angeion*, vessel + *oma*, tumor

THE CLINIC

DISEASES OF THE BLOOD VESSELS

The blood vessels are often involved in systemic diseases or diseases that originate in other organ systems. Some diseases, however, do originate primarily in the cardiovascular system of arteries and veins.

Developmental and Hereditary Defects

Although the blood vessels tend to be quite consistent in their anatomy, developmental abnormalities may occur. Such vessels are called **anomalous** or **aberrant vessels** and are usually only significant in the arterial system where they cause compression of vital structures. Improper development of the embryonic branchial arterial arches, for example, can result in absence of portions of the aortic arches and abnormal masses and pulsations in the neck and upper extremities. Aberrant renal arteries crossing and blocking the ureter may cause damage to the obstructed kidney. **Congenital cerebral aneurysms** (see Clinical Note: Aneurysm) may cause persistent headaches, neurologic defects, or cerebral hemorrhage if they rupture. **Angiomas** are localized vascular lesions of the skin and subcutaneous tissues that may cause unsightly "port wine stains" on the skin or swelling of an extremity.

Marfan's syndrome is an interesting hereditary disease that affects the medial layer of the ascending aorta. It results in aneurysm of the aorta and sometimes sudden death following rupture. These unfortunate people are very tall with excessively long extremities and poor vision due to dislocated lenses. If, because of their stature, they become athletes, they risk sudden death from overexertion.

Infections

Infections of the vascular tree may result directly from cuts or other wounds to a blood vessel that become infected. Bits of infected emboli in the bloodstream may also indirectly infect the blood vessels. **Phlebitis,** or inflammation of veins, may result from intravenous needles or catheters. The venereal disease syphilis can cause infection of the thoracic aorta with resultant aneurysm.

Trauma

Lacerations to blood vessels can result in external hemorrhage that can be considerable, particularly when arteries are involved. Excessive blood loss can result in **circulatory shock** with decreased circulating volume and blood pressure. Untreated, shock can result in unconsciousness, circulatory collapse, and death by cardiac arrest (see Clinical Note: Shock). Blunt trauma can result in internal bleeding into body cavities or into tissues producing masses of blood around the vessel called **hematomas.** Fractures can result in massive, hidden blood loss. Rare injuries to adjacent arteries and veins may develop into an abnormal interconnection between the vessels called **arteriovenous fistulas,** which produce a localized pulsating mass and increased cardiac output.

Immunologic and Allergic Disorders

Arteries are sometimes affected by nonspecific inflammations of undetermined cause described as **angitis** or **arteritis. Thromboangiitis** (thrŏm″bō·ăn″jē·ī′tĭs) **obliterans** (Buerger's disease) is an inflammatory disease of small and medium-sized arteries that results in ischemic symptoms, usually in the legs. It usually occurs in men 20 to 40 years of age who are cigarette smokers. **Temporal arteritis** affects the temporal arteries in the elderly, causing a characteristic headache and possibly resulting in blindness or stroke. It can be controlled by treatment with cortisone.

Acute allergic reactions sometimes affect small veins and capillaries causing them to leak small amounts of blood into the tissues. In the skin this may cause a rash of pinhead-sized blood spots called **petechiae** (pē·tē′kē·ē).

Spasm of the small arterioles of the digits and nose can cause **Raynaud's phenomenon** in which the affected digits turn white when exposed to cold and then turn deep red when rewarmed. Raynaud's phenomenon occurs principally in women. It may also occur secondary to connective tissue diseases and result in atrophy and ulceration of the skin.

Neoplasms

Primary tumors of the vascular system are usually noncancerous and result in masses of entangled blood vessels and fibrous tissues called **hemangiomas** (hē·măn″jē·ō′măz). They usually bleed excessively if traumatized and can cause problems if in a vital organ. **Telangiectases** (těl·ăn″jē·ěk·tā′sēs) are groups of dilated capillaries that resemble spider webs, usually occurring on the skin of the face and thighs.

Aging

The most significant problem of aging in the cardiovascular system is development of **atherosclerosis** (see Chapter 22, The Clinic: Diseases of the Heart). It is estimated that 50% of deaths in the elderly in the United States are due to atherosclerosis, either through coronary artery disease or strokes. Atherosclerosis is the occlusion of medium-sized and large arteries by atherosclerotic plaques (see photograph on next page). Its symptoms include ischemia and pain in the muscles of the extremities after exercise (similar to the symptoms of angina pectoris). As the blockage progresses, the person may have pain even when resting. Peripheral arterial pulsations may be reduced or absent. Atrophic changes may begin in the distal extremity with dry, scaling skin and nails and the absence of hair growth. The signs of arterial insufficiency are pallor of the extremity when it is elevated followed by redness when it is lowered, coldness, and delayed venous filling time. Severe disease may result in numbness of the extremity, skin ulcerations, and possibly even gangrene.

Acute ischemia may result from an embolism or from acute thrombosis blocking a vessel (see Clinical Note:

(continued)

THE CLINIC (continued)

DISEASES OF THE BLOOD VESSELS

Thrombosis and Embolism). The symptoms include severe pain, coldness, numbness, and pallor. The usual treatment of acute arterial occlusion is surgical removal of the obstruction or bypass grafting.

Atherosclerosis may also produce an aneurysm of the abdominal aorta with massive enlargement of this vessel. This may cause obstruction of the major branches of abdominal arteries, or if it ruptures, it may cause sudden death by massive abdominal hemorrhage. Surgical excision of the aneurysm and replacement with a Teflon graft can prevent the inevitable disaster.

Diseases of the smaller arteries may be due to diabetes mellitus or other generalized diseases.

Venous Diseases

The most common disease of the venous system is **venous thrombosis,** which is the presence of a thrombus (or clot) in a vein (see Clinical Note: Thrombosis and Embolism). Persistent venous obstruction may lead to skin changes in the lower extremities called **stasis dermatitis.** The treatment of venous thrombosis is through prevention or, once established, with anticoagulant drugs to prevent extension of the thrombus and anti-inflammatory drugs.

Varicose veins are dilated, tortuous superficial veins that occur in the legs because of increased hydrostatic pressure. Obesity, pregnancy, and abdominal tumors all contribute to the development of varicose veins (see Clinical Note: Varicose Veins).

Hemorrhoids are varicose veins of the venous plexus in the anorectal area most commonly caused by constipation. They may be complicated by inflammation, thrombosis, or bleeding. Usually, improved anal hygiene and stool softeners to relieve strain are sufficient to correct the problem. Severe cases may require surgical excision, injection with sclerosing agents, or banding with elastic bands.

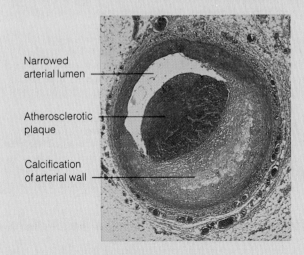

Coronary artery blocked by atherosclerosis.

STUDY OUTLINE

Blood Vessels (pp. 585–587)

Three main kinds of blood vessels exist: *arteries, veins,* and *capillaries.*

Arteries Arteries have thick walls and conduct blood from the heart to tissues. The wall of an artery has three layers or *tunics.* The *tunica intima,* a thin layer of endothelial cells, lines the cavity of the artery. The *tunica media* consists of a thick cylinder of smooth muscle and elastic connective tissues. The *tunica externa* consists of fibrous connective tissue.

Arterioles Arterioles are arteries less than 0.5 mm in diameter.

Capillaries Capillaries are only 0.01 mm in diameter, and their walls consist of only an endothelial layer.

Venules Venules are formed when several capillaries join together. At the point where the tunica externa appears, the venules become veins.

Veins Veins conduct blood from the venules to the heart. In a vein the tunica media and tunica externa are much thinner than in an artery. Veins possess *valves* that prevent the backward flow of blood.

Anatomy of the Circulatory Pathways (pp. 587–602)

Pulmonary Circulation Pulmonary circulation starts at the right ventricle and conducts blood through the *pulmonary trunk, pulmonary arteries,* and *lobar arteries* leading to the lobes of the lungs. Oxygenated blood is returned to the left atrium through *pulmonary veins.*

Systemic Circulation: Arteries Systemic circulation is that part of the vascular system that carries blood from the left ventricle to the tissues and returns it to the right atrium.

The *aorta* is the major artery of the systemic circuit. It begins at the superior medial border of the left ventricle, proceeds upward for a short distance, executes a sharp U turn, and extends down the length of the thorax and abdomen.

Major arteries of the head and neck are the *brachiocephalic, common carotids, subclavians,* and *vertebrals.* The *circle of Willis* supplies the brain.

Major arteries supplying the upper extremities are the *axillary, brachial, radial,* and *ulnar.*

Major arteries of the thorax are the *posterior intercostals, superior phrenics, bronchials,* and *esophageals.*

Major arteries of the abdomen are the *inferior phrenics, lumbars, middle sacral, celiac, left gastric, splenic, hepatic, superior mesenteric, inferior mesenteric, adrenals, renals, gonadals (testiculars* or *ovarians), common iliacs, internal iliacs,* and *external iliacs.*

Major arteries of the lower extremities are the *femoral, deep femoral, medial femoral circumflex, lateral femoral circumflex,*

anaphylaxis: (Gr.) *ana,* away + *phylaxis,* protection
glomerulonephritis: (L.) *glomerulus,* little ball + (Gr.) *nephros,* kidney + *itis,* inflammation

Anaphylaxis (ăn″ă·fĭ·lăk′sĭs) is an acute, often explosive systemic reaction occurring with hives, respiratory distress, shock, and gastrointestinal symptoms that occurs in a previously sensitized person when he or she is again exposed to the antigen. This is a severe, life-threatening emergency condition (see Chapter 23: Clinical Note: Shock). It can happen with exposure to drugs (for example, an allergic reaction to penicillin) and at times following insect stings. **Anaphylactoid reactions** are similar to anaphylaxis but occur after the first exposure to a drug (for example, histamine, morphine, or X-ray contrast media) and are dose related. Immediate treatment with an epinephrine injection can be life saving.

Physical allergy is a condition in which allergic symptoms occur upon exposure to cold, sunlight, heat, or mild trauma. **Serum sickness** is an allergic reaction involving fever, joint pain, skin rash, and swelling of lymph nodes that occurs 7 to 10 days after the administration of serums or certain drugs.

Autoimmune disorders are conditions in which the immune system produces **autoantibodies** to a natural antigen, with consequent injury to the tissues. Many examples of this type of disease have been described in clinical notes in previous chapters. A few examples include **glomerulonephritis** (an inflammatory disease affecting the glomeruli of the kidneys), acquired hemolytic anemia, rheumatoid arthritis, and systemic lupus erythematosus.

Organ Transplantation

A healthy immune system will, of course, react violently to any foreign tissue or organ and **reject** it. Thus, the ability to perform organ transplantation surgery has been dependent upon the development of drugs and techniques to control the immune system.

The **HLA system** is a group of genetically determined tissue antigens responsible for most of the problems of graft rejection. HLA stands for human leukocyte group A, or histocompatibility antigens (see text). The genes that control the production of these antigens have been found at four different loci. Prior to transplantation, the recipient's lymphocytes are typed, and a donor organ of the closest possible type is selected. It has also been discovered that some diseases seem to be associated with specific HLA antigens. The reverse may also be true in that some diseases appear to be associated with the absence of certain HLA antigens, such as rheumatic fever.

Several drugs have been developed to suppress the immune system and are given to transplant patients to prevent or alleviate **graft-versus-host** reaction. Of course, they suppress the entire immune system and also the metabolism of rapidly dividing cells so that the patient must be protected from infection.

Tumor Immunology

The immunology of tumors is a very intriguing study. It has been discovered that transplanted tumors may result in the production of tumor-specific antigens, or **tumor-specific transplantation antigen (TSTA).** Many attempts have been made to use TSTA to destroy cancers, either directly or by attaching radioactive or toxic components to the antigens. This is called **immunotherapy.** It has been used successfully in conjunction with chemotherapy to treat three types of cancer: choriocarcinoma, Burkitt's lymphoma, and neuroblastomas.

Immunology has also been helpful in diagnosing or following the course of some tumors. **Carcinoembryonic antigen** (CEA) is a protein polysaccharide found in adult colon cancers and in the intestines of normal fetuses. It has been used to test for the presence of colon cancer and as a test for recurrence after treatment.

STUDY OUTLINE

Lymphatic System (pp. 619–624)

The *lymphatic system* returns *lymph* (or *tissue fluid*) to systemic veins. It also transports fatty materials from the digestive system and produces *lymphocytes* and *antibodies*.

Lymphatic Vessels *Lymph capillaries* are closed at one end and contain valves that prevent backward flow of lymph. They collect lymph and pass it through the *lymphatics* to the lymph nodes on its way to the venous system. The major vessels are the *cisterna chyli, thoracic duct,* and *right lymphatic duct*.

Lymph Nodes *Lymph nodes* are small masses of lymphatic tissue distributed along the lymphatics. Each node consists of densely packed lymphocytes forming *lymph nodules,* which surround numerous interior sinuses. Dead cells and pathogens are filtered out of the lymph by lymphocytes and macrophages.

Lymph Circulation Once within lymphatic vessels, lymph flows slowly from the tissue level through progressively larger lymphatics into lymph nodes. Lacking a pump, the lymphatic vessels, like veins, make use of the squeezing action of contracting muscles to move lymph. About 2 to 4 L of lymph passes into the venous system every day.

Lymphatic Organs The *spleen* is the largest lymph organ. It contains *red pulp* (venous sinuses) and *white pulp* (splenic nodules) that are masses of lymphocytes. The spleen produces lymphocytes and antibodies and destroys aging erythrocytes.

Tonsils are masses of lymphatic tissue embedded in the mucous membrane of the pharynx. They function as lymph nodes.

The *thymus gland* is a double-lobed mass of lymphatic tissue in the mediastinum. The thymus produces and processes lymphocytes.

Nonspecific Defense Against Disease (pp. 624–627)

Nonspecific response is a defense against disease and infection that includes generalized mechanical and chemical means of warding off invaders.

Skin and Mucous Membrane The skin discourages infection with its layers of keratinized cells and the secretion of sweat and sebum. Mucus traps microbes and is moved away by ciliated epithelium.

Mechanical Processes Saliva and tears physically remove microbes.

Chemical Processes *Lysozyme* present in tears and acids in sweat and gastric juice kill and detoxify most microbes.

Nonspecific Chemicals *Interferon* inhibits reproduction of viruses. *Complement* and *properdin* promote inflammation, destroy microbes, and attract and stimulate the phagocytic activity of macrophages.

Phagocytosis *Microphages* and *macrophages* recognize, ingest, and digest bacteria in a process called *phagocytosis*.

Inflammation *Inflammation* is a coordinated localized response to a limited infection, which involves vasodilation and attraction of phagocytes to the area.

Specific Defense Against Disease (pp. 627–629)

Specific response is a type of defense that involves the production of specific antibodies that recognize and destroy specific antigens and the cells upon which they are carried.

Antigens *Antigens* are chemical substances that provoke a specific immune reaction when introduced into the body. They are large molecules containing specific *antigenic determinant sites* that are recognized by antibodies.

Antibodies *Antibodies* are *immunoglobulins* (specialized proteins) produced in response to a specific antigen. The molecule has a Y shape. The forked end of the Y contains *variable regions* that recognize and combine with the determinant sites on antigens. The complex that forms accumulates as insoluble aggregations that are phagocytized by leukocytes.

Cellular and Humoral Immunity (pp. 629–634)

In *cellular* (or *cell-mediated*) *immunity,* lymphocytes directly attack pathogens, while in *humoral immunity,* they utilize antibodies in the attack. Lymphocytes develop from *stem cells* in the red bone marrow. Three types are known: T cells, B cells, and natural killer cells.

Cellular Immunity and T Cells *T cells* accomplish cellular immunity. They are produced in the red marrow, mature in the thymus, and migrate to lymphoid tissue throughout the body, where they remain as a lifetime source of newly produced T cells.

When presented with macrophage-bound antigens, T cells divide, producing a clone of several subclasses of cells. *Effector T cells* destroy microbes directly and secrete chemicals that attract and activate macrophages in the area. *Helper T cells* increase the ability of B cells to secrete antibodies during the humoral response. *Suppressor T cells* inhibit the activities of other T and B cells. *Memory T cells* remain in the body for use against future infections. *Amplifier T cells* increase the activities of helper T cells and suppressor T cells. *Delayed hypersensitivity T cells* are active in allergic responses.

Natural killer cells identify and destroy tumor cells.

Humoral Immunity and B Cells *B cells* accomplish humoral immunity. They are produced in the red marrow and mature in other lymphoid tissue where they accumulate and remain as a source of newly produced B cells.

There are thousands of kinds of B cells covered with antibodies that recognize bacteria, viruses, and other invaders. Should a pathogen penetrate the body, B cells combine with it and divide repeatedly to produce a clone of two kinds of cells. *Plasma cells* produce antibodies that destroy the invader. *Memory cells* remain in the body for future use.

Organ Transplants The immune system is able to tolerate *self-antigens* (*histocompatibility antigens*), and rejects *non-self-antigens*. Organ transplants may be rejected if the number of donor non-self-antigens is sufficiently large.

Immunity and Vaccination (pp. 634–635)

Induced Immunity The second exposure to an antigen induces a faster, stronger *secondary response. Vaccination* is the injection of dead or inactivated microbes or antigens to induce the *primary response*. If the individual is exposed to a virulent member of the strain in the future, the secondary response overwhelms the invader.

Natural Immunity Many dangerous microorganisms are controlled by *natural immunities* present at birth.

Active and Passive Immunity Induced and natural immunity are forms of *active immunity*. *Passive immunity* occurs when one accepts antibodies from an external source.

Monoclonal Antibodies *Monoclonal antibodies* are molecules produced from a single B cell clone. The B cell is fused to a tumor cell producing a *hybridoma*. The antibodies produced are specific to a single antigen or line of cells.

Blood Groups (pp. 636–637)

At least 14 different blood groups have been identified in humans. They are due to *agglutinogens* (*isoantigens*) carried on the surfaces of erythrocytes and *agglutinins* (*isoantibodies*) in the plasma.

ABO Blood Groups Four *blood types*—A, B, AB, and O—are produced by the presence or absence of two agglutinogens, A and B. Type A people possess anti-B antibody, type B people have anti-A antibody, type AB people have neither antibody, and type O people have both antibodies.

Transfusions If type A blood is given to a type B person, or vice versa, clumping will occur in the recipient's blood. Type O blood can be given to other types without causing *agglutination* (*universal donor*). People with type AB blood can receive all the other types in transfusions (*universal recipient*).

Rh Blood Group People who possess the Rh antigen are Rh^+, and those who lack it are Rh^-. If Rh^+ blood is transfused into an Rh^- person (or leaks across the placenta into an Rh^- pregnant mother), the recipient of the cells will produce anti-Rh antibodies that destroy the erythrocytes.

SELF-TEST OF CHAPTER OBJECTIVES

True-False Questions

1. Afferent lymphatics enter a lymph node at a shallow indentation called the hilus.
2. Once an invading bacterium enters a tissue, it may be phagocytized by a defensive cell called a macrophage.
3. Most of the humoral antibodies are found in the serum protein fraction of the blood consisting of γ-globulins.
4. Lymphokine production is a primary function of B cell lymphocytes.
5. T cells lymphocytes are active in the rejection of transplanted human organs.
6. The differences among antibodies is due to differences in the structure of the variable section of both the heavy and light chains that compose the molecule.
7. The lymphatic system returns tissue fluid to the aorta, where it mixes with the blood.
8. The right lymphatic duct drains lymph fluid only from the upper right quadrant of the body.
9. T cells are so named because they mature in the thyroid gland after their original formation.
10. Active immunity against a disease is acquired when an individual receives a vaccination containing antibodies produced in another animal.
11. B cells are so named because they mature in the bursa of Fabricius in chickens.
12. Interferon inhibits the ability of viruses to reproduce within a cell.
13. B cells are noted for their ability to phagocytize bacteria.
14. T cells are active in humoral immunity and are noted for their ability to secrete antibodies.
15. Helper T cells do not themselves secrete antibodies, however, they stimulate other cells to produce them.
16. Natural killer cells are able to destroy tumor cells without prior sensitization to their antigens.
17. Like T cells, B cells produce clones with numerous subclasses of cells, including helper B cells and suppressor B cells.
18. Histocompatibility antigens are found on the surfaces of all body cells except mature erythrocytes.
19. The vaccine for smallpox is an example of a natural immunity in humans.
20. People with type AB blood possess both anti-A and anti-B antibodies.

Matching Question

Match the component on the left with the type of immunity with which it is associated on the right:

21. B cells
22. T cells   a. humoral immunity
23. immunoglobins   b. cellular immunity
24. plasma cells

Multiple-Choice Questions

25. Invading cells that escape phagocytosis may be immobilized and destroyed by combination with antibodies. The humoral antibodies that are involved are produced by
    a. reticulocytes   c. macrophages
    b. B cells   d. T cells

26. When immunity is induced in a person, initial exposure to a disease-causing bacterium is usually followed by a latent period. During this latent period, which of the following probably does *not* happen?
    a. B cells are proliferating
    b. the bacteria are proliferating
    c. antibodies against the bacteria are killing them
    d. disease symptoms due to the bacterium are being expressed
27. In induced immunity the more rapid secondary response is probably due to
    a. preselection of T cells induced in the primary response
    b. preselection of B cells induced in the primary response
    c. an undiminished concentration of antibodies remaining from the primary response
28. Parvovirus causes an often fatal illness in puppies. The puppies can be protected by injection of killed viruses, but humans do not need to be inoculated because they do not become ill from the virus. Puppies and humans, respectively, are displaying examples of
    a. induced and natural immunity
    b. induced and passive immunity
    c. natural and passive immunity
    d. passive and induced immunity
29. Which of the following is known to be present in stomach secretions?
    a. IgG   c. IgA
    b. IgM   d. IgD
30. Antibodies naturally present in a blood type AB person include
    a. anti-A antibodies but not anti-B antibodies
    b. anti-B antibodies but not anti-A antibodies
    c. both anti-A and anti-B antibodies
    d. neither anti-A nor anti-B antibodies

Essay Questions
31. Describe the major functions of the lymphatic system. Trace a drop of lymph from a lymphatic vessel in the right leg to the superior vena cava.
32. Discuss the complement system. Describe the four main ways in which complement enhances the body's ability to fight microbial infection.
33. Describe the steps involved in phagocytosis.
34. Discuss the importance of inflammation as a localized response to a limited infection.
35. Describe the steps involved in formation of an antigen-antibody complex.
36. Compare and contrast cellular immunity and humoral immunity.
37. List six subclasses of cells found in a T cell clone and describe the function of each.
38. Distinguish between a polyclonal antibody and a monoclonal antibody.
39. Distinguish between natural and induced immunity and discuss the advantages and disadvantages of passive and active immunity.
40. Discuss the antigens and antibodies that cause the ABO blood groups.

Clinical Application Questions
41. How does a radical mastectomy differ from a modified radical mastectomy?
42. Discuss hypochondriasis. Describe some of the nonmedical factors that may predispose an individual to this condition.
43. Define the term AIDS. Name the agent that causes the disease. Do victims with the disease usually die from the direct effects of the virus or from some other cause?
44. Discuss hemolytic disease of the newborn. How can it be prevented?
45. Compare and contrast primary specific immunodeficiency disorders with primary nonspecific immunodeficiency disorders.
46. Describe anaphylaxis. Why is it such a dangerous reaction?

UNIT V CASE STUDY

HEART PATIENT

Mr. A. S. is a 52-year-old businessman who began seeing his doctor two years ago when he first complained of fatigue and shortness of breath whenever he exerted himself. About one year ago his symptoms became so severe that he was unable to work at previous levels and had to curtail his favorite hobby, gardening. During the past year he experienced several episodes of crushing chest pain upon exertion. On three occasions he lost consciousness when digging in his garden. One event in his history is significant: at age 14 he was bedridden for several months with prolonged fever, migrating joint pain, weakness, and weight loss. After recovering from this illness, his doctor noted that he had developed a heart murmur. From that time up to two years ago, he enjoyed good health and an active life.

On his initial physical examination, his internist noted that he appeared older than his age of 52 years and was fatigued and mildly depressed. The cardiovascular examination revealed a loud, rough systolic murmur heard over the left second intercostal space which was transmitted to his neck vessels. A strong vibration (thrill) could be felt over this same area. His pulse rate was normal, but his blood pressure was mildly elevated (150/110) with a narrow pulse pressure. His chest X ray showed a normal sized heart, but calcium deposits were observed in the area of the aortic valve. A routine electrocardiogram (EKG) showed evidence of thickening of the left ventricle, as well as some nonspecific T wave changes. His internist diagnosed his problem as **aortic valve stenosis** (narrowing of the aortic valve), which apparently resulted from the rheumatic fever he had contracted in his youth. A catheter was inserted into the left side of his heart to evaluate the degree of valve stenosis. It revealed an increase in the left ventricular pressure, which then decreased as the catheter was manipulated through the narrowed valve. The valve area was calculated at less than 40% of its original size. On the basis of these findings, it was advised that Mr. A. S. undergo a surgical replacement of his defective aortic valve. In view of the 80% mortality rate within two years in patients with this diagnosis and a less than 10% mortality rate from the surgery, the patient agreed to the surgery.

The operation began with Mr. A. S. being anesthetized and placed on a heart-lung machine to support his circulation while the heart was opened. His defective valve was then replaced with a prosthetic (artificial) plastic and steel valve (see figure). As suspected, his aortic valve was severely deformed and calcified and was less than 50% of its original size. Mr. A. S. recovered well from his surgery. He is now being maintained on anticoagulants for several months to prevent embolisms (blood clots traveling from the new valve). He is expected to make an excellent recovery, and he can look forward to a long, active life.

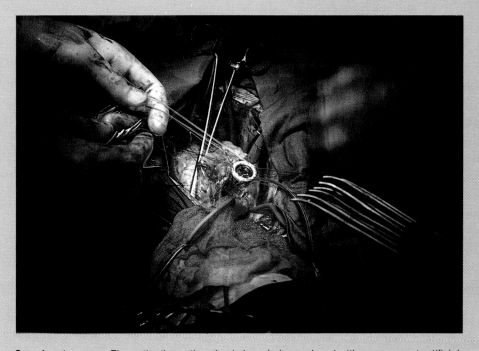

Open heart surgery. The patient's aortic valve is here being replaced with a permanent artificial valve.

Unit VI

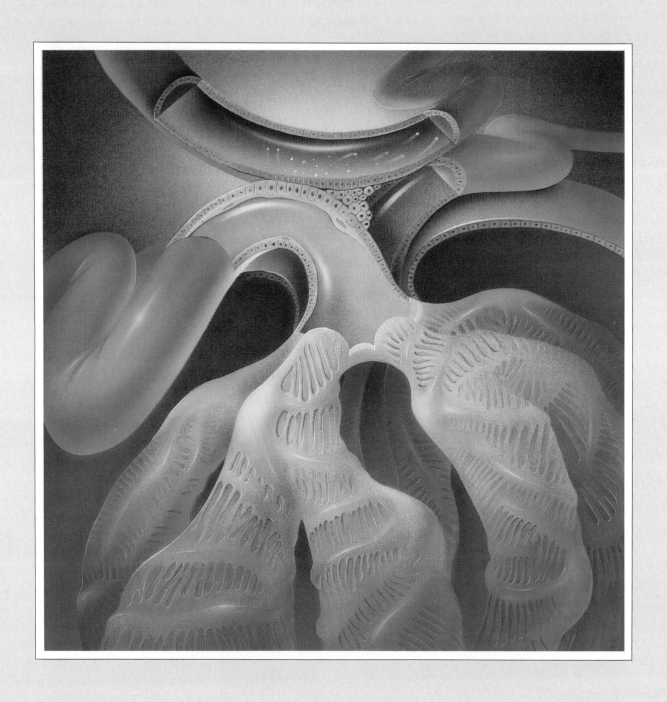

Homeostatic Control of Body Fluids

Unit VI describes the homeostatic mechanisms that neutralize and remove toxic materials that are constantly being added to the blood as a result of the metabolic activities of cells. We begin with an examination of the kidneys, organs that remove excess water, dissolved materials, wastes, and toxic substances from the blood during the formation of urine. The unit ends with a study of several homeostatic mechanisms that regulate the composition of the blood and tissue fluids.

25
The Urinary System

26
Composition and Control of Body Fluids

25
The Urinary System

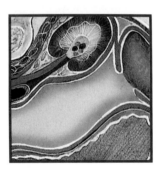

The urinary system plays enormously important and diverse roles in maintaining homeostasis. It is responsible for (1) removal of waste products from the blood, particularly nitrogen-containing products from the breakdown of amino acids; (2) elimination of many foreign substances, such as certain drugs and toxins produced by infecting bacteria; (3) maintenance of the extracellular fluid volume, including correct mineral levels in the body and correct amounts of total body water; (4) adjustment of the pH of blood; (5) regulation of blood pressure; (6) stimulation of the production of blood; and (7) maintenance of healthy bones through the activation of vitamin D.

Most of these functions are performed by the **kidneys,** a pair of organs located on either side of the vertebral column about even with the junction of the abdominal and thoracic cavities. The kidneys produce **urine,** an aqueous solution of waste products that passes from the kidneys through the ureters to the urinary bladder located in the lower abdominal cavity. The bladder stores the urine until it can be eliminated, at which time it flows to the outside of the body through the urethra. Collectively, these structures comprise the **urinary system** (Figure 25-1).

In this chapter we will examine

1. the structure of the kidneys,
2. the organization of the nephrons, the functional units of the kidneys,
3. the processes involved in the production, storage, and elimination of urine, and
4. accessory functions performed by the kidneys.

retroperitoneal: (L.) *retro*, backward + (Gr.) *peritonaion*, peritonium

Anatomy of the Kidneys

After studying this section, you should be able to:

1. Describe the location and external anatomy of the kidneys.
2. Explain how blood is delivered to and removed from the kidneys.
3. Describe the macro- and microscopic internal organization of the kidneys.

As the fetal urinary system develops, it moves posteriorly and comes to lie between the parietal peritoneum and the wall of the abdominal cavity (see the box on Embryonic Development of the Urinary System). Hence, the kidneys, ureters, and urinary bladder are **retroperitoneal** (rē″trō·pĕr″ĭ·tō·nē′ăl) **organs,** that is, they are outside the peritoneal cavity. Also, as the kidneys move upward, they rotate until the region where blood vessels attach to the kidney faces the vertebral column (Figure 25-2). They normally lie on either side of the vertebral column, with the upper end of each kidney nearly level with the 12th thoracic vertebra and the lower end about even with the 3rd lumbar vertebra. The left kidney is usually slightly higher, probably because the right one is pushed down by the liver.

External Structure

Each kidney is about 12 cm long, 7.5 cm wide, and 3 cm thick (Figure 25-3). It is shaped like a kidney bean (hence, the name of the bean), but much larger, with the concave surface facing the vertebral column. This concave depression, the **renal hilum,** is a region where the **renal artery,** the **renal vein,** and the ureter attach to the kidney. The kidney is surrounded by a thin **renal capsule,** which in turn is surrounded by a 2- to 3-cm-thick layer of **adipose capsule.** The fatty tissue of the adipose capsule cushions the kidney and protects it from damage from physical shock. The kidney and its capsules are enclosed in a fibrous sheath called the

Figure 25-1 Organization of the urinary system. The urinary system consists of the kidneys, ureters, bladder, and urethra.

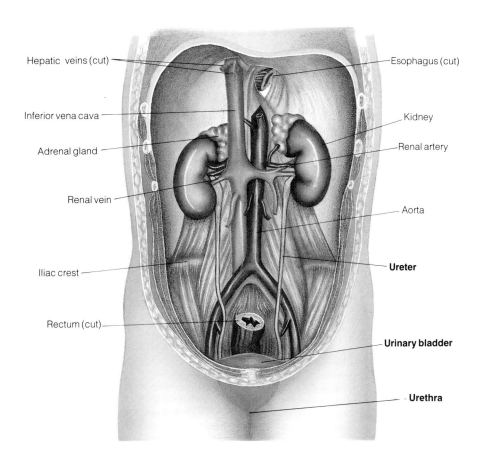

calyx: (Gr.) *kalyx*, cup
cribrosa: (L.) *cribum*, a sieve

CLINICAL NOTE

URINARY TRACT IMAGING TECHNIQUES

There are several techniques for producing images of the urinary tract. A simple X-ray film of the abdomen or of the kidney, ureters, and bladder may show the size of the kidneys, their position, and the presence of radiopaque stones. However, a more useful view is produced by an **intravenous pyelogram (IVP)**, performed by injecting a radiopaque dye into a vein. Multiple X rays will then reveal the structural anatomy of the kidneys, ureters, and bladder as the dye passes through the urinary tract.

A **retrograde pyelogram**, especially useful for the detection of urinary tract obstructions is performed by injecting the dye directly into the ureters while a cystoscopic examination of the bladder is performed. Cystoscopy is also a useful method for finding tumors for biopsy or for removing polyps.

A **voiding cystogram** is obtained by injecting a dye into the bladder via a catheter and taking X rays while the patient urinates. This test is used to evaluate the condition of the bladder and urethra and the mechanics of voiding.

The deep structures of the urinary system may also be examined with radioisotopes and sound waves. **Scintiphotography** shows the distribution of a radioisotope in an organ following injection of the isotope. **Ultrasonography** gives an image of internal organs by measuring the reflection of continuous high-frequency sound waves (see Chapter 28 Clinical Note: Ultrasonography of the Uterus).

In some cases a renal biopsy must be obtained to determine if a pathologic kidney condition exists. An **open biopsy** involves an incision that allows direct visualization of the kidneys and the removal of a large segment of kidney tissue. A **closed biopsy**, in contrast, involves the removal of a small piece of kidney tissue with a needle. The closed biopsy requires little recovery time and reduces the chance for infection to occur.

renal fascia (făsh′ē·ă), which attaches the entire mass to the abdominal wall. This attachment is flexible, allowing the kidneys to move with the diaphragm during inspiration and expiration.

Internal Organization

A coronal section of the kidney (Figure 25-4a) reveals three loosely defined renal areas: the **cortex** in the outer region, the **medulla** occupying most of the middle region, and the **renal sinus**, a cavity that opens at the hilum. The cortex and medulla form the **parenchyma** (păr·ĕn′kĭ·mă) of the kidney.

The cortex is an uninterrupted layer around the periphery of the organ, extending inward through the medulla to the edge of the renal sinus at several roughly regular intervals called **renal columns**. These columns divide the medulla into 8 to 18 conical regions called the **renal pyramids**. These pyramids are oriented such that tips of two and sometimes three pyramids fuse to form **renal papillae** (pă·pĭl′ē).

The renal sinus contains several different structures, including branches of the renal artery and vein, loose connective tissue, fat, and the enlarged opening of the ureter forming the **renal pelvis**. Two or three extensions of the renal pelvis radiate from the main chamber of the pelvis to form **major calyces**, each of which is subdivided further into two or three **minor calyces**. The tip of each papilla is perforated by 16 to 20 minute pores in the **area cribrosa** (ă′rē·ă crĭb·rō′să), which open into a minor calyx. Urine

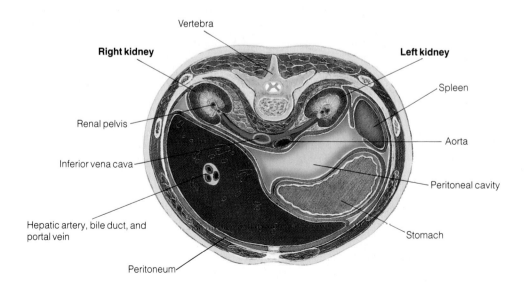

Figure 25-2 **A transverse section through the abdominal cavity showing the location and orientation of the kidneys in relation to other structures in the cavity.** Note the location of the kidneys outside the peritoneum and on either side of the vertebral column.

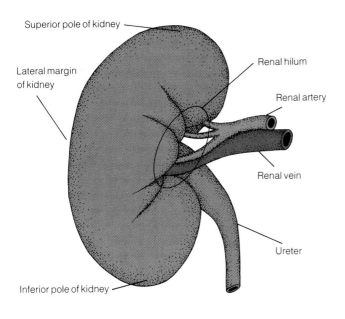

Figure 25-3 External view of the ventral surface of the right kidney. The renal hilum is the depression in the side of the kidney through which the renal artery and vein pass and from which the ureter emerges.

flows through these pores into the minor calyces, then into the major calyces and renal pelvis, and finally through the ureter to the bladder (Figure 25-4b).

Microscopic Anatomy

Under low magnification, the cortex can be seen to be filled with thousands of tiny granules separated by columns of tissue that extend from the outermost region of the cortex to the surface of the medulla (Figure 25-5). The granular regions are called **cortical labyrinths,** and the regions that separate them from one another are **medullary rays.** A single medullary ray and the cortical labyrinth it surrounds together form a functional unit of the kidney called a **lobule.**

The granular region contains many round capillary-filled structures intermixed with numerous highly twisted and convoluted tubules, veins, and arteries. The medullary rays contain long, relatively straight tubules that increase in diameter as more and more smaller tubules combine into larger ones, much like the tributaries of a river. These lead into the medullary pyramids, which contain tubules of still larger diameter formed by additional merging of smaller collecting tubules. After several fusions, the tubules in the pyramids form **papillary** (păp′ĭ·lăr·ē) **ducts,** the final tubules in the sequence,

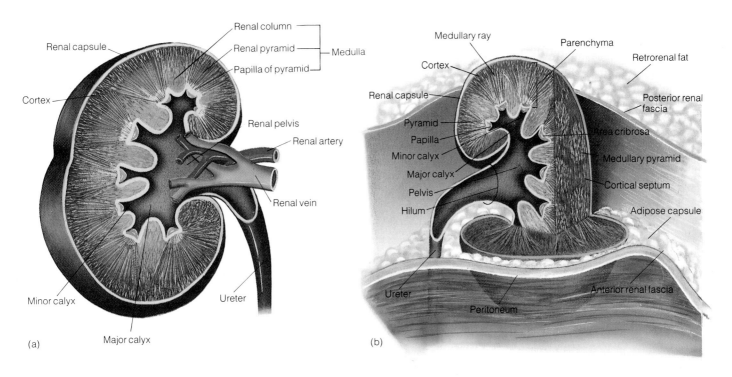

Figure 25-4 Internal organization of the kidney. (a) Coronal section. (b) Perspective view through the kidney.

Embryonic Development of the Urinary System

By the third week of development, embryonic mesoderm has differentiated into several subdivisions, including **intermediate mesoderm**, which are paired columns lying along the posterior half of the dorsal side of the embryo. Each column gives rise to a kidney in three successive stages, proceeding in a cranial to caudal direction.

The **pronephros** (prō·nĕf′rŏs) is the first kidney to form (Figure 25-A). It develops in the cranial end of the intermediate mesoderm as a series of vesicles that soon degenerate before ever becoming functional. Although nonfunctional, the pronephros is attached to a **pronephric duct** that connects to the cloaca. The **cloaca** (meaning "sewer") is an early embryonic sac that represents a common receptacle for the digestive, urinary, and reproductive systems. Although functional in other vertebrates, the cloaca soon disappears in the human embryo.

As the pronephros degenerates, the **mesonephros** devellops in a more caudal position. It connects to the pronephric duct, which is now called the **mesonephric duct.** As the mesonephros grows caudally, its cranial end begins to degenerate. As with the pronephros, there is no evidence that the mesonephros ever functions in the embryo, and by the eighth week it has completely disappeared.

During the fifth week, while the mesonephros is still growing, an outgrowth called the **ureteric** (ū″rē·tĕr′ik) **bud** begins to grow from the base of each mesonephric duct. The bud grows cranially toward a mass of intermediate mesoderm called the **metanephros** which is located near the posterior end of the mesonephros. The ureteric bud becomes encapsulated by the metanephros, and this combined structure develops into the adult kidney. The expanded portion of the ureteric bud gives rise to the collecting tubules, calyces, and pelvis of the kidney, while the unexpanded portion, the **metanephric duct**, becomes the ureter. The intermediate mesoderm of the metanephros produces the nephrons, the active filtering units of the kidney.

The cloaca divides into the **urogenital sinus** and rectum. In the embryo both urinary and reproductive ducts empty into the urogenital sinus, and as development proceeds, the urinary bladder develops from the sinus. The short duct leading from the urinary bladder to the sinus becomes the urethra in both sexes. The **vestibule** in females, an area into which both urinary and reproductive ducts empty, also develops from the urogenital sinus.

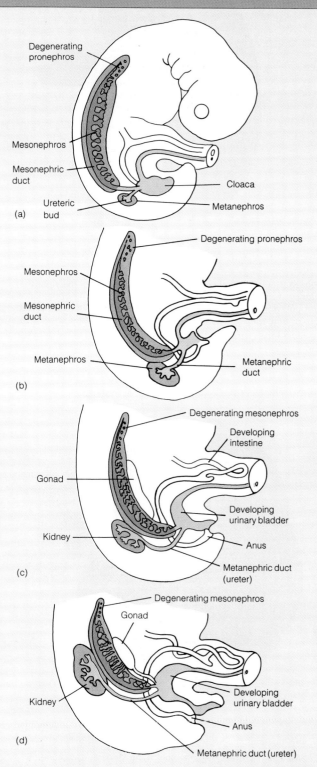

Figure 25-A Embryonic development of the urinary system. (a) During the fifth week of development, the urinary system consists of the pronephros and mesonephros. During the (b) sixth, (c) seventh, and (d) eighth weeks, the pronephros and mesonephros degenerate and the kidney begins to take its adult form.

nephron: (Gr.) *nephros*, kidney

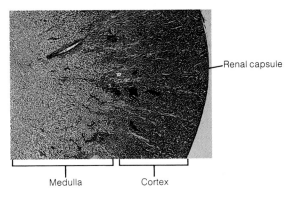

Figure 25-5 Photomicrograph of the renal cortex. Note its granular appearance.

The Nephron

After studying this section, you should be able to:
1. Describe the structure and organization of a glomerulus in relation to Bowman's capsule.
2. List and describe the structure through which urine passes as it is formed.
3. List and describe the arteries, veins, and capillaries involved in supplying blood to a nephron.

The functional unit of the kidney is the **nephron** (něf′rŏn), a structure that begins as a spherical body in the cortical labyrinth and continues as an unbranched tubule that eventually connects with a collecting tubule in an adjacent medullary ray (Figure 25-6). A nephron is responsible for most kidney functions, including the removal of wastes; production of urine; regulation of mineral, water, and hydrogen ion excretion; and control of blood pressure.

Renal Corpuscle

The spherical end of the nephron is a **renal corpuscle.** Averaging about 0.2 mm in diameter, it consists of two parts:

which open into the minor calyces at the pores of the area cribosa. The diameter of the tubules ranges from about 40 μm in the medullary rays to over 200 μm where they open into a minor calyx. Collectively, the convoluted tubules in the granular region of the cortex and the straight tubules into which they lead comprise a system of **collecting tubules.** Urine passes through this system as it flows from the granular region to a minor calyx.

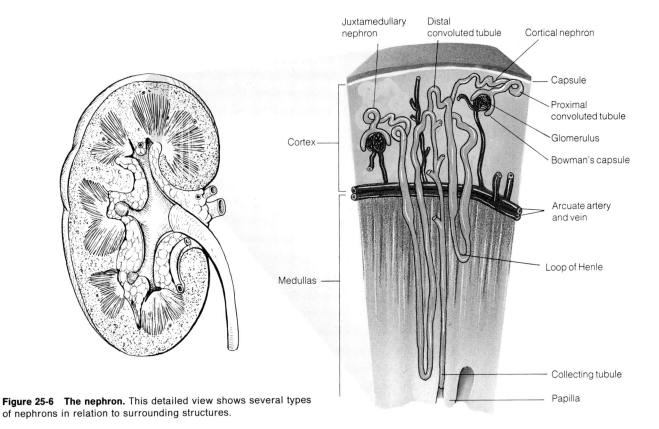

Figure 25-6 The nephron. This detailed view shows several types of nephrons in relation to surrounding structures.

glomerulus: (L.) *glomerulus*, little ball
podocyte: (Gr.) *podos*, foot + *kytos*, cell

a central mass of blood capillaries called the **glomerulus** (glō·měr′ū·lŭs) and a surrounding, double-walled, cup-shaped structure called the **Bowman's** (or **glomerular**) **capsule** (Figure 25-7). About 1 to 3 million nephrons are normally present in each kidney, and their renal corpuscles are largely responsible for the granular appearance of the cortical labyrinth.

Bowman's Capsule

The walls of a Bowman's capsule consist of a layer of epithelial cells that invaginates to form a cavity surrounded by two cell layers. The outer wall forms the **parietal layer**, which is separated from the inner wall, the **visceral layer**, by the **urinary space** (Figure 25-7). The folding back of the visceral layer to form the parietal layer leaves an opening into the interior of the cavity. This opening is the **vascular pole** of the corpuscle. This pole provides a passageway for a blood vessel that passes into the glomerulus and a second one that comes out. The parietal wall continues around the capsule to a point on the opposite side where it continues as the wall of the tabular portion of the nephron. The point where the urinary space opens into the lumen of the tubular part is the **urinary pole** of the corpuscle, located at the neck of the nephron.

Epithelial cells of the parietal layer are relatively uniform in structure, resting on a thin basal lamina and covered by a thin layer of connective tissue. The visceral layer, in contrast, is more elaborate, consisting of cells with highly branched processes that extend from the body of each cell to tightly cover the surface of the glomerular capillaries. Each process has numerous secondary extensions that create a featherlike pattern. These cells are **podocytes** (pŏd′ō·sīts). They rest on and conform to the shape of the **glomerular basement membrane**, a relatively thick, folded membrane that conforms to the surface of the glomerulus that it surrounds. Extensions of neighboring podocytes are interleaved to form narrow **filtration slits**. The processes that extend from the cell body of a podocyte are **foot processes**, or **pedicels** (pěd·ī·sěls′). Filtration slits range in width from 0.02 to 0.04 μm (Figure 25-8).

Glomerulus

The glomerular portion of the renal corpuscle is a mass of endothelial cells that branch and merge repeatedly to form a network of passageways. The cytoplasm of these cells is particularly thin where it lies against the basement membrane and is perforated with numerous pores that allow glo-

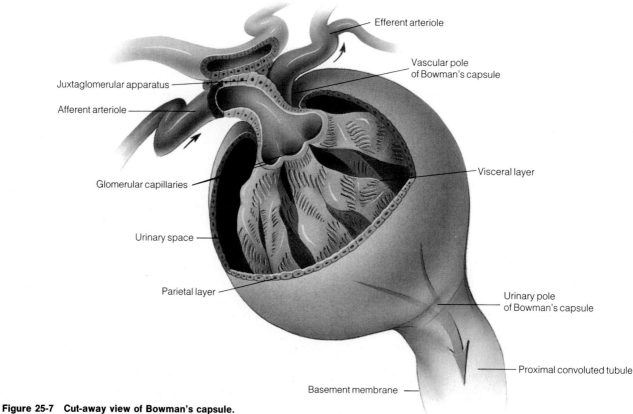

Figure 25-7 Cut-away view of Bowman's capsule.

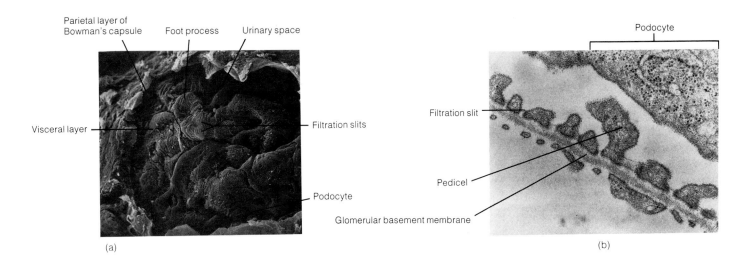

Figure 25-8 Internal structure of Bowman's capsule.
(a) Scanning electron micrograph of podocytes surrounding glomerular capillaries (mag. 328×). (b) Electron micrograph showing higher magnification view of filtration slits (mag. 21,528×). Filtration occurs at the visceral layer, as plasma is forced through gaps in the endothelial cells, through the basement membrane, and through the slits between the pedicels of the podocytes.

merular blood to make contact with the underlying basement membrane. These pores have an average diameter of about 0.08 μm and are thus small enough to prevent passage of blood cells and platelets.

Glomerular Basement Membrane

The basement membrane between the endothelial cells and the podocytes is about 0.08 to 0.12 μm thick and is composed of fibrils suspended in a matrix of glycoprotein. Spaces between the fibrils are especially narrow, about 0.004 to 0.014 μm wide. Although pores in the endothelial cells are small enough to prevent the passage of platelets and blood cells, they are not small enough to interfere with the passage of proteins and other macromolecules in the blood. The pores through the membrane, in contrast, are small enough to prevent passage of all but the smallest macromolecules. As a result, only water, small solute molecules, and ions can pass through the membrane into the urinary space, leaving most macromolecules, cells, and platelets in the blood. The blood is thus filtered as it passes through the glomerulus, with a portion of the plasma passing into the urinary space.

Tubular Portion of a Nephron

The tubular portion of a nephron begins at the urinary pole of the corpuscle and ends where it merges with a collecting duct. The end closer to Bowman's capsule is the **proximal** (or **initial**) portion. This part is subdivided into a highly twisted segment followed by a relatively straight portion that leads to the **thin segment.** The thin segment leads into the **distal** portion of the tubule. Like the proximal portion, the distal part is subdivided into straight and twisted regions. The twisted region connects with a collecting duct. The twisted regions at both ends of the tubule are referred to as the **proximal** and **distal convoluted tubules,** respectively (Figure 25-9).

The region of the tubule that includes the proximal and distal straight regions and the thin segment that lies between them comprise a functional unit of the nephron called the **loop of Henle.** This loop is a hairpin-shaped region that extends into the medulla for a variable distance, depending on the nephron. The proximal straight region is called the **thick descending limb** of the loop of Henle. It is followed by the **thin descending limb,** which reverses direction at its innermost point in the kidney and becomes the **thin ascending limb** of the loop of Henle. The thin ascending limb is followed by a relatively thick-walled section, the **thick ascending limb,** which continues on to become the distal convoluted tubule.

Location and Dimensions of Nephrons

Nephrons are arranged more or less radially in the kidney, with the renal corpuscles located in the cortical region and the loop of Henle extending toward or into the medulla.

juxtamedullary: (L.) *juxta*, near + *medulla*, marrow

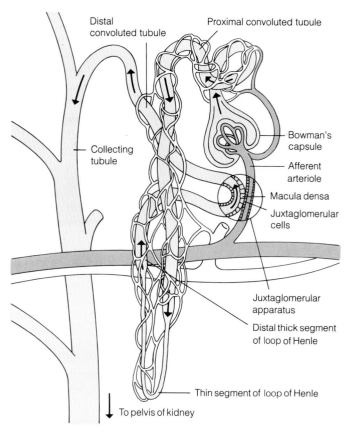

Figure 25-9 The tubular portion of the nephron. This schematic drawing shows how the tubular portion begins at the urinary pole of the renal corpuscle as a highly twisted tube that becomes a straight portion looping downward and then back up to the region of the corpuscle, where it again becomes twisted.

The depth to which the loop reaches into the medulla depends on the location of the capsule in the cortex: some capsules are located close to the surface of the kidney and their nephrons extend little or no distance into the medulla. This type of nephron is a **cortical nephron**. **Juxtamedullary** (jŭks″tă·mĕ′dŭl·lă·rē) **nephrons** are situated with their capsules deep in the renal columns and frequently have tubular portions that extend all the way to the papilla of a pyramid.

Typically, the proximal convoluted tubule is about 14 mm long, although it is so compact because of its convolutions that it occupies a spherical volume of only about 0.05 to 0.06 mm diameter. The distal convoluted tubule is less twisted and only about one-third as long as the proximal convoluted tubule. It occupies a spherical volume with a diameter of about 0.02 to 0.05 mm. Differences in the overall lengths of nephrons result primarily from differences in lengths of the thick descending limb and the thin portions of the loop of Henle.

Blood Supply to a Nephron

Blood is carried to the kidneys in the renal arteries, two large trunks that branch from the aorta at about the level of the first and second lumbar vertebrae. Each artery divides to produce four or five branches before entering the kidney, and some of the branches provide blood to neighboring organs and tissues. Most of the blood, however, is carried through the hilum and distributed through a succession of arterial branches until it reaches a glomerulus (Figure 25-10).

Distribution Within the Kidney

The primary branches of the renal artery within the kidney are **interlobar** (ĭn″tĕr·lō′băr) **arteries.** These pass between the calyces and extend outward from the renal sinus until they reach the interface between the medulla and cortex. From that level they branch further to produce **arcuate** arteries, which lie within the interface between the cortex and the medullary pyramids. **Interlobular arteries** radiate outward from the arcuate arteries into the cortical labyrinths, where they branch again to produce small, short branches, the **afferent glomerular arterioles.**

Passage of Blood Through a Glomerulus

Each afferent glomerular arteriole travels to a renal corpuscle, passing into a Bowman's capsule through a vascular pole. Immediately within the capsule, the afferent arteriole branches into four or five capillaries, each of which branches further to produce the network of passageways that makes up the glomerulus. Blood emerges from the glomerulus in an **efferent glomerular arteriole** that exits through the vascular pole, where it lies alongside the incoming afferent arteriole. The efferent arteriole is considerably narrower than the afferent arteriole, hence, it carries considerably less blood from the glomerulus than the afferent arteriole carried in. The difference in volume corresponds to the amount of fluid that has been filtered from the blood during its transit through the glomerulus.

Juxtaglomerular Apparatus

The afferent and efferent arterioles come into close association with a specialized portion of the distal convoluted tubule where they emerge from the vascular pole of the Bowman's capsule. The association of the arterioles and the tubule at this point, along with a group of cells that lies between the arterioles, comprise the **juxtaglomerular** (jŭks″tă·glō·mĕr′

extraglomerular: (L.) *extra*, outside + *glomerulus*, little ball
mesangium: (Gr.) *mesos*, middle + *angeion*, vessel
macula densa: (L.) *macula*, spot + *densus*, thick

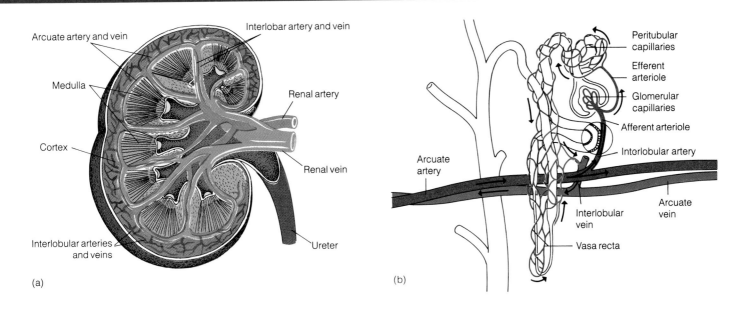

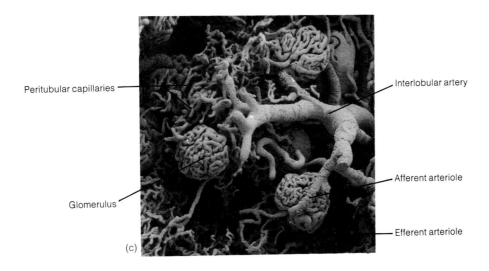

Figure 25-10 Blood supply to the nephrons. (a) Schematic drawing showing distribution of blood vessels to kidney. (b) Schematic drawing showing blood supply to the nephron. Arrows show direction of blood flow. (c) Electron micrograph of blood vessels associated with the glomerulus (mag. 100×).

ŭ·lăr) **apparatus** (Figure 25-11). The additional group of cells lying between the arterioles is an **extraglomerular mesangium** (měs·ăn′jē·ŭm). Cells in this region of the wall of the afferent arteriole, and sometimes the efferent arteriole as well, are modified smooth muscle cells called **juxtaglomerular cells.** Juxtaglomerular cells produce **renin** (rĕn′ĭn), a protein that helps in regulating blood pressure. Juxtaglomerular cells are also thought to help control the filtration rate in renal corpuscles. Tubule cells in juxtaglomerular apparatus appear narrower and denser than cells in other regions of the wall. These cells comprise the **macula densa** (măk′ū·lă děn′să) of the juxtaglomerular apparatus and are also involved in regulating the filtration rate in renal corpuscles.

Blood Distribution After Leaving a Glomerulus

The pattern of branching that the efferent arteriole undergoes after it leaves the renal corpuscle depends on the type of nephron involved. In cortical nephrons, the arteriole subdivides immediately into many small capillaries that extend to cortical tissues and lie near collecting tubules. This network of capillaries is the **peritubular plexus.** In juxtamedullary nephrons, the efferent arteriole travels toward the

vasa recta: (L.) *vas*, vessel + *recta*, straight

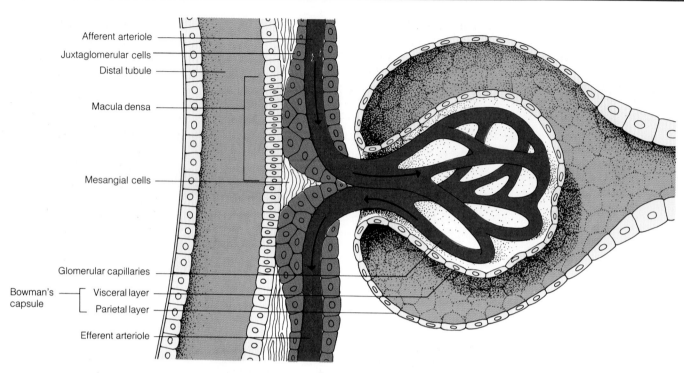

Figure 25-11 **Juxtaglomerular apparatus.**

medulla and then branches into several relatively long capillaries that extend to the tip of the renal pyramids and then return to the region of the corpuscle, running roughly parallel with the descending and ascending portions of the loop of Henle. This system of capillaries is called the **vasa recta** (vā′să rĕk′tă) (Figure 25-12).

After passing through capillaries of the peritubular plexus or the vasa recta, the blood flows successively into **interlobular**, **arcuate**, and **interlobar veins**, all of which parallel the corresponding vessels of the renal arterial system. Interlobar veins merge to form the renal vein, which delivers blood to the ascending vena cava.

Nerve and Lymphatic Supply to a Nephron

Most nerves that lead to the kidney are sympathetic nerves and emanate from the celiac or aortic ganglia. Some sensory nerves are also present and are responsible for the sensation of renal pain. Sympathetic nerves generally parallel branches of the arterial supply to the kidney and terminate at the walls of renal blood vessels outside the renal corpuscle. Consequently, most motor nerve function in the kidney appears to control vasoconstriction.

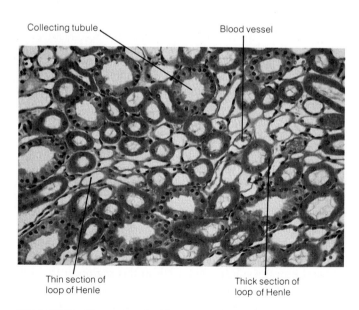

Figure 25-12 **The vasa recta.** The vasa recta consists of vessels that loop down into the medulla paralleling the loop of Henle. This cross sectional view shows how the tubules of the loops of Henle are intermixed with the blood vessels.

Kidneys are supplied with lymphatic vessels that also parallel the blood vessels from the interlobular level to the level of the renal artery and vein. There is also a highly branched network of lymphatic vessels that originates in the capsular region surrounding the kidney. Vessels that originate within the kidney converge in the sinus to produce a number of lymphatic trunks that exit the kidney at the hilum and pass to lymph nodes that lie alongside the vena cava and aorta. Those that originate in the capsular region communicate with lymphatic vessels present in neighboring tissues and organs.

Physiology of the Kidneys

After studying this section, you should be able to:

1. Describe the process of filtration that occurs in the nephron.
2. List and explain the factors that regulate filtration rate.
3. Describe where and how substances are reabsorbed from urine.
4. Explain the countercurrent mechanism for concentrating urine.
5. Explain the role of the vasa recta.
6. List and describe the function of hormones involved in regulating urine production.

In the course of any given day, about 25% of the blood pumped by the heart, about 1800 L, passes through the kidneys. Since only between 5 and 6 L of blood are in the body, it means blood must be filtered between 300 and 400 times a day. About 180 L of fluid is filtered from the blood, and of that, 99% is recovered. The remaining 1%, about 1.5 L, is excreted as urine.

The work of the kidneys begins in a glomerulus as the blood flows in and is filtered. The filtrate enters the urinary space of Bowman's capsule and then proceeds through the tubular portion of the nephron to the collecting ducts. During its transit through the nephron and collecting ducts, beneficial components and most of the water are recovered, leaving an aqueous solution of metabolic wastes that passes through the rest of the urinary tract and is excreted.

Filtration

Glomerular filtration results from the difference in pressure between blood in glomerular capillaries and the fluid in a urinary space. Since pores and slits in the visceral layer are too small to permit the passage of anything larger than the smallest proteins, the fluid that passes through is an aqueous solution of low molecular weight plasma components. In fact, it is very similar to plasma in composition. This fluid is **primary urine.**

The rate of filtration is termed the **glomerular filtration rate (GFR).** The GFR depends on a person's size, and in a healthy person it corresponds to about 70 mL of blood filtered each minute per square meter of body surface. An average-sized adult female has about 1.5 m^2 of body surface area and normally filters about 105 mL of fluid per minute. An adult male with a body surface area of about 2 m^2 normally filters about 140 mL of fluid per minute.

Glomerular Blood Pressure

Hydrostatic pressure is the force per unit area that an aqueous solution exerts on the container that holds it. Filtration occurs because the hydrostatic pressure of blood in the glomerulus is greater than the hydrostatic pressure of the fluid in the urinary space. The difference in pressure causes the fluid in the blood to flow from the region of higher pressure (the blood) to the region of lower pressure (the urinary space). There are several factors involved in determining this pressure difference: (1) differences in structure of afferent and efferent arterioles, (2) degree of constriction of renal arteries and glomerular arterioles, (3) hydrostatic pressure in the urinary space, and (4) resistance to water movement due to osmotic attraction of water by the blood.

Two key differences in the structure of the afferent and efferent arterioles exist. One is the greater diameter of the afferent arteriole over that of the efferent arteriole. The larger diameter of the afferent arteriole allows more blood to enter than leaves. As a result, blood pressure in the glomerulus is higher than is normal in a capillary. A similar effect can be demonstrated by connecting a large diameter hose and a small diameter hose to a plastic sphere and then pumping water in through the large hose. Because of the smaller diameter of the hose carrying water out, pressure develops in the sphere. A second difference in the arterioles is that the wall of afferent arterioles is thicker and stronger than that of efferent arterioles. The thicker wall prevents swelling in response to elevated pressure, which would decrease the pressure of the blood in the glomerulus.

Vasoconstriction increases blood pressure (see Chapter 23). Sympathetic signals sent to arteries and arterioles in the kidney cause the vessels to contract, decreasing their diameters and the rate of blood flow into the glomeruli. With decreased blood flow there is also a decrease in filtration. This effect can be caused by several factors, including decreased oxygen and increased levels of catecholamines in

necrosis: (Gr.) *nekrosis*, deadness

> **CLINICAL NOTE**
>
> ### RENAL FAILURE
>
> **Renal failure** is the loss of the kidneys' ability to respond to the constantly changing physiologic conditions within the body. When the kidneys fail, volume regulation suffers, electrolyte and acid-base balance is impaired, and waste products are retained.
>
> Renal failure can be acute or chronic. **Acute renal failure** is the abrupt cessation of renal function. Initially, it most often results in scanty urine, an increase in the level of urea nitrogen and creatinine (both nitrogenous waste products) in the blood, and perhaps pain. The severity of acute renal failure varies widely from a mild reduction of renal function to severely compromised function and uremia (see Clinical Note: Syndromes Associated with Renal Disease). However, it is usually a self-limiting disorder.
>
> Three types of acute renal failure exist: prerenal, intrarenal, and postrenal. **Prerenal renal failure** is generally caused by poor blood flow to the kidney as a result of shock; excessive treatment of hypertension; loss of fluid through diarrhea and bleeding; excessive urinary loss because of osmotic diuresis, diabetes insipidus, or excessive use of diuretics; and sequestering of fluid in burn spaces. In short, prerenal failure may occur because of decreased blood pressure or loss of circulating blood volume.
>
> **Intrarenal renal failure** results from actual renal damage because of glomerulonephritis (see The Clinic: Urinary System Disorders), tubular damage, infections, or vascular damage in the kidney. The most common cause of intrarenal renal failure is **acute tubular necrosis (ATN)** caused predominantly by agents that are directly toxic to the tubules, such as heavy metals, X-ray dyes, and tetracyclines.
>
> **Postrenal renal failure** is caused by obstruction of the urinary tract by prostatic hypotrophy, renal stones, or cancers. Pressure reflected back to the kidney causes damage.
>
> The treatment of acute renal failure is to support the patient and remove or modify the cause of the failure. A period of reduced urine production is followed by a diuretic phase as the kidneys heal. Eventually function returns to normal.
>
> **Chronic renal failure,** also known as renal insufficiency, is the slow, steady loss of functioning nephrons with an increasing solute load for the remaining functioning nephrons. Chronic renal failure is manifested by decreased glomular filtration rates, increased nitrogenous wastes, anemia, electrolyte disturbances, and acid-base imbalances. Treatment includes dietary and fluid restrictions, dialysis, and renal transplant. Causes of chronic renal failure are many, including connective tissue disorders, metabolic disorders, toxic nephropathies, and hypertensive vascular diseases.

the blood. Catecholamines, for example, cause vasoconstriction to occur primarily in the afferent arteriole and interlobular arteries and reduce renal blood flow.

Blood pressure in glomerular capillaries (**capillary hydrostatic pressure**), is offset by the pressure of the fluid in the urinary space (**tubular hydrostatic pressure**). Normally, capillary hydrostatic pressure is about 45 mm Hg and tubular hydrostatic pressure about 10 mm Hg.

Proteins unable to pass through the visceral wall have an osmotic effect that impedes the movement of water. This phenomenon, called **colloid osmotic pressure,** has the same effect as a pressure working in the opposite direction to the capillary hydrostatic pressure. Colloid osmotic pressure becomes greater toward the efferent arteriole end of the glomerulus as more and more water is filtered from the blood. In experimental animals, colloid osmotic pressure is about 20 mm Hg at the afferent end, before much water has been removed, and about 35 mm Hg at the efferent end where the blood is more concentrated.

The net effect of the contrasting forces on water is called the **net filtration pressure**, which is the capillary hydrostatic pressure minus the tubular hydrostatic pressure and colloid osmotic pressure (Figure 25-13). Net filtration pressure is generally about 15 mm Hg at the afferent arteriole end of the glomerulus ($45 - 10 - 20 = 15$) and declines to 0 mm Hg before the efferent arteriole end is reached ($45 - 10 - 35 = 0$).

Measurement of Glomerular Filtration Rate

A valuable concept for understanding how well glomeruli are working is **clearance**, which is defined as the volume of plasma that would have to pass through the kidneys within a given time period to produce the amount of a solute present in the urine. Clearance is usually expressed in milliliters per minute.

In a test to determine clearance, a known amount of a substance is injected intravenously over a period of time to maintain a steady concentration as the substance is simultaneously filtered by the kidneys. Urine is collected over the same period of time. The amount of the test substance present in the urine is then determined, as was its concentration in the blood as the urine was collected. Blood concentration is determined from a plasma sample collected midway in the test. Clearance is determined from the relationship

$$c = \frac{uv}{p}$$

Physiology of the Kidneys

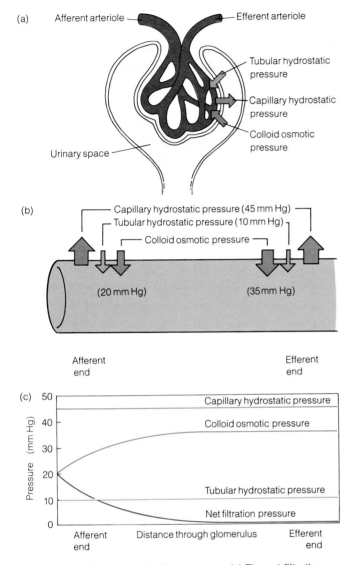

Figure 25-13 Glomerular filtration pressure. (a) The net filtration pressure is the result of pressure forcing fluid out of the glomerular capillaries into the urinary space minus the pressures resisting that movement due to pressure already in the urinary space and the attractive effect of proteins in the blood. (b) Schematic representation showing relative magnitudes of the different pressures in relation to distance through the glomerulus. (c) Graph of relationships diagrammed in (b). Note that increased colloid pressure reduces net filtration pressure to zero before the end of the glomerulus is reached.

where c represents the clearance rate of the substance, u represents concentration of the substance in the urine, v the volume of urine excreted per minute, and p the concentration of the substance in the plasma. For substances not reabsorbed from the urine or secreted from the tubules, the clearance rate for a substance equals the glomerular filtration rate.

A substance sometimes used in clearance tests is **inulin** (ĭn´ū·lĭn), an inert polysaccharide. Inulin is useful because it is small enough to pass through the glomerular filter and is not reabsorbed or secreted by the nephron or collecting ducts. Even though a person might produce a normal amount of urine, the effectiveness of removal of wastes might be diminished due to disease of the glomeruli. An inulin clearance of about 130 mL/min is normal, so if a patient's clearance rate were about 30 mL/min, it would indicate a significant loss of filtration capacity and improperly functioning renal corpuscles even though a normal volume of urine were being produced.

Creatinine, a waste product of muscle activity described in Chapter 9, is also commonly used to measure kidney clearance rates. Since creatinine is secreted from the tubule walls (see next section), it does not provide as accurate a measure of the GFR as does a substance that is not produced by metabolic reactions and that enters the urine solely through the glomerulus. Unlike substances that must be injected, however, tests of creatinine clearance only require the patient to collect the urine released in 24 hours and to supply a sample of blood. Results obtained in creatinine clearance tests are accurate enough for most clinical purposes.

Function of the Tubular Portion of a Nephron

The fluid that leaves the urinary space of Bowman's capsule and enters the tubular portion of the nephron is a dilute solution similar in concentration to plasma. It contains glucose, amino acids, nitrogen-containing waste materials, minerals, hormones, breakdown products from blood, and a host of other compounds. Yet when it leaves the tubular portion, it is a concentrated solution of metabolic wastes and excess minerals. The function of the tubular portion of the nephron is to accomplish this transformation.

Urine production is accomplished through several mechanisms, each of which occurs in a specific region of the nephron. These functions include reabsorption of nutrients and certain mineral ions; secretion of wastes, drugs, and toxins; concentration of urine; maintenance of body water balance; regulation of sodium content of the body; and regulation of body pH.

Reabsorption

Reabsorption is a process by which a nephron recovers nutrients and certain other substances that have been filtered from the blood in Bowman's capsule. Primary urine leaves the urinary space through the urinary pole and enters the proximal convoluted tubule. By the time it leaves the proximal convoluted tubule, its volume has been reduced by about two-thirds, and essentially all the carbohydrates

azotemia: (Gr.) *a*, not + *zoe*, life + *haima*, blood
uremia: (Gr.) *ouron*, urine + *haima*, blood

> **CLINICAL NOTE**
>
> ## SYNDROMES ASSOCIATED WITH RENAL DISEASE
>
> Renal disease is often accompanied by several characteristic syndromes, or groups of related symptoms and signs. The most common of these are azotemia, uremia, and the nephrotic syndrome.
>
> **Azotemia** is an accumulation of nitrogenous wastes—urea, creatinine, and uric acid—in the blood. It occurs if the wastes are not delivered to the kidney, as would happen with circulatory failure. Azotemia also occurs if the kidney is failing.
>
> **Uremia** is a syndrome that results when the biochemical and metabolic changes caused by renal failure become grossly symptomatic. The impaired volume regulation, electrolyte and acid base balance, and retained nitrogenous wastes lead to signs and symptoms in all organs. The degree of uremia will vary according to the severity of renal failure, and if severe and progressive, it can lead to death. Azotemia can be used as a measure of the severity of uremia, but the systemic toxicity that accompanies uremia may actually be unrelated to these particular waste products.
>
> The **nephrotic syndrome** is a term for complex signs and symptoms related to severe protein loss via the kidney. Many diseases may cause protein loss. An increase in the permeability of the glomerulus is usually a part of the cause. This syndrome includes **proteinuria** (presence of protein in the urine), widespread edema, slight elevation in blood pressure, and often headache. In children who develop nephrotic syndrome, almost all recover. Adults may develop complications such as congestive heart failure, acute renal failure, or hypertension.

and amino acids have been reabsorbed. Other substances recovered in the proximal convoluted tubule include sodium ion (Na^+), potassium ion (K^+), phosphate ion (PO_4^{3-}), sulfate ion (SO_4^{2+}), ascorbic acid, and even some substances considered waste products, such as creatinine and urea. Water recovery in this part of the nephron is the result of osmotic effects (colloid osmotic pressure) caused by the proteins present in the cells that line the tubule and by the active transport of sugars, amino acids, small proteins, and other substances recovered from the primary urine. As such, movement of water from the proximal convoluted tubule is a passive process, resulting from the change in osmotic concentrations brought about by active transport of solutes in the primary urine. Consequently, although the volume of the primary urine is reduced considerably by the time the fluid has reached the distal end of the proximal convoluted tubule, its overall concentration of dissolved materials is still the same as it was at the proximal end.

Not all substances are removed from primary urine with equal effectiveness, some being removed completely and others only partially. The effectiveness of removal of a substance from the fluid depends on the number and activity of active transport carriers in the membranes of the cells that line the lumen of the tubule. Glucose carriers are so concentrated and so active, for example, that they remove virtually all of the glucose present in the filtrate.

Some substances, such as urea, are so small that they diffuse unimpeded along with the actively transported materials. Anywhere from 30 to 80% of some substances may be reabsorbed, depending on their size and concentration in the primary urine, even though they may be waste products.

Cells that make up the wall of the proximal portion of the nephron have a surface that makes them efficient at reabsorption (Figure 25-14). Like cells of the duodenum, they are covered by a brush border made up of countless tiny villi that greatly increase the surface area exposed to the fluid in the tubule. The opposite side of the cells is marked by deep infoldings in which are located long, narrow mitochondria. Water and solutes diffuse through the brush border and travel down a concentration gradient that exists between the cell interior and the lumen of the tubule. Once within the cells, the substances diffuse across the cell to the opposite side. Active transport systems located in the membrane on the opposite side of the cell actively transport the substances outside the cell.

Water and solutes pass through a basement membrane that surrounds the tubule and enter fluid-filled spaces between the tubules. These tubules are **interstitial spaces.** The transported substances pass back into the blood in the capillaries that also lie among the tubules, as does water, which follows passively. The solutes and water are carried by the blood back to the general circulation.

The efficiency with which a substance is recovered from primary urine depends on the availability of carrier proteins specialized for transporting that substance. At elevated concentrations of a substance, its transport system may become overloaded and fail to remove the substance completely from the urine. The plasma concentration of a substance at which it begins to appear in the urine is the substance's **renal threshold.** The maximum amount of a substance that can be reabsorbed in a unit of time is the **transport maximum (Tm)** for the substance. Many genetic and metabolic disorders either cause the concentration of a substance in the blood to be raised past its renal threshold or cause its transport maximum to be lowered. In diabetes mellitus, for example, the inability of tissue cells to absorb glucose can result in plasma glucose levels that exceed the renal threshold for glucose (see Chapter 17 The Clinic: Diabetes Mellitus). As a result, not all the glucose in the urine can be reabsorbed, and some is excreted. When this happens, glucose is said to be "spilled" into the urine.

cystinuria: (Gr.) *kystis,* bladder + *ouron,* urine

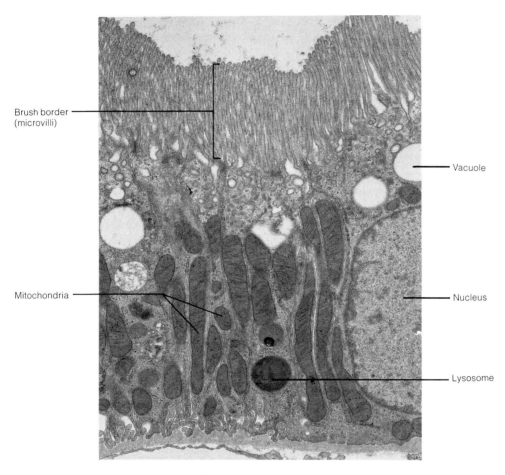

Figure 25-14 Electron micrograph showing brush border of cells of the proximal convoluted tubule (mag. 8760×).

Cystinuria (sĭs″tĭ·nū′rē·ă) is an inherited deficiency in the kidney's ability to reabsorb the amino acid cystine. Because of a defective transport system for this amino acid, its transport maximum is exceeded by normal levels in the urine, and the amino acid is excreted. The damage done by this disease occurs when the concentration of cystine in the urine becomes too high and the amino acid begins to precipitate in the tubules. As the precipitate accumulates, it blocks the flow of urine and causes progressive damage to the kidneys that can lead to kidney failure.

Tubular reabsorption also occurs in the loop of Henle, the distal convoluted tubule, and collecting tubules. About 60% of sodium reabsorption occurs in the proximal convoluted tubule, about 25% in the loop of Henle, about 8 to 9% in the distal convoluted tubule, and about 1% in the collecting ducts. Nearly 90% of bicarbonate ion (HCO_3^-) and carbonic acid (H_2CO_3) is reabsorbed in the proximal convoluted tubule and about 10% in the distal convoluted tubule. Chloride ion is also reabsorbed in the loop of Henle, where it accompanies the active transport of Na^+. These reactions are important to the role of the nephron in maintaining homeostasis and in concentrating urine, subjects which will be discussed shortly.

Secretion

Not all substances secreted in the urine are obtained solely by glomerular filtration; some are also secreted by the tubules themselves. Hydrogen ions, creatinine, and potassium ions are examples of substances that enter the urine by filtration and tubular secretion. Penicillin is also removed from the blood by both mechanisms, as is para-aminohippuric acid, a drug used to measure renal flow rate. The region of the nephron where a substance is secreted depends largely on the substance. Some hydrogen ion secretion occurs in the proximal convoluted tubule, but most occurs in the distal convoluted tubule. Potassium ions are also secreted in the distal convoluted tubule in conjunction with absorption of sodium ions. Organic acids and bases and urea are secreted primarily by the straight portion of the proximal tubule.

Concentrating the Urine

One requirement of the kidneys is that they be able to rid the body of wastes independent of the volume of water that must also be eliminated. During periods when the body is relatively dehydrated, for example, following loss due to perspiration, diarrhea, or excessive bleeding, it becomes essential to conserve water. During such periods the kidneys produce a concentrated urine that contains the waste materials in as little water as possible. In contrast, during periods when the body contains a surplus of water, the kidneys produce a relatively dilute urine that contains both the waste products and the excess water that is also being eliminated.

Because individual solutes in urine may vary widely in concentration, it is generally more useful to consider the total solute concentration than the individual solute concentrations. Total solute concentration is usually expressed in terms of milliosmoles per liter (mOsm/L). An **osmole** is a unit of measurement of the total number of osmotically active particles in a solution, regardless of whether they are individual atoms, molecules, or ions. The solute concentration of urine and other biological solutions is generally low enough that a **milliosmole**, 1/1000 of an osmole, is a more convenient unit of measurement. The solute concentration of plasma, for example, normally ranges from 281 to 295 mOsm/L. Urine normally has a concentration of about 700 mOsm/L, but it can range from 30 to 1400 mOsm/L depending on how much water must also be eliminated (or conserved) as the wastes are eliminated.

By the time urine has passed through the proximal convoluted tubule, virtually all the organic nutrients and about two-thirds of the mineral ions have been reabsorbed. About two-thirds of the water has also been reabsorbed passively as it entered surrounding interstitial tissues made hypertonic by mineral and nutrient reabsorption. Consequently, even though the volume of fluid entering the loop of Henle is considerably less than the amount of fluid originally filtered, its osmolarity is unchanged, remaining about 300 mOsm/L.

From this point on, the fluid is subjected to a regulated process that produces a final urine containing wastes that must be eliminated in a volume of water determined by the water needs of the body. Urine production in subsequent portions of the nephron and in neighboring collecting ducts involves a **countercurrent concentrating mechanism.**

Countercurrent Concentrating Mechanism

To understand how the countercurrent concentrating mechanism operates, it is necessary to understand the placement of juxtamedullary nephrons in the kidneys and the properties of various regions along their tubules and the collecting ducts to which they connect. The proximal tubule becomes the thin descending limb of the loop of Henle, which continues deep into the pyramid, reversing direction close to the tip of the pyramid and proceeding upward as the thin ascending limb. As the tubule reaches the level of the outer border of the medulla, it becomes the thick ascending limb. This limb leads into the distal convoluted tubule in a region close to Bowman's capsule at the innermost portion of the cortex. The distal convoluted tubule then connects to a collecting duct, which passes downward to the tip of the papilla and joins other collecting ducts to form a papillary duct. Fluid in these tubules thus reverses direction twice as it proceeds from the proximal tubule to the papillary ducts, once in the loop of Henle and once again where the distal convoluted tubule meets a collecting duct.

Solute concentrations in different regions of the tubular system and surrounding tissues are shown in Figure 25-15. As fluid enters the descending limb of the loop of Henle, it has a solute concentration approximately equal to plasma, about 300 mOsm/L, which increases to a maximum of nearly 1200 mOsm/L as the fluid approaches the tip of the loop, a fourfold increase in solute concentration. As the fluid flows through the ascending limb, its solute concentration diminishes still further. As it passes through the distal convoluted tubule, it once again increases in concentration, entering the collecting duct at nearly its original concentration, about 300 mOsm/L. Final concentration of urine occurs in the collecting duct, where it becomes progressively more concentrated until it reaches a final concentration of about 1200 mOsm/L. It is at this concentration that it passes into the renal pelvis.

The amount of fluid diminishes progressively as it passes through these tubules. About 33% remains of the primary

Figure 25-15 The countercurrent multiplier system for concentrating urine. (a) A drawing of a cortical (upper) and a juxtamedullary (lower) nephron showing solute concentration in the urine and surrounding tissues. (b) Regions of movement through the nephron wall. Relative degree of movement is indicated by the weight of the arrow. Part (b) adapted from Robert M. Berne and Matthew N. Levy, *Physiology* (St. Louis: C. V. Mosby Company, 1983).

Physiology of the Kidneys 663

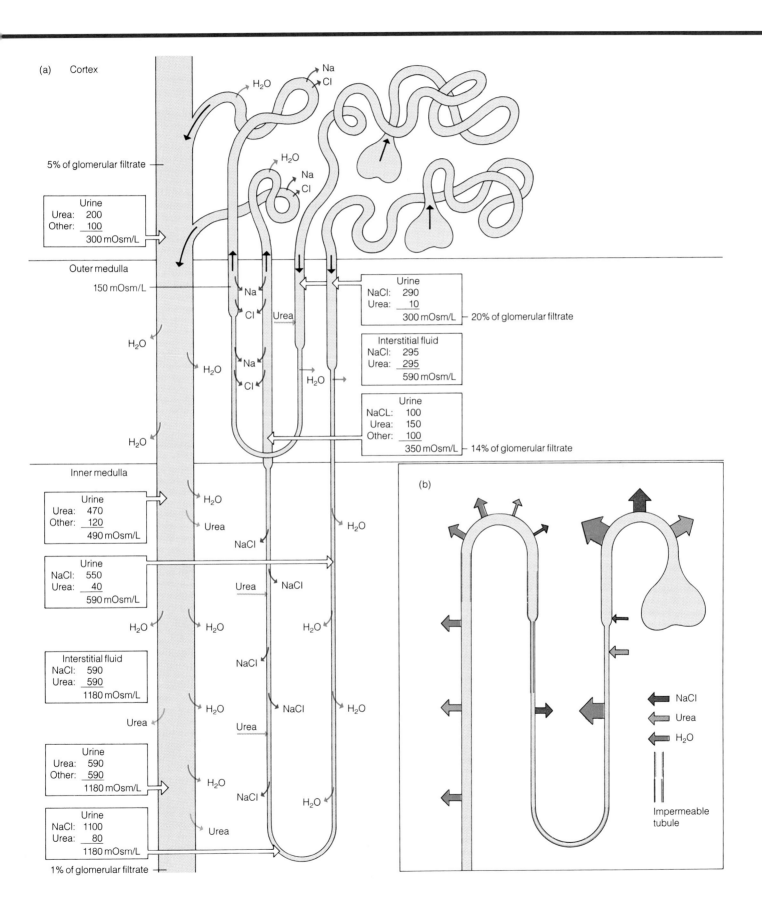

filtrate after passing through the proximal tubule. This amount is further reduced in the descending limb to about 15% of the original volume. No further water reabsorption occurs until the distal convoluted tubule is reached, where another 10% is removed. Additional water reabsorption in the collecting duct reduces the volume further to 1% or less of its original volume.

The solute concentration in the surrounding tissues plays an important role in this process. Interstitial fluids in the cortex where Bowman's capsules and distal tubules are located have a concentration of nearly 300 mOsm/L, increasing toward the inner medulla to a maximum of about 1200 mOsm/L. The same concentrations are found in corresponding regions of the descending limbs and collecting ducts.

Whenever exchange occurs between adjacent fluids moving in opposite directions, it is referred to as **countercurrent exchange.** In the kidney countercurrent exchange occurs between different regions of the filtrate as it flows through the descending and ascending limbs of the loop of Henle and again through the collecting ducts. This exchange results in increased solute concentration in the filtrate and surrounding interstitial fluids. It is therefore referred to as a **countercurrent multiplier mechanism.** Two solutes are primarily used in this mechanism, NaCl and urea, so the mechanism is also referred to as a **two-solute model** for the countercurrent multiplier.

The ability of the kidney to concentrate urine depends on its being able to establish a concentration gradient between solutes in the interstitial fluids and solutes in the filtrate. This concentration gradient causes water to flow by osmosis from the tubule, where the fluid is hypotonic, into the hypertonic interstitial fluid.

The countercurrent multiplier mechanism works as follows. The thin descending limb of the loop of Henle is highly permeable to water, only slightly permeable to urea, and impermeable to salt. As urine flows through the descending limb, the higher solute concentration in the surrounding interstitial fluid causes a net flow of water into the interstitial tissue. The thin ascending limb is relatively impermeable to water but relatively permeable to salt. Salt passes from the urine into the interstitial fluids in this region. In the next section, the thick ascending limb, active transport of Na^+ from the tubular fluid to the interstitial fluid decreases the solute concentration in the urine and increases the solute concentration in the interstitial fluid. Cl^- also leaves the tubule in this region, attracted by the flow of the positively charged sodium ions. This leaves behind a solution low in Na^+ and Cl^- but high in urea.

The fluid flows into the distal convoluted tubule and on to a collecting duct. Movement of water out of the distal convoluted tubule and the collecting duct depends on antidiuretic hormone (ADH), which controls the permeability of the tubules to water. Through the action of ADH, the kidney controls the amount of water reabsorbed.

As fluid flows from the distal convoluted tubule into a collecting duct, it enters a region where the walls become especially permeable to urea. Urea diffuses from the filtrate into the interstitial region. NaCl is already in high concentration in interstitial fluid, having come from the fluid in the ascending thin limb of the loop of Henle, but it is unable to diffuse into a collecting duct because the wall of the duct is impermeable to sodium. Consequently, NaCl and urea each contribute to the total osmolarity of the interstitial fluid. The combined osmotic effects of these two solutes cause additional water to diffuse from the descending thin limb of the loop of Henle, making the contents of that part of the tubule even more concentrated. Rounding the loop of Henle, the fluid (now containing a salt concentration greater than that of the interstitial fluid) moves into a tubular region where the wall is permeable to salt but not to water. Salt diffuses from the tubular fluid into the interstitial fluid, raising the total solute concentration of the interstitial fluid even more. The tubular fluid continues to the descending ducts, where some of the urea in it diffuses into the interstitial tissue. This process continues and produces an interstitial fluid with a solute concentration that gets progressively more concentrated from the cortex to the innermost portion of the medulla.

Role of the Vasa Recta

Like other tissues, the medulla must be provided with oxygen and nutrients to live. In addition, to maintain the high solute concentration in the medulla, water reabsorbed from the loop of Henle and collecting ducts must be removed without also removing the solutes in the interstitium. These functions are provided by the vasa recta (Figure 25-10b).

The vasa recta consists of looping capillaries that dip into the medulla, forming a countercurrent system in parallel with that of the tubules and bathed by the same hypertonic interstitial fluid. Blood in the descending portion of these capillaries gains ions from the interstitial fluid and loses water to it because of concentration differences between the plasma and the interstitial fluid. As it loses water and gains solutes, the blood thickens and becomes sluggish, allowing more time for exchange. In the ascending portion, blood (now hypertonic to interstitial fluids) loses those same solutes it picked up on the way down and regains water that had diffused out. In addition, plasma proteins exert a colloid osmotic force that causes additional water to enter the blood and be carried away. Water and solute exchange is facilitated in the ascending portion by a thin, porous capillary wall that allows efficient reabsorption of water and diffusion of solutes into the interstitium. As a consequence of the capillary arrangement, blood leaves the vasa recta carrying a few

diuresis: (Gr.) *diourein*, to urinate

more solutes than when it entered but also carrying the water reabsorbed as the urine was concentrated.

Regulation of Urine Concentration and Water Excretion

If a person with normal kidneys were to drink a large volume of water very rapidly, within about 15 min the kidneys would begin to produce a greater volume of an increasingly more dilute urine. After about 40 min, production would slow down and the urine would become increasingly concentrated until its rate of production and concentration returned to what they had been before the water was drunk (Figure 25-16). Production of an increased amount of a more dilute urine following water consumption is called **diuresis** (dī″ū·rē′sĭs). Diuresis results from a decrease in the rate of water reabsorption in the distal convoluted tubules and collecting ducts rather than from an increase in glomerular filtration rate in the renal corpuscle.

Water reabsorption is also affected by dehydration. When a person does not drink enough water to replace water lost by evaporation, diarrhea, or hemorrhage, increased amounts of water are recovered from the tubular fluid, resulting in the production of a smaller volume of a more concentrated urine.

Reabsorption is controlled by antidiuretic hormone (ADH), released by the hypothalamus to the posterior pituitary gland, which in turn releases the hormone into the blood (see Chapter 17). Carried by blood to the kidneys, ADH increases the permeability of distal convoluted tubules and collecting ducts to water. The fluid in these tubules is hypotonic to the surrounding fluids, so water flows out of the tubules in response to ADH (Figure 25-17).

ADH release is regulated by the concentration of solutes in blood and by blood pressure, both of which operate in a negative feedback mechanism. During periods of dehydration, the concentration of solutes in the blood increases, raising its osmotic pressure. This increase is detected by osmoreceptors in the hypothalamus that transmit signals to other hypothalamic regions where ADH is produced. ADH is released in response to these signals (see Chapter 17). After the ADH has had its effect on the kidney and additional water has been reabsorbed, the osmotic pressure of the blood diminishes and causes a decrease in the frequency of signals to the hypothalamus. The hypothalamus produces less ADH in response to this decreased rate of stimulation, less hormone is released to the blood, and the tubules in the kidney become less permeable to water.

During diuresis, ADH production is inhibited by the decrease in the osmolarity of blood that follows the ingestion of water. As a result, distal convoluted tubules and collecting ducts become less permeable to water. Water reabsorption

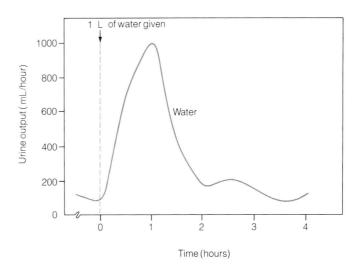

Figure 25-16 Diuresis. This graph shows the effect of rapidly drinking 1 L of water on urine output in a healthy adult male. Urine output begins to increase almost immediately, reaching a maximum 1 hour after drinking the water. Most of the water is eliminated within 2 hours.

diminishes, and water that would otherwise be reabsorbed remains in the tubules.

ADH release is also controlled by blood pressure. When water is ingested and absorbed, the extra fluid volume of the blood causes its pressure to rise. This increase in pressure is sensed by baroreceptors in the arteries, which respond by increasing the rate at which they send signals to the hypothalamus. This increase has an opposite effect to that of an increase in signal rate from osmoreceptors and results in a decrease in the rate of ADH release and water reabsorption.

Homeostatic Functions of the Kidneys

After studying this section, you should be able to:

1. Describe how the kidneys control the secretion of sodium ions.
2. Explain how kidneys are involved in regulating blood pressure.
3. Describe how kidneys help regulate the pH of body tissues.
4. Describe the mechanism used by kidneys to regulate potassium levels in the blood.
5. Define erythropoiesis and describe how the kidneys and liver are involved in controlling the rate of red blood cell production.
6. Describe the role of the kidneys in activating vitamin D.

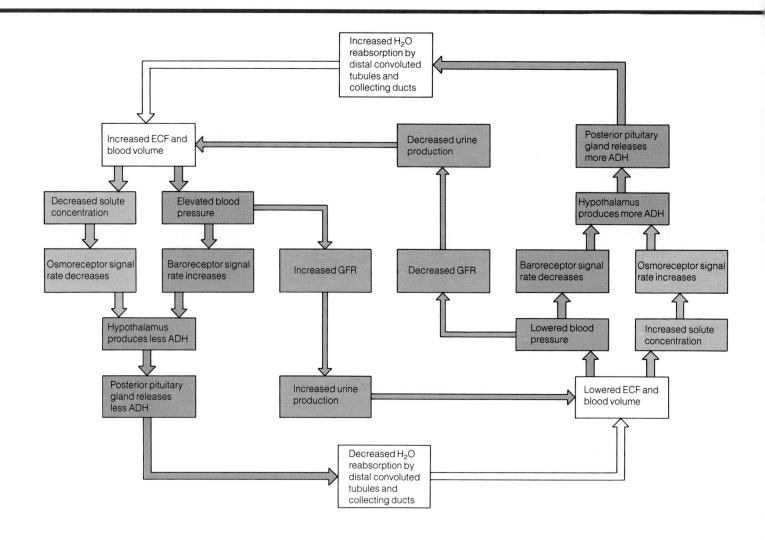

Figure 25-17 Regulation of extracellular fluid (ECF) and blood volumes by the kidneys. GFR is the glomerular filtration rate.

Renal Regulation of Sodium Content

Control of Na$^+$ excretion is an important mechanism for regulating body water. Na$^+$ is actively absorbed throughout the tubular portion of the nephron, but in the distal convoluted tubule its absorption is controlled by aldosterone, a hormone produced by the adrenal cortex that stimulates Na$^+$ absorption. By regulating the amount of aldosterone in the blood, the distal convoluted tubule is regulated in its reabsorption of Na$^+$. The amount of water reabsorbed depends in turn on the amount of Na$^+$ reabsorbed because the water follows the Na$^+$ passively as the ion is transported from the urine.

The regulatory system that controls aldosterone production is termed the **renin-angiotensin** (rĕn´ĭn ăn˝jē·ō·tĕn´ sĭn) **mechanism.** As fluid passes through the distal tubule in the region of the juxtaglomerular apparatus, cells of the apparatus release renin. Renin is an enzyme that cleaves **angiotensin I** from **angiotensinogen** (ăn˝jē·ō·tĕn·sĭn´ō·jĕn), a plasma protein produced in the liver. Angiotensin I is cleaved further to produce **angiotensin II** as it is transported through the circulatory system, with most conversion occurring in the lungs. Angiotensin II is carried to the adrenal cortex and converted to **angiotensin III.** This substance stimulates production of aldosterone and a subsequent increase in the rate of Na$^+$ absorption in the kidney. The effect of such reabsorption is that more water is recovered from the urine and retained by the tissues. This extra fluid load in the tissues results in increased blood pressure.

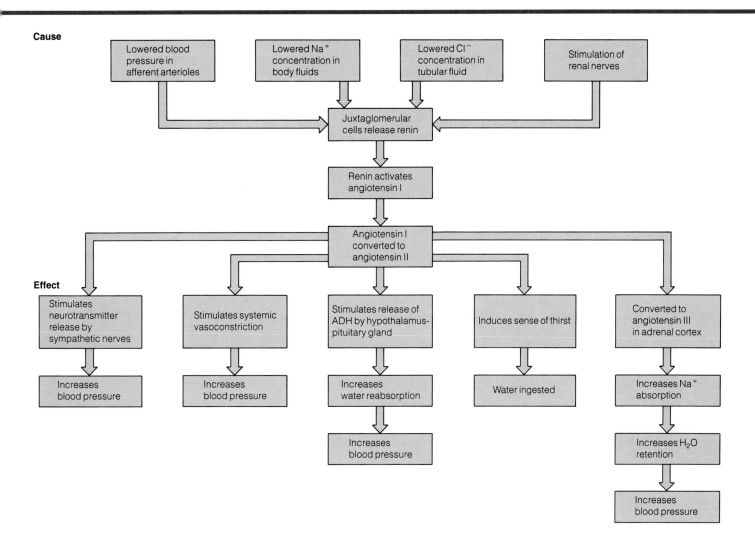

Figure 25-18 The renin-angiotensin mechanism for controlling blood pressure. Angiotensin is directly responsible for controlling blood pressure and fluid volume through several mechanisms, but its production is controlled by renin, produced by the juxtaglomerular apparatus. The system operates by negative feedback: as water retention and blood pressure increases, the stimulus for renin release decreases.

Control of Blood Pressure

Control of the Na^+ concentration in body fluids is only one aspect of the role played by angiotensin in controlling extracellular fluid (ECF) volume and blood pressure. Other effects of angiotensin include constriction of systemic arteries, increased secretion of transmitter substances in the sympathetic nervous system, increased production of ADH by the hypothalamus, and, by stimulating appropriate centers in the brain, a sense of thirst. All these effects increase blood pressure directly by causing vasoconstriction (see Chapter 23) or indirectly by increasing extracellular fluid volume (Figure 25-18).

Production of angiotensin II is controlled by mechanisms that result in the production of renin. Its production is stimulated, for example, by lowered blood pressure in afferent arterioles, lowered Na^+ concentration in blood plasma, or lowered Cl^- concentration in distal tubular fluid, and also by direct sympathetic nervous stimulation of juxtaglomerular cells in response to signals from baroreceptors reacting to lowered blood pressure.

Renal Regulation of pH

Hydrogen ions are produced by many metabolic reactions, and their concentration must be controlled. The kidneys participate in this control by secreting excess H^+ and by absorbing HCO_3^- from tubular fluid (Figure 25-19).

The amount of H^+ secreted into urine depends on the acidity of the extracellular fluid and blood, which depends in turn on diet and certain other physiologic factors. When there is an excess of H^+, it is secreted into the tubular fluid where much of it combines with phosphate, ammonia, or bicarbonate. Phosphate ions and ammonia serve as important pH buffers and allow the urine to carry more H^+ than it would be able to in their absence.

Some H^+ secretion is linked directly to Na^+ reabsorption, with H^+ secreted into the tubular fluid at the same time Na^+ is absorbed. H^+ secretion also contributes to HCO_3^- recovery by combining with HCO_3^- in the tubular fluid and converting it to carbonic acid. Dissociation of carbonic acid yields H_2O and CO_2, which diffuse more readily into the cells of the tubule wall than HCO_3^- does. Once within the cell, CO_2 reacts with water to produce carbonic acid. Ionization of the carbonic acid then yields HCO_3^- and H^+. The H^+ is again pumped from the cell into the tubule, perhaps in conjunction with the absorption of more Na^+, and the HCO_3^- diffuses to the blood, which carries it away. More than 99% of the HCO_3^- ions present in the glomerular filtrate can be recovered in this way if the body's HCO_3^- reserves are low.

Potassium Absorption and Secretion

A considerable amount of potassium is consumed in the diet each day, and the kidney is the primary organ for regulating potassium levels in the body. Small amounts are excreted in feces and perspiration, but the major amount leaves in urine. About 70% of the K^+ in glomerular filtrate is reabsorbed in the proximal tubule where it is recovered both by active transport and passive diffusion. Another 25% is reabsorbed as the fluid passes through the loop of Henle, so only about 5% reaches the distal tubule. Nevertheless, K^+ concentration in urine is at least twice as great as in plasma even when dietary K^+ is minimal, so it is apparent that kidneys secrete more K^+ than they absorb.

K^+ secretion occurs primarily in the distal tubule where it is linked to Na^+ reabsorption; secretion of each potassium ion accompanies the absorption of each sodium ion. H^+ secretion is also linked to Na^+ reabsorption, and the amount of K^+ secreted depends on how much H^+ is also secreted. K^+ secretion diminishes and H^+ secretion increases when plasma and ECF are relatively acid. Conversely, when body fluids are relatively alkaline, K^+ secretion is enhanced and H^+ secretion is reduced.

Regulation of Red Blood Cell Production

The kidney is also involved in controlling **erythropoiesis**, the production of red blood cells (see Chapter 21). When the number of red blood cells declines, for example, due to disease or excessive bleeding, the rate of hemoglobin synthesis and erythropoiesis goes up, compensating for the loss. The chemical that stimulates this increase in red cell produc-

Figure 25-19 Relationship of Na^+, H^+, HCO_3^-, K^+ exchange across the tubular cell membrane. Transport of H^+ and K^+ is linked to Na^+ reabsorption, with H^+ exchange taking precedence when plasma acidity is high. H^+ can also be recycled through the membrane as HCO_3^- is recovered. Excess H^+ is absorbed by NH_3 and PO_4^{3-} as it enters the urine and is excreted in combination with those buffers.

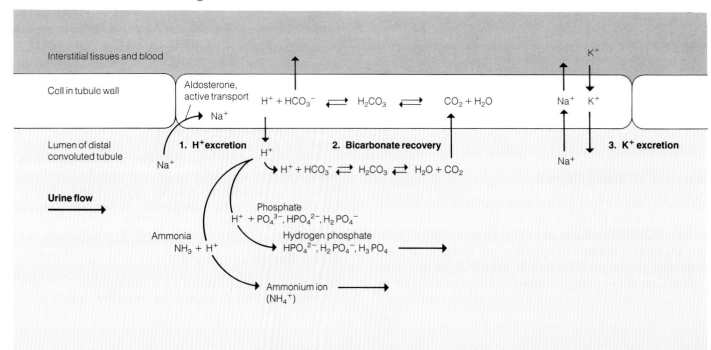

tion is **erythropoietin** (ĕ·rĭth″rō·poy´ĕ·tĭn), a glycoprotein that circulates in the blood. Erythropoietin is derived from an inactive protein produced by the liver and is activated by a renal product called **renal erythropoietic factor (REF)**. REF is an enzyme that catalyzes the conversion of the erythropoietin precursor into erythropoietin.

Vitamin D Activation

Another kidney function is the conversion of vitamin D to the chemically active form called 1,25-dihydroxycholecalciferol. It is in this form that the vitamin is active in stimulating calcium absorption from the small intestine.

Conduction, Storage, and Elimination of Urine

After studying this section, you should be able to:

1. List the structures through which urine passes after it leaves the nephrons.
2. Describe how urine is propelled through the urinary tract after accumulating in the renal pelvis.
3. Describe the structure of the urinary bladder and explain how release from the bladder is controlled.
4. Compare the structure of the urethra in females and males.
5. Define micturition and describe the neural mechanisms involved.
6. List and describe the components of urine.

Conduction Through the Ureters

As urine reaches the ends of papillary ducts, it passes through the pores of the area cribosa and enters the calyces of the renal pelvis. From the calyces it collects in the main chamber of the pelvis and passes into the **ureter** (ū´rĕ·tĕr) (Figure 25-20a). Each ureter is 20 to 35 cm in length and serves as a channel for conducting urine to the bladder. The ureters pass behind the urinary bladder and enter it near its lower end, passing through the bladder wall at an oblique angle (Figure 25-20b). Entering at an angle increases the length of the ureter that is in the wall, which contributes to the ability of the bladder to close the opening into the ureter when the bladder is emptying.

The ureter wall is three layers thick. The inner lining, the mucosa, is composed of transitional epithelium and lacks glands. The middle region is a muscular layer which, in the upper two-thirds of each ureter, consists of two sublayers of

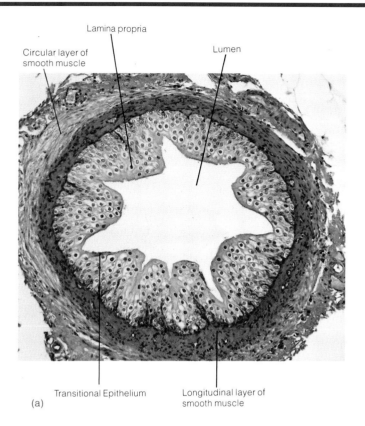

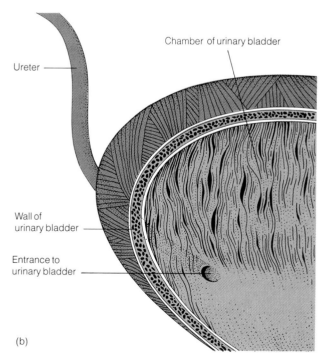

Figure 25-20 Ureters. (a) Photomicrograph of the ureter wall, showing the muscle layers responsible for peristalsis (mag. 654×). (b) Entry of the ureters into the bladder.

> **CLINICAL NOTE**
>
> **URINARY CALCULI**
>
> The formation of **urinary calculi** (kăl´kū·lī), or stones, is a common problem of the urinary system. It is calculated that 1 in every 1000 adults may be hospitalized for treatment of urinary calculi annually. The causes of stone formation may be related to (1) factors that cause an increase in the urinary concentration of stone crystalloids, such as excessive calcium intake, dehydration, or metabolic abnormalities; and (2) factors that may encourage stone formation at normal concentrations of urinary crystalloids, such as stasis, infection, changes in urinary pH, and foreign bodies in the urinary tract. The chemical composition of the stones varies according to the cause. Stones are most frequently composed of calcium salts.
>
> The symptoms of urinary calculi include severe intermittent pain (**renal colic**), hematuria (blood in the urine), and frequent painful urination (dysuria). Treatment for acute urinary calculi is to relieve the pain, treat the infection if present, increase hydration, and give antispasmodic medications. Treatment of recurrent stones involves eliminating the causative factors and adjusting the urinary pH to prevent further stone formation.
>
> Some stones are so large that they cannot pass through the urinary system and thus require surgical removal. A new method of dealing with large, multiple stones has been developed called **lithotripsy**, which utilizes ultrasound waves that are focused on the stone while the patient is immersed in a water bath. This technique fractures the large stones into smaller pieces, which can then be passed. Experiments have also been performed using laser light beams to fracture the stones.

smooth muscle. The inner of these consists of cells oriented lengthwise, and the outer sublayer consists of cells oriented around the circumference of the ureter. Another longitudinal sublayer in the middle region is also present in the distal end of the tube. The outermost layer consists of fibrous tissue continuous with the fibrous renal capsule.

Urine is forced into the ureter by contractions of muscle in the pelvic wall. As it enters the ureter, stretching of the ureter wall causes the circular muscle in the wall to contract. This effectively closes the opening from the pelvis and forces the urine farther down the ureter. Distension of the ureter wall also intitiates peristaltic waves that force the urine the rest of the way through the ureter and into the bladder.

Urinary Bladder

The lower wall of the **urinary bladder** has three openings: those from the two ureters and an **internal urethral orifice.** These three orifices form a triangular area called a **trigone** (trī´gōn) (Figure 25-21a). The trigone remains in position whether the bladder is empty or full. During filling, the upper portion of the bladder expands upward. As it fills, expansion of the bladder causes the wall to squeeze down on each ureter where it passes through the wall. This automatically closes the ureter as urine accumulates in the bladder, preventing backflow of urine. This effect is important because it prevents pressure from damaging the delicate kidney tubules.

The wall of the bladder has four tissue layers (Figure 25-21b). An inner tunica mucosa is continuous with that of the ureters, although the transitional epithelium is somewhat thicker than that of a ureter. Unlike the lining of the ureters, mucous glands are present in the bladder lining. A submucosal layer of loose (areolar) connective tissue lies external to the mucosa. A muscular layer (tunica muscularis) outside the submucosa consists of several sublayers of smooth muscle oriented in numerous directions, with a distinct external longitudinal layer. The muscular layer is the **detrusor** (dē·trū´sōr) **muscle.** A portion of it surrounds the upper end of the urethra where it connects to the bladder, forming the **internal urethral sphincter.** The bladder is covered by a fibrous coat of connective tissue continuous with that of the ureters.

Urethra

Urine enters the **urethra** (ū·rē´thră) through the internal urethral orifice and is excreted at the other end through the **external urethral orifice.** In females the urethra is relatively short (4 cm long) and lies immediately behind the **symphysis pubis** (sĭm´fĭ·sĭs pū´bĭs) in the wall of the vagina. The external urethral orifice lies within the vestibule of the vagina, just in front of the vaginal orifice (Figure 25-22a).

The urethra is considerably longer in males (20 cm long). There are three different regions in the male urethra (Figure 25-22b), corresponding in name to the region of the body through which each passes. The **prostatic urethra** (3 cm long) extends from the internal urethral sphincter to the floor of the pelvic cavity and, as the name implies, passes through the prostate gland. This portion is slightly larger in diameter than the other two regions. It has openings into ducts from the prostate gland and the two ejaculatory ducts. In males the urethra not only conducts urine, but also serves as the passageway for sperm and secretions of male sex glands. The next part of the male urethra, the **membranous urethra** (1 to 2 cm long), passes through the pelvic floor. Voluntary muscles surround the urethra in this region, forming the **external urethral sphincter.** The **cavernous (penile) urethra** is about 15 cm long and extends from the external urethral sphincter to the tip of the penis, passing through a column of erectile tissue called the **corpus spongiosum** (cōr´pŭs

Conduction, Storage, and Elimination of Urine 671

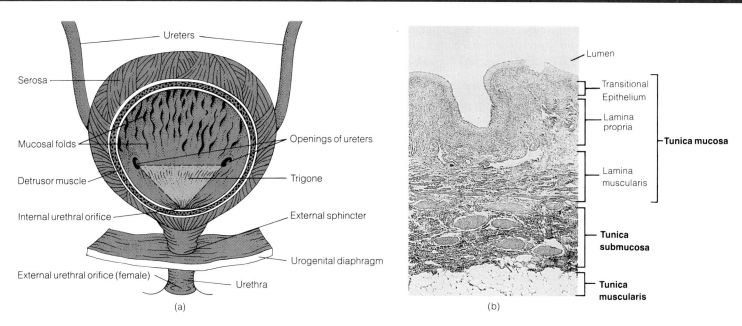

Figure 25-21 The urinary bladder. (a) Internal and external views of the urinary bladder (shown here for a female). (b) Photomicrograph of the bladder wall (mag. 124×).

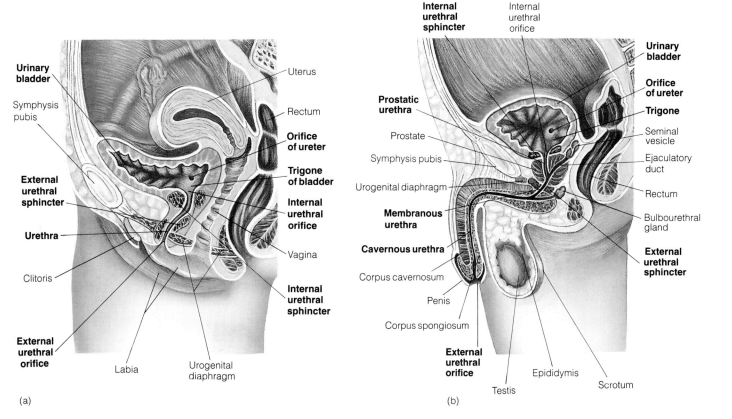

Figure 25-22 Sagittal views of the urethras. (a) Female. (b) Male.

spŭn′jē·ō·sŭm). Bulbourethral glands empty into the upper end of the cavernous urethra (see Chapter 27).

Micturition

Micturition (mĭk·tū·rĭ′shŭn), or **urination,** is the process by which urine is moved from the urinary bladder through the urethra to the outside. Essentially, it is a reflex activity initiated when urine accumulates in the bladder. The bladder is capable of expanding to a capacity of about 800 mL of urine, yet it rarely does. Usually when 200 to 300 mL of urine are in the bladder, the detrusor muscle is stretched sufficiently to stimulate proprioceptors located in it. Impulses generated by these proprioceptors are transmitted to reflex centers in the sacral region of the spinal cord, which in turn transmit parasympathetic motor impulses to the detrusor muscle and cause it to begin to contract rhythmically. Impulses are simultaneously sent to the internal urethral sphincter causing it to relax. Impulses are also transmitted to the cerebral cortex, making one aware of the increasing need to urinate (Figure 25-23). In an infant a reflex action takes over at this point, causing the internal urethral sphincter to relax and the detrusor muscle to contract, and the bladder is emptied. Older children and adults can override the micturition reflex through the use of voluntary centers in the cerebral cortex (see Chapter 14). These centers act through pyramidal tracts of the medulla oblongata, suppressing the micturition reflex and voluntarily maintaining the external urethral sphincter in a contracted condition. Urination can thus be delayed until a more convenient time. (The main reason that toilet training is usually unsuccessful in children less than 2 years of age is that the nerve connections to the external urethral sphincter have not developed.) Once the bladder has emptied, the detrusor muscle relaxes, the internal sphincter closes, and urine begins to collect again.

Figure 25-23 Micturition reflex. The detrusor muscle contracts, forcing urine into the urethra. Parasympathetic afferent neurons in the urethra are stimulated to send impulses to the spinal cord that inhibit impulses to the external urethral sphincter, causing it to relax and allow urine to pass into the urethra.

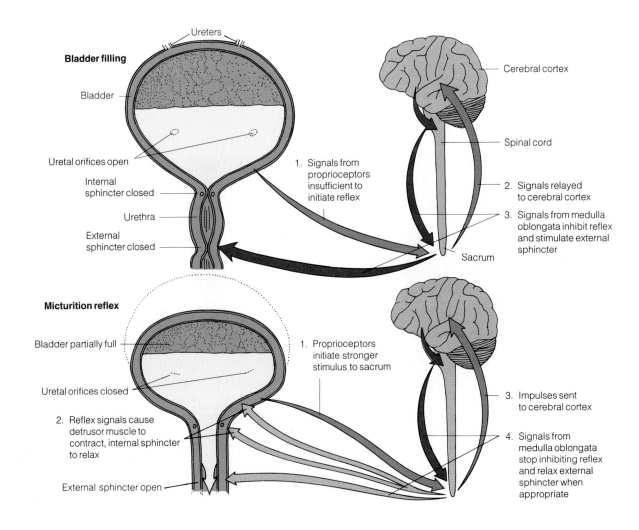

Composition of Urine

After completing this section, you should be able to:

1. Describe the components and characteristics of urine.
2. Describe several abnormal components of urine and their cause.

Excreted urine varies considerably in concentration and composition of dissolved and suspended substances depending on diet, state of health, and even time of day (see Clinical Note: Urinalysis). It usually is a light amber color that becomes darker with increased concentration. It is usually transparent, but may become turbid due to the presence of salts of uric acid, calcium or magnesium phosphate, bacteria, or pus cells resulting from an infection. The content of dissolved and suspended substances is usually measured with a hydrometer, a device that measures specific gravity. Urine

CLINICAL NOTE

URINALYSIS

A standard medical examination has several focuses. In addition to gathering a medical history and ascertaining whether a patient has suffered recent minor illnesses, or has current complaints about ill health, a physician may carry out a series of probes, including (but not limited to) a superficial visual estimate of the patient's present condition, palpation (touch and pressure tests), auscultation (listening to internal sounds with a stethoscope), X rays, an electrocardiogram, blood tests, and urinalysis. Information gathered during these tests may suggest further, more invasive tests to determine the meaning of any questionable or negative results.

The **urinalysis** is a particularly informative test because urine contains products from almost every organ in the body. In addition to yielding information about the urinary tract, a urinalysis also indicates the presence of abnormal levels of standard cellular products. It can also provide the first indication of infections that may not even involve the urinary tract. Presented below is a list of the standard and more specific tests performed in a urinalysis, the expected quantities of substances normally present, and the clinical implications of deviations from the expected quantities. In addition to monitoring these standard substances, the presence of parasites, erythrocytes, leukocytes, and fibrous or fatty solids (called **casts**) associated with infection or disease may also be noted in a microscopic examination of the urine specimen.

Test	Normal Quantities	Clinical Implications
Acetone and acetoacetate	0	Quantities increase in diabetic acidosis
Albumin	0 to trace	Quantities increase in kidney disease, hypertension, and heart failure
Ammonia	20 to 70 mEq/L	Quantities increase in diabetes mellitus and liver diseases
Bacterial count	Under 10,000/mL	Quantities increase in urinary tract infections
Bile and bilirubin	0	Quantities increase in melanoma and obstructions of gallbladder
Calcium	Under 250 mg/24 hr	Quantities increase in hyperparathyroidism and decrease in hypoparathyroidism
Creatinine	1 to 2 g/24 hr	Quantities increase in infections and decrease in anemia, muscular atrophy, leukemia, and kidney diseases
Creatinine clearance	70 to 130 mL/min	Quantities increase in kidney disease
Glucose	0	Quantities increase in diabetes mellitus and pituitary gland disorders
17-Hydroxycorticosteroids	2 to 10 mg/24 hr	Quantities increase in Cushing's syndrome and decrease in Addison's disease
Phenylpyruvic acid	0	Quantities increase in phenylketonuria
Urea clearance	Over 40 mL blood cleared of urea/min	Quantities increase in kidney disease
Urobilinogen	0 to 4 mg/24 hr	Quantities increase in liver disease and hemolytic anemia; a decrease occurs in severe diarrhea and obstruction of gallbladder.
Urea	25 to 35 g/24 hr	Quantities increase with excessive protein breakdown; a decrease indicates impaired kidney function
Uric acid (as urate)	0.6 to 1.0 g/24 hr as urate	Quantities increase in gout and decrease in several kidney diseases

glycosuria: (Gr.) *glykys*, sweet + *ouron*, urine
hematuria: (Gr.) *haima*, blood + *ouron*, urine

Table 25-1 QUANTITIES OF SELECTED SOLUTES EXCRETED IN URINE IN A TYPICAL DAY (ADULTS)

Solute	Millimoles		Weight	
	Amount Excreted (mmol)	Amount Relative to Urea (%)	Amount Excreted (mg)	Amount Relative to Urea (%)
Na^+	40 to 220	32	920 to 5060	12
K^+	25 to 150	21	975 to 5850	14
Ca^{2+}	1.25 to 3.75	<1	50 to 150	<1
Mg^{2+}	1 to 10.5	1	24 to 255	<1
Cl^-	110 to 250	44	3850 to 8750	25
HCO_3^-	14	2	854	4
Inorganic phosphorus	9.7 to 42	6	300 to 1300	3
Glucose	0	0	0	0
Urea	250 to 570	100	15,000 to 34,000	100
Uric acid	1.5 to 4.5	<1	250 to 750	2
Creatinine	8.8 to 17.7	13	994 to 2000	6
Protein	—	—	50 to 150	<1

produced by healthy people varies in specific gravity from 1.010 to 1.025. Its odor is described as aromatic or ammoniacal and its taste as bitter and disagreeable. Table 25-1 shows the amounts of several solutes excreted in urine in the course of a day.

Abnormal constituents appear in urine as a result of disease or damage to the urinary tract. The presence of the plasma protein albumin, for example, is an indication that glomeruli are not filtering plasma properly, and it may result from fever or inflammation of nephrons or other parts of the urinary tract. Glucose, normally reabsorbed in the proximal tubule, may be present in the urine of diabetics and in certain orther cases. The presence of glucose in urine is termed **glycosuria** (glī″kō·sū′rē·ă). Blood cells may also be found in the urine, a condition called **hematuria** (hē″mă·tū′rē·ă), as a result of urinary tract bleeding caused by infection, tumors, or stones produced by the accretion of precipitated salts. **Renal (urinary) casts** are formed from mucoprotein or other substances that form in renal tubules where they are molded in the shape of the tubule. Casts are formed as the protein, a normal component of urine, precipitates from solution (Figure 25-24). **Urinary stones** (or **calculi**) are precipitated minerals, often consisting of calcium oxalate or other calcium salts that may develop in any part of the urinary tract. Stones may be large enough to cause damage to renal tissue or even to block the flow of urine from the kidney. They can be extremely painful and a source of secondary infection of the damaged tissue (see Clinical Note: Urinary Calculi).

The kidneys play an extremely important role in regulating the amount and composition of body fluids. They eliminate the bulk of water consumed or produced in the body, they eliminate nitrogenous and other wastes, they eliminate excess minerals, they play a major role in maintaining pH, and they help regulate the production of blood. If the kidneys are functioning properly, an appropriate amount of water and solutes are retained for efficient cellular activity. The fluid that remains is a complex mixture that is distributed within cells, in the fluid surrounding the cells, and in blood. The composition and properties of those fluids are the subject of Chapter 26.

Figure 25-24 Common types of sediment found in urine. (a) Blood cells in infected urine (mag. 292 ×). (b) Renal casts (mag. 37 ×). (c) Yeast cells (mag. 1460 ×). (d) Uric acid crystals (mag. 234 ×).

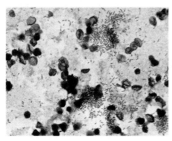
(a)
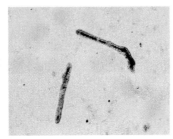
(b)
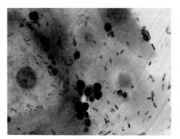
(c)
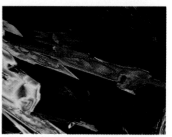
(d)

cystitis: (Gr.) *kystis*, bladder + *itis*, inflammation
pyelonephritis: (Gr.) *pyelos*, pelvis + *nephros*, kidney + *itis*, inflammation

THE CLINIC

URINARY SYSTEM DISORDERS

Many disorders and diseases affect the urinary system, especially the main functional organ of the system, the kidney. With its diverse and vital functions, the kidney is particularly susceptible to many disorders that adversely affect the body's ability to function.

Developmental and Hereditary Diseases

On rare occasions, the kidneys may completely fail to develop. The absence of both kidneys **(bilateral renal agenesis)** is fatal. **Unilateral renal agenesis** (one kidney missing) is a fairly common abnormality and may be entirely without symptoms because the single kidney enlarges to perform the functions of the missing organ. Duplication anomalies may also occur with double kidneys and/or ureters on one or both sides. Fusion anomalies may occur with an isthmus of renal tissue across the midline resulting in one large **horseshoe kidney.**

Polycystic renal disease is an inherited congenital disorder in which multiple bilateral cysts occur in the kidneys that enlarge the total renal mass and may cause reduced renal function by compression of the renal tissue (see figure). People with this disease may survive into old age before renal failure occurs. Kidney transplantation and/or renal dialysis are necessary if the kidneys cease to function.

Medullary sponge kidney is a congenital cystic dilation of the collecting tubules of the kidney that may result in formation of urinary calculi (stones) (see Clinical Note: Urinary Calculi).

Infectious Diseases

Bacterial infections of the urinary tract are a common medical problem. Infection may be limited to a specific part of the system or ascend to involve the entire system. **Urethritis** involves the urethra alone and is a comon problem in males as a result of sexually transmitted diseases such as gonorrhea. **Cystitis** is an infection confined to the urinary bladder. The person voids small amounts of urine frequently with varying degrees of discomfort, burning, and urgency. The urine is cloudy and when examined microscopically shows evidence of infection. Women, presumably because of the short urethra, are more susceptible to cystitis. Cystitis is often caused by coliform organisms. Simple precautions for prevention include emptying the bladder immediately after sexual intercourse, wearing cotton underwear, avoiding colored or perfumed toilet paper, drinking ample fluids, and urinating as soon as the urge is felt. Pregnant women are especially susceptible to cystitis, as are people with diabetes mellitus.

If the infection ascends to involve the kidney, it is known as **pyelonephritis.** The infecting bacteria is usually *E. coli*.

Polycystic kidneys. (a) An IVP of normal kidneys (for comparison). (b) An IVP of kidneys affected by polycystic renal disease. The cysts are the dark, round areas that have deformed the collecting system of the kidney.

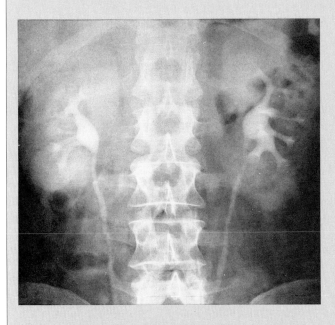
(a)

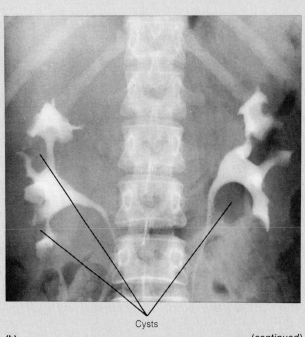
(b)

Cysts

(continued)

glomerulonephritis: (L.) *glomerulus*, little ball + (Gr.) *nephros*, kidney + *itis*, inflammation

THE CLINIC (continued)

URINARY SYSTEM DISORDERS

The person will experience fever, chills, headache, and pain in the lower back. Treatment is through use of antibiotics. Chronic pyelonephritis can lead to chronic renal failure.

Obstruction to the normal urinary flow anywhere along the urinary system usually results in infection above the obstruction. It also increases pressure in the urinary tract and causes dilation of the pelvis and the calyces of the kidney. It may eventually lead to renal damage and failure. Common causes of obstructed urinary flow are kidney stones (renal calculi), which can obstruct the ureters or urethra; tumors; prostatic hypertrophy; inflammatory damage, such as from sexually transmitted diseases, which can lead to urethral strictures; and pregnancy. The obstruction should be removed if possible.

Trauma

Injuries to the urinary system are quite common today due to our high speed automobile travel and increased participation in contact sports. Severe motor vehicle accidents in which riders are ejected from the car frequently result in trauma to the lower abdomen or fractures of the pelvis. These injuries may result in rupture of the urethra and urinary bladder. Symptoms of these injuries include hematuria, inability to void, difficulty in passing a catheter into the bladder, and release of urine into the surrounding soft tissues. Diagnosis is made by performing an intravenous pyelogram (IVP) (see Clinical Note: Urinary Tract Imaging Techniques). Treatment is by surgical repair.

Injuries may occur to the kidneys through blows to the kidney area. This is a fairly frequent injury in contact sports, particularly in football. Such injuries have caused football leagues to rule against "blocking from behind." The kidney may sustain a contusion characterized by severe local pain and temporary hematuria. More severe injuries can result in rupture of the kidney or even complete **avulsion** (forcibly tearing away) of the kidney. The kidney is a retroperitoneal organ, and therefore bleeding can be hidden, but an IVP will usually delineate the extent of the injury. Surgical repair or even excision of the kidney may be required to stop severe hemorrhage.

Immunologic and Allergic Diseases

Acute glomerulonephritis is a classic example of an immune complex disease in which antigens to streptococcal infections combine with antibodies, and the complex is deposited in the glomerular capillary walls. These complexes activate the complement system, liberating chemotactic factors that attract polymorphonuclear leukocytes to the area (see Chapter 24). The streptococcal infection may be in the respiratory tract ("strep throat") or on the skin but not in the kidney itself. The inflammatory reaction in the glomerulus results in protein, blood, and red blood cell casts in the urine giving the urine a brown and cloudy appearance. Fortunately 85 to 95% of patients recover in a matter of weeks if the infection is treated with antibiotics. A few patients may progress to chronic nephritis and subsequent renal failure.

Goodpasture's syndrome is a rare condition in which the circulating antibody is less specific and may damage the pulmonary alveolar basement membrane as well as cause both renal and pulmonary hemorrhages. It may be rapidly fatal. Treatment is by administration of corticosteroid drugs.

Renal function may be adversely affected by many medications, either by direct toxic effect or by allergic reactions to prolonged use. These include some analgesics and anti-inflammatory drugs as well as some antibiotics.

Cancers

Renal malignancies are common in both children and adults. The **Wilms' tumor** is an adenocarcinoma that occurs in the fetal kidney and may lie dormant for many years. Prompt treatment by surgery, chemotherapy, and radiation can result in better than a 50% five-year survival. It is second only to **neuroblastomas** in frequency of solid tumors in childhood. Neuroblastoma arises from the adrenal gland or sympathetic chain. Treatment is less successful than for Wilms' tumor.

Carcinoma of the kidney accounts for 1 to 2% of adult cancers. Two-thirds of the patients are males. Five year survival is about 45%. **Transitional cell carcinomas** can occur in the renal pelvis, ureters, or bladder. Hematuria is a sign, and treatment is surgical excision and chemotherapy.

Aging

Renal function gradually declines with age. The usual cause is progressive arteriosclerosis and hypertension. The glomerular filtration rate at age 70 is about 50% of what it is at age 40. Renal tubular function also declines with a decrease in concentrating ability. Acid-base regulation is usually retained.

A common problem of aging males is obstruction due to **benign prostatic hypertrophy,** which is an increase in the size of the prostate. It is treated by surgical removal through the urethra of the enlarged central portion of the prostate.

STUDY OUTLINE

Anatomy of the Kidneys (pp. 647–651)
The kidneys lie on either side of the vertebral column at a level between the 12th thoracic and 3rd lumbar vertebrae.

External Structure The concave surface of a kidney is the *renal hilum*. Kidneys are surrounded by *fibrous* and *adipose capsules* and a *renal fascia*.

Internal Organization Three regions of a kidney are the *cortex, medulla,* and *renal sinus. Renal columns* from the cortex divide the medulla into *renal pyramids. Renal papillae* are formed where the tips of the pyramids come together. The *renal pelvis* is an enlarged opening in the kidney opposite the hilum. *Calyces* are regions of the pelvis that extend from the main chamber. An *area cribrosa* at the base of each papilla opens into a calyx.

Microscopic Anatomy *Cortical labyrinths* are granular regions of the cortex separated from one another by *medullary rays*. Medullary rays and the enclosed cortical labyrinths comprise a *lobule*. Cortical labyrinths contain portions of nephrons; medullary rays contain *collecting tubules*.

The Nephron (pp. 651–657)
The functional unit of the kidney is the *nephron*, consisting of a spherical, capillary-filled end and an attached tubule that leads to a collecting duct.

Renal Corpuscle The *renal corpuscle* consists of a glomerulus and the surrounding *Bowman's capsule*. The Bowman's capsule is a double-walled spherical structure consisting of *parietal* and *visceral layers* separated by the *urinary space*, a *vascular pole* that opens into the interior, and a *urinary pole* that drains the urinary space. *Podocytes* on the wall of the visceral layer form filtration slits through which substances are filtered.

The *glomerulus* is a network of passageways within Bowman's capsule through which blood passes. Pores in the endothelial cells lining the passageways allow blood to make contact with an underlying basement membrane at the inner (glomerular) wall of the visceral layer.

The *glomerular basement membrane* lies between the endothelial cells of the glomerulus and the podocytes on the urinary space side of the visceral layer. Spaces between the fibrils in the membrane prevent the passage of proteins and other macromolecules.

Tubular Portion of a Nephron Beginning at the urinary pole, the tubular portion of a nephron consists of a *proximal convoluted tubule,* a *loop of Henle,* and a *distal convoluted tubule*. The loop of Henle contains a *descending limb* that reverses direction and becomes an *ascending limb*.

Location and Dimensions of Nephrons *Cortical nephrons* have loops of Henle that are largely confined to the cortex. *Juxtamedullary nephrons* extend into the medulla pyramid.

Blood Supply to a Nephron In its course through a kidney, blood passes from the renal artery successively through *interlobar, arcuate,* and *interlobular arteries;* an *afferent glomerular arteriole;* the glomerulus; an *efferent glomerular arteriole;* the *peritubular plexus; interlobular, arcuate,* and *interlobar veins;* and out through the renal vein.

The *juxtaglomerular apparatus* is a region where an afferent and efferent arteriole lie close to one another at the vascular pole of Bowman's capsule. Cells lying between the arterioles form an *extraglomerular mesangium,* which is involved in the production of *renin,* a protein that helps control blood pressure.

Nerve and Lymphatic Supply to a Nephron Most of the nerves leading into a kidney are sympathetic, originating in the celiac or aortic ganglia. Some sensory nerves are present. Most motor nerves probably control vasoconstriction. Lymphatic vessels originate in the capsule surrounding the kidney and within the kidney.

Physiology of the Kidneys (pp. 657–665)
Filtration Filtration through the visceral wall of Bowman's capsule results from a difference in glomerular blood pressure and pressure in the urinary space. The rate is termed *glomerular filtration rate. Capillary hydrostatic pressure* in the glomerulus is offset by *tubular hydrostatic pressure* and *colloid osmotic pressure. Net filtration pressure* is the resulting pressure. Filtration rate is determined from *clearance,* the volume of plasma that would have to pass through the kidneys in a given time to produce an amount of solute in urine.

Function of the Tubular Portion of a Nephron Primary urine is similar to plasma in composition, lacking dissolved proteins. Nutrients, water, and minerals are reabsorbed in various portions of the nephron. Some substances are secreted into the filtrate through the walls of the tubular portion of the nephron.

Concentrating the Urine As the glomerular filtrate passes through the proximal convoluted tubule, water and solute removal reduces its volume without changing concentration.

Countercurrent Concentrating Mechanism The filtrate is concentrated in subsequent portions of the nephron and collecting ducts by *countercurrent exchange*. Dissolved NaCl and urea makes the surrounding fluids hypertonic to the fluid in the tubules, resulting in osmotic flow of water from the tubules. Water is carried from the tissues by blood in the *vasa recta*.

Regulation of Urine Concentration and Water Excretion *Diuresis* is the increase in urine production that follows water consumption. The amount of urine produced is controlled by antidiuretic hormone (ADH) released from the posterior pituitary gland in response to hypothalamic osmoreceptors sensitive to osmotic concentration of blood and to blood pressure. ADH increases permeability of the distal convoluted tubules and collecting ducts, allowing more water to pass into surrounding tissues.

Homeostatic Functions of the Kidneys (pp. 665–669)

Renal Regulation of Sodium Content Na$^+$ is reabsorbed throughout the nephron tubules, but in the distal convoluted tubule its reabsorption is controlled by aldosterone. Na$^+$ reabsorption results in increased water absorption. Aldosterone production is controlled by renin and *angiotensin*.

Control of Blood Pressure Angiotensin causes an increase in blood pressure by increasing vasoconstriction and extracellular fluid volume.

Renal Regulation of pH Excess H$^+$ is secreted into renal tubules. It is linked to Na$^+$ absorption and helps recover HCO$_3^-$.

Potassium Absorption and Secretion Kidneys are the primary organs for secretion of excess K$^+$. K$^+$ is secreted primarily in the distal convoluted tubule in conjunction with sodium reabsorption, the amount depending on pH of the blood.

Regulation of Red Blood Cell Production Kidneys produce *renal erythropoietic factor*, which activates *erythropoietin*. Erythropoietin is a hormone that stimulates erythrocyte production.

Vitamin D Activation The kidney converts vitamin D to dihydroxycholecalciferol, the active form of the vitamin.

Conduction, Storage, and Elimination of Urine (pp. 669–672)

Conduction through the Ureters Urine passes from calyces into the renal pelvis. From there it is forced into *ureters* by contraction of the pelvis and peristalsis of the ureters. It passes down to the *urinary bladder*, where it is stored.

Urinary Bladder The ureter openings into the bladder and the *internal urethral orifice* comprise the *trigone*. Contraction of the *detrusor muscle* forces urine through the *internal urethral orifice*, past the *internal urethral sphincter*, and into the urethra.

Urethra Adult females have a *urethra* about 4 cm long. Adult males have a urethra about 20 cm long, including *prostatic, membranous,* and *cavernous* portions. The *external urethral sphincter* is in the membranous portion.

Micturition Micturition is the removal of urine from the bladder. Filling of the bladder stimulates proprioceptors in the bladder wall. Proprioceptors send signals to the spinal cord, which sends signals to the detrusor muscle in the bladder wall and to the cerebral cortex. Voluntary signals from the cerebral cortex can cause the external urethral orifice to remain closed for a time. When the external urethral orifice relaxes, the bladder contracts, forcing urine out through the urethra.

Composition of Urine (pp. 673–674)

Urine is variable in composition, depending on a number of conditions. Abnormal constituents can include glucose, blood, or protein. Precipitates may form *renal casts* and stones.

SELF-TEST OF CHAPTER OBJECTIVES

True-False Questions

1. Angiotensin II is a compound that stimulates the production of aldosterone.
2. The specialized cells of the afferent arteriole that produce renin are juxtaglomerular cells.
3. Podocytes form spaces that filter blood in the glomerulus.
4. All the waste products that ultimately appear in the urine first appear in the glomerular filtrate.
5. The vasa recta is associated with juxtamedullary nephrons as opposed to cortical nephrons.
6. The walls of an ascending portion of the loop of Henle are relatively impermeable to water.
7. Renin is the chemical product of the kidney that is responsible for stimulating the production of blood.
8. The rate at which water is reabsorbed is controlled primarily by aldosterone.
9. Sodium reabsorption is hormonally regulated in the distal convoluted tubule.
10. The high permeability of a collecting duct to NaCl contributes to the high solute concentration in interstitial tissues surrounding the loop of Henle.

Matching Questions

Match the structure on the left to its primary function on the right:

11. glomerulus
12. Bowman's capsule
13. collecting duct
14. distal convoluted tubule
15. proximal convoluted tubule

a. final concentration of urine
b. recovers glucose and amino acids
c. controls water removal
d. filters the blood
e. collects primary filtrate

Matching the substance on the left with the substance on the right with which it is most closely related in function:

16. renin
17. aldosterone
18. REF
19. ADH

a. Na$^+$
b. water
c. angiotensinase
d. erythropoietin

Multiple-Choice Questions

20. Which one of the following is not a function of the kidney?
    a. controls blood pressure
    b. controls body water volume
    c. produces erythropoietin
    d. converts vitamin D to an active form
21. Which of the following is involved in the pH regulation function performed by the kidney?
    a. bicarbonate ions
    b. phosphate ions
    c. hydrogen ions
    d. all of the above

22. Of the following, which item is most directly involved in the control of body water volume by the kidney?
    a. sodium ions
    b. potassium ions
    c. phosphate ions
    d. bicarbonate ions
23. Which of the following lists structures in the order through which urine passes?
    a. kidney, urethra, ureter, urinary bladder
    b. kidney, ureter, urinary bladder, urethra
    c. kidney, urethra, urinary bladder, ureter
    d. kidney, ureter, urethra, urinary bladder
24. Which of the following lists structures of the kidney in the order through which urine passes?
    a. collecting tubule, distal convoluted tubule, loop of Henle, proximal convoluted tubule
    b. distal convoluted tubule, collecting tubule, loop of Henle, proximal convoluted tubule
    c. loop of Henle, distal convoluted tubule, collecting tubule, proximal convoluted tubule
    d. proximal convoluted tubule, loop of Henle, distal convoluted tubule, collecting tubule
25. Water is primarily reabsorbed from the urine in all but which one of the following structures?
    a. proximal convoluted tubule
    b. descending arm of the loop of Henle
    c. ascending arm of the loop of Henle
    d. distal convoluted tubule
26. K^+ is secreted from the nephron primarily in the
    a. Bowman's capsule
    b. proximal convoluted tubule
    c. loop of Henle
    d. distal convoluted tubule
27. K^+ is absorbed from the glomerular filtrate primarily in the
    a. proximal convoluted tubule
    b. loop of Henle
    c. distal convoluted tubule
    d. collecting ducts
28. Aldosterone is a hormone that causes the kidney to
    a. absorb Na^+ in the distal convoluted tubule
    b. absorb Na^+ in the proximal convoluted tubule
    c. secrete Na^+ in the distal convoluted tubule
    d. secrete Na^+ in the proximal convoluted tubule
29. The primary receptors sensitive to the osmotic pressure of blood are found in
    a. the cortex of the kidney
    b. the medulla of the kidney
    c. the hypothalamus
    d. the juxtaglomerular apparatus
30. Water recovered from the glomerular filtrate is delivered first to the
    a. afferent arterioles
    b. efferent arterioles
    c. glomerulus
    d. vasa recta
31. Which of the following reflects the role of the kidney in the production of red blood cells?
    a. The kidney produces a hormone that directly stimulates red blood cell production.
    b. The kidney produces a substance that activates a hormone that stimulates red blood cell production.
    c. The kidney itself produces red blood cells in response to anemia.
    d. The kidney removes excess red blood cells.
32. A trigone is located in
    a. a nephron
    b. the urinary bladder
    c. the renal pelvis
    d. the urethra
33. Which of the following is not normally found in urine?
    a. K^+
    b. H^+
    c. glucose
    d. creatinine
34. Which of the following does not occur during a micturition reflex?
    a. The internal urethral sphincter relaxes.
    b. Proprioceptors in the detrusor muscle are stimulated.
    c. The detrusor muscle contracts.
    d. Stimulatory impulses are sent to the bladder from the medulla.
35. Which of following is most distal to the bladder in a male?
    a. cavernous urethra
    b. membranous urethra
    c. prostate urethra
    d. internal urethral orifice

Essay Questions
36. List at least five functions performed by the kidney.
37. List the organs of the urinary system in the order of production and passage of urine.
38. Describe the microscopic anatomy of the nephron.
39. Describe how filtration occurs in the renal corpuscle.
40. Explain how nutrients in the filtrate are recovered as urine is produced.
41. Describe the two-solute countercurrent multiplier mechanism for concentrating urine.
42. Explain how water volume and concentrations of Na^+, K^+, and H^+ in the body are controlled by the kidneys.
43. Compare the roles of the different hormones that are produced by the kidney and that regulate kidney function.
44. Describe the structures through which urine passes after its production in the kidneys and the mechanisms involved in urination.
45. List several normal and abnormal components of urine.

Clinical Application Questions
46. List and describe techniques used to produce images of the kidneys.
47. Distinguish among prerenal, renal renal, and postrenal acute renal failure and chronic renal failure.
48. Describe some of the factors that can lead to the production of renal calculi. How are renal calculi treated?
49. List and describe examples of developmental, infectious, traumatic, immunologic, and cancerous diseases that affect the kidneys.

26
Composition and Control of Body Fluids

Water is the most abundant substance in the body. It is the chief solvent for minerals and molecules, a component of many chemical reactions, and helps maintain homeostatic conditions necessary for efficient cell function. Water bathes cells and is a major component of the interior of a typical cell. Blood plasma, which carries blood cells from one part of the body to another, is mostly water. Water helps to maintain a relatively constant body temperature by transporting heat from metabolically active tissues, such as in muscles, to less active tissues. Evaporation of water from the skin and lungs helps cool the body, preventing overheating during hot weather and strenuous activity. In short, water plays an extremely important role in the reactions and functions that characterize life.

In this chapter we consider the function of water and its solutes, examining

1. the composition of water inside and outside cells,
2. the regulatory mechanisms that control water balance, and
3. the mechanisms used to control the pH of body fluids.

Regulating Water Content

After studying this section, you should be able to:

1. Compare and describe the composition of intra- and extracellular fluids.
2. List three sources of body water.
3. Compare carbohydrate, fat, and protein as sources of metabolic water.
4. Describe how water intake is regulated.
5. List three ways in which water is lost.

Table 26-1 DAILY INTAKE AND OUTPUT OF WATER

	Amounts (mL/day)
Intake	
Drink	1300
Food	850
Metabolic water	350
Total	2500
Output	
Urine	1500
Expired air	400
Skin	500
Feces	100
Total	2500
Secretions of the Digestive Tract (reabsorbed)	
Saliva	500–1500
Gastric juice	2000–3000
Pancreatic juice	300–1500
Bile	250–1100
Intestinal secretions	about 3000

Not all people have the same amount of water in their tissues, since differences in water content correspond primarily to age and the amount of fatty tissue present. Because fatty tissue contains almost no water, a lean person tends to have a greater proportion of water to total body weight than an overweight person. Because most females have a thicker layer of subcutaneous fat than males, the tissues of an average adult male contain more water (about 60%) than those of an average adult female (about 50%). Infants have the highest proportion of water to body weight, and the proportion decreases with age.

Extracellular and Intracellular Fluids

About two-thirds of the fluid in the body is within cells, where it forms the **intracellular fluid (ICF).** Water and dissolved substances outside the cells is **extracellular fluid (ECF).** Extracellular fluid includes synovial fluid, interstitial fluid, fluids of the eyes, peritoneal and pleural fluids, cerebrospinal fluid, pericardial fluid, fluid within the urinary tract, lymph, and plasma.

Water moves back and forth between the ECF and ICF by osmosis. Movement of substances dissolved in the water, however, is controlled by cells and their membranes, and this movement often involves active and facilitated transport as well as simple diffusion.

Water Intake

We obtain water from three sources: we drink water and other fluids (about 1300 mL/day); we obtain water in food that we eat (about 850 mL/day); and we produce water as a byproduct of internal chemical reactions, called **metabolic water** (about 350 mL/day) (Table 26-1). Water obtained by ingestion is **preformed water.** The amount obtained from each source varies with the environment and the size, weight, and physical activity of a person. How much is drunk depends on how hot and dry the weather is because these factors affect the amount of evaporative loss. Diet influences the other two categories: some foods contain more water than others and some kinds of nutrients produce more metabolic water than others. For example, the water content of an orange is obviously greater than that of grain. On an equal weight basis, about twice as much water is produced by fat catabolism than by starch catabolism, and protein catabolism yields still less. Moreover, in contrast to the breakdown of fats and carbohydrates, which yield carbon dioxide as a waste product, the breakdown of protein produces urea as a waste product, and urea can only be excreted in solution. Consequently, the modest amount of water gained through protein breakdown is offset by the amount required to remove urea.

Water that is drunk is absorbed throughout the digestive tract, the last of it passing into the tissues of the colon. Fluid is also added to this water in salivary, gastric, and intestinal secretions, and water and minerals present in these secretions are also absorbed. Once absorbed, water enters the circulation and is distributed throughout the body. Pressure and osmotic effects that result from increased fluid volume induce urine production, and excess water is removed within a few hours.

The main regulator for the control of water intake is **thirst.** The sensation of thirst is initiated by a region of the hypothalamus called the **thirst center** (see Chapter 12), which is stimulated by a loss of intracellular fluid that results when fluid volume diminishes (Figure 26-1). As water is lost from the body, the concentration of dissolved substances in the ECF rises and cells of the thirst center, which may have been in osmotic equilibrium with the ECF, lose water by

electrolyte: (Gr.) *elektron*, amber + *lytos*, soluble

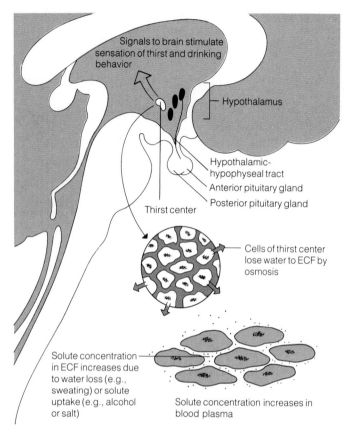

Figure 26-1 A lateral view of the hypothalamus and pituitary gland. The thirst center is located in the hypothalamus, just above the pituitary gland.

osmosis. The thirst center responds by sending signals to higher brain centers, which induce one to drink. A feeling of thirst results when only about 1% of the water is lost from thirst center cells, so the thirst center provides a very delicate indicator of water needs.

Control of thirst does not appear to be exercised exclusively by the thirst center, however, because this thirst can be quenched by drinking water long before the concentration of solutes in the ICF is raised (see Chapter 12). In fact, when a thirsty person drinks water, the sensation of thirst is lost almost immediately, although the return to a normal ICF concentration in the cells of the thirst center may take several minutes.

Water Output

Under normal circumstances, water uptake and water loss are balanced. In an adult, about 2500 mL of water is taken in and lost each day. Of the water lost, about 1500 mL is excreted as urine, about 100 mL is lost in the feces, and about 900 mL evaporates from the skin and lungs (Table 26-1). This does not include sweat lost due to heat or strenuous physical activity; if one is very active on a hot day, the rate of water loss increases dramatically and may reach 10 L or more. Water loss can be increased by fever, vomiting, and diarrhea, which can lead to dangerous degrees of dehydration if not compensated by increased intake. Sometimes, when water is lost from excessive vomiting, fluids must be administered intravenously.

Water excretion is controlled largely by factors that stimulate or retard urine production. Regulation is complex, however, and several mechanisms are used, particularly involving the hypothalamus and pituitary gland. The rate of excretion depends on the concentration of minerals in the blood and the amount of blood pumped by the heart. Control of the volume of water excreted is accomplished by the renal mechanisms described in Chapter 25.

Composition of Body Fluids

After studying this section, you should be able to:

1. Define electrolytes and identify the major ions of the ECF and ICF.
2. Describe the source and distribution of bicarbonate ion in body fluids.
3. Compare the electrolyte composition of the ECF with that of the ICF.
4. Explain the role of sodium, potassium, calcium, magnesium, and other cations, and describe the mechanisms used to maintain appropriate levels of each in body fluid.
5. Describe the role of chloride, bicarbonate, phosphate, sulfate, and other anions in body fluids.

Electrolytes

Electrolytes are compounds capable of dissolving in the watery medium of body fluid and dissociating into ions. Positively charged ions are **cations**, and negatively charged ones are **anions**. Concentrations of electrolytes are commonly expressed in **equivalents** per liter (Eq/L), a term that relates to the potential chemical reactivity of the ions of an element. A **chemical equivalent** of a substance is the weight in grams of the substance that will react with or produce 1 mol of hydrogen ions or 1 mol of electrons. The concentrations of electrolytes in body fluids are generally so low that it is more convenient to refer to **milliequivalents** per liter (mEq/L), equal to 1/1000 of an equivalent per liter.

For ions that have a single charge, the weight of a chemical equivalent is the same as its molecular or atomic weight. For ions with multiple charges, however, equivalent weight is equal to the molecular weight of the substance divided by its electric charge. Since 1 mol of Mg^{2+} can absorb 2 mol of electrons (two electrons per ion), there are two equivalents in each mole of magnesium ions. The atomic weight of a mole of Mg^{2+} is 24.3 g, therefore the weight of an equivalent of Mg^{2+} is 12.15 g.

The concentration of electrolytes in body fluids depends on the balance between how much is present in food, how efficiently electrolytes are absorbed, and how much is excreted. There is a distinct difference in the concentration of each of the cations and anions in the ECF and ICF (Figure 26-2). Sodium is predominantly a cation of the ECF and potassium is the major cation of the ICF. Magnesium is also largely an intracellular ion, although its concentration inside the cell is considerably less than that of potassium. A dramatic difference also exists in the anion composition of the two fluids. Chloride is the major anion of the ECF, where it is nearly 40 times more concentrated than in the ICF. In contrast, no single anion predominates in the ICF, where the major anions are phosphate, sulfate, and negatively charged organic ions.

Differences in composition between the ECF and the ICF are maintained primarily by active transport. Cell membranes are largely impermeable to ions in body fluids, so energy must be expended to actively move most ions from one side of a membrane to the other.

Major Cations in Body Fluid

Sodium

Sodium (Na^+) is the major cation of extracellular fluid, accounting for about 90% of the cations in the ECF. It is essential to the transmission of nerve impulses throughout the nervous system. It is also the major substance used to regulate ECF volume. Because of its importance, disorders that affect sodium concentration generally result in a deficiency or an excess of water in the ECF. Changes in ECF volume, in turn, cause changes in ICF volume that impair proper cell function. For example, changes in ICF volume in neurons of the central nervous system, brought on by a sodium imbalance in the ECF that bathes the cells, can cause seizures, coma, or even death. Loss of sodium can result in a diminished plasma volume, which in turn causes rapid heartbeat and unusually low blood pressure. Retention of too much sodium causes the retention of too much water, with a resulting puffiness and swelling of the tissue. This condition is

Figure 26-2 Electrolytes. Ions are distributed in three separate compartments of body fluids: blood plasma, interstitial fluid, and intracellular fluid. The values given are milliequivalents of ion per liter (meq/L) of solution.

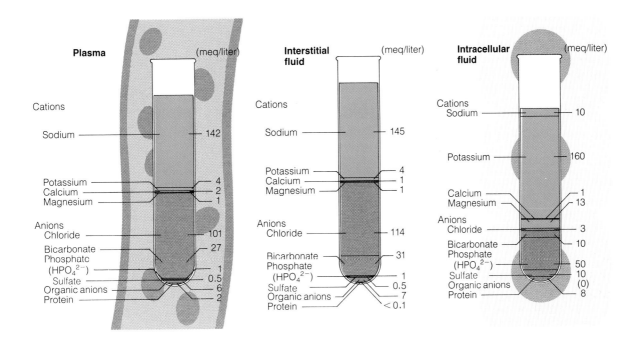

edema: (Gr.) *oidema*, swelling

CLINICAL NOTE

BODY FLUID AND ELECTROLYTE IMBALANCE

One of the most common problems in clinical medicine is an imbalance of fluids and electrolytes. The history of the illness will provide clues to the nature of the imbalance and its magnitude. High fever, persistent vomiting, diarrhea, profuse sweating, and lack of oral fluid intake should alert the physician to suspect dehydration. The magnitude of the problem can be estimated by a careful analysis of the amount of vomit and diarrhea, the duration of the illness, and the amount of fluid retained. A good clue to the extent of dehydration is the length of time between urinations, since with severe dehydration the kidneys will attempt to conserve body fluids and urination will become infrequent. Other complicating factors may also be significant, such as the presence of renal disease or diabetes.

During the physical exam, the physician will note the height, weight, temperature, blood pressure, and pulse rate; this will indicate whether vascular collapse (shock) has occurred. Evaluation of the moisture content of the mucous membranes, skin dryness, and skin turgor (resistance of skin to being grasped between the fingers) also give valuable information about the state of hydration.

When fluid and electrolyte imbalance is suspected, a few laboratory tests will help to confirm the magnitude of the problem. The **urine specific gravity** is a good indication of dehydration. Normal kidneys will concentrate urine at a specific gravity above 1.010, usually at about 1.020. If the specific gravity rises above 1.020, and there are no abnormal solutes in the urine (glucose, for example), dehydration may be suspected. A significantly elevated hematocrit (packed cell volume of whole blood) also suggests a reduced plasma volume and, thus, dehydration. The **serum electrolytes** are a measurement of the serum sodium (Na^+), serum potassium (K^+), serum chloride (Cl^-), and serum CO_2. The serum CO_2 combining power indicates the presence of acidosis or alkalosis.

Additional tests may be ordered, if indicated, such as serum calcium, serum phosphorus, and serum magnesium. Usually the complete blood count, urinalysis, and serum electrolyte tests will provide enough information to begin treatment. These tests may be repeated at intervals to help evaluate the progress of the treatment plan.

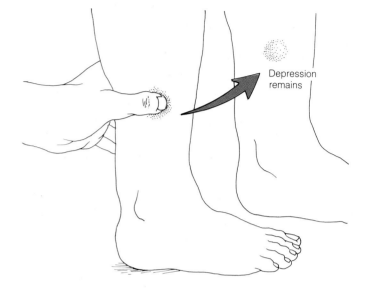

Figure 26-3 A simple test for edema that physicians sometimes apply is to hold the thumb firmly against the calf for a few seconds. If a depression remains after the thumb is removed, it indicates that fluid has collected in the tissues and that edema is present.

called **edema** (ĕ·dē′mă), which can lead to heart, kidney, and pulmonary failure and death (Figure 26-3).

Usually, sufficient sodium is present in food to meet body requirements. Surplus sodium is excreted if the kidneys are functioning properly, thus, chronic sodium imbalance is seldom the result of occasional poor eating habits. Instead, sodium imbalance is almost always the result of kidney malfunction brought on by disease of the kidney itself or by impairment of hormonal mechanisms that control the kidney.

Assuming adequate sodium intake from the diet, homeostatic control of the level of sodium in the blood is accomplished by a negative feedback mechanism involving aldosterone, which is secreted from the adrenal cortex (see Chapter 17). Under normal circumstances, the level of aldosterone is sufficient to maintain adequate amounts of sodium in the blood. Several factors will cause an increase in the production of aldosterone, including a drop in the sodium concentration of the ECF, stress, a reduction in the arterial blood pressure, and an increase in the concentration of potassium in the ECF. Aldosterone stimulates distal convoluted tubules of the kidneys to reabsorb sodium from the glomerular filtrate, and the sodium is then returned to the blood. This causes the level of sodium in the blood to rise, and the stimulus for the production of aldosterone subsides.

Significant sodium loss sometimes occurs during periods of excess sweating, as when one engages in strenuous activity in hot weather. As a result of **acclimatization** (ă·klī·mă·tĭ·zā′shŭn), however, the body adapts to these conditions and reduces the amount of sodium lost in sweat.

Potassium

Over 95% of the body's dissolved potassium (K^+) is in the ICF, and most of that is in muscles. Potassium is important to the transmission of nerve impulses and muscle contraction. A normal diet usually provides sufficient potassium, but when deficiencies do occur, they lead initially to muscle

gastroenteritis: (Gr.) *gaster*, stomach + *enteron*, intestine + *itis*, inflammation

---

CLINICAL NOTE

REPAIR OF FLUID AND ELECTROLYTE IMBALANCE: AN EXAMPLE

There are many formulas for calculating fluid and electrolyte losses and developing protocols for treatment. While these formulas may be useful in treating very complex cases, such as serious burns, diabetic ketoacidosis, and kidney failure, most physicians acknowledge that a healthy kidney can repair simple imbalances on its own without much intervention. Therefore, physicians provide the body with the fluids and electrolytes estimated to be lost and allow the kidneys to retain those needed.

Intravenous solutions are supplied prepackaged and are adjusted to be isotonic with the blood so that they do not cause hemolysis. Common examples are 5% dextrose in water, used for hydration, normal saline solution (0.9% sodium chloride), and 5% dextrose in normal saline. Special solutions can also be prepared by adding small amounts of concentrated electrolyte solutions, such as potassium chloride or sodium bicarbonate, to the stock bottles.

In one typical example, a patient shows symptoms of **gastroenteritis,** inflammation of the stomach and intestines. He has been ill for about 12 hours, has not been able to keep down any food, and has not urinated in the past 4 hours. It is estimated that the patient has lost about 1000 mL of fluid through vomiting and diarrhea. Since vomit and feces contain large amounts of sodium, chloride, and potassium, the physician starts the patient on 1000 mL of 5% dextrose in normal saline solution intravenously and plans to follow it with 2000 mL of 5% dextrose in water for hydration. Potassium chloride is then added to the solution. This electrolyte must be watched closely since excess or deficiency can lead to arrhymias.

Two hours following the initial treatment the laboratory reports that the serum electrolytes confirm the suspected depression of sodium, chloride, and potassium. The patient has urinated a good volume at this time, so the physician continues the addition of potassium to the intravenous solutions. Four hours later the patient is much improved, is retaining small amounts of water by mouth, and has had no further diarrhea.

Fluid and electrolyte problems can be much more complicated than this example, particularly in children and the elderly with chronic disease. Intravenous fluids may be required for several days. These patients require strict recording of their fluid intake and output as well as an estimation of fluids lost through the lungs and perspiration. Daily adjustments of intravenous fluids are calculated on the basis of these figures and on serial analysis of the serum electrolytes.

---

weakness. When the level of potassium in the serum falls too low, the heart begins to beat erratically. In severe cases there is muscle twitching, lowered blood pressure, paralysis, respiratory failure, and death. Too much potassium in the serum usually brings on rapid changes in heart function and may lead to sudden cardiac arrest.

Like sodium, potassium levels are controlled primarily by the kidney. Under normal circumstances the amount of potassium excreted in the urine equals the amount ingested with food, and the potassium concentration remains in balance. An increase of potassium in the ECF causes an increase in the production of aldosterone, which causes increased filtration of potassium from the blood by the kidney. This homeostatic mechanism reduces the concentration of potassium in the ECF, which is followed by a corresponding decrease in aldosterone production and a reduction of potassium filtration. Potassium imbalance is often the result of kidney malfunction, and individuals with insufficient kidney function must be monitored carefully and their intake of potassium restricted to prevent accumulation of excessive amounts of the ion (see Chapter 25 The Clinic: Urinary System Disorders).

Calcium

Calcium (Ca^{2+}) is essential for several reasons. It is involved in the formation of bones, it is necessary for blood coagulation, and it is essential for proper nerve and muscle activity. Its principle dietary sources are beans, dark green leafy vegetables, and dairy products, although significant quantities can be obtained from other foods, such as sardines and salmon. Normally, there is about three times as much calcium in the ECF as in the ICF. Calcium in the ECF is not equally divided between blood plasma and interstitial fluid; the concentration of calcium is about twice as great in plasma as in interstitial fluid. Calcium is present in three different forms. In plasma, about half is bound to proteins, a little less than half is present as free ions, and about 5% is bound to phosphate or citrate anions. This distribution depends directly on the pH of the blood and is altered when pH is lower or higher than normal. Increased acidity causes the calcium bound to protein to be released, increasing the concentration of free calcium ions. Increased alkalinity generally has the opposite effect and lowers the concentration of calcium ions in the fluid.

hypocalcemia: (Gr.) *hypo*, under + (L.) *calx*, lime + (Gr.) *haima*, blood

Only a small portion of the calcium in a normal diet is usually absorbed; most is passed through the digestive tract and eliminated in feces. The amount absorbed often depends on the availability of vitamin D in the small intestine. This vitamin enhances the ability of carrier proteins in intestinal epithelial cells to bind and transport calcium into the tissue. Calcium deficiency diseases sometimes result from inadequate quantities of vitamin D or the carrier protein rather than from inadequate quantities of the mineral itself.

Calcium concentration is regulated primarily through a homeostatic mechanism involving parathyroid hormone (PTH) and calcitonin (see Chapter 17). The overall effect of PTH is to increase the level of calcium ions and to decrease the level of phosphate ions in blood. PTH controls calcium concentration in three ways: (1) it causes calcium ions to be released from bones where 99% of the body's calcium is located; (2) it regulates the rate at which calcium is excreted by the kidneys; and (3) it increases the efficiency of calcium absorption in the intestines. The latter effect is probably due to increased conversion of vitamin D from an inactive to an active form by the kidney and is consequently an indirect effect of the hormone.

The action of calcitonin is not as well understood as that of PTH. Calcitonin inhibits the loss of calcium from bone, so its effect is opposite that of PTH. It also acts more quickly. A change in blood calcium concentration causes a change in calcitonin levels in less than 1 hour, as opposed to several hours required for PTH levels to change. These responses provide a feedback mechanism in which hormone secretion by the thyroid and parathyroid glands is regulated by the calcium level in the blood (Figure 26-4).

Symptoms of depressed or elevated levels of calcium largely involve the nervous and muscular systems. Calcium deficiency, **hypocalcemia** (hī″pō·kăl·sē′mē·ă), is characterized by mental depression, psychotic behavior, and deteriorated mental ability. Severe deficiency can cause muscle spasms, especially of laryngeal muscles, generalized convulsion, tetany, and death. Excess calcium has similar effects on behavior and can also cause nausea, vomiting, and kidney failure. It, too, can cause death.

Magnesium

Magnesium (Mg^{2+}) is the fourth most common cation in the body. It is located in two major areas: about 50% in bones and about 50% dissolved in body fluid. Almost all dissolved magnesium ions are in the ICF, with less than 3% of the body's total in the ECF. Magnesium is necessary for the production of bones and teeth. It is necessary for conduction of nerve impulses and is also involved in muscle contraction. Magnesium plays an important role in many metabolic reactions in which it acts as a cofactor to cellular enzymes.

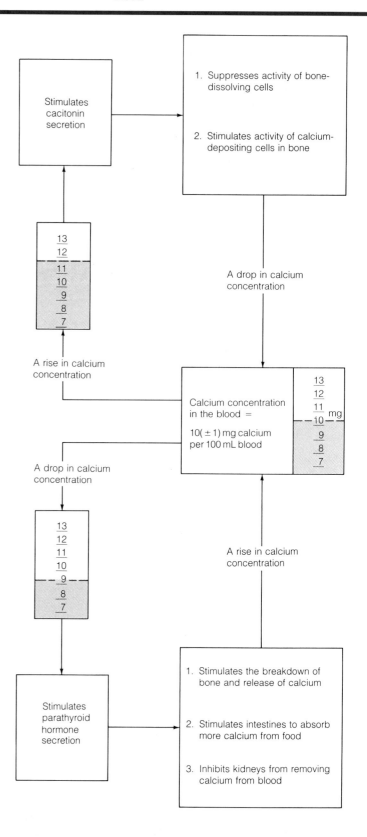

Figure 26-4 The concentration of calcium ions in the blood is under the control of parathyroid hormone from the parathyroid glands and of calcitonin from the thyroid gland.

It appears that the level of magnesium in the ECF is also regulated by aldosterone. A decrease in the level of magnesium causes increased secretion of aldosterone, which causes the kidney to reabsorb magnesium ions from the glomerular filtrate. Decreased levels of serum magnesium, a condition called **hypomagnesia,** can result in muscle disorder, lethargy and weakness, and seizures. Excessive magnesium levels, **hypermagnesia,** can cause generalized impairment of impulse transmission at neuromuscular junctions, decreased blood pressure and respiratory rate, and loss of sensation.

Other Cations

Body fluid contains at least 14 cations in addition to those just given (see Table 19-3). Each is present in low quantities, however, and collectively they are referred to as **trace elements** (see Chapter 19). Deficiency symptoms for most trace elements are unusual because amounts sufficient to meet the body's needs are very small compared to other cations and are usually present in sufficient quantities in food. Dangers due to overconsumption of trace elements are probably greater than those due to underconsumption. Toxic effects are often seen in miners exposed to dust rich in particular elements or in workers in industries that use the elements.

Major Anions in Body Fluids

The major anions of the ECF are chloride (Cl^-) and bicarbonate (HCO_3^-). The most abundant anions of the ICF are phosphate (PO_4^{3-}) and sulfate (SO_4^{2-}), in addition to many negatively charged proteins within the interior of the cells. Chloride distribution is closely related to sodium distribution. Chloride ions leave when sodium is transported from a cell because they are repelled by the negatively charged environment that remains after sodium has been removed and because they are attracted by the positive charge resulting from the excess of sodium ions outside the cell.

Significant quantities of bicarbonate ion are present in the ECF and ICF because it is derived from carbon dioxide produced by respiring cells. Upon leaving a cell, carbon dioxide enters the ECF where it reacts with water to form carbonic acid:

$$CO_2 + H_2O \rightleftharpoons H_2CO_3$$

Carbonic acid dissociates in solution to produce hydrogen ions and bicarbonate ions:

$$H_2CO_3 \rightleftharpoons H^+ + HCO_3^-$$

Phosphate is present both as free phosphate ion and in combination with calcium. It is used in energy metabolism as a component of ATP and related compounds. Dietary deficiencies of phosphate do not occur because it is present in virtually all food, unless one is existing on a very unusual diet, such as pure carbohydrate. There is generally enough phosphate in the body to withstand long periods of inadequate phosphate intake, during which time problems due to other deficiencies will occur first.

Acid-Base Balance

After studying this section, you should be able to:

1. List three ways in which pH is controlled in body fluids.
2. Define the term *pH buffer* and explain in general what pH buffers do.
3. List and describe four buffer systems that operate in body fluid.
4. Describe how the lungs and kidneys help control pH.

Metabolic reactions in the body are influenced by many factors. One important factor is the pH of the fluid in which a reaction occurs. Most enzymatic reactions occur at a pH that is near neutral, and it is critical to the body that acid-base conditions be maintained as close to neutral as possible. Under normal circumstances the pH of extracellular body fluids is close to 7.4, which is slightly basic. This control is accomplished by three interrelated mechanisms: intracellular and extracellular buffers, respiratory ventilation, and renal control.

pH Buffers

Mixtures of chemicals that maintain a relatively constant pH in spite of the addition of acids or bases to the solution are called **pH buffers.** Acids are substances that release H^+ and bases are substances that release OH^-. Strong acids and bases dissociate completely, whereas weak acids only partially dissociate (see Figure 2-10 in Chapter 2). Consequently, a strong acid or base is capable of causing greater changes in pH than the same amount of a weak acid or base. Four principal buffer mechanisms operate within the body: the carbonic acid–bicarbonate, phosphate, hemoglobin-oxyhemoglobin, and protein buffer systems (Figure 26-5).

Carbonic Acid–Bicarbonate Buffer

The **carbonic acid–bicarbonate buffer system** involves a weak acid (carbonic acid) and a weak base (sodium bicarbonate). Both chemicals are in solution in body fluids where they dissociate into ions. Carbonic acid (H_2CO_3) and sodium bicarbonate ($NaHCO_3$) will dissociate in water as follows:

$$H_2CO_3 \rightleftharpoons H^+ + HCO_3^-$$
$$NaHCO_3 \rightleftharpoons Na^+ + HCO_3^-$$

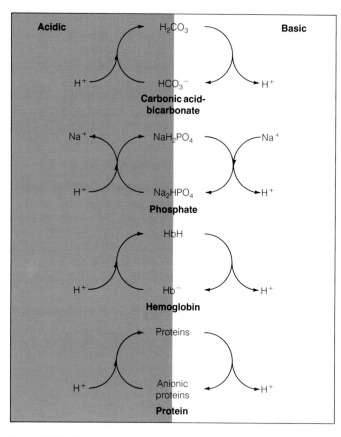

Figure 26-5 A summary of the major pH buffers in body fluids.

If a strong acid, such as hydrochloric acid (HCl), is added to the solution, excess hydrogen ions will react with the bicarbonate ions to produce carbonic acid:

$$HCl \longrightarrow H^+ + Cl^-$$
$$H^+ + Cl^- + Na^+ + HCO_3^- \rightleftharpoons H_2CO_3 + Na^+ + Cl^-$$

If a strong base, such as sodium hydroxide, is added to the solution, excess hydroxide ions combine with hydrogen ions released by carbonic acid to produce water:

$$NaOH \longrightarrow Na^+ + OH^-$$
$$H_2CO_3 + Na^+ + OH^- \rightleftharpoons H_2O + Na^+ + HCO_3^-$$

Thus, the carbonic acid–bicarbonate system stabilizes the pH of body fluids by removing excess hydrogen ions and hydroxide ions that may be produced by normal metabolic activities of the cells.

Phosphate Buffers

The **phosphate buffer system** is particularly important within cells, primarily cells of the kidney tubules and erythrocytes (red blood cells). It operates through sodium dihydrogen phosphate (NaH_2PO_4), a weak acid, and sodium monohydrogen phosphate (Na_2HPO_4), a weak base. Excess hydrogen ions are removed from the solution in the following way:

$$H^+ + Na_2HPO_4 \rightleftharpoons NaH_2PO_4 + Na^+$$

If excess hydroxide ions are introduced into the system, they are removed by combining with a hydrogen ion from NaH_2PO_4:

$$OH^- + Na^+ + NaH_2PO_4 \rightleftharpoons H_2O + Na_2HPO_4$$

Hemoglobin-Oxyhemoglobin Buffer

The **hemoglobin-oxyhemoglobin buffer system** operates primarily in the blood. As an erythrocyte passes through a capillary, some of the hemoglobin molecules contained within the cell release the oxygen molecules they are carrying. The oxygen molecules diffuse out of the erythrocyte and eventually enter the cells of the surrounding tissues. At the same time, some of the carbon dioxide produced in the tissue diffuses into the erythrocyte, where it combines with water to produce carbonic acid. Carbonic acid subsequently dissociates to produce hydrogen ions. The reactions are as follows:

$$CO_2 + H_2O \rightleftharpoons H_2CO_3$$
$$H_2CO_3 \rightleftharpoons H^+ + HCO_3^-$$

When hemoglobin (Hb) gives up its oxygen to tissues, it undergoes a change in conformation that causes certain of its amino acids to attract H^+. In combination with H^+ the deoxygenated hemoglobin molecule becomes a weak acid:

$$H^+ + Hb^- \rightleftharpoons HbH$$

As a consequence, deoxygenated hemoglobin acts as a buffer, absorbing H^+ when there is excess in the surrounding fluid and releasing that H^+ when its concentration declines.

Protein Buffer

The **protein buffer system** primarily involves plasma proteins. Under physiological conditions, proteins normally exist as anions, and the overall negative charge associated with these large molecules causes them to attract and form loose associations with hydrogen ions. This helps to lower the concentration of free hydrogen ions in body fluids, thus contributing to the control of pH. In addition, certain of the amino acids present in proteins act as weak acids or bases, releasing or absorbing hydrogen ions as the pH of the surrounding fluid changes.

acidosis: (L.) *acidum;* acid + (Gr.) *osis,* a condition
alkalosis: (Arab.) *algaliy,* ashes of salt wort plant + (Gr.) *osis,* a condition

Respiratory Ventilation

As cells respire, they release carbon dioxide into the surrounding fluids. The CO_2 combines with water to produce carbonic acid, which dissociates in a reversible reaction to produce hydrogen ions and bicarbonate ions. Addition or removal of one of the reactants controls the direction in which the reaction proceeds. Addition of carbon dioxide causes a buildup of hydrogen ions. Removal of carbon dioxide reduces the free hydrogen ion concentration.

Respiratory ventilation (breathing) is important in the control of acid-base balance because, with each breath exhaled from the lungs, CO_2 from the blood is discharged into the atmosphere (see Chapter 20). The amount of CO_2 discharged depends on how rapidly and deeply one is breathing—the more rapid and deeper the breaths, the more effective the ventilation and elimination of CO_2. Removing CO_2 from the blood shifts both reactions to the left and the H^+ concentration is reduced.

Breathing rate is controlled by the respiratory center in the medulla oblongata. This center is sensitive to H^+ in the blood carried to it and responds to elevated concentrations of H^+ by increasing the rate and depth of breathing. As a result, CO_2 is eliminated more rapidly and the blood becomes more basic. As the pH of the blood rises, the stimulus to the respiratory center is reduced and there is a corresponding decrease in the center's stimulation of breathing. As breathing slows down, CO_2 again begins to accumulate in the blood. The resulting increase in dissolved CO_2 drives the carbonic acid reaction to the right and increases the H^+ concentration again. Thus the respiratory center and lungs participate in a feedback mechanism, regulating CO_2 concentration and alternately raising and lowering the pH in response to the H^+ concentration in the blood. This homeostatic mechanism is very effective and exerts more control over acid-base balance than all the buffer systems combined.

It is important to remember that the buffer systems and the respiratory ventilation mechanism do not eliminate excess hydrogen ions, but merely maintain an appropriate concentration of the ion in body fluids by absorbing or releasing it in response to pH. While it is true that these mechanisms are indispensible in regulating the acid-base balance of the body, it is the kidney that is responsible for the eventual elimination of excess hydrogen ions.

Kidney Control

As described in Chapter 25, the kidneys provide a major mechanism for controlling the acid-base balance of body fluids. Blood is carried into the kidneys and filtered to remove waste products and excess ions, including H^+ and HCO_3^- (bicarbonate ions). In the course of this purification, the pH of the blood is adjusted through the kidney's control of the amount of H^+ that is excreted. The rate of excretion is dictated by the acidity of body fluids and other factors (see section on Homeostatic Functions of the Kidneys in Chapter 25).

Acid-Base Imbalance

After studying this section, you should be able to:

1. **Define acidosis and distinguish between metabolic and respiratory causes of acidosis.**
2. **Define alkalosis and distinguish between metabolic and respiratory causes of alkalosis.**

A number of conditions and diseases can lead to fluctuations in the pH of body fluids, as can the ingestion of excessive amounts of certain foods and drugs. It is important that such fluctuations be held within bounds because of the profound effect that deviations from the normal pH have on the body. As previously mentioned, the pH of body fluids averages about 7.4, slightly on the alkaline side. Lowered pH of body fluids is referred to as **acidosis** (ăs″ĭ·dō′sĭs) and usually involves pH values in the range of about 7.35 to 7.0. Higher than normal pH is referred to as **alkalosis** (ăl″ka·lō′sĭs) and usually involves pH values of about 7.45 to 7.7 (Figure 26-6). A pH value above 7.7 or below 7.0 is lethal.

Acidosis

Metabolic acidosis arises primarily from a decrease in the amount of bicarbonate in the blood. This might be caused, for example, by excessive diarrhea. Loss of bicarbonate reduces the body's buffering ability, and the concentration of hydrogen ions in the body fluids rises. Metabolic acidosis can also be caused by direct addition of acids to body fluids. In diabetes, for example, the inability of cells to use glucose causes the body's metabolic apparatus to switch over to using stored fat. One of the products of fat metabolism is acetoacetic acid, a substance that releases H^+, and directly lowers pH. Alcohol, shock to the vascular system, extensive and rigorous exercise, and anything else that reduces the availability of oxygen to muscle tissues all cause the rate of lactic acid production to increase. Lactic acid releases H^+ and thus contributes to acidosis.

Respiratory acidosis occurs as a result of hypoventilation, which is impaired elimination of CO_2 from the lungs (see Chapter 20 Clinical Note: Hyperventilation and Hypoventilation). When CO_2 cannot be removed properly, it accumulates in the body fluids and drives the carbonic acid–bicarbonate reaction to the right, causing an accumulation

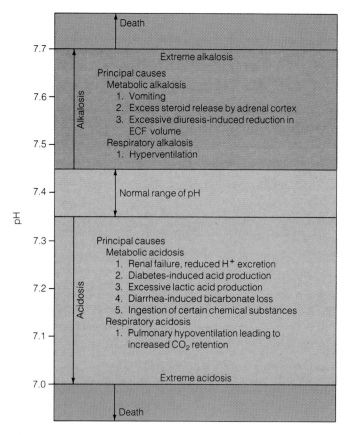

Figure 26-6 Metabolic pH. Normal pH ranges from 7.35 to 7.45. Alkalosis exists at pH values greater than 7.45 and acidosis exists at pH values less than 7.35.

of hydrogen ions and a lowering of the pH. Hypoventilation can be caused by injury to the lungs, by emphysema, and by some drugs that induce shallow, troubled breathing.

Alkalosis

Metabolic alkalosis results from excessive loss of hydrogen ions or increased presence of bicarbonate ions. Increased HCO_3^- levels can be caused simply by eating too much of it. Sodium bicarbonate is a leading antacid readily available in markets and drugstores. When excessive quantities are taken, so much so that one's kidneys are not able to excrete the extra amount rapidly enough, metabolic alkalosis can result. Normally, the kidney is very effective at eliminating excess HCO_3^-, but if this ability is impaired, pH rises.

Excessive vomiting, either caused by illness or the binge-and-purge behavior of bulimics, is another cause of metabolic alkalosis (see Chapter 19 Clinical Note: Anorexia Nevosa). The stomach walls secrete considerable amounts of H^+ into the stomach where acidic conditions are necessary for efficient enzyme action and digestion. Vomiting of the gastric contents eliminates most of this H^+, so less HCO_3^- is required to neutralize the chyme in the duodenum. As a result, HCO_3^- concentrations in body fluids rise higher than they normally would and alkalosis results.

Respiratory alkalosis results from **hyperventilation,** the rapid and deep breathing that occurs when the respiratory center is stimulated. When hyperventilation occurs, more CO_2 is eliminated than necessary, and pH rises. Mountain sickness, which involves headache, dizziness, and a general malaise that sometimes comes with excessive physical exertion at high altitudes, is one of the causes of hyperventilation, and respiratory alkalosis is one of the effects associated with it (see Chapter 20 Clinical Note: Altitude Sickness).

In Part VI we have examined the organs and mechanisms used to maintain appropriate amounts of body fluids and electrolytes and to maintain the pH of those fluids within close tolerances. The systems involved are closely interrelated and are also responsible for the elimination of wastes as they control the elimination of electrolytes and hydrogen ions. As such, these systems play an extremely important homeostatic role, and when functioning properly, ensure that conditions are right for body chemistries to occur with maximum efficiency.

In Part VII we go on to discuss organs and processes that are not so concerned with maintaining and providing for the individual's metabolic needs. Instead, we deal with those structures and processes involved in maintaining the species, that is, the production of a new person.

STUDY OUTLINE

Regulating Water Content (pp. 681–682)
Human bodies usually consist of 50 to 60% water, depending on age, sex, and amount of body fat. On an equal weight basis, fatty tissue generally contains less water than do other kinds of tissues.

Extracellular and Intracellular Fluids Body water is distributed between two components. *Extracellular fluid* (ECF) includes all the fluids outside cell membranes. *Intracellular fluid* (ICF) is contained within cells.

Water Intake Water is obtained by drinking, eating, and metabolism. Water obtained by drinking and eating is *preformed water* and is absorbed from the digestive tract. *Metabolic water* is produced primarily by electron transport reactions in cellular respiration.

Water intake is regulated by the sensation of *thirst* controlled by the *thirst center* in the hypothalamus, which responds to cellular conditions that result from osmotic water

hyperthermia: (Gr.) *hyper*, above + *therme*, heat

THE CLINIC

DISEASES AFFECTING THE BODY FLUIDS

Conditions that cause minor changes in fluid and electrolyte balance are quite common. Physicians must be alert to these changes and correct them in their patients who have serious disease. Here is a compilation of several conditions that we have studied in other chapters in which fluid and electrolyte problems are of primary concern.

Diabetes mellitus that is out of control is a common and urgent problem (see Chapter 17, The Clinic: Diabetes Mellitus). Radical changes in diet, failure to take or adjust insulin requirements, and acute infections can tip a diabetic patient into diabetic ketoacidosis, which can result in coma and death if not corrected. The lack of sufficient insulin allows the blood sugar to rise to very high levels. The kidney cannot remove such a large load of sugar and the osmotic effects of the sugar interfere with water recovery; as a result, large amounts of water are excreted with resulting severe dehydration. In addition, with decreased insulin, the body switches to metabolizing fats, which produce acetoacetic and β-hydroxybutyric acids. These cause acidosis. These people require large amounts of intravenous fluids to combat the dehydration, insulin to reverse the process, and addition of HCO_3^- and sometimes K^+ to restore the fluid and electrolyte balance.

ADH-sensitive **diabetes insipidus** is a rare condition in which the pituitary gland fails to produce sufficient antidiuretic hormone, resulting in the kidneys producing large amounts of very dilute urine (see Chapter 17 Clinical Note: Diabetes Insipidus). These patients produce enormous amounts of urine daily and need to ingest huge quantities of water to compensate for the loss.

Primary aldosteronism is a condition due to a tumor of the adrenal cortex that causes production of excessive amounts of aldosterone (see Chapter 17 Clinical Note: Aldosteronism). Aldosterone causes Na^+ retention and K^+ loss by the kidney with resultant hypertension, fluid retention, and **hypokalemia** (hī″pō·kă·lē′mē·ă), a condition of very low K^+ levels in the blood manifested by periodic weakness, transient paralysis, and tetany. Treatment of aldosteronism is by surgical removal of the tumor.

Hyperthermia, or heat sickness, is a common problem in athletes working out in conditions of high temperature and high humidity as is often experienced in late summer (see Chapter 19 Clinical Note: Heat Stroke). Excessive sweating can result in the loss of 10 to 15 lbs during a single practice session in an athlete who is not acclimated to the heat. Severe salt and water loss can result in **heat stroke** with absence of sweating, uncontrolled body temperature elevation to 106°F, brain damage, and death. Heat sickness is such a common problem that many athletic teams have strict rules about monitoring weather conditions, fluid intake, and practice schedules during hot weather.

---

loss to hypertonic extracellular fluids. Signals from the thirst center induce drinking behavior.

Water Output Most water is usually lost in urine, with additional amounts lost by evaporation from the lungs and skin. Some water is also lost in the feces. During illness, vomiting and diarrhea can cause dehydration due to excessive water loss.

Water output is regulated by the kidneys, the primary organs involved in the control of water loss. They regulate the concentration of water in urine to compensate for loss by evaporation and other causes.

Composition of Body Fluids (pp. 682–687)

Electrolytes Compounds that dissolve in aqueous fluids and produce ions are *electrolytes*. Positively charged ions are collectively referred to as *cations*, negatively charged ones as *anions*. The electrolyte composition of ECF and ICF differ in the concentrations of specific ions that are present.

Major Cations in Body Fluid The major cations of the ECF and ICF are sodium, potassium, calcium, and magnesium, with numerous cationic trace elements.

Sodium is the major cation of the EFC. It is the primary ion used to maintain body fluid volume, and it is used for nerve and muscle function. Its concentration is controlled by aldosterone activity in the kidneys.

More than 95% of the potassium in body fluid is in the ICF. It is active in nerve and muscle function. Potassium concentration is controlled by the kidney in response to aldosterone.

Primarily in the ECF, calcium is involved in blood clotting, formation of bones, muscle contraction, and nerve function. Calcium is transported bound to proteins, as free ions, or bound to phosphate or citrate. Calcium concentration is controlled by parathyroid hormone and calcitonin.

Almost all magnesium in body fluid is present in the ICF. Much of the body's magnesium is bound in solid form in bones. In cytoplasm, magnesium ion serves as a cofactor for many enzymes. Magnesium is necessary for the production of bones and teeth, impulse conduction in nerves, and muscle contraction.

At least 14 other cations are also necessary for life and collectively comprise the *trace elements*. Amounts required are generally so low that sufficient amounts are present in normal diets, making trace element deficiencies uncommon.

Major Anions in Body Fluids The major anion of the ECF is chloride. Major anions of the ICF are phosphate and sulfate. Bicarbonate ions are common in both the ICF and ECF. Chloride distribution is generally linked to sodium ion distribution.

Acid-Base Balance (pp. 687–689)

Proper fluid pH is critical to the functioning of the body. pH variation is kept to a minimum by pH buffers, respiratory ventilation, and renal control mechanisms.

pH Buffers pH buffers are mixtures of chemicals that minimize the effect of addition of hydrogen ions or hydroxide ions by absorbing or releasing hydrogen ions in response to changes in pH. The primary buffer systems are the *carbonic acid–bicarbonate, phosphate, hemoglobin-oxyhemoglobin,* and *protein buffer systems.*

Respiratory Ventilation Carbon dioxide exhaled from the lungs is produced from carbonic acid in the blood, leaving water as a product of the reaction. Loss of carbonic acid removes a source of hydrogen ions, raising the pH of the blood and, subsequently, other body fluids. The effectiveness of respiratory ventilation in removing hydrogen from solution depends on the depth and rate of breathing, both of which are controlled by the respiratory center.

Kidney Control Kidneys secrete hydrogen when blood pH is low.

Acid-Base Imbalance (pp. 689–690)

Normal body fluid pH is 7.4; pH less than that is *acidosis.* Elevated body fluid pH, greater than 7.4, is *alkalosis.*

Acidosis Metabolic acidosis results from metabolic reactions that produce acids, for example, acetoacetic acid produced from fat catabolism and lactic acid produced from muscle activity under anaerobic conditions. Respiratory acidosis results when lungs do not eliminate carbon dioxide adequately and excess carbonic acid remains.

Alkalosis Metabolic alkalosis results from excessive loss of hydrogen, resulting from ingesting bicarbonate or another base or from excessive vomiting. Respiratory alkalosis results from excessive loss of carbon dioxide from the lungs due to rapid and deep breathing.

SELF-TEST OF CHAPTER OBJECTIVES

True-False Questions

1. Of the three sources of water available to humans, the smallest volume is provided by metabolic water.
2. The thirst center is located in the hypothalamus.
3. If the extracellular fluid is hypotonic to the cells of the thirst center, one is likely to be thirsty.
4. The most common cation of the ECF is K^+.
5. The most common anion of ICF is Cl^-.
6. The single most important ion used in regulating the volume of fluid in the body is Na^+.
7. Aldosterone is important in regulating the concentration of Na^+ ions in the body.
8. Magnesium is more highly concentrated in the ECF than in the ICF.
9. Calcitonin causes the release of calcium from the bones.
10. Parathyroid hormone increases the efficiency of calcium absorption from the intestine.

Matching Questions

Match the following disorders with their possible causes:

11. metabolic alkalosis
12. metabolic acidosis
13. respiratory alkalosis
14. respiratory acidosis

   a. excessive fat metabolism
   b. lung damage leading to hypoventilation
   c. altitude sickness
   d. kidney damage leading to inadequate bicarbonate excretion

Multiple-Choice Questions

15. Which of the following is not part of the ECF?
    a. vitreous humor of the eyes
    b. acetylcholine-containing vesicles of a neuron
    c. lymphatic fluid
    d. glomerular filtrate
16. Which of the following lists sources of water in decreasing order of importance?
    a. drinking, metabolism, food
    b. drinking, food, metabolism
    c. metabolism, food, drinking
    d. food, metabolism, drinking
17. Which organ is responsible for regulating water loss to the greatest degree?
    a. lung   c. kidney
    b. skin   d. colon
18. Which of the following lists sources of water loss in increasing order of importance?
    a. feces, skin and lung evaporation, urine
    b. feces, urine, skin and lung evaporation
    c. skin and lung evaporation, feces, urine
    d. urine, feces, skin and lung evaporation
19. Metabolic water is produced principally from which of the following?
    a. glycolysis   c. electron transport
    b. Krebs cycle   d. gluconeogenesis
20. Which of the following lists compounds in the order of increasing yield of water upon chemical breakdown?
    a. carbohydrates, fats, proteins
    b. proteins, carbohydrates, fats
    c. fats, proteins, carbohydrates
    d. fats, carbohydrates, proteins
21. Kidney disease patients are monitored especially carefully for potassium levels in the plasma. This is because
    a. too low a level can cause a drop in blood pressure
    b. too low a level can cause respiratory failure
    c. too high a level can cause cardiac arrest
    d. all of the above
22. The bulk of the body's calcium is found in
    a. the ICF   c. the plasma
    b. the ECF   d. the bones

23. Which of the following anions is in the highest concentration in the ECF?
    a. bicarbonate   c. phosphate
    b. chloride   d. proteins
24. Which of the following anions is in the highest concentration in the ICF?
    a. bicarbonate   c. phosphate
    b. chloride   d. proteins
25. Ions of which of the following elements are involved to the greatest degree in regulating the volume of body fluid?
    a. sodium   c. calcium
    b. potassium d. magnesium
26. Excessive ion excretion of which of the following elements would probably have the greatest effect on bone structure?
    a. sodium   c. calcium
    b. potassium d. magnesium
27. Deficiency of which of the following would probably have the greatest effect on the ability of enzymes to catalyze reactions?
    a. sodium   c. calcium
    b. potassium d. magnesium
28. Which of the following is produced most directly by a reaction between carbon dioxide and water?
    a. bicarbonate ion   c. carbonate ion
    b. carbonic acid   d. sodium bicarbonate
29. The organ most directly involved in eliminating excess hydrogen ions is the
    a. lung   c. kidney
    b. skin   d. colon
30. Which organ is involved to the greatest degree in regulating pH?
    a. lung   c. kidney
    b. skin   d. colon
31. ECF pH is controlled to the greatest degree by
    a. hemoglobin buffering
    b. plasma protein buffering
    c. sodium phosphate buffering
    d. carbonic acid–sodium bicarbonate buffering
32. Which of the following compounds would help remove excess hydrogen ions most effectively?
    a. sodium dihydrogen phosphate   c. oxyhemoglobin
    b. carbonic acid   d. hemoglobin

Essay Questions
33. Describe the differences in content of intra- and extracellular fluid and explain how water passes from one to the other.
34. List three sources of body water, describing factors that affect the volume of each.
35. Explain why different amounts of metabolic water are produced as a result of the catabolism of carbohydrates, fats, and proteins.
36. Explain how loss of water through excessive sweating can stimulate the thirst center.
37. List three ways water is lost from the body and describe factors that affect the rate of loss for each.
38. Describe the major electrolytes of the ECF and ICF, citing functions performed by each.
39. Describe how bicarbonate ion is formed from carbon dioxide and explain how it is involved in regulating fluid pH.
40. List and compare the mechanism of action of four different buffer systems involved in regulating fluid pH.
41. Compare the functions performed by the lungs and kidneys in regulating body fluid pH.
42. List and compare the effect on body fluid pH of substances that can cause acidosis or alkalosis when ingested. Describe how the acidosis or alkalosis caused by such substances differs from changes in pH caused by hypo- or hyperventilation.

Clinical Application Questions
43. Describe why the concentration of solutes in urine (specific gravity) is an indicator of the degree of dehydration of the body.
44. Explain why dehydration can result in an elevated hematocrit.
45. List and describe four ailments that can cause imbalances in electrolyte concentration and body fluid volume.
46. Explain how intravenously administered solutions can be used to adjust electrolyte imbalances in body fluids.
47. List several causes of dehydration and how they can be prevented or controlled.

UNIT VI CASE STUDY

DIABETES AND FLUID IMBALANCE

Proper fluid and electrolyte balance is extremely important for the normal function of body systems, and imbalances can have profound effects. One of the most common and serious conditions that can cause an imbalance is diabetes mellitus, a condition that affects more than 10 million Americans and is the cause of death of over 40,000 Americans per year. The following is an account of one such case and its management.

A 17-year-old high school girl was brought to the emergency room of the local hospital unconscious. Her distraught mother told the emergency room doctor that Mary had been in good health, but that she had been acting a little strange lately in that she seemed to have developed an insatiable thirst and appetite. "I can't seem to fill her up, she eats and drinks soda all the time," her mother said. "She also seems to spend a lot of time in the bathroom. Then last night she seemed a little tired, did her homework, had a large snack of cake and ice cream, and went to bed early. When she didn't appear for breakfast this morning, I went up to her room but couldn't awaken her."

In the examining room Mary appeared to be in a deep sleep, a little pale perhaps, and her skin felt dry and cool. The nurse reported that her blood pressure was 100/70, pulse 88 and regular, and her respiration rate was 12 per minute, very slow and deep. The nurse also noted that Mary had a strange "fruity" odor to her breath. The doctor started some intravenous fluids in Mary's left arm and ordered some immediate blood tests. A thorough physical exam revealed a rather thin young girl who appeared to be in a deep sleep but aroused somewhat to painful stimuli, such as starting the IV or pinching her heel cords. Her physical and neurologic examination was otherwise normal.

Then the phone rang. It was the laboratory technician reporting that Mary's blood sugar was extremely high (625 mg/100 mL), her hemoglobin was high (16 g/100 mL), and her white blood count was only slightly elevated. Mary was in diabetic coma!

The doctor ordered insulin for Mary, 25 units intramuscularly and 10 units to be added to the IV. He then arranged for her admission to the hospital. The doctor talked with Mary's mother, telling her that Mary had diabetes and was in diabetic acidosis. He then explained that Mary would require hospitalization for several days, and that she would have to take insulin every day for the rest of her life. Her mother was relieved and related that her father had diabetes, took insulin, and lived a long life.

In the hospital Mary was given large amounts of intravenous fluids to combat the dehydration caused by her body's attempt to flush out the excessive sugar in her bloodstream. She was given a steady supply of insulin intravenously while her blood glucose levels were monitored hourly. She soon regained consciousness and began to take food and fluids by mouth. While she was recovering in the hospital, Mary and her mother attended classes to learn how to test Mary's urine and blood sugars, how to give her twice daily injections of insulin, and all about her diet.

Mary was released from the hospital on a diet suggested by the American Diabetic Association consisting of 250 g of carbohydrate, 100 g of protein, 130 g of fat, and providing 2600 kilocalories per day. Everyday she injected herself with 25 units of long-acting insulin and 15 units of short-acting insulin before breakfast and a second dose of 10 units of short-acting insulin before supper. Mary did very well on this program, but she found it difficult to stick with her diet when her friends were enjoying burgers and fries after school. She learned to reduce her insulin dose on days that she played field hockey and carried a supply of hard candy with her should she start to feel weak during the games.

A few years later, in college, she continued to be well. On one occasion, however, the morning of a big exam, she took her usual dose of insulin but did not eat breakfast. During the exam, she began to feel weak and confused, the exam paper became blurry, she began to perspire, and she lost consciousness. In the emergency room of the college hospital the doctor again started intravenous fluids, but this time her blood sugar was very low (45 mg/100 mL). Mary was in insulin shock. The doctor immediately injected 50 mL of 50% glucose solution directly through the IV tubing in Mary's arm. Her eyelids fluttered and she immediately regained consciousness.

This young diabetic can maintain a normal lifestyle through daily self-injections of insulin.

Following college, Mary married and at age 26 was pregnant with her first child. She was watched closely throughout the pregnancy, and her weight and blood sugar were controlled meticulously. Her doctor made plans to deliver the baby four weeks early in an attempt to spare the fetus the known late complications of pregnancy in a diabetic mother: excessive fetal size, fluid accumulation, and 15% chance of perinatal death (see Chapter 28).

The size and maturity of the fetus were monitored by ultrasound examinations as well as maternal urine tests for urinary estriol (estrogenic hormone) three times a week after the 34th week of gestation. A falling estriol level suggests the need for termination of the pregnancy. An amniocentesis (withdrawing fluid from the uterus) was done to check on the maturity of the fetal lungs (see Chapter 28 Clinical Note: Amniocentesis). Mary's labor was induced at the 37th week and she delivered a healthy 7 lb 6 oz boy.

Mary continued having good health for the next 20 years, except for increased susceptibility to infections. Then at age 50 she began to notice deterioration in her vision. Her eye doctor noted progressive diabetic retinopathy with many microaneurysms and hemorrhages in the retina (see Chapter 16 Clinical Note: The Eye: Window to the Body). She received several laser treatments, which stabilized her vision. At age 60 she began to notice loss of sensation in her feet, sharp pains in her legs and arms, and vague digestive problems diagnosed by her doctor as diabetic neuropathies. Her doctor also noticed that her kidney function was starting to deteriorate. Her diet was adjusted to reduce the protein and salt load on her kidneys.

Mary died of a stroke at age 72, probably related to atherosclerosis of her cerebral arteries from the abnormalities of fat metabolism that also accompany diabetes mellitus. Mary had lived a long and productive life in spite of her long struggle with diabetes.

Unit VII

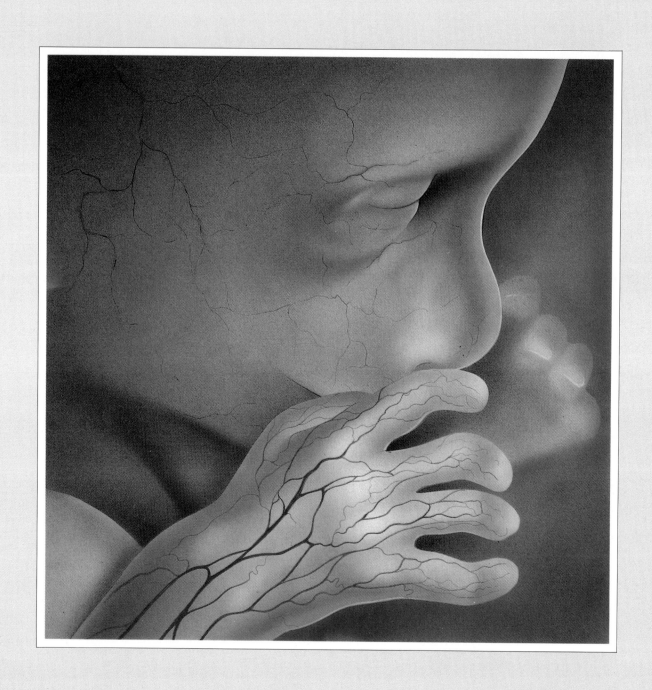

Reproduction: The Continuity of Life

Unit VII describes the anatomy and physiology of human reproduction. We begin with an examination of the reproductive systems that produce specialized sex cells. These cells combine to form a fertilized egg, which develops into an embryo that is nurtured within its mother's uterus. The characteristics of any individual are determined by the genes that person receives from his or her parents. The unit ends with a study of the patterns of development and inheritance, information that allows one to predict which characteristics might be present in a child.

27
Reproductive Organs and Their Functions

28
Development and Inheritance

27
Reproductive Organs and Their Functions

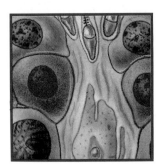

Every system we have studied so far is virtually the same in both males and females. In contrast, the reproductive system differs substantially between the two sexes, which is fundamentally due to genetic differences between males and females. Although these differences are the primary criteria by which one is identified as male or female, they are not the only ones. Chromosomal and psychologic criteria are also used, and occasionally an individual is considered a male according to one set of criteria and a female according to another.

One other difference between reproductive systems and other organ systems is the role they play in maintaining the body. Whereas muscles, nerves, blood, and digestive organs, for example, are all essential to life and the maintenance of homeostatic conditions, reproductive organs are not vital to life nor are they generally involved in maintaining homeostasis. Instead, the reproductive systems have the related functions of producing sex hormones responsible for male and female characteristics and producing sex cells used to produce offspring. Rather than being essential to the survival of an *individual*, the reproductive systems are essential to the survival of the *species*.

In this chapter we will consider

1. the anatomy of both the male and female reproductive systems,
2. the production, storage, and ejaculation of sperm,
3. the production, storage, and release of eggs,
4. the hormonal control of both processes,
5. the physiology of sexual intercourse, and
6. the mechanism of fertilization.

albuginea: (L.) *albus*, white
vaginalis: (L.) *vagina*, sheath
sustentacular: (Gr.) *sustentaculum*, support

Anatomy and Function of the Male Reproductive Tract

After studying this section, you should be able to:

1. List the organs of the male reproductive tract, citing the role each organ plays in reproduction.
2. Describe the process of sperm production and maturation.
3. List the hormones involved in sperm production, citing the source and specific function of each.
4. Describe the composition of semen and explain how it is produced.
5. Describe the mechanisms involved in erection.

The adult male reproductive tract consists of four general regions (Figure 27-1). The testes are paired organs in which sperm (male reproductive cells) develop. The *testes* are carried in a small sac, the *scrotum*. Sperm leave the testes and are stored in a network of tubules through which they pass when they are expelled from the penis. *Accessory glands* produce most of the fluid that carries the sperm and comprise the third major region. The fourth major part of the male reproductive system is the *penis*, an organ through which sperm are delivered to a female's reproductive system during sexual intercourse, or coitus.

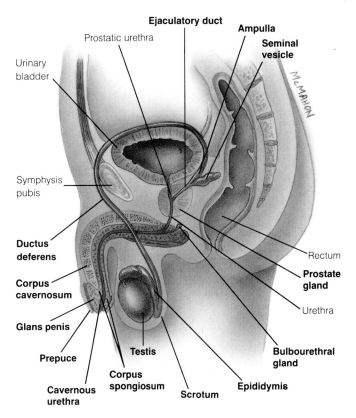

Figure 27-1 The reproductive system of an adult male.

Scrotum

The **scrotum** (skrō'tŭm) is divided into two halves exteriorly by a median ridge of tissue, the **scrotal raphe** (skrō'tăl rā'fē). Internally, the scrotum is partitioned into two lateral compartments by a septum of connective tissue. Each compartment contains one testis. The scrotal skin covers subcutaneous tissue that contains a layer of smooth muscle, the **tunica dartos** (dăr'tōs). The muscle extends into the septum that divides the scrotum and enables the wall of the scrotum to contract and thicken in response to low temperature. This provides an important mechanism for temperature control in the testes. Sperm development requires a temperature lower than that of the body, which is achieved by suspension of the testes outside the body in the scrotum. When the external temperature is too cold for optimal sperm development, the tunica dartos contracts, thickening the scrotal wall and bringing the testes closer to the body where the temperature is warmer. Another small muscle, the **cremasteric** (krē·măs'tĕr), assists in this process. It is located in the inguinal region above the testes. In warmer temperatures these muscles relax, lowering the testes and allowing the scrotal wall to become thin. Richly endowed with sweat glands, the scrotum also secretes sweat that cools the testes by evaporation.

Testes

Adult **testes** (tĕs'tēs), or **testicles**, are egg-shaped structures. Each testis is about 25 by 50 mm and is divided into 200 to 300 **lobules** (Figure 27-2). Lobular walls are extensions of a thick, fibrous covering called the **tunica albuginea** (ăl·bū·jĭn'ē·ă) that surrounds each testis. External to the tunica albuginea is a thinner membrane, the **tunica vaginalis** (văj·ĭn·ăl'ĭs).

Each lobule in the testis consists of a collection of highly coiled **seminiferous** (sĕm·ĭn·ĭf'ĕr·ŭs) **tubules.** Located between the seminiferous tubules are **Leydig** (lī'dĭg) **cells,** also called **interstitial endocrinocytes** or **interstitial cells of Leydig** (Figure 27-3a). These cells secrete hormones important to sperm production (see section on Hormonal Control of Sperm Production). Microscopic examination of the inner wall of a seminiferous tubule reveals sperm in various stages of development. The periphery of each tubule is marked by a basement membrane. Sperm are produced by the division of cells that lie just inside this membrane, with progressively more mature sperm lying nearer the lumen. Also embedded in the wall of a seminiferous tubule are **Sertoli's** (sĕr·tō'lēz) **cells,** also known as **sustentacular**

Embryonic Development of the Reproductive System

Embryos of both sexes develop identically for the first eight weeks, a period referred to as the **indifferent stage.** Paired gonads differentiate from intermediate mesoderm during this period and develop near the embryonic kidneys and mesonephric ducts (Figure 27-A). An additional pair of ducts, the **paramesonephric** (păr″ă·měs″ō·něf′rĭk) **ducts,** also appear by the sixth week of development.

After the eighth week, development proceeds along different lines in the two sexes. In males, seminiferous tubules begin to develop in the interior of each gonad, initiating its differentiation into a testis. These tubules soon connect with the adjacent mesonephric duct, which undergoes further development and produces the efferent ducts, epididymis, ductus deferens, ejaculatory duct, and seminal vesicles. The prostate and bulbourethral glands develop as outgrowths from the urethra. Simultaneous with these developments, the paramesonephric duct degenerates and does not contribute further to the male reproductive system.

In females, follicles begin to develop in the exterior of each gonad, which now differentiates into an ovary. The distal ends of the paramesonephric ducts fuse, forming the uterus and vagina, while the proximal ends develop into the uterine (fallopian) tubes. The mesonephric duct degenerates without further contributions to the female reproductive system. Bartholin's glands and the vestibular glands develop from the vestibule, a remnant of the urogenital sinus.

It is not until the ninth week of development that external differences between males and females are evident (Figure

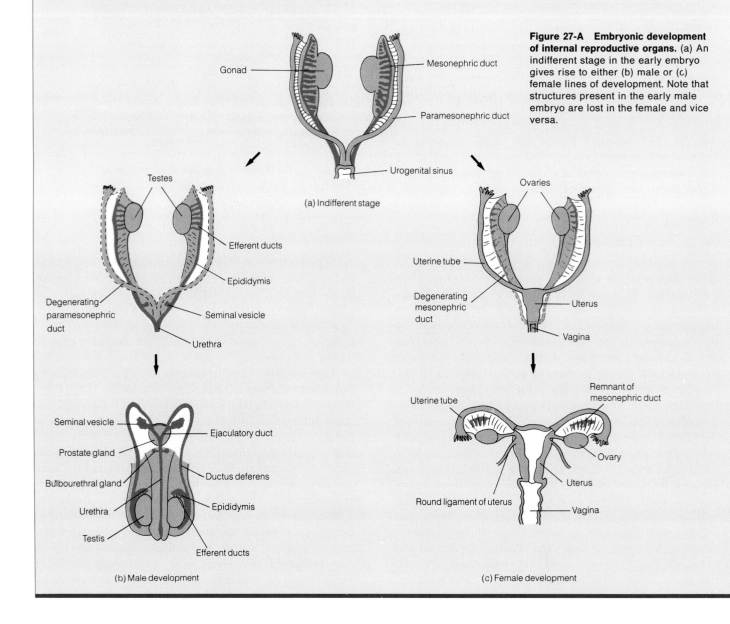

Figure 27-A Embryonic development of internal reproductive organs. (a) An indifferent stage in the early embryo gives rise to either (b) male or (c) female lines of development. Note that structures present in the early male embryo are lost in the female and vice versa.

27-B). In the undifferentiated embryo, a **genital tubercle**, consisting of paired **urethral folds** and **labioscrotal** (lā″bē·ō·skrō′tăl) **swellings**, protrudes between the tail and umbilical cord. The **urethral groove** between the folds is the opening into the urogenital sinus.

In males, the genital tubercle elongates to form the penis. The labioscrotal swellings enlarge into the scrotum, into which the testes later descend. The urethral folds fuse, enclosing the urethra. The line of fusion produces the scrotal raphe on the scrotum, and the urethral raphe on the penis.

In females, the external genitalia are produced as the labioscrotal swellings enlarge to produce the labia majora and the urethral folds develop into the labia minora. The genital tubercle develops into the clitoris, while the vestibule is produced by the urethral groove. The establishment of a distinctively male or female reproductive tract is completed by about the end of the first three months of gestation. During the remaining six months of fetal growth, the external genitalia enlarge and the internal portions of the tract continue to develop. Guided by a fibromuscular cord, the **gubernaculum** (gū″bĕr·năk′ū·lŭm), the testes descend into the scrotum during the eighth fetal month. Occasionally one or both of the testes fails to descend, remaining in the abdominal cavity or lodged in the inguinal canal; this condition is called cryptorchidism (see The Clinic: Disorders of the Reproductive System).

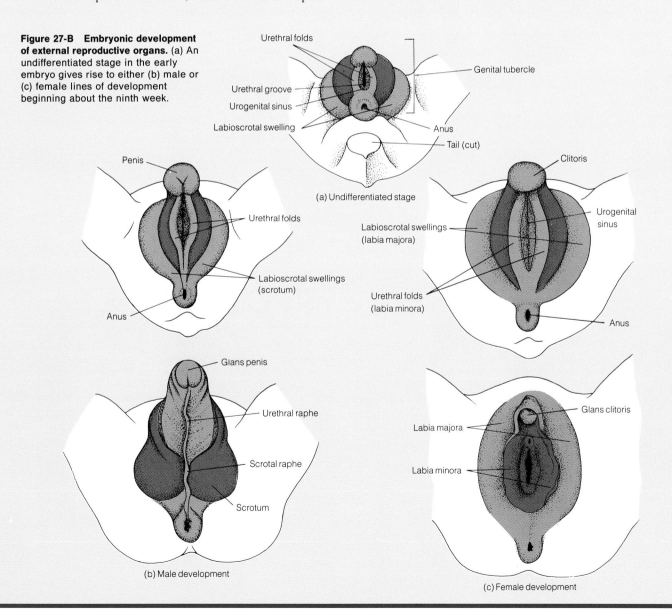

Figure 27-B Embryonic development of external reproductive organs. (a) An undifferentiated stage in the early embryo gives rise to either (b) male or (c) female lines of development beginning about the ninth week.

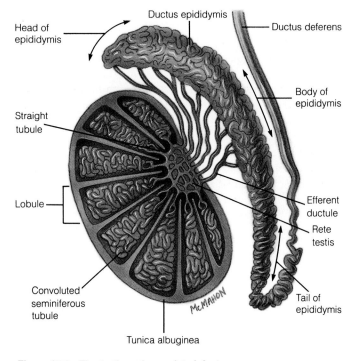

Figure 27-2 The testis and associated ducts.

(sŭs″tĕn·tăk´ū·lŭr) or **nurse cells,** that provide chemical assistance to developing sperm.

Seminiferous tubules connect and fuse at one side of each testis to form another network of tubules, the **rete testis** (rē´tē těs´tĭs). From the rete testis emerges a smaller collection of larger diameter tubules, the **efferent ductules.** These empty into a single tubular structure called the **epididymis** (ĕp″ĭ·dĭd´ĭ·mĭs), located outside of the testis but within the scrotum.

Epididymis

The epididymis is formed by the fusion of efferent testicular ductules into the **ductus epididymis,** a tightly coiled tube. An epididymis lies alongside each testis, curved in shape to conform to the testis (Figure 27-4). The upper end is the head, the middle section the body, and the lower end the tail. Although each epididymis is only about 4 cm long externally, the ductus epididymis included in it is nearly 6 m long. The duct is only 1 mm in diameter and very tightly coiled. The wall contains pseudostratified columnar epithelial cells. Long, branching microvilli extend into the lumen of the tube. The epididymis is where sperm undergo final maturation and where mature sperm are stored.

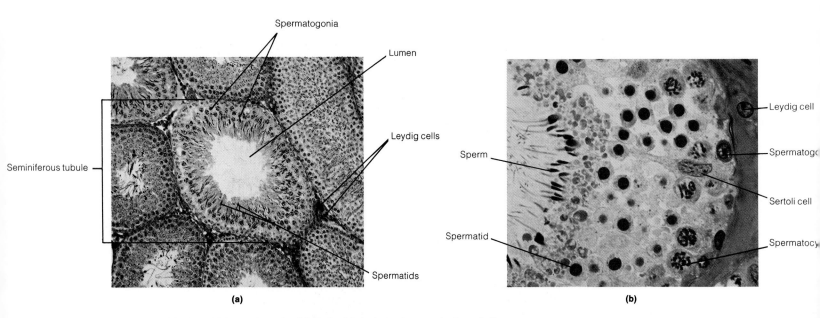

Figure 27-3 Histology of the testis and epididymis. (a) A photomicrograph of seminiferous tubules (mag. 524×). (b) A higher magnification photomicrograph of a seminiferous tubule showing a variety of stages of sperm development (see Figure 27-5) (mag. 655×)

ductus deferens: (L.) *ducere*, to lead + *deferens*, carrying away
inguinal: (L.) *inguinalis*, pertaining to groin

Ductus Deferens and Ejaculatory Duct

The tail of each epididymis leads to the **ductus deferens** (dŭk'tŭs dĕf'ĕr·ĕns) (or **vas deferens**), a tube about 45 cm long. The ductus deferens from each testis passes out of the scrotum through the **inguinal** (ĭng'gwĭ·năl) **canal,** a narrow opening in the abdominal wall, and then enters the pelvic cavity. Each duct is accompanied by testicular arteries, veins, nerves, and lymphatic vessels. The entire assemblage, the **spermatic cord,** is surrounded by the cremasteric muscle and connective tissue sheaths. Once inside the pelvic cavity, the ductus deferens loops over the ureter and passes behind the urinary bladder (Figure 27-4). The terminal portion is enlarged in diameter, forming the **ampulla.** The lining of the ductus deferens is similar to that of the ductus epididymis, except that the pseudostratified epithelium is surrounded by three layers of muscle. Peristaltic contractions of these muscles move sperm through the tube and into the urethra (see Clinical Note: Sterilization).

The ductus deferens receives fluids from a seminal vesicle, an accessory gland. The union of the ductus deferens and a duct from the seminal vesicle produces the **ejaculatory duct.** This short duct, only about 2 cm long, conducts sperm into the urethra.

Urethra

The urethra is a single tube extending from the urinary bladder to the tip of the penis (Figures 27-1 and 27-4). In males the urethra conducts urine from the bladder during urination and also provides a path for sperm during sexual activity (see Chapter 25). The initial portion of the urethra is the **prostatic urethra,** a section about 3 cm long that passes through the prostate gland. Prostatic secretions enter the urethra in this section. At the base of the prostate gland, the **membranous urethra** passes through the urogenital membrane. This portion is less than 2 cm long and leads directly to the **cavernous** (or **penile**) **urethra,** which passes through the corpus spongiosum of the penis. The penile urethra, approximately 16 cm long, terminates at the **urethral orifice.**

Production and Ejaculation of Sperm

Sperm are produced in the wall of seminiferous tubules by a process called **spermatogenesis** (sper″măt·ō·jĕn′ĕ·sĭs) (Figure 27-5). These tubules are surrounded by a thin basement membrane. Attached to this membrane is the **seminiferous** (or **germinal**) **epithelium,** populations of cells in the

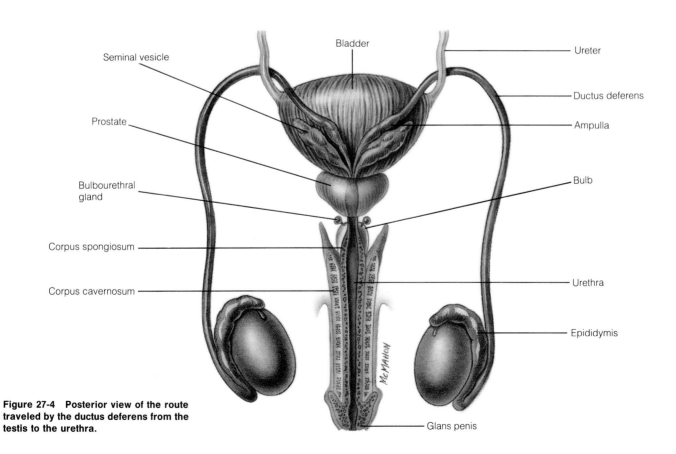

Figure 27-4 Posterior view of the route traveled by the ductus deferens from the testis to the urethra.

vasectomy: (L.) *vas*, vessel + (Gr.) *ektome*, excision
laparoscopy: (Gr.) *lapara*, loin + *skopein*, to examine
spermatogonium: (Gr.) *spermatos*, seed + *gone*, generation
spermatocyte: (Gr.) *spermatos*, seed + *kytos*, cell

CLINICAL NOTE

STERILIZATION

Sterilization can be effective in preventing pregnancy and can be used by both men and women. In men, a small incision is made on either side of the scrotum. The ductus (or vas) deferens is then cut and tied off, which blocks the passage of spermatazoa out of the testes. The procedure, called a **vasectomy** (văs·ĕk′tō·mē), is done in the physician's office (Figure 1a). It is 100% effective and is virtually irreversible. Vasectomy has no other physiological effect on the man; it does not affect sexual response or virility. Sperm continue to be manufactured, but they are unable to get past the tied off ductus deferens and are absorbed. Men continue to have a normal ejaculation, but the seminal fluid contains no sperm.

Women can be sterilized in many ways, but the simplest for birth control purposes is through the use of **laparoscopy** (lăp·ăr·ŏs′kō·pē). The woman receives a general anesthetic in the operating room, and a small incision is made beneath the umbilicus. An instrument with a light called a **laparoscope** is inserted through this incision, and carbon dioxide is pumped in so that the abdominal wall is lifted. The surgeon locates the fallopian tube on each side and cuts and ties off each tube, a procedure known as a **tubal ligation** (Figure 1b). A more recent and simpler technique involves passing an electric current through the laparoscope for about 3 to 5 sec, thereby coagulating the tubal tissue and sealing the tube. Either method prevents ova from passing down the tube and sperm from moving up it. New techniques are being developed that may allow this sterilization procedure to be reversed. Note that in contrast to a vasectomy, which is an office procedure, this surgery does require general anesthesia and must be done in the hospital (usually an overnight stay is not necessary). Also, there is some discomfort for some days as the carbon dioxide is removed.

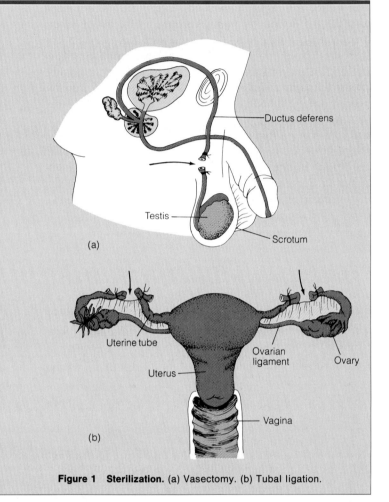

Figure 1 Sterilization. (a) Vasectomy. (b) Tubal ligation.

---

process of dividing and differentiating into sperm, and cells that support and regulate sperm differentiation. Sperm production requires continuous reproduction of **spermatogonia** (sper″măt·ō·gō′nē·ă), cells located within the seminiferous epithelium. As spermatogonia proliferate, some are triggered to differentiate into **primary spermatocytes** (sper·măt′ō·sīts). Primary spermatocytes undergo meiosis, a two-step division sequence that produces four cell products (see Chapter 28). The first division of this sequence produces two **secondary spermatocytes,** each of which divides again to produce two **spermatids** (sper′mă·tĭds).

In the final stage of spermatogenesis, **spermiogenesis,** the spermatids undergo a dramatic change in shape and size to become **sperm** (or **spermatozoa**). During this phase of the process, a cap develops on the nucleus and spreads over 50 to 70% of the nuclear surface. This structure is an **acrosome** (ăk′rō·sōm). About the same time, a long, thin flagellum extends from the cytoplasm on the side opposite the acrosome, cytoplasm is lost, chromatin condenses, and the transition from spermatid to sperm is complete (Figure 27-6). The mature sperm is essentially an acrosome-covered nucleus with a flagellum, at the base of which is an ATP-driven mechanism for operating the flagellum.

During spermiogenesis, spermatids are surrounded by cytoplasmic extensions of neighboring Sertoli's cells, which regulate the differentiation of spermatids into sperm. Sertoli's cells are sometimes referred to as nurse cells in reference to the support and chemical assistance they give to differentiating spermatids. Mature sperm are released into the central cavity of the seminiferous tubules and carried by peristaltic contractions and by a flow of fluid released into the tubules from Sertoli's cells.

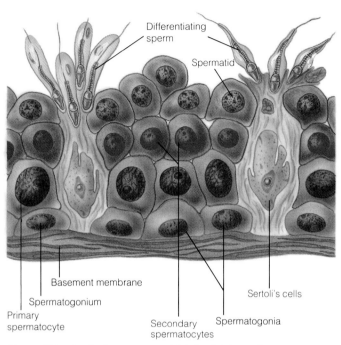

Figure 27-5 Production of sperm. Meiotic divisions of spermatogonia that line the walls of the seminiferous tubules produce spermatids. Under the influence of neighboring Sertoli's cells, these differentiate into sperm.

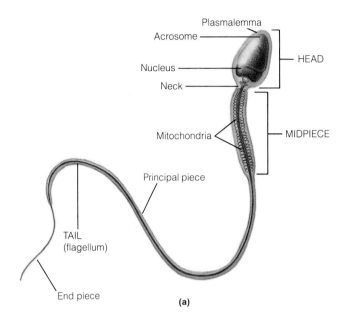

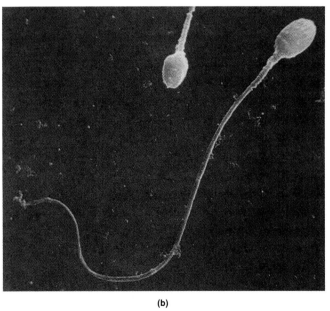

Figure 27-6 Sperm. (a) Drawing of a mature sperm. (b) Scanning electron micrograph of sperm (mag. 3220×).

After sperm pass from seminiferous tubules, they are carried into the epididymis and the ductus deferens where they are stored in a nonmotile state for about three weeks. While stored, the sperm undergo additional biochemical and structural changes that make them motile.

Sperm are discharged from the penis during sexual activity by **ejaculation** (ē·jăk″ū·lā′shŭn). Sperm not ejaculated degenerate and are absorbed in the epididymis.

Hormonal Control of Sperm Production

Sperm production is controlled by several hormones. Gonadotropin-releasing hormone (GRH) produced in the hypothalamus is carried by the hypophyseal portal veins to the anterior pituitary lobe (see Chapter 17). The anterior lobe is stimulated to release gonadotropins by GRH, which are carried to the testes where they stimulate sperm production.

At least two gonadotropins are involved in stimulating sperm production. Follicle-stimulating hormone (FSH), named for its role in the female reproductive process, stimulates the differentiation of epithelial cells in seminiferous tubules into primary spermatocytes. Luteinizing hormone (LH), also named for its role in females, stimulates interstitial cells that lie between neighboring seminiferous tubules (Leydig cells) to produce testosterone.

Testosterone

Testosterone (tĕs·tŏs′tĕr·ōn) is the primary male sex hormone responsible for widespread effects on development, including the characteristic male body form and musculature, development of body hair, deepening of voice quality, and other features generally regarded as secondary male sexual characteristics. Testosterone also stimulates differentiation of spermatids into mature sperm.

bulbourethral: (L.) *bulbus*, bulbous root + (Gr.) *ourethra*, urethra

The amount of testosterone produced by interstitial cells depends on how much LH is secreted. LH and FSH levels are affected, in turn, by the amount of testosterone present in the blood, elevated levels inhibiting GRH production by the hypothalamus. Thus, elevated levels of testosterone indirectly cause a decrease in the amount of FSH and LH released. As the concentration of FSH and LH falls, stimulation of interstitial cells to produce testosterone declines. With lowered testosterone production, inhibition of the hypothalamus is relaxed and GRH production again begins to rise, completing the cycle.

Inhibin

In addition to the feedback mechanism involving testosterone just described, a control mechanism may exist involving a substance produced by Sertoli's cells. Increased blood levels of FSH stimulate Sertoli's cells to produce **inhibin,** a polypeptide that appears to inhibit FSH production. Decreased FSH then results in decreased inhibin production by Sertoli's cells. Interplay between these two hormones is thought to help regulate the rate of sperm production.

Accessory Glands

In addition to fluid produced in the testes, fluid is added to the mixture of sperm by cells of sacs in the ampulla wall of the ductus deferens. Once sperm have passed through the ductus deferens, they are carried in fluid produced by the **accessory glands,** which include the seminal vesicles, prostate gland, and bulbourethral glands.

The paired **seminal vesicles** (sĕm′ĭ·năl vĕs′ĭ·kăls) are small pouches connected by short ducts to the junction of the ampulla and the ejaculatory duct. Each seminal vesicle, which is from 5 to 10 cm long, consists of a tightly twisted and convoluted tubule (Figure 27-7). Each tubule is lined with secretory cells that release an alkaline fluid rich in fructose, a sugar used by the sperm for energy. The alkaline nature of the secretions helps neutralize the acidity of urine remaining in the urethra. The mixture of sperm and secretions is **semen** (sē′mĕn), or **seminal fluid.** Seminal vesicles contribute slightly more than half the seminal fluid volume.

The **prostate** (prŏs′tāt) **gland** lies below the bladder, where it surrounds the urethra and the two ejaculatory ducts. Shaped like a cored apple, the prostate gland is divided into several branched compartments, each of which opens through a duct directly into the urethra. The compartments are separated from one another by walls of smooth muscle and connective tissue. The entire organ is surrounded by a jacket of connective tissue.

The prostate gland produces 1 to 2 mL of fluid per day, which collects and is released into the urethra just prior to ejaculation. Prostatic fluid adds about 20 to 30% to the volume of semen and provides several enzymes that aid in activating the swimming movement of sperm, thus helping them to reach and fertilize an egg. Prostatic fluid is alkaline and helps maintain the proper pH for sperm.

The paired **bulbourethral** (bŭl″bō·ū·rē′thrăl) **glands** (or **Cowper's glands**) are located below the prostate gland (Figure 27-7). They are about pea-size and empty into the cavernous urethra through ducts located close to the base of the penis. The release of bulbourethral secretions is triggered by erotic activity and precedes ejaculation. These secretions provide lubricants for sexual intercourse and nutrients that sperm require for motility, and they also help neutralize the acidity of any urine remaining in the urethra.

Numerous bulges occur in the walls of the cavernous urethra, some of which are branched chambers connected to the urethra by short ducts. These structures, called **urethral glands,** produce mucus that serves as a lubricant to semen.

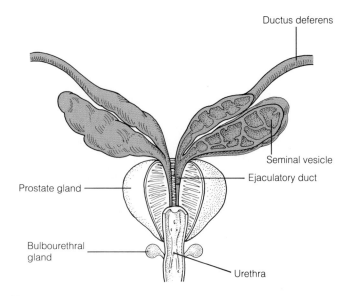

Figure 27-7 The seminal vesicles and prostate gland.

Composition of Semen

Semen is a complex mixture of sperm, water, dissolved nutrients, enzymes, and proteins. The average ejaculate contains about 3 mL of fluid, with sperm contributing less than 5% of the volume. The concentration of sperm in semen varies from 50 to 150 million per mL. Semen also contains citric acid (0.1 to 1%), fructose (0.1 to 0.5%), ascorbic acid (0.01%), and lesser amounts of enzymes and other nutrients. Included among the enzymes is **hyaluronidase** (hī″ă·lūr·ŏn′ĭ·dās), an enzyme that digests mucus in the cervix of the female

corpora cavernosa: (L.) *corpus*, body + *caverna*, hollow
corpus spongiosum: (L.) *corpus*, body + (Gr.) *sphongos*, sponge
pudendum: (L.) *pudenda*, external genitals

reproductive tract and makes it possible for sperm to pass through. Hyaluronidase also helps remove protective cells that surround an ovum, making the ovum accessible to sperm for fertilization.

Penis

The **penis** consists of three cylindrical bodies of spony tissue surrounded by sheaths of connective tissue and skin (Figure 27-8). Two of the spongy bodies comprise the **corpora cavernosa** (kōr′pōr·ă kăv·ĕr·nō′să). These paired structures lie parallel to one another in the superior portion of the penis. The third body, the **corpus spongiosum** (spŭn·jē·ō′sŭm), lies inferior to the corpora cavernosum, extending past the distal ends of the corpus cavernosum into an enlarged tip, the **glans penis**. The urethra enters the corpus spongiosum at its base and terminates at the glans penis in a slitlike opening, the **urethral orifice**. The glans penis is covered by a loose extension of the skin of the penis called the **prepuce** (prē′pūs), or **foreskin**. The prepuce is often surgically removed in infancy by **circumcision** (sŭr″kŭm·sĭ′shŭn).

The two spongy cylinders of the corpora cavernosa are separated from one another for most of their length by a tough, inelastic sheath called the tunica albuginea, which also surrounds the testes. The corpus spongiosum is also surrounded by a sheath, but one that is more elastic. The three bodies in turn are surrounded by the **fascia penis**, a sheath that lies just below the skin.

Blood is supplied to the penis by the **internal pudendal** (pū·dĕn′dăl) **arteries**, which branch from the internal iliac arteries. Each pudendal artery divides further to lead to capillaries and **helicine** (hĕl′ĭ·sĭn) **arteries**. Helicine arteries line the walls of chambers in the spongy bodies and empty directly into the chambers. These arteries derive their name from their highly coiled state in a flaccid penis. When the penis is flaccid, much of the blood bypasses the helicine arteries, passing instead through an arteriovenous shunt directly to the venous system (Figure 27-9).

Walls of the helicine arteries, as well as the arteries from which they branch, possess elongated, longitudinally oriented smooth muscular bundles. In a flaccid penis these bundles are contracted, bulging into the interior of the artery and impeding the flow of blood through the arteries.

In response to psychologic, tactile, or other stimuli, parasympathetic impulses cause the muscles in the arterial walls to relax. This in turn removes the barriers to blood flow and allows blood to flow into the spony tissue. At the same time, the arteriovenous shunt, which lies outside the erectile bodies, narrows. This impedes the flow of blood through the shunt and causes more blood to flow into the spony tissue. Accumulation of blood in the spongy tissue causes the spongy bodies to swell. Their enlargement exerts pressure on the efferent veins that drain the tissues, impeding

Figure 27-8 Anatomy of a penis. This cutaway view shows the relationship of the spongy bodies, major arteries, urethra, and surrounding tissues.

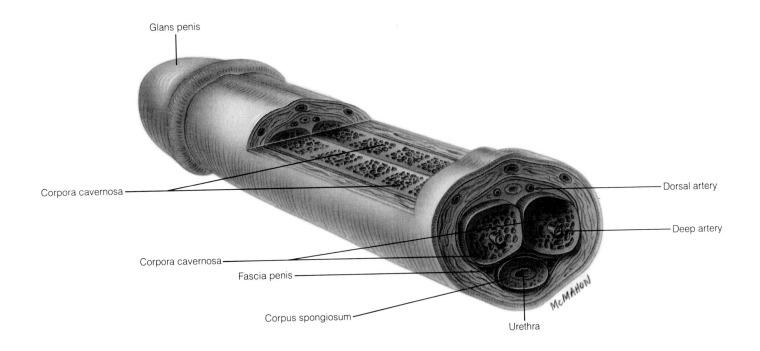

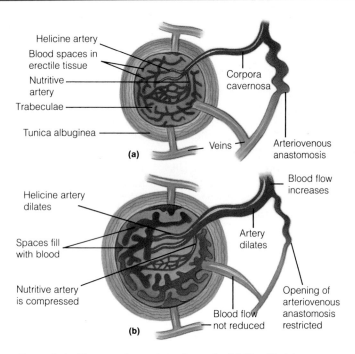

Figure 27-9 The vascular system of a penis. (a) Flaccid state. (b) During sexual excitement, blood is shunted into the spongy tissues from which its exit is blocked. The pressure that builds up as a result of the pumping of blood into the spongy tissues causes the penis to enlarge during an erection.

blood flow through them and causing still more blood to accumulate in the spony tissues.

As a result of these coordinated events, the corpora cavernosa and corpus spongiosum enlarge. The tough, inelastic tunica albuginea around them restricts their enlargement, however, and pressure develops in the bodies as more and more blood enters. The somewhat more elastic tunic that surrounds the corpus spongiosum allows that structure to enlarge in such a way that the urethra is also enlarged, rather than being forced closed by the pressure. As a result, the penis is transformed from a flaccid organ ranging in length from about 7 to 12 cm to a firm, erect organ about 14 to 18 cm long.

If tactile stimulation continues, as during sexual intercourse, sympathetic impulses increase in intensity until a series of intense nervous, glandular, and muscular reactions occur. In response to the nervous impulses, the muscles in the walls of the epididymis, ductus deferens, and seminal vesicles contract in a series of peristaltic contractions. At the same time, the urethral sphincter closes, preventing the backflow of sperm into the bladder and passage of urine into the urethra. Driven by contractions of the muscles surrounding the tubules, semen is forced into the urethra. This phenomenon is called **emission**.

As semen enters the prostatic region of the urethra, additional nervous impulses are sent to the **bulbocavernosus** (bŭl″bō·kăv″ĕr·nō·sŭs) and **ischiocavernosus** (ĭs″kē·ō·kă″vĕr·nō′sŭs) muscles at the base of the penis, causing them to contract forcefully and spasmodically. This forces the semen through the remaining length of the urethra, and it leaves the urethral orifice in a series of spurts. As semen is ejaculated, there is a dramatic increase in breathing rate, heart rate, and blood pressure, as well as widespread muscle contractions, all of which are accompanied by pleasurable sensations. This complex pattern of reactions, which is distinct from but triggered by ejaculation, is referred to as an **orgasm**.

Following orgasm, erection subsides over a period of a few to several minutes as the helicine arteries contract and blood leaves the spongy tissue. The penis then becomes flaccid and insensitive to the same stimuli that induced erection a short time before. The length of time before erection can occur again, and consequently the frequency with which a male can achieve orgasm, depends largely on health, psychologic state, and physical condition.

Anatomy and Function of the Female Reproductive Tract

After studying this section, you should be able to:

1. List the organs of the female reproductive tract and cite the function of each organ.
2. Describe the anatomy of an ovary and the process of oocyte production and release.
3. Describe the path followed by an oocyte as it passes through a uterine tube.
4. Draw and label a uterus, illustrating the structure of the uterine wall.
5. Describe the function of mucus-secreting glands in the female reproductive tract.
6. Describe the structure of a mammary gland and hormonal control of milk production.

An adult female reproductive tract consists of five general regions (Figure 27-10): two *ovaries* in which eggs, or ova, are produced; two tubular *oviducts* through which eggs pass after release; a saclike *uterus* in which an embryo develops; a vagina, which leads from the uterus to the exterior; and a *vulva*, a collective term for the external genitalia. Mammary glands used to provide milk for a newborn infant are also considered part of the female reproductive system.

mesovarium:	(Gr.) *mesos*, middle + (L.) *ovarium*, ovary	fimbriae:	(L.) *fimbria*, a fringe
oocyte:	(Gr.) *oon*, egg + *kytos*, cell	infundibulum:	(L.) *infundibulum*, a funnel
atresia:	(Gr.) *a*, not + *tresis*, a perforation		

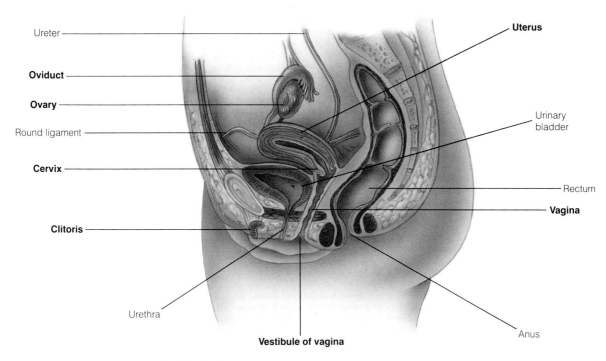

Figure 27-10 Anatomy of the female reproductive tract.

Ovaries

Ovaries are almond-shaped, paired organs about 2 to 5 cm long located on either side of the lower portion of the abdominal cavity, where they are held in place by the **mesovarium** (měs″ō·vā′rē·ŭm). This ligament in turn connects to and supports other organs of the female reproductive tract. Medially, ovaries are connected to the uterus by the **ovarian ligament;** laterally, they are attached to the abdominal wall by the **suspensory ligament.**

Histologically, each ovary consists of an inner **medulla**, a **cortex** surrounding the medulla, and a single layer of cells, a modified **mesothelium,** that covers the cortex (Figure 27-11). Nerves, lymph, and blood vessels lie within the medulla and extend into the cortical region. The cortex and mesothelium are surrounded by a layer of connective tissue, the **tunica albuginea.** Embedded within the cortex are 300,000 to 400,000 undeveloped but potential egg cells. In the course of a woman's reproductive years, only about 400 to 500 of these cells will complete development and be released from the ovary. A mature egg cell, an **ovum,** is a female gamete capable of fertilization.

Several other structures are also visible in the ovary. Potential egg cells in the cortex of the ovary of a young woman are surrounded by a single layer of cuboidal or columnar cells enclosed within a basement membrane. The cell at this stage is a **primary oocyte** (ō′ō·sīt). The oocyte and the cells surrounding it form an **ovarian follicle.** Numerous follicles in various stages of development are present, as well as some that failed to develop completely and have degenerated. Follicles that cease developing and subsequently degenerate are said to undergo **atresia** (ă·trē′zē·ă), and their remains are referred to as **atretic follicles.**

In addition to producing eggs, the ovaries are important sources of **estrogens,** steroid hormones responsible for developing the body form and other features characteristic of a female. Although small quantities of estrogens are produced by the adrenal cortex, most are produced by developing follicles and associated interstitial cells.

Oviducts

The **oviducts,** or **uterine tubes,** are about 10 to 12 cm long and extend from the superior surface of the uterus (see Figure 27-13). The end of the oviduct distal to the uterus and adjacent to the ovary consists of a mass of highly convoluted fingerlike projections called **fimbriae** (fĭm′brē·ē). Collectively, fimbria form a funnel-shaped opening to the tube, the **infundibulum** (ĭn″fŭn·dĭb′ū·lŭm). In cross section, the oviduct consists of a maze of narrow passageways formed by the fimbriae and convolutions of the wall of the infundibulum (Figure 27-12).

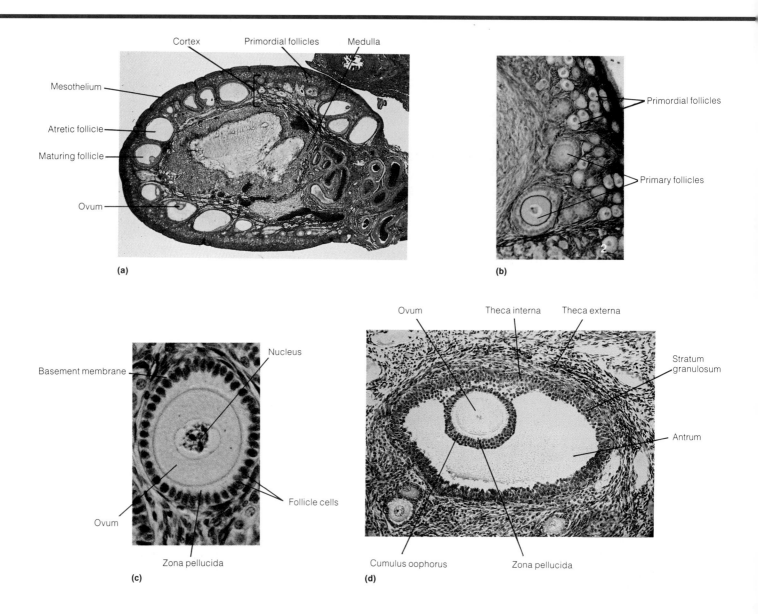

Figure 27-11 The ovary. (a) Photomicrograph showing a longitudinal section through an ovary. (b) A higher magnification of the cortex showing primordial and primary follicles (mag. 118×). (c) Photomicrograph of a primary follicle (mag. 413×). (d) Photomicrograph of a graafian follicle (mag. 472×).

The internal surface of an oviduct consists of two cell types: ciliated simple columnar cells and secretory simple columnar cells. Movements of the cilia create a current that helps carry an ovum from an ovary into the infundibulum; the secretory cells provide nutrients for the ovum as it passes through the oviduct. External to the columnar cells is a layer of involuntary muscle, the **muscularis.** Peristaltic contractions of the muscularies also help move the ovum to the uterus. The outermost layer of the oviduct is the **serosa** (Figure 27-12).

The infundibulum leads directly into the **ampulla,** an enlarged portion occupying about one-third to one-half the length of the oviduct. The interior wall of the ampulla is highly convoluted. It is in this region that fertilization normally occurs.

The remainder of the oviduct includes the **isthmus,** a portion that leads from the ampulla to the uterus, and an

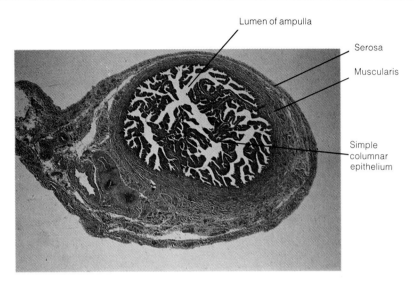

Figure 27-12 Photomicrograph of a transverse section through an oviduct.

interstitial segment that penetrates the uterine wall. There is a progressive decrease in the degree of infolding of the wall and the percentage of ciliated cells from the infundibulum to the interstitial segment. At the end of the uterine tube closest to the uterus, the infoldings are reduced to several longitudinal ridges.

Uterus

The **uterus** is a hollow, pear-shaped, thick-walled sac that rests on the floor of the abdominopelvic cavity, located between the urinary bladder and the rectum (Figures 27-10 and 27-13). It is held loosely in place by **broad ligaments** on either side, **uterosacral ligaments** that connect it to the sacrum, **lateral cervical ligaments** that connect the cervix to the pelvic diaphragm, and **round ligaments** that are anchored in the tissues beneath the labia majora. Some of the ligaments also carry uterine blood vessels and nerves. The bladder, rectum, and other adjacent organs also help position the uterus, and because of its flexibility, the uterus can assume a number of different forms and orientations as the bladder and rectum fill and empty.

The normal uterus in a nonpregnant woman is about 6 cm high, 4.5 cm wide, and 2.5 cm deep. The hollow interior of the uterus, the **uterine cavity**, connects to the outside through **cervical** and **vaginal canals.** Regions within the uterus include the **fundus,** the domelike cap on the body (or **corpus**) of the uterus; the **isthmus,** a region below the corpus where the uterus narrows; and the **cervix** (sĕr´vĭks), a narrow, necklike extension that protrudes into the vagina. The junction of the uterine cavity and the cervical canal forms a narrow opening, the **internal os** (ŏs) (see Figure 27-16). The opposite end of the cervical canal, where it opens into the vagina, is the **external os.** The uterus is usually bent in **anteflexion,** a position in which the body of the uterus projects anteriorly over the urinary bladder, while the cervix projects posteriorly, entering the upper end of the vagina at nearly a right angle.

The wall of the body of the uterus consists of an inner **endometrium** (ĕn·dō·mē´trē·ŭm), a complex layer of epithelial cells, glands, and blood vessels; a middle layer, the **myometrium** (mī˝ō·mē´trē·ŭm), within which lie layers of smooth muscle, connective tissue, and many large blood and lymphatic vessels; and an outermost **perimetrium** (pĕr·ĭ·mē´trē·ŭm). The perimetrium is a serosa continuous with the broad ligaments that support the uterus and uterine tubes.

The endometrium consists of two functionally and structurally distinct regions, the **stratum functionalis** and **stratum basalis** (Figure 27-14). Prior to menstruation these two regions together may be as thick as 6 or 7 mm, but during menstruation the cells of the stratum functionalis die and slough off, reducing the endometrium essentially to the stratum basalis, a layer only about 1 mm thick (see Clinical Note: Endometriosis).

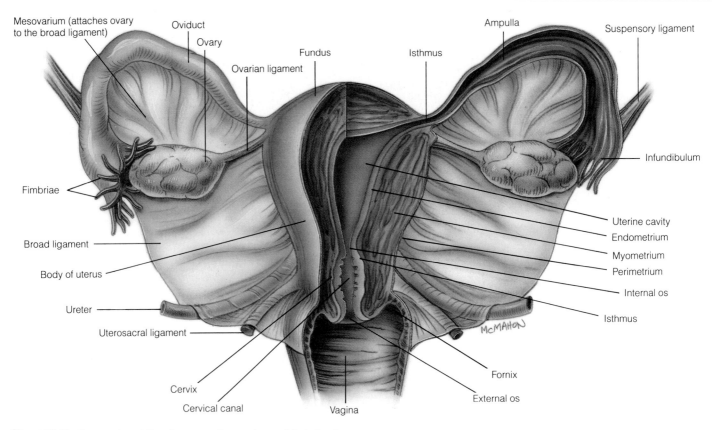

Figure 27-13 Suspension of the uterus, ovaries, and associated structures.

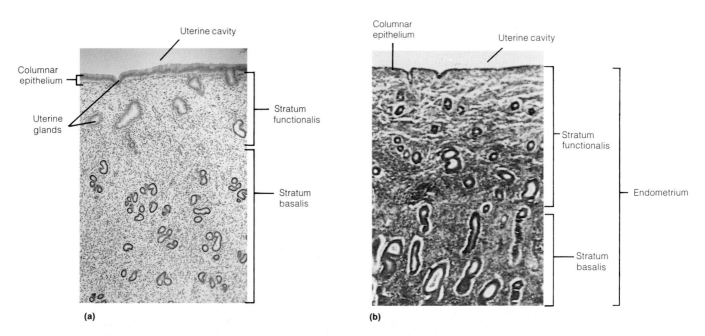

Figure 27-14 Histology of the uterine wall. (a) Photomicrograph of a section of the uterine wall in a preovulatory stage. Because the stratum functionalis and stratum basalis are defined more on the basis of function than structure, the dividing line between them is only approximate.
(b) Photomicrograph of a section of the uterine wall in a postovulatory state. Note the increased development of the stratum functionalis.

endometriosis: (Gr.) *endos*, within + *metra*, uterus + *osis*, condition

> **CLINICAL NOTE**
>
> **ENDOMETRIOSIS**
>
> One of the causes of severe dysmenorrhea (menstrual cramps), sterility, and dyspareunia (painful intercourse) is **endometriosis** (ĕn″dō·mē″trē·ō′sĭs). This condition involves abnormal implantations of endometrial tissue in the uterine wall, ovaries, or other extragenital sites. These implants mimic the changes of the normal endometrium during the menstrual cycle. Severe cases may require surgical removal of the implants and prolonged treatment with hormones to suppress ovulation.

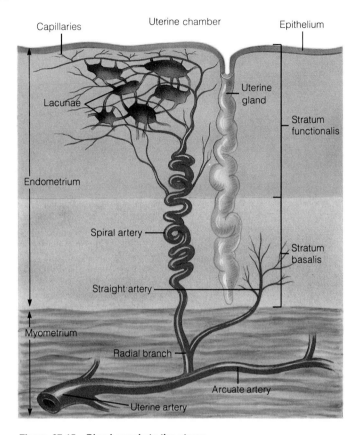

Figure 27-15 Blood supply to the uterus.

The endometrial surface consists largely of simple columnar epithelium underlain by cells of the **endometrial stroma**. The stroma attaches directly to the myometrium. Many tubular uterine glands extend from the base of the endometrium, communicating with the uterine cavity through openings in the epithelium.

Branches from the internal iliac artery called **uterine arteries** supply blood to the uterus. **Arcuate branches** give rise to **radial branches** that traverse the myometrium and extend into the endometrium to form **straight arteries** and **spiral arteries**. Straight arteries supply the stratum basalis. Spiral arteries branch again and extend through the stratum functionalis to terminate in capillaries that supply the epithelium (Figure 27-15).

Cervix

Although the cervix is part of the uterus, the structure and function of its walls are quite distinct (Figure 27-16). The cervical wall is relatively thick, and its inner surface is covered by an elaborate network of ridges and valleys, the **plicae palmatae**. The elaborate folding produces a large surface area that secretes mucus that fills the cervical canal (see Clinical Note: Cancer of the Cervix).

Vagina

The vagina (vă·jĭ′nă), about 10 to 15 cm long, provides a passage from the uterus to the outside of the body. The vagina is largely a fibromuscular tube and has a wall consisting of three tissue layers. Outermost lies a thick layer of connective tissue covering a middle layer of muscle. Blood vessels and nerve bundles lie in this wall. The innermost layer of the vagina is a folded mucosa, with the folds oriented transversely. These folds are **rugae** (roo′jē) and are located primarily in the lower two-thirds of the vagina. The mucosal surface is lined with several layers of stratified squamous

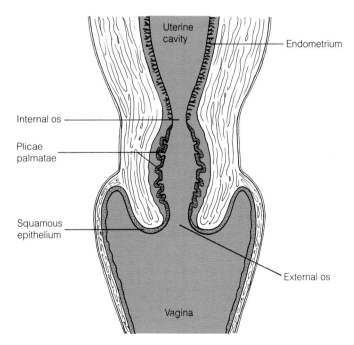

Figure 27-16 Anatomy of the cervix.

colposcopy: (Gr.) *kolon*, colon + *skopein*, to examine
labia majora: (L.) *labium*, lip; *major*, larger

> **CLINICAL NOTE**
>
> **CANCER OF THE CERVIX**
>
> **Cancer of the cervix** is ranked after breast, colon, and rectal cancer as having the highest incidence in women. It occurs most often in women 40 to 49 years of age. As with other cancers, the earlier the detection, the better the prognosis.
>
> A **pap smear** is a screening measure for cervical cancer. It is not a diagnostic test; therefore, if the cells obtained from the external os, the endocervix, or the vaginal pool are abnormal, then further diagnostic tests are necessary. Malignant cells have a decreased attachment to one another and thus exfoliate into the vagina where they can be gathered and examined for their degree of likeness to normal cells. The test is named for George Papanicolaou, a physician who developed the system for classifying cells shed from the cervix into Class I (normal) through Class V (definitely malignant). This system is rarely used today and has been replaced by descriptive reports that include the next steps to be taken in treatment if the cells are suspicious. The pap smear for detecting cervical cancer is 90 to 95% accurate, so it is a very reliable screening tool. It is now recommended that women over 20 years of age (or younger if sexually active) should have a pap smear once every year.
>
> A relatively new technique in the diagnosis of cancer of the cervix is **colposcopy** (kŏl·pŏs´kō·pē). The colposcope is a binocular microscope that magnifies the cervix and vaginal surfaces to allow detailed inspection. In those patients with abnormal pap smears, the colposcope allows the experienced operator to make directed biopsies of suspicious areas of the cervix and thus refine the diagnosis of cervical cancer. The colposcope is also used to search for small areas of **condylomata accuminata** (kŏn˝dĭ·lō´mă tă ăk·ū·mĭ´nă´tă) (venereal warts), which are believed to be implicated in the development of malignant changes in the cervix.

cells that produce secretions during sexual activity. Insertion of a penis into the vagina is made easier by the lubricating effect of these secretions.

The connection of the vagina to the cervix forms the **vaginal fornix** (fōr´nĭks), an enlarged portion of the vagina in which sperm collect during sexual intercourse. Deposited sperm are thus in close proximity to the external os, enhancing the likelihood of sperm continuing into the cervix.

External Genitalia (Vulva)

The vagina leads inferiorly and anteriorly from the cervix and opens into the **vulva**, or **pudendum** (pū·dĕn´dŭm), the external genitalia (Figure 27-17). This opening is surrounded by a membranous ring of tissue, the **hymen** (hī´mĕn), which consists of relatively fine and highly vascularized tissue. Normally an opening exists through the hymen, but sometimes the hymen covers the opening of the vagina completely. In such cases, the hymen must be opened surgically once menstruation begins to permit menstrual flow to escape.

The external genitalia consist largely of modified skin and subcutaneous tissues. The **mons pubis** (mŏns pū´bĭs) is a rather thick layer of fatty tissue that lies over the **symphysis pubis** (sĭm´fĭ·sĭs pū´bĭs), the junction of the two pubic bones that lie at the base of the abdominopelvic cavity. The mons pubis divides into two thick fatty pads that proceed posteriorly to join again in the space between the openings of the vagina and the anus. These two folds of fatty tissue are the **labia majora** (lă´bē·ă mă·jŏr´ă), which enclose the remainder of the external genitalia in the **pudendal cleft.** Just inside the two labia major are two smaller folds of mucous membrane, the **labia minora.** These labia join at the anterior end of the pudendal cleft to form a hoodlike structure that covers the **clitoris** (klĭ´tō·rĭs) and surrounds the **vestibule** (vĕs´tĭ·būl) of the vagina, the cleft between the labia minora. As puberty is reached, the mons pubis and the labia majora become covered with a thick mat of coarse **pubic hair.** Hair does not develop on the labia minora, which are kept moist by secretions of sebaceous glands. The entire region enclosed within a roughly diamond shaped area from the clitoris anteriorly to the coccyx posteriorly, and laterally to the region external to the ischial tuberosity of the coxal bone, is the **perineum** (pĕr˝ĭ·nē´ŭm). (Some authorities, however, restrict their definition of the perineum to only the small area between the anus and the vulva.) The triangular anterior half of the perineum forms the **urogenital triangle;** the posterior triangle of the perineum is the **anal triangle.**

From a developmental standpoint, the clitoris and the penis are equivalent because they are derived from the same embryonic structure. The clitoris, however, is much smaller and does not contain the urethra. Like a penis, it contains erectile tissue that becomes engorged with blood during periods of sexual stimulation, causing the clitoris to become erect. The spongy tissue responsible for erection of the clitoris is the corpus cavernosum. A corpus spongiosum is absent. The clitoris terminates in a **glans clitoridis** (glănz klĭ˝tō·rĭd´ĭs) that is nearly covered by the folds of the labia minora. The folds covering the glans comprise the **female prepuce.**

The vestibule also contains the opening of the urethra and several mucus-secreting glands that provide lubrication during sexual intercourse. **Paraurethral** (păr˝ă·ū·rē´thrăl) **glands** open into the area surrounding the urethra; a pair of small, round **Bartholin's** (băr´tō·lĭnz) **glands** are posterior to the base of each of the labia minora and open through ducts on either side of the vaginal orifice. In addition to these major glands, the vestibule contains many smaller glands that also contribute mucus.

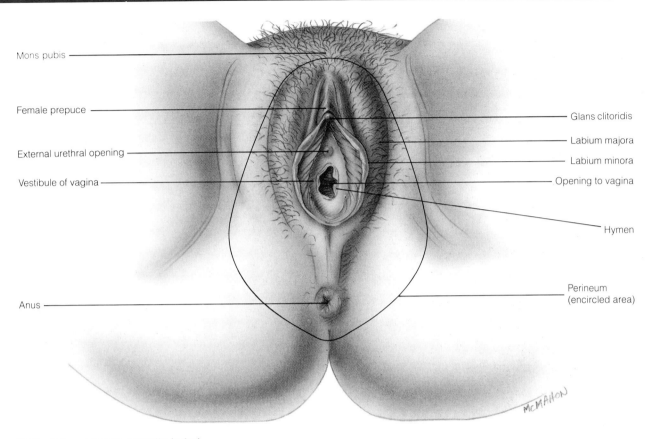

Figure 27-17 External female genitalia (vulva).

Mammary Glands

Mammary glands are considered part of the reproductive system because of the important role they play in nurturing an infant. Mammary glands begin to secrete milk within two or three days after a baby is born. The milk is rich in nutrients and immunoglobulins that convey disease resistance to the suckling infant. It is a mixture consisting of about 88% water, 7% lactose, 4% fat, 1% protein, and minute amounts of other valuable nutrients.

Mammary glands are approximately the same size in male and female infants and do not really begin to grow in females until the onset of puberty. The glands do not achieve full development in a woman, however, until she has had a child and the child is allowed to nurse. Although milk-producing tissues develop during pregnancy under the influence of estrogen and progesterone, high levels of those hormones inhibit milk production. The inhibition persists for a few days after birth, during which time the glands secrete a watery solution called **colostrum** (kō·lŏs′trŭm). Following birth of a child, however, estrogen and progesterone levels decline and the glands are no longer inhibited. Milk production is stimulated by prolactin, an anterior pituitary gland hormone, the production of which is stimulated by the suckling infant. The pituitary hormone oxytocin stimulates milk secretion.

A mature, lactating mammary gland consists of 15 to 20 compartments, or **lobes,** separated by fatty and connective tissues (Figure 27-18). The lobes consist of subdivisions **(lobules)** made up of numerous grapelike clusters of **alveoli** that produce the milk. A duct emerges from each cluster of alveoli and fuses with others to form **lactiferous** (or mammary) **ducts,** each one carrying the milk produced by the alveoli of a lobe. Each lactiferous duct has an enlarged region, an **ampulla,** that lies just beneath the nipple. Each ampulla leads into a short continuation of the lactiferous duct that opens at the nipple.

As milk is produced, it accumulates in each breast, collecting in ampullae and the extensive duct system. This system opens externally in the **nipple.** The nipple is surrounded by a pigmented ring of tissue called the **areola** (ă·rē′ō·lă). Oxytocin causes smooth muscle cells in alveoli walls to contract and force milk from the alveoli into the ducts. An infant's suckling draws this milk out through openings of the lactiferous ducts in the nipple, relieving the pressure caused by the collected milk.

CLINICAL NOTE

BREAST FEEDING

Although normal newborn infants can be bottle fed with artificial formula, **breast feeding** is much more desirable because breast milk contains nutritional substances ideal for the human infant. Breast milk contains monosaturated fatty acids that are easily absorbed and do not combine with calcium to form unabsorbable soaps. It has the highest lactose concentration of all mammalian milks as an energy source. It has a calcium to phosphorus ratio of 2:1 (cow's milk has the reverse ratio), helping to avoid calcium-deficiency tetany. In addition, breast milk provides a favorable pH and bacterial flora in the infant's intestinal tract preventing bacterial diarrheas. All infectious diseases are less common in breast-fed than in bottle-fed infants.

Nursing may begin immediately following delivery. Early breast milk is a high caloric, high protein, thin, yellow, fluid called **colostrum** that also contains antibodies, lymphocytes, and macrophages, which are useful to the newborn in fighting infections. Colostrum may be produced for several days following birth.

One problem that occurs with breast feeding is that many drugs taken by the lactating mother may be excreted in the breast milk and affect the infant. Drugs such as tetracyclines, atropine, anticoagulents, antithyroid drugs, antimetabolites, and narcotics should be avoided by nursing mothers. In fact, all unnecessary drugs should be avoided by the lactating mother, just as they are throughout pregnancy.

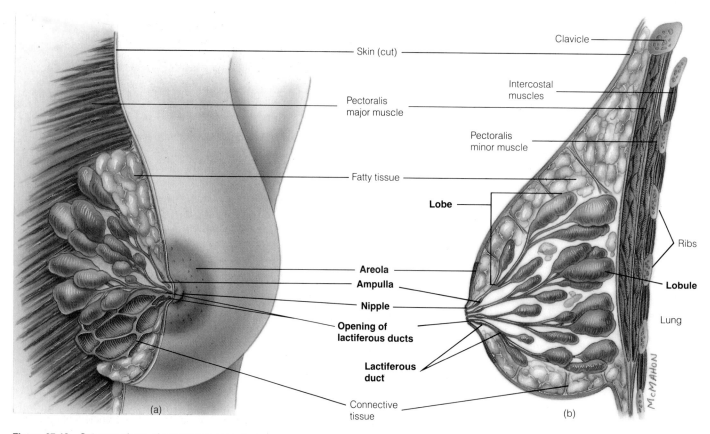

Figure 27-18 Cutaway views of a mammary gland.

Production and Release of Eggs

After studying this section, you should be able to:

1. Describe follicles as they appear before they begin developing.
2. Describe the changes that occur in a follicle as it develops prior to ovulation.
3. Distinguish between primordial, primary, secondary, graafian, and atretic follicles.
4. Explain the process by which an egg is released from an ovary and how it is carried into an oviduct.
5. Describe the changes that occur in a follicle after an egg is released.

Development of Follicles

When a baby girl is born, her ovaries already contain the cells that will produce eggs after she reaches sexual maturity. These cells are contained in hundreds of thousands of resting **primordial follicles** that lie in the cortex of her ovaries. Each primordial follicle consists of a single primary oocyte surrounded by a layer of flattened **follicular cells.** As the girl matures, and throughout her reproductive years, primordial follicles will develop into larger, more complex structures that will play an essential role in her reproductive processes (Figure 27-19).

Even before birth, a few such follicles continuously leave their resting state and begin to grow into **primary follicles.** This transition is marked by an increase in the size of the oocyte, division and growth of the follicular cells into a layer of cuboidal cells called the **stratum granulosum,** and the appearance of a thick membrane called the **zona pellucida** between the follicular cells and the oocyte.

Before sexual maturity is attained, developing follicles reach a point in their development and then stop developing, degenerating instead into **atretic follicles.** As the girl matures, follicles proceed further in their development until at some point, usually when the girl is in her early teens, one follicle completes development and releases a female reproductive cell, an **ovum** (or egg). Release of the ovum, **ovulation,** marks the attainment of sexual maturity, or **puberty.**

During its development, the primary follicle continues to grow as follicular cells proliferate and the oocyte increases in size. The zona pellucida also continues to thicken. Follicular cells immediately adjacent to the zona pellucida maintain contact with the oocyte through cytoplasmic extensions that pass through the membrane to the oocyte cytoplasm.

As growth continues, small vesicles begin to appear within the stratum granulosum, increasing in size until they coalesce to form a single fluid-filled cavity, an **antrum** (an'trum). At this point, the follicle is a **secondary or vesicular follicle.**

In a secondary follicle, the oocyte has reached its full size, and the antrum is surrounded by a multilayered stratum granulosum. Cells of the stratum granulosum are bounded by a basement membrane, outside of which lies the **theca folliculi** (thē'kă fō·lĭk'ū·lī), a layer of cortical cells oriented circumferentially about the follicle. In a mature follicle, the theca folliculi consists of two layers, an innermost layer of glandular and vascular tissue, the **theca interna,** and an outer layer of connective tissue, the **theca externa.** The oocyte lies to the side of the antrum surrounded by a body of follicular cells, the **cumulus oophorus** (kū'mū·lŭs ō·ŏf'ō·rŭs), which is connected to the stratum granulosum by a **stalk of follicular cells** (Figure 27-19).

Initially, growing primary follicles migrate into the ovary as they develop, but as they become larger, development of surrounding cortical cells forces them toward the ovarian surface. By this point, most of the follicles that began developing together have ceased developing, and some may have begun to degenerate and form atretic follicles. In a sexually mature female, one follicle continues to enlarge and become a fully mature **graafian** (grăf'ē·ăn) **follicle** capable of ovulation. The graafian follicle continues to migrate to the ovary surface until it pushes against the surrounding tunica albuginea, forming a bulge, or **stigma,** on the smooth surface of the ovary.

By this time, the oocyte has begun meiosis and has divided into two cells, one a large **secondary oocyte** and the second a small **polar body.** Both cells are still enclosed within the surrounding zona pellucida, with the polar body pushed to one side by the much larger secondary oocyte. The secondary oocyte immediately begins the second division of meiosis, but the process is arrested midway and the cell fails to complete meiosis unless it is subsequently fertilized (see Chapter 28).

Within a few hours the membrane ruptures and antral fluid oozes out, carrying the oocyte and polar body out, still surrounded by the zona pellucida and the cumulus oophorus. The remaining cells of the cumulus oophorus now comprise the **corona radiata** (kō·rō'nă ra·dē·ă'tă) (Figure 27-20).

The mass of cells adheres to the ovarian surface until movements of fimbriae at the end of the neighboring infundibulum sweep the cells into the oviduct. Currents caused by cilia in the infundibulum then send the cells on their way down the oviduct.

Fate of Follicles

Following ovulation, the walls of the follicle collapse and fill the antrum. Blood is released into this cavity where it

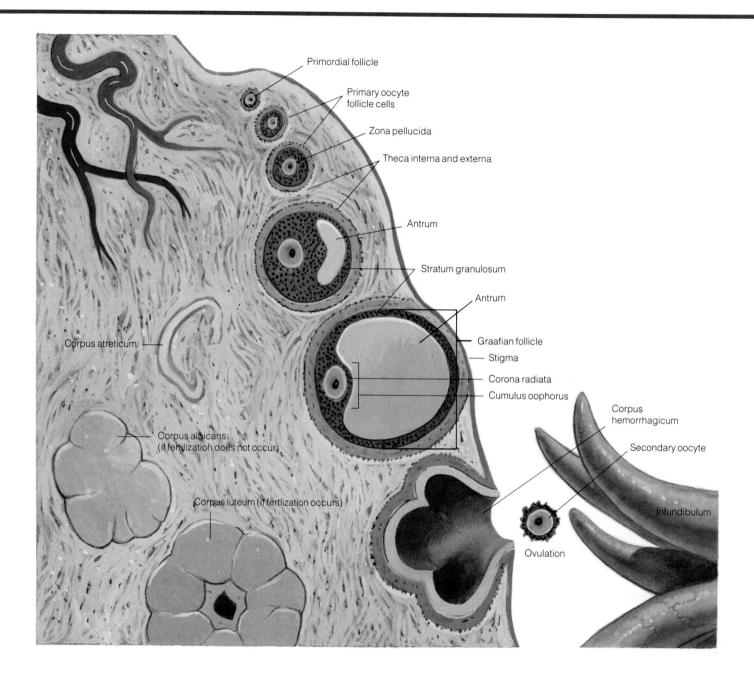

Figure 27-19 The ovarian cycle. This diagram depicts the chronological stages in the development of a follicle in the ovary.

coagulates and mixes with cells remaining in the cavity. The follicle is now a **corpus hemorrhagicum** (hĕm·ō·răj´ĭ·kŭm). Cells of the stratum granulosum that remain begin to divide and fill the cavity. These cells produce and accumulate a yellowish lipid material that gives the structure a yellow color. The ruptured follicle is now a **corpus luteum** (lū´tē·ŭm) (Figure 27-21).

If fertilization occurs, the corpus luteum continues to enlarge. It will also produce hormones important in maintaining the uterine lining and in postponing additional ovulations during the woman's pregnancy. If fertilization does not occur, however, the corpus luteum will degenerate and develop into a scarlike **corpus albicans** (ăl´bĭ·kănz), which may persist on the surface of the ovary for several months.

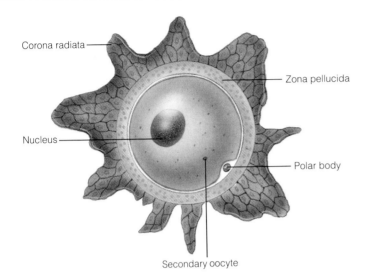

Figure 27-20 The ovum and corona radiata.

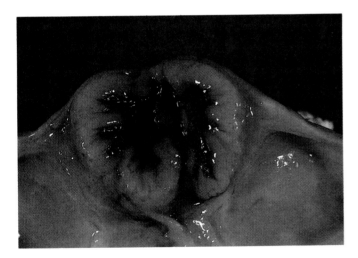

Figure 27-21 Photograph of a corpus luteum.

Hormonal Control of Egg Production

After studying this section, you should be able to:

1. List and cite the role of hormones that control ovulation.
2. List three phases of the menstrual cycle and describe the hormonal and structural features that characterize each phase.
3. Explain why the menstrual cycle stops during pregnancy.

Conversion of primordial follicles to primary follicles probably goes on continuously from before birth until the supply of primordial follicles is exhausted. Continued growth of a follicle, which is controlled by follicle-stimulating hormone (FSH), results from a rapid increase in sensitivity of the developing follicle to FSH in the blood. As the follicle continues to develop, the theca interna begins to secrete increased amounts of estrogen. As a result, there is a rapid rise in blood levels of estrogen beginning about five days prior to ovulation (Figure 27-22). The effect of estrogen is twofold: it stimulates cell proliferation in the endometrium, and it stimulates the hypothalamus to produce gonadotropin-releasing hormone (GRH). GRH is carried to the neighboring anterior pituitary gland where it stimulates the production of additional FSH and a second glycoprotein, luteinizing hormone (LH). The result is a rapid rise in blood levels of these two hormones, peaking about 16 hours before ovulation. The effect of the rapid increase in FSH is to stimulate the growing follicle to complete its development to a graafian follicle. The rapid rise in LH causes the follicular wall to rupture and ovulation to occur.

Following ovulation, the ruptured follicle enters a new phase. During this period, there is a rapid return of LH and FSH concentrations to the levels that existed prior to their estrogen-induced increase. At the same time, and under the influence of the remaining LH, the ruptured follicle becomes a corpus luteum. As the corpus luteum develops, it begins again to produce estrogen and a second steroid hormone, **progesterone**, which contributes to further development and maintenance of the endometrium. However, in this phase of the process, increased levels of the hormones exert a negative feedback effect on the hypothalamus and anterior pituitary gland that results in a further decrease in the level of FSH. The lowered level of FSH that results removes the stimulus to follicular development, so no new follicles begin developing.

Coincident with the negative feedback effect that estrogen and progesterone have on anterior pituitary secretion, the corpus luteum begins regressing and becomes a corpus albicans. Although the cause of this transition is unclear, it is known that it can be prevented by LH. As it regresses, the corpus luteum stops producing progesterone and estrogen, and blood levels of those hormones decline (see Figure 27-22).

Menstrual Cycle

The **menstrual cycle** is a series of changes occurring in the endometrium that prepares the uterus to receive and nurture a developing embryo. The onset of the first menstrual cycle, **menarche** (měn·ăr′kē), precedes the first ovulation, but once begun, menstrual cycles occur periodically for the remainder of a woman's reproductive life, broken only during periods of pregnancy, lactation, illness, and emotional stress. The cycles repeat themselves about once a month, but the

dysmenorrhea: (Gr.) *dys*, difficult + *men*, month + *rhein*, to flow

CLINICAL NOTE

PREMENSTRUAL TENSION SYNDROME

The **premenstrual tension syndrome (PMS)** is a combination of many symptoms, including irritability, depression, headache, light-headedness, tiredness, fluid retention, and sore breasts, some or all of which can affect women 7 to 10 days prior to the onset of menstruation. It usually disappears within hours of the onset of the menstrual flow. Recent research on PMS shows that it is caused by imbalances in levels of estrogen and progesterone during the last half of the menstrual cycle. Symptoms are worsened by stress and poor health habits. Self-treatment of PMS includes a nutritious (low sugar, low salt) diet, a good vitamin program, regular exercise, and reduction of stress. If this "natural" approach isn't effective, the woman should consult a supportive physician who is willing to work with her to find effective treatment. Treatment can be symptomatic, with diuretics, analgesics, and sedatives, but comprehensive treatment with progesterone seems to be the most promising.

CLINICAL NOTE

DYSMENORRHEA

Dysmenorrhea (dĭs″měn·ō·rē′ă), or painful menstruation, is a mild to severe cramping pain in the lower abdomen that begins with the onset of menstruation. It is often associated with nausea and vomiting. Often referred to as "cramps" by affected women, it should not be confused with premenstrual tension syndrome (see Clinical Note). Dysmenorrhea does not occur prior to menstruation. The development of dysmenorrhea in older women is often associated with pelvic pathology. The symptoms are frequently relieved with the use of nonsteroid, anti-inflammatory drugs such as ibuprofen or Naprosyn, which block the production of **prostaglandins.** Prostaglandins (Chapter 17) are known to increase at the time of menstruation and cause uterine muscle spasms.

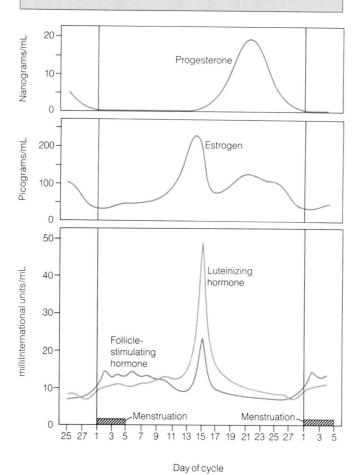

Figure 27-22 Reproductive hormone concentrations during a 28-day menstrual cycle. Source: A. R. Midgley et al., "Human Reproductive Endocrinology," *Human Reproduction*, E. S. E. Hafey and T. N. Evans, eds. (Hagerstown, MD: Harper & Row, 1973).

period with which they occur is quite variable, even in the same woman from period to period. The average length of time between onsets of menstrual flow is 28 days, but this value commonly varies from 24 to 32 days and can be even longer or shorter in some women.

The menstrual cycle consists of the **menstrual, preovulatory** (proliferative), and **postovulatory** (secretory) **phases.** By convention, the first day of the menstrual phase is considered the beginning of a menstrual cycle. However, to show the correlation between ovulation and the menstrual cycle, we will begin with the preovulatory phase (Figure 27-23).

The preovulatory phase is initiated by estrogens secreted by developing follicles and usually lasts about 10 days. This phase is marked by a buildup of the stratum functionalis. The endometrium thickens considerably as new cells are produced and enlarge. Blood vessels invade the thickened tissue, and straight tubular endometrial glands grow into the stratum functionalis. The end of this phase coincides with ovulation.

The postovulatory phase, which lasts about 13 days, completes the preparation of the uterus for receiving an embryo. It is initiated by progesterone from the corpus luteum, which causes blood vessels in the endometrium to continue enlarging. Endometrial glands begin to secrete a nutrient-rich mucus. The endometrium continues to thicken, largely due to swelling of the tissue rather than additional cell production.

If an egg is not fertilized, the corpus luteum disintegrates and ceases producing hormones. With the source of estrogen and progesterone removed, the newly formed lining of the uterus begins to degenerate and slough into the uterine cavity. This material, which includes blood from the vessels in the endometrium, is discharged through the vagina as

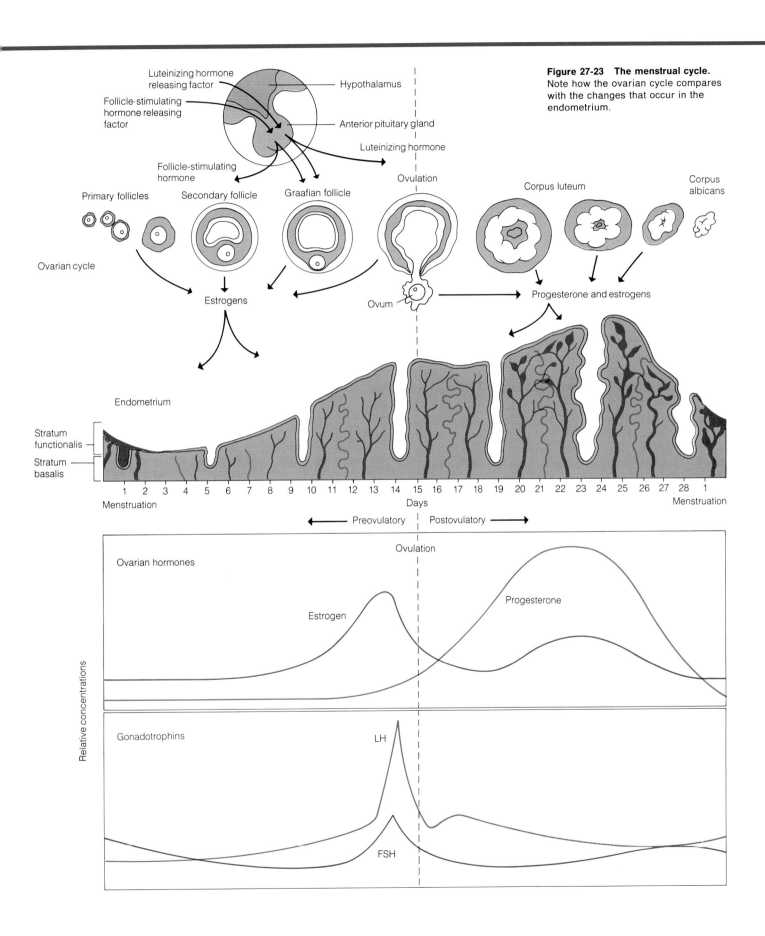

Figure 27-23 The menstrual cycle. Note how the ovarian cycle compares with the changes that occur in the endometrium.

menstrual flow. This starts the menstrual phase, or **menses** (měn´sēz), which usually lasts about 5 days. By the time menses draws to a close, the stratum functionalis is virtually gone, leaving the underlying stratum basalis to provide tissues for another preovulatory phase.

Menstrual cycles stop during pregnancy because the embryo, after becoming implanted in the uterine wall, begins secreting a hormone that prevents the corpus luteum from degenerating. This hormone, human chorionic gonadotropin (HCG), enables the corpus luteum to continue secreting estrogen and progesterone. Under the influence of these three hormones, the endometrium remains healthy for the growth of the embryo, and release of FSH and LH from the anterior pituitary gland is inhibited.

Menopause

Menstrual cycles end as the supply of ovarian follicles is depleted, generally about 35 to 40 years after onset of menstruation. This termination, called **menopause,** causes hormonal and other physiological imbalances that produce occasional feelings of warmth, flushed skin, and perspiration. Headaches, muscular cramps, and general emotional upset may also occur (see The Clinic: Disorders of the Reproductive System).

Coitus

After studying this section, you should be able to:

1. Define coitus.
2. Identify four different stages of sexual response.
3. Describe differences in the stages of sexual response as they occur in males and females.

Coitus (kō´ĭ·tŭs), or **copulation,** is the act of sexual intercourse in which semen is ejaculated into a woman's reproductive tract. It usually involves several discrete phases in both men and women, during which various sensory and psychologic stimuli initiate physiologic responses.

Male Sexual Response

Arousal and Erection

Thoughts, sights, sounds, odors, and touch can act through sensory and parasympathetic nerves as sexual stimuli. In a male, such stimuli can induce nervous impulses that allow the rate of blood flow into the penis to increase while constricting the rate of blood flow out of the penis. As a result, the spongy tissues become engorged with blood, causing the penis to enlarge and stiffen. This reaction is an **erection.** As the erection occurs, muscles of the dartos contract, causing the scrotum to lift and become more compact. Within the scrotum, the **cremaster muscle,** a band of skeletal muscle lying adjacent to the ductus deferens, also contracts, lifting the testes still more. When it is in this state, the penis can serve as a copulatory organ.

Plateau

Upon insertion of the penis into the vagina, reactions are intensified and accompanied by flushed skin, increased blood pressure, and accelerated heart and respiratory rates. Semen moves into the urethra from the ductus deferens and is mixed with prostatic and seminal vesicle secretions. This period, termed the **plateau,** may last from a few seconds to several minutes, depending on the individual and his physiologic and psychologic states.

Orgasm and Ejaculation

As sexual excitement grows in the plateau stage, the skin becomes more flushed, and blood pressure, heart rate, and respiratory rate increase further. Continued friction on the glans and shaft of the penis from rhythmic pelvic thrusts transmits sensory stimuli to the central nervous system where they are summed. When a threshold is reached, orgasm begins and cannot be prevented voluntarily. In both sexes, orgasm is associated with widespread nervous and muscular reactions and intense pleasure as the pent-up muscular contractions are suddenly released.

Resolution

Orgasm in both women and men is followed by a **resolution** phase, during which reactions that occurred during the previous stages decline and the body returns to its "prearousal" state. In a man, the penis loses its erection as muscles that prevented blood from leaving it relax. During resolution, the same stimuli that aroused him may lose their effect and the man may enter a **refractory** period during which greater stimulation is required for arousal. The length of the refractory period depends on a number of factors, including health, fatigue, and psychologic state.

Female Sexual Response

Arousal

As in a male, arousal in a female is triggered by certain stimuli and initially results in physiologic changes involving the breasts, vagina, and external genitalia. The nipples be-

> **CLINICAL NOTE**
>
> ## SEXUALLY TRANSMITTED DISEASES
>
> A number of disorders are transmitted almost entirely through intimate sexual contact. These are collectively known as **sexually transmitted diseases (STDs)**, once referred to as "venereal diseases." The incidence of STDs seems to be increasing, although true figures are hard to obtain because many cases are not reported. These diseases are caused by organisms as diverse as viruses, bacteria, protozoa, and fungi. Two of the most common have already been discussed: genital herpes (see Chapter 5 Clinical Note: Common Skin Problems) and AIDS (see Chapter 24 Clinical Note: AIDS). Three other important STDs are gonorrhea, syphilis, and chlamydial infection.
>
> **Gonorrhea** is caused by the bacterium *Neisseria gonorrhoeae*. The incidence of this disorder seems to have declined, although infections by a strain resistant to penicillin seem to be on the increase. The bacteria grow best in a warm, moist environment and evoke inflammatory reactions resulting in a discharge of pus. The organism can enter the body through the genitourinary tract, eyes, mouth and throat, rectum, or skin. One can infect one's own eyes by accident. Children born of mothers who are infected may develop gonorrheal infection of the conjunctiva, which may result in blindness if not treated. Because of the danger of this type of infection, newborns are routinely treated with an antibacterial agent applied to the eye immediately upon delivery.
>
> While men are more likely to display symptoms than are women, both may be without symptoms and therefore transmit the disease unwittingly. Men usually notice a creamy yellow or bloody discharge and may experience urethral pain. Women may notice unusual vaginal or urinary discharge, pain on voiding, pain during sexual intercourse, unusual vaginal bleeding, abdominal tenderness, and fever. If the infection is untreated in women, it will spread up the reproductive tract into the oviducts leading to infections of these tubes with scarring and, ultimately, sterility.
>
> The causative organism of **syphilis** is *Treponema pallidum*. This spirochete is readily killed by soap and water but is easily transmitted by intimate sexual contact. This organism can also easily spread from mother to infant during pregnancy, leading to congenital defects in the child as well as active disease.
>
> Syphilis has three clinical stages. **Primary syphilis** first appears as a painless, buttonlike sore or **chancre** at the site of exposure within 3 weeks to 3 months of exposure. This disease is very contagious at this point. If untreated, the chancre will heal in about 3 to 12 weeks. The fact that the lesion goes away, is probably why many people do not seek medical advice and therefore do not receive treatment.
>
> **Secondary syphilis** is manifested as long as 6 months later by a rash often appearing on the hands and feets, fever, sore throat, and nausea. These symptoms may appear and disappear for a year or more. The person may also lose hair and develop elevated reddish brown lesions in the genital area. These may ulcerate and produce a particularly foul discharge that is highly contagious.
>
> The delayed response of the untreated disorder can occur as long as 20 years later and is known as **tertiary syphilis.** Only about one-third of people with untreated syphilis go on to this stage of the disease. This stage takes on one of three forms. Cardiovascular changes result from scarring to the aorta and the development of aortic aneurysms and may lead to aortic valve insufficiency. **Tabes dorsalis** refers to the sensory loss that occurs as a result of lesions of syphilis which affect the central nervous system. Other effects on the central nervous system include dementia and blindness.
>
> A leading cause of infant blindness in underdeveloped countries is **inclusion conjunctivitis,** an ocular disease in newborns caused by infection by a parasitic microorganism called *Chlamydia trachomatis*. In these countries the organism is mainly spread by flies and nonsexual personal contact. In developed countries the disease is spread by sexual contact. The **chlamydial** (klă·mĭd´ē·ăl) **infection** results in a wide variety of genital and urinary infections in men and women. Infants become infected during birth when they come into contact with their mother's infected tissues. This disease has probably twice the incidence of gonorrhea, whose signs and symptoms it closely resembles.
>
> Except for AIDS, most sexually transmitted diseases can be treated if the person seeks help from health practitioners. More importantly, all of these diseases can be prevented by good sex education at an early age, careful screening of sexual partners, and the consistent use of condoms during intercourse.

come erect as a result of smooth muscle contraction and the engorgement of the nipples and surrounding areola with blood. The glans of the clitoris enlarges, and the vestibular glands and cells of the vaginal wall secrete fluid. This fluid will lubricate the vulva and vagina, facilitating penetration by a man's penis. The vaginal fluid also neutralizes the pH of the vagina, a necessary condition for proper functioning of sperm. Simultaneously, the labia become congested with blood and swell, and the vagina lengthens and dilates as the uterus is pulled upward. All these reactions increase the probability of sperm being successfully delivered to the reproductive tract, should coitus occur.

Plateau

As in a male, a female moves from the initial flush of arousal to a plateau stage. Her reactions may also include a rush of blood to the skin along with increases in blood pressure,

> **CLINICAL NOTE**
>
> **SAFE SEX**
>
> The recent AIDS epidemic and spread of other sexually transmitted diseases has resulted in a reexamination of sexual habits. The following are useful guidelines for sexually active people.
>
> 1. *Make careful choices about sexual activity* and negotiate with your sexual partner for safer sexual practices. For those who are sexually active, a long-term, mutually monogamous relationship prior to which both have been safe from sexually transmitted disease is nearly risk free.
> 2. *Know your sexual partner* well before having sex. It is a good idea to ask a prospective partner about his or her health, sexual history, and awareness of safety precautions.
> 3. *Use condoms during sex.* The proper use of condoms greatly reduces the chance of transmitting the AIDS virus, but it has not been proven that condoms eliminate the risk of AIDS.
> 4. *Carefully avoid any injury* to the body tissues during sex.
> 5. *Do not mix alcohol or other drugs with sexual encounters;* they may cloud your judgment and weaken your immune system.

heart rate, and respiratory rate. The plateau phase is also marked by increased secretions of fluid from the numerous glands active in the reproductive tract.

Orgasm

Continued stimulation of the clitoris and vagina causes a woman to move from the plateau stage to the orgasmic stage. This can be enhanced by pressure exerted on the mons pubis, which creates traction on the prepuce and stimulates the shaft of the clitoris. Additional traction on the labia by the penis as it moves in and out increases the level of sexual excitement until a peak of stimulation is reached, and a sudden release from the pent-up muscle tension is experienced. This reaction can also involve extensive body responses, including spasmodic contractions of the vagina, uterus, and various skeletal muscles.

Resolution

During resolution in a woman, the labia return to their prearousal state, the vagina and uterus return to their original position, and the nipples and clitoris decrease in size. Unlike men, women are thought not to experience a refractory period, or at least not one that is as pronounced as in males. At least some women are able to continue responding to sexual stimuli and to experience several orgasms before entering a refractory period.

Fertilization

After studying this section, you should be able to:

1. Describe the mechanism used by sperm in reaching and binding to the zona pellucida.
2. Explain the mechanism that usually allows only one sperm to fertilize an egg.
3. Describe the events occurring in an egg following sperm and egg fusion that lead to nuclear fusion.

Fertilization (or **conception**) normally occurs in the ampulla of the oviduct within 12 hours after ovulation and sexual intercourse. After being deposited in the vagina, sperm are transported to the egg by whiplike swimming movements of their flagellae and by wavelike contractions of the walls of the uterus and oviducts. Such movements are stimulated by prostaglandins in seminal fluid.

Sperm usually live as long as 48 hours, but have sufficient viability to fertilize an egg for only about 24 hours after release. An ovum can be fertilized for approximately 12 hours after ovulation, but the sooner it is fertilized, the better its chances for normal development. Most women, therefore, have a period of about 36 hours each month during which sexual intercourse can result in fertilization (24 hours before ovulation plus 12 hours after ovulation).

Capacitation

While traveling toward the egg, the sperm come into contact with secretions from the wall of the uterus and oviduct that make them subsequently able to fertilize an egg. The changes that occur in the sperm are collectively called **capacitation.** Although the details of capacitation are not fully understood, it is essential because sperm that do not come into contact with the walls of the uterus and oviduct or secretions from these tissues cannot fertilize an egg.

Once capacitated sperm reach the egg, digestive enzymes are released from the acrosome. These enzymes weaken the corona radiata, allowing sperm to reach the zona pellucida (Figure 27-24). Upon reaching the zona pellucida, sperm undergo **attachment,** in which they form a loose association with the membrane. After forming this loose attachment, the zona pellucida and the sperm plasma membrane attach to one another, **binding** the sperm tightly to the zona pellucida.

spermicide: (Gr.) *spermatos,* seed + (L.) *caedere,* to kill

CLINICAL NOTE

CONTRACEPTIVE DEVICES

Attempts to prevent or plan conception go far back in history, and some of these ancient techniques are still in use today, such as the rhythm method and breast feeding. Not until recently, however, has the advent of effective, consistent birth control techniques resulted in a marked change in our society, especially in the lives of women. Modern contraceptive devices allow us the ability to prevent unwanted children or to put off having children until a more appropriate time. Here are some of the main methods of contraception that are now available.

Oral contraceptives, or "birth control pills," are composed of synthetic estrogen and a small amount of synthetic progesterone (see figure). They are taken by the female from day 5 through day 25 of the menstrual cycle. The estrogen suppresses the gonadotrophic hormones of the pituitary and prevents ovulation. Progesterone probably increases the viscosity of cervical mucus (which limits the access of sperm) and also interferes with the maturation of the endometrium so that implantation is unlikely. The actual effectiveness of oral contraceptives is statistically between 96 and 100%. It is only because women forget to take them that the effectiveness is not 100%. These pills have many side effects, including nausea, weight gain, headache, breast tenderness, spotting, and vaginal infections. Levels of the hormones taken in the pills, however, can be adjusted to avoid many of these effects. Women who smoke, are over 35 years of age, are overweight, have high blood pressure, or are at risk for heart attack should not take oral contraceptives.

A **diaphragm** is a circular latex disk that fits over the cervix and acts as a barrier against the entrance of spermatozoa to the uterus. It is prescribed and fitted by a physician or nurse practitioner and should be refitted after pregnancy, therapeutic abortion, or the gain or loss of more than 20 lbs. Before intercourse, the woman coats the diaphragm with contraceptive jelly and then inserts it in the vagina over the cervix. Since spermatozoa are still alive for up to 6 hours following ejaculation, the woman should leave the diaphragm in place for 6 hours. Although the diaphragm can be inserted 2 hours before intercourse, some women dislike the diaphragm because it decreases spontaneity. If checked and fitted properly, the diaphragm is effective as a contraceptive.

Two variations on the diaphragm are available. The **cervical cap** is a smaller version of the diaphragm that is placed over just the cervix and may be left in place for several days. The **vaginal sponge** is a small, disk-shaped sponge saturated with spermicide that is placed in the posterior vagina prior to intercourse. The sponge does not require refitting and is readily available.

Spermicides are chemical substances placed in the vagina that kill spermatozoa. They may be in the form of jellies, creams, impregnated sponges, tablets, or vaginal suppositories. This method of contraception, while inexpensive and requiring no prescription, is only about 80% effective and is normally used in conjunction with another method.

Condoms are sheaths made of rubber or animal membranes that are placed over the erect penis before sexual intercourse. The spermatozoa are therefore deposited in the sheath rather than in the vagina, and pregnancy is usually prevented. This is an inexpensive method of birth control requiring no prescription. Condoms are not entirely effective in preventing pregnancy, but their efficiency can be improved by also using spermicide.

Condoms must be used properly to be effective. They must be placed on the penis before penile-vulvar contact since preejaculation fluid may contain sperm. They must also be carefully held in place as the penis is withdrawn promptly after ejaculation so that no sperm leaks from the sheath, which may loosen as the penis becomes flaccid.

The condom was first used as a contraceptive device by the Chinese more than 2000 years ago. It has recently become an important technique in preventing the spread of sexually transmitted diseases, particularly the AIDS virus.

Intrauterine devices, or **IUDs,** have been used by many women as a convenient form of birth control. An IUD is a small device inserted into the uterus by a physician that can remain in place for several years. Through a mechanism that remains unclear, its presence in the uterus prevents implantation of the embryo on the uterine wall. Unfortunately, IUDs have harmful side effects in some women, including chronic infections, increased risk of miscarriage, infertility, and ectopic pregnancy.

There are presently many other possibilities being pursued to prevent contraception (see Clinical Note: Sterilization). The search continues, however, for the ideal contraceptive: completely safe, having no side effects, inexpensive, easily available, and acceptable to the user.

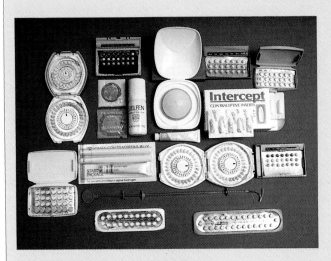

A variety of available contraceptive devices.

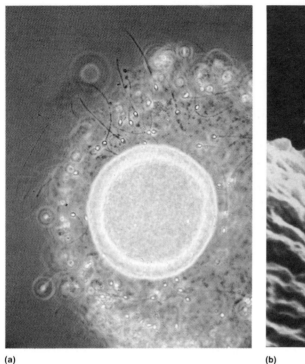
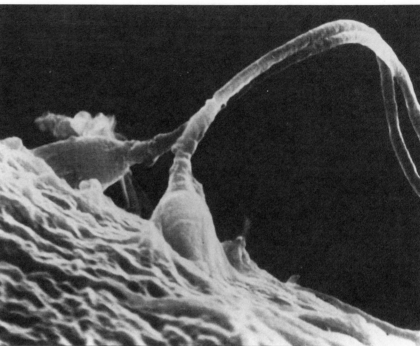

Figure 27-24 Fertilization. (a) Photomicrograph of sperm clustered around the ovum. (b) Scanning electron micrograph of attached and penetrating spermatozoa on the surface of the zona pellucida.

Following binding, sperm pass through the zona pellucida, probably by using acrosomal enzymes that digest the membrane in advance of the sperm. Several sperm can penetrate the zona pellucida, each leaving a tiny channel in the membrane with a diameter about the size of the sperm head, but relatively few sperm make it all the way through the membrane.

Fusion of Sperm and Egg

Fertilization occurs after a sperm emerges from the zona pellucida and enters the space between that membrane and the egg plasma membrane, the **perivitelline** (pĕr″ĭ·vī·tĕl´ĕn) **space,** and fuses with the egg plasma membrane. Fusion of the membrane of one sperm (normally) with the vitellin membrane probably induces a rapid depolarization of the vitellin membrane and prevents any more sperm from fusing. This is a **fast block** to the fusion of more than one sperm with the egg **(polyspermy).** Following the fast block, chemical changes occur in the zona pellucida that make it more difficult for sperm to attach or penetrate. This constitutes a **slow block** to polyspermy.

After the sperm and vitellin membranes fuse, the sperm is drawn into the egg cytoplasm. This stimulates the egg nucleus to complete meiosis and to produce a **female pronucleus.** The sperm nucleus enlarges and becomes a **male pronucleus.** Fertilization is complete when the male and female pronuclei fuse, and the resulting cell is a **zygote.** The zygote travels down the oviduct, swept along by currents from the moving cilia and nourished by secretions from cells in the wall. As it travels, the zygote divides, beginning the process of forming a new human being.

A zygote is a cell that combines, for the first time, a particular set of genetic information, half of which was in the female pronucleus and the other half in the male nucleus. In the subsequent months of development, that information will be used to produce the structures from which a new human being will develop, as well as the structures required for protection and nourishment during that time. If all goes well, after about 280 days an infant will be born. In Chapter 28 we discuss those developmental changes and the mechanisms that determine the genetic potential of that new human being.

STUDY OUTLINE

Anatomy and Function of the Male Reproductive Tract (pp. 699–708)

The major divisions are the testes, scrotum, accessory glands, and penis.

Scrotum The *scrotum* is a sac that contains the testes and related ducts and tubules. Divided by the *scrotal raphe*, the scrotum maintains the testes at a temperature necessary for sperm production.

Testes Contained within the scrotum are the testes and epididymis. *Testes* are lobular structures in which sperm are produced. *Seminiferous tubules* lead to the *rete testis*, from which emerges the *efferent ductules*.

Epididymis Efferent ductules empty in an *epididymis*, where sperm are stored.

Ductus Deferens and Ejaculatory Duct *Ductus deferens* are tubes that lead out of the scrotum through the *inguinal canal*. The ductus deferens and associated nerves, arteries, veins, and lymphatic vessels form the *spermatic cord*. The ductus deferens combines with a duct from a seminal vesicle to form an *ejaculatory duct*.

Urethra The tube leading from the urinary bladder to the end of the penis is the urethra, which consists of *prostatic*, *membranous*, and *cavernous* portions.

Production and Ejaculation of Sperm Sperm are produced in the wall of seminiferous tubules by *spermatogenesis*. *Spermatogonia* divide by meiosis to produce *spermatids*, which differentiate into sperm (spermiogenesis). The mature sperm is an acrosome-covered nucleus with a flagellum. Spermiogenesis is stimulated by *Sertoli's* (or *sustentacular* or *nurse*) *cells*. Stored in the epididymis and ductus deferens, sperm are discharged from the penis by *ejaculation*.

Hormonal Control of Sperm Production Hypothalamic gonadotropin-releasing hormone (GRH) stimulates production of follicle-stimulating and luteinizing hormones (FSH and LH) by the anterior pituitary gland. FSH stimulates differentiation of seminiferous tube epithelial cells into primary spermatocytes. LH stimulates interstitial (Leydig) cells to produce testosterone.

Testosterone stimulates development of secondary sexual characteristics in males and differentiation of spermatids into sperm. GRH (and thus FSH and LH) levels are regulated by testosterone by negative feedback.

Inhibin is a polypeptide produced by Sertoli's cells which may inhibit FSH production and vice versa. Interaction may regulate the rate of sperm production.

Accessory Glands *Accessory glands* contribute fluid to production of semen. They include *seminal vesicles, prostate gland, bulbourethral glands*, and *urethral glands*.

Composition of Semen *Semen* is an aqueous mixture of sperm, citric acid, fructose, ascorbic acid, various enzymes, and other nutrients.

Penis The male copulatory organ is the *penis*, which consists of three cylindrical spongy bodies, two *corpora cavernosa* and one *corpus spongiosum*, enclosed in sheaths called the *tunica albuginea*. The urethra passes through the corpus spongiosum and ends in the urethral orifice, an opening in the *glans penis*. A loose extension of skin is the *foreskin*, or *prepuce*. The *fascia penis* is a sheath surrounding the entire penis below the skin. An erection results when blood enters the penis and is prevented from flowing out. Sympathetic impulses cause sperm to be forced through the urethra and ejaculated.

Anatomy and Function of the Female Reproductive Tract (pp. 708–716)

Ovaries The *ovaries* are paired organs located in the lower portion of the abdominal cavity and are held in place by the *mesovarium, ovarian ligament*, and *suspensory ligament*. Internally, they consist of the *medulla* and *cortex* surrounded by the *mesothelium*. The *tunica albuginea* forms the outer covering. Hundreds of thousands of *follicles* are present in the cortex, each containing a *primary oocyte*. Ovaries produce *estrogens*, hormones responsible for the development of secondary female characteristics.

Oviducts The tubes extending from the ovaries to the uterus are the *oviducts*. The end closest to ovary contains *fimbriae*, which form the *infundibulum*. The infundibulum leads into the *ampulla, isthmus*, and *interstitial segment* at the junction with the uterus.

Uterus The *uterus* is a thick-walled sac lying on the floor of the abdominal and pelvic cavity. It is held in place by the *broad, uterosacral, lateral cervical*, and *round ligaments*. The uterine cavity connects to the outside through *cervical* and *vaginal canals*. Regions of the uterus include the *fundus, corpus, isthmus*, and *cervix*. The *internal os* is the junction of the uterine cavity and cervical canal.

The *external os* is at the junction of the cervical and vaginal canals. The wall of the uterine cavity includes the *endometrium, myometrium*, and *perimetrium*. Two subdivisions of the endometrium are the *stratum functionalis* and *stratum basalis*.

Cervix Although part of the uterus, the *cervix* has a different structure. It is thick walled, with ridges and valleys on its inner surface called the *plicae palmatae*.

Vagina The *vagina* leads from the cervix to the outside. Its wall is three layers thick, the innermost layer being a folded mucosa. An enlarged region at the end closest to the cervix is the *vaginal fornix*.

External Genitalia (Vulva) At the opening of the vagina is the external genitalia, or *vulva*, surrounded by the *hymen*. The *mons pubis* divides to form *labia majora*. The separation between the labia majora forms the *pudendal cleft*. The *labia minora* are within this cleft. The *clitoris* is located at the anterior end of pudendal cleft. The *perineum* is a diamond-shaped region enclosing the vulva. The clitoris terminates in the *glans clitoris*, which is covered by the *female prepuce*. Paraurethral and *Bartholin's glands* and smaller glands are present in the vulva.

hypospadias: (Gr.) *hypo*, under + *span*, to draw
hermaphroditism: (Gr.) *Hermaphroditos*, mythological son of Hermes and Aphrodite, who was man and woman

---

THE CLINIC

DISORDERS OF THE REPRODUCTIVE SYSTEM

Developmental Problems

Although there is much more fusion of the genital folds in males than in females, normal development of the reproductive system involves some fusion in both (see Figure 27-B in box on embryonic development). Failure of the proper amount of union of the folds results in abnormal development in both sexes.

In the male, fusion failure can result in displacement of the terminal urethral orifice. The orifice can develop on the dorsum of the penis proximal to the glans, a condition called **epispadias** (ĕp″ĭ·spā′dē·ăs), or more commonly, the orifice may occur on the ventral side of the penis, called **hypospadias**. Hypospadias is often associated with a fibrosis of the corpus spongiosum, resulting in a ventral curvature of the penis called **chordee**. These abnormalities can be repaired by plastic surgery. More severe fusion failure may result in defects that cause confusion as to the true sex of the infant; this is called **pseudohermaphroditism** (sū″dō·hĕr·măf′rō·dīt″ĭzm). Surgical exploration may be necessary to identify which organs are present, ovaries or testes. Very rarely, both ovaries and testes may be present, resulting in **true hermaphroditism**.

Fusion failures in the female can result in a double vagina, double cervix, and sometimes a partial or complete double uterus. Childbearing in these women is often possible but complicated. Plastic surgery is often required.

Another common developmental defect in the male is failure of one or both testes to descend into the scrotum, called **cryptorchidism** (krĭpt·or′kĭd·ĭzm). This condition can result in an increased incidence of testicular tumors and reduced fertility. It is easily corrected surgically or sometimes by administering male hormones.

Congenital or inguinal hernias result from a failure of fusion of the tract developed by the descending testes through the peritoneum and inguinal canal. A partial closure failure of the peritoneal tract results in a cystic mass in the scrotum called a **hydrocele** (hī′drō·sĕl). A common condition is a reduplication of the veins of the spermatic cord, causing an enlargement in the scrotum resembling "a bag of worms" called a **varicocele** (văr′ĭ·kō·sĕl). A varicocele in the left scrotum is not a problem, but in the right scrotum it is suggestive of obstruction of the veins by an intraabdominal tumor and must always be investigated. Occasionally, the testis is not fixed in the scrotum and can rotate on the spermatic cord, shutting off the blood supply to the testis. This **testicular torsion** produces a sudden onset of severe pain and swelling of the involved testis and must be corrected by immediate surgery or the testis may be destroyed.

Infection

Diseases that are spread during sexual intercourse, orogenital, or anogenital relations are known as sexually transmitted diseases (STDs) (see Clinical Note: Sexually Transmitted Diseases). STDs are the most common infectious diseases in the world. Recent figures estimate that 250 million people are infected with STDs annually. There were close to 3 million cases of gonorrhea and 400,000 cases of syphilis requiring treatment in the United States in a single year. Even more prevalent are cases of nonspecific urethritis, trichomoniasis, genital candidiasis, chlamydia, yeast infections, herpes genitalis, warts, scabies, pubic lice, and molloscum contagiosum. In addition the AIDS epidemic is causing great concern throughout the world (see Chapter 24 Clinical Note: AIDS).

Mammary Glands In mature women, the *mammary glands* consist of *lobes*, subdivided into *lobules* of *alveoli*. Milk is produced in the alveoli. The *lactiferous duct* carries milk to the *ampulla* region where milk is stored until secreted. Milk passes from the ampulla through a short section of the duct to an opening in the surface of the nipple. Prolactin causes milk to be produced, and oxytocin causes milk to be forced into ducts (secreted).

Production and Release of Eggs (pp. 717–718)

Development of Follicles *Primordial follicles* are present before birth. Each one contains a *primary oocyte*, a cell capable of producing an *ovum*. Many develop into *primary follicles*, accompanied by enlargement of the oocyte and division of follicular cells into *stratum granulosum*. Follicles that do not develop to maturity degenerate into *atretic follicles*. Follicles that develop to maturity release an oocyte at ovulation.

As the follicle matures, vesicles form a single fluid-filled *antrum*, and *theca folliculi* form outside the stratum granulosum. Meiosis begins and produces a *secondary oocyte* and a *polar body*. The follicle matures into a *graafian follicle*, which moves to the surface of the ovary where it ruptures and releases the oocyte and *cumulus oophorus*, now the *corona radiata*, which are swept into infundibulum.

Fate of Follicles A ruptured follicle produces a *corpus hemorrhagicum* and then a *corpus luteum*. A corpus luteum will be maintained through pregnancy if fertilization occurs, otherwise it will degenerate into a *corpus albicans*.

Hormonal Control of Egg Production (pp. 719–722)

Development of follicle results from increased sensitivity to follicle-stimulating hormone (FSH). During development, the follicle releases estrogen, which stimulates cell proliferation in the endometrium and synthesis of gonadotropin-releasing hor-

cryptorchidism: (Gr.) *kryptos,* hidden + *orchis,* testis + *ismos,* condition

Trauma
Injury to the external genitalia may occur in straddle-type falls, motor vehicle accidents, and athletic activity. Lacerations of the vagina sometimes occur during sexual intercourse. The majority of these injuries are contusions and lacerations that are amenable to surgical repair.

Tumors and Cancer
Cancers of the female genital tract are common. They account for one of six diagnosed cancers and one of ten cancer deaths in women in the United States. Cancer of the uterine cervix occurs in 15,000 women annually and results in 6,800 deaths (see Clinical Note: Cancer of the Cervix). Preinvasive cervical cancer may be detected by the pap smear (a simple smear of cells scraped from the cervix and examined under a microscope) ten years or more prior to its becoming invasive. Risk factors for cervical cancer include early age of first coitus and multiple sexual partners. Cancer of the uterine body is even more prevalent: 37,000 new cases are reported each year resulting in 2,900 deaths. Ovarian cancer occurs in 19,000 women annually in the United States causing 11,600 deaths.

Cancer of the female breast is the most common malignant tumor in women (124,000 cases annually) in the Western world. Self-breast examination and X-ray examination (mammography) are effective measures for early diagnosis. Treatment includes surgical excision of the tumor and irradiation of local lymph nodes (see Chapter 24 Clinical Note: Mastectomy).

Cancer of the testis is a relatively rare tumor but is often highly malignant and rapidly fatal. Recent advances in treatment have markedly increased the cure rate. Testicular self-examination is important to early detection of the tumor.

Allergic and Hypersensitivity Diseases
There are only a few hypersensitivity diseases that affect the reproductive system. Contact dermatitis may cause uncomfortable rashes in the area of the external genitalia. On rare occasions an allergic reaction may occur between sperm and the female cervical and vaginal area, which may result in reduced fertility.

Aging
Aging has its most profound effects in the female with the onset of menopause. **Natural menopause** is a result of declining ovarian function that usually occurs between the ages of 40 and 50. The ovary becomes atrophic and gradually fails to respond to gonadotrophic stimuli. The remaining follicles become atretic and menstruation ceases, signaling the end of the reproductive life of the woman. **Premature menopause** may occur before the age of 40. **Artificial menopause** may occur at any age following the surgical removal of the ovaries or following irradiation for cancer. Declining estrogen levels may result in uncomfortable symptoms of depression, vasomotor instability resulting in sudden attacks of flushing and sweating (hot flashes), and dryness and atrophy of the external genitalia which can result in painful intercourse. Menopause may be accompanied by reduced calcification of bones (osteoporosis) resulting in spontaneous fractures of the vertebra and hips (see Chapter 6, The Clinic: Diseases of the Skeletal System).

mone (GRH), which stimulates production of FSH and LH. FSH stimulates the final development of the follicle; LH induces ovulation. FSH and LH levels decline after ovulation as the corpus luteum begins to synthesize estrogen and progesterone, which inhibit production of FSH. In the absence of pregnancy, the corpus luteum regresses, and estrogen and progesterone levels decline, releasing the anterior pituitary gland from inhibition.

Menstrual Cycle The *menstrual cycle* consists of *menstrual, preovulatory,* and *postovulatory phases.* The preovulatory phase is marked by a buildup of the endometrium and ends at the time of ovulation. During the postovulatory phase, the endometrium develops further in preparation for nurturing an embryo if fertilization occurs. If fertilization does not occur, the stratum functionalis disintegrates and is discharged through the vagina as menstrual flow. If fertilization does occur, the embryo produces human chorionic gonadotropin, which maintains the corpus luteum, and progesterone and estrogen levels remain elevated.

Menopause Menopause is the cessation of menstrual periods resulting from the exhaustion of the supply of oocytes.

Coitus (pp. 722–724)
Sexual intercourse, or *coitus,* has four stages in both males and females.

Male Sexual Response The *arousal stage* in males is initiated by stimuli. The *plateau stage* is accompanied by flushed skin, increased heart and respiratory rates, and elevated blood pressure. The *orgasm stage* is associated with widespread nervous and muscular reactions and sudden release of muscle tension, accompanied by ejaculation in the male. This is followed by a *resolution stage,* which includes a *refractory stage* in males.

Female Sexual Response The stages of coitus for females are basically the same as outlined above for males, except that orgasm is accompanied by vaginal and uterine contractions in females. Also, females are thought not to experience a refractory stage.

Fertilization (pp. 724–726)

In *fertilization*, sperm deposited in the vagina swim and are carried through the uterus and up the oviduct.

Capacitation Sperm must be *capacitated* to fertilize an egg. The corona radiata disintegrates due to the action of digestive enzymes from sperm, and sperm bind to the zona pellucida. Sperm digest and swim through the zona pellucida and emerge in the *perivitelline space*.

Fusion of Sperm and Egg Fusion of the membrane of a single sperm with the egg membrane probably causes an action potential that prevents fusion of a second sperm with the membrane (*fast block*). This is followed by chemical changes that cause a *slow block* to sperm. Fusion of the sperm and egg is followed by completion of meiosis in the egg, producing a *female pronucleus*. The nucleus from the sperm forms a *male pronucleus*. Fusion of pronuclei produces a *zygote*.

SELF-TEST OF CHAPTER OBJECTIVES

True-False Questions
1. Leydig cells secrete testosterone.
2. Sperm are capable of fertilizing the egg immediately after they are produced.
3. The urethral raphe is a distinctly male structure.
4. A prepuce is found in both male and female external genitalia.
5. A sperm is produced from a spermatid by cell division.
6. Helicine arteries are found in the stratum functionalis of the endometrium.
7. Follicle-stimulating hormone (FSH) is important only to the female reproductive process.
8. Luteinizing hormone stimulates interstitial cells in the testes to produce testosterone.
9. The onset of a menstrual cycle is measured from the time of ovulation.
10. Menstruation is the indirect result of the decrease in GRH production by the hypothalamus.

Multiple-Choice Questions
11. Secretions from which of the following glands do sperm first encounter as they pass through the male reproductive tract?
    a. seminal vesicles   c. urethral glands
    b. bulbourethral glands   d. prostate gland
12. At the time an egg is fertilized, it is most likely in the
    a. ovary   c. isthmus of the oviduct
    b. ampulla of the oviduct   d. uterus
13. Of the following tubules, which is the first where a mature sperm is likely to be found?
    a. rete testis   c. epididymis
    b. urethra   d. ejaculatory duct
14. Which is the largest organizational unit in the mammary gland?
    a. lobe   c. alveolus
    b. lobule   d. ampulla
15. In a male, the urethra passes through the
    a. corpus spongiosum   c. tunica albuginea
    b. corpus cavernosum   d. infundibulum
16. Helicine arteries perform which of these functions?
    a. supply blood to the penis
    b. supply blood to the ovaries
    c. are located in spongy tissue of the penis, where they contribute to erection
    d. supply blood to the uterus
17. Sertoli's cells are most important to which of the following processes?
    a. spermiogenesis   c. oogenesis
    b. spermatogenesis   d. ovulation
18. After ovulation, the follicle develops into a
    a. corpus hemorrhagicum   c. corpus luteum
    b. corpus albicans   d. corpus atretica
19. Which of the following hormones most directly influences milk production (as opposed to secretion)?
    a. FSH   c. prolactin
    b. GRH   d. oxytocin
20. The reason that ovulation and menstruation do not occur in a pregnant woman is that
    a. the oviducts are blocked by a fetus growing in the uterus
    b. hormone released by the growing embryo or fetus inhibits GRH production
    c. estrogen prevents the development of additional follicles
    d. an accumulation of prolactin blocks FSH production
21. The portion of the uterine wall that undergoes a cyclic buildup and disintegration is the
    a. myometrium   c. serosa
    b. endometrium   d. muscularis
22. Menstrual flow follows most directly in time
    a. a decrease in FSH concentration in the blood
    b. a decrease in estrogen concentration in the blood
    c. a rise in progesterone concentration in the blood
    d. a rise in LH concentration in the blood
23. During follicle development, several follicles begin to develop, but usually only one follicle develops to maturity and ovulates. The remaining ones
    a. are arrested in development and resume developing during the next menstrual cycle
    b. regress to their original condition and begin developing again in the next menstrual cycle
    c. degenerate to form atretic follicles
    d. release oocytes that have not yet matured
24. An endometrium is found in
    a. a testis   c. a uterus
    b. a ductus deferens   d. a uterine tube

25. The role of Sertoli's cells is to
    a. produce testosterone
    b. produce inhibin
    c. stimulate sperm maturation
    d. produce semen
26. Which of the following usually contributes the greatest amount of fluid to semen?
    a. testes
    b. seminal vesicles
    c. prostate gland
    d. bulbourethral gland
27. The function of an oviduct is to
    a. transport sperm to an egg
    b. nourish a young embryo
    c. transport a young embryo
    d. all of the above
28. What is an infundibulum?
    a. The end of a uterine tube adjacent to an ovary.
    b. The end of a uterine tube that opens into the uterus.
    c. An enlarged portion of an oviduct in which fertilization usually occurs.
    d. A supporting ligament for an ovary.
29. What is an ampulla?
    a. The end of a uterine tube adjacent to an ovary.
    b. The end of a uterine tube that opens into the uterus.
    c. An enlarged portion of an oviduct in which fertilization usually occurs.
    d. A supporting ligament for an ovary.
30. The narrow, necklike extension at the inferior end of a uterus is the
    a. fundus
    b. corpus
    c. isthmus
    d. cervix
31. Following ovulation, blood fills the cavity formed by the follicle where it coagulates and mixes with cells in the cavity. At this point the structure is a
    a. corpus hemorrhagicum
    b. corpus luteum
    c. corpus albicans
    d. cumulus oophorus
32. Following ovulation, a ruptured follicle fills with a yellowish lipid material. At this point, the structure is a
    a. corpus hemorrhagicum
    b. corpus luteum
    c. corpus albicans
    d. cumulus oophorus
33. Which of the following hormones is produced by a corpus luteum in a pregnant woman?
    a. progesterone
    b. FSH
    c. LH
    d. oxytocin
34. Which of the following hormones cause a corpus luteum to develop?
    a. progesterone
    b. FSH
    c. LH
    d. oxytocin
35. One function of the enzyme hyaluronidase is to
    a. capacitate sperm
    b. dissolve mucus in the cervix
    c. stimulate sperm production
    d. stimulate sperm maturation

Essay Questions
36. List the organs through which sperm pass from the time they are produced until ejaculation.
37. List the organs of the female reproductive tract through which an egg or embryo passes from the time the egg is released.
38. List the hormones involved in reproduction in males and females and describe the source and role of each hormone.
39. Describe why a single oocyte is usually released from an ovary even though many follicles begin developing at the same time.
40. Describe the structure of a penis and explain the events associated with an erection.
41. Describe the changes that occur in the uterine wall during a menstrual cycle in relation to the development of an ovarian follicle.
42. List the components and sources of semen.
43. Describe the changes that occur in a follicle following ovulation and the role played by the resulting structure.
44. List and describe the physiologic events associated with coitus in females and males.
45. Describe how a sperm penetrates an ovum.

Clinical Application Questions
46. Describe how surgical sterilization is performed in men and women as a means of preventing pregnancy.
47. Describe how the results of a pap smear are used to help detect cervical cancer.
48. Describe the composition of human milk and explain how its components benefit a newborn baby.
49. Describe the symptoms associated with premenstrual tension syndrome versus dysmenorrhea.
50. Define an STD and compare the symptoms of gonorrhea, syphilis, and inclusion conjunctivitis.
51. List four preventive measures to reduce the spread of sexually transmitted diseases.
52. List and describe the mechanisms by which contraceptive chemicals and devices prevent fertilization.
53. List and compare factors that can interfere with proper functioning of the reproductive system.

28
Development and Inheritance

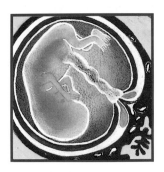

Fertilization produces a cell that contains a unique combination of genetic information, half provided by the sperm and the other half by the egg. Division of the zygote begins **development**, a process that uses this information to produce a new human being. Development involves three coordinated processes: **growth**, the production and enlargement of new cells; **morphogenesis**, the movement of masses of cells within the embryo to form embryonic structures; and **cell differentiation**, the accumulation of biochemical and structural changes that enable cells to perform specific functions.

Because of the manner in which DNA is duplicated and subsequently transmitted by mitosis, genetic information will be transmitted with high fidelity to all the body cells produced throughout life. When a person begins to produce reproductive cells, information will be included in those cells that is nearly identical to that received from his or her parents. To prevent doubling the information in the next generation, only half as much DNA is included in sex cells as is present in body cells. Combination of sperm and egg then produces a zygote with the appropriate amount of DNA. Meiosis is the process that produces cells having half the DNA of body cells. Because of the way information is distributed at meiosis, traits controlled by that information are usually inherited in predictable ways.

In this chapter we will examine

1. the developmental process that follows division of a zygote and leads to the birth of a baby,
2. meiosis, the mechanism used to transmit DNA to sex cells,
3. the principles of inheritance that derive from that mechanism and that govern the transmission of genes from parents to their offspring, and
4. how changes in the number and organization of chromosomes can affect development.

Embryonic Development 733

morphogenesis: (Gr.) *morphe*, form + *genesis*, production
blastomere: (Gr.) *blastos*, germ + *meros*, a part of
blastocyst: (Gr.) *blastos*, germ + *kystis*, bag
trophoblast: (Gr.) *trophe*, nourishment + *blastos*, germ

Embryonic Development

After studying this section, you should be able to:

1. Describe the divisions that occur as an embryo travels through an oviduct.
2. Explain how an embryo becomes embedded in the uterine wall.
3. Define gastrulation and explain how germ layers are produced.
4. List three primary embryonic membranes and structures produced from them in a developing embryo.
5. Describe the structure of a placenta and umbilical cord, and explain how nutrients and waste products are exchanged between mother and developing child.
6. Distinguish between an embryo and fetus.
7. Explain how fraternal and identical twins are produced.

Following fertilization, development and growth occurs over a period of about 280 days. Division of the zygote begins in the oviduct, and the mass of cells produced travels to the uterus where it becomes implanted in the uterine wall.

Figure 28-1 Cleavage. (a) Photomicrograph of two-cell stage (mag. 69,153×). (b) Photomicrograph of four-cell stage (mag. 77,829×). (c) Scanning electron micrograph of morula. Note that the zona pellucida has been removed in part (c). Along the bottom are generalized drawings of each stage of cleavage.

Cleavage

The zygote, still surrounded by a zona pellucida (see Chapter 27), begins to undergo a series of cell divisions in a process called **cleavage**. This process leads to the formation of a ball of cells, a **morula** (mōr′ū·lă), contained in the zona pellucida (Figure 28-1). Each cell of the morula is a **blastomere** (blăs′tō·mēr). The first cleavage division occurs during the first day following fertilization. The second division happens in the second day, and a morula forms by the end of the third day. As it enlarges, the morula becomes spherical and develops a fluid-filled space in its interior, a **blastocoel** (blăs′tō·sēl). At this point, the sphere is called a **blastocyst** (blăs′tō·sĭst). A blastocyst consists of a thin outer layer of cells, the **trophoblast** (trŏf′ō·blăst) (or **trophectoderm**), and a group of cells lying to one side of the interior, the **inner cell mass** (Figure 28-2). Within three days to a week following fertilization, the blastocyst passes into the uterus, where it remains free in the uterine cavity for several days. During this time the trophoblast absorbs nutrients from the secretions present within the uterus. The zona pellucida disintegrates, and the blastocyst increases in size.

Implantation

After two to three days the blastocyst adheres to the endometrium. Cells of the trophoblast secrete enzymes that dissolve adjacent endometrial tissue, and the trophoblast undergoes **implantation** as its cells grow into the space created

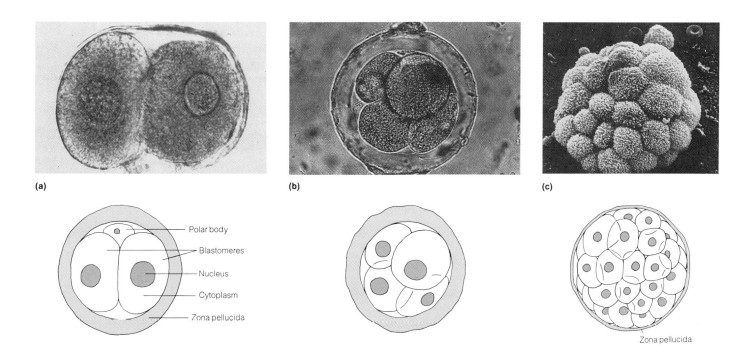

in vitro: (L.) in glass

> ### CLINICAL NOTE
>
> ### ARTIFICIAL REPRODUCTION
>
> Scientists have made great strides in manipulating the process of reproduction. The National Center for Health Statistics reports that about one in five couples may be infertile. Whether the origin of the infertility problem is male or female, infertile couples and single people not only have new opportunities for creating babies, they also have options among a variety of reproductive techniques. (see Unit VII Case Study: A Problem of Reproductive Failure).
>
> #### Artificial Insemination
>
> If the male is infertile, one solution is the injection of anonymous donor sperm (at an average cost of $100 per insemination) into a woman's uterus without coitus, which is called **artificial insemination.** This method can also be used by women who by choice are without partners. The development of liquid nitrogen refrigeration techniques for freezing and storing sperm **(cryopreservation)** has greatly increased the use of artificial insemination (Figure 1). People can now return to the same sperm bank donor for their second baby, thus creating true siblings. Sperm banks rigorously screen the quality of the donated sperm for infectious and genetic diseases (Figure 2).
>
> #### *In Vitro* Fertilization
>
> Once known as test tube fertilization, the technique of *in vitro* **fertilization (IVF)** is suggested for women who have fallopian tube blockage. At the present time more than 100 clinics offer this procedure in the United States, and it costs an average of $6,000 per attempt. The woman is placed on hormone therapy to stimulate egg production. Her eggs are then removed from the ovaries by **laparoscopy** and placed in a petri dish with the partner's or donor's sperm. When the eggs have incubated between 42 and 72 hours and are of the proper size, the now fertilized eggs are placed directly into the woman's uterus to be gestated. Louise Brown, born
>
>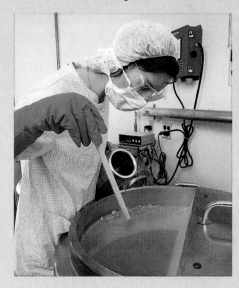
>
> **Figure 1 Cryopreservation.** Here frozen embryos are being placed in liquid nitrogen.

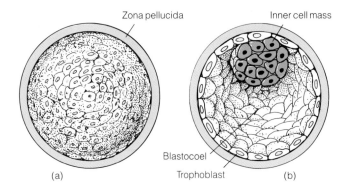

Figure 28-2 Blastocyst. (a) Exterior view showing the sphere formed by a single layer of cells. (b) Interior view showing an inner cell mass located to one side of the interior of the sphere.

by the disintegration of endometrial tissue. Endometrial tissues grow around the blastocyst, eventually surrounding it (Figure 28-3). By the second week following fertilization, the only visible evidence of the implanted blastocyst is a pimple-like bump on the surface of the endometrium. Enzymes produced by the trophoblast continue to dissolve the uterine wall, releasing nutrients used by the embryo for several weeks.

Primary Germ Layers

Shortly after implantation, two layers of cells in the inner cell mass separate from adjacent trophoblast cells and form an **embryonic disk** (Figure 28-4). These cells give rise to the embryo. The space produced between the embryonic disk and the trophoblast is the **amniotic** (ăm·nē·ŏt´ĭk) **cavity.** The cell layer adjacent to the amniotic cavity com-

in vivo: (L.) in the living body
gastrula (L.) *gastrula,* little belly

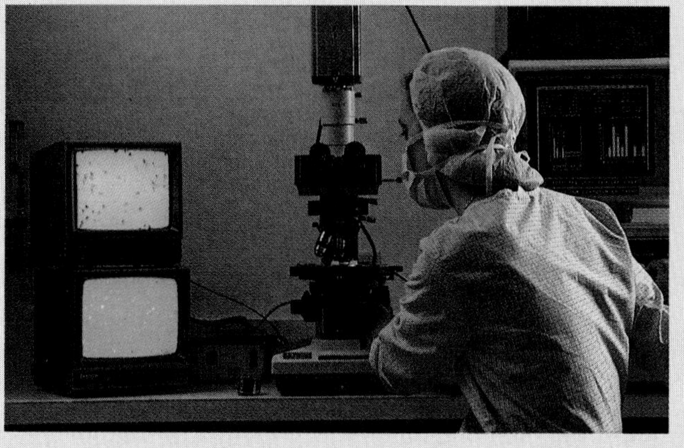

Figure 2 An *in vitro* fertilization embryologist examines semen on a computer analyzer.

in England in 1978, was the world's first "test tube baby." The current IVF rate is 20 to 30% successful pregnancies, which is the same percentage as natural reproduction. IVF results in no increase in the incidence of birth defects but can result in multiple pregnancies.

In Vivo Fertilization and Embryo Transfer
When a woman cannot produce eggs, she can still give birth by means of *in vivo* **fertilization** and embryo transfer. In this process, an ovum of a fertile woman is fertilized through artificial insemination. The embryo that develops is then transferred from the uterus of the fertile woman into the uterus of the infertile woman where the infant completes its development.

Gamete Interfallopian Transfer
Sometimes the exact cause of a couple's infertility is unknown. In this situation, one possibility is **gamete interfallopian transfer (GIFT).** In this process, the eggs and sperm are brought into contact outside the body, then replaced in the fallopian tube together. This creates a more natural environment for conception than the *in vitro* method. The success rate of this method has been about 40% or better.

Surrogacy or Contract Motherhood
If the female partner is infertile or cannot carry a baby due to health reasons, another woman can be contracted (for approximately $25,000) to be artificially inseminated with the male partner's sperm and carry the baby to birth. Upon delivery of the baby, the couple keeps and raises the child as their own and the surrogate mother's responsibilities are ended. There have been approximately 600 surrogate births in this country. Recently, the case of Baby M and her surrogate mother Mary Beth Whitehead has raised many ethical and legal questions that will determine the fate of surrogacy as a means of artificial reproduction.

---

prises the embryonic **ectoderm** (ĕk′tō·dĕrm), and the layer adjacent to the blastocoel is the embryonic **endoderm** (ĕn′dō·dĕrm). The blastocoel is now referred to as the **extraembryonic coelum.** By the beginning of the third week, a third layer of cells, a **mesoderm** (mĕs′ō·dĕrm), develops between the ectoderm and endoderm (Figure 28-5). Each of these layers gives rise to a group of adult structures (Table 28-1). The ectoderm, mesoderm, and endoderm together comprise the **primary germ layers.** Development of the three germ layers is called **gastrulation,** and the embryo is called a **gastrula** (găs′trū·lă) at this stage of its development.

Embryonic Membranes

Simultaneous with development of the primary germ layers, four membranes develop from neighboring tissues: the amnion, yolk sac, allantois, and chorion. These help to nourish and protect the embryo.

Amnion

The **amnion** (ăm′nē·ŏn) develops from cells surrounding the amniotic cavity (Figure 28-4). Early in development, it is continuous with the embryonic ectoderm (Figure 28-6). However, as development proceeds, the amnion expands and surrounds the embryo in a fluid-filled bag. **Amniotic fluid** in the bag forms a shock absorber and helps to maintain an appropriate temperature for the embryo (see Clinical Note: Amniocentesis). The amnion usually ruptures just prior to birth of the child.

Yolk Sac

The **yolk sac** develops as a cavity on the opposite side of the embryonic disk from the amnion (Figures 28-4 and 28-6). Its tissues are continuous with the embryonic endoderm. Only a small amount of yolk is present, and the yolk sac

736 Chapter 28 Development and Inheritance

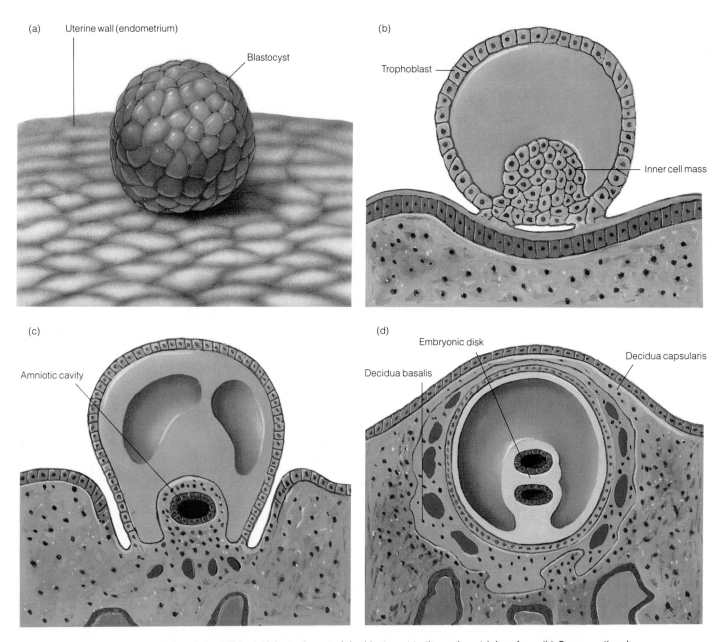

Figure 28-3 Implantation. (a) External view of the initial attachment of the blastocyst to the endometrial surface. (b) Cross-sectional view of the attached blastocyst. (c) The blastocyst begins to digest its way into the endometrium. (d) The endometrium grows over the surface of the blastocyst, completely enclosing it. Source: Adapted from J. E. Crouch, *Functional Human Anatomy*, 2nd ed. (Philadelphia: Lea & Fibeger, 1972).

Embryonic Development   737

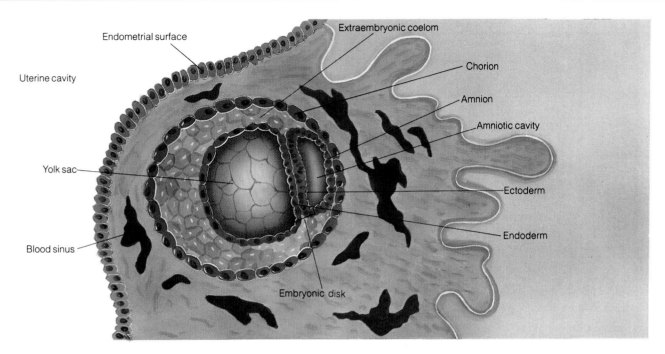

Figure 28-4   Detailed diagram of the embryonic disk and surrounding structures.

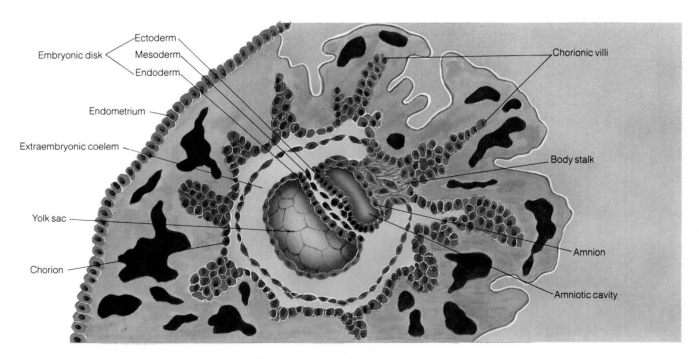

Figure 28-5   Germ layers. This detailed view shows the development of the three primary germ layers in the embryonic disk that occurs during gastrulation.

amniocentesis: (Gr.) *amnion*, lamb + *kentesis*, puncture
allantois: (Gr.) *allantos*, sausage + *eidos*, resemblance
chorion: (Gr.) *choroeides*, resembling a membrane

Table 28-1 EMBRYONIC GERM LAYER DERIVATIVES

Germ Layer	Tissues and Organs
Ectoderm	Adrenal medulla
	Brain, spinal cord, and peripheral nerves
	Epidermis, hair, nails, and glands of the skin
	Lining of the mouth, nasal cavity, and paranasal sinuses
	Salivary glands
	Sense organs
	Lining of the anus
Mesoderm	Bone, bone marrow, and cartilage
	Connective tissue
	Blood and cardiovascular organs
	Dermis
	Tunics of the eyes
	Skeletal and most smooth muscle
	Kidneys, ureters, and urinary bladder
	Peritoneum and mesentery
	Reproductive organs
	Lymphoid tissue
Endoderm	Digestive organs, liver, and gallbladder
	Lungs
	Urinary bladder epithelium
	Lining of the middle ear
	Thyroid, parathyroid, and thymus glands

CLINICAL NOTE

AMNIOCENTESIS

Amniocentesis (ăm″nē·ō·sĕn·tē′sĭs) is a technique of obtaining a sample of the amniotic fluid from the uterus of a pregnant woman. A needle is inserted into the amniotic sac, usually guided with the use of ultrasonography, and 10 to 15 mL of fluid is removed. This fluid contains a high percentage of fetal cells, which can be cultured and a **chromosomal analysis** performed. Over 50 metabolic diseases can be diagnosed by this technique; the sex of the embryo can also be determined. Amniocentesis is usually performed between the 14th and 15th week of gestation so that the diagnosis can be made while a therapeutic abortion is still an option. Amniocentesis is routinely performed on pregnant women 35 years of age and older to check for chromosomal abnormalities in developing fetuses, the likelihood of which may increase dramatically with the age of the mother.

Later in pregnancy, amniocentesis is used to help determine the maturity of the fetal lungs by analyzing the lethicin to sphingomyelin ratio in the fluid. A ratio greater than 2:1 ensures maturity of the fetal lung. Such tests help the obstetrician determine maturity when an early delivery is planned, such as occurs for mothers with diabetes.

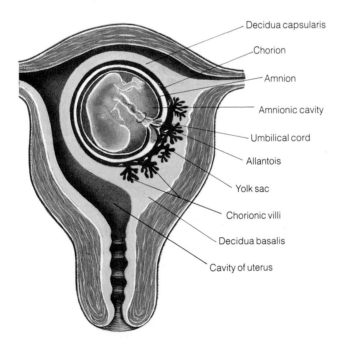

Figure 28-6 Fetal membranes.

provides no known nutritive function. In an early embryo it provides primordial germ cells that will migrate into the ovary or testis of the embryo, and it is where blood cells form prior to development of blood-forming tissues within the embryo. Eventually, the yolk sac is incorporated into the placenta, and some of its cells also contribute to the formation of the digestive tube.

Allantois

The **allantois** (ă·lăn′tō·ĭs) is a small, highly vascularized sac that develops from endoderm in an early embryo (Figure 28-6). Its blood vessels become the umbilical arteries and vein, and it produces blood cells during the early life of an embryo.

Chorion

The **chorion** (kō′rē·ŏn) develops from the trophoblast and mesoderm. As an amnion enlarges, it fuses to the chorion, forming a double-layered protective sac around the embryo (Figure 28-5). The embryo, amnion, and yolk sac are connected to the chorion by the **body stalk.** With time, the body stalk lengthens and forms the **umbilical cord** (Figure 28-6).

Placenta and Umbilical Cord

The **placenta** and umbilical cord attach an embryo to its mother (Figure 28-7). Nourishment and other necessities are delivered to the embryo and wastes are removed through these structures. A placenta consists of structures derived from fetal tissue, the **fetal portion,** and structures derived

decidua: (L.) *deciduus*, falling off

> **CLINICAL NOTE**
>
> ## ULTRASONOGRAPHY OF THE UTERUS
>
> The use of **ultrasonography** (or ultrasound) for diagnosis is of particular interest to the obstetrician. It is a painless, noninvasive diagnostic examination that can be repeated as needed with no apparent harm to the mother or fetus. Ultrasonography is particularly useful for defining fluid-filled structures such as the pregnant uterus. A pregnancy can be visualized as early as the fifth week of gestation (see figure). The technique is also useful in differentiating a normal intrauterine pregnancy from an ectopic or tubal pregnancy. Serial scans are used to determine the fetal growth rate by measuring the diameter of the head and the chest. At times, even the sex of the fetus can be determined if the genitals are visualized. The position of the placenta can be determined, which is useful in predicting problems with a low lying placenta and is necessary if amniocentesis is planned.
>
> Ultrasonography is especially useful in detecting abnormalities of the fetal heart, kidneys, and neural tubes. This technique, along with amniocentesis and **fetoscopy** (using an optical device made of fiberoptics to visualize the fetus directly), has allowed surgeons to operate on the fetus in the uterus and repair abnormalities before birth. In another example of its use, a child was recently found to have an undeveloped left ventricle of the heart while still in the uterus. Ultrasonography allowed the physicians to deliver the child by cesarean section as soon as a suitable donor heart was available and to proceed with an immediate heart transplant, which saved the infant's life.
>
> **Ultrasound image of a fetus showing fetal profile.** The EKG record across the bottom shows the fetal heartbeat.
>
>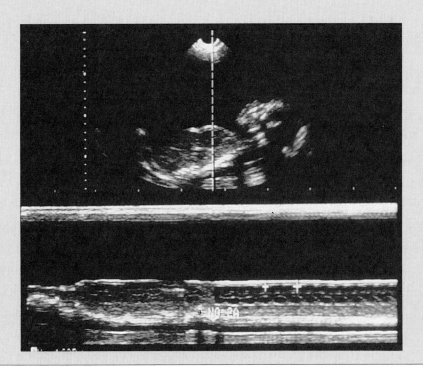

from the uterine endometrium, which comprise the **maternal portion** (Figures 28-8 and 28-9).

As previously discussed, during implantation the trophoblast sinks into the endometrium. The outer layer of endometrium, regularly lost during menses, is called the **decidua** (dē·sĭd´ū·ă). The portion that grows over the trophoblast is the **decidua capsularis** (căp·sū·lăr´ŭs). The deep portion of the endometrium, the **decidua basalis** (bā·săl´ŭs), is responsible for forming the maternal part of the placenta.

About three weeks after fertilization, the outer layer of the chorion produces many **chorionic villi** (kō·rĕ·ŏn´ĭk vĭl´ī), fingerlike projections that penetrate the decidua basalis. Enzymes secreted from the trophoblast cells that help form the chorion digest neighboring endometrial cells, creating **intervillous spaces** that fill with maternal blood. The chorionic villi enlarge as the placenta increases in size and complexity. (Figure 28-8). Umbilical arteries of a fetus carry deoxygenated blood into the placenta and branch into capil-

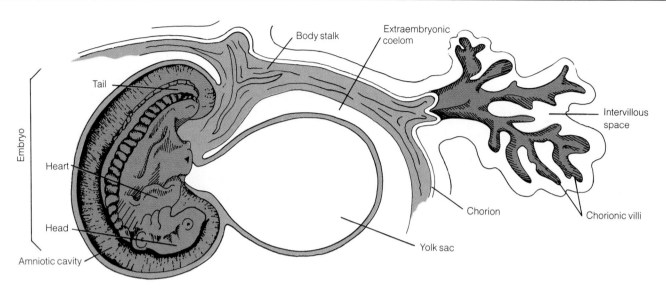

Figure 28-7 Placenta and associated structures.

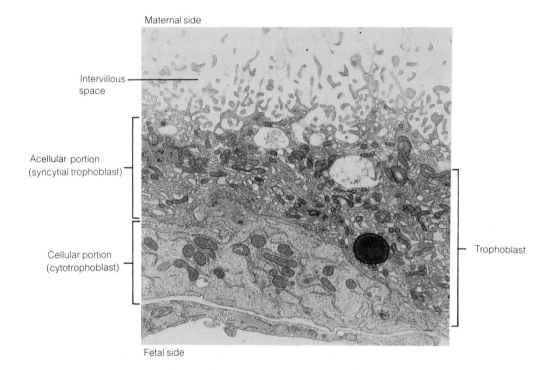

Figure 28-8 Transmission electron micrograph of a portion of a placenta at full term. Nutrients in maternal blood traveling through intervillus spaces are absorbed by a modified acellular portion of the trophoblast that lies on the villus surface (mag. 8260×).

laries of the villi. The few layers of cells that separate fetal blood from maternal blood comprise a **placental membrane** that prevents mixing of the blood. Oxygen, nutrients, water, hormones, vitamins, minerals, and antibodies in maternal blood diffuse across the placental membrane into capillaries of the chorionic villi. At the same time metabolic wastes, mainly carbon dioxide and urea, diffuse from fetal blood into the intervillous blood. In this way, a mother supplies her fetus with materials essential for development and removes waste materials produced by it.

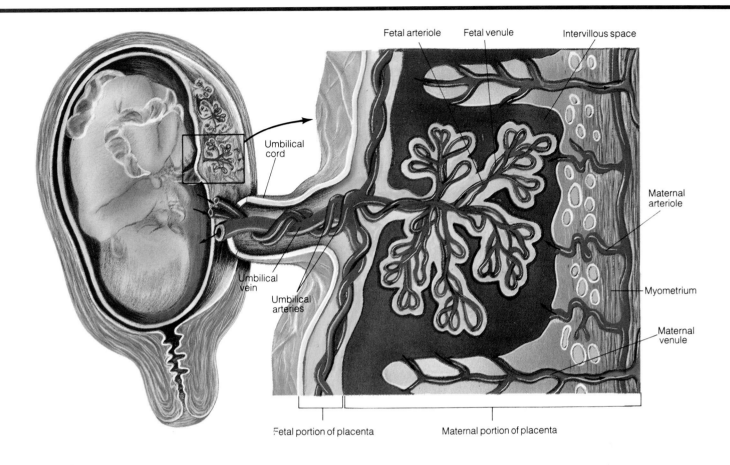

Figure 28-9 Structure of the placenta and umbilical cord.

The umbilical cord (Figure 28-9) is a ropelike connection between the fetus and placenta, which becomes about 35 to 50 cm long by the end of pregnancy. The outer surface is stratified epithelium derived from the inner wall of the amnion as the fetus becomes surrounded by the amniotic cavity. Within the cord, two **umbilical arteries** carry fetal blood to the placenta, and a single **umbilical vein** returns blood from the placenta to the fetus. The vessels are surrounded by connective tissue that produces large amounts of a gelatinous material called **Wharton's** (hwar′tŏnz) **jelly.**

Fetal Development

During the first 8 weeks of embryonic life, the primary germ layers begin to produce the major organs. The sex of the embryo is apparent, and subsequent development refines the basic patterns established during the first 8 weeks. There is a rapid increase in size, and many organs become functional. From this stage onward in its development, the embryo is referred to as a **fetus.** Growth and development of the embryo and fetus is referred to as **gestation.** This period involves an average of 280 days. A summary of fetal development is presented in Figure 28-10.

Twins

Twinning is the simultaneous development of two or more fetuses. It occurs in about 1 out of 85 pregnancies. The most common type occurs when graafian follicles develop in both ovaries simultaneously and each releases an oocyte. Fertilization of both oocytes produces **fraternal** or **dizygotic** twins. **Identical** or **monozygotic** twins result from a single fertilized egg. This usually occurs from a failure of the two cells produced by the first cleavage division to stick together. Only rarely does an embryo split later in cleavage. After separating, each cell develops into an embryo. Because both embryos originate from the same zygote, they are genetically identical. Fraternal twins, in contrast, develop from different zygotes.

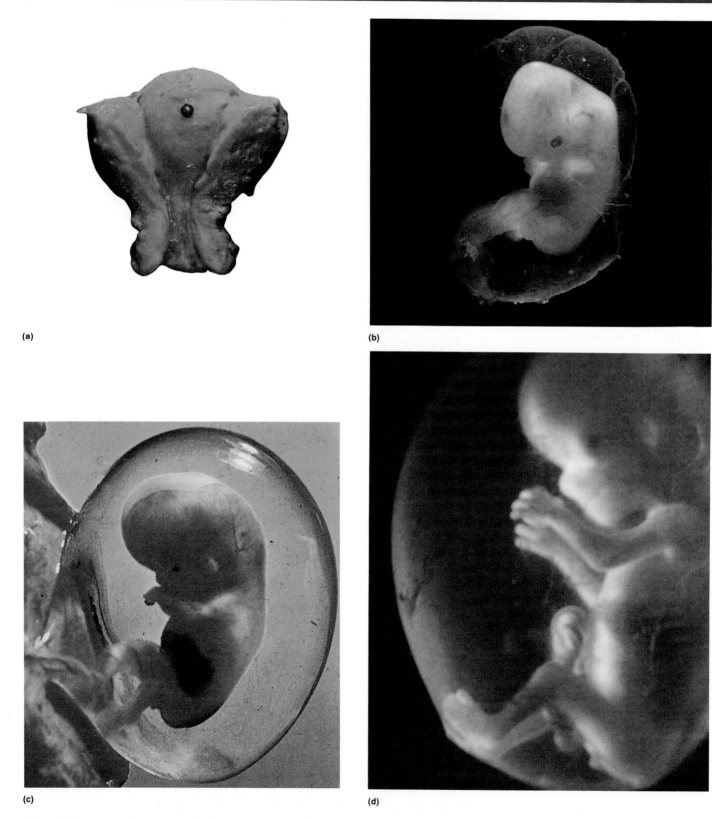

Figure 28-10 Summary of fetal development. (a) Embryo implanted on uterine wall. (b) Embryo at 5 to 6 weeks development. (c) Very young fetus, about 2 months old. (d) Fetus about 4 months old.

Pregnancy and Childbirth 743

> **CLINICAL NOTE**
>
> **ABORTION**
>
> A **therapeutic abortion** is an interruption of pregnancy before the fetus can survive outside the uterus. Other names for a therapeutic abortion are induced, medical, or planned; these differentiate it from a spontaneous abortion, which occurs from natural causes. A therapeutic abortion can be done for a variety of reasons: if the woman's life is endangered by a continued pregnancy; if the fetus is found upon amniocentesis to have a chromosomal defect; or if the woman does not wish to have a child at that time in her life. Therapeutic abortions are an option for any woman in the United States as long as the pregnancy has progressed less than 12 weeks. After 12 weeks abortions are allowed in some states but not in others.
>
> During the first trimester the most common abortion method is vacuum aspiration, which removes the fetus by suction from the uterine wall. The uterus is then scraped with surgical knives called curettes. Another process called "D and C" (dilation and curettage) may also be used in the first trimester. If the abortion is performed in the second trimester (which is less common), a method called "D and E," or dilation and evacuation, is used. Another method involves the injection of a salt solution into the amniotic fluid surrounding the fetus. The solution kills the fetus and causes it to be expelled from the mother's body. Still another method, only performed in the second trimester, is through an injection of prostaglandin drugs that cause muscle contractions to abort the fetus.
>
> Abortion raises several volatile issues in our society. Under what circumstances, if any, should the law permit a woman to have an abortion? Should there be laws that protect the unborn fetus and to what extent? Supporters of abortion believe that legalized abortion eliminates many illegal, unprofessional, and unsanitary abortions, as well as preventing many unwanted births to people who cannot afford or are unable to raise children. Opponents, on the other hand, believe that an abortion is tantamount to murder of an unborn child and that legal abortion creates a sexually irresponsible society. In the 1960s most states prohibited abortions except when the health of the mother was in jeopardy. In 1973, however, the Supreme Court ruled that a woman's right to privacy was protected by the Constitution and that the states could not forbid an abortion in the first trimester of a woman's pregnancy.

Pregnancy and Childbirth

After studying this section, you should be able to:

1. List and describe several indicators of pregnancy.
2. Describe how human chorionic gonadotrophin is used to determine whether or not a woman is pregnant.
3. Explain how a birth date is predicted.
4. Describe the physiological changes that occur in a woman during pregnancy.
5. Describe the three stages of childbirth.

During pregnancy, a woman's body undergoes numerous physiological and anatomical changes. In addition to providing nutrients to the fetus, the placenta secretes hormones that affect the mother's metabolism. The physiological changes that result help nourish the fetus as it grows and develops.

Diagnosis of Pregnancy

There are several **presumptive signs** of pregnancy. The first is usually **amenorrhea**, or the absence of menstruation. However, because other factors can have the same effect, missed menses can only be regarded as a possible indicator of pregnancy. A second presumptive sign is fatigue. Even before a missed menstruation, a pregnant woman often has episodes of fatigue and drowsiness. About one-third of pregnant women experience nausea that often leads to vomiting. Because this experience often occurs early in the day, it is referred to as "morning sickness." The nausea usually disappears by midday and may leave the woman especially hungry, so that she is motivated to eat and meet the additional nutritional requirements that her pregnancy imposes. Other early signs that indicate possible pregnancy include more frequent urination, changes in color of vaginal and cervical tissue (as a result of increased blood flow to those areas), darkening of the areola and elevation of small glands around the nipples, and tingling sensations in the breasts. Fetal movement may be felt by the woman after 18 to 20 weeks.

More reliable than presumptive signs of pregnancy are the **probable signs** of pregnancy. Pelvic examination will reveal enlargement and softening of the uterus after about 6 weeks following conception. An enlarged abdomen and uterus become apparent after 8 to 12 weeks, and there may be intermittent and painless uterine contractions. By about the 24th week, the fetal outline can be felt and fetal movements are considerably more discernible.

Positive signs of pregnancy include fetal heart sounds heard with a stethoscope after about 18 to 20 weeks or with

> **CLINICAL NOTE**
>
> ### COMPLICATIONS DURING PREGNANCY
>
> The first trimester of pregnancy is particularly hazardous for a developing fetus. Such substances as alcohol, medication, and drugs as well as infection can cause damage to a fetus early in its development. Pregnant women are advised to avoid the use of alcohol, drugs, and any unnecessary medications. Maternal rubella in the first trimester can result in fetal defects that will lead to deafness and congenital heart disease after the baby is born.
>
> First trimester bleeding can be an innocent occurrence, but when combined with cramping may suggest spontaneous abortion. **Spontaneous abortion,** or **miscarriage,** is the loss of a fetus and placenta before the 20th week of pregnancy. About 30 to 40% of pregnant women experience bleeding and cramping in the first trimester; about 20% go on to abort spontaneously. In about 90% of spontaneous abortions the fetus is absent or grossly deformed and an additional 5% show chromosomal abnormalities.
>
> Another life-threatening cause of early bleeding and cramping is an **ectopic pregnancy,** which results when an embryo implants outside the uterine cavity. This usually occurs in the oviduct, although ectopic pregnancies can also occur on an ovary, in the cervix, or in the abdominal cavity. Aside from a cessation of menstruation, a woman is symptomless until the area where the embryo is growing becomes distended. Ectopic pregnancy can rupture into the abdomen and cause massive intraabdominal bleeding and shock. Emergency surgery is required to stop the bleeding. Tubal pregnancies have been associated with the use of intrauterine devices (see Chapter 27 Clinical Note: Contraceptive Devices).
>
> The first trimester may also be complicated by **pernicious vomiting.** This is defined as nausea and vomiting that is either prolonged past the first 12 weeks of pregnancy or of such a serious nature during the first 12 weeks that dehydration, improper fat metabolism, and significant weight loss occurs. In some cases, it requires hospitalization to treat dehydration and acidosis.
>
> The second trimester is usually the most problem-free period, during which the developing fetus increases in size.
>
> A rare complication of pregnancy that may start in this period is **pre-eclampsia.** This is actually a toxic reaction to pregnancy involving hypertension, headaches, albuminuria, and edema during the second trimester, which may persist until a week after delivery. Rarely this condition leads to true **eclampsia,** which results in convulsions or coma. Treatment consists of bedrest, control of the blood pressure, and prevention of convulsions.
>
> The third trimester is a period of rapid growth of the fetus and may be made uncomfortable by the enlarging pregnancy, which can cause obstruction to venous return from the lower extremities, resulting in ankle edema, varicose veins, and hemorrhoid problems. Vaginal bleeding during the third trimester suggests a different and perhaps more serious group of problems. **Abruptio placentae** is a premature separation of the placenta from the uterine wall. Bleeding occurs behind the placenta and may be concealed or it may pass behind the membranes and appear vaginally. Major bleeding will cause fetal distress and possibly death. Shock and serious complications for the mother can also occur. If the bleeding persists or if the fetus shows signs of distress, the fetus should be delivered as soon as possible either vaginally or by cesarean section. **Placenta previa** is low implantation of the placenta near or over the internal os of the cervix. When the cervix dilates in early labor, bleeding can occur. If the placenta is completely covering the os, cesarean section may be necessary.
>
> Premature labor is another problem of the third trimester, which can result in the delivery of a premature infant. The size and maturity of the infant is critical. Every attempt to delay delivery until the fetus is viable should be made.
>
> Maternal diabetes mellitus (see Chapter 17 The Clinic: Diabetes Mellitus) is a serious complication of pregnancy that frequently results in a very large baby and excessive amniotic fluid in the uterus. Diabetic mothers are controlled very carefully and plans should be made for early delivery before the excessive size of the baby complicates delivery. In spite of the best care, there is a 5% chance of fetal mortality between the 36th week and full term.

other sound detection instruments as early as 10 to 12 weeks. Fetal movement can be felt or heard by a physician after about 24 weeks. X-ray examination, although rarely used, can provide positive evidence of pregnancy after about 14 weeks. Proof of pregnancy may also be obtained by ultrasound examination which reveals fetal movement and heartbeat.

The most definitive early test for pregnancy is based on the production of **human chorionic gonadotrophin (HCG),** a hormone produced by the trophoblast. HCG is a glycoprotein that consists of two polypeptide subunits, designated α (alpha) and β (beta). The α-subunit is identical to polypeptides in thyroid-stimulating hormone, luteinizing hormone, and follicle-stimulating hormone. The β-subunit, however, is distinctive and allows the production of antibodies specific for it. Within about 9 days following conception, HCG can be detected in the mother's urine and plasma on the basis of a test using antibody that recognizes the β-subunit. Urine and serum pregnancy tests are about 95% accurate within the first 2 or 3 weeks of pregnancy and virtually 100% accurate after about 6 weeks.

Duration of Pregnancy

A full-term pregnancy averages about 266 days from the time of conception. If the woman's menstrual periods were regularly 28 days apart before pregnancy, it is customary to date the pregnancy from the first day of the last menstrual period, which will normally have occurred 2 weeks before conception. Based on this, a normal pregnancy is counted as 280 days, or approximately 9 months. It is customary to divide this into three 3-month periods, or **trimesters.**

The date of delivery, or estimated date of confinement (EDC), is predicted with Nägele's rule by subtracting 3 months from the date of onset of the last menstrual period and adding 7 days. While only 10% or fewer patients will deliver on the predicted date, 50% will deliver within 1 week and 74 to 88% within 2 weeks. Consequently, birth 2 weeks before or after an estimated date is within the normal range of error in predicting the precise date of delivery.

Changes in Physiology During Pregnancy

In addition to the signs of pregnancy noted above, many other changes occur in the physiology of a pregnant woman and affect every organ system in her body (Table 28-2). These changes generally reflect increased demands on the mother's organs that result from the added requirements of the growing fetus. Most of the changes, however, revert to normal following delivery of the baby.

Cardiovascular Changes

Some of the most dramatic changes occur in cardiac output and blood volume, as the uterine and placental blood vessels develop. Starting at about the 6th week of pregnancy, the amount of blood pumped by the heart increases until, by about the 24th week, it may be 50% higher than it was before pregnancy. Blood volume may increase another 30% during labor. Heart rate may increase from a normal of about 70 beats per minute to 80 or 90, and there is a corresponding increase in blood volume pumped with each beat.

Increased blood volume parallels increased cardiac output, which may be 50% greater than it was before the woman became pregnant. Red blood cells do not increase proportionately, however, (typically rising about 25%), thus blood hemoglobin concentration becomes more dilute. Increased red blood cells require increased iron, as do the placenta and fetus, and anemia may result if iron supplements are not taken. Leukocytes also increase, especially during labor and the first few days following birth of the baby. Plasma concentrations of cholesterol and triglycerides increase, and some of these compounds are used to produce hormones and milk.

Pulmonary Changes

Although vital capacity does not change, tidal volume, respiratory rate, and minute volume all increase during pregnancy. Inspiratory and expiratory reserve volumes, residual volume, and residual capacity all decrease. Carbon dioxide

Table 28-2 COMMON PHYSIOLOGICAL CHANGES IN A PREGNANT WOMAN

First Trimester
Increased urination because of hormonal changes and the pressure of the enlarging uterus on the bladder
Enlarged breasts as milk glands develop
Darkening of the nipples and the area around them
Nausea or vomiting, particularly in the morning
Fatigue
Increased vaginal secretions
Pinching of the sciatic nerve as the pelvis bones widen and begin to separate
Irregular bowel movements

Second Trimester
Thickening of the waist as the uterus grows
Weight gain
Increase in total blood volume
Slight increase and change in position of the heart
Darkening of the pigment around the nipple and from the navel to the pubic region; darkening of face
Increased salivation and perspiration
Secretion of colostrum from breasts
Indigestion, constipation, and hemorrhoids
Varicose veins

Third Trimester
Increased urination because of pressure from the uterus
Tightening of the uterine muscles (Braxton-Hicks contractions)
Breathlessness because of increased pressure from uterus on lungs and diaphragm
Heartburn and indigestion
Trouble sleeping because of baby's movements or need to urinate often
Navel pushed out

output and oxygen uptake increase as the mother's blood provides for the needs of the developing fetus. Respiratory tract edema may occur and contribute to nasal stuffiness and blockage of the eustachian tubes.

Renal Changes

Like the cardiovascular and pulmonary systems, greater demands are placed on the kidneys during pregnancy. There is increased flow of blood into the kidneys, and glomerular filtration rate increases by as much as 50% by the 24th week. As the kidneys work more effectively, nitrogenous wastes in the blood diminish. The ureters dilate, and the woman feels a need to urinate more frequently as a result of the pressure on the urinary bladder caused by the enlarged uterus. This is especially pronounced when the woman is lying in a supine position.

Digestive System Changes

As the uterus enlarges, it compresses the rectum and colon. This may result in constipation. Elevated progesterone levels that occur during pregnancy inhibit smooth muscle activity so gastric motility diminishes and stomach contents take

CLINICAL NOTE

NATURAL CHILDBIRTH PREPARATION

Natural childbirth is the delivery of an infant without the use of analgesics, sedatives, or anesthesia. It is called "natural" because this has been the approach used throughout history before the development of modern obstetrics. Today, natural childbirth is accomplished by the pregnant woman and her partner attending a series of classes in which they are educated about childbirth and pregnancy. Thus, natural childbirth could more accurately be called **prepared childbirth.** The couple shares their feelings and concerns about pregnancy and childbirth with other couples attending the classes. They do exercises to help them relax, and the mother and her partner learn breathing exercises that help the mother to have control over her discomfort during the actual birth. The mother's partner acts as a coach to assist her in relaxing and breathing appropriately during labor and delivery (see figure). The mother and her partner may see movies of births and deliveries and have an opportunity to visit the maternity unit of a hospital or birthing centers. If it is known that the woman's baby will be delivered by cesarean section, classes to prepare for this event are also available.

Natural childbirth. (a) Father "coaching" mother in breathing exercises during labor. (b) The infant's head is emerging with physician's guidance. (c) The infant takes its first breath; the umbilical cord is still attached. (d) Mother and father comforting their newborn infant.

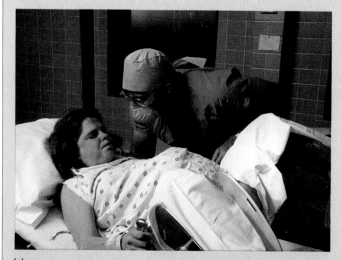
(a)

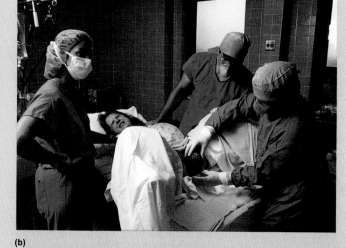
(b)

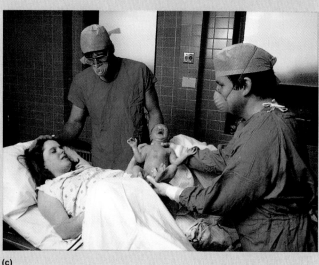
(c)

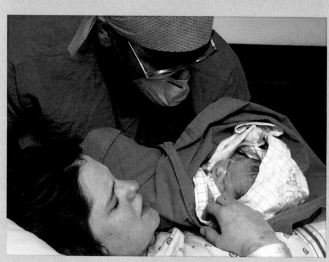
(d)

parturition: (L.) *parturitio*, act of giving birth

> **CLINICAL NOTE**
>
> **COMPLICATIONS OF LABOR AND DELIVERY**
>
> During the first stage of labor it is of great importance for the attending physician to discern the presenting part of the baby (head, buttocks, foot, or hand), its position, and at what level it is in the pelvis. Abnormal fetal presentation and position may make vaginal delivery unlikely and cause undue stress on the fetus, resulting in a **cesarean** delivery. It is equally important to monitor the condition of the fetus via the fetal heart rate. A slowing of the rate suggests stress on the fetus.
>
> **Fetopelvic disproportion** is a complication that can occur during the second stage of labor as the baby passes through the narrowest part of the pelvis. This occurs when the presenting part, such as the buttocks, is too large for the maternal pelvis to accommodate, causing prolonged labor and excessive stress on the fetus and mother. Cesarean section is necessary for a safe delivery.
>
> Complications of the third stage, which is the actual emergence of the baby, can mean lacerations of the maternal perineum. Most physicians elect to make an **episiotomy** (ĕ·pĭs´ē·ŏt´ō·mē), an incision in the lower vagina and perineum to provide more room and to avoid trauma to the fetus. The placenta is normally delivered into the vagina several contractions after the delivery of the baby and is lifted out. On occasions, the placenta may adhere to the uterine wall or may be torn and result in excessive bleeding, requiring surgical removal.
>
> In a normal, uncomplicated birth, the baby presents its head first with the face aimed posteriorly toward the anus and slightly to the left or right. Vaginal examination at the time of admission to the delivery room allows the physician to determine the baby's presentation, the position, and the station (the degree of descent through the pelvis). Presentations other than head first, face down are usually due to an abnormality of the pelvis or of the presenting part and complicate the delivery. At this time, almost any abnormal presentation is considered an indication for a cesarean section to avoid excessive trauma to the mother and baby.
>
> In times past when cesarean sections were more dangerous, the outcome of an abnormal presentation depended on the skill of the physician. In cases of **breech** (buttocks) presentation, the physician might attempt to turn the infant in the uterus to a cephalic presentation or deliver it in the breech position. A breech delivery consists of reaching into the uterus and sweeping down each foot in turn, putting traction on the ankles that usually delivers the buttocks and torso, then turning the baby to one side and sweeping down and delivering the lower arm. A similar maneuver to the opposite side allows delivery of the remaining arm. The head may be delivered by cradling the torso over the physician's forearm while he or she controls the head by inserting a finger into the baby's mouth; the head is then delivered by extending the entire torso up over the mother's abdomen. The head may also be delivered by a specially designed instrument (Piper forceps) that is inserted between the vaginal walls and the baby's buttocks cheeks while the physician gently extracts the head.
>
> Some abnormal presentations, such as a posterior face with the neck extended or an impacted shoulder, are considered to be impossible situations for vaginal delivery so the cesarean section is elected. Improved methods of anesthesia and the use of antibiotics have today reduced the complications of cesarean section.

longer to move into the small intestine. Perhaps due to this, belching and heartburn are common in pregnant women.

Itchiness often develops in the second or third trimester as a result of decreased bile output, the cause of which may be hormonal imbalances. Jaundice may occur subsequently and be accompanied by unusually dark urine. These symptoms disappear after the baby is born.

Endocrine Changes

With pregnancy, the placenta begins to produce gonadotrophins. As a result, progesterone and estrogen levels remain elevated in the blood throughout the pregnancy. Prolactin concentration also rises throughout pregnancy, as does **human chorionic somatomammotrophin (HCS).** HCS is unique to pregnancy and increases the delivery of glucose and other nutrients to the developing fetus.

Most other hormone levels are changed in a pregnant woman as well. Insulin production increases in the second trimester, reaches a maximum, and then declines in the third trimester. Aldosterone concentration rises continuously during pregnancy and at its peak may be six to eight times as high as it was prior to pregnancy. Cortical steroid hormone concentrations rise and may be responsible for increased adipose tissue and breast development. Parathyroid hormone concentration increases, probably in response to the mother's loss of calcium to fetal bone growth. Thyroid hormone levels also increase and result in increased metabolic rate and heart rate. Human growth hormone levels decline, probably due to the activity of the more potent HCS, and LH and FSH secretion is inhibited due to the sustained high levels of progesterone and estrogen.

Labor and Delivery

The birth of an infant is referred to as **parturition** (păr·tū·rĭsh´ŭn) (Figure 28-11). The process is initiated by wavelike

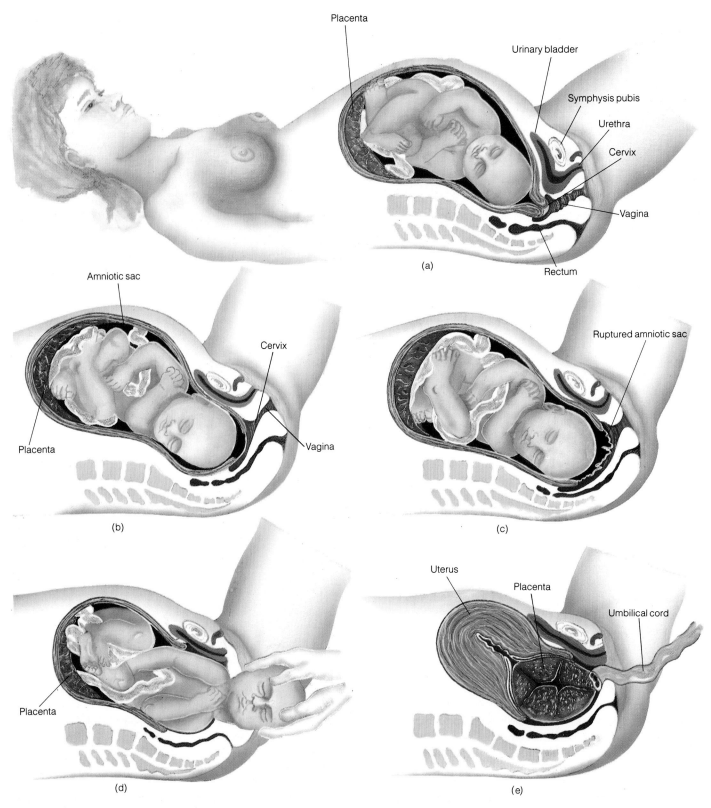

Figure 28-11 Parturition. (a) Initial stage of childbirth. Contractions have begun but the cervix has not yet begun to dilate. (b) The cervix has dilated and the baby's head has begun to push into the vagina. (c) The amniotic sac ruptures. (d) The baby's head emerges from the birth canal. Here, the baby's head has become slightly misshaped from being forced through the narrow open. Normal shape will return within about two weeks. (e) Continued contractions after the baby is born separate the placenta from the uterus and force it into the lower part of the uterus or vagina.

meiosis: (Gr.) *meiosis,* diminution

muscular contractions in the uterus stimulated by an increase in the blood level of oxytocin. These contractions, collectively referred to as **labor,** force the baby from the uterus and through the cervix and vagina. Labor and delivery are divided into three stages. The first stage begins with uterine contractions and dilation of the cervix. Contractions begin sporadically and progress to a regularity that can be timed, usually about 3 minutes apart in effective labor. Contractions start at the upper end of the uterus, and like peristaltic contractions, work their way toward the cervix. This pushes the amniotic sac into the birth canal, where it usually ruptures and releases the amniotic fluid. This is usually a sign that birth is about to occur. During the last weeks of pregnancy and during early labor, the cervix becomes flattened and begins to dilate. The frequency of contractions and the degree of cervical dilation are indications of the progress of labor.

The second stage of labor begins when the cervix is fully dilated (at about 10 cm in diameter) and the fetus begins to descend in the uterus through the narrowest part of the pelvis. Usually, the baby's head is directed downward so that the baby will emerge head first (but see Clinical Note: Complications of Labor and Delivery).

The third stage of labor begins with the appearance of the baby at the opening to the birth canal. Emergence and expulsion usually takes only a few minutes once the infant's head has entered the birth canal. Shortly after the baby is born, additional uterine contractions expel the placenta and umbilical cord as afterbirth. The cord is usually severed by the mother or attending physician near the point where it attaches to the infant. The stub shrivels and forms a circular depression, the **umbilicus** (ŭm·bĭ·lĭ′kŭs) or navel. Over the next few weeks, continued contractions of uterine muscles reduce the uterus to near its prepregnancy size.

Meiosis

After studying this section, you should be able to:

1. Define homologous chromosomes and describe their behavior and appearance at the beginning of meiosis.
2. List the steps of the first division of meiosis and explain how the two cell products differ in their chromosome complements from the cell that produced them.
3. List the steps of the second division of meiosis and describe what happens to chromosomes at each step.
4. Describe the differences in meiosis in males and females.

The role of a sperm and an egg is to carry genetic information contributed by each parent to the production of a new individual. This information is encoded in DNA, organized into discrete units known as **genes** that lie along the length of the DNA molecule present in each chromosome (Figure 28-12). With the important exception of genes on chromosomes involved in determining sex, similar genes are generally carried in eggs and sperm. The fusion of an egg and a sperm thus produces a zygote in which each gene is represented twice, one from the mother and one from the father. As DNA replicates and cells divide, genes are passed to successive cell generations as the individual grows and develops. When he or she reaches adulthood, spermatogonia or oogonia are produced that contain essentially the same genes as the zygote from which they descended.

The production of eggs and sperm begins with **meiosis,** a process involving two successive cell divisions rather than one as in mitosis. The first division produces two cell products, and the second division produces two more from each of those. As a result, four cells are produced by a meiotic division (Figure 28-13).

In the first division, **meiosis I,** the 46 chromosomes of a cell are distributed equally to two new cells, each cell receiving 23 chromosomes. In the second division, **meiosis II,** each chromosome splits lengthwise and chromatids are distributed to opposite sides of the cell.

Meiosis in Males

In males, meiosis occurs in the walls of seminiferous tubules (Figure 28-14). Here, spermatogonia differentiate into primary spermatocytes. As a primary spermatocyte begins meiosis, its chromosomes shorten and thicken sufficiently to enable them to be seen with a microscope. Initially, the chromosomes appear as a tangled mass of long, thin threads. Shortening and thickening continues until the chromosomes are untangled and free of one another. At the same time, they come to lie alongside one another in **homologous** (hō·mŏl′ō·gŭs) **pairs.** Members of a pair are alike in size and shape and together from a **bivalent** (bī·vā′lĕnt). By the time the chromosomes finish pairing, each one has also divided lengthwise into pairs of sister chromatids, which remain connected at the centromere (see Chapter 3). At this point, each pair of homologous chromosomes consists of a **tetrad** of four chromatids. As the chromosomes shorten and pair, the nuclear membrane disintegrates, a spindle forms, and the tetrads migrate to a central region of the cell in a plane midway between the ends of the spindle. This part of meiosis is **prophase I.**

The time spent in this central region constitutes **metaphase I.** Following metaphase I, members of each homologous pair of chromosomes separate and move to opposite ends of the spindle. The period during which they move

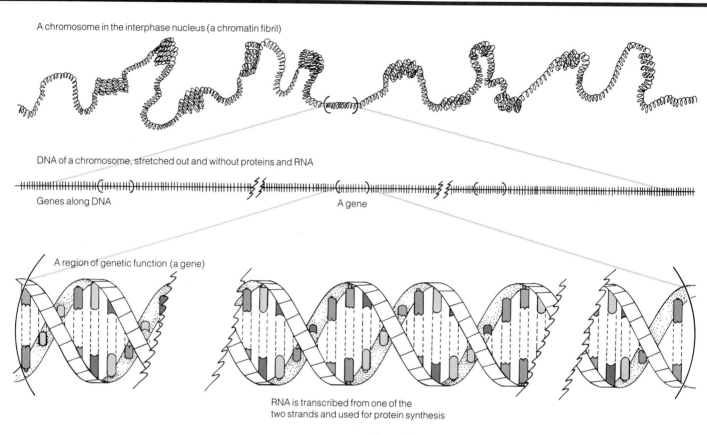

Figure 28-12 (above) Organization of genes along the chromosome. Each chromosome contains a single DNA molecule associated with proteins and RNA. Genes are regions along the length of DNA that are responsible for the production of specific RNA molecules used in protein synthesis (see Chapter 3).

Figure 28-13 (left) A schematic comparison of mitosis and meiosis. (a) In mitosis, a single division of a cell produces two new cells, each with the same number of chromosomes as the original cell. (b) In meiosis, two successive divisions (meiosis I and II) produce four cells, each with half as many chromosomes as the original cell. (For simplication this figure does not show the division of the chromosomes into chromatids in the two cells produced by the first division.)

is **anaphase I**. The final stage of meiosis I, **telophase I**, is marked by the chromosomes clumping together at opposite ends of the spindle. Nuclear membranes reform around each group of chromosomes and the cytoplasm divides. The two cells that result are **secondary spermatocytes**.

The second division of meiosis, **meiosis II**, is similar to mitosis in somatic cells in that sister chromatids are separated and distributed to new cells. Each chromosome, still consisting of two sister chromatids, migrates to the center of the cell. The nuclear membrane disintegrates and a new spindle forms. After remaining a short time midway between

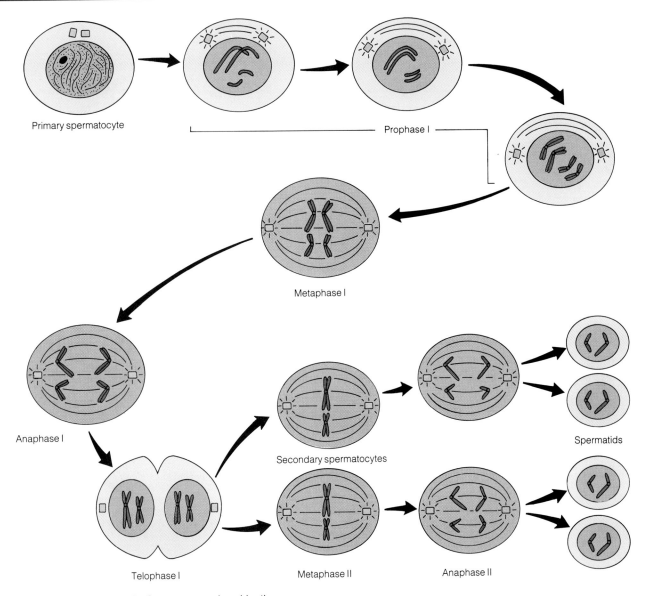

Figure 28-14 Meiosis in the testis. Sperm are produced by the differentiation of a spermatogonium into a primary spermatocyte and subsequent meiosis. By the end of prophase I, each chomosome consists of two longitudinal halves called chromatids. The four meiotic products differentiate into sperm.

the ends of the spindle during **metaphase II,** the sister chromatids move to opposite ends of the spindle. This stage is **anaphase II.** The chromatids, now chromosomes, clump together at opposite ends of the spindle **(telophase II)** and then relax as the spindle disintegrates, nuclear membranes form, and the cytoplasm divides. Four cells, spermatids, have been produced from a single primary spermatocyte. With the assistance of Sertoli cells, the spermatids differentiate into sperm (see Chapter 27).

Meiosis in Females

Meiosis begins in females before birth when primary oocytes, present in primordial follicles, enter prophase I. Meiosis continues through prophase and stops with the chromosomes in metaphase I. The cells are held in this stage until just before ovulation, at which time division resumes and anaphase and telophase I are completed (Figure 28-15). The subsequent division produces a secondary oocyte and a polar body

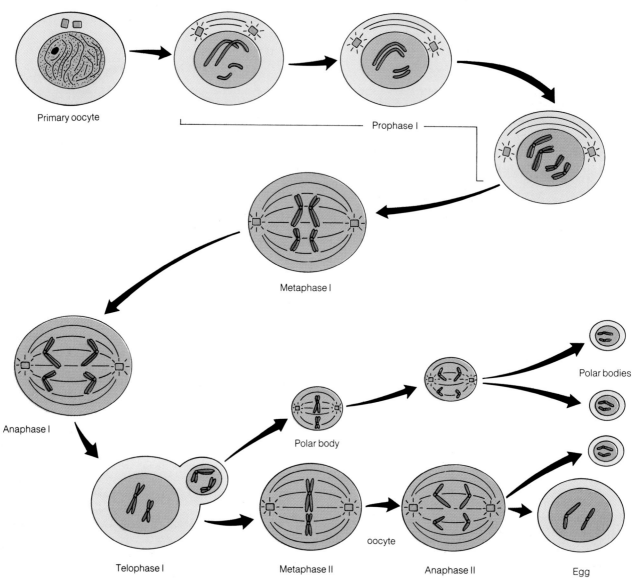

Figure 28-15 Meiosis in the ovary. This process differs from meiosis in a male in two ways: division is arrested from before birth until ovulation and meiosis does not go to completion unless the egg is fertilized. The cell products also differ in that they include one large cell and two to three smaller cells called polar bodies, depending on whether or not the first polar body also divides.

(see Chapter 27). The secondary oocyte then begins the second division of meiosis. Meiosis II proceeds only to metaphase II, however (unless fertilization occurs), and the oocyte remains in metaphase II until it disintegrates.

If fertilization occurs, the nucleus of the secondary oocyte divides again, producing two nuclei. One nucleus enlarges and becomes the female pronucleus. The other is extruded with a small amount of cytoplasm and becomes a **secondary polar body.** The first polar body may divide again at this point, and if it does, there will be four meiotic products, three small polar bodies and a large ovum (egg), each containing 23 chromosomes (Figure 28-15). Because fertilization has occurred, the ovum also contains the male pronucleus. Fusion of the two pronuclei produces a zygote.

genotype:	(Gr.) *genus*, race + *typos*, type
phenotype:	(Gr.) *phainein*, to show + *typos*, type
homozygous:	(Gr.) *homos*, same + *zygon*, pair
heterozygous:	(Gr.) *heteros*, different + *zygon*, pair

Inheritance

After studying this section, you should be able to:

1. Distinguish between genotype and phenotype.
2. Explain what alleles are and name the different combinations in which pairs of alleles can exist.
3. Distinguish between complete dominance, incomplete dominance, and codominance.
4. Define and compare the phenomena of segregation, independent assortment, and linkage.
5. Describe crossing-over and explain its importance.
6. Describe how restriction fragment analysis is used to study the organization of genes on chromosomes.
7. Explain how and why the inheritance of genes on sex chromosomes differs from that on autosomes.

DNA is organized into chromosomes, each of which consists of a long, slender nucleoprotein fibril formed by the association of a single DNA molecule with many protein molecules. The information in DNA is organized into thousands of discrete units, **genes,** distributed along the length of the DNA of each chromosome.

Genotype and Phenotype

Although genes are usually quite stable and retain their structure through many generations, changes sometimes occur that result in alternative forms called **alleles** (ă·lēls′). The different alleles in the human population are the basis for much of its diversity.

Unless they are identical twins, no two people have precisely the same genetic makeup. As a result, when a man and a woman have a child, each contributes a unique set of alleles to the child. The specific combination of alleles present in the child comprise its **genotype** (jēn′ō·tīp). The way those alleles are expressed comprise its **phenotype** (fē′nō·tīp).

Distribution of Genes

To understand these concepts, let's consider a gene that controls production of growth hormone. Growth hormone is a protein produced by the pituitary gland that enables a person to grow to normal height. In the absence of this hormone, a person will not achieve normal height and will be a **dwarf.** Dwarfism that results from insufficient growth hormone is **pituitary dwarfism** (see Chapter 17). The most common allele of the gene for growth hormone is one that specifies a normal, functional hormone molecule. An alternative allele exists, however, that lacks this information, specifying instead either a defective hormone or none at all. We can designate an allele that specifies growth hormone as *A* and one that specifies defective hormone as *a*. An individual with two normal alleles (*AA*) will be of normal height. A person with two defective alleles (*aa*) will be a dwarf.

When a person's cells have two identical alleles of a gene, as the *AA* and *aa* individuals do, the person is said to be **homozygous** (hō″mō·zī′gŭs) for the allele. A third combination, *Aa*, is also possible, in which case a person is **heterozygous** (hĕt″ĕr·ō·zī′gŭs) with respect to that gene. The phenotype that results from a heterozygous condition depends on the gene involved. In our example of pituitary dwarfism, heterozygous individuals achieve normal height because the normal allele produces enough hormone to overcome the effect of the defective allele. When one allele overcomes the effect of a second allele in this way, it is **dominant.** The allele whose expression is masked in the heterozygote is **recessive.** Likewise, the corresponding phenotypes, normal height and dwarfism, are dominant and recessive traits, respectively.

Suppose two people of normal height, each of whom is heterozygous for pituitary dwarfism (*Aa*), produce a child. Because of the way chromosomes are distributed at meiosis, each parent will produce gametes that contain the *A* allele or the *a* allele in equal frequency. That means that half the man's sperm and half the woman's ova will carry the *A* allele and the other half will carry the *a* allele.

Whether their child is a dwarf or not depends on which kind of egg and sperm join at fertilization. If either or both gametes carry the *A* allele, the child will not be dwarf, since he or she will either be *AA* or *Aa* in genotype. Only if both the egg and the sperm carry the *a* allele will the child be a dwarf. Since *A* and *a* sperm are equally likely, as are *A* and *a* eggs, there are four equally likely ways in which eggs and sperm can combine (Figure 28-16). One combination produces an *AA* zygote, two combinations produce an *Aa* zygote, and one combination produces an *aa* zygote. Thus, one out of four combinations results in a dwarf. The remaining three combinations produce children of normal height, one of whom is homozygous and two of whom are heterozygous.

Figure 28-17 shows all possible combinations of parental genotypes and the genotype and phenotype frequencies expected in their children. In cases (a) and (b), only one kind of combination of egg and sperm is possible and all offspring will have the same genotype and phenotype. In (a) the offspring will be *AA* and normal; in (b) they will be *aa* and dwarf. Case (c) shows the result if one parent is homozygous dominant and the other parent is homozygous recessive. All children will be heterozygous (*Aa*) and of normal height. Cases (d) and (e) show combinations of parental genotypes that can produce two different genotypic results; in (d) one out of two possible outcomes is a child homozygous for the

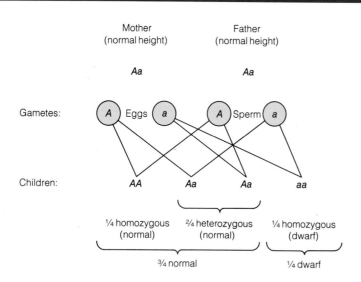

Figure 28-16 Inheritance of pituitary dwarfism in heterozygous parents. If both parents are heterozygous, 3/4 of the offspring will be phenotypically normal and 1/4 dwarf.

dominant gene (*AA*), and the other is a child who is heterozygous, both of whom will have normal height. In (e), half the children will be heterozygous *Aa* (normal height) and half will be homozygous *aa* (dwarf). Example (f) is the example described in Figure 28-16 in which both parents were heterozygous. In this case, one-fourth of the offspring will be homozygous for the dominant gene (*AA*), one-half heterozygous (*Aa*), and one-fourth homozygous for the recessive gene (*aa*). Thus, on the average, three out of four children can be expected to have normal height and one out of four to be a dwarf.

Many traits in humans are inherited in this way, although sometimes the normal allele is the recessive one. For example, **achondroplasia** (ă·kŏn″drō·plā′sē·ă) is a form of dwarfism in which the allele that causes dwarfism is dominant to the allele for normal height and heterozygous individuals are dwarfs.

In other cases, there is no clear dominance of one allele over another, and the heterozygous individual has a phenotype intermediate between the two homozygous genotypes.

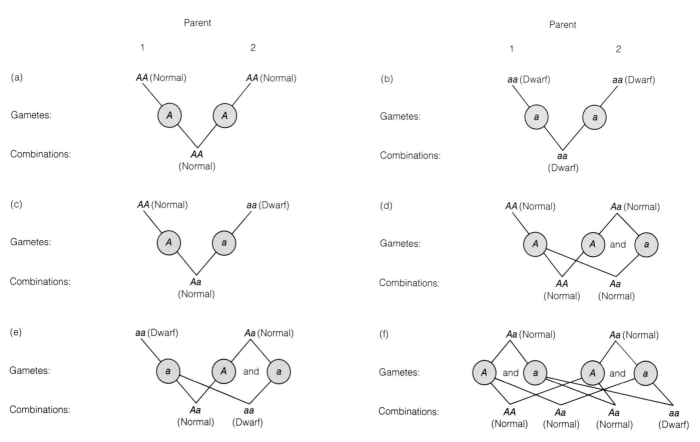

Figure 28-17 All possible outcomes of crosses involving pituitary dwarfism.

This is **incomplete dominance.** Tay-Sachs disease is an example of a trait that shows incomplete dominance, at least at the biochemical level. Children who are homozygous for the allele responsible for this disease lack an enzyme called hexosaminidase A that is necessary for production of certain central nervous system lipids. They show clear evidence of retardation by about 1 year of age, although they appear normal at birth. In the enusing months, there is increased loss of motor skills, accompanied by blindness and deafness. They die between the ages of 3 and 5 years. Children who are heterozygous for the Tay-Sachs gene show normal development, so the trait is a recessive one with respect to the disease. Although heterozygous individuals produce the enzyme, they produce roughly half as much as people who are homozygous for the normal allele. Thus, Tay-Sachs disease is a recessive trait, but hexosaminidase A production is an incompletely dominant trait.

Multiple Alleles

There is no particular limit to the number of different allelic forms of a gene in a population, and many different alleles of some genes are known. When more than two alleles of a gene are known, they comprise a family of **multiple alleles.** Some of the genes that govern the production of certain proteins found in the membrane of blood cells, for example, exist in the human population in dozens of forms, although no more than two different alleles of a particular gene can exist in one person.

The ABO system of blood classification in humans is an example of a multiple allelic genetic system (see Chapter 24). In this system, a gene determines the presence or absence of an antigen in the erythrocyte membrane. The antigen exists in two different forms, designated A and B. If a person's blood contains the A antigen alone, it is classified as type A blood. If the B antigen is present, the blood is classified as type B. If both antigens are present, the blood is classified as type AB, and if neither antigen is present, the blood is classified as type O. The gene responsible for these antigens exists in three allelic forms: I^A causes the production of A antigen; I^B causes the production of B antigen, and i fails to cause the production of either antigen. The different combinations that are possible for these alleles and their associated blood types are shown in Table 28-3.

Studies on the inheritance of blood type show that i is recessive to both I^A and I^B, since heterozygous ($I^A i$) individuals have type A blood. Likewise, heterozygous $I^B i$ individuals have type B blood, as do people who are homozygous for the I^B allele.

In the $I^A I^B$ combination, we see a new phenomenon, that of **codominance.** The heterozygous $I^A I^B$ combination

Table 28-3 GENETIC BASIS FOR HUMAN ABO BLOOD GROUPS

Phenotype (Bloodtype)	Genotype(s)	Antigen(s) in Blood
A	$I^A I^A$, $I^A i$	A
B	$I^B I^B$, $I^B i$	B
AB	$I^A I^B$	A and B
O	ii	Neither

Note: I^A and I^B are codominant. I^A and I^B are dominant to i.

shows both phenotypes simultaneously, having both A and B antigens in the cell membranes; this person's blood is classified as type AB. Only in the homozygous ii person is the antigen absent.

The story is even more complex than indicated here because chemical tests show that there are different forms of A and B antigens as well. Over the years several chemical variants of the A antigen and a few of the B antigen have been found, each presumably produced by a corresponding allele.

Segregation and Independent Assortment of Genes

Because genes are carried on chromosomes, their distribution to gametes depends directly on the way the chromosomes are distributed during meiosis. When we consider only one gene pair, the distribution is relatively simple to follow. At metaphase I, the pairs of homologous chromosomes line up midway between the two ends of the spindle, separating and moving to opposite ends during anaphase I. If a particular pair of chromosomes is carrying a heterozygous pair of alleles, (e.g., Aa), then the two alleles will be carried to opposite ends of the spindle and be present in different cells at the end of meiosis I. At the conclusion of meiosis II, two of the four meiotic products will have a chromosome that carries A and two will have a chromosome that carries a. The two alleles are said to **segregate** as they are distributed to the different cells during meiosis (Figure 28-18).

When two pairs of alleles are involved that are carried on nonhomologous chromosomes, segregation of one pair is independent of the segregation of the other pair because the chromosomes on which they are located segregate independently. For example, in cells of genotype $AaBb$ where A and a are on one pair of chromosomes and B and b are on another pair, the chromosomes line up in some cells such that A and B are carried in one direction and a and b are carried in the other direction. This results in two kinds of

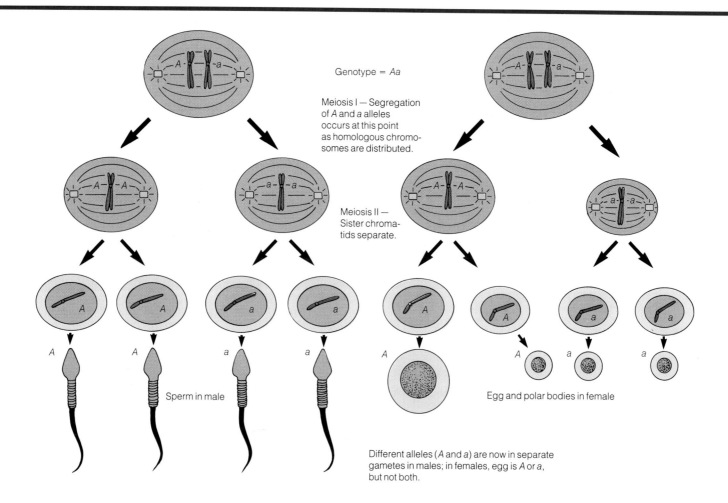

Figure 28-18 The chromosomal basis for segregation. As homologous chromosomes separate from one another at anaphase I, the different alleles for a gene in a heterozygous person are carried with them. As a result, no cell produced by meiosis will contain both alleles of the gene.

meiotic products, one with an *AB* genotype and one with an *ab* genotype.

Equally likely, however, is a chromosome alignment that results in *A* and *b* being carried in one direction and *a* and *B* in the other. When this occurs, half the cells that are produced have an *Ab* genotype and the other half have an *aB*.

Since both chromosome alignments are equally likely, half the meiotic divisions produce *AB* and *ab* gametes and the other half produce *Ab* and *aB* gametes. Consequently, there are equal proportions of all four allelic combinations in the population of gametes produced. This phenomenon is referred to as **independent assortment** (Figure 28-19). Independent assortment is important because it provides a way in which genes can be mixed to provided new combinations.

Chromosomal Linkage

Genes segregate independently of one another when they are on chromosomes that segregate independently. Chromosomes can carry hundreds of genes, however, and if two genes are on the same chromosome, the genes may not show independent assortment. For example, if one chromosome of a pair carries the genes *A* and *B* and the other member of the pair carries alleles *a* and *b*, then *A* and *B* will be carried in one direction at anaphase I and *a* and *b* will be carried in the other direction. As a result, half of the gametes that are produced will be *AB* in genotype and the other half will be *ab*, and no gametes of the genotypes *Ab* or *aB* will be produced. This phenomenon is referred to as **chromosomal linkage** (Figure 28-20).

In practice, chromosomal linkage is not usually so complete that only two gametic genotypes are produced. Instead, meiotic cells usually produce gametes of all four genotypic combinations, although not necessarily in equal frequencies. For example, if genes *A* and *B* are on one chromosome and

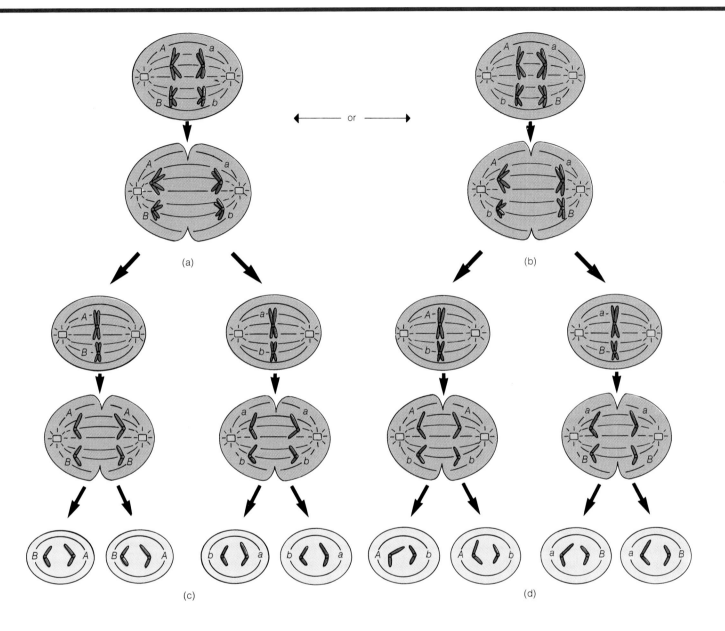

Figure 28-19 The chromosomal basis for independent assortment. The segregation of a pair of alleles depends on the metaphase I orientation of the chromosomes on which they are carried. (a) and (b) show alternative orientations. (c) and (d) show the outcome of the orientations shown in (a) and (b), respectively. Since the orientation of one pair of chromosomes is independent of the orientation of other pairs of chromosomes, segregation of alleles on one pair of chromosomes is independent of the segregation of alleles on any other pair and each of the different combinations of nonallelic genes is equally likely to occur.

their alleles *a* and *b* are on the other, then a majority of gametes may be produced of genotypes *AB* and *ab*, with a smaller proportion of *Ab* and *aB* gametes. This phenomenon is referred to as **partial linkage.**

Partial linkage presents a problem because if *A* and *B* are on one chromosome and *a* and *b* are on the other, it is difficult to understand (from what we have learned so far) how chromosomes with the genotype *Ab* or *aB* can be produced. The answer to the problem lies in **crossing-over,** a phenomenon that occurs early in meiosis I. During this stage, homologous chromosomes come to lie alongside one another in such a way that portions of chromatids from each of the homologous chromosomes come into intimate contact with one another. When this occurs, each chromatid breaks at almost exactly the same point, the broken ends rearrange, and the breaks heal. The result is a reciprocal exchange of the chromatid ends (Figure 28-21). If the break and rearrangement occurs in the region between genes *A* and *B*, then the

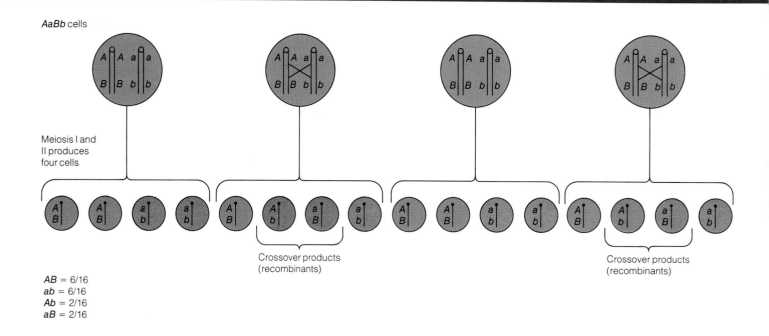

AB = 6/16
ab = 6/16
Ab = 2/16
aB = 2/16

new arrangements *Ab* and *aB* will be present on the chromatids involved in the crossover. As the chromosomes are pulled apart at anaphase I, each chromosome has a chromatid that has been rearranged by the crossover. These chromatids are subsequently distributed to cells in the second meiotic division.

At the completion of meiosis, two of the four cells produced will contain these chromosomes and be *Ab* and *aB* in genotype. These cells are referred to as **recombinants.** The remaining two cells will carry the chromosomes that were not involved in the crossover and will be *AB* and *ab* in genotype. These cells are referred to as **parental types.**

The likelihood of a crossover occurring between two genes depends on how far apart the genes are on the chromosome. If they are very close together, crossovers may be so uncommon that recombinants are rarely if ever produced. On the other hand, if the genes are relatively far apart, a crossover between them may be so common that recombinants are produced in virtually every meiotic division. If this happens, there will be as many recombinant gametes as there are parental type gametes, which is similar to the results obtained when the genes are not on the same chromosome. The frequency of recombinant gametes can be used to determine whether or not two genes are on the same chromosome and, if they are, their proximity to one another. Determination of the chromosomal location of genes through this technique is called **mapping.**

Figure 28-20 Linkage. If all genes were inherited independently of one another, there would be equal numbers of each combination of alleles in the gametes. In linkage, a predominance of two of the combinations over the other two occurs, altering the genetic ratio in the offspring. Here 50% of the cells have a crossover producing 25% recombinant cells and 75% nonrecombinant cells.

RFLP-Based Mapping of Chromosomes

With the development of new techniques of DNA analysis made possible by recombinant DNA technology, new methods of chromosome mapping have become possible (see The Clinic: Advances in Molecular Genetics). The techniques involved are based on the use of **restriction endonucleases,** bacterial enzymes that cleave DNA at specific nucleotide sequences. The techniques also make use of the ability to identify fragments of DNA with particular sequences through the use of DNA probes. For example, one enzyme, which has been named **EcoRI,** cuts DNA only at the sequence G-A-A-T-T-C. Human DNA has millions of such "restriction" sites, each located at a precise point in the human genome. Consequently, when a person's DNA is isolated and subjected to the action of EcoRI, it is cut into millions of fragments of specific sizes. The fragments vary in length from a few nucleotides to perhaps thousands, depending on how far apart any two sites might be located in a DNA

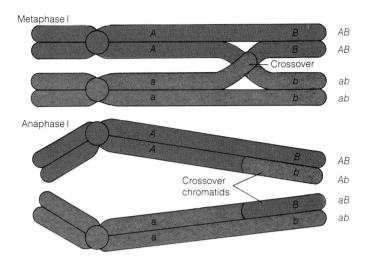

Figure 28-21 The break and rejoin model of crossing-over. Breaks occur at the same point in each of two nonsister chromatids in a pair of homologous chromosomes. The segments of the chromatids are exchanged as the broken ends rejoin in a new pattern and the chromosomes are pulled apart.

molecule. At present, about 200 different restriction endonucleases have been discovered that collectively cleave DNA at more than 60 different sequences.

Because each enzyme is specific in the sequence it recognizes, any deviation from the sequence at a site—for example, as a result of chemical change or misincorporation during DNA synthesis—will cause the enzyme to be unable to cut the DNA at that site. Conversely, mistakes can generate sites where one normally does not exist and provide a new point of cleavage. The presence or absence of a given site will then result in the production of two small fragments or one large one from the region surrounding the site. The fragments that are produced can be distinguished from the millions of other fragments produced by the rest of the DNA through a separation process called electrophoresis. They are then identified on the basis of their ability to combine with a suitably labeled molecule of complementary DNA (a DNA probe) (see figure in The Clinic).

When the DNA of some individuals in a population possesses a particular site and the DNA of other individuals lacks it, the corresponding fragments produced from their DNA will differ in length. The detection of different lengths of a fragment, as identified by a DNA probe, is referred to as **restriction fragment length polymorphism (RFLP).**

If suitable DNA probes are available, the presence or absence of a restriction site (as identified on the basis of the length of the restriction fragments produced) can be used like any other genetic marker. The site can be mapped on the basis of its tendency to be inherited jointly with other restriction sites or even entire genes that may be located nearby on the chromosome. When any site of known location (or absence of a site) shows a tendency to be inherited along with a particular genetic trait, the approximate location of the gene for that trait is also revealed. Thousands of sites are now known, involving all the chromosomes, and knowledge of the chromosomal location of genes is developing rapidly.

Sex Linkage

The members of each of the 23 pairs of homologous chromosomes in the nucleus of a typical female cell each have a distinctive size, shape, and staining pattern that makes their identification possible. In a male cell, only 22 pairs of homologous chromosomes are present, however. The remaining pair consists of one relatively large and one relatively small chromosome. The large member of this pair is an **X chromosome.** The small member is a **Y chromosome.** X and Y chromosomes are **sex chromosomes.** The remaining 22 pairs are **autosomes** (ăw′tō·sōms) (Figure 28-22).

In addition to visible differences in size, X and Y chromosomes differ from one another in the genes they carry. The X chromosome probably carries several hundred genes, more than a hundred of which have been identified. None of the genes identified on the X chromosome, however, have been found on the Y chromosome. Genes located on the X chromosome are said to be **X-linked,** whereas genes on the Y chromosome are said to be **Y-linked,** or **holandric** (hŏl·ăn′drĭk).

The difference in genetic content of X and Y chromosomes causes genes on these chromosomes to be inherited differently than genes on autosomes. Y-linked genes, for example, are necessarily transmitted only from father to son because the Y chromosome, which can only come from the father, causes the child to be a male. For this reason, genes on the Y chromosome will only be expressed in males. In fact, few if any genes have been identified on the Y chromosome, although genetic information must be present since the presence of a Y chromosome causes a person to be a male. The only genes that have been conclusively shown to be on the Y chromosome are one responsible for a protein called **H-Y antigen** and one responsible for a presumed "testis-determining factor" (TDF). H-Y antigen is a substance that is produced very early in embryonic development, but its function is unknown. The TDF gene appears to be responsible for causing the embryo to develop into a male, but the product of its activity is unknown.

hemizygous: (Gr.) *hemi*, half + *zygon*, pair

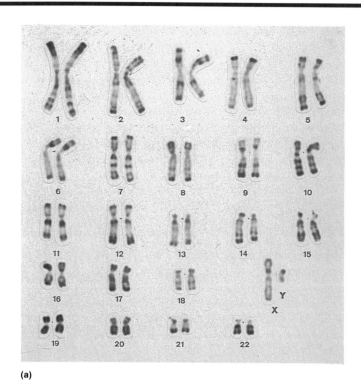
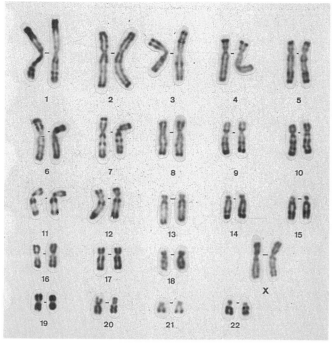

Figure 28-22 The metaphase chromosomes of (a) a human male and (b) a female. Chromosomes from both are similar, except that the Y chromosome is replaced with another X chromosome in a female.

Female cells, with two X chromosomes, possess two copies of each X-linked gene. In females, these genes are expressed and distributed much like autosomal genes. In male cells, however, only one X-linked allele of a gene can be present because male cells have only one X chromosome. Because they are present only once, the concepts of homozygosity, heterozygosity, dominance, and recessiveness do not apply to the genotype and phenotype of X-linked genes in the male. Instead, a male is **hemizygous** (hĕm·ē·zī′gŭs) for X-linked genes, and the genes are expressed regardless of whether they would be dominant or recessive in a female cell.

An example of an X-linked gene is one responsible for the production of factor VIII, a protein involved in blood clotting (see Chapter 21). Factor VIII catalyzes a step in the production of fibrin, a filamentous protein laid down over a wound as blood clots. If factor VIII cannot be made, clotting cannot occur properly, and there is danger of bleeding to death from an otherwise minor injury. This is an example of a group of inherited diseases called **hemophilias**. Several forms of hemophilia are known, each one resulting from an inability to make one of the components of the clotting process. The specific type of hemophilia that results when factor VIII is missing is **hemophilia A**. People with hemophilia A are unable to make factor VIII because they do not have an allele that specifies the normal structure of the factor, carrying instead a recessive allele that lacks the correct information.

Table 28-4 shows the genotypes and phenotypes expected among progeny from matings involving the hemophilia A allele and illustrates the pattern that characterizes inheritance of X-linked genes in general. Men who carry the allele are hemizygous and will be afflicted because they have no second allele to mask its effects. Since the allele is recessive, women must be homozygous to have hemophilia A. Moreover, all the male children of a homozygous female will also be afflicted because they receive their X chromosome from their mother. Although heterozygous women do not have hemophilia, half their male children will because half the eggs the women produce will have an X chromosome that carries the hemophilia allele.

It is characteristic of X-linked traits that they are passed by fathers only to daughters. Fathers cannot pass an X-linked gene to their sons because they do not pass an X chromo-

Table 28-4 PROGENY PHENOTYPES FROM UNIONS INVOLVING HEMOPHILIA A

Parental genotypes and phenotypes	Father × Mother H–, normal HH, normal	Father × Mother H–, normal Hh, normal	Father × Mother H–, normal hh, hemophiliac
Offspring genotypes and phenotypes	1/2 HH daughters (normal) 1/2 H– sons (normal)	1/4 HH daughters (normal) 1/4 Hh daughters (normal) 1/4 H– sons (normal) 1/4 h– sons (hemophiliac)	1/2 Hh daughters (normal) 1/2 h– sons (hemophiliac)
Parental genotypes and phenotypes	Father × Mother h–, hemophiliac HH, normal	Father × Mother h–, hemophiliac Hh, normal	Father × Mother h–, hemophiliac hh, hemophiliac
Offspring genotypes and phenotypes	1/2 Hh daughters (normal) 1/2 H– sons (normal)	1/4 Hh daughters (normal) 1/4 hh daughters (hemophiliac) 1/4 H– sons (normal) 1/4 h– sons (hemophiliac)	1/2 hh daughters (hemophiliac) 1/2 h– sons (hemophiliac)

Note: The dash in H– and h– stands for the missing allele in hemizygous males.

some to them. Consequently, all X-linked traits expressed in males have been transmitted to them by their mothers, who either also showed the trait or were recessive carriers of the allele for the trait.

Genetic Disorders due to Chromosomal Abnormalities

After studying this section, you should be able to:

1. Describe the distribution of chromosomes when nondisjunction occurs.
2. Describe the effect on the chromosome composition of cells produced from a meiotic division in which nondisjunction occurred in anaphase I.
3. List several syndromes that result from an imbalance in chromosome number.
4. List three ways in which chromosome structure may change and the consequences of such changes.
5. Describe techniques for the prenatal detection of chromosomal and genetic disorders.

Chromosomal Nondisjunction

Although separation of chromosomes at anaphase usually occurs accurately, sometimes homologous chromosomes fail to separate properly and both chromosomes go to the same side of the cell. As a result, one side of the cell receives both chromosomes and the other side receives neither. This phenomenon is **chromosomal nondisjunction**. When nondisjunction occurs during anaphase I, the gametes that result either have an extra chromosome or lack one (Figure 28-23).

The fate of a cell that has such a chromosome imbalance depends on which chromosome is involved. Absence or extra copies of certain chromosomes can cause a gamete to be unable to function, either because it dies or because it has lost the ability to produce substances necessary for fertilization. Imbalances of other chromosomes may permit gametes to function, but the embryo that results may not survive, again depending on which chromosome is involved. If a gamete that lacks a chromosome is fertilized by a normal gamete, the resulting zygote and all the cells descended from it will have one fewer chromosome than normal. Conversely, if a gamete has two of a particular chromosome, fertilization by a normal gamete will produce a zygote with three of the chromosomes. In general, cells that have one less chromosome than normal are **monosomics**. Those that have an extra chromosome are **trisomics**.

In humans, monosomy is usually lethal; the only cases that allow survival occur in people whose cells possess a single X chromosome and no Y chromosome. These people are invariably female and possess traits referred to as **Turner syndrome**. Monosomy involving a single Y chromosome (and no X chromosome) is a lethal condition because at least one X chromosome is essential for survival. Trisomy for sex chromosomes involves three X chromo-

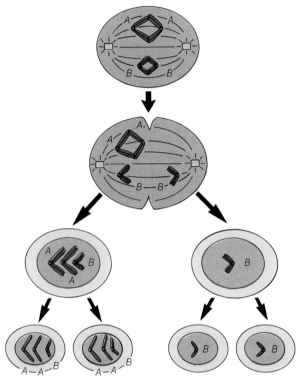

Figure 28-23 Chromosomal nondisjunction as a basis for monosomy and trisomy. If the chromosomes do not separate from one another at anaphase, gametes may be produced that lack a chromosome or that have an extra chromosome.

somes, two X and one Y chromosome, and one X and two Y chromosomes. People with three X chromosomes are female and usually normal. Individuals with one X and two Y chromosomes are male, many of whom are sterile. As a group they tend to be taller than average and have slightly lower than normal intelligence.

Individuals with two X and one Y chromosome are more severely affected. These people have a set of characters called **Klinefelter syndrome.** They are invariably infertile males who often have impaired intelligence, undeveloped testes, and reduced testosterone levels, in addition to certain other characteristics.

Down Syndrome

Nondisjunction can involve autosomes, but only a few of the possible chromosomal imbalances that result allow an embryo or fetus to survive. Probably the best known and most thoroughly studied of the autosomal chromosome disorders is **Down syndrome.** Cells of most people afflicted with Down syndrome have three chromosome 21s (Figure 28-24). The genetic imbalance that results causes extensive alteration in development. People with Down syndrome are often easily identified by folds of skin about the eyes, a face characteristically flat in profile, and ear pinnae that are often small and round. In Down syndrome the tongue is relatively large and flat, protruding slightly from the mouth, and causing the mouth to be held partially open. Body stature is short and stubby, as are hands, fingers, and feet. People with Down syndrome are mentally retarded, generally having IQ's that range from 0 to 50, although some afflicted people have an IQ as high as about 70 (Figure 28-25).

People with Down syndrome are prone to heart disease and infection, and before the use of antibiotics, they often died at an early age, especially from tuberculosis. Now, with the use of antibiotics, many live into middle or old age. Nevertheless, mortality rates remain high and many die within the first five years of life from infection and heart disease. The overall incidence of Down syndrome is about one in 700 live births.

The parent who has contributed the extra chromosome can often be determined by comparing the parents' and child's chromosomes after staining them with special dyes.

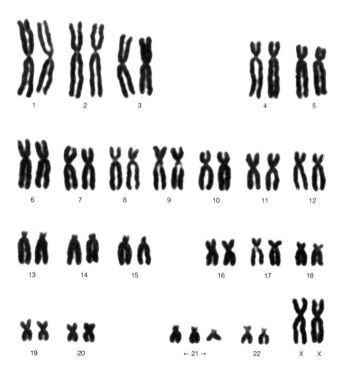

Figure 28-24 Karyotype of a person with Down syndrome. There are 47 chromosomes in the karyotype, including 3 of the chromosome designated number 21.

Figure 28-25 A child with Down syndrome.

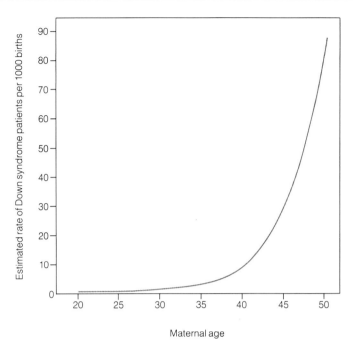

Figure 28-26 This graph shows the relationship between the frequency of Down syndrome and the age of the mother.

On this basis, about 80% of the cases of Down syndrome are attributable to an extra maternal chromosome and about 20% to an extra paternal chromosome.

There is a dramatic correlation between the likelihood of a child being born with Down syndrome and the mother's age (Figure 28-26). Although women over 40 years of age give birth to only about 4% of babies born each year, those babies include more than 50% of children born with Down syndrome. A child born to a 45-year-old woman is 60 times more likely to have Down syndrome than one born to a 20-year-old woman (see Clinical Note: Amniocentesis).

Other Examples of Autosomal Nondisjunction

Two other examples of autosomal trisomy involve chromosomes 13 and 18. Trisomy for chromosome 13 causes **Patau syndrome** in about one child in 5000 births. Babies with Pateau syndrome are often blind and deaf and may have a severe cleft palate, a small head, extra fingers and toes, and other abnormalities. About 70% of afflicted babies die within 6 months and 80% within the first year.

Edwards syndrome is associated with trisomy for chromosome 18. Like Patau syndrome, Edwards syndrome is lethal. Newborn infants have a skull that bulges outward in the back, a small mouth, tightly clenched hands, characteristic clubfeet, and fused fingers and toes. A cleft lip or palate often occurs. Numerous developmental anomalies occur in the heart, kidneys, lungs, ureters, and other organs. About three-fourths of babies born with this syndrome are female and about 90% of them die within a year.

Other Chromosomal Abnormalities

Genetic imbalances are also caused by changes in chromosome structure. These include translocations, deletions, and duplications. A **translocation** occurs when a portion of one chromosome is transferred to another chromosome. **Deletions** involve loss of a portion of a chromosome, and **duplications** involve a repeat of a portion of a chromosome, usually on the same chromosome. Most of these chromosomal abnormalities are lethal before birth, and only a few very specific examples are found in newborn infants, where they usually lead to early death. **Cat's cry** (or cri du chat) **syndrome,** for example (Figure 28-27), is a rare phenomenon that results from a deletion of a portion of chromosome number 5. Newborn babies make a high-pitched mewing sound like the cry of a kitten. In time, that characteristic goes away, but several other characteristics remain, including

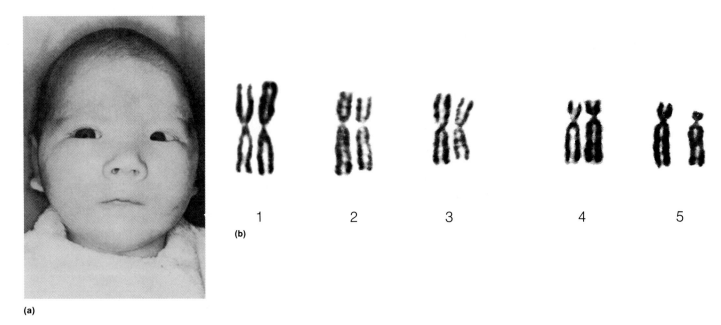

Figure 28-27 Cat's cry (or cri du chat) syndrome. (a) An infant with cat's cry syndrome. (b) Partial karyotype of an infant with cat's cry syndrome, showing a portion of chromosome number 5 missing (the short one in the last pair).

a small head, round face, and heart defects. Some children survive to adulthood, but mental and physical development is severely retarded in those who do.

Deletions of portions of chromosomes 4, 13, 18, 21, and 22 have also been reported and generally cause severe mental and physical retardation and early death. One exception is a deletion in which the short arm of chromosome 21 (measured from the centromere) is missing. About 1% of the population has this abnormality, and these people appear to be normal in spite of the missing chromosome arm.

Few duplications have been reported, the most notable example being a duplication of a part of chromosome 21 and its translocation to another chromosome. People who have the translocated chromosome and two normal chromosomes have a type of Down syndrome called **translocation Down syndrome.**

Translocation of a piece of chromosome 22 to another chromosome is responsible for about 80% of the cases of **chronic myelogenous leukemia,** a cancer of the tissue that produces white blood cells. The shortened chromosome 22 that results is called a **Philadelphia chromosome,** named in reference to the city in which the anomoly was first discovered. There is no evidence for any net loss or duplication of chromosomal material in people with this condition, and the leukemia may result from the genetic material of chromosome 22 being moved to a location where it cannot function properly.

Prenatal Detection of Genetic and Chromosomal Abnormalities

Many genetic and chromosomal abnormalities can be detected before the birth of a baby. The primary technique, **amniocentesis** (see Clinical Note: Amniocentesis) involves removing fluid from the amnion through a needle inserted through the abdominal wall. Cells in the fluid are cultured and examined in mitosis for chromosomal aberrations, or else they are examined chemically for evidence of certain genetic conditions. Disadvantages of amniocentesis are that it cannot be performed until about the fifteenth week of pregnancy, and results of the examination require two to three weeks to obtain.

A more recent technique is **chorionic villus sampling (CVS).** In this technique, a tissue biopsy is taken from the chorion with a device inserted through the cervix, and DNA is obtained from the tissue. RFLP analysis of the DNA is then performed to determine whether or not alleles for certain genetic diseases are present. In the case of recessive disorders, analysis can reveal whether or not the fetus is homozygous and therefore whether or not it will be afflicted after birth.

Chorionic villus sampling can be performed as early as the eighth to tenth week of pregnancy, and results can be obtained within two days. Consequently, CVS is potentially suitable for identifying genetic disorders earlier in pregnancy.

Offsetting this advantage, however, is an increased risk of infection and miscarriage resulting from the procedure as compared with amniocentesis.

With this chapter we conclude our discussion of the anatomy and physiology of the human body. We have discussed the processes by which a new human being develops and the manner in which a pregnant woman's physiology changes to accommodate that development. We have also seen how the genetic potential of that new person is determined through the distribution of chromosomes, the carriers of genes, to the gametes. Fusion of the gametes then produces a zygote with a combination of genes that probably has never occurred before. As the person grows and develops, he or she will use the potentials represented by those genes to produce the structures and chemicals necessary for life and, perhaps, to repeat the cycle.

STUDY OUTLINE

Embryonic Development (pp. 733–742)

Cleavage Division of the zygote produces a *morula*, a ball of cells contained in the zona pellucida. Each cell is a *blastomere*. A fluid-filled *blastocoel* develops, and the morula becomes a *blastocyst*. The outer layer of cells is the *trophoblast*. A group of cells inside is the *inner cell mass*.

Implantation The blastocyst adheres to the endometrium and digests adjacent endometrial tissue in *implantation*. The surrounding endometrium grows over the blastocyst.

Primary Germ Layers The inner cell mass gives rise to an *embryonic disk*. Ectoderm, endoderm, and mesoderm develop from the embryonic disk during *gastrulation*.

Embryonic Membranes An *amnion* develops and surrounds the developing embryo and a bag containing *amniotic fluid*. A *yolk sac* develops and provides primordial germ cells that migrate to a developing ovary or testis. The *allantois* produces blood and forms vessels that will become umbilical arteries and vein. The *chorion* surrounds the amnion, fusing with it to form a protective sac.

Placenta and Umbilical Cord The *placenta* and *umbilical cord* develop from the chorion and body stalk, respectively. Fingerlike extensions of the chorion (*chorionic villi*) penetrate the *decidua basalis*, creating *intervillous spaces* that fill with blood from the mother. Exchange of nutrients and wastes between mother and embryo occurs across the *placental membrane* that separates intervillous blood from capillary blood in the chorionic villi.

Fetal Development By about eight weeks of *gestation*, development of major organs has occurred and the embryo becomes a *fetus*.

Twins *Identical twins* develop from the same fertilized egg, usually following separation of the two cells produced by division of the zygote. *Fraternal twins* develop from separate fertilized eggs.

Pregnancy and Childbirth (pp. 743–749)

Diagnosis of Pregnancy There are several presumptive, probable, and positive indications of pregnancy. One of the earliest is *amenorrhea*, a presumptive indicator. Fetal heart sounds and ultrasound are positive signs. Increased *human chorionic gonadotrophin (HCG) levels* are usually accurate indicators of pregnancy.

Duration of Pregnancy Full-term pregnancy usually lasts 266 days from conception, plus or minus 2 weeks. Predicted date of birth is calculated by subtracting 3 months from the onset of the previous menstrual period and adding 7 days. This is correct to within 2 weeks about 80% of the time.

Changes in Physiology During Pregnancy Pregnancy causes changes in virtually all organ systems of the pregnant woman. Cardiovascular, respiratory, renal, digestive, and endocrine activities generally increase.

Labor and Delivery Birth of an infant is referred to as *parturition*. Contractions of uterine muscles are induced by oxytocin, and force the baby out during labor. Expulsion of the placenta occurs subsequently, and the placenta emerges as afterbirth. The umbilical cord is cut and leaves a scar called an *umbilicus*, or navel.

Meiosis (pp. 749–752)

Meiosis is a division process that produces the cells that will differentiate into eggs and sperm. Its two subdivisions are *meiosis I* and *meiosis II*.

Meiosis in Males In males, meiosis occurs in the walls of seminiferous tubules. Spermatogonia differentiate into primary spermatocytes which undergo meiosis. The first division is marked by the pairing of *homologous* chromosomes to form *bivalents*, doubling of each chromosome into sister chromatids, which converts each bivalent to a *tetrad*, and distribution of the chromosomes of each pair to two new secondary spermatocytes. In the second division, sister chromatids are distributed to new cells. The four spermatids from each primary spermatocyte differentiate into sperm.

Meiosis in Females Meiosis begins before birth in females. Primary oocytes in primordial follicles begin meiosis, but the process is halted in the first division until ovulation. The products of the first division are a secondary oocyte and a polar body. The secondary oocyte initiates the second division of meiosis, but the process halts until fertilization occurs. If fertilization occurs, nuclear division in the secondary oocyte produces a female pronucleus and a second nucleus that is extruded as a *secondary polar body*. Upon fusion of the female and male pronuclei, the cell becomes a zygote.

THE CLINIC

ADVANCES IN MOLECULAR GENETICS*

The speed with which the new DNA technology is being moved from the laboratory bench to the care of patients is without precedent in the history of medicine. A few of the main acts in this rapidly developing drama are highlighted below.

Isolation of Human Genes
New techniques for three important processes have been crucial in the rapid advancement of DNA technology: purification of DNA and RNA, separation of the two polynucleotide strands of the double helix, and understanding the triplet code for translating the sequence of bases in the genetic material to the corresponding sequence of amino acids in polypeptide chains. For example, the gene for human insulin was synthesized by first purifying the two polypeptide chains of the hormone itself. The sequence of amino acids yielded the sequence of bases that coded them, and the appropriate bases were chemically linked together in the correct order *in vitro*. The genes for the α- and β-hemoglobin polypeptide chains were synthesized from purified messenger RNA (mRNA). The specific RNA was relatively easy to purify from human erythroid precursor cells because about half of their total mRNA is for hemoglobin production. Transcription from an RNA template to DNA occurs thanks to the enzyme reverse transcriptase.

Several additional methods for purifying and analyzing human genes have been developed. It turns out that a gene is not simply a segment of DNA coding for the amino acid sequence of a polypeptide chain. Instead, it consists of coding sequences called exons that are interspersed by noncoding intervening sequences known as introns. The introns are transcribed initially but are spliced out of the mRNA before it leaves the cell nucleus. The function of the introns is unknown, but many of the mutations that cause β-thalassemia, for example, lie within them. In addition, there are flanking sequences on both sides of structural genes that regulate their activity and provide start and stop signals for their transcription.

Restriction Endonucleases
Restriction endonucleases are bacterial enzymes that break the genetic material into thousands of pieces, but each specific endonuclease breaks the DNA or RNA only when there is a very specific sequence of bases. For example, one endonuclease might recognize the sequence G-C-C-T-A-A and break the chain between the T and the A (the four bases that make up the DNA are cytosine, guanine, thymidine, and adenine). This sequence will occur by

Southern blot technique. DNA is cleaved, placed in a thin layer of gel, and subjected to an electric field. In electrophoresis, individual fragments are drawn through the gel by electrical attraction, with the smaller fragments traveling faster. The DNA is then transferred to filter paper and a radioactive or other suitably labeled probe is allowed to hybridize with it. The location of the labeled probe on the filter reveals the location, and thus the size, of DNA fragments complementary to the probe.

DNA extracted from human cells

DNA fragments after incubation with restriction endonucleases

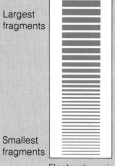
Electrophoresis (Largest fragments / Smallest fragments)

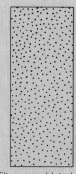
Filter paper blot; dots represent labeled probe

Autoradiography showing specific fragment identified by probe

chance over and over again in the human genome and create the many fragments of varying lengths (see figure).

Gene Cloning

Cloning the genes is the vital step in which a piece of human DNA (be it a purified gene or not) is inserted into a cloning vehicle, usually a bacterial plasmid (a circular bit of DNA present naturally in bacteria). Then the bacterium is cultured. Each newly produced plasmid contains the human DNA, and literally billions of copies of the human gene or DNA segment can be produced in this bacterial factory. The circular plasmids can be opened for insertion of foreign DNA using restriction endonucleases. The combination of bacterial plasmid and a human DNA segment is one example of recombinant DNA.

Gene Probes

Any segment of DNA or RNA will bind to its complementary sequence. That is how the double helix is formed from a single strand in the first place—G always binds to C, and A to T. Thus, if one takes the many copies on one of the cloned genes described above and adds a radioactive label to each copy, one has a labeled probe. The probe will seek out its complementary segment of DNA and can then be found by autoradiography. A probe added to a chromosome spread can actually locate a gene in a specific chromosome or use it to identify a piece of DNA in **Southern blot** procedure (see figure).

The Southern blot, a procedure named after the investigator who devised it, is the cornerstone of the application of recombinant DNA to the prevention of genetically determined disease. Briefly, DNA is extracted from cells of the patient and fragmented by one of the restriction endonucleases. The fragments are electrophoresed, which separates them primarily by size, the smaller ones moving more rapidly through the pores of the gel. The fragments are then blotted onto nitrocellulose filter paper and overlaid with a labeled probe. The probe will bind only to its complementary sequence and thus identify the gene of interest as a band or bands on the paper.

Restriction Fragment Length Polymorphisms The various available restriction endonucleases have revealed a remarkable amount of variability in the human genome. If just one base substitution occurs in the sequence recognized by the endonuclease, the DNA will not break at that site, and instead of 32 small fragments, there will be one blot procedure and the probe will identify the different-sized fragments. Since these variants are so common, they have provided us with an almost unending series of markers known as **restriction fragment length polymorphisms (RFLPS).** If the base substitution occurs within the gene in question and coincidentally at the site of the mutation that causes the disease (as is the case for sickle cell anemia), the marker will identify an individual's genotype without family studies. If the variant flanks the gene in question but is very close to it (as in Huntington's disease), family studies are required to determine which sized fragment is associated with the Huntington's disease gene in a given family.

Medical Applications These techniques have already led to extraordinary advances in mapping the human genome. The next step will likely be the development of probes for most human diseases resulting from single gene mutations, even though the nature of the genetic defect is unknown. This will permit precise identification of genotype, including heterozygotes and homozygotes, and provide the means for diagnosis before birth and before the development of detectable clinical signs in such conditions as Huntington's disease. Genes whose function is unknown will soon be isolated and sequenced and their products deduced, which could lead to understanding of the basic defect and eventually to treatment approaches. Already reported are probes for the X-linked genes causing the hemophilias, Lesch-Nyhan syndrome, and Duchenne muscular dystrophy. Probes are also available for other medically important conditions due to autosomal mutant genes, including hemoglobinopathies, cystic fibrosis, and Huntington's disease. If the markers turn out to be very close to the genes, they will soon be made available for pre- and postnatal diagnosis.

Additional applications include the use of recombinant DNA techniques in the production of scarce human replacement products, including growth hormone and pure factor VIII, both of which are currently nearing commercial availability. The practicality of actual gene replacement therapy, the ultimate genetic engineering, still remains a distant possibility. We know very little about the nature or even the location of the crucial regulatory genes. We have no idea as to how to direct genes to specific sites in the genome, and in the wrong locale, they could have serious deleterious effects.

Possibly the most exciting discoveries to date have been the markers for several genetically influenced conditions that are both common and represent serious medical problems. These include the specific HLA associations with specific forms of arthritis and diabetes, and most recently the uncovering of a RFLP flanking the human insulin gene. One form of this RFLP appears to identify individuals at high risk of developing arteriosclerotic heart disease, whether or not they actually have diabetes. Early identification of these high-risk subgroups may permit interventions that will delay or even prevent the effects of the disease.

*Source: Adapted from *The Merck Manual*, 15th ed. (Rahway N.J.: Merck Sharp & Dohme Research Laboratories, a division of Merck & Co., Inc.), 2153–56.

Inheritance (pp. 753–761)

Genotype and Phenotype The alleles in the cells of a person constitutes the person's *genotype*. The way the alleles are expressed is the *phenotype*.

Distribution of Genes Each *gene* is represented twice in body cells and only once in sex cells. Genes may exist in more than one form. Different forms of the same gene are *alleles*. In body cells of *homozygous* individuals, the same allele is represented twice. In *heterozygous* individuals two different alleles of a gene are present. If only one of the two alleles of a heterozygote is expressed, that allele is *dominant* and the other is *recessive*. *Incomplete dominance* and *codominance* occur when both alleles are expressed simultaneously.

Multiple Alleles Although only two different alleles of a gene may be present in an individual (because he or she has two sets of genes), a gene may be represented by more than two alleles in a population. Genes that exist in more than two allelic forms are *multiple allelic*.

Segregation and Independent Assortment of Genes Distribution of alleles corresponds to the distribution of the chromosomes on which they are located. The two alleles of a gene are located on the members of a *homologous pair* of chromosomes. Chromosome distribution at meiosis distributes the alleles to different cells. The orientation of one pair of homologous chromosomes at metaphase I is independent of the orientation of other pairs. Consequently, *segregation* of a pair of alleles on one chromosome pair occurs *independently* of the segregation of alleles located on other chromosome pairs.

Chromosomal Linkage Alleles of different genes located on the same chromosomes will segregate together unless crossing-over causes them to recombine. If no crossing-over occurs, there will be no recombinants, and *chromosomal linkage* exists. In crossing-over, nonsister chromatids of a tetrad break and exchange their ends. Alleles distal to the break will no longer be associated with alleles proximal to the break and *recombinants* will result. This is *partial linkage*. The likelihood of crossing-over between linked genes depends on how far apart they are. If crossing-over occurs between them in virtually every meiotic division, recombinants will be as common as parental types and no evidence of linkage will exist.

RFLP-Based Mapping of Chromosomes. Cutting DNA with *restriction endonucleases* can reveal the presence or absence of specific restriction sites in the DNA on the basis of the length of *restriction fragments* created by the enzyme. The tendency of a genetic trait and a restriction fragment to be inherited together can be used to identify the chromosomal location of the gene responsible for that trait.

Sex Linkage Females have 23 homologous pairs of chromosomes in their body cells. In males, 44 chromosomes are present as 22 homologous pairs. The remaining two chromosomes are dissimilar. They are *sex chromosomes*, designated X and Y. The remainder are *autosomes*. Males have one X and one Y chromosome; females have two X chromosomes. Since genes on an X chromosome are different from genes on a Y chromosome, inheritance of traits due to X- and Y-borne genes differs from inheritance of autosomal traits. *X-linked* traits in males are inherited from their mothers. *Y-linked* traits can only occur in males and are inherited from their fathers. X-linked alleles in males are necessarily transmitted to their daughters. Whether or not they are expressed in a daughter depends on the corresponding allele inherited from the mother.

Genetic Disorders due to Chromosomal Abnormalities (pp. 761–765)

Chromosomal Nondisjunction This is the failure of homologous chromosomes or chromatids to separate at anaphase I or II, respectively. When *chromosomal nondisjunction* occurs, cell products have both members of a chromosome pair or neither, instead of one. The resulting genetic imbalance usually impairs development and is often lethal. Chromosomal imbalance is tolerated best when it involves sex chromosomes. Autosomal imbalances are usually lethal. *Down syndrome* is the most common autosomal trisomy, involving chromosome 21.

Other Chromosomal Abnormalities Chromosomes can be rearranged in *translocations*, material can be lost due to *deletions (cat's cry syndrome)*, or regions can be *duplicated*. Most such changes in structure are lethal before birth.

Prenatal Detection of Genetic and Chromosomal Abnormalities. Two techniques can detect genetic and chromosomal abnormalities before birth. *Amniocentesis* involves the removal and examination of amnionic fluid. In *chorionic villus sampling (CVS)*, a tissue biopsy is taken from the chorion. DNA in the tissue is analyzed by RFLP to determine if genetic diseases are present and whether the baby will be afflicted.

SELF-TEST OF CHAPTER OBJECTIVES

True-False Questions

1. Cleavage is a series of cellular divisions that transforms a unicellar zygote into a multicellular embryo.
2. The blastocoel is a cavity that develops within the zygote.
3. The second division of meiosis is the division responsible for reducing the chromosome number of a cell.
4. There are generally as many DNA molecules in a chromosome as there are genes in it.
5. The members of a homologous pair of chromosomes always carry identical genes.

6. An individual who is heterozygous for a particular gene will generally produce gametes carrying one allele or the other in equal numbers.
7. There can never be more than two allelic forms of a gene in a person.
8. The genes for the ABO blood typing system represent a multiple allelic series.
9. If there is independent assortment in the cross $AaBb \times aabb$, one-fourth of the offspring on the average will be heterozygous for both gene pairs.
10. It is apparent from examining the chromosomes of various people that a Y chromosome imparts maleness, regardless of the number of X chromosomes present.

Matching Questions
Sickle cell anemia is a blood disease that is due to a single recessive gene (s). Match the proportion of children expected with the corresponding phenotypic combinations from marriages between people who are heterozygous for sickle cell anemia (Ss) and for blood type A ($I^A i$):

11. normal blood, type O
12. normal blood, type A
13. sickle cell anemia, type O
14. sickle cell anemia, type A

a. 9/16
b. 3/16
c. 1/16

Match the following chromosome complements with the correct syndrome:

15. 45 chromosomes, including one X and no Y
16. 47 chromosomes, including two X and one Y
17. 47 chromosomes, trisomic for number 13
18. 47 chromosomes, trisomic for number 21

a. Klinefelter syndrome
b. Down syndrome
c. Turner syndrome
d. Patau syndrome

Multiple-Choice Questions
19. The outer layer of cells surrounding the blastocyst is the
    a. trophoblast   c. blastomere
    b. blastocoel   d. endometrium
20. An inner cell mass gives rise directly to which of the following?
    a. primary germ layer   c. blastocyst
    b. morula   d. placental villi
21. Which of the following embryonic membranes enlarges until it forms a sac around a developing embryo?
    a. amnion   c. allantois
    b. yolk sac   d. chorion
22. Which of the following is a source of blood early in embryonic development?
    a. amnion   c. allantois
    b. yolk sac   d. chorion
23. During implantation, which structure digests its way into the uterine wall?
    a. morula   c. inner cell mass
    b. blastocyst   d. decidua basalis
24. Nutrients are transferred from maternal blood to the fetal circulation by
    a. capillaries in chorionic villi
    b. blood in umbilical arteries
    c. blood in the umbilical vein
    d. blood in intervillous spaces
25. Wharton's jelly is located in the _____, surrounding the _____.
    a. amnion, fetus
    b. yolk sac, embryo
    c. umbilical cord, umbilical arteries and vein
    d. endometrium, blastocyst
26. Which of the following is considered a positive sign of pregnancy?
    a. amenorrhea
    b. enlargement and softening of the uterus
    c. morning sickness
    d. elevated human chorionic gonadotrophin levels in the blood and urine.
27. Which of the following hormone levels usually decreases during pregnancy?
    a. prolactin
    b. follicle-stimulating hormone
    c. thyroid hormone
    d. cortical steroid hormones
28. The average length of gestation is about
    a. 215 days   c. 290 days
    b. 266 days   d. 310 days
29. The hormone that induces labor is
    a. human chorionic gonadotrophin
    b. melatonin
    c. oxytocin
    d. parturition inducing hormone
30. Human somatic cells have 46 chromosomes. How many bivalents are expected in prophase I of meiosis?
    a. 23   c. 69
    b. 46   d. 92
31. How many chromatids are present in a human secondary spermatocyte?
    a. 23   c. 69
    b. 46   d. 92
32. In what stage of meiosis does crossing-over occur?
    a. prophase I   c. anaphase I
    b. metaphase I   d. metaphase II
33. How many sperm will be produced by 10 secondary spermatocytes?
    a. 10   c. 30
    b. 20   d. 40
34. How many ova will be produced by 10 secondary oocytes?
    a. 10   c. 30
    b. 20   d. 40
35. How many different gamete genotypes could a person of genotype $AaBBCcddEeff$ produce, assuming these genes assort independently of one another?
    a. 6   c. 12
    b. 8   d. 64

36. In humans, albinism (lack of pigment) is a recessive, autosomal trait. Two people with normal pigmentation, who each had a parent who was an albino, have children. What proportion of their children would you expect to be albino?
    a. none of them
    b. one-fourth
    c. one-half
    d. all of them
37. Ignoring crossing-over, what proportion of a man's sperm would have only maternal chromosomes, that is, chromosomes descended from those contributed by his mother?
    a. $1/2$
    b. $(1/2)^{23}$
    c. $2 \times (1/2)^{23}$
    d. $(1/2)^{46}$
38. The genes for Rh and the Duffy blood group are on the same chromosome, but they are so far apart that at least one crossover happens between them in virtually every meiotic cell. What would the relative number of recombinants to nonrecombinants be in the gametes of a man heterozygous for both of the genes?
    a. All sperm would be nonrecombinant.
    b. There would be an equal number of recombinant and nonrecombinant sperm.
    c. All sperm would be recombinant.
    d. The nonrecombinant sperm would outnumber the recombinant ones.
39. Red-green color blindness is a recessive sex-linked trait in humans. If a woman with normal vision whose father was colorblind marries a colorblind man, what proportion (by sex) of their children are likely to be colorblind?
    a. half of the boys and half of the girls
    b. half of the boys and none of the girls
    c. all of the boys and none of the girls
    d. all of the boys and half of the girls
40. A pair of twins appears, by all chemical tests, to be identical, yet one member of the pair is a boy and the other is a girl with Turner's syndrome. Which of the following is the most likely explanation?
    a. They are fraternal (two-egg) twins who happen to have gotten identical sets of genes from their parents.
    b. They are identical twins, but one resulted from a cell in which nondisjunction occurred after the zygote had cleaved.

Essay Questions
41. Describe the changes that occur in an embryo from the time of fertilization until it enters the uterine cavity.
42. List the three primary germ layers and the structures that develop from them.
43. Identical twins are genetically identical, but fraternal twins are no more similar genetically than other siblings. Explain why.
44. Describe the distribution of chromosomes during meiosis, and explain how this distribution is the basis for the distribution of genes from parents to progeny.
45. Explain the principle of segregation of alleles and the chromosomal basis for it.
46. Define multiple alleles.
47. Explain the principle of independent assortment and the chromosomal basis for it.
48. Define the term *linkage* and explain its chromosomal basis.
49. Distinguish between the inheritance of genes on sex chromosomes and autosomes.
50. Explain how imbalances in genetic information can occur and list several examples of syndromes caused by chromosomal abnormalities.

Clinical Application Questions
51. Explain how amniocentesis can provide information about an infant before it is born.
52. Explain the difference between therapeutic and spontaneous abortion. How and when are therapeutic abortions performed?
53. Describe how an ectopic pregnancy occurs.
54. Describe some of the problems associated with childbirth when the infant is not positioned to emerge head first. What measures might a physician use to help avoid birth by cesarean section?
55. What are restriction endonucleases and why are they useful for cleaving DNA at specific sites?
56. What are restriction fragment length polymorphisms, and how might they be used to determine when one is at risk for a genetic disease that has not yet been expressed?

A PROBLEM OF REPRODUCTIVE FAILURE

Mrs. M. is a 26-year-old wife who has been married for three years and has failed to become pregnant. Her past gynecologic history is significant in that she has been sexually active since entering college at age 18. Early in her freshman year she started taking birth control pills and was seen by a physician on multiple occasions for a recurring bladder infection, usually caused by an *E. coli* organism. She was successfully treated with appropriate antibiotics. She was also plagued with recurrent bouts of vaginitis usually caused by the yeast organism *Monilia*. In her junior year she decided to have an intrauterine contraceptive device (IUD) placed in her uterus and discontinued the birth control pills in attempt to alleviate her problems with frequent infections (see Chapter 27 Clinical Note: Contraceptive Devices). In her senior year she was hospitalized with serious pelvic inflammatory disease (PID) caused by the *Chlamydia* microorganism. She was treated with intravenous antibiotics, the IUD was removed, and she made a good recovery. Since that time she has used a diaphragm for contraception.

Mrs. M. has had no medical problems during her marriage until a year ago when the couple decided to start a family. She has thus far been unsuccessful in becoming pregnant. Urological examination and sperm counts on her husband have proved to be normal. An examination of Mrs. M. has revealed normal ovulation, but a **uterosalpingogram** (special X rays that allow direct visualization of the uterus and fallopian tubes) showed a persistent blockage of both fallopian tubes probably due to her previous bout with PID.

The couple was very determined to have a child and wanted to be referred to a clinic to be evaluated for *in vitro* fertilization (see Clinical Note: Artificial Reproduction). Although this process is expensive ($5,000 to $10,000), it is encouraging to know that health insurance companies in many states are now required to offer artificial fertilization services as part of their contracts.

Appointments were made for Mr. and Mrs. M. for a prolonged and detailed discussion about the possibility of referral to an *in vitro* fertilization clinic in a nearby city. Following this complete discussion concerning the options available and the possible complications, the couple agreed to pursue *in vitro* fertilization and arrangements for the referral were completed. At the clinic, the medical records were reviewed, complete physical and psychological examinations performed, and plans for an attempt at *in vitro* fertilization were completed. Mrs M. was given a prescription for **pergonal** (to hyperstimulate her ovaries) with instructions concerning the time to take the medication during her next menstrual cycle. The following month the couple made frequent visits to the clinic for hormone testing and repeated ultrasonography to determine the precise time of expected ovulation. A few days prior to her expected ovulation time, Mrs. M. was admitted to the clinic. Under local anesthesia, five mature ova were identified using laparoscopy and "harvested" with a gentle suction device (see figure). The ova were examined and appeared normal. A solution of washed and concentrated sperm from her husband was added to the solution containing the ova and the specimens were placed in an incubator. Several hours later examination revealed four developing embryos; these were inserted into Mrs. M.'s uterus.

Two weeks later, serial tests for human chorionic gonadotrophin (HCG) showed a marked increase in concentration, suggesting a successful implantation. At 4 weeks, uterine ultrasonography identified two developing embryos—a twin pregnancy. The couple was delighted with the success and expectation of two babies. Mrs. M's pregnancy progressed without any serious complications. At the 38th week of gestation, she went into labor spontaneously and delivered a 4 lb 6 oz girl and a 3 lb 12 oz boy. Both babies were healthy and vigorous; they gained weight rapidly and were sent home to their new parents when they reached 5 lb in weight.

Only a few short years ago this happy resolution to such a difficult reproductive problem would not have been possible. Because of the rapidly advancing technology in this fascinating field of obstetrics, the possibility of having a baby (or even two) has gone from a dream to a reality for many infertile couples.

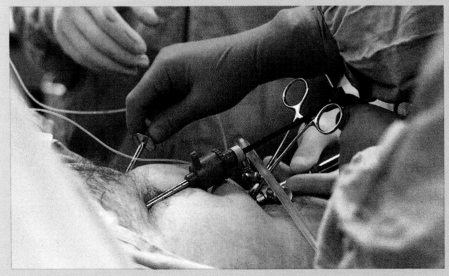

***In vitro* fertilization.** Ova are being removed from this woman's ovaries by laparoscopy, an abdominal exploratory procedure using a type of endoscope called a laparoscope.

Appendix I
Answers to Self-Tests of Chapter Objectives

CHAPTER 1

1. T	6. F	11. c	16. b	21. c
2. F	7. T	12. d	17. d	22. d
3. F	8. F	13. b	18. c	23. b
4. T	9. T	14. a	19. a	24. a
5. F	10. T	15. e	20. b	25. d

CHAPTER 2

1. T	10. T	19. a	28. b
2. F	11. d	20. c	29. b
3. F	12. b	21. a	30. c
4. F	13. a	22. b	31. c
5. F	14. b	23. pH buffer	32. b
6. T	15. bonds	24. a	33. c
7. F	16. covalent bond	25. d	34. c
8. F	17. ionic bond	26. b	35. d
9. F	18. a	27. b	

CHAPTER 3

1. F	8. T	15. b, d, g, e	22. b	29. c
2. F	9. T	16. f, g	23. d	30. b
3. F	10. T	17. f, g	24. c	31. d
4. T	11. b	18. b	25. b	32. b
5. F	12. a	19. a	26. a	33. a
6. F	13. c	20. a	27. c	34. c
7. F	14. c, f, g	21. b	28. b	35. b

CHAPTER 4

1. F	7. T	13. e	19. a	25. a
2. F	8. F	14. d	20. e	26. a
3. T	9. F	15. b	21. d	27. a
4. F	10. T	16. b	22. b	28. c
5. T	11. c	17. d	23. a	29. b
6. F	12. a	18. c	24. c	30. d

CHAPTER 5

1. F	7. T	13. e	19. b	25. d
2. T	8. F	14. b	20. a	26. b
3. T	9. T	15. d	21. b	27. e
4. T	10. F	16. b	22. e	28. c
5. T	11. c	17. a	23. a	29. a
6. F	12. a	18. b	24. a	30. e

CHAPTER 6

1. T	7. T	13. b	19. a	25. b
2. T	8. T	14. a	20. b	26. d
3. T	9. F	15. a	21. b	27. a
4. T	10. F	16. b	22. d	28. c
5. F	11. a	17. a	23. a	29. b
6. F	12. c	18. b	24. c	30. d

CHAPTER 7

1. axial
2. pelvic
3. pectoral, pelvic
4. pectoral
5. sphenoid
6. occipital
7. ethmoid
8. sagittal
9. sesamoid
10. vertebral foramen
11. atlas, axis
12. manubrium
13. sternum, seventh costal cartilage
14. head, tubercle
15. medial, lateral
16. b
17. d
18. c
19. a
20. c
21. c
22. a
23. a
24. b
25. b
26. a
27. c
28. a
29. b
30. a
31. c
32. a
33. a
34. b
35. a
36. c
37. a
38. b
39. c
40. b
41. c
42. b
43. c
44. a
45. a
46. b
47. a
48. a
49. b
50. d

CHAPTER 8

1. structural
2. functional
3. Abduction
4. Flexion
5. plantar flexion
6. circumduction
7. protraction
8. hinge joint
9. synovial joint
10. menisci
11. c
12. c
13. a
14. c
15. b
16. b
17. c
18. c
19. a
20. e
21. d
22. e
23. e
24. c
25. b
26. a
27. b
28. e
29. e
30. e
31. c
32. d
33. a
34. d
35. b

CHAPTER 9

1. F
2. F
3. T
4. F
5. T
6. T
7. T
8. T
9. T
10. T
11. b
12. d
13. d
14. a
15. c
16. d
17. b
18. c
19. c
20. c
21. c
22. b
23. b
24. a
25. b
26. a
27. b
28. a
29. a
30. b
31. d
32. a
33. c
34. c
35. a

CHAPTER 10

1. T
2. F
3. T
4. T
5. T
6. T
7. T
8. T
9. T
10. F
11. F
12. T
13. F
14. T
15. T
16. b
17. e
18. a
19. c
20. d
21. b
22. b
23. d
24. d
25. a
26. c
27. b
28. a
29. d
30. c
31. d
32. a
33. e
34. c
35. b

CHAPTER 11

1. T
2. T
3. F
4. F
5. T
6. F
7. F
8. F
9. F
10. T
11. b
12. c
13. d
14. d
15. a
16. d
17. b
18. d
19. a
20. b
21. a
22. b
23. b
24. b
25. c
26. b
27. c
28. c
29. c
30. a
31. c
32. d
33. c
34. a
35. c

CHAPTER 12

1. T
2. F
3. F
4. T
5. F
6. F
7. F
8. F
9. T
10. T
11. F
12. T
13. F
14. T
15. T
16. d
17. a
18. d
19. b
20. e
21. c
22. b
23. a
24. d
25. a
26. c
27. b
28. a
29. d
30. c
31. d
32. b
33. a
34. b
35. e

CHAPTER 13

1. T
2. F
3. F
4. F
5. T
6. F
7. F
8. T
9. T
10. T
11. T
12. T
13. F
14. T
15. b
16. e
17. a
18. c
19. d
20. b
21. c
22. d
23. a
24. b
25. a
26. b
27. d
28. a
29. c

CHAPTER 14

1. F
2. T
3. T
4. T
5. F
6. T
7. T
8. T
9. F
10. F
11. b
12. d
13. c
14. a
15. e
16. d
17. b
18. b
19. c
20. a
21. b
22. a
23. b
24. d
25. a

CHAPTER 15

1. F
2. T
3. F
4. F
5. T
6. T
7. T
8. F
9. T
10. F
11. T
12. T
13. F
14. T
15. T
16. T
17. F
18. T
19. F
20. F
21. d
22. b
23. a
24. b
25. c
26. a
27. d
28. a
29. c
30. a

CHAPTER 16

1. F	9. T	17. F	25. a	33. b
2. F	10. F	18. T	26. d	34. c
3. F	11. F	19. T	27. a	35. b
4. F	12. T	20. T	28. c	36. d
5. T	13. F	21. c	29. e	37. a
6. T	14. T	22. b	30. d	38. a
7. T	15. T	23. e	31. a	39. c
8. F	16. F	24. d	32. d	

CHAPTER 17

1. F	10. F	19. b	28. f	37. a
2. F	11. c	20. a	29. e	38. a
3. T	12. b	21. a	30. c	39. a
4. T	13. a	22. a	31. b	40. b
5. T	14. d	23. b	32. c	41. a
6. F	15. a	24. d	33. b	42. d
7. T	16. c	25. b	34. d	43. c
8. T	17. b	26. g	35. c	
9. F	18. b	27. a	36. c	

CHAPTER 18

1. F	12. T	23. d	34. a	45. a
2. F	13. F	24. c	35. c	46. c
3. T	14. F	25. b	36. e	47. b
4. F	15. T	26. a	37. c	48. c
5. F	16. F	27. d	38. a	49. c
6. T	17. F	28. e	39. c	50. c
7. T	18. c, d	29. a	40. c	51. d
8. F	19. a, b, d	30. a	41. a	52. a
9. F	20. d	31. c	42. b	53. d
10. T	21. d	32. d	43. c	54. b
11. T	22. e	33. e	44. c	

CHAPTER 19

1. T	8. T	15. d	22. d	29. c
2. T	9. F	16. b	23. c	30. d
3. F	10. T	17. b	24. b	31. a
4. F	11. a	18. a	25. b	32. a
5. F	12. d	19. b	26. a	33. c
6. F	13. b	20. a	27. d	34. a
7. F	14. a	21. a	28. d	

CHAPTER 20

1. F	8. F	15. a	22. c	29. a
2. F	9. F	16. b	23. a	30. a
3. T	10. F	17. d	24. b	31. e
4. T	11. a	18. c	25. c	32. c
5. F	12. b	19. d	26. a	33. d
6. T	13. d	20. c	27. a	34. a
7. T	14. c	21. b	28. d	35. a

CHAPTER 21

1. T	8. T	15. F	22. e	29. e
2. T	9. F	16. T	23. a	30. b
3. F	10. T	17. T	24. d	31. d
4. T	11. F	18. F	25. d	32. b
5. F	12. F	19. T	26. b	33. e
6. T	13. T	20. b	27. a	34. a
7. F	14. T	21. c	28. c	

CHAPTER 22

1. F	9. F	17. T	25. h	33. b
2. T	10. T	18. F	26. e	34. d
3. F	11. T	19. F	27. d	35. c
4. T	12. F	20. T	28. g	36. b
5. T	13. T	21. c	29. c	37. b
6. F	14. T	22. a	30. d	38. c
7. F	15. F	23. f	31. d	39. a
8. T	16. F	24. b	32. d	40. b

CHAPTER 23

1. T	8. F	15. F	22. d	29. d
2. T	9. T	16. T	23. a	30. a
3. T	10. F	17. F	24. e	31. a
4. T	11. F	18. T	25. c	32. d
5. F	12. T	19. T	26. a	33. a
6. F	13. T	20. F	27. c	34. d
7. T	14. F	21. b	28. a	35. a

CHAPTER 24

1. F	7. F	13. F	19. F	25. b
2. T	8. T	14. F	20. F	26. d
3. F	9. F	15. T	21. a	27. b
4. F	10. F	16. T	22. b	28. a
5. F	11. T	17. F	23. a	29. c
6. T	12. T	18. T	24. a	30. d

Answers to Self-Tests of Chapter Objectives

CHAPTER 25

1. F	8. T	15. b	22. a	29. c
2. T	9. T	16. c	23. b	30. d
3. T	10. F	17. a	24. d	31. b
4. F	11. d	18. d	25. c	32. b
5. T	12. e	19. b	26. d	33. c
6. T	13. a	20. c	27. a	34. d
7. F	14. c	21. d	28. a	35. a

CHAPTER 26

1. T	8. T	15. b	21. d	27. d
2. T	9. F	16. b	22. d	28. b
3. F	10. T	17. c	23. b	29. c
4. F	11. d	18. c	24. c	30. a
5. F	12. a	19. c	25. a	31. d
6. T	13. c	20. b	26. c	32. d
7. T	14. b			

CHAPTER 27

1. T	8. T	15. b	22. b	29. c
2. F	9. F	16. c	23. c	30. d
3. T	10. F	17. a	24. c	31. a
4. T	11. a	18. c	25. c	32. b
5. F	12. b	19. c	26. b	33. a
6. F	13. a	20. c	27. d	34. c
7. F	14. a	21. b	28. a	35. b

CHAPTER 28

1. T	9. T	17. d	25. c	33. b
2. F	10. T	18. b	26. d	34. a
3. F	11. b	19. a	27. b	35. b
4. F	12. a	20. a	28. b	36. b
5. F	13. c	21. a	29. c	37. b
6. T	14. b	22. b	30. a	38. b
7. F	15. c	23. b	31. b	39. a
8. T	16. a	24. d	32. a	40. b

Appendix II
Allied Health Careers

The topics discussed in this text are presented in a way that emphasizes the scientific approach to the study of human anatomy and physiology. It has been our goal to show the relationship between human structure and function and how together they produce a smoothly functioning person. Even though this scientific study is interesting for its own sake, we are aware that most students enrolled in an anatomy/physiology class are doing so to satisfy the requirements of a professional program leading to a career in medicine or nursing, or an allied health field, such as radiography or medical technology. Admission to one of these programs usually requires successful completion of introductory courses in biology and chemistry, as well as human anatomy and physiology. Moreover, once admitted, success in the program demands a knowledge of the names, locations, and functions of the body tissues and organs studied in this text.

In recognition of this interest in careers in allied health fields, and to help you acquire more information about these careers, we include here a brief description of 13 of the major allied health careers (in alphabetic order) and addresses to which you can write for additional details. This material comes from the *Allied Health Education Directory* to which you can refer for more information:

Allied Health Education Directory, 16th ed., edited by William R. Burrows (Chicago: American Medical Association, 1988), 32–169.

For information on careers in physical therapy (not included in the *Allied Health Education Directory*), write to:

American Physical Therapy Association
1111 Fairfax Street
Alexandria, Virginia 22314

For information on careers in dietetics (also not in the directory), please write to:

American Dietetic Association
208 South LaSalle Street
Suite 1100
Chicago, Illinois 60604

This appendix includes only careers considered to be in the *allied* health fields. If instead you are thinking of a career in medicine, an excellent resource book to consult is:

T. Donald Rucker and Martin D. Keller, *Planning Your Medical Career: Traditional and Alternative Opportunities* (Garrett Park, Maryland: Garrett Park Press, 1986).

Questions concerning career opportunities in nursing can be directed to:

American Nurses' Association, Inc.
Attn: Publication Orders
2420 Pershing Road
Kansas City, Missouri 64108

Emergency Medical Technician-Paramedic

Occupational Description Emergency medical technician-paramedics (EMT-paramedics), working under the direction of a physician (often through radio communication), recognize, assess, and manage medical emergencies of acutely ill or injured patients in pre-hospital care settings. EMT-paramedics work principally in advanced life support units and ambulance services which are under medical supervision and direction.

Job Description EMT-paramedics are competent in recognizing a medical emergency, in assessing the situation, and in managing the emergency care. This includes the ability to recognize the patients's condition and to initiate appropriate invasive and non-invasive treatments for a variety of surgical and medical emergencies, including airway and respiratory problems, cardiac dysrhythmias and standstills, and psychological crises. In addition, EMT-paramedics are able to assess the response of the patient to the treatment received and to modify that therapy as required through the direction of a designated physician or other authorized individuals.

EMT-paramedics maintain written records and dictate details relating to a patient's emergency care and a description of the incident which led to that care. They also direct the maintenance and preparation of emergency care equipment and supplies, as well as direct and coordinate the transport of patients. EMT-paramedics are responsible for exercising personal judgment when communication failures interrupt contact with medical direction or, in cases of immediate life threatening conditions, when such emergency care has been specifically authorized in advance.

Employment Characteristics Variations in geographic, sociologic, and economic factors impact on emergency medical services and subsequently on the type of practice engaged in by EMT-paramedics. Some EMT-paramedics are employed by community fire and police departments and have related responsibilities in those fields, or they serve as community volunteers. Not only are these individuals being employed in the pre-hospital phase of acute care provided by fire departments, police departments, public services, and private purveyors, but there is also an increased demand for their skills in hospital emergency departments and private industry.

According to a 1987 survey of program directors, entry level salaries average $18,200 per year for emergency medical technicians-paramedics.

Careers Inquiries regarding careers should be addressed to:

National Association of Emergency Medical Technicians
9140 Ward Parkway
Kansas City, MO 64114
(816) 444-3500

Medical Assistant

Occupational Description The medical assistant is a professional, multi-skilled person dedicated to assisting in patient care management. This practitioner performs administrative and clinical duties and may manage emergency situations, facilities, and/or personnel. Competence in the field also requires that a medical assistant display professionalism, communicate effectively, and provide instruction to patients.

Job Description Medical assistants are broadly defined as individuals who assist physicians in their offices or other medical settings, performing those administrative and/or clinical duties delegated in relation to the degree of training and in accordance with respective state laws governing such actions and activities. Medical assistants have a wide range of duties in many aspects of the physician's practice.

Their business-administrative duties include scheduling and receiving patients; obtaining patient's data; maintaining medical records; typing and medical transcription; handling telephone calls, correspondence, reports, and manuscripts; and assuming responsibility for office care, insurance matters, office accounts, fees and collections.

Their clinical duties may include preparing the patient for examination, obtaining vital signs, taking medical histories, assisting with examinations and treatments, performing routine office laboratory procedures and electrocardiograms, sterilizing instruments and equipment for office procedures, and instructing patients in preparation for x-ray and laboratory examinations.

Employment Characteristics More medical assistants are employed by practicing physicians than any other type of allied health personnel. Medical assistants are usually employed in physicians' offices where they perform a variety of administrative and clinical tasks to facilitate the work of the physician. The responsibilities of medical assistants may vary depending on whether they work in a clinic, hospital, large group practice, or small private office. With a demand from more than 200,000 physicians, there are and will probably continue to be almost unlimited opportunities for formally educated medical assistants. Salaries vary depending on the employer and geographic location.

According to a 1987 survey of program directors, entry level salaries average $11,700 per year for medical assistants.

Careers Inquiries regarding careers or accredited programs should be addressed to:

American Association of Medical Assistants
20 N Wacker Dr, Ste 1575
Chicago, IL 60606

Medical Laboratory Technician (Associate Degree)

Occupational Description Laboratory tests play an important role in the detection, diagnosis, and treatment of many diseases. Medical laboratory workers perform these tests under the supervision or direction of pathologists (physicians who diagnose the causes and nature of disease) and other physicians, or scientists who specialize in clinical chemistry, microbiology, or the other biological sciences. Medical laboratory workers develop data on the blood, tissues, and fluids in the human body by using a variety of precision instruments.

Job Description Medical laboratory technicians (associate degree) perform all of the routine tests in an up-to-date medical laboratory and can demonstrate discrimination between closely similar items and correction of errors by use of pre-set strategies. The technician has knowledge of specific techniques and instruments and is able to recognize factors which directly affect procedures and results. For confirmation of results, the technician conducts more than one test for each specialty area. The technician also monitors quality control programs within predetermined parameters.

Employment Characteristics Most medical laboratory personnel work in hospital laboratories, averaging a 40-hour week. Salaries vary depending on the employer and geographic location.

According to a 1987 survey of program directors, salaries of associate degree level medical laboratory personnel average $15,200.

Careers *See* Medical Technologist.

Medical Laboratory Technician (Certificate)

Occupational Description Laboratory tests play an important role in the detection, diagnosis, and treatment of many diseases. Medical laboratory workers perform these tests under the supervision or direction of pathologists (physicians who diagnose the causes and nature of disease) and other physicians, or scientists who specialize in clinical chemistry, microbiology, or biological sciences. Medical laboratory technicians (certificate) perform many routine procedures in the clinical laboratory under the direction of a qualified physician and/or medical technologist.

Job Descriptions Medical laboratory technicians (certificate) perform routine, uncomplicated procedures in the areas of hematology, serology, blood banking, urinalysis, microbiology, and clinical chemistry. These procedures involve the use of common laboratory instruments in processes where discrimination is clear, errors are few and easily corrected, and results of the procedures can be confirmed with a reference test or source within the working area.

Employment Characteristics Most medical laboratory personnel work in hospital laboratories, averaging a 40-hour week.

According to a 1987 survey of program directors, salaries of certificate level medical laboratory personnel average $14,300.

Careers *See* Medical Technologist.

Medical Technologist

Occupational Description Laboratory tests play an important role in the detection, diagnosis, and treatment of many diseases. Medical technologists perform these tests in conjunction with pathologists (physicians who diagnose the causes and nature of disease) and other physicians, or scientists who specialize in clinical chemistry, microbiology, or the other biological sciences. Medical technologists develop data on the blood, tissues, and fluids in the human body by using a variety of precision instruments.

Job Description In addition to the skills possessed by medical laboratory technicians, medical technologists perform complex analyses, fine line discrimination and correction of errors. They are able to recognize interdependency of tests and have knowledge of physiological conditions affecting test results in order to confirm these results and to develop data which may be used by a physician in determining the presence, extent, and, as far as possible, the cause of disease.

Medical technologists assume responsibility for, and are held accountable for, accurate results. They establish and monitor quality control programs and design or modify procedures as necessary. Tests and procedures performed or supervised by medical technologists in the clinical laboratory center on major areas of hematology, microbiology, immunohematology, immunology, clinical chemistry, and urinalysis.

Employment Characteristics Most medical technologists are employed in hospital laboratories. The remainder are chiefly employed in physicians' private laboratories and clinics, by the armed forces, by city, state, and federal health agencies, in industrial medical laboratories, in pharmaceutical houses, in numerous public and private research programs dedicated to the study of specific diseases, and as faculty of accredited programs preparing medical laboratory personnel. Salaries vary depending on the employer and geographic location.

According to a 1987 survey of program directors, salaries of entry level medical technologists average $19,800.

Careers Inquiries regarding careers and curriculum should be addressed to:

American Society for Medical Technology
2021 L St, NW, Ste 400
Washington, DC 20056
(202) 785-3311

Nuclear Medicine Technologist

Occupational Description Nuclear medicine is the medical specialty that utilizes the nuclear properties of radioactive and stable nuclides to make diagnostic evaluations of the anatomic or physiologic conditions of the body and to provide therapy with unsealed radioactive sources. The skills of the nuclear medicine technologist complement those of the nuclear medicine physician and of other professionals in the field.

Job Description Nuclear medicine technologists perform a number of tasks in the areas of patient care, technical skills, and administration. When caring for patients, they acquire adequate knowledge of the patients' medical history to understand and relate to their illness and pending diagnostic procedures for therapy; instruct the patient prior to and during procedures; evaluate the satisfactory preparation of the patient prior to commencing a procedure; and recognize emergency patient conditions and initiate lifesaving first aid when appropriate.

Technically, they apply their knowledge of radiation physics and safety regulations to limit radiation exposure; prepare and administer radiopharmaceuticals; use radiation detection devices and other kinds of laboratory equipment that measure the quantity and distribution of radionuclides deposited in the patient or in a patient specimen; perform in-vivo and in-vitro diagnostic procedures; utilize quality control techniques as part of a quality assurance program covering all procedures and products in the laboratory; and participate in research activities.

Administrative functions may include supervising other nuclear medicine technologists, students, laboratory assistants, and other personnel; participating in procuring supplies and equipment; documenting laboratory operations; participating in departmental inspections conducted by various licensing, regulatory, and accrediting agencies; and participating in scheduling patient examinations.

Employment Characteristics The employment outlook in nuclear medicine technology is very bright. Opportunities may be found both in major medical centers and in smaller hospitals. Opportunities are also available for obtaining positions in clinical research, education, and administration. Salaries vary depending on the employer and geographic location.

According to a 1987 survey of program directors, salaries of entry level nuclear medicine technologists average $20,200 per year.

Careers Inquiries regarding careers and curriculum should be addressed to:

> Joint Review Committee on Educational Programs in
> Nuclear Medicine Technology
> 445 S 300 E
> Salt Lake City, UT 84111

Occupational Therapist

Occupational Description Occupational therapy is the application of purposeful, goal-oriented activity in the evaluation, diagnosis, and/or treatment of persons whose function is impaired by physical illness or injury, emotional disorder, congenital or developmental disability, or the aging process, in order to achieve optimum functioning, to prevent disability, or to maintain health. Occupational therapists provide services to those individuals whose abilities to cope with tasks of living are threatened or impaired by developmental deficits, the aging process, poverty and cultural differences, physical injury or illness, or psychologic and social disability. Individuals are helped to attain the highest possible functional level, to become self-reliant, and to balance work and leisure in their lives through selected educational vocational and rehabilitation activities.

Job Description Specific occupational therapy services include, but are not limited to, education and training in activities of daily living (ADL); the design, fabrication, and application of orthoses (splints); guidance in the selection and use of adaptive equipment; therapeutic activities to enhance functional performance; pre-vocational evaluation and training; and consultation concerning the adaptation of physical environments for the handicapped. These services are provided to individuals or groups, and to both in-patients and out-patients.

Employment Characteristics Occupational therapists work in hospitals, schools, and mental health and community agencies. The wide population served by occupational therapists is located in a variety of settings such as clinics, rehabilitation facilities, long-term facilities, extended care facilities, schools and camps, and private homes. Occupational therapists both receive referrals from and make referrals to the appropriate health, educational, or medical specialists.

According to a 1987 survey of program directors, salaries of entry level occupational therapists average $22,500.

Careers Inquiries regarding careers and curriculum should be addressed to:

> American Occupational Therapy Certification Board
> Division of Credentialing
> 1383 Piccard Dr, Ste 300
> Rockville, MD 20850
> (301) 948-9626

Physician Assistant

Occupational Description The physician assistant is academically and clinically prepared to provide health care services with the direction and responsible supervision of a doctor of medicine or osteopathy. Physician assistants are accountable for their own actions, as well as being accountable to their supervising physician. The functions of the physician assistant include performing diagnostic, therapeutic, preventive and health maintenance services in any setting in which the physician renders care, in order to allow more effective and focused application of the physician's particular knowledge and skills.

Job Description The role of the physician assistant demands intelligence, sound judgment, intellectual honesty, the ability to relate effectively with people and the capacity to react to emergencies in a calm and reasoned manner. An attitude of respect for self and others, adherence to the concept of privilege and confidentiality in communicating with patients, and a commitment to the patient's welfare are essential attributes.

Physician assistants are educated in those areas of basic medical science and clinical disciplines which prepare them to function as generalists. Services performed by physician assistants include but are not limited to the following:

1. *Evaluation.* Initially approaching a patient of any age in any setting to elicit a detailed and accurate history, perform an appropriate physical examination, delineate health problems, and record and present the data.
2. *Monitoring.* Assisting the physician in conducting rounds in acute and long-term inpatient care settings, developing and implementing patient management plans, recording progress notes, and assisting in the provision of continuity of care in office-based and other ambulatory care settings.
3. *Diagnostics.* Performing and/or interpreting, at least to the point of recognizing deviations from the norm, common laboratory, radiologic, cardiographic, and other routine diagnostic procedures used to identify pathophysiologic processes.
4. *Therapeutics.* Performing routine procedures such as injections, immunizations, suturing and wound care, managing simple conditions produced by infection or trauma, assisting in the

management of more complex illness and injury, and taking initiative in performing evaluation and therapeutic procedures in response to life-threatening situations.

5. *Counseling.* Instructing and counseling patients regarding compliance with prescribed therapeutic regimens, normal growth and development, family planning, emotional problems of daily living, and health maintenance.
6. *Referral.* Facilitating the referral of patients to the community's health and social service agencies when appropriate.

Employment Characteristics A study published in 1987 indicated that just over half of physician assistant graduates were working with family physicians and internists. Almost thirty-eight percent worked in a private solo practice or private partnership practice. Similarly, thirty-eight percent were working in one of a variety of hospital settings, including county, city, and private hospitals, Veterans Administration hospitals, academic medical centers, and the like. The rest worked in such diverse settings as health maintenance organizations, industrial health clinics, military facilities, and prisons.

The normal work week for a majority of PAs exceeds 45 hours. Almost half devote more than 40 hours to direct patient contact, with one in five working over 50 hours a week. Similarly, about half report spending some additional hours of the week on call. It is anticipated that the demand for PAs will continue to increase in the coming decade.

Salaries vary depending on the experience and education of the individual, the economy of a given region, and the nature of the practice. According to a 1987 survey of program directors, salaries of entry level physician assistants average $25,600. Experienced PAs commonly earn in the $40,000–$50,000 range.

Careers Inquiries regarding careers and curriculum should be addressed to:

Association of Physician Assistant Programs
1117 North 19th St
Arlington, VA 22209
(703) 525-4200

Radiographer

Occupational Description Radiographers provide patient services using imaging modalities, as directed by physicians qualified to order and/or perform radiologic procedures. When providing patient services, they continually strive to provide quality patient care and are particularly concerned with limiting radiation exposure to patients, self, and others. Radiographers exercise independent judgment in the technical performance of medical imaging procedures by adopting variable technical parameters of the procedure to the condition of the patient and by initiating life-saving first aid and basic life support procedures as necessary during medical emergencies.

Job Description Professional competence requires that radiographers apply knowledge of anatomy, physiology, positioning, and radiographic technique in the performance of their duties. They must also be able to communicate effectively with patients, other health professionals, and the public. Additional duties may include processing of film, evaluating radiologic equipment, managing a radiographic quality assurance program, and providing patient education relevant to specific imaging procedures. The radiographer displays personal attributes of compassion, courtesy and concern in meeting the special needs of the patient.

Employment Characteristics Most radiographers are employed in hospitals. However, there are also positions open to qualified professionals in specialized imaging centers, urgent care clinics, private physician offices, industry, and civil service and public health service facilities. Radiographers who are employed fulltime usually work 40 hours per week. Salaries and benefits vary according to experience, ability, and geographic location, but are generally competitive with those of professions requiring comparable educational preparation. Employment opportunities are available throughout the nation, but may vary geographically.

According to a 1987 survey of program directors, salaries of entry level radiographers average $17,200.

Careers Inquiries regarding careers and curriculum should be addressed to:

American Society of Radiologic Technologists
15000 Central Ave, SE
Albuquerque, NM 87123
(505) 298-4500

Respiratory Therapist

Occupational Description The respiratory therapist applies scientific knowledge and theory to practical clinical problems of respiratory care. The respiratory therapist is qualified to assume primary responsibility for all respiratory care modalities, including the supervision of respiratory therapy technician functions. The respiratory therapist may be required to exercise considerable independent clinical judgment, under the supervision of a physician, in the respiratory care of patients.

Job Description In fulfillment of the therapist role the respiratory therapist may:

1. Review, collect, and recommend obtaining additional data. The therapist evaluates all data to determine the appropriateness of the prescribed respiratory care, and participates in the development of the respiratory care plan.
2. Select, assemble, and check all equipment used in providing respiratory care.
3. Initiate and conduct therapeutic procedures, and modify prescribed therapeutic procedures to achieve one or more specific objectives.
4. Maintain patient records and communicate relevant information to other members of the health care team.
5. Assist the physician in performing special procedures in a clinical laboratory, procedure room, or operating room.

Employment Characteristics Respiratory therapy personnel are employed in hospitals, nursing care facilities, clinics, physicians' offices, companies providing emergency oxygen services, and municipal organizations. Salaries vary depending on employer and geographic location.

According to a 1987 survey of program directors, salaries of entry level respiratory therapists average $19,300.

Careers Inquiries regarding careers and curriculum should be addressed to:

> American Association for Respiratory Care
> 1720 Regal Row, Ste 125
> Dallas, TX 75235

Respiratory Therapy Technician

Occupational Description The respiratory therapy technician administers general respiratory care. Technicians may assume clinical responsibility for specified respiratory care modalities involving the application of well defined therapeutic techniques under the supervision of a respiratory therapist and a physician.

Job Description In fulfillment of the technician role the respiratory therapy technician may:

1. Review clinical data, history, and respiratory therapy orders.
2. Collect clinical data by interview and examination of the patient. This will include portions of the data by inspection, palpation, percussion, and auscultation of the patient.
3. Recommend and/or perform and review additional bedside procedures, x-rays, and laboratory tests.
4. Evaluate data to determine the appropriateness of the prescribed respiratory care.
5. Assemble and maintain equipment used in respiratory care.
6. Assure cleanliness and sterility by the selection and/or performance of appropriate disinfecting techniques, and monitoring their effectiveness.
7. Initiate, conduct, and modify prescribed therapeutic procedures.

Employment Characteristics Respiratory therapy personnel are employed in hospitals, nursing care facilities, clinics, doctor's offices, companies providing emergency oxygen services, and municipal organizations. Salaries vary depending on employer and geographic location.

According to a 1987 survey of program directors, salaries of entry level respiratory therapy technicians average $16,000.

Careers *See* Respiratory Therapist.

Surgeon's Assistant

Occupational Description The purpose of the physician assistant in surgery is to help surgeons provide personal health services to patients under their care. They perform a number of functions and tasks formerly done only by surgeons in the operating room, in the pre- and post-operative care of the hospitalized patient, in the emergency room, in the surgeon's office practice, and in other settings.

Job Description Surgeon's assistants perform selected diagnostic and therapeutic functions and tasks in order to allow surgeons to extend their services through more effective use of their knowledge, skills, and abilities. Surgeon's assistants gather the data necessary for the surgeon to reach a decision and help in implementing the therapeutic plan for the patient. Tasks performed by assistants include transmission and execution of the surgeon's orders and such diagnostic and therapeutic procedures as may be delegated by the supervising surgeon. Such tasks and functions include but need not be limited to the following:

1. Eliciting a detailed and accurate health history, performing an appropriate physical examination and recording and presenting data to the supervising surgeon;
2. Performing or assisting in routine laboratory work and related diagnostic studies;
3. Performing therapeutic procedures common to the surgical practice;
4. Helping the supervising surgeon in making rounds of hospitalized patients, recording patient progress notes, executing standing and other orders, and preparing comprehensive discharge summaries; and
5. Assisting in operative procedures and in the provision of care to patients in the surgical office.

Employment Characteristics Surgeon's assistants are prepared to help in any medical setting in which the surgeon functions and may therefore be involved with patients in any medical setting. They may work in an area of surgery which requires a wide variety of procedures, such as those performed by general surgeons in community hospitals or in the surgical specialty of their supervising physician.

Surgeon's assistants commonly work in both a hospital and a surgical office practice. A normal work week often ranges between 45 and 55 hours. In addition, they may be expected to assume responsibility for additional hours on call.

Salaries vary depending on the experience and education of the individual, the economy of a given region, the nature of the surgical practice, and the time demands of the work schedule and responsibilities. Starting salaries begin around $30,000 and progress to $50,000 and higher. It is anticipated that the demand for surgeon's assistants will continue to exceed the supply during the immediate future.

Careers Inquiries regarding careers and curriculum should be addressed to:

> American Association of Surgeon's Assistants
> 11800 Sunrise Valley Drive, Ste 808
> Reston, VA 22091
> (703) 434-7311

Surgical Technologist

Occupational Description Surgical technologists work principally in the operating room performing functions and tasks that provide for a safe environment for surgical care, contribute to the efficiency of the operating room team, and support the operating surgeons and others involved in operative procedures. Surgical technologists also work in other patient service settings which call for special knowledge of asepsis.

Job Description Because of the marked variations in practice resulting from geographic, sociologic and economic factors, the role of surgical technologists is not rigidly defined. Surgical technologists perform functions and tasks which contribute to patient care and safety, to the efficiency of the operating team, and to the cleanliness of the surgical care environment. Knowledge of and experience with aseptic surgical techniques qualify surgical technologists to prepare instruments and materials for use at the operating table and elsewhere and to assist in the use of these materials.

Employment Characteristics A majority of surgical technologists work in hospitals, principally in operating rooms and occasionally in emergency rooms and other settings which call for knowledge of and ability in maintaining asepsis. A much smaller number work in a wide variety of settings and arrangements including out-patient surgicenters, either under the employ of physicians or as self-employed technologists.

Those who work in hospital and other institutional settings are usually expected to work rotating shifts or to accommodate on-call assignments to assure adequate staffing for emergency surgical procedures which need to be done during evening, night, weekend, and holiday hours. Otherwise, surgical technologists follow a standard hospital work day.

Salaries vary depending upon the experience and education of the individual, the economy of a given region, the responsibilities of the position, and the working hours. Starting annual salaries begin in the range of $14,000, with experienced technologists earning incomes of $20,000 and higher. Demand for technologists varies among communities and geographic regions. Prospective students are advised to assess the market for graduates within the region in which they would like to work before matriculating in an educational program. Such information is likely to be available through local employment offices, local accredited programs, and hospital councils or hospitals.

According to a 1987 survey of program directors, salaries of entry level surgical technologists average $14,200.

Careers Inquiries regarding careers and curriculum should be addressed to:

Association of Surgical Technologists
8307 Shaffer Parkway
Littleton, CO 80127

Appendix III
Word Parts for Building Anatomical and Medical Terminology

Prefix	Meaning (and Example)
a-, ab-	away from, opposite (*abnormal*)
a-, an-	negative, not, without (*anaerobic*)
abdomin-	pertaining to abdomen (*abdominal*)
ad-	toward, adhere to (*adduction*)
aer-	pertaining to air (*aerobic*)
af-	movement toward a point (*afferent nerve*)
algi-	pain (*algesic*)
all-	other, different (*allele*)
amb-	both, on both sides (*ambidextrous*)
amph-	on both sides (*amphiarthrosis*)
ana-	up (*anaphylaxis*)
angio-	pertaining to blood or lymph vessels (*angiogram*)
ante-	before (*antebrachium*)
anti-	against (*antibiotic*)
apo-	from, opposed to (*apocrine*)
aqua-	water (*aqueous*)
arche(i)-	first (*archetype*)
arthr-	joint (*arthritis*)
aud-	pertaining to hearing (*audiology*)
auto-	pertaining to self (*autoimmunity*)
bi-	double (*binocular*)
bio-	life, living (*biofeedback*)
blast-	germinal, bud (*blastocyst*)
brachi-	arm (*brachium*)
brachy-	short (*brachycephalic*)
brady-	slow (*bradycardia*)
bucc-	cheek (*buccinator muscle*)
cac-	bad, evil (*cachexia*)
calc-	pertaining to calcium, lime, or stone (*calcification*)
carcin-	cancer (*carcinogen*)
cardi-	heart (*cardiac*)
cata-	down, lower (*catatonia*)
caud-	tail (*caudal*)
centri(o)-	center, middle (*centrilobular*)
cephal-	pertaining to head (*cephalomeningitis*)
cerebro-	brain (*cerebrovascular accident*)
chol-	bile (*cholangiography*)
chondr-	cartilage (*chondrocranium*)
chromo-	color (*chromosome*)
circum-	around (*circumcision*)
co-	together (*coarctation*)
coel-	hollow cavity (*coelom*)
com-	together (*communicable*)
con-	together, with (*congested*)
contra-	against (*contraceptive*)
corn-	hardened, horny (*cornea*)
corp-	body (*corpus cavernosum*)
cryo-	pertaining to cold (*cryopreservation*)
crypt-	hidden (*cryptocephalus*)
cyan-	blue in color (*cyanotic*)
cyst-	bladder, baglike (*cystocarcinoma*)
cyto-	cell (*cytoplasm*)
dacry-	tears, lacrimal (*dacryostenosis*)
dactyl-	pertaining to fingers (*dactylospasm*)
de-	from, not, down (*decongestant*)
deci-	tenth (*decibel*)
dent-	pertaining to teeth (*dentition*)
derm-	skin (*dermatitus*)
di-	two, double (*digastric*)
dia-	through, between, separate (*diastema*)
dipl-	double (*diplococcus*)
dis-	apart, away from, negative (*dislocation*)
duct-	to lead or conduct (*ductus arteriosus*)
dys-	difficult, bad, painful (*dysmenorrhea*)
ecto-	external, outside (*ectoderm*)
ef-	out (*effusion*)
endo-	internal, inside (*endoderm*)
entero-	internal, inside (*enteroscope*)
ento-	within (*entocyte*)
epi-	upon, over, in addition to (*epicardium*)
erythro-	red (*erythrocyte*)
eu-	well, healthy, normal (*euphoria*)

Prefix	Meaning (and Example)
ex-	from, out of (*excrement*)
exo-	outside (*exoskeleton*)
extra-	beyond, outside of (*extraembryonic*)
fasci-	band (*fasciculus*)
febr-	pertaining to fever (*febricide*)
feto-	pertaining to a fetus (*fetoscope*)
fibr-	fibrous (*fibroma*)
for-	opening (*foramen rotundum*)
fore-	in front of (*forebrain*)
galact-	pertaining to milk (*galactosemia*)
gastr-	pertaining to the stomach (*gastroenteritis*)
gloss-	tongue (*glossectomy*)
gluco-	pertaining to sugar (*glucogenesis*)
glyco-	pertaining to sugar (*glycosemia*)
gonado-	pertaining to the gonads (*gonadotropin*)
gran-	particle, grain (*granulocyte*)
gravi-	heavy (*gravimetric*)
gyn-	female sex (*gynecologist*)
haplo-	simple, single (*haploid*)
hema(o)-	blood (*hemapoiesis*)
hemato-	blood (*hematologist*)
hemi-	half (*hemisphere*)
hepat-	liver (*hepatitis*)
hetero-	different, other (*heterosexual*)
histo-	tissue (*histocompatibility*)
holo-	whole (*holocrine*)
homo-	same (*homosexual*)
hydro-	water (*hydrocephalic*)
hyper-	excessive, above (*hypertension*)
hypo-	under, below (*hypoglycemia*)
idio-	individual, distinct (*idiopathic*)
ilei(o)-	pertaining to the ileum (*ileitis*)
ilio-	pertaining to ilium (*iliosacral*)
in-	in, within, negative (*incontinence*)
infra-	beneath (*infraorbital*)
inter-	between, among (*intercostal*)
intra-	within, inside (*intramuscular*)
irid-	pertaining to the iris (*iridectomy*)
iso-	like, equal (*isometric*)
karyo-	nucleus (*karyotype*)
kerato-	pertaining to tough substances, cornea (*keratosis*)
keto-	pertaining to ketone bodies (*ketosis*)
kine-	movement (*kinesiology*)
kypho-	humped (*kyphosis*)
labi-	lip (*labia majora*)
lacri-	tears (*lacrimal duct*)
lacto-	pertaining to milk (*lactogenic*)

Prefix	Meaning (and Example)
laparo-	pertaining to loin or abdomen (*laparoscopy*)
laryngi(o)-	pertaining to larynx (*laryngoscope*)
later-	side (*lateral*)
leuk-	white (*leukocyte*)
lipo-	pertaining to fat (*lipocyte*)
lith-	stone (*lithodialysis*)
lymph-	pertaining to lymphatic system (*lymphoma*)
macro-	large (*macrophage*)
mal-	abnormal, bad, ill (*malnutrition*)
mamm-	pertaining to breast (*mammography*)
mast-	pertaining to breast (*mastectomy*)
medi-	middle (*mediastinum*)
mega-	large (*megakaryocyte*)
mela-	black (*melanoma*)
meningo-	meninges (*meningopathy*)
meso-	middle (*mesoderm*)
meta-	beyond, after (*metacarpal*)
metra(o)	pertaining to uterus (*metrocystosis*)
micro-	small (*microcephaly*)
mio-	less, smaller (*mioplasmia*)
mito-	thread (*mitosome*)
mono-	one, alone (*mononucleosis*)
morph-	shape (*morphogenesis*)
muco-	pertaining to mucous (*mucoenteritis*)
multi-	many (*multicellular*)
myelo-	pertaining to bone marrow (*myeloma*)
myo-	muscle (*myocardium*)
narc-	numbness, stupor (*narcolepsy*)
naso-	nose (*nasoscope*)
necro-	dead, corpse (*necrosis*)
neo-	young, new (*neonatal*)
nephro-	kidney (*nephropathy*)
neuro-	nerve (*neurosurgery*)
nucleo-	nucleus (*nucleotide*)
ob-	against, in front of (*obstruction*)
oc-	against (*occlusion*)
oculo-	eye (*oculomotor*)
odont-	tooth (*odontoblast*)
oligo-	small, few (*oligospermia*)
omo-	shoulder (*omoclavicular*)
oo-	egg (*oocyte*)
oophoro-	ovary (*oophorotomy*)
opisth-	backward, behind (*opisthotic*)
orchi-	testis (*orchidectomy*)
ortho-	straight, normal (*orthopedics*)
osteo-	bone (*osteocyte*)
oto-	ear (*otopathy*)
ovi(o)-	egg (*oviduct*)
pachy-	thick, heavy (*pachyderma*)
pan-	all (*panarthritis*)
para-	beside, near (*paranasal*)

Word Parts for Building Anatomical and Medical Terminology

Prefix	Meaning (and Example)
path-	disease (*pathology*)
pedia-	children (*pediatrics*)
per-	through (*perfusion*)
peri-	surrounding, near (*periosteum*)
phag-	pertaining to eating (*phagocyte*)
pharmaco-	drugs, medicine (*pharmacotherapy*)
phlebi(o)-	vein (*phlebitis*)
photo-	pertaining to light (*photoreceptor*)
platy-	flat, broad (*platycephaly*)
pleur-	pertaining to pleura (*pleurisy*)
pneumo-	air, lung (*pneumonia*)
pod-	foot (*podiatry*)
poly-	much, many (*polydactyly*)
post-	after, behind (*postmortem*)
pre-	before (*prenatal*)
pro-	before (*prochondral*)
proct-	anus (*proctology*)
proto-	first (*protoplasm*)
pseudo-	false (*pseudogout*)
psycho-	mental (*psychogenesis*)
pyo-	pus (*pyodermatitis*)
quad-	pertaining to four (*quadriplegia*)
re-	again, back (*reabsorb*)
reno-	pertaining to the kidney (*renography*)
rete(i)-	network (*reticulocyte*)
retro-	behind, backward (*retroperitoneal*)
rhino-	nose (*rhinopathy*)
saccharo-	sugar (*saccharuria*)
sarco-	flesh (*sarcolemma*)
sclero-	hard (*scleroderma*)
semi-	half (*semilunar*)
steno-	narrow (*stenosis*)
sub-	under, below, beneath (*subcutaneous*)
super-	beyond, above, upper (*supervirulent*)
supra-	over, above, (*supramandibular*)
sym-	joined together (*symphysis*)
syn-	joined together (*synarthrosis*)
tachy-	fast (*tachycardia*)
tele-	far (*telencephalon*)
tetra-	four (*tetraploid*)
therm-	pertaining to heat (*thermogenesis*)
thorac-	chest (*thoracic cavity*)
thrombo-	clot, lump (*thrombosis*)
tox-	poison (*toxicology*)
trans-	over, across (*transfuse*)
tri-	three (*triceps brachii*)
tropho-	pertaining to nourishment (*trophoblast*)
ultra-	excess, beyond (*ultrasonogram*)
uni-	one (*unicellular*)
uro-	urinary organs, urine (*urogenital*)

Prefix	Meaning (and Example)
vas-	vessel (*vasodilation*)
viscer-	internal organs (*visceroreceptor*)
vit-	life (*vitality*)
xanth-	pertaining to yellow (*xanthoderma*)
zoo-	animal (*zoology*)
zygo-	union, join (*zygote*)

Suffix	Meaning (and Example)
-able	capable of (*digestable*)
-a(e)sthesia	sensation (*anesthesia*)
-algia	pain (*gastralgia*)
-ase	enzyme (*lactase*)
-asis	condition (*homeostasis*)
-cele	tumor, cyst (*hydrocele*)
-cide	destroy (*fungicide*)
-coele	enlarged cavity, swelling (*blastocoele*)
-cyte	cell (*erythrocyte*)
-dynia	pain (*ophthalmodynia*)
-ectomy	surgical removal (*appendectomy*)
-emesis	vomiting (*hyperemesis*)
-emia	pertaining to the blood (*glycemia*)
-ferent	moving, carrying (*afferent*)
-form	shape (*penniform*)
-fuge	drive away (*centrifuge*)
-gen	causative agent (*pathogen*)
-genic	causing (*carcinogenic*)
-gram	recording, record (*electrocardiogram*)
-ia	condition (*uremia*)
-iatrics	medical specialties (*geriatrics*)
-ism	condition (*rheumatism*)
-itis	inflammation (*arthritis*)
-kinesis	movement (*orthokinesis*)

Suffix	Meaning (and Example)	Suffix	Meaning (and Example)
-lith	stone (*otolith*)	-phylaxis	protection (*prophylaxis*)
-logy	study of (*osteology*)	-plasty	to reconstruct (*rhinoplasty*)
-lysis	dissolve, break apart (*hemolysis*)	-plegia	stroke, paralysis (*paraplegia*)
		-pnea	pertaining to breathing (*apnea*)
		-poiesis	formation of (*hemapoiesis*)
-megaly	enlarged (*acromegaly*)		
		-rrhage	overflow (*hemorrhage*)
		-rrhea	discharge, flow (*menorrhea*)
-oid	resembling (*nephroid*)		
-oma	tumor (*melanoma*)	-sclerosis	hardness, dryness (*arteriosclerosis*)
-ory	pertaining to (*sensory*)	-sis	action, process (*dialysis*)
-ose	full of (*adipose*)		
-osis	condition (*scoliosis*)		
		-tomy	cut, incision (*tracheostomy*)
		-trophy	pertaining to nutrition (*hypertrophy*)
-pathy	disease, abnormality (*colonopathy*)	-tropic	changing, influencing (*gonadotropic*)
-penia	deficiency (*leukopenia*)		
-phil	attraction to (*neutrophil*)		
-phobia	abnormal fear (*hematophobia*)	-uria	pertaining to urine (*glycosuria*)

Glossary

A

abdomen the portion of the body between the thoracic cavity and the pelvis.

abdominopelvic (ăb·dŏm″ĭ·nō·pĕl′vĭk) pertaining to the abdomen and pelvis.

abduction (ăb·dŭk′shŭn) movement of a body part away from the midline of the body.

ABO blood group one of several blood typing systems, based on the presence or the absence of A and B antigens in the membrane of erythrocytes and certain other cells.

abortion spontaneous or induced termination of pregnancy before the fetus is viable.

abscess a collection of pus in a localized cavity or space.

absolute refractory period the period during an action potential when an additional action potential cannot be induced.

absorption the movement of substances or energy into body fluids or tissues.

absorptive state a period when digested nutrients are absorbed from the digestive tract.

acclimatization (ă·klī″m·tĭ·za′shŭn) physiological changes that occur as one adapts to new environmental conditions.

accommodation (ă·kŏm″ō·dā′shŭn) adaptation to meet new needs.

acetabulum (ăs″ĕ·tăb′ū·lŭm) the socket in the pelvic bone with which the head of the femur articulates.

acetoacetic (ăs″ĕ·tō·ă·sē′tĭk) **acid** a four-carbon ketone body formed when fats are incompletely catabolized; found in abnormal quantities in the blood and urine of diabetics.

acetone (ăs′ĕ·tōn) a two-carbon ketone body formed when fats are incompletely catabolized, especially in diabetics.

acetylcholine (ăs″ĕ·tĭl·kō′lēn) a neurotransmitter in synapses of motor and parasympathetic nerves and in neuromyal junctions.

acetylcholinesterase (ăs″ĕ·tĭl·kō″lĭn·ĕs′tĕr·ās) an enzyme that destroys acetylcholine.

achalasia (ăk″ă·lā′zē·ă) failure of normally contracted sphincter muscles to relax.

Achilles (ă·kĭl′ēz) **tendon** the tendon that attaches the calf muscle to the calcaneus bone in the heel.

achondroplasia (ă·kŏn″drō·plā′sē·ă) a dominant, hereditary dwarfism resulting from a defect in epiphyseal growth of long bones.

acid a substance that lowers the pH of a solution; a solution with a pH less than 7.0.

acidosis (ăs″ĭ·dō·sĭs) excessive lowering of the pH of the body fluids.

acini (ăs′ĭ·nī) a group of secretory cells in a gland that surround and empty into a cavity.

acne (ăk′nē) chronic inflammation of the sebaceous glands and hair follicles in the skin, especially on the face.

acoustic (ă·koos′tĭk) having to do with the sense of hearing.

acromegaly (ăk″rō·mĕg′ă·lē) a disorder characterized by enlargement and elongation of the bones of the extremities and the frontal bone of the skull and jaws, and an enlargement of the nose, lips, and soft tissues of the face.

acromion (ă·krō′mē·ŏn) portion of the scapula that articulates with the clavicle and projects from the spine of the scapula.

acrosome (ăk′rō·sōm) an enzyme-filled structure at the anterior end of a sperm.

actin (ăk′tĭn) a protein found in the thin filaments of muscle and in the cytoskeleton.

action potential a rapid change in the polarity of the membrane of a neuron or a muscle fiber; used to transmit an impulse.

active immunity immunity established by prior exposure to an antigen.

active transport transport of materials across a cell membrane against a concentration gradient and using ATP as a source of energy.

acupuncture a procedure involving the insertion of fine needles into specific points in the skin to relieve pain or to induce localized anesthesia.

acute occurring rapidly and subsiding quickly after a short period.

acute mastoiditis (măs·toyd·ī′tĭs) bacterial infection of the mastoid process.

acute mountain sickness (AMS) headache, nausea, fatigue, and accelerated heartbeat resulting from diminished oxygen at high altitude.

adaptation the adjustment of the pupil of the eye to changes in light intensity; change in the intensity of a sensation with continuous stimulation; decreased frequency of firing of a neuron in response to a continuing stimulus of constant intensity.

Addison's disease a disease that results from deficient production of adrenal cortex hormones characterized by increased skin pigmentation, weight loss, fatigue, and hypotension.

adduction (ă·dŭk′shŭn) movement toward the main axis of the body.

adductor a muscle that draws a limb toward the medial line of the body.

adenohypophysis (ăd″ē·nō·hī·pŏf´ĭ·sĭs) the anterior lobe of the pituitary gland.

adenoma a benign tumor of glandular tissue.

adenoid (ăd´ē·noyd) lymphatic tissue located in the nasopharynx.

adenosine diphosphate (ADP) (ă·děn´ō·sēn dī·fos´fāt) a molecule formed from ATP when ATP is used as a source of energy.

adenosine triphosphate (ATP) a molecule used as a source of energy in most of the energy-requiring reactions of metabolism; it is formed by inorganic phosphate combining with ADP.

adenyl cyclase (ăd″ē·nĭl sī´klās) an enzyme involved in the production of cyclic AMP from ATP following the combination of a hormone with a membrane-bound receptor.

ADH *see antidiuretic hormone*

ADH-resistant diabetes insipidus a disorder resulting from loss or absence of ADH receptors in the kidney that results in the production of excessive amounts of urine.

ADH-sensitive diabetes insipidus a disorder resulting from deficient production of ADH by the posterior pituitary gland and characterized by the production of excessive urine.

adhesion an abnormal joining together of parts by fibrous tissue; an attraction of dissimilar substances to one another.

adipocyte (ăd″ĭ·pō·sīt) a fat-storing cell in fatty tissue; a lipocyte.

adipose tissue fatty tissue.

adrenal cortex (ăd·rē´năl kor´těks) the outer portion of the adrenal gland and the source of corticosteroid hormones.

adrenal glands endocrine glands located at the superior end of each kidney.

adrenal medulla (ăd·rē´năl mě·dū·lă) the inner portion of the adrenal glands and the source of catecholamines.

adrenal virilism (ăd·rē´năl vĭr´ĭl·ĭzm) development of secondary male characteristics in a woman due to excess production of adrenal androgens.

adrenaline (ă·drěn´ă·lěn) **(epinephrine)** a hormone produced by the adrenal medulla and certain other tissues.

adrenergic nerves nerves that use epinephrin in their synapses, including most of the sympathetic postganglionic fibers (except those that stimulate sweat glands).

adrenocorticotrophic hormone (ACTH) (ăd·rē″nō·kor″tĭ·kō·trō´fĭk) an anterior pituitary hormone involved in stimulating growth.

adventitia (ăd″věn·tĭsh´ē·ă) the loose connective tissue that forms the outermost covering of an organ.

aerobic (ā·ěr·ō´bĭk) using oxygen.

afferent carrying impulses or blood inward toward a center.

afferent arteriole the blood vessel that carries blood into the glomerulus in the kidney.

afferent neuron a nerve cell that carries an impulse toward the central nervous system.

affinity a chemical attraction between substances.

afterbirth the placenta and associated structures expelled from the uterus following the birth of a baby.

agglutination (ă·glū″tĭ·nā´shŭn) the clumping of red blood cells following the mixing of mismatched blood types.

agglutinin (ă·glū´tĭ·nĭn) a substance that can cause red blood cells to agglutinate.

agglutinogen (ă·glū·tĭn´ō·jěn) a substance that stimulates the production of an agglutinin.

agonist (ăg´ŏn·ĭst) a muscle that produces a desired action when it contracts.

agranulocytes (ă·grăn´ū·lō·sīts) nongranular leukocytes, including lymphocytes and monocytes.

air contrast enema a technique for X raying the colon in which barium salts are injected into the colon followed by inflation of the colon with air.

akinetic seizure a brief and generalized epileptic seizure in children.

ala (ā´lă) a winglike structure or process.

albinism (ăl·bĭn´ĭzm) a genetic condition that prevents the production of pigment in the skin, hair, and eyes.

Albright's syndrome a childhood condition resulting from abnormal bone development due to hyperparathyroidism and characterized by softening and loss of calcium in the bones.

albuginea (ăl·bū·jĭn´ē·ă) a layer of tough, whitish fibrous tissue that surrounds an organ or other structure.

albumin (ăl·bū´mĭn) a common protein found in blood plasma.

albuminuria (ăl·bū·mĭ·nū´rē·ă) the presence of albumin in the urine.

aldosterone (ăl·dōs´těr·ōn) a steroid hormone of the adrenal cortex that is primarily involved in regulating body fluid volume by controlling sodium chloride and potassium excretion.

aldosteronism a condition in which there are abnormally high levels of aldosterone.

alimentary canal the digestive tube from the mouth to the anus.

alkaline an aqueous solution with a pH greater than 7.

alkalosis (ăl″kă·lō´sĭs) a condition in which the pH of body fluids is higher than normal.

all-or-none effect in muscles, the phenomenon that muscle fibers in a motor unit contract to their full extent independent of the magnitude of the stimulus as long as a threshold is reached; in neuron physiology, the phenomenon that the magnitude of an action potential is independent of the strength of the stimulus as long as the threshold is reached.

allantois (ă·lăn´tō·ĭs) the membrane between the chorion and amnion in a fetus.

allele (ă·lēl´) one of two or more forms of a gene.

allergen (ăl´ěr·jěn) a substance that causes an allergic reaction.

allergy (ăl´ěr·jē) a hypersensitive reaction of body tissues to a specific substance or allergen.

alpha-blocking agents any substance that interferes with the interaction between neurotransmitters and alpha-receptors in sympathetic synapses.

alpha cell a cell type in the islets of Langerhans responsible for producing glucagon.

altitude sickness a response to low oxygen content associated with high altitudes, characterized by nausea and dizziness.

alveolar duct an extension of a respiratory bronchiole that communicates with alveoli and alveolar sacs.

alveolar macrophage a cell type found in the walls of the alveoli in the lungs that is responsible for removing foreign matter; also called a "dust cell."

alveolar sac a group of alveoli that have a common opening.

alveolus (ăl·vē´ō·lŭs) a small, hollow space; a tooth socket; the smallest air space in lung tissue.

Alzheimer's disease progressive decline in intellectual ability due to loss of cells of the cerebral cortex typically occurring in the age range of 40 to 60 years.

amblyopia (ăm″blē·ō´pē·ă) dimness of vision not associated with any visible defect.

amenorrhea (ă·mĕn″ō·rē′ă) cessation or absence of menstruation.
amino acid subunit of protein containing a carboxyl group and an amino group separated by a carbon atom to which a variable side group is attached.
amnesia (ăm·nē′zē·ă) the temporary loss of memory usually caused by an impact or blow to the head.
amniocentesis (ăm″nē·ō·sĕn·tē′sĭs) puncturing the amniotic sac of a pregnant woman to remove amniotic fluid, usually with a needle or syringe.
amnion (ăm′nē·ŏn) the embryonic sac that contains the developing embryo.
amphiarthrosis (ăm″fē·ăr·thrō′sĭs) an articulation in which the bony surfaces move slightly in any direction with respect to one another.
ampulla (ăm·pŭl′lă) a flasklike tubular structure.
amygdaloid (ă·mĭg′dă·loyd) shaped like an almond.
amylase (ăm′ĭ·lās) a salivary and pancreatic enzyme that breaks down starch into maltose units and other carbohydrate residues.
anabolic steroids steroid hormones that stimulate anabolism.
anabolism (ă·năb′ō·lĭzm) a metabolic process in which large compounds are formed through the assembly of smaller subunits.
anaerobic respiration cellular respiration in the absence of oxygen in which glucose is converted to lactic acid.
anal canal the terminal portion of the large intestine.
anal column longitudinal folds in the membrane of the anal canal.
anal fissure a painful ulcer in a cleft at the margin of the anus.
analgesia (ăn·ăl·jē′zē·ă) absence of a normal sense of pain.
anal triangle the posterior half of the male or female perineum.
anaphase (ăn′ă·fāz) the period of mitosis or meiosis following metaphase during which chromosomes move to opposite poles of the cell.
anaphylaxis (ăn″ă·fĭ·lăk′sĭs) a sudden and especially severe reaction to an allergen.
anastomosis (ă·năs″tō·mō′sĭs) a connection between two vessels; end-to-end joining of elongated structures such as blood vessels, nerves, or lymphatic vessels.
anatomical position a conventional positioning of the body used in descriptive anatomy; the body is erect facing the viewer with the arms hanging to the side of the body and the palms facing forward.
anatomy the study of the structure of the body.
anconeus (ăn·kō′nē·ŭs) the forearm extensor muscle located on the back of the elbow.
androgen (ăn′drō·jĕn) any substance that stimulates the development of male characteristics.
anemia (ă·nē′mē·ă) reduction in the number of circulating erythrocytes or hemoglobin in the blood.
anencephaly (ăn·ĕn·sĕf′ă·lē) absence of a brain or spinal cord.
anesthesia (ăn″ĕs·thē′zē·ă) partial or complete loss of sensation or feeling.
aneurysm (ăn·ū·rĭzm) a localized sac in the wall of an artery at a point where the wall has become weakened.
angina pectoris (ăn·jī′nă pĕk·tō′rĭs) pain in the area of the heart possibly radiating to the shoulder and down the left arm; caused by inadequate oxygen supply to the heart.
angiography (ăn″jē·ŏg′ră·fē) X-ray examination of blood vessels following injection of an opaque substance into the circulatory system.
angiotensin (ăn″jē·ō·tĕn′sĭn) **I** a polypeptide formed from angiotensin by the action of renin; angiotensin I is then converted to angiotensin II.
angiotensin II a polypeptide concerned with the control of blood pressure.
anion (ăn′ī·ŏn) a negatively charged ion.
ankylosing spondylitis (ăng″kĭ·lō′sĭng spŏn″dĭ·lī′tĭs) a systemic joint disease characterized by inflammation of the vertebral, costovertebral, and sacroiliac joints.
annulus fibrosus (ăn′u·lŭs fī′brō·sŭs) a ring of fibrous material lying in the outer edge of an intervertebral disk.
anorectal (ă″nō·rĕk′tăl) **abscesses** abscesses in the region of the anus and rectum resulting from bacterial invasion of anal and rectal glands and surrounding soft tissues.
anorexia (ăn·ō·rĕk′sē·ă) abnormal loss of appetite.
anorexia nervosa (nĕr·vō′să) a psychological disorder usually of young women, involving an extreme loss of appetite for food and resulting in excessive weight loss.
anosmia (ăn·ŏz′mē·ă) the loss of the ability to smell.
anoxia (ăn·ŏk′sē·ă) inadequate supply of oxygen.
ansa (ăn′să) a looplike structure.
antagonist a muscle that resists the action of an agonist.
antebrachium (ăn″tē·brā′kē·ŭm) the forearm.
antepartum (ăn′tē·păr′tŭm) occurring before birth.
anterior referring to the ventral surface of the body or located in front of an organ or structure.
anterior pituitary gland (adenohypophysis) the portion of the pituitary gland that secretes the hormones FSH, LH, ACTH, TSH, GH, and prolactin in response to hypothalamic hormones.
anterior root the nervous tissue consisting of axons of motor or efferent fibers that project anteriorly from the spinal cord; it extends laterally to combine with a posterior root and form a spinal nerve.
anthropometry (ăn·thrō·pŏm′ĕt·rē) measurement of the human body and its parts.
antibody a protein produced in response to an antigen; antibodies react with the antigens that stimulated their production and aid in the removal of the antigen, which is the basis for immunity.
anticholinesterase (ăn″tĭ·kō·lĭn·ĕs′tĕr·ās) any substance that inhibits the activity of cholinesterase in catalyzing the breakdown of acetylcholine.
anticoagulant (ăn″tĭ·kō·ăg′ū·lănt) any substance that prevents the coagulation of blood.
anticodon (ăn″tĭ·kō·dŏn) the triplet of bases in a transfer RNA molecule that is complementary to a codon.
antidiuretic (ăn″tĭ·dī·ū·rĕt′ĭk) any substance that reduces the production of urine.
antidiuretic hormone (ADH) a hypothalamic hormone secreted by the posterior lobe of the pituitary gland that causes reabsorption of water in the kidney and reduced urine output; also called vasopressin.
antigen (ăn′tĭ·jĕn) any substance either formed in the body or introduced into the body, such as toxins, bacteria, or foreign blood cells, that stimulates the production of antibodies.
antiserum (ăn′tĭ·sē′rŭm) serum that contains antibodies against particular antigens.
antrum (ăn′trŭm) a cavity or chamber especially in a bone.
anus outlet of the colon.
anvil (ăn′vĭl) middle ossicle of the ear; also called the incus.
aorta (ā·or′tă) the main artery that carries blood from the heart to the systemic circulation.

aortic body a structure in the aortic arch sensitive to oxygen pressure, carbon dioxide pressure, and pH in the blood; also involved in regulating blood pressure and breathing rate.
aortic hiatus an opening in the diaphragm through which the aorta passes.
aphasia (ă·fā′zē·ă) inability to express oneself through speech.
aplastic (ă·plăs′tĭk) **anemia** a reduction in red blood cells or hemoglobin resulting from destruction of bone marrow.
apnea (ăp·nē′ă) cessation of breathing.
apneusis (ăp·nū′sĭs) sustained inspiration uninterrupted by expiration.
apneustic center a portion of the respiratory center in the brain stem that helps regulate the rhythmicity of breathing.
apocrine (ăp′ō·krēn) characterized by a granular secretion involving the loss of a portion of the secreting cell containing the secretion product.
apocrine gland sweat glands that occur in hairy areas of the body.
aponeurosis (ăp″ō·nū·rō′sĭs) a flattened, fibrous sheet of connective tissue that attaches a muscle to a bone or other tissue.
apoprotein (ăp″ō·prō′tēn) the protein to which a coenzyme attaches to form an active enzyme.
appendage a structure attached to a larger structure.
appendicitis (ă·pěn″dĭ·sī′tĭs) inflammation of the vermiform appendix.
appendix an appendage attached to a main and larger structure; the vermiform appendix.
apposition (ăp″ō·zĭ′shŭn) growth of successive layers; lying side by side or next to each other.
aqueous humor the fluid in the anterior and posterior chambers of the eye.
arachnoid (ă·răk′noyd) weblike.
arachnoid membrane the middle of the three membranes (meninges) that cover the brain and spinal cord.
arbor vitae (ăr′bor vī′tē) the treelike arrangement of white matter occurring in sections of the cerebellum.
arch of the aorta the bend of the aorta as it undergoes a transition from the ascending to descending portions.
areola (ă·rē′ō·lă) any tiny space within a tissue; a circular area surrounding a central point and of different color than the point; the area surrounding a nipple.
arm the portion of the upper extremity that extends from the shoulder to the elbow.
arrector pili the smooth muscle of a hair follicle; contraction causes a "goose bump" and the hair to stand erect.
arrhythmia (ă·rĭth′mē·ă) irregular heartbeat.
arteriole (ăr·tē′rē·ōl) the smallest subdivision of an artery, from which blood flows into a capillary.
arteriosclerosis (ăr·tě″rē·ō·″sklē·rō′sĭs) a circulatory system disease or condition characterized by a thickening, hardening, and loss of elasticity of the walls of arteries.
artery a blood vessel that carries blood away from the heart.
arthritis (ăr·thrī′tĭs) inflammation of the joints.
arthrogram (ăr′thrō·grăm) an X ray of a joint that has been infused with an opaque dye.
arthrogryposis (ăr″thrō·grī·pō′sĭs) an abnormal condition in which a joint is fixed in position.
arthrogryposis multiplex congenita a birth defect characterized by fixed joints.
arthroscope a device for the optical examination of the interior of a joint.
articular cartilage hyaline cartilage that covers the bony surfaces in synovial joints.
articulate to form a joint.
articulation a joint formed by two bones that move.
arytenoid (ăr″ĭ·tē′noyd) **cartilages** a pair of cartilages at the superior end of the larynx.
ascending colon the portion of the colon that lies between the cecum and the hepatic flexure.
ascites (ă·sī′tēz) accumulation of excess fluid in the peritoneum.
aseptic meningitis inflammation of the meninges with an absence of evidence of bacterial origin.
asphyxia (ăs·fĭk′sē·ă) unconsciousness due to inadequate oxygen supply.
aspirate to remove by suction.
association area portions of the cerebral cortex that integrate certain motor and sensory functions.
association neuron a nerve cell that lies within the central nervous system between afferent and efferent neurons.
asthma (ăz′mă) a condition in which there are attacks of shortness of breath and difficulty of breathing due to spasm or swelling of the bronchial tubes.
astigmatism (ă·stĭg′mă·tĭzm) an irregular or asymmetrical curvature of the lens or cornea that causes portions of an image to be out of focus.
astrocyte (ăs′trō·sīt) a neuroglial cell with many processes, which provides nutrients, support, and insulation for central nervous system neurons.
ataxia (ă·tăk·sē·ă) lack of voluntary muscular coordination.
atelectasis (ăt′ě·lěk′tă·sĭs) a collapsed lung.
atherosclerosis (ăth″ěr·ō·sklě·rō′sĭs) a form of arteriosclerosis characterized by local deposits of lipid materials in the walls of blood vessels.
atlas the first cervical vertebra.
atom the smallest particle of an element, consisting of a nucleus which contains the protons and neutrons and surrounding orbitals of electrons.
atomic number the number of protons in the atoms of an element.
atopic dermatitis itching dermatitis of unknown origin.
atrial (ā′trē·ăl) **fibrillation** rapid unproductive contraction of the atria due to spontaneous contraction of individual fibers.
atrioventricular (ā″trē·ō·věn·trĭk′ū·lăr) **bundle** fibers in the heart that carry impulses from the atrioventricular node through the interventricular septum to the muscles that form the two ventricles; also called bundle of His.
atrioventricular (AV) node a mass of cardiac fibers lying in the cardiac septum that separates the two atria and part of the conducting system of the heart.
atrioventricular valve a valve located between an atrium and a ventricle.
atrium (ā′trē·ŭm) a chamber that provides entrance to a second, larger chamber; one of the two chambers of the heart that receives blood from the vena cava and the pulmonary veins.
atrophy (ăt′rō·fē) decrease in size of an organ or muscle resulting from lack of use or disease.
atropine (ăt′rō·pēn) a drug that counteracts the effects of parasympathetic stimulation.
attention deficit disorder a developmental neuropsychiatric disorder characterized by limited attention span, impulsive behavior, and hyperactivity.
audiometer an instrument for testing hearing.
auditory ossicle any of the three small bones of the middle ear.

auditory tube a tube that extends from the middle ear to the nasopharynx; also called eustachian tube.

aura (aw′ră) awareness of an approaching onset of an epileptic attack or similar disorder.

auricle (aw′rĭ·kl) the pinna or external ear; an appendage in each of the cardiac atria.

auscultation (aws″kūl·tā′shŭn) examination of the body based on sounds produced in certain body cavities such as the stomach and thoracic cavities.

autoantibody antibodies formed in response to molecules in one's own body acting as antigens.

autoimmune response production of antibodies or T cells against one's own tissues.

autonomic ganglion a structure containing autonomic cell bodies outside the central nervous system.

autonomic nervous system the portion of the nervous system that controls smooth muscle, cardiac muscle, and glands.

autopsy examination of the organs and tissues of a dead body to determine the cause of death.

autosome (aw′tō·sōm) a chromosome other than the X or Y chromosome.

axial located in or related to the axis of an organ or other structure.

axilla (ăk·sĭl′ă) the armpit.

axis the second cervical vertebra; an imaginary line running vertically through the center of the body.

axoaxonic (ăk″sō·ăk·sŏn′ĭk) a synaptic junction formed between the axon of one neuron and the axon of a second neuron.

axodendritic a synaptic junction formed between the axon of one neuron and a dendrite of another neuron.

axon (ăk′sŏn) an extension of a neuron that carries an impulse away from the cell body.

axosomatic a synaptic junction between the axon of one neuron and the cell body of a second neuron.

azotemia (ăz·ō·tē′mē·ă) presence of abnormal amounts of nitrogen-containing substances in the blood.

azygous (ăz′ĭ·gŭs) a single vein that passes from the abdomen through the aortic hiatus, proceeding along the right side of the vertebral column to the level of the fourth thoracic vertebra where it bends and connects to the vena cava.

B

Babinski's (bă·bĭn′skēz) **reflex** a normal reflex in newborns in which the large toe extends instead of flexing and the toes spread out upon stroking the sole of the foot; after infancy, it is a possible indicator of organically based central nervous system malfunction.

Bainbridge reflex acceleration of heart rate resulting from pressure in or distension of the veins of the right atrium.

ball-and-socket joint a joint in which the rounded head of one bone moves in a concave socket of another bone.

barbiturates (băr·bĭt′ū·rāts) a group of chemical agents that depress the central nervous system.

baroreceptor pressure-sensitive receptor organs of the nervous system.

barotrauma pain and damage to the paranasal sinuses due to a difference between the pressure in the sinuses and atmospheric pressure.

basal ganglia masses of neuronal cell bodies in the gray matter of each cerebral hemisphere, including caudate, lentiform, and amygdaloid nuclei and the claustrum; also called cerebral nuclei.

basal metabolic rate (BMR) the rate of energy use by the body under controlled resting conditions.

base a substance that causes the pH of a solution to rise; a solution with a pH greater than 7.0; the nitrogen-containing ring structure of a nucleotide found in DNA and RNA.

basement membrane a thin, collagenous layer on which epithelial tissue rests.

basilar (băs′ĭ·lăr) **membrane** membrane in the cochlea of the inner ear that lies between the cochlear duct and the scala tympani.

basophil (bă′sō·fĭl) a granular leukocyte that reacts readily with alkaline dyes.

B cell lymphocyte that differentiates into an antibody-producing plasma cell.

Bell's palsy sudden occurrence of unilateral facial pain and paralysis due to inflammation of the facial nerve or surrounding tissue.

belly the abdomen; the fleshy part of a muscle where its diameter is greatest.

benign a mild or nonprogressive disorder.

beriberi (běr′ē·běr′ē) a disease due to deficiency of vitamin B_1 characterized by central and peripheral nervous system abnormalities, cardiac and vascular changes, and tissue edema.

beta-blocking agent a substance that blocks beta-adrenergic receptors in the sympathetic nervous system.

beta cell the insulin-secreting cells in the islets of Langerhans in the pancreas.

bicuspid having two projections or cusps.

bicuspid teeth the teeth between the molars and the canines.

bicuspid valve the valve that lies between the left atrium and ventricle; also called the mitral valve.

bile (bīl) a fluid secreted by the liver and stored in the gallbladder that aids digestion by emulsifying fats.

biliary colic (bĭl′ē·ăr·ē kōl′ĭk) spasm occurring in the bile ducts, often due to the presence of gallstones.

bilirubin (bĭl·ĭ·rū′bĭn) one of the components of bile.

biopsy (bī′ŏp·sē) the surgical removal of a small piece of tissue for clinical examination.

Biot's (bē·ōz′) **respiration** a breathing pattern characterized by irregular periods of apnea followed by a few short breaths; due to head injury or other cerebral lesion.

bipolar neuron a neuron that has two processes.

bitemporal hemianopia (bī·těm′pō·răl hěm″ē·ă·nō′pē·ă) a defect or loss of the temporal half of each visual field due to lesions of the optic nerve.

bivalent paired homologous chromosomes in prophase I of meiosis.

blastocoel (blăs′tō·sěl) the cavity within a blastula.

blastocyst (blăs′tō·sĭst) a stage in embryonic development after the blastocoel develops.

blastomere (blăs′tō·měr) one of the cells of a embryo after the division of the zygote.

blastula early stage in an embryo's development, consisting of a hollow sphere of cells around a central cavity.

blind spot an area in the retina located where the optic nerve enters the eye, characterized by an absence of rods and cones.

blood fluid that circulates through the heart, arteries, veins, and capillaries and which carries nutrients, oxygen, cellular products, and waste materials.

blood-brain barrier the barrier that separates blood from the cells of the central nervous system formed by the capillary walls and surrounding neuroglia.

blood pressure force exerted by blood on the walls of the blood vessels.
bolus (bō′lŭs) the wad of food matter that passes from the mouth to the stomach.
bone dense connective tissue consisting of a mineralized matrix surrounding living osteocytes.
bony labyrinth a network of passages within the temporal bone that forms the vestibule, cochlea, and semicircular canals of the inner ear.
botulism (bŏt′ū·lĭzm) a dangerous form of food poisoning due to toxins produced in food contaminated by the bacterium *Clostridium botulinum* growing in anaerobic conditions.
Bowman's capsule the cup at the end of a nephron that contains the glomerulus.
brachium (brā′kē·ŭm) the upper arm extending from the shoulder to the elbow.
bradycardia (brăd″ē·kăr′dē·ă) slow heart rate.
bradykinin (brăd′ē·kī′nĭn) a plasma polypeptide that causes relaxation of smooth muscle.
brain portion of the central nervous system contained within the skull and the primary organ for regulating and coordinating body activities.
brain abscess an encapsulated collection of pus in the cerebrum.
brain death loss of reflex motor functions above the neck, including loss of respiration and absence of brain waves for specified periods of time.
brain sand *see corpora arenacea.*
brain stem the base of the brain that contains the medulla, the pons, and the midbrain.
Broca's (brō′kăs) **area** the portion of the frontal lobe of the brain responsible for speech.
bronchial asthma narrowing of the bronchi due to smooth muscle spasm, inflammation, edema, or reaction of the bronchial mucosa to foreign agents.
bronchiole (brŏng′kē·ōl) the successive branches of the bronchi.
bronchitis (brŏng·kī′tĭs) inflammation of the bronchi.
bronchus (brŏng′kŭs) the major branches of the trachea.
brush border *see microvilli.*
buccal (bŭk′ăl) pertaining to the cheek or mouth.
bulbourethral (bŭl″bō·ū·rē′thrăl) **glands** two small glands on either side of the prostate gland that produce a clear, viscid secretion that acts as a lubricant and contributes to the formation of seminal fluid.
bulimia (bū·lĭm′ē·ă) abnormally excessive appetite; a psychological disorder usually of young women characterized by binging followed by induced vomiting.
bundle branch one of the branches of the bundle of His.
bundle branch block a failure of one of the bundle branches to conduct electrical impulses.
bundle of His a bundle of fibers that conduct signals from the atrioventricular node into the ventricles.
bunion (bŭn′yŭn) inflammation and swelling of the bursa of the large toe.
bursa (bŭr′să) a small fluid-filled cavity usually located in joints where it reduces friction between structures that rub against one another.
bursitis (bŭr·sī′tĭs) inflammation of a bursa.
butterfly rash a skin rash involving the cheeks and the skin above the nose, often seen in patients with systemic lupus erythematosus and certain skin conditions.
buttocks the fleshy masses posterior to the hip formed by the gluteal muscles.

bouton (bū·tŏn′) **terminaux** the enlarged tip of a nerve cell where it forms a synapse with a muscle fiber.

C

calcaneus (kăl·kā′nē·ŭs) the heel bone.
calcitonin (kăl″sĭ·tō′nĭn) a thyroid hormone that helps regulate calcium and phosphate levels in the blood.
calcium channel blocker a substance that blocks calcium ion channels in cell membranes.
calculus a mass of precipitated or crystallized salts or other material formed within a chamber of the body such as the kidneys, ureters, urinary bladder, or gallbladder; also called a stone.
callus (kăl′ŭs) a thickened, horny layer of skin.
calmodulin (kăl·mŏd′u·lĭn) a calcium receptor protein involved in cyclic-AMP mediated hormone responses.
calorie the amount of heat required to raise 1 g of water 1°C from 14.5°C to 15.5°C; a large calorie (or kilocalorie) is 1000 "small" calories.
calyx (kā′lĭx) any cuplike cavity or organ; the chambers in the pelvis of the kidney.
canal of Schlemm (shlĕm) a venus sinus at the junction of the sclera and the cornea of the eye that drains the aqueous humor from the anterior chamber of the eye.
cancer a malignant growth of tissue that has the capability of spreading to other parts of the body; a carcinoma.
canine tooth the cuspid tooth located next to the lateral incisors.
canker sore an ulcerated region in or about the mouth and lips.
canthus (kăn′thŭs) the angle formed by the upper and lower eyelids on each side of the eye.
capacitation the process by which sperm become capable of fertilizing an egg as a result of exposure to the substances within the female reproductive tract.
capillary microscopic vessels of the circulatory system through which exchange of materials between the blood and surrounding tissues occurs.
capitate (kăp′ĭ·tāt) shaped like a head.
carbaminohemoglobin (kăr·băm″ĭ·nō·hē″mō·glō′bĭn) the combination of hemoglobin and carbon dioxide.
carbohydrate organic substances that contain carbon, hydrogen, and oxygen in the approximate ratio of 1:2:1.
carbonic anhydrase an enzyme that catalyzes the reversible reaction between carbon dioxide and water to produce carbonic acid.
carboxypeptidase (kăr·bŏk·sĭ·pĕp′tĭ·dās) a pancreatic enzyme involved in the breakdown of peptides.
carcinogen a substances that causes cancer.
carcinoma a cancer that originates in epithelial tissue
cardiac muscle the type of muscle tissue found in the heart and in the major arteries where they leave the heart.
cardiac output (CO) the volume of blood that is pumped from the left ventricle in one minute.
cardiac reserve the capacity of the heart to increase cardiac output over the resting value.
cardiopulmonary resuscitation (CPR) bringing one back to consciousness through external cardiac massage and externally forced respiration.
caries (kăr′ēz) tooth or bone decay.
carotene (kăr′ō·tēn) a yellowish pigment that is a precursor to vitamin A.

carotid body small masses of tissue at the bifurcation of the common carotid artery that detect changes in oxygen, carbon dioxide, and pH levels in the blood.

carotid sinus a dilation of the common carotid artery where it bifurcates to form the internal and external carotid arteries; it contains receptors that help regulate blood pressure.

carpal tunnel syndrome pain or numbness in the thumb or hand resulting from pressure on the median nerve as it passes through the tunnel formed by the flexor retinaculum and the carpal bones.

carpus (kăr′pŭs) the eight bones of the wrist.

cartilage (kăr′tĭ·lĭj) dense connective tissue consisting of cells imbedded in a tough but flexible matrix.

caruncle (kār′ŭng·kl) a small, fleshy protrusion.

cast molded material produced in a cavity or tubule and discharged from the structure.

catabolism metabolic reactions involving the breakdown of larger molecules into smaller ones.

catalyst any substance that speeds up a chemical reaction without itself being used up in the reaction, such as enzymes that catalyze biochemical reactions.

cataract increased opacity of the lens of the eye, its capsule, or both.

catecholamines (kăt″ĕ·kŏl′ă·mēns) a group of hormones that includes epinephrine, norepinephrine, and dopamine, characterized by the presence of a catechol group in the molecule.

catheter a tubular instrument for withdrawing fluids from body cavities.

cation a positively charged ion.

cauda equina (kaw′dă ē′kwīn·ă) the base of the spinal cord and the associated roots of the spinal nerves that emanate below the first lumbar nerve.

caudate pertaining to the tail or the lower portion of the body.

cecum (sē′kŭm) a pouch at the beginning of the large intestine.

cell-mediated immunity immunity based on the actions of T-cell lymphocytes.

cellular respiration the biochemical processes involved in the oxidation of an energy source and the use of the released energy for the production of ATP.

cementum (sē·měn′tŭm) the thin layer of calcified tissue that covers the dentine of the root of a tooth and helps to anchor the tooth in its bony socket.

central canal a tube that runs through the gray commissure throughout the length of the spinal cord.

central fovea a cuplike depression that lies in the middle of the macula lutea in the retina.

central lacteal a tubule that projects into a villus in the small intestine that receives fatty materials that have been absorbed from the chyme in the lumen of the intestine.

central nervous system the brain and spinal cord.

centriole (sĕn′trē·ōl) a tiny cylinder of microtubules located near the nucleus of a cell; it divides in two during mitosis and migrates to opposite sides of the nucleus and helps form the spindle.

centromere (sĕn′trō·mēr) a constricted region of the chromosome observed during mitosis or meiosis where the chromatids are attached to one another and to which spindle fibers attach.

cephalalgia (sĕf′ă·lăl′jē·ă) headache.

cerebellar peduncle (sĕr·ĕ·bĕl′ăr pē·dŭn′kl) a group of nerve fibers that connects the brain stem and the cerebellum.

cerebellum (sĕr′ĕ·bĕl′ŭm) a portion of the hindbrain responsible for coordinating movement.

cerebral (sĕr′ĕ·brăl) **peduncles** a pair of bundles of nerve fibers located on the ventral surface of the midbrain that conduct signals between the cerebral hemispheres and the pons.

cerebral aqueduct a channel between the third and fourth ventricles of the brain.

cerebral death damage to the cerebrum without simultaneous damage to the diencephalon and brain stem, resulting in loss of awareness without loss of autonomic or motor functions.

cerebral edema an abnormal accumulation of fluid in the cerebrum.

cerebral infarction inhibited neurological function due to blockage of blood flow to a localized region of the brain.

cerebral insufficiency a condition in which blood flow to the cerebrum is inadequate to supply the necessary amount of oxygen.

cerebritis (sĕr″ĕ·brī′tĭs) inflammation of the cerebrum.

cerebrospinal fluid (CSF) the fluid located in the ventricles in the subarachnoid space around the brain and spinal cord.

cerebrovascular accident a general term for the clinical syndromes that result from vascular insufficiency or hemorrhage; a stroke.

cerebrovascular disease a general term for disorders involving the vasculature of the brain, such as cerebral insufficiency.

cerebrum (sĕr′ĕ·brŭm) the largest part of the brain, consisting of left and right hemispheres; receives conscious sensation and controls voluntary motor activity.

cerumen (sĕ·roo′mĕn) ear wax.

cervical pertaining to the neck or a necklike region within an organ.

cervix part of an organ resembling a neck, especially the neck of the uterus.

Charley horse pain and muscular spasm due to inflammation of the muscles in the thigh.

chemiosmotic theory the mechanism by which ATP is thought to be produced during electron transport in cellular respiration

chemoreceptor a sensory cell of the nervous system that responds to specific chemical substances.

chemotaxis (kē″mō·tăk′sĭs) movement of a cell or part of a cell in response to a chemical substance.

chemotherapy the use of chemical substances to treat disease.

Cheyne-Stokes respiration difficulty of breathing characterized by regularly alternating periods of apnea and hyperpnea.

chiasma (kī·ăz′mă) an X-shaped structure, such as formed by the optic nerves or as a result of crossing over in chromosomes during meiosis.

chickenpox see *varicella*.

chloride shift the movement of chloride ions through a red cell membrane into the plasma as a result of movement of bicarbonate ion into the red blood cell from the plasma.

cholecystectomy (kō″lē·sĭs·tĕk′tō·mē) removal of the gallbladder by surgery.

cholecystogram (kō″lē·sĭs′tō·grăm) an X-ray photograph of the gallbladder.

cholecystokinin (ko″lē·sĭs′tō·kī′nĭn) a duodenal hormone that stimulates gallbladder contraction and pancreatic secretion into the small intestine; also called pancreozymin.

cholelithiasis (kō″lē·lĭ·thī′ă·sĭs) presence of stones in the gallbladder or the ducts that drain the gallbladder.

cholesterol (kō·lĕs′tĕr·ŏl) a widely distributed steroid found in bile cell membranes and other lipid-containing structures; the primary component of gallstones and a precursor in the synthesis of many steroid hormones.

cholinergic (kō″lĭn·ĕr′jĭk) neurons that use acetylcholine as a neurotransmitter.

chondral fracture a cartilaginous fracture.

chondrin (kŏn′drĭn) a gelatinous protein in cartilage.

chondroblast (kŏn′drō·blăst) a cell that forms cartilage.

chondrocyte (kŏn′drō·sīt) a cartilage cell.

chondromalacia (kŏn·drō·măl·ă′shē·ă) softening of cartilage in a joint, especially at the patella.

chordae tendineae (kor′dē tĕn·dĭn′ē·ē) small cords in the heart that connect the edges of the atrioventricular valves to the papillary muscles.

chorion (kō′rē·ŏn) the extraembryonic membrane that forms the outer wall of the blastocyst during the early stages of development.

chorionic villi extensions of the chorion into the endometrium that give rise to the placenta.

chorionic villus sampling (CVS) a prenatal diagnostic procedure in which a biopsy of the chorion is taken at 8 to 10 weeks and used for DNA and chomosome analysis to detect genetic abnormalities.

choroid (kō′royd) the vascular region of an eye between the sclera and retina.

chromaffin (krō·măf′ĭn) **cells** cells in the adrenal medulla that contain brightly staining granules.

chromatid (krō′mă·tĭd) one of the pair of subunits of chromosomes seen during mitosis or the first division of meiosis and which become chromosomes upon separation.

chromatin (krō′mă·tĭn) the material in the cell nucleus that contains granular DNA and protein and that appears during interphase; it later condenses into discrete chromosomes during prophase.

chromophore the source of color in a pigmented cell.

chromosome one of the 46 structures of the cell nucleus that contain DNA and protein seen as darkly staining bodies during mitosis and meiosis.

chronic long lasting, such as an ongoing disease.

chylomicron (kī″lō·mī′krŏn) the small, spherical assemblage of triglyceride, phospholipid, cholesterol, and protein that passes into a lacteal when nutrients are absorbed from the digestive tract.

chyme (kīm) the milky white fluid produced by digestion of food matter in the stomach.

chymotrypsin (kī″mō·trĭp′sĭn) a pancreatic peptide-digesting enzyme produced from chymotrypsinogen, a precursor molecule.

cilia (sĭl′ē·ă) hairlike projections of a cell that are usually present in large numbers; movement of cilia is often used to transport materials over the surface of tissues.

ciliary bodies part of the choroid layer of the eye from which the aqueous humor is secreted and in which the ciliary muscle is located.

circadian (sĭr″kā′·dē·ăn) pertaining to a period of about one day.

circadian dysrhythmia (dĭs·rĭth′mē·ă) *see jetlag.*

circle of Willis the circular union of the anterior and posterior cerebral arteries where they surround the pituitary gland at the base of the brain.

circumcision surgical removal of the foreskin of the penis.

circumduction the swinging of an arm or leg in a cone-shaped figure with the joint of the limb at the base of the cone.

circumflex a long, narrow structure such as a nerve or blood vessel that winds around the surrounding tissue as it passes through it.

cirrhosis (sĭ·rō′sĭs) disease of the liver that results in the replacement of functional liver cells with nonfunctional connective and fatty tissues.

claustrum (klŏs′trŭm) one of the basal ganglia; a layer of nervous tissue that lies between the external capsule and the isle of Reil.

clavicle (klăv′ĭ·kl) the collarbone.

cleavage the cell divisions that follow the fertilization of an egg.

cleavage reaction a chemical reaction in which a large molecule is divided into two or more smaller molecules.

cleft lip a congenital fissure of the upper lip.

cleft palate a congenital fissure in the roof of the mouth.

climax the period of greatest intensity during sexual stimulation; orgasm.

clitoris (klĭ′tō·rĭs) an erectile structure beneath the anterior labial commissure in the vulva of a female.

cloacal (klō·ā′kăl) **membrane** a membrane formed from the outer and inner germ layers of an embryo that covers the hindgut.

clonic spasm (klŏn′ĭk spăzm) abnormal rapid and rhythmic contractions and relaxations of a muscle.

clot a mass of coagulated blood or lymph resulting from the formation of fibrin molecules.

clot retraction pulling together of surrounding tissues associated with the contraction of the fibrin in a clot.

clubfoot a congenital abnormality of the foot, characterized by unusual twisting and mishapen development; talipes.

cluster headache a suddenly occurring, severe sequence of headaches occurring over days or weeks, involving the area around the eyes and nose, each attack usually lasting about an hour.

coagulation (kō·ăg″ū·lā′shŭn) a clumping together of blood or lymph.

cobalamin (kō·băl′ă·mĭn) a cobalt-containing portion of the vitamin B_{12} group.

coccyx (kŏk′sĭks) a small bone inferior to the sacrum formed by the union of four or more vertebrae.

cochlea (kŏk′lē·ă) a spirally wound tube in the inner ear essential to hearing.

codon (kō′dŏn) the sequence of three nucleotides and a messenger RNA molecule that specifies a particular amino acid to be incorporated into a peptide translated from the messenger RNA or which serves as a signal to terminate the translation.

coelom (sē′lŏm) the embryonic cavity that develops between the mesoderm layers; it forms the adult pleural, peritoneal, and pericardial cavities.

coenzyme (kō·ĕn′zīm) a generally low molecular weight substance necessary to impart enzymatic activity to an otherwise inactive enzyme.

coitus (kō′ĭ·tŭs) sexual intercourse between a man and a woman.

cold sores skin vesicles around the mouth and nares caused by herpes simplex virus.

collagen (kŏl′ă·jĕn) the primary protein in skin, tendons, bones, cartilage, and connective tissue.

colloid (kŏl′oyd) a gelatinous substance, such as egg white, containing dispersed particles that remain uniformly distributed, yet fail to form a true solution.

colloid osmotic pressure osmotic pressure of a solution that results from high molecular weight compounds suspended in the fluids, such as proteins present in plasma; also called oncotic pressure.

colon (kō′lŏn) section of the large intestine from the cecum to the rectum.

colostomy (kō·lŏs′tō·mē) formation of an artificial opening from the colon to the exterior of the body.

colostrum (kŏ·lŏs′trŭm) fluid secreted by mammary glands immediately after childbirth and preceding the secretion of milk.

colposcopy (kŏl·pŏs′kō·pē) use of an optical device for examining the vagina and cervix.

coma (kō′mă) a prolonged state of unconsciousness resulting from injury, poisoning, or disease.

comminuted (kom′ĭ·nū″tĕd) **fracture** a fracture in which the bone is shattered into many small pieces at the point of the break.

commissure (kŏm′ĭ·shŭr) a junction formed between two anatomical structures.

complement a group of inactive plasma proteins which when activated combine with antibodies and destroy bacteria.

compliance (kŏm·plī′ăns) a measure of the ability of the lungs to change volume in response to changes in pressure.

compound fracture a bone fracture in which the fractured ends of a bone have penetrated the skin.

computerized axial tomography (CAT) a diagnostic technique using an instrument that directs X rays through a particular part of the body and absorbs them with detectors that relay the signals to a computer. The computer uses the information to generate a photographic image of the internal structure of a specific slice of the body part.

conarium (kō·nā′rē·ūm) the pineal body.

conception union of an egg and sperm to form a zygote.

concussion (kŏn·kŭsh′ŭn) a sudden jar or shock to an organ; in a brain concussion there is loss of consciousness due to a blow to the head.

condyle (kŏn′dīl) a rounded projection on a bone.

cone the light-sensitive receptor in the retina that is responsible for color vision.

congenital metatarsus varus an abnormality of the foot in which the foot is twisted such that the inner border of the sole is raised and the person walks on the outer border.

congestive heart failure a decrease in cardiac output to the point where the heart is unable to meet the body's needs.

conjunctiva (kŏn″jŭnk·tī′vă) the mucous membrane lining the eyelids.

conjunctivitis (kŏn·jŭnk″tī·vī′tĭs) inflammation of the conjunctiva.

connective tissue the type of tissue used to connect and support organs and other tissues, such as fibrous tissue, cartilage, and bone.

constipation infrequent and difficult defecation.

contact dermatitis an allergic reaction of the skin caused by contact with a substance.

contact lens a glass or plastic lens placed in direct contact with the cornea to correct defects in vision.

continuous positive airway edema (CPAE) spontaneous breathing in a person who is receiving ventilation by artificial means.

contractility (kŏn·trăk·tĭl′ĭ·tē) the ability to shorten in length, as in certain filaments and muscle fibers.

contralateral involving opposite sides of the body.

contrast radiography X-ray photography in which contrast in the image is enhanced through the use of opaque dyes.

convergence the turning of the eyes inward to focus on a point as it approaches the viewer; where the axons of several neurons form synapses with a single or few postsynaptic neurons.

copulation sexual intercourse.

coracoid (kor′ă·koyd) **process** a bony projection on the upper anterior surface of a scapula.

corium (kō′rē·ŭm) the layer of skin immediately beneath the epidermis.

cornea the transparent anterior portion of an eyeball.

corona radiata (kō·rō′nă rā·dē·ā′tă) the mass of follicle cells that remain associated with the zona pellucida following ovulation.

coronary artery one of a pair of arteries that supply blood directly to the heart.

coronary sinus a vein that lies in the groove between the left atrium and ventricle on the posterior surface of the heart which collects blood that is passed through the heart muscle and delivers it to the right atrium.

coronoid (kor′ō·noyd) **process** a bony projection on the proximal end of an ulna; a bony projection on the upper anterior end of a mandible where the temporalis muscle attaches.

corpora arenacea granular bodies found in the pineal body of the brain; also called brain sand.

corpora cavernosa (kăv″ĕr·nō′să) the paired cylinders of erectile tissue in a penis.

corpora quadrigemina (kor′pō·ră kwod″rĭ·jĕm′ĭn·ă) two pairs of rounded structures that lie on the superior portion of the midbrain.

corpus albicans (kor′pŭs ăl′bĭ·kăns) the whitish structure that remains after a corpus luteum disappears.

corpus callosum (kă·lō′sŭm) a band of white matter at the base of the longitudinal fissure that connects the two cerebral hemispheres.

corpus luteum (lū′tē·ŭm) a yellowish endocrine structure that develops within a ruptured ovarian follicle following ovulation.

corpus spongiosum (spŭn″jē·ō′sŭm) the erectile tissue in a penis through which the urethra passes.

cortex (kor′tĕks) the outer layer of an organ usually surrounding an inner medulla.

corticobulbar (kor″tĭ·kō·bŭl′băr) pertaining to the upper region of the brainstem and the cerebral cortex.

corticosteroid (kor″tĭ·kō·stēr′oyd) **hormone** the steroid hormones produced by the cortex of the adrenal gland.

corticotrophin (kor″tĭ·kō·trō′fĭn) the hormone produced by the anterior lobe of the pituitary gland that stimulates corticosteroid production by the adrenal cortex.

corticotrophin-releasing factor (CRF) hypothalamic hormone that stimulates secretion of ACTH by the anterior lobe of the pituitary gland.

cortisol (kor′tĭ·sŏl) an adrenocortical hormone; also known as hydrocortisone.

coryza (kŏ·rī′ză) the common cold.

costal (kŏs′tăl) pertaining to a rib.

costal cartilage a cartilage that connects the end of a rib to the sternum or to the cartilage associated with a rib above.

costocervical (kŏs″tō·sĕr′vĭ·kăl) pertaining to the ribs and the neck region of the spine.

countercurrent multiplier mechanism the mechanism used by a kidney to produce a concentrated urine.

coxa (kŏk′să) the hip joint or the hip.

cramp a spasmodic and usually painful muscle contraction.

craniosynostosis (krā″nē·ō·sĭn″ŏs·tō′sĭs) premature closing of the sutures in the skull.

craniotabes (krā″nē·ō·tā′bēz) an abnormal softening of the skull due to loss of minerals.

creatinine (krē·ăt′ĭn·ĭn) a nitrogenous waste product formed from creatine and urea and often used to measure glomerular filtration rate in kidneys.

crenation (krē·nā′shŭn) the conversion of red blood cells to

angular knobbed forms as a result of loss of water to a hypertonic solution.
cretinism (krē'tĭn·ĭzm) dwarfism in children due to a deficiency of thyroid hormone production.
crib death a sudden, unexplained death of an infant occurring during sleep; sudden infant death syndrome.
cribriform (krĭb'rĭ·form) perforated like a sieve.
cricoid (krī'koyd) a ring-shaped cartilage in the lower larynx.
crista (krĭs'tă) a crest or a ridge; the projections of the inner membrane of a mitochondrion into the interior.
Crohn's (krōnz) **disease** a chronic inflammation of the alimentary canal, especially occurring in the ileum and colon.
cumulus oophorus (kū'mū·lŭs ō·ŏf'ō·rŭs) a mass of cells that project from the wall of the graafian follicle into the antrum and surrounding the ovum.
cuneate (kū'nē·āt) shaped like a wedge.
cupula (kū'pū·lă) a receptor in the ampulla of a semicircular canal.
curettage (kū″rĕ·tăzh′) scraping of the inner surface of a hollow organ.
Cushing's syndrome a set of characteristics that result from excess production of glucocorticoid hormones by the adrenal cortex, characterized by edema, excess hair growth, fatigue, and osteoporosis.
cusp a projection on the chewing surface of a tooth; one of the divisions in a heart valve.
cuticle (kū'tĭ·kl) the epidermis.
cyclic-AMP a nucleotide formed by the action of adenylcyclase on ATP in response to the combination of certain hormones with their receptors in a cell membrane.
cyst (sĭst) a closed sac lined by an epithelium and containing a liquid or semisolid material.
cystic duct the tube through which bile travels from the gallbladder to the common bile duct.
cystic fibrosis (sĭs'tĭk fī·brō'sĭs) a genetic disease of infants and young people that is characterized by excessive electrolyte concentration in sweat, pancreatic deficiency, and symptoms of pulmonary disease.
cystinuria (sĭs″tĭ·nū'rē·ă) the occurrence of cystine or certain other amino acids in the urine; a hereditary metabolic disorder that results in the excretion of cystine and other amino acids in the urine.
cystitis (sĭs·tī'tĭs) inflammation of the urinary bladder.
cystoscope (sĭst'ō·skōp) an optic device for the visual examination of the interior of the urinary bladder.
cytochrome a group of mitochondrial proteins involved in the transport of electrons released in the Krebs cycle.
cytokinesis (sī″tō·kĭ·nē'sĭs) the portion of mitosis or meiosis during which the cytoplasm divides.
cytoplasm (sī'tō·plăzm) that portion of a cell contained within the plasma membrane, exclusive of the nucleus.
cytoskeleton the network of microfilaments and microtubules found in the cytoplasm of a cell.

D

deamination (dē″ăm·ĭ·nā'shŭn) the removal of an amino group from an organic compound, especially from an amino acid.
decibel (dĕs'ĭ·bĕl) a unit of measure used to describe electrical power or sound intensity.
decidua basalis (dē·sĭd'ū·ă bă·săl'ĭs) the part of the endometrium in a pregnant woman that combines with the chorion to form a placenta.
deciduous teeth the 20 temporary teeth that are shed and replaced by permanent teeth in a child (also called "baby" teeth or "milk" teeth).
decompression sickness pain and neurological effects that occur from rapid reduction in the pressure surrounding a person, such as when ascending too quickly from deep underwater, caused by the formation of gas bubbles in the blood or surrounding tissues; also called the "bends."
decussation an X-shaped crossing of two similar structures, such as in the optic chiasma.
deep fascia the layer of connective tissue that surrounds a muscle.
defecation the elimination of feces through the anus.
defecation reflex the pattern of neural and musculature activity stimulated by distention of the rectum that results in a relaxation of the anal sphincter and expulsion of feces.
defibrillation (dē·fĭb″rĭ·lā'shŭn) stopping a pattern of rapid and spontaneous muscular contractions of the heart through electrical shock or by chemical means.
degenerative arthritis a chronic disease that effects the bones at the joints, characterized by loss of cartilage and underlying bony tissue and growth of spurs and other structures on the bony surface; osteoarthritis.
deglutition (dē″glū·tĭsh'ŭn) swallowing.
delayed hypersensitivity a type of T cell mediated immunity in which an allergic response occurs several hours or days following exposure to an antigen.
delta cells cells in the islets of Langerhans that produce somatostatin, a hormone involved in controlling growth.
dendrites (dĕn'drĭts) processes that emanate from the body of a neuron that carry impulses in the direction of the body.
dentate serrated; toothed.
dentin (dĕn'tĭn) hard, calcified tisssue that forms the major part of a tooth.
deoxyribonucleic (dē·ŏk″sē·rī″bō·nū·klē'ĭk) **acid (DNA)** the genetic material of cells, composed of two chains of nucleotides wrapped around one another in a double helix.
depolarization the reduction or reversal of a membrane potential.
de Quervain's disease a painful condition resulting from inflammation of the tendon sheath of the abductor pollicis longus and the extensor pollicis brevis, the muscles that move the thumb.
dermatology (dĕr″mă·tŏl'ō·jē) the study of the skin and its disorders.
dermatome (dĕr'mă·tōm) an area of the skin innervated by a particular spinal nerve.
dermatomyositis (dĕr″mă·tō·mī″ō·sī'tĭs) a systemic disease of connective tissue characterized by edema and inflammation of the skin and underlying muscles.
dermatophytosis (dĕr″mă·tō·fī·tō'sĭs) a fungal infection of the hands and feet; athlete's foot.
dermis the layer of skin below the epidermis; also called the corium or the true skin.
descending colon the part of the colon that extends down the left side of the abdomen from the splenic flexure to the sigmoid colon.
desmosome (dĕs'mō·sōm) a type of cell-to-cell junction that binds adjacent cells to one another.
dextrinase (dĕks'trĭn·ās) a duodenal enzyme that catalyzes the hydrolysis of dextrins.
dextrins short lengths of polysaccharides that remain as carbo-

hydrate residue following amylase-catalyzed digestion of starch and glycogen.

diabetes insipidus (dī″ă·bē′tēz ĭn·sĭp′ĭ·dŭs) a disorder that results from insufficient antidiuretic hormone production or use, characterized by excretion of excessive volumes of dilute urine.

diabetes mellitus (mĕ·lī′tŭs) an inherited carbohydrate metabolism disorder that results from inadequate production or use of insulin.

diagnostic laparotomy (lăp·ăr·ŏt′ō·mē) a surgical incision that opens the abdomen for the purpose of determining the cause of a disorder.

dialysis (dī·ăl′ĭ·sĭs) the technique of separating relatively small molecules from relatively large ones in an aqueous solution by diffusion through a membrane permeable to the smaller molecules and impermeable to the larger ones; *see hemodialysis.*

diapedesis (dī″ă·pĕd·ē′sĭs) squeezing of leukocytes through the unruptured wall of a capillary.

diaphragm (dī′ă·frăm) the muscular partition that separates the thoracic cavity from the abdomen and that contributes to the filling and emptying of the lungs through its contraction; a contraceptive device fitted over the cervical opening to prevent the passage of sperm into the cervix.

diaphysis (dī·ăf′ĭ·sĭs) the shaft of a long bone.

diarrhea (dī·ă·rē′ă) increased frequency of defecation of feces of abnormally fluid consistency.

diarthroses (dī·ăr·thrō′sĭs) a freely moveable, or synovial, joint.

diastole (dī·ăs·tō′lē) the period when a chamber of the heart is relaxed and filling with blood.

diastolic blood pressure the blood pressure that exists during ventricular diastole.

diencephalon (dī″ĕn·sĕf′ă·lŏn) the portion of the brain lying between the mesencephalon and the telencephalon, connecting the midbrain and the cerebral hemispheres, and including the thalamus, hypothalamus, and pituitary gland.

differentiation the chemical and structural changes that occur in a cell as it undergoes transition to its mature functional form.

diffusion the spontaneous movement of molecules or ions from a region of high concentration to a region of low concentration.

digestion the chemical and mechanical processes involved in the breakdown of food into component molecules that can be absorbed.

dehydration abnormally low body fluid volume due to excessive loss of water.

dipeptidase an enzyme produced in the small intestine that degrades pairs of amino acids joined by a peptide bond into individual amino acids.

diphtheria (dĭf·thē′rē·ă) a contagious disease caused by the bacillus *Corynebacterium diphtheriae* that results in the formation of fibrous membranes over the mucous tissues lining the throat and other air passages.

diploid (dĭp′loyd) having two sets of chromosomes in somatic cell nuclei.

diplopia (dĭp·lō′pē·ă) perceiving two images of an object; double vision.

disaccharide a carbohydrate molecule composed of two monosaccharide subunits joined together by a covalent bond.

displaced fracture a fracture in which the broken ends of the bone are shifted in position with respect to each other.

distal farther, or farthest, from the origin of a structure or the midline of the body.

diuresis (dī″ū·rē′sĭs) increased or excessive excretion of urine.

divergence an arrangement of neurons in which one or a few neurons form synapses with a significantly greater number of postsynaptic neurons.

diverticulitis (dī″vĕr·tĭk″ū·lī′tĭs) inflammation of a diverticulum.

diverticulosis (dī″vĕr·tĭk″ū·lō′sĭs) the presence of diverticula in the colon but not necessarily accompanied by discomfort or inflammation.

diverticulum (dī″vĕr·tĭk″ū·lŭm) a sac or pouch that forms in the wall of a hollow organ.

dizygotic twins twins produced following the fertilization of two separate eggs.

DNA cloning the procedure by which specific nucleotide sequences of DNA are placed in a bacterium or other microorganism and replicated to produce multiple copies of DNA with the same sequence.

dominant allele an allele that is expressed when heterozygous with a second, recessive allele of the same gene.

dopamine a neurotransmitter of certain peripheral nerves and of many central nervous system neurons.

dorsal pertaining to the back or posterior part of an organ or body; posterior.

dorsal root ganglion groups of sensory cell bodies that form an enlarged region in the dorsal roots of spinal nerves; also called a spinal ganglion.

dorsiflexion movement of a body part, such as the hand or foot, at a joint in such a way as to bend the part in the direction of its dorsum.

Down syndrome a combination of characteristics that result in most cases from trisomy for chromosome number 21. Characteristic symptoms include mental retardation and developmental abnormalities.

ductus arteriosus (dŭk′tŭs ăr·tĕr″ē·ō′sŭs) a blood vessel that forms a shunt between the left pulmonary artery and the aorta in a fetus.

ductus deferens (dĕf′ĕr·ĕnz) the duct in the scrotum that transports sperm from the epididymis to the ejaculatory duct; also called vas deferens.

ductus venosus (vĕ·nō′sŭs) one of the two branches of the umbilical vein that occurs where the umbilical vein enters the abdomen.

duodenal papilla a small projection where the common bile duct and the pancreatic duct empty into the lumen of the duodenum.

duodenal ulcer an inflamed lesion on the internal surface of the duodenum caused by irritation from stomach fluids before they can be neutralized.

duodenum (dū″ō·dē′nŭm) the first 10 to 12 inches of the small intestine where liver and pancreatic secretions are mixed with food being digested.

dura mater the membrane that forms the outermost covering of the brain and spinal cord.

dwarfism abnormally short in height; *see also pituitary dwarfism.*

dysentery any of several infections of the colon and other parts of the lower digestive tract characterized by diarrhea, inflammation of the wall of the intestine, pain, and fever.

dysgraphia (dĭs·grăf′ē·ă) a disorder of higher brain function in which a person has trouble expressing himself or herself in writing.

dyslexia (dĭs·lĕk′sē·ă) impaired ability to read, particularly in children, in spite of an apparently adequate intellectual capability.

dysmenorrhea (dĭs″mĕn·ō·rē´ă) difficult or painful menstruation.
dysmetria (dĭs·mē´trē·ă) inability to control the range of movement of body parts, usually resulting from a disorder of the cerebellum.
dysphagia (dĭs·fā´jē·ă) difficulty in swallowing.
dyspnea (dĭsp·nē´ă) difficult or labored breathing.
dystrophy (dĭs´trō·fē) defective or abnormal development or degeneration of a tissue or organ, especially muscle tissue.

E

eardrum the tympanic membrane of the ear, which vibrates in response to changes in air pressure associated with sound and moves bones of the middle ear to stimulate the sensory organ of the inner ear.
eccrine (ĕk´rĭn) **gland** a gland in the skin that secretes a clear fluid that evaporates and cools the body; a sweat gland.
ectoderm the outermost of the three primary germ layers in an embryo and the source of epidermis and epidermal tissues, the nervous system, external sense organs, and the mucous membrane that lines the mouth and anus.
ectopic out of place; situated in an unusual or abnormal location.
eczema (ĕk´zĕ·mă) a generalized noncontagious inflammation of the skin characterized by itchiness, redness, and the presence of vesicles and pustules that may be dry or moist.
edema (ĕ·dē´mă) abnormal swelling of body tissues due to an accumulation of tissue fluid.
effector a muscle that is stimulated to contract or a gland that is stimulated to secrete a product in response to nerve impulses.
efferent to be directed away from a point of reference.
efferent arteriole an arteriole that carries blood away from a glomerulus in a nephron.
efferent nerve nerves that carry signals away from the central nervous system.
efferent neuron a neuron that carries impulses in a direction away from the central nervous system toward muscles or glands; a motor neuron.
ejaculation the sudden and forceful discharge of fluid, especially the discharge of semen from the penis in response to sexual excitement.
ejaculatory duct the terminal portion of the seminal tract formed by the convergence of an excretory duct from a seminal vesicle and the ductus deferens.
elastin the principal protein of elastic tissue.
electrocardiogram (EKG or **ECG)** a recording of the electrical activity of the heart showing characteristic peaks designated P, Q, R, S, and T; the relative timing, magnitude, frequency, and shapes of these peaks on the electrocardiogram are used to diagnose heart function.
electroencephalogram (EEG) a graph showing tracings of electropotentials at the skull that arise from brain activity.
electrolyte ionized solute molecules present in body fluids.
electromagnetic radiation radiation consisting of waves of energy from any part of the electromagnetic spectrum.
electromagnetic spectrum the range of electromagnetic waves ranging from low frequency radio waves to high frequency cosmic rays and X rays.
electromyography the recording and study of muscle twitches in response to electric shock.
electron a negatively charged particle that travels in an orbit around an atomic nucleus.
electron transport chain a series of oxidation-reduction reactions occurring in mitochondria that release energy used in the synthesis of ATP.
electroneurography the recording and study of the response of nervous tissue to electric stimulation.
eleidin (ĕ·lē´ĭ·dĭn) a keratin-like protein present in cells of the stratum lucidum of the skin.
embolism a sudden block of an artery by a blood clot or foreign body.
embolus (ĕm´bō·lŭs) the mass of clotted blood or other substance responsible for an embolism.
embryo the stage in development of a human between the zygote and the fetus, that is, from the time of conception to the eighth week of development.
emesis (ĕm´ĕ·sĭs) vomiting.
eminence a projecting portion of a structure, particularly of a bone.
emmetropia (ĕm″ĕ·trō´pē·ă) normal vision, in which light rays are focused precisely on the retina without special effort.
emphysema (ĕm″fĭ·sē´mă) a diseased condition of the lungs characterized by degradation of the alveolar walls and accumulation of fibrous tissue.
emulsification the conversion of large fat droplets into smaller fat droplets of a size that can remain dispersed in the fluid.
enamel the hard outer covering of the exposed part of a tooth.
encephalitis (ĕn·sĕf″ă·lī´tĭs) inflammation of the brain.
encephalomyelitis (ĕn·sĕf″ă·lō·mī·ĕl·ī´tĭs) an acute inflammation of the brain and spinal cord.
endocarditis (ĕn″dō·kăr·dī´tĭs) inflammation of the endocardium.
endocardium (ĕn″dō·kar·dē´ŭm) the membrane that lines the interior of the heart.
endochondral (ĕn″dō·kŏn´drăl) contained within or related to cartilage.
endocrine gland any gland that secretes its products directly into the blood; a ductless gland.
endocrine system the glands and tissues responsible for the production and secretion of hormones.
endocytosis (ĕn″dō·sī·tō´sĭs) uptake of a substance by a cell surrounding the substance in cytoplasm.
endoderm the innermost layer of the three primary cell layers in an embryo.
endolymph the fluid contained in the membranous labyrinth of the ear.
endometrium (ĕn″dō·mē´trē·ŭm) the layer of mucous membrane that lines the inner surface of the uterus.
endomysium (ĕn″dō·mĭs´ē·ŭm) the connective tissue in a muscle that lies between the fascicles.
endoneurium (ĕn″dō·nū´rē·ŭm) the connective tissue in a nerve that separates the individual nerve fascicles from one another.
endoplasmic reticulum (ĕn″dō·plaz´mĭk rĕ·tĭk´ū·lŭm) a network of membranes in the cytoplasm of a cell.
endorphin (ĕn·dor´fĭn) a class of central nervous system polypeptide neurotransmitters involved in the relief of pain and the control of mood and behavior.
endoscope an optical device used to examine the interior of hollow organs.
endosteum (ĕn·dŏs´tē·ŭm) the layer of connective tissue that lines the medullary cavity of a bone.
endothelium (ĕn″dō·thē´lē·ŭm) the layer of squamous epithelium that lines the heart, blood vessels, and lymph vessels.
enkephalin (ĕn·kĕf´ă·lĭn) a group of central nervous system

peptide neurotransmitters similar to endorphins that are involved in the control of pain.

enteroceptive impulses nerve impulses that originate in internal organs and travel toward the central nervous system.

enterogastric response the increased gastric motility and secretion of hydrochloric acid and digestive enzymes that occur in the stomach in response to distension caused by the ingestion of food.

enterokinase (ĕn″tĕr·ō·kī′nās) a duodenal enzyme that catalyzes the conversion of trypsinogen to trypsin.

entrapment neuropathy a disorder of a nerve due to compression or other mechanical effect.

enzyme a protein catalyst of metabolic reactions.

eosinophil (ē″ō·sĭn′ō·fĭl) a granulocyte of the blood or bone marrow containing granules that stain red with eosin or other acidic dyes.

ependyma (ĕp·ĕn′dĭ·mă) the type of neuroglial cell that lines the cavities of the brain and spinal cord.

ephedrine (ĕ·fĕd′rĭn) a synthetic or naturally occurring drug used as a vasoconstrictor in the treatment of asthma and allergies.

epicardium (ĕp″ĭ·kărd′ē·ŭm) the inner layer of the pericardium that forms a layer of serous tissue on the surface of the heart.

epicondylitis (ĕp″ĭ·kŏn″dĭ·lī′tĭs) pain near the lateral epicondyle of the humerus due to a strain of the lateral muscles of the forearm; also called "tennis elbow."

epicranium (ĕp″ĭ·krā′nē·ŭm) the skin, muscles, and other soft tissue that cover the cranium.

epidermis (ĕp″ĭ·dĕr′mĭs) the outermost portion of the skin, consisting of the stratum corneum, stratum lucidum (if present), stratum granulosum, stratum spinosum, and stratum basale.

epididymis (ĕp″ĭ·dĭd′ĭ·mĭs) the long, convoluted tubule on the posterior surface of a testis in which sperm are stored.

epidural hematoma an abnormal mass of blood accumulated in the epidural space outside the dura mater of the brain where it can create pressure on the brain.

epiglottis (ĕp″ĭ·glŏt″ĭs) a cartilaginous structure covered with mucous membrane at the superior end of the larynx which covers the glottis when swallowing.

epilepsy (ĕp″ĭ·lĕp′sē) a neurological disorder characterized by recurrent and temporary attacks involving motor, sensory, and psychic malfunctions.

epimysium (ĕp″ĭ·mĭz″ē·ŭm) a sheath of connective tissue that surrounds a muscle.

epinephrine (ĕp″ĭ·nĕf′rĭn) a hormone produced by the adrenal medulla that acts by stimulating the sympathetic division of the autonomic nervous system.

epiphyseal plate a thin layer of cartilage that lies between the epiphysis and newly forming tissue in a growing long bone.

epiphysis (ĕ·pĭf′ĭ·sĭs) the end of a long bone which is either composed of cartilage or is separated from the shaft of the bone by a cartilaginous disk; a part of a bone that develops as a secondary center of ossification by cartilage and that joins and becomes part of the larger bone during development.

epiploic (ĕp″ĭ·plō″ĭk) pertaining to the omentum, a portion of the peritoneum the covers the intestines.

epiploic foramen a connection between the greater and lesser cavities of the peritoneum.

epithalamus (ĕp″ĭ·thăl″ă·mŭs) the part of the interbrain superior and posterior to the thalamus that includes the pineal body, the habenula, the trigonum habenulae, and the habenular commissure.

epithelium a layer of cells supported by a basement membrane that forms the epidermis of the skin, the surface layer of mucous and serous membranes of hollow organs, and the passageways of the respiratory, digestive, and genital urinary tracts. Depending on the shape, organization and function of the cells, it is classified into columnar, cuboidal, squamous, simple, pseudostratified, stratified, protective, sensory, glandular, or secreting epithelium.

eponychium (ĕp″ō·nĭk″ē·ŭm) the band of epidermal tissue that surrounds a nail; horny epidermal tissue from which a toenail or fingernail develops in a fetus.

Epstein-Barr virus a virus thought to be responsible for infectious mononucleosis and often associated with cancer.

equilibrium a state in which opposing forces or activities are occurring at rates that equalize one another so that there is no apparent change in condition.

erection the enlargement and stiffening of a penis or a clitoris that occur during sexual excitement.

eructation (ĕ·rŭk·tā′shŭn) belching.

erythema (er″ĭ·thē′mă) diffuse redness of the skin caused by dilation of skin capillaries in response to an external stimulus, inflammation, or nervous response.

erythema marginatum a rash associated with rheumatic fever.

erythema multiforme a disease characterized by the appearance of dark red spots or tubercles in the skin, usually on the face and neck, forearms, legs, and posterior surfaces of the hands and feet.

erythema nodosum an inflammatory bacterial disease of the skin that forms painful nodules on the leg.

erythrocyte (ĕ·rĭth′rō·sīt) a red blood cell.

erythropoiesis (ĕ·rĭth″rō·poy·ē′sĭs) the process involved in the production of erythrocytes.

erythropoietin a hormone that stimulates the production of erythrocytes.

esophagus (ē·sŏf′ă·gŭs) the tubular structure that connects the pharynx to the stomach.

esotropia (ĕs·ō·trō′pē·ă) a disorder of the eyes in which the visual axis of one eye inclines toward the visual axis of the other eye; cross-eyed.

essential amino acids several necessary amino acids that cannot by synthesized in sufficient amounts by the body and therefore must be obtained in the diet.

estrogen (ĕs′trō·jĕn) the group of hormones responsible for the development of secondary sexual characteristics in a female and the development of endometrium in the menstrual cycle; it is also produced by Sertoli cells in the testes in response to FSH stimulation.

etiology the study of the cause of a disease; the cause of a disease itself.

eupnea (ūp·nē′ă) normal breathing.

eversion turning outward of an organ or structure in relation to its normal position.

excitability ability of a cell or a tissue to be stimulated by an internal or external event or condition.

excitatory postsynaptic potential (EPSP) a transient reduction in the polarity of the membrane of a postsynaptic neuron in response to an action potential in the presynaptic cell.

excretion the discharging of waste products from the body.

exhalation *see expiration.*

exocrine (ĕks′ō·krĭn) **gland** a gland that secretes its products onto an epithelial surface either through ducts or directly to the surface.

exocytosis (ĕks″o·sī·to·sĭs) the process by which cells release substances that are too big to diffuse or to be transported directly through the membrane; usually enclosed in vesicles that are released upon fusion of the vesicle membrane with the cell membrane.

exophthalmos (ĕks″ŏf·thal′mŏs) an abnormal protrusion of an eyeball from its socket.

expiration the breathing outward phase of the respiratory cycle; exhalation.

expiratory reserve volume (ERV) the amount of air that can be forced from the lungs by maximum expiratory effort in excess of the total volume.

extension a straightening out, especially of the members on either side of a flexed joint.

extensor a category of muscle that extends or stretches a limb or other body part.

external acoustic meatus the canal in the temporal bone that leads to the middle ear; also called external auditory meatus.

external ear the portion of the ear external to the tympanic membrane, consisting of the external acoustic meatus and the pinna.

external nares the two openings into the nasal cavity in the nose; the nostrils.

exteroceptive pertaining to a stimulus received from outside the body.

exteroceptive reflex a reflex initiated in response to stimulation of an exteroceptor.

exteroceptor a sensory structure in the skin or a mucous membrane that responds to stimuli from outside the body.

extracellular fluid body fluid that is not contained within cells.

F

face the anterior surface of the head.

facet a small flat surface, especially on a bone or other hard surface.

facilitated diffusion diffusion through a cell membrane involving specialized carrier molecules within the membrane; net flow is from the side of high concentration to the side of low concentration and does not require ATP or other energy source.

falciform (făl′sĭ·form) **ligament of the liver** the ligament on the anterior surface of the liver that extends from the diaphragm to the umbilicus.

fallopian tube an oviduct.

false vocal cords the ventricular folds that lie above the true vocal cords in the larynx.

falx (fălks) a sickle-shaped structure.

falx cerebelli a fold in the dura mater that lies between the two cerebellar hemispheres.

falx cerebri a fold of the dura mater that lies in the longitudinal fissure between the two cerebral hemispheres.

fascia (făsh′e·ă) the fibrous membrane that forms sheaths of the muscles, nerves, and certain blood vessels; the areolar tissue that lies under the skin.

fascicle a bundle of muscle fibers contained within a muscle or a bundle of nerve fibers contained within a nerve.

fasciculation (fă·sĭk″u·lā′shŭn) the spontaneous contraction of a single group of muscle fibers innervated by a single motor neuron, often visible as a spontaneous twitching of the fibers under the skin.

fat adipose tissue; a triglyceride composed of a molecule of glycerol and three molecules of fatty acid.

fauces (fŏ′sēz) the constricted space between the mouth and the oropharynx surrounded by the soft palate, the palatoglossal and palatopharyngeal arches, and the base of the tongue.

febrile (fĕ′brĭl) having a fever.

feces the excretory material discharged from the large intestine through the anus.

femur the thigh bone.

fenestra (fĕn·ĕs′tră) any small opening.

fenestra cochlea an opening between the middle and inner ear covered by the second tympanic membrane; the round window.

fenestra vestibuli (vĕs·tĭb′u·lī) an opening on the inner wall of the middle ear into which the base of the stapes fits; the oval window.

fertilization conception; the fusion of male and female gametes.

fetus a developing offspring from the beginning of the ninth week of development following fertilization until birth.

fever body temperature elevated above normal, in humans above 37°C or 98.6°F.

fibrillation rapid and spontaneous contraction of individual muscle fibers in a larger muscle.

fibrin an insoluble fibrous protein formed by the action of thrombin on fibrinogen during the formation of a blood clot.

fibrinogen (fī·brĭn′o·jĕn) a globulin protein present in blood plasma and the precursor to fibrin molecules laid down during the formation of a blood clot.

fibrinolysis the digestion or degradation of fibrin.

fibroblast a connective tissue cell that differentiates into various specialized forms responsible for producing fibrous tissues in developing or healing tissues.

fibrocartilage a dense fibrous type of connective tissue in which small masses of cartilage have been formed between the fibers, characteristically found in intervertebral disks.

fibromyalgia (fī″bro·mī·ăl′jē·ă) pain in the fibrous connective tissue associated with muscles, ligaments, and tendons occurring, for example, in the lower back (lumbago) and in the thigh (charley horse).

fibromyositis (fī″bro·mī″o·sī′tĭs) a group of muscle and tendon disorders characterized by pain, tenderness, and stiffness of tendons, muscles, and neighboring tissues; the source of lumbago in the lower back and charley horse in the thigh.

fibrosis (fī·bro′sĭs) an abnormal formation of scar tissue or other fibrous tissue.

fibrous indicating the presence of fibers in a tissue or organ.

fibrous dysplasia replacement of bony tissue by fibrous tissue in a bone.

fibula the smaller bone that extends from the knee to the ankle; also called the "shinbone."

fight-or-flight response the set of effects of the sympathetic division of the autonomic nervous system on the heart and circulatory system, adrenal hormone production, and other organs in response to stress, anxiety, or fright.

filiform having a threadlike appearance or structure.

filiform papilla (fĭl′ĭ·form pă·pĭl′ă) one of the narrow papillae at the base of the tongue.

filtration slits the narrow spaces between extensions of the processes of podocytes in a glomerulus.

filum (fī′lŭm) a long, slender filament.

filum terminalae the long slender filaments of nervous tissue that occur at the inferior end of the spinal cord.

fimbriae (fĭm′brē·ē) fringed with long, slender, branched processes, as found at the superior opening of the oviduct.

fissure a deep furrow in the surface of an organ.

fistulo in ano an opening in the skin near the anus that may extend into the anal canal; an anal fistula.
flaccid (flăk′sĭd) soft and limp, lacking muscle tone or firmness.
flagellum (flă·jĕl′ŭm) a long, slender tail or whiplike process that contains a central filament surrounded by a thin layer of cytoplasm, used for locomotion by sperm.
flatfoot an abnormally flat sole of the foot; also called pes planus.
flatulence excess gas in the digestive tract.
flavin adenine dinucleotide (FAD) the electron receptor in the oxidation of succinic acid to fumaric acid in the Krebs cycle.
flexion the bending of a joint.
flexor a muscle that flexes a limb or a part.
flexor reflex a reflex arc that withdraws a limb from the source of stimulation.
flexure a bend or fold in a structure.
focal length the distance from the lens at which parallel rays entering the lens converge to a point.
focal point the point of convergence of rays behind a lens.
folium the individual leaflets of the arbor vitae in the gray matter of the cerebellum.
follicle a small cavity or sac with a secretory function.
follicle-stimulating hormone (FSH) an anterior pituitary gland hormone that stimulates the growth of follicles in an ovary and spermatogenesis in a testis.
fontanel (fŏn″tă·nĕl′) a membrane-covered space that lies between cranial bones of the skull of a newborn infant or fetus.
foramen (for·ā′mĕn) an opening that serves as a passage between two neighboring structures.
foramen ovale an opening in the wall between the two atria of a fetus that allows the blood to travel directly from the right atrium to the left atrium.
forebrain the part of the brain comprising the diencephalon and telencephalon; the anterior part of the three primary brain vesicles in a embryo that later produces the diencephalon and telencephalon.
foregut the first part of the embryonic digestive tube from which the pharnyx, esophagus, stomach, and duodenum arise.
forensic medicine the branch of medicine that deals with the relationship between medicine and law.
foreskin a fold of skin that covers the glans penis; the male prepuce.
fornix a structure with an arched or vaulted shape.
fossa a low pit or depression.
fourth ventricle a brain cavity that lies between the cerebellum and the brain stem.
fovea (fō′vē·ă) a small pit or depression.
fovea centralis a small pit in the middle of the macula lutea in the retina.
fracture a bone injury involving breakage of bone tissue.
freckles small spots, or macules, in the skin resulting from a localized concentration of melanin.
frenulum (frĕn′ū·lŭm) a small fold of mucous membrane that lies between two structures and acts like a bridle in limiting the movement of one of those structures.
frontal pertaining to the anterior portion of an organ or the body; pertaining to the bone in the forehead.
fructose a six-carbon sugar produced in a phosphorylated form from glucose in glycolysis.
fulcrum the pivot point on which a lever rotates.
fulminating hepatitis a rare but serious form of hepatitis that develops quickly and involves massive destruction of large areas of the liver, caused either by viral infection or by drug injury.
fumaric acid a component in the Krebs cycle that lies between succinic acid and malic acid.
functional residual volume (FRV) the volume of air in the lungs at the end of a normal exhalation, it is the sum of the expiratory reserve volume and the residual volume.
fundus the largest part of a hollow organ; the part farthest removed from the opening of a hollow organ.
fungiform papilla small, round structures on the anterior dorsum and sides of the tongue.
fusiform (fū′zĭ·form) spindle shaped; enlarged in the middle and tapering at each end.

G

GABA gamma-aminobutyric acid; an inhibitory neurotransmitter of the central nervous system.
gallbladder the sac on the underside of the right lobe of the liver in which bile is concentrated and stored.
gallstone stones formed from precipitated cholesterol or calcium salts in the gallbladder and bile ducts.
gamete a reproductive cell; an egg or sperm.
gametogenesis (găm″ĕt·ō·jĕn′ĕ·sĭs) the process by which gametes are produced.
gamma-aminobutyric acid see GABA.
gamma rays a type of electromagnetic radiation produced by the decay of various radioactive isotopes. Gamma rays have sufficient energy to penetrate and damage or destroy body cells and tissues and are used to treat certain diseases.
ganglion (găng′lē·ŏn) a mass of nerve cell bodies localized in structures that lie outside the central nervous system. May consist of somatic sensory neurons or motor neurons of the autonomic nervous system.
gangrene (găng′grēn) tissue death that results from inadequate blood supply.
gap junction a type of cell junction characterized by a narrow intercellular gap between the membranes of two adjacent cells; it allows direct communication between the interior of the two cells, also called a nexus.
gas embolism blockage of a blood vessel by the formation of gas bubbles in the blood, usually resulting from overinflation of the lungs that occurs as a result of a rapid decrease in surrounding pressure. The bubbles are thought to travel from the lungs to the cerebral circulation where they damage tissues, sometimes causing coma or death.
gas gangrene a type of gangrene resulting from infection by a bacterium, usually *Clostridium perfringens*.
gastric pertaining to the stomach.
gastric ulcer a lesion of the stomach mucosa brought on by irritation of the stomach wall by gastric juices; see *peptic ulcer*.
gastrin a hormone secreted by the mucosa of the stomach and carried in the blood to other digestive organs, including the stomach, gallbladder, pancreas, and small intestine. Gastrin stimulates secretion of products of these organs.
gastritis (găs·trī′tĭs) inflammation of the stomach.
gastroenteritis (găs″trō·ĕn·tĕr·ī′tĭs) inflammation of the stomach and intestinal mucosa.
gastrointestinal (GI) tract the portion of the digestive system that includes the stomach and the intestines.
gastrula (găs·trū′lă) a stage in development following invagination of the blastula when the embryo consists of three layers of cells.

gates specialized structures in cell membranes that regulate the passage of ions in response to changes in membrane potential and to the action of chemical transmitter substances.

gene the fundamental functional hereditary unit; composed of DNA and responsible for the production of RNA used in the synthesis of proteins.

genetic engineering the technology involved in producing recombinant DNA molecules and introducing those molecules into a bacterium or other microorganism for the production of substances encoded in the recombinant DNA molecule.

geniculate (jĕn·ĭk′ū·lāt) sharply bent, for example, as occurs at a knee.

genitalia (jĕn·ĭ·tāl′ē·ă) the external sex organs.

genome (jē′nōm) a haploid set of genes or chromosomes.

genotype the specific set of genes in the cells of an organism.

gestation the period of development from conception to birth.

gibbous (gĭb′ŭs) a hump on a body part.

gigantism growth to abnormally large size and height.

gingiva (jĭn·jī′vă) the tissues that cover the neck of a tooth; the gum.

gingivitis (jĭn·jĭ·vī′tĭs) inflammation of the gums.

ginglymus (jĭng′lĭ·mŭs) a hinged joint; a joint that has only forward and backward motion; diarthrosis.

glabella (glă·bĕl′ă) the portion of the frontal bone that lies directly above the nose.

gland any organ that secretes a substance that is used elsewhere in the body.

glans clitoridis the distal end of the clitoris.

glans penis the enlarged distal end of the penis.

glaucoma (glaw·kō′mă) a disorder of the eye associated with high intraocular pressure; effects can range from slightly impaired vision to total blindness.

glenoid (glē′noyd) a depression that forms a socket in which another structure rests.

glia (glī′ă) nervous tissue in the central and peripheral nervous systems exclusive of the neurons; neuroglia.

gliding joint a synovial joint that allows sliding of one surface over another.

globin the protein parts of hemoglobin and myoglobin.

globulin a category of proteins characterized by solubility in salt solutions and certain other physical characteristics.

glomerular filtrate the primary urine; the fluid that passes from the glomerulus into the urinary space of a Bowman's capsule.

glomerular filtration rate (GFR) the average total volume of fluid that passes from the glomeruli into the urinary space of the Bowman's capsules in the kidneys in one minute.

glomerulonephritis (glō·mĕr″ū·lō·nĕ·frī′tĭs) a kidney disorder resulting from inflammation of the glomeruli, possibly from bacterial infection or autoimmune response, which results in blood and protein in the urine and decreased kidney function.

glomerulus a small, round structure; the body of capillaries contained within a Bowman's capsule.

glossopalatine (glŏs″ō·păl′ă·tīn) related to the tongue and palate.

glossopharyngeal (glŏs″ō·fă·rĭn·jē·ăl) related to the tongue and pharynx.

glottis (glŏt′ĭs) the true vocal cords and the space between them.

glucagon a hormone produced by the alpha cells of the pancreas that stimulates the degradation of glycogen and the release of glucose from the liver.

glucocorticoid hormone a class of hormones produced by the adrenal cortex that stimulate carbohydrate metabolism.

gluconeogenesis (glū″kō·nē″ō·jĕn′ĕ·sis) the formation of glucose in the liver from noncarbohydrate molecules.

glucose a six-carbon monosaccharide that serves as a primary source of energy and material for anabolic reactions in a cell.

glucosuria the presence of glucose in the urine.

glycerol a three-carbon compound present in glycerides, phosphatides, and certain other lipids, to which fatty acids, phosphate groups, and other compounds are bound.

glycocalyx (glī″kō·kăl′ĭks) the network of polysaccharides on the exterior surface of a cell membrane.

glycogen a polysaccharide formed by joining glucose units; a primary storage form for carbohydrates in the liver and muscle tissue.

glycogenesis the formation of glycogen from glucose.

glycogenolysis (glī″kō·jĕn·ōl′ĭ·sĭs) breakdown of glycogen into glucose in the liver and other tissues.

glycolysis the series of reactions by which glucose is converted to pyruvic acid or lactic acid with the concomitant production of ATP.

goblet cell the unicellular mucus-secreting gland in the epithelium of the respiratory and digestive tracts.

goiter an enlargement of the thyroid gland, due to iodine deficiency or a thyroid gland dysfunction.

Golgi apparatus an array of closely packed parallel membranes in the cytoplasm of a cell responsible for enclosing materials in vesicles that will be secreted by the cell.

gomphosis (gŏm·fō′sĭs) a type of joint in which a conical structure is inserted into a socket, such as a tooth held in an alveolus.

gonad (gō′năd) the organ that produces gametes; the ovary in females and the testis in males; the developing sex gland in an embryo before it can be identified as an ovary or testis.

gonadotrophin (gŏn″ă·dō·trō′fĭn) hormones that stimulate gonadal functions, including follicle-stimulating hormone and luteinizing hormone produced by the anterior pituitary gland and human chorionic gonadotrophin produced by the chorion.

gonococcus (gŏn″ō·kŏk′ŭs) the bacterium that causes gonorrhea, *Neisseria gonorrhoea*.

gout arthritis and inflammation of the joints resulting from excessive uric acid crystals accumulating in the joints.

graafian (grăf′ē·ăn) **follicle** a fully developed ovarian follicle prior to rupture and release of the ovum.

graded response a nerve or muscle response that varies with the strength of a stimulus.

grand mal seizure a type of epileptic seizure often preceeded by an aura, which is characterized by strong involuntary muscle spasm and coma; also called primary generalized tonic-clonic seizure.

granulocyte (grăn′ū·lō·sīt″) a leukocyte containing characteristic cytoplasmic granules that take up certain dyes.

granulosa the layer of cells that surround the oocyte in a developing ovarian follicle.

Graves' disease hyperthyroidism characterized by goiter, exophthalmos, myxedema, increased metabolic rate, and other symptoms, possibly due to an autoimmune response.

gray matter the portion of the central nervous system that lacks myelinated nerve fibers and has a gray appearance as opposed to the glistening white appearance of white matter.

greater omentum the portion of the double fold of peritoneum that covers the intestines and is suspended from the greater

curvature of the stomach. It is also attached to the transverse colon.
greenstick fracture a type of bone fracture where the bone is broken on one side and bent on the other side.
growth hormone anterior pituitary hormone that stimulates skeletal and soft tissue growth before maturity and effects numerous metabolic reactions in adults.
gubernaculum (gū″bĕr·năk′ū·lŭm) a guiding structure; a cord in the fetus that extends from the end of the testis through the inguinal canal to the scrotal swelling where it guides the descent of a testis during development.
Guillain-Barré syndrome an acute inflammation of multiple nerves that results in muscular weakness and mild loss of sensation, of unknown cause.
gustation (gŭs·tā′shŭn) the sense of taste.
gut the intestine; an embryonic digestive tube.
gyrus one of the convolutions on the surface of the cerebral hemisphere, delineated by sulci or fissures.

H

habenula stalk attached to the pineal body of the brain.
hair a keratinized filament that grows from a hair follicle in the skin.
hallucinogen a substance that induces hallucination; an imagined sensory perception of a stimulus that is not actually present.
hamate shaped like a hook.
hand the organ at the end of the arm that contains the carpals metacarpals, and phalanges.
haploid pertaining to the number of chromosomes normally present in an egg or sperm.
hapten the portion of an antigen recognized by an antibody; a determinant site.
hard palate the anterior, hard portion of the roof of the mouth.
haustrum (haw′strŭm) one of the major pouches formed by constrictions in the wall of the colon.
Haversian canal one of the tiny vascular canals present in bone.
Haversian system a unit of bone that contains a Haversian canal and the surrounding osteocytes and matrix; an osteon.
hay fever an allergic reaction involving the mucous passages of the upper respiratory tract caused by pollen or other allergens.
head the portion of the body superior to the neck; the proximal end of a bone.
heart the four-chambered muscular structure that pumps blood to the lungs and to the systematic circulatory system.
heart attack a general term for heart failure or for the symptoms that result from blockage of the coronary arteries.
heart block arrhythmia of the heart due to improper conduction of impulses from the atrium to the ventricles.
heart murmur an abnormal sound of the heart due to incomplete closing of a valve during contraction of an atrium or ventricle.
heart sounds the "lubb" and "dupp" associated with the closure of the atrioventricular and semilunar valves, respectively.
heat exhaustion weakness, dizziness, and nausea occurring acutely in response to overexposure to heat. Associated with cold clammy skin, dilated pupils, cramps, and possible fainting.
heat stroke an acute reaction to heat characterized by elevated body temperature, absence of sweat, elevated pulse and respiratory rates, resulting from an inability of the body's temperature regulating mechanisms to counteract the effect of external heat.
helicine (hĕl′ĭ·sĭn) **arteries** coiled arteries in the penis, clitoris, and the wall of the uterus.
helicotrema (hĕl″ĭ·kō·trē′mă) the opening at the end of the cochlear canal where the scala vestibuli joins the scala tympani.
helper T cells T lymphocytes that help stimulate antibody production by B lymphocytes in response to antigens.
hematocrit (hē·măt′ō·krĭt) the percentage of whole blood volume due to erythrocytes.
hematocytoblast (hĕm″ă·tō·sī′tō·blăst) a stem cell of the bone marrow from which all blood cells are believed to be produced.
hematology (hē″mă·tŏl′ō·jē) the study of blood and blood-forming tissues.
hematoma (hē″mă·tō′mă) a local mass of blood, usually clotted, that has entered an organ, space, or tissue from a ruptured blood vessel.
hematopoiesis (hē″mă·tō·poy·ē′sĭs) the formation and development of blood cells in the bone marrow.
hematuria (hē″mă·tū′rē·ă) the presence of bloods cells in the urine.
heme (hēm) an iron-containing nonproteinaceous portion of a hemoglobin molecule.
hemiplegia (hĕm·ē·plē′jē·ă) paralysis restricted to one side of the body
hemivertebra (hĕm″ē·vĕr′tĕ·bră) an abnormality of development in which one side of a vertebra is incompletely developed.
hemizygous (hĕm″ē·zī′gŭs) the condition of having a single allele of a gene rather than the normal two alleles of a gene, especially pertaining to genes on the X chromosome in a male.
hemodialysis (hē″mō·dī·ăl′ĭ·sĭs) selective removal of waste products and other substances from blood by diffusion through selectively permeable membranes as the blood circulates through a machine outside the body.
hemoglobin (hē″mō·glō′bĭn) the iron-containing protein of red blood cells that is primarily responsible for carrying oxygen to the tissues and, to a lesser extent, carbon dioxide from the tissues.
hemolysis (hē·mŏl′ĭ·sĭs) destruction of red blood cells.
hemolytic anemia lowered red blood cell count that results from excessive destruction of red blood cells.
hemophilia (hē″mō·fĭl′ē·ă) a sex-linked recessive genetic disorder characterized by reduced ability of blood to clot.
hemopoiesis (hē″mō·poy·ē′sĭs) the processes that result in the formation of blood.
hemorrhage (hĕm′ĕ·rĭj) bleeding; an unchecked escape of blood from blood vessels.
hemorrhoids (hĕm′ō·royds) masses of distended, knotted veins around the anus and distal end of the rectum.
hemostasis (hē·mŏs′tă·sĭs) the stopping of a flow of blood from a tissue.
heparin (hĕp′ă·rĭn) a naturally occurring mucopolysaccharide that delays blood clotting.
hepatic pertaining to the liver.
hepatic artery one of the three branches of the celiac trunk that supplies blood to the organs of the abdominal cavity.
hepatic duct the duct that receives bile from the liver and joins the cystic duct to form the common bile duct.
hepatic portal vein the vein that carries blood that has absorbed nutrients in capillaries of the small intestine to the liver.

hepatitis (hĕp″ă·tī′tĭs) inflammation of the liver.
Hering-Breuer (hĕr′ĭng-broy′ĕr) **reflex** a reflex in which distension of the pulmonary airways stimulates stretch receptors that inhibit further inspiration and that stimulate expiration.
hernia an abnormal protrusion of an organ or part of an organ through the wall of the body cavity in which it is contained.
herniated disk a protrusion or rupture of the nucleus pulposus of an intervertebral disk through the surrounding fibrocartilage; also called a ruptured disk.
herpes genitalis herpes simplex infection of the genitals, usually transmitted by sexual contact and characterized by painful blisters.
herpes simplex an infectious viral disease characterized by the appearance of small fluid-filled vesicles in the skin or mucous membrane.
herpes zoster a viral infection involving the dorsal root ganglia, characterized by inflammation of a few ganglia and the sensation of pain and appearance of vesicles in the area of the skin supplied by the sensory nerves that arise in the infected ganglia. Caused by the same virus that causes chickenpox; also called shingles.
heterochromatin one of two states of chromatin in which the chromatin is particularly condensed.
heterocrine gland a gland that has both endocrine and exocrine functions.
heterozygous a genotype consisting of two allelic forms of a gene.
hiatal hernia a hernia involving the protrusion of a portion of the abdomen through the esophageal hiatus of the diaphragm into the thoracic cavity.
high altitude cerebral edema a high altitude illness characterized by headache, clumsiness, confusion, and double vision and associated with the accumulation of fluid in the brain.
high altitude pulmonary edema a high altitude illness characterized by shortness of breath, weakness, staggering, and poor balance, along with an accumulation of fluid in the lungs.
high density lipoprotein (HDL) one of the forms in which lipids circulate in plasma, characterized by relatively small micelles and relatively high concentrations of protein and phospholipid. Cholesterol that leaves the cells is absorbed into HDL synthesized in the liver and intestine and carried in HDL micelles.
hillock a raised area at the base of an axon at which a nerve impulse usually originates.
hilus (hī′lŭs) a recess or opening in an organ through which nerves, muscles, or ducts pass into the organ; also called a hilum.
hindgut the distal end of the embryonic digestive tract that produces the small and large intestines, urinary bladder, and urogenital openings.
hinged joint a synovial joint in which a convex surface of one bone fits into a concave surface of another bone; also called a ginglymus joint.
hippocampus (hĭp″ō·kăm′pŭs) a curved, raised area in the floor of the inferior horn of the lateral ventricle.
histamine a naturally occurring product derived from the amino acid histidine which causes a red flush and a wheal around damaged or irritated skin.
histology the study of the cellular organization of tissues and organs.
histone basic proteins which, in conjunction with DNA, form nucleosomes.
hives a skin condition characterized by itchy eruptions caused by an allergen or a psychological reaction; also called urticaria.

Hodgkin's disease a disease of unknown cause that affects the lymphoid tissue, causing enlargement of the lymph glands, spleen, liver, and other tissues. Symptoms include itching, fever, weight loss, and night sweats.
Hoffman's reflex an abnormal reflex induced by striking the nail of the ring, middle, or index finger that causes flexion of the thumb and of the second and third phalanx of one of the other fingers.
holandric (hŏl·ăn′drĭk) related to genes on the Y chromosome.
holocrine (hŏl′ō·krĭn) pertaining to glands in which the secretion contains altered or degraded cells of the gland.
holocrine gland a type of gland in which the secretions consist of cells of the gland that contain the secretory products.
homeostasis the maintenance of a relatively constant internal environment.
homologous chromosomes chromosomes that pair during prophase I of meiosis and that carry similar distributions of genes, although not necessarily in the same allelic forms.
horizontal plane a plane that divides the body or organs into superior and inferior divisions; a transverse plane.
hormone a regulatory substance produced in a gland or other organ that is carried in the blood to another part of the body where it stimulates specific biochemical responses.
human chorionic gonadotrophin (HCG) a hormone produced by the chorionic villi of the placenta that helps the mother's ovaries to maintain the corpus luteum and that stimulates secretion of progesterone and estrogens by the corpus luteum until the placenta begins to produce its own hormones.
human chorionic somatomammotrophin (HCS) a protein produced by the placenta that regulates maternal metabolism so as to provide nutrients to the developing fetus.
humerus (hū′mĕr·ŭs) the bone of the upper arm extending from the elbow to the shoulder.
humoral pertaining to body fluids or substances carried in them.
humoral immunity part of the immune system involving B lymphocyte production of antibodies.
Huntington's disease a dominant genetic disease that begins to apppear in the fourth or fifth decade and results in progressive loss of intellectual capability, personality changes, and various motor disturbances; formerly called Huntington's chorea.
hyaline (hī′ă·lĭn) clear, glassy, or translucent in appearance.
hyaluronic acid a mucopolysaccharide found in connective tissue where it helps to bind the matrix.
hyaluronidase (hī″ă·lūr·ŏn′ĭ·dās) one of the enzymes in semen and sperm that hydrolyses hyaluronic acid and disperses the cells of the corona radiata that surround an ovum.
hybridoma a cell produced by artificial fusion of nuclei from two different cells; used to produce monoclonal antibodies.
hydrocarbon a chemical substance that contains only carbon and hydrogen atoms.
hydrocele (hī′drō·sēl) an accumulation of fluid in a sac or cavity, particularly in the tunica vaginalis of a testis.
hydrocephalus (hī·drō·sĕf′ă·lŭs) an accumulation of cerebrospinal fluid within the skull.
hydrochloric acid a gastric acid produced by parietal cells in gastric glands involved in digestion; a strong acid that dissociates to hydrogen and chloride ions.
hydrolysis a chemical reaction in which the bond between two parts of a molecule is broken, with the addition of the OH group of a water molecule to one side of the broken bond and the remaining hydrogen atom of the water molecule to the other side of the bond.

hydrophilic the tendency of a substance to absorb water or readily dissolve in water.

hydrophobic the tendency of a substance to be insoluble in water and to repel water.

hydrostatic pressure the pressure produced by a fluid in a container.

hymen (hī′měn) a membrane that partially covers the entrance to the vagina.

hyoid bone (hī′oyd) a horseshoe-shaped bone located at the base of the tongue suspended from the styloid processes of the temporal bones and located above the larynx.

hypercapnia (hī″pěr·kăp′nē·ă) an abnormally high concentration of carbon dioxide in the blood.

hyperemia (hī″pěr·ē′mē·ă) increased blood in a body part.

hyperextension overextension of a limb or other body part.

hyperglycemia (hī″pěr·glī·sē′mē·ă) an abnormally high concentration of glucose in the blood, often associated with diabetes mellitus.

hyperinsulinism (hī″pěr·ĭn′sū·lĭn·ĭzm) abnormally high levels of insulin in the blood, resulting in hypoglycemic-like symptoms of weakness, sweating, double vision, and possible coma and death.

hyperkalemia (hī″pěr·kă·lē′mē·ă) abnormally high concentrations of potassium in the blood, often due to kidney failure, and resulting in muscle weakness or cardiac arrest.

hypermagnesia (hī″pěr·măg·nē·zē·ă) elevated magnesium ion concentration in the blood and other body fluids, often associated with kidney failure or ingestion of excessive amounts of antacids. Can cause irregular heartbeat and breathing.

hypermetropia (hī″pěr·mě·trō′pē·ă) a condition of defective vision in which the eye focuses an image behind the retina; also called farsightedness or hyperopia.

hyperplasia (hī″pěr·plā′zē·ă) production of an excessive number of cells in an organ or tissue.

hyperpolarized an increased level of polarization of a cell membrane.

hypersensitivity an unusually high sensitivity of the skin or other tissue to a stimulus.

hypertension a chronic or acute condition in which blood pressure is higher than normal.

hyperthermia abnormally high body temperature.

hyperthyroidism excess production of thyroid hormone, usually resulting in high calcium ion concentration and low phosphate ion concentration in the blood, loss of calcium from the bone, and increased basal metabolic rate.

hypertonic a solution that has a higher concentration of osmotically active solute than a second solution with which it is compared.

hypertrophy increase in size of an organ resulting from cell enlargement or from the production of cell products rather than from an increase in cell number.

hyperventilation increased rate and depth of breathing, sometimes leading to abnormally low carbon dioxide levels that result in dizziness, anxiety, localized numbness, and other symptoms.

hypervitaminosis a condition caused by the ingestion of excessive amounts of a vitamin.

hypocalcemia (hī″pō·kăl·sē′mē·ă) lower than normal concentrations of calcium ion resulting from a number of causes including deficiency or inability to use vitamin D, kidney malfunction, inflammation of the pancreas, or deficiency of parathyroid hormone production.

hypochlorhydria (hī″pō·klor·hī′drē·ă) release of abnormally low amounts of hydrochloric acid into the stomach by the cells in the stomach wall.

hypodermis (hī″pō·děr′mĭs) subcutaneous tissues.

hypoesthesia (hī″pō·ěs·thē′zē·ă) abnormally low sensitivity of the skin to stimuli.

hypogastric (hī″pō·găs′trĭk) pertaining to the hypogastrium, the region below the stomach in the lower portion of the abdomen.

hypoglossal (hī″pō·glŏs′ăl) pertaining to the region beneath the tongue.

hypoglycemia (hī″pō·glī·sē′mē·ă) abnormally low concentrations of glucose in the blood, usually resulting from excess production of insulin or inadequate intake of food.

hypomagnesemia (hī″pō·măg″ně·sē′mē·ă) abnormally low concentrations of magnesium ion in the blood, most commonly due to impaired absorption, protein deficiency, and parathyroid disease, and characterized by a broad spectrum of symptoms involving personality and neuromuscular disorders.

hyponatremia (hī″pō·nă·trē′mē·ă) abnormally low concentration of sodium ions in the blood, often due to excessive water intake and deficient water loss.

hyponychium (hī·pō·nĭk′ē·ŭm) the thickened epidermal layer that lies under the free edge of a nail.

hypoparathyroidism (hī″pō·păr·ă·thī′royd·ĭzm) abnormally low secretion of parathyroid hormone resulting in reduced serum phosphorus levels.

hypophysis (hī·pŏf′ĭ·sĭs) the pituitary gland.

hyposecretion abnormally low production of the secretion products of a gland.

hypothalamic-hypophyseal tract a group of nerve fibers that have their cell bodies situated in the hypothalamus and extend to the posterior pituitary gland where they secrete antidiuretic hormone and oxytocin.

hypothalamus (hī″pō·thăl′ă·mŭs) part of the diencephalon, including a portion of the ventral wall of the third ventricle and the ventricular floor. It produces hormones that are transported to and released from the posterior lobe of the pituitary gland (ADH and oxytocin) and other hormones that control the release of anterior pituitary lobe hormones.

hypothenar (hī·pŏth′ě·năr) the fleshy ridge on the ulnar side of the palm that includes the muscles that control the small finger.

hypothermia (hī″pō·thěr′mē·ă) abnormally low body temperature.

hypotonic a situation in which a solution has a lower concentration of osmotically active solutes than a second solution with which it is compared.

hypoventilation abnormally low exchange of air resulting from shallow breathing or reduced breathing rate.

hypoxemia (hī·pŏks·ē′mē·ă) abnormally low oxygen levels in the blood; hypoxia.

hypoxia (hī·pŏks′ē·ă) insufficient respiratory oxygen intake.

hysterectomy (hĭs·těr·ěk′tō·mē) surgical removal of the uterus.

I

ichthyosis (ĭk″thē·ō′sĭs) a skin disease in which the skin becomes dry and scaly, resembling fish scales.

idiopathic (ĭd″ē·ō·păth′ĭk) an abnormal or diseased state for which the cause is not known.

ileocecal (ĭl″ē·ō·sē′kăl) pertaining to the junction of the ileum and cecum.

ileocecal valve the sphincter located where the ileum empties into the cecum of the large intestine.

ileum (ĭl´ē·ŭm) the distal two-thirds of the small intestine where bile salts and vitamin B_{12} are reabsorbed.

ilium (ĭl´ē·ŭm) one of the bones located in each half of the pelvis.

immunity condition of being resistent to a disease or protected from infection by disease.

immunoglobulins a family of plasma proteins that contains antibodies; plasma proteins that confer immunity.

immunosuppression inhibition of an immune response with drugs or other agents.

impetigo (ĭm·pē·tī´gō) a contagious skin disease in which the skin is inflamed and marked by itching blisters that break and become encrusted. The blisters usually form on the face or around the mouth and nostrils and spread to neighboring regions.

implantation the attachment and digestion into the uterine wall by a blastocyst.

impotence inability of a male to produce or maintain an erection, due possibly to fatigue, stress, poor health, drugs, psychologic factors, or neurologic dysfunction.

incisor one of the eight teeth that lie between the canines in each jaw and are used for cutting.

incontinence loss of sphincter control that results in the inability to retain urine, feces, or semen.

incus (ĭng´kŭs) one of the three small bones of the middle ear; also called the anvil.

infant respiratory distress syndrome a disorder that results in diffuse lung collapse in infants as a result of deficient production of pulmonary surfactant; also called hyaline membrane disease.

infarction (ĭn·fărk´shŭn) an area of tissue that is dead or dying from lack of oxygen, usually resulting from a blockage of blood flow caused by an embolus.

infectious mononucleosis an acute, infectious viral disease that primarily affects lymphoid tissue and is marked by sore throat, fever, tender lymph nodes, an enlarged spleen, and an increase of abnormal white cells in the blood. Possibly caused by the Epstein-Barr virus.

inferior a term relating to a position below a structure or on the underside of a structure.

inferior vena cava the large vein that delivers blood from the lower part of the body to the right atrium.

infertile a condition in which a woman is unable to conceive or a male is unable to cause conception due to physiologic, anatomic, or psychologic causes.

inflammation reddening of a tissue caused by dilatation of the blood vessels in an area in response to injury or infection.

inflation reflex a neural reflex stimulated by receptors in the lung airways that inhibits excessive inflation of the lung; also called the Hering-Breuer reflex.

influenza a contagious virus-caused infection with symptoms that include sore throat, muscular achiness, weakness, cough, and fever. Onset usually occurs very quickly with chills and fever.

infundibulum (ĭn″fŭn·dĭb´ū·lŭm) any funnel-shaped passageway, as in the stalk of the pituitary gland, the opening into the oviduct, the opening at the upper end of the cochlear canal, or the tube that connects the frontal sinus to the middle nasal meatus.

ingest to take food, liquid, or other substances into the alimentary canal by way of the mouth.

inguinal (ĭng´gwĭ·năl) **canal** the opening from the abdominal cavity that carries the round ligament in a female or the spermatic cord in a male and through which the testes descend in a developing male fetus.

inhalation see inspiration.

inheritance transmission of genes from parents to progeny.

inhibin a hormone produced by seminiferous tubules that inhibits the production of FSH by the anterior pituitary gland.

inhibitory postsynaptic potential (IPSP) a change in the membrane potential of a neuron that makes it more negative and thus requires a greater stimulus to achieve the threshold membrane potential.

inner cell mass a group of cells in the blastocyst that gives rise to the cell layers from which the embryo and associated structures develop.

innervation the group of nerves that enter or leave a particular body part.

insertion point of attachment of a muscle that moves when the muscle contracts.

insomnia a condition in which one is unable to fall asleep or awakes prematurely and is often unable to go back to sleep. It can be caused by a variety of factors including emotional stress, pain, and neurological problems.

inspiration the act of breathing in or drawing air into the lungs; inhalation.

inspiratory capacity the total volume of air that can be inspired; the total of the tidal volume and the inspiratory reserve volume.

inspiratory reserve volume (IRV) the amount of air that can be brought into the lungs by forced inspiration over and above the tidal volume.

insula the triangular area of the cerebral cortex that lies on the floor of the lateral fissure; the central lobe of the cerebral hemisphere; also called the isle of Reil.

insulin a hormone produced by beta cells of the islets of Langerhans of the pancreas that enables cells to take in glucose, among many other effects.

insulin-dependent diabetes mellitus (IDDM) a form of diabetes mellitus in which the patient must receive insulin from an external source to regulate glucose metabolism.

integument the skin.

interatrial (ĭn″tĕr·ā´trē·ăl) pertaining to the area between the atria.

intercalated disk the convoluted pair of cell membranes that forms a junction between the ends of cardiac muscle fibers.

intercostal (ĭn·tĕr·kŏs´tăl) pertaining to the areas between the ribs.

interferon (ĭn·tĕr·fĕr´ŏn) a protein produced by cells in response to viral infection that interferes with reproduction of the virus.

interleukin (ĭn·ter·lū´kĭn) a substance produced by cells in response to bacterial toxins that stimulates the immune system and also stimulates production of a fever by its action on the preoptic area of the hypothalamus.

interlobar pertaining to the area lying between the lobes of an organ.

internal ear the innermost chamber of the ear, containing the vestibule, cochlea, and semicircular canal.

internal nares the openings in the posterior end of each of the nasal cavities that open into the nasopharynx; also called choanae.

internal respiration gas exchange that goes on between the blood and capillaries and the surrounding tissues in the body.

internuncial (ĭn″tĕr·nŭn′shē·ăl) **neuron** a neuron that lies between two other neurons in a neural pathway; association neuron.

interoceptor a receptor that is sensitive to internal stimuli.

interphase the period in the cell cycle between mitotic divisions.

interstitial cell one of the cells that lies between the seminiferous tubules in the testes, responsible for testosterone secretion.

interstitial fluid extracellular fluid located in intercellular spaces.

interventricular pertaining to the region between the ventricles.

intervertebral disk one of the cartilaginous disks that lies between vertebrae.

intervertebral foramina the spaces between the pedicles of successive vertebrae through which spinal nerves pass.

intestine the part of the digestive tract that extends from the stomach to the anus, consisting of the small and large intestines.

intraarticular fracture a fracture of the articular surface of a bone.

intracellular pertaining to substances or events occurring within a cell.

intracellular fluid the fluid that is contained within a cell.

intrafusal fiber an encapsulated striated muscle fiber contained within a muscle stretch receptor.

intrapleural pressure the air pressure in the space between the visceral and parietal pleurae.

intrarenal failure *see renal renal failure.*

intrauterine device (IUD) a plastic or metal device inserted and maintained in the uterus to prevent conception by preventing implantation of a blastocyst in the uterine wall.

intrinsic factor a glycoprotein produced by parietal cells of the gastric mucosa that aids in the absorption of vitamin B_{12}.

intrinsic muscle muscle in which both the insertion and origin are located in the same structure.

inulin a polysaccharide sometimes used in clearance tests to determine kidney function.

in utero within the uterus.

inversion to turn an extremity or other organ upside down or inside out from the normal orientation.

in vitro occurring or present in a test tube or other artificial container.

in vivo present or occurring in living tissue.

involuntary muscle muscle that is not subject to voluntary control, usually innervated by the autonomic nervous system; smooth muscle.

ion an atom or molecule that has an electric charge due to a difference in the number of protons and electrons present.

ionic bond the bond formed by the electrostatic attraction between ions of opposite charge.

ipsilateral (ĭp″sĭ·lăt′ĕr·ăl) affecting the same side of the body.

iris the contractile membrane that lies between the lens and the cornea which regulates the amount of light that enters the eye.

irritability capability of a cell to respond to a stimulus.

irritable bowel syndrome involuntary contractions of the muscles of the small and large intestines, often associated with abdominal pain and often caused by emotional stress; also called spastic colon.

ischemia (ĭs·kē′mē·ă) a condition in which a localized region receives inadequate blood due to localized blockage of the blood vessels.

ischium (ĭs′kē·ŭm) the lower portion of the hip bone.

islet of Langerhans one of many clusters of endocrine cells in the pancreas that secrete insulin, glucagon, and somatostatin.

isoantibody an antibody produced in response to an isoantigen.

isoantigen a substance that acts as an antigen in other members of the same species but not in the individual in which it is contained.

isoenzyme enzymes that catalyze the same biochemical reaction but have different amino acid sequences.

isometric having constant dimension.

isometric contraction a muscle contraction in which the muscle is prevented from shortening.

isotonic relating to two solutions with equal concentrations of osmotically active solutes.

isotonic contraction a muscle contraction in which the muscle shortens and maintains a constant tension.

isotope one of the forms of the atoms of an element differing from the rest in the number of neutrons present in the nucleus.

isthmus a narrow passageway between two cavities; a constricted region within a tubular organ.

J

Jacksonian seizure a form of epilepsy characterized by rapid contraction and relaxing of a group of muscles and then proceeding from one group to another as a result of the movement of the epileptic activity through the motor cortex of the brain.

jaundice yellow color of the skin that results from an accumulation of bile pigments in the blood.

jejunum (jē·jū′nŭm) the second portion of the small intestine, comprising about 40% of its length, that lies between the duodenum and ileum, responsible for absorbing organic nutrients.

jet lag disruption of daily physiological cycles caused by flying across several time zones; also called circadian dysrhythmia.

joint the junction between two bones; an articulation.

joint fracture a fracture involving the articular surfaces of the bony structures in a joint.

jugular pertaining to the throat or neck.

junction a point of attachment of the cell membranes of adjacent cells.

juxtaglomerular (juks″tă·glō·mĕr′ŭ·lăr) located near a glomerulus.

juxtaglomerular apparatus a renin-secreting structure that consists of the macula densa of the distal convoluted tubule and juxtaglomerular cells of the afferent arteriole in a nephron.

juxtamedullary (jŭks″tă·mĕd′ū·lăr·ē) adjacent to the medulla.

juxtamedullary nephron a nephron that originates in the inner portion of the cortex and extends relatively deep into the renal medulla.

K

Kallmann's syndrome a set of conditions resulting from an endocrine disorder in which there is a deficiency of gonadotrophin-releasing hormone production by the hypothalamus. This results in reduced secretion of luteinizing hormone and follicle-stimulating hormone, and the activity of the gonads is reduced.

karyokinesis the stage of mitotic division involving the division of the nucleus.

karyotype a photographic or diagrammatic display of prophase or metaphase chromosomes that have been aligned according to the size, centromere location, and staining pattern.

keloid (kē´loyd) an elevated and irregularly shaped overgrown scar that results from the production of excessive amounts of fibrous tissue in the area of a wound.

keratin an especially tough protein found as the principle component of skin, nails, hair, and tooth enamel.

keratohyalin (kĕr˝ă·tō·hī´ă·lĭn) a form of hyalin found in the epidermis.

ketogenesis the metabolic production of ketone bodies, often occurring in diabetes mellitus or as a result of excessive fat catabolism.

ketone body a substance produced as a result of fat digestion, including acetone, acetoacetic acid, and beta-hydroxybutyric acid. Ketone bodies are often found in the urine of diabetics.

ketosis the accumulation of ketone bodies in body fluids as a result of incomplete catabolism of fatty acids, causing increased acidity of the fluids. Ketosis sometimes occurs following digestion of large amounts of fats and often in diabetes mellitus.

kidney one of a pair of bean-shaped organs that lie on each side of the spine and are involved in waste elimination, control of body pH and electrolyte concentration, blood pressure, blood formation, and vitamin D production.

kilocalorie the amount of heat required to raise the temperature of 1 kg of water from 14.5°C to 15.5°C; equal to 1000 calories. "Calories" referred to in dietetics and nutrition are actually kilocalories (*see calorie*).

kinesiology the study of muscular and skeletal mechanics of body movement.

kinetic energy the energy of motion.

kinins a group of polypeptides that influence smooth muscle contraction and thereby increase blood flow and reduce blood pressure as a result of arterial dilation.

Klinefelter syndrome the syndrome of characters present in a person who has one or more extra X chromosome(s).

Koebner phenomenon the appearance of skin lesions that occur in psoriasis and certain other skin conditions.

Krebs cycle a sequence of metabolic reactions by which acetate is degraded to carbon dioxide. ATP is produced and electrons are released to an electron transport chain as a result of reactions of the Krebs cycle.

Kupffer cells cells found in the sinusoids of the liver that phagocytize bacteria and other small foreign proteins that are carried through the liver in the blood.

Kussmaul (kūs´mowl) **breathing** a pattern of breathing characterized by deep rapid breaths; associated with diabetic acidosis.

kyphosis (kī·fō´sĭs) an abnormally convex curvature of the spine when seen in the lateral view.

L

labia (lā´bē·ă) fleshy borders surrounding an opening; the lips about the mouth.

labia majora two long folds of fatty tissue on either side of the vaginal opening forming the borders to the vulva.

labia minora two thin folds of skin that lie between the labia majora and the opening to the vagina.

labial frenulum (lā´bē·ăl frĕn´ū·lŭm) a fold of mucous membrane that connects the middle portion of the upper and lower lips to the upper and lower gums.

labyrinth (lăb´ĭ·rĭnth) any network or system of cavities or canals that connect with each other through passages; the inner ear consisting of the vestibule, three semicircular canals, and the cochlea, which form the osseous labyrinth.

lacrimal apparatus structures in the eye that are involved with the production and conduction of tears, including the lacrimal gland and ducts.

lactase an enzyme produced in the lining of the small intestine that catalyzes the conversion of lactose to glucose and galactose.

lacteal (lăk´tē·ăl) pertaining to the production of milk (*see central lacteal*).

lactic acid an acid produced from pyruvic acid when glycolysis occurs during periods of oxygen absence or deficiency; it is produced by active muscles when oxygen is being used faster than it can be delivered.

lactiferous ducts *see mammary ducts.*

lactogenic the ability of a substance to stimulate the production and secretion of milk.

lactose intolerance an inability to digest lactose resulting from the loss of ability to produce lactase.

lacuna (lă·kū´nă) a small chamber in bone or cartilage tissue that contains an osteocyte or chondrocyte, respectively.

lamella (lă·mĕl´ă) a thin plate or leaf of material.

lamina (lăm´ĭ·nă) a thin and relatively flat layer of tissue.

lamina propria (prō´prē·ă) the layer of connective tissue that lies beneath the epithelium of mucous membranes.

large intestine the part of the digestive tract through which material passes after leaving the small intestine and consisting of the cecum, colon, rectum, anal canal, and anus.

laryngitis (lăr·ĭn·jī´tĭs) inflammation of the larynx.

laryngopharynx (lăr·ĭn˝gō·făr´ĭnks) pertaining to the area around the larynx and pharynx.

larynx (lăr´ĭnks) the structure at the superior end of the trachea that contains the vocal cords and produces sound used in speech.

lateral pertaining to the side of the body or body part.

lateral ventricle a cavity located in each hemisphere of the cerebrum.

leg the part of the lower extremity between the knee and the ankle, including the tibia, fibula, and patella.

leiomyoma (lī˝ō·mī·ō´mă) a nonmalignant tumor that develops in smooth muscle, especially in the uterus, stomach, esophagus, or small intestine.

lemniscus (lĕm·nĭs´kŭs) a general term for bundles of sensory fibers located in the central nervous system, particularly in the medulla and pons.

lens a fluid-filled transparent structure in the anterior of the eye used to focus light rays on the retina.

lens placode an ectoderm-derived platelike thickening that forms the lens of an eye during embryological development.

lentiform (lĕnt´ĭ·form) shaped like a lens.

lesion an injured, infected, or otherwise damaged area of tissue.

lesser omentum a portion of the peritoneum attached to the lesser curvature of the stomach and extending to the transverse fissure of the liver.

leukemia (lū·kē´mē·ă) a cancer of tissues responsible for producing blood that results in excessive numbers of white blood cells in the blood and bone marrow.

leukocyte (lū´kō·sīt) any of several forms of white blood cells.

leukocytosis (lū˝kō·sī·tō´sĭs) the appearance of an abnor-

mally high number of white blood cells (leukocytes) in the blood, as often occurs following or during a bacterial infection.
leukopenia (lū″kō·pē′nē·ă)   an abnormally low number of white cells in the blood.
levator   a general term for a muscle that raises a body part.
ligament   a band or sheet of tough, flexible, white, fibrous tissue that connects bones together in a joint and allows for movement of the bones about the joint; a thickened membrane that suspends and supports organs within the abdominal cavity.
limbic system   a group of structures in the midbrain that are associated with certain autonomic functions and various emotions including pleasure, fear, and happiness.
lingual   pertaining to the tongue.
lingual frenulum (frĕn′ū·lŭm)   the longitudinal fold of mucous membrane extending from the midline of the inferior surface of the tongue to the floor of the mouth.
lingual tonsil   lymphatic tissue in the mucous membrane near the base of the tongue.
linitis (lĭn·ī′tĭs)   inflammation of the stomach lining.
linitis plastica   thickening of the walls of the stomach due to the formation of new tissue.
lipase   a primarily pancreatic enzyme that catalyzes the degradation of fat in the small intestine. Gastric lipase is produced by children but not by adults.
lipid   organic compounds including fat, steroids, and phospholipids that are generally insoluble in water and soluble in chloroform and certain other organic solvents.
lipocyte   a fat storage cell in adipose tissue; an adipocyte.
lipofuscin (lĭp″ō·fŭs′sĭn)   a golden brown pigment that collects in granules in cells that tend to live a long time, such as neurons and cardiac muscle cells; believed to be the residue of intracellular digestion that accumulates over a long period of time.
lipogenesis (lĭp″ŏ·jĕn′ĕ·sĭs)   the metabolic reactions that produce fat.
lipolysis (lĭp·ŏl′ĭ·sĭs)   the metabolic reactions that hydrolyze fat into free fatty acids and glycerol.
lipoma   a benign tumor of fatty tissue.
lipoprotein   a protein molecule to which one or more lipid molecules are covalently bound.
liver   the largest organ of the body, located in the abdominal cavity where it is responsible for numerous functions: storage of glucose absorbed in the digestive tract, synthesis and degradation of amino acids, production of many serum proteins, removal of waste products from blood, production of blood-clotting agents, removal of bacteria and other foreign agents from blood, production and storage of vitamin B_{12} and certain other vitamins, and production of bile.
liver spots   a common term for brownish spots in the skin of the face and hands.
lobe   a structural subunit of many organs defined by fissures, membranes, furrows, or other boundary-forming structures.
lobule   a subunit of a lobe.
lordosis   an abnormal degree of anterior curvature of the spine.
low density lipoprotein   a complex of plasma proteins and lipids in which the bulk of dietary cholesterol and saturated fatty acids are carried. Elevated levels of low density lipoprotein are thought to increase the risk of coronary heart disease.
lumbago (lŭm·bā′gō)   a general term for pain in the lower part of the back.
lumbar   pertaining to the lower area of the back.
lumbar plexus   the network of nerves formed by the division of the anterior branches of spinal nerves LI, LII, LIII, and part of LIV.
lumbosacral   pertaining to the region around the junction of the lumbar and sacral portions of the spine.
lumen   the space within a hollow organ or tubular structure.
lunate   one of the bones of the wrist.
lung   one of the pair of spongy organs in the chest involved in respiration. Uptake of oxygen and release of carbon dioxide occurs primarily in the alveoli of the lungs.
lunula (lū′nū·lă)   the lightly colored arch at the base of a nail.
luteinizing hormone   an anterior pituitary hormone that stimulates the production of the corpus luteum in an ovary following ovulation and the secretion of testosterone by interstitial cells in the testes.
luteinizing hormone-releasing hormone   a hypothalamic hormone that controls the production and release of luteinizing hormone by the anterior pituitary gland.
lymph (lĭmf)   a transparent, slightly yellowish fluid that circulates in the lymphatic vessels and is filtered at lymph nodes. Lymph is similar to blood plasma except it has very few cells and its protein content is lower.
lymph node   one of many small, roundish bodies composed of lymphatic tissue and lying at various places along the lymphatic system where they serve as filters that remove bacteria cells and other debris from the lymph.
lymphatic duct   a vessel that transports lymph.
lymphocyte (lĭm′fō·sīt)   a type of white blood cell involved in maintaining immunity by producing antibodies or by attacking and removing foreign protein. Two types of lymphocytes are B cells and T cells.
lymphokines (lĭm′fō·kīnz)   chemical substances produced by activated T lymphocytes that affect and stimulate other cells in the immune system by attracting them to a site of infection, stimulating proliferation of certain lymphocytes and promoting antibody production, among other functions.
lysosome (lī′sō·sōm)   a cellular organelle that contains enzymes used to digest phagocytized substances; especially prominent in cells that are active in removing bacteria and other substances from the blood, such as certain leukocytes and Kupffer cells in the liver.
lysozyme (lī′sō·zīm)   an enzyme that is effective in causing the lysis of many bacteria and is found in saliva, tears, sweat, and monocytes and granulocytes.

M

macromolecules   molecules made up of an especially large number of subunits, such as glycogen, protein, and nucleic acid molecules.
macronutrients   those chemical elements that are needed in relatively large quantities by the body as compared to micronutrients. They include carbon, hydrogen, oxygen, nitrogen, potassium, sodium, calcium, chloride, magnesium, phosphorus, and sulfur.
macrophage   any of several types of cells in the body's defense mechanism that have the ability to remove cells and other particulate matter by phagocytosis.
macula densa   a densely nucleated spot on the distal convoluted tubule of a nephron that lies between the efferent and afferent arterioles. The macula densa is in contact with the vascular pole of the glomerulus and the juxtaglomerular cells of the afferent arteriole.
Madelung's deformity   a condition in which the hand deviates

from its normal position as a result of excessive growth of the distal ulna or undergrowth of the radius.

malabsorption syndrome impaired absorption of nutrients by the small intestine due to certain diseases, insufficient mixing of the contents of the small intestine, deficient pancreatic function, abnormal structures, or other causes.

malic acid a component of the Krebs cycle produced from fumaric acid and oxidized to oxaloacetic acid.

malignant worsening to the point of being life-threatening; with cancer, advanced and spreading.

malleus (măl′ē·ŭs) one of the three bones in the middle ear, lying between the eardrum and incus.

Mallory-Weiss syndrome laceration of the distal esophagus where it connects to the stomach due to excessive vomiting, often associated with alcoholism, and resulting in gastrointestinal bleeding.

maltase an enzyme produced by the small intestine that catalyzes the conversion of maltose to glucose.

mammary ducts ducts of the mammary glands through which milk flows.

mammary gland one of the two milk-producing glands on the chest of an adult female.

mammillary (măm′ĭ·lār·ē) shaped like a nipple.

mandible the lower jawbone.

mandibulofacial dysostosis (măn·dĭb″ū·lō·fā′shăl dĭs″ŏs·tō′sĭs) an inherited malformation of the skull and face that occurs during embryonic development; also called Treacher Collins syndrome.

manubrium (mă·nū′brē·ŭm) the upper bone of the sternum that articulates with the clavicle and the first two pairs of ribs.

manus (mā′nŭs) the hand.

marrow the soft tissue that occupies the cavities in the interior of the shafts of long bones where blood is produced.

mass number the total number of protons and neutrons in the nucleus of an atom.

masseter (măs·sē′tĕr) the thick muscle on each side of the face that closes the jaw.

mast cell certain cells in connective tissue that release histamine and heparin when tissue is injured to aid in combating infection and inducing blood clotting.

mastication chewing.

mastoid shaped like a nipple.

mastoiditis inflammation of the cells of the mastoid process.

mastoid process a nipple-shaped extension of the portion of the temporal bone that lies behind the external ear and below the line of the temple.

matrix a substance that serves as a binding agent for tissues or tissue secretions.

maxilla (măks′ĭl·ă) the upper jawbone.

meatus an opening or passageway through an organ.

mechanoreceptors receptors of the sensory nervous system that are sensitive to physical stimuli, such as pressure or touch.

medial pertaining to the middle of the body or a part of the body.

mediastinum (mē″dē·ăs·tī′nŭm) a space between the pleura surrounding the lungs that contains the heart and associated arteries and veins.

medulla the innermost part of an organ or structure.

medulla oblongata (ŏb″lŏng·gă′tă) the lowest portion of the brain stem lying below the pons and above the spinal cord just within the opening into the skull. It is involved in the regulation of respiration, circulation, and certain special senses.

megakaryocyte (mĕg″ă·kăr′ē·ō·sīt″) a large cell in bone marrow which is a source of blood platelets.

megaloblast (mĕg′ă·lō·blăst) an abnormally large, incompletely differentiated nonfunctional red blood cell, characteristically found in the blood of patients with pernicious anemia.

meibomian gland (mī·bō′mē·ăn) a sebaceous gland that lies in the eyelid and releases sebum through ducts that open at the outer end of each eyelid.

meiosis (mī·ō′sĭs) a cellular division process in animals that produces the cells that differentiate into eggs or sperm. The number of chromosomes in a cell is reduced by one-half during meiosis.

Meissner's corpuscle sensory receptors in the skin that are sensitive to touch.

melanin a brown pigment responsible for the color of hair, skin, and the iris of the eyes.

melanocyte a cell that produces melanin.

melatonin a hormone produced by the pineal gland, production of which is rhythmic, with a greater amount being produced during the night than during the day. It may be involved in the control of other hormones and in the production of skin color, although its functions in humans are not well understood.

membrane potential the electrical potential difference between the fluids on the two sides of a cell membrane that results from the unequal distribution of ions in those two regions.

membranous bone bone that develops within and from connective tissue, as opposed to cartilaginous bone.

menarche (mĕn·ăr′kē) the beginning of menstrual cycles in the life of a female, usually occurring about age 13.

Ménière's (mān″ē·ārz′) **disease** a disorder of the inner ear that causes dizziness, deafness, and ringing in the ear.

meninges (mĕn·ĭn′jēz) the three membranes that surround the spinal cord and brain, including the dura mater, arachnoid, and pia mater.

meningitis (mĕn·ĭn·jī′tĭs) inflammation of the meninges, the membranes that surround the brain and spinal cord.

meniscus (mĕn·ĭs′kŭs) the curved cartilaginous material found in the knee and other joints.

menopause the period in a woman's life when menses diminish or stop completely, usually completed by the age of 55.

menses the period of the menstrual cycle during which blood and uterine tissues flow from the uterus.

menstrual cycle the alternating period of buildup and degradation of the uterine wall as a result of the interaction of ovarian and pituitary hormones. Ovulation occurs about midway between the end of one menses and the beginning of the next.

merocrine gland a type of gland in which the cell that secretes the glandular product remains intact upon release of the secretory product. The salivary and pancreatic glands are examples of merocrine glands.

mesangium (mĕs·ăn′jē·ŭm) a thin membrane that supports the capillaries within a glomerulus.

mesencephalon (mĕs·ĕn·sĕf′ă·lŏn) the midbrain.

mesenchyme (mĕs′ĕn·kīm) a network of cells in the embryonic mesoderm that produces connective tissue, the lymphatic system, and the blood vessels of the circulatory system.

mesentery a portion of the peritoneum that surrounds a portion of the small intestine and connects to the back of the abdominal wall.

mesocolon (mĕs″ō·kō′lŏn) the mesentery that connects the colon to the posterior portion of the abdominal wall.

mesoderm the middle of the three primary germ layers of an embryo from which are produced connective tissue, bone and cartilage, muscle, blood and blood vessels, peritoneum, kidney, gonads, and other structures.

mesothelium (měs″ō·thē′lē·ŭm) a layer of cells from the mesoderm that lines body cavities in an embryo and becomes the epithelium that covers the surface of the serous membranes.

mesovarium (měs″ō·vā′rē·ŭm) a portion of the peritoneal fold that attaches an ovary to the broad ligament of the uterus.

messenger RNA (mRNA) a class of RNA molecules that carries the information specifying the order of amino acids to be incorporated into a protein as it is synthesized. Each amino acid is specified by a sequence of three nucleotides called a codon (*see ribonucleic acid*).

metabolic rate the amount of energy used by the body in a specific amount of time.

metabolism the total of all the chemical reactions that go on in the body.

metacarpus the palm of the hand.

metaphase the stage of meiotic or mitotic division when the chromosomes line up midway between the poles of the spindle.

metaphysis (mě·tǎf′ĭ·sĭs) a region in a long bone where the shaft meets the epiphysis and where growth occurs.

metastasis (mě·tǎs′tǎ·sĭs) the movement of bacteria or cancer cells from one point of infection or growth in the body to another point.

metastatic growth growth of a tumor in new regions resulting from release and distribution of individual tumor cells.

metatarsus the part of the foot between the tarsus (the ankle and its seven bones) and the toes.

metencephalon (mět″ěn·sěf′ǎ·lŏn) the portion of the brain consisting of the cerebellum and the pons.

micelle (mī′sěl′) tiny spherical particles formed by the aggregation of lipids in chyme.

microangiopathy (mī″krō·ǎn″jē·ŏp′ǎ·thē) a disease of small blood vessels, for example, the capillaries and arterioles, resulting from trauma, elevated blood pressure, or emboli.

microfilament a network of filaments in a cell's cytoplasm that gives the cell shape.

microglia (mī·krŏg′lē·ǎ) a category of nonconducting cells of the central nervous system that phagocytize and remove foreign matter.

micronutrient substances necessary for nutrition in small amounts, including vitamins, iodine, zinc, and several other electrolytes.

microphage a small white blood cell that can phagocytize bacteria and foreign matter in blood and other body fluids.

microtubule tubules in the cytoplasm of a cell involved in the transport of organelles from one part of the cell to another.

microvilli numerous fine, parallel processes that cover the surface of a cell membrane, providing increased surface area for cell absorption; together, the microvilli form a brush border.

micturition (mĭk·tū·rĭ′shun) *see urination*.

micturition reflex the reflex that occurs as a result of increased pressure within the urinary bladder that causes the bladder wall to contract and the urinary sphincter to relax and allow urine to pass into the urethra.

midbrain a portion of the brain stem that lies between the pons and the diencephalon involved in controlling posture, movement, vision, and auditory reflexes; also called the mesencephalon.

midgut the middle portion of the embryonic gut from which arise the ileum and jejunum.

migraine headache an extremely severe throbbing headache that may last for hours or days, of uncertain cause, but associated with widening of the blood vessels in the head.

mineral corticoids a class of hormones produced by the adrenal cortex that are involved in regulating electrolyte concentrations in body fluids.

minerals inorganic ions in the intracellular and extracellular fluids that serve numerous functions including, for example, as cofactors necessary for enzyme activity.

miosis (mī·ō′sĭs) abnormal narrowing of the pupil of the eye, resulting in pinpoint openings.

mitochondria (mĭt″ō·kŏn′drē·ǎ) membranous cytoplasmic organelles that contain the enzymes used in the Krebs cycle and the components of the electron transport chain; they are the primary source of ATP.

mitosis a process of cell division that distributes replicated chromosomes equally to each of two new cells produced by the division.

mitral valve the valve that lies between the left atrium and left ventricle of the heart and controls the flow of blood from the atrium to the ventricle; also called the bicuspid valve or the left atrioventricular valve.

modiolus (mō·dī′ō·lŭs) the central pillar of the cochlea in the inner ear.

molar a tooth used to grind food when chewing; in chemistry, a unit of concentration consisting of one mole of solute per liter of solution.

mole a unit of measure in chemistry consisting of 6.02×10^{23} atoms or molecules; the amount of a chemical compound whose mass in grams is the same as its molecular weight; an elevated pigmented spot on the skin.

molecule a unit of matter consisting of two or more atoms held together by covalent bonds.

monochromat a rare condition in which a person is completely colorblind.

monoclonal antibody an antibody produced *in vitro* by a population of genetically identical B-lymphocytes; hybridoma.

mononeuritis (mŏn″ō·nū·rī′tĭs) a disease that affects a single nerve.

monosaccharide a type of simple sugar such as glucose, fructose, or galactose, in which the molecules are not composed of sugar subunits.

monosomic an abnormal condition in which an individual's cells have only one representative of a particular chromosome instead of the usual two.

monozygotic pertaining to twins who have developed from a single fertilized egg, that is, identical twins.

mons pubis an elevated region in the pubic area of a female caused by a layer of fatty tissue that lies above the pubic symphysis.

Moro reflex the reflex of a baby to a sudden loud noise characterized by the infant pulling its arms in toward its chest in an embracing attitude.

morphogenesis a general term referring to the developmental processes that produce a body and its structures.

morula (mor′ū·lǎ) the stage in the development of an embryo following cleavage of the zygote when the embryo is a solid mass of cells.

motor area a region of the cerebral cortex from which nerve impulses are sent to muscles or glands.

motor nerve a nerve that contains only motor neurons.

motor neuron a nerve cell that carries signals away from the central nervous system to a muscle and gland.

motor unit a group of muscle cells controlled by a single motor neuron.

mucin (mū′sĭn) a glycoprotein present in mucus, saliva, connective tissue, and other tissues. In solution it creates a slippery material that has a protective or lubricating function.

mucosa a type of membrane containing mucous glands. The membrane typically consists of a thin sheet of cells that line or cover various structures, including the airways of the lungs, the alimentary canal, and the genital and urinary ducts.

multicellular consisting of many cells.

multiple sclerosis a disorder of the central nervous system characterized by patches of demyelinated nervous tissue. The disease is chronic and progresses slowly, resulting in a complex set of symptoms involving emotion, muscle control, and nervous control.

multipolar neuron a neuron that has multiple processes.

muralium (mū·rǎ′lē·ŭm) an interconnected network of chambers formed by spaces between cells in an organ, as in the adrenal medulla.

murmur an abnormal sound in the body, such as a heart murmur.

muscle a type of tissue that consists of cells capable of contracting; an organ that consists primarily of muscle tissue used to effect movement in the body.

muscle fibers the contractile elements of muscle tissue; muscle cells.

muscle spindle a group of muscle fibers enclosed in a capsule and contained within a skeletal muscle where it serves as a stretch receptor.

muscle stretch reflex a reflex contraction of a muscle stimulated by stretching of the muscle.

muscle tone a sustained state of partial contraction of skeletal muscle and smooth muscle.

muscle twitch a single rapid contraction and relaxation of a muscle in response to a single stimulus.

muscular dystrophy a progressive wasting away of the muscles due to a defect in metabolism.

muscularis the layer of smooth muscle in the walls of many internal organs.

musculotension headache a headache due to emotional stress and chronic overwork.

mutation a change in the structure of the genetic material.

myalgia (mī·ǎl′jē·ǎ) achiness and pain in specific muscles.

myasthenia gravis (mī·ǎs·thē′nē·ǎ grǎ′vĭs) a disease characterized by a wasting away of muscles due to disuse caused by impulse transmission across myoneural synapses.

myelencephalon (mī″ĕl·ĕn·sĕf′ǎ·lŏn) the portion of the brain that contains the medulla oblongata and the lower part of the fourth ventricle; the hindbrain in a developing embryo.

myelin (mī′ĕ·lĭn) a fatty material produced by Schwann cells and oligodendrocytes and incorporated in sheaths wrapped around the axons of neighboring neurons.

myelin sheath a sheath of myelin-containing membrane formed by Schwann cells and oligodendrocytes around the axons of neurons. The sheath increases the speed of an action potential along the axon and simultaneously insulates the axon and reduces leakage of electrolytes through the axonal membrane.

myelitis (mī·ĕ·lī′tĭs) inflammation of the spinal cord or of bone marrow.

myelogram an X-ray photograph of the spinal cord after injection of an opaque dye.

myenteric (mī″ĕn·tĕr′ĭk) pertaining to the myenteron, the layer of smooth muscle in the intestinal wall.

myocardial infarction development of regions of dead tissue in the cardiac muscle as a result of blockage of blood flow to those tissues.

myocardium the middle layer of tissue in the wall of the heart consisting of cardiac muscle.

myofibril a longitudinally oriented subunit of a skeletal or cardiac muscle fiber set off by the sarcoplasmic reticulum and containing numerous thick and thin filaments.

myofilaments the filaments of a muscle cell that are responsible for contraction, consisting of both the thin and thick filaments.

myoglobin an oxygen-binding substance in muscle tissue used for oxygen storage.

myoglobinuria (mī·ō·glō″bĭn·ū′rē·ǎ) the presence of myoglobin in the urine, usually caused by muscle injury or excessive exercise.

myoma (mī·ō′mǎ) a tumor of muscle tissue.

myometrium (mī·ō·mē′trē·ŭm) the layer of smooth muscle tissue in the wall of the uterus.

myoneural junction a junction between a motor neuron and a muscle cell that includes the synapse.

myopia (mī·ō′pē·ǎ) a defect in the shape of the lens that results in a focusing of the light in front of the retina; also called nearsightedness.

myosin (mī′ō·sĭn) the protein subunit that forms the thick filaments involved in the sliding filament mechanism of muscle contraction.

myositis (mī·ō·sī′tĭs) inflammation of a muscle, especially skeletal muscle.

myositis ossificans (ŏs·ĭf′ĭ·kǎns) myositis characterized by the development of bony material in the muscles.

myotome (mī′ō·tōm) a group of muscles innervated by a single spinal nerve segment.

myotonia atrophica an autosomal dominant genetic disorder characterized by slow relaxation of voluntary muscles following contraction and atrophy of the muscles. Numerous other abnormalities may also occur that may ultimately lead to mental retardation and death.

myotonia congenita an autosomal dominant genetic condition resulting in muscle stiffness and slow relaxation times.

myxedema (mĭks″ĕ·dē′mǎ) an abnormal condition resulting from deficient production of thyroid hormone in adults, characterized by swollen facial features that result from the accumulation of mucin in the underlying tissue.

N

nail the rounded, horny, translucent structure at the end of each finger or toe, consisting largely of keratin.

narcolepsy (nǎr′kō·lĕp″sē) an abnormal condition of unknown cause characterized by frequent and uncontrollable desire to sleep, loss of muscle tone, and sometimes associated with auditory or visual hallucinations.

narcotic an often abused drug or other substance that depresses the central nervous system, relieving pain and sometimes inducing sleep or stupor.

nares (nǎr′ēz) openings in the anterior and posterior ends of the nasal cavity through which air passes during breathing. The external nares are also called nostrils.

nasal cavity a chamber that lies between the mouth and the cranial cavity through which air passes as it is breathed through the nose.

nasal conchae projections from the lateral walls of the nasal fossae that extend into the cavities and provide a surface area of epithelium over which air must pass as it is breathed in and out through the nose.

nasal fossae the two halves of the nasal cavity formed by the nasal septum.

nasal septum a thin perpendicular wall that divides the nasal cavity into two nasal fossae, composed of the perpendicular plate of the ethmoid bone, the vomer, and cartilage.

nasolacrimal duct A channel that carries lacrimal secretions (tears) from the lacrimal sac into the nasal cavity.

nasopharynx (na″zō·făr′ĭnks) the uppermost portion of the pharynx, lying above the soft pallet.

navel *see umbilicus.*

navicular (nă·vĭk′ū·lăr) shaped like a boat; one of the bones of the wrist.

neck a constricted portion of an organ or other body structure; the part of the body that connects the head to the trunk.

necrosis (nĕ·krō′sĭs) death of a tissue in a localized region due to injury or disease.

negative feedback mechanism a physiological regulatory process in which a product acts to remove the stimulus for its production.

neonatal the four weeks of life following birth.

neoplasm (nē′ō·plăzm) a newly formed mass of abnormal tissue, such as a tumor.

nephritis (nĕf·rī′tĭs) inflammation of the kidney or kidney tissues.

nephron (nĕf′rŏn) the functional unit of a kidney, consisting of a Bowman's capsule with an enclosed glomerulus, proximal convoluted tubule, loop of Henle, and distal convoluted tubule.

nephropathy (nĕ·frŏp′ă·thē) a general term for a disorder involving inflammatory or degenerative kidney disease.

nerve a cordlike collection of nerve cells and associated connective and circulatory tissues involved in the transmission of nerve impulses. Motor nerves carry impulses away from the central nervous system, sensory nerves carry signals toward the central nervous system, and mixed nerves carry impulses in both directions.

nerve impulse an action potential that is usually initiated in an axon hillock and propagated the length of the axon.

neural tube defect an abnormality resulting from incomplete fusion of neural folds during embryonic development of the brain and spinal cord resulting, for example, in the absence of cerebral hemispheres (anencephaly) or a protrusion of the spinal cord through an opening in the spinal column (spina bifida).

neuralgia (nū·răl′jē·ă) pain associated with one or more nerves, distinguished by the cause of the pain or the part that is affected.

neurilemma (nū′rĭ·lĕm″mă) a thin membrane that covers the myelin sheaths of myelinated neurons in the peripheral nervous system.

neuritis (nū·rī′tĭs) inflammation of a nerve.

neuroblast (nū′rō·blăst) an embryonic cell that develops into a neuron.

neurofibromatosis (nū″rō·fī·brō″mă·tō′sĭs) an autosomal dominant disorder characterized by the presence of tumors on the skin and peripheral nerves.

neurofilament numerous intermediate filaments that lie in the cell body of an axon of a neuron where they provide structural support.

neuroglandular junction the synaptic region between a neuron and glandular tissue.

neuroglia (nū·rŏg′lē·ă) nonconductive cells of nervous tissue, including the Schwann cells of the peripheral nervous system and the astrocytes, ependyma, oligodendrocytes, and microglia of the central nervous system.

neurohypophysis (nū″rō·hī·pŏf′ĭs·ĭs) the posterior lobe of the pituitary gland and the source of antidiuretic hormone (ADH) and oxytocin.

neuromuscular junction the junction between neurons and muscle cells where synapses are formed.

neuron the impulse-transmitting cell of nervous tissue, consisting of a cell body, dendrites, and an axon.

neurophysins hypothalamic peptides to which antidiuretic hormone and oxytocin bind as they are transported in axons that extend from the hypothalamus to the posterior pituitary gland.

neurophysiology the branch of physiology that deals with the function of the nervous system.

neurotransmitter one of several chemical substances released by a presynaptic neuron into a synapse and which then stimulate or inhibit the production of an action potential in the postsynaptic cell.

neutron an uncharged particle in an atomic nucleus. Isotopes of an element differ in the number of neutrons in their respective atoms.

neutrophil (nū′trō·fĭl) a leukocyte that stains readily with neutral dyes as opposed to acidic or basic dyes; they destroy and remove bacteria, dead cells, and cell debris from the blood.

nexus (nĕk′sŭs) a type of junction between the membranes of adjacent cells that allows for communication between the cells; also called a gap junction.

niacin (nī′ă·sĭn) a water soluble vitamin of the B group that plays an important role in oxidation-reduction reactions of cell metabolism as a component of NAD and NADP. Niacin deficiency causes pellagra; also called nicotinic acid.

nicotinamide adenine dinucleotide (NAD) an organic compound that receives electrons in most of the oxidation-reduction reactions of cellular respiration. In the presence of oxygen the electrons are passed on to the respiratory chain to produce water. Energy released as electrons pass through the chain is used to form ATP. In the absence of oxygen, reduced NAD provides electrons for the conversion of pyruvic acid to lactic acid.

nicotinic acid *see niacin.*

nipple a cylindrical, rounded protuberance on each breast through which milk is secreted in a lactating female.

Nissl (nĭs′l) **bodies** brightly staining granules in the cytoplasm of nerve cell bodies formed by aggregates of ribosomes.

nitrogen narcosis an intoxicated state caused by elevated concentrations of nitrogen gas in the blood that can occur in divers breathing compressed air at great depths.

nociceptor (nō sĭ·sĕp′tor) a free sensory nerve ending sensitive to tissue damage and responsible for painful sensation.

node a small, rounded mass of tissue, such as a lymph node.

node of Ranvier a gap in the myelin sheath of a myelinated nerve fiber that allows for saltatory conduction of a nerve impulse.

nodular silicosis (sĭl·ĭ·kō′sĭs) a respiratory lung disease that occurs among workers and others exposed to silica-based dust, which leads to fibrous nodules in the lung tissues.

nonarticular rheumatism a group of disorders characterized by pain and tenderness in muscles, tendons, and surrounding tissues, but not involving the joints.

nondisjunction failure of paired chromosomes or of chromatids to separate from one another during the anaphase stage of cell division, with both structures being pulled to the same end of the spindle fiber.

noninsulin-dependent diabetes mellitus (NIDDM) a type of diabetes mellitus in which patients do not need exogenously supplied insulin to survive.

noradrenalin see *norepinephrine*.

norepinephrine (nor·ĕp″ĭ·nĕf′rĭn) a hormone produced by the adrenal medulla that increases vasoconstriction and blood pressure without affecting cardiac output. It serves as a transmitter substance in peripheral sympathetic nerves and certain synapses of the central nervous system; also called noradrenalin.

nose the organ of olfaction and the entrance for most of the air breathed into or out of the respiratory tract. Consists of the external nose, which projects from the face, and the internal nose, or nasal cavity. Air is warmed, filtered, and moisturized as it passes through the nose.

nostril one of the two openings in the external nose leading into a nasal fossa; also called an external naris.

nuclear magnetic resonance imaging (NMRI) a technique that uses the effect hydrogen nuclei have on certain radio waves when the nuclei are maintained in a magnetic field. The body or body part is placed in an intense magnetic field, and radio waves are transmitted through it; perturbations of the radio waves are then detected and analyzed by a computer to produce a visual image.

nucleic acid a macromolecule formed by the polymerization of nucleotides, each of which contains a five-carbon sugar, a phosphate group, and a nitrogen-containing organic base (see *deoxyribonucleic acid* and *ribonucleic acid*).

nucleolus (nū″klē′ō·lŭs) a darkly staining, nonmembrane-bound structure within a cell nucleus in which ribosomal subunits are formed.

nucleoplasm the contents of a cell nucleus contained within the nuclear membrane.

nucleosome the fundamental structural unit of chromatin, consisting of a core of eight histone molecules around which is wrapped a length of DNA of about 200 base pairs held in place by a ninth histone molecule.

nucleus a cell organelle that contains most of the cell's genetic material; a group of neuronal cell bodies in the central nervous system with processes that extend into neighboring nervous tissue; the central portion of an atom that contains protons and neutrons and around which electrons orbit.

nucleus pulposus (pŭl·pō′sŭs) the inner pliable portion of an intervertebral disk surrounded by a tougher fibrous outer portion called the annulus fibrosis.

nutriment any food or other material that supplies nutrition to the body.

nystagmus (nĭs·tăg′mŭs) involuntary, constant, rapid, and random movement of an eyeball.

O

obesity an excessive accumulation of body fat, that may be harmful to one's health; a body weight 20% over the normal weight indicated in standard height-to-weight tables.

occipital (ŏk·sĭp′ĭ·tăl) pertaining to the back of the head.

occlusion a closing or blocking of an opening to a duct or other passageway; the fitting together of the teeth of the upper and lower jaws when the jaws are closed.

oculomotor pertaining to the muscles that move the eyes.

olecranon (ō·lĕk′răn·ŏn) a process at the superior end of the ulna that projects backward and forms the bony prominence of the elbow.

olfactory (ŏl·făk′tō·rē) pertaining to the sense of smell.

olfactory bulb a bulbous enlargement at the anterior end of each olfactory tract that receives signals initiated by olfactory receptor cells and transmitted to the bulbs by olfactory nerves.

olfactory placode a platelike thickening of the embryonic ectoderm from which the nasal cavity develops.

olfactory tract the tract of sensory nerve fibers that carries signals received by the olfactory bulbs to the olfactory regions of the cerebrum.

oligodendrocyte (ŏl″ĭ·gō·dĕn′drō·sīt) a type of central nervous system neuroglia that forms the latticework that supports and suspends neurons by forming myelin sheaths around the axons of many neurons.

omentum (ō·mĕn′tum) a portion of the peritoneum that hangs down over the small intestine (the greater omentum) and lies between the lesser curvature of the stomach and the liver (the lesser omentum).

oncology (ŏng·kŏl′ō·jē) a branch of medicine that deals with the study of tumors.

oncotic pressure see *colloid osmotic pressure*.

oocyte cells in the ovary from which eggs may be produced (see *primary oocyte* and *secondary oocyte*).

oogenesis the cellular and chemical events associated with the production and release of an ovum from the ovary.

oogonium an ovarian cell which differentiates into a primary oocyte.

ophthalmic (ŏf′thăl′mĭk) pertaining to the eye.

ophthalmologist a physician who specializes in disorders of the eye.

ophthalmoscope an instrument used to examine the interior of the eye.

opiate chemical substances that ease pain and induce sleep.

opiod a narcotic drug that has properties similar to opium in numbing pain and inducing sleep.

opsin (ŏp′sĭn) the protein in the rods and cones of the retina that combines with the visual pigments involved in light absorption and vision.

opsonin (ŏp·sō′nĭn) serum peptides that combine with antibodies on the surface of microorganisms and other foreign agents and thereby increase the efficiency of phagocytosis by certain leukocytes.

optic pertaining to the eye.

optic chiasma the region in the brain where optic nerve fibers from each of the eyes cross over one another.

optic disk the region of the retina that marks the point of attachment of the optic nerve; also called the blind spot.

optic neuritis an inflammation of the portion of the optic nerve visible within the eyeball or immediately behind the eyeball.

optic tract bundles of optic nerve fibers posterior to the optic chiasma that carry impulses from the eyes to the thalamus and other portions of the brain.

ora serrata the irregularly serrated anterior margin of the retina.

oral pertaining to the mouth.

oral contraceptive steroid drugs taken orally that mimic the effect of naturally occurring estrogens and progesterone in inhibiting the synthesis of gonadotropin-releasing hormone by the hypothalamus. In the absence of these hormones, follicle-stimulating hormone and luteinizing hormone are not released by the pituitary gland and follicles do not develop.

orbit the bony cavity in which the eyeball rests; one of the volumes within which an electron travels about an atomic nucleus.

organ a structure that contains two or more different cell types that are responsible for a specific function in the body.

organ of Corti the functional unit of hearing in the cochlea of the inner ear which contains hair cells that convert the mechanical energy of sound into electrical energy that in turn stimulates nerve endings in the organ.

organelles structural subunits of a cell that are responsible for specific biochemical functions, for example, mitochondria, nucleus, and ribosomes.

organic compounds chemical compounds that contain carbon atoms.

orgasm (or′găzm) sexual climax, involving strong muscular contractions accompanied by intense pleasurable sensations.

orifice (or′ĭ·fĭs) an opening into or out of a hollow structure or organ.

origin the end of a skeletal muscle attached to the bone that remains stationary when the muscle contracts.

oropharynx (or″ō·făr′ĭnks) the portion of the pharynx that lies at the posterior end of the oral cavity.

os the mouth or opening into a hollow structure; bone.

osmoreceptor a structure containing sensory cells that are stimulated by changes in the osmotic pressure of the surrounding fluid.

osmosis the diffusion of water or other solvent through a membrane that is selectively permeable to solutes in the solvent.

osmotic pressure the hydrostatic pressure that develops in a closed container in which there is a solution that is hypertonic to the surrounding fluid. Pressure develops as a result of a net flow of solvent into the container. Equilibrium is reached when osmotic pressure is great enough to cause the flow of solvent into the container to match the flow of solvent out of the container.

osseous (ŏs′ē·ŭs) **tissue** bony tissue.

ossicle a very small bone, especially the auditory ossicles of the middle ear, including the malleus, incus, and stapes.

ossification the growth of bone or bony material, including abnormal growth in soft tissues.

osteitis (ŏs·tē·ī′tĭs) **deformans** a bone disorder of unknown cause that occurs in adults, characterized by replacement of localized areas of metabolically active bone with fibrous tissue that leaves the area soft and weakened; also called Paget's disease of bone.

osteoarthritis (ŏ″tē·ō·ăr·thrī′tĭs) a joint disease characterized by the breakdown of cartilage, growth of bony spikes, swelling of the surrounding membrane, stiffness, and tenderness of the joint; also called degenerative arthritis.

osteoblast (ŏs′tē·ō·blăst) a cell that develops from a fibroblast and is involved with the production of bone.

osteochondroma (ŏs″tē·ō·kŏn·drō′mă) a benign bone tumor usually occurring within the marrow of a long bone; a congenital condition characterized by abnormal fragility of the bones such that an infant may be born with multiple fractures, soft skull, and physical deformities.

osteoclasis (ŏs′tē·ō·klăs″ĭs) the destruction and absorption of bony tissue by osteoclasts that can occur during the healing of broken bones; the destruction of bony tissue that occurs during certain bone diseases.

osteoclast large cells within bone that remove bony tissue surrounding the medullary cavity, enabling the cavity to enlarge as the bone enlarges.

osteocyte a bone cell surrounded by a matrix of hard bony material which communicates with other osteocytes through narrow cytoplasmic processes that extend through the canals of the bony matrix.

osteogenesis the processes involved in the production and growth of bony tissue.

osteogenic sarcoma a metastasizing, malignant bone tumor.

osteomalacia (ŏs″tē·ō·măl·ā′shē·ă) a disorder of the bones marked by increasing softness and deformation. It is one result of vitamin D deficiency in adults.

osteomyelitis (ŏs″tē·ō·mī″el·ī′tĭs) inflammation of a bone caused by infection by bacteria or other microorganisms.

osteon (ŏs′tē·ŏn) the structural unit of bone consisting of a haversian canal and several layers of hard bony material called lamellae.

osteoporosis (ŏs″tē·ō·por·ō′sĭs) increased porosity of bone due to resorption of bone tissue mass associated with aging, hormonal disorders, or inadequate calcium absorption in the intestine.

ostium (ŏs′tē·ŭm) a small opening into a tubular or otherwise hollow organ.

otic pertaining to the ear.

otic placode a platelike thickening of the embryological epithelium that gives rise to the inner ear.

otitis media inflammation of the middle ear.

otoconium (ō″tō·kō′nē·ŭm) a tiny calcium-containing granule in a gelatinous substance that forms the otolithic membrane in the vestibule of the inner ear. Movement of the otolithic membrane provides sensory information on the position and movement of the head in space; also called otolith.

otolith see *otoconium*.

otosclerosis (ō″tō·sklē·rō′sĭs) a progressive condition in which new bone develops within the otic chambers and leads to progressive loss of hearing.

ototoxicity the property of many drugs to impair the function of the inner ear.

oval window an oval-shaped aperture in the wall between the middle and inner ears into which fits the base of the stapes.

ovarian cycle the approximately monthly cycle of endometrial buildup, follicle development, ovulation, endometrial disintegration, and menstrual flow.

ovarian follicle a spherical structure in the ovary in which an oogonium develops into an oocyte. Migration of the mature follicle to the surface of the ovary and rupture releases the oocyte at ovulation.

ovarian ligament a fibrous band that connects the ovary to the uterus.

ovary one of two organs of the female reproductive tract in which ova form and which is responsible for the production of hormones that cause secondary sexual characteristics to develop.

overweight a condition not considered to be harmful to one's health in which one's body weight exceeds the weight indicated in standard height-to-weight tables by less than about 20% (see *obesity*).

oviduct one of a pair of tubes in the female reproductive system through which an egg passes from the ovary to the uterus; also called a fallopian tube.

ovulation a process in which an ovarian follicle develops and, upon rupturing, releases an ovum enclosed within a surrounding mass of cells.

ovum the female reproductive cell, capable of being fertilized. In humans the ovum is a secondary oocyte that is arrested at metaphase of the second meiotic division. Meiosis II is completed only upon fertilization.

oxidation the removal of electrons from an atom or molecule.

oxidation-reduction reaction a class of chemical reactions in which electrons are transferred from one atom or molecule to another atom or molecule. The component of the reaction that loses the electron is said to be oxidized, and the component that receives the electron is said to be reduced.

oxidative phosphorylation the production of ATP from ADP and inorganic phosphate using energy released as electrons travel through the respiratory chain in cell respiration.

oxidizing agent any atom or molecule that receives one or more electrons in an oxidation-reduction reaction.

oxygen debt the phenomenon of continued above-normal oxygen utilization by tissues after a period of strenuous exercise is over.

oxyhemoglobin (ŏk″sē·hē′mō·glō′bĭn) a hemoglobin molecule that has bound one or more oxygen molecules.

oxyphil cells in general, cells that stain readily with acid-based dyes; specifically, cells of the parathyroid glands of unknown function, although they may be derived from chief cells, which produce parathyroid hormone.

oxytocin (ŏk″sē·tō′sĭn) a hormone produced in the hypothalamus and stored in and released from the posterior pituitary gland. The hormone stimulates contractions of the uterus during labor and contractions of mammary gland alveoli during lactation.

P

pacemaker the sinoatrial node; a specialized mass of nervous tissue at the junction of the superior vena cava and the right atrium that spontaneously initiates impulses that spread to other parts of the heart and cause coordinated contraction of the atria and ventricles; an artificial electrical device implanted under the skin that stimulates and controls heart rate by sending electrical stimuli to the heart.

pacinian (pă·sĭn′ē·an) **corpuscle** a pressure-sensitive, encapsulated sensory nerve ending located in the skin, in certain connective tissues of the hands and soles of the feet, and in joints.

palate (păl′ăt) the roof of the mouth, consisting of an anterior hard palate and a posterior soft palate.

palatine (păl′ă·tīn) relating to or associated with the palate.

palatine tonsil one of a pair of masses of lymphatic tissue that lie in the fauces immediately posterior to the palatoglossal arch.

palatoglossal (păl″ă·tō·glŏs′ăl) **arch** a vault-shaped fold of tissue that extends from either side of the soft palate to the tongue, from which is suspended the uvula and which forms the isthmus of the fauces.

palatopharyngeal (păl″ă·tō·fă·rĭn′jē·ăl) **arch** a vaulted fold of tissue that lies posterior to the fauces and the palatine tonsils in the mouth.

palmar (păl′măr) relating to the palm of the hand, the region between the wrist and the fingers.

palmar reflex a reflex that causes the fingers to curl inward when the palm is lightly stroked or tickled.

palpable (păl′pă·bl) capable of being felt with the hands or fingers through an overlying tissue.

palpation an examination technique in which the size and shape of a structure is determined by feel.

palpebra (păl′pă·bră) an eyelid.

palpebral aperture the space between the upper and lower eyelids.

pancreas (păn′krē·ăs) a mixed gland that lies below the stomach and toward the back of the abdominal cavity. It secretes bicarbonate and digestive enzymes to the small intestine where they are used in digestion. It also secretes, from islets of Langerhans, glucagon and insulin used to control carbohydrate metabolism.

pancreatic duct the duct that leads from the pancreas to the small intestine through which digestive enzymes and bicarbonate ions are secreted.

pancreatic enzymes digestive enzymes that are secreted into the small intestine from the pancreas.

pancreatitis (păn″krē·ă·tī′tĭs) inflammation of the pancreas.

pancreozymin (păn″krē·ō·zī′mĕn) see *cholecystokinin*.

pantothenic (păn·tō·thĕn′ĭk) **acid** a member of the vitamin B complex that is widely distributed in food. It is a component of coenzyme A, which carries acetyl groups into the reactions of the Krebs cycle, as well as serving in many other metabolic reactions.

papilla (pă·pĭl′ă) a small cone or nipple-shaped projection.

papillary muscle muscles in the ventricles of the heart attached to the chordae tendineae which are involved in opening and closing heart valves.

papilledema (păp″ĭl·ĕ·dē′mă) a swelling of the optic papilla due to brain tumor or abscess, trauma to the brain, hemorrhage, or other causes. Eventually will lead to loss of vision if the pressure is not relieved.

papovavirus (păp″ō·vă·vī′rŭs) a group of viruses that cause benign tumors (warts) in the skin.

pap smear a test for cervical cancer involving staining and microscopic examination of cells; Papanicolaou test.

papular (păp′ū·lăr) **lesion** a red, solid, circular, elevated area of the skin symptomatic of several diseases, such as measles and smallpox.

parafollicular (păr″ă·fŏ·lĭk′ū·lăr) **cells** cells of the thyroid gland that lie beside the thyroid follicles and produce calcitonin, a hormone that reduces blood calcium levels; also called C cells.

paralysis agitans see *Parkinson disease*.

paranasal sinuses mucous membrane-lined chambers in the frontal, maxillary, ethmoid, and sphenoid bones that drain into the nasal cavity.

paraplegia (păr·ă·plē′jē·ă) paralysis of the legs and lower portion of the trunk.

parasympathetic nervous system the division of the autonomic nervous system responsible for such actions as increasing glandular secretion (not including the sweat glands), slowing heart rate, constriction of smooth muscle of the digestive tract and bronchioles, and constriction of the pupils.

parasympatholytic (păr″ă·sĭm″pă·thō·lĭt′ĭk) **drugs** a substance that specifically inhibits sympathetic nerve fibers.

parasympathomimetic (păr″ă·sĭm″pă·thō·mĭm·ĕt′ĭk) **drugs** drugs that produce effects similar to the effects of parasympathetic nerve impulses.

parathormone (păr″ă·thor′mōn) see *parathyroid hormone*.

parathyroid (păr·ă·thī′royd) **gland** one of four or more approximately pea-sized glands embedded in the posterior surface of the thyroid gland which secrete parathyroid hormone.

parathyroid hormone a polypeptide hormone secreted by the parathyroid gland that causes calcium to be released from the bones and causes a rise in calcium levels in the blood. The hormone also causes calcium to be reabsorbed from urine in the kidneys and increases kidney secretion of phosphate ions, as well as stimulating the production in the kidney of 1,25-dihydroxycholecalciferol, an active form of vitamin D.

paraurethral (păr″ă·u·rē′thrăl) **glands** glands that lie on either side of the urethral opening in the vestibule of the female genitalia which secrete fluid that moistens and lubricates the vestibule.

paraventricular nuclei (păr″ă·vĕn·trĭk′u·lar) groups of neuron cell bodies that lie in the hypothalamus. Axons of these nuclei form the hypothalamus-hypophyseal tract that extends to the posterior pituitary gland. Oxytocin is produced in the nuclei and transported through the tract to the posterior pituitary gland for storage and secretion.

parenchyma (păr·ĕn′kī·mă) a general term used to indicate the functional part of an organ as opposed to the structural framework (stroma) of the organ.

paresthesia (păr″ĕs·thē′zē·ă) an abnormal sensation without apparent cause, such as numbness, tingling, or burning feeling. Sometimes due to central and peripheral nervous system lesions.

parietal (pă·rī′ĕ·tăl) pertaining to the wall of a cavity.

parietal bone one of the pair of bones that form the side of the skull.

parietal lobe a lobe that lies on the lateral surface of a cerebral hemisphere posterior to the central sulcus.

parietal pleurae the peripheral part of the pleural membranes in the thoracic cavity that covers the ribs, the diaphragm, and the pericardium.

Parkinson disease a progressive disorder of the central nervous system characterized by a tremor of resting muscles (especially of the hand), muscular rigidity, and slowness of movement. It appears to be due to loss of cells in the midbrain that communicate with the caudate nucleus and putamen and the resulting depletion of dopamine, a substance that serves as a neurotransmitter in those regions; also called shaking palsy or paralysis agitans.

parotid glands (pă·rŏt′ĭd) the largest of the salivary glands, lying under the skin anterior and inferior to the external ear in both sides of the head.

pars nervosa (părz nĕr·vō′să) an alternative name for the posterior lobe of the pituitary gland.

parturition (păr·tu·rĭsh′ŭn) the process of childbirth.

passive immunity a temporary immunity conferred by antibodies that enter the fetus through the placenta or are ingested by an infant in colostrum as it nurses; it can also be conferred by direct injection of a solution that contains antibodies against a particular disease.

patch test a skin test used to identify substances that cause allergic reactions. The substance is placed on a patch that is held against the patient's skin for one or two days; the skin is then examined for a sensitivity reaction.

patella (pă·tĕl′ă) a more or less flat bone about 5 cm in diameter that lies in the front of the knee joint.

pathogen any substance or microorganism that has a potential for causing disease.

pathologic fracture a bone fracture that results from a relatively minor injury because of an abnormal bone condition due to an illness such as a tumor or osteoporosis.

pathologic reflex an abnormal reflex that results from illness or other abnormal condition.

pathologist a person who is a specialist in diagnosing diseases on the basis of changes in the structure and function of tissues and organs.

pectoral (pĕk′tō·răl) pertaining to the chest.

pedicel (pĕd′ĭ·sĕl) the footlike process of a podocyte in a Bowman's capsule that intermeshes with pedicels of neighboring podocytes to produce filtration silts.

pedicle (pĕd′ĭ·kl) the root of the vertebral arch of a vertebra, to which the lamina connect.

peduncle (pē·dŭn′kl) one of several bands of nerve fibers that connects various portions of the brain to one another.

pelvis the bony structure at the base of the trunk. Each half consists of an ilium, ischium, and pubis bones.

pelvic girdle a general term for the hip bones, not including the associated vertebrae.

penis the male copulatory and urinary excretion organ.

pennate (pĕn′āt) shaped like or looking like a feather; penniform.

pepsin (pĕp′sĭn) a gastric enzyme that catalyzes the cleavage of proteins.

pepsinogen (pĕp·sĭn′ō·jĕn) an inactive precursor to pepsin that is secreted into the stomach by chief cells of the gastric glands. It is converted to its active form, pepsin, by the acidic conditions of the stomach content.

peptic ulcer a lesion in the wall of the esophagus, stomach, or duodenum caused by digestive fluids.

peptide bond the covalent bond that joins the amino group of one amino acid to the carboxyl group of a second amino acid.

perception the process of receiving stimuli by the sensory elements of the nervous system.

periarteritis (pĕr″ē·ăr·tĕr·ĭ′tĭs) **nodosa** a disorder characterized by widespread inflammation of medium- and small-sized arteries. Their swelling blocks the flow of blood and causes necrosis in the organs and tissues supplied by the affected arteries.

pericardial space the space between the inner and outer membrane of the pericardium. It contains a fluid that lubricates the two surfaces as they slide over one another as the heart beats.

pericarditis (pĕr·ĭ·kăr·dī′tĭs) inflammation of the pericardium.

pericardium (pĕr·ĭ·kăr′dē·ŭm) a double-layered membranous sac that surrounds the heart. The innermost layer is composed of serous tissue and lies closely appressed to the heart, while the outermost layer is composed of fibrous tissue.

perichondrium (pĕr·ĭ·kŏn′drē·ŭm) a layer of fibrous connective tissue that surrounds cartilage.

perikaryon (pĕr″ĭ·kăr′ē·ŏn) the cytoplasmic portion of the cell body of a neuron.

perilymph (pĕr′ĭ·lĭmf) a clear fluid contained in the space between the osseous labyrinth and the membranous labyrinth of the inner ear.

perilymphatic space the space in which perilymph is contained in the inner ear, lying between the osseous and membranous labyrinths.

perimetrium (pĕr·ĭ·mē′trē·ŭm) a serous membrane that surrounds the uterus.

perimysium (pĕr″ĭ·mīs′e·ŭm) the connective tissue within a skeletal muscle that divides the muscle into fasciculi.

perineum (pĕr″ĭ·nē′ŭm) a roughly diamond-shaped area at the inferior surface of the body trunk bounded posteriorly by the coccyx, anteriorly by the pubic symphysis, and the ischial tuberosities to either side. The anus, and in a female, the vulva, are included. (According to some authorities, the perineum is the area between the anus and the vulva in a female and between the anus and scrotum in a male.)

perineurium (pĕr″ĭ·nū′rē·ŭm) the connective tissue layers in a nerve that divide the nerve into fascicles.

periodontal ligament a layer of connective tissue that lines the alveolar process in which the root of a tooth is imbedded and which helps hold the tooth in its socket.

periodontium (pĕr·ē·ō·dŏn′shē·ŭm) the tissues that surround and support the teeth.

periosteum (pĕr·ē·ŏs′tē·ŭm) a layer of fibrous connective tissue that covers bones at all but their articular surfaces. It serves as a support for blood vessels and nerves and for the attachment of tendons and ligaments.

periostitis (per″ē·ŏs·tī′tĭs) inflammation of the periosteum.

peripheral nervous system that portion of the nervous system lying outside the brain and spinal cord.

peripheral neuropathy any disorder that occurs in nerves outside the central nervous system.

peristalsis (pĕr·ĭ·stăl′sĭs) a mechanism of contraction of a tubular structure that moves contents in the lumen from one end to the other. It is characterized by a coordinated constriction of circular and longitudinal muscle layers that advances along the length of the tube.

peritoneum (pĕr″ĭ·tō·nē′ŭm) a serous membrane that lines the abdomen and surrounds many of the abdominal organs.

peritonitis (pĕr·ĭ·tō·nī′tĭs) inflammation of the peritoneum.

peritubular plexus a network of capillaries that branch from efferent arterioles of cortical nephrons.

perivitelline membrane (pĕr″ĭ·vī·tĕl′ĕn) A matrix that surrounds the egg membrane and separates from the egg in response to penetration by a sperm, thereby preventing penetration by additional sperm.

perivitelline space the space that lies between the perivitelline membrane and the zona pellucida in an ovum.

permeability the ability of a substance to pass through a cell membrane.

pernicious anemia a deficiency of red blood cells due to vitamin B_{12} deficiency, resulting from inadequate secretion of intrinsic factor by the gastric mucosa.

peroneal pertaining to the fibula or to the outer part of the leg.

peroxisome (pĕ·rŏks′ĭ·sōm) a membrane-bound vesicle in a cell cytoplasm that contains the enzymes peroxidase, catalase, and oxidase. Peroxisomes appear to be involved in the formation of glucose from noncarbohydrate molecules and in degrading hydrogen peroxide into molecular oxygen and water.

perspiration the fluid produced by the sweat glands that is used to cool the body and rid it of certain waste products.

pes pertaining to the foot.

pes planus a foot deformity characterized by lowering of the arch of the sole of the foot; also called flatfoot.

petit mal seizure a form of epilepsy characterized by a brief period of unconsciousness, and perhaps mild muscular fluttering, and an absence of the strong convulsions that characterize certain other forms of epilepsy; also called absence seizure.

petrous a descriptive term meaning hard, or like a rock.

Peyer's patches small nodules or groups of nodules of lymphatic tissue found in the mucosa and submucosa of the small intestine, particularly toward the distal end of the ileum.

pH a unit used to indicate the hydrogen ion concentration of a solution. The pH of a solution is the negative value of the logarithm of the hydrogen ion concentration of the solution expressed in moles per liter.

pH buffer a solution that contains chemicals that resist a change in the hydrogen ion concentration of a solution.

phagocyte (făg′ō·sīt) any of several cells that have the capability of ingesting microorganisms or particulate matter by phagocytosis.

phagocytosis (făg″ō·sī·tō′sĭs) a cellular process by which certain cells ingest microorganisms, cell debris, or other particulate matter by engulfing it in cytoplasm.

phagolysosome (făg″ō·lī′sō·sōm) lysosomes within a phagocytic cell that contain enzymes used to digest materials that have been taken into the cell by phagocytosis.

phalanges (fă·lăn′jēz) any of the bones present in the fingers and toes.

pharynx (făr′ĭnks) the part of the digestive tract that lies between the mouth and the esophagus. It consists of three parts: the nasopharynx, oropharynx, and laryngopharynx.

phenothiazines (fē″nō·thī′ă·zēn) a class of chemical substances used as antipsychotic drugs in the treatment of schizophrenia, manic-depressive psychosis, and certain other psychotic conditions.

phenotype the genetic traits of an individual that can be measured or observed, as opposed to the genotype, which consists of the genes that underlie phenotypic expression.

phenylephrine hydrochloride a decongestant used to relieve hay fever or other allergies.

pheochromocytoma (fē·ō·krō″mō·sī·tō′mă) a usually benign tumor of the adrenal medulla that causes the chromaffin cells to produce excessive amounts of epinephrine and norepinephrine.

photophobia (fō tō·fō′bē·ă) an abnormal condition characterized by an intolerance of light. Occurs in albinos, in such diseases as measles and rubella, and when the eyes are inflamed.

photoreceptors sensory cells and structures that are stimulated by light, including the rods and cones in the retina of an eye.

phrenic (frĕn′ĭk) pertaining to the diaphragm.

phrenic nerve a nerve that emanates from the cervical plexus and innervates the pleura, pericardium, diaphragm, peritoneum, and sympathetic plexuses.

physiology the study of the processes and functions performed by the body and its parts.

pia mater the innermost of the meninges covering the brain and spinal cord.

pickwickian syndrome (pĭk·wĭk′ē·ăn) a condition associated with obesity, characterized by frequent hypoventilation and the resulting effects of increased carbon dioxide in the blood and increased red blood cell counts.

pigeon toe a deformity of the foot that causes the feet to turn inward; also called pes varus.

pigmentation the presence of colored matter in a tissue, hair, or hairlike structure.

pillar (pĭl′ĕr) an upright column or support of tissue.

pillars of Corti structures in the organ of Corti of the ear that enclose the inner tunnel and support the tectorial membrane.

pillars of the fauces folds of soft tissue that form the sides of the glossopalatine and pharyngopalatine arches in the mouth.

pineal (pĭn′ē·ăl) **body** a small endocrine gland suspended from the roof of the third ventricle of the brain which releases melatonin.

pinealocyte (pĭn′ē·ă·lō′sīt) the primary cell type in the pineal body and probably the source of melatonin.

pinna (pĭn′ă) the external, visible flap of the ear.

pinocytosis (pi″no·si·to′sis) a form of endocytosis involving engulfment of liquid by enclosing the liquid in cell membrane.

pisiform (pi′si·form) resembling a pea in size and shape; one of the small bones lying on the ulnar side of the wrist.

pituitary dwarfism abnormally short stature, usually caused by a deficiency in the production of growth hormone and characterized by normal body proportions and normal mental development.

pituitary gland a small, roundish gland suspended from the floor of the third ventricle of the brain by the infundibular stalk. The pituitary gland consists of an anterior lobe, which releases growth hormone, adrenocorticotrophic hormone, thyrotrophic hormones, follicle-stimulating hormone, and luteinizing hormone, and a posterior lobe which releases oxytocin and antidiuretic hormone.

pivot joint a type of joint where the head of one bone rotates in a cuplike surface of another bone.

placebo (plă·se′bo) an inactive substance given to a patient who is under the impression that it is a drug with therapeutic properties. Placebos are used as controls in drug studies designed to test the effectiveness of substances as medication.

placenta (plă·sĕn′tă) a structure in the uterine wall of a pregnant female through which nutrients and oxygen are delivered to the fetal circulation and fetal waste products are removed from the fetal circulation.

placode (plăk′od) a platelike thickening in an embryonic epithelium that gives rise to particular organs or structures as the embryo and fetus develop.

plantar (plăn′tăr) pertaining to the bottom surface of the foot.

plantar flexion extension of the foot at the ankle.

plaque (plăk) a patch of lipid material, usually rich in cholesterol, that collects in an arterial wall and slowly obstructs circulation or induces emboli as it protrudes into the artery.

plasma the fluid portion of the blood, exclusive of cells and platelets.

plasma cell a cell of bone marrow and connective tissue derived from B-lymphocytes that is responsible for the production of humoral antibodies.

plasmin an enzyme that degrades the fibrin formed in a blood clot and thereby prevents excessive growth of fibrin molecules and helps control the formation of a clot.

plasminogen a precursor to plasmin.

platelet plug a plug of platelets that forms at a wound in a blood vessel and helps stem the flow of blood.

platelets fragments of megakaryocytes present in red bone marrow that are involved in initiating blood clotting reactions; also called thrombocytes.

platybasia (plat″e·bă′se·ă) a developmental abnormality in which the base of the skull is flatter than normal. Can result in nervous disorders from pressure on the lower brain stem, cervical spinal cord, cerebellum, and the related cranial and spinal nerves.

pleocytosis (ple″o·si·to′sis) a condition in which there are an abnormally high number of lymphocytes in the cerebrospinal fluid.

pleura (plu′ră) the serous membrane that surrounds the lungs and folds back to cover the inner surface of the thoracic cavity and diaphragm.

pleural cavity the space, or potential space, that lies between the parietal and visceral pleurae in the thoracic cavity.

pleurisy (plu′ris·e) swelling and inflammation of the pleura, usually associated with pain when breathing and coughing.

plexus (plĕks′ŭs) a general term relating to a network of nerves, blood vessels, or lymphatic vessels.

plicae circularis (pli′ke sir·ku·lăr′is) permanent mucosal and submucosal folds that extend from the wall of the small intestine into the lumen.

plicae palmatae (păl′mă·te) folds of the mucosal lining that extend into the lumen of the uterine cervix.

pneumonia (nu·mo′ne·ă) a lung disorder involving an inflammation of lung tissues responsible for gas exchange. Inflammation may be due to infection by viruses, bacteria, fungi, protozoa, or physical agents.

pneumotaxic center (nu″mo·tăk′sik) an area in the pons that regulates the activity of inspiratory neurons in the medulla oblongata.

pneumothorax (nu·mo·tho′răks) the presence of air in the space between the visceral and parietal pleurae, and the usual cause of a collapsed lung.

podocyte (pŏd′o·sit) cells that cover and form the visceral layer of a Bowman's capsule.

polar body one of the small cells produced during meiosis in a female which does not serve as an egg.

polarity pertaining to the unequal distribution of ions on either side of a cell membrane and responsible for the membrane potential.

pollex (pŏl′ĕks) the thumb.

pollicis pertaining to the thumb.

polycystic renal disease an inherited disorder of the kidney in which multiple cysts occur in both kidneys. As they enlarge, the cysts compress and damage the renal tissue and cause reduced renal function.

polymer a macromolecule formed by linking a few to many thousands of similar or identical subunits. The subunits of a polymer are referred to as monomers.

polymorphonuclear leukocyte a white blood cell that contains a nucleus composed of two or more lobes; a granulocyte, including basophils, eosinophils, and neutrophils.

polymyositis (pŏl″e·mi″o·si′tis) an inflammation of several muscles at the same time, usually of unknown cause, although possibly due to an autoimmune reaction.

polypeptide a chain of a few to many amino acids held together by peptide bonds.

polyribosome (pŏl″e·ri′bo·som) a group of ribosomes associated with a single messenger RNA molecule, each of which is involved with the production of a polypeptide.

polysaccharide (pŏl″e·săk′ă·rid) a polymer of monosaccharides including, for example, starch and glycogen.

polysome (pŏl″e·som) see *polyribosome*.

polysynaptic pertaining to a nerve impulse pathway involving more than one synapse.

polyunsaturated fat a fat molecule in which there are numerous double bonds binding carbon atoms together in the fatty acid portions of the molecule.

polyuria (pŏl″e·u′re·ă) production and excretion of an abnormally large volume of urine.

pons in general, a bridge of tissue that connects two parts of an organ; specifically, the part of the brain stem that lies above the medulla oblongata and anterior to the cerebellum which contains nuclei of various nerves and controls several voluntary and involuntary functions.

popliteal (pŏp″lit·e′ăl) pertaining to the posterior side of the knee joint.

portal vein a vein that carries blood from the stomach and other abdominal organs to the liver, or in general, from one capillary bed to another.

positive feedback mechanism a response to a stimulus that has the effect of increasing the level of the stimulus that caused the response in the first place.

positron emission tomography (PET) a technique using a device that forms images of internal structures based on gamma rays released by radioactive substances ingested by the patient. A detector measures the pattern and intensity of gamma emission and provides information to a computer which constructs an image of the structure involved.

postabsorptive state the period following absorption of nutrients from the digestive tract during which the nutrients are used for metabolism.

posterior in humans, a term that pertains to the back or dorsal surface of the body or body parts.

posterior pituitary gland the posterior lobe of the pituitary gland, the source of oxytocin and antidiuretic hormone.

posterior root a cord that contains the axons of sensory neurons extending from a dorsal root ganglion to the spinal cord.

posterior root ganglion a bulbous structure on the dorsal root of a spinal nerve that contains the cell bodies of sensory neurons entering the spine.

postictal (pōst·ĭk'tăl) **state** a condition experienced by a person following seizure or other attack, often involving headache, muscle achiness, or sleep.

postpartum (pŏst·pär'tŭm) the period of time immediately following the birth of a baby.

postrenal renal failure a form of acute renal failure resulting from obstruction of the urinary tract by overgrowth of the prostate gland, the presence of stones, or cancer. The pressure that develops is reflected back to the kidney where is causes damage.

postsynaptic inhibition inhibition of impulse initiation in a neuron by increasing the magnitude of the membrane potential of the postsynaptic membrane.

postsynaptic membrane the membrane that receives the stimulus delivered to a synapse by a presynaptic neuron.

precursor any substance that is converted to a product in a chemical reaction; a substance, condition, or behavior that is the predecessor to some other substance or condition.

prefrontal lobotomy a surgical procedure in which the connection between the hypothalamic area and the prefrontal cortex is severed. The procedure is used to relieve the symptoms of certain psychiatric disorders.

preganglionic neuron (prē"găng·lē·ŏn'ĭk) a ganglion in an autonomic pathway that carries impulses away from the brain or spinal cord. Its cell body is in the central nervous system and its axon extends to a ganglion where it forms a synapse with the cell body or dendrites of a postganglionic neuron.

pregnancy the growth and development of an embryo and fetus inside a uterus, lasting from about 266 to 280 days.

premenstrual syndrome (PSM) a condition that occurs about a week or more before menstruation characterized by nervousness and emotional stress, irritability, depression, headache, and fluid retention.

premolar teeth the four teeth on the upper and lower jaws, two on each side, that lie between the canines and the molar teeth.

prenatal the period of time preceding the birth of a baby, in reference both to the pregnant woman and the developing offspring.

prepuce (prē'pūs) a fold of skin that covers the glans penis in the male and the clitoris in the female; also called foreskin.

prerenal renal failure a form of acute renal failure that results from a blockage of blood to the kidney as might occur following trauma, dehydration, excessive treatment of hypertension, or other source of diminished blood volume or pressure.

presenile dementia (prē·sē'nīl dě·měn'shă) any neurologic disorder involving progressive decline of intellectual capability that is not due to the mental deterioration that sometimes accompanies old age.

presentation the orientation of a fetus as it enters the birth canal.

pressor (prěs'or) a chemical substance that will induce an increase in blood pressure.

pressoreceptor a nerve ending or sensory organ that is sensitive to changes in blood pressure.

pressure palsy paralysis of a limb or tissues caused by the mechanical effects of compression of the nerve that supplies the organ or tissue.

presynaptic inhibition decrease in the amount of neurotransmitter released into a synaptic cleft as a result of impulses from a neuron that forms synapses with the presynaptic axon.

presynaptic membrane the membrane in a synapse that releases neurotransmitters into the synaptic cleft in response to the arrival of an impulse.

presynaptic neuron the neuron that delivers an impulse to a synapse.

primary aldosteronism a condition caused by a tumor of the adrenal cortex that causes production of excess aldosterone. The effect is to increase sodium retention and potassium loss by the kidney, which in turn causes fluid retention, muscle weakness, and tetany.

primary germ layers the ectoderm, mesoderm, and endoderm of an embryo, that give rise to organs and tissues as the embryo develops.

primary growth the initial growth of a metastasizing tumor which releases cells that cause growths elsewhere in the body.

primary oocyte an ovarian cell that initiates meiosis. In humans, primary oocytes divide prior to birth (*see also secondary oocyte*).

procarboxypeptidase (prō"kăr·bŏk"sē·pěp'tī·dās) a precursor protein secreted by the pancreas into the small intestine that is converted to carboxypeptidase, an enzyme used in the digestion of polypeptides.

proctitis (prŏk·tī'tĭs) inflammation of the anus and rectum.

proctodeum (prŏk·tō·dē'ŭm) a depression in the anal region of the hindgut in an embryo formed by invagination of the ectoderm. It gives rise to the cloacal membrane, which is involved in the formation of the anal canal.

progeny (prŏj'ě·nē) the offspring produced by parents.

progesterone (prō·jěs'těr·ōn) a steroid hormone produced by the corpus luteum, adrenal glands, and placenta that stimulates the growth and maintenance of the uterine endometrium.

proglucagon (prō·glū'kă·gŏn) a protein precursor to glucagon, a hormone produced by alpha cells in the islets of Langerhans in the pancreas that increases glucose levels in the blood, among other effects.

progressive systemic sclerosis a chronic progressive disease of the skin and internal organs involving hardening and thickening of connective tissue.

prolactin (prō·lăk'tĭn) a hormone released by the anterior pituitary gland that stimulates secretion of milk in a woman after the birth of her baby.

prolapse (prō'lăps') downward movement an organ from its usual position.

proliferation a rapid increase in the number of cells or other structures in a tissue.

proliferative phase the phase of the menstrual cycle following menses during which the endometrial lining of the uterus thickens.

prominence a portion of a structure that juts or projects outward.

pronation (prō·nā′shŭn) lying flat with one's belly side down; turning the forearm so that the palm faces backward or downward on a surface.

pronator a muscle involved in pronation.

pronephric duct a tube that connects the pronephros to the cloaca in the embryo.

pronephros (prō·něf′rŏs) the embryonic kidney.

pronucleus a nucleus from an egg or sperm in a zygote prior to fusing.

pro-opiomelanocorticotrophin (prō·ŏp″ē·ō·měl·ăn′ō·kor·tĭ·kō·trō′phĭn) **(POMC)** a protein produced by the anterior lobe of the pituitary gland that serves as a precursor to melanocyte-stimulating hormone (MSH) and β-endorphin.

prophase the early stages of mitosis, meiosis I, and meiosis II during which the chromosomes become visible microscopically, prior to metaphase.

proprioception (prō″prē·ō·sĕp′shŭn) the process of sensing movement and position of the body and body parts.

proprioceptors sensory nerve structures that provide information to the central nervous system concerning the movement of the body and body parts.

prosencephalon (prŏs″ĕn·sĕf′ă·lŏn) the part of the embryonic brain that gives rise to the telencephalon and diencephalon, the embryonic forebrain.

prostacyclin (prŏs″tă·sī′klĭn) a prostaglandin that acts to inhibit blood clotting by interfering with the attachment of platelets to the inner wall of intact blood vessels.

prostaglandins (prŏs″tă·glănd′ĭns) a class of hormones derived from fatty acid that are involved in the regulation of blood clotting, inflammation, and numerous functions normally controlled by the autonomic nervous system.

prostate gland a gland in the male that surrounds the urethra where it emerges from the urinary bladder and which contributes to the formation of seminal fluid.

prosthesis (prŏs′thē·sĭs) an artificial replacement for a missing body part, or a device used to enhance the function of a body part.

protein a polypeptide of particular form and function, usually consisting of several dozen to hundreds of amino acids.

proteinuria (prō″tē·ĭn·ū′rē·ă) an abnormal condition characterized by the presence of a protein, usually albumin, in the urine.

prothrombin (prō·thrŏm′bĭn) an inactive plasma protein that is converted by the action of substances on the surfaces of platelets or tissue cells to thrombin in the formation of a blood clot.

proton the positively charged subatomic particle in the nucleus of an atom.

protoplasm the metabolically active portion of the cell, including the cytoplasm and the nucleus.

protraction the movement of a part forward in relation to the body.

protractor a muscle that moves a part forward.

proximal pertaining to the portion of a part that is relatively close to the midline or the origin of a structure, as opposed to distal.

prune belly syndrome the congenital absence of certain portions of the abdominal musculature. The name comes from the protruding and wrinkled appearance of the abdomen.

pruritus (prū·rī′tŭs) a sensation of itchiness resulting, for example, from an allergic reaction or a skin infection.

pseudogout (sū′dō·gowt″) a goutlike joint disease caused by the accumulation of calcium phosphate crystals in synovial fluid and calcification and degeneration of the associated cartilage; also called chondrocalcinosis.

pseudohypertrophic muscular dystrophy a sex-linked recessive disease associated with progressive weakness and atrophy of the muscles; also called Duchenne muscular dystrophy.

pseudopodia (sū″dō·pō′dē·ă) the protoplasmic extensions formed by motile cells, such as certain leukocytes, as they travel or when they ingest particles or foreign substances by phagocytosis.

pseudostratified epithelium a type of epithelium in which one end of all of the cells rests on a membrane, but the other end may or may not extend to the surface of the epithelium, resulting in the appearance of successive cell layers in the tissue.

psoriasis (sō·rī′ă·sĭs) a skin disorder characterized by red patches covered with thick, dry scales of extra skin.

psychogenic headache headaches resulting from hysteria, anxiety, and emotional disturbance. The pain is typically long-lasting and is not associated with any discernible physical problem.

psychomotor seizure a form of epilepsy associated with damage to the temporal lobe of the cerebrum.

puberty the attainment of sexual maturity in a young male or female.

pubic pertaining to the pubic bone, which lies in the lower anterior portion of the hip bone.

pudendum (pū·děn′dŭm) a term used to refer to all of the external genitalia of a female.

pulmonary embolism a blood clot carried into the lungs, where it blocks the flow of blood to the pulmonary tissue.

pulmonary emphysema a lung disorder characterized by increased size of alveoli and other chambers involved in gas exchange with a resulting decrease in efficiency.

pulmonary pertaining to the lungs and external respiration.

pulp cavity the cavity of a tooth that contains the soft tissue, including the nerves and blood vessels.

pulse the rhythmic expansion and contraction of arteries that occurs in response to increased and decreased blood pressure.

punctum (pŭnk′tŭm) a general term relating to a very small spot or point.

pupil the opening in the iris that controls the amount of light that enters the eye.

purine a class of nitrogen-containing chemical compounds that includes the adenine and guanine organic bases of DNA and RNA.

Purkinje (pŭr·kĭn′jē) **fibers** specialized muscle fibers in the heart that branch from the bundle branches into the myocardium. Impulses traveling to the Purkinje fibers cause a coordinated contraction of the ventricles that forces blood up and out through the semilunar valves into the pulmonary trunk and aorta.

purulent (pūr′ū·lĕnt) pertaining to a tissue that contains or is producing pus.

pus a collection of fluid produced at a point of infection that contains suspended white blood cells and materials produced by the infected tissue.

pustular lesion any sore in the surface of the skin that contains pus.

putamen (pū·tā′mĕn) a portion of the lentiform nucleus in the basal ganglia of the cerebral cortex which is involved with the subconscious control of certain skeletal muscles.

P wave the sharp spike in an electrocardiogram that results from depolarization of the atria.

pyelonephritis (pī″ĕ·lō·nĕ·frī′tĭs) a bacterial infection of the kidney that causes inflammation of the tubules, glomeruli, blood vessels, and pelvis and is associated with the production of pus.

pyloric glands mucous-secreting glands in the wall of the pyloric portion of the stomach.

pyloric sphincter the muscular constriction at the base of the stomach that controls the flow of chyme into the small intestine.

pyloric stenosis a decrease in diameter of the pylorus due to excessive growth of the muscle or mucosal layers in the pyloric wall. The growth may be so great as to block the flow of stomach contents into the small intestine.

pylorus (pī·lor′ŭs) the distal part of the stomach including the opening into the small intestine.

pyoderma (pī·ō·dĕr′mă) a general term for any skin disease involving inflammation and pus formation.

pyramidal tract one of the three descending tracts in the spinal cord.

pyridoxine (pĭ·rĭ·dŏks′ēn) a member of the vitamin B_6 group of compounds involved in the formation and degradation of amino acids and in the conversion of glycogen to glucose.

pyrimidine (pĭ·rĭm′ĭd·ēn) a class of nitrogen-containing compounds that includes the thymine, cytosine, and uracil bases of nucleic acids.

pyrogen (pī′rō·jĕn) any substance that induces a fever, including naturally occurring cell products and bacterial toxins.

pyuria (pī·ū′rē·ă) the presence of leukocytes and pus in urine.

Q

quadrate (kwŏd′rāt) having four equal sides.

quadriceps (kwŏd′rĭ·sĕps) having four heads.

quadriceps reflex a reflex in which the leg is extended following a tap on the patellar tendon; also called patellar reflex.

quadriplegia (kwŏd″rĭ·plē′jē·ă) paralysis that involves all four limbs.

Quinsy an acute bacterial infection of the area around the tonsils, usually causing fever and pain when swallowing.

QRS wave the complex of deflections in an electrocardiogram produced by depolarization of the ventricles.

Q wave the short, sharp downward spike of an electrocardiogram that follows the P wave and precedes the spike associated with the R wave.

R

rabies (rā′bēz) an infectious viral disease of the central nervous system transmitted by the bite of infected dogs and other mammals. It travels from the peripheral nerves to the central nerves where it causes hemorrhaging in the meninges and brain, leading progressively to spasm and paralysis of the laryngeal and pharyngeal muscles and eventually death; also called hydrophobia.

rachitic (ră·kĭt′ĭk) **rosary** swelling of the junction of the ribs with the cartilage seen in infants who are suffering from vitamin D deficiency.

radiograph an X-ray photograph.

radiologist a specialist in the diagnosis of disease and other disorders on the basis of radiant energy imaging techniques.

radionucleotide scanning the analysis of the function and structure of an organ on the basis of the detection of radioactive isotopes preferentially accumulated in the organ.

radius the outermost bone of the forearm.

ramus (rā′mŭs) a general term for a smaller structure that branches off a larger structure, such as a branch of a nerve or a blood vessel.

raphe (rā′fē) a line that indicates where two halves of a structure joined, as the raphe of the penis.

Raynaud's phenomenon (rā·nōz′) a disorder of the peripheral arterioles characterized by spasm in the smooth muscle of the arterioles of the fingers and toes, and sometimes of the tongue, nose, and ears. Characteristically brought on by cold or emotional upset. The effect is a pallor or discoloration of the skin due to the presence of poorly oxygenated hemoglobin.

receptor-mediated endocytosis a mechanism for the transport of substances into a cell utilizing receptors that aggregate in specific pits in the cell membrane.

receptors specialized sensory nerve structures at the end of sensory nerves that perceive stimuli of specific types; molecules in the membrane or cytoplasm of a cell that bind with specific substances to initiate a response.

recessive allele a form of a gene that is not expressed when in combination with an alternative, dominant form of the same gene; an allele that is expressed only when in the homozygous or hemizygous state (see dominant allele).

recessive trait a phenotypic trait expressed by a recessive allele.

recombinant DNA DNA molecules that contain sequences derived from two or more different sources that may be naturally occurring or synthesized in the laboratory.

rectum the lower end of the large intestine lying between the sigmoid flexure and the anal canal where the feces are stored prior to defecation.

red marrow the tissue in the medullary cavity of long bones from which blood cells are produced.

red nucleus a highly vascularized region of neuronal cell bodies located in the upper region of the midbrain.

reduced a state in which a chemical substance possesses electrons that can be donated in an oxidation-reduction reaction.

reducing agent a chemical substance that can readily donate electrons in an oxidation-reduction reaction.

reduction the acquiring of electrons by an atom or molecule in an oxidation-reduction reaction.

referred pain a sense of pain in an area of the body other than the location of the source of the pain.

reflex an involuntary response to a stimulus involving a neural pathway in an effector organ.

reflex arc the neural pathway involved in a reflex, usually consisting of at least a sensory neuron that delivers signals to the central nervous system and a motor neuron that delivers the signal to an effector organ.

refraction the bending of light rays as they pass from a medium of one density into a medium of a different density, as they do when they pass into the lens of the eye and then into the interior of the eye.

refractory period a period of time immediately after the induction of an action potential in a neuron or muscle cell during which the membrane is incapable of restimulation.

relaxin (rē·lăk′sĭn) a polypeptide hormone produced by the corpus luteum in a pregnant woman and perhaps by the uterus and placenta which appears to relax pelvic ligaments, soften the cervix, and inhibit contraction of the uterine wall and thus to reduce the chance of spontaneous abortion.

REM sleep a sleep pattern characterized by rapid eye movements and dreaming.

renal pertaining to the kidney.

renal calculus a hard, stonelike structure that occurs in the kidney pelvis or ureter where it may block the flow of urine.

renal calyx (kā′lĭx) branches of the renal pelvis that receive urine flowing from the collecting ducts.

renal clearance rate a kidney function test used to determine the volume of plasma from which a substance is removed in the kidneys in one minute.

renal corpuscle the portion of a nephron that consists of the glomerulus and the Bowman's capsule.

renal cortex the outermost portion of the body of a kidney.

renal dialysis removal of waste products from the blood by dialysis across artificial membranes as the blood is led through a dialysis machine connected to the patient.

renal medulla the innermost portion of the body of a kidney, lying between the cortex and the renal pelvis.

renal pelvis the large inner cavity of the kidney from which urine passes into the ureter.

renal pyramids organizational subunits of the renal medulla consisting of the loops of Henle, collecting ducts, and the capillaries of the vasa recta.

renal renal failure loss of a kidney's ability to function as a result of damage to the nephrons and associated tubules and blood vessels, usually caused by infection or the effect of drugs or toxic substances.

renin (rĕn′ĭn) a proteolytic enzyme produced by the kidney that catalyzes the conversion of angiotensin to an active form that increases blood pressure.

repolarization the return of a membrane potential to a negative value in the latter part of an action potential.

reproduction the process by which offspring are produced.

residual volume the volume of gas in the lungs at the end of a forced expiration.

respiration the process by which gases are exchanged between the body and the atmosphere. External respiration involves the exchange that occurs in the lungs, and internal respiration the exchange that occurs in the tissue capillaries (see *cellular respiration*).

respiratory acidosis an acidic condition of the blood due to an accumulation of CO_2 that results from inadequate elimination by the lungs.

respiratory alkalosis an elevated blood pH resulting from an abnormally low concentration of CO_2 following hyperventilation.

respiratory centers localized areas within the brain stem that regulate the rate and depth of breathing.

respiratory distress syndrome a disorder of the lungs most often found in premature infants resulting from a deficiency in the production of surfactant, which decreases the lungs' ability to exchange gases between the alveoli and blood; also called hyaline membrane disease.

resting potential the electric potential across the membrane of a nerve cell or muscle cell when an action potential is not occurring.

restriction endonuclease one of many commercially available enzymes produced by bacteria or other microorganisms that cleaves DNA molecules at specific nucleotide sequences. Their use makes possible the isolation of specific regions of DNA for use in genetic engineering.

restriction fragment length polymorphism (RFLP) variable lengths of DNA generated by endonuclease digestion of the DNA. The presence of a particular fragment may be correlated with a particular genetic allele, and thus may be used to predict the presence of that allele in individuals in which it is not expressed.

rete testis (rē′tē tĕs′tĭs) a plexus of tubules in the testis that receives sperm produced in the seminiferous tubules.

reticular (rĕ·tĭk′ū·lăr) netlike in form or appearance.

reticular formation a network of cells distributed in the brain stem.

reticulocyte (rĕ·tĭk′ū·lō·sīt) an immature red blood cell normally found in the bone marrow where the cells are being produced.

retina (rĕt′ĭ·nă) a light-sensitive structure on the inner surface of the eye that contains light-sensitive receptors.

retinal detachment separation of the layer of the retina that contains the nervous tissue, or a portion of it, from the underlying layer that contains the pigment epithelium, usually as a result of fluid collecting between the two layers.

retinene (rĕt′ĭ·nēn) a pigment in the retina derived from vitamin A.

retraction the act of drawing back, as in clot retraction when a clot pulls away from the wall of a blood vessel.

retrograde amnesia loss of memory concerning the periods immediately before a concussion or other trauma to the head or the onset of certain brain disorders.

retrograde pyelogram (pī′ĕ·lō·grăm) an X-ray examination of the kidneys in which an opaque dye is introduced into the kidney through the ureter.

retroperitoneal (rĕ″trō·pĕr″ĭ·tō·nē′ăl) lying behind the peritoneum.

rheumatism (rū′mă·tĭzm) a general term for pain and stiffness in the joints and associated muscles.

rheumatoid arthritis chronic and nonspecific inflammation of peripheral joints, resulting in pain, stiffness, and potential damage to the joints.

Rh factor a genetically determined antigen in the red cell membrane of persons who have an Rh-positive blood type.

rhinencephalon (rī″nĕn·sĕf′ă·lŏn) several structures of the limbic system involved in emotion. It also contains structures involved in olfaction, the sense of smell.

rhodopsin (rō·dŏp′sĭn) the visual pigment present in the rods of the retina.

rhombencephalon (rŏm″bĕn·sĕf′ă·lŏn) the part of the brain that consists of the cerebellum, pons, and medulla oblongata; the posterior of the three brain vesicles in the embryo; also called the hindbrain.

rhythmicity area a group of neurons in the medulla oblongata that control breathing by alternately stimulating inspiration and expiration.

riboflavin a member of the vitamin B complex used as a cofactor in the oxidative breakdown of glucose and other fuels.

ribonucleic acid (RNA) a nucleic acid that consists of adenine, uracil, guanine, and cytosine-containing nucleotides. Nucleotides also contain phosphate and ribose, a five-carbon sugar. Several forms of RNA exist, the most common of which are ribosomal RNA, messenger RNA, and transfer RNA.

ribosome a cytoplasmic organelle that consists of a large and small subunit, each of which is composed of RNA and protein molecules; used in combination with messenger RNA, transfer RNA, and amino acids for the synthesis of proteins.

rickets a vitamin D deficiency disease characterized by abnormalities in the shape and structure of bones.

rigor mortis (rĭg′or mor′tĭs) a condition that develops in the muscles of the body shortly after death in which muscles are unable to relax due to ATP deficiency.

Rinne's test (rĭn′nēz) a clinical test of hearing ability in which a tuning fork is used to produce sounds and the ability to hear the sound by air conduction is compared with the ability to hear the sound by bone conduction.

rod a form of photoreceptor located in the retina of the eye and primarily involved in the reception of dim light, but not sensitive to the detection of colors.

roentgenogram (rĕnt·gĕn′ō·grăm) a photographic image produced by X rays (*see X rays*).

Romberg's sign swaying of the body when one is standing with the eyes closed and the feet close together. The positive sign is an indication of damage to the spinal cord.

root canal a central cavity in a tooth that is filled with soft tissue. It is continuous with a pore at the bottom of the tooth through which pass the nerves and blood vessels that enter the tooth.

root hair plexus a network of nerve cell endings in and around the hair follicle that are stimulated by movement of the hair.

rooting reflex a reflex observed in a newborn infant in which stimulating the baby's cheek or lip causes it to turn its head in the direction of the source of the stimulus.

rotation the turning of a structure about an axis.

rotator a type of muscle used to make a part revolve on its axis.

round window a membrane-covered aperture that lies in the wall between the middle and inner ears at the end of the cochlea; also called the fenestra cochlea or secondary tympanic membrane.

rubella (rū·bĕl′ă) an infectious viral disease with some of the symptoms of measles, except that it lasts for only a few days and is accompanied by a milder rash; also called German measles.

rubeola (rū·bē′ō·lă) a very contagious viral disease characterized by fever, cough, characteristic skin rash and spots on the cheeks and lips; also called measles.

rubriblast (rū′brĭ·blăst) one of the stages in the development of a red blood cell.

rubricyte (rū′brĭ·sīt) a nucleated immature red blood cell at a stage when hemoglobin has first started to appear.

rubrospinal tract a bundle of nerve fibers that arises in the red nucleus of the midbrain and terminates in the anterior horn of gray matter.

rugae (rū′gē) the folds of mucous membrane on the inner surface of the stomach.

ruptured disk a tear of one of the disks of cartilaginous material that lies between the vertebrae, resulting in compression or irritation of an associated nerve root or the spinal cord, with resulting pain and impaired muscle and nerve functions; also called herniated disk.

S

saccule (săk′ūl) a small sac.

sacral plexus (sā′krăl plĕks′ŭs) a network of nerve fibers that emanate from the fourth lumbar spinal nerve to the third sacral spinal nerve and fuse to form a nerve path that innervates the muscles in the hip, thigh, leg, and foot.

sacrum (sā·krŭm) a large triangular bone that lies at the base of the vertebral column above the coccyx. It is formed by the fusion of five sacral vertebrae.

sagittal (săj′ĭ·tăl) **plane** an anterior-ventral longitudinal plane through a body when in the anatomic position that would divide the body into two left and right halves.

saliva a complex fluid secreted by the salivary glands in the mouth that helps lubricate the food as it is chewed and swallowed, begins the digestion of starch, and helps combat the ingestion of bacteria.

salivary amylase an enzyme in saliva that catalyzes the breakdown of starch.

saltatory conduction the pattern of conduction along a myelinated axon in which the impulses jump from one node of Ranvier to the next.

sarcolemma (săr″kō·lĕm′a) the outer membrane of a muscle cell.

sarcoma (săr·kō′mă) a cancerous tumor of connective tissue, especially muscle and bone.

sarcomere (săr′kō·mēr) the fundamental unit of contraction of a skeletal or cardiac muscle cell, delineated by Z lines and containing A, I, and M bands produced by the overlapping of the thin and thick filaments.

sarcoplasm (săr′kō·plăzm) the fluid material in the cytoplasm of a muscle cell.

sarcoplasmic reticulum (săr″kō·plas′mĭk) the network of membranes that surround the myofibrils in a muscle cell.

satellite cells a type of peripheral neuroglial cell associated with nerve cell bodies in the peripheral nervous system, especially in ganglia.

saturated fat fat molecules in which there are no carbon-to-carbon double bonds in the fatty acids.

scala media (skā′lă) a helical chamber between the scala tympani and the scala vestibuli in the inner ear where the organ of Corti is located; the sensory hearing apparatus.

scala tympani (tĭm′pă·nē) the spiral, perilymph-filled chamber of the cochlea bordered by the basilar membrane and ending at the round window.

scala vestibuli (vĕs·tĭb′ū·lī) a spiral canal in the cochlea that lies above the scala media, extending from the vestibuli to the tip of the cochlea, where it communicates with the scala tympani through the helicotrema.

scapula (skăp′ū·lă) the large triangular bone that forms an eminence on the posterior aspect of each shoulder; also called the shoulder blade.

scar tissue dense fibrous tissue formed at the site of an injury.

Scheuermann's (shoy′ĕr·mănz) **disease** an abnormal condition affecting spinal development in adolescents, characterized by a misalignment of vertebrae that causes a round-shouldered posture and a persistent, mild, low backache.

Schwann (shwŏn) **cell** a neuroglial cell that forms sheaths around the axons of peripheral nerves. In the larger nerve fibers, the sheaths contain myelin that forms an insulating cover around the axon except at the nodes of Ranvier.

sciatic (sī·ăt·ĭk) **nerve** a large nerve that extends from the sacral plexus to the thigh, where it divides to form the tibial and peroneal nerves.

sciatica (sī·ăt′ĭ·kă) low back and shooting buttock and leg pain, usually resulting from compression of the sciatic nerve by a herniated intervertebral disk or spinal tumor.

scintiphotography (sĭn″tĭ·fō·tŏg′ră·fē) a technique used to study the structure and function of internal organs by exposing photographic film to the decay products of radioactive substances injected into the body.

sclera (sklĕ′ră) the tough, white membrane of the surface of the eyeball.

sclerosis (sklĕ·rō′sĭs) a condition in which tissues have hardened due to the accumulation of fibrous tissue.

sclerotic plaque a localized region in which soft tissue has become hardened due to the production of fibrous tissue.

scoliosis (skō′lē·ō′sĭs) a deformity of the spine in which the vertebral column curves laterally.

scotopsin (skō·tŏp′sĭn) the protein portion of rhodopsin, the visual pigment in the rods of the retina.

scrotum (skrō′tŭm) the external pouch in a male that contains the testes and associated structures.

scurvy a vitamin C deficiency disorder characterized by abnormal bone and teeth production and maintenance.

sebaceous (sē·bā′shŭs) **gland** holocrine glands in the skin that secrete an oily substance called sebum.

seborrheic keratoses (sĕb″ō·rē′ĭk kĕr′ă·tō′sĭs) pigmented epithelial warts that occur most often after middle age.

sebum (sē′bŭm) the secretion product of a sebaceous gland, containing a fatty liquid and the cellular debris resulting from the breakdown of the cells involved in its secretion.

secondary growth cancerous growths that develop from cells that have been released by a metastatic tumor.

secondary oocyte one of the two products of division of a primary oocyte, the other product being a polar body; it completes the meiotic division only upon penetration by a sperm (*see oocyte*).

second messenger a molecule or ion that induces a specific biochemical effect after it is released into the cytoplasm in response to combination of a hormone with a receptor in the cell membrane. Cyclic-AMP is an example of a second messenger induced by peptide hormones.

secretin a hormone released from the duodenal mucosa that stimulates pepsinogen release from the gastric mucosa.

sedative a drug used to calm anxiety.

segregate the process by which two different alleles of a heterozygous pair are distributed to separate cells during meiosis.

seizure a sudden loss of nervous muscle control, often characterized by involuntary muscle contractions and relaxations, sometimes with the loss of consciousness.

sella turcica (sĕl′ă tŭr′sĭ·kă) a concave depression in the sphenoid bone that contains the pituitary gland.

semen (sē′mĕn) a complex mixture containing enzymes, electrolytes, and sperm that is ejaculated by the male at orgasm.

semicircular canal tubules in the inner ear that emanate from the vestibule and contain the structures that signal the position and movement of the head.

semilunar valves valves located at the junction of the left and right ventricles and the aorta and pulmonary arteries, respectively.

seminal vesicle a pair of elongated sacs that contribute fluid to semen located behind and below the urinary bladder and opening into the ductus deferens at the junction of the ampulla and ejaculatory duct.

seminiferous (sĕm·ĭn·ĭf′ĕr·ŭs) **tubules** network of sperm-producing tubules in a testis.

semipermeable membrane a membrane that allows the passage of some molecules but not others, usually on the basis of size.

senile dementia a progressive loss of various aspects of intellectual function that sometimes develops in old age.

sensation an awareness of a stimulus as a result of its perception by sensory receptors.

sensory area an area in the cerebral cortex that is responsible for receiving and interpreting impulses initiated in sensory receptors.

septum (sĕp′tŭm) a division between two cavities or chambers.

serosa (sē·rō′să) a relatively thin, delicate membrane that lacks glands and secretes a lymphlike fluid. Serous membranes are found lining the peritoneal, pleural, and pericardial cavities.

serous otitis media an accumulation of fluid in the middle ear resulting from a middle ear infection or blockage of a eustachian tube.

Sertoli's (sĕr·tō·lēz) **cells** cells in the lining of the seminiferous tubules that nourish and support spermatids as they differentiate into sperm; also called sustentacular cells.

serum the fluid portion of blood that remains after the formed elements and clotting elements have been removed; the fluid secreted by a serous membrane.

serum sickness an allergic reaction to the transfusion of serum from a foreign source, as when an antibody-containing serum from an animal is administered. Certain drugs will also cause the reaction, which is characterized by a skin rash, fever, and swelling of the lymph glands.

sesamoid bone (sĕs′ă·moyd) a small bone embedded in a tendon or in an articular capsule. The patella is the largest sesamoid bone.

sex chromosomes the X and Y chromosomes. Females usually have two X chromosomes per nucleus, whereas males usually have one X and one Y chromosome.

sex linkage the characterisic pattern of inheritance of traits controlled by genes carried on the X chromosome. Since the male has only one X chromosome, recessive alleles carried on this chromosome will be expressed as if they were in the homozygous condition.

sexually transmitted disease (STD) any of several diseases that are transmitted during sexual contact.

shin splint strain of the flexor digitorum longus muscle in the lower leg marked by pain along the shin bone.

shingles eruption of fluid-filled vesicles and itchiness on the skin along the course of a peripheral nerve due to a herpes zoster infection in the ganglia and peripheral nerves that supply the effected area. Caused by the same virus that causes chickenpox.

shunt a passageway that develops or is placed between two conducting vessels. An arteriovenous shunt is a plastic tube inserted between a vein and an artery, often in the forearm, used to provide a structure into which a hemodialysis machine can be connected.

sickle cell anemia a fatal form of anemia caused by an autosomal recessive trait that affects the structure of hemoglobin and causes red blood cells to assume a sickle shape and lyse under conditions of low oxygen tension.

sigmoid (sĭg·moyd) **colon** the portion of the colon that extends from the lower end of the descending colon to the rectum.

simple fracture a fracture in which the ends of the broken bone do not penetrate the skin.

sinoatrial node (sīn″ō·ā′trē·ăl) specialized muscle tissue in the wall of the right atrium from which rhythmic impulses originate that cause the heart to contract; also called pacemaker of the heart.

sinus a general term for a cavity or chamber, such as the paranasal sinuses.

sinusitis (sī·nūs·ī·tĭs) inflammation and swelling of one or more of the paranasal sinuses caused by infection that blocks sinus drainage.

skeletal muscle striated (or voluntary) muscle generally attached to and responsible for the movement of bones.

sliding filament mechanism the mechanism by which muscle cells contract, involving the sequential and stepwise movement of thin filaments along thick filaments in a sarcomere. Movement of the thin filaments pulls the Z disks to which they are anchored, and the sarcomere shortens.

small intestine the portion of the digestive tract between the stomach and the large intestine. It is involved in chemical digestion and absorption of nutrients and water.

smooth muscle a type of muscle found in the viscera consisting of uninucleate cells that are free of striations.

sodium-potassium-ATPase pump the mechanism used in the linked active transport of sodium ions to the outside of a membrane and potassium ions to the inside of a membrane. ATP is used in the process.

soft palate the posterior portion of the roof of the mouth.

solute any substance dissolved in a solvent.

solution a mixture of substances dissolved in a fluid.

soma (sō′mă) an alternative term for the cell body of a neuron.

somatic pertaining to the body.

somatic capillaries capillaries in tissues other than the lungs.

somatic nervous system the portion of the nervous system other than the brain and spinal column; the peripheral nervous system.

somatomammotrophin (sō·mă″tō·mă″mō·trō′fĭn) a hormone uniquely produced by the chorion during pregnancy that stimulates several functions that have in common the effect of increasing nutrient supply to a developing fetus; also called human chorionic somatomammotrophin (HCS).

somatomedin (sō″mă·tō·mē′dĭn) hormones produced in the liver that mediate the effect of human growth hormone.

somatostatin (sō·măt′ō·stăt″ĭn) a hormone produced by the hypothalamus and delta cells of the islets of Langerhans. The hormone inhibits growth hormone secretion from the pituitary gland and the release of insulin and glucagon from the pancreas.

somatotrophin (sō″mat·ō·trō′fĭn) a polypeptide hormone produced by the anterior pituitary gland in response to the action of hypothalamic growth hormone-releasing hormone (GHRH). In general, somatotrophin stimulates growth, particularly of long bones. It is also known as human growth hormone.

somite (sō′mīt) one of several pairs of clumps of mesodermic tissue that lie alongside the neural tube of an embryo from which the vertebral column is formed.

sonogram (sō′nō·grăm) an image of an internal structure formed from reflected high frequency sound waves.

spasm (spăzm) a sudden involuntary tightening of one or more muscles.

sperm the male gamete or sex cell.

spermatic cord a tubular structure in the male reproductive system that consists of the ductus deferens and associated nerves, vessels, and connective tissue.

spermatid the product of meiosis in the seminiferous tubules that differentiates into a sperm.

spermatocyte (spĕr·măt′ō·sīt) a cell in the seminiferous tubules about to undergo meiosis (primary spermatocyte) or following the first division of meiosis (secondary spermatocyte).

spermatogenesis (spĕr″măt·ō·jĕn′ĕ·sĭs) the process by which sperm are produced.

spermatogonium a testicular cell that differentiates into a primary spermatocyte.

spermatozoan (spĕr″măt·ō·zō′ăn) a sperm cell.

spermiogenesis (spĕr″mē·ō·jĕn′ĕ·sĭs) the process by which spermatids differentiate into sperm.

sphenoid (sfē′noyd) **bone** one of the eight bones of the skull, lying at the base of the skull.

sphincter (sfĭnk′tĕr) a circular muscle that surrounds an opening into or out of a hollow organ. Constriction of the sphincter regulates flow through the opening.

sphincter of Oddi (ŏd′ē) a muscular band that surrounds and controls the opening of the common bile and pancreatic ducts into the small intestine.

sphygmomanometer (sfĭg″mō·măn·ŏm′ĕt·ĕr) a device used to measure blood pressure.

spina bifida (spī′nă bĭf′ĭ·dă) a neurologic defect that results from the failure of the vertebral column to close during development. In the most severe form, the spine is completely open and death results. In spina bifida cystica, a sac protrudes from the back of the baby that can include the meninges and the spinal cord. Spina bifida occulta is less severe and there is no protrusion of the cord or meninges.

spinal cord the portion of the central nervous system that extends through the vertebral canal.

spinal nerve a member of one of the pairs of nerves that emanate from the spinal cord and extend into the periphery.

spinal shock the loss of all sensation and reflex activity as well as paralysis in the regions of the body innervated by nerves originating below the level of a trauma to the spinal cord.

spinal tap the removal of cerebral spinal fluid through a lumbar puncture. Performed for diagnostic purposes.

spindle the network of fibers present in a cell undergoing mitosis or meiosis that are involved in the movement of chromosomes to opposite sides of the cell.

spine the spinal column, consisting of the 33 vertebrae and their associated intervertebral disks and ligaments.

spiral fracture a fracture of the bone involving a twisting apart of the two broken ends.

spirometer a device used to measure lung volumes as a person exhales or inhales into a chamber that has a bellows or other apparatus that moves as the volume of the chamber changes.

splanchnic (splăngk′nĭk) pertaining to the organs in the abdominal cavity.

spleen a lymphatic organ located in the upper left portion of the abdominal cavity. It serves to store and filter blood.

spondylolisthesis (spŏn″dĭ·lō·lĭs″thē′sĭs) an anterior displacement of a vertebrae, usually of the fifth lumbar vertebra as it is pushed over the sacrum.

spondylolysis (spŏn″dĭ·lŏl′ĭ·sĭs) a loss of bony material from a vertebra due to dissolution.

spondylosis (spŏn″dĭ·lō′sĭs) stiffness of the spine.

sputum (spū′tŭm) a viscous fluid coughed up from the lungs, bronchi, and trachea.

squamous (skwā′mŭs) **epithelium** a type of epithelium characterized by one or more layers of flattened cells. In simple squamous epithelium there is a single layer of cells. In stratified squamous epithelium there are two or more layers of cells, with the cells in the uppermost layers flattened.

stapes (stā′pēz) the innermost of the three small bones of the middle ear.

Starling's law of the heart a mechanism by which stroke volume of a heart is regulated by the degree to which the muscle fibers in the myocardium are stretched at the end of diastole. The more they are stretched, the more strongly they contract, thus allowing the heart to adjust to changing demands independent of any nerve impulses; also called Frank-Starling law of the heart.

stasis (stā′sĭs) the stopping of flow of a fluid such as blood.

statoconia (stăt″ō·kō′nē·ă) *see otoconium.*

status epilepticus (ĕp″ĭ·lĕp′tĭ·kŭs) a series of epileptic seizures in an unconscious person.

sternum (stĕr′nŭm) the large, flat bone at the anterior of the thoracic cavity to which most of the ribs attach through costal cartilage.

steroids a category of lipid molecules that include cholesterol and serve important functions in cell membranes and as hormones.

stethoscope an instrument used to listen to sounds that emanate from the body.

stigma (stĭg′mă) a dimple that develops on the surface of an ovary that marks the point where ovulation will occur.

stimulant any substance that increases nervous activity or speeds up a body process.

stomach an organ of the digestive tract between the esophagus and the small intestine where the initial digestion of proteins occurs.

stomodeum (stō″mō·dē′ŭm) a region in the ectoderm of an embryo from which the mouth is derived.

strabismus (stră·bĭz′mŭs) an abnormality of the eye in which the line of sight of one eye is not parallel to the sight of the other eye when viewing distant objects. It may be caused by a lesion of the nerves that control the oculomotor muscles; also called cross-eye.

stratified epithelium epithelial tissue that consists of two or more layers of cells.

stratum basalis (bă·săl′ĭs) a layer in the wall of the uterus that underlies and produces the stratum functionalis, the portion of the wall that undergoes repeated cycles of buildup and degeneration during a menstrual cycle.

stratum corneum (kor′nē·um) the outermost layer of the skin.

stratum functionalis (fŭng·shun·al′ĭs) the innermost layer of the uterine wall. It undergoes alternate cycles of thickening and degradation corresponding to alternating levels of estrogen in the blood and is the source of menstrual flow as it sloughs off the wall of the uterus.

stratum granulosum (grăn″ū·lō′sŭm) a layer of cells in the skin that lies between the stratum germinativum and the stratum lucidum; a layer of cells that line the cavity of a matured ovarian follicle.

stress fracture a fracture of a bone caused by repeated stress to the bone.

stretch receptor specialized nerve endings in the walls of the airways of the lungs that send impulses to the respiratory center in response to stretching of the airways, thereby inhibiting the inspiration phase and preventing overinflation of the lungs.

stretch reflex a reflex arc involving a sensory and motor neuron induced by stretching specialized receptors in a muscle. The impulse travels to the central nervous system and back out to the muscle that has been stretched, inducing contraction.

striated muscle skeletal muscle; named for the appearance of microscopic striations formed by sarcomeres present in the muscle fibers.

stroke blocking of blood flow to the brain by an embolus in a cerebral blood vessel.

stroke volume the volume of blood pumped from a ventricle as it contracts.

stroma (strō′mă) the supporting tissue of an organ as opposed to its functional tissue, which is called the parenchyma.

subarachnoid (sŭb″ă·răk′noyd) **space** a space between the arachnoid and pia mater layers of the meninges. It is filled with cerebrospinal fluid.

subcutaneous (sŭb″kū·tā′nē·us) lying below the skin.

subdural hematoma a clot or other mass of blood in the space between the dura mater and the brain.

subdural space the space between the arachnoid and dura mater membranes of the meninges.

sublingual (sŭb·lĭng′wăl) pertaining to the region under the tongue.

sublingual gland a salivary gland located below the tongue.

submucosa (sub·mū·kō′să) a layer of loosely packed tissue that lies underneath a mucous membrane.

substrate the starting material of an enzyme-catalyzed reaction.

sucking reflex a reflex observed in an infant in which the infant goes through sucking motions in response to light touch to the lips.

sucrase (sū′krās) an enzyme secreted by the duodenum that catalyzes the breakdown of sucrose (table sugar) to glucose and fructose.

sudden infant death syndrome (SIDS) the sudden unexpected death of an otherwise healthy infant in its sleep; also called crib death.

sudoriferous glands (sū·dor·ĭf′er·ŭs) sweat glands.

sulcus (sŭl′kŭs) a furrow or groove, especially on the surface of the brain where it marks the boundary of a gyrus.

summation induction of an impulse in a neuron through the accumulated effects of multiple stimuli arriving from presynaptic neurons over a period of time or at several synaptic junctions.

sunstroke a reaction to excessive heat resulting from inadequate response of the body's heat regulating mechanism. Characterized by red, dry skin with little sweating, rapid pulse, and elevated temperature.

superficial relating to the surface of an organ or tissue.

superior located above a structure.

supination to be in a supine position; to rotate the forearm so that the palm faces forward when hanging at the side of the body.

supine a body position where one lies on his or her back.

suppressor T cells a type of T lymphocyte that helps regulate antibody production by inhibiting the activity of B lymphocytes in producing antibodies.

supraoptic (sū″pră·ŏp′tĭk) **nucleus** a group of neuronal cell bodies in the hypothalamus responsible for the production of antidiuretic hormone and oxytocin, which are carried by neurophysin in the axons of the neurons to the posterior pituitary gland from which they are released.

surfactant (sur·făk′tănt) a substance in the fluid that lines the wall of the alveolus that enhances diffusion of gases by lowering the surface tension of the liquid.

sustentacular (sŭs″tĕn·tăk′ū·lăr) **cells** cells in the wall of seminiferous tubules that aid in the differentiation of spermatids to sperm; also called Sertoli's cells.

suture (sū′chūr) a line that marks the junction of two bones that form an immovable joint, such as are visible on the skull.

sweat a colorless, salty fluid secreted by sudoriferous glands in the skin and used for the evaporative cooling of the body and the elimination of certain waste materials.

sympathetic nervous system one of the divisions of the autonomic nervous system, emanating from the thoracic and upper lumbar regions of the spinal cord.

sympathomimetic (sĭm″pă·thō·mĭm·ĕt′ĭk) having an effect similar to that which results from stimulating the sympathetic division of the autonomic nervous system.

symphysis pubis (sĭm′fĭ·sĭs pū′bĭs) the junction of the two pubic bones in the anterior of the pelvis, where they are held together by cartilage and ligaments.

symptomatic the presence of conditions that suggest the existence of a specific disease or disorder.

synapse (sĭn′ăps) the functional junction between a terminal branch of the axon of a nerve cell and a point on the surface of another nerve cell or a gland or muscle cell. Neurotransmitter is secreted into this narrow space between the two membranes.

synaptic cleft the neural gap that lies between the presynaptic and postsynaptic membranes of a synapse.

synaptic delay the relatively long time it takes for a nerve impulse to cross a synapse compared to the time it takes for the impulse to travel along an axon.

synaptic vesicle one of the numerous vesicles in the cytoplasm of a presynaptic neuron that fuses with the membrane and releases neurotransmitter into the synaptic cleft in response to the arrival of a nerve impulse.

synarthrosis (sĭn″ăr·thrō′sĭs) a type of bone junction in which the bones are immovable with respect to one another. The sutures visible on the surface of the skull are examples of synarthrotic joints.

synchondrosis (sĭn″kŏn·drō′sĭs) a rigid joint between two bones in which the bones are held in place by an intervening layer of cartilage.

syndesmosis (sĭn″dĕs·mō′sĭs) a type of joint in which the bones are held together by ligaments and are able to move to a slight extent with respect to one another.

syndrome a collection of traits and symptoms that collectively indicate the presence of a particular disorder or abnormal condition.

syneresis (sĭn·ĕr′ĕ·sĭs) the separation of a liquid from a gel caused by the shrinking of the gel, for example, the release of fluid that occurs when the fibrin molecules in a clot shrink.

synergist (sĭn′ĕr·jĭst) an organ, such as a muscle, that works cooperatively with another organ.

synostosis (sĭn″ŏs·tō′sĭs) a condition in which the connective tissue that lies between two bones in a synarthrotic joint is replaced by bone so that the bones on either side of the joint are firmly fused.

synovial (sĭn·ō′vē·al) **cavity** a fluid-filled space that lies between the bones in a synovial joint.

synovial joint a type of joint in which the bones are freely movable with respect to each other, limited only by associated ligaments, tendons, and other tissues.

synovial membrane a membrane that lines the inner surface of the articular capsule that encloses a synovial joint and which secretes synovial fluid.

synovitis (sĭn″ō·vī′tĭs) inflammation of a synovial membrane.

syphilis (sĭf′ĭ·lĭs) a disease caused by the bacterium *Treponema pallidum* that is transmitted through sexual contact or is congenital in babies born of mothers with the disease. It affects nearly all tissues of the body and is lethal if not controlled.

systemic pertaining to the entire body.

systemic circulation the portion of the circulatory system that carries blood from the left ventricle to the right atrium.

systemic lupus erythematosus (ĕr·ĭ″thē·mă·tō′sŭs) a chronic and progressive disease of the immune system that affects virtually all the organs of the body. It causes fatigue, widespread inflammation, and possibly kidney failure and nerve disorders.

systole (sĭs′tō·lē) the period when a heart chamber is contracting, as in arterial systole and ventricular systole.

systolic blood pressure the hydrodynamic pressure of the blood in the arteries that results from ventricular systole. The first and higher of the two numbers indicating blood pressure, as for example 120/80.

T

tachycardia (tăk″ē·kăr′dē·ă) an abnormally rapid heart rate.

tactile (tăk′tĭl) pertaining to the sense of touch.

taenia coli (tē′nē·ă kō′lī) the three longitudinal bands of smooth muscle on the surface of the large intestine.

talipes (tăl′ĭ·pēz) any congenital deformity of the foot and ankle.

talus (tā′lŭs) a bone in the ankle that articulates with the tibia and fibula of the leg.

target organ the specific organ or organs that respond to a particular hormone.

tarsus (tăr′sŭs) the bones of the ankle; connective tissue in an eyelid.

taste bud one of many specialized structures on the surface of the tongue that is stimulated by certain chemicals substances to produce a sensation of taste.

Tay-Sachs (tā·săks′) **disease** a recessive genetic disorder characterized by an inability to metabolize certain lipids, resulting in a progressive loss of neurological function, with death occurring usually by the age of four.

T cell a type of lymphocyte involved in cellular immunity, some of which help regulate antibody production by B-lymphocytes.

tectorial membrane of the cochlear duct a membrane in a cochlea of the inner ear that lies above the organ of Corti; also called membrane of Corti.

tectum (tĕk′tŭm) a rooflike covering over a structure or cavity.

telencephalon (tĕl·ĕn·sĕf′ă·lŏn) the portion of an embryo from which the cerebral hemispheres, corpora striata, and rhinencephalon arise.

telophase (tĕl′ē·ō·fāz) that period during mitosis or meiosis when chromosomes have passed to the ends of the spindle fibers and clumped together.

temporal summation a mechanism for producing an impulse in a postsynaptic neuron in which impulses are delivered to the synapse more rapidly than the neurotransmitter can be removed. The resulting accumulation of neurotransmitter eventually reaches a level sufficient to stimulate the postsynaptic cell.

tendon a band of connective tissue through which a skeletal muscle is attached to a bone or other structure.

tendonitis (tĕn″dŏn·ī′tĭs) inflammation of a tendon.

tendon sheath membranes around tendons that form a synovial cavity where tendons cross over bones, cushioning the tendon at that point.

tenesmus (tĕ·nĕz′mŭs) a painful but unsuccessful urge to empty the urinary bladder or rectum.

tennis elbow a strain of muscles in the forearm that originate on the lateral epicondyle of the humerus; also called lateral humeral epicondylitis.

tenosynovitis (tĕn″o·sĭn″o·vī′tĭs) inflammation of the sheath of fibrous material that surrounds a tendon.

tensor (tĕn′sor) a type of muscle that tightens a membrane or other structure.

tentorium cerebelli (tĕn·tō′rē·ŭm sĕr·ĕ·bel′ē) a portion of the duramater that lies between the cerebellum and the cerebral hemispheres.

teres (tē′rēz) a muscle that is cylindrical in shape.

testicle (tĕs′tĭ·kl) see *testis*.

testis (tĕs′tēs) one of a pair of structures normally contained in the scrotum in which sperm and testosterone are produced; also called testicle.

testosterone (tĕs·tŏs′tĕr·ōn) a hormone produced by cells between the seminiferous tubules in the testis which is responsible for the production and maintenance of secondary male sex characteristics, including facial and body hair, deepened voice, and the musculature and bone development characteristic of the male body.

tetanus (tĕt′ă·nŭs) a condition in which a muscle is held in constant contraction; an infectious disease caused by the bacterium *Clostridium tetani* growing in a wound that receives inadequate oxygenation. Toxins produced by the microorganism induce muscle spasms and hypersensitivity; also called lockjaw.

tetrad a group of four chromatids produced by the pairing of homologous chromosomes in prophase I of meiosis.

thalamus (thăl′ă·mŭs) a pair of oval masses of gray matter that form the lateral walls of the third ventricle of the brain and a portion of the diencephalon of the forebrain. It integrates and relays sensory information to other parts of the brain.

theca folliculi (thē′ka fō·lĭk′ū·lī) ovarian cells that surround a developing follicle.

thenar (thē′năr) pertaining to the palm of the hand or the sole of the foot; the fleshy mass of tissue on the palmar aspect at the base of the thumb.

thermoreceptors sensory nerve endings that are stimulated by heat.

thiamine a vitamin that serves as a cofactor in carbohydrate metabolism; also called vitamin B_1.

thirst center a region in the hypothalamus that contains cells stimulated by the increased osmotic pressure of surrounding fluids that occurs when the body becomes dehydrated. Activity of these cells induces the sensation of thirst.

thoracic cavity the space contained within the rib cage and above the diaphragm.

thoracic duct the major lymphatic vessel in the body. It begins in the upper portion of the abdominal cavity, passes through the thoracic cavity alongside the aorta, and then joins the left subclavian vein.

thoracolumbar (thō″răk·o·lŭm′băr) pertaining to the thoracic and lumbar portions of the spine.

threshold the membrane potential that must be reached for an impulse to be initiated in a nerve or muscle membrane.

thrombin (thrŏm′bĭn) an enzyme that catalyzes the conversion of fibrinogen to fibrin in the clotting of blood.

thrombocyte (thrŏm′bō·sīt) an alternative term for a blood platelet.

thrombocytopenia (thrŏm″bō·sī″tō·pē′nē·ă) an abnormally low number of platelets in the blood.

thromboplastin (thrŏm″bō·plăs′tĭn) a blood-clotting factor involved in the conversion of prothrombin to thrombin.

thrombosis (thrŏm·bō′sĭs) the formation of a blood clot.

thromboxanes (thrŏm·boks′āns) a group of compounds derived from prostaglandins that are active as vasoconstrictors and in causing platelet aggregation.

thrombus (thrŏm′bus) a blood clot that forms in a blood vessel and blocks the flow of blood.

thymic (thī′mĭk) **corpuscle** roughly spherical bodies of unknown function present in the medulla of the thymus gland.

thymoma (thī·mō′mă) a tumor that develops in the epithelial or lymphoid tissue of a thymus gland.

thymosin (thī·mō′sĭn) a substance produced by the thymus gland that appears to stimulate the maturation of T lymphocytes and thereby to increase their ability to make antibodies.

thymus gland a lymphatic gland located above the heart in the mediastinal cavity that is important in the development of immune response in children, although it diminishes in size and importance in adults.

thyroglobulin (thī″ro·glŏb′ū·lĭn) an iodine-containing protein stored in the follicles of the thyroid gland from which thyroid hormone is derived.

thyroid cartilage a large cartilaginous structure that forms the anterior surface of the larynx.

thyroid follicle one of many spherical structures in a thyroid gland formed by follicle cells and containing colloid in which thyroxin (T4) and triiodothyronine (T3) are produced.

thyroid hormone thyroxin (T4) and triiodothyronine (T3) produced upon degradation of thyroglobulin in the colloid of thyroid follicles; it plays an important role in regulating metabolic rate in all tissues.

thyroid-stimulating hormone (TSH) a hormone produced by the anterior pituitary gland that stimulates thyroid hormone production by the gland; also called thyrotrophin.

thyroxin (thī·rŏks′ĭn) one of two primary thyroid hormones; also designated T4.

tic spasmodic twitching of muscles, especially of the face.

tidal volume the volume of air that moves into the lungs with each inspiration when at rest.

tinea circinata (tĭn′ē·ă sĭr·sĭ·nă′tă) a fungal infection that occurs in the body's skin; also called ringworm of the body.

tinea cruris (krū′rĭs) a fungal infection that involves the skin in the upper inner thigh; also called jock itch.

tinea pedia (pĕd·ē′ă) a fungal infection in the skin of the foot; also called ringworm of the feet.

tinnitus (tĭn·ī·tus) a sensation of a noise in the ears, such as a ringing or buzzing sound.

tissue any population of cells of related structure and function.

tone see *muscle tone*.

tonic pertaining to muscle contraction.

tonic neck reflex a reflex seen in newborn infants in which turning of the head causes the arm on the side to which the head is turned to extend outward and the arm and leg on the other side to be pulled in.

tonic spasm an involuntary and painful muscle contraction in which the muscle contracts and remains contracted and rigid for an extended period.

tonsillitis acute infection of the tonsils.

tonsils masses of lymphoid tissue on the walls of the oral cavity behind the constriction of the fauces.

torticollis (tor′tĭ·kŏl′ĭs) stiff neck caused by a spasm of the neck muscles.

total lung capacity the total potential gaseous volume of the lungs, consisting of the inspiratory reserve volume, the tidal volume, the expiratory reserve volume, and the residual volume.

toxin any poisonous substance produced by the metabolic activity of an organism.

trabeculae (tră·běk´u·lē) a general term for supporting cords or sheaths of fibrous tissue that forms septa within an organ.

trachea (trā´kē·ă) the tube bounded by C-shaped bands of cartilage that conducts inspired air from the larynx to the bronchi.

tract a bundle of nerve fibers that forms a pathway in the central nervous system.

trait in genetics, the manner in which one or more genes are expressed.

tranquilizer a drug used to induce a calming effect in a person.

transamination a reaction largely performed in the liver by which nitrogen is removed from amino acids.

transcortin (trăns·kor´tĭn) a protein to which corticosteroid hormones are bound as they circulate in plasma.

transcription the process by which RNA is synthesized using DNA as a template.

transferrin (trăns·fěr´ĭn) a protein to which iron ions are bound as they circulate in plasma.

translation the process of using the information in a messenger RNA molecule for the synthesis of a protein.

translocation an abnormal alteration in chromosome structure involving the transfer of a portion of one chromosome to another, nonhomologous, chromosome.

transmitter substance any of several chemicals released into a synaptic cleft from a presynaptic cell that stimulates or inhibits the production of an impulse in the postsynaptic cell. The postsynaptic cell may be another neuron, muscle cell, or gland cell.

transport maximum the maximum plasma concentration of a substance that can be completely absorbed by nephrons as the glomerular filtrate passes through the nephrons.

transpulmonary pressure the pressure difference between the gas in the alveoli and the gas in the interpleural space. When it is positive, the lungs expand; when it is negative, they contract.

trapezium (tră·pē´zē·ŭm) an irregularly shaped four-sided figure.

trauma (traw´mă) a wound or injury to an organ or tissue or to one's mental state.

treppe (trěp´ē) a phenomenon of muscle contraction in which the strength of contraction increases as the muscle is stimulated rapidly and the muscle is not allowed to relax between stimuli.

triad an arrangement of membranes and transverse tubules at the junction of two sarcomeres in striated muscle.

trichinosis (trĭk˝ĭn·ō´sĭs) a rare disease caused by eating a parasite in uncooked pork, *Trichinella spiralis*.

tricuspid valve a valve that has three divisions, or cusps, that controls the flow of blood from the right atrium into the right ventricle; also called the right atrioventricular valve.

trigeminal neuralgia (trī·jěm´ĭn·ăl nū·răl´jē·ă) sudden and intense pain associated with the trigeminal nerve (fifth cranial nerve); also called tic douloureux.

trigone (trī´gōn) a triangular area formed at the base of the urinary bladder by the orifices of the ureters and the urethra.

triiodothyronine (trī˝ī·o˝dō·thī´rō·nēn) one of two primary thyroid hormones; also designated T3.

trisomy (trī·sōm´ē) an abnormal condition in which the cells of a person have three chromosomes of a particular type instead of the normal two.

trochlea (trŏk´lē·ă) a ring of fibrous material through which the tendon of a muscle passes that enables a muscle to pull a structure in a direction other than its own orientation, much like a pulley enables a force in one direction to effect movement of a structure in another direction. The superior oblique muscle of the eye turns the eyeball in this way.

trophic hormones hormones produced by the anterior pituitary gland that stimulate another endocrine gland to produce its hormone.

trophoblast (trŏf´ō·blăst) the thin shell of cells that surround the inner cell mass in a blastocyst.

tropomyosin (trō˝pō·mī´ō·sĭn) a protein associated with actin units in the thin filaments of a sarcomere and involved in the sliding filament mechanism of muscle contraction.

troponin (trō´pō·nĭn) a protein that occupies active sites in the thin filaments of a sarcomere. Movement of the troponin molecules in response to calcium ions exposes the active site and allows the filaments to slide over one another.

true vocal cords the lower of two pairs of folds of mucous tissue in the larynx used to produce sound by controlling their tension.

trunk the chest and abdominal portions of the body.

trypsin (trĭp´sĭn) a pancreatic enzyme involved in the digestion of protein in the small intestine.

trypsinogen a precursor of trypsin, which is activated in the small intestine.

T tubule a membranous tubule continuous with the sarcoplasmic reticulum that extends down into the muscle fiber; also called a transverse tubule.

tubal pregnancy pregnancy in which the embryo has attached and is developing in the wall of the oviduct instead of the uterus.

tubercle (tū´běr·kl) a small nodule caused by infection with *Mycobacterium tuberculosis*; a small bump on a bone.

tuberculosis an infectious disease usually caused by *Mycobacterium tuberculosis*. In humans it usually begins as a lung infection and can spread to other tissues in the body. The disease is characterized by the formation of tubercles in the infected tissues with a resulting tissue necrosis.

tubular reabsorption the process by which solutes in the glomerular filtrate are recovered as the filtrate passes through the nephron.

tubular secretion the process by which solutes are transported through the walls of a nephron into the glomerular filtrate as the filtrate passes through the tubular portion of the nephron.

tumor an abnormal growth of tissue, which may be either benign or cancerous.

tunica a membrane that envelopes or covers an organ.

tunica adventitia the outermost layer of the wall of an artery or vein, composed primarily of connective tissue; also called the tunica externa.

tunica albuginea a layer of tough, dense, fibrous tissue that surrounds each ovary in a female and each testis and corpus cavernosum in a male.

tunica intima an innermost coat of endothelial cells that lines the inner walls of arteries and veins.

tunica media the middle layer in an artery or vein, which contains smooth muscle cells associated with elastic and collagenous fibers.

turbinate (tŭr´bĭ·nāt) shaped like a cone or a child's top.

Turner syndrome a genetic disorder of development in females resulting from a single X chromosome and no second X or Y chromosome in the cells, thus there are only 45 chromosomes per nucleus, instead of the usual 46.

T wave a deflection in the waves of an electrocardiogram that corresponds to repolarization of the ventricle musculature.

twitch a single brief contraction of a muscle, muscle fiber, or group of fibers in response to a single stimulus.

tympanic membrane the membrane that separates the external acoustic meatus from the tympanic cavity of the middle ear; also called the eardrum.

tyrosinase (tī·rō′sĭn·ās) an enzyme that catalyzes the conversion of tyrosine to the pigment melanin.

tyrosine (tī′rō·sēn) one of the 20 amino acids found in protein, which serves as a precursor to melanin.

U

ubiquinone (ū·bĭk′wĭ·nōn) an alternative name of coenzyme Q, the component of the respiratory chain that receives electrons from reduced NAD or FAD in cellular respiration.

ulcer any open sore or other lesion in the surface of a mucous membrane or the skin.

ulcerative colitis a chronic, often fatal disease of the colon, of unknown cause, characterized by inflammation, ulceration of the mucosal membrane, bloody diarrhea, and microbial infection.

ulna the larger of the two bones of the forearm, located on the same side of the arm as the little finger.

ultrasound scan a technique for visualizing internal organs on the basis of their reflection of high frequency sound. The sound is detected and converted to a visual image for analysis.

umbilical cord a tube that contains the arteries and veins that conduct fetal blood to and from the placenta.

umbilicus (ŭm·bĭ·lī′kŭs) the scar on the surface of the abdomen that marks the point where the umbilical cord was attached to the fetus; also called the navel.

unicellular consisting of one cell.

unipennate muscle a type of muscle in which the fasciculi insert diagonally along the length of one side of a tendon.

unipolar neuron a type of neuron characteristic of sensory ganglia in which there is a single process that divides into two branches, one terminating in a receptor in a peripheral sensory organ and the other usually extending to the central nervous system. Impulses initiated in the sensory end thus pass through both branches.

universal donor a person who has type O blood, which lacks antigens in the ABO blood-typing series, and thus fails to cause a reaction with anti-A and anti-B antibodies in a potential recipient.

universal recipient a person who has type AB blood in the ABO system, so named because the person's blood lacks antibodies against either the A or B antigens. Consequently, he or she will not have an allergic reaction to A or B antigens that might be received in a blood transfusion.

unsaturated fat a fat molecule in which one or more of the component fatty acids has pairs of carbon atoms joined by double bonds.

urea a waste product formed by the combination of two amino groups and a CO_2 molecule in the removal of excess nitrogen; a major waste product in urine.

uremia (ū·rē′mē·ă) an abnormally high concentration of nitrogen-containing waste products in the blood that are normally excreted by the kidneys.

ureter (ū′rē·ter) one of the two tubes that conduct urine from the kidneys to the urinary bladder.

ureteric bud an outgrowth of the mesonephros of the embryo that begins to grow from the base of each mesonephric duct about the fifth week of development and eventually gives rise to the collecting tubules, calyces, and pelvis of the kidney and the ureter.

urethra (ū·rē′thră) the tube through which urine passes as it leaves the urinary bladder. It also serves as a conducting channel for semen in a male.

urethritis (ū″rē·thrī′tĭs) inflammation of the urethra.

urinalysis (ū″rĭ·năl′ĭ·sĭs) a clinical examination of urine.

urinary calculi stones formed in various portions of the urinary tract consisting of cystine or calcium salts of uric, oxalic, carbonic, or phosphoric acid.

urinary cast gels of protein formed in renal tubules that are flushed out with the urine.

urinary pole the opening from the Bowman's capsule that enters into the proximal tubule of the nephron.

urinary reflex a reflex stimulated by distension of the urinary bladder. It results in contraction of the bladder and relaxation of the internal sphincter.

urinary system the organs involved in the production, storage, and excretion of urine, consisting of the two kidneys, the two ureters, the urinary bladder, and the urethra.

urination the process by which the urinary bladder is emptied of urine; also called micturition.

urine a fluid formed in the kidneys, stored in the bladder, and released through the urethra. In a healthy person, it consists of water, nitrogenous wastes, and various electrolytes.

urobilinogen (ū″rō·bī·lĭn′ō·jĕn) a derivative of bilirubin formed in the large intestine by the action of microorganisms that live there.

urogenital sinus a chamber in the embryo that gives rise to the urinary bladder.

urogenital triangle the anterior division of the perineum that extends from the symphysis pubis anteriorly to a line extending between the two ischial tuberosities. In the female it includes the external genitalia.

uterine tube a tube that opens at an ovary and connects to the uterus. It is also called the oviduct or fallopian tube.

uterosacral (ū″tĕr·ō·sā′krăl) pertaining to the region around the uterus and the sacrum.

uterus (ū′tĕr·ŭs) a hollow, muscular structure in the female in which an embryo becomes implanted and in which the fetus develops during pregnancy.

utricle (ū′trĭk′l) the larger of two sacs in the membranous labyrinth of the inner ear containing a sensory region that senses the position and movement of the head.

uvea (ū′vē·ă) the vascular tunic of the eye lying just below the sclera and consisting of the choroid, ciliary body, and iris.

uvula (ū′vū·lă) a small, fleshy mass of tissue that hangs down from the palatoglossal arch over the base of the tongue.

V

vaccination an injection or other introduction of antigens that will cause an immune reaction and the production of antibodies against a disease-causing virus or microorganism. The antigens may be present on inactivated microbes, may be extracted from the infective microbes, or may be taken from a different organism but cause the same response.

vaccine a solution used in vaccination containing antigens that will cause an immune response that confers immunity to a specific disease-causing microbe.

vagina (vă·jī′nă) the canal in the female genitalia that extends from the external opening to the cervix.

vagus nerve the tenth cranial nerve, which is widespread in distribution, innervating most thoracic and abdominal organs.

valgus (văl′gŭs) a term that indicates a position or orientation in which a structure is bent outward.

vallate (văl′āt) **papilla** one of a group of taste buds at the posterior portion of the dorsal surface of the tongue.

valve a structure that controls the passage of material from one chamber to another, as in the valves of the heart, which control the flow of blood from the atria to the ventricles and from the ventricles to the major arteries.

varicella (văr″ĭ·sĕl′ă) a contagious viral disease characterized by inflamed eruptions causing small red spots on the skin and mucous membranes; also called chickenpox.

vas deferens (văs def′er·ens) *see ductus deferens*.

vasa recta (vā′să rĕk′tă) a network of long, looping capillaries that extend from the efferent arterioles of juxtamedullary nephrons, looping down to the tip of the renal pyramid and back toward the corpuscle. Blood in these vessels carries off water and solutes that have been reabsorbed from the nephrons.

vasa vasorum (vā′să văs·ō′rŭm) tiny arteries and veins that supply nutrients and remove waste products from the tissues in the walls of the larger blood vessels.

vascular pertaining to the blood and lymph vessels.

vasectomy (văs·ĕk′tō·mē) the cutting and sealing off of the cut ends of the vas (ductus) deferens as a means of preventing the addition of sperm to semen, used as a contraceptive technique in males.

vasoconstriction a decrease in the diameter of blood vessels, particularly of arterioles.

vasodilation (văs″ō·dī·lā′shŭn) an increase in the diameter of blood vessels, especially arterioles.

vasodilator any substance that causes dilation of blood vessels.

vasopressin (văs″ō·prĕs′ĭn) *see antidiuretic hormone*.

vasospasm spontaneous contraction of the muscles in blood vessels, causing them to constrict in diameter.

vein a vessel that carries blood in the direction of the heart.

vena cava the major inferior and superior veins that deliver blood to the right atrium.

ventilation a general term for the exchange of gases that occurs between alveolar gases and blood in the lungs.

ventral pertaining to the forward part of the body; in humans it means the same as anterior.

ventricle (vĕn′trĭk·l) either of the two large muscular chambers that pump blood out of the heart; one of several cavities in the brain that contains cerebrospinal fluid.

ventricular fibrillation spontaneous rapid contractions of the ventricles.

venule a small vein.

vermiform (vĕr′mĭ·form) **appendix** a long, narrow sac, about 7 to 15 cm in length, that extends from the tip of the cecum of the large intestine.

vermis (vĕr′mĭs) wormlike in appearance or form.

vertebra (vĕr″tĕ·bră) one of the 33 bones in the spinal column.

vertebral canal the hollow column formed by the vertebral foramina through which passes the spinal cord.

vertebral column the column of vertebrae that forms the spine.

vertigo (vĕr′tĭ·gō) a false sensation of spinning; dizziness.

very low density lipoprotein (VLDL) a lipid micelle rich in triglycerides and containing smaller amounts of cholesterol and phospholipids.

vesicle (vĕs′ĭ·kl) small, membrane-bound, spherical organelles in the cytoplasm of a cell.

vesicular lesion a small fluid-filled sac on the skin.

vestibular membrane the membrane in the inner ear that separates the scala vestibuli from the scala media; also called Reissner's membrane.

vestibule (vĕs′tĭ·būl) a partially enclosed area that serves as an entrance to another chamber, as in the vestibule of the vagina.

villi (vĭl′ī) tiny projections of cells that line the inner surface of certain tubules, such as the small intestine, where they absorb nutrients and fluids.

villous pigmented synovitis inflammation of a synovial membrane in a joint characterized by the production of small brown nodes.

virilism (vĭr′ĭl·ĭzm) the development of secondary male sexual characteristics in a female, for example, as a result of the effect of testosterone released from the adrenal cortex.

virulent the capacity of a microorganism or other agent to cause disease.

visceral (vĭs′ĕr·ăl) pertaining to the organs of the abdominal cavity.

visceral peritoneum the portion of the peritoneum that covers the anterior surface of the abdominal organs.

visceral pleura the portion of the pleurae that covers the lungs and lies opposite the parietal pleura, with the pleural cavity lying between them.

visceroceptors sensory receptors in the walls of the larger blood vessels and thoracic and abdominal organs. They monitor such phenomena as pH, carbon dioxide level, pressure, and stretching of smooth muscle. They are also called interoceptors.

vital capacity the gaseous volume of the lungs exclusive of the residual volume; that is, the total of the inspiratory reserve volume, the tidal volume, and the expiratory reserve volume.

vitamin any of a group of organic substances that serve as cofactors in chemical reactions, thereby helping regulate metabolism.

vitelline (vī′tĕl′ēn) **sac** an extraembryonic sac that develops by the second week after conception. It is a source of blood for the early embryo, the primitive gut is derived from it, and it produces cells that migrate to the gonads where they give rise eventually to spermatogonia and oogonia; also called yolk sac.

vitreous (vĭt′rē·ŭs) **humor** the viscous, transparent gel that lies between the lens of the eye and the retina.

volley a pattern of impulses arriving at different times at a point on a nerve some distance away from the point of stimulation. The impulses arrive at different times because the nerve contains neurons that transmit impulses at different speeds.

voluntary muscle an alternative name for skeletal muscle, reflecting that it is largely under voluntary control.

vomer (vō′mĕr) a thin, flat vertical bone that forms the posterior portion of the nasal septum.

vulva (vŭl′vă) a general term for the external female genitalia including the labia majora, labia minora, clitoris, and vestibule of the vagina.

W

wart an elevated, sometimes pigmented, growth on the skin.

Weber's test a hearing test in which a vibrating tuning fork is touched to the forehead.

Wharton's jelly a gelatinous material that fills large intercellular spaces in the connective tissue of an umbilical cord.

whiplash injury that occurs to the spine and spinal cord at the fourth and fifth cervical vertebrae as a result of a rapid accel-

eration or deceleration of the head with respect to the vertebral column.
white matter portions of the central nervous system that have a glistening white appearance due to the presence of the myelinated axons of nerve cells.
wisdom tooth one of the third molars, located behind the first and second molars in both jaws. They do not appear in some people.
withdrawal reflex a reflex in which one quickly withdraws an extremity from a stimulus.

X

X chromosome one of the sex chromosomes, present twice per nucleus in most females and once per nucleus in most males.
xeroderma pigmentosum (zē″rō·dĕr′mă pĭg″měn·tō′sŭm) a genetic disorder resulting from a deficient DNA repair system that makes the DNA susceptible to damage by ultraviolet irradiation. Areas of the skin lose their pigmentation or become excessively pigmented; scarring of the skin also occurs and skin cancers often develop.
X ray a portion of the electromagnetic spectrum of sufficient energy and wavelength to pass through soft tissue and used to expose photographic film for examination of internal structures of the body; the photograph made by exposing X rays to film; also called radiograph.
xiphoid process (zĭf′oyd) a bony protuberance that projects downward at the posterior end of the sternum and to which certain ligaments and muscles are attached.

Y

Y chromosome a sex chromosome for which few if any genes are known, except that its presence is normally necessary and sufficient to produce maleness.
yolk sac *see vitelline sac.*

Z

Z line the dark line in an electron micrograph of striated muscle that marks the borders of the sarcomere; also called Z disk.
zona fasciculata (zō′nă făs·sĭk′ū·lă·tă) the middle zone of the adrenal cortex, responsible primarily for the production of glucocorticoid hormones, which control carbohydrate and protein metabolism.
zona glomerulosa (glŏ″měr·ū·lō′să) the outermost zone of the adrenal cortex, just within the capsule, responsible for the production of mineralocorticoid hormones.
zona pellucida (pěl·ū′·sĭ·dă) a glycoprotein-rich covering that surrounds a primary oocyte.
zona reticularis (rĕ·tĭk″ū·lăr′ĭs) the innermost zone of the adrenal cortex, which produces sex hormones that contribute to the development of secondary sexual characteristics.
zygote (zī′gōt) the cell produced by the fusion of an egg and sperm.
zymogenic cell (zī″mō·jěn′ĭk) cells in the gastric glands that secrete pepsinogen, the precursor to pepsin in the stomach juices.

Copyrights and Acknowledgments

Illustrations

Unit Openers Pages 2, 98, 258, 438, 536, and 644: Keith Kasnot; p. 696: Keith Kasnot, source: adapted from © Lennart Nilsson, *Behold Man*, Little, Brown and Co.

Chapter 1 Page 4, Figures 1-2, 1-3, 1-4, 1-5, 1-6, p. 14 (top), and p. 15 (top): David Fischer; 1-7: Art Tom.

Chapter 2 Page 18, Figures 2-1, 2-2, 2-3, 2-4, 2-5, 2-6, 2-7, 2-8, 2-9, 2-10, 2-11, 2-12, 2-13, 2-15, 2-16, 2-17, and 2-19: David Fischer; 2-14 and 2-18: Art Tom.

Chapter 3 Page 46, Figures 3-1a, 3-3, 3-5, 3-6, 3-7, 3-8, 3-9, 3-10, 3-11, 3-12, 3-13, 3-14b, 3-15b and c, 3-16b and c, 3-17b and c, 3-19, 3-21a, b, and c, 3-24b and c, 3-25, 3-26c, 3-27, 3-28, and 3-29: David Fischer; 3-18b and c: Art Tom.

Chapter 4 Figures 4-1a, 4-2, 4-3, 4-4, 4-5, 4-8, and 4-9: David Fischer.

Chapter 5 Page 98, Figures 5-1, 5-3, 5-4, 5-5, and p. 108: David Fischer.

Chapter 6 Page 114, Figures 6-1, 6-2, 6-4, 6-5, and 6-6b: David Fischer.

Chapter 7 Page 130, Figures 7-1, 7-2, 7-3a, 7-4, 7-5, 7-6, 7-7, 7-8, 7-9a, 7-10, 7-11, 7-12, 7-13, 7-14, 7-15a, 7-16, 7-17, 7-18, 7-19, 7-20, 7-21, 7-22, 7-23, 7-24, 7-25, 7-26a and b, 7-27a, 7-28a, 7-29, 7-30, 7-31a, b, d, and e, 7-32a and b, 7-33a, and 7-34: Darwen and Vally Hennings.

Chapter 8 Page 164, Figures 8-1, 8-2, 8-3, 8-4, 8-5, 8-6, 8-7, and 8-8: Darwen and Vally Hennings.

Chapter 9 Page 180, Figures 9-4a, 9-8, 9-9, 9-17c, and 9-18b: Ed Zilberts; 9-1 and 9-2: Keith Kasnot; p. 182, 9-3b and d, 9-5b, 9-6b, 9-7, 9-11, and 9-12: David Fischer; 9-10, 9-13, 9-14, and 9-15: Art Tom; 9-16: R. Margaria (Ed.), *Exercise at Altitude*, 1967, Amsterdam: Exerpta Medica; 9-19: Sandra McMahon.

Chapter 10 Page 206, Figures 10-1, 10-2, 10-3, 10-4, 10-5, 10-6, 10-7, 10-8, 10-9, 10-10, 10-11, 10-12, 10-13, 10-14, 10-15, 10-16, 10-17, 10-18, 10-19, 10-20, 10-21, 10-22, 10-23, and 10-24: Sandra McMahon.

Chapter 11 Page 260 and Figure 11-3a: Keith Kasnot; 11-1, 11-2, 11-4, 11-5, 11-6, 11-8, 11-9, 11-11, 11-12, 11-13, 11-15a, 11-17, and 11-19: David Fischer; 11-10 and 11-18: Art Tom; 11-14: Bell, Emslie-Smith, and Paterson, *Textbook of Physiology and Biochemistry*, Churchill Livingstone, 1976; 11-16: Jim Cherry; 11-20: Art Tom, source: from *Molecular Biology of the Cell* by Bruce Alberts, Dennis Bray, Julian Lewis, Martin Raff, Keith Roberts, and James D. Watson. Garland Publ., New York, 1983, p. 1048.

Chapter 12 Page 290, Figures 12-1, 12-2, 12-3, 12-4, 12-5, 12-6, 12-9, 12-11, 12-12, and 12-16: Darwen and Vally Hennings; p. 292, 293, 12-8, 12-10, 12-14, 12-17, 12-18, and 12-19: Christine Bondante; 12-7, 12-13, and 12-15: Keith Kasnot.

Chapter 13 Page 322, Figure 13-1, p. 324, 13-2, 13-3, 13-4, 13-5, 13-6, 13-7, 13-8, 13-9, 13-10, 13-12, 13-13, 13-15, and 13-16: Christine Bondante; 13-11, 13-14, 13-17, 13-18, and 13-19: David Fischer; p. 336: Art Tom, source: adapted from *Medical Times*, June 1980, p. 95.

Chapter 14 Page 342, Figures 14-1 and 14-3: Christine Bondante; 14-2: David Fischer.

Chapter 15 Page 356, Figures 15-1, 15-2, 15-3, 15-4, 15-5, 15-6, and 15-7: Christine Bondante; 15-10a, b, and c: David Fischer.

Chapter 16 Page 376, Figures 16-4, 16-7, 16-16, and 16-19: Ed Alexander and Cynthia Turner; 16-1, p. 378, 16-2a, p. 392, 16-14, p. 394, 16-21, and 16-22: Christine Bondante; 16-3, 16-5a, 16-8, 16-9, p. 389 (left), 16-10, 16-12, 16-13, 16-17, 16-18, 16-20, and 16-23: David Fischer; p. 389 (right): Art Tom; p. 394: Keith Kasnot; p. 403: © Copyright 1970, CIBA-GEIGY Corporation. Reproduced with permission from *Clinical Symposia* by Frank H. Netter, M.D. All rights reserved.

Chapter 17 Page 406, Figures 17-1, 17-5, and 17-17a: Jim Cherry; 17-2, 17-3, 17-6, 17-7, and 17-18: David Fischer; 17-4: Art Tom; 17-8a, 17-11a, 17-12, 17-13a and b, 17-14, 17-15a, and 17-16: Christine Bondante.

Chapter 18 Page 440, Figures 18-1, 18-2, 18-12a, 18-15a, 18-18, 18-20, 18-21, 18-22, and 18-26a: Keith Kasnot; p. 442, 18-5, 18-6a and b, 18-9, 18-11, 18-14, 18-16a and b, 18-17, and 18-19: David Fischer; 18-13: David Fischer, source: adapted from A. Hopkins, *Journal of Physiology*, 182, 1966, pp. 142–149; 18-3, 18-4a, 18-8a, 18-10, 18-23, 18-25, and 18-28: Christine Bondante.

Chapter 19 Figures 19-1, 19-2, 19-3, 19-4, 19-5, 19-6, 19-7, 19-8, 19-9, 19-11, 19-12, and 19-14: David Fischer.

Chapter 20 Page 504 and Figure 20-1: Keith Kasnot; p. 506, 20-11, 20-12a, 20-13, 20-16, 20-17, and 20-20: David Fischer; 20-2, 20-3, 20-4, 20-5, 20-6, 20-7, 20-8, and 20-9: Christine Bondante; 20-14: Bell, Emslie-Smith, and Paterson, *Textbook of Physiology and Biochemistry*, Churchill Livingstone, 1976; 20-15, 20-18, 20-19, and 20-21: Art Tom.

Chapter 21 Page 538, Figures 21-3, 21-6, 21-7, 21-8, and 21-10: Keith Kasnot; 21-1 and 21-2: Art Tom; 21-5: David Fischer.

Chapter 22 Page 558, Figures 22-1, 22-3a and b, 22-4, 22-5, 22-6, 22-7, and 22-10: Keith Kasnot; p. 560, 22-2, 22-8, 22-9b, and 22-13: David Fischer; 22-11: Art Tom; 22-14b, 22-15: Christine Bondante.

Chapter 23 Page 584, Figures 23-1, 23-2, 23-7, 23-9, 23-16, 23-20, and 23-22: Sandra McMahon; 23-3a, 23-4, 23-5, 23-8, 23-23, and 23-26a: David Fischer; 23-6 and 23-15: Keith Kasnot; 23-10, 23-11, 23-12, 23-13, 23-14, 23-17, 23-18, 23-19, 23-21, 23-24, and 23-27: Christine Bondante; 23-25: Art Tom.

Chapter 24 Figures 24-1, 24-5a, 24-6, 24-7, 24-8, 24-9, 24-10, 24-11, and 24-14: David Fischer; 24-2, 24-3a, and 24-4a: Christine Bondante; p. 622(left): Christine Bondante, source: American Cancer Society, Inc.; 24-13: Art Tom.

Chapter 25 Page 646, Figures 25-1, 25-2, 25-3, 25-4, p. 650, 25-10a and b, 25-20b, 25-21, and 25-22: Christine Bondante; 25-6, 25-7, and 25-23: Keith Kasnot; 25-9, 25-11, 25-13, and 25-19: David Fischer; 25-15: David Fischer; source: (a) adapted by permission from Brian R. Duling, "Integrated Nephron Function," in *Physiology*, Berne/Levy, editors, 2/E, 1988, fig. 47-5, p. 788; (b) adapted by permission from Brian R. Duling, "Integrated Nephron Function," in *Physiology*, Berne/Levy, editors, 2/E, 1988, fig. 47-8, p. 791; 25-16: Bell, Emslie-Smith, and Paterson, *Textbook of Physiology and Biochemistry*, Churchill Livingstone, 1976; 25-17 and 25-18: Art Tom.

Chapter 26 Page 680, Figures 26-1, and 26-3: David Fischer; 26-2, 26-5, and 26-6: Art Tom.

Chapter 27 Page 698, Figures 27-1, 27-2, 27-4, 27-5, 27-6a, 27-7, 27-8, 27-9, 27-10, 27-13, 27-15, 27-16, 27-17, 27-18, and 27-20: Sandra McMahon; pp. 700, 701, and 704: Christine Bondante; 27-19: Keith Kasnot; 27-22: David Fischer, source: adapted from Midgley, A. R., in *Human Reproduction*, Hafez, E. S. E., and Evans, T. N., eds., 1973, fig. 23-26. Reproduced with permission by J. B. Lippincott Co./Harper & Row, Publ.; 27-23: David Fischer.

Chapter 28 Page 732, Figure 28-1, and p. 766: David Fischer; 28-2, 28-3, 28-12, 28-13, 28-14, 28-15, 28-19, and 28-20: Keith Kasnot; 28-4, 28-5, 28-6, 28-7, 28-9, and 28-11: Christine Bondante; 28-16, 28-17, 28-18, 28-21, 28-23, and 28-26: Art Tom; p. 766: Adapted from *The Merck Manual of Diagnosis and Therapy*, 15th Edition, p. 2154, edited by Robert Berkow. Copyright © 1987 by Merck & Co., Inc. Used with permission.

Photographs

Chapter 1 Figures 1-1a and 1-1b: © Phototake; p. 14 (bottom) and p. 15 (bottom): © Dan McCoy/Rainbow.

Chapter 2 Page 24 (top): © Hank Morgan/Rainbow; p. 24 (bottom): © Dan McCoy/Rainbow.

Chapter 3 Figure 3-1b: © Biophoto Associates/Photo Researchers; 3-2, 3-4c, and 3-21d: © Dennis Kunkel/Phototake; 3-4a and b: Courtesy of Daniel S. Friend, M.D., University of California, San Francisco; p. 52: © Richard Hutchings/Photo Researchers; 3-14a, 3-15a, 3-16a, 3-17a, and 3-24a: © Lennart Nilsson, *The Body Victorious*, Delacorte Press, Dell Publishing Co.; 3-18a: Reproduced from *An Atlas of Immunofluorescence in Cultured Cells* by M. C. Willingham and I. Pastan; 3-20: Courtesy of Keith R. Porter, University of Maryland; 3-22: © Martin M. Rotker/Phototake; 3-23, 3-26a and b, and p. 75: © Phototake.

Chapter 4 Page 80 and Figure 4-7d: © Robert Knauft/Photo Researchers; 4-1b: © James A. Dennis/Phototake; p. 84: © Robert Caputo/Stock, Boston; 4-6a and b, 4-7c and g: © Dr. Mary Notter/Phototake; 4-7a: © Manfred Kage/Peter Arnold, Inc.; 4-7b: © Ed Reschke/Peter Arnold, Inc.; 4-7e: © Michael Abbey/Photo Researchers; 4-7f: © Harry J. Przekop/Medichrome; p. 96: Scripps Memorial Hospital and staff.

Chapter 5 Page 102 (a and b): © Zeva Oelbaum/Peter Arnold, Inc.; p. 102 (c): © John Radcliffe/Photo Researchers; 5-2: © Dr. Mary Notter/Phototake; p. 104 (a and b): © Phototake; p. 105: © James Stevenson/Photo Researchers.

Chapter 6 Pages 116 (a), p. 117 (f and h), and Figure 6-6: Courtesy of Mission Bay Hospital Radiology Department; p. 116 (b), p. 119 (a), 6-3, 6-7, and p. 127 (left): © Phototake; pp. 116 (c), 119 (b), and 120 (bottom left): © SIU/Peter Arnold, Inc.; p. 116 (d): Courtesy of Mission Bay Hospital Radiology Department and Nikki Kelly; p. 117 (e): © Donald C. Wetter/Medichrome; p. 117 (g): Dr. Olcay S. Cigtay in Rosse, Cornelius, and Clawson, D. Kay, *The Musculoskeletal System in Health and Disease*, 1980, Harper & Row, Publ., fig. 6-1, p. 91; p. 126: © Biophoto Associates/Photo Researchers; p. 127 (right): © Jacques M. Chenet/Woodfin Camp & Associates.

Chapter 7 Figure 7-3b: Courtesy of Mission Bay Hospital Radiology Department and Nikki Kelly; 7-9b: © AFIP/Photo Researchers; 7-15b, 7-30c, 7-33b, and p. 158 (top): © Phototake; 7-26c and p. 159: © Martin M. Rotker/Taurus Photos, Inc.; 7-27b: Courtesy of Mission Bay Hospital Radiology Department; 7-28b: © Walter A. Reiter III/Phototake; 7-31c: Courtesy of San Diego Diagnostic Radiology Medical Group, Inc.; 7-32c: © Dr. John Denton/Phototake; p. 156: © Biophoto Associates/Photo Researchers; p. 158 (bottom): © Susan Leavines/Photo Researchers.

Chapter 8 Pages 168 (left) and 170 (bottom): © Phototake; p. 170 (top): © James Stevenson/Photo Researchers.

Chapter 9 Figure 9-1a: © Eric Grave/Photo Researchers; 9-1b and c, 9-4b, 9-5a, 9-6a (top), and 9-18a: © Biophoto Associates/Photo Researchers; 9-3a and 9-17b: © Don Fawcett/Photo Researchers; 9-3c: Courtesy of D. E. Kelly, USC School of Medicine, in Rosse, Cornelius, and Clawson, D. Kay, *The Musculoskeletal System in Health and Disease*, 1980, Harper & Row, Publ.; 9-6a (left, center, and right): Reprinted from F. Pepe, *Biological Macromolecules Series: Subunits in Biological Systems*. Marcel Dekker, Inc., N.Y., 1971; 9-17a: © Michael Abbey/Photo Researchers.

Chapter 10 Page 256: © Tony Duffy/Allsport.

Chapter 11 Figures 11-3b and 11-15c: © Scott/Phototake; 11-7: Courtesy of D. E. Kelly, USC School of Medicine. From *Bailey's Textbook of Microscopic Anatomy*, 18th edition, Kelly, D. E., Wood, R. L., and Enders, A. E., editors. Williams & Wilkins Co., Baltimore, 1984, p. 354; 11-15b: © Dennis Kunkel/Phototake.

Chapter 12 Figure 12-2d: © Martin M. Rotker/Phototake; p. 298 (left and right): © Dan McCoy/Rainbow; p. 304 (all): © Dr. David Rosenbaum/Phototake; p. 315: © Nubar Alexanian/Stock, Boston.

Chapter 14 Page 350 (left): © F. B. Grunzweig/Photo Researchers; p. 350 (right): © 1986, M. H. English/Georgia Crime Lab/Photo Researchers.

Chapter 15 Page 364: © Paul Biddle and Tim Malyon/Photo Researchers; 15-8: © Tony Duffy/Allsport; 15-9: © Dr. David Rosenbaum/Phototake; 15-10d: © Dennis Kunkel/Phototake.

Chapter 16 Figure 16-2b: © Lennart Nilsson, *Behold Man*, Little, Brown and Co.; 16-5b: From *Tissues and Organs: A Text-Atlas of Scanning Electron Microscopy* by Richard G. Kessel and Randy H. Kardon. Copyright © 1979 W. H. Freeman and Co. Reprinted with permission; 16-6: © F. R. Masini/Phototake; p. 384 (all): © Margaret Cubberly/Phototake; 16-11: HBJ Photo/Larry Hoagland, Professional Photographic Services. Courtesy Jerry Young and Colin Young; 16-15: HBJ Photo/Larry Hoagland, Professional Photographic Services. Courtesy Colin Young.

Chapter 17 Figure 17-8b and 17-9: © Martin M. Rotker/Phototake; 17-10: © John Paul Kay/Peter Arnold, Inc.; 17-11b: © Ida Wyman/Phototake; 17-13c and d: © Phototake; 17-15b: © Biophoto Associates/Photo Researchers; 17-17b and c: © Dr. Mary Notter/Phototake; p. 437 (left): Courtesy of University of California at San Diego Medical Center; p. 437 (right): © Richard Wood/Taurus Photos, Inc.

Chapter 18 Figure 18-4b: © Margaret G. Cubberly/Phototake; 18-4c: © Omikron/Photo Researchers; p. 446 (left and right) and 18-7: Dr. R. P. Renner, Professor, SUNY at Stony Brook, School of Dental Medicine; p. 446 (center) and 18-8b: © Biophoto Associates/Photo Researchers; 18-6c (all): © Dr. S. L. Gibbs/Peter Arnold, Inc.; p. 452: Scripps Memorial Hospital and staff; 18-12b, p. 456 (center and right), 18-15b, and 18-26c: © Phototake; p. 456 (left): Courtesy of Olympus Corp.; 18-15c: © Fawcett/Ito/Photo Researchers; 18-16c: Massimo Piccin, M.D., Piccin Nuova Libraria, S.p.A., Padua, Italy; 18-24a, 18-24b, and 18-26b: From *Tissues and Organs: A Text-Atlas of Scanning Electron Microscopy* by Richard G. Kessel and Randy H. Kardon. Copyright © 1979 W. H. Freeman and Co. Reprinted with permission; 18-27: © Don Fawcett/Photo Researchers.

Chapter 19 Page 478 and Figure 19-10: From *Tissues and Organs: A Text-Atlas of Scanning Electron Microscopy* by Richard G. Kessel and Randy H. Kardon. Copyright © 1979 W. H. Freeman and Co. Reprinted with permission; p. 489: © Phototake; 19-13: © Dr. R. P. Clark and M. Goff/Photo Researchers.

Chapter 20 Figure 20-10: © Science Photo Library/Photo Researchers; p. 514 (right) and 20-12b: © Phototake; p. 530: © Biophoto Associates/Photo Researchers; p. 531: © Martin M. Rotker/Phototake; p. 535: Scripps Memorial Hospital and staff.

Chapter 21 Figure 21-4: © Dr. W. Reinhart/Phototake; p. 544: © Phototake; p. 546: Scripps Memorial Hospital and staff; 21-9: © Dennis Kunkel/Phototake; p. 555 (1a): © Southern Illinois University Photo Researchers; p. 555 (1b): Scripps Memorial Hospital and staff.

Chapter 22 Figure 22-3c and page 580: © Martin M. Rotker/Phototake; 22-9a: © Michael Abbey/Photo Researchers; 22-14 and p. 575: © Phototake; p. 581: Scripps Memorial Hospital and staff.

Chapter 23 Figure 23-3b: Reproduced with permission from *Atlas of Radiologic Anatomy* by Lothar Wicke, copyright 1987, Urban & Schwarzenberg, Baltimore/Munich; 23-26b: From *Tissues and Organs: A Text-Atlas of Scanning Electron Microscopy* by Richard G. Kessel and Randy H. Kardon. Copyright © 1979 W. H. Freeman and Co. Reprinted with permission; p. 614: © Biophoto Associates/Photo Researchers.

Chapter 24 Page 618 and Figure 4-12: © Boehringer Ingelheim International GmbH. Photo: Lennart Nilsson, *The Body Victorious*, Dell Publishing Co., p. 191; 24-3b: © Ed Reschke/Peter Arnold, Inc.; 24-5b: © Phototake; p. 630 (top): © Chuck Nacke/Picture Group; p. 630 (bottom): Boehringer Ingelheim International GmbH. Photo: Lennart Nilsson; 24-15a and b: Lester V. Bergman, N.Y.; 643: © Paolo Koch/Photo Researchers.

Chapter 25 Figures 25-5, 25-8a: © Manfred Kage/Peter Arnold, Inc.; 25-8b: Courtesy Richard L. Wood, from *Bailey's Textbook of Microscopic Anatomy*, 18th edition, Kelly, D. E., Wood, R. L., and Enders, A. E., editors. Williams & Wilkins Co., Baltimore, 1984; 25-10c: From *Tissues and Organs: A Text-Atlas of Scanning Electron Microscopy* by Richard G. Kessel and Randy H. Kardon. Copyright © 1979 W. H. Freeman and Co. Reprinted with permission; 25-12: © Biophoto Associates/Photo Researchers; 25-14: From Rhodin, *Histology: A Text and Atlas*, Oxford University Press, New York, 1974, fig. 32-14; 25-20a: © Ed Reschke/Peter Arnold, Inc.; 25-21b: © Dr. Mary Notter/Phototake; p. 674 (a and b): Lester V. Bergman, N.Y.; p. 674 (c): © Phototake; p. 674 (d): © Margaret Cubberly/Phototake; p. 675 (a): Scripps Memorial Hospital and staff; p. 675 (b): © Biophoto Associates/Photo Researchers.

Chapter 26 Page 694: © Biophoto Associates/Photo Researchers.

Chapter 27 Figures 27-3a and b, 27-11c and d, and 27-14a: © Ed Reschke/Peter Arnold, Inc.; 27-6b; © Phototake; 27-11a and b, 27-12, and 27-14b: © Biophoto Associates/Photo Researchers; 27-21; © Martin M. Rotker/Phototake; p. 725: © Ray Ellis/Photo Researchers; 27-24a: © Lennart Nilsson, *A Child is Born*, Dell Publishing Co.; 27-24b: © Dr. Sundstroem/Gamma-Liaison.

Chapter 28 Figure 28-1a: © Omikron/Photo Researchers; 28-1b, 28-10b and d: © Petit Format/Nestle/Photo Researchers; 28-1c: © Institut Pasteur/Phototake; p. 734 (top) and 735: © Yoav/Phototake; p. 739: © Siemens-Lutheran Hospital/Peter Arnold, Inc.; 28-8: Courtesy Dr. A. E. Enders, from *Bailey's Textbook of Microscopic Anatomy*, 18th edition, Kelly, D. E., Wood, R. L., and Enders, A. E., editors. Williams & Wilkins Co., Baltimore, 1984; 28-10a: © Donald Yeager, 1984, for Camera M. D. Studios; 28-10c: © Marjorie Jacobson, 1973, for Camera M. D. Studios; p. 746 (all): © SIU/Peter Arnold, Inc.; 28-22a and b: © Phototake; 28-24: © Biophoto Associates/Photo Researchers; 28-25: © Michael and Elvan Habicht/Taurus Photos, Inc.; 28-27: Thompson and Thompson, *Genetics in Medicine*, W. B. Saunders Co., Philadelphia, 1980. Reprinted by permission; p. 771: © Joel Landau/Phototake.

Index

Page numbers set in boldface indicate pages where terms are first introduced and defined. Page numbers followed by the letter F indicate structures and concepts illustrated in figures.

A bands, **184**, 202
Abcesses, anorectal, 473
Abdomen
  arteries of, 594
  quadrants of, **9–10**
  regions of, **10**
Abdominal aorta, 594
Abdominal cavity
  membranes of, 471–72
  spatial regions of, 10F
  veins of, 601
Abdominal ganglia, **344**
Abdominal oblique muscle, external and internal, **226**
Abdominal wall, muscles of, 226–27, 227F
Abdominopelvic cavities, **9**, 16
Abducens nerves, 305, **329**, 329F, 340
Abduction, 173, **177**, 178, 209
Abductor pollicis longus, **238**
Aberrant blood vessels, **613**
ABO blood groups, 636–37
  genetic basis for, 755
Abortion, 743
Abruptio placentae, **744**
Abscess, **90**
  of the brain, 286
Absorption, in the small intestine, 460–63
Absorptive state, **479**
Accessory glands of scrotum, **706**
Accessory hemiazygos vein, **600**
Accessory nerves, **332**, 332F, 340
Accessory structures of the eyeball, **380**
Acclimatization, **684**
Accommodation
  pupillary, **390**
  visual, **387**
Acetabulum, **152**
Acetazolamide, **529**
Acetoacetic acid, **489**
Acetone, **489**
Acetylcholine, **189**, 202, 283, 285, 353
  release into neuromuscular cleft, 189F
  removal from neuromuscular cleft, 189–90
Acetyl group, **482**
A chains, **426**
Achalasia, **472**

Achilles' tendon, **248**
Achondroplasia, 125, 126, **754**
Acid-base balance, in body fluid, 687–89
Acidosis, **689**, 689–90
  metabolic, 689
Acids, **31**, 32F, 42
Acne, **111**
Acoustic meatus, internal, 136
Acoustic nerves, 329
Acquired immune deficiency syndrome (AIDS), **630**
Acromegaly, **417**
Acromial extremity, **148**
Acromion, **149**
Acrosome, **704**
Actin, **186**, 202
Action of a muscle, **209**
Action potential, **269**, 269–70, 285
  propagating, 270–74, 272F
Active immunity, **635**
Active sites, **30**
Active transport, **54**, 54F, 76
Acupuncture, **364**
Acute epiglottitis, **530**
Acute gastritis, **454**
Acute glaucoma, **387**
Acute glomerulonephritis, **676**
Acute ischemia, **613**
Acute lymphoblastic leukemia (ALL), 555
Acute mastoiditis, 134
Acute monoblastic leukemia, 555
Acute mountain sickness (AMS), **529**
Acute renal failure, **658**
Acute rheumatic fever, **581**
Acute sinusitis, **508**
Acute tonsilitis, **530**
Acute tubular necrosis (ATN), **658**
Adam's apple, **509**
Adaptation
  light and dark, **392**
  of sensory neurons, **358**
Addison's disease, **425**
Addition reaction, 27F, **28**, 42
Adduction, 173, **178**, 209
Adductor brevis, **242**
Adductor longus, **242**
Adductor magnus, **242**
Adenine, **39**, 43
Adenohypophysis, **412**
Adenoids, **509**, 624
Adenomas, **74**
Adenosine, **28**, 43

Adenosine diphosphate (ADP), **28**
Adenosine triphosphate (ATP), **28**, 40
Adenylate cyclase, **408**
Adenylate kinase, **195**
Adequate stimulus, **357**
Adhering junctions, **50**, 73
Adhesions, **84**
ADH-resistant diabetes insipidus, **415**
ADH-sensitive diabetes insipidus, **415**
Adipocytes, **487**, 488F
  anabolism of, 490–91
Adipose capsule of kidney, **647**
Adipose tissue, **88**, 94, 480, 488
Adrenal androgens, **424**
Adrenal arteries, **594**
Adrenal cortex, 422–26
  diseases of, 425
Adrenal glands, 422, 423F
Adrenaline, **283**
Adrenal medulla, **348**, 424–26
Adrenal virilism, **425**
Adrenergic nerves, **283**
Adrenocorticotrophic hormone (ACTH), **417**
Adrenocorticotrophin, **417**
Adult connective tissue, 87–90, 94
Advanced life support (ALS), **305**
Afferent conduction pathways, **361**
Afferent glomerular arterioles, **654**
Afferent lymphatics, **620**
Afferent portion of the central nervous system, **261**
Afferent portion of the peripheral nervous system, 284
Afterimage, **358**
Agglutinate, **636**
Agglutination reaction, 637F
Agglutinins, **546**, 636
Agglutinogens, **546**, 636
Aging
  Alzheimer's disease, 287, 298
  blood vessel defects and, 613
  central nervous system and, 287
  heart rate and, 578
  joints and, 177
  muscles and, 253
  renal function and, 676
  respiratory disease and, 531
  skeletal system and, 127
  of the skin, 111
  urinary system disorders and, 676
Agonist, **209**, 254
Agranulocytes, **547**

837

Agranulocytosis, **554**
AIDS. *See* Acquired immune deficiency syndrome
Air, composition of, 520
Air cells, **134**
Air contrast enema, **456**
Akinetic seizures, **304**
Ala, **146**
Alanine, **37**
Albinism, **103**
Albright's syndrome, **121**
Albumin, **448**, 541
Alcohol abuse, 470
Aldosterone, **423**, 666
    regulatory system of, 666
Aldosteronism, **424**, 607
Alimentary canal, **440**
Alkaline phosphatase, **120**, 128
Alkalosis, metabolic, 424, **689**, 690
Allantois, **738**
Alleles, **753**
    multiple, 755
Allergens, **111**
Allergic disorders, 638–39
Allergic reactions, respiratory system and, 530–31
Allergy shots, **111**
All-or-none effect, **190**, 190F, 202, 271
α-apoprotein, **487**
Alpha-blocking agents, **352**
Alpha cells, **426**
α-globin, **523**
Alpha helix, **37**
α-ketoglutaric acid, **483**
Alpha-receptors, **284**, 347, 353
Alpha waves, **367**
Altitude illness, **529**
Alveolar clusters, **470**
Alveolar ducts, **514**
Alveolar glands, **86**
Alveolar process, **140**, 142
Alveolar sac, **514**
Alveolar wall, 514–16, 515F
Alveoli, **444**, 514
    gas exchange in, 521–24
    of mammary glands, 715
Alzheimer's disease, **287**, 298
Amblyopia, **390**
Amenorrhea, **743**
Amines, 282–83
Amino acid derivatives, 282–83
Amino acids, **37**, 43, 282–83
    absorption by the small intestine, 462
    alanine, 37
    amino group of, 37
    aspartic acid, 283, 285
    cysteine, 38
    gamma-aminobutyric acid, 283, 285
    glutamic acid, 283, 285, 491
    glycine, 37, 283, 285
    heavy chains of, 628
    light chains of, 628
    metabolism of, 491–92
    molecular structure, 37F
    thymine, 39
    tyrosine, 105
Amino group, **37**
Amino peptidase, **460**
Ammonium ion, **492**

Amnesia, retrograde, **287**
Amniocentesis, **738**, **764**
Amnion, **735**
Amniotic cavity, **734**
Amniotic fluid, **735**
Amphiarthroses, **165**, 176
Amplifier T cells, **632**
Amplitude, **398**
Ampulla, **396**
    of ductus deferens, **703**
    of the ear, 401F
    of mammary duct, **715**
    of oviduct, **710**
    of Vater, **457**
Ampullary crests, **396**
Amygdaloid nucleus, **297**
Anabolic processes, **12**, 16
Anabolic steroids, **201**
Anabolism, **29**, 42
    of adipocytes, 490–91
Anal canal, **465**, 465F
    disorders of, 473
Anal columns, **465**
Anal fissures, **473**
Anal sinuses, **465**
Anal sphincter, **228**
    external, 466
Anal triangle, **714**
Anaphase I, **750**
Anaphase II, **751**
Anaphase of mitosis, 66F, **68**, 77
Anaphylactic shock, **352**, 609
Anaphylactoid reactions, **639**
Anaphylaxis, **531**, 639
Anastomoses, **566**
Anatomical terms, 12
Anatomic neck, **149**
Anatomy, defined, **5**
Anconeus, **234**
Androgens, 407
    adrenal, 424
Anemia, **545**, 546
    aplastic, 555
    hemolytic, 546
    hemorrhagic, 546
    Mediterranean, 554
    pernicious, 473, 546
    sickle cell, 554, 555F
Anencephaly, **286**
Anesthesia, **318**, 329
Anesthesiologists, **329**
Aneurysm, **607**
    cerebral, 613
Angina pectoris, **567**, 580
Angiodynography, **593**
Angiography, **575**, 593
    digital subtraction, 593
Angiomas, **613**
Angiotensin I, **666**
Angiotensin II, **424**, 609, 666
Angiotensin III, **666**
Angiotensinogen, **666**
Angitis, **613**
Anions, **31**, 682
    in body fluid, 687
Ankle, bones of, 120F
Ankylosing spondylitis, **177**
Annulus fibrosis, **159**
Anomalous blood vessels, **613**

Anorectal abscesses, **473**
Anorexia nervosa, **489**
Anosmia, **140**, 379
ANS. *See* Autonomic nervous system.
Ansa cervicalis, **336**
Antagonist, **209**
Anteflexion of uterus, **711**
Anterior canal of ear, **396**
Anterior cavity, **7–8**, 385
Anterior chamber, **385**
Anterior commissure, **297**
Anterior corticospinal tract, **365**
Anterior cruciate ligaments, **176**
Anterior ducts of ear, **396**
Anterior horn, **307**
Anterior interventricular artery, **567**
Anterior lateral sulcus, **312**
Anterior lobe of pituitary gland, **412**
Anterior median fissure, **312**
Anterior pillar, **444**
Anterior pituitary, **412**
Anterior (ventral) portion of the body, **10**
Anterior spinothalamic fasciculus, **313**
Anterior tibial artery, **596**
Anterior tibial veins, **602**
Anthropometry, **9**
Antibodies, 93, **547**, 618, 628
    structure of, 628F
Anticholinergics, **353**
Anticholinesterases, **353**
Anticoagulant, **547**
Anticodons, **73**, 77
Antidiuretic hormone (ADH), **413**, 413–14, 665
Antigen-antibody complex, **628**
Antigenic determinant sites, **627**
Antigens, 88, **546**, 627, 628F
    complete, 627
Antithrombin III, **553**
Antrum of stratum granulosum, **717**
Anus, **465**
Anvil, **395**
Aorta, **591**
    abdominal, 594
    ascending and descending, 591
    coarctation of, 579
    thoracic, 594
Aortic arch, **591**
Aortic bodies, **527**
Aortic receptors, **527**
Aortic reflex, **577**
Aortic semilunar valve, **566**, 566
Aortic valve stenosis, **643**
Aphasia, **314**
Apical foramen, **444**
Aplastic anemia, **555**
Apnea, **527**
Apneusis, **526**
Apneustic center, **526**
Apneustic nuclei, **305**
Apocrine glands, **85**, 93
Aponeuroses, **90**, 94
Appendages, **7**, 16
Appendicitis, **465**
Appendicular skeleton, **131**, 132F, 160
Appendix, **464**
Appositional growth, **119**, 128
Aqueduct of Sylvius, **307**
Aqueous humor, **385**

Arachnoid, **310**, 319
Arachnoid villi, **307**
Arbor vitae, **303**
Arches of the foot, 156–57, 161
Arcuate arteries, **654**
Arcuate popliteal ligament, **176**
Arcuate veins of nephron, **656**
Area cribrosa, **649**
Areola, **715**
Areolar tissue, **88**, 94
Arrector pili muscle, **108**, 111
Arrhythmia, cardiac, 579
Arterial circulation, 605–10
Arteries, **585**, 585–86, 589–96. *See also* Aorta
  abdominal, 594
  adrenal, 594
  afferent glomerular, 654
  anterior interventricular, 567
  anterior tibial, 596
  arcuate, 654
  arterial circulation to the brain, 592F
  axillary, 593
  basilar artery, 593
  brachial artery, 594
  bronchial, 594
  carotid, 591
  celiac artery, 594, 595F
  circumflex artery, 567
  common carotid, 591
  common iliac, 594
  coronary, 566
  costocervical, 593
  deep femoral, 596
  digital, 594, 596
  dorsalis pedis, 596
  elastic, 585
  esophageal, 594
  external carotid, 591
  external iliac, 594
  femoral, 596
  fistulas in, 613
  gastric, 594
  gonadal, 594
  of the head, 591–93
  helicine, 707
  hepatic, 594
  hypogastric, 594
  iliac, 594
  inferior mesenteric, **594**
  inferior phrenic, **594**
  interlobar, 654
  interlobular, 654
  internal carotid, 591
  internal iliac, 594
  internal pudendal, 707
  internal thoracic, 593
  interventricular, 567
  lateral femoral circumflex, 596
  left common carotid, 591
  left coronary, 566
  left gastric, 594
  left subclavian, 591
  lobar, 589
  of the lower extremity, 594–96
  lumbar, 594
  marginal, 567
  medial femoral circumflex, 596
  medial plantar, 596
  mesenteric, 596F

middle sacral, 594
of the neck, 591–93
neural control of pressure in, 609F
ophthalmic, 591
ovarian, 594
parietal, 594
peroneal, 596
plantar, 596
popliteal, 596
posterior cerebral, 593
posterior communication, 593
posterior intercostal, 594
posterior interventricular, 567
posterior tibial, 596
pulmonary, 589
radial, 594
renal, 594, 647
right common carotid, 591
right coronary, 566
right subclavian, 591
spermatic, 594
spiral, of endometrium, 713
splenic, 594
straight, of endometrium, 713
superior mesenteric, 594
superior phrenic, 594
suprarenal, 594
systemic, 590F
testicular, 594
of the thorax, 594, 595F
thyrocervical, 593
tibial, 596
truncus arteriosus, 560
tunics of, 585
ulnar, 594
umbilical, 605, 741
of the upper extremity, 593–94
uterine, 713
vertebral, 591
visceral, 594
Arteriogram, 573F
Arterioles, **585**, 586, 586F
  glomerular, 654
Arteriovenous fistulas, **613**
Arteritis, **613**
Arthritis, **177**
  gout, 166
  rheumatoid, **170**
Arthrodial joints, **168**
Arthrogram, **15**
Arthrogryposis, 252
Arthrogryposis multiplex congenita, **177**
Arthrosclerotic disease, **314**
Arthroscope, **119**
Arthroscopy, **174**
Articular cartilage, **123**, 167
Articular disks, **168**
Articular facets, **147**, 148, 155
Articular processes, 161
Articulations, **115**, 164–78, 176. *See also* Joints
  classification of, 165
Artificial heart, **564**
Artificial insemination, **734**
Artificial menopause, **729**
Asbestosis, **531**
Ascending aorta, **591**
Ascending colon, **464**
Ascending fasciculi, **313**
Ascites, **470**

Aseptic meningitis, **286**
Aspartic acid, 283, 285
Association areas (cerebral cortex), **296**, 319
Association fibers, **297**
Association neurons, **312**
Association tracts, **297**
Asthma
  bronchial, 518, 531
  extrinsic, 531
Astigmatism, **389**
Astrocyte, 264F, **265**, 284
Atelectasis, **515**, 530
Atheromas, **580**
Atherosclerosis, **432**, 580, 613
Atherosclerotic plaques, **287**
Athlete's foot, 111
Atlas vertebra, **144**, 145F, 161
Atom, 7, 16, **19**, 41
  structure of, 21F
Atomic mass, **21**, 41
Atomic number, 41
Atomic weight, **21**, 41
Atopic dermatitis, **105**
Atopic diseases, **638**
ATPase, **484**
Atresia, **709**
Atretic follicles, **709**
Atrial fibrillation, 571, **579**
Atrial flutter, **579**
Atrial septal defects, **579**
Atrial septum, **560**
Atrioventricular block, **579**
Atrioventricular (AV) bundle, **569**
Atrioventricular valves, **561**
Atrium of heart, **197**, 560, 561
  left, 564–65, **594**
  right, 563, **563**
Atrophy, **199**
Atropine, **353**
Attachment of sperm, **724**
Attention deficit disorder (ADD), **373**
Attenuated polio virus, **634**
Audiogram, conventional, **403**
Audiology analysis, **403**
Audiometers, **403**
Auditory meatus, internal, 136
Auditory nerves, **329**
Auditory tube, **395**
Aura, **299**
Auricle of ear, **393**
Auricles of heart, **563**
Auricular surfaces, **146**, 153
Auscultation, **574**
Autoantibodies, **639**
Autoimmune disorders, 93, **627**, 638, 639
Autonomic hyperreflexia, **315**
Autonomic nervous system (ANS), **199**, 261, 284, 342
  anatomy of, 343–47
  conscious control of, 349–51
  distribution of neurons in, 343
  drugs that influence, 352–53
  efferent fibers of, 345F
  embryonic development of, 343
  physiological effects of, 348
  physiology of, 347–51
  sensory input into, 349
Autopsy, **12**
Autoregulation, **609**

Autosomes, **759**
Avogadro's number, **29**, 42
Axial skeleton, **131**, 132F, 160
Axilla, **109**
Axillary artery, **593**
Axillary vein, **600**
Axis vertebra, **145**, 145F, 161
Axoaxonic synapses, **274**, 285
Axodendritic synapses, **274**, 285
Axoneme, **63**, 76
Axons, **92**, 262, 284
Axon terminals, **262**, 264, 284
Axosomatic synapses, **274**, 285
Azotemia, **660**
Azygos vein, **600**

Babinski's sign (reflex), **316**, 318
Baby teeth, 445
Backwater fever, **544**
Bacterial endocarditis, **581**
Bainbridge reflex, **577**
Balance, physiology of, 400–402
Ball-and-socket joints, **169**, 169–70, 178
"Bamboo spine," 177
Barbiturates, **350**
Barnard, Christiaan, 564
Baroreceptors, **357**
Barotrauma, **525**
Bartholin's glands, **714**
Basal body, **63**, 76
Basal cell carcinoma, **106**
Basal ganglia, **291**, **297**, 299F, 319
Basal layer, **83**, 92–93
Basal metabolic rate (BMR), **497**
  standards of, 498
Basal nuclei, **291**, **297**
Basement membrane, **81**
Bases, **31**, 32F, 42
  nitrogenous, 43
Basilar artery, **593**
Basilar membrane, **397**
Basilic veins, **600**
Basophils, **547**
B cell lymphocytes, **632**
B cells, formation of, 629F
B chains, **426**
Bell's palsy, **330**
Belt desmosomes, **50**
Bends, 525
Benign prostatic hypertrophy, **676**
Benign tumors, **74**
Beriberi, 41
Berylliosis, **531**
β-apoprotein, **487**
Beta-blockers, **284**, 353
Beta cells, **426**
β-globin, **523**
β-hydroxybutyric acid, **489**
β-oxidation, **488**
Beta-receptors, **284**, 347, 353
Biaxial joints, **170**
Biaxial movement, **169**
Biceps brachii, **207**, **234**
Biceps femoris, **246**
Bicipital groove, **149**
Bicuspids, **445**
Bicuspid valve, **566**
Bilateral renal agenesis, **675**
Bile, **459**

Bile canaliculi, **468**
Bile ducts, **468**
Bile salts, **460**
Biliary colic, **460**
Bilirubin, **459**, 545
Binding of sperm, **724**
Binocular vision, **390**
Biofeedback, **351**, 354
Biologic death, **305**
Biopsy, **96**
  closed, 649
Biotin, **493**
Biot's respiration, **527**
Bipedal lymphangiogram, **97**
Bipennate fascicles, **207**
Bipenniform tendons, 207
Bipolar neurons, **264**, 284
Birth control, 725
Bitemporal hemianopia, **326**
Bivalent chromosomes, **749**
Black lung disease, **531**
Bladder. *See* Urinary bladder
Blastocoel, **733**
Blastocyst, **733**, 734F
Blastomere, **733**
Bleeding time, **553**
Blindness
  color, **393**
  night, **392**
Blind spot, **383**
Blood, **90**, 94, **538**, 538–55. *See also* Vascular tissue
  calcium ions in, 686F
  cancer of, 555
  composition of, 539–41
  diseases of, 554–55
  erythrocytes, 543–46
  formed elements of, 539, 541
  functions of, 539
  hemostasis, 550–55
  leukocytes, 546–48
  oxygen transport in, 521–23
  thrombocytes, 549–50
  types of, **636**
  viscosity of, 607
  volume of, 606
Blood-brain barrier (BBB), **311**, 320
Blood calcium, 13–16
Blood cells
  origin and development of, 542
  production of, 541–43
Blood clot, **549**, 550–55
  formation of, **550**, 551F
Blood count, 548. *See also* Complete blood count
Blood groups, 546, 636–37. *See also* ABO blood groups; Blood types
  immunity and, 636–37
Blood hydrostatic pressure (BHP), **610**
Blood osmotic pressure (BOP), **611**
Blood pressure
  control of, 607–10
  factors affecting, 606–7
  nervous control of arterial, 608F
  renin-angiotensin mechanism for controlling, 667F
Blood reservoirs, **612**
Blood transfusions, 637
Blood types, **636**

Blood vessels, **584**, 585F, 585–87. *See also* Arteries; Arterioles; Capillaries; Veins
  aberrant, 613
  anatomy of circulatory pathways, 587–602
  anomalous, 613
  arteries, 585–96
  capillary exchange, 610–11
  diseases of, 613–14
  disorders of, 593
  veins, 596–602
  venous circulation, 611–12
Body
  anatomic regions and positions, 7–10, 16
  anterior portion of, 10
  chemical composition of, 19
  coronal section of, 10
  distribution of heat in, 494F
  frontal section of, 10
  major cavities of, 10F
  major muscles of, 212F
  measurement of, 9
  mechanisms for cooling, 496
  mechanisms for warming, 495–96
  midsagittal section of, 10
  minerals required by, 492
  most abundant chemical elements in, 19
  movements of, 172F
  oblique section of, 10
  organ systems of, 9F
  parasagittal section of, 10
  structural complexity of, 6F
  structural plan of, 7, 16
  superior portion of, 10
  survey of joints in, 171
  temperature regulation of, 494–96
  transverse section of, 10
  ventral portion of, 10
Body building, 200, **201**
Body fluids
  acid-base balance in, 687–89
  acid-base imbalance in, 689–90
  anions in, 687
  composition of, 682–87
  diseases affecting, 691
  imbalance of, 684
  regulating water content, 681–82
  repair of imbalance of, 685
Body of stomach, **451**
Body planes, **10**
Body sections, **10**
Body stalk, **738**
Body temperature, 494–98
  regulation of, 494–96, 495F
Bolus, 449
Bonds, 41. *See also* Chemical bonds
Bone, **90**, 90–91, 94, **115**. *See also* Osseous tissue
  cancellous, 118
  compact, 118, 128
  development of, 119–23, 128
  floating, 142
  histology of, 115–18, 128
  spongy, 118, 128
  trabecular, 118, 128
  tumors of, 126
  ulcers of, 124
Bones. *See also* Humerus; Joints; Skeletal system; Skeleton; Skull
  of ankle, 120F

anvil, 395
atlas vertebra 144, 145F, 161
axis vertebra, 145, 161
carpals, 150, 151F, 161
cervical vertebrae, 143–45, 161
of cheek, 444
clavicles, 148, 148F, 161, 174
coccygeal vertebrae, 143, 146F, 147, 161
compact bone, 118, 128
coxal, 152
embryonic development of, 120–23
ethmoid, 139–40, 160
fibula, 155, 161
hyoid, 43, 142, 142F, 160, 219F
ilium, 153, 161
incus, 395
intramembranous versus endochondral, 120F
ischium, 153, 161
lesser multangular carpal, 150, 151F
lumbar vertebrae, 143, 145, 146F, 161
malleus, 395
mandible, 142, 160
maxillae, 140, 141F, 160
medial pterygoid, 218
metacarpals, 151F, 152, 161
metatarsals, 156, 161
nasals, 140, 160
nasal turbinates, 507
navicular carpal, 150, 151F
navicular tarsal, 156
neural arch, 143
number in the body, 131
occipital, 136–38, 160
ossicles, 395
palatine, 140, 141
parietals, 133, 160
patella, 154F, 155, 161
pectoral girdle, 49, 131, 148, 148F, 161
pelvic, 152, 152F
phalanges of foot, 156, 161
phalanges of hand, 151F, 152, 161
pisiform carpal, 150, 151F
radius, 150, 151F, 161
ribs, 147–48, 148F
sacral vertebrae, 145
sacrum, 143, 145–46, 146F, 161
scapulae, 148–49, 149F, 161
sesamoid, 115, 142
shinbone, 155, 161
short, 115
shoulder girdle, 148–49, 161
sphenoid, 138, 139F, 160
stapes, 395
stirrup, 395
talus, 156
tarsal, 156, 161
temporal, 133–36, 136F, 160
thoracic vertebrae, 143, 145, 145F, 161
tibia, 155, 155F, 161
trapezoid carpal, 150, 151F
ulcers of, 124
ulna, 150, 151F, 161
vertebrae, 143–45, 160, 161
vertebral ribs, 147
wormian, 142, 160
Bony labyrinth, **396**
Botulism, **286**
Bouton, **276**

Bowman's capsule, **652**
Boyle's law, **519**
Brachial artery, **594**
Brachialis, **234**
Brachial plexus, **336**, 337F, 336–38, 340
Brachial vein, **600**
Brachiocephalic vein, **600**
Brachioradialis, **234**
Bradycardia, **579**
Brain, **261**, 290. See also Cerebellum; Cerebrum; Medulla oblongata
    arterial circulation to, 592F
    association areas of, 296
    cavernous sinus of, 597
    embryonic development of, 293, 293F
    forebrain, 293
    formation of brain vesicles, 293F
    hypothalamus, 300–301, 319, 412–13, 415, 416F, 494–95
    integrative functions of, 367–72
    lateral ventricles, 307, 319
    major regions of, 291F
    medial sagittal section, 302F
    meninges of, **309**, 309–10, 310F
    motor areas of, 296
    relationship of embryonic and adult brain regions, 293, 293F
    respiratory center of the, 526F
    sensory areas of, 295–96
    thalamus, 299–300, 319
    venous sinuses of, 599F
    ventricles of the, **307**, 309F, 319
    visual pathways to, 393
Brain abcesses, **286**
Brain cells, electrical activity of, 368F
Brain death, **305**
"Brain sand," 431
Brain stem, 291F
Branchial grooves, **394**
Breast feeding, 716
Breathing, **504**. See also Respiratory system
    blood movement and, 612
    chemical regulation of, 527
    mechanics of, 516–19
    neural control of, 526
Breathing therapy, **351**
Breech birth, 177, **747**
Brim of the pelvis, **152**
Broad ligaments of uterus, **711**
Broca, Paul, 296
Broca's area, **296**
Bronchi, 511F
Bronchial arteries, **594**
Bronchial asthma, **518**, 531
Bronchial veins, **600**
Bronchioles, **513**
Bronchitis, **530**
Bronchogenic carcinoma, **511**
Brown atrophy, **287**
Brucellosis, 626
Brunner's glands, **457**
Brush border, **457**
Buccal cavity, **443**
Buccal glands, **447**
Buccal stage, **449**
Buccinator, **214**
Buffers, 32–33, 42, 539
Bulbar conjunctiva, **380**
Bulbocavernosus muscle, **708**

Bulbourethral glands, **706**
Bulimia, **489**
Bundle branch block, **580**
Bundle branches of heart, **569**
Bundle of His, **569**
Bunion, **156**
Burns, 109
Bursae, **168**
Bursa of Fabricius, 632
Bursitis, **168**
Butterfly rash, **93**

Cachexia, **74**
Calcaneal tendon, **248**
Calcaneus, **156**
Calcitonin, **13**, 123, 128, 420
Calcium
    in the blood, 13–16
    in body fluid, 685–66
    homeostatic regulation of concentration in blood plasma, 13F
Calcium channel blockers, **191**
Calcium deficiency, 686
Calculi in urine, **674**
Calculus, **446**
Calluses, **103**, 117F
Calmodulin, **199**, 203
Caloric testing, **402**
Calories, **494**
Calvarium, **133**
cAMP, **409**
cAMP-induced mechanism of hormone action, 409–10
Canaliculi, **118**
Canal of Schlemm, **385**
Cancellous bone, **118**
Cancer, 74–75. See also Carcinoma; Tumors
    acute lymphoblastic leukemia, 555
    acute monoblastic leukemia, 555
    adenomas, 74
    angiomas, 613
    atheromas, 580
    basal cell carcinoma, 106
    benign tumors, 74
    of the blood, 555
    of the cervix, **714**
    of the colon, **473**
    Hodgkin's disease, 96
    hybridoma, 635
    Kaposi's sarcoma, 630
    leukemia, 555, 764
    lipomas, 74
    lung cancer, 511, 531
    malignant melanoma, 106
    malignant myeloma, 126
    malignant tumors, 74
    metastasis, **74**
    metastatic disease, 97
    metastatic growths, 287
    multiple neurofibroma, 74
    muscle tumors, 253
    myomas, 253
    neuroblastomas, 676
    oncologists, 75
    osteochondroma, 126
    osteogenic sarcoma, 126
    of reproductive organs, 729
    sarcomas, **74**, 253
    skin tumors, 111

Cancer (*continued*)
squamous cell carcinoma, 106, 472
thymoma, 253
of thyroid, 74
transitional cell carcinomas, 676
Canines, **445**
Canker sores, **286**
Cannon, Walter, 12
Capacitation, **724**
Capillaries, **585**, 586
Capillary hydrostatic pressure, **658**
Capitate carpal, **150**, 151F
Capitulum, **149**
Capsule of adrenal gland, **422**
Capsule of liver, **468**
Capsule of lymph node, **621**
Capsule of thymus gland, **431**
Capsules, **90**, 94
Carbaminohemoglobin, **544**
Carbohydrate metabolism, 479–86
overview of, 486F
Carbohydrates, **34**, 42
absorption by the small intestine, 461–62
digestion of, 459–60
molecular structure, 34F
Carbon dioxide transport, 524
Carbonic acid-bicarbonate buffer system, **687**
Carboxyl group, 42
Carboxypeptidase, **459**
Carcinoembryonic antigen (CEA), **639**
Carcinogenic, defined, 74
Carcinoma, **74**. *See also* Cancer
basal cell, 106
basal cell and squamous cell, **106**
bronchogenic, 511
of the kidney, 676
Cardiac arrhythmia, **579**
Cardiac catheterization, **575**
Cardiac center, **306**
Cardiac contraction, 568–70
Cardiac cycle, **572**, 572–74
Cardiac diagnostic procedures, 575
Cardiac glands, **452**
Cardiac intrinsic blood supply, 566–67
Cardiac muscle, **92**, 94, 181, 197–98, 198F, 203
Cardiac muscle tissue, histology of, 568F
Cardiac orifice, **451**
Cardiac output, 574–76
Cardiac pacemaker, **580**
Cardiac patient, office evaluation of, 574
Cardiac plexus, **347**
Cardiac reserve, **576**
Cardiac veins, **567**
Cardioaccelerator center, **577**
Cardiogenic shock, **609**
Cardioinhibitor center, **577**
Cardiology, **6**
Cardiopulmonary resuscitation (CPR), **305**
Cardiovascular system, **558**. *See also* Blood vessels; Heart
Cardioversion, **580**
Carotene, **103**, 110
Carotid arteries, **591**
Carotid bodies, **527**
Carotid chemoreceptors, **527**
Carotid foramen, **136**
Carotid plexuses, **344**
Carotid sinus reflex, **577**
Carpals, **150**, 151F, 161

Carpal tunnel syndrome, **339**
Carpus, **150**, 161
Cartilage, **90**, 94, 128
articular, **123**, 167
development of, 119, 128
hyaline, 90, 94
torn, 119
Cartilaginous joint, **165**, 176
Caruncle, **386**
Casts in urine, **673**
Catabolic processes, **12**, 16
Catabolism, **29**, 42
of fats, 488–90
Catalysis, **30**
Cataracts, **387**
Cat's cry syndrome, **763**
Catecholamines, **281**, 283, 285, 424
Catheter, **575**
Catheterization, cardiac, 575
Cations, **31**, 682
in body fluid, 683–87
CAT scan, **5**, 5F
apparatus for, 14F
Cauda equina, **312**
Caudate lobe of liver, **468**
Caudate nucleus, **297**
Cavernous sinus of brain, **597**
Cavernous urethra, **670**, 703
Cavities, dental, 446
C cells, **418**
Cecum, **464**
Celiac artery, **594**, 595F
Celiac ganglion, **344**
Celiac plexus, **347**
Cell differentiation, **732**
Cell division, **65**, 66–68, 76–77. *See also* Mitosis
Cell-mediated immunity, **629**
Cell membrane, 47
fluid mosaic structure of, 49F
junctions in, 50F
movement of materials through, 50–56, 55F, 73–74
Cell physiology, **5**
Cells, **7**, 16, 46. *See also* Neurons; Nucleus of a cell; Platelets; Red blood cells
adipocytes, 487–88, 490–91
agranulocytes, 547
alpha cells, 426
amplifier T cells, 632
astrocytes, 264F, 265, 284
axons, 92, 262, 284
basophils, 547
B cell lymphocytes, 632
B cells, 629F
beta cells, 426
blood, 541–43
brain, 368F
C cells, 418
chief cells, 420, 453
chondroblasts, 119, 128
chondrocytes, 90, 119
chromaffin, 424
columnar, 456
cones, 382, 391–92
daughter cells, 68
delayed hypersensitivity T cells, 632
delta cells, 426
effector T cells, 631

enteroendocrine, 457
eosinophils, 546
erythrocytes, 543–46
fat cells, 88, 94
fibroblasts, 84, 87, 94
follicular, 717
glandular, 7
goblet, 83, 456
granulocytes, 546
gustatory, 379–80
helper T cells, 632
hematocytoblasts, 541
intact, 632
interstitial cells of Leydig, 429, 699
interstitial endocrinocytes, 699
juxtaglomerular, 655
killer T cells, 631
Kupffer cells, 55
leukocytes, 546–48
lipocytes, 88, 94
lymphoblasts, 543
lymphocytes, 431, 547, 618
macrophages, 88, 94, 545, 626, 632
mast cells, 88, 94, 547
megakaryoblasts, 543
megakaryocytes, 549
megaloblasts, 546
meiosis, 749–52, 750F
Meissner's corpuscles, **359**
melanocytes, 103, 110
memory cells of B-cell clone, 633
memory T cells, 632
mesenchymal, 87, 94
metarubricyte, 544
microphages, 626
mitochondria, 59, 61F, 62F, 70
mitosis, 66F, 67–68, 77, 750F
monoblasts, 543
monocytes, 547
motor neurons, 359F
mucous neck cells, 452
multipolar neurons, 264, 284
myeloblasts, 543
natural killer cells, 632
neural crest cells, 292
neuroblasts, 324
neutrophils, 546
nurse, 702
olfactory, 377
oligodendrocytes, 264F, 265, 265F, 284
organelles in, 7, 57, 58F
osteoblasts, 120, 128
osteocytes, 90–91, 115, 120, 128
osteoprogenitor, 120, 128
ovarian follicles, 709
ovum, 709, 717, 719F
oxyphil, 420
pacinian corpuscles, **360**
packed cell volume, 548
pancreatic acinar, 471F
parafollicular, 418
parietal, 453
phagocytes, 547, 626
photoreceptive, 382
plasma, 88, 94, 133
podocytes, 652
polymorphonuclear white blood cells, 548
primary oocytes, 709
primary spermatocytes, 704

primordial follicles, 717
renal corpuscles, 651, 652F
respiration of, **480**, 485
reticular, 88
reticulocytes, 544
rods, 382, 391–92
rubriblasts, 543
rubricytes, 544
satellite, 264, 265, 266F, 284
Schwann cells, 264, 266, 284
secondary oocytes, 717
secondary spermatocytes, 704, 750
sensory, 260
septal, 515
Sertoli's, 429, 699
spermatozoa, 704
stem cells, 629
structure of, 48F
supporting, 377
suppressor T cells, 632
sustentacular, 377, 702
tactile corpuscles, **359**
T cell lymphocytes, 629, 629F, 632F
thrombocytes, 549–50
thymic corpuscles, 431, 624
trophoblast, 733
zymogenic, 453
Cellular immune response, 631–32, **629**
Cellulose, 35, 35F
Cementum, **445**
Central canal, **306**
Central nervous system (CNS), **261**, 284, 290. *See also* Brain; Spinal cord
    afferent portion of, 261
    corticospinal tracts of, 365F
    diencephalon, 299–302
    mesencephalon, 302–3
    metencephalon, 303–5
    myelencephalon, 305–6
    neuroglia of, 265
    telencephalon, 291–98
    tumors of, 287
    ventricles and meninges of, 307–11
Central tendon of diaphragm, **224**, 516
Centromere, **67**, 77
Cephalalgia, **299**
Cephalic phase, **454**
Cephalic veins, **600**
Cerebellar hemispheres, 290, **303**
Cerebellar peduncles, **303**, 319
Cerebellum, 291F, 302F, **303**, 303–5
Cerebral aqueduct, **302**, 307, 319
Cerebral cortex, 291–95, **291**, 319
    functional areas of, 295–96, 295F
    pathway for auditory stimuli to, 400F
Cerebral death, **305**
Cerebral edema, **287**
Cerebral hemispheres, 290, **291**, 294F, 319
Cerebral infarction, **314**
Cerebral insufficiency, **314**
Cerebral malaria, **544**
Cerebral nuclei, **297**
Cerebral peduncles, **303**, 319
Cerebral vascular accident, **588**
Cerebritis, **286**
Cerebrospinal accidents (CVAs), **314**
Cerebrospinal fluid (CSF), **140**, 291, 307, 319
    circulation of, 308F
Cerebrovascular accident (CVA), 554

Cerebrovascular disease, **287**
Cerebrum, 291–98, **291**, 291F, 319
    gray matter of, 297–98
    white matter of, 296–97, 297F
Cerumen, **109**, 393
Ceruminous glands, **109**, 111, 393
Cervical canals of uterus, **711**
Cervical cancer, 714
    test for, 75
Cervical cap, **725**
Cervical curve, **143**, 161
Cervical enlargement, **312**
Cervical plexus, **334**, 334–36, 337F, 340
Cervical veins, **600**
Cervical vertebrae, **143**, 143–45, 161
Cervix, **711**, 713F
Cesarean section, 747
Chancre of syphilis, **723**
Channels, 269
Charles' law, **520**
Charley horse, **200**
Chediak-Higashi syndrome, **638**
Cheek bone, **141**
    fracture of, 141
Cheeks, **444**
Chemical bonds, 23, 41
Chemical elements, **19**, 41
Chemical equivalent of a substance, **682**
Chemical isomers, **28**
Chemical reactions, **26**, 26–28, 42
Chemical symbols, **19**, 41
Chemical theory of olfactory cell stimulation, **377**
Chemiosmotic hypothesis, **484**
Chemoreceptors, **357**, 377, 608
    medullary, 528
    peripheral, 527–28, 528
Chemotaxis, **626**
Chewing, muscles of, 218
Cheyne-Stokes respiration, **527**
Chicken pox, 104
Chief cells, **420**, 453
Childbirth, 747–49, 748F
    complications of, 747
    natural, 746
    prepared, 746
Children, rheumatoid arthritis in, 170
Chlamydial infection, **723**
Chloride shift, **524**
Chloroquine, 544
Choanae, **506**
Cholecystogram, **456**
Cholecystokinin-pancreozymin (CCK-PZ), **458**
Cholelithiasis, **460**
Cholesterol, **37**
    molecular structure, 37F
Choline esterase, **189**, 202
Choline esters, 353
Cholinergic alkaloids, **353**
Cholinergic nerves, **283**
Cholinergics, **353**
Cholinesterase inhibitors, 190
Chondral fracture, **177**
Chondrin, **90**
Chondroblasts, **119**, 128
Chondrocalcinosis, **166**
Chondrocytes, **90**, 119
Chondrodystrophies, **125**

Chordae tendineae, **563**
Chordee, **728**
Chorion, **738**
Chorionic villi, **739**
Chorionic villus sampling (CVS), 764
Choroid, **381**
Choroid fissure, **381**
Choroid plexus, **307**
Chromaffin cells, **424**
Chromatids, **67**, 77
Chromatin, **65**, 65F, 76
Chromophils, **415**
Chromosomal abnormalities, 761–65
Chromosomal analysis, **738**
Chromosomal linkage, **756**, 756–58
Chromosomal nondisjunction, **761**, 762F
Chromosome arms, **67**
Chromosomes, **67**, 67F
    bivalent, 749
    chromosomal basis for independent assortment, 757F
    chromosomal basis for segregation, 756F
    male, 760F
    RFLP-based mapping of, 758
Chronic active hepatitis, **471**
Chronic gastritis, **454**
Chronic glaucoma, **387**
Chronic granulocytic leukemia (CGL), 555
Chronic lymphocytic leukemia (CLL), 555
Chronic myelogenous leukemia, **764**
Chronic renal failure, **658**
Chronic sinusitis, **508**
Chylomicrons, **463**
Chyme, **452**
    neutralization of, 458
Chymotrypsin, **459**
Chymotrypsinogen, **459**
Cigarette smoking, 74, 75
Cilia, 7, **63**, 63F, 76
Ciliary body, **381**
Ciliary ganglion, **327**, 346
Ciliary glands, **386**
Cimetidine, 452
Cingulate gyrus, **302**
Circadian dysrhythmia, **367**
Circadian rhythm, **367**
Circle of Willis, **591**
Circular fascicles, **207**
Circular folds, **456**
Circulation. *See also* Blood vessels; Heart
    arterial, 589–96, 605–10
    fetal, 603–5, 604F
    major circulatory routes, 588F
    pulmonary, 589
Circulatory shock, **613**
Circulatory system, **7**
Circumcision, **707**
Circumduction, **174**, 178
Circumflex artery, **567**
Cirrhosis, **470**
Cistern, **188**, 202
Cisterna chyli, **619**
Citric acid, **483**
Citric acid cycle, **482**
Class I lever, **209**
Class II lever, **210**
Class III lever, **211**
Claustrum, **298**
Clavicles, **148**, 148F, 161, 174

Clavicular notches, **147**
Clearance in kidneys, **658**
Cleavage, **733**, 733F
Cleavage reaction, **28**, 42
Cleft lip, **125**, 472
Cleft palate, **125**, 140, 472
Clinical death, **305**
Clitoris, **714**
Cloaca, **648**
Cloacal membranes, **442**
Clonic spasm, **193**
Cloning, DNA, 70
Clonus, **318**
Closed biopsy, **649**
Closed fracture, **116**
Clot formation, enhancement of, 553
Clot retraction, **553**
Clotting mechanism, 552F
Clotting time, **553**
Clubfoot, **156**
Cluster headaches, **299**
CNS. *See* Central nervous system
Coagulation, **550**
Coarctation of aorta, **579**
Cobalamin, **493**
Coccygeal vertebrae, **143**, 147, 161
Coccygeus, **228**
Coccyx, **143**, 146F, 147, 161
Cochlea, **396**, 397F
  functions of, 398–99
Cochlear canals, **396**
Cochlear duct, **396**
Cochlear implants, **403**
Cochlear nerve, **330**, 340
Cochlear organ, **394**
Codominance, **755**
Codons, **73**
Coenzyme A (CoA), **482**
Cofactors, defined, 492
Coitus, **722**, 722–24
Cold sores, **102**, 102F
Colic, biliary, 460
Collagenous fibers, **87**, 94
Collagen vascular diseases, **93**
Collarbones, **148**, 161
Collateral branches, **264**, 284
Collateral circulation, **566**
Collateral ganglia, **344**
Collecting tubules, **651**
Colles' fracture, 117F
Colliculi, 319
Colloid osmotic pressure, **658**
Colon, **464**
  ascending, 464
  cancer of, 473
  descending, 465
  disorders of, 473
Color blindness, **393**
Colostrum, **715**, 716
Colposcopy, **714**
Columnar cells, **456**
Columns of gray matter, **312**
Coma, **367**
Combined gas law, **520**
Comminuted fracture, 116F
Commissural fibers, **297**
Commissural interneurons, **317**
Commissural tracts, **297**
Commissurotomy, **564**

Common bile duct, **470**
Common carotid arteries, **591**
Common cold, **507**
Common hepatic duct, **470**
Common iliac arteries, **594**
Common iliac veins, **601**
Communicating canals, **118**
Communicating hydrocephalus, **307**
Communicating junctions, **49**, 73
Communicating rami, **344**, 352
Compact bone, **118**, 128
Complement, **626**
Complete antigens, **627**
Complete blood count (CBC), **548**
Complex molecules, **29**
Compliance of lungs, **518**
Compound action potential, **274**
Compound alveolar glands, **86**
Compound fracture, 116F
Compound tubuloalveolar glands, **86**
Computerized axial tomogram (CAT scan), 5, **15**
Concentration, **29**
  defined, 42
Conception, **724**
Concussion, **287**
Conditioning, **372**
Condoms, **725**
Conducting division, **513**
Conduction system of heart, **568**, 569F
Conductive hearing loss, **403**
Conductive heat loss, **496**
Condyles, of the femur, **155**
Condyloid joints, **168**, 168–69
Condyloid process, **142**
Condylomata acuminata, **714**
Cones, **382**, 392–93
Conformational change, **26**, 27F
Conformational change hypothesis, **55**
Congenital cerebral aneurysms, **613**
Congenital deformities, of the face, **125**
Congenital heart disease, **579**
Congenital neutropenia, **638**
Congestive heart failure (CHF), **576**
Conjugate base of an acid, **33**
Conjunctival fornix, **386**
Conjunctivitis, **386**, 387
  inclusion, 723
Connective tissue, 86–91, 90, 93–94
  adult, 87–90, 94
  blood, 90
  bone, 90–91
  cartilage, 90
  dense, 87, 90, 94
  embryonic, 88F
  embryonic versus adult, 87, 94
  loose, 87, 87F, 88, 94
  types of adult, 89F
Connexons, **49**
Consciousness, **367**
  versus unconsciousness, 367–69
Constant region of antibody, **629**
Constipation, **467**
Contact dermatitis, 105
Contact lenses, **385**
Continuous positive airway pressure (CPAP), **530**
Contraception, **725**
Contralateral reflex, **317**

Contrast enema, **456**
Contrast radiography, **15**
Conus medullaris, **312**
Conventional audiogram, **403**
Convergence, **279**, 280F, 285
Convergent fascicles, **207**
Convulsive spasm, **193**
Copulation, **722**. *See also* Sexuality
Coracobrachialis, **232**
Coracoid process, **149**
Coracoid (conoid) tubercle, **148**
Cords, **338**
Core temperature, **494**
Corium, **105**, 110
Cornea, **380**
  transplantation of, 382
Cornua, **143**
Coronal section of the body, **10**
Coronal suture, **142**, 160
Corona radiata, **717**, 719F
Coronary arteries, **566**
Coronary artery disease, **580**
Coronary bypass operations, **564**, 581
Coronary circulation, **566**
Coronary ligaments, **176**, 468
Coronary sinus, **563**, 597
Coronary thrombosis, **554**
Coronary trigon, **561**
Coroners, **12**
Coronoid fossa, **150**
Coronoid process, **142**, 150
Corpora arenacea, **431**
Corpora cavernosa, **707**
Corpora quadrigemina, **303**, 319
Corpus albicans, **718**
Corpus callosum, **297**
Corpuscles. *See* Cells
Corpus hemorrhagicum, **718**
Corpus of uterus, **711**
Corpus luteum, **418**, 718, 719F
Corpus spongiosum, **670**, 707
Corpus striatum, **298**
Corrugator supercilii, **214**
Cortex, adrenal, **422**
Cortex (hair), **107**
Cortex of kidney, **649**
Cortex of lymph node, **621**
Cortex of ovary, **709**
Cortex of thymus gland, **431**
Cortical labyrinths, **650**
Cortical nephron, **654**
Cortical sinuses of lymph node, **621**
Corticobulbar tract, **365**
Corticoids, mineral, 407
Corticospinal fasciculi, **314**
Corticospinal pathways, **365**
Corticospinal tracts, **365**, 365F
Corticosteroids, **186**, 423
Corticotrophin-releasing factor (CRF), **416**
Corticotrophs, **415**
Cortisol, **424**
Coryza, **507**
Costal cartilages, **147**, 161
Costal groove, **148**
Costal pleura, **512**
Costocervical arteries, **593**
Countercurrent concentrating mechanism, **662**
Countercurrent exchange, **664**

Countercurrent multiplier mechanism, **664**
Countercurrent multiplier system, 663F
Covalent bonds, **23**, 25F, 41
Cowper's glands, **706**
Coxal bones, **152**, 152F
CPR. *See* Cardiopulmonary resuscitation
Cranial cavity, **9**
Cranial group, **131**, 160
Cranial nerves, 323–33, 323F, 325, 340
Craniofacial abnormalities, 125
Craniosacral nervous system, **344**, 352
Craniosynostosis, **133**
Craniotabes, **126**
Creatine phosphate, **195**
Creatinine, **197**, 203
Creatinine clearance test, 659
Cremasteric muscle, **699**, **722**
Cretinism, **419**
Crib death, **526**
Cribriform plate, **139**
   fracture of, 140
Cricoarytenoid muscles, **509**
Cricoid cartilage, **509**
Cricothyroid muscles, **510**
Cri du chat syndrome, 763
Cristae, **61**, 76, 401
Crista galli, **139**
Crohn's disease, **473**
Cross bridges, **187**, 202
Crossed dominance, **373**
Crossed extensor reflex, **318**
Crossed eyes, 390
Crossing-over, **757**
   break and rejoin model of, 759F
Crown, **444**
Cruciate ligaments, 176, **176**
Crude touch, **359**
Cryopreservation, **734**
Cryptorchidism, **701**, **728**
Crypts of Lieberkuhn, **457**
Crystalline lattice, **23**
Cuboid tarsal, **156**
Cumulus oophorus, **717**
Cuneiforms, **156**
Cupula, **401**
Curettage, **134**
Cushing's syndrome, **425**
Cuspids, **445**
Cusps, **445**
Cutaneous nerves, **334**, 340
Cutaneous veins, **597**
Cuticle (hair), **107**
Cyclic AMP, **409**
Cysteine, **38**
Cystic duct, **468**, 470
Cystic fibrosis, **530**
Cystinuria, **661**
Cystitis, **675**
Cytochromes, **484**
Cytokinesis, **66**, 76
Cytology, **47**
Cytoplasm, **47**
   contents of, 57–63, 76
Cytosine, **43**
Cytoskeleton, **61**, 76
Cytosol phase, **57**, 76

Dalton, **21**, 41
Dalton, John, 19
Dark adaptation, **392**
Daughter cells, **68**
Deamination, **491**
Death, 305
Decibels, **398**
Decidua, **739**
Decidua basalis, **739**
Decidua capsularis, **739**
Deciduous teeth, **445**
Decompression sickness, **525**
Decussation of pyramids, **306**
Deep fascia, **182**, **202**
Deep femoral artery, **596**
Deep palmar arches, **594**
Deep tendon reflex, **316**, 318
Deep veins, **597**
   of arm, **600**
Defecation, **467**
Defecation reflex, **467**
Defibrillation, **193**
Degenerative arthritis, **177**
Deglutition, **449**
Delayed hypersensitivity T cells, **632**
Deletions of chromosome portions, **763**
Delta cells, **426**
δ-globin, **523**
Deltoid muscle, 210F, **232**
Deltoid tuberosity, **149**
Demifacets, **145**
Dendrites, **92**, **262**, 284
Dense bone, **118**
Dense connective tissue, **87**, 90, 94
Dental caries, **446**
Dental plaque, **446**
Dentate nucleus, **303**
Dentin, **444**
Deoxyribonucleases, **459**
Deoxyribonucleic acid (DNA), **39**, 40F, 43
   nucleotides of, 39F
   recombinant, **70**
Deoxyribonucleotides, 39F
Deoxyribose, **39**
Dependency, **350**
Dephosphorylation, **199**
Depolarized membrane, **269**, 285
Depressor labii inferioris, **214**
Depth perception, **390**, 391
de Quervain's disease, **186**
Dermatitis, **105**
   types of, 105
Dermatology, **111**
Dermatome, **182**
Dermatomyositis, **93**, 253
Dermatophytoses, **111**
Dermis, **100**, 101F, 105–6, 110
Descending aorta, **591**
Descending colon, **465**
Descending fasciculi, **314**
Descending limb of loop of Henle, **653**
Desmosomes, **50**, 73
Detoxification, **350**
Detrusor muscle, **670**
Deuterium, **20**
Developmental anatomy, **5**
Developmental dyslexia, **373**
Dextrinase, **461**
Dextrins, **459**
Diabetes insipidus, **415**, 691
Diabetes mellitus, **428**, **432**, 691

retinal changes resulting from, 384
Diagnostic laparotomy, **97**
Diagnostic techniques
   angiodynography, 593
   angiography, 575, 593
   arteriogram, 573F
   arthrogram, 15
   audiogram, 403
   bipedal lymphangiogram, 97
   cardiac, 574, 575
   cholecystogram, 456
   computerized axial tomography (CAT) scan, 5, 14, 15
   digital subtraction angiography, 593
   echocardiography, 575
   electrocardiogram, 31, 570
   electrocardiography, 570–71
   electrodiagnosis, 336
   electroencephalogram, 304F, 367, 368F
   electromyography, 336
   electroneurography, 336
   gastrointestinal, 456
   intravenous pyelogram (IVP), 649, 676
   laparoscopy, 704, 734
   laparotomy, 97
   lymphangiogram, 97
   myelogram, 15
   myogram, 193F, 194F
   ophthalmoscope, 384
   papanicolaou test (pap smear), 75
   positron emission tomography (PET), 24
   pyelograms, 649, 676
   radioisotopes, 22
   respirometer, 517
   roentgenology, 14
   single photon emission computed tomography (SPECT), 24
   sonogram, 15
   spirometer, 517
   for testing gustation, 379–80
   tomogram, 14
   tomography, 5, 15
   ultrasonography, 575, 593, 69, 739
   ultrasound scans, 456
   urinary tract imaging techniques, 648
   uterosalpingogram, 771
   voiding cystogram, 649
   xeroradiograph, 14
   X rays, 5, 16
Dialysate, **52**
Diapedesis, **547**, 547F, 586
Diaphragm, **9**, **224**, **725**
   central tendon of, 224, 516
Diaphragma sellae, **309**
Diaphragmatic hernia, 224
Diaphragmatic pleura, **512**
Diaphysis, **121**
Diarrhea, **467**
Diarthroses, **165**, 176
Diastole, **572**
Dicrotic notch, **574**
Diencephalon, 291F, **293**, 299–302, 319
   sagittal section, 301F
Diets, 498–99
Differential blood count, **548**
Diffusion, **51**, **52**, 73
   facilitated, **54**, 76
Digastric muscle, **220**
DiGeorge's syndrome, **638**

Digestion
  in large intestine, 466
  in the mouth, 448
  peptide, 460
  in the small intestine, 460
Digestion phase of phagocytosis, **626**
Digestive system, 440–72, 441F
  accessory structures, 468–72
  embryonic development of, 442
  histology, 441–43, 443F
  large intestine, 464–67
  mouth, 443–48
  pharynx and esophagus, 449–50
  small intestine, 455–63
  stomach, 451–55
Digital arteries, **594**, 596
Digitalis, **576**
Digital subtraction angiography (DSA), **593**
Dihydroxyacetone phosphate (DHAP), **481**
Dilation and curettage (D and C), **743**
Dilation and evacuation (D and E), **743**
Dipeptidase, **460**
Diphtheria, **286**
Diplopia, 253
Disaccharides, **34**, 42
Disease
  defenses against, 624–29
  infectious, 111
Disks
  intercalated, 92, 197, 203
  intervertebral, 143, 159, 161
  optic, 383
  ruptured, 159, 338, **338**
Displaced fracture, 116F
Dissecting aneurysm, **607**
Distal convoluted tubules of nephron, **653**
Distal portion of nephron, **653**
Distemper, **634**
Disulfide bonds, **628**
Diuresis, **413**, **665**, 665F
Divergence, **279**, 280F, 285
Diverticulosis, **473**
Diving, illness associated with, 525
Dizygotic twins, **741**
DNA. *See* Deoxyribonucleic acid
DNA cloning, **70**
DNA function, 68–73, 77
DNA polymerase, **68**, 77
DNA replication, **68**, 69F, 77
Dominant alleles, **753**
Dopamine, **281**, 352
Dorsal cavity, **7**
Dorsalis pedis artery, **596**
Dorsal portion of the body, **10**
Dorsal respiratory group (DRG), **526**
Dorsiflexion, 173
Dowager's hump, 127, 127F, **158**
Down syndrome, **286**, **762**
  age of mother and, 763F
Drug abuse, 350–51
  AIDS and, 630
Drug-induced neuromuscular junction block, 190
Drugs
  abuse of, 350–51, 630
  influence on the autonomic nervous system, 352–53
Dual innervation, **349**, 353
Duchenne's muscular dystrophy, 252
Ducted glands, **407**

Ductless glands, **407**
Duct of Santorini, **457**
Duct of Wirsung, **470**
Ducts
  alveolar, 514
  anterior ear, 396
  aqueduct of Sylvius, 307
  bile, 468, 470
  cerebral aqueduct, 302, 307, 319
  cochlear, 396
  cystic, 468, 470
  ductus deferens, 703
  ejaculatory, 703
  endolymphatic, 396
  of exocrine glands, **407**
  hepatic, 470
  interlobular, 470–71
  lacrimal, 140, 386
  lactiferous, 715
  lateral ear, 396
  mammary, 715
  mesonephric, 648
  metanephric, 648
  nasolacrimal, 387
  oviducts, 709–11, 711F
  pancreatic, 470
  papillary, 650
  paramesonephric, 700
  parotid, 447
  patent ductus arteriosus, 579
  posterior ear, 396
  pronephric, 648
  right lymphatic, 619
  semicircular, 396
  Stensen's duct, 447
  submandibular, 447
  thoracic, 619
  Wharton's ducts, 447
Ductus arteriosus, **605**
Ductus deferens, **703**
Ductus epididymis, **702**
Ductus venosus, **605**
Duodenal papilla, **457**
Duodenal ulcers, **452**, 473
Duodenitis, **473**
Duodenum, **455**
Duplications of chromosome portions, **763**
Dural sinuses, **309**, 597
Dura mater, **309**, 319
Dwarf, **753**
Dwarfism
  inheritance of, 754F
Dynamic equilibrium, **400**, 401–2
Dynamic spatial reconstructor (DSR), **15**
Dynamic stability, **176**
Dysentery, **467**
Dyslexia, 373
  developmental, **373**
Dysmenorrhea, **720**
Dysphagia, **93**, 224, 472
Dyspnea, **531**
Dysrhythmia, circadian, 367

Ear. *See also* Equilibrium; Hearing
  anatomy of, 393–98
  anterior canal and ducts of, 396
  auricle of, 393
  conductive hearing loss, 403
  embryonic development of, 394
  external or outer, **393**, 395F

  inner or internal, **396**
  innervation of, 398
  internal, 396
  middle, **393**, 395F
  pinna, 393
  posterior ducts of, 396
Eardrum, **393**
Ear ossicles, 131, 395
Earwax, **393**
Eccrine glands, **108**
Echocardiography, **575**
Eclampsia, **744**
EcoRI, **758**
Ectoderm, **182**, 292, 735
Ectopic pregnancy, **744**
Eczema, **105**, 105F
Edema, **124**, 621, 683
  high-altitude, 529
  test for, 684F
Edwards syndrome, **763**
Effector organs, **260**
Effector T cells, **631**
Efferent ductules, **702**
Efferent glomerular arteriole, **654**
Efferent lymphatics, **620**
Efferent portion of the central nervous system, **261**
Efferent portion of the peripheral nervous system, 284
Effort, **209**
Effusion, **167**
Ejaculation, **705**
Ejaculatory duct, **703**
Elastic arteries, **585**
Elastic cartilage, **90**, 94
Elastic fibers, **87**, 94
Elastic tissue, 90
Electrical synapses, 278
Electric charge (protons and electrons), **19**
Electric potential, **266**, 266–67, 284
Electrocardiogram (EKG or ECG), **31**, 570
  abnormal, 571F
  normal, 570F
Electrocardiography, **570**, 570–71
Electrodiagnosis, 336
Electroencephalogram (EEG), **367**, 368F
  normal versus epileptic, 304F
Electrolyte imbalance, 684
Electrolytes, **682**, 683F
  repair of imbalance in, 685
Electromagnetic radiation, **20**
Electromyography (EMG), **336**
Electron acceptor, **28**
Electron donor, **28**
Electroneurography (ENG), **336**
Electron orbitals, 21–23, 22F, 41
Electrons, **19**, 41
Eleiden, **101**
Elements, **19**, 41
Elephant Man disease, 74
Elimination diets, **499**
Ellipsoidal joints, **168**, 168–69, 178
Elongation phase of translation, **73**, 77
Embolus, **314**, **517**, 554, 588
Embryonic connective tissue, 87, 94
Embryonic development, 733–42
  of the autonomic nervous system, 343
  of bones, 120–23
  of the brain, 293, 293F
  of circulation, 603–5

of connective tissue, 87, 88F
of the digestive system, 442
of the ear, 394
of the eye, 381
fetal circulation, 603–5
genetic disorders due to chromosomal abnormalities, 761–65
of the heart, 560
immunity during, 633
indifferent stage of, 700
inheritance, 753–61
meiosis, 749–52
of muscle, 182
of the nervous system, 292–93
of olfactory organs, 378
of peripheral nerves, 324
pregnancy and childbirth, 743–49
of the reproductive system, 700–701
of the respiratory system, 506
of the urinary system, 648
Embryonic disk, **734**, 737F
Embryonic germ layers, **734**–35, 737F
derivatives of, 738
Embryonic membranes, 735–38
Embryo transfer, 735
Emission, **708**
Emmetropia, **389**
Enamel, **444**
Encephalitis, **286**
Encephalomyelitis, **286**
Encoding, 280–81, **281**, 281F
End-diastolic volume (EDV) of heart, **575**
Endemic goiter, **419**
Endocardial tubes, **560**
Endocarditis, bacterial, 581
Endocardium, **561**
Endochondral ossification, **120**, 121–23, 122F, 128
Endocrine glands, **84**, **93**, **407**, 409F
Endocrine system, **16**, 260, 406–33. *See also* Hormones
 adrenal cortex, 422–26
 gonads, 429
 hypothalamus, 412–13
 pancreas, 426–29
 pineal gland, 431
 pituitary gland, 413–18
 thymus gland, 429–31, 430F
 thyroid and parathyroid glands, 418–21
Endocrinocytes, interstitial, 699
Endocytosis, **55**, 56F, 76
 receptor-mediated, **56**
Endoderm, **182**, 292, 735
Endogenous pyrogens, **627**
Endolymph, **396**
Endolymphatic duct, **396**
Endometrial stroma, **713**
Endometriosis, **713**
Endometrium, **711**
Endomysium, **183**, 202
Endoneurium, **261**, 284
Endoplasmic reticulum, **57**, 59F, 76
End organ of Ruffini, 359
Endorphins, **281**
Endoscopy, **456**
Endosomes, **56**
Endosteum, **118**, 128
Endothelium, 82F
End-systolic volume of heart, **575**
Endurance, **200**, 203

Energy level, **21**
Energy levels of electrons, 41
Enkephalins, **281**
Enteroceptive neurons, 354
Enteroceptive sensory neurons, **349**
Enteroceptors, **357**
Enteroendocrine cells, **457**
Enterogastric reflex, **454**
Enterokinase, **459**
Entrapment neuropathies, **339**
Enzymes, **29**, 29–30, 42
 adenylate cyclase, 408
 adenylate kinase, 195
 amino peptidase, 460
 amylase, 459
 anticholinesterase, 353
 ATPase, 484
 choline esterase, 189, 202
 cholinesterase inhibitors, 190
 chymotrypsin, 459
 coenzyme A, 482
 cofactors of, 492
 deoxyribonucleases, 459
 dextrinase, 461
 dipeptidase, 34, 42
 DNA polymerase, 68, 77
 enterokinase, 459
 hyaluronidase, 706
 intestinal, 460
 lactase, 460
 lipases, 459, 460
 lysozyme, 386, 448, 625
 maltase, 460
 pancreatic, 458–59, 458–60
 pepsin, 453
 permeases, 55, 76
 procarboxypeptidase, 459
 propionyl CoA, 489
 protein kinases, 409
 renin, 655
 restriction endonucleases, **758**, 766–67
 ribonucleases, 459
 saliva, 446, 448
 of the small intestine brush border, 461
 streptokinase, 581
 succinyl CoA, 483
 sucrase, 460
 telangiectases, 613
 tyrosinase, 105
Eosinophils, **546**
Ependyma, 264F, **265**, 284
Ephedrine, 352
Epicardium, **559**
Epicondyles, **155**
Epicranius, **214**
Epidermal derivatives, 106–7, 110–11
Epidermal layers, 101–3
Epidermis, **100**, 101, 101F, 110
Epididymis, **702**
Epidural hematoma, **287**
Epiglottis, 449, **509**
Epiglottitis, acute, 530
Epilepsy, **304**
Epimysium, 202
Epinephrine, **281**, 283, 352, 353, 424–26, 480
 control of glycogen synthesis and breakdown, 481
Epineurium, **261**, 284
Epiphyseal fractures, 123
Epiphyseal line, **123**

Epiphyseal plate, 122F, **123**, 128
Epiphyses, **121**, 431
Epiploic appendages, **464**
Episiotomy, **747**
Epispadias, **728**
Epithalamus, **301**, 319
Epithelial tissues, 81F, 81–86, 92–93
 glandular, 84–86
 pseudostratified, 83
 simple, 82–83
 stratified, 83–84
Epithelium, **81**, 441
 germinal, 703
 glandular, 84–86, 93
Eponychium, **107**
Epstein-Barr virus, **75**, 530
Equilibrium, **33**, **400**, 400–402
Equilibrium concentration, 42
Equilibrium constant, **33**, 42
Equivalents, in electrolyte concentration, **682**
Erection, **722**
Erector spinae, **222**
Erythema marginatum, **111**
Erythema multiforme, **111**
Erythema nodosum, **111**
Erythroblastosis fetalis, **636**
Erythrocytes, **543**, 543–46
 control of production of, 545
Erythrocyte sedimentation rate (ESR), **551**
Erythropoiesis, **544**, 668
Erythropoietin, **545**, 669
Esophageal artery, **594**
Esophageal plexus, **347**
Esophageal sphincter, lower, 449
Esophageal stage, **449**
Esophageal vein, **600**
Esophagus, **449**
 disorders of, 472–73
Esotropia, **390**
Essential fatty acids, **490**
Estrogens, **407**, **429**, 709
Ethmoid bone, **139**, 139F, 139–40, 160
Ethmoid sinuses, **508**
Etiology, **304**
Evaporative heat loss, **496**
Eversion, 174, 178
Exchange reaction, **28**
Excitability, **269**
Excitatory neurotransmitters, **278**, 285
Excitatory postsynaptic potential (EPSP), **287**, 278, 279F
Excretion. *See also* Urinary system
 sweat glands and, 110
Exhalation, **516**
Exocrine glands, **84**, **93**, **407**
 functional classification of, 84–85, 85F
 structural classification of, 85–86
Exocytosis, **55**, 56F, 76
Exophthalmos, **419**
Expiration, **516**
Expiratory reserve volume, **517**
Extended wear contact lenses, **385**
Extension, 173, 178
Extensor carpi radialis brevis, **238**
Extensor carpi radialis longus, **238**
Extensor carpi ulnaris, **238**
Extensor digiti minimi, **238**
Extensor digitorum brevis, **250**
Extensor digitorum communis, **238**

Extensor digitorum longus, **248**
Extensor hallucis longus, **248**
Extensor indicis, **238**
Extensor pollicis brevis, **238**
Extensor pollicis longus, **238**
Extensors, of the wrist and hand, 238–39
External acoustic (auditory) meatus, **134**, 393
External anal sphincter, **466**
External auditory canal, **393**
External carotid artery, **591**
External ear, **393**, 395F
External iliac arteries, **594**
External iliac veins, **601**
External jugular vein, **597**
External nares, **506**
External occipital protuberance, **138**
External os, **711**
External respiration, **504**
External root sheath (hair), **107**
External urethral orifice, **670**
External urethral sphincter, **670**
Exteroceptive neurons, **354**
Exteroceptive sensory neurons, **349**
Exteroceptors, **357**
Extracellular fluid (ECF), **12**, 539, 681
Extracorporeal circulation pump, **564**
Extracorporeal hemodialysis, **52**
Extraembryonic coelum, **735**
Extraglomerular mesangium, **655**
Extraocular motions, **327**
Extrapyramidal disorders, 366
Extrapyramidal pathways, **365**, 366
Extrapyramidal tracts, **296**
Extrinsic asthma, **531**
Extrinsic autonomic plexuses, **344**
Extrinsic pathway of prothrombin activator synthesis, **552**
Eye, 380–87. *See also* Vision
    accessory structures of, **380**, 385–87, 386F
    anatomy of, 380–85
    crossed, 390
    diseases of, 387
    embryonic development of, 381
    light adaptation of, 391–92
    optics of, 389F
    sagittal section, 382F
Eyeball, **380**
    extrinsic muscles of, 217F
    muscles that move, 216–17
Eyebrows, **385**
Eye examinations, 384
Eyelashes, **385**
Eyelid folds, **381**
Eyelids, **385**
Eye sockets, 133

Facets, **145**
Facial deformities, congenital, 125
Facial expression, muscles of, 214–15, 215
Facial group, **131**, 160
Facial nerves, **305**, **329**, 330F, 340
Facilitated diffusion, **54**, 54F, 76
Fainting, 570
Falciform ligament, **468**
Falciparum malaria, 544
False pelvis, **152**
False ribs, **147**, 161
Falx cerebelli, **309**, 311F
Falx cerebri, **309**, 311F

Familial periodic paralysis, 252–53
Farmer's lung disease, **531**
Far point of vision, **388**
Farsightedness, 389
Fascia lata, **242**
Fascia penis, **707**
Fascicles, **182**, 202, 261, 284
    patterns of, 207
Fasciculations, **193**
Fasciculi, **313**, 320
    ascending, 313
Fasciculus cuneatus, **313**, 362
Fasciculus gracilis, **313**, 362
Fast block to sperm/egg fusion, **726**
Fast twitch muscle fibers (Type ll), **195**, 195–96, 202
Fat
    absorption by the small intestine, 462–63
    digestion of, 459, 460
    obesity, 488
Fat cells, **88**, 94
Fatiguing of a muscle, **194**, 195F
Fats
    catabolism of, 488–90, 490F
    metabolism of, 488
    neutral, 35, 42
    storage of, 487–88
Fat-soluble vitamins, **492**
Fatty acids, **35**, 36F, 42
Feces, **466**
Feedback, negative, **411**
Feedback mechanisms, **12**, 13, 16
Female genitalia, external, **714**, 715F
Female prepuce, **714**
Female pronucleus, **726**
Female sexual response, 722–24
Femoral artery, **596**
Femoral veins, **602**
Femur, **154**, 154F, 154–55, 161
Fenestra ovalis, **395**
Fenestra rotundum, **395**
Fertilization, **724**. *See also* Reproduction
Fe-S protein, **484**
Fetal circulation, 603–5, 604F
Fetal development. *See* Embryonic development; Reproduction
Fetal membranes, 738
Fetal portion of placenta, **738**
Fetopelvic disproportion, **747**
Fetoscopy, **739**
Fetus, **741**
Fever, **497**, 497
Fiberoptics, **456**
Fibers, **262**, 284
Fibrillation, **193**
Fibrin, **550**
Fibrinogen, **551**
Fibrinolysis, **553**
Fibrin stabilizing factor, **552**
Fibroblasts, **84**, 87, 94
Fibrocartilage, **90**, 94
Fibrocystic disease, **622**
Fibromyalgia, **253**
Fibromyositis, **253**
Fibrosis, **93**
Fibrous capsule of kidney, **647**
Fibrous dysplasia, **121**
Fibrous joints, **165**, 165–67, 176
Fibrous pericardium, **561**

Fibrous tunic, **380**
Fibula, **155**, 155F, 161
Fibular collateral ligament, **176**
Filaments, **61**, 62F
    arrangement in a myofibril, 186F
    arrangement of thick and thin, 187F
    molecular organization of, 186–87
Filiform papillae, **444**
Filtration slits of Bowman's capsule, **652**
Fimbriae, **709**
Fine touch, **359**
Fingernails, **106**
    structure of, 107F
First class lever, **209**, 254
First-degree burns, **109**
First-order neurons, **362**
Fissures, **295**
Fistula in ano, **473**
Fistulas, arteriovenous, 613
Fixator, **209**, 254
Fixed pore hypothesis, **55**
Flagella, **63**, 63F, 76
Flat bones, **115**
Flatfoot, **156**, 157
Flavin adenine dinucleotide (FAD), **483**
Flexion, **173**, 178
Flexor carpi radialis, **236**
Flexor carpi ulnaris, **207**, **236**
Flexor digitorum, 207
Flexor digitorum longus, **248**
Flexor digitorum profundus, **236**
Flexor digitorum superficialis, **236**
Flexor hallucis longus, **248**
Flexor pollicis longus, **237**
Flexor reflex, **317**
Flexors, of the wrist and hand, 236–37
Floating bone, **142**
Floating ribs, **147**, 161
Flu, **507**
Fluid mosaic structure of a cell membrane, **47**, 49F
Fluids. *See also* Body fluids
    extracellular and intracellular, 681
Focal seizures, **304**
Folia cerebelli, **303**
Folic acid, **493**
Follicles, **418**, 717–18
Follicle-stimulating hormone (FSH), **417**, 705
Follicular cells, **717**
Fontanels, **133**, 160
Foot, 156–57, 161
    abnormalities of, 125, 156
    arches of, 156–57, 157F, 161
    bones of, 120F, 157F
    hyperpronation of, 156
    intrinsic muscles of, 250–51, 251F
    muscles that move, 248–49, 249F
Foot processes of podocytes, **652**
Foramen
    apical, 444
    carotid, 136
    incisive, 140
    infraorbital, 140
    interventricular, 307
    intervertebral, 143
    jugular, 136
    mandibular, 142
    mental, 142
    obturator, 153

optic, 138
stylomastoid, 136
supraorbital, 133
vertebral, 143, 161
zygomaticofacial, 142
Foramen lacerum, **138**
Foramen magnum, **136**
Foramen of Luschka, **309**
Foramen of Magendie, **309**
Foramen of Monro, **307**
Foramen ovale, **138**, 563
Foramen rotundum, **138**
Foramen spinosum, **138**
Foramina, **133**, 134
palatine, 141
sacral, 146
of the skull, 134
transverse, 144, 161
Forearm
extensors of the wrist and hand, 238–39
flexors of the wrist and hand, 236–37
muscles that move, 234–35, 235F
Forebrain, **293**
Foregut, **442**
Forensic medicine, **192**
Foreskin, **707**
Formed elements of blood, **539**, 541
Fourth ventricle, **305**, 307, 319
Fovea centralis, **154**, 383
Fractures, 116–17
chondral, 177
of the joints, **177**
reduction of, 140
of the spine, 159–60
Fraternal twins, **741**
Freckles, **103**
Free nerve endings, **359**
Frequency, **398**
Frontal bone, **133**
Frontalis, **214**
Frontal lobe, **295**, 302F, 319
Frontal plane, 16
Frontal process, **140**, 142, 160
Frontal (coronal) section of the body, **10**
Frontal sinuses, **508**
Frontal squama, 133
Fructose, **34**
Fructose diphosphate, **481**
Fulcrum, **209**
Fumaric acid, **483**
Fundus, **451**, 711
Fungiform papillae, **444**
Funiculi, **313**
Fusiform muscles, **207**

GABA. *See* Gamma-aminobutyric acid
Galactose, **34**
Galea aponeurotica, **214**
Gallbladder, 470
Gallium citrate, 22
Gallstones, **460**
Gamete interfallopian transfer (GIFT), **735**
Gamma-aminobutyric acid (GABA), **283**, 285
Ganglion (ganglia), **265**
abdominal, 344
basal, 291, 297, 319
celiac, 344
ciliary, 327, 346
collateral, 344

Gasserian, 328
geniculate, 329
inferior jugular, 331
inferior mesenteric, **344**
inferior nodose, 331
otic, 331, 346
paravertebral, 344, 352
posterior root, 334, 340
pterygopalatine, 346
retinal, 383
semilunar, 328
spiral, 330
submandibular, 346
superior jugular, 331
superior mesenteric, 344
vestibular, 330
Ganglion neurons of the retina, **383**
Gap junction, **49**
Gas embolism, **525**
Gases
ideal gas laws, 519–20
solubility in liquids, 520–21
Gas exchange, in alveoli, 521–24
Gas pressure, **519**
Gasserian ganglion, **328**
Gastric glands, **452**, 453F
control of secretion from, 454
Gastric-inhibitory peptide, 431
Gastric inhibitory protein (GIP), **455**
Gastric juice, **452**, 452–54
Gastric phase, **454**
Gastric pits, **452**
Gastric ulcers, **452**
Gastrin, 420, **453**
Gastritis, **454**
chronic, 454
Gastrocnemius, **248**
Gastroenteritis, **685**
Gastrointestinal (GI) tract, **440**
Gastrointestinal diagnostic techniques, 456
Gastrula, **735**
Gastrulation, **735**
Gemellus inferior, **242**
Gemellus superior, **242**
Gene probes, 767
General anesthesia, **329**
Generalized seizures, **304**
General senses, **357**
General sensory area (cerebral cortex), **295**
General sensory neurons, **334**
Generator potential, **357**
Genes, **749**
distribution of, 753–55
organization along a chromosome, 750F
segregation and independent assortment of, 755–56
Genetic disorders, 761–65
Genetic engineering, **70**
Genetics, molecular, 766–67
Geniculate ganglion, **329**
Genioglossus, **219**
Genitalia. *See also* Reproduction
external female, 714, 715F
Genital tubercle, **701**
Genotype, **753**
German measles, 104
Germinal center of lymph nodule, **621**
Germinal epithelium, **703**
Gestation, **741**

Gibbous, **158**
Gigantism, **417**
Gingivae, **444**
Gingivitis, **446**
Ginglymus joints, **168**
Glabella, **133**
Glands, **84**, 93
accessory glands of scrotum, **706**
adrenal, 422, 423F
alveolar, 86
apocrine, 85, 93
Bartholin's, 714
Brunner's, **457**
buccal, 447
bulbourethral, 706
cardiac, **452**
ceruminous, 109, 111, 393
ciliary, 386
classification of types found in the body, 85
compound alveolar, 86
compound tubuloalveolar, 86
Cowper's, 706
ducted and ductless, **407**
eccrine, 108
endocrine, 84, 93, **407**, 409F
exocrine, 84–86, 93, **407**
gastric, **452**, 454
glandular epithelial tissue, 84–86
heterocrine, **408**
holocrine, 85, 93
intestinal, **457**
islets of Langerhans, 426
lacrimal, 386
mammary, 109, 715–16, 716F
meibomian, 386
merocrine, 85, 93
multicellular exocrine, 85, 86F
oil, 108, 111
pancreas, 426–29, 427F, 470–71
parathyroid, 13, **418**, 420–21
paraurethral, 714
parotid, **447**
pineal, 301, 319, **431**
pituitary, 300, **412**, 413–18
prostate, 706, 706F
pyloric, **453**
salivary, 447, 448F
sebaceous, 108, 111
simple alveolar, 86
simple tubular, 85
of the skin, 108–9, 111
sublingual, **447**
submandibular, **447**
submaxillary, **447**
sudoriferous, 108, 111
sweat, 108, 110, 111
tarsal, 386
thymus, **429**, 429–31, 430F, 624, 624F
thyroid, 13, **418**, 418–20, 419F
tubercular, 85, 86
tubuloalveolar, 86
unicellular exocrine, 85
urethral, 706
Glandular cells, 7
Glandular epithelium, 84–86, 93
Glans clitoridis, **714**
Glans penis, **707**
Glaucoma, **387**, 387

Glenoid fossa, **149**
Glia, **262**, 284
Gliding joints, **168**, 176
Globulin, **448**, 541
Globus pallidus, **298**
Glomerular arterioles, afferent, 654
Glomerular basement membrane, **652**, 653
Glomerular blood pressure, 657–58
Glomerular capsule, **652**
Glomerular filtration, 657
Glomerular filtration pressure, 659F
Glomerular filtration rate (GFR), **657**, 658–59
Glomerulonephritis, **639**
   acute, 676
Glomerulus, **652**, 652–53
Glossopalatine arch, **444**
Glossopharyngeal nerves, 306, **330**, 330–31, 331F, 340
Glottis, **509**
Glucagon, **428**
Glucocorticoids, 407, **424**
Gluconeogenesis, **479**
Glucose, **34**, 480–81
   oxidation of, 486
   as a source of carbon, 485–86
   sparing of, **479**
Glucose phosphate, **28**, 195, 480
Glutamic acid, 283, 285, **491**
Gluteal lines, **153**
Gluteal tuberosity, **155**
Gluteus maximus, **242**
Gluteus medius, **242**
Gluteus minimus, **242**
Glycerides, 36F
Glycerol, **35**
Glycine, **37**, 283, 285
Glycocalyx, **460**
Glycogen, 35F
Glycogenesis, **480**
Glycogenolysis, **480**
Glycogen storage diseases, 252
Glycogen synthesis, 480
Glycolysis, **481**, 481–82
   reaction of, 482F
Glycosaminoglycans, **120**, 128
Glycosuria, **674**
Goblet cells, **83**, 456
Goiter, **419**, 420F
Golgi apparatus, **57**, 60F, 76
Golgi tendon organs, **360**
Gomphosis, **166**
Gonadal arteries, **594**
Gonadal veins, **601**
Gonadocorticoids, **424**
Gonadotrophin-releasing hormone (GnRH), **416**
Gonadotrophins, **417**
Gonadotrophs, **415**
Gonads, **429**
Gonococcus, **177**
Gonorrhea, **723**
Goodpasture's syndrome, **676**
Gout, **166**
G protein, **408**
Graafian follicle, **717**
Gracilis pectineus, **242**
Graft-versus-host reaction, **639**
Grand mal seizures, **304**
Granulocytes, **546**

Graves' disease, **419**
Gray commissure, **312**
Gray communicating rami, **344**
Gray matter, **291**, 319, 297–98
   columns of, **312**
   of the spinal cord, 312
Great cardiac vein, **567**
Greater cornua, **143**
Greater curvature, **451**
Greater multangular carpal, **150**, 151F
Greater omentum, **472**
Greater palatine foramina, **141**
Greater pelvis, **152**
Greater sciatic notch, **153**
Greater trochanter, **154**
Greater tubercles, **149**
Greater wings (sphenoid bone), **138**
Great saphenous vein, **602**
Greenstick fracture, 116F
Groin pulls, 246
Gross anatomy, **5**
Gross obesity, **488**
Ground substance, 62F, **63**, 86, 94
Growth, **732**
Growth abnormalities, 417
Growth hormone release inhibitory hormone (GHRIH), **416**
Guanine, **39**, 43
Guanosine, 43
Guanosine diphosphate (GDP), **483**
Guanosine monophosphate (GMP), 40
Guanosine triphosphate (GTP), **483**
Gubernaculum, **701**
Guillain-Barre syndrome, **287**, 339
Gum disorders, 446
Gums, **444**
Gustation, **379**, 379–80
   testing, 380
Gustatory cells, **379**
Gustatory pathway, 380
Gustatory pore, **379**
Gyri, **291**, 319

Hair, 107–8, 107F, 111
   olfactory, **377**
Hair bulb, **107**
Hair follicles, **107**, 108F, 111
Hair root, **107**
Hair shaft, **107**
Hallucinogenic drugs, 351
Hallux, **156**
Hamate carpal, **150**, 151F
Hammer, **395**
Hamstring, 246
Hand, 150–52, 151F, 161
   forearm extensors of, 238–39, 239F
   forearm flexors of, 236–37, 237F
   intrinsic muscles of, 240–41, 241F
Haptens, **627**
Hard corneal contact lenses, **385**
Hard palate, **140**, 444
Haustra, **464**
Haversian canals, **118**
Haversian system, **118**
Hay fever, **638**
Head, 7, 16
   arteries of, 591–93
   muscles of, 214–19, 215F
   trauma to, 287

Headache
   musculotension, **214**, 299
   types of, 299
Hearing, 398–400
   disorders and tests of, 403
   sensory pathways for, 399–400
Heart, **558**
   anatomy of, 559–67, 565F
   artificial, 564
   atria of, 563–65
   auricles of, 563
   bundle branches of, 569
   cardiac cycle, 572–74, 572F
   cardiac output, 574–76
   cardiac physiology, 567–70
   coronary vessels, 567F
   detail of, 562F
   diseases of, 579–81
   electrocardiography, 570–71
   embryonic development of, 560
   end-diastolic and end-systolic volume of, 575
   fibrous skeleton of, 563F
   muscles of, 92
   pain in, 567, 580
   physiologic control of heart rate, 576–78
   surgery of, 564
Heart attack, 567, **580**
   pain from, 360
Heartburn, 450
Heart disease
   congenital, 579
   congestive heart failure, 576
Heart-lung machine, 564
Heart murmurs, **574**
Heart rate, 576–78
   aging and, 578
   nerve pathways that control, 577F
Heart sounds, 574
Heart transplant operation, **564**
Heat stroke, **496**, 691
Heavy chains of amino acids, **628**
Heel bone, 156
Helicine arteries, **707**
Helicotrema, **396**
Helper T cells, **632**
Hemangiomas, **613**
Hemapoiesis, **115**
Hematocrit, **548**
Hematocytoblasts, **541**
Hematoma, **117**
Hematomas, 117, 287, **613**
Hematopoiesis, 115, **541**
Hematuria, **674**
Heme, **523**
Hemiazygos vein, **600**
Hemiplegia, **314**
Hemisection (spinal cord), **315**
Hemivertebra, **159**
Hemizygous persons, **760**
Hemoglobin (Hgb), **521**, 523–24, 543, 548
   oxygen dissociation curve of, 523F
   reuse of, from erythrocytes, 545F
Hemoglobin A, 523
Hemoglobin A2, 523
Hemoglobin molecule, 523F, 543F
Hemoglobinopathies, **554**
Hemoglobin-oxyhemoglobin buffer system, **688**

Hemolysis, **544**
Hemolytic anemias, **546**
Hemolytic disease of the newborn, 636
Hemophiliacs, AIDS and, 630
Hemophilias, **554**, 760
    A type, **760**, 761
Hemopoiesis, 128, **541**, 542F
Hemorrhage, **554**
Hemorrhagic anemia, **546**
Hemorrhoids, **473**, 614
Hemostasis, **550**, 550–55
Heparin, 88, **547**
Hepatic artery, **594**
Hepatic ducts, **470**
Hepatic flexure, **464**
Hepatic portal system, **601**
Hepatic portal vein, **601**
Hepatic veins, **601**
Hepatitis, **471**
Hereditary disease, 70
Hering-Breuer reflex, **527**
Hermaphroditism, true, **728**
Hernia
    diaphragmatic, 224
    herniated disk, 338
    hiatus, 224, 472
Herniated intervertebral disk, **159**
Herniation, **224**
Herpes genitalis, **102**
Herpes simplex, **102**, 102F, 634
Herring bodies, **413**
Heterocrine glands, **408**
Heterozygous persons, **753**
Hexoses, **34**
Hiatus, **224**
Hiatus hernia, **224**, 472
Hiccups, 193
High-altitude cerebral edema (HACE), **529**
High-altitude pulmonary edema (HAPE), **529**
High density lipoprotein (HDL), **487**
Hillock, **264**, 284
Hilum, **620**
Hilus, **620**
Hindbrain, **293**
Hindgut, **442**
Hinge joints, **168**, 176
Hip, muscles of, 245F
Hipbones, 152, 152F
Hippocampal gyrus, **302**
Hip prosthesis, **169**
Histamine, 88, **102**, 547
Histocompatibility antigens, **633**
Histology, **5**, 115
Histones, **65**, 76
Hives, **102**, 102F
HLA system, **639**
Hodgkin's disease, **96**
Hoffman's reflex, **318**
Holandric genes, **759**
Holocrine glands, **85**, 93
Homeostasis, **12**, 16
Homologous pairs of chromosomes, **749**
Homosexuals
    AIDS and, 630
    hepatitis and, 471
Homozygous persons, **753**
Horizontal plate (ethmoid bone), **139**
Hormones, **84**, 260, 407–12
    adrenal, 423–24

    adrenal androgens, 424
    adrenaline, 283
    adrenocorticotrophic (ACTH), 417
    aldosterone, 423, 666
    anabolic steroids, 201
    androgens, 407
    angiotensins, 424, 609, 666
    antidiuretic (ADH), 413–14, 665
    antithrombin III, 553
    chemical classification of, 408
    controlling factors for secretion of, 412
    corticosteroids, 423
    corticotrophin-releasing factor, 416
    cortisol, 424
    egg production and, 719–22
    estrogens, 407, 429, 709
    follicle-stimulating (FSH), 417, 705
    gonadotrophin-releasing (GnRH), 416
    growth hormone release inhibitory (GHRIH), 416
    histones, 65, 76
    human growth (HGH), 417
    lactogenic, 416
    luteinizing, 417
    melanocyte-stimulating, 417
    parathormone, 124, 128, 420
    parathyroid (PTH), 420
    peptide, 407
    pituitary, 413–18
    progesterone, 429, 719
    prolactin-inhibiting, 416
    prolactin-releasing, 416
    protein, 407
    regulation of activity, 411–12
    release-inhibiting, 415
    releasing, 415
    sperm production and, 705–6
    steroid, 36, 42, 201, 407, 410–11
    sympathomimetic function of, 426
    testosterone, 429, 705
    thyroid, 418–20
    thyroid-stimulating, 417
    thyrotrophin-releasing, 416
    trophic, 415
    water-soluble, 407, 408
Horseshoe kidney, **675**
HTLV (human T cell lymphotrophic virus) III, **630**
Human chorionic gonadotrophin (HCG), 431, **744**
Human chorionic somatomammotrophin (HCS), **747**
Human development, **732**
Human growth hormone (HGH), **417**
Humerus, **149**, 149–50, 150F, 161
    muscles that move, 232, 233F
Humoral immunity, **629**, 632–33, 633F
Huntington's disease (chorea), **283**, 286
Hyaline cartilage, **90**, 94
Hyaluromic acid, **167**
Hyaluronidase, **706**
H-Y antigen, **759**
Hybridoma, **635**
Hydrocele, **728**
Hydrocephalus, **286**, 307
    communicating, 307
Hydrogen bond, **25**, 42
    in water, 26F
Hydrogen ion concentration, 30–33, 42

Hydrogen ion transport, 524
Hydrogen transport protein complexes, **484**
Hydrostatic pressure, **610**, 657
Hydroxyl group, **28**
Hydroxyl ions, **31**
Hymen, **714**
Hyoglossus, **219**
Hyoid bone, 43, **142**, 142F, 160, 219F
Hyperextension, **167**, 173, 178
Hyperglycemia, **432**
Hyperinsulinism, **427**
Hyperlordosis, **158**
Hypermagnesia, **687**
Hyperopia, **389**
Hyperparathyroidism, **421**
Hyperpolarized membrane, **278**
Hyperpronation of the foot, **156**
Hyperreflexia, autonomic, 315
Hypersensitivity disorders, **638**
Hypersplenism, **623**
Hypertension, **314**, 581, 607
    biofeedback techniques and, 351
    retinal changes resulting from, 384F
Hyperthermia, **691**
Hyperthyroidism, **419**, 497
Hypertonic dialysate, **52**
Hypertonic solution, **52**, 76
Hyperventilation, **527**, 690
Hypocalcemia, **686**
Hypochlorhydria, **454**
Hypochondriasis, **625**
Hypodermis, **106**
Hypoesthesia, **318**
Hypogastric arteries, **594**
Hypoglossal canals, **138**
Hypoglossal nerves, 306, **332**, 332F, 340
Hypoglycemia, **428**
Hypokalemia, **691**
Hypomagnesia, **687**
Hyponychium, **107**
Hypoparathyroidism, **421**
Hypopharyngeal sphincter, **449**
Hypophyseal stalk, **412**
Hypophysis, **300**, 412
Hypospadias, **728**
Hypothalamic nuclei, **300**
Hypothalamus, **300**, 301, 319, 412–13, 415
    control of anterior pituitary hormone production, 416F
    control of body temperature, 494–95
Hypothenar eminence, **240**
Hypothyroidism, **419**, 497
Hypotonic solution, **52**, 76
Hypoventilation, **527**
Hypovolemic shock, **609**
Hypoxemia, **529**
H zone, **184**, 202

I bands, **184**, 202
Ichthyosis, **111**
Ideal gas laws, **519**, 519–20
Identical twins, **741**
Idiopathic epilepsy, **304**
Ig antibodies, **628**
Ileocecal valve, **457**
Ileum, **455**
Iliac crest, **153**
Iliac fossa, **153**
Iliac spines, **153**

Iliac tuberosity, **153**
Iliacus, **242**
Iliopsoas, **242**
Iliotibial band, **242**
Iliotibial band friction syndrome, 242
Ilium, **153**, 161
Immune system, **627**
Immunity, **627**
  active, 635
  blood groups and, 636–37
  cell-mediated, 629
  cellular and humoral, 629–34
  immunologic disorders, 638–39
  nonspecific defense against disease, 624–27
  specific defense against disease, 627–29
  vaccination and, 634–35
Immunodeficiency diseases, **638**
Immunodeficiency with thymoma, **638**
Immunoglobulins (Ig), **628**
Immunotherapy, **75**, 639
Impaction, **447**
Impermeable junctions, **50**, 73
Impetigo, **111**
Implantation, **733**, 736F
Impulses
  initiating, 271
  speed of, 271–74
Incisive foramen (canal), **140**
Incisors, **445**
Inclusion conjunctivitis, **723**
Incomplete dominance, **755**
Incus, **395**
Independent assortment
  of alleles, **756**
  chromosomal basis for, 757F
Indifferent stage of embryonic development, **700**
Infarction
  cerebral, 314
  myocardial, 567, 580, 588
Infectious colitis, **473**
Infectious mononucleosis, **530**
Inferior anterior nucleus, **300**
Inferior articular processes, **143**
Inferior colliculus, **303**
Inferior conjunctival fornix, **386**
Inferior horn, **307**
Inferior jugular ganglia, **331**
Inferior lateral nucleus, **300**
Inferior mesenteric artery, **594**
Inferior mesenteric ganglion, **344**
Inferior mesenteric plexus, **347**
Inferior mesenteric vein, **601**–602
Inferior nasal conchae, **140**
Inferior nodose ganglion, **331**
Inferior nuchal lines, **138**
Inferior phrenic arteries, **594**
Inferior portion of the body, **10**
Inferior posterior nucleus, **300**
Inferior temporal lines, **133**
Inferior trunk, **336**
Inferior vena cava, **563**, 597
Infertility, **771**
Infiltration anesthesia, **329**
Inflammation, **627**
Influenza, **507**
Infrahyoid muscles, **220**
Infraorbital foramen, **140**

Infraspinatus, **232**
Infraspinous fossa, **149**
Infundibulum, **300**, **412**
  of oviduct, **709**
Ingestion phase of phagocytosis, **626**
Inguinal canal, **703**
Inhalation, **516**
Inheritance, 753–61
Inhibin, **429**, 706
Inhibitory neurotransmitters, **278**, 285
Inhibitory postsynaptic potential (IPSP), **278**, 278–79, 279F
Initial portion of nephron, **653**
Initiation phase of translation, **73**, 77
Inner cell mass, **733**
Inner ear, **396**
Inorganic phosphate (Pi), **482**
Insemination, artificial, **734**
Insensible perspiration, **497**
Insertion of a muscle, **209**, 254
Insomnia, **369**
Inspiration, **516**
Inspiratory capacity, **517**
Inspiratory reserve volume, **517**
Insula, **295**
Insulin, **426**
Insulin-dependent diabetes mellitus (IDDM), **432**
Insulin shock, **609**
Intact cells, **632**
Integument, **100**
Integumentary system, 100–12
  dermis, 105–6
  epidermal derivatives, 106–9
  epidermis, 101–5
  functions of the skin, 109–10
Internal ear, **396**
Interalveolar septa, **514**
Interatrial septum, **561**
Intercalated disks, **92**, 197, 203
Intercondylar eminence, **155**
Intercondylar fossa, **155**
Intercostal nerves, **334**
Intercostals, **224**
Intercostal veins, **600**
Interferon (IFN), **75**, 625
Interleukin 2, **632**
Interlobar arteries, **654**
Interlobar veins of nephron, **656**
Interlobular arteries, **654**
Interlobular ducts, **470**–71
Interlobular veins of nephron, **656**
Intermediate filaments, **61**
Intermediate mesoderm, **648**
Internal acoustic (auditory) meatus, **136**
Internal anal sphincter, **465**
Internal capsule (white matter), **298**
Internal carotid artery, **591**
Internal iliac arteries, **594**
Internal iliac veins, **601**
Internal nares, **506**
Internal os, **711**
Internal pudendal arteries, **707**
Internal respiration, **504**
Internal root sheath (hair), **107**
Internal thoracic artery, **593**
Internal urethral orifice, **670**
Internal urethral sphincter, **670**
Internuncial neurons, **312**

Interosseous borders, **150**
Interphase of mitosis, 66F, **67**, 77
Interstitial cells of Leydig, **429**, 699
Interstitial endocrinocytes, **699**
Interstitial fluid, **539**, 621
Interstitial fluid hydrostatic pressure (IFHP), **611**
Interstitial fluid osmotic pressure (IFOP), **611**
Interstitial growth, **119**, 128
Interstitial segment of oviduct, **711**
Interstitial spaces in nephron, **660**
Intertrochanteric crest, **154**
Intertrochanteric fossa, **154**
Intertrochanteric line, **154**
Intertubercular sulcus, **149**
Interventricular arteries, **567**
Interventricular foramen, **307**
Interventricular septum, **561**
Intervertebral disks, **143**, 159, 161
  herniated, 159
Intervertebral foramina, **143**
Intervillous spaces in chorion, **739**
Intestinal enzymes, **460**
Intestinal flora, 466–67
Intestinal gastrin, **454**
Intestinal glands, **457**
Intestinal lipase, **460**
Intestinal phase, **454**
Intraarticular fractures, **177**
Intraarticular ligaments, **176**
Intracellular fluid (ICF), **681**
Intrafusal fibers, **361**
Intramembranous ossification, **120**, 120–21, 128
Intraocular pressure, **385**
Intrauterine devices (IUDs), **725**
Intravenous pyelogram (IVP), **649**, 676
Intrinsic asthma, **531**
Intrinsic autonomic plexuses, **344**
Intrinsic factor, **453**
Intrinsic pathway of prothrombin activator synthesis, **552**
Inulin, **659**
Invasive cardiac diagnostic procedures, **575**
Inversion, **174**, 178
In vitro fertilization (IVF), **734**, 771
In vivo fertilization, **735**
Involuntary muscle, **91**, 94
Iodide, **418**
Ionic bond, **23**, 25F, 41
Ions, **23**, 41
  hydroxyl, 31
Ipsilateral reflex, **316**
Iris, **381**
Irregular bones, **115**
Irregularly arranged dense connective tissue, **90**, 94
Irritability, **16**
Irritable bowel syndrome, **473**
Ischemia, **93**
  acute, 613
  myocardial, 580
Ischial spine, **153**
Ischial tuberosity, **153**
Ischiocavernosus muscle, **708**
Ischium, **153**, 161
Islets of Langerhans, **426**
Isoantibodies, **636**
Isoantigens, **636**
Isocitric acid, **483**

Isoimmunization, **636**
Isomerization, 27F, **28**
Isomers, **28**
Isometric contraction, **200**, 203
  versus isotonic contraction, 201F
Isoniazid, **514**
Isoproterenol, **352**
Isotonic contraction, **200**, 203
Isotonic solution, **52**, 76
Isotopes, **20**, 20–21, 41
Isthmus of oviduct, **710**
Isthmus of the fauces, **509**
Isthmus of thyroid gland, **418**
Isthmus of uterus, **711**

Jacksonian seizures, **304**
Jejunum, **455**
Jenner, Edward, 635
Jet lag, 367
Jock itch, 111
Joint cavity, **165**, 176
Joints, **164**, 176. *See also* Articulations
  acetabulum, 152
  arthrodial, 168
  ball-and-socket, 169–70, 178
  biaxial, 170
  bursae, 168
  cartilaginous, 165, 176
  condyloid, 168–69
  disorders of, 177
  ellipsoidal, 168–69, 178
  fibrous, 165–67, 176
  fractures of, 116–17
  gliding, 168, 176
  hinge, 168, 176
  knee, 168, 174–76, 178
  patellofemoral, 174, 178
  pivot, 169, 178
  saddle, 170, 178
  sellaris, 170
  stability of, 176
  survey of joints in the body, 171
  synovial, 165, 168–70, 176–78
  tibiofemoral, 174, 178
Jugular foramen, **136**
Jugular notch, **147**
Jugular vein, 597
Junctions, **47**, 50F, 73
  adhering, **50**, 50F, 73
  communicating, 49, 73
  impermeable, 50, 73
  neuromuscular, 285
Juxtacapillary receptors, **527**
Juxtaglomerular apparatus, **654–55**
Juxtaglomerular cells, **655**
Juxtamedullary nephrons, **654**

Kallmann's syndrome, 379
Kaposi's sarcoma, **630**
Karyokinesis, **66**, 77
Keratin, **83**, 93, 101, 110
Keratinized squamous epithelium, **83**
Keratohyalin, **101**, 110
Keto acid, **491**
Ketone bodies, **489**
Ketosis, **489**
Kidneys, **646**, 647–51, 648F, 649F
  adipose capsule of, 647
  cancer of, 676

clearance in, 658
cortex of, 649
homeostatic functions of, 665–69
nephron, 651–57
physiology of, 657–65
serum creatinine, 197
Killer T cells, **631**
Kilocalorie, **494**
Kilogram calorie, **494**
Kinesiology, **6**
Kinetic energy, movement through cell
  membranes and, **51**, 73
Kinocilium, **400**
Klinefelter's syndrome, **762**
Knee, **161**, 168, 174–76, 178
  iliotibial band friction syndrome, 242
  locking of, **119**
  pain in, 360
  torn cartilage in, 174
Kneecap. *See* Patella
Knee jerk reflex, **317**
Knee joint, 168F, 174, 175F
Knock-knees, 155
Koebner phenomenon, **102**
Krebs cycle, **482**
  reactions of the, 483F
Kupffer cells, **55**
Kussmaul breathing, **527**
Kyphosis, **127**, 158

Labia, **444**
Labial frenulum, **444**
Labia majora, **714**
Labia minora, **714**
Labioscrotal swellings, **701**
Labor, **749**
Lacrimal apparatus, **386**
Lacrimal canal, **140**, 387
Lacrimal duct, **140**, 386
Lacrimal gland, **386**
Lacrimal papilla, **387**
Lacrimal punctum, **387**
Lacrimals, **140**, 160
Lacrimal sac, **387**
Lactase, **460**
Lacteals, **456**, 619
Lactic acid, **197**, 203, 479, 482
  formation of, 482F
Lactiferous duct, **715**
Lactogenic hormone, **416**
Lactose, **34**
Lactose intolerance, **460**
Lacunae, **120**
Lambdoidal suture, **142**, 160
Lamellae, **118**, 128
Lamellar bone, **118**, 128
Laminae, **143**
Lamina propria, **377**, 441
Language area (cerebral cortex), **296**
Laparoscope, **704**
Laparoscopy, **704**, 734
Laparotomy, **97**
Large intestine, **464**, 464F, 464–67
  connection with small intestine, 465F
  digestion in, 466
  histology of, 466F
Laryngeal aperture, **509**
Laryngeal prominence, **509**
Laryngitis, **530**

Laryngopharynx, **449**, 509
Laryngotracheal bud, **506**
Larynx, **509**, 510F
Latent period, **193**, 202
Latent period of immune response, **634**
Lateral canal of ear, **396**
Lateral canthus, **386**
Lateral cervical ligaments of uterus, **711**
Lateral cord, 338
Lateral corticospinal tract, **365**
Lateral ducts of ear, **396**
Lateral epicondyle, **149**
Lateral femoral circumflex artery, **596**
Lateral fissure, **295**
Lateral geniculate nucleus, **300**
Lateral malleolus, **155**
Lateral masses (ethmoid bone), **139**
Lateral pterygoid, **218**
Lateral sacral crest, **146**
Lateral spinothalamic fasciculus, **313**
Lateral ventricles, **319**
Lateral ventricles (brain), **307**
Latissimus dorsi, **232**
Learning, 372
Left atrioventricular valve, **566**
Left atrium of heart, **564**
Left common carotid artery, **591**
Left coronary artery, **566**
Left gastric artery, **594**
Left lobe of liver, **468**
Left subclavian artery, **591**
Left ventricle of heart, **566**
Leg, muscles that move the, 244F, 245F, 246–
  47, 247F, 248F
Leiomyomas, 253
Lens, **383**
Lens placode, **381**
Lentiform nucleus, **298**
Lesions, **111**
Lesser cornua, **143**
Lesser curvature, **451**
Lesser multangular carpal, **150**, 151F
Lesser omentum, **468**
Lesser palatine foramina, **141**
Lesser pelvis, **152**
Lesser sciatic notch, **153**
Lesser trochanter, **154**
Lesser tubercles, **149**
Lesser wings (sphenoid bone), **138**
Leukemia, **555**, 555F
  chronic myelogenous, 764
Leukocytes, **546**, 546–48
Leukocytosis, **548**
Leukopenia, **548**
Levator ani, **228**
Levator labii superioris, **214**
Levator palpebrae superioris, **385**
Levator scapulae, 207, **230**
Lever arm, **210**
Levers, **209**, 209–11, 211F, 254
  in moving parts of the body, 211F
Leydig cells, **699**
Life
  characteristics of, 10–16
  homeostasis, 12–16
  irritability, 16
  metabolism, 10–12
  organizational complexity, 10
  reproduction, 16

Ligaments, **90**, 91, 94, 167
   arcuate popliteal, 176
   fibular collateral, 176
   of the knee, 174–76
   patellar, 174, 246
   periodontal, 166, 445
   popliteal, 176
   posterior cruciate, 176
   round, 468
   round, of uterus, 711
   suspensory, of eye, **381**
   suspensory, of ovary, 709
   tibial collateral ligament, 176
   transverse ligament, 176
   triangular, 468
   vocal, 510
Ligamentum arteriosum, **605**
Ligamentum teres, **468**
Light, adaptation of the eye to, 391–92, **392**
Light chains of amino acids, **628**
Limbic system, **301**, 301F, 301–302, 319
Linea aspera, **154**
Lingual frenulum, **444**
Lingual tonsil, **444**, 624
Linitis plastica, **471**
Linkage, 758F
   chromosomal, **756**, 756–58
   partial, **757**
   sex, 759–61
Lipid bilayer, **47**
Lipids, **35**, 42
   metabolism of, 487–91
   transport of, 487
Lipocytes, **88**, 94
Lipofuscin, **287**
Lipomas, **74**
Lipoprotein, **487**
   high-density, 487
Lipoprotein micelles, 487
Lips, **444**
Lithotripsy, **670**
Liver, 468F, 468–70, 469F
   caudate lobe of, 468
   cirrhosis of, 470
   left lobe of, 468
Liver spots, **111**
Lobar arteries, **589**
Lobes of lung, **512**
Lobes of mammary glands, **715**
Lobotomy, **296**
Lobules, **303**, 514F
   of kidney, **650**
   of liver, **468**
   of mammary glands, **715**
   of thymus gland, **431**
   of thyroid gland, **418**
Localization of sound, **400**
Long bones, **115**, 118F
Longissimus capitis, **220**
Long-term memory, **371**
Longus colli, **223**
Loop of Henle, **653**
Loose connective tissue, **87**, 87F, 88, 94
Low calorie diet, **498**
Low density lipoprotein (LDL), **487**
Lower esophageal sphincter, **449**
Lower extremity, 154–60, 161
   arteries of, 594–96, 597F
   muscles of, 242–51
   veins of, 602, 603F
Lower labial frenulum, **444**
Lower motor neuron, **365**
Lower respiratory tract, **505**, 510–11
Low-sodium diets, **499**
Lumbar arteries, **594**
Lumbar curve, **143**, 161
Lumbar enlargement, **312**
Lumbar plexus, **338**
Lumbar veins, **601**
Lumbar vertebrae, **143**, 145, 146F, 161
Lumbosacral plexus, **338**, 339F, 340
Luminal phase, **57**, 76
Lunate carpal, **150**, 151F
Lung cancer, **511**, 531
Lungs, **511**, 511–12, 512F
   compliance of, 518
   tumors of, 531
Lunula, **107**
Lupus erythematosus, **177**
Luteinizing hormone (LH), **417**, 705
Lymph, **618**
Lymphangiogram, **97**
Lymphatic fluid, **618**
Lymphatic organs, 622–24
Lymphatics, **619**
   afferent, 620
Lymphatic system, 618–24. *See also* Immunity
Lymphatic vessels, 619–20, 620F
Lymph capillaries, **619**, 619F
Lymph circulation, 621–22
Lymph nodes, **620**, 620–21
   cortex of, 621
   internal structure of, 621F
Lymph nodules, **621**
Lymphoblasts, **543**
Lymphocytes, **431**, 547, 618
Lymphoepithelial organ, **429–30**
Lymphokines, **631**
Lysosomes, **57**, 76, **626**
Lysozyme, **386**, 448, 625

Macroangiopathy, **432**
Macromolecules, **25**, 42
Macronutrients, **41**
Macrophage activating factor, **632**
Macrophage migration inhibitory fraction (MIF), **632**
Macrophages, **88**, 94, 545, 626
Macula, **396**
Macula densa, **655**
Maculae, **400**
Macula lutea, **383**
Macular degeneration, **387**
Macular lesions, **111**
Madelung's deformity, **177**
Magnesium, in body fluid, 686–87
Major calyces, **649**
Major elements, **492**
Malabsorption syndromes, **460**, 473
Malaria, **544**
Male pronucleus, **726**
Male sexual response, **722**
Malic acid, **483**
Malignant myeloma, **126**
Malignant tumors, **74**
Malleolus, **155**
Malleus, **395**
Mallory-Weiss syndrome, **473**
Malnutrition, **489**
Malonyl CoA, **490**
Maltase, **460**
Maltose, **34**
Maltotriose, **459**
Mammary ducts, **715**
Mammary glands, **109**, 715–16, 716F
   alveoli of, 715
Mammillary bodies, **300**
Mammography, **622**
Mammotrophs, **415**
Mandible, **142**, 160
Mandibular condyle, **142**
Mandibular foramen, **142**
Mandibular fossa, **134**
Mandibular nerve, **328**, 340
Mandibular symphysis, **142**
Mandibulofacial dystocia, **125**
Manubrium, **147**, 161
Manus, **150**, 161
Marey's law of the heart, **577**
Marfan's syndrome, **613**
Marginal artery, **567**
Marrow, **118**, 128
   primary marrow spaces, **121**
   secondary marrow cavity, **123**
Massa intermedia, **300**
Masseter, **218**
Mass number, 41
Mast cells, **88**, 94, 547
Mastectomy, **622**
Mastication, **448**
   muscles of, 218, 218F
Mastoiditis, **396**
   acute, 134
Mastoid portion of the temporals, **134**
Mastoid process, **134**
Mastoid sinuses, **134**
Maternal portion of placenta, **739**
Matrix, **61**, 76, 86, 94
Matrix (in hair bulb), **107**
Matter, **19**, 41
   nature of, 19–26, 41–42
Maxillae, **140**, 141F, 160
Maxillary nerve, **328**, 340
Maxillary process, **142**
Maxillary sinus, **140**, 508
Maximal voluntary ventilation, **518**
McArdle's disease, 252
McBurney's point, 465
Measles, 104
Meatuses, **507**
Mechanoreceptors, **357**, 359–61
Medial canthus, **386**
Medial cord, **338**
Medial cubital vein, **600**
Medial epicondyle, **149**
Medial femoral circumflex artery, **596**
Medial geniculate nucleus, **300**
Medial lemniscus, **303**, 362
Medial malleolus, **155**
Medial nuchal lines, **138**
Medial plantar artery, **596**
Medial pterygoid, **218**
Median sacral crest, **146**
Mediastinal pleura, **512**
Mediastinum, **9**, 511
Medical examiner, **12**

Meditation, 351, 354
Mediterranean anemia, 554
Medulla
    adrenal, **422**
    of hair, **107**
    of kidney, **649**
    of lymph node, **621**
    of ovary, **709**
    of thymus gland, **431**
Medulla oblongata, 302F, **305**, 305–6, 319
Medullary chemoreceptors, **527**
Medullary cords of lymph node, **621**
Medullary rays, **650**
Medullary rhythmicity area, **306**
Medullary sinuses of lymph node, **621**
Medullary sponge kidney, **675**
Megakaryoblasts, **543**
Megakaryocytes, **549**
Megaloblasts, **546**
Meibomian glands, **386**
Meiosis
    compared with mitosis, 750F
    in females, 751–52
    in males, 749–51
    in the testis, 751F
Meiosis I, **749**
Meiosis II, **749**, 750
Meissner's corpuscles, **359**
Melanin, **103**, 110
Melanocytes, **103**, 110
Melanocyte-stimulating hormone (MSH), **417**
Melanoma, malignant, **106**
Melatonin, **431**
Membrane potential, **266**, 266–69, 268F
Membrane receptors, **347**, 347–48, 353
Membranous labyrinth, **396**
Membranous urethra, **670**, **703**
Memory, **370**, 370–71
    long-term, 317, **371**
    short-term, **370**
Memory cells of B-cell clone, **633**
Memory T cells, **632**
Menarche, **719**
Meniere's disease, **402**
Meninges, 319
    of the brain, **309**, 310F
Meningitis, **286**
Menisci, **119**, 168, 176, 178
Menopause, **722**
    types of, **729**
Menses, **722**
Menstrual cramps, 713, 720
Menstrual cycle, **719**, 719–22
Menstrual flow, **722**
Menstrual phase of menstrual cycle, **720**
Mental foramen, **142**
Merkel's disks, **359**
Merocrine glands, **85**, 93
Mesencephalon, **293**, 302–3, 319
Mesenchymal cells, **87**, 94
Mesenchyme, **94**
Mesenchyme tissue, **87**
Mesenteric arteries, 596F
Mesentery, **472**
Mesocolon, **472**
Mesoderm, **182**, 292, **735**
    intermediate, 648
Mesonephric duct, **648**

Mesonephros, **648**
Mesothelium, 82F
    of ovary, **709**
Mesovarium, **709**
Messenger RNA (m-RNA), **70**, 77
Metabolator, **497**
Metabolic acidosis, **689**
Metabolic alkalosis, **424**, 690
Metabolic pathways, 28
Metabolic processes, 12
    citric acid cycle, 482
Metabolic reactions, 27–30, 42
Metabolic water, **681**
Metabolism, 10, 16, 478–98
    basal metabolic rate, 497, 498
    of carbohydrate, 479–86
    cofactors in, 492–93
    heat as a by-product of, 493–98
    Krebs cycle, 482, 483F
    of lipids, 487–91
    pathways of, 28
    of protein and amino acids, 491–92
Metacarpals, 151F, **152**, 161
Metacarpus, **152**, 161
Metanephric duct, **648**
Metanephros, **648**
Metaphase I, **749**
Metaphase II, **751**
Metaphase of mitosis, 66F, **67**, 77
Metaphysis, **123**
Metarubricyte, **544**
Metastasis, **74**
Metastatic calcifications, **126**
Metastatic disease, **97**
Metastatic growths, **287**
Metatarsals, **156**, 161
Metatarsus, **156**, 161
Metatarsus varus, **156**
Metencephalon, **293**, 303–5, 319
Micelles, **462**
    lipoprotein, 487
Microangiopathy, **432**
Microfilaments, **63**, 76
Microglia, 264F, **265**, 284
Micronutrients, **41**
Microphages, **626**
Microscope, operative, **403**
Microscopic anatomy, **5**
Microtrabecular lattice, 62F, 63
Microtubules, **61**, 62F, 76
Microvilli, **457**, 457F
Micturition, **672**
Micturition reflex, 672F
Midbrain, **293**
Middle cardiac vein, **567**
Middle ear, **393**, 395F
Middle nasal conchae, **140**
Middle sacral artery, **594**
Middle trunk, 336
Midgut, **442**
Midpalmar muscles, **240**
Midsagittal section of the body, **10**
Migraine headaches, **299**
Military spine, **158**
Milliequivalents, **683**
Milliosmole, **662**
Millivolts, **267**
Mineral corticoids, **407**, **423**
Minerals, **492**

    absorption by the small intestine, 463
Minimal brain dysfunction (MBD), 373
Minor calyces, **649**
Minor tranquilizers, **350**
Miosis, **190**
Miscarriage, **743**
Mitochondria, **59**, 61F, 76
    movement along a microtubule, 62F
Mitosis, 66F, **67**. *See also* Cell division
    compared with meiosis, 750F
    phases of, 67–68, 77
Mitral valve, **566**
Mixed micelle, **462**
Mixed nerves, **261**, 284, 323, 340
Mixing waves, **452**
M line, **184**, 202
Modified radical mastectomy, **622**
Modiolus, **397**
Molars, **445**
Mole, **29**
Molecular genetics, 766–67
Molecular weight, **29**, 42
Molecules, 7, 16, **25**, 42
    covalent and ionic bonds in, **23**, 25F, 41
Monoblasts, **543**
Monochromats, **393**
Monoclonal antibodies, **75**, 635
Monocytes, **547**
Mononeuritis, **339**
Mononucleosis, 530
Monosaccharides, **34**, 42
Monosomic cells, **761**
Monosynaptic reflex, **314**, 320
Monozygotic twins, **741**
Mons pubis, **714**
Moro's reflex, **316**
Morphogenesis, **732**
Morula, **733**
Motion sickness, **401**, 402
Motor area for speech (cerebral cortex), **296**, 319
Motor end plate, **183**, 184F, 202
Motor nerves, **261**, 284, 323, 340
Motor neurons, 359F
Motor units, **183**, 202
Mountain sickness, acute, 529
Mouth, **443**, 443–48
    digestion in the, 448
    disorders of, 472
Movements of the body, 172F
Mucin, **448**
Mucous neck cells, **452**
Mucous tissue, **87**, 94
Multicellular exocrine glands, **85**, 86F
Multiple alleles, **755**
Multiple neurofibroma, **74**
Multiple sclerosis, **265**
Multipolar neurons, **264**, 284
Multiunit smooth muscle, **198**, 203
Muralium, **424**
Muscarinic receptors, **347**, 353
Muscle, **91**. *See also* Muscle contraction
    action of, 209
    cardiac, 92, 94, 181, 197–198, 198F, 203
    chemical waste products of, 197
    effects of training on, 199–202, 203
    embryonic development of, 182, 182F
    involuntary, 91
    motor units, 183

Muscle (continued)
  multiunit smooth muscle, 198, 203
  organization of skeletal muscle, 181–88
  origin of, 209, 254
  production of heat, 197
  smooth, 91, 94, 181, 198–99, 199F, 203
  steps in contraction and relaxation, 192
  striated, 181, 185F, 202
  trumpeter's, 214
  types of, 181F
  visceral, 91
  visceral smooth, 198, 203
  voluntary, 91, 181, 202
Muscle contraction, 188–92, 202
  abnormal, 193
  energy supply for, 195–196
  initiation of contraction, 189–90
  mechanism of contraction, 190–92
  physiology of, 192–97, 202–3
  sliding filament model of, 191F
Muscle fatigue, 195F
Muscle fibers, **182**, 202
Muscle relaxation, 192, 202
Muscles, **180**, 180–203. *See also* Skeletal muscles
  abdominal, 226, 227F
  abductors, 238
  adductors, 173, 178, 209, 242
  biceps brachii, 207, 234
  biceps femoris, 246
  brachialis, 234
  buccinator, 214
  of chewing, 218
  cremaster, 722
  cricoarytenoid, 509
  cricothyroid, 510
  deltoid muscle, 210F, 232
  detrusor, 670
  diaphragm, 9, 224, 725
  digastric, 220
  effect on movement of venous blood, 612F
  extensors, 238–39, 248, 250
  external anal sphincter, 466
  external urethral sphincter, 670
  of the eye, 216–17
  flexors of the wrist and hand, 236–37, 237F
  of the foot, 248, 249F, 250–51
  of the forearm, 234–35, 235F, 236–37, 237F, 238–39, 239F
  of the hand, 236–41, 237F, 239F, 241F
  of the head, 214–19
  of the heart, 561
  iliopsoas, 242
  infrahyoid, 220
  infraspinatus, 232
  insertion of, 254
  internal anal sphincter, 465
  internal urethral sphincter, 670
  levator ani, 228
  levator labii superioris, 214
  levator palpebrae superioris, 385
  levator scapulae, 207, 230
  longissimus capitis, 220
  lower esophageal sphincter, 449
  of lower extremity, 242–51
  major, 212F, 213F
  masseter, 218
  of mastication, 218, 218F
  midpalmar, 240
  movement of lymph and, 622F
  movements of, 207–13
  musculi pectinati, 563
  naming, 207, 254
  of the neck, 215F, 220–21
  oblique, 216
  orbicularis oculi, 207, 214
  orbicularis oris, 214
  overuse syndromes, 256–57
  of the pectoral girdle, 230–41
  pectoralis major, 207, 232
  pectoralis minor, 207, 230
  penniform, 207
  popliteus, 248
  pronators, 236
  psoas major and minor, 223, 242
  pterygoid, 218
  quadratus femoris, 242
  quadratus lumborum, 223
  quadriceps femoris, 207, 246
  quadriceps group, 246
  radiate, 207
  rectus, 216
  rectus abdominis, 226
  rectus capitis, 223
  rectus femoris, 246
  sartorius, 246
  serratus anterior, 230
  serratus posterior, 224
  soleus, 248
  steps in contraction and relaxation of, 192
  sternocleidomastoid, 220
  suprahyoid, 220
  supraspinatus, 232
  temporalis, 207, 208
  tensor fasciae latae, 242
  teres major and minor, 232
  that move the scapula, 230, 231F, 233F
  of the thigh, 242–45, 244F, 247F
  of thoracic wall, 224–25, 225F
  thyroarytenoid, 510
  tibialis anterior, 207, 248
  tibialis posterior, 207, 248
  of the tongue, 219, 219F
  trachealis, 511
  trapezius, 207, 230
  triceps brachii, 234
  of the trunk, 222–29, 231F
  of upper extremity, 230–41
  of the vertebral column, 222–23, 223F
  wrist extensors, 238–39
  wrist flexors, 236–37
  zygomaticus, 207
  zygomaticus major and minor, 214
Muscle spindle, **316**, 360
Muscle stretch reflex, **316**
Muscle tears, 246
Muscle tension headaches, **299**
Muscle tissue, 91F, 91–92, 94
Muscle tone, **197**, 203
Muscle trunk, **182**, 202
Muscle tumors, 253
Muscular atrophies, **199**, 203
Muscular diseases, 252–53
Muscular dystrophies, **200**, 203, 252
Muscularis mucosa, **441**
Muscularis muscle layer, **710**
Muscular system, **7**
Musculi pectinati, **563**
Musculotension headache, **214**
Myalgia, **252**
Myasthenia gravis, **200**, 253
Myelencephalon, **293**, 305–6, 319
Myelin, **265**
Myelinated neurons, 263F, **265**, 284
Myelin sheath, 284
Myeloblasts, **543**
Myelogram, **15**
Myenteric plexus, **454**
Myocardial infarction (MI), **567**, 580, 588
Myocardial ischemia, 580
Myocarditis, **581**
Myocardium, **561**, 568
Myofibril, **184**, 202
  arrangement of filaments in, 186F
Myoglobin, **196**, 202
Myoglobinuria, **196**
Myogram, 193F
  of a muscle twitch, 194F
Myomas, 253
Myometrium, **711**
Myoneural cleft, **183**
Myoneural junction, **183**, 202, 274
Myopathies, myotonic, 252
Myopia, **389**
Myosin, **186**, 202
Myositis, **253**
Myositis ossificans, **121**
Myotome, **182**
Myotonia atrophica, **252**
Myotonia congenita, **252**
Myotonic myopathies, 252
Myxedema, **419**, 420F

NADH, **481**
NADPH, **491**
Nail bed, **107**, 110
Nail matrix, 106, 111
Nail plate, **107**, 111
Nails, 106–7, 110–11
  structure of, 107F
Nail wall, **107**
Narcolepsy, **369**
Narcotics, **350**
Nares, internal, 506
Nasal cavity, **506**, 507–8
Nasal conchae, 140, 160, **507**
Nasal fossae, **507**
Nasal organs, **378**
Nasals, **140**, 160
Nasal sacs, **378**
Nasal septum, **139**, 507
Nasal turbinates, **507**
Nasolacrimal duct, **387**
Nasomaxillary suture, **142**
Nasopharynx, **508**
Natural childbirth, **746**
Natural hallucinogens, **351**
Natural immunities, **634**
Natural killer (NK) cells, **632**
Natural menopause, **729**
Navicular carpal, **150**, 151F
Navicular tarsal, **156**
Near point of vision, **388**
Nearsightedness, 389
Neck, **444**
  arteries of, 591–93
  muscles of, 215F, 220–21, 221F

Negative charge, **19**
Negative feedback, **411**
Negative feedback mechanism, **13**
Neoplasms, **74**
Neovascularization, **387**
Nephrons, **651**
    arcuate veins of, 656
    blood supply to, 654–56, 655F
    cortical, 654
    location and dimensions of, 653–54
    nerve and lymphatic supply to, 656–57
    tubular portion of, 653, 654F, 659
Nephropathy, **432**
Nephrotic syndrome, **660**
Nerve endings, 358F
Nerve impulse, **269**
Nerves, **261**. *See also* Nervous system; Neurons; Peripheral nerves; Peripheral nervous system (PNS); Spinal cord
    abducens, 305, 329, 329F
    accessory, 332, 340
    acoustic, 329
    adrenergic, 283
    auditory, 329
    cholinergic, 283
    cochlear, 330, 340
    cranial, 306, 323–33, 340
    cutaneous, 334, 340
    facial, 305, 329, 330F, 340
    glossopharyngeal, 330–31, 340
    hypoglossal, 332, 332F, 340
    intercostal, 334
    mandibular, 328, 340
    maxillary, 328, 340
    mixed, 261, 284, 323, 340
    motor, 261, 284, 323, 340
    oculomotor, 326–27, 327F, 340
    olfactory, 324–25, 326F, 340, **377**, 377–78
    ophthalmic, 328, 340
    optic, 325–26, 326F, 340
    organization of, 261
    pelvic splanchnic, 347
    phrenic, 334–35, 340
    sciatic, 338, 340
    sensory, 261, 284, 323
    spinal, 333F, 333–39, 335F, 340
    spinal accessory, 332
    splanchnic, 344, 352
    splanchnic, of pelvis, 347
    trigeminal, 305, 328–29, 328F, 340
    trochlear, 327–28, 327F, 340
    vagus, 331, 332F, 340
    vestibular, 330, 340
    vestibulocochlear, 305, 329–30, 331F, 340
Nervous system, **16**, 260. *See also* Central nervous system (CNS); Nerves; Peripheral nervous system (PNS)
    action potential, 269–70
    cells of the, 262–65
    craniosacral, 352
    diseases of, 286–87
    embryonic development of, 292F, 292–93
    examination of, 318
    functional aspects of, 356–72
    general sense receptors, 359–61
    integrating stimuli in, 279–81
    membrane potential, 266–69
    motor pathways, 364–67
    neurotransmitters, 281–84
    organization of, 261, 261F
    parasympathetic, 352
    postsynaptic membrane potentials, 278–79
    sensory pathways, 361–64
    sensory reception, 357–59
    somatic, 261, 284
    sympathetic, 343, 344, 346F, 352
    thoracolumbar, 343, 352
Nervous tissue, **92**, 94
Nervous tunic, **382**
Net filtration pressure, **658**
Neural arch, **143**
Neural canal, **143**
Neural crest cells, **292**
Neural folds, **292**
Neuralgia, 102, **339**
    trigeminal, 328
Neural plate, **292**
Neural tube, **292**
Neural tube defects (NTDs), **286**
Neuritis, **339**
Neuroblastomas, **676**
Neuroblasts, **324**
Neurofibromatosis, **159**
Neurofilaments, **263**, 284
Neurogenic shock, **609**
Neuroglandular junction, **274**, 285
Neuroglia, **92**, 262, 264–65, 284
Neurohypophysis, **412**
Neuromuscular cleft, **183**, 202
Neuromuscular junction, **183**, 202, 274, 285
Neuromuscular junction block, drug-induced, 190
Neurons, **92**, 92F, 262, 284. *See also* Receptors; Synapses
    association, 312
    in the autonomic nervous system, 343
    bipolar, 264, 284
    enteroceptive, 349, 354
    exteroceptive, 349, 354
    first-order, **362**
    general sensory, 334
    internuncial, 312
    major ions involved in function of, 266
    motor, 359F
    myelinated, 263F, 265, 284
    nonmyelinated, 265
    olfactory, **378**
    postganglionic, 343, 352
    postsynaptic, 274, 285, 343
    preganglionic, 343, 352
    presynaptic, 274, 285, 343
    pseudounipolar sensory, 358
    retinal ganglion, **383**
    second-order, **362**
    sensory, 349, 359F
    somatic motor, 263F, 333
    somatic sensory, 349, 354
    third-order, **362**
    threshold value of, 271
    touch receptors, 359
    transmission between, 274–78
    types of, 264F
    unipolar, 264, 284
    upper motor, 365
    visceral efferent, 343, 352
    visceral sensory, 349, 354
    visceroreceptors, 357
Neuropathies, entrapment, 339
Neurophysins, **413**
Neurophysiology, **6**
Neurotransmitters, **276**, 277–78, 281–85, 347, 353
    inhibitory, 278, 285
    release into a synaptic cleft, 277F
Neurotransmitters, inhibitory, 278, 285
Neurulation, **292**
Neutral fats, **35**, 42
Neutral solution, 42
Neutrons, **19**, 41
Neutrophils, **546**
Newborn, hemolytic disease of, 636
Niacin, **493**
Nicotine adenine dinucleotide (NAD+), **481**
Nicotinic receptors, **347**, 353
Night blindness, **392**
Nipple, **715**
Nissl bodies, **262**, 284
Nitrogen narcosis, **525**
Nitrogenous bases, 43
Nociceptors, **357**, 360
Nodes of Ranvier, **265**, 284
Nodular sclerosis, **96**
Non-A/non-B hepatitis, **471**
Nonarticular rheumatism, **253**
Noncommunicating hydrocephalus, **307**
Nonhistone chromosomal proteins (NHCP), **65**, 76
Non-insulin-dependent diabetes mellitus (NIDDM), **432**
Noninvasive cardiac diagnostic procedures, **575**
Nonkeratinized squamous epithelium, **83**
Nonmyelinated neurons, **265**
Nonrapid eye movement (NREM) sleep, **370**
Non-self-antigens, **633**
Nonspecific response to foreign intrusions, **624**
Nonunion fracture, 117F
Norepinephrine, **199**, 281, 284, 352, 353, 426
Nose, **506**, 507F
    fracture of, 140
Nostrils, **506**
Nuchal lines, medial, 138
Nuclear envelope, **47**, 64, 76
Nuclear magnetic resonance, **15**, 15F
Nucleic acids, **39**, 43. *See also* Deoxyribonucleic acid (DNA); Ribonucleic acid (RNA)
Nucleolus, **57**, 64, 76
Nucleolus organizer, **65**
Nucleoplasm, **47**
Nucleosomes, **65**
Nucleotides, **39**, 43
Nucleus cuneatus, **306**
Nucleus gracilis, **306**
Nucleus of a cell, **47**, 64F
    contents of, 64–65, 76
Nucleus of an atom, **19**, 41
Nucleus pulposus, **159**
Nurse cells, **702**
Nutriment, **41**
Nystagmus, **402**

Obesity, **488**
Oblique muscles, 216
Oblique popliteal ligament, **176**

Oblique section of the body, **10**
Obstruction of urinary flow, **676**
Obturator externus, **242**
Obturator foramen, **153**
Obturator internus, **242**
Occipital bone, **136**, 136–38, 160
Occipital condyles, **138**
Occipitalis, **214**
Occipital lobe, **295**, 302F, 319
Octet rule, **23**
Oculomotor nerves, **326**, 326–27, 327F, 340
Odontoid process, **145**
Oil glands (skin), **108**, 111
Olecranon fossa, **150**
Olecranon process, **150**
Oleic acid, **35**
Olfaction, 377–79
Olfactory adaptation, **378**
Olfactory bulb, **325**, 378
Olfactory cells, **377**
Olfactory cell stimulation, 377
Olfactory hairs, **377**
Olfactory nerves, **324**, 324–25, 326F, 340, 377
Olfactory neurons, **378**
Olfactory organs, **378**
 embryonic development of, 378
Olfactory pathway, 378–79
Olfactory placodes, **378**
Olfactory receptors, 377F
Olfactory tracts, **379**
Oligodendrocytes, 264F, **265**, 265F, 284
Olive, **306**
Oncologists, **75**
Open biopsy, **649**
Open fracture, 116F
Open heart surgery, **564**, 643
Operative microscope, **403**
Ophthalmic artery, **591**
Ophthalmic nerve, **328**, 340
Ophthalmologists, **384**
Ophthalmoscope, **384**
Opiates, **350**
Opioid, **281**
Opportunistic infections, **630**
Opsin, **391**
Opsonins, **626**
Opsonization, **626**
Optic chiasma, **300**, 325, 326F
Optic cup, **381**
Optic disk, **383**
Optic foramen, **138**
Optic nerves, **325**, 325–26, 326F, 340
Optic neuritis, **327**
Optic stalk, **381**
Optic tracts, **325**
Oral cavity, **443**, 444F
Oral contraceptives, **725**
Ora serrata, **382**
Orbicularis oculi, 207, **214**
Orbicularis oris, **214**
Orbital bones, 135F
Orbital process, **142**
Orbitals, **21**, 41
Orbits, **133**
Organelles, **7**, **57**, 58F
Organic bases, **39**
Organic compounds, **34**, 42–43
 categories of, 33–40

Organisms, characteristics of, 10–16
Organ of Corti, **398**
Organs, **7**, 16
Organ transplants, 633–34
Orgasm, **708**
 female, 724
 male, 722
Origin of a muscle, **209**, 254
Oropharynx, **449**, 509
Orton-Gillingham system, **373**
Oscillation frequency (wavelength), **20**
Os coxae, **152**, 152F, 152–53, 161
Osmole, **662**
Osmosis, **51**, 52, 53F, 76
Osmotic pressure, **610**
Osseous fracture, **177**
Osseous labyrinth, **396**
Osseous tissue, **90**, 90–91, 94. *See also* Bone
Ossicles, 131, **395**
Ossification, 120–23
 endochondral, 128
Osteitis deformans, **127**
Osteoarthritis, **177**
Osteoblasts, **120**, 128
Osteochondral fracture, **177**
Osteochondroma, **126**
Osteoclasis, **123**, 123–24, 128
Osteoclasts, **123**, 124F, 128
Osteocytes, **90–91**, 115, 120, 128
Osteogenesis, **123**, 123–24, 128
Osteogenesis imperfecta, **125**
Osteogenic sarcoma, **126**
Osteomalacia, **126**
Osteomyelitis, **124**
Osteon, **118**
Osteonal canals, **118**
Osteoporosis, **127**
Osteoprogenitor cells, **120**, 128
Otic ganglion, **331**, 346
Otic placodes, **394**
Otoconia, **400**
Otocyst, **394**
Otoliths, **400**
Otosclerosis, **403**
Outer ear, **393**
Oval window, **395**
Ovarian arteries, **594**
Ovarian cycle, 718F
Ovarian follicle, **709**
Ovarian ligament, **709**
Ovaries, **429**, 709, 710F
 cortex of, 709
 meiosis in, 752F
Overeaters Anonymous (OA), **488**, 489
Overuse syndromes, 256–57
Overweight, **488**
Oviducts, **709**, 709–11, 711F
 ampulla of, 710
Ovulation, **418**, 717
Ovum, **709**, 717, 719F
Oxaloacetic acid, **483**
Oxidation of glucose, **481**
Oxidation-reduction reactions, **28**, 42
Oxidative phosphorylation, **484**
Oxidized electron donor, **28**
Oxygen debt, **196**, 196F, 203
Oxygen transport, 521–23
Oxyhemoglobin, **523**, 543

 oxygen release from, 524
Oxyphil cells, **420**
Oxytocin, **413**, 414–15

PABA, **106**
Pacemaker (node) of heart, **182**, 568
Pacemaker potential, **568**
Pacinian corpuscles, **360**
Packed cell volume, **548**
Paget's disease, **127**, 160
Pain
 control of, 364
 pain-blocking substances, 281
 receptors of, 360
 referred, **159**, 360
 sensory pathway for, 362F, 363–64
 somatic, 360
 visceral, 360
Palate, **140**, 444
Palatine bones (palatines), **141**, 160
Palatine process, **140**
Palatine tonsils, **444**, 624
Palatoglossal arch, **509**
Palatopharyngeal arch, **444**
Palmar arches, **594**, 600
Palmar grasp, **316**
Palmaris longus, **236**
Palpebrae, **385**
Palpebral conjunctiva, **386**
Palpebral fissure, **386**
Pancreas, 426–29, 427F, 470–71
Pancreatic acinar cell, 471F
Pancreatic acini, **470**
Pancreatic amylase, **459**
Pancreatic duct, **470**
Pancreatic enzymes, 458–59
Pancreatic lipase, **459**
Pancytopenia, **638**
Pantothenic acid, **493**
Papanicolaou test (pap smear), **75**
Papillae, **106**, 107, 444
Papillary ducts, **650**
Papillary muscles, **561**
Papillary region, **106**, 110
Papilledema, **327**
Papovaviruses, **102**
Pap smear, **714**
Papular lesions, **111**
Paradoxical sleep, **370**
Parafollicular cells, **418**
Parallel fascicles, **207**
Paralysis, **437**
 familial periodic, 252–53
Paramesonephric ducts, **700**
Paranasal sinuses, **508**
Paraplegia, **160**, 315
Parasagittal section of the body, **10**
Parasympathetic (craniosacral) nervous system, **344**, 344–47, 352
Parasympatholytic (anticholinergic) drugs, **353**
Parasympathomimetic (cholinergic) drugs, **353**
Parathormone, **420**
Parathyroid glands, **13**, 418, 420–21
Parathyroid hormones, **13**, 124, 128, 420
Paraurethral glands, **714**
Paraventricular nuclei, **413**

Index

Paravertebral ganglia, **344**, 352
Parenchyma, **86**, 649
Parental types, **758**
Paresthesia, **339**
Parietal arteries, **594**
Parietal cells, **453**
Parietal layer of Bowman's capsule, **652**
Parietal lobe, **295**, 302F, 319
Parietal pericardium, **561**
Parietal peritoneum, **471**
Parietal pleura, **512**
Parietals, **133**, 160
Parkinsonism, **366**
Parotid duct, **447**
Parotid glands, **447**
Paroxysmal atrial tachycardia (PAT), **579**
Pars distalis, **412**
Pars intermedia, **412**
Pars nervosa, **412**
Pars tuberalis, **412**
Partial linkage, **757**
Partial pressure (P), **519**, 522F
Parturition, **747**. *See also* Childbirth
Passive depolarization, **271**, 285
Passive immunity, **635**
Patau's syndrome, **763**
Patch testing, **111**
Patella, 154F, **155**, 161, 174, 178, 246, 317
Patellar ligament, **174**, 246
Patellar reflex, **317**
Patellar retinacula, **176**
Patellar surface, **155**
Patellar tendon, **246**
Patellofemoral joint, **174**, 178
Patent ductus arteriosus, **579**
Pathologic fracture, 117F
Pathologic reflexes, **318**
Pathologists, **12**, 96
Pavlov, Ivan, 372
Pectoral girdle, 49, **131**, 148, 148F, 161
    muscles of, 230–41
Pectoralis major, 207, **232**
Pectoralis minor, 207, **230**
Pedicels of podocytes, **652**
Pedicles, **143**
Pellagra, 41, **41**
Pelvic bones, 152, 152F
    male versus female, 154
Pelvic cavity, **9**
Pelvic floor, muscles of, 228–29, 229F
Pelvic girdle, **131**, 152–54, 153F, 161
Pelvic inlet, **153**
Pelvic outlet, **153**
Pelvic splanchnic nerves, **347**
Pelvis, **152**, 153F, 161
Penile urethra, **670**, 703
Penis, **707**, 707F, 707–8
    vascular system of, 708F
Pennate fascicles, **207**
Penniform muscles, **207**
Pentose, **34**
Pepsin, **453**
Pepsinogen, **453**
Peptic ulcer disease (PUD), **452**
Peptide bonds, **37**, 38F, 43
Peptide digestion, **460**
Peptide hormones, **407**
Perception, **357**

Perceptive hearing loss, **403**
Perennial allergic rhinitis, **638**
Perforating canals, **118**
Pergonal, **771**
Pericardial cavity, **9**, 561
Pericardial fluid, **561**
Pericardial sac, **559**
Pericardial vein, **600**
Pericarditis, **561**, 581
Pericardium, **559**, 559–61, 561F
Perichondrium, **90**, 119, 121, 128
Perikaryon, **262**, 284
Perilymph, **396**
Perilymphatic labyrinth, **396**
Perimetrium, **711**
Perimysium, **182**, 202
Perineum, **714**
Perineurium, **261**, 284
Perinuclear space, **64**, 64F, 76
Periodontal disease, **446**
Periodontal ligament, **166**, 445
Periodontium, **446**
Periosteum, **118**, 121, 128
Periostitis, **256**
Peripheral chemoreceptors, **527**
Peripheral fiber, **365**
Peripheral nerves, 262F
    embryonic development of, 324
Peripheral nervous system (PNS), **261**, 284, 322
    afferent portion of, 284
    cranial nerves, 323–33
    neuroglia of, 265
    spinal nerves, 333–39
Peripheral neuropathy, **339**
Peripheral resistance, **606**
Peristalsis, **301**, 450F
Peristaltic wave, **449**
Peritoneal cavity, **471**
Peritoneal dialysis, **52**
Peritoneal membrane, **52**
Peritoneum, **471**, 472F
Peritubular plexus, **656**
Perivitelline space, **726**
Permanent teeth, **445**
Permeases, **55**, 76
Pernicious anemia, **473**, 546
Pernicious vomiting, **744**
Peroneal artery, **596**
Peroneal veins, **602**
Peroneus brevis, **248**
Peroneus longus, **248**
Peroneus tertius, **248**
Peroxisomes, **59**, 76
Perpendicular plate (ethmoid bone), **139**
Perspiration, **497**
Pes, **156**, 161. *See also* Foot
Petechiae, **613**
Petit mal attacks, **304**
Petrosal sinus of brain, **597**
Petrous portion of the temporals, **134**
Peyer's patches, **457**, 623
pH, 42
    of common fluids, 32
    metabolic, 690
    relationship to concentration of hydrogen ion, 31
    renal regulation of, 667–68

Phagocytes, **547**
    pseudopodia of, 626
Phagocytic vacuole, **626**
Phagocytosis, **56**, 76, 547, 626, 626F
    ingestion phase of, 626
Phagolysosome, **626**
Phalanges
    of the foot, **156**, 161
    of the hand, 151F, **152**, 161
Pharyngeal apertures, **509**
Pharyngeal pouches, **394**
Pharyngeal stage, **449**
Pharyngeal tonsils, **509**, 624
Pharyngeal tympanic tube, **395**
Pharyngitis, **530**
Pharynx, **449**, 508–9, 509F
Phasic receptors, **358**
pH buffer, **32**, 687
    in body fluid, 688F
Phenothiazine, **366**
Phenotype, **753**
Phenylephrine, **352**
Pheochromocytoma, **426**
Philadelphia chromosome, **764**
Phlebitis, **613**
Phosphate buffer system, **688**
Phosphoglyceraldehyde (PGAl), **481**
Phosphoglyceric acid (PGA), **482**
Phospholipids, **36**, 42
Phosphorylation, **199**
Photon imaging, **20**, 97
Photons, **20**
Photophobia, **214**
Photoreceptive cells, **382**
Photoreceptors, **357**
Phrenic nerve, **334–35**, 340
Phrenic veins, **601**
Physical allergy, **639**
Physical theory of olfactory cell stimulation, 377
Physical therapy, **176**
Physiological dead space, **518**
Physiologic shunt, **521**
Physiology, **5**, 16
Pia mater, **309**, 319
Pickwickian syndrome, **527**
Pigeon toe, **156**
Pigmentation, 103–5
Pineal body (gland), **301**, 319, 431
Pinna, **393**
Pinocytosis, **56**, 76
Piriformis, **242**
Piriformis syndrome, **242**
Pisiform carpal, **150**, 151F
Pituitary dwarfism, **417**, 753
Pituitary gland, **300**, 412–18
    anterior lobe of, 412
Pituitary hormones, 413–18
Pituitary tumors, 326
Pivot joints, **169**, 178
Placebo effect, **364**
Placenta, **603**, 738
    structure of, 741F
Placental corticotrophin, **431**
Placental membrane, **740**
Placental thyrotrophin, **431**
Placenta previa, **744**
Planes of the body, 7–10, 16

Plantar arch, **596**, 602
Plantar artery, **596**
Plantar flexion, 173
Plantaris muscle, **248**, 248
Plaques, **553**
Plasma, **90**, 539–41
   major components of, 540
Plasma cells, **88**, 94, 633
Plasma coagulation factors, **551**
Plasmalemma, **47**, 47–50, 49F, 73
   flow of water through, 53F
Plasmin, **553**
Plasminogen, **553**
Plateau of coitus, **722**
Platelet coagulation factors, **551**
Platelet plug formation, 550
Platelets, **549**. *See also* Thrombocytes
   internal structure of, 550F
Platybasia, **160**
Platysma, **214**
Pleocytosis, **286**
Pleurae, **511**, 513F
Pleural cavities, **9**
Plexus, **334**, 340, 352
   primary capillary, **415**
   root hair, **359**
   secondary capillary, **415**
Plicae circulares, **456**
Plicae palmatae of cervix, **713**
Pneumocystis carinii pneumonia, 630
Pneumonia, **513**
Pneumotaxic center, **526**
Pneumotaxic nuclei, **305**
Podocytes, **652**
Polar body, **717**
Polarized membranes, **267**
Poliomyelitis, **159**
Polio virus, **634**
Pollex, **152**
Polyarteritis nodosa, **93**
Polyclonal antibody, **635**
Polycystic renal disease, **675**
Polycythemia, **548**
Polymerization, **25**, 42
Polymers, **25**
Polymorphonuclear white blood cells, **548**
Polymyositis, **93**
Polyneuritis, **339**
Polyneuropathy, **287**
Polynucleotides, **39**, 39F
Polypeptide chains, 38F
Polysaccharides, **34**, 35F, 42
Polysome, **73**, 77
Polyspermy, **726**
Polysynaptic reflexes, **317**, 317F
Pompe's disease, **252**
Pons, **319**
   pons varolii, **305**
Popliteal artery, **596**
Popliteal ligaments, **176**
Popliteal veins, **602**
Popliteus, **248**
Pores, Na$^+$-specific, **269**
Portal systems, 601–2
Portal vessels, **415**
Positive charge, **19**
Positive feedback mechanisms, **13**
Positive signs of pregnancy, **743**
Positron emission tomography (PET), **24**

Postabsorptive state, **479**
Posterior canal of ear, **396**
Posterior cavity, **7**, 385
Posterior cerebral arteries, **593**
Posterior chamber, **385**
Posterior commissure, **301**, 319
Posterior communication arteries, **593**
Posterior cord, **338**
Posterior cruciate ligaments, **176**
Posterior ducts of ear, **396**
Posterior horn, **307**
Posterior intercostal arteries, **594**
Posterior intermediate sulcus, **312**
Posterior interventricular artery, **567**
Posterior lateral sulcus, **312**
Posterior lobe of pituitary gland, **412**
Posterior median sulcus, **312**
Posterior pillar, **444**
Posterior pituitary, **412**
Posterior (dorsal) portion of the body, **10**
Posterior root ganglion, **334**, 340
Posterior tibial artery, **596**
Posterior tibial veins, **602**
Postganglionic neuron, **343**, 352
Postictal state, **304**
Postmortem examination, **12**
Postovulatory phase of menstrual cycle, **720**
Postrenal renal failure, **658**
Postsynaptic membrane, **274**
Postsynaptic membrane potentials, 278–79
Postsynaptic neuron, **274**, 285, 343
Postsynaptic potential, inhibitory, **278**, 279F
Potassium, **31**
   absorption and secretion of, 668
   in body fluid, 684–85
Precapillary sphincter, **586**
Pre-eclampsia, **744**
Preformed water, **681**
Prefrontal lobotomy, **296**
Preganglionic neuron, **343**, 352
Pregnancy, 743–47
   complications during, 744
   diagnosis of, 743–44
   duration of, 744–45
   miscarriage, 743
   physiological changes during, 745–47
Premature menopause, **729**
Premature ventricular contractions (PVCs), **580**
Premenstrual tension syndrome (PMS), **720**
Premolars, **445**
Premotor area (cerebral cortex), **296**, 319
Preovulatory phase of menstrual cycle, **720**
Prepared childbirth, **746**
Prepuce, **707**
Prerenal failure, **658**
Presenile dementia, **298**
Pressoreceptors, **349**, 608
Pressure
   afferent pathways for, 363F
   sensory pathway for, 362F
Pressure palsies, **339**
Pressure receptors, **359**, 360
Presumptive signs of pregnancy, **743**
Presynaptic inhibition, **279**, 285
Presynaptic membrane, **274**
Presynaptic neuron, **274**, 285, 343
Primary aldosteronism, **691**
Primary auditory area (cerebral cortex), **295**, 319

Primary bronchi, **510**
Primary capillary plexus, **415**
Primary follicles, **717**
Primary gustatory area (cerebral cortex), **296**, 319
Primary hypertension, **607**
Primary immune response, **634**
Primary immunodeficiency diseases, **638**
Primary lobule of lung, **514**
Primary marrow spaces, **121**
Primary motor area (cerebral cortex), **296**, 319
Primary olfactory area (cerebral cortex), **295**, 319
Primary oocyte, **709**
Primary ossification center, **121**, 128
Primary peristaltic waves, **449**
Primary sensory area (cerebral cortex), **295**, 319
Primary spermatocytes, **704**
Primary syphilis, **723**
Primary urine, **657**
Primary visual area (cerebral cortex), **295**, 319
Prime mover, **209**, 254
Primordial follicles, **717**
P-R interval, **570**
Probable signs of pregnancy, **743**
Procarboxypeptidase, **459**
Processes, **262**, 284
   articular, 161
Proctitis, **473**
Proctodeum, **442**
Products, **32**
Progesterone, **429**, 719
Progestogens, **407**
Proglucagon, **428**
Progression, **350**
Progressive systemic sclerosis, 93
Projection, 358–59, **359**
Projection fibers, **296**
Projection tracts, **296**
Prolactin, 416–17
Prolactin-inhibiting hormone (PIH), **416**
Prolactin-releasing factors (PRF), **416**
Prolactin-releasing hormone (PRH), **416**
Pronation, **174**
Pronator quadratus, **236**
Pronator teres, **236**
Pronephric duct, **648**
Pronephros, **648**
Proopiomelanocortin (POMC), **415**
Properdin, **626**
Prophase I, **749**
Prophase of mitosis, **67**, 66F, 77
Propionic acid, **489**
Propionyl CoA, **489**
Proprioception, **295**, 319
   sensory pathway for, 362–63
   tracts for, 363F
Proprioceptors, **357**, 360–61
Prosencephalon, **293**
Prostacyclins, **431**
Prostaglandins, **431**, 431–33, 720
Prostate gland, 706F, **706**
Prostatic hypertrophy, **676**
Prostatic urethra, **670**, 703
Prosthesis, of the hip, **169**
Protein, **37**, 43
   digestion of, 459
   metabolism of, 491–92

molecular structure of, 38F
primary and secondary structures of, **37**, 43
quaternary structure of, **39**, 43
tertiary structure of, **38**, 43
Protein buffer system, **688**
Protein hormones, 407
Protein kinases, **409**
Proteinuria, **660**
Prothrombin, **551**
Prothrombin activator, **551**
Prothrombin time test, **553**
Protons, **19**, 41
Protoporphyrin, **523**
Protraction, **174**, 178
Proximal convoluted tubules of nephron, **653**
Proximal portion of nephron, **653**
Prune belly syndrome, **252**
Pseudogout, **166**
Pseudohermaphroditism, **728**
Pseudohypertrophic muscular dystrophy, **252**
Pseudopodia, **55**
Pseudopodia of phagocyte, **626**
Pseudostratified ciliated columnar epithelium, 82F
Pseudostratified columnar epithelium, **83**
Pseudostratified epithelium, **81**, 92
Pseudounipolar sensory neuron, **358**
Psoas major, **223**, 242
Psoas minor, **223**, 242
Psoriasis, **102**, 102F
Psychoactive drugs, 296
Psychogenic headaches, **299**
Psychomotor attacks, **304**
Pterygoid, lateral, 218
Pterygoid muscles, 218
Pterygoid processes, **138**
Pterygopalatine ganglion, **346**
Puberty, **717**
Pubic crest, **153**
Pubic hair, **714**
Pubic symphysis, **153**
Pubis, **153**, 161
Pudendal cleft, **714**
Pudendal plexus, **338**
Pudendum, **714**
Pulmonary arteries, **589**
Pulmonary circulation, **563**, 589
Pulmonary embolism, **517**
Pulmonary embolus, **588**
Pulmonary emphysema, **518**, 531
Pulmonary hilum, **512**
Pulmonary irritant receptors, **527**
Pulmonary pleura, **512**
Pulmonary plexus, **347**
Pulmonary semilunar valve, **564**
Pulmonary stretch receptors, **527**
Pulmonary tree, **512**, 513F
Pulmonary trunk, **564**, 589
Pulmonary veins, **564**, 589
Pulmonary ventilation, **504**
Pulp, **444**
Pulp cavity, **444**
Pupil, **381**
Pupillary accommodation, **390**
Purkinje fibers, **569**
Pus, **547**, 627
Pustular lesions, **111**

Putamen, **298**
P wave, **570**
Pyelogram
  intravenous, **649**, 676
  retrograde, 649
Pyelonephritis, **675**
Pyloric antrum, **451**
Pyloric glands, **453**
Pyloric orifice, **451**
Pyloric pump, **452**
Pyloric sphincter, **451**
Pyloric stenosis, **473**
Pyodermas, **111**
Pyramidal fiber, **365**
Pyramidal pathways, **365**
Pyramidal tracts, **296**
Pyramids, **306**
Pyridoxine, **493**
Pyrogens, **497**
Pyrosis, **450**
Pyruvic acid, **481**, 482

QRS complex, **570**
Quadrate lobe of liver, **468**
Quadratus femoris, **242**
Quadratus lumborum, **223**, 226
Quadriceps femoris, **207**, 246
Quadriceps group, 246
Quadriplegia, **160**, 315
Quinsy, **530**

Rabies, **286**
Rachitic rosary, **126**
Radial artery, **594**
Radial branches, **468**
  of uterine arteries, **713**
Radial fossa, **150**
Radial notch, **150**
Radial tuberosity, **150**
Radial vein, **600**
Radiate muscles, **207**
Radiation, **21**
Radiative heat loss, **496**
Radical mastectomy, **622**
Radioactive elements, 41
Radiographic anatomy, **5**, 14
Radiographs, **5**, 16
Radiography, contrast, 15
Radioisotopes, **22**
Radioisotope scanning, 22
Radiologists, **14**
Radionucleotide scanning, **97**
Radius, **150**, 151F, 161
Rami, **142**, 153, 340
  anterior and posterior, **334**
  communicating, **344**, 352
  gray communicating, 344
  superior and inferior, **153**
Rapid eye movement (REM) sleep, **370**
Rash, butterfly, **93**
Rautidine, **452**
Raynaud's phenomenon, **93**, 613
Reabsorption, **659**
Reactants, **32**
Reaction rate, **29**
Rearrangement reaction, **26**, 42
Receptor activation, **411**
Receptor-mediated endocytosis, **56**
Receptor potential, **357**, 357–58

Receptors, **276**, 285, 357
  general sense, 359–61
  olfactory, 377–78
  peripheral, 526–27
  phasic, **358**
  pressure, **359**
  stimulation of, 357–59
  taste, 379–80
  tonic, **358**
  touch, **359**
Recessive alleles, **753**
Reciprocal innervation, **318**
Recognition and adherence phase of phagocytosis, **626**
Recombinant DNA, **70**
Recombinants, **758**
Rectum, **465**, 465F
  disorders of, 473
Rectus abdominis, **226**
Rectus capitis, **223**
Rectus femoris, **246**
Rectus muscles, **216**
Red blood cells, **543**. *See also* Erythrocytes
  regulation of production of, 668–69
Red cell indices, **548**
Red marrow, **115**
Red measles, **104**
Red nucleus, **298**, **303**
Red pulp of spleen, **623**
Reduced electron acceptor, **28**
Referred pain, **159**, 360, 361F
Reflexes, 316
  Babinski's, **316**, 318
  Bainbridge, **577**
  contralateral, **317**
  crossed extensor, **318**
  deep tendon, **316**, 318
  defecation, **467**
  enterogastric, **454**
  Hering-Breuer, **527**
  Hoffman's, **318**
  ipsilateral, **316**
  micturition, 672F
  monosynaptic, **320**
  Moro's, **316**
  muscle stretch, **316**
  patellar, **317**
  right atrial, **577**
  rooting, **316**
  of the spinal cord, **311**, 314–18, 320
  startle, **316**
  stretch, **316**, 316F
  sucking, **316**
  tonic neck, **316**
  withdrawal, **317**
Reflex motor pathways, 364
Refraction, **388**
Refractory period, **194**, 202
Refractory period of cardiac muscle, **569**
Refractory period of coitus, **722**
Regional anatomy, **5**
Regional anesthesia, **329**
Regional ileitis, **473**
Regularly arranged dense connective tissue, **90**, 94
Regularly arranged elastic tissue, **90**
Rejection of organ transplant, **639**
Relaxation pressure, **518**
Release-inhibiting hormones, **415**

Releasing hormones, **415**
Renal arteries, **594**, 647
Renal capsule, 653F
Renal casts, **674**
Renal colic, **670**
Renal columns, **649**
Renal corpuscles, **651**, 652F
Renal cortex, 651F
Renal dialysis, **52**
Renal disease syndromes, 658, 660, 675
Renal erythropoietic factor (REF), **669**
Renal failure, **658**
Renal fascia, **647–49**
Renal hilum, **647**
Renal malignancies, 676
Renal papillae, **649**
Renal pelvis, **649**
Renal physiology, **5**
Renal pyramids, **649**
Renal sinus, **649**
Renal threshold, **660**
Renal veins, **601**, 647
Renin, **655**
Renin-angiotensin mechanism, **666**, 667F
Renin-angiotensin system, **609**
Replacement (exchange) reaction, **28**
Replication, DNA, 68, 69F, 77
Replication bubbles, **68**, 77
Replication fork, 68
Repolarized membrane, **269**, 285
Reproduction, **16**. *See also* Childbirth; Embryonic development; Pregnancy
  artificial, 734–35
  coitus, 722–24
  disturbances of the reproductive system, 728–29
  embryonic development of reproductive system, 700–701
  female reproductive tract, 708–16, 709F
  fertilization, 724–26, 726F
  fetal development, 741
  hormonal control of egg production, 719–22
  male reproductive tract, 699F, 699–708
  production and release of eggs, 717–18
  reproductive failure, 771
Residual volume of lung, **518**
Resistance, **209**, 254
Resistance arm, **210**
Resistance to disease, **624**
Resolution of coitus, **722**
Respiration
  cellular, 480, 485
    Cheyne-Stokes, 527
    external, 504
    internal, 504
Respiratory acidosis, **689**
Respiratory alkalosis, **690**
Respiratory bronchioles, **514**
Respiratory center, **526**, 526F
Respiratory chain, **483**, 483–84
  production of ATP, 485F
  steps in, 484F
Respiratory dead space, **518**
Respiratory distress syndrome (RDS), **515**, 530
Respiratory division, **513**
Respiratory physiology, **6**
Respiratory quotient (RQ), **521**

Respiratory system, 7, 504–28. *See also* Breathing; Lungs
  diseases of, 530–31
  embryonic development of, 506
  gas exchange in alveoli, 521–24
  mechanics of breathing, 516–19
  organization of, 505F, 505–16
  properties of gases, 519–21
  regulation of respiration, 525–28
Respiratory tract, **504**
  lower, 505, 510–11
Respiratory ventilation, 689
Respiratory volumes, 517–19, 518F
Respirometer, **517**
Resting cardiac output, **575**
Resting potential, 267–69, **269**, 284
Resting stroke volume of heart, **575**
Restricted diet, **499**
Restriction endonucleases, **758**, 766–67
Restriction fragment length polymorphism (RFLP), **758**, 767
Rete testis, **702**
Reticular activating system (RAS), **306**
Reticular cells, **88**
Reticular epithelium, **431**
Reticular fibers, **87**, 94
Reticular formation, **306**, 306F, 319
  sleep and, 369
Reticular region, **106**, 110
Reticular tissue, **88**, 94
Reticulocyte, **544**
Reticulocyte counts, 544
Retina, **382**, 382–83, 383F
Retinal, **391**
Retinaldehyde, **391**
Retinal detachment, **387**
Retinal ganglia, **383**
Retinal-rhodopsin cycle, 392F
Retinene, **391**
Retinopathy, **432**
Retraction, **174**, 178
Retrograde amnesia, **287**
Retrograde pyelogram, **649**
Retroperitoneal organs, **472**, 647
Retrovirus, **630**
Reverberating circuits hypothesis, 371F
Reverberating neuron circuits, **370**
Reversible reaction, **32**
Rh antigen, **637**
Rheumatic fever, **530**
Rheumatism, nonarticular, 253
Rheumatoid arthritis, 170
Rh group, **637**
Rh-negative (Rh$^-$) individuals, **637**
Rhodopsin, **391**
Rhombencephalon, **293**
Rhomboideus, 230
Rh-positive (Rh$^+$) individuals, **637**
Rib cage, 147, 147F, 161
Riboflavin, **493**
Ribonucleases, 459
Ribonucleic acid (RNA), **39**, 43
  messenger RNA (m-RNA), transfer RNA (t-RNA), and ribosomal RNA (r-RNA), **70**
  nucleotides of, 39F
Ribonucleotides, **39**, 39F
Ribose, **39**
Ribosomal RNA (r-RNA), **70**, 77

Ribosomes, **57**, 60F, 76
Ribs, 147–48, 148F, 161
  anterior view of a typical rib, 148F
  vertebral, 147
Rickets, **126**, 126F
Right atrial reflex, **577**
Right atrioventricular valve, **563**
Right atrium of heart, **563**
Right common carotid artery, **591**
Right coronary artery, **566**
Right lobe of liver, **468**
Right lymphatic duct, **619**
Right subclavian artery, **591**
Right ventricle of heart, **563**
Rigor mortis, **192**
Ringworm, 111
Rinne test, **403**
RNA polymerase, **70**
RNA processing, **70**
Rods, **382**, 391–92
Roentgenology, **14**
Romberg's sign, **402**
Root canals, **444**
Root hair plexuses, **359**
Rooting reflex, **316**
Root of lung, **512**
Roots, **444**
Rotation, 173–74, 178
Rotator cuff, **232**
Rough ER, **57**
Round ligament, **468**
  of uterus, **711**
Round window, **395**
Route deferens, 703F
Routine hospital diets, **499**
Rubella, **104**
Rubeola, **104**
Rubriblasts, **543**
Rubricytes, **544**
Rubrospinal fasciculi, **314**
Rubrospinal tract, **303**, 366
Ruffini, end organ of, **359**
Rugae, **451**
  of vagina, **713**
"Rule of nines," 109
Ruptured disk, 159, **338**

Saccule, **396**
Sacral canal, **146**, 161
Sacral curve, **143**, 161
Sacral foramina, **146**
Sacral hiatus, **146**
Sacral plexus, **338**
Sacral promontory, **146**
Sacral vertebrae, **145**
Sacrum, **143**, 145–46, 146F, 161
Saddle joints, **170**, 178
Sagittal plane, **16**
Sagittal sections of the body, **10**
Sagittal sinus of brain, **597**
Sagittal suture, **142**, 160
Saliva, 446, 448
Salivary glands, 447, 448F
Saltatory conduction, **273**, 273F, 285
Salts, **32**, 32F, 42
Sarcolemma, **183**, 202
Sarcomas, **74**, 253
Sarcomere, **184**, 184–86, 202
  membranes of, 187–88

organization of, 188F
Sarcoplasm, **183**, 202
Sarcoplasmic reticulum, **184**, 202
Sartorius, 246
Satellite cells, 264, **265**, 266F, 284
Saturated fats, 35
Saturated fatty acids, **35**
Saturated oxyhemoglobin, **523**
Scala media, 396
Scala tympani, **396**
Scala vestibuli, **396**
Scaphoid carpal, **150**, 151F
Scapula, **148**, 148–49, 149F, 161
    muscles that move, 230–31, 231F, 233F
Scapular notch, **148**
Scar tissue, 84
Scheuermann's disease, **158**
Schleiden, Matthias, 47
Schwann, Theodor, 47
Schwann cells, 264, **265**, 266F, 284
Sciatica, **159**, 242, 338
Sciatic nerve, **338**, 340
Sciatic notch, lesser, 153
Scintiphotography, **649**
Sclera, **380**
Scleroderma, **93**, 472
Sclerotic plaques, **265**
Sclerotome, **182**
Scoliosis, **158**
Scotopsin, **391**
Scratch testing, **111**
Scrotal raphe, **699**
Scrotum, **699**
    accessory glands of, 706
Scurvy, **41**
Sebaceous glands, **108**, 111
Seborrheic keratoses, **111**
Sebum, **108**, 111
Secondary capillary plexus, **415**
Secondary follicle, **717**
Secondary immune response, **634**
Secondary immunodeficiency diseases, **638**
Secondary marrow cavity, 123
Secondary oocyte, **717**
Secondary ossification centers, **121**, 128
Secondary peristaltic waves, **449**
Secondary polar body, **752**
Secondary reconstruction, **91**
Secondary spermatocytes, **704**, 750
Secondary syphilis, **723**
Second class lever, **210**, 254
Second-degree burns, **109**
Second messenger, **410**
Second-order neurons, **362**
Secretin, 431, **458**
Sedatives, **350**
Sedimentation rate, 551
Segmentation, **457**
Segregation
    of alleles, **755**
    chromosomal basis for, 756F
Seizures, 304
Selective IgA deficiency, **638**
Selectively permeable membrane, **47**, 76
Self-antigens, **633**
Sellaris joints, **170**
Sella turcica, **138**
Semen, **706**
    composition of, 706–7

Semicircular canals, **396**
Semicircular ducts, **396**
Semilunar ganglion, **328**
Semilunar notch, **150**
Semimembranosus, **246**
Seminal fluid, **706**
Seminal vesicles, **706**, 706F
Seminiferous epithelium, **703**
Seminiferous tubules, **699**
Semispinalis capitis, **220**
Semitendinosus, **246**
Senile dementia, **298**
Sensation, **357**
Sense organs, 376–402
    ear, 393–402
    eye, 380–93
    gustatory cells, 379–80
    olfactory receptors, 377–79
Senses, **357**
Sensible perspiration, **497**
Sensory cells, **260**
Sensory nerves, **261**, 284, 323
Sensory neurons, **349**, 359F
Sensory pathways, **361**, 361–64
Sensory reception, 357–59
Sensory receptors, **271**, 285
Septa, interalveolar, 514
Septal cells, **515**
Septicemia, **554**
Septum, interatrial, 561
Septum pellucidum, **307**
Serosa, **710**
Serous otitis media, **403**
Serous pericardium, **561**
Serratus anterior, **230**
Serratus posterior, **224**
Sertoli's cells, **429**, 699
Serum, **553**
Serum creatinine, **197**
Serum electrolytes, **31**, 684
Serum sickness, **177**, 639
Sesamoid bones, **115**, 142
Severe, fulminating hepatitis, **471**
Sex chromosomes, **759**
Sex linkage, 759–61
Sexuality. *See also* Coitus; Reproduction
    safe sex, 724
Sexually transmitted diseases (STDs), **723**
Shallow water blackout, **525**
Sharpey's fibers, **118**
Shells, **21**, 41
Shell temperature, **494**
Shinbone, 155, 161
Shingles, **102**
Shivering, **197**, 203, 496
Shock, **609**
    anaphylactic, 352, 609
    cardiogenic, 609
    circulatory, 613
    hypovolemic, 609
    neurogenic, 609
    spinal, 315
    vasogenic, 609
Short bones, **115**
Short-term memory, **370**
Shoulder blades, 148–49, 161
Shoulder girdle, 148
Shoulder impingement syndrome, **256**
Sickle cell anemia, **554**, 555F

Sigmoid colon, **465**
Silicosis, **531**
Simple alveolar glands, **86**
Simple columnar epithelium, 82F, **83**, 92
Simple cuboidal epithelium, 82F, **83**, 92
Simple epithelial tissue, 92
Simple epithelium, **81**
Simple fracture, **116**
Simple mastectomy, **622**
Simple squamous epithelium, **82**, 82F, 92
Simple tubular glands, **85**
Singing, 509–10
Single photon emission computed
    tomography (SPECT), **24**
Sinoatrial (SA) node, **568**
Sinus arrhythmia, **579**
Sinus bradycardia, **579**
Sinuses, **139**, 508F
    adrenal, 422
    frontal, 508
    mastoid, 134
    maxillary, 508
    paranasal, 508
    renal, 649
    sagittal, 597
    urogenital, 648
Sinusitis, 508
Sinusoids, **468**, 527, 597
Sinus venosus, **560**
Skeletal muscles, 91, 94, 181, 207
    fiber structure of, 183–84
    organization of, 181–88, 183F, 202
    oxygen debt in, 196
    shapes of, 208F
Skeletal system, **7**, 114–28. *See also* Bone;
    Skeleton; Vertebral column
    anatomy of, 130–61
    development of, 128
    diseases of, 125–27
    functions of, 115, 128
    physiological maintenance of, 123–25, 128
    remodeling of, 124–25, 128
Skeleton
    appendicular, 131, 132F, 160
    axial, 132F
    development of, 119–23
    divisions of, 131, 160
    movements of, 170–74, 178
    terms used for surface features of, 131
Skin, **100**. *See also* Dermis; Epidermis;
    Integumentary system
    functions of, 109–10, 112
    glands of, 108–9, 111
    hypodermis, 106
    immunity and, 625
    structure of, 103F
    subcutaneous layer, 88, 106
    viral infections of, 104
Skin problems, 102
Skin tumors, 111
Skull, **131**, 131–42, 160
    anterior view, 135F
    cranial group of skull bones, 133–40
    facial group of skull bones, 140–42
    fetal, 133F
    flat bones of the, 121F
    lateral view, 135F
    major foramina of, 134
    sagittal section, 138F

Skull (continued)
  sinuses of the, 141F
  superior and inferior views, 137F
  sutures and wormian bones, 142
Sleep, 369, 369–70
Sliding filament model of muscle contraction, 191F, 202
Slow block to sperm/egg fusion, 726
Slow twitch muscle fibers (Type I), 196, 202
Slow wave sleep, 370
Small intestine, 455, 455–63
  digestion in, 460
  disorders of, 473
  enzymes of the brush border of, 461
  functions of, 463F
  segmentation of, 458F
Smallpox, 635
Small saphenous vein, 602
Smell. See Olfaction
Smooth ER, 57
Smooth muscle, 91, 94, 181, 198–99, 199F, 203
Social death, 305
Sockets, 444
Sodium
  in body fluid, 683–84
  renal regulation of, 666
Sodium-potassium-ATPase pump, 267, 267F
Soft hydrophilic contact lenses, 385
Soft palate, 444
Solenoids, 65
Soleus, 248
Solubility coefficient, 520
Soma, 262, 284
Somatic nervous system, 261, 284
Somatic (somatomotor) neurons, 263F, 333
Somatic pain, 360
Somatic sensory (exteroceptive) neurons, 349, 354
Somatic symptoms, 625
Somatomammotrophin, 431
Somatomedin, 417
Somatostatin, 416, 429, 431
Somatotrophs, 415
Somites, 182, 182F
Sonogram, 15
Sound
  localization of, 400
  transmission to the ear, 398, 399F
Southern blot technique, 767
Sparing of glucose, 479
Sparing of proteins, 428
Spasm, 193
Spatial summation, 278, 285
Special senses, 357
Special test diets, 499
Specific compliance of lungs, 518
Specific learning disability (SLD), 373
Specific response to foreign instrusions, 624
SPECT (single photon emission tomography), 24
Speech, motor area for, 296
Sperm, 704
  attachment and binding of, 724
  hormonal control of production of, 705–6
  production and ejaculation of, 703–5
Spermacides, 725
Spermatic arteries, 594
Spermatic cord, 703
Spermatids, 704

Spermatogenesis, 703
Spermatogonia, 704
Spermatozoa, 704
Spermiogenesis, 704
Sphenofrontal suture, 142
Sphenoid bone, 138, 139F, 160
Sphenoid sinus, 508
Sphincter
  of Oddi, 470
  pyloric, 451
Spina bifida, 286
Spina bifida cystica, 286
Spina bifida occulta, 286
Spinal accessory nerves, 306, 332
Spinal cord, 261, 290, 302F, 311–18, 312F, 320
  anatomy of, 311–14
  hemisection of, 315
  major tracts in, 314
  meninges of, 309–10
  organization of, 313
  reflexes of, 314–18
  with spinal nerves, 334F
Spinal cord injury, 315
Spinal nerves, 333F, 333–39, 340
  adult, 333–34
  distribution of, 335F
  peripheral regions innervated by, 334
Spinal reflex arc, 316
Spinal reflexes, 311, 314, 320
Spinal shock, 315
Spinal tap, 311
Spindle, 67, 77
Spine, 143, 160–61
  disorders of the, 158–60
Spinocerebellar fasciculi, 313
Spinothalamic fasciculus, anterior, 313
Spinous process, 143, 161
Spiral arteries of endometrium, 713
Spiral fracture, 116F
Spiral ganglion, 330
Spiral lamina, 397
Spirometer, 517
Splanchnic nerves, 344, 352
  pelvic, 347
Spleen, 623, 623F, 623–24
  removal of, 623
Splenectomy, 623
Splenic artery, 594
Splenic deficiency syndromes, 638
Splenic flexure, 465
Splenic nodules, 623
Splenic vein, 601
Splenius capitis, 220, 223
Splenius cervicis, 223
Spondylitis, ankylosing, 177
Spondylolisthesis, 159
Spondylolysis, 159
Spongy bone, 118, 128
Spontaneous abortion, 743, 744
Sports injuries, 174
Spot desmosomes, 50
Sputum, 511
Squama, 136
Squamosal suture, 142, 160
Squamous cell carcinoma, 106, 472
Squamous epithelium
  keratinized and nonkeratinized, 83
  simple, 82
Squamous portion of the temporals, 134

Staging of a disease, 96
Staircase effect, 194, 195F
Stalk of follicular cells, 717
Stapes, 395
Staphylococcus, 177
Staphylococcus aureus, 124
Starch, 35, 35F
Starling's law of the heart, 576
Startle reflex, 316
Stasis dermatitis, 614
Static equilibrium, 400
Static stability, 176
Statoconia, 400
Status epilepticus, 304
Steinert's disease, 252
Stem cells, 629
Stenosis
  of cardiac valves, 579
  pyloric, 473
Stensen's duct, 447
Stereocilia, 400
Sterilization, 704
Sternal extremity, 148
Sternocleidomastoid, 220
Sternum, 147, 161
Steroid hormone-binding globulin (SHBG), 429
Steroid hormones, 407, 410–11
Steroids, 36, 42, 201
Stethoscope, 574
Stiff neck, 220
Stigma of ovary, 717
Stimulants, 351
Stimulus, adequate, 357
Stirrup, 395
Stomach, 451, 451F, 451–55
  body of, 451
  disorders of, 473
Stomodeum, 442
Strabismus, 390
Straight arteries of endometrium, 713
Straight sinus of brain, 597
Stratified columnar epithelium, 82F, 83, 93
Stratified cuboidal epithelium, 82F, 83, 93
Stratified epithelium, 81, 83, 92
Stratified squamous epithelium, 82F, 83, 92–93
Stratum basale, 101, 110
Stratum basalis, 711
Stratum corneum, 103, 110
Stratum functionalis, 711
Stratum germinativum, 101
Stratum granulosum, 101, 110, 717
Stratum lucidum, 103, 110
Stratum spinosum, 101, 110
Streptokinase, 581
Streptomycin, 402
Stress fracture, 116
Stress testing of cardiac patient, 575
Stress ulcers, 473
Stretch reflex, 316, 316F
Striated muscle, 181, 202
Striated muscle fiber, 185F
Striations, 91
Strokes, 287, 314, 554
Stroma, 86
Styloglossus, 219
Styloid process, 134, 150
Stylomastoid foramen, 136
Subarachnoid space, 310

Subclavius, 230
Subcutaneous layer, 88, 106
Subdural hematoma, chronic, 287
Sublingual glands, 447
Submandibular ducts, 447
Submandibular ganglia, 346
Submandibular glands, 447
Submaxillary glands, 447
Subscapular fossa, 149
Subscapularis, 232
Subshells, 21
Substance abuse, 350
Substance P, 194
Substantia nigra, 298, 303
Substrate, 24
Substrate molecules, 30
Subthalamic nucleus, 298
Succinic acid, 483
Succinyl CoA, 483
Sucking reflex, 316
Sucrase, 460
Sucrose, 34
Sudden cardiac death (SCD), 581
Sudden infant death syndrome (SIDS), 526
Sudoriferous glands, 108, 111
Sulcus, 291, 319
 anterior lateral, 312
Sulfhydryl group, 38
Summation, 193, 194, 194F, 202
Sun protection factor (SPF), 106
Suntanning, 105, 106
Superciliary arches, 133
Superficial fascia, 106
Superficial palmar arches, 594
Superficial veins, 597
Superficial veins of arm, 600
Superior articular processes, 143, 146
Superior colliculus, 303
Superior conjunctival fornix, 386
Superior jugular ganglion, 331
Superior mesenteric artery, 594
Superior mesenteric ganglion, 344
Superior mesenteric plexus, 347
Superior mesenteric vein, 601
Superior nasal conchae, 140
Superior nuchal lines, 138
Superior orbital fissure, 138
Superior phrenic arteries, 594
Superior portion of the body, 10
Superior temporal lines, 133
Superior trunk, 336
Superior vena cava, 563, 597
Supination, 174
Supporting cells, 377
Suppressor T cells, 632
Suprahyoid muscles, 220
Supraoptic nuclei, 413
Supraorbital foramen, 133
Supraorbital ridges, 133
Suprarenal arteries, 594
Suprarenal veins, 601
Supraspinatus, 232
Supraspinous fossa, 149
Surfactant, 515
Surgical neck, 149
Surrogate motherhood, 735
Susceptibility to disease, 625
Suspensory ligament
 of eye, 381
 of ovary, 709
Sustentacular cells, 377, 702
Sutures, 133, 142, 160, 165, 166F
Swallowing, 449–50
Swallowing center, 449
Swallowing receptor areas, 449
Swayback, 158
Sweat glands, 108, 110, 111
Sympathetic (thoracolumbar) nervous
 system, 343, 344, 352
 motor routes of, 346F
Sympathetic trunk (chain), 344
Sympatholytic drugs, 352
Sympathomimetic (adrenergic) drugs, 352
Sympathomimetic function of hormones,
 426
Symphysis, 166, 167F, 176
Symphysis pubis, 670, 714
Symptomatic epilepsy, 304
Synapses, 274, 275F, 285
 axosomatic, 274, 285
 communication across, 276–77
 electrical, 278
 types of, 276F
Synaptic cleft, 274
Synaptic delay, 277, 285
Synaptic fatigue, 277, 285
Synaptic knob, 276
Synarthroses, 165, 176
Synchondroses, 166, 176
Syncopal attack, 570
Syndesmosis, 165, 166F
Syneresis, 553
Synergists, 209, 254
Synostosis, 165
Synovial cavities, 165
Synovial fluid, 167
Synovial joints, 165, 167F, 169F, 176
 movement of, 168–70, 176–78
Synovial membrane, 167, 176
Synovitis, 167
Synthetic hallucinogens, 351
Synthetic nonopiates, 350
Syphilis, 111, 723
System, 7, 16
Systemic anatomy, 5
Systemic arteries, 590F
Systemic circulation, 563, 589
Systemic lupus erythematosus (SLE), 93, 638
Systole, 572

Tabes dorsalis, 723
Tachycardia, 579
Tactile corpuscles, 359
Taenia coli, 464
Talipes, 156
Talus, 156
Target tissues and organs, 406
Tarier's disease, 252
Tarsal bones, 156, 161
Tarsal glands, 386
Tarsal plate, 385–86
Tarsus, 156, 161
Taste, 379. *See also* Gustation
Taste bud, 379, 379F
Tay-Sachs disease, 286, 755
T cell (lymphocytes), 629, 632F
 formation of, 629F
Tears, 386
Tectorial membrane, 398
Tectospinal fasciculi, 314
Tectospinal tracts, 366
Tectum, 303
Teeth, 444, 444–47
 anatomy of, 444
 articulation of, 165–66
 decay of, 446
 deciduous (baby), 445
 disorders of, 446
 periodontal disease, 446
 permanent, 445
 types of, 444–45, 447F
 wisdom, 445
Telangiectases, 613
Telencephalon, 291, 319, 293
Telophase I, 750
Telophase II, 751
Telophase of mitosis, 66F, 68, 77
Temperature, 494. *See also* Body temperature
 sensory pathway for, 362F, 363–64
Template, 68, 77
Temporal arteritis, 613
Temporal bones (temporals), 133, 133–36,
 136F, 160
Temporalis, 207, 218
Temporal lobe, 295, 302F, 319
Temporal process, 142
Temporal summation, 278, 285
Tendonitis, 186
Tendons, 90, 94, 168, 181, 202
 Achilles', 248
 bipenniform, 207
 calcaneal, 248
 central, of diaphragm, 224, 516
 patellar, 246
 unipenniform, 207
Tendon sheaths, 168
Tenesmus, 467
Tenosynovitis, 186, 256
Tensor fasciae latae, 242
Tentorium cerebelli, 309
Terbutaline, 352
Teres major and minor, 232
Terminal bronchioles, 513
Termination phase of translation, 73, 77
Tertiary syphilis, 723
Testes, 429, 699, 702F
 meiosis in, 751F
Testicles, 699
Testicular arteries, 594
Testicular torsion, 728
Testis-determining factor (TDF), 759
Testosterone, 429, 705
Tetanus, 194, 195F, 202, 286
Tetrad of chromatids, 749
Tetralogy of Fallot, 579
Thalamic nuclei, 300, 300F
Thalamus, 299, 299–300, 319
Thalassemia, 554
Theca externa, 717
Theca folliculi, 717
Theca interna, 717
Thenar eminence, 240
Therapeutic abortion, 743
Therapeutic diets, 499
Thermography, 622
Thermoreceptors, 357, 360
Thermostatic center, 495

Thiamine, **493**
Thick ascending limb of loop of Henle, **653**
Thick filaments, **186**, 202
Thigh
   individual muscles of, 247F
   muscles that move, 242–45, 244F
Thin ascending limb of loop of Henle, **653**
Thin descending limb of loop of Henle, **653**
Thin filaments, **186**, 202
Thin segment of nephron, **653**
Third class lever, **211**, 254
Third-degree burns, **109**
Third-order neurons, **362**
Third ventricle, **307**, 319
Thirst, **681**
Thirst center, **681**, 682F
Thomsen's disease, 252
Thoracic aorta, **594**
Thoracic area, X ray of, 5
Thoracic cavity, 16
Thoracic curve, **143**, 161
Thoracic duct, **619**
Thoracic vertebrae, **143**, 145, 145F, 161
Thoracic wall, muscles of, 224–25, 225F
Thoracolumbar nervous system, **343**, 352
Thorax, **147**, 147–48, 161
   arteries of, 594, 595F
   veins of, 600–601
Threshold, 202
Threshold value, 285
Threshold value of a neuron, **271**
Thrombin, **551**
Thromboangiitis obliterans, **613**
Thrombocytes, **549**, 549–50
Thrombosis, **314**, 553, 588
   coronary, 554
Thromboxanes, **431**
Thrombus, **553**, 588
Thymic corpuscles, **431**, 624
Thymine, **39**
Thymoma, 253
Thymosins, **431**
Thymus gland, **429**, 429–31, 430F, 624
   in a child, 624F
   cortex of, 731
Thyroarytenoid muscles, **510**
Thyrocervical artery, **593**
Thyroglobulin, **418**
Thyroid, tumors of, 74
Thyroid cartilage, **509**
Thyroid follicles, **418**
Thyroid gland, **13**, 418, 419F
   isthmus of, 418
Thyroid hormones, 418–20
Thyroid-stimulating hormone (TSH), **417**
Thyrotrophin, **417**
Thyrotrophin-releasing hormone (TRH), **416**
Thyrotrophs, **415**
Thyroxine, **407**, 418
Tibia, **155**, 155F, 161
Tibial artery, **596**
Tibial collateral ligament, **176**
Tibialis anterior, 207, **248**
Tibialis posterior, 207, **248**
Tibial tuberosity, **155**
Tibial veins, 602
Tibiofemoral joint, **174**, 178
Tic, **193**
Tidal volume, **517**

Tight junctions, **50**, 73
Timed vital capacity, **518**
Tinea circinata, **111**
Tinea cruris, **111**
Tinea pedis, **111**
Tinel's sign, **339**
Tinnitus, **402**
Tissue, **7**, 16, **80**, 80–94, 92
   connective, 86–91, 93–94
   dense connective, 87, 90, 94
   epithelial, 81–86, 92–93
   mucous, 87, 94
   muscle, 91–92, 94
   nervous, 92, 94
   scar, 84
   types of, 82F
   vascular, 90, 94
   white adipose, 488
Tissue fluid, **12**, 539, 618
Tissue plasma activator (TPA), **581**
Tissue thromboplastin, **553**
Tissue typing, 633
Toes, 156
   muscles that move, 248–49, 249F
Tolerance, **350**
Tomogram, **14**
Tomography, 5, 15
Tongue, **444**, 445F
   muscles, 219, 219F
   taste regions, 380F
Tonic neck reflex, **316**
Tonic receptors, **358**
Tonic spasm, **193**
Tonsillitis, acute, 530
Tonsils, **444**, 624
   lingual, 444, 624
   pharyngeal, 509, 624
Tooth decay, **446**
Torticollis, **220**
Total body scanning, **22**
Total lung capacity, **518**
Total mastectomy, **622**
Touch, **359**
   afferent pathways for, 363F
   crude, **359**
   fine, **359**
   sensory pathway for, 362
Touch receptors, **359**
Trabeculae, **118**, 128, 310
Trabeculae carneae, **561**
Trabeculae of lymph node, **621**
Trabecular bone, **118**, 128
Trace elements, **687**
Trachea, **510**, 511F
Trachealis, **511**
Tracheitis, **530**
Tracts, **261**
Transamination, **491**
Transcortin, **424**
Transcription, **70**, 71F, 77
Transection (spinal cord), **315**
Transfer reaction, **27**F, **28**, 42
Transferrin, **545**
Transfer RNA (t-RNA), **70**, 77
Transfusions, **637**
Transient ischemic attacks (TIAs), **314**
Transitional cell carcinomas, **676**
Transitional epithelium, 82F, 84, 93
Translation, **70**, 72F, 77

Translocation, **763**
Translocation Down syndrome, **764**
Transmitter substances, **188**
Transplantation, of cornea, 382
Transport maximum (Tm), **660**
Transverse colon, **464**
Transverse foramina, **144**, 161
Transverse ligament, **176**
Transverse lines, **145**
Transverse plane, 16
Transverse processes, **143**, 161
Transverse section of the body, **10**
Transverse sinuses of brain, **597**
Transverse thoracis, **224**
Transversus abdominis, **226**
Trapezium carpal, **150**, 151F
Trapezius, 207, **230**
Trapezoid carpal, **150**, 151F
Treppe, **194**, 195F, 202
Triad, **188**, 202
Triangular ligaments, **468**
Tricarboxylic acid cycle, **482**
Triceps brachii, **234**
Trichinosis, **253**
Tricuspid valve, **563**
Trigeminal nerves, 305, **328**, 328F, 328–29, 340
Trigeminal neuralgia, **328**
Triglycerides, 42
Trigone of urinary bladder, **670**
Triiodothyronine, **418**
Trimesters, **745**
Triose, 34
Triquetrum carpal, **150**, 151F
Trisomic cells, **761**
Trisomies, **286**
Trochanter, greater and lesser, **154**
Trochlea, **149**
Trochlear nerves, **327**, 327F, 327–28, 340
Trochlear notch, **150**
Trochoid joints, **169**
Trophectoderm, **733**
Trophic hormones, **415**
Trophoblast, **733**
Tropomyosin, **186**, 202
Troponin, **186**, 202
True pelvis, **152**
True ribs, **147**, 161
Trumpeter's muscle, **214**
Truncus arteriosus, **560**
Trunk, **7**, 16
   muscles of, 222–29, 231F
Trunks of nerves, 336
Trypsin, **459**
Trypsinogen, **459**
T tubules, **188**, 202
Tubal pregnancy, **744**
Tuber cinereum, **300**
Tubercles, **148**, 445, 514
   lesser, 149
Tuberculosis, **514**
Tubular glands, **85**, 86
Tubular hydrostatic pressure, 658
Tubular necrosis, acute, 658
Tubuloalveolar glands, **86**
Tularemia, 626
Tumors, **74**
   of the adrenal medulla, 426
   benign, 74

of the central nervous system, 287
of the joint, 177
of the pituitary, 326
Tumor-specific transplantation antigen (TSTA), **639**
Tunica adventitia, **443**, 585
Tunica albuginea, **699**, 709
Tunica dartos, **699**
Tunica externa, **585**
Tunica intima, **585**
Tunica media, **585**
Tunica mucosa, **441**
Tunica muscularis, **443**
Tunica serosa, **443**
Tunica submucosa, **443**
Tunica vaginalis, **699**
Tunics, **441**
   of arteries, 585
Turner syndrome, **177**, 761
T wave, **570**
Twinning, **741**
Twins, 741
Twitches, **193**, 193–94, 202
Two-point discrimination test, 359
Two-solute model for countercurrent multiplier, 664
Tympanic antrum, **395**
Tympanic membrane, **393**
Tympanum, **393**
Tyrosinase, **105**
Tyrosine, **105**

Ubiquinone, **484**
UDP-glucose, **480**
Ulcerative colitis, **467**, 473
Ulcers, **452**, 473
   of bone, 124
   gastric, 452
   stress and, 473
Ulna, **150**, 151F, 161
Ulnar artery, **594**
Ulnar notch, **150**
Ulnar tuberosity, **150**
Ulnar vein, **600**
Ultrasonography, **575**, 593, 649, 739
Ultrasound scans, **456**
Umbilical arteries, **605**, 741
Umbilical cord, **603**, 738
   structure of, 741F
Umbilical vein, **605**, 741
Umbilicus, **749**
Unconsciousness, 367–69
Unicellular exocrine glands, **85**
Unilateral renal agenesis, **675**
Unipennate fascicles, **207**
Unipenniform tendons, **207**
Unipolar neurons, **264**, 284
Universal blood donor, **637**
Universal blood recipient, **637**
Unsaturated fatty acids, **35**
Upper extremity, 149–52, 161
   arteries of, 593–94
   muscles of, 230–41
   veins of, 600
Upper labial frenulum, **444**
Upper motor neuron, **365**
Upper respiratory infection, 507, **530**
Upper respiratory tract, **505**, 506–7
Uracil, **39**, 43

Uremia, **660**
Ureteric bud, **648**
Ureters, **669**, 669F
Urethra, **670**
   cavernous, 670, 703
   male, 703
   male and female, 671F
Urethral folds, **701**
Urethral glands, **706**
Urethral groove, **701**
Urethral orifice, **703**, 707
   internal, 670
Urethritis, **675**
Uridine triphosphate (UTP), **480**
Urinalysis, **673**
Urinary bladder, **670**, 671F
Urinary calculi, **670**
Urinary casts, **674**
Urinary pole, **652**
Urinary space, **652**
Urinary stones, **674**
Urinary system, **646**, 647F, 646–74
   composition of urine, 673–74
   conduction, storage, and elimination of urine, 669–72
   disorders, 675–76
   embryonic development of, 650
   homeostatic functions of the kidneys, 665–69
   kidneys, 647–51
   nephron, 651–57
   physiology of the kidneys, 657–65
Urinary tract imaging techniques, 648
Urination, **672**
Urine, **646**
   calculi in, 674
   casts in, 673
   composition of, 673–74
   concentration of, 662–64, 665
   conduction, storage, and elimination of, 669–72
   flow obstruction, 676
   myoglobinuria, 196
   primary, 657
   solutes excreted in, 674
Urine specific gravity, **684**
Urogenital sinus, **648**
Urogenital triangle, **714**
Urothelium, **84**, 93
Urticaria, **102**, 102F
Uterine arteries, **713**
Uterine cavity, **711**
Uterine tubes, **709**
Uterosacral ligaments, **711**
Uterosalpingogram, **771**
Uterus, **711**, 711–13
   anteflexion of, 711
   blood supply to, 713F
   histology of uterine wall, 712F
   isthmus of, 711
   ultrasonography of, 739
Utricle, **396**
Uvula, **444**

Vaccination, **634**, 634–35
Vaccine, **634**
Vacuum sinusitis, **508**
Vagina, 713–14
Vaginal canals, **711**

Vaginal fornix, **714**
Vaginal sponge, **725**
Vagus nerves, 306, **331**, 332F, 340
Valgus, **177**
Vallate papillae, **444**
Valve of the vena cava, **605**
Valves of veins, **587**
Valvular heart disease, **581**
Variable region of antibody, **629**
Varicella, 104
Varices, **611**
Varicocele, **728**
Varicose veins, **611**, 614
Variola, **635**
Vasa recta, **656**, 656F
   role of, 664–65
Vasa vasorum, **585**
Vascular constriction, 550
Vascular pole, **652**
Vascular tissue, **90**, 94. *See also* Blood
Vascular tunic, **380**
Vas deferens, **703**
Vasectomy, **704**
Vasoconstriction, **585**, 608. *See also* Blood pressure
Vasodilation, **586**
Vasodilator, **547**
Vasogenic shock, **609**
Vasomotor center, **306**, 607
Vasospasm, **193**
Vastus intermedius, **246**
Vastus lateralis, **246**
Vastus medialis, **246**
Veins, **585**, 587, 596–602, 598F
   abdominal, 601
   accessory hemiazygos, 600
   anterior tibial, 602
   arcuate, 656
   axillary, 600
   azygos, 600
   basilic, 600
   brachial, 600
   brachiocephalic, 600
   bronchial, 600
   cardiac, 567
   cephalic, 600
   cervical, 600
   common iliac, 601
   cutaneous, 597
   deep, 597
   esophageal, 600
   external iliac, 601
   external jugular, 597
   femoral, 602
   fistulas in, 613
   gonadal, 601
   great cardiac, 567
   great saphenous, 602
   of the head and neck, 597–600
   hemiazygos, 600
   of the hepatic portal system, 601–2, 602F
   iliac, 601
   inferior mesenteric, **601**–2
   inferior vena cava, 563, 597
   intercostal, 600
   interlobar, 656
   interlobular, 656
   internal iliac, 601
   jugular, 597

Veins (continued)
　of the lower extremity, 602
　lumbar, 601
　medial cubital, 600
　middle cardiac, 567
　pericardial, 600
　peroneal, 602
　phrenic, 601
　popliteal, 602
　portal, 601
　posterior tibial, 602
　pulmonary, 564, 589
　radial, 600
　renal, 601, 647
　small saphenous, 602
　splenic, 601
　superficial, 597, 600
　superior mesenteric, 601
　superior vena cava, 563, 597
　suprarenal, 601
　of the thorax, 600–601
　tibial, 602
　ulnar, 600
　umbilical, 605, 741
　of the upper extremity, 600
　valves of, 587, 587F, 605
　varicose, 611, 614
　vena cava, superior and inferior, 563
　venous circulation, 611–12
　venous sinuses of the brain, 599F
　vertebral, 597
Vena cava, superior and inferior, **563**
Venous thrombosis, **614**
Ventral cavity, **9**
Ventral portion of the body, **10**
Ventral respiratory group (VRG), **526**
Ventricles, brain, **307**, 309F, 319
Ventricles, heart, **197**, 560, 561
　left, **566**
　right, **563**
Ventricular fibrillation, 193, 571, **580**
Ventricular septal defects, **579**
Ventricular septum, **560**
Ventricular tachycardias, **580**
Venules, **585**, 586–87
Vermiform appendix, **464**
Vermis, **303**, 319
Verrucae, **102**
Vertebrae, **143**, 143–45, 160, 161. *See also* Bones
Vertebral arch, **143**, 161
Vertebral artery, **591**
Vertebral body, **143**
Vertebral canal, **311**
Vertebral cavity, **9**
Vertebral column, **143**, 143–47, 144F, 160–61, 312F. *See also* Bones
　muscles of, 222–23, 223F
　types of vertebrae, 143–45
Vertebral foramen, **143**, 161
Vertebral notches, **143**
Vertebral plexus, **344**
Vertebral ribs, **147**
Vertebral veins, **597**
Vertebra prominens, **145**
Vertigo, **402**

Very low density lipoprotein (VLDL), **487**
Vesicles, **55**
Vesicular follicle, **717**
Vesicular lesions, **111**
Vestibular folds, **509**
Vestibular ganglion, **330**
Vestibular nerves, **330**, 340
Vestibular organ, **394**
Vestibule, **396**, 444
　of larynx, **509**
　of nose, **506**
　urogenital, **648**
　of vagina, **714**
Vestibulocochlear nerves, 305, 306, **329**, 329–30, 331F, 340
Vestibulospinal fasciculi, **314**
Vestibulospinal tracts, **366**
Villi, **456**, 457F
　absorption of nutrients into, 461F
　arachnoid, 307
Villous pigmented synovitis, **167**
Viral skin infections, 104
Virilism, adrenal, 425
Visceral arches, **125**
Visceral arteries, **594**
Visceral efferent neurons, **343**, 352
Visceral layer of Bowman's capsule, **652**
Visceral muscle, 91
Visceral pain, **360**
Visceral pericardium, **559**
Visceral peritoneum, **471**
Visceral pleura, **512**
Visceral sensory (enteroceptive) neurons, 354
Visceral sensory neurons, **349**
Visceral smooth muscle, **198**, 203
Visceroreceptors, **357**
Vision, 387–93
　binocular, 390–91
　biochemistry of, 391–93
　near and far points of, **388**
　visual pathways to the brain, 393
Visual accommodation, **387**, 388F
Visual disorders, 389
Visual purple, **391**
Vital capacity, **517**
Vital signs of cardiac patient, **574**
Vitamin B complex, **492**
Vitamin D, 123, **492**
　activation by the kidneys, 669
　hypervitaminosis D, **126**
　production by skin, 110, 112
Vitamins, 41, 492–93
　vitamins A, C, E, K, **492**
　vitamins B_1, B_2, B_6, B_{12}, D_3, K_2, **493**
　water- and fat-soluble, **492**, 493
Vitelline sac, **442**
Vitreous humor, **385**
Vocal folds, **509**
Vocal ligaments, **510**
Voice, 509–10
Voiding cystogram, **649**
Volkmann's canals, **118**
Volkmann's ischemic paralysis, **339**
Volley, **274**, 274F, 285
Voltage dependent channels, 270F, 285
Voltage independent channels, **270**

Volts, **266**
Voluntary motor pathways, **364**, 364–67
Voluntary muscle, 91, **181**, 202
Vomer, **140**, 160
von Recklinghausen's disease, 74
Vulva, **714**

Warts, **102**
Water
　absorption by the small intestine, 463
　intake and output of, 681–82
　regulating content of, 681–82
Water excretion, 665
Water-soluble hormones, 407, 408
Water-soluble vitamins, **492**
Wavelength, **20**
Weber's test, **403**
Wharton's ducts, **447**
Wharton's jelly, 87, **741**
Wheals, **111**
Whiplash, **144**
White adipose tissue, **488**
White blood cells, **546**. *See also* Leukocytes
　polymorphonuclear, 548
White commissure, **313**
White communicating rami, **344**
White fibrous connective tissue, **90**
White matter, **291**, 319, 296–97, 297F
　of the spinal cord, 313
White pulp of spleen, **623**
Wilms tumor, **676**
Wisdom teeth, **445**
Wiskott-Aldrich syndrome, **638**
Withdrawal reflex, **317**
Wormian bones, **142**, 160
Wrist, 150, 151F
　forearm extensors of, 238–39, 239F
　forearm flexors of, 236–37, 237F

X-chromosome, **759**
Xeroderma pigmentosum, **111**
Xeroradiograph, **14**
Xiphoid process, **147**, 161
X-linked genes, **759**
X rays, 5, 16

Y-chromosome, **759**
Y-linked genes, **759**
Yoga, **351**, 354
Yolk sac, **735**

Z disk, **184**
Z lines, **184**, 202
Zona fasciculata, **423**, 424
Zona glomerulosa, **422**, 423–24
Zona pellucida, **717**
Zona reticularis, **423**, 424
Zygomatic arch, **134**, 142
Zygomatic fracture, 141
Zygomaticofacial foramen, **142**
Zygomatic processes, **133**, 140, 142
Zygomatics, **141**, 141–42, 160
Zygomaticus, **207**
　major and minor, **214**
Zygote, **66**, 726
Zymogenic cells, **453**

CLINICAL ABBREVIATIONS

The following abbreviations are commonly used in clinical situations. Many of them have been used in this text. Those that have not are included because of their relatively common use in medical, pharmacological, and health-related contexts.

a.c.	before eating
abs. feb.	without fever
ACTH	adrenocorticotrophic hormone
ad effect.	until effective
ad lib.	as desired
ad us.	as customary
ad us. ext.	for external use
ADH	antidiuretic hormone
ADP	adenosine diphosphate
adst. feb.	when fever is present
ag. feb.	when fever increases
AIDS	acquired immunodeficiency syndrome
alt. dieb.	every other day
alt. hor.	every other hour
AQ	aqueous
ARC	AIDS-related complex
ARDS	adult respiratory distress syndrome
ARF	acute respiratory failure; acute renal failure
AS	aortic stenosis
atm	atmosphere
ATP	adenosine triphosphate
AV	atrioventricular
bid	twice daily
BP	blood pressure
BBB	bundle branch block; blood-brain barrier
bis in 7d	twice a week
BMR	basal metabolic rate
BP	blood pressure
BSA	body surface area
BUN	blood urea nitrogen
C	celsius; complement; gallon
$\bar{c}$	with
C_{Cr}	renal creatinine clearance
C1, etc.	first cervical vertebra, etc.
ca	about
CAD	coronary artery disease
cal	calorie
Cal	kilocalorie
cAMP	cyclic AMP
CAPD	continuous ambulatory peritoneal dialysis
CAT	computerized axial tomography
CBC	complete blood count
cc	cubic centimeter (cm^3)
CCK, CCK-PZ	cholecystokinin-pancreozymin
CDC	Centers for Disease Control
CF	complement fixation; cystic fibrosis
CHD	congenital heart disease
CHF	congestive heart failure
cm	centimeter
CNS	central nervous system
CO	cardiac output
CoA	coenzyme A
COAD	chronic obstructive airways disease
comp	compound
contra	against
COPD	chronic obstructive pulmonary disease
CPAP	continuous positive airway pressure
CPR	cardiopulmonary resuscitation
Cr	creatinine
CRH	corticotrophin-releasing hormone
CRF	chronic renal failure
CSF	cerebrospinal fluid
CT	computed tomography
CV	cardiovascular
CVA	cerebrovascular accident
CVP	central venous pressure
CVS	chorionic villus sampling
d	dextro (right); day
D & C	dilation and curettage
da	day
DBP	diastolic blood pressure
def	defecation
DH	delayed hypersensitivity
DHAP	dihydroxyacetone phosphate
dL	deciliter
DNA	deoxyribonucleic acid
dr	dram
DTP	diphtheria-tetanus-pertussis
ECF	extracellular fluid
ECG	electrocardiogram
ECT	electroconvulsive therapy
EDV	end-diastolic volume
EEG	electroencephalogram
EM	electron micrograph
EMG	electromyogram
ENT	ear, nose, and throat
EPSP	excitatory postsynaptic potential
ER	endoplasmic reticulum
ESR	erythrocyte sedimentation rate
ESRD	end-stage renal disease
ESV	end-systolic volume
F	Fahrenheit
F_1, F_2	first, second filial generation
FDA	Food and Drug Administration
FEV	forced expiratory volume
fl oz	fluid ounce
FSH	follicle-stimulating hormone
FUO	fever, unknown origin
G6PD	glucose 6-phosphate dehydrogenase
GABA	gamma-amino butyric acid
GFR	glomerular filtration rate
GI	gastrointestinal
GnRH	gonadotrophin-releasing hormone
gr	grain
GTP	guanosine triphosphate
gtt	drops
GU	genitourinary
Hb	hemoglobin
Hcy	hematocrit
HD	Huntington's disease
HDL	high density lipoprotein
hgb	hemoglobin
HGH	human growth hormone
HIV	human immunodeficiency virus
HLA	human leukocyte group A
HTLV-III	human T-lymphotrophic virus Type III
hypo	hypodermic
Hz	hertz (cycles per second)
IBS	irritable bowel syndrome
ICF	intracellular fluid
IDL	intermediate density lipoprotein
Ig	immunoglobulin
IM	intramuscular
inf	infusion
inhal	inhalation
inj	injection
IPPB	inspiratory positive pressure breathing
IPSP	inhibitory postsynaptic potential
IU	international unit
IUD	intrauterine device
IV	intravenous
IVP	intravenous pyelogram
$\bar{s}$	without
kcal	kilocalorie
kg	kilogram
l	laevo (left)
L	liter
L1, etc.	first lumbar vertebra, etc.
LD_{50}	median lethal dose
LDH	lactic dehydrogenase
LDL	low density lipoprotein
LE	lupus erythematosus
LH	luteinizing hormone